Dave Sobecki | Brian Mercer

MATH
In Our World

Fifth Edition

T0175673

McGraw Hill

MATH IN OUR WORLD

Published by McGraw Hill LLC, 1325 Avenue of the Americas, New York, NY 10019. Copyright ©2023 by McGraw Hill LLC. All rights reserved. Printed in the United States of America. No part of this publication may be reproduced or distributed in any form or by any means, or stored in a database or retrieval system, without the prior written consent of McGraw Hill LLC, including, but not limited to, in any network or other electronic storage or transmission, or broadcast for distance learning.

Some ancillaries, including electronic and print components, may not be available to customers outside the United States.

This book is printed on acid-free paper.

1 2 3 4 5 6 7 8 9 LWI 27 26 25 24 23 22

ISBN 978-1-265-14527-9
MHID 1-265-14527-X

Cover Image: *Rawpixel.com/Shutterstock*

All credits appearing on page or at the end of the book are considered to be an extension of the copyright page.

The Internet addresses listed in the text were accurate at the time of publication. The inclusion of a website does not indicate an endorsement by the authors or McGraw Hill LLC, and McGraw Hill LLC does not guarantee the accuracy of the information presented at these sites.

Table of Contents

James P. Blair/Photodisc/Getty Images

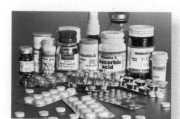

Gennadiy Poznyakov/123RF

Digital Vision/SuperStock

clintscholz/iStock/Getty Images

CHAPTER 8

Measurement 448

David Sobecki/McGraw Hill

CHAPTER 9

Geometry 480

Steve Allen/Stockbyte/Getty Images

CHAPTER 10

Probability and Counting Techniques 556

Helen H. Richardson/The Denver
Post/Getty Images

Peter Joneleit/Zuma Wire Service/
ZUMA Press/Alamy Stock Photo

John Lund/Tiffany Schoepp/Blend
Images

About the Authors

Courtesy of David Sobecki

Dave Sobecki

Dave was born and raised in Cleveland, and in spite of starting college with a major in creative writing, he somehow ended up with a doctorate degree in math. Go figure. Dave spent two years at Franklin and Marshall College in Lancaster, Pennsylvania, followed by 22 years at the Hamilton campus of Miami University in southwest Ohio. He recently relocated to North Port, Florida. Dave has won a number of teaching awards in his career and has written or co-authored either nine or twenty-one books, depending on how you count them. He's also working on his first novel. When not teaching or writing, Dave's passions include Ohio State and Cleveland professional sports teams, heavy metal music, travel, golf, collecting fine art, and most importantly, spending time with his wife Cat and a cast of characters at Heron Creek Golf and Country Club.

Courtesy of Brian Mercer

Brian Mercer

Brian is a tenured professor at Parkland College in Champaign, Illinois, where he has taught developmental and transfer math courses for 25 years. He began writing in 1999 and has currently co-authored eight textbooks, with others in the planning stages. Outside of the classroom and away from the computer, Brian is kept educated, entertained, and ever-busy by his wonderful wife Nikki, their two children, Charlotte (15) and Jake (14), and dog Molly. He is an avid St. Louis Cardinals fan and enjoys playing softball and golf in the summertime with colleagues and friends.

Letter from the Authors

This is the story behind *Math in Our World*. Liberal Arts Math is different from the other classes we typically teach to first- and second-year students, and we think that it requires a different approach. Many of the students have had negative experiences in algebra, and they come into any math course thinking it's going to be the same old thing again—finding *x*. Liberal Arts Math provides a great opportunity to show students that math isn't just an abstract subject studied by high-level intellectuals. In this course, we have the opportunity to really teach students about reasoning and thinking, rather than train them to mimic procedures. Who wouldn't look forward to that?

Math in Our World has a different style than you'll find in most college math texts. Both the structure of the chapters and the style of writing are designed to make the students think "Wow, this isn't what I expected … I can actually read and understand this!" We like to call it "teaching backwards": rather than first learning the math and then studying how it can be applied, every topic is introduced from a conceptual, applied point of view. The goal is to engage students at the beginning of each topic, helping them to avoid falling into the old "Why do I have to know this—I'm never going to use it" trap. I (Dave) have been passionate about writing as far back as I can remember, and I switched my major from creative writing to math education when I got a first taste as a tutor in college. I was captivated then by the "Aha!" moment when struggling students GET IT—and I still feel that way today. This book is special to me because it really allowed my two passions to come together into a unique product.

No one has ever become stronger by watching someone else lift weights, and our students aren't going to be any better at thinking and problem solving unless we encourage them to practice it. *Math in Our World* includes a veritable cornucopia of applications for students to hone their skills. The exercises for this edition were carefully evaluated to ensure that they are engaging for students and apply to fields of study that are common for Liberal Arts Math students. Additionally, we've continued to incorporate key ideas from the growing quantitative reasoning movement. You'll see more open-ended questions, more discovery learning, and more critical thinking. The goal is to help students develop into problem solvers and thinkers beyond the halls of academia. We've made a special effort toward diversity in this edition, carefully evaluating all aspects to make sure that *Math in Our World* provides an inviting atmosphere for students and faculty from the diverse world in which we live.

While no book can prevent students entering the course underprepared, we believe the digital component of the *Math in Our World* program (ALEKS) will help engage students and encourage them to develop their own questions about the world. ALEKS can play an important role in your class by providing opportunities to practice and master the computational as well as conceptual aspects of this course. In addition, we've added a corequisite guide to our supplements, recognizing that many schools are working on creative ways to help students fulfill their math requirements expediently.

We're confident that this book offers a fantastic vehicle to drive your classes to higher pass rates because of the pedagogical elements, writing style, interesting problem sets, and digital components. We hope you and your students enjoy using *Math in Our World* as much as our team enjoyed creating this program together. Good luck, and please don't hesitate to reach out with comments, questions or suggestions. We love hearing from instructors and students at davesobecki@gmail.com and bmercer51@gmail.com.

—*Dave and Brian*

Engage

Highly relevant **Application Exercises and Examples** drawn from the experiences and research of the authors further emphasize the importance that *Math in Our World* places upon students' ability to form a distinct connection with the mathematical content. The new edition brings many brand new and updated application exercises to students in each chapter, with topics ranging from credit card usage, college degree majors, elections, and relevant business decisions to scenarios involving popular statistics.

Chapter Openers engage student interest by immediately tying mathematical concepts to their everyday lives. These vignettes introduce concepts by referencing popular topics familiar to a wide variety of students—travel, demographics, the economy, television, and even college football.

Used to clarify concepts and emphasize particularly important points, **Math Notes** provide suggestions for students to keep in mind as they progress through the chapter.

Sidelights highlight relevant interdisciplinary connections within math to encourage and motivate students who have a variety of interests. These include biographic vignettes as well as other interesting facts that emphasize the importance of math in areas like weather, photography, music, and health.

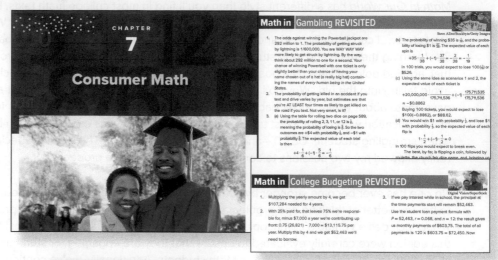

Math Note

When finding percent increase or decrease, make sure you divide by the *original* amount, not the ending amount.

Sidelight   Unlikely Events Are Very Likely to Occur

When the probability of some event is low, like say .05, it's common and reasonable to guess that event most likely will not occur. But if you look at it a certain way, events that are incredibly unlikely happen all the time. Let's use poker as an example. When you're dealt five cards, there's just a 1 in 13 chance that the first card you are dealt will be a 5, because there are 13 denominations. That's a probability of about 0.077. Now think about this: regardless of what denomination your first card ends up being, there was the same probability—just 0.077—of getting it. But of course the first card has to be *something*, so every time that first card is dealt, something with probability 0.077 is *guaranteed to happen*.

Now let's extend that out to all five cards. As we saw in Example 2, there are 2,598,960 possible hands you can be dealt. That means the probability of getting any particular hand is 0.00000038. Yet every single time that you are dealt five cards,

you have to get one possible hand, so something with that spectacularly low probability will occur with probability 1. Wild!

Finally, let's talk about the most unlikely lottery win ever, which happened to you even if you don't realize it. That's the genetic lottery, which resulted in you being born as you. Genetics tells us that the number of possible DNA combinations, which determine all of your characteristics and make you who you are, is on the order of 3 followed by 614 zeros. This is far larger than the total number of stars in the universe. Of all those possible combinations, only one of them could have resulted in you, an event so unlikely that the probability is for all intents and purposes zero. And yet somehow that happened. And continues to happen thousands of times across the world every single day. Next time you're feeling a little down, think about how incredibly lucky you were to have ever been born, and maybe you'll feel just a little special.

Practice

Carefully chosen questions help students to form a connection between relevant examples and the mathematical concepts of the chapter. Using the engaging writing style characteristic of the text, the author supports concepts through abundant examples, helpful practice problems, and rich exercise sets.

● **Worked Example Problems** with detailed solutions help students master and in some cases discover key concepts. Solutions demonstrate a logical, orderly approach to solving problems. Each example is titled to help you clearly identify the relevant learning objective.

● As one of our hallmark features, **Try This One** practice exercises provide immediate reinforcement for students. Designed to follow each example, these practice exercises ask students to solve a similar problem, actively involving them in the learning process. All answers to Try This One exercises can be found immediately preceding the end-of-section exercise sets for students to complete the problem-solving process by confirming their solutions.

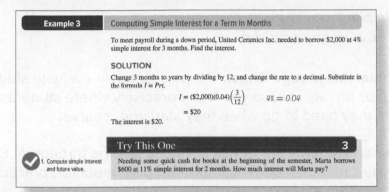

Calculator Guide

Computing log base 10 of 1.25 with a calculator is really simple:

Typical Scientific or Online Calculator:
1.25 [LOG]

Typical Graphing Calculator:
[LOG] 1.25 [ENTER]

● With an increased emphasis on using technology appropriately to allow more focus on interpretation and understanding, **Using Technology** boxes and **Calculator Guides** show students how to perform important calculations using spreadsheets and calculators (both graphing and scientific).

Using Technology

Using a Spreadsheet for Trial and Error

One of the things that spreadsheets do best is carry out repetitive calculations, which is a pretty good description of what we did in Example 3. Here's a screenshot of a spreadsheet I put together when writing that problem. I knew

● The rich variety of problem material found in the **End-of-Section Exercise Sets** helps students to test their understanding in a variety of ways to cater to their varying interests and educational backgrounds. In each end-of-section exercise set, there are **Writing Exercises, Computational Exercises, Applications in Our World, and Critical Thinking Exercises.**

● Exercises and activities located in the **End-of-Chapter Material** provide opportunities for students to prepare for success on quizzes or tests. At the conclusion of each chapter, students find critical summary information that helps them pull together each concept learned while moving through the chapter. In each end-of-chapter segment, students and instructors will also find Review Exercises, and a Chapter Test.

● Also featured are the **Chapter Projects,** which encourage more in-depth investigation for students working to summarize key concepts from the entire chapter. These projects are valuable assets for instructors looking for ways that students can work collaboratively.

Students start your course with varying levels of preparedness. Some will get it quickly. Some won't. ALEKS is a course assistant that helps you meet each student where they are and provide the necessary building blocks to get them where they need to go. You determine the assignments and the content, and ALEKS will deliver customized practice until your students truly get it.

Experience the ALEKS Difference

Easily Identify Knowledge Gaps

Gain greater visibility into student performance so you know immediately if your lessons clicked.

- **ALEKS's "Initial Knowledge Check"** helps accurately evaluate student levels and gaps on day one, so you know precisely where students are at and where they need to go when they start your course.

- **You know when students are at risk of falling behind** through ALEKS Insights so you can remediate—be it through prep modules, practice questions, or written explanations of video tutorials.

- **Students always know where they are,** know how they are doing, and can track their own progress easily.

Gain More Flexibility and Engagement

Teach your course your way, with best-in-class content and tools to immerse students and keep them on track.

- **ALEKS gives you flexibility** to assign homework, share a vast library of curated content including videos, review progress, and provide student support, anytime anywhere.

- **You save time** otherwise spent performing tedious tasks while having more control over and impact on your students' learning process.

- **Students gain a deeper level of understanding** through interactive and hands-on assignments that go beyond multiple-choice questions.

with ALEKS® Constructive Learning Paths.

**Narrow the
Equity Gap**

Efficiently and effectively create individual pathways for students without leaving anyone behind.

- **ALEKS creates an equitable experience for all students,** making sure they get the support they need to successfully finish the courses they start.

- **You help reduce attrition,** falling enrollment, and further widening of the learning gap.

- **Student success rates improve**—not just better grades, but better learning.

**Count on
Hands-on Support**

A dedicated Implementation Manager will work with you to build your course exactly the way you want it and your students need it.

- **An ALEKS Implementation Manager** is with you every step of the way through the life of your course.

- **You never have to figure it out on your own** or be your student's customer service. We believe in a consultative approach and take care of all of that for you, so you can focus on your class.

- **Your students benefit** from more meaningful moments with you, while ALEKS—directed by you—does the rest.

Already benefitting from ALEKS?

Check out our New Enhancements:
mheducation.com/highered/aleks/new-releases.html

Supplements

McGraw Hill ALEKS is a course assistant that helps math and chemistry instructors forge Constructive Learning Paths for their students—blending personalized modules with instructor-driven assignments to ensure every student always has another block to build on their knowledge base. ALEKS gives instructors the flexibility to assign homework, share curated content such as videos, review student progress, and provide student support anytime, anywhere. Instructors can identify what their students know and don't know and, from there, ALEKS will deliver customized practice until they truly get it. In other words, with ALEKS by your side, learning more is a given.

Create
Your Book, Your Way
McGraw Hill's Content Collections Powered by Create® is a self-service website that enables instructors to create custom course materials—print and eBooks—by drawing upon McGraw Hill's comprehensive, cross-disciplinary content. Choose what you want from our high-quality textbooks, articles, and cases. Combine it with your own content quickly and easily, and tap into other rights-secured, third-party content such as readings, cases, and articles. Content can be arranged in a way that makes the most sense for your course, and you can include the course name and information as well. Choose the best format for your course: color print, black-and-white print, or eBook. The eBook can be included in your Connect course and is available on the free ReadAnywhere app for smartphone or tablet access as well. When you are finished customizing, you will receive a free digital copy to review in just minutes! Visit McGraw Hill Create®—www.mcgrawhillcreate.com—today and begin building!

Corequisite Workbook

This resource provides corequisite remediation of the necessary skills for a liberal arts math course. The included topics are arranged by chapter and section, so that the skills needed for upcoming material are studied in a just-in-time manner. This is available through McGraw Hill Create.

TestGen
Computerized Test Bank Online
Among the supplements is a computerized test bank using the algorithm-based testing software TestGen® to create customized exams quickly. Hundreds of text-specific, open-ended, and multiple-choice questions are included in the question bank.

Lecture and Exercise Videos

These videos introduce concepts, definitions, formulas, and problem-solving procedures to help students comprehend topics.

What's New in This Edition?

Here's What is New in This Edition

- In response to numerous requests, the subsection on fractals in Chapter 9 has been expanded again, this time with a new beginning based on observations in geography that directly led to the development of the topic.

- We've always tried to base as many applications as possible on real, current data. That's a good thing. (Math in Our World, remember?) The downside is having to update or replace all of those problems with each revision. It's a lot of work, but well worth it, especially for Information Age students, many of whom feel like anything that happened before 2015 was recorded in spiral-bound notebooks accompanied by sepia-tinged photographs.

- As always, we reevaluated every single passage in the book, looking for opportunities to improve readability and tone. Of course, we never pass up an opportunity to add a little more humor—learning is supposed to be fun, right?

- Exercises and videos throughout this text are available online in ALEKS Homework, Test, and Quiz assignments. With flexibility to create video-only assignments or include videos mixed in with textbook exercises and questions you author, you can ensure that students' work in ALEKS supports and aligns with your course.

- We believe in unlocking the potential of every learner at every stage of life. To accomplish that, we are dedicated to creating products that reflect, and are accessible to, all the diverse, global customers we serve. Within McGraw Hill, we foster a culture of belonging, and we work with partners who share our commitment to equity, inclusion, and diversity in all forms.

The fifth edition of this text received an extensive and thorough audit to ensure that the examples, applications, and topics referenced throughout support a welcoming and sensitive experience for all learners and uphold McGraw Hill's commitment to equity, inclusion, and diversity.

Acknowledgments

The McGraw Hill math team and authors would like to thank the following instructors past and present who participated in reviews of the content, both print and digital. This feedback was used to guide the direction of the final text and digital product.

Maria T. Alzugaray, PhD, *Suffolk County Community College, New York*

Holly Ashton, *Pikes Peak Community College*

Thomas Beatty, *Florida Gulf Coast University*

Andrew Beiderman, *Community College of Baltimore County*

Lisa Benson, *Olney Central College*

LaVerne Blagmon-Earl, *University of the District of Columbia*

Steve Brick, *University of South Alabama*

Barbara Broughton, *Ivy Tech Community College, Terre Haute*

C. Allen Brown, *Wabash Valley College*

Keisha Brown, *Perimeter College at Georgia State University*

Sam Buckner, *North Greenville University*

Barbara Burke, *Hawai'i Pacific University*

Paul Canepa, *Associate Professor of Mathematics, Delaware County Community College*

Kyle Carter, *University of West Georgia*

Eun J. Cha, *College of Southern Nevada*

Jerry J. Chen, *Suffolk County Community College*

Ivette Chuca, *El Paso Community College*

Ray E. Collings, *Georgia Perimeter College-Clarkston*

Erin Cooke Church, *Georgia State University-Perimeter College*

Jacob Dasinger, Ph.D, *University of South Alabama*

Michelle Davis, *Northeast Iowa Community College*

Shari Davis, *Old Dominion University*

Professor Tuan Dean, *Triton College*

Dr. Scott Demsky, *Broward College*

Carrie Elledge, *San Juan College*

Emily Elrod, *Valencia College*

Shurron Farmer, *University of District of Columbia*

Robert H. Fay, *St. Petersburg College*

Dion Fleitas, *Dallas Baptist University*

Frederick Fritz, *Central Carolina Community College*

Alicia Frost, *Santiago Canyon College*

Peter Georgakis, *Santa Barbara City College*

Dr. Larry Green, *Lake Tahoe Community College*

Jane Golden, *Hillsborough Community College*

Sheryl Griffith, *Iowa Central Community College*

John M. Hansen, *Iowa Central Community College*

Ryan M. T. Harper, *Spartanburg Community College*

Mahshid Hassani, *Hillsborough Community College*

Mary Beth Headlee, *State College of Florida*

Sonja Hensler, *St. Petersburg College Mathematics Department*

Dr. Abushieba Ibrahim, *Broward College*

Kelly Jackson, *Camden County College*

Gretchen Jordan, *Ivy Tech Community College*

Joann Kakascik-Dye, *Hillsborough Community College-Dale Mabry Campus*

Najam Khaja, *Centennial College*

David Keller, *Kirkwood Community College*

Angie Kenrick, *Washtenaw Community College*

Elizabeth Kiedaisch, *College of DuPage*

Harriet Kiser, *Georgia Highlands College*

Shannon Kratzmeyer, *Dallas College Mathematics*

Scott Krise, *Valencia College*

Rachel Lamp, *North Iowa Area Community College*

Dr. Jennifer Lawhon, *Valencia College*

Julia Ledet, *Louisiana State University*

James Lee, *College of Southern Nevada*

Kathy Lewis, *California State University-Fullerton*

Qiana T. Lewis, *Richard J. Daley College*

Lorraine Lopez, *San Antonio College*

Jackie MacLaughlin, *Central Piedmont Community College*

Antonio Magliaro, *Quinnipiac University*

Joanne E. Manville, *Bunker Hill Community College*

Rich Marchand, *Slippery Rock University*

Jeffery Marsh, *Northeast Wisconsin Technical College*

Monica Meissen, *University of Dubuque*

Dennis Monbrod, *South Suburban College*

Carla A. Monticelli, Ed.D., *Camden County College*

Kathy Mowers, *Owensboro Community and Technical College*

Rinav Mehta, *Central Piedmont Community College*

Tamara Miller, *Ivy Tech Community College of Indiana, Columbus Campus*

Shauna Mullins, *Murray State University*

Martha Nega, *Georgia Perimeter College*

Cornelius Nelan, *Quinnipiac University*

Shai Neumann, *Eastern Florida State College*

Stanley Perrine, *Georgia Gwinnett College*

Betty Peterson, *Mercer County Community College*

Alice Pollock, *Lone Star College-Montgomery*

Chrystal Portier, *Nicholls State University*

Glenn Preston, *George Mason University*

Elena Rakova, *Middlesex County College*

Dr. Traci M. Reed, *St. Johns River State College*

Vic Roeske, *Ivy Tech Community College-Indianapolis*

Lisa Rombes, *Washtenaw Community College*

Jason Rosenberry, *Harrisburg Area Community College*

Robin Rufatto, *Ball State University*

Tracy Saltwick, *Bergen Community College*

Jason Samuels, *City University of New York*

Jennifer Sanchez, M.S., *Bunker Hill Community College*

Ioana Sancira, *Olive Harvey College*

Cyrus Screwvala, *Ivy Tech Community College-Columbus*

Edith Silver, *Mercer County Community College*

Zeph Smith, *Salt Lake Community College*

Kristen L Soots, *Ivy Tech Community College, Richmond*

Susanna Sotomayor, M.Ed., *Montgomery County Community College*

Mike Spencer, *Ivy Tech Community College-Columbus*

Leslie Sterrett, *Indian River State College*

Rebecca Steward *Texas A&M University-Commerce*

Jim Stewart, *Jefferson Community and Technical College*

Edward Stumpf, *Central Carolina Community College*

Linda Tansil, *Southeast Missouri State University*

Cindy Vanderlaan, *Purdue University*

Sasha Verkhovtseva, *Anoka-Ramsey Community College*

Camilla Walker, *Indian Hills Community College*

Sister Marcella Louise Wallowicz, *Holy Family University*

Carol Weideman, *St. Petersburg College-Gibbs Campus*

Amelia Jo Weston, *Ivy Tech Community College*

Greg Wisloski, *Indiana University of Pennsylvania*

Fred Worth, *Henderson State University*

We would like to send out thanks to the many kind folks who aided in the continuing evolution of *Math in Our World*. The lifeblood of a successful revision is feedback from users and reviewers, and over the last several years, we've received valuable suggestions both electronically and in person from far too many instructors to count.

At McGraw Hill, thanks to my team: Portfolio Manager Marisa Dobbeleare; Product Developer Megan Platt; Marketing Manager Claire McLemore; and Project Manager Vicki Krug. Finally, thanks to the wonderful national sales and marketing force. It's a pleasure to work with you, my friends.

Index of Applications

Sports, Leisure, and Hobbies

Statistics and Demographics

Design Element Number Image: ©pixeldreams.eu/Shutterstock RF.

Problem Solving

Photo: CBS Photo Archive/Contributor

Outline

Math in | Criminal Justice

In traditional cops-and-robbers movies, crime fighters use guns, fists, and high-speed chases to catch criminals. But in real life, most often it's brain power that brings the bad guys to justice. That's why the TV show *CSI: Las Vegas* marked a revolution of sorts when it debuted in October 2000: for the first time, a series featured *scientists* fighting crime, not tough guys. Solving a case is intimately tied to the process of problem solving: investigators gather and organize as much relevant information as they can, then use logic and intuition to formulate a plan. Hopefully, this will lead them to a suspect.

This same strategy is the essence of problem solving in many walks of life other than criminal justice. Students in math classes often ask, "When am I going to use what I learn?" The best answer to that question is, "Every day, if you're doing it right." Many people think that math classes are just about facts and formulas, and that belief is a big part of why so many people struggle with math. If you don't understand the value of learning something it's a lot harder to learn it. At their very core, math classes are about exercising your mind, training your brain to think logically, and learning effective strategies for solving problems. And not just math problems. Every day of our lives, we face a wide variety of problems: they pop up in our jobs, in school, and in our personal lives. Which smartphone should you buy? What should you do when your car starts making an awful noise? What would be a good topic for a research paper? What's the best approach to telling your roommate that you spilled a whole bottle of vitamin water on their laptop? How can you get all your work done in time to go to that party Friday night?

Chapter 1 of this book is dedicated to the most important topic we'll cover: an introduction to some of the classic techniques of problem solving. These techniques will prove to be useful tools that you can apply in the rest of your education. But more importantly, they can be applied just as well to situations outside the classroom.

And this brings us back to our friends from *CSI*. The logic and reasoning that they use to identify suspects and prove their guilt are largely based on problem-solving skills we'll study in this chapter. By the time you've finished the chapter, you should be able to evaluate the situations below, all based on episodes of *CSI: Las Vegas*. In each case, you should identify the type of reasoning, inductive or deductive, that was used and decide whether the conclusion would stand up as proof in a court of law.

1. After a violent crime, the investigators identify a recently paroled suspect living in the area who had previously committed three very similar crimes.
2. A homeless man is found dead from exposure after being roughed up. His wrists look like he had been handcuffed, and fingerprints on his ID lead them to a local police officer.
3. A murder victim grabbed a pager from the killer while being attacked and threw it under the couch. With the suspect identified, the investigators found that his DNA matched DNA found under the victim's fingernails.
4. A series of five bodies are found posed like mannequins in public places. The lead suspect is an artist that is found to have sketches matching the poses of all five victims.

For answers, see Math in Criminal Justice Revisited on page 38

Section 1-1   The Nature of Mathematical Reasoning

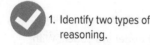

LEARNING OBJECTIVES

1. Identify two types of reasoning.
2. Use inductive reasoning to make conjectures.
3. Find a counterexample to disprove a conjecture.
4. Explain the difference between inductive and deductive reasoning.
5. Use deductive reasoning to prove a conjecture.

A big part of being an adult is making decisions on your own—every day is full of them, from the simple, like what to eat for breakfast, to the critical, like a choice of major, career, job, or mate. How far do you think you'll get in life if you flip a coin to make every decision? I'm going to say, "Not very." Instead, it's important to be able to analyze a situation based on logical thinking. What are the possible outcomes of making that decision? How likely is it that each choice will have positive or negative consequences? We call the process of logical thinking **reasoning**. It doesn't take a lot of imagination to understand how important reasoning is in everyone's life.

foodandmore/123RF

You may not realize it, but every day in your life, you use two types of reasoning to make decisions and solve problems: *inductive reasoning* or *induction*, and *deductive reasoning* or *deduction*.

> **Inductive reasoning** is the process of reasoning that arrives at a general conclusion based on the observation of specific examples.

For example, suppose that your instructor gives a surprise quiz every Friday for the first four weeks of your math class. At this point, you might make a **conjecture**, or educated guess, that you'll have a surprise quiz the next Friday as well. As a result, you'd probably study before that class.

This is an example of inductive reasoning. By observing certain events for four *specific* Fridays, you arrive at a general conclusion. Inductive reasoning is useful in everyday life, and it is also useful as a problem-solving tool in math, as shown in Example 1.

1. Identify two types of reasoning.

Example 1   Using Inductive Reasoning to Find a Pattern

A game show host provides the following string of numbers:

$$1, 2, 4, 5, 7, 8, 10, \_\_\_, \_\_\_, \_\_\_$$

The first contestant who can continue the pattern and fill in the three blanks correctly will win $4,000 and an 18-month supply of Rice-A-Roni (the San Francisco treat). Use inductive reasoning to find a correct answer. How confident are you in your answer? 100%? Or less than that?

SOLUTION

To find patterns in strings of numbers, it's often helpful to think about operations that can turn a number into the next one. In this case, we can use addition to find a regular pattern:

$$1 \underset{+1}{\frown} 2 \underset{+2}{\frown} 4 \underset{+1}{\frown} 5 \underset{+2}{\frown} 7 \underset{+1}{\frown} 8 \underset{+2}{\frown} 10 \underset{+1}{\frown} \_\_ \underset{+2}{\frown} \_\_ \underset{+1}{\frown} \_\_$$

The pattern seems to be to add 1, then add 2, then add 1, then add 2, etc. So a reasonable conjecture for the next three numbers is 11, 13, and 14.

Now let's talk about confidence. I like this answer a lot, but I'm not 100% confident. I made an educated guess based on just seven numbers. There's no guarantee that this is the ONLY correct answer. And, in fact, there are many other potential answers. If you can find one, use it to impress your instructor.

Try This One   1

Use inductive reasoning to find a pattern and make a reasonable conjecture for the next three numbers by using that pattern.

$$1, 4, 2, 5, 3, 6, 4, 7, 5, \underline{\quad}, \underline{\quad}, \underline{\quad}$$

Example 2   Using Inductive Reasoning to Find a Pattern

Make a reasonable conjecture for the next figure in the sequence.

SOLUTION

In the first four figures, the flat part goes from facing up to right, down, then left. There's also a solid circle • in each figure. The sequence then repeats with an open circle ○ in each figure, so in the next one, the flat part should face left and have an open circle:

Photodisc/Getty Images

Recognizing, describing, and creating patterns are important in many fields. For example, many types of patterns are used in music.

Try This One   2

Make a reasonable conjecture for the next figure in the sequence.

Example 3   Using Inductive Reasoning to Make a Conjecture

a. When two odd numbers are added, will the result always be an even number? Use inductive reasoning to make a conjecture.
b. How many pairs of numbers would you need to try in order to be CERTAIN that your conjecture is true?

SOLUTION

a. First, let's try several specific examples of adding two odd numbers:

$$3 + 7 = 10 \qquad 25 + 5 = 30$$
$$5 + 9 = 14 \qquad 1 + 27 = 28$$
$$19 + 9 = 28 \qquad 21 + 33 = 54$$

Since all the answers are even, it seems reasonable to conclude that the sum of two odd numbers will be an even number.

b. Did we really slip in a trick question this early in the book? You bet. This is a very important point about inductive reasoning: you can try specific examples all day and always get an even sum, but that can never guarantee that it will ALWAYS happen. For that, we're going to need deductive reasoning, which we'll study in just a bit.

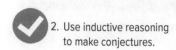

2. Use inductive reasoning to make conjectures.

Try This One   3

If two odd numbers are multiplied, is the result always odd, always even, or sometimes odd and sometimes even? Use inductive reasoning to answer.

One number is *divisible* by another number if the remainder is zero after dividing. For example, 16 is divisible by 8 because $16 \div 8$ has remainder zero, but 17 is not divisible by 8 because $17 \div 8$ has remainder 1.

Example 4   Using Inductive Reasoning to Test a Conjecture

Math Note

Once again, inductive reasoning can't tell us for certain that the conjecture is true. On the other hand, in this case it's at least *possible* to try every four-digit number. But who has time for that? I surely do not.

Use inductive reasoning to decide if the following conjecture is likely to be true: any four-digit number is divisible by 11 if the difference between the sum of the first and third digits and the sum of the second and fourth digits is divisible by 11.

SOLUTION

Let's make up a few examples. For 1,738, the sum of the first and third digits is $1 + 3 = 4$, and the sum of the second and fourth digits is $7 + 8 = 15$. The difference is $15 - 4 = 11$, so if the conjecture is true, 1,738 should be divisible by 11. To check: $1{,}738 \div 11 = 158$ (with no remainder).

For 9,273, $9 + 7 = 16$, $2 + 3 = 5$, and $16 - 5 = 11$. So if the conjecture is true, 9,273 should be divisible by 11. To check: $9{,}273 \div 11 = 843$ (with no remainder).

Let's look at one more example. For 7,161, $7 + 6 = 13$, $1 + 1 = 2$, and $13 - 2 = 11$. Also $7{,}161 \div 11 = 651$ (with no remainder), so the conjecture is true for this example as well. While we can't be positive based on three examples, inductive reasoning indicates that the conjecture is likely to be true.

Try This One   4

Use inductive reasoning to decide if the following conjecture is likely to be true: if the sum of the digits of a number is divisible by 3, then the number itself is divisible by 3.

Inductive reasoning can definitely be a useful tool in decision making, and we use it very often in our lives. But we've talked about a pretty serious drawback: because you can very seldom verify conclusions for every possible case, you can't be positive that the conclusions you're drawing are correct. In the example of the class in which a quiz is given on four consecutive Fridays, even if that continues for 10 more weeks, there's still a chance that there won't be a quiz the following Friday. And if there's even one Friday on which a quiz is not given, then the conjecture that there will be a quiz every Friday proves to be false.

This is a useful observation: while it's not often easy to prove that a conjecture is true, it's much simpler to prove that one is false. All you need is to find one specific example that contradicts the conjecture. This is known as a **counterexample**. In the quiz example, just one Friday without a quiz serves as a counterexample: it proves that your conjecture that there would be a quiz every Friday is false. In Example 5, we'll use this idea to prove that a conjecture is false.

| **Example 5** | **Finding a Counterexample** |

Find a counterexample that proves the conjecture below is false.

Conjecture: A number is divisible by 3 if the last two digits are divisible by 3.

SOLUTION

We'll pick a few numbers at random whose last two digits are divisible by 3, then divide the original number by 3, and see if there's a remainder.

1,527: Last two digits, 27, divisible by 3; $1{,}527 \div 3 = 509$
11,745: Last two digits, 45, divisible by 3; $11{,}745 \div 3 = 3{,}915$

At this point, you might start to suspect that the conjecture is true, but you shouldn't! We've only checked two cases, and there are literally infinitely many possibilities.

1,136: Last two digits, 36, divisible by 3; $1{,}136 \div 3 = 378\frac{2}{3}$

This counterexample shows that the conjecture is false.

Try This One 5

3. Find a counterexample to disprove a conjecture.

Find a counterexample to disprove the conjecture that the name of every month in English contains either the letter y or the letter r.

| **CAUTION** | *Remember:* One counterexample is enough to show that a conjecture is false. But one positive example is *never* enough to show that a conjecture is true. In fact, in most cases even 100 positive examples wouldn't do it. That's why we'll need a different type of reasoning. |

| **Example 6** | **Making and Testing a Conjecture** |

Use inductive reasoning to make a conjecture about the number of sections a circle is divided into when there are a given number of points on a circle, and every pair of points is connected by a chord. (A chord is a line connecting two points on a circle.) Then test the conjecture with one further example.

SOLUTION

This is a really good example of a common technique in math and other sciences: start trying some examples to see if it leads you somewhere useful. We'll draw several circles, connect the points with chords, and then number the sections to make it easy to count them.

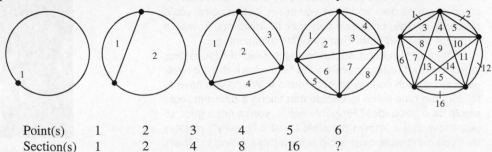

Point(s)	1	2	3	4	5	6
Section(s)	1	2	4	8	16	?

Looking at the pattern in the number of sections, we see that a logical guess for the next number is 32. It looks like the number of sections keeps doubling. In fact, the number of sections appears to be 2 raised to the power of 1 less than the number of points. This will be our conjecture. Let's see how we did by checking with six points:

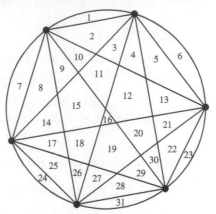

Uh oh . . . there are 31 sections! It looks like our conjecture is not true.

Math Note

Example 6 illustrates what is, in some sense, the very essence of math and science: we have an idea that something might be true, and we use inductive reasoning to test it. But the result of the example shows why inductive reasoning can't be used to *prove* results: what appears to be true after looking at several examples can still turn out to be false.

Try This One   6

Given that there are 31 sections with 6 points, guess how many there will be with 7, and then check your answer.

The other method of reasoning that we will study is called *deductive reasoning*, or *deduction*.

Deductive reasoning is the process of reasoning that arrives at a conclusion based on previously accepted general statements. It's based on overall rules, NOT specific examples.

Here's an example of deductive reasoning. At many colleges, a student has to be registered for at least 12 hours to be considered full-time. So we accept the statement

Sidelight   Of Fuzzy Dogs and Inductive Reasoning

Here's a clever way to understand the difference between inductive and deductive reasoning. Suppose a friend invites you over to their new apartment, and while walking over there, you come across a house with a loose dog in the yard. As you pass, the dog runs over and bites you on the ankle. Ouch! If you're walking back to your friend's place again a week later, you might decide to take the same path—hey, it's the quickest way to get there, and the dog bite was an isolated incident. But as you're passing by the same house, the same dog runs over and bites you again. At that point, you'd probably decide, based on two specific incidents, to walk a different way next time. That's inductive reasoning.

On the other hand, suppose that instead of a loose dog, in front of that house there's a big hole where the sidewalk used to be, with no way to get around it. Would you have to fall into the hole twice to decide that taking a different route would be a good idea? Probably not . . . you're not a goof, so you know that a known principle (I call it "gravity") dictates that you can't walk over a big hole without falling in. That's deductive reasoning.

Ingram Publishing/SuperStock

4. Explain the difference between inductive and deductive reasoning.

"A student is full-time if they are registered for at least 12 hours" as true. If we're then told that a particular student has full-time status, we can conclude that the student is registered for at least 12 hours. The key is that we can be *positive* that this is true. On the other hand, you might have 10 friends who are full-time students, and all of them might be taking 15 or more hours of classes. If you concluded that "full-time" means more than 14 hours, you'd be using inductive reasoning, not deductive. And in this case, you'd probably be wrong.

So that's what separates deductive reasoning from inductive reasoning. In Example 7, we'll look at an incredibly important aspect of these types of reasoning. We'll use inductive reasoning to MAKE a conjecture, and deductive reasoning to PROVE it.

| Example 7 | Using Deductive Reasoning to Prove a Conjecture |

Math Note

Here's another way to think about the difference between inductive and deductive reasoning: inductive reasoning begins with *specific* examples, and leads to a more *general* conclusion. Deductive reasoning starts with a *general* principle, and uses it to draw conclusions about *specific* examples.

Consider the following problem: think of any number. Multiply that number by 2, then add 6, and divide the result by 2. Next subtract the original number. What is the result?

(a) Use inductive reasoning to make a conjecture for the answer.
(b) Use deductive reasoning to prove your conjecture.

SOLUTION

(a) Let's begin by picking a few specific numbers randomly, and performing the described operations to see what the result looks like.

Number:	12	5	43
Multiply by 2:	$2 \times 12 = 24$	$2 \times 5 = 10$	$2 \times 43 = 86$
Add 6:	$24 + 6 = 30$	$10 + 6 = 16$	$86 + 6 = 92$
Divide by 2:	$30 \div 2 = 15$	$16 \div 2 = 8$	$92 \div 2 = 46$
Subtract the original number:	$15 - 12 = 3$	$8 - 5 = 3$	$46 - 43 = 3$
Result:	3	3	3

At this point, you may be tempted to conclude that the result is always 3. But this is just a conjecture: we've tried only three of infinitely many possible numbers! Not to mention the fact that we've chosen only whole numbers. What about decimals? Fractions? Negative numbers? As usual when using inductive reasoning, we can't be completely sure that our conjecture is always true. But at this point, it seems like it would at least be worth the effort to see if we can prove that our conjecture is true.

(b) The problem with the inductive approach is that it requires using specific numbers, and we know that we can't check every possible number. Instead, we'll choose an *arbitrary* number and call it a. Think of that as standing for "any old number." If we can show that the result is 3 in this case, that will tell us that this is the result for *every* number. Remember, we'll be doing the exact same operations, just on an arbitrary number a. This will require some basic algebra skills.

Number:	a
Multiply by 2:	$2a$
Add 6:	$2a + 6$
Divide by 2:	$\dfrac{2a + 6}{2} = \dfrac{2a}{2} + \dfrac{6}{2} = a + 3$
Subtract the original number:	$a + 3 - a = 3$

Now we know for sure that the result will always be 3, and our conjecture is proved.

| Try This One | 7 |

5. Use deductive reasoning to prove a conjecture.

Consider the following problem: think of any number. Multiply that number by 3, then add 30, and divide the result by 3. Next subtract the original number. What is the result?

(a) Use inductive reasoning to make a conjecture for the answer.
(b) Use deductive reasoning to prove your conjecture.

Sidelight   **Making an Arbitrary Point**

In common usage, the word *arbitrary* is often misinterpreted as a synonym for *random*. When reaching into a bag of potato chips, you make a random selection, and some people would also call this an arbitrary selection. But in math, the word *arbitrary* means something very different. When randomly selecting that chip, you have still chosen a specific chip—it is probably not representative of every chip in the bag. Some chips will be bigger, and others smaller. Maybe you got that one gross little burned chip. Yuck.

When we use *arbitrary* in math, we're referring to a non-specific item that is able to represent *all* such items. In the series of calculations we looked at above, we could never

be sure that the result will always be 3 by choosing specific numbers. Why? Because we'd have to try it for *every* number, which is, of course, impossible. You have better things to do than spend the rest of your life testing number after number. The value of performing the calculation on an *arbitrary* number *a* is that this one calculation proves what will happen for *every* number you choose. It is absolutely crucial in the study of mathematics to understand that choosing specific numbers can almost never *prove* a result, because you can't try every number. Instead, we'll rely on using arbitrary numbers and deductive reasoning. Now ponder this deep question: is there such a thing as an arbitrary potato chip?

Let's try another example. Try to focus on the difference between inductive and deductive reasoning, and the fact that inductive reasoning is great for giving you an idea about what the truth might be for a given situation, but deductive reasoning is needed for proof.

Example 8   Using Deductive Reasoning to Prove a Conjecture

Calculator Guide

If you use a calculator to try the repeated operations in Example 8, you'll need to press [=] (scientific/online calculator) or [ENTER] (graphing calculator) after every operation, or you'll get the wrong result. Suppose you simply enter the whole string:

Typical Scientific or Online Calculator

12 [+] 50 [×] 2 [−] 12 [−]

Typical Graphing Calculator

12 [+] 50 [×] 2 [−] 12 [ENTER]

The result is 100, which is incorrect. In Chapter 5, we'll find out why when we study the order of operations.

Use inductive reasoning to arrive at a general conclusion, and then prove your conclusion is true by using deductive reasoning.

Pick a number:
Add 50:
Multiply by 2:
Subtract the original number:
Result:

SOLUTION

Approach: Inductive Reasoning

Try a couple different numbers and make a conjecture.

Original number:	12	50
Add 50:	$12 + 50 = 62$	$50 + 50 = 100$
Multiply by 2:	$62 \times 2 = 124$	$100 \times 2 = 200$
Subtract the original number:	$124 - 12 = 112$	$200 - 50 = 150$
Result:	112	150

A reasonable conjecture is that the final answer is 100 more than the original number.

Approach:	Deductive Reasoning
Pick an arbitrary number:	a
Add 50:	$a + 50$
Multiply by 2:	$2(a + 50) = 2a + 100$
Subtract the original number:	$2a + 100 - a$
Result:	$a + 100$

Our conjecture was right: the final answer is always 100 more than the original number.

Math Note

You probably recognized the deductive approach to proving our conjecture in Example 8 as algebra, where we use a symbol (in this case the letter *a*) to represent an arbitrary number. If you need some help brushing up on algebra, we have you covered: Chapter 6 reviews some key elements of algebra, and there are online resources as well.

Try This One   8

Arrive at a conclusion by using inductive reasoning, and then try to prove your conclusion by using deductive reasoning.

Pick a number:
Add 16:
Multiply by 3:
Add 2:
Subtract twice the original number:
Subtract 50:
Result:

Now that we've seen how inductive and deductive reasoning can be used, let's review by distinguishing between the two types of reasoning in some real situations.

Example 9   Comparing Inductive and Deductive Reasoning

The last six times we played our archrival in football, we won, so I know we're going to win on Saturday. Did I use inductive or deductive reasoning?

SOLUTION

This conclusion is based on six specific occurrences, not a general rule that we know to be true. (No team wins *every* game!). I used inductive reasoning.

Try This One   9

There is no mail delivery on holidays. Tomorrow is Labor Day so I know my student loan check won't show up. Did I use inductive or deductive reasoning?

Example 10   Comparing Inductive and Deductive Reasoning

The syllabus states that any final average between 80 and 90% will result in a B. If I get 78% on my final, my overall average will be 80.1%, so I'll get a B. Did I use inductive or deductive reasoning?

SOLUTION

Although we're talking about a specific person's grade, the conclusion that I'll get a B is based on a general rule: all scores in the 80s earn a B. So this is deductive reasoning.

Try This One   10

Everyone I know in my sorority got at least a 2.5 GPA last semester, so I'm sure I'll get at least a 2.5 this semester. Did I use inductive or deductive reasoning?

Remember that both inductive reasoning and deductive reasoning are useful tools for problem solving. But the biggest difference between them is that conclusions drawn from inductive reasoning, no matter how reasonable, are still at least somewhat uncertain. In Problems 67–72, we'll distinguish between *weak* and *strong* inductive arguments. But conclusions drawn by using deductive reasoning can be considered definitely true, as long as the general rules used to draw the conclusion are known to be true.

And finally, here's one of the biggest takeaways from this section: to disprove a conjecture, you only need to find *one specific example* for which it's not true. But to prove a conjecture, you have to show that it's true in *every* possible case.

Answers to Try This One

1   Pattern: every entry is 1 more than the one that comes two spots before it. The next three numbers are 8, 6, 9.

2

3   Always odd

4   True

5   June has neither a y nor an r.

6   There are 57.

7   (a) 10
   (b) Pick an arbitrary number: a
   Multiply by 3: $3a$
   Add 30: $3a + 30$
   Divide by 3: $(3a + 30)/3 = a + 10$
   Subtract the original number: $a + 10 - a$
   Result: 10

8   (a) Conclusion: The result is the original number.
   (b) Pick an arbitrary number: a
   Add 16: $a + 16$
   Multiply by 3: $3(a + 16) = 3a + 48$
   Add 2: $3a + 48 + 2 = 3a + 50$
   Subtract twice the original number: $3a + 50 - 2a = a + 50$
   Subtract 50: $a + 50 - 50$
   Result: a

9   Deductive

10 Inductive

Exercise Set   1-1

Writing Exercises

1. Explain the difference between inductive and deductive reasoning.
2. What is meant by the term *conjecture*?
3. Give an example of a decision you made based on inductive reasoning that turned out well, and one that turned out poorly.
4. What is a counterexample? What are counterexamples used for?
5. Explain why you can never be sure that a conclusion you arrived at using inductive reasoning is true.
6. Explain the difference between an arbitrary number and a number selected at random.
7. Take another look at the opener for Chapter 1 on page 3. How do the terms inductive and deductive reasoning apply to evidence in court?
8. Describe the difference between being confident that a conjecture is true and being CERTAIN that it's true.

Computational Exercises

For Exercises 9–18, use inductive reasoning to find a pattern, and then make a reasonable conjecture for the next number or item in the sequence.

9. 1 2 4 7 11 16 22 29 ____
10. 6 10 22 58 166 490 ____
11. 10 20 11 18 12 16 13 14 14 12 15 ____
12. 2 3 8 63 3,968 ____
13. 100 99 97 94 90 85 79 ____
14. 9 12 11 14 13 16 15 18 ____
15.
16.

17.

18.

For Exercises 19–22, find a counterexample to show that each statement is false.

19. The sum of any three odd numbers is even.
20. When an even number is added to the product of two odd numbers, the result will be even.
21. When an odd number is squared and divided by 2, the result will be a whole number.

22. When any number is multiplied by 6 and the digits of the answer are added, the sum will be divisible by 6.

For Exercises 23–26, use inductive reasoning to make a conjecture about a rule that relates the number you selected to the final answer. Try to prove your conjecture by using deductive reasoning.

23. Pick a number:
    Double it:
    Subtract 20 from the answer:
    Divide by 2:
    Subtract the original number:
    Result:

24. Pick a number:
    Multiply it by 9:
    Add 21:
    Divide by 3:
    Subtract three times the original number:
    Result:

25. Pick a number:
    Add 6:
    Multiply the answer by 9:
    Divide the answer by 3:
    Subtract 3 times the original number:
    Result:

26. Pick an even number:
    Multiply it by 4:
    Add 8 to the product:
    Divide the answer by 2:
    Subtract 2 times the original number:
    Result:

For Exercises 27–36, use inductive reasoning to find a pattern for the answers. Then use the pattern to guess the result of the final calculation, and perform the operation to see if your answer is correct.

27. $12,345,679 \times 9 = 111,111,111$
    $12,345,679 \times 18 = 222,222,222$
    $12,345,679 \times 27 = 333,333,333$
    $\vdots$
    $12,345,679 \times 72 = ?$

28. $0^2 + 1 = 1$
    $1^2 + 3 = 2^2$
    $2^2 + 5 = 3^2$
    $3^2 + 7 = 4^2$
    $4^2 + 9 = 5^2$
    $5^2 + 11 = ?$

29. $999,999 \times 1 = 0,999,999$
    $999,999 \times 2 = 1,999,998$
    $999,999 \times 3 = 2,999,997$
    $\vdots$
    $999,999 \times 9 = ?$

30. $1 = 1^2$
    $1 + 2 + 1 = 2^2$
    $1 + 2 + 3 + 2 + 1 = 3^2$
    $\vdots$
    $1 + 2 + 3 + 4 + 5 + 6 + 7 + 6 + 5 + 4 + 3 + 2 + 1 = ?$

31. $9 \times 9 = 81$
    $99 \times 99 = 9,801$
    $999 \times 999 = 998,001$
    $9,999 \times 9,999 = 99,980,001$
    $99,999 \times 99,999 = ?$

32. $1 \times 8 + 1 = 9$
    $12 \times 8 + 2 = 98$
    $123 \times 8 + 3 = 987$
    $1,234 \times 8 + 4 = 9,876$
    $12,345 \times 8 + 5 = ?$

33. $1 \cdot 1 = 1$
    $11 \cdot 11 = 121$
    $111 \cdot 111 = 12,321$
    $1,111 \cdot 1,111 = 1,234,321$
    $11,111 \cdot 11,111 = ?$

34. $9 \cdot 91 = 819$
    $8 \cdot 91 = 728$
    $7 \cdot 91 = 637$
    $6 \cdot 91 = 546$
    $5 \cdot 91 = ?$

35. Explain what happens when the number 142,857 is multiplied by the numbers 2 through 8.

36. A Greek mathematician named Pythagoras is said to have discovered the following number pattern. Find the next three sums by using inductive reasoning. Don't just add!
    $1 = 1$
    $1 + 3 = 4$
    $1 + 3 + 5 = 9$
    $1 + 3 + 5 + 7 = 16$
    $1 + 3 + 5 + 7 + 9 = ?$
    $1 + 3 + 5 + 7 + 9 + 11 = ?$
    $1 + 3 + 5 + 7 + 9 + 11 + 13 = ?$

37. Use inductive reasoning to make a conjecture about the next three sums, and then perform the calculations to verify that your conjecture is true.

$$1 + \frac{1}{2} = \frac{3}{2}$$

$$1 + \frac{1}{2} + \frac{1}{2 \cdot 3} = \frac{5}{3}$$

$$1 + \frac{1}{2} + \frac{1}{2 \cdot 3} + \frac{1}{3 \cdot 4} = \frac{7}{4}$$

$$1 + \frac{1}{2} + \frac{1}{2 \cdot 3} + \frac{1}{3 \cdot 4} + \frac{1}{4 \cdot 5} = ?$$

$$1 + \frac{1}{2} + \frac{1}{2 \cdot 3} + \frac{1}{3 \cdot 4} + \frac{1}{4 \cdot 5} + \frac{1}{5 \cdot 6} = ?$$

$$1 + \frac{1}{2} + \frac{1}{2 \cdot 3} + \frac{1}{3 \cdot 4} + \frac{1}{4 \cdot 5} + \frac{1}{5 \cdot 6} + \frac{1}{6 \cdot 7} = ?$$

38. Use inductive reasoning to find the unknown sum, then perform the calculation to verify your answer.

$$2 = 1(2)$$
$$2 + 4 = 2(3)$$
$$2 + 4 + 6 = 3(4)$$
$$2 + 4 + 6 + 8 = 4(5)$$
$$2 + 4 + 6 + 8 + 10 + 12 + 14 = ?$$

In Exercises 39–42, use inductive reasoning to find a pattern, then make a reasonable conjecture for the next three items in the pattern. You may have to think outside the box on some of them. That's a good thing.

39. d b e c f d ___ ___ ___
40. a b c e d f i g h o ___ ___ ___
41. J F M A ___ ___ ___
42. D N O S A ___ ___ ___

Applications in Our World

In Exercises 43–58, determine whether the type of reasoning used is inductive or deductive reasoning.

43. The last four congressional representatives from this district were all Republicans. I don't know why the Democratic candidate is even bothering to run this year.

44. I know I'll have to work a double shift today because I have a migraine and every time I have a migraine I get stuck pulling a double.

45. If class is canceled, I go to the beach with my friends. I didn't go to the beach with my friends yesterday; so class wasn't canceled.

46. On Christmas Day, movie theaters and Chinese restaurants are always open, so this Christmas Day we can go to a movie and get some Chinese takeout.

47. For the first three games this year, the parking lot was packed with tailgaters, so we'll have to leave extra early to find a spot this week.

48. Every time Beth sold back her textbooks, she got about 10% of what she paid for them; so this semester she decided to not bother selling her books back.

49. Experts say that opening email attachments that come from unknown senders is the easiest way to get a virus on your computer. My mom constantly opens attachments from random people, so she will probably end up with a computer virus.

50. Whenever Omar was set up by friends on a date, it turned out to be a disaster. Next time a friend offers a fix-up, Omar plans to decline, and maybe move to a new state.

51. Dr. Spalsbury's policy is that any student who gets a text notification on their phone during class will be asked to leave. So when Mateo forgot to turn off the ringer and got a text . . . no more Mateo in class that day.

52. While experimenting on learning in mice, a biology student was able to successfully train six different mice to finish a maze, so she was really surprised when the next one was unable to learn the maze.

53. Since Phil ate a diet consisting largely of pizza and other foods high in saturated fat, he was not surprised when his doctor said his cholesterol levels were too high.

54. In the past, even when Chris followed a recipe, the meal was either burned or underdone. Now party guests know to eat before they attend Chris's dinners so they won't starve all evening.

55. Working as a nurse in a hospital requires at least a two-year degree in this state, so when I was in the emergency room last week I asked the nurse where he went to college.

56. Marathon runners should eat extra carbs before a big race, and since I didn't eat enough carbs before a recent race, I felt sluggish the entire time. (*Note:* This question is hypothetical. I don't actually run marathons. In fact, I get tired if I drive 26 miles.)

57. Organizing chapter contents in your own words before the test will decrease the amount of study you have to do before a test. Aziz was pleasantly surprised at how fast test prep went when trying this method.

58. The last several network dramas I've followed have been canceled just when I started getting into them. So I'm not going to bother watching the new one they're advertising even though it looks good, because I don't want to be disappointed when it gets canceled.

Critical Thinking

59. Do a Google search for the string "studies texting while driving." Suppose that you've driven while texting 10 times in the past without any incident. How likely would you be to text while driving if you use (a) inductive reasoning, and (b) deductive reasoning based on your Google search? Describe your reasoning in each case.

60. Just about everyone had a conversation like this with their parents or caregivers at some point in their childhood: "But all my friends are doing it!" "If your friends jumped off a bridge, would you jump, too?" Describe how arguments like this apply to inductive and deductive reasoning. Specifically, what type of reasoning is each person using, and who in your opinion makes a stronger argument?

61. (a) Find a likely candidate for the next two numbers in the following sequence: 2, 4, 8, . . .

    (b) Was your answer 16 and 32? How did you get that answer? Can you find a formula with variable n that provides the numbers in your sequence?

    (c) My answer is 14 and 22. How did I get that answer? Can you find a formula with variable n that provides the numbers in my sequence? (*Note:* The last question is NOT easy!)

(d) Fill in the following table by substituting the given values of n into the formula. Can you answer the last question in part (c) now? Based on all parts of this problem, what can you conclude about finding a pattern when you have just a bit of information to use?

n	1	2	3	4	5
$n^2 - n + 2$					

62. (a) Find a likely candidate for the next two numbers in the following sequence: 3, 9, 27, . . .

(b) Was your answer 81 and 243? How did you get that answer? Can you find a formula with variable n that provides the numbers in your sequence?

(c) My answer is 57 and 99. How did I get that answer? (*Hint:* Find differences between the first two pairs of terms and look for a pattern.) Can you find a formula with variable n that provides the numbers in my sequence? (*Note:* This is NOT easy!)

(d) Fill in the following table by substituting the given values of n into the formula. Can you answer the last question in part (c) now? Based on all parts of this problem, what can you conclude about finding a pattern when you have just a bit of information to use?

n	1	2	3	4	5
$6n^2 - 12n + 9$					

63. (a) In several of the problems in this section, you looked at a string of numbers then decided what the next number would be. This time, write a string of five numbers with a pattern so that the next number in the string would be 10.

(b) Next, write a string of five numbers with a pattern so that the next two numbers in the string would be 10 and 13.

(c) Finally, write a string of five numbers with a pattern so that the next three numbers in the string would be 10, 13, and 17.

64. Refer to Problem 63.

(a) Write a string of three numbers so that the next number in the string would be $\frac{4}{81}$.

(b) Next, write a string of three numbers so that the next two numbers in the string would be $\frac{4}{81}$ and $-\frac{5}{243}$.

(c) Find a formula with variable n that provides the numbers in your sequence.

Problems 65 and 66 use the formula average speed = distance/time.

65. Suppose that you drive a certain distance at 20 miles per hour, then turn around and drive back the same distance at 60 miles per hour.

(a) Choose at least four different distances and find the average speed for the whole trip. Then use

inductive reasoning to make a conjecture as to what the average speed is in general.

(b) Use algebra to prove your conjecture from part (a).

66. Suppose that you drive a certain distance at 40 miles per hour, then turn around and drive twice as far at 20 miles per hour.

(a) Choose at least four different distances and find the average speed for the whole trip. Then use inductive reasoning to make a conjecture as to what the average speed is in general.

(b) Use algebra to prove your conjecture from part (a).

*All conclusions drawn from inductive reasoning aren't created equal. We can distinguish between a **weak inductive argument**, where a conclusion is drawn from just a few specific instances, and a **strong inductive argument**, where a conclusion is drawn from a large amount of observations. In Problems 67–72, classify each argument as weak or strong induction, and discuss your reasoning.*

67. The Cubs lost the first six games of the season. They have NO chance tonight.

68. There are 52 people on my flight to the Bahamas, and 40 of them brought carry-on bags to avoid paying baggage fees. So I'm thinking that about 80% of air travelers bring carry-ons to avoid fees.

69. NPR and Marist College conducted a poll of 1,115 Americans in July 2021 and reported that two-thirds of them feel that American democracy is under threat. I conclude that a majority of Americans feel like our democracy is threatened.

70. I played the first nine holes of a local golf course last night, and the grass was brown on all nine of them. I imagine all nine holes on the back nine are burned out, too.

71. There were 109,000 people at the Ohio State-Michigan State game last year. Almost all of them cheered when Ohio State scored, so the game must have been played in Ohio.

72. All three high school math teachers I had were kind of nerdy. I guess that all math teachers are nerds.

73. The numbers 1, 3, 6, 10, 15, . . . are called *triangular numbers* since they can be displayed as shown.

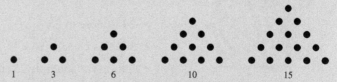

The numbers 1, 4, 9, 16, 25, . . . are called *square numbers* since they can be displayed as shown.

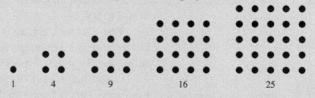

The numbers 1, 5, 12, 22, 35, . . . are called *pentagonal numbers* since they can be displayed as shown.

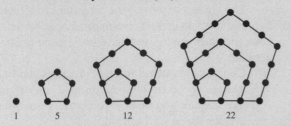

1   5   12   22

(a) Using inductive reasoning, find the next three triangular numbers.

(b) Using inductive reasoning, find the next three square numbers.

(c) Using inductive reasoning, find the next three pentagonal numbers.

(d) Using inductive reasoning, find the first four hexagonal numbers.

74. Refer to Exercise 73. The formula for finding triangular numbers is $\dfrac{n[1n - (-1)]}{2}$ or $\dfrac{n(n + 1)}{2}$. The formula for finding square numbers is $\dfrac{n(2n - 0)}{2}$ or n^2. The formula for pentagonal numbers is $\dfrac{n(3n - 1)}{2}$. Find the formula for hexagonal numbers, using inductive reasoning.

Section 1-2   Estimation and Interpreting Graphs

  LEARNING OBJECTIVES

1. Identify some uses for estimation.

2. Round numbers to a given level of accuracy.

3. Estimate the answers to problems in our world.

4. Use estimation to obtain information from graphs.

imagehitevo/123RF

Everyone likes buying items on sale, so we should all be familiar with the idea of finding a rough approximation for a sale price. If you're looking at a pair of shoes that normally sells for $70 and the store has a 40% off sale, you might figure that the shoes are a little more than half price, which would be $35, so they're probably around $40. We will call the process of finding an approximate answer to a math problem **estimation**. Chances are that you use estimation a lot more than you realize, unless you use your cell phone calculator a whole lot more than pretty much everyone.

Estimation comes in handy in a wide variety of settings. When the auto repair shop technicians look over your car to see what's wrong, they can't know for sure what the exact cost will be until they've made the repairs, so they will give you an estimate. When you go to the grocery store and have only $20 to spend, you'll probably keep a rough estimate of the total as you add items to the cart. (Imagine buying a week's worth of groceries and keeping track of every price to the penny on your smartphone. Who has time for that?) If you plan on buying carpet for a room, you'd most likely measure the square footage and then estimate the total cost as you looked at different styles of carpet. You could find the exact cost if you really needed to, but often an estimate is good enough for you to make a sound buying decision.

Estimation is also a really useful tool in checking answers to math problems, particularly word problems. Let's say you're planning an outing for a student group you belong to, and lunch is included at $3.95 per person. If 24 people signed up and you were billed $94.80, you could use estimation to quickly figure that this is a reasonable bill.

Here's a look at what a sound thought process would look like in that case.

- 24 is close to 25.

- $3.95 is close to $4.

- It's easy to multiply 25 and $4 to get $100.

- So the bill should be somewhere close to $100.

See? Anyone can do this kind of basic reasoning, but many people are, for some reason, afraid to try.

You also use estimation when rounding numbers for simplicity. If someone asks your height and age, you might say 5′11″ and 36, even if you're actually 5′10½″ (everyone fudges a little) and 36 years, 4 months, and 6 days old.

Since the process of estimating uses rounding, we'll start with a brief review of rounding numbers. This is based on the concept of place value. The **place value** of a digit in a number tells the value of the digit in terms of ones, tens, hundreds, etc. For example, in the number

  1. Identify some uses for estimation.

325, the 3 means 3 hundreds or 300 since its place value is hundreds. The 2 means 2 tens, or 20, and the 5 means 5 ones. A place value chart is shown here, along with instructions for rounding numbers.

Math Note

We often use the word *nearest* to describe the place value to round to. Instead of saying, "Round to the hundreds place," we might say, "Round to the nearest hundred."

Steps for Rounding Numbers

1. Locate the place-value digit of the number that is being rounded. Here is the place-value chart for whole numbers and decimals:

8,	9	8	5,	7	3	0,	2	6	1	.	2	3	5	6	7	8
billions	hundred-millions	ten-millions	millions	hundred-thousands	ten-thousands	thousands	hundreds	tens	ones		tenths	hundredths	thousandths	ten-thousandths	hundred-thousandths	millionths

2a. If the digit to the right of the place-value digit is 0 through 4, don't change the place-value digit.

2b. If the digit to the right of the place-value digit is 5 through 9, add 1 to the place-value digit.

Note: When you round whole numbers, replace all digits to the right of the digit being rounded with zeros. When you round decimal numbers, drop all digits to the right of the digit that is being rounded.

Example 1

Rounding Numbers

Round each value as requested.

(a) $147.38 to the nearest 10 dollars
(b) According to Wikipedia, an average person's nails grow at the rate of 0.1181 inches per month. (That makes me well above average I guess. Yay me.) Round that length to the nearest hundredth of an inch.
(c) According to worldometers.info, as of August 4, 2021, 4,263,688 people have died world-wide from COVID-19. Round this terrifying number to the nearest ten thousand.
(d) Is $1 million a lot of money? It depends who you ask. That amount is 0.13374% of the amount the federal government spent in 2020— in ONE HOUR. Seriously. Round that percentage to the nearest thousandth of a percent.

SOLUTION

(a) Rounding to the nearest ten dollars is the tens place, and the digit there is 4. The digit to the right of that is 7, which is more than 4. So we round the 4 up to 5, and replace the 7 with a zero, to get $150. You could include zeros in the two places after the decimal if it means a lot to you, but you can just leave them off as well.
(b) The digit in the hundredths place is 1, and it's followed by an 8. Again, this is more than 4, so we round up the 1 to a 2 to get 0.12 inch.
(c) The ten-thousands place is the third digit here, which is 6. The following digit (3) is less than 5, so we leave the 6 as is and replace the digits after it with zeros. The rounded result is 4,260,000 deaths.
(d) Thousandths is the third digit after the decimal point, which in this case is 3. The following digit (7) is more than 4, so we round the 3 in the thousandths place up to 4, giving us 0.134%.

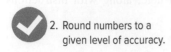

2. Round numbers to a given level of accuracy.

Try This One   1

Round each value as requested.

(a) The average distance from Earth to the moon is 238,856 miles. Round to the nearest hundred miles.
(b) The moon is moving away from Earth at the rate of 1.488189 inches per year. Round to the nearest ten-thousandth of an inch.
(c) $8.93 to the nearest 10 cents.

Estimation

When we use estimation to simplify numerical calculations, we use two steps:

1. Round the numbers being used to numbers that make the calculation simple.
2. Perform the operation or operations involved.

Example 2   **Estimating Total Cost When Shopping**

The owner of an apartment complex needs to buy six refrigerators for a new building. The chosen model costs $579.99 per refrigerator. Estimate the total cost of all six and decide if your answer is an **overestimate** (more than the actual value) or an **underestimate** (less than the actual value).

SOLUTION

Step 1 Round the cost of the refrigerators. In this case, rounding up to $600 will make the calculation easy.

Step 2 Perform the calculation: $600 × 6 = $3,600. Our estimated cost is $3,600.

The actual cost will be a little less than $3,600—since we rounded the price up, our estimate will be high.

Try This One   2

At one ballpark, large frosty beverages cost $7.25 each. Estimate the cost of buying one for each member of a group consisting of four couples. Is the actual cost more or less than your answer?

Students often wonder, "How do I know what digit to round to?" It would be nice if there were an exact answer to that question, but there isn't—it depends on the individual numbers. In Example 2, the cost of the refrigerators could have been rounded from $579.99 to $580. Then the cost estimate would be $580 × 6 = $3,480. This is a much closer estimate because we rounded the cost to the nearest dollar, rather than the nearest $100. But the calculation is harder.

Deciding on how much to round is really a trade-off: ease of calculation versus accuracy. In most cases you'll get a more accurate result if you round less, but the calculation will be a little harder. Since there's no exact rule, it's important to evaluate the situation and use good old-fashioned common sense. And remember, when you're estimating, there is no one correct answer.

Sidelight   Just How Big Are Big Numbers?

Back when we were still living in caves, really large numbers probably weren't of much use. Early humans could use their fingers and toes to count their families and possessions, and I imagine that was good enough. How the world has changed! In the 21st century, we're bombarded with large numbers from every direction, and being able to have some perspective on the size of those numbers is a useful skill. The truth is that most people have absolutely no idea how big a number like a million actually is.

One million is a one followed by six zeros (1,000,000). If you wanted to count to a million and you counted one number each second with no time off to eat or sleep, it would take you just about $11\frac{1}{2}$ days. Wow. A stack of one million pennies would be almost a mile high; a pile of one million dollar bills would weigh almost a ton. As of this writing, the federal minimum wage is $7.25 per hour. If you worked 40 hours a week at a minimum wage job, it would take you over 66 years to make one million dollars—before taxes!

Add three more zeros to the end of a million and you get one billion (1,000,000,000); that makes a billion equal to 1,000 million. In 2020, the federal government spent almost 18 billion dollars—per DAY. So how big is a billion?

Counting to one billion by ones would take you about 32 years (no rest or sleep, of course). A billion pennies would make a stack almost 1,000 miles high; a pile of one billion dollar bills would be about the size of a medium-sized office building and weigh over a thousand tons. And guess what? From some perspectives, a billion isn't even that much.

In 2020, the federal government spent 345 billion dollars just paying *interest* on the national debt. (And guess where that money is coming from?) Here's the actual amount of money spent by our government in 2020: $6,550,000,000,000. I would be willing to bet that 80% of the population can't even read that number out loud, let alone have any perspective on just how big it is. For the record, that's 6 trillion, 550 billion dollars. It would take over 100,000 YEARS to count that high. Everyone knows that a person making a million dollars a year is wealthy, but if you were lucky enough to reach that lofty goal, you'd still have to work for over six and a half million years to make the amount of money the feds spend in one year.

Once you get past a trillion, things get just plain silly. A quadrillion is one followed by 15 zeros. Eighteen zeros gives you a quintillion, and 21 zeros a sextillion. The entire Earth weighs about 6 sextillion, 570 quintillion tons; the weight of all the people on Earth is a mere 525 million tons. The largest number with a name ending in -illion that I know of is the vigintillion, which has 63 zeros.

So is that the biggest number of all? Not even close. A nine-year-old girl came up with the name "googol" in 1938 to describe one followed by 100 zeros. Scientists think this is more than the total number of protons in our universe. But if you want to make a googol look small, go up to a googolplex, which is one followed by a googol of zeros. It's almost impossible to imagine how large a number this is, but writing it out would require a piece of paper far longer than the known universe.

Finally, here's an easy way to show that there IS no biggest number: give me any number, and I can give you a bigger one by adding one to it. Maybe a million isn't so big after all.

Example 3	Estimating the Cost of a Cell Phone

Sam Edwards/Glow Images

You're considering a new cell phone plan where you have to pay $179 up front for the latest phone, but the monthly charge of $39.99 includes unlimited minutes, data, and messaging. Estimate the cost of the phone for 1 year if there are no additional charges. If you were trying to decide if you could afford this phone in your budget, would you want an overestimate or an underestimate?

SOLUTION

Step 1 We can round the cost of the phone to $180 and the monthly charge to $40.

Step 2 The monthly cost estimate for 1 year will be $40 × 12 = $480; add the estimated cost of the phone to get an estimated total cost of $480 + $180 = $660.

When building a budget, you'd want an overestimate to make sure that you don't end up spending more than you were planning on.

Try This One   3

A rental car company charges a rate of $178 per week to rent an economy car. For an up-front fee of $52, you can upgrade to a midsize car. Estimate the cost of renting a midsize car for 3 weeks. When planning a vacation, would you want an overestimate or an underestimate of the cost?

| Example 4 | Estimating Remodeling Costs |

Math Note

While rounding up or down to intentionally get either an over- or underestimate can be useful, for calculations involving addition and multiplication, rounding one quantity up and the other down can reduce the total error.

The Osbueño family plans on remodeling the living room. They will be replacing 21 square yards of carpet at a cost of $23 per square yard (installed), and they also need to have 26 linear feet of crown molding installed. They'd like to keep the total cost around $1,000. Estimate the cost per foot of crown molding that they can afford.

SOLUTION

First, we'll estimate how much is going to be spent on carpet: we can round the 21 square yards down to 20, and the $23 per square yard up to $25, giving us 20 × $25 = $500. So the Osbueños will have about $500 left to spend on crown molding.

We can round the 26 linear feet down to 25, and use division to estimate the price per foot: $500 ÷ 25 = $20, so they should be looking for crown molding that costs no more than $20 per linear foot installed.

| Try This One | 4 |

3. Estimate the answers to problems in our world.

Next up for the Osbueños: a bedroom remodel. They will have 28 linear feet of wall painted at $12 per foot, and need 19 square yards of carpet. If the budget for the bedroom is $900, estimate the price per square yard of carpet they can afford.

In our world, useful or interesting information is often displayed in graphical form. In Examples 5–7, we'll illustrate how estimation applies to interpreting graphical information.

A **bar graph** is used to compare amounts or percentages using either vertical or horizontal bars of various lengths; the lengths correspond to the amounts or percentages, with longer bars representing larger amounts.

| Example 5 | Getting Information from a Bar Graph |

Suicide is a serious problem in our society. If you or someone you know may be in danger of self-harm, please call 1-800-273-TALK. Every life is important.

The graph shows the number of deaths by suicide for every 100,000 people in the United States in 2015, broken down by both age and sex. Use the graph to find about how many deaths by suicide there were per 100,000 males in the 15–24 age range, and also to estimate the largest discrepancy between males and females in any age range. How confident are you in these estimates?

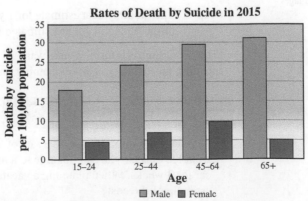

Source: Centers for Disease Control and Prevention

SOLUTION

The blue bars illustrate the rate for males, so we need to estimate the height of the first blue bar (the first pair of bars corresponds to the 15–24 age group). That bar appears to be a little more than halfway between 15 and 20, so I'd estimate that height to be 18, and say that there were about 18 deaths by suicide for every 100,000 males in that age range.

The biggest discrepancy between males and females is in the 65 and over age range. The blue bar looks to be about height 31, while the female is right on 5, so there were about 26 more deaths by suicide for every 100,000 males than females in that age range.

Confidence in an estimate is in the eye of the beholder, but because the grid only has lines marked every 5 units, it's hard to be much more precise than to the nearest whole number. So if we're interested in accuracy greater than that, I wouldn't be terribly confident.

Try This One 5

Which difference in the rate of deaths by suicide is greater for females: between the 15–24 and 25–44 groups, or between the 45–64 and 65+ age groups? By how much?

A **pie chart**, also called a **circle graph**, is constructed by drawing a circle and dividing it into parts called sectors, according to the size of the percentage of each portion in relation to the whole.

Example 6 — Getting Information from a Pie Chart

Math Note

To change a percent to decimal form, move the decimal point two places to the left and drop the percent sign: 18 percent means "18 per hundred," which is 18/100, or 0.18. We'll study percents in detail in Section 7-1.

The pie chart shown represents military spending by country for 2020. The total spent worldwide was \$1,981 billion. Estimate how much was spent by the United States and how much by Russia.

Military Spending by Country

Russia 3.1%
United Kingdom 3.0%
India 3.7%
China 12.7%
Rest of the world 38.2%
United States 39.3%

Source: Stockholm International Peace Research Institute

SOLUTION

The sector labeled "United States" indicates that 39.3% of worldwide military spending was done by the U.S. So we need to find 39.3% of \$1,981 billion. To find a percentage, we multiply the percentage in decimal form by the full amount. In this case, we get $0.393 \times 1,981 = 778.533$. This tells us that the U.S. spent about \$778.5 billion on military spending in 2020. The Russia slice accounts for 3.1%, so we get $0.031 \times 1,981 = 61.411$, so Russia spent about \$61.4 billion.

Try This One 6

Using the pie chart shown in Example 6, find the approximate amount of military spending by China and the United Kingdom.

A **time series graph** or **line graph** shows how the value of some variable quantity changes over a specific time period.

Example 7	**Estimating Information from a Line Graph**

The graphs below each illustrate the number of Americans that were living in poverty according to federal guidelines that determine the poverty level.

(a) Find the approximate number of Americans living in poverty in 2015.
(b) Find the average rate at which the number of folks living in poverty changed between 1995 and 2000.
(c) Estimate the year in which the number of Americans living in poverty first topped 40 million.
(d) Which of the two graphs do you think gives a more accurate view of the poverty rate?

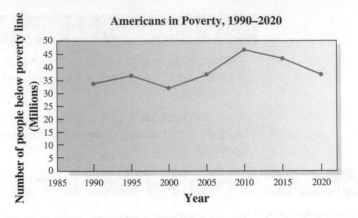

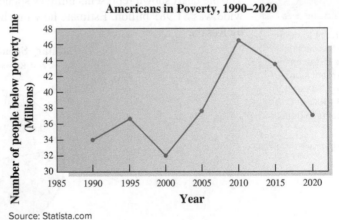

Source: Statista.com

SOLUTION

(a) First, find the year 2015 along the horizontal axis, and move straight upward until hitting the graph. Then move horizontally to the left to find the height on the vertical axis as shown.

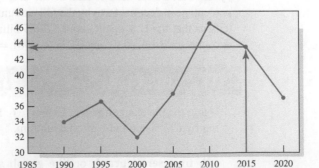

Source: Statista.com

The height is a bit below 44, so we estimate that there were about 43.5 million people living in poverty in 2015.

(b) The point corresponding to 1995 is almost halfway between heights 36 and 38, so we can estimate about 37 million people in poverty in 1995. The height of the point for 2000 is almost right on height 32, so we can estimate 32 million people in 2000. Subtracting, we find that the approximate change in number of people in poverty is 32 million − 37 million, or −5 million (the negative represents a decrease in amount). This occurred over a span of 5 years, so the rate of change is

$$\frac{-5 \text{ million people}}{5 \text{ years}} = -1 \text{ million people per year}$$

In other words, between 1995 and 2000, the number of people living in poverty decreased by an average of about a million people per year.

(c) This question is basically the opposite of part (a): we're given a number of people (a height on the graph), and are asked to find the corresponding year. So we locate height 40 on the vertical axis (which corresponds to 40 million people), and move across to the graph, then move down to find the year on the horizontal axis.

Americans in Poverty, 1990–2020

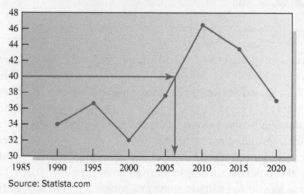

Source: Statista.com

It certainly happened after 2005. Either 2006 or 2007 would be good guesses, but I'd probably go with 2006.

(d) This question brings up a really interesting point about the scale for graphs. Notice that the vertical axis on the first graph starts at zero, resulting in a lot of "dead space" near the bottom of the graph. The second starts at 30. That makes it a little easier to read the graph precisely, which is good. But the first graph provides a more accurate picture of the poverty situation: it shows that the number in poverty fluctuated over 25 years, but not THAT much. The second graph makes the change look much more drastic than it actually was.

Try This One 7

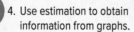

4. Use estimation to obtain information from graphs.

(a) About how many people lived in poverty in 2010?
(b) Find the average rate of change in the number of people living in poverty over the entire span from 1990 to 2020.
(c) Do you think your answer to part (b) provides a complete picture of the change in poverty over that time span? Why or why not?
(d) Estimate the year or years in which there were about 44 million people living in poverty.

Answers to Try This One

1 (a) 238,900 miles
   (b) 1.4882 inches
   (c) $8.90

2 About $56; actual cost is more

3 About $590; you'd want an overestimate

4 About $30 per square yard

5 The difference between 45–64 and 65+ is about 2 or 2.5 deaths by suicide per 100,000 people greater than the difference between 15–24 and 25–44.

6 China: About 251.6 billion. United Kingdom: About 59.4 billion.

7 (a) About 46.5 million
   (b) About 100,000 more people per year
   (c) It provides an incomplete picture, because it totally ignores all of the ups and downs in between those two years.
   (d) Around 2008 and 2014

Exercise Set   1-2

Writing Exercises

1. Think of three situations in our world where you could use estimation.
2. Explain why an exact answer to a math problem isn't always necessary.
3. How can estimation be used as a quick check to see if the answer to a math problem is reasonable?
4. Describe the rules for rounding numbers to a given place.
5. Explain why there is never a single, correct answer to a question that asks you to estimate some quantity.
6. Explain how to estimate the size of a quantity from a bar graph.
7. How is information described in a pie chart? What sort of information works well with pie charts?
8. How can you tell when a quantity is getting larger over time from looking at a time-series graph?
9. Think of a situation where you'd most likely want an overestimate, and one where you'd most likely want an underestimate.
10. When information is presented in the form of a bar graph or time-series graph, you could get more exact values if all of the data were just listed out in table form. Then why not always do that? Why bother with graphs?

Computational Exercises

For Exercises 11–30, round the number to the place value given.

11. 2,861 (hundreds)
12. 732.6498 (thousandths)
13. 3,261,437 (ten-thousands)
14. 9,347 (tens)
15. 62.67 (ones)
16. 45,371,999 (millions)
17. 218,763 (hundred-thousands)
18. 923 (hundreds)
19. 3.671 (hundredths)
20. 56.3 (ones)
21. 327.146 (tenths)
22. 83,261,000 (millions)
23. 5,462,371 (ten-thousands)
24. 7.8662 (thousandths)
25. 272,341 (hundred-thousands)
26. 63.715 (tenths)
27. 264.97348 (ten-thousandths)
28. 1,655,432 (thousands)
29. 482.6002 (hundredths)
30. 426.861356 (hundred-thousandths)

For Exercises 31–34, estimate the result of the computation by rounding the numbers involved, then use a calculator to find the exact value and find the percent error. (Note: Percent error is the amount of error divided by the exact value, written in percentage form.)

31. $-4.21(7.38 + 3.51)$
32. $10.24(-8.93 + 2.77)$
33. $\dfrac{\sqrt{9.36}}{7.423 - 9.1}$
34. $\dfrac{47.256 - 9.90}{\sqrt{24.501}}$

Applications in Our World

For Exercises 35–50, supply the requested estimate, then describe your answer as an overestimate or underestimate.

35. Estimate the total cost of eight high-intensity LED lightbulbs on sale for $16.99 each.
36. Estimate the cost of five months of HD cable at $39.95 per month.
37. Estimate the time it would take you to drive 237 miles at 37 miles per hour.
38. Estimate the distance you can travel in 3 hours 25 minutes if you drive on average 42 miles per hour.
39. Estimate the sale price of a futon you saw on eBay that costs $178.99 and is now on sale for 60% off.
40. Estimate the sale price of a sweater that costs $42.99, on sale for 15% off.
41. Estimate the total cost of the following meal at McDonald's:

Quarter pounder with cheese	$3.89
Large fries	$1.89
Small shamrock shake	$1.29

42. Estimate the total cost of the following items for a dorm room:

Loft bed	$159.95
Beanbag chair	$49.95
Storage cubes	$29.95
Lava lamp	$19.95

43. A group of five architecture students enters a design for an eco-friendly building in a contest, and wins third place, with a prize of $950. Estimate how much each student will get.
44. A biology lab houses 47 rats for experiments, and they go through about 105 pounds of food each week. Estimate how much food the average rat eats per week.
45. If Erin earns $48,300.00 per year, estimate how much she earns per hour. Assume that she works 40 hours per week and 50 weeks per year.
46. If Jamaal earns $8.75 per hour, estimate how much he would earn per year. Assume that he works 40 hours per week and 50 weeks per year.
47. Estimate the cost of putting up a decorative border in your family room if the room is 24 feet long and 18 feet wide and the border costs $5.95 every 10 feet.
48. Estimate the cost of painting a concrete patio if it's a 12 foot by 16 foot rectangle, and a quart of paint that covers 53 square feet costs $11.99.
49. The Green Party at a large university plans to line both sides of a 30-foot-long hallway with posters endorsing a candidate for state senate. Each of the posters costs them $4.95, they're 2 feet wide, and there will be 5 feet between posters. Estimate how much this will cost.
50. Estimate your cost to share an apartment for 1 year if your share of the rent is $365.00 per month and utilities are $62.00 per tenant per month.

Use the information shown in the bar graph for Exercises 51–54. The graph shows the number of people (in millions) that are native speakers of the top five most common languages in the world.

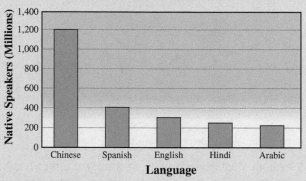

Most Common Native Languages Spoken

Source: Ethnologue

51. Estimate the number of native English speakers in the world.
52. Estimate the number of native Chinese speakers in the world.
53. Estimate the difference in the number of native speakers between the first and fifth most common languages.
54. Which has more native speakers: Chinese, or the next four languages combined?

Use the information shown in the graph for Exercises 55–58. The graph represents a survey of 1,385 office workers and shows the percent of people who indicated what time of day is most productive for them.

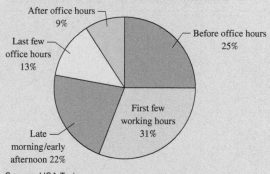

Most Productive Time of Day

Source: USA Today

55. Estimate the number of people who feel they are most productive outside normal office hours.
56. Estimate the number of people who feel they are most productive before late morning.
57. How many more people feel they're most productive in the first few working hours compared to those that feel they're most productive in the last few office hours?
58. How many times more people are most productive before office hours compared to after?

The next graph covers Exercises 59–62. It shows the percentage of Americans surveyed in June 2021 who gave a series of responses to their attitude toward getting the COVID-19 vaccine.

Covid Vaccination Plans as of June 2021

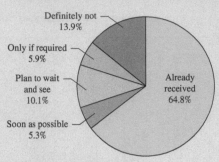

Source: Kaiser Family Foundation

59. About what percent of those surveyed either had been vaccinated or were at least willing to consider getting vaccinated? Round to the nearest full percent.
60. About what percent of those surveyed had not been vaccinated at that point? Round to the nearest full percent.
61. If these results are typical for all Americans, if there were 410 people who said they would definitely not get vaccinated in a similar survey, about how many people were surveyed?
62. In this particular survey, if 114 people said that they planned to wait and see, how many people were surveyed?

Use the line graph shown for Exercises 63–68. The graph shows annual cigarette consumption (in billions) for the United States for the years 1900 to 2016.

Cigarette Consumption in the United States

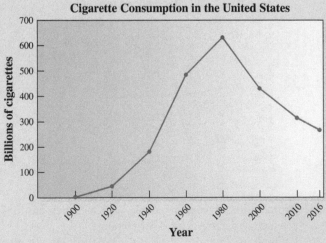

Source: www.infoplease.com

63. Estimate the number of cigarettes smoked in 1950.
64. Estimate the number of cigarettes smoked in 1985.
65. Estimate the year or years in which 200 billion cigarettes were smoked.
66. Estimate the year or years in which 400 billion cigarettes were smoked.

67. Find the average rate of change in cigarette consumption for the years shown when consumption was increasing.
68. Find the average rate of change in cigarette consumption for the years shown when consumption was decreasing.

The next two graphs describe Internet access by continent as of 2018. The first shows the NUMBER of people that have Internet access in the place where they live, in millions. The second shows Internet penetration, which is the PERCENTAGE of people in each region that have Internet access where they live. Use the graphs to answer Exercises 69–74.

Internet Users in Millions

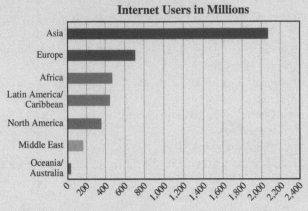

Source: Internet World Stats- www.internetworldstats.com/stats.html

Internet Penetration

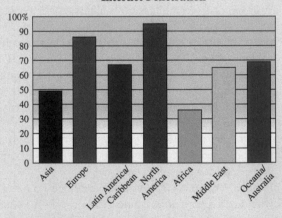

Source: www.internetworldstats.com

69. Which region has the most people with Internet access at home? The least?
70. If you picked a person at random from each region, on which region would it be most likely that the person has Internet access at home? Least likely?
71. Estimate the number of Internet users in Europe and the Internet penetration in Asia.
72. Estimate the Internet penetration in Africa and the number of Internet users in the Middle East.
73. What do you think accounts for the fact that North America has the longest bar on the second graph, but is middle of the pack on the first one?
74. Does Asia have more Internet users than the rest of the world combined? Justify your answer.

The next graph displays growth in Internet penetration from 1998 to 2018 in the developed world compared to growth in the developing world. Use this graph to answer Exercises 75–78.

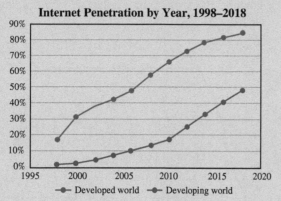

Internet Penetration by Year, 1998–2018

Source: International Telecommunication Union

75. By how much did the percentage change from 1998 to 2018 in the developed world? What about the developing world?

76. Is the gap between the developed and developing worlds getting bigger or smaller? Discuss.

77. Which had a more significant increase between 1998 and 2018, the developed or developing world? Explain how you decided.

78. Which of the two trends shown on the graph is likely to continue for the next 5–10 years? Explain how you decided.

79. In Tyreke's math class, he scored 84, 92, 79, and 86 on the first four tests. He needs at least an 80% overall test average to earn the B he's shooting for. Without doing any computation at all, make a guess as to the score you think he would need on the final (which is worth two test grades) to have an average of at least 80%. Then find what his average would be if he got the score you guessed. How did you do?

80. Entering a postseason basketball tournament, Marta's goal is to average at least 15 points per game for the tournament. (Marta is not the greatest team player in the world.) After scoring 16, 12, 9, and 18 in the first four games, how many points do you think Marta will need to score in the remaining two games to reach that goal? First, make a guess without doing any computation. Then find what her average would be if Marta scored the number of points you guessed. How did you do?

81. One of the most valuable uses of estimation is to roughly keep track of the total cost of items when shopping. Use rounding to estimate the total cost of the following items at a grocery store: 4 cans of green beans at 79 cents each; 8 cups of yogurt at 49 cents each; a 29-ounce steak at $5.80 per pound; 4 energy drinks at $1.29 each; and 100 ounces of mineral water at $3.08 per gallon. (*Hint:* You may need to look up conversions for units of weight and volume.)

82. Refer to Problem 81. In getting my swimming pool ready for summer, I usually buy the following supplies. Use rounding to estimate the total cost: 8 boxes of baking soda at 89 cents per box; 20 gallons of bleach at $1.29 for

96 ounces; 8 pounds of chlorine stabilizer at $11.99 for a 4-pound bottle; and four 24-can cases of refreshments at 60 cents per can.

83. The following graph appeared in a *New York Times* article on August 4, 2021.
    (a) What was the difference between the two types of counties in the daily death rate on May 1 and the daily death rate on August 1?
    (b) Describe trends in the daily death rate for counties with high vaccination rates and for counties with low vaccination rates.
    (c) What does this graph tell you about vaccination?

Daily Deaths by U.S. County Vaccination Rate

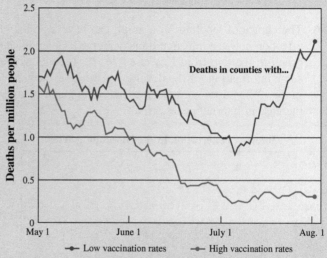

Source: The New York Times

84. This line graph appeared in the Morning Brew mailing list in March 2020.

Streaming Minutes per Week (Billions)
February 24 to March 16, 2020

Source: Nielsen/Axios

(a) Which streaming service had the biggest increase in minutes from the week of March 1 to the week of March 15? Which had the smallest?
(b) All of the services had a significant increase in streaming minutes in early March 2020. What do you think accounts for this? (A bit of online research on what was happening in the world in that time frame will likely help.)

Critical Thinking

85. Sometimes graphs are drawn in such a way as to support a conclusion that may or may not be true. Look at the graph and see if you can find anything misleading.

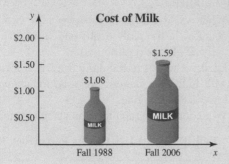

Cost of Milk

86. The choice of labeling on a graph can have a profound effect on how the information is perceived. Compare the graph to the one from Exercises 75–78. It contains the same information for the developing world as the earlier graph. Why does it appear to show a much sharper increase in Internet penetration?

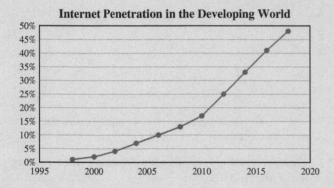

Internet Penetration in the Developing World

Recall that the average speed of an object over some time period is calculated using the formula: speed = distance/time. The following graph describes an overnight road trip to go to an off-campus party at another college: the horizontal axis is hours driven, and the vertical axis is miles from the home campus. Use the graph to answer Exercises 87–92.

87. Estimate the average speed over the first hour of the trip, then the first 2 hours, the first 3 hours, and the first 4 hours. Based on your results, write a description of the trip.

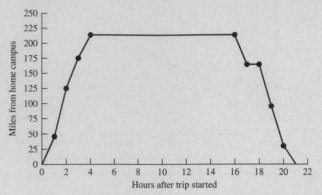

88. How long did the road-trippers stay at the other school? How far away was it from their own campus?

89. Estimate the average speed from time 16 hours to 17 hours, then from 17 hours to 18 hours. What is the significance of the sign in the first answer? What happened between 17 and 18 hours?

90. Without doing any calculations, how can you tell if the average speed on the trip back was more or less than the average speed on the trip there?

91. Using your results from Exercises 87–90, what feature of the graph do you think corresponds to the average speed over a portion of the trip?

92. Find the average speed for the whole trip. Why is your answer nonsense? What can you conclude about using the average speed formula?

Section 1-3   **Problem-Solving Strategies**

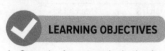

LEARNING OBJECTIVES

1. State the four steps in the basic problem-solving procedure.

2. Solve problems using a diagram.

3. Solve problems using trial and error.

4. Solve problems involving money.

5. Solve problems using calculation.

Here's an idea that can help you understand your math classes better: once in a while, think about what some of the math words you take for granted actually mean in English. For example, have you ever thought about why we call math problems "problems?" In real life, when a problem confronts you, you probably think about different strategies you might use to overcome an obstacle and then decide on the best course of action. So why not use that same strategy in solving math problems? This is probably the single biggest reason why taking math classes is useful to *anyone*: math is all about learning and practicing problem-solving strategies.

A Hungarian mathematician named George Polya did some important research on the nature of problem solving in the first half of the 20th century. His biggest contribution to the field was an attempt to identify a series of steps that were fundamental to problem-solving strategies used by great thinkers throughout human history. One of his books, published in

1945 (and still a big seller on Amazon!), set forth these basic steps. *How to Solve It* is so widely read that it has been translated into at least 17 languages.

Polya's strategy isn't necessarily earth-shattering: its brilliance lies in its simplicity. It provides four basic steps that can be used as a framework for problem solving in any area, from math to personal issues, and everything in between.

1. State the four steps in the basic problem-solving procedure.

Polya's Four-Step Problem-Solving Procedure

Step 1 *Understand the problem.* The best way to start any problem is to write down information that's provided as you come to it. Especially with longer word problems, if you read the whole thing all at once and don't DO anything, it's easy to get overwhelmed. If you read the problem slowly and carefully, writing down information as it's provided, you'll always at least have a start on the problem. Another great idea: carefully identify and *write down* what it is they're asking you to find; this almost always helps you to devise a strategy.

Step 2 *Devise a plan to solve the problem.* This is where problem solving is at least as much art as science—there are many, many ways to solve problems. Some common strategies: making a list of possible outcomes; drawing a diagram; trial and error; finding a similar problem that you already know how to solve; and using arithmetic, algebra, or geometry.

Step 3 *Carry out your plan to solve the problem.* After you've made a plan, try it out. If it doesn't work, try a different strategy! There are many different ways to attack problems. Be persistent!

Step 4 *Check your answer.* It's always a good idea to think about whether or not your answer is reasonable, and in many cases you'll be able to use math to check your answer and see if it's exactly correct. If not, don't forget what we learned about estimation in Section 1-2—that can be a big help in deciding if an answer is reasonable.

In each of the examples in this section, we'll illustrate the use of Polya's four-step procedure.

Example 1   **Solving a Problem Using a Diagram**

Math Note

Sometimes problems will contain extraneous information. In Example 1, the height of the plants is immaterial to the distance between them, so you can (and should) ignore that information. Students sometimes get into trouble by trying to incorporate *all* the information provided without considering relevance.

A gardener is asked to plant eight tomato plants that are 18 inches tall in a straight line with 2 feet between each plant.

(a) How much space is needed between the first plant and the last one?
(b) Can you devise a formula to find the length needed for ANY number of plants?

SOLUTION

(a) Be careful—what seems like an obvious solution is not always correct! You might be tempted to just multiply 8 by 2, but instead we'll use Polya's method.

Step 1 *Understand the problem.* In this case, the key information given is that there will be eight plants in a line, with 2 feet between each. We're asked to find the total distance from the first to the last.

Step 2 *Devise a plan to solve the problem.* This sounds a lot like a situation where drawing a diagram would be a big help, so we'll start there.

Step 3 *Carry out the plan to solve the problem.* The diagram would look like this:

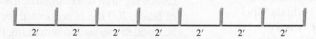

Now we can use the picture to add up the distances:

$$2 + 2 + 2 + 2 + 2 + 2 + 2 = 14 \text{ feet}$$

Step 4 *Check the answer.* There are eight plants, but only seven spaces of 2 feet between them. So $7 \times 2 = 14$ feet is right.

(b) Our drawing was designed to solve a specific problem, but we can actually use it for the more general case. It shows us that for eight plants, there were seven 2-foot gaps in between. We can deduce that for n plants, there would be $n - 1$ gaps, so the total length would be $2(n - 1)$ feet.

Try This One   1

Suppose you want to cut a 4-foot-long party sub into 20 pieces by first cutting it lengthwise down the middle, then making a series of crosscuts.

(a) How many cuts in all would you need?
(b) How many cuts would you need to cut it into n pieces?

Sidelight   **The Story Behind "Story Problems"**

Want to make a math professor cringe? Use the term *story problem*. Elementary school teachers use this term to try to give kids a positive attitude toward word problems. I would ask them, "How's that working out for you?" I've never seen a math problem that reads "Once upon a time, there was a beautiful princess who was on a train leaving Chicago going west at 45 miles per hour." Math problems don't begin with "Once upon a time," and they only have happy endings if you find the right answer.

A more adult name is *word problems*, but even that isn't great because it's not really descriptive. My favorite term is *application problems*, which explains why many of the problems in this book are called *Applications in Our World*. That term is very descriptive of WHY you should care about solving word problems: it's the way you APPLY the skills and knowledge you're acquiring to more than just number crunching or finding what *x* is. So next time you're tempted to complain about having to do word problems, think of them as *applications*, and consider that they're not a burden or an afterthought. Applying book knowledge to solve problems is one of the most valuable skills you'll acquire from a college education. And besides: if all you can do is crunch numbers and find *x*, doesn't that kind of make you a nerd?

Example 2   Solving a Perimeter Problem

Alinari Archives/Corbis/Getty Images

Scientists and inventors often use sketches to organize their thoughts as Leonardo da Vinci did. He wrote backward in Latin to protect his work. I guess industrial espionage is older than I thought.

A campus group is setting up a rectangular area for a tailgate bash. They have 100 feet between two roads to use as width and 440 feet of fence to use. What length will use up the total amount of fence and enclose the biggest space?

SOLUTION

Step 1 *Understand the problem.* We're asked to consider a rectangular area, so there will be four sides. We're told that the width is 100 feet and that the four sides add up to 440 feet. (That is, the perimeter is 440 feet.) We're asked to find the length.

Step 2 *Devise a plan to solve the problem.* This is another classic example of a problem where a diagram will be useful. This should help us to figure out all the dimensions.

Step 3 *Carry out the plan to solve the problem.* Our diagram looks like this:

100 feet ▭ 100 feet

(Since the area is rectangular, the opposite sides have the same length.) Of the 440 feet of fence, 200 feet is accounted for in our diagram. That leaves $440 - 200 = 240$ feet to be divided among the remaining two sides. Each has length 120 feet.

Step 4 *Check the answer.* If there are two sides with width 100 feet and two others with length 120 feet, the perimeter is $100 + 100 + 120 + 120 = 440$ feet.

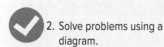

2. Solve problems using a diagram.

Try This One   2

A rectangular poster promoting the tailgate bash in Example 2 is restricted to 10 inches in width to fit bulletin board restrictions. The printer suggests a perimeter of 56 inches for maximum savings in terms of setup. How tall will the posters be?

| Example 3 | Solving a Problem Using Trial and Error |

Math Note

The trial and error method works nicely in Example 3, but for many problems, there are just too many possible answers to try. It's useful for some problems, and can be a *great* way to help understand a problem, but you shouldn't rely on it for every problem.

As part of your duties as the new hire at a job, you're put in charge of buying 12 door prizes for an end-of-year staff meeting. You've got $110 to spend, and the boss is thinking that insulated drink cups and smartphone stands would make nice gifts. If the cups cost $11 and the stands cost $8 each, how many of each should you buy? (The corporate world is kind of funny about budgeting: it's considered a good thing to spend as much of a budgeted amount as you can. Seriously.)

SOLUTION

Step 1 *Understand the problem.* The key information: total of 12 items purchased, $110 to spend, $11 for each cup, and $8 for each stand. We're asked to find how many cups and how many stands will result in a cost of $110.

Step 2 *Devise a plan to solve the problem.* I can easily figure out the total cost if I have a specific number of each item, so let's try a combination at random and see if that helps. If you buy one cup and 11 stands, the cost is $1 \times \$11 + 11 \times \$8 = \$99$. This isn't right, but now we have a strategy.

Cups	Stands	Total cost
1	11	$99

Step 3 *Carry out the plan.* Let's just keep trying combinations until we find the right one.

Cups	Stands	Total cost
1	11	$99
2	10	$102
3	9	$105
4	8	$108
5	7	$111

At this point we can stop. Since the cups cost more, as we keep adding more of them the cost will keep going up, and we're already past our $110 target. The last row is out because that costs more than what we budgeted, so the best we can do is to buy 4 insulated cups and 8 smartphone stands at a cost of $108. Then you can use the leftover $2 for some Reese's cups because they're awesome.

Step 4 *Check your answer.* One of the nice things about using trial and error is that the strategy basically IS checking your answer . . . we already know we found the best choice.

Try This One   3

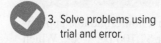

3. Solve problems using trial and error.

Michelle shipped some packages for her boss and can't find the receipt but needs an itemized list to get reimbursed. There were definitely 12 packages total, split between small flat-rate boxes ($5 each) and medium flat-rate boxes ($10.50 each). The total cost was about $110. How many of each type were shipped?

Using Technology

Using a Spreadsheet for Trial and Error

One of the things that spreadsheets do best is carry out repetitive calculations, which is a pretty good description of what we did in Example 3. Here's a screenshot of a spreadsheet I put together when writing that problem. I knew that the number of items had to add to 12, so after entering the number of cups in column A, I used the formula " = 12−A2" in cell B2 to get the corresponding number of stands. That formula was then copied down the rest of column B.

To calculate the total cost in cell C2, I used the formula " =11*A2+8*B2", which multiplies the number of cups by 11, the number of stands by 8, and adds the result. Copying that formula down then allowed me to calculate all the combinations very quickly.

See Section 1-3 Using Tech video in online resources for further information.

	A	B	C
1	# cups	# stands	total cost
2	1	11	99
3	2	10	102
4	3	9	105
5	4	8	108
6	5	7	111
7	6	6	114
8	7	5	117
9	8	4	120
10	9	3	123
11	10	2	126
12	11	1	129

C2 | fx =11*A2+8*B2

Source: Microsoft Corporation

Example 4   **Solving a Problem Involving Salary**

So you've graduated from college and you're ready for that first real job. In fact, you have two offers! One pays an hourly wage of $19.20 per hour, with a 40-hour work week. You work for 50 weeks and get 2 weeks' paid vacation. The second offer is a salaried position, offering $41,000 per year. Which job will pay more?

SOLUTION

Step 1 *Understand the problem.* The important information is that the hourly job pays $19.20 per hour for 40 hours each week, and that you will be paid for 52 weeks per year. We are asked to decide if that will work out to be more or less than $41,000 per year.

Step 2 *Devise a plan to solve the problem.* We can use multiplication to figure out how much you would be paid each week and then multiply by 52 to get the yearly amount. Then we can compare to the salaried position.

Step 3 *Carry out the plan to solve the problem.* Multiply the hourly wage by 40 hours; this shows that the weekly earnings will be $19.20 × 40 = $768. Now we multiply by 52 weeks: $768 × 52 = $39,936. The salaried position, at $41,000 per year, pays more.

Step 4 *Check the answer.* We can figure out the hourly wage of the job that pays $41,000 per year. We divide by 52 to get a weekly salary of $788.46. Then we divide by 40 to get an hourly wage of $19.71. Again, this job pays more.

Try This One   4

4. Solve problems involving money.

A condo in Myrtle Beach can be rented for $280 per day, with a nonrefundable application fee of $50. Another one down the beach can be rented for $2,100 per week. Which condo costs less for a week's stay?

Example 5	Solving a Problem Using Calculation

Ron Chapple/Thinkstock Images/
Getty Images

Nutritionists often say that you need to burn 3,500 calories while exercising to shed one pound of excess body fat. A general rule of thumb on exercise is that an average-sized person can burn about 100 calories while walking a mile at a fairly brisk pace. How many miles per day would an average person have to walk in order to shed a pound of body fat in a week?

SOLUTION

Step 1 *Understand the problem.* We're told that a person needs to burn 3,500 calories in 7 days to shed a pound of body fat, and that they'll burn 100 calories while walking a mile. We're asked to find the number of miles they need to walk each day.

Step 2 *Devise a plan.* We'll calculate how many calories need to be burned each day, then divide by 100 to see how many miles need to be walked.

Step 3 *Carry out the plan.* Since 3,500 calories need to be burned in 7 days, divide 3,500 by 7 to get 500 calories per day. Then divide 500 calories by 100 calories per mile to get 5 miles. An average person would need to walk 5 miles per day to lose one pound of body fat in a week.

Step 4 *Check the answer.* First, 5 miles per day times 100 calories is 500 calories per day. Multiply that by 7 days, and we get the 3,500 calories we need.

Try This One   5

A rental car gets 28 miles per gallon on long trips. The driver leaves home, bound for a friend's house 370 miles away. If the drive takes 6 hours, how many gallons of gas per hour is the car burning?

Sidelight   **A Legendary Problem Solver**

Archimedes is considered by some to be the greatest mathematician of the ancient world. But he is best known for a famous problem he solved in the bathtub.

King Hieron II of Syracuse ordered a solid gold crown from a goldsmith sometime around 250 BCE. He suspected that the crown was not really solid gold, but he didn't want to have it melted down to check, because of course that would ruin it. He asked his friend Archimedes if he could figure out a way to determine whether the crown really was pure gold without destroying it.

After some serious thought, Archimedes decided to relax by taking a bath. While sitting in the tub, he saw that the water level rose in proportion to his weight. Suddenly, he realized that he had discovered the solution to his problem. By placing the crown in a full tub of water, he could tell how much of it was gold by weighing the amount of water that overflowed. Legend has it that Archimedes was so excited by this burst of inspiration that he forgot to clothe himself and ran naked through the streets of Syracuse, shouting "Eureka" ("I have found it!"), a performance that would most likely get him arrested, or maybe institutionalized, today.

De Agostini/Getty Images

It turns out that the crown was not solid gold, and the goldsmith either was placed in jail or had his head removed from his shoulders, depending on which account you believe. The moral of this story is that you never know where inspiration in solving a problem might come from. But please put your clothes on before sharing your latest brilliant insight with the world.

The world is an imperfect place, my friends, and not every problem can be solved. A big part of problem solving is being able to recognize when there is no solution.

Example 6	Recognizing a Problem with No Solution

The grade in Jie's history class will be determined completely by three tests, each worth 100 points. She scored 78 and 84 on the first two tests, but still hopes to get an A, which would require an average of 92. What's the minimum score she needs on the third test?

SOLUTION

Step 1 *Understand the problem.* We're given two test scores of 78 and 84, and asked about the average for three tests. Specifically, we want it to be at least 92.

Step 2 *Devise a plan.* We'll start by seeing what the average would be if Jie scores 100 on the last test, then decide how much lower she can go and still have an average of at least 92.

Step 3 *Carry out the plan.* With a third test score of 100, Jie's average would be

$$\frac{78 + 84 + 100}{3} = \frac{262}{3} = 87.3$$

Uh oh: bad news for Jie. Even with a perfect score, her average will be only 87.3, so it's not possible for her to get an A.

Step 4 *Check the answer.* In this case, we already checked our answer as part of the plan, so we know there's no solution to the problem.

Try This One	6

✓ 5. Solve problems using calculation.

A student teacher wants to divide a sixth grade class into groups to work on a project. The teacher would like to have somewhere between 3 and 6 students in each group and wants every group to have the same number of students. Find the group sizes that will accommodate a class of 26 students.

Math Note

One of the hardest things to do as a math instructor is come up with good examples to use as practice for problem solving. We try to write problems that are realistic, but sometimes it's really helpful to practice problem-solving strategies on problems that don't sound worthwhile (Exercises 7–10 are good examples of this).

So try not to get too wrapped up in whether or not a problem sounds like something a real person might need to solve: focus on the four-step strategy and remember that exercise for your mind is just as important as exercise for your body.

Strategies for Understanding Problems

One reason that problem solving is challenging is that every problem can be a little different, so you can't just memorize a procedure and mimic it over and over. You actually have to think on your own! But here are some suggestions for helping to understand a problem and devise a strategy for solving it.

- If the problem describes something that can be diagrammed, a drawing almost always helps.

- If you find that you read through a problem and feel like you have no idea where to start, you should consider NOT reading the entire problem at first. This can sometimes lead to feeling overwhelmed or intimidated. Instead . . .

- Write down all the numeric information in the problem as you read it. This often helps to organize your thoughts. But maybe more importantly, it's a guaranteed way to get a START on every problem, so you avoid that moment of panic when you feel like you have no clue what to do.

- Sometimes making a chart to organize information is helpful, especially when you're trying to use trial and error.

- Don't expect that you'll know exactly how to solve a problem after reading it once! Most often you'll need to read through a problem several times, and even then you may need to just try some approaches before finding one that works. Don't be afraid to try! Nobody ever became a good problem solver by being afraid to make a mistake.

Answers to Try This One

1   Ten; $n/2$

2   18 inches

3   Three small and nine medium

4   The condo for $280 per day costs less for a week.

5   About 2.2 gallons per hour

6   None of those group sizes will work.

Exercise Set   1-3

Writing Exercises

1. List and describe the four steps in problem solving.
2. Discuss what you should do first when given an application problem to solve.
3. Discuss different ways you might be able to check your answer to a problem.
4. Why is trial and error not always a good problem-solving strategy?
5. Why is "application problem" a better name than "word problem"?
6. Think about a problem you've had to solve outside of school, and outline how Polya's procedure could be used to solve it.

Computational Exercises

7. One number is 6 more than another number, and their sum is 22. Find the numbers.
8. One number is 7 more than another number. Their sum is 23. Find the numbers.
9. If 24 is added to a number, it will be 3 times as large as it was originally. Find the number.
10. If the smaller of two numbers is one-half of the larger number and the sum of the two numbers is 57, find the numbers.
11. The sum of the digits of a two-digit number is 7. If 9 is subtracted from the number, the answer will be the number with the digits reversed. Find the number.
12. If the sum of the digits of a two-digit number is 7 and the tens digit is one more than the ones digit, find the number.
13. When the mortgage is completely paid off for Isiah and Kiara's house, it will be 5 times as old as it is now. If they have 28 years left on the mortgage, how old is the house right now?
14. Hoang has worked as a nurse at Springfield General Hospital for 5 years longer than her friend Bill. Four years ago, she had been at the hospital for twice as long. How long has each been at the hospital?
15. A retired police officer passes away, leaving $140,000 to be divided among two children and three grandchildren. The will specifies that each child is to get twice as much as each grandchild. How much does each get?
16. At the dog park, there are several dogs with their owners. Counting heads, there are 12; counting legs, there are 38. How many dogs and owners are there? (Assume that there are no one-legged humans or three-legged dogs.)
17. Arrange the digits 1, 2, and 3 to form two numbers divisible by 6.
18. Nine athletes from a co-ed track team stop for lunch. The men eat four pieces of pizza each, the women eat two pieces each, and the coach eats three pieces of pizza. If the team bought four pizzas, each cut into eight pieces, and there were five pieces left over (mmm, doggy bag), how many men and women were on the team?

Applications in Our World

19. Barney and Betty break into a parking meter with $5.05 in dimes and quarters in it (legal disclaimer: don't do this), and agree that Barney will get all the dimes, and Betty will get all the quarters. (Barney isn't terribly bright.) Barney ends up with five more coins than Betty. How much money did each get?
20. A tip jar contains twice as many quarters as dollar bills. If it has a total of $12 in it, how many quarters and how many dollar bills does it contain?
21. A fraternity charged $2.00 admission for dudes and $1.00 admission for ladies to their finals week bash. The fraternity made $75 and sold 55 tickets. How many ladies attended the party?
22. While reviewing the previous day's arrest report, a police sergeant notices that nine suspects were arrested, all of whom had either one or two previous arrests. Including yesterday's arrests, there were 19 total arrests among them. How many suspects had fewer than two prior arrests?
23. Mae receives $87 for working one 8-hour day. One day she had to stop after working 5 hours because of a doctor's appointment. How much did she make that day?

24. The manager of the new campus Internet café wants to put six PCs on each table with 3 feet between the PCs and 2 feet on each end. The PCs measure $1\frac{1}{2}$ feet wide and 21 inches high. What length of tables should the Internet café order?

25. A contractor is building a deck that is 32 feet long, and posts that are 6 inches square will be used to build the railing. If the posts are placed 4 feet apart with one on each end, how many posts will the contractor use?

26. A decorator hangs 10 pictures that each measure $8\frac{1}{2}$ inches wide on a wall. They're spaced 6 inches apart with 2 inches on each end, so that they fit perfectly on the wall. How wide is the wall?

27. You want to cut a width of mat board to frame pictures so you can fit six pictures that have a width of 6 inches each with 2 inches between them and a 1-inch border. How wide should you cut the mat board?

28. A physical therapy facility is building a new pool that is 60 feet long and 5 feet deep. They have ordered enough tile for a 220-foot-long border around the edge. How wide should the pool be to ensure that all tiles are used?

29. A landscape architect is planning a new nature area in the middle of an urban campus. The plan is for the length to be twice the width, and there will be a 3-foot-high retaining wall around the perimeter. There will be 300 total feet of wall installed. How wide will this area be?

30. A standard tube of silicone caulk will make a 3/16″ bead of caulk 50 feet long. Manuel plans to seal all the window casings in the family home; there are three windows that measure 2 feet by 3 feet, three that measure 4 feet by 5 feet, and two that measure 6 feet by 4 feet. How many tubes of caulk will Manuel need?

31. Zahra is interning with a wedding planner, and is given the task of decorating a reception hall. She's chosen some decorative lights that come in boxes of 12 lights, with each string having a length of 20 feet. The room is 70 feet long and twice as wide. How many boxes of lights should Zahra buy so that the strings go all the way around the perimeter of the room once? How many lights will there be?

32. Two students are paid a total of $60 for leading a campus tour at freshman orientation. The tour lasts approximately 2 hours. If Mikael works for $1\frac{1}{2}$ hours, Pete works $\frac{1}{2}$ hour, and each makes the same hourly wage, how much does each receive?

33. You have one-half of an energy drink left. If you drink one-half of that, how much of your energy drink do you have left?

34. A small beverage company has 832 bottles of water to ship. If there are 6 bottles per case, how many cases are needed and how many bottles will be left over?

35. Kam's monthly budget includes $256 for food, $125 for gasoline, and $150 for utilities. If Kam earns $1,624 per month after taxes, how much money is left for other expenses?

36. Cheryl is training for a half-marathon, and is supplementing her diet with protein bars the week before the race. Each bar has 20 grams of protein and 15 grams of carbs. Cheryl's personal trainer would like for her to take in an extra 300 grams of carbs and 350 grams of protein during that week. How many protein bars should Cheryl buy?

37. A physical education teacher plans to divide the seventh graders at Wilson Middle School into teams of equal size for a year-ending mock Olympic event. Each team will have between 5 and 9 students, and all teams need to have the same number of students. The seventh grade at Wilson consists of three classes; one with 24 students, one with 26, and one with 21. How many students should be on each team?

38. Bart is taking a pass-fail math class, and needs to average 70% to get a passing grade. The grade is determined by five 50-point tests, and Bart's doing really well, scoring 48, 45, 45, and 47. The week of the final test, Bart has a really important chemistry test to focus on but doesn't want to risk failing math. Find the range of scores Bart can get on test 5 that will result in failing the class.

39. If a family borrows $12,381 for an addition to their home, and the loan is to be paid off in monthly payments over a period of 5 years, how much should each payment be? (Interest has been included in the total amount borrowed.)

40. On the way to the airport on December 8, 2011, my wife's car hit a huge hole in a construction area, damaging the front right wheel. So that was cool. The temporary tire forced us to decrease our average speed by 15 miles per hour. If the 39-mile drive usually takes us 45 minutes, how much time should we have budgeted for the drive home from the airport?

41. Four friends decide to rent an apartment. Because each will be using it for different lengths of time, Mary will pay $\frac{1}{2}$ of the monthly rent, Ji-Woo will pay $\frac{1}{4}$ of the monthly rent, Claire will pay $\frac{1}{8}$ of the monthly rent, and Margie will pay the rest. If the monthly rent is $2,375, how much will each person pay?

42. Harry fills up his Jeep with gasoline and notes that the odometer reading is 23,568.7 miles. The next time he fills up his Jeep, he pays for 12.6 gallons of gasoline. He notes his odometer reading is 23,706.3 miles. How many miles per gallon did he get?

43. A clerk earns $9.50 per hour and is paid time and a half for any hours worked over 40. Find the pay if the clerk worked 46 hours during a specific week.

44. A cell phone company charges 35 cents per minute during the daytime and 10 cents per minute during the evening for those who go over their allotted monthly minutes. If a customer went 32 minutes over their monthly minutes, find out how much they'd save if they went 32 minutes over their monthly minutes during the evening compared to during the daytime.

Last week at Chili's, my wife and I played a game on a tabletop video screen. The object is to move your frog so that it hits every lily pad exactly once. The frog cannot move diagonally and cannot go back in the direction it came from. Lily pads disappear after the frog jumps off them, and the frog can't jump over an existing lily pad. In Exercises 45 and 46, find a path that wins each game.

45.

46.

Critical Thinking

47. A king decided to pay a knight one piece of gold for each day's protection on a 6-day trip. The king took a gold bar 6 inches long and paid the knight at the end of each day; however, he made only two cuts. How did he do this?

48. At the finals of the campus-wide Brainbowl League Tournament, your team was asked to measure out exactly 1 gallon of water using only a 5-gallon and a 3-gallon container, without wasting any water. Your team completed the task. How did they do it?

49. What is the smallest number of cars that can tailgate in a straight line at a football game so that one car is in front of two cars, one car is behind two cars, and one car is between two cars?

50. To purchase a computer for the Student Activities office, the freshman class decides to raise $\frac{1}{3}$ of the money, and the sophomore class decides to raise $\frac{1}{2}$ of the money. The Student Government Association agrees to contribute the rest, which amounted to $400. What was the cost of the computer?

51. An important message about troop movements has to be delivered in person across a stretch of desert. The Jeeps available to make the trip can carry enough fuel to get them halfway across the desert, and there are no fueling stations along the way. There are a number of identical Jeeps, all of which can transfer their fuel to any other, but there are no fuel containers available or ropes to tow extra Jeeps. Be a hero and devise a strategy to get the message delivered.

52. How many total triangles are in this figure?

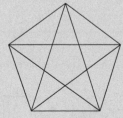

53. A professional driver completes one lap around the track at Daytona International Speedway at 80 miles per hour. How fast would the second lap have to be driven to make the overall average 160 miles per hour?

54. If you run around a quarter-mile track at some average speed *s*, at what average speed would you need to run a second lap to make your overall average twice as much as it was for the first lap?

Almost everyone thinks at first that the next two problems can't be solved, but I promise you they can be!

55. A math professor writes down two consecutive whole numbers from the range from 1 through 10. She tells one of the numbers to Maurice, and the other to Hani. Maurice says to Hani, "I don't know your number." Hani replies, "And I don't know your number." Maurice then says, "Now I know what your number is." What are the two numbers? (There are four different solutions to this problem!)

56. Two women playing golf have the following conversation.

    Jasmine: How old are your kids?
    Judi: I have three kids, and the product of their ages is 36.
    Jasmine: That's not enough information to figure out their ages.
    Judi: The sum of their ages is exactly the number of holes we have played so far today.
    Jasmine: Still not enough information.
    Judi: My oldest child almost always wears a red hat.
    Jasmine: Okay, now I know how old they are.
    How old are the three kids?

57. TECHNOLOGY. Recreate the spreadsheet used to solve the problem in Example 3 of Section 1-3. The spreadsheet (with a description) is in the Technology box on page 32. Make sure you use formulas to calculate the total cost!

58. TECHNOLOGY. A group of friends at a ballgame sends two gofers to get hot dogs, beers, and sodas for everyone. Hot dogs are $3.50 each, beers are $8, and each soda is $5. Eight beverages total were purchased, and the gofers brought back $20.50 from $100 that was collected. They tipped the vendor $5. How many hot dogs, beers, and sodas were purchased? Create a spreadsheet similar to the one on page 32 to solve this problem using trial and error. There's a template in the online resources for this section to get you started.

Section	Important Terms	Important Ideas
1-1	Inductive reasoning Deductive reasoning Conjecture Counterexample	**In math** (and in life!), we can use two types of reasoning: inductive and deductive. Inductive reasoning is the process of arriving at a general conclusion based on observing specific examples. Deductive reasoning is the process of arriving at a conclusion based on known rules and principles. Remember: inductive goes from specific to general, and deductive goes from general to specific.
1-2	Estimation Place Value Bar graph Pie chart Time series graph Overestimate Underestimate	**In many** cases, it isn't necessary to find the exact answer to a problem. When only an approximate answer is needed, you can use estimation. This is often accomplished by rounding the numbers used in the problem and then performing the necessary operation or operations.
1-3	Polya's four-step problem- solving procedure	**A mathematician** named George Polya devised a procedure to solve mathematical problems. The steps of his procedure are (1) understand the problem, (2) devise a plan to solve the problem, (3) carry out the plan to solve the problem, and (4) check the answer.

Math in Criminal Justice REVISITED

CBS Photo Archive/Contributor

1. The suspect was identified by specific incidents in the past, which makes this inductive reasoning that would not hold up as proof of guilt in court without further evidence.

2. Fingerprints positively identify the officer as having had contact with the victim. This is deductive reasoning and would be useful in court.

3. Like fingerprints, DNA can positively show that the suspect had physical contact with the victim. This

evidence, based on deductive reasoning, would hurt the suspect badly in court.

4. While this is compelling evidence, it's based on assuming that those five drawings indicate the artist is the killer. While unlikely, it could be a coincidence based on five drawings, so this is inductive reasoning. It might impress a jury to some extent, but wouldn't be sufficient for a conviction.

Review Exercises

Section 1-1

For Exercises 1–4, make reasonable conjectures for the next three numbers or letters in the sequence.

1. 3   4   6   7   9   10   12   13   15   16   ___ ___ ___
2. 2   7   4   9   6   11   8   13   ___ ___ ___
3. 4   z   16   w   64   t   256   ___ ___ ___
4. 20   A   18   C   15   F   11   J   ___ ___ ___

For Exercises 5 and 6, make a reasonable conjecture and draw the next figure.

5.

6.

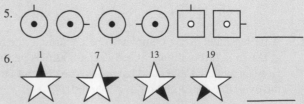

For Exercises 7 and 8, find a counterexample to show that each statement is false.

7. The product of three odd numbers will always be even.
8. The sum of three multiples of 5 will always end in a 5.

For Exercises 9 and 10, use inductive reasoning to find a rule that relates the number selected to the final answer, and then try to prove your conjecture, using deductive reasoning.

9. Pick an even number.

   Add 6:
   Divide the answer by 2:
   Add 10:
   Result:

10. Pick a number.

    Multiply it by 9:
    Add 18 to the number:
    Divide by 3:
    Subtract 6:
    Result:

In Exercises 11 and 12, use inductive reasoning to find the next two equations in the pattern:

11. $337 \times 3 = 1{,}011$
    $337 \times 6 = 2{,}022$
    $337 \times 9 = 3{,}033$

12. $33 \times 33 = 1{,}089$
    $333 \times 333 = 110{,}889$
    $3{,}333 \times 3{,}333 = 11{,}108{,}889$

13. Use inductive reasoning to solve the last equation.

    $\sqrt{1} = 1$
    $\sqrt{1+3} = 2$
    $\sqrt{1+3+5} = 3$
    $\sqrt{1+3+5+7} = 4$
    $\sqrt{1+3+5+7+9+11+13+15+17} = ?$

14. One of the statements is true, and one is not. Use inductive reasoning to decide which is which.

    (a) If a number is divisible by 3, then its square is divisible by 9.
    (b) The square of a two-digit number has three digits.

In Exercises 15–18, decide whether inductive or deductive reasoning was used.

15. My professor has given extra credit to all students for contributing canned goods to a food pantry during finals week for the last 5 years, so I know I'll get a chance for some easy extra credit on my final.
16. A GPA of 3.5 is required to make the dean's list. I checked with all my teachers to see what my final grades will be, and my GPA works out to be 3.72, so I'll be on the dean's list this semester.
17. To qualify for bowl games, college football teams have to win at least six games. Our team finished 5–7, so they won't be playing in a bowl game this year.
18. The fastest time that I've ever made it to class from my apartment is 8 minutes, and class starts in 7 minutes, so I'll be late today.

Section 1-2

For Exercises 19–23, round each number to the place value given.

19. 132,356 (thousands)
20. 186.75 (ones)
21. 14.63157 (ten-thousandths)
22. 0.6314 (tenths)
23. 3,725.63 (tens)
24. Estimate the cost of four lawn mowers if each one costs $329.95.
25. Estimate the cost of five textbooks if they cost $115.60, $89.95, $29.95, $62.50, and $43.10.
26. According to the trip computer on my car, I averaged 19.7 miles per gallon on my last tank of gas and drove 364 miles. Estimate the size of the gas tank.
27. A family of six consists of four people older than 12 and two people 12 or under. Tickets into an amusement park are $57.95 for those over 12 and $53.95 for those 12 and under. Estimate how much it would cost the family to go to the amusement park.
28. At M.T. Wallatts University, it costs a student $689 per credit-hour to attend.

    (a) Estimate the cost for a student to attend one semester if the student registers for 9 credit-hours.
    (b) If a student makes $11 an hour at her part-time job (after taxes) and works 30 hours a week, approximately how many weeks will that student have to work to afford one semester with 9 credit-hours?

29. Isn't it nice that the holiday season and fall semester book buyback come at the same time? Looks like everyone on your list is getting college logo merchandise this year! Suppose you get $130 for selling back your books (yeah, right). If college logo T-shirts are $19.75 each and sweatpants are $17.15 each, estimate how many of each you could buy with your book money. (*Hint:* There are many different combinations!)

The following pie chart was published in USA Today. It shows what one thousand respondents to a survey plan to do with the worst holiday gift they receive. Use the chart to answer Exercises 30–33.

Plans for the Worst Gift You Get This Year

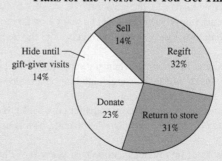

30. How many people plan to either regift or sell their worst gift?
31. How many more people plan to return the gift to a store than donate the gift?
32. How many people will not end up keeping the gift?
33. What can you conclude from adding up all of the percentages in the chart?

Use the information shown in the graph for Exercises 34–38.
The graph shows the average weekly salary (in dollars) for U.S.
production workers from 1970 to 2015.

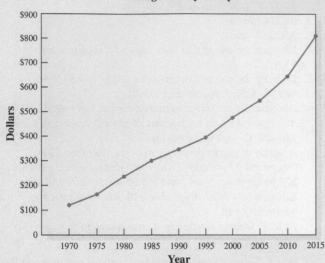

Average Weekly Salary

Source: *The World Almanac and Book of Facts*

34. Estimate the weekly salary in 1988.
35. Estimate the year in which weekly salary went over $400.
36. Find the average rate of change in salary between 1970 and 2010.
37. Just looking at the graph, would you guess that the average rate of change was greater from 1985 to 1995 or from 2000 to 2010? Explain how you made your choice, then find each rate of change to see if you guessed correctly.
38. Use the graph to make an estimate for average weekly salary in 2017, then use the Internet to find how accurate your guess was. The U.S. Bureau of Labor website is a good place to start.

Section 1-3

39. Ximena has 32 flyers about a campus symposium on environmental issues that she is organizing. She gave away all but 9. How many did she have left?
40. A tennis team played 40 matches. The team won 20 more matches than they lost. How many matches did the team lose?
41. If Iesha weighs 110 pounds when standing on one foot on a scale, how much will Iesha weigh standing on a scale with both feet?
42. A small mocha latte and biscotti together cost $3.40. If the mocha latte cost $0.40 more than the biscotti, how much did each cost?
43. I put a new floor on my deck a couple of years ago. The deck is rectangular, and measures 10 feet by 40 feet. The deck boards come in 12-foot lengths, and are $5\frac{1}{2}$ inches wide. I planned to leave a half-inch gap between boards that are lying next to each other side-to-side; boards that meet end-to-end were to be placed with no gap in between. Find the smallest number of boards I could have used for the job.

44. Mary got $80 in tips last night for her waitressing job at the campus café. She spent $8.00 downloading songs online to her phone and then spent $\frac{1}{3}$ of the remainder on tickets for a comedy club. How much did she have left?
45. Your science textbook and its accompanying lab packet cost $120. If the textbook costs twice as much as the lab packet, how much did the lab packet cost?
46. Fill in the squares with digits to complete the problem. There are several correct answers.

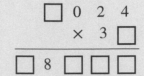

47. In 10 years, my house will be 5 times as old as it was 10 years ago. How old is it now?
48. At times during the summer in Alaska, the day is 18 hours longer than the night. How long is each?
49. Using +, −, and ×, make a true equation. Do not change the order of the digits.

$$2 \quad 9 \quad 6 \quad 7 \quad = \quad 17$$

50. Tina and Joe are doing homework problems together for their math class. Joe says to Tina, "If I do one more problem, then we'll have both done the same number of problems." Tina says to Joe, "If I do one more problem, then I will have done twice the number you have!" How many problems has each one done so far?
51. Can you divide a pie into 11 pieces with four straight cuts? The cuts must go from rim to rim but not necessarily through the center. The pieces need not be identical.
52. How many triangles are in the figure shown here?

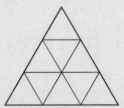

53. This one's a classic: two trains are 200 miles apart, traveling toward each other, each going 20 miles per hour. A really speedy fly takes off from one train and flies directly toward the other at 60 miles per hour. When it reaches the train, it bounces off and flies back to the first train. The fly repeats the trip until the trains collide and the poor little guy gets squashed. How far did the fly fly?
54. The sum of two numbers is 120 and the difference is 15. Find the numbers.
55. A health food store charges $2.00 per pound for high-protein nature mix and $2.75 for low-carb soy medley. If 10 pounds of the two items mixed together costs $24.50, find the amount of each type in the mixture.
56. Taylor had $1,000 to invest. One part was invested at 8% and the other part at 6%. If the total simple interest was $76.00, find how much was invested for 1 year at each rate. (Simple interest is the amount invested times the interest rate.)

Practice Test

1. Make a reasonable conjecture for the next three numbers on the list, then describe how confident you are in your answer.
   2 5 4 8 6 11 8 ___ ___ ___

2. What state logically goes next on this list?
   Kansas, Louisiana, Maine, Nebraska, Ohio

3. Use inductive reasoning to find the solution to the problem; then check it by performing the calculation.
   $0 \cdot 9 + 8 = 8$
   $9 \cdot 9 + 7 = 88$
   $98 \cdot 9 + 6 = 888$
   $987 \cdot 9 + 5 = 8{,}888$
   $\vdots$
   $9{,}876{,}543 \cdot 9 + 1 = ?$

4. Use inductive reasoning to find the solution to the problem; then check it by performing the calculation.
   $6 \times 7 = 42$
   $66 \times 67 = 4{,}422$
   $666 \times 667 = 444{,}222$
   $6{,}666 \times 6{,}667 = ?$

5. Use inductive reasoning to find a rule that relates the number you selected to the final answer, and try to prove your conjecture using deductive reasoning.

   Pick a number:
   Add 10 to the number:
   Multiply the answer by 5:
   Add 15 to the answer:
   Divide the answer by 5:
   Result:

6. There were 12 students in line to register for the Underwater Basketweaving 101 course. All but 2 changed their minds. How many remained in line?

7. An eccentric business owner decides to give holiday bonuses this year by giving some money for each of the 12 days of Christmas; the amount doubles each day. On the 12th day, all employees get $204.80.
   (a) On which day did they get close to $25?
   (b) How much did they get on the first day?
   (c) What was the total amount of the bonuses?

8. What are the next two letters in the sequence, T, T, F, F, S, S, . . . ? (*Hint:* It has something to do with numbers.)

9. By moving just one coin, make two lines, each three coins long. There are two solutions!

10. A ship is docked in harbor with a rope ladder hanging over the edge, and 9 feet of the ladder is above the waterline. The tide is rising at 8 inches per hour. After 6 hours, how much of the ladder remains above the waterline?

11. This problem was written by the famous mathematician Diophantus. Can you find the solution?
    The boyhood of a man lasted $\frac{1}{6}$ of his life; his beard grew after $\frac{1}{12}$ more; after $\frac{1}{7}$ more he married; 5 years later his son was born; the son lived to one-half the father's age; and the father died 4 years after the son. How old was the father when he died?

12. A number divided by 3 less than itself gives a quotient of $\frac{8}{5}$. Find the number.

13. The sum of $\frac{1}{2}$ of a number and $\frac{1}{3}$ of the same number is 10. Find the number.

14. Add five lines to the square to make three squares and two triangles.

15. One person works for 3 hours and another person works for 2 hours. They are given a total of $60.00. How should it be divided up so that each person receives a fair share?

16. The sum of the reciprocals of two numbers is $\frac{5}{6}$ and the difference is $\frac{1}{6}$. Find the numbers. (*Hint:* The reciprocal of a number n is $\frac{1}{n}$.)

17. Sam scored 72 and 78 on her first two 100-point tests.
    (a) What score does she need on the third test to bring her average up to 80%?
    (b) If she gets that score, what's the lowest score she can get on the last exam to get an A−, which is assigned to average scores between 90% and 92%?

18. Denali is about 20,300 feet above sea level, and Death Valley is 280 feet below sea level. Find the vertical distance from the top of Denali to the bottom of Death Valley, and the average rate of change in height if an adventurer travels from the summit of Denali to the deepest part of Death Valley in a cross-country eco challenge.

19. Mark's mother is 32 years older than Mark. The sum of their ages is 66 years. How old is each?

20. Round 1,674,253 to the nearest hundred-thousand.

21. Round 1.3752 to the nearest hundredth.

22. Estimate the cost of Stuart's new wardrobe for a job interview if a blazer costs $69.95, a new tie costs $32.54, and new pants cost $42.99.

23. The graph shows the percentage of waste paper that was recycled over the years from 1950 to 2020.

(a) Use the graph to estimate the percentage of waste paper that was recycled in 1960 and in 2000.

(b) Estimate the first year in which more than half of all waste paper was recycled.

(c) Find the average rate of change in the recycling percentage for the period from 1970 to 2010.

24. Using the pie chart, of the 3,646 students surveyed at seven residential universities, estimate the number of students surveyed who do not live in a residence hall, and how many more live at home than in a frat or sorority house.

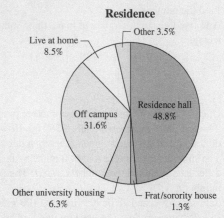

Residence

Other 3.5%
Live at home 8.5%
Off campus 31.6%
Residence hall 48.8%
Other university housing 6.3%
Frat/sorority house 1.3%

Source: http://www.acha-ncha.org/data/DEMOGF06.html

The bar graph shows the number of COVID-19 deaths per 100,000 residents for the five largest U.S. states as of August 2021; the table shows the population of each state in that year. Use this information in Exercises 25–27.

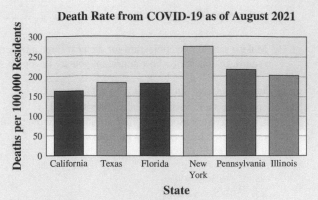

Death Rate from COVID-19 as of August 2021

Deaths per 100,000 Residents

State

Source: Statista.com

State	Population in 2021 (Millions)
New York	19.3
California	39.6
Florida	21.9
Texas	29.7
Pennsylvania	12.8
Illinois	12.6

25. The rates for Texas and Florida are very similar. Which of those two do you think had more deaths from COVID?

26. What state had the most total COVID deaths?

27. Ohio had 20,530 COVID deaths at that point, and a population of 11.6 million. Make a copy of the bar graph and draw in a seventh bar corresponding to Ohio's COVID death rate.

Projects

1. One of my students asked me an intriguing question this past semester: he was really bothered by the fact that his chemistry professor claimed that "science can never really prove anything," and wanted me to try and explain what he meant. In fact, this issue is ALL about inductive and deductive reasoning. Do an Internet search for the string "can science prove anything," and research the argument. Then discuss it with group members or friends, and write a paper that provides your opinion on whether science can or cannot prove anything. One of the main focuses of your paper should be what you've learned about inductive and deductive reasoning.

2. Many of the problems in this book are based on real data, and the fact is that I know all of that stuff off the top of my head. Just kidding—I get most of it from the Internet. One good source of data is the website Infoplease.com. Your mission is to write your own problems based on estimation from graphs. Look through the website, and find some data on a topic you find interesting. Draw a bar graph corresponding to the data, then write at least six questions for students in another group based on the graph. Do the same for a pie chart, and a time-series graph (use different topics for each type of chart). When all groups have completed their questions, you can exchange information and answer the questions. Try to shoot for a range of difficulties: some simple problems and some challenging ones. Looking at the problems in Section 1-2 would probably help—they're brilliant.

3. One reason that Polya's method has stood the test of time is that it is used not just in math, but also for any problem needing to be solved. For each of the problems listed, write a paragraph explaining how each of the steps in Polya's method could be applied.

(a) A family has a small set of stairs leading from a back door down to a featureless backyard. They would like to have a place for outdoor parties during the summer.

(b) Your car has been making a really funny noise whenever you go over 50 miles per hour.

(c) You've balanced your checkbook and found out that you won't have enough money to pay for tuition and books next semester.

(d) One of your friends heard something you said about him behind his back, and now he's very upset with you.

(e) A couple that you hang out with has broken up, and you're still friends with each of them. It's going to be tough to set up days and nights out together in your group of friends.

(f) You saw your boss do something really unethical, maybe even illegal, at work, and are trying to decide what to do about it.

Design elements: Front matter, Chapter Opener, Summary and End Matter header design (random numbers background illustration): ©pixeldreams.eu/Shutterstock RF

Set Theory

Photo: John Lund/Blend Images

Outline

Demography is the statistical study of the changing characteristics of people in a population. If there's one thing we can guarantee for certain in this field, it's that the racial makeup of the U.S. population has undergone a radical shift in recent years, and that shift is only expected to speed up.

The bar graph below shows the percentage of the total population that fell into four distinct categories in 1960, 2005, and 2019, and what those percentages are projected to be in 2050 by the Pew Research Center, an organization that ranks among the world leaders in population statistics.

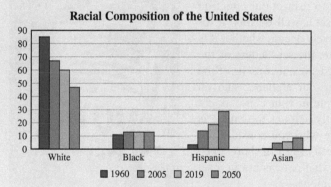

Racial Composition of the United States

There's a lot going on here: The percentage of whites decreases dramatically, with Asians and Hispanics both showing significant gains, and Blacks staying more or less the same.

No matter how you slice it, our society is becoming more and more diverse every year, to the point where even defining what we mean by "race" isn't so easy anymore. The most common racial groups referred to in population statistics are white, Black, Asian, and Hispanic, but of course there are many people that fall into more than one category (including the 44th president of the United States, and the 46th vice president!). Actually, it's even more confusing than that: "Hispanic" isn't really a race, but rather an ethnic group, and many Hispanics also report themselves as either white or Black. (For the data in the bar graph, "white" means "non-Hispanic white.")

Sorting it all out to get any sort of meaningful picture of what we as Americans look like isn't easy, and the techniques of set theory are very useful tools In trying to do so. In this chapter, we'll define what we mean by sets and study sets and how they can be used to organize information in an increasingly complex world. The concept of sets has been used extensively since people began studying mathematics, but it wasn't until the late 1800s that the theory of sets was studied as a specific branch of math. One of the major tools that we will use to study sets, the Venn diagram, was introduced in an 1880 paper by a man named John Venn. These diagrams enable us to picture complicated relationships between sets of objects, like people of certain races.

Because race and ethnicity are self-reported in a variety of different ways, it's very difficult to find detailed data on the breakdown of races, but there are some reasonable estimates out there. The following estimates were cobbled together from a number of different sources. In a group of 1,000 randomly selected Americans, 727 would self-report as white, 134 as Black, and 185 as Hispanic. In addition, 18 would self-report as Black and white, 16 as Black and Hispanic, and 119 as white and Hispanic. Finally, 11 would self-report as all three. Based on these estimates, after completing this chapter, you should be able to answer the following questions:

1. How many of the original 1,000 report as white only, Black only, and Hispanic only?

2. How many report as Hispanic and Black, but not white?

3. How many report as either Hispanic or Black?

4. How many report as none of white, Black, or Hispanic?

5. Based on the bar graph to the left, which set will have more elements in 2050: the set of Americans reporting as white, or those reporting as nonwhite?

For answers, see Math in Diversity Revisited on page 96

Section 2-1 | Introduction to Set Theory

LEARNING OBJECTIVES

1. Define set.
2. Write sets three different ways.
3. Define the empty set.
4. Find the cardinality of a set.
5. Classify sets as finite or infinite.
6. Decide if two sets are equal or equivalent.

Have you ever thought about the role that grouping things plays in everyday life? Think of all the subgroups just within the people you know: You have a group of close friends, a group of social media friends, a group of family members, a group of casual acquaintances, a group of classmates, a group of professors, a group of coworkers. . . . You also have a group of keys, groups of clothes, electronics, foods, TV shows, and many others. Our whole world is divided into groups of things, or what we call *sets*. So studying sets from a mathematical standpoint is a good opportunity to study how math is used in our world. Our entire study of sets will be set up by the topics in this section. Set up—get it? You don't get that quality of bad humor in most textbooks.

Basic Concepts

Let's begin with a basic definition of sets.

> A **set** is a collection of objects.

 1. Define set.

In our study of sets, we'll want to restrict our attention to sets that are well-defined. A set is **well-defined** if for any given object, we can objectively decide whether it is or is not in the set. For example, the set "letters of the English alphabet" is well-defined since it consists of the 26 symbols we use to make up our alphabet, and no other objects. The set "tall people in your class" is not well-defined because who exactly belongs to that set is open to interpretation; someone who you consider tall may not look tall to me. (I'm pretty big.) In short, to be well-defined, the definition of what is or is not in a set has to be based on facts, not opinions.

Each object in a set is called an **element** or a **member** of the set. One method of designating a set is called the **roster method**, in which elements are listed between braces, with commas between the elements. The order in which we list elements isn't important: {2, 5, 7} and {5, 2, 7} are the same set. Often, we'll name sets by using a capital letter.

Frontpage/Shutterstock

Laws that specify who can vote in a specific election determine a well-defined set of people. If that set were not well-defined, it would be almost impossible to enforce the law.

Example 1 | Writing a Set Using the Roster Method

Write the set of months of the year that begin with the letter M. Is this set well-defined? Why or why not?

Math Note

The commas between elements make it clear that the elements of the set are the names, not the individual letters.

SOLUTION

The months that begin with M are March and May. So, the answer can be written in set notation as

$$M = \{\text{March, May}\}$$

Each element in the set is separated by a comma. This is a well-defined set, because every month either begins with M or it does not; there's no opinion involved. However, if you really wanted to nitpick, you could claim that this set is not well-defined because it didn't specify a language for the names of months.

Try This One   1

Write the set of months that end with the letter *y*.

In math, the set of *counting numbers* or **natural numbers** is defined as $N = \{1, 2, 3, 4, \ldots\}$. (When we are designating sets, the three dots, or *ellipsis,* mean that the list of elements continues indefinitely in the same pattern.) The set $E = \{2, 4, 6, 8, \ldots\}$ is the set of **even natural numbers** and the set $O = \{1, 3, 5, 7, \ldots\}$ is the set of **odd natural numbers**.

| **Example 2** | **Writing Sets Using the Roster Method** |

Math Note

You can list an element of a set more than once if it means a lot to you, but there's no particularly good reason to list repeats. For example, the set of letters in the word *letters* is written as {l, e, t, r, s}.

Use the roster method to do the following:

(a) Write the set of natural numbers less than 6.
(b) Write the set of odd natural numbers greater than 4.
(c) Can you think of another way to describe each set in words?

SOLUTION

(a) $\{1, 2, 3, 4, 5\}$
(b) $\{5, 7, 9, 11, \ldots\}$
(c) The first set could be described as the set of natural numbers less than or equal to 5, or between 0 and 6. The second set could be described as the set of odd natural numbers greater than 3, or greater than or equal to 5.

Try This One 2

Write each set, using the roster method, then write at least one alternate description for each set.

(a) The set of even natural numbers from 80 to 90.
(b) The set of odd natural numbers greater than 10.

CAUTION

Students often wonder how many elements of a set to write before ending with an ellipsis. The correct answer is "seven." Just kidding—there is no set rule. Just make sure to include enough initial numbers that the pattern is clear. For Example 2(b), just writing {5, 7, ...} would leave any number of possible interpretations: {5, 7, 8, 10, 11, 13, 14, 16, ...}, and {5, 7, 10, 14, ...} are two that come to mind.

The symbol $\in$ is used to show that an object is a member or element of a set. For example, if A is the set of days of the week, we could write Monday $\in A,$ and read this as "Monday is an element of set $A.$" Likewise, we could write Friday $\in A.$

When an object is not a member of a set, we use the symbol $\notin$. Since "Icecreamday" is not a day of the week (although it probably should be), we can write Icecreamday $\notin A,$ and read this as "Icecreamday is not an element of $A.$"

| **Example 3** | **Understanding Set Notation** |

Decide whether each statement is true or false.

(a) Oregon $\in A,$ where A is the set of states west of the Mississippi River.
(b) $27 \in \{1, 5, 9, 13, 17, \ldots\}$
(c) $z \notin \{v, w, x, y, z\}$

Math Note

Set notation can be a little fussy, and you have to be careful that you're writing what you mean. For example, it's wrong to write {6} ∈ {2,4,6}. The symbol {6} represents a set containing 6 because of the brackets, which is not an element of {2,4,6}. Instead, you should write 6 ∈ {2,4,6}.

SOLUTION

(a) Oregon is west of the Mississippi, so Oregon is an element of A. The statement is true.
(b) The pattern shows that each element is 4 more than the previous element. So the next three elements are 21, 25, and 29; 27 got skipped, so it's not in the set. The statement is false.
(c) The letter z is an element of the set, so this statement is also false.

Try This One   3

Decide whether each statement is true or false.

(a) July ∈ A, where A is the set of months between Memorial Day and Labor Day.
(b) $21 \in \{2, 5, 8, 11, \ldots\}$
(c) map $\notin \{m, a, p\}$

There are three common ways to designate sets:

1. The *list* or *roster* method.

2. The *descriptive* method.

3. *Set-builder* notation.

We already know a lot about using the list or roster method; the elements of the set are listed in braces and are separated by commas, as in Examples 1 through 3.

The **descriptive method** uses a short verbal statement to describe the set.

Example 4   Describing a Set Using the Descriptive Method

Use the descriptive method to describe the set B containing 2, 4, 6, 8, 10, and 12 in two different ways.

SOLUTION

All of the elements in the set are even natural numbers, and all are less than 14, so B is the set of even natural numbers less than 14. There are plenty of other ways the set could be described. Another is the set of natural numbers between 1 and 15 that are divisible by 2.

Try This One   4

Use the descriptive method to describe the set A containing $-3, -2, -1, 0, 1, 2, 3$ in two different ways.

Math Note

When you hear *variable*, you might automatically think *letter*, like x or y. But you should think about what the word *variable* really means: something that can change, or vary. A variable is just a symbol that represents some number or object that can change.

The third (and fanciest) method of designating a set is **set-builder notation**, and this method uses *variables*.

A **variable** is a symbol (usually a letter) that can represent different elements of a set.

Set-builder notation uses a variable, braces, and a vertical bar | that is read as "such that." For example, the set $\{1, 2, 3, 4, 5, 6\}$ can be written in set-builder notation as

$$\{x \mid x \in N \quad \text{and} \quad x < 7\}$$

This is read as "the set of elements x such that x is a natural number and x is less than 7." We can use any letter or symbol for the variable, but it's common to use x. (If you need a review of inequality symbols, see the online algebra review resources.)

| Example 5 | **Writing a Set Using Set-Builder Notation** |

Use set-builder notation to designate each set, then write how your answer would be read aloud.

(a) The set R contains the elements 2, 4, and 6.
(b) The set W contains the elements red, yellow, and blue.

SOLUTION

(a) $R = \{x \mid x \in E \text{ and } x < 7\}$, the set of all x such that x is an even natural number and x is less than 7.
(b) $W = \{x \mid x \text{ is a primary color}\}$, the set of all x such that x is a primary color.

| **Math Note** |

You might have noticed that one of the cool things about set notation is that there's often more than one way to write a given set. In Example 5, we could have written $W = \{x \mid x \text{ is a color in the flag of Colombia}\}$ if we wanted to be wise guys. Or were from Colombia, I guess.

Try This One 5

Use set-builder notation to designate each set, then write how your answer would be read aloud.

(a) The set K contains the elements 10, 12, 14, 16, 18.
(b) The set W contains the elements Democrat and Republican.

| Example 6 | **Using Different Set Notations** |

Designate the set S with elements 32, 33, 34, 35, . . . using

(a) The roster method.
(b) The descriptive method.
(c) Set-builder notation.

SOLUTION

(a) $\{32, 33, 34, 35, \ldots\}$
(b) The set S is the set of natural numbers greater than 31.
(c) $\{x \mid x \in N \text{ and } x > 31\}$

| **Math Note** |

I'll be the first to admit that sets consisting of letters and numbers are abstract and not particularly interesting. But hang in there—we're using simple examples to introduce terminology and notation so that we can work with relevant sets more easily later on.

Try This One 6

Designate the set with elements 11, 13, 15, 17, . . . using

(a) The roster method.
(b) The descriptive method.
(c) Set-builder notation.

If a set contains many elements, we can again use an ellipsis to represent the missing elements as long as we illustrate a clear pattern. For example, the set $\{1, 2, 3, \ldots, 99, 100\}$ includes all the natural numbers from 1 to 100. Likewise, the set $\{a, b, c, \ldots, x, y, z\}$ includes all the letters of the alphabet.

| Example 7 | **Writing a Set Using an Ellipsis** |

Using the roster method, write the set containing all even natural numbers between 99 and 201.

SOLUTION

$\{100, 102, 104, \ldots, 198, 200\}$

2. Write sets three different ways.

Try This One   7

Using the roster method, write the set of odd natural numbers between 50 and 500.

There are some situations in which it's necessary to define a set with no elements. For example, the set of female players in the National Football League would contain no people, so it has no elements (at least as of this writing).

A set with no elements is called an **empty set** or **null set**. The symbols used to represent the empty set are { } or Ø.

| **Example 8** | Identifying Empty Sets |

Which of the following sets are empty?

(a) The set of woolly mammoth fossils in museums
(b) $\{x \mid x$ is a living woolly mammoth$\}$
(c) $\{\emptyset\}$
(d) $\{x \mid x$ is a natural number between 1 and 2$\}$

SOLUTION

(a) There is certainly at least one woolly mammoth fossil in a museum somewhere, so the set is not empty.
(b) Woolly mammoths have been extinct for almost 8,000 years, so this set is most definitely empty.
(c) This one's tricky. Each of { } and Ø represent the empty set, but $\{\emptyset\}$ is the set *containing* the empty set, which has one element. It's a goofy set, mind you, but it does have one element, so it is not empty.
(d) This set is empty because there are no natural numbers between 1 and 2.

Math Note

In December 2011, a group of scientists from Japan and Russia announced that it hoped to clone a woolly mammoth from long-frozen DNA found in Siberia within 5 years. It's now 2021 and we're still waiting, although there are also teams from both Penn State University and Harvard working to de-extinct mammoths. I think we can all agree that would be super-cool. So I reserve the right to change my answer to Example 8(b).

Try This One   8

Which of the following sets are empty?

(a) $\{x \mid x$ is a natural number divisible by 7$\}$
(b) $\{x \mid x$ is a human being living on Mars$\}$
(c) $\{\{ \ \}\}$
(d) The set Z consists of the living people on Earth who are over 120 years old.

CAUTION

Make sure you don't write the empty set as {Ø}; the brackets indicate a set containing what's inside, so that symbol represents a set containing one element: the empty set.

3. Define the empty set.

Cardinal Number of a Set

The number of elements in a set is called the *cardinal number* of a set. For example, the set $R = \{2, 4, 6, 8, 10\}$ has a cardinal number of 5 since it has 5 elements. This could also be stated by saying the **cardinality** of set R is 5. Formally defined,

The **cardinal number** of a set is the number of elements in the set. For a set A the symbol for the cardinality is $n(A)$, which is read as "n of A."

| Example 9 | Finding the Cardinality of a Set |

Find the cardinal number of each set.

(a) $A = \{5, 10, 15, 20, 25, 30\}$
(b) $B = \{x \mid x \in N \text{ and } x < 16\}$
(c) $C = \{16\}$
(d) $\emptyset$

SOLUTION

(a) $n(A) = 6$ since set A has 6 elements
(b) B is the set $\{1, 2, 3, 4, \ldots, 14, 15\}$, which has 15 elements. So $n(B) = 15$.
(c) $n(C) = 1$ since set C has 1 element
(d) $n(\emptyset) = 0$ since there are no elements in an empty set

Try This One   9

4. Find the cardinality of a set.

Find the cardinal number of each set.

(a) $A = \{z, y, x, w, v\}$
(b) $B = \{x \mid x \in E \text{ and } x \text{ is between 15 and 31}\}$
(c) $C = \{\text{Chevrolet}\}$

Finite and Infinite Sets

Sets can be classified as *finite* or *infinite*.

A set is called **finite** if it has no elements, or has cardinality that is a natural number. A set that is not finite is called an **infinite set**.

The set $\{p, q, r, s\}$ is a finite set since it has four members: p, q, r, and s. The set $\{10, 20, 30, \ldots\}$ is an infinite set since it has an unlimited number of elements: the natural numbers that are multiples of 10.

| Example 10 | Classifying Sets as Finite or Infinite |

Math Note

If you're wondering how to describe the cardinality of an infinite set, you're going to love Section 2-5.

Classify each set as finite or infinite.

(a) $\{x \mid x \in N \text{ and } x < 100\}$
(b) Set R is the set of letters used to make Roman numerals.
(c) $\{100, 102, 104, 106, \ldots\}$
(d) Set M is the set of people who went to your elementary school.
(e) Set S is the set of songs that can be written.

SOLUTION

(a) The set is finite since there are 99 natural numbers that are less than 100.
(b) The set is finite since the letters used are C, D, I, L, M, V, and X.
(c) The set is infinite since it consists of an unlimited number of elements.
(d) The set is finite since there is a specific number of people who went to any given school. You might have an awful time finding what that number is, but it's certainly finite.
(e) The set is infinite because an unlimited number of songs can be written.

5. Classify sets as finite or infinite.

Classify each set as finite or infinite.

(a) Set P is the set of numbers that are multiples of 6.
(b) $\{x \mid x$ is a member of the U.S. Senate$\}$
(c) $\{3, 6, 9, \ldots, 24\}$
(d) The set of all possible computer passwords

Equal and Equivalent Sets

When studying set theory, we'll need to understand the difference between two key concepts: *equal sets* and *equivalent sets*.

> Two sets A and B are **equal** (written $A = B$) if they have exactly the same members or elements. Two finite sets A and B are said to be **equivalent** (written $A \cong B$) if they have the same number of elements: that is, $n(A) = n(B)$.

Math Note

All equal sets are equivalent since both sets will have the same number of members, but not all equivalent sets are equal.

For example, the two sets $\{a, b, c\}$ and $\{c, b, a\}$ are equal since they have exactly the same members, a, b, and c. Also the set $\{4, 5, 6\}$ is equal to the set $\{4, 4, 5, 6\}$ since 4 doesn't have to be written twice in the second set. The set of all names of students in your class and the set of their student ID numbers are equivalent sets because they have the same number of elements, but they have different elements, so the sets are not equal.

Example 11 | Deciding If Sets Are Equal or Equivalent

Math Note

Can you think of two sets that are equal but not equivalent? What about the other way around? What can you conclude?

State whether each pair of sets is equal, equivalent, or neither.

(a) $\{p, q, r, s\}$; $\{a, b, c, d\}$
(b) $\{8, 10, 12\}$; $\{12, 8, 10\}$
(c) $\{213\}$; $\{2, 1, 3\}$
(d) $\{1, 2, 10, 20\}$; $\{2, 1, 20, 11\}$
(e) $\{$even natural numbers less than 10$\}$; $\{2, 4, 6, 8\}$

SOLUTION

(a) Equivalent
(b) Equal and equivalent
(c) Neither
(d) Equivalent
(e) Equal and equivalent

State whether each pair of sets is equal, equivalent, or neither.

(a) $\{d, o, g\}$; $\{c, a, t\}$
(b) $\{run\}$; $\{r, u, n\}$
(c) $\{t, o, p\}$; $\{p, o, t\}$
(d) $\{10, 20, 30\}$; $\{1, 3, 5\}$

When two sets have a relatively small number of elements, the simplest way to decide if sets are equivalent is to count the number of elements, as I imagine you've already realized. But when sets are really big, or infinite, there's a clever way to recognize equivalent sets: It's called putting them in one-to-one correspondence. This is going to come in really handy when we study infinite sets in Section 2–5.

Two sets have a **one-to-one correspondence** of elements if each element in the first set can be paired with exactly one element of the second set and each element of the second set can be paired with exactly one element of the first set.

Example 12	Putting Sets in One-to-One Correspondence

Show that (a) the sets {8, 16, 24, 32} and {s, t, u, v} have a one-to-one correspondence and (b) the sets {x, y, z} and {5, 10} do not have a one-to-one correspondence. Then draw a conclusion about what one-to-one correspondence has to do with equivalence of sets.

SOLUTION

(a) We need to demonstrate that each element of one set can be paired with one and only one element of the second set. One possible way to show a one-to-one correspondence is this:

$$\{8, \quad 16, \quad 24, \quad 32\}$$
$$\updownarrow \quad \updownarrow \quad \updownarrow \quad \updownarrow$$
$$\{s, \quad t, \quad u, \quad v\}$$

(b) The elements of the sets {x, y, z} and {5, 10} can't be put in one-to-one correspondence. No matter how we try, there will be an element in the first set that doesn't get matched up with any element in the second set.

What can we conclude? The sets that *could* be put into one-to-one correspondence had the same number of elements, while the two that could not had a different number of elements. Conclusion? Two sets are equivalent exactly when they can be put into one-to-one correspondence.

Mitrofanov Alexander/Shutterstock

Two sets of basketball teams on the court have a one-to-one correspondence (assuming each has five healthy players).

6. Decide if two sets are equal or equivalent.

Try This One   12

Show that the sets {North, South, East, West} and {sun, rain, snow, sleet} have a one-to-one correspondence.

Correspondence and Equivalent Sets

Two sets are

- Equivalent if you can put their elements in one-to-one correspondence.
- Not equivalent if you cannot put their elements in one-to-one correspondence.

Answers to Try This One

1 {January, February, May, July}

2 (a) {80, 82, 84, 86, 88, 90}; (Alternate descriptions can vary)
   (b) {11, 13, 15, 17, . . .}

3 (a) True   (b) False   (c) True

4 The set of integers from −3 to 3 and the set of the first four whole numbers and their negatives are two possibilities.

5 (a) $K = \{x | x \in E, x > 9,$ and $x < 19\}$, the set of all x such that x is an even natural number, x is greater than 9, and x is less than 19.
   (b) $W = \{x | x$ is a major American political party$\}$, the set of all x such that x is a major American political party.

6 (a) {11, 13, 15, 17, . . .}
   (b) The set of odd natural numbers greater than 10
   (c) $\{x | x \in N, x$ is odd, and $x > 10\}$

7 {51, 53, 55, . . . , 497, 499}

8 (b) and (d)

9 (a) 5   (b) 8   (c) 1

10 (a) Infinite   (b) Finite   (c) Finite   (d) Infinite

11 (a) Equivalent   (c) Equal and equivalent
   (b) Neither   (d) Equivalent

12 North South East West
   $\updownarrow$ $\updownarrow$ $\updownarrow$ $\updownarrow$
   Sun   Rain   Snow Sleet

Exercise Set   2-1

Writing Exercises

1. Explain what a set is.
2. What does it mean for a set to be well-defined?
3. Write an example of a set that is well-defined, and one that is not. (No stealing examples from the book!)
4. List and describe three ways to write sets.
5. What is the difference between equal and equivalent sets?
6. Explain the difference between a finite and an infinite set.
7. What is meant by "one-to-one correspondence between two sets"?
8. Define the empty set, and give two examples of an empty set.

Computational Exercises

For Exercises 9–22, write each set using the roster method. Pay attention to repeated elements, and THINK about why you don't need to list the same element more than once. You may have to do a bit of Internet research for some problems.

9. T is the set of letters in the word *thinking*.
10. A is the set of letters in the word *Alabama*.
11. P is the set of natural numbers between 50 and 60.
12. R is the set of even natural numbers between 10 and 40.
13. $C = \{x \mid x \in N \text{ and } x < 9\}$
14. $F = \{x \mid x \in N \text{ and } x > 100\}$
15. $G = \{x \mid x \in N \text{ and } x > 10\}$
16. B is the set of natural numbers greater than 100.
17. Y is the set of natural numbers between 2,000 and 3,000.
18. $Z = \{x \mid x \in N \text{ and } 500 < x < 6,000\}$
19. C is the set of colors in the flags of the states that begin with O.
20. S is the set of current U.S. senators from states that begin with A.
21. L is the set of ligaments in the human knee.
22. A is the set of capitals of the seven mainland countries in Central America.

For Exercises 23–28, decide if the statement is true or false.

23. $5 \in \{1, 3, 5, 7\}$
24. $8 \notin \{2, 4, 6, \ldots\}$
25. $\frac{1}{2} \notin N$
26. $0.6 \in N$
27. $\{x \mid x \text{ is a living stegosaurus}\}$ is an empty set.
28. Cleveland $\in \{x \mid x \text{ is one of the United States}\}$

For Exercises 29–36, write each set, using the descriptive method.

29. $\{5, 10, 15, 20, \ldots\}$
30. $\{4, 8, 12, 16\}$
31. $\{13, 26, 39, 52\}$
32. $\{7, 14, 21, 28, \ldots\}$
33. $\{s, t, e, v, n\}$
34. $\{a, u, g, s, t\}$
35. $\{100, 101, 102, \ldots, 199\}$
36. $\{21, 22, 23, \ldots, 29, 30\}$

For Exercises 37–42, write each set using set-builder notation, then write an alternate description for each set.

37. $\{10, 20, 30, 40, \ldots\}$
38. $\{3, 6, 9, 12, \ldots\}$
39. X is the set of odd natural numbers less than 16.
40. Z is the set of natural numbers between 70 and 76.
41. $\{red, white, blue\}$
42. $\{black, white, gray\}$

For Exercises 43–48, list the elements in each set.

43. H is the set of natural numbers less than 0.
44. $\{x \mid x \in N \text{ and } 70 < x < 80\}$
45. $\{x \mid x \text{ is a season of the year}\}$
46. R is the set of letters that can be both a vowel and a consonant.
47. $\{x \mid x \text{ is an even natural number between 100 and 120}\}$
48. $\{x \mid x \text{ is an odd natural number between 90 and 100}\}$

For Exercises 49–54, state whether each collection is well-defined or not well-defined.

49. L is the set of contestants that have won *Survivor*.
50. $\{I \mid I \text{ is a death row inmate in Texas}\}$
51. $\{NBA \text{ players that had awesome dunks last week}\}$
52. N is the set of patients that deserve a heart transplant.
53. $B = \{x \mid x \text{ is a large number}\}$
54. $C = \{x \mid x \text{ is a number greater than the number of people in the United States}\}$

For Exercises 55–60, decide if the statement is true or false.

Let $A =$ the set of U.S. state capitals
   $B = \{10, 20, 30, 40, \ldots\}$
   $C =$ the set of presidents of the United States

55. $35 \in B$
56. Benjamin Franklin $\in C$
57. Philadelphia $\notin A$
58. $350 \in B$
59. Cheyenne $\in A$
60. James Madison $\in C$

For Exercises 61–68, state whether each set is infinite or finite.

61. $\{x \mid x \in N \text{ and } x \text{ is even}\}$

62. $\{1, 2, 3, \ldots, 999, 1,000\}$
63. K is the set of letters of the English alphabet.
64. $\{x \mid x \in$ years in which the past presidents of the United States were born$\}$
65. $\{x \mid x \in N$ and x is a number whose last digit is zero$\}$
66. $\varnothing$
67. $\{x \mid x$ is a current television program$\}$
68. $\{x \mid x$ is a fraction$\}$

For Exercises 69–74, state whether each pair of sets is equal, equivalent, or neither.

69. $\{s, t, u, v, w\}$ and $\{t, v, w, s, u\}$
70. $\{1, 2, 3, 4, 5\}$ and $\{10, 20, 30, 40, 50\}$
71. $\{2, 4, 6, 8\}$ and $\{2, 4, 6, 8, \ldots\}$
72. $\{three\}$ and $\{t, h, r, e, e\}$
73. $\{3\}$ and $\{\varnothing\}$
74. $\{x \mid x \in$ months with exactly 30 days$\}$ and $\{$April, June, September, November$\}$

For Exercises 75–78, show that each pair of sets is equivalent by using a one-to-one correspondence.

75. $\{10, 20, 30, 40\}$ and $\{40, 10, 20, 30\}$
76. $\{w, x, y, z\}$ and $\{1, 2, 3, 4\}$

77. $\{x \mid x \in N\}$ and $\{x \mid x$ is a multiple of 4$\}$
78. $\{x \mid x$ is an odd natural number less than 11$\}$ and $\{x \mid x$ is an even natural number less than 12$\}$

For Exercises 79–86, find the cardinal number for each set.

79. $A = \{63, 72, 51, 44\}$
80. $B = \{10, 11, 12, \ldots, 20\}$
81. $C = \{x \mid x$ is a day of the week$\}$
82. $D = \{x \mid x$ is a month of the year$\}$
83. $E = \{three\}$
84. $F = \{t, h, r, e, e\}$
85. $G = \{x \mid x \in N$ and x is negative$\}$
86. $H = \varnothing$

For Exercises 87–92, determine whether each statement is true or false.

87. All equal sets are equivalent.
88. No equivalent sets are equal.
89. $n(\{\varnothing\}) = 0$
90. $E = \{2, 4, 6, 8, \ldots\}$ is a finite set
91. $\{1, 2, 3, 4, \ldots\}$ is equivalent to $\{10, 20, 30, 40, \ldots\}$
92. $n(\{\ \}) = 0$

Applications in Our World

93. The table shows the top 10 states in number of immigrants granted permanent resident status in 2018.

State	Number of Immigrants	% of Total Immigrants to United States
California	200,897	18.3%
New York	134,839	12.3%
Florida	130,405	11.9%
Texas	104,515	9.5%
New Jersey	54,424	5.0%
Illinois	38,287	3.5%
Massachusetts	33,174	3.0%
Virginia	27,426	2.5%
Georgia	26,725	2.4%
Pennsylvania	26,078	2.4%

Source: U.S. Dept. of Homeland Security

(a) List the set of states with more than 100,000 immigrants.
(b) List the set of states in the top 10 with fewer than 50,000 immigrants.
(c) List $\{x \mid x$ is a state with at least 4% of the immigrant total$\}$.
(d) List $\{x \mid x$ is a state with between 3% and 10% of the immigrant total$\}$.

94. This table shows the top 10 states in terms of undocumented immigrant population (estimated) in 2018. This problem uses this table and the one from Exercise 93.

State	Number of Undocumented Immigrants	% of Total Undocumented Immigrant Population
California	2,625,000	23.9%
Texas	1,730,000	15.8%
New York	866,000	7.9%
Florida	732,000	6.7%
Illinois	437,000	4.0%
New Jersey	425,000	3.9%
Georgia	330,000	3.0%
North Carolina	298,000	2.7%
Arizona	281,000	2.6%
Virginia	251,000	2.3%

Source: Statista

(a) List the set of states that rank in the top 5 in both documented and undocumented immigration.
(b) List the set of states in both of the top 10 that have a higher percentage of documented immigrants than undocumented.
(c) Write a description of a set of states that is empty based on the data in the two tables.
(d) Using the information in the two tables, write a verbal description of the set $\{$California, New York, Texas, Florida$\}$.

95. Excessive alcohol consumption by those aged 18–24 affects nearly all U.S. college students, whether they choose to drink or not. Some consequences of excessive drinking are listed in the next table.

Consequence	Average Number of College Students Aged 18–24 Affected per Year
Death	1,825
Injury	599,000
Assault	696,000
Sexual abuse	97,000
Unsafe sex	400,000
Health problems	150,000
Drunk driving	3,360,000

Source: http://www.collegedrinkingprevention.gov/StatsSummaries/snapshot.aspx

(a) List the set of the three consequences with the most students affected by excessive alcohol consumption.

(b) List the set of consequences that affect between 100,000 and 600,000 college students each year.

(c) Find the set $\{x \mid x$ is a consequence of which over a half million students are affected$\}$.

(d) Find the set $\{x \mid x$ is the average number of college students affected by sexual abuse, death, or health problems$\}$.

(e) If A is the set of students affected by health problems, injury, or drunk driving, can you find $n(A)$? Why or why not?

96. The number of bachelor's degrees awarded in the United States in the top 10 majors for 2018 is listed in the table, with data for 2000 and 2010 as well.

Major	2000	2010	2018
Business	263,515	365,133	390,564
Health professions and related programs	75,933	143,463	251,355
Social sciences and history	128,036	177,169	160,628
Engineering	58,209	76,356	126,687
Biological and biomedical sciences	60,576	89,984	121,191
Psychology	73,645	100,906	116,536
Communication/Journalism/ related programs	58,013	83,231	92,528
Visual and performing arts	61,148	93,939	89,730
Computer and information sciences	44,142	43,066	88,633
Education	105,458	104,008	83,946

Source: Consumer Sentinel Network

(a) List the set of majors that increased in popularity every year listed.

(b) List the set of majors that did not increase in popularity from 2010 to 2018.

(c) List the set of majors that had between 90,000 and 120,000 bachelor's degrees awarded in 2018.

(d) Find the set $\{x \mid x$ decreased in popularity between 2000 and 2010$\}$.

(e) To find the percent increase P between an original amount O and a new amount N, use the following formula: $P = (N - O)/O$. Calculate the percent increase for any major that saw an increase in degrees awarded between 2010 and 2018. List the set of majors that increased at least 30%.

97. Identity theft is now one of the most costly crimes in the United States, and younger adults are often affected. In 2018 alone, almost half a million people under the age of 30 were victims of financial fraud. The following charts show types of identity theft fraud reported in 2018 and the percentage of victims by age.

Types of Identity Theft Fraud Reported in 2018

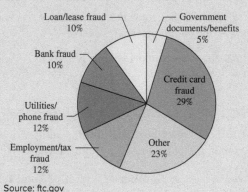

Source: ftc.gov

Percentage of Victims by Age

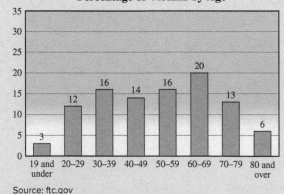

Source: ftc.gov

(a) List the set of the three types of identity fraud with the lowest percentage of reported crimes in 2018.

(b) List the set of age groups that are above 18%.

(c) List the set of identity fraud types that make up more than 15% of reported crimes.

(d) Find the set $\{x \mid x$ is a percentage of those 40 and over who are victims of identity fraud$\}$.

(e) Find the set $\{x \mid x$ is a type of fraud that makes up between 9% and 20% of reported crimes$\}$.

(f) Add up the percentages in the pie chart. What do you notice? Can you think of possible explanations?

98. In the last several years, revenues from music in the form of physical media (like CDs and LPs) and in digital downloads have been dwarfed by revenue from online streaming. The chart displays the annual revenue in millions of dollars for each of these types of distribution for the years from 2015 to 2020.

Music Industry Revenue by Medium

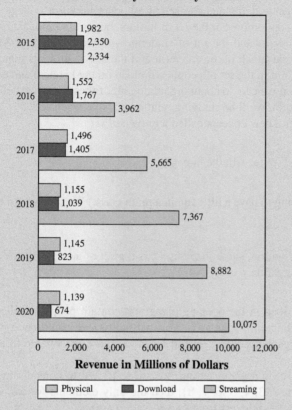

Revenue in Millions of Dollars

☐ Physical   ■ Download   ☐ Streaming

(a) List the set of years in which the value of digital downloads exceeded the value of physical sales.

(b) List the set of years in which the value of physical sales was more than the value of streaming revenue.

(c) List the set $\{x \mid x$ is a year in which streaming revenue exceeded the combined total for digital downloads and physical sales$\}$.

(d) List the set $\{x \mid x$ is a distribution medium that exceeded a billion dollars in sales in 2018$\}$.

99. The median prices of existing homes in the United States for the years from 2004 to 2021 are shown in the line graph below.

Median Price of Existing Homes in the United States

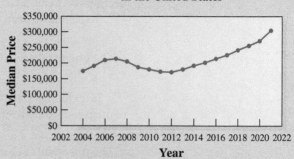

Source: National Association of Realtors

(a) List the set of years in which the median price was above $200,000.

(b) List the set of years in which the median price was between $160,000 and $200,000.

(c) Find $\{x \mid x$ is a year in which the median price increased from the year before$\}$.

(d) Find $\{x \mid x$ is a year in which the median price decreased from the year before$\}$.

Critical Thinking

100. If $A \cong B$ and $A \cong C$, is $B \cong C$? Explain your answer.

101. Is $\{0\}$ equivalent to $\varnothing$? Explain your answer.

102. Write two sets that are equivalent but not equal. Why is it not possible to write two sets that are equal but not equivalent?

103. We know that two sets are equivalent if we can match up their elements in a one-to-one correspondence.
    (a) Which set has more elements: $A = \{1, 2, 3, 4, 5, 6, \ldots\}$ or $B = \{2, 4, 6, 8, 10, \ldots\}$?
    (b) Write out a correspondence between A and B where every element in A gets matched with its double in B. Does this change your mind about your answer to part (a)? (If you find this problem interesting, you'll like Section 2-5 very much.)

104. (a) List all of the different sets you can form using only the elements in the set $\{2, 4, 6\}$.
    (b) There are eight sets that can be formed in part (a). Did you find seven of them? If so, can you figure out why you missed one?

105. Explain why each of the following sets is not well-defined.
    (a) The set of all Americans
    (b) The set of luxury cars in the 2021 model year
    (c) The set of all colleges with a legitimate chance to win the NCAA basketball tournament (There are at least two reasons!)
    (d) The set of all jobs that pay over $50,000 per year
    (e) The set of mothers

106. Sometimes when a set is not well-defined, you can provide a better description that makes it well-defined. For example, the set of really great movies from 2016 is not well-defined. But if we change our description to the set of movies that were released in 2016 and have at least a 90/100 critics rating on Rotten Tomatoes, we've defined what we mean by "really great," and now the set is well-defined. For each of the sets in Exercise 105, write an improved description that makes the set well-defined.

Section 2-2	**Subsets and Set Operations**

✓ **LEARNING OBJECTIVES**

1. Define the complement of a set.
2. Find all subsets of a set.
3. Use subset notation.
4. Find the number of subsets for a set.
5. Find intersections, unions, and differences of sets.

We've seen that set theory is about identifying relationships between things that are grouped together for some reason. Taking that idea a little further, sets often have relationships with other sets, and that's when things get a bit complicated. In that case, a system for displaying and studying those relationships will come in handy, which ultimately is kind of the main point of studying set theory. So far, we've just scratched the surface. For example, you are a member of both the set of college students and the set of students taking a college math course. You could be in the set of sophomores or the set of juniors, but not in both. You might be in the set of nontraditional students and the set of students who work full time. Maybe you're in the set of students who eat lunch in the cafeteria and the set of students that think the french fries are too soggy, but not in the set of people who load on the ketchup and eat the lousy things anyway. Let's see what we can do about organizing all of these complicated connections between sets. In this section, we'll be studying relationships between sets.

To begin, we need to consider a new concept called a *universal set.*

> The **universal set** is the set of all objects in the universe.

Just kidding. (No harm in trying to have a little fun in a math class.) We now return to our regularly scheduled definition:

> The **universal set** for a given situation, symbolized by U, is the set of all objects that are reasonable to consider in that situation.

For example, all of the sets described in the opening paragraph contain people, so theoretically we could use the set of all human beings as U. For that matter, we could choose U to be the set of all carbon-based organisms. But it would be much more reasonable to assign $U = \{$College students$\}$.

Once we define a universal set in a given setting, we are restricted to considering only elements from that set. If $U = \{1, 2, 3, 4, 5, 6, 7, 8\}$, then the only elements we can use to define other sets in this setting are the integers from 1 to 8.

In the remainder of this chapter, we'll use a clever method for visualizing sets and their relationships called a *Venn diagram* (so named because it was developed by a man named John Venn in the 1800s). Figure 2-1 shows an example.

You can get a lot of information from this simple diagram. A set called A is being defined. The universal set from which elements of A can be chosen is $U = \{1, 2, 3, 4, 5, 6, 7, 8\}$. The set A is $\{2, 4, 6, 8\}$, and the elements not in A are $\{1, 3, 5, 7\}$. We will call the elements in U that are not in A the *complement* of $A,$ and denote it A'.

> The **complement** of a set $A,$ symbolized A', is the set of elements in the universal set that are *not* in A. Using set-builder notation, the complement of A is $A' = \{x \mid x \in U \text{ and } x \notin A\}$.

In a Venn diagram, the complement of a set A is all the things inside the rectangle that are not inside the circle representing set A. This is shown in Figure 2-2.

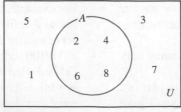

Figure 2-1

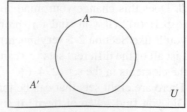

Figure 2-2

Example 1	**Finding the Complement of a Set**

(a) Let $U = \{v, w, x, y, z\}$ and $A = \{w, y, z\}$. Find A' and draw a Venn diagram that illustrates these sets.

(b) What is the complement of the universal set for a given situation?

SOLUTION

(a) Using the list of elements in U, we just have to cross out the ones that are also in A. The elements left over are in A'.

$$U = \{v, \cancel{w}, x, \cancel{y}, \cancel{z}\} \qquad A' = \{v, x\}$$

The Venn diagram is shown in Figure 2-3.

Figure 2-3

(b) There are no elements in a universal set that are not in the universal set, so according to the definition of complement, there are no elements in the complement of a universal set, which means the complement is the empty set.

Try This One   1

(a) Let $U = \{10, 20, 30, 40, 50, 60, 70, 80, 90\}$ and $A = \{10, 30, 50\}$. Find A' and draw a Venn diagram that illustrates these sets.

(b) What is the complement of the empty set?

1. Define the complement of a set.

Subsets

At the beginning of the section, we pointed out that you're in both the set of college students and the set of students taking a college math course. Notice that everyone in the second set is automatically in the first one. (Obviously, if you're taking a college math course, you have to be a college student.) We could say that the set of students taking a college math course is contained in the set of all college students. When one set is contained in a second set, we call the smaller set a *subset* of the larger one.

> If every element of a set A is also an element of a set B, then A is called a **subset** of B. The symbol $\subseteq$ is used to designate a subset; in this case, we write $A \subseteq B$.

An alternate definition is that A is a subset of B if there are no elements in A that are not also in B. Here are a couple of observations about subsets.

- Every set is a subset of itself. Every element of a set A is obviously an element of set A, so $A \subseteq A$.

- The empty set is a subset of every set. The empty set has no elements, so for any set A, you can't find an element of $\emptyset$ that is not also in A.

If we start with the set $\{x, y, z\}$, let's look at how many subsets we can form:

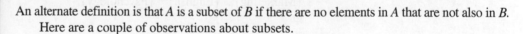

Number of Elements in Subset	**Subsets with That Number of Elements**	
3	$\{x, y, z\}$	(One subset)
2	$\{x, y\}, \{x, z\}, \{y, z\}$	(Three subsets)
1	$\{x\}, \{y\}, \{z\}$	(Three subsets)
0	$\emptyset$	(One subset)

So for a set with three elements, we can form eight subsets.

Barry Barker/McGraw Hill

There are many subsets of this set of spring breakers: the subset of female students, the subset of guys fruitlessly trying to impress those female students, the subset of students who had their fake I.D. confiscated by the police, and so on.

| **Example 2** | Finding All Subsets of a Set |

Find all subsets of $A = \{\text{Cold, Flu}\}$.

SOLUTION

The subsets are

> {Cold, Flu}
> {Cold}
> {Flu}
> Ø

Note that a set with two elements has four subsets.

Try This One   2

Find all subsets of $B = \{\text{Verizon, T-Mobile, AT\&T}\}$.

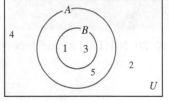

Figure 2-4 $B \subset A$

To indicate that a set is not a subset of another set, the symbol $\not\subseteq$ is used. For example, $\{1, 3\} \not\subseteq \{0, 3, 5, 7\}$ since $1 \notin \{0, 3, 5, 7\}$.

Of the four subsets in Example 2, only one is equal to the original set. We will call the remaining three *proper subsets* of A. The Venn diagram for a proper subset is shown in Figure 2-4. Notice that the blue circle is entirely inside the red one, and there's an element inside the red circle that is not in the blue one. In this case, $U = \{1, 2, 3, 4, 5\}$, $A = \{1, 3, 5\}$, and $B = \{1, 3\}$.

> If a set A is a subset of a set B and is not equal to B, then we call A a **proper subset** of B, and write $A \subset B$. That is, $A \subseteq B$ and $A \neq B$.

| **Example 3** | Finding Proper Subsets of a Set |

Find all proper subsets of {Marketing, English, Psychology}

SOLUTION

> {Marketing, English}   {Marketing, Psychology}   {English, Psychology}
> {Marketing}   {English}   {Psychology}
> Ø

Try This One   3

Find all proper subsets of {Spring, Summer, Fall, Winter}.

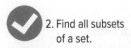

2. Find all subsets of a set.

The symbol $\not\subset$ is used to indicate that the set is not a proper subset. For example, $\{1, 3\} \subset \{1, 3, 5\}$, but $\{1, 3, 5\} \not\subset \{1, 3, 5\}$.

| **Example 4** | Understanding Subset Notation |

Decide if each statement is true or false.

(a) $\{1, 3, 5\} \subseteq \{1, 3, 5, 7\}$
(b) $\{a, b\} \subset \{a, b\}$
(c) $\{x \mid x \in E \text{ and } x > 10\} \subset N$

Math Note

It's important not to confuse the concept of subsets with the concept of elements. For example, the statement $6 \in \{2, 4, 6\}$ is true since 6 is an element of the set $\{2, 4, 6\}$, but the statement $\{6\} \in \{2, 4, 6\}$ is false since it states that the set containing the element 6 is an element of the set containing 2, 4, and 6. However, it *is* correct to say that $\{6\} \subseteq \{2, 4, 6\}$, or that $\{6\} \subset \{2, 4, 6\}$.

(d) $\{r, s, t\} \not\subset \{t, s, r\}$

(e) $\{$Lake Erie, Lake Huron$\} \not\subset$ The set of Great Lakes

(f) $\varnothing \subset \{5, 10, 15\}$

(g) $\{u, v, w, x\} \subseteq \{x, w, u\}$

(h) $\{0\} \subseteq \varnothing$

SOLUTION

(a) All of 1, 3, and 5 are in the second set, so $\{1, 3, 5\}$ is a subset of $\{1, 3, 5, 7\}$. The statement is true.

(b) Even though $\{a, b\}$ is a subset of $\{a, b\}$, it is not a proper subset, so the statement is false.

(c) Every element in the first set is a natural number, but not all natural numbers are in the set, so that set is a proper subset of the natural numbers. The statement is true.

(d) The two sets are identical, so $\{r, s, t\}$ is not a proper subset of $\{t, s, r\}$. The statement is true.

(e) Lake Erie and Lake Huron are both Great Lakes, so the set $\{$Lake Erie, Lake Huron$\}$ is a subset of the set of Great Lakes. The statement is false.

(f) True: The empty set is a proper subset of every set except itself.

(g) False: v is an element of $\{u, v, w, x\}$ but not $\{x, w, u\}$.

(h) The set on the left has one element, 0. The empty set has no elements, so the statement is false.

Try This One   4

Decide if each statement is true or false.

(a) $\{8\} \subseteq \{x | x$ is an even natural number$\}$

(b) $\{6\} \subseteq \{1, 3, 5, 7, \ldots\}$

(c) $\{2, 3\} \subseteq \{x | x \in N\}$

(d) $\{a, b, c\} \subset \{$letters of the alphabet$\}$

(e) $\varnothing \in \{x, y, z\}$

(f) $\varnothing \subseteq \{$red, yellow, blue$\}$

(g) $\{100, 200, 300, 400\} \subset \{200, 300, 400\}$

(h) $\{\varnothing\} \subseteq \varnothing$

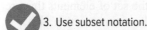 3. Use subset notation.

A set with one element has two subsets—itself and the empty set. We have seen that if a set has two elements, there are four subsets, and if a set has three elements, there are eight subsets. This is an excellent opportunity to use the inductive reasoning that we practiced in Chapter 1!

Number of elements	0	1	2	3
Number of subsets	1	2	4	8

Math Note

In Exercise 111, we'll look at an alternate approach to developing the formula for the number of subsets. It uses deductive reasoning rather than inductive reasoning.

It's not too hard to see that the number of subsets doubles every time you add one more element to the original set. It takes a little more thought to see that, so far at least, in every case, the number of subsets is 2 raised to the number of elements. So we might use inductive reasoning to conjecture that the number of subsets for any set follows that same pattern, and if we did so, we would be correct. Also, the number of proper subsets is always one less, so we get the following formulas:

The Number of Subsets for a Finite Set

If a finite set has n elements, then the set has 2^n subsets and $2^n - 1$ proper subsets.

Example 5	Finding the Number of Subsets of a Set

(a) Find the number of subsets and proper subsets of the set $\{$Ace, King, Queen, Jack, Ten, Nine$\}$.

(b) Explain why the number of proper subsets for a set is always one less than the total number of subsets.

SOLUTION

(a) The set has $n = 6$ elements, so there are 2^n, or $2^6 = 64$, subsets. Of these, $2^n - 1$, or 63, are proper. (Recall that 2^6 means $2 \cdot 2 \cdot 2 \cdot 2 \cdot 2 \cdot 2$, which is 64.)

(b) When finding proper subsets, there's only one subset excluded from the list of ALL subsets: the original set itself. So there will always be one fewer proper subsets than total subsets.

Try This One   5

Find the number of subsets and proper subsets of the set {OSU, USC, KSU, MSU, UND, PSU, UT, FSU}.

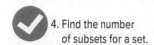

4. Find the number of subsets for a set.

Intersection and Union of Sets

At the beginning of the section, we pointed out that you might be in both the set of nontraditional students and the set of students who work full time. We will identify objects that are common to two or more sets by using the term *intersection*.

> The **intersection** of two sets A and B, symbolized by $A \cap B$, is the set of all elements that are in both sets. In set-builder notation, $A \cap B = \{x \mid x \in A \text{ and } x \in B\}$.

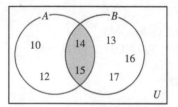

Figure 2-5 $A \cap B = \{14, 15\}$

For example, if $A = \{10, 12, 14, 15\}$ and $B = \{13, 14, 15, 16, 17\}$, then the intersection $A \cap B = \{14, 15\}$, since 14 and 15 are the only elements that are common to both sets. The Venn diagram for $A \cap B$ is shown in Figure 2-5. Notice that the elements of A are placed inside the circle for set A, and the elements of B are inside the circle for set B. The elements in the intersection are placed into the portion where the circles overlap: $A \cap B$ is the shaded portion. This makes perfect sense because that's exactly the region that is inside both circles.

Intersection is an example of a **set operation**—a rule for combining two or more sets to form a new set. The intersection of three or more sets consists of the set of elements that are in every single set. Note that the word *and* is sometimes used to indicate intersection; $A \cap B$ is the set of elements in A and B.

Example 6   **Finding Intersections**

Three experimental medications are being evaluated for safety. Each has a list of side effects that has been reported by at least 1% of the people trying the medication. This a blind trial, so the medications are simply labeled A, B, and C. The side effects for each are listed below.

$A = \{$nausea, night sweats, nervousness, dry mouth, swollen feet$\}$
$B = \{$weight gain, nausea, nervousness, blurry vision, fever, trouble sleeping$\}$
$C = \{$dry mouth, nausea, blurry vision, fever, weight loss, eczema$\}$

Find each requested set.

(a) $A \cap B$ (b) $B \cap C$ (c) $A \cap B \cap C$

SOLUTION

(a) There are two side effects listed for both A and B: nausea and nervousness. So $A \cap B = \{$nausea, nervousness$\}$.

(b) There are three side effects common to drugs B and C: nausea, blurry vision, and fever. So $B \cap C = \{$nausea, blurry vision, fever$\}$.

(c) This example indicates that it makes perfect sense to find the intersection of more than two sets: you just find elements that are in EVERY set. In this case, only one of the side effects is listed for all three drugs: nausea. So $A \cap B \cap C = \{$nausea$\}$.

Try This One 6

If A = {Cleveland, Indianapolis, Chicago, Des Moines, Detroit}, B = {New York, Los Angeles, Chicago, Detroit}, and C = {Seattle, Los Angeles, San Diego}, find $A \cap B$, $B \cap C$, and $A \cap B \cap C$.

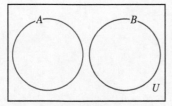

Figure 2-6 $A \cap B = \emptyset$

When the intersection of two sets is the empty set, the sets are said to be **disjoint**. For example, the set of students who stop attending class midway through a term and the set of students earning A's are disjoint, because you can't be a member of both sets. The Venn diagram for a pair of disjoint sets A and B is shown in Figure 2-6. If the sets have no elements in common, the circles representing them don't overlap at all.

Another way of combining sets to form a new set is called *union*.

The **union** of two sets A and B, symbolized by $A \cup B$, is the set of all elements that are in either set A or set B (or both). In set-builder notation,

$$A \cup B = \{x \mid x \in A \text{ or } x \in B\}$$

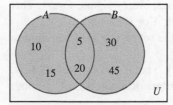

Figure 2-7 $A \cup B$

For example, if A = {5, 10, 15, 20} and B = {5, 20, 30, 45}, then the union $A \cup B$ = {5, 10, 15, 20, 30, 45}. Even though 5 and 20 are in both sets, we list them only once in the union. The Venn diagram for $A \cup B$ is shown in Figure 2-7. The set $A \cup B$ is the shaded area consisting of all elements in either set.

Example 7	Finding Unions

For the sets in Example 6, find each of the following, then describe verbally what each set represents.

(a) $A \cup B$
(b) $A \cup C$
(c) $A \cup B \cup C$

SOLUTION

To find a union, just make a list of all the elements from each set without writing repeats.

(a) $A \cup B$ = {nausea, night sweats, nervousness, dry mouth, swollen feet, weight gain, blurry vision, fever, trouble sleeping}. This is the set of side effects that were reported by more than 1% of subjects taking either medication A or B.
(b) $A \cup C$ = {nausea, night sweats, nervousness, dry mouth, swollen feet, blurry vision, fever, weight loss, eczema}. This is the set of side effects reported by more than 1% of subjects taking either medication A or C.
(c) $A \cup B \cup C$ = {nausea, night sweats, nervousness, dry mouth, swollen feet, weight gain, blurry vision, fever, trouble sleeping, weight loss, eczema}. This is the set of side effects reported by more than 1% of subjects taking any of the three medications.

Try This One 7

For the sets in Try This One 6, find $A \cup B$, $B \cup C$, and $A \cup B \cup C$.

What about operations involving more than two sets and more than one operation? Just like with operations involving numbers, we use parentheses to indicate an order of operations. This is illustrated in Example 8.

| **Example 8** | Performing Set Operations |

Once again using the sets in Example 6, find each requested set and write a verbal description of what each set represents.

(a) $(A \cup B) \cap C$
(b) $A \cap (B \cup C)$
(c) $(A \cap B) \cup C$

SOLUTION

The key is to perform the operation in parentheses first.

(a) First, find $A \cup B$: this is the set we found in part (a) of Example 7. Now find elements common to that set and C: $(A \cup B) \cap C = \{$dry mouth, nausea, blurry vision, fever$\}$. This is the set of side effects common to medication C and either A or B (or both).
(b) $A \cap (B \cup C) = \{$nausea, dry mouth, nervousness$\}$. This is the set of side effects common to medication A and either B or C (or both).
(c) $(A \cap B) \cup C = \{$nausea, nervousness, dry mouth, blurry vision, fever, weight loss, eczema$\}$. This is the set of side effects reported either by users of medication C, or both users of A and users of B.

| Try This One | 8 |

For the sets in Try This One 6, find $A \cup (B \cap C)$, $(A \cap B) \cup C$, and $A \cap (B \cup C)$.

| **CAUTION** | When combining union and intersection with complements as we will in Example 9, we'll have to be extra careful. Pay particular attention to the parentheses and to whether the complement symbol is inside or outside the parentheses. |

| **Example 9** | Performing Set Operations |

Math Note

Don't forget the importance of the universal set when finding complements: the complement of a set A is all of the elements in *the universal set* that are not in A, not all of the objects in the universe that are not in A.

Going back to our sets of side effects, recall that these were reported by at least 1% of users. The universal set below is the set of all side effects reported by ANY users. Use this to find the following sets.

$U = \{$nausea, night sweats, nervousness, dry mouth, swollen feet, weight gain, blurry vision, fever, trouble sleeping, weight loss, eczema, motor mouth, darting eyes, uncontrollable falling down$\}$

(a) $A' \cap C'$ (b) $(A \cap B)' \cap C$ (c) $B' \cup (A \cap C')$

SOLUTION

(a) First, find A', which is all of the items in the universal set that are not in A: $A' = \{$weight gain, blurry vision, fever, trouble sleeping, weight loss, eczema, motor mouth, darting eyes, uncontrollable falling down$\}$.

Next, find C': $C' = \{$night sweats, nervousness, swollen feet, weight gain, trouble sleeping, motor mouth, darting eyes, uncontrollable falling down$\}$.

Now $A' \cap C'$ is the elements common to A' and C': $A' \cap C' = \{$weight gain, trouble sleeping, motor mouth, darting eyes, uncontrollable falling down$\}$.

(b) The parentheses tell us that we should find $A \cap B$ first: $A \cap B = \{$nausea, nervousness$\}$. Next, find the complement: $(A \cap B)' = \{$night sweats, dry mouth, swollen feet, weight gain, blurry vision, fever, trouble sleeping, weight loss, eczema, motor mouth, darting eyes, uncontrollable falling down$\}$. The elements of this set that are also in set C are what we're looking for. This is $(A \cap B)' \cap C = \{$dry mouth, blurry vision, fever, weight loss, eczema$\}$.

(c) First, find $A \cap C'$: $C' = \{$night sweats, nervousness, swollen feet, weight gain, trouble sleeping, motor mouth, darting eyes, uncontrollable falling down$\}$, so $A \cap C' = \{$night sweats, nervousness, swollen feet$\}$.

Next, note that $B' = \{$night sweats, dry mouth, swollen feet, weight loss, eczema, motor mouth, darting eyes, uncontrollable falling down$\}$.

The union we're looking for is all of the stuff in B' along with the stuff in $A \cap C'$ that aren't already listed in B':

$B' \cup (A \cap C') = \{$night sweats, dry mouth, swollen feet, weight loss, eczema, motor mouth, darting eyes, uncontrollable falling down, nervousness$\}$.

Try This One   9

The universal set for the sets in Try This One 6 is $U = \{$Cleveland, Indianapolis, Chicago, Des Moines, Detroit, New York, Los Angeles, Seattle, San Diego, Phoenix, Boston$\}$. Find each set.

(a) C' (b) $(A \cup B)'$ (c) $A' \cap C'$ (d) $(A \cup B) \cap C'$

The union and intersection of sets are commonly used in real life—it's just that you might not have thought of it in those terms. For example, the intersection of the set of U.S. citizens older than 17 and the set of U.S. citizens who are not convicted felons makes up the set of those eligible to vote in national elections. The union of the set of your mother's parents and your father's parents forms the set of your grandparents.

Set Subtraction

The third set operation we'll study is called the *difference* of sets. We also call it *set subtraction* and use a minus sign to represent it.

Momentum Creative Group/
Alamy Stock Photo

> The **difference** of set A and set B is the set of elements in set A that are *not* in set B. In set-builder notation, $A - B = \{x \mid x \in A \text{ and } x \notin B\}$.

| **Example 10** | Finding the Difference of Two Sets |

Math Note

Sometimes operations can be written in terms of other operations. For example, $3 - 5$ is also $3 + (-5)$. Can you think of a way to write $A - B$ using intersection and complement? Drawing a Venn diagram might help.

Once more, we're going to use the sets from Example 6. You should be pretty well-acquainted with them by now.

Find each set.

(a) $A - B$ (b) $B - C$ (c) $(A - B) - C$

SOLUTION

(a) Start with the elements of A, then take out anything in B that's also in A. The common elements are nausea and nervousness, so $A - B = \{$night sweats, dry mouth, swollen feet$\}$.

(b) This time, start with set B, then throw out the things in C that are also in B. The common elements are nausea, blurry vision, and fever, so $B - C = \{$weight gain, nervousness, trouble sleeping$\}$.

(c) We already know that $A - B = \{$night sweats, dry mouth, swollen feet$\}$; now we need to find any elements that are also in C and throw them out. Only dry mouth is also in C, so $(A - B) - C = \{$night sweats, swollen feet$\}$.

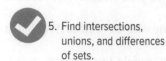

5. Find intersections, unions, and differences of sets.

Try This One   10

For the sets in Try This One 6, find each set.

(a) $A - B$ (b) $B - C$ (c) $(B - C) - A$

So far, we've used Venn diagrams as a way to picture certain sets. In the next two sections we'll study how these diagrams can be used to study set operations in greater depth and in a variety of applied settings.

Answers to Try This One

1 (a) $A' = \{20, 40, 60, 70, 80, 90\}$

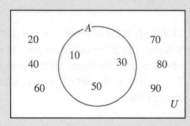

(b) The complement of the empty set is the universal set.

2 $\{$Verizon, T-Mobile, AT&T$\}$, $\{$Verizon, T-Mobile$\}$, $\{$Verizon, AT&T$\}$, $\{$T-Mobile, AT&T$\}$, $\{$Verizon$\}$, $\{$T-Mobile$\}$, $\{$AT&T$\}$, $\emptyset$

3 $\{$Spring, Summer, Fall$\}$, $\{$Spring, Summer, Winter$\}$, $\{$Spring, Fall, Winter$\}$, $\{$Summer, Fall, Winter$\}$, $\{$Spring, Summer$\}$, $\{$Spring, Fall$\}$, $\{$Spring, Winter$\}$, $\{$Summer, Fall$\}$, $\{$Summer, Winter$\}$, $\{$Fall, Winter$\}$, $\{$Spring$\}$, $\{$Summer$\}$, $\{$Fall$\}$, $\{$Winter$\}$, $\emptyset$

4 (a) True (c) True (e) False (g) False
    (b) False (d) True (f) True   (h) False

5 Subsets: $2^8 = 256$; proper subsets: 255

6 $A \cap B = \{$Chicago, Detroit$\}$;
    $B \cap C = \{$Los Angeles$\}$; $A \cap B \cap C = \emptyset$

7 $A \cup B = \{$Cleveland, Indianapolis, Chicago, Des Moines, Detroit, New York, Los Angeles$\}$
    $A \cup C = \{$Cleveland, Indianapolis, Chicago, Des Moines, Detroit, Seattle, Los Angeles, San Diego$\}$
    $A \cup B \cup C = \{$Cleveland, Indianapolis, Chicago, Des Moines, Detroit, New York, Los Angeles, Seattle, San Diego$\}$

8 $A \cup (B \cap C) = \{$Cleveland, Indianapolis, Chicago, Des Moines, Detroit, Los Angeles$\}$
    $(A \cap B) \cup C = \{$Chicago, Detroit, Seattle, Los Angeles, San Diego$\}$
    $A \cap (B \cup C) = \{$Chicago, Detroit$\}$

9 (a) $C' = \{$Cleveland, Indianapolis, Chicago, Des Moines, Detroit, New York, Phoenix, Boston$\}$
    (b) $(A \cup B)' = \{$Seattle, San Diego, Phoenix, Boston$\}$
    (c) $A' \cap C' = \{$New York, Phoenix, Boston$\}$
    (d) $(A \cup B) \cap C' = \{$Cleveland, Indianapolis, Chicago, Des Moines, Detroit, New York$\}$

10 (a) $A - B = \{$Cleveland, Indianapolis, Des Moines$\}$
     (b) $B - C = \{$New York, Chicago, Detroit$\}$
     (c) $(B - C) - A = \{$New York$\}$

Exercise Set 2-2

Writing Exercises

1. What is a subset?
2. Explain the difference between a subset and a proper subset.
3. Explain the difference between a subset and an element of a set.
4. Explain why the empty set is a subset, but not a proper subset, of itself.
5. Explain the difference between the union and intersection of two sets.
6. When are two sets said to be disjoint?
7. What is a universal set?
8. What is the complement of a set?
9. Write an example from real life that represents the union of sets and explain why it represents union. Then do the same for intersection.
10. Write an example from real life that represents the difference of sets and explain why it represents difference.

Computational Exercises

For Exercises 11–14, let $U = \{2, 3, 5, 7, 11, 13, 17, 19\}$, $A = \{5, 7, 11, 13\}$, $B = \{2\}$, $C = \{13, 17, 19\}$, and $D = \{2, 3, 5\}$. Find each set.

11. A'
12. B'
13. C'
14. D'
15. If $U =$ the set of natural numbers and $A = \{4, 6, 8, 10, 12, \ldots\}$, find A'.
16. If $U =$ the set of odd natural numbers and $B = \{13, 15, 17, 19, 21, 23, \ldots\}$, find B'.

For Exercises 17–24, find all subsets and all proper subsets of each set.

17. {OVI, theft, fraud}
18. {assault, manslaughter, battery}
19. {radio, TV}
20. {online, print}
21. Ø
22. { }
23. {fever, chills, nausea, headache}
24. {seizures, numbness, paralysis, pain}

For Exercises 25–34, state whether each is true or false.

25. $\{3\} \subseteq \{1, 3, 5\}$
26. $\{a, b, c\} \subset \{c, b, a\}$
27. $\{1, 2, 3\} \subseteq \{123\}$
28. $\varnothing \subset \varnothing$
29. $\varnothing \in \{\ \}$
30. $\{x \mid x \in E \text{ and } x > 100\} \subset \{x \mid x \in N \text{ and } x > 52\}$
31. $\{3\} \in \{1, 3, 5, 7, \ldots\}$
32. $\{x \mid x \in N \text{ and } x > 10\} \subseteq \{x \mid x \in N \text{ and } x \geq 10\}$
33. $\varnothing \subset \{a, b, c\}$
34. $\{7, 11, 13, 17\} \subseteq \{17, 13, 11\}$

For Exercises 35–40, find the number of subsets and proper subsets each set has. Do not list the subsets.

35. {25, 50, 75}
36. {a, b, c, d, \ldots, z}
37. Ø
38. {0}
39. {x, y}
40. {2, 4, 6, 8, 10, \ldots, 30}

For Exercises 41–50, use the Venn diagram to find the elements in each set.

41. U
42. A
43. B
44. $A \cap B$
45. $A \cup B$
46. A'
47. B'
48. $(A \cup B)'$
49. $(A \cap B)'$
50. $A \cap B'$

For Exercises 51–60, let

$U = \{11, 12, 13, 14, 15, 16, 17, 18, 19, 20\}$
$A = \{14, 15, 16, 17\}$
$B = \{11, 13, 15, 17, 19\}$
$C = \{12, 14, 15, 19, 20\}$

Find each set.

51. $A \cup C$
52. $A \cap B$
53. A'
54. $(A \cap B) \cup C$
55. $A' \cap (B \cup C)$
56. $(A \cap B) \cap C$
57. $(A \cup B)' \cap C$
58. $A \cap B'$
59. $(B \cup C) \cap A'$
60. $(A' \cup B)' \cup C'$

For Exercises 61–70, let

$U = \{x \mid x \in N \text{ and } x < 25\}$
$W = \{x \mid x \in N \text{ and } 5 < x < 15\}$
$X = \{x \mid x \in \text{ even natural numbers less than } 10\}$
$Y = \{x \mid x \in N \text{ and } 20 < x < 25\}$
$Z = \{x \mid x \in \text{ odd natural numbers less than } 13\}$

Find each set.

61. $W \cap Y$
62. $X \cup Z$
63. $W \cup X$
64. $(X \cap Y) \cap Z$

65. $W \cap X$
66. $(Y \cup Z)'$
67. $(X \cup Y) \cap Z$

68. $(Z \cap Y) \cup W$
69. $W' \cap X'$
70. $(Z \cup X)' \cap Y$

For Exercises 71–74, let

$U = \{1, 2, 3, \dots\}$
$A = \{3, 6, 9, 12, \dots\}$
$B = \{9, 18, 27, 36, \dots\}$
$C = \{2, 4, 6, 8, \dots\}$

Find each set.

71. $A \cap B$
72. $A' \cap C$

73. $A \cap (B \cup C')$
74. $A \cup B$

For Exercises 75–80, let

$U = \{p, q, r, s, t, u, v, w\}$
$A = \{p, q, r, s, t\}$
$B = \{r, s, t, u, v\}$
$C = \{p, r, t, v\}$

Find each set.

75. $C - B$
76. $A - C$
77. $B - C$

78. $B - A$
79. $B \cap C'$
80. $C \cap A'$

For Exercises 81–84, let

$D = \{11, 12, 13, 14, 15, \dots\}$
$M = \{x \mid x \in E \text{ and } x > 10\}$
$T = \{x \mid x \in N \text{ and } x < 100\} \cup \{x \mid x \in O \text{ and } x > 100\}$

Find each set.

81. $D - M$
82. $T - D$

83. $(D - M) - T$
84. $(T - D) - M$

For Exercises 85–88, use the Venn diagram to write each set in terms of A, B, and/or U.

85. $\{1, 2, 3, 4\}$
86. $\{2, 3, 5, 6, 7, 8, 9\}$
87. $\{2, 3, 6, 7, 8\}$
88. $\{1, 4\}$

Applications in Our World

89. A student can have a tablet, a smartphone, and a laptop while hanging out on campus between classes. List all the sets of different communication options a student can select, considering all, some, or none of these technologies.

90. If a person is dealt five cards and has a chance of discarding any number including 0, how many choices will the person have?

91. A first-year college student can choose one, some, or all of the following classes for their first semester: an English class, a math class, a foreign language class, a science class, a philosophy class, a physical education class, and a history class. How many different possible schedules could be chosen?

92. Since the student union is being remodeled, there is a limited choice of foods and drinks a student can buy for a snack between classes. Students can choose none, some, or all of these items: pizza, fries, big soft pretzels, Coke, Diet Coke, and Hawaiian Punch. How many different selections can be made?

93. Suzie is buying a new laptop for school and can select none, some, or all of the following choices of peripherals: a laser mouse, a DVD burner, a Web cam, or a jump drive. How many different selections of peripherals are possible for the laptop?

94. To integrate aerobics into an exercise program, you could pick one, some, or all of these machines: treadmill, cycle, and stair stepper. List all possibilities for the aerobics selection.

95. The want ad at the top of the next column is looking for a person who's in the intersection of three sets. What are those three sets?

96. The following ad actually appeared on Craigslist: "I need a nemesis. I'm willing to pay $350 up front for your services

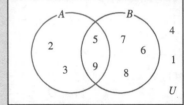

Help Wanted

Accounts Manager, requires strong management skills, good at working as part of a team and has at least 5-years experience with a similar project. P. O. Box No. 5474

⋯⋯ Manager strong

as an archenemy over the next six months. Nothing crazy. Steal my parking space, knock my coffee over, trip me when I'm running to catch the BART and occasionally whisper in my ear, "Ahha, we meet again." That kind of thing. Just keep me on my toes. Complacency will be the death of me. You need to have an evil streak and be blessed with innate guile and cunning. You should also be adept at inconspicuous pursuit. Evil laugh or British accent required. Send me a photo and a brief explanation why you would be a good nemesis." Write a statement involving five sets, intersection, and union that describes this person's requirements for a nemesis.

Exercises 97–100 use the following sets:

U = the set of all people who have been charged with a felony
A = the set of people who are on trial or awaiting trial on felony charges
B = the set of people who have been convicted of a felony
C = the set of people who have been convicted of a felony and have been released from prison
D = the set of people who were charged with a felony and found not guilty
E = the set of people who were charged with a felony and had charges dropped before standing trial

Write a verbal description of each set.

97. (a) $B \cup C$ (b) $C \cup D$ (c) $D \cup E$
98. (a) A' (b) C' (c) E'

99. (a) $B \cap C$ (b) $A \cap B$ (c) $C \cap B'$
100. (a) $(A \cup B)'$ (b) $(B \cup D)'$ (c) $A - (B \cap C)$

Critical Thinking

*The **Cartesian product** or **cross product** is a set operation that we haven't studied yet. When elements from two sets are paired together in a set of parentheses, like (flu, chills), we call that an **ordered pair**. The cross product of two sets A and B, denoted A × B, is the set of all ordered pairs with first entry an element from A, and second entry from B. This operation is used to pair up things that go together naturally, like ailments and symptoms. For example, if A = {cold, flu} and B = {fever, chills}, A × B = {(cold, fever), (cold, chills), (flu, fever), (flu, chills)}. For each pair of sets in Exercises 101 and 102, find the cross product and write a verbal description of what its significance is.*

101. A = {chocolate, yellow, red velvet}, B = {chocolate icing, cream cheese icing}
102. A = {guilty, not guilty}, B = {possession with intent, DUI, assault}
103. If $n(A) = n$ and $n(B) = m$, what is $n(A \times B)$? Explain how you got your answer.
104. What's the connection between the Cartesian product and regular multiplication?
105. Can you find two sets whose union and intersection are the same set?
106. Pick three medications and find a resource on the Internet that lists the possible side effects of each. Find the intersection of the sets.

107. (a) Make up two sets A and B with somewhere between 4 and 8 elements in each so that $A \cap B$ is nonempty. Find each of $n(A)$, $n(B)$, $n(A \cup B)$, and $n(A \cap B)$.
     (b) Repeat part (a) with two completely different sets A and B.
     (c) Use the results of parts (a) and (b) to make a conjecture about a formula for finding the cardinality of a union of two sets.
108. (a) Write two sets A and B for which $n(A \cup B) > n(A \cap B)$.
     (b) Write two sets A and B for which $n(A \cup B) = n(A \cap B)$.
     (c) Can you write two sets A and B for which $n(A \cup B) < n(A \cap B)$? Use a Venn diagram to illustrate why you can or cannot.
109. Think about two sets A and B. What has to happen in order to have $A \cap B = A$? What about $A \cap B = B$?
110. For any two sets A and B, what has to happen for BOTH of the things in Exercise 109 to occur?
111. Here's an alternate approach to developing the formula for the number of subsets for a set with n elements. If a set has two elements, when forming a subset, there are two choices for each element: it's either in the subset or it's not. If we multiply two choices for the first element by two choices for the second element, we get four choices of subset. (This illustrates an important idea we'll encounter in Chapter 10 called the fundamental counting principle.) Generalize this idea to derive the formula for the number of subsets.

Section 2-3	**Using Venn Diagrams to Study Set Operations**

 LEARNING OBJECTIVES

1. Illustrate set statements involving two sets with Venn diagrams.

2. Illustrate set statements involving three sets with Venn diagrams.

3. Use De Morgan's laws.

4. Use Venn diagrams to decide if two sets are equal.

5. Use the formula to find the cardinality of a union of two sets.

Have you ever wanted to post something on Twitter or Instagram, then decided not to because it may not be something you want everyone you know to see? Some people are social media friends only with their closest buddies: more people are friends with hundreds of people, from their Mom to casual work acquaintances. My closest friends and I have a secret private group on Facebook for exactly this reason: the set of things I feel comfortable saying to my college buddies is most definitely not equal, or even equivalent, to the set of things I feel comfortable saying to my Mom or my golf pro. Most of us have many distinct circles of friends and contacts, and when those worlds collide the results can be unpredictable, and maybe unintentionally hilarious.

One good way to get a handle on complicated interplay between varying groups is with diagrams. And can you guess what kind of diagrams we'll choose to accomplish this? If you didn't say "Venn," please go back and reread the previous section. We'll wait here for you.

In this section, we'll develop a method for drawing Venn diagrams that will help us to illustrate set operations. We'll start with diagrams involving interactions between two sets, as in Figure 2-8. Notice that there are four distinct regions in a Venn diagram illustrating two sets A and B. We'll want to number the regions for reference; we use Roman numerals so that we don't confuse the number of the region with elements in the set or the cardinality of the set.

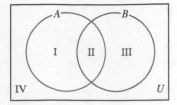

Figure 2-8

The procedure that we will use to illustrate set statements, found in the box below, is demonstrated in Examples 1 and 2. For now, we're going to be working with strictly abstract sets, but hang in there. The skills we learn will serve us well in solving real problems later on. Trust me, I'm a doctor.

Illustrating a Set Statement with a Venn Diagram

Step 1 Draw a diagram for the sets, with Roman numerals in each region.

Step 2 Using those Roman numerals, list the regions described by each set.

Step 3 Find the set of numerals that correspond to the set given in the set statement.

Step 4 Shade the area corresponding to the set of numerals found in step 3.

Region I represents the elements in set A that are not in set B.
Region II represents the elements in both sets A and B.
Region III represents the elements in set B that are not in set A.
Region IV represents the elements in the universal set that are in neither set A nor set B.

Example 1   Drawing a Venn Diagram

Draw a Venn diagram to illustrate the set $(A \cup B)'$.

SOLUTION

Step 1 Draw the diagram and label each area with a Roman numeral.

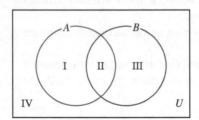

Step 2 From the diagram, list the regions that make up each set.

$U = \{I, II, III, IV\}$
$A = \{I, II\}$
$B = \{II, III\}$

Step 3 Using the sets in step 2, find $(A \cup B)'$.
First, all of I, II, and III are in either A or B, so $A \cup B = \{I, II, III\}$. The only region not in $A \cup B$ is IV, so the complement is $(A \cup B)' = \{IV\}$.

Step 4 Shade region IV to illustrate $(A \cup B)'$.

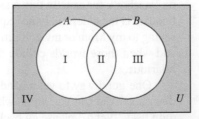

Try This One   **1**

Draw a Venn diagram to illustrate the set $A' \cap B$.

| Example 2 | Drawing a Venn Diagram |

Draw a Venn diagram to illustrate the set $A \cap B'$.

SOLUTION

Step 1 Draw the diagram and label each area. This will be the same diagram as in Step 1 of Example 1.

Step 2 From the diagram, list the regions that make up each set.

$U = \{$I, II, III, IV$\}$
$A = \{$I, II$\}$
$B = \{$II, III$\}$

Step 3 Using the sets in step 2, find $A \cap B'$.
First, regions I and IV are outside of set B, so $B' = \{$I, IV$\}$. Of these two regions, I is also in set A, so $A \cap B' = \{$I$\}$.

Step 4 Shade region I to illustrate $A \cap B'$.

> **Math Note**
>
> In any problem where we're asked to illustrate a set statement involving two sets, steps 1 and 2 will be exactly the same.

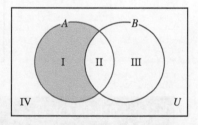

 1. Illustrate set statements involving two sets with Venn diagrams.

Try This One   2

Draw a Venn diagram to illustrate the set $A' \cup B$.

In the opener for this chapter, we asked you to solve a more complicated problem involving three sets of people: those self-reporting as white, Black, or Hispanic. Venn diagrams are great for sorting out information of this nature, and we'll want some experience with Venn diagrams involving three sets before we move on to solving those problems.

Fortunately, the procedure we used for two sets can be used for three sets as well: you just get a more complicated diagram (see Figure 2-9).

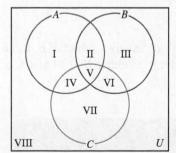

Figure 2-9

Region I represents the elements in set A but not in set B or set C.
Region II represents the elements in set A and set B but not in set C.
Region III represents the elements in set B but not in set A or set C.
Region IV represents the elements in sets A and C but not in set B.
Region V represents the elements in sets A, B, and C.
Region VI represents the elements in sets B and C but not in set A.
Region VII represents the elements in set C but not in set A or set B.
Region VIII represents the elements in the universal set U, but not in set A, B, or C.

| Example 3 | Drawing a Venn Diagram with Three Sets |

Draw a Venn diagram to illustrate the set $A \cap (B \cap C)'$.

SOLUTION

Step 1 Draw and label the diagram as in Figure 2-9.

Math Note

When illustrating complicated sets like $A \cap (B \cap C)'$, don't forget to find the set in parentheses first. That's why the parentheses are there!

Step 2 From the diagram, list the regions that make up each set.

$U = \{I, II, III, IV, V, VI, VII, VIII\}$
$A = \{I, II, IV, V\}$
$B = \{II, III, V, VI\}$
$C = \{IV, V, VI, VII\}$

Step 3 Using the sets in step 2, find $A \cap (B \cap C)'$.
First, find $B \cap C$: $B \cap C = \{V, VI\}$. The complement is $(B \cap C)' = \{I, II, III, IV, VII, VIII\}$. Regions I, II, and IV are also part of A, so $A \cap (B \cap C)' = \{I, II, IV\}$.

Step 4 Shade regions I, II, and IV to illustrate $A \cap (B \cap C)'$.

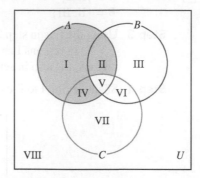

Try This One   3

Draw a Venn diagram to illustrate the set $(A \cap B') \cup C$.

To get even better at working with Venn diagrams, it's helpful to turn the process around, starting with a shaded diagram and figuring out what set it represents, as in Example 4.

Example 4 | Finding a Set Corresponding to a Venn Diagram

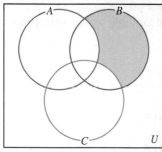

Figure 2-10

Write the set illustrated by the Venn diagram in Figure 2-10.

SOLUTION

The shaded portion is completely inside the circle for B, so it's definitely a subset of B. But it doesn't include anything in either A or C, so we could write it as either $B - (A \cup C)$, or $B \cap (A \cup C)'$.

Try This One   4

Write the set illustrated by the Venn diagram in Figure 2-11.

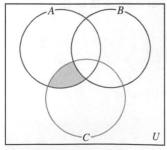

Figure 2-11

De Morgan's Laws

There are two very well-known formulas that are useful in simplifying some set operations. They're named in honor of a 19th-century mathematician named Augustus De Morgan.

First, we'll write the formulas and illustrate each with an example. Then we'll see how Venn diagrams can be used to prove the formulas.

De Morgan's Laws

For any two sets A and B,

$(A \cup B)' = A' \cap B'$

$(A \cap B)' = A' \cup B'$

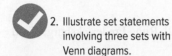

2. Illustrate set statements involving three sets with Venn diagrams.

The first law states that the complement of the union of two sets will always be equal to the intersection of the complements of each set.

Example 5 | **Using De Morgan's Laws**

If U = {Antietam, Bull Run, Gettysburg, Shiloh, Fredericksburg, Cold Harbor, Chancellorsville}, A = {Antietam, Bull Run, Gettysburg, Shiloh}, and B = {Gettysburg, Shiloh, Fredericksburg, Cold Harbor}, find $(A \cup B)'$ and $A' \cap B'$. What can we observe about these two sets?

SOLUTION

$A \cup B$ = {Antietam, Bull Run, Gettysburg, Shiloh, Fredericksburg, Cold Harbor} and
$(A \cup B)'$ = {Chancellorsville}

A' = {Fredericksburg, Cold Harbor, Chancellorsville}
B' = {Antietam, Bull Run, Chancellorsville}
$A' \cap B'$ = {Chancellorsville}

These two sets are the same, which matches the first of De Morgan's laws.

Try This One | **5**

If U = {ABC, NBC, CBS, Fox, USA, TBS, TNT, MTV}, A = {NBC, Fox, USA, TBS}, and B = {ABC, NBC, CBS, Fox}, find $(A \cup B)'$ and $A' \cap B'$.

Sidelight **Now and Venn**

Venn diagrams are generally credited to the British mathematician John Venn, who introduced them in an 1880 paper as they are used today. That makes it sound like a pretty old concept, but the general idea can be traced back much further. The great mathematician Leonhard Euler used similar diagrams in the 1700s, and other figures like them can be traced as far back as the 1200s!

While it's absolutely true that John Venn was a classic academic—at one time writing or lecturing in each of

morality, mathematics, logic, probability theory, philosophy, metaphysics, and history—he had a somewhat surprising hobby: building machines. In particular, he was best known for building a machine for bowling cricket balls (which is roughly analogous to pitching in baseball). His machine was so good that in 1909 it "clean bowled" one of the top cricket players of the time on four occasions, which is kind of like a turn-of-the-century pitching machine striking out Babe Ruth.

The second of De Morgan's laws states that the complement of the intersection of two sets will equal the union of the complements of the sets.

Example 6

Using De Morgan's Laws

Math Note

In Examples 5 and 6, we're looking at specific examples, so we're using inductive reasoning to conclude that De Morgan's laws are likely to be true. In Example 7, we'll use deductive reasoning to *prove* them.

If U = {Antietam, Bull Run, Gettysburg, Shiloh, Fredericksburg, Cold Harbor, Chancellorsville}, A = {Antietam, Bull Run, Gettysburg, Shiloh}, and B = {Gettysburg, Shiloh, Fredericksburg, Cold Harbor}, find $(A \cap B)'$ and $A' \cup B'$. What can we observe about these two sets?

SOLUTION

$A \cap B$ = {Gettysburg, Shiloh} and $(A \cap B)'$ = {Antietam, Bull Run, Fredericksburg, Cold Harbor, Chancellorsville}

A' = {Fredericksburg, Cold Harbor, Chancellorsville}
B' = {Antietam, Bull Run, Chancellorsville}
$A' \cup B'$ = {Fredericksburg, Cold Harbor, Chancellorsville, Antietam, Bull Run}

Even though they're listed in different orders (which we know doesn't matter), these two sets are the same, which matches the second of De Morgan's laws.

Try This One   6

3. Use De Morgan's laws.

If U = {ABC, NBC, CBS, Fox, USA, TBS, TNT, MTV}, A = {NBC, Fox, USA, TBS}, and B = {ABC, NBC, CBS, Fox}, find $(A \cap B)'$ and $A' \cup B'$.

Now that we know how to illustrate sets with Venn diagrams, we can use them to show that two sets that look different are actually the same. In Example 7, we'll illustrate the procedure by proving the first of De Morgan's laws. We'll leave the second one for you to try.

Example 7

Using Venn Diagrams to Show Equality of Sets

Use Venn diagrams to show that $(A \cup B)' = A' \cap B'$, proving the first of De Morgan's laws.

SOLUTION

Start by drawing the Venn diagram for $(A \cup B)'$.

Step 1 Draw the figure (as shown in Step 4).

Step 2 Set U contains regions I, II, III, and IV. Set A contains regions I and II, and B contains regions II and III.

Step 3 $A \cup B$ = {I, II, III}, so $(A \cup B)'$ = {IV}.

Step 4 Shade region IV to illustrate $(A \cup B)'$.

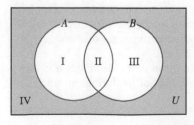

Next draw the Venn diagram for $A' \cap B'$. Steps 1 and 2 are the same as above.

Step 3 A' = {III, IV} and B' = {I, IV}, so $A' \cap B'$ = {IV}.

Step 4 Shade region IV to illustrate $A' \cap B'$.

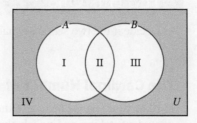

Since the diagrams for each side of the equation are identical, we use deductive reasoning to conclude that $(A \cup B)' = A' \cap B'$.

Try This One 7

Use Venn diagrams to show that $(A \cap B)' = A' \cup B'$.

Here's an example using three sets.

Example 8	Using Venn Diagrams to Decide If Two Sets Are Equal

Math Note

As you get more comfortable working with Venn diagrams, you'll probably be able to shade the regions illustrated by a set without formally going through our four-step process. This is what we're doing in Example 8.

Decide if these two sets are equal using Venn diagrams: $(A \cup B) \cap C$ and $(A \cap C) \cup (B \cap C)$.

SOLUTION

The set $A \cup B$ consists of regions I through VI. Of these, IV, V, and VI are also in C, so $(A \cup B) \cap C$ consists of regions IV, V, and VI.

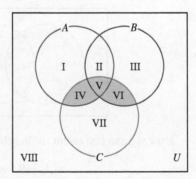

The set $A \cap C$ consists of regions IV and V, and the set $B \cap C$ consists of regions V and VI. Their union is regions IV, V, and VI.

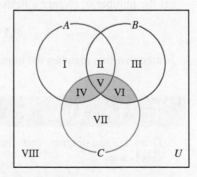

Since the shaded areas are the same, the two sets are equal.

4. Use Venn diagrams to decide if two sets are equal.

Try This One   8

Decide if the two sets are equal using Venn diagrams: $B \cup (A \cap C)$ and $(A \cup B) \cap (B \cup C)$.

The Cardinal Number of a Union

If 10 of your friends belong to the set of students taking a math class, and 14 belong to the set of students taking an English class, how many are in the union of those two sets? If your first instinct is 24, you're not alone—that's sort of the standard guess. And it might actually be right, but only if none of your friends are taking both a math and an English class. If any of them are in both classes, you'd be counting them twice by just adding the number of friends in each set. Venn diagrams can be used to analyze this situation.

Example 9   Finding the Cardinality of a Union

Draw a Venn diagram illustrating the sets below, then use the diagram to find the cardinality of A, B, $A \cap B$, and $A \cup B$.

$$A = \{1, 2, 3, 4, 5, 6, 7, 8, 9\} \quad B = \{6, 7, 8, 9, 10, 11, 12\}$$

Use the result to develop a formula for the cardinality of a union.

SOLUTION

First, notice that 6, 7, 8, and 9 are in both sets, so we'll begin our Venn diagram by putting those elements in the intersection portion of the diagram. Then we put the remaining elements in A inside the circle for A but outside the intersection, and do the same for the remaining elements in B.

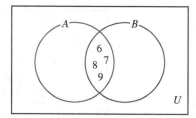 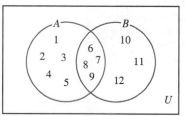

Now we can just count to find the cardinalities in question.

$$n(A) = 9 \quad n(B) = 7 \quad n(A \cap B) = 4 \quad n(A \cup B) = 12$$

From the diagram, we can see that if you simply add the number of elements in A and B, you'll be adding the elements in the intersection twice. So to account for that we can subtract off the number of elements in the intersection, giving us the formula

$$n(A \cup B) = n(A) + n(B) - n(A \cap B).$$

Notice that this matches the numbers for the sets in this example: $9 + 7 - 4 = 12$.

Try This One   9

Draw a Venn diagram that illustrates our formula for the cardinality of a union using the sets below.
$$A = \{a, c, f, g, l, k, m, n, p\} \quad B = \{g, l, m, o, q, r, t, z\}$$

Math Note

In words, the formula to the right says that to find the number of elements in the union of A and B, you add the number of elements in A and B and then subtract the number of elements in the intersection of A and B.

The Cardinality of a Union

If $n(A)$ represents the cardinal number of set A, then for any two finite sets A and B,

$$n(A \cup B) = n(A) + n(B) - n(A \cap B).$$

Next, we'll see how this formula can be used in an applied situation.

Example 10

Using the Formula for Cardinality of a Union

Digital Vision/Getty Images

In a survey of 100 randomly selected first-year students walking across campus, it turns out that 42 are taking a math class, 51 are taking an English class, and 12 are taking both. How many students are taking either a math class or an English class?

SOLUTION

If we call the set of students taking a math class A and the set of students taking an English class B, we're asked to find $n(A \cup B)$. We're told that $n(A) = 42$, $n(B) = 51$, and $n(A \cap B) = 12$. So,

$$n(A \cup B) = n(A) + n(B) - n(A \cap B) = 42 + 51 - 12 = 81$$

Try This One   10

5. Use the formula to find the cardinality of a union of two sets.

A poll of 200 doctors across the nation found that 112 were assisted in their office by registered nurses, 83 were assisted by licensed practical nurses, and 21 were assisted by both. How many were assisted by at least one type of nurse?

In this section, we saw how Venn diagrams can be used to illustrate sets, prove the equality of two sets, and solve problems. We'll explore the problem-solving aspect of Venn diagrams further in Section 2-4 and learn how to solve problems like the one in the chapter opener.

Answers to Try This One

1

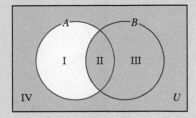

2

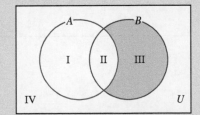

3

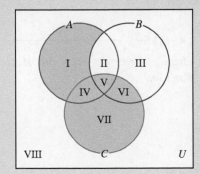

4 $(A \cap C) - B$ or $(A \cap C) \cap B'$

5   Both are {TNT, MTV}.

6   Both are {ABC, CBS, USA, TBS, TNT, MTV}.

7   Both diagrams are

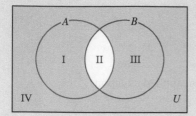

8   Both diagrams are

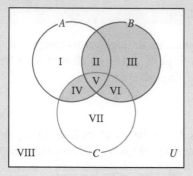

9

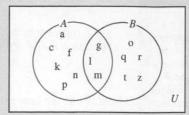

10   174

Exercise Set   2-3

Writing Exercises

1. One of your buddies is paging through your textbook and sees some Venn diagrams, and asks, "What's the point of these pictures?" How would you answer?
2. Explain in your own words how to draw a Venn diagram representing the set $A \cup B$.
3. Explain in your own words how to draw a Venn diagram representing the set $A \cap B$.
4. How can we use Venn diagrams to decide if two sets that look different are actually equal?
5. Describe in your own words what De Morgan's laws say.
6. Describe in your own words how to find the cardinal number of the union of two sets.

Computational Exercises

For Exercises 7–30, draw a Venn diagram and shade the sections representing each set.

7. $A \cup B'$
8. $(A \cup B)'$
9. $A' \cup B'$
10. $A' \cup B$
11. $A' \cap B'$
12. $A \cap B'$
13. $A \cup (B \cap C)$
14. $A \cap (B \cup C)$
15. $(A \cup B) \cup (A \cap C)$
16. $(A \cup B) \cap C$
17. $(A \cup B) \cap (A \cup C)$
18. $(A \cap B) \cup C$
19. $(A \cap B)' \cup C$
20. $(A \cup B) \cup C'$
21. $A \cap (B \cup C)'$
22. $A' \cap (B' \cup C')$
23. $(A' \cup B') \cap C$
24. $A \cap (B \cap C)'$
25. $(A \cup B)' \cap (A \cup C)$
26. $(B \cup C) \cup C'$
27. $A' \cap (B' \cap C')$
28. $(A \cup B)' \cap C'$
29. $A' \cap (B \cup C)'$
30. $(A \cup B) \cap (A \cap C)$

For Exercises 31–38, use Venn diagrams to decide if the two sets are equal.

31. $(A \cap B)'$ and $A' \cup B'$
32. $(A \cup B)'$ and $A' \cup B'$
33. $(A \cup B) \cup C$ and $A \cup (B \cup C)$
34. $A \cap (B \cup C)$ and $(A \cap B) \cup (A \cap C)$
35. $A' \cup (B \cap C')$ and $(A' \cup B) \cap C'$
36. $(A \cap B) \cup C'$ and $(A \cap B) \cup (B \cap C')$
37. $(A \cap B)' \cup C$ and $(A' \cup B') \cap C$
38. $(A' \cup B') \cup C$ and $(A \cap B)' \cap C'$

For Exercises 39–50, use the following Venn diagram to find the cardinality of each set.

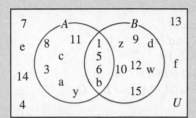

39. $n(A)$
40. $n(B)$
41. $n(A \cap B)$
42. $n(A \cup B)$
43. $n(A')$
44. $n(B')$
45. $n(A' \cap B')$
46. $n(A' \cup B')$
47. $n(A - B)$
48. $n(B - A)$
49. $n(A \cap (B - A))$
50. $n(B' \cup (B - A))$

For Exercises 51–60, use the following information:

$$U = \{x \mid x \text{ is a natural number less than } 20\}$$
$$A = \{x \mid x \text{ is an odd natural number less than } 16\}$$
$$B = \{x \mid x \text{ is a prime number greater than } 5\}$$

(Note: The prime numbers less than 20 are 2, 3, 5, 7, 11, 13, 17, and 19.) Find the cardinality of each set.

51. $n(A)$
52. $n(B)$
53. $n(A \cap B)$
54. $n(A \cup B)$
55. $n(A \cap B')$
56. $n(A' \cup B)$
57. $n(A')$
58. $n(B')$
59. $n(A - B)$
60. $n(B' - A)$

Applications in Our World

In Exercises 61–64, $A = \{$people who drive an SUV$\}$ and $B = \{$people who drive a hybrid vehicle$\}$. Draw a Venn diagram of the following, and write a sentence describing what the set represents.

61. $A \cup B$
62. $A \cap B$
63. A'
64. $(A \cap B)'$

In Exercises 65–68, $O = \{$students in online courses$\}$, $B = \{$students in blended courses$\}$, and $T = \{$students in traditional courses$\}$. Draw a Venn diagram of the following, and write a sentence describing what the set represents.

65. $O \cap (T \cup B)$
66. $B \cup (O \cap T)$
67. $B \cap O \cap T$
68. $(B \cup O) \cap (T \cup O)$

In Exercises 69–72, $D = \{$students voting Democrat$\}$, $R = \{$students voting Republican$\}$, and $I = \{$students voting Independent$\}$. Draw a Venn diagram of the following, and write a sentence describing what the set represents.

69. $D' \cup R$
70. $D' \cap I'$
71. $(D \cup R) \cap I'$
72. $I - (D \cup R)$

In Exercises 73–76, $G = \{$people who regularly use Google$\}$, $Y = \{$people who regularly use Yahoo!$\}$, and $B = \{$people who regularly use Bing$\}$. Draw a Venn diagram of the following, and write a sentence describing what the set represents.

73. $G - Y$
74. $G - (Y \cap B)$
75. $G' \cap Y' \cap B'$
76. $(Y \cap B) \cup (Y \cap G)$

The table and Venn diagram below are to be used for Exercises 77–82. The table shows the set of football teams that made the playoffs in the American Football Conference (AFC) from 2018 to 2020. For each exercise, write the region(s) of the Venn diagram that would include the team listed.

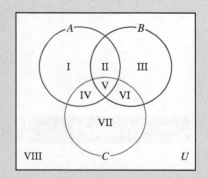

2018	2019	2020
Kansas City	Baltimore	Kansas City
New England	Kansas City	Buffalo
Houston	New England	Pittsburgh
Baltimore	Houston	Tennessee
Los Angeles	Buffalo	Baltimore
Indianapolis	Tennessee	Cleveland
		Indianapolis

Note: Set A represents 2018 playoff teams, set B represents 2019 playoff teams, and set C represents 2020 playoff teams.

77. New England
78. Indianapolis

79. Los Angeles
80. Cleveland
81. Buffalo
82. Cincinnati

Critical Thinking

83. For two finite sets A and B, is $n(A - B)$ equal to $n(A) - n(B)$? If not, can you find a formula for $n(A - B)$?

84. Can you find a formula for $n(A \cap B)$ in terms of only $n(A)$ and $n(B)$? Why or why not? See if you can find a formula for $n(A \cap B)$ using any sets you like.

85. Make a conjecture about another form for the set $(A \cup B \cup C)'$ based on the first of De Morgan's laws. Check out your conjecture by using a Venn diagram.

86. Make a conjecture about another form for the set $(A \cap B \cap C)'$ based on the second of De Morgan's laws. Check out your conjecture by using a Venn diagram.

In Exercises 87–92, (a) use a Venn diagram to show that the two sets are not equal in general; (b) try to find specific sets A, B (and C if necessary) for which the two sets are equal; and (c) try to find a general condition under which the two sets are always equal. Recall that U represents the universal set.

87. $A \cap B$ and B
88. $A - B$ and A
89. $(A \cap B)'$ and U
90. $(A \cap B)'$ and A'
91. $(A - C) \cap B$ and $B \cap A$
92. $(A - C) \cup (B - A)$ and $B - C$

Section 2-4

Using Sets to Solve Problems

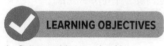

1. Solve problems using Venn diagrams.

Communication in our society is becoming cheaper, easier, and more effective all the time. In the age of smartphones and 24/7 Internet connectivity, businesses are finding it simpler than ever to contact people for their opinions, and more and more people are finding out that companies are willing to pay to hear what they have to say. There are literally hundreds of companies in the United States today whose main function is to gather opinions on everything from political candidates to potato chips. In fact, the amount spent on

yurakrasil/Shutterstock

market research in the United States each year increased from $13 billion in 2009 to over $47 billion in 2019. That's billion. With a b. Maybe you'll think twice the next time somebody asks you for your opinion for free.

With all the money at stake, not surprisingly it's important to organize all of the information that gets gathered. That sure sounds a lot like what we've been using set theory for! We've learned a lot about working with Venn diagrams so far. In my opinion, this knowledge is a great way to organize information gathered from surveys and other sources. (Oh, and no charge—you can have that opinion for free.)

When things are classified into two distinct sets, we can use a two-set Venn diagram to interpret the information. This is illustrated in Example 1.

Example 1

Solving a Problem Using a Venn Diagram

In 2021, there were 41 states that had some form of casino gambling in the state, 45 states that sold lottery tickets of some kind, and 38 states that had both casinos and lotteries. Draw a Venn diagram to represent the survey results, and find how many states have only casino gambling, how many states have only lotteries, and how many states have neither.

SOLUTION

Math Note

The first piece of information we're given is that 41 states have casino gambling, so it's tempting to begin by putting 41 in region I.

But this isn't right—region I represents states that have casino gambling but NOT lotteries, and we don't know that number yet. If we know the number in the intersection, that's where we'll always start.

Step 1 Draw a Venn diagram with circles for casino gambling (C) and lotteries (L), labeling the regions with Roman numerals as usual.

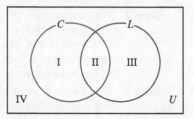

Step 2 Thirty-eight states have both, so put 38 in the intersection of C and L, which is region II.

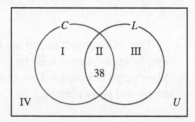

Step 3 Since 41 states have casino gambling and 38 have both, there must be 3 that have only casino gambling. Put 3 in region I. Since 45 states have lotteries and 38 have both, there are 7 that have only lotteries. Put 7 in region III.

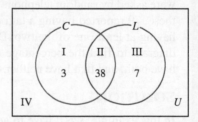

Step 4 Now 48 states are accounted for, so there must be 2 left to put in region IV.

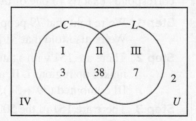

Now we can answer the questions easily. There are 3 states that have casino gambling but not lotteries (region I), 7 that have lotteries but not casinos (region III), and just 2 that have neither (region IV). In case you're wondering, those two states are Hawaii and Utah.

Try This One   1

In an average year, Columbus, Ohio, has 163 days with some rain, 63 days with some snow, and 24 days with both. Draw a Venn diagram to represent these averages, and find how many days have only rain, only snow, and neither.

We can use the results of Example 1 to write a general procedure for using a Venn diagram to interpret information that can be divided into two sets.

Using Venn Diagrams with Two Sets

Step 1   Find the number of elements that are common to both sets and write that number in region II.

Step 2   Find the number of elements that are in set A and not set B by subtracting the number in region II from the total number of elements in A. Then write that number in region I. Repeat for the elements in B but not in region II, and write in region III.

Step 3   Find the number of elements in U that are not in either A or B, and write it in region IV.

Step 4   Use the diagram to answer specific questions about the situation.

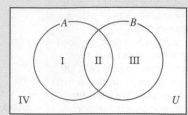

Quick Vote

Are you worse off financially than you were in recent years?
○ Yes
○ No
○ About the same

[VOTE] or see results

Many news websites include daily surveys, like this one from cnn.com.

One of the most useful applications of Venn diagrams is for studying the results of surveys. Whether they are for business-related research or just to satisfy curiosity, surveys seem to be everywhere these days, especially online. Example 2 analyzes the results of a survey on tattoos and body piercings.

Example 2   Solving a Survey Problem Using a Venn Diagram

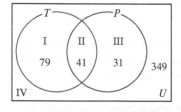

Neil Marriott/Digital Vision/ Getty Images

In a survey published in the *Journal of the American Academy of Dermatologists,* 500 people were asked by random telephone dialing whether they have a tattoo and/or a body piercing. Of these, 79 reported having a tattoo only, 31 reported having a piercing only, and 151 reported having at least one of the two. Draw a Venn diagram to represent these results and use your diagram to find the percentage of respondents that have a tattoo, that have a piercing, that have both, and that have neither.

SOLUTION

In this example, we'll have to adapt the procedure from Example 1 because we don't know the number that have both. The key is to begin by putting in information we're given that corresponds exactly to one of the regions in our Venn diagram.

Step 1   We're told that 79 people have only a tattoo, which means we can put 79 in region I. We're also told that 31 have only a piercing, so that goes in region III.

Step 2   There are 151 with a tattoo, a piercing, or both. This is the union of sets T and P, which makes up regions I, II, and III. We already know there are 110 people in regions I and III combined (79 + 31), so there must be 151 − 110 = 41 people in region II.

Step 3   There are 151 of the 500 accounted for so far, so region IV must contain 500 − 151 = 349 people.

Step 4   There is a total of 120 people in the regions that make up set T, so 120 people have tattoos; 120/500 = 0.24, so 24% have tattoos. Seventy-two have piercings (14.4%), 41 have both (8.2%), and 349 have neither (69.8%).

Try This One   **2**

According to an online survey on Howstuffworks.com, 12,595 people gave their thoughts on Coke vs. Pepsi. Of these, 5,786 drink only Coke, 3,763 drink only Pepsi, and 11,405 drink at least one. Draw a Venn diagram to represent these results and use your diagram to find the percentage of respondents that drink Coke, that drink Pepsi, that drink both, and that drink neither.

Sidelight   The Lighter Side of Polls

Most of us have gotten pretty accustomed to public opinion polls that deal with weighty issues—politics, climate change, the economy—but not every survey is quite that serious. Let's have a look at some odd facts that came out of recent surveys.

- Just about 50% of lost remote controls are found in the cushions of furniture. About 4% end up in the fridge or freezer, and about 2% are found outside or in a car.

- 29% of Americans think that "cloud computing" involves actual clouds in the sky.

- Almost 10% of American college graduates believe that Judge Judy is a member of the United States Supreme Court.

- Over 60% of self-proclaimed vegetarians admitted to eating meat in the previous 24 hours.

- 47% of Americans don't put a single penny of their paycheck into long-term savings.

- 52% of Americans sing in the shower, 47% do something in the shower that is more traditionally done in a different bathroom facility, and 7% claim to never bathe at all.

- (This is my favorite.) 2% of voters polled during the 2012 election season thought that Republican presidential nominee Mitt Romney's full first name was "Mittens."

When a classification problem or a survey consists of three sets, a similar procedure is followed, using a Venn diagram with three sets. We just have more work to do since there are now eight regions instead of four.

| Example 3 | Solving a Problem Using a Three-Set Venn Diagram |

A criminal justice major is studying the frequency of certain types of crimes in a nearby county. He studies the arrest records of 300 inmates at the county jail, specifically asking about drug-related offenses, domestic violence, and theft of some sort. He finds that 194 had been arrested for theft, 210 for drug offenses, and 170 for domestic violence. In addition, 142 had arrests for both theft and drugs, 111 for both drugs and domestic violence, 91 for both theft and domestic violence, and 45 had been arrested for all three. Draw a Venn diagram to represent these results, and find the number of inmates that had been arrested for

(a) Only drug-related offenses.
(b) Theft and domestic violence but not drugs.
(c) Theft or drugs.
(d) None of these offenses.

SOLUTION

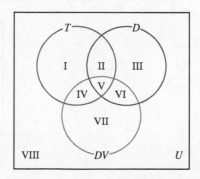

Step 1 The only region we know for sure from the given information is region V—the number of inmates arrested for all three offenses. So we begin by putting 45 in region V.

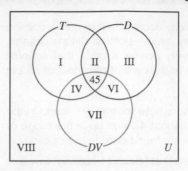

Step 2 There are 142 inmates with arrests for both theft and drugs, but we have to subtract the number arrested for all three offenses to find the number in region II: $142 - 45 = 97$. In the same way, we get $91 - 45 = 46$ in region IV (both theft and domestic violence) and $111 - 45 = 66$ in region VI (both drugs and domestic violence).

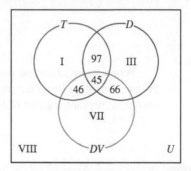

Step 3 Now we can find the number of elements in regions I, III, and VII. There were 194 inmates arrested for theft, but $97 + 45 + 46 = 188$ are already accounted for in the diagram, so that leaves 6 in region I. Of the 210 inmates with drug arrests, $97 + 45 + 66 = 208$ are already accounted for, leaving just 2 in region III. There were 170 inmates arrested for domestic violence, with $46 + 45 + 66 = 157$ already accounted for. This leaves 13 in region VII.

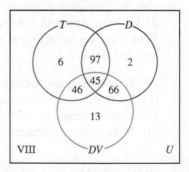

Step 4 Adding up all the numbers in the diagram so far, we get 275. That leaves 25 in region VIII.

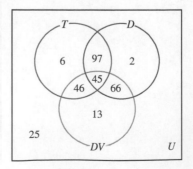

Math Note

Notice that in filling in the Venn diagram in Example 3, we started with the number of elements in the innermost region and worked our way outward.

Step 5 Now that we have the diagram completed, we turn our attention to the questions.

(a) Inmates arrested only for drug-related offenses are in region III: there were just 2.

(b) Theft and domestic violence with no drug arrests is region IV, so there were 46 inmates.

(c) Those arrested for either theft or drugs are in all regions except VII and VIII. So there are only $13 + 25 = 38$ that weren't arrested for at least one of those, and $300 - 38 = 262$ that were.

(d) Only 25 inmates (outside of all circles) haven't been arrested for any of those offenses.

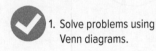

1. Solve problems using Venn diagrams.

Try This One   3

An online music service surveyed 500 customers and found that 270 listen to hip-hop music, 320 listen to rock, and 160 listen to country. In addition, 140 listen to both rock and hip-hop, 120 listen to rock and country, and 80 listen to hip-hop and country. Finally, 50 listen to all three. Draw a Venn diagram to represent the results of the survey and find the number of customers who

(a) Listen to only hip-hop.

(b) Listen to rock and country but not hip-hop.

(c) Don't listen to any of these three types of music.

(d) Don't listen to country music.

Instead of writing a general procedure for solving problems using a three-circle Venn diagram, we'll solve one more example. This time, the information provided is a little bit different. The key, which is really the key to all of these problems, is finding information that applies exactly to certain regions in the diagram, then using subtraction to find other regions one at a time.

Example 4   Solving a Problem Using a Three-Set Venn Diagram

Three of the most dangerous risk factors for heart attack are high blood pressure, high cholesterol, and smoking. In a survey of 690 heart attack survivors, 62 had only high cholesterol among those three risk factors; 36 had only smoking; and 93 had only high blood pressure. There were 370 total with high cholesterol, 159 with high blood pressure and cholesterol that didn't smoke, and 23 that smoked and had high cholesterol but not high blood pressure. Finally, 585 had at least one risk factor. Draw a Venn diagram representing this information, and use it to answer the following questions.

(a) How many survivors had all three risk factors?

(b) How many had exactly two of the three risk factors?

(c) How many had none?

(d) What percentage were smokers?

SOLUTION

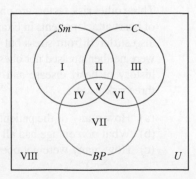

Step 1   This problem's a little different than the previous because we're not given the number of people in the intersection of all three sets. The good news, though, is that we're actually given the exact number in five different regions: 62 only high cholesterol (region III), 36 only smoking (region I), 93 only high blood pressure (region VII), 159 with high blood pressure and cholesterol but no smoking (region VI), and 23 with high cholesterol and smoking but not high blood pressure (region II).

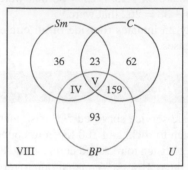   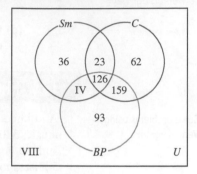

Step 2   There were 370 total with high cholesterol, and we have 23 + 62 + 159 = 244 accounted for so far, so region V must contain 370 − 244 = 126 survivors.

Step 3   The last piece of information we have is that 585 had at least one risk factor. This will allow us to find the remaining two regions. All of the numbers currently in the diagram add up to 499, so region IV must contain 585 − 499 = 86 survivors. Also, if 585 patients had at least one risk factor, that leaves 690 − 585 = 105 in region VIII.

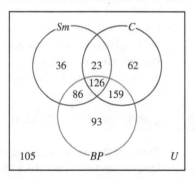

Step 4   Now we can answer a whole bunch of questions about the study.

(a) The intersection of all three risk factors contains 126 survivors.
(b) Regions II, IV, and VI are patients that had exactly two of the risk factors; this is 86 + 23 + 159 = 268 survivors.
(c) From region VIII we see that 105 patients had none of the risk factors.
(d) The total number inside the smoker circle is 36 + 23 + 126 + 86 = 271; this represents 271/690, or 39.3% of the survivors.

Try This One   4

Three other risk factors are obesity, family history of heart disease, and stress. Of the group of heart attack patients in Example 4, 213 had a family history of heart disease, 47 of those also suffered from stress but not obesity, 60 were obese but had no stress issues, and 12 were neither stressed nor obese. Stress was a factor for 170 total, 8 of whom had no family history of heart disease and weren't obese. There were 396 patients with none of these three risk factors.

(a) How many of the patients were obese?
(b) What percentage had all three of these risk factors?
(c) How many were not obese and did not have a family history of heart disease?

We've come a pretty long way from simply defining what sets and elements are! In this section, we've seen that Venn diagrams can be effectively used to sort out some pretty complicated situations in our world. And the better you are at interpreting information, the better equipped you'll be to survive and thrive in the information age.

Answers to Try This One

1 Only rain: 139; only snow: 39; neither: 163 (assuming it's not leap year!)

2 Coke: 60.7%, Pepsi: 44.6% both: 14.7%, neither: 9.4%

3 (a) 100   (c) 40
   (b) 70   (d) 340

4 (a) 227   (b) 13.6%   (c) 404

Exercise Set   **2-4**

Applications in Our World

1. In a survey of 85 college students, 72 use Instagram, 31 use Twitter, and 21 use both.
   (a) How many use Twitter only?
   (b) How many use Instagram only?
   (c) How many use neither?

2. In a class of 25 students, 18 were math majors, 12 were computer science majors, and 7 were dual majors in math and computer science.
   (a) How many students were majoring in math only?
   (b) How many students were not majoring in computer science?
   (c) How many students were not math or computer science majors?

3. A court record search of 250 incoming first-year students at a state university shows that 26 had been arrested at some point for underage drinking but not drug possession, and 12 had been arrested for possession but not underage drinking. There were 202 that had been arrested for neither.
   (a) How many of the students had been arrested for drug possession?
   (b) How many had been arrested for underage drinking?

4. Twenty-five mice were involved in a biology experiment involving exposure to chemicals found in cigarette smoke. Fifteen developed at least one tumor, nine suffered from respiratory failure, and four developed tumors and had respiratory failure.
   (a) How many only got tumors?
   (b) How many didn't get a tumor?
   (c) How many suffered from at least one of these effects?

5. Out of 20 students taking a midterm psychology exam, 15 answered the first of two bonus questions, 13 answered the second bonus question, and 2 didn't bother with either one.
   (a) What percentage of students tried both questions?
   (b) What percentage tried at least one question?

6. In a study of 400 entrées served at 75 campus cafeterias, 70 had less than 10 grams of fat but not less than 350 calories; 48 had less than 350 calories but not less than 10 grams of fat; 140 had over 350 calories and over 10 grams of fat.
   (a) What percentage of entrées had less than 10 grams of fat?
   (b) What percentage of entrées had less than 350 calories?

7. The financial aid department at a college surveyed 70 students, asking if they receive any type of financial aid. The results of the survey are summarized in the table below.

Financial Aid	Number of Students
Scholarships	16
Student loans	24
Private grants	20
Scholarships and loans	9
Loans and grants	11
Scholarships and grants	7
Scholarships, loans, and grants	2

   (a) How many students got only scholarships?
   (b) How many got loans and private grants but not scholarships?
   (c) How many didn't get any of these types of financial aid?

8. The manager of a campus gym is planning the schedule of fitness classes for a new school year, and will decide how often to hold certain classes based on the interests of students. She polls 47 students at various times of day, asking what type of classes they'd be interested in attending. The results are summarized in the table below.

Type of Class	Students Interested
Yoga	17
Pilates	13
Spinning	12
Yoga and Pilates	9
Pilates and spinning	3
Yoga and spinning	5
All three	2

   (a) How many students are interested in yoga or spinning, but not Pilates?
   (b) How many are interested in exactly two of the three classes?
   (c) How many are interested in yoga but not Pilates?

9. One semester in a chemistry class, 14 students failed due to poor attendance, 23 failed due to not studying, 15 failed because they did not turn in assignments, 9 failed because of poor attendance and not studying, 8 failed because of not studying and not turning in assignments, 5 failed because of poor attendance and not turning in assignments, and 2 failed because of all three of these reasons.

   (a) How many failed for exactly two of the three reasons?
   (b) How many failed because of poor attendance and not studying but not because of not turning in assignments?
   (c) How many failed because of exactly one of the three reasons?
   (d) How many failed because of poor attendance and not turning in assignments but not because of not studying?

10. According to a survey conducted by the National Pizza Foundation that I just now made up, out of 109 customers surveyed, 32 prefer pizzas with just pepperoni, 40 with just sausage, and 18 with only onions. Thirteen big-time carnivores like pepperoni and sausage, 10 customers prefer sausage and onions, 9 customers like pepperoni and onions; in each case, the third item could be included as well. Seven go all out, ordering all three.

   (a) How many customers prefer pepperoni, or sausage, or pepperoni and sausage with no onions?
   (b) What about sausage, or onions, or sausage and onions with no pepperoni?
   (c) How many go the boring route—none of those toppings?

11. Two hundred patients suffering from depression enrolled in a clinical trial to test the effects of various antidepressants. Zoloft was given to 27% of patients, Lexapro was given to 30%, and Prozac was given to 27%. Thirteen percent were treated with at least Zoloft and Lexapro, 11.5% were given at least Lexapro and Prozac, 7% were given at least Zoloft and Prozac, and 4% were treated with all three drugs.

   (a) How many patients in the trial were given at most two of the three drugs?
   (b) How many patients were treated with Zoloft and Prozac but not Lexapro?
   (c) How many patients were given a placebo containing none of the three drugs?

12. A survey of 96 students on campus showed that 29 read the *Campus Observer* student newspaper that morning, 24 read the news via the Internet that morning, and 20 read the local city paper that morning. Eight read the *Campus Observer* and the Internet news that morning, while four read the Internet news and the local paper, seven read the *Campus Observer* and the local city paper, and one person read the *Campus Observer,* the Internet news, and the local paper.

   (a) How many read the Internet news or local paper but not both?
   (b) How many read the Internet news and the local paper but not the *Campus Observer*?
   (c) How many read the Internet news or the *Campus Observer* or both?

13. Of the 50 largest cities in the United States, 11 have a team in the National Basketball Association but not a major league baseball team; 9 have a major league baseball team but not a team in the NBA; 12 have neither.

   (a) How many cities have both a major league baseball team and a team in the NBA?
   (b) Chicago, New York, and Los Angeles have two baseball teams, but Los Angeles is the only city with two basketball teams. Each of those cities has teams in both leagues. How many teams are there in each league?

14. One hundred new books are released nationally over a busy 3-day stretch in December. Eight had an e-book version available only on Amazon, 5 were available only on Google books, and 18 were available only on iTunes. There were 26 total available on Google, 7 that could be found on both Amazon and Google but not iTunes, and 4 that could be found on both iTunes and Google but not Amazon. Draw a Venn diagram representing this information, and use it to answer the following questions.

   (a) How many books were available on all three services?
   (b) Explain why you can't find the number of books that were not available on any of the three services.

(c) If every book released was available as an e-book on at least one of Amazon, Google, or iTunes, how many were available on Amazon and iTunes but not Google?

(d) In that case, how many were available on exactly two of those three services?

15. A marketing firm is hired to conduct research into the listening habits of drivers in a large urban area. On the first day, 121 drivers were surveyed: 26 listen to FM radio while driving, 4 of whom listen to only FM. Eight more listen to FM and AM only, while 4 listen to FM and satellite only. There were 6 that listen to only AM, 22 that listen to only satellite, and 69 that listen to at least one of the three.

(a) Are there more people who listen to satellite radio, or more who listen to none of the three types in the survey?

(b) How many more listen to AM radio than FM?

(c) How many listen to some form of radio, but not AM?

16. The arts communities in 230 cities across the country were rated according to whether or not they have an art museum, a symphony orchestra, and a ballet company. There are 119 cities with an art museum; 20 of those also have ballet but no orchestra, and 41 have an orchestra but no ballet; 30 have neither. Of the 75 cities with a ballet company, 10 have an orchestra as well, but lack an art museum. Twenty-two cities have only an orchestra.

(a) What percentage of the cities has an orchestra?

(b) How many more cities are there with none of these three than with all of them?

(c) If you pick a city at random from this list to travel to and really want to go either to an art museum or an orchestral concert, what is the percent chance that you'll end up disappointed?

Critical Thinking

17. A researcher was hired to examine the drinking habits of energy drink consumers. Explain why he was fired when he published the results below, from a survey of 40 such consumers:

23 said they drink Red Bull.
18 said they drink Monster.
19 said they drink G2.
12 said they drink Red Bull and Monster.
6 said they drink Monster and G2.
7 said they drink Red Bull and G2.
2 said they drink all three (not at the same time, hopefully).
2 said they don't drink any of the three brands.

18. The marketing research firm of OUWant12 designed and sent three spam advertisements to 40 e-mail accounts. The first one was an ad for hair removal cream, the second was an ad for Botox treatments, and the third was an ad for a new all lima bean diet. Explain why, when the following results occurred, the sponsors discontinued their services.

23 recipients deleted the ad for hair removal cream without looking at it.
18 recipients deleted the ad for Botox treatments.
19 recipients deleted the ad for the all lima bean diet.
12 recipients deleted the ads for hair removal cream and Botox treatments.
6 recipients deleted the ads for Botox treatments and the all lima bean diet.
7 recipients deleted the ads for the hair removal cream and the all lima bean diet.
2 recipients deleted all three ads.

19. A TV network considering new contracts to televise pro sports hires a marketing consultant to conduct a survey of randomly selected TV viewers asking them which of football, baseball, and basketball they go out of their way to watch on TV. Of those surveyed, 35 watch baseball, 235 watch basketball, 295 watch football, 90 watch basketball and football, and 560 watch none of the three.

(a) Explain why this is not enough information to find the total number of people surveyed.

(b) Upon looking at the results more carefully, a member of the consultant team discovers that every single person who watches baseball also watches basketball, and none of those people watch football. Now can you find the number of people surveyed?

(c) How many people watch only football? How many watch only basketball?

20. (a) How many distinct regions are there in a two-circle Venn diagram? What about a three-circle? Use this to make a conjecture on the number of regions that would be needed for a four-circle Venn diagram.

(b) Make a few attempts at drawing a Venn diagram with four circles, then explain why it can't be done.

(c) Use the Internet to find a Venn diagram for four sets, and use it to check your conjecture from part (a).

Section 2-5   Infinite Sets

LEARNING OBJECTIVES

1. Formally define infinite sets.
2. Show that a set is infinite.
3. Find a general term for an infinite set.
4. Define countable and uncountable sets.

John Carleton/Moment/Getty Images

The night sky may look infinite, but have you ever thought about what that really means?

Infinity is a concept that's tremendously difficult for us human beings to wrap our minds around. Because our thoughts are shaped by experiences in a physical world with finite dimensions, things that are infinitely large always seem just out of our grasp. Whether you're a proponent of the Big Bang theory or creationism, you probably think that one or the other must be true because the alternative is too far beyond our experience: that time didn't have a beginning at all, but extends infinitely far in each direction. Some philosophers feel that human beings are fundamentally incapable of grasping the concept of something infinitely large at all!

The study of infinity and infinite sets from a mathematical standpoint is a relatively young one compared to the history of math in general. For at least a couple of thousand years, the nature of infinity so confounded the greatest human minds that they chose to not deal with it at all. And yet in working with a set as simple as the natural numbers, we deal with infinite sets in math all the time. It's an interesting paradox.

A Definition of Infinite Sets

Recall from Section 2-1 that a set is considered to be finite if the number of elements is either zero or a natural number. Otherwise, it is considered to be an infinite set. For example, the set $\{10, 20, 30, 40\}$ is finite because the number of elements (four) is a natural number. But the set $\{10, 20, 30, 40, \ldots\}$ is infinite because the number of elements is unlimited, and therefore not a natural number.

You might recognize an infinite set when you see one, but it's not necessarily easy to make a precise definition of what it means for a set to be infinite (other than the obvious definition, "not finite"). The German mathematician Georg Cantor, widely regarded as the father of set theory, is famous for his 19th-century study of infinite sets. Cantor's simple and elegant definition of an infinite set is as follows:

> A set is **infinite** if it can be placed into a one-to-one correspondence with a proper subset of itself.

1. Formally define infinite sets.

Math Note

Think about this for a second: which set is bigger, $\{1, 2, 3, 4, 5, \ldots\}$ or $\{2, 4, 6, 8, 10, \ldots\}$? Almost any reasonable person would say the first is bigger, right? But the discussion to the right shows that they're actually the same size. Mind blown!

First, notice that a finite set definitely does not meet the condition in this definition: if a set has some finite number of elements, let's say 10, then any proper subset has at most 9 elements, and an attempt at one-to-one correspondence will always leave out at least one member.

The trickier thing is to understand how an infinite set can meet this definition. We'll illustrate with an infinite set we know well, the set of natural numbers $\{1, 2, 3, 4, \ldots\}$. The set of even natural numbers $\{2, 4, 6, 8, \ldots\}$ is a proper subset: every even number is also a natural number, but there are natural numbers that are not even numbers. Now we demonstrate a clever way to put these two sets into a one-to-one correspondence: match each natural number with its double.

$$1 \leftrightarrow 2, \qquad 2 \leftrightarrow 4, \qquad 3 \leftrightarrow 6, \qquad 4 \leftrightarrow 8, \ldots$$

In general, we can define our correspondence as matching any n from the set of natural numbers with a corresponding even number $2n$. This is a one-to-one correspondence because every natural number has a match (its double), and every even number has a match (its half). So we've put the natural numbers into one-to-one correspondence with a proper subset, proving that they are an infinite set.

Let's try another example.

Sidelight   The Infinite Hotel

Suppose a hotel in some far-off galaxy was so immense that it actually had infinitely many rooms, numbered 1, 2, 3, 4, ... There's a big convention of creepy alien creatures in town, so every room is filled. A weary traveler drags into the lobby and asks for a room, and when informed that the hotel is full, they protest that the hotel can most definitely accommodate them. Do you agree? Can they find a room without kicking someone out?

People tend to be split on this question about half and half: half think they can't accommodate our traveler because all the rooms are full, and half think they can because there are infinitely many rooms. In fact, the traveler is correct—it just takes inconveniencing every other guest! If the manager asks every guest to move into the room whose number is 1 higher than their current room, everyone that was originally in a room still has one, and our traveler can rest comfortably in room 1.

Naufal MQ/Getty Images

This clever little mind exercise is a consequence of the fact that the natural numbers form an infinite set—they can be put in one-to-one correspondence with a proper subset of themselves by corresponding any n with $n + 1$.

Example 1	Showing That a Set Is Infinite

Show that the set $\{5, 10, 15, 20, 25, \ldots\}$ is an infinite set.

SOLUTION

A simple way to put this set in correspondence with a proper subset of itself is to match every element n with its double $2n$:

$$\{5, \quad 10, \quad 15, \quad 20, \quad 25, \ldots\}$$
$$\updownarrow \quad \updownarrow \quad \updownarrow \quad \updownarrow \quad \updownarrow$$
$$\{10, \quad 20, \quad 30, \quad 40, \quad 50, \ldots\}$$

The second set, $\{10, 20, 30, 40, 50, \ldots\}$ is a proper subset of the first, and the two are in one-to-one correspondence, so $\{5, 10, 15, 20, 25, \ldots\}$ is an infinite set.

Try This One   1

Show that the set $\{-1, -2, -3, -4, -5, \ldots\}$ is an infinite set.

2. Show that a set is infinite.

A General Term for an Infinite Set

One consequence of the way we showed that the set of natural numbers is infinite is that we can find a generic formula for the set of even numbers: $2n$, where n is the set $\{1, 2, 3, 4, \ldots\}$. We will call $2n$ in this case a **general term** of the set of even numbers. Notice that we said "a general term," not "the general term." There are other general terms we could write for this set: $2n - 6$, where n is the set $\{4, 5, 6, 7, \ldots\}$, is another possibility. But in most cases the simplest general term is the one where the first listed number is obtained from substituting in 1 for n, and that's the one we'll typically find.

Example 2   Finding a General Term for an Infinite Set

Find a general term for the set $\{4, 7, 10, 13, 16, \ldots\}$.

SOLUTION

We should always begin by trying to recognize a pattern in the numbers of the set. In this case, the pattern is that the numbers increase by 3. When this is the case, $3n$ is a good choice, because as n increases by 1, $3n$ increases by 3. But simply using $3n$ will give us the set $\{3, 6, 9, 12, \ldots\}$, which is not quite what we want. We remedy that by adding 1 to our general term, to get $3n + 1$. (We encourage you to check that answer by substituting in $1, 2, 3, \ldots$ for n to see that it generates the set $\{4, 7, 10, 13, 16, \ldots\}$.)

Math Note

Finding a general term for a set is not always easy. In some cases, it can be very difficult or even impossible. You may need to do some trial and error before finding a formula that works.

3. Find a general term for an infinite set.

Try This One   2

Find a general term for the set $\{2, 8, 14, 20, 26, \ldots\}$.

Different Kinds of Infinity?

Quick, which set is bigger, the set of natural numbers or the set of real numbers? You probably answered the set of real numbers. But both sets are infinitely large, so aren't they the same size? Cantor attacked this problem in the late 1800s. He defined a set to be **countable** if it is finite or can be placed into one-to-one correspondence with the natural numbers and an infinite set to be **uncountable** if it cannot. He used the symbol $\aleph_0$, pronounced aleph-null or aleph-naught (aleph is the first letter of the Hebrew alphabet), to represent the cardinality of a countable set.

Example 3   Showing That a Set Is Countable

Show that the set of integers is countable.

SOLUTION

If a set is finite, it's automatically countable, so that's at least worth considering. But the integers are not a finite set, so we need to find a way to put them in one-to-one correspondence with the natural numbers. We could match up 0 with 1, 1 with 2, 2 with 3, and so on, but that would leave out the negatives. So let's get fancier:

Natural numbers	1	2	3	4	5	6	7	8	9	...
Integers	0	1	−1	2	−2	3	−3	4	−4	...

This works because we can see that every integer will eventually get matched with a natural number, so this defines a one-to-one correspondence. The proof would be stronger, though, if we can define a formula for the correspondence. For every natural number n,

$$n \to \begin{cases} \dfrac{n}{2} & \text{if } n \text{ is even} \\[2mm] -\dfrac{n-1}{2} & \text{if } n \text{ is odd} \end{cases}$$

defines a one-to-one correspondence.

Sidelight   The Sad Case of the Man Who Was Too Insightful

Whoever coined the phrase "true genius is never recognized in its own time" would have LOVED Georg Cantor. Simply being known as the founder of what is now a major branch of mathematics (set theory) is a tremendous accomplishment, but it was Cantor's unique ability to formalize a study of the infinite that was both his greatest achievement, and ultimately his greatest curse.

For centuries, the study of math essentially ignored the fact that its most basic building blocks—the natural and real numbers—were infinite sets; infinity simply wasn't studied. For the most part, this was attributed to religious beliefs—infinity was thought to be the realm of God alone, and attempts to study it scientifically were considered inappropriate by some, and outright heresy by others. To say the very least, Cantor's work was not celebrated at the time it was produced. Two of Cantor's greatest critics were also among the most celebrated mathematicians of the late 1800s, Henri Poincaré and Leopold Kronecker. Poincaré referred to Cantor's work as "a grave disease infecting mathematics." Kronecker evidently preferred personal attacks, labeling Cantor a "scientific charlatan," a "renegade," and a "corrupter of youth." And this was from mathematicians—the religious philosophers felt that he should be imprisoned, or worse.

This criticism weighed heavily on poor Georg, who was first hospitalized for severe depression in 1884 after 10 years of founding work on set theory. Although he remained active as a mathematician until 1913, he was in and out of institutions, and spent the last 5 years of his life in a sanatorium, dying in 1918.

But believe it or not, it could have been worse. Some of the ideas that we now accept about infinity can be traced back to the Italian philosopher, mathematician, and astronomer Giordano Bruno, who was rewarded for his groundbreaking understanding of the infinite nature of the universe by being burned at the stake on February 17, 1600.

Try This One   3

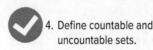

4. Define countable and uncountable sets.

Show that the set of positive rational numbers with denominators 2 or 3 is a countable set. (Rational numbers are fractions with integers in the numerator and denominator.)

Now the obvious question is this: What kinds of sets are uncountable? And how do you prove this? That, friends, is not an easy question at all. In fact, one of Cantor's greatest achievements was showing that the set of real numbers is not countable. So if you guessed that there are more real numbers than natural numbers, you were right. But the study of infinite sets is a strange and interesting one, with unexpected results at nearly every turn. For example, it can be shown that the cardinality of the set of numbers just between 0 and 1 is exactly the same as the cardinality of the entire set of real numbers! If you find these ideas interesting, you'll get a big kick out of Projects 3 and 4 at the end of this chapter.

Answers to Try This One

1   Can be done in many ways: one choice is to correspond -1 with -2, -2 with -4, and in general $-n$ with $-2n$.

2 $6n - 4$

3   One possibility:

$$
\begin{array}{cccccc}
1 & 2 & 3 & 4 & 5 & 6\ldots \\
\updownarrow & \updownarrow & \updownarrow & \updownarrow & \updownarrow & \updownarrow \\
\frac{1}{2} & \frac{1}{3} & \frac{2}{2} & \frac{2}{3} & \frac{3}{2} & \frac{3}{3}\ldots
\end{array}
$$

$$
n \to \begin{cases} \dfrac{(n+1)/2}{2} & \text{if } n \text{ is odd} \\[2ex] \dfrac{n/2}{3} & \text{if } n \text{ is even} \end{cases} \quad \text{or} \quad n \to \begin{cases} \dfrac{(n+1)}{4} & \text{if } n \text{ is odd} \\[2ex] \dfrac{n}{6} & \text{if } n \text{ is even} \end{cases}
$$

Exercise Set   2-5

Writing Exercises

1. Define an infinite set, both in your own words and by using Cantor's definition.
2. What is meant by a general term for an infinite set?

3. What does it mean for a set to be countable?
4. Explain how you can tell that the set of natural numbers and the set of even numbers have the same cardinality.

Computational Exercises

For Exercises 5–20, find a general term for the set.

5. $\{7, 14, 21, 28, 35, \ldots\}$
6. $\{1, 8, 27, 64, 125, \ldots\}$
7. $\{4, 16, 64, 256, 1{,}024, \ldots\}$
8. $\{1, 4, 9, 16, 25, \ldots\}$
9. $\{-3, -6, -9, -12, -15, \ldots\}$
10. $\{22, 44, 66, 88, 110, \ldots\}$
11. $\{\frac{1}{4}, \frac{1}{2}, \frac{3}{4}, 1, \frac{5}{4}, \frac{3}{2}, \frac{7}{4}, \ldots\}$
12. $\{\frac{1}{6}, \frac{1}{3}, \frac{1}{2}, \frac{2}{3}, \frac{5}{6}, 1, \ldots\}$
13. $\{2, 6, 10, 14, 18, \ldots\}$
14. $\{1, 4, 7, 10, 13, \ldots\}$
15. $\{\frac{2}{3}, \frac{3}{4}, \frac{4}{5}, \frac{5}{6}, \frac{6}{7}, \ldots\}$
16. $\{1, \frac{1}{8}, \frac{1}{27}, \frac{1}{64}, \frac{1}{125}, \ldots\}$
17. $\{100, 200, 300, 400, 500, \ldots\}$
18. $\{50, 100, 150, 200, 250, \ldots\}$
19. $\{-4, -7, -10, -13, -16, \ldots\}$
20. $\{-3, -5, -7, -9, -11, \ldots\}$

For Exercises 21–30, show each set is an infinite set.

21. $\{3, 6, 9, 12, 15, \ldots\}$
22. $\{10, 15, 20, 25, 30, \ldots\}$
23. $\{9, 18, 27, 36, 45, \ldots\}$
24. $\{4, 10, 16, 22, 28, \ldots\}$
25. $\{2, 5, 8, 11, 14, \ldots\}$
26. $\{20, 24, 28, 32, 36, \ldots\}$
27. $\{10, 100, 1{,}000, 10{,}000, \ldots\}$
28. $\{100, 200, 300, 400, 500, \ldots\}$
29. $\{\frac{5}{1}, \frac{5}{2}, \frac{5}{3}, \frac{5}{4}, \frac{5}{5}, \ldots\}$
30. $\{\frac{1}{2}, \frac{1}{4}, \frac{1}{8}, \frac{1}{16}, \ldots\}$

For Exercises 31–34, show that the given set is countable. (See Example 3 for guidance.)

31. $\{5, 10, 15, 20, 25, \ldots\}$
32. $\{-3, -6, -9, -12, -15, -18, \ldots\}$
33. The set of numbers whose square root is a whole number
34. The set of negative rational numbers with denominators 5 and 7

Critical Thinking

35. The set of rational numbers is the set of all possible fractions that have integer numerators and denominators. Intuitively, do you think there are more rational numbers than natural numbers? Why? Do you think that the set of rational numbers is countable?
36. Can you think of any set of tangible objects that is infinite? Why or why not?
37. Study Example 3 carefully, then compare it to Example 2. What did we actually prove in Example 2 without even realizing it?
38. True or false:
    (a) A subset of an infinite set is infinite.
    (b) If set A has a subset that is infinite, then A has to be infinite as well.

Exercises 39 and 40 use the fact that the cardinality of the set of natural numbers $\{1, 2, 3, 4, \ldots\}$ is $\aleph_0$.

39. (a) Define a one-to-one correspondence between the set of natural numbers and the set $\{0, 1, 2, 3, 4, \ldots\}$.

(b) Write an arithmetic problem involving $\aleph_0$ that is illustrated by part (a). (*Hint:* How many more elements than the natural numbers does the set $\{0, 1, 2, 3, 4, \ldots\}$ have?

40. (a) Define a one-to-one correspondence between the set of natural numbers and the set of all integers excluding zero.

(b) Write an arithmetic problem involving $\aleph_0$ that is illustrated by part (a).

In Exercises 41–46, find the cardinality of the given set. You may find the ideas in Exercises 39 and 40 helpful.

41. $\{10, 11, 12, 13, 14, \ldots\}$
42. $\{-1, -2, -3, -4, -5, \ldots\}$
43. $\{1, 3, 5, 7, 9, \ldots, 29\}$
44. $\{2, 4, 6, 8, 10, \ldots, 24\}$
45. The set of odd natural numbers.
46. The set of even negative integers.

CHAPTER 2 Summary

Section	Important Terms	Important Ideas
2-1	Set Roster method Element Well-defined Natural numbers Descriptive method Set-builder notation Variable Finite set Infinite set Cardinal number Empty set Equal sets Equivalent sets One-to-one correspondence	**A set** is a collection of objects; a set is well-defined if any object can be objectively determined to be either in the set or not in the set. Each object is called an element or member of the set. We use three ways to identify sets: the roster method, the descriptive method, and set-builder notation. A finite set contains a specific number of elements, while an infinite set contains an unlimited number of elements. If a set has no elements, it is called an empty set or a null set. Two sets are equal if they have the same elements, and two finite sets are equivalent if they have the same number of elements. Two sets are said to be in one-to-one correspondence if it's possible to pair the elements so that each element in the first set has exactly one match in the second set, and vice versa.
2-2	Universal set Complement Subset Proper subset Intersection Union Subtraction Cartesian product	**The universal** set is the set of all elements used for a specific problem or situation. The complement of a specific set is a set that consists of all elements in the universal set that are not in the specific set. A set A is called a subset of another set B if every element in A is also in B. We call A a proper subset of B if there's at least one element in B that's not in A. The union of two sets is the set of all elements that are in at least one of the sets. The intersection is the set of all elements in both sets. The difference of set A and set B, denoted $A - B$, is the set of elements in set A but not in set B. The Cartesian product of two sets A and B is $A \times B = \{(x, y) \mid x \in A \text{ and } y \in B\}$.
2-3	Venn diagram	**A mathematician** named John Venn devised a way to represent sets pictorially. His method uses overlapping circles to represent the sets. Items in the intersection of the sets are placed where the circles overlap. De Morgan's laws for two sets A and B are $(A \cup B)' = A' \cap B'$ and $(A \cap B)' = A' \cup B'$. For any two finite sets A and B, $n(A \cup B) = n(A) + n(B) - n(A \cap B)$.
2-4		**Venn diagrams** can be used to solve problems in our world involving surveys and classifications.
2-5	Infinite set General term Countable set Uncountable set	**An infinite** set can be placed in a one-to-one correspondence with a proper subset of itself. A set is called countable if it is finite or if there is a one-to-one correspondence between the set and the set of natural numbers. A set is called uncountable if it is not countable. The natural numbers are an example of an infinite set that is countable, and the real numbers are an uncountable set.

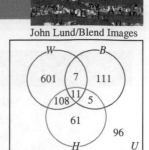

John Lund/Blend Images

Math in Diversity REVISITED

The Venn diagram shown to the right is based on the given demographic estimates:

1. White only: 601; Black only: 111; Hispanic only: 61
2. Hispanic and Black, but not white: 5
3. Hispanic or Black: 303
4. None of white, Black, or Hispanic: 96

5. This is simple if you think about it: the percentage for whites is less than 50, so there will be more nonwhites.

Review Exercises

Section 2-1

For Exercises 1–8, write each set in roster notation.

1. The set D is the set of even numbers between 50 and 60.
2. The set F is the set of odd numbers between 3 and 40.
3. The set L is the set of letters in the word *letter*.
4. The set A is the set of letters in the word *Arkansas*.
5. The set B is $\{x \mid x \in N \text{ and } x > 500\}$.
6. The set C is the set of natural numbers between 5 and 12.
7. M is the set of men that have walked on the moon.
8. W is the set of women that have walked on the moon.

For Exercises 9–12, write each set using set-builder notation.

9. $\{18, 20, 22, 24\}$
10. $\{5, 10, 15, 20\}$
11. $\{101, 103, 105, 107, \ldots\}$
12. $\{8, 16, 24, \ldots 72\}$

For Exercises 13–20, state whether the set is finite or infinite.

13. $\{x \mid x \in N \text{ and } x \geq 9\}$
14. $\{4, 8, 12, 16, \ldots\}$
15. {annoying commercials}
16. $\{3, 7, 9, 12\}$
17. Ø
18. {people who have red hair}
19. {10 digit numbers}
20. Which of the sets in Exercises 13–19 are not well-defined?

Section 2-2

For Exercises 21–24, decide if the statement is true or false.

21. $\{80, 100, 120, \ldots\} \subseteq \{40, 80, 120, \ldots\}$
22. $\{6\} \subset \{6, 12, 18\}$
23. $\{5, 6, 7\} \subseteq \{5, 7\}$
24. $\{a, b, c\} \subset \{a, b, c\}$
25. Find all subsets of $\{r, s, t\}$.
26. How many subsets and proper subsets does the set $\{a, e, i, o, u, y\}$ have?

An auto testing website chose 11 new vehicles for three types of safety tests: airbag deployment, rollover strength, and 20-MPH crash test. The vehicles that rated exceptional in airbag deployment are listed in set A. Set B is those that rated exceptional in rollover strength, and set C is the ones rated exceptional in crash test. Use these sets for Exercises 27–38.

$U = \{$Chevy, Ford, BMW, Mercedes, Toyota, Honda, Lexus, Acura, Hyundai, Tesla, Dodge$\}$
$A = \{$Chevy, BMW, Toyota, Honda, Lexus$\}$
$B = \{$Toyota, Honda, Lexus, Hyundai, Tesla$\}$
$C = \{$Mercedes, Acura, Dodge$\}$

Find each set.

27. $A \cap B$
28. $B \cup C$
29. $(A \cap B) \cap C$
30. B'
31. $A - B$
32. $B - A$
33. $(A \cup B)' \cap C$
34. $B' \cap C'$
35. $(B \cup C) \cap A'$
36. $(A \cup B) \cap C'$
37. $(B' \cap C') \cup A'$
38. $(A' \cap B) \cup C$

39. If $K = \{x \mid x \in N, x > 25\}$ and $L = \{x \mid x \in E, x > 10\}$, find $K \cap L$, $K \cup L$, and $L - K$.
40. For each exercise number listed below, write a verbal description of what the set in that exercise represents.

   (a) 27   (d) 30   (g) 34
   (b) 28   (e) 31   (h) 35
   (c) 29   (f) 32

Section 2-3

For Exercises 41–46, use the Venn diagram below. Describe the region or regions provided in each problem, using set operations on A and B. There may be more than one right answer.

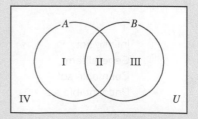

41. Region I
42. Region II
43. Region III
44. Region IV
45. Regions I and III
46. Regions I and IV

For Exercises 47–50, draw a Venn diagram and shade the appropriate area for each.

47. $A' \cap B$
48. $(A \cup B)'$
49. $(A' \cap B') \cup C$
50. $A \cap (B \cup C)'$
51. If $n(A) = 15$, $n(B) = 9$, and $n(A \cap B) = 4$, find $n(A \cup B)$.
52. If $n(A) = 24$, $n(B) = 20$, and $n(A \cap B) = 14$, find $n(A \cup B)$.

The table and Venn diagram below are to be used for Exercises 53–56. The table shows the bottom five states in 2021 in terms of percentage of the population that have reached at least a certain level of education. For each question, write the region of the Venn diagram that would include the state listed.

High School Graduate	Bachelor's Degree	Advanced Degree
California	West Virginia	North Dakota
Texas	Mississippi	West Virginia
Mississippi	Arkansas	Arkansas
Louisiana	Louisiana	Louisiana
New Mexico	Kentucky	Mississippi

Source: Wallethub

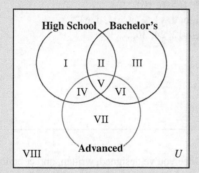

53. West Virginia
54. Mississippi
55. North Dakota
56. Ohio

Section 2-4

57. On the eve of the 2020 Iowa Democratic caucuses, 250 likely voters were polled: 155 were supporters of Joe Biden,

140 supported Bernie Sanders, and 120 were supporters of both candidates.
   (a) How many of those polled supported neither of those candidates?
   (b) How many supported only Bernie Sanders?

58. A hearing specialist conducts a study on hearing loss at certain frequencies among a group of patients in a retirement community. Of 94 residents tested, 10 had significant hearing loss at low frequencies but not high, 40 had significant loss at high frequencies but not low, and 26 showed no significant hearing loss at all.
   (a) How many residents had hearing loss at both low and high frequencies?
   (b) What percentage suffer from hearing loss at high frequencies?

59. Fifty-three callers to a campus radio station were asked what they usually listened to while driving to school. Of those asked, 22 listened to a local radio station, 18 listened to satellite radio, 33 listened to online music, 8 listened to a local radio station and satellite radio, 13 listened to satellite radio and online music, 11 listened to a local radio station and online music, and 6 listened to all three.
   (a) How many listened to only satellite radio?
   (b) How many listened to local radio stations and online music but not satellite radio?
   (c) How many listened to none of these?

60. The manager of a campus bookstore finds that in the last hour before closing, 41 students bought two or more textbooks. Of these, 4 used cash only, 5 used a financial aid voucher only, and 5 used a debit card only. Seven used a debit card and financial aid voucher and no cash; three used cash and a debit card with no financial aid assistance. Sixteen students total used a debit card for at least part of their purchase, while 9 used none of those forms of payment.
   (a) How many students used all of cash, debit card, and financial aid voucher?
   (b) Did more students use cash or financial aid vouchers?
   (c) What percentage of students did not use a financial aid voucher?

Section 2-5

61. Find a general term for the set $\{-5, -7, -9, -11, -13, \ldots\}$
62. Show that the set $\{12, 24, 36, 48, 60, \ldots\}$ is an infinite set.
63. Show that the set in Exercise 62 is countable.

Chapter Test

For Exercises 1–4, write each set in roster notation.

1. The set P is the set of even natural numbers between 90 and 100.
2. The set K is the set of letters in the word *envelope*.
3. $X = \{x \mid x \in N \text{ and } x < 80\}$
4. The set J is the set of months of the year that begin with the letter J.

For Exercises 5 and 6, write each set using set-builder notation.

5. $\{12, 14, 16, 18\}$
6. $\{4, 8, 16, \ldots 128\}$

For Exercises 7–11, state whether the set is finite or infinite.

7. $\{x \mid x \in N \text{ and } x \text{ is a multiple of } 6\}$
8. $\{a, b, c, \ldots s, t\}$

9. The set V is the set of people with awesome hair.

10. Explain why the set in Exercise 9 is not well-defined. Your answer should be written so that someone who has no idea what "well-defined" means would understand.

11. Find all subsets and all proper subsets of the set of states that border California. How do you know how many subsets you're looking for?

For Exercises 12–16, let $U = \{a, b, c, d, e, f, g, h, i, j, k\}$, $A = \{a, b, d, e, f\}$, $B = \{a, g, i, j, k\}$, and $C = \{e, h, j\}$. Find each.

12. $(A \cap B) \cup C$

13. $(A \cup B)'$

14. $A - B$

15. $(A - B) - C$

16. Draw and shade a separate Venn diagram for each set: $B - A$, $B' \cup A$, $A \cup B \cup C$

17. Find both Cartesian products that can be formed using the set in Exercise 11 and set C in Exercises 12–16.

For Exercises 18–20, draw a Venn diagram for each set.

18. $A' \cap B$

19. $(A \cap B)'$

20. $(A' \cup B') \cap C'$

21. If $n(A) = 1,500$, $n(B) = 1,150$, and $n(A \cap B) = 350$, find $n(A \cup B)$.

22. A student studying for a master's degree in sports management is working on a thesis about the prevalence of women's college sports since Title IX mandated equal access to women. He compiles data on 119 schools with football teams that compete in the NCAA FBS (which many fans still call Division 1-A), and finds that 69 have a women's golf team, 63 have a field hockey team, and 83 have a women's swimming team. There are 28 schools that field teams in all three sports. Forty-six have golf and women's swimming teams, 40 have women's swimming and field hockey teams, and 47 have women's golf and field hockey teams.

   (a) How many schools have a women's golf team, but no women's swimming or field hockey?

   (b) What percentage of teams have at least two of the three sports?

   (c) If you pick one of the schools in the study at random, what's the percent chance that it has none of those three sports?

23. Find a general term for the set $\{15, 30, 45, 60, 75, \ldots\}$.

24. Show that the set $\{1, 2, 3, 4, \ldots\} \cup \{-1, -2, -3, -4, \ldots\}$ is countably infinite. (*Hint:* There are actually two separate questions to answer!)

For Exercises 25–30, state whether each is true or false.

25. $\{s, e, s, a, m, e\}$ is equivalent to $\{s, a, m, e\}$

26. $\{4, 8, 12, 16, \ldots\} \subseteq \{2, 4, 6, 8, \ldots\}$

27. $\{15\} \subset \{3, 6, 9, 12, \ldots\}$

28. $9 \notin \{2, 4, 5, 6, 10\}$

29. $\{a, e, i, o, u, y\} \subseteq \{a, e, i, o, u\}$

30. $\{12\} \in \{12, 24, 36, \ldots\}$

Projects

1. Have the students in your class fill out this questionnaire:

   A. Sex: Male _____ Female _____
   B. Age: Under 21 _____ 21 or older _____
   C. Work: Yes _____ No _____

   Draw a Venn diagram, and from the information answer these questions:
   (a) How many students are female?
   (b) How many students are under 21?
   (c) How many students work?
   (d) How many students are under 21 and work?
   (e) How many students are males and do not work?
   (f) How many students are 21 or older and work?
   (g) How many students are female, work, and are under 21?

2. If you've had a pulse for this entire chapter, you know that surveys play a big role in the way that Venn diagrams are used to organize information. Now it's time for you to design a survey of your very own. You should make the topic of the survey something you find interesting, and design the survey so that the results can be summarized and studied using a Venn diagram with three sets. You can accomplish this by asking three separate questions, or by designing a single-question poll where responders can choose any, all, or none of the responses.

   Once you've designed, written, and conducted your survey, organize the results with a Venn diagram, then write a report on your findings. Don't just include the raw numbers—the most important part of a survey is often summarizing and interpreting the meaning of survey results. You may want to keep this requirement in mind when designing your survey—it's pretty hard to write an intelligent interpretation if the questions are totally unconnected.

3. You may have wondered why we stated in Section 2-5 that the set of real numbers is uncountable, but didn't back that up. The short answer is that it's a pretty involved process to prove that result. It's not all that hard, but it's VERY clever and kind of subtle. The good news is you can find the proof on about ten thousand different Web pages by searching for "Cantor diagonal." Find a page that describes Cantor's diagonal argument in a way that you can understand, then put together a demonstration for classmates to help them understand why the real numbers are uncountable. *Extra credit:* Include a discussion of the fact that the set of real numbers has the same cardinality of just the real numbers between zero and one.

4. If you're really interested by the ideas of counting infinite sets, you'll find this project extra interesting. One of the strangest, most fascinating sets ever introduced is known as

the Cantor set. It starts with the set of all real numbers between 0 and 1 and throws away a sequence of portions. Do an online or library search (they still have actual libraries, right?) for the Cantor set, and answer the following questions about it:

(a) How can you be sure that there's at least SOMETHING in the Cantor set? (You'll find that at first glance, it looks like the set might be empty.)

(b) How do you know that the Cantor set is not only nonempty, but infinite?

(c) Is the Cantor set countable or uncountable? How do you know that?

(d) What's the total length of all intervals that are discarded when defining the Cantor set? Why is that such a shocking result?

Design elements: Front matter, Chapter Opener, Summary and End Matter header design (random numbers background illustration): ©pixeldreams.eu/Shutterstock RF

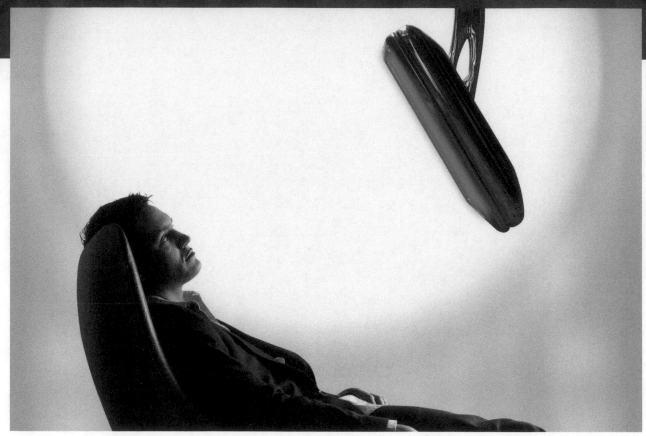

Photo: Colin Anderson/Blend Images

Outline

Math in Mind Control

The term *mind control* is an absolute favorite of conspiracy theorists and science fiction writers, but is it real? That largely depends on what you mean by the term, but I can guarantee you one thing: there are a LOT of people and organizations out there that are trying to control the way you think each and every day. Everywhere you turn in modern society, someone is trying to convince you of something. "Buy my product!" "My candidate is great, and yours is an evil dunce!" "You can't afford NOT to lease a car from our dealership!" "Bailing out banks is a horrible idea!" "You should go out with me, and soon!" "Pledge our fraternity—all the other ones stink!" Should I go on?

Logic is sometimes defined as correct thinking or correct reasoning, but many folks refer to logic by a more casual name: *common sense*. Regardless of what you call it, the ability to think logically is crucial for all of us because our lives are inundated daily with advertisements, contracts, product and service warranties, political debates, and news commentaries, to name just a few. People often have problems processing these things because of misinterpretation, misunderstanding, and faulty logic.

You can look up the truth or falseness of a fact on the Internet (and absolutely should), but that won't help you in analyzing whether a certain claim is logically valid. The term *common sense* is really misleading, because evaluating logical arguments involves skills that are anything but common—but we're here to help! Forget what you've seen about mind control in the movies: the best way to control someone's mind is to bias their thinking by using their emotions and opinions against them. And that's where this chapter comes in. To decide whether or not an argument makes sense, our goal will be to study the structure of the argument without thinking about the particular topic: to make that possible, we'll use variables to stand in for sentences so that we won't *know* what the exact topic is. This allows us to focus simply on whether you can logically draw a conclusion based on certain types of statements.

To that end, we've written some claims below. To be honest, some of them are a little controversial, touching on hot-button issues like religion, politics, and the most contentious of all, college football. We sincerely apologize if any of these statements offends you, but our aim is to illustrate how difficult it can be to decide if an argument is logically sound when you're distracted by emotion and strongly held beliefs. Your job is to evaluate each claim as a logical argument, and decide if the argument is valid: that is, if the conclusion can be logically drawn from a set of statements. The skills you learn in this chapter wll help you to do so, but making your best guess now might prove to be very enlightening later on.

1. Where there's smoke, there's fire.
2. People with lots of money are always happy. My neighbors are so happy you just want to smack them sometimes, so I guess they have a lot of money.
3. Every team in the SEC is good enough to play in a bowl game. Florida State isn't in the SEC, so it doesn't belong in a bowl game.
4. Scripture is the word of God. I know this because it says so in the Bible.
5. If a presidential candidate wins the popular vote, they should win the election. The Republican candidate lost the popular vote and shouldn't have been elected.
6. The world will end before a woman of color is elected vice president. Wait, what? That actually happened? I guess I better start preparing for the end.

For answers, see Math in Mind Control Revisited on page 153

Section 3-1 | Statements and Quantifiers

✓ **LEARNING OBJECTIVES**

1. Define and identify statements.
2. Define the logical connectives.
3. Write the negation of a statement.
4. Write statements symbolically.

For dozens of years, surveys done by marketing companies have shown that the primary concern for a majority of American parents is taking care of their families. So it's no coincidence that a lot of the household products that are traditionally bought by caregivers are advertised in a way that plays on that concern. The implication, sometimes overt and sometimes subconscious, is that if you're a good parent who cares about your family, you'll buy this particular product. It's easy to say that people can't possibly fall for stunts that blatant, but here's all the evidence you need that it works: advertisers keep doing it.

This is exactly the sort of mind control we talked about in the chapter opener: the advertiser doesn't want you to think about the fact that their actual argument is silly, so they try to tug at your heartstrings. But today, you begin to fight back by beginning your study of **symbolic logic**, which uses letters to represent statements and special symbols to represent words like *and*, *or*, and *not*. This will allow us to remove our emotional bias from an argument so that we can analytically evaluate the logic behind it.

Jamie Marcial/Purestock/SuperStock

Statements

In the English language there are many types of sentences; a few of the types are

- Factual statements (You have to pass the bar exam to practice law.)
- Commands (Get out of my face!)
- Opinions (Chocolate cake with cream cheese icing is the best dessert EVER.)
- Questions ('Sup with you?)
- Exclamations (Holy cow!)

In the objective study of logic, we will use only factual statements—it's pretty hard to decide if "Get out of my face" is true or false. And by "pretty hard" I mean "completely impossible."

> A **statement** is a declarative sentence that can be objectively determined to be either true or false, but not both.

Sidelight   **When True or False Fails**

Did it strike you as odd that the definition of statement requires a sentence to be "true or false, but not both"? Isn't something either true or false? How can it be both? Glad you asked.

Consider this fun example: "This sentence isn't true." If the sentence is in fact true, it means that it isn't true. But if it's false, it means that it is in fact true. Isn't that bizarre? This is a classic example of what's known as a **paradox**: something that appears to be a statement, but contradicts itself. It's literally impossible to decide if a paradox is true or false, because it's either both or neither depending on whether you're a "glass half full" type of person. An Internet search for the string "paradox example" is hours of fun (but not while you're

supposed to be doing your homework). Here are some examples I liked. See if you can figure out why they're paradoxical.

- There's an exception to every rule.
- I know nothing at all.
- Moderation in all things, including moderation.
- What would happen if Pinocchio said, "My nose is about to grow"?
- Does the set of all sets that do not contain themselves contain itself?

Example 1   Recognizing Statements

Decide which of the following are statements and which are not.

(a) Most scientists agree that climate change is a threat to the environment.
(b) Is that your laptop?
(c) Man, that hurts!
(d) $432 + 8 \div 1.3 = \sqrt{115,000}$
(e) This book is about database management.
(f) Watching reality shows turns your brain to mush.

SOLUTION

Parts (a), (d), and (e) are statements because they can be judged as true or false in a nonsubjective manner.

Part (b) is not a statement because it is a question.

Part (c) is not a statement because it is an exclamation.

Part (f) is not a statement because it requires an opinion (unless there's a scientific study I'm unaware of).

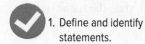

1. Define and identify statements.

Try This One 1

Decide which of the following are statements and which are not.

(a) Those pants rock!
(b) $12 - 8 = 5$
(c) Give me a bottle of anything and a glazed donut—to go.
(d) Cat can send animated gifs with her cell phone.
(e) When does the party start?
(f) History is interesting.

Simple and Compound Statements

Statements can be classified as *simple* or *compound*. A **simple statement** contains only one idea. Each of these statements is an example of a simple statement.

> Your jeans are torn.
> My room has a virtual mountain of dirty socks in it.
> Daytona Beach is in Florida.

A statement like "I'm taking chemistry this semester and I'm going to get an A" is called a *compound statement* since it's formed from more than one simple statement.

A **compound statement** is a statement formed by joining two or more simple statements with a **connective**. There are four basic connectives used in logic: and (the **conjunction**), or (**disjunction**), if . . . then (**conditional**), and if and only if (**biconditional**).

Here are some examples of compound statements using connectives.

> John studied for 5 hours, and he got an A.   (conjunction)
>
> I'm going to pass this class or I'm going to change my major.   (disjunction)
>
> If I get 80% of the questions on the LSAT right, then I will get into law school.   (conditional)
>
> We will win the game if and only if we score more points than the other team.   (biconditional)

Math Note

In standard usage, the word *then* is often omitted from a conditional statement; instead of "If it snows, then I will go skiing," you'd probably just say, "If it snows, I'll go skiing."

| **Example 2** | Classifying Statements as Simple or Compound |

Math Note

Technically, we've given the names *conjunction*, *disjunction*, *conditional*, and *biconditional* to the connectives, but from now on, we'll refer to whole statements using these connectives by those names. For example, we would call the compound statement in Example 2d a disjunction.

Classify each statement as simple or compound. If it is compound, state the name of the connective used.

(a) Our school mascot is a moose.
(b) If you register for WiFi service, you will get three days of free access.
(c) Tomorrow is the last day to register for classes.
(d) In the interest of saving the planet, I plan to buy either a hybrid or a motorcycle.

SOLUTION

(a) There are no connectives involved, so this is a simple statement.
(b) This if . . . then statement is compound and uses a conditional connective.
(c) This is a simple statement.
(d) Ultimately, this statement could be restated as "I will buy a hybrid, or I will buy a motorcycle," which makes it a compound statement: specifically, a disjunction.

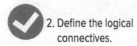
2. Define the logical connectives.

| Try This One | 2 |

Classify each statement as simple or compound. If it is compound, state the name of the connective used.

(a) My jacket is both warm and light.
(b) This is an informative website on STIs.
(c) If it doesn't rain tomorrow, I'm going windsurfing.
(d) I'm going to eat at Taco Bell or Wendy's today.
(e) Yesterday was the deadline to withdraw from a class.

Math Note

We'll worry later about determining whether statements involving quantifiers and connectives are true or false. For now, focus on learning and understanding the terms.

Quantified Statements

Quantified statements involve terms such as *all, each, every, no, none, some, there exists,* and *at least one.* The first five (*all, each, every, no, none*) are called *universal quantifiers* because they either include or exclude every element of the universal set. The latter three (*some, there exists, at least one*) are called *existential quantifiers* because they claim the existence of something, but don't include the entire universal set. Here are some examples of quantified statements:

> Every student taking philosophy this semester will pass.
>
> No nursing student is also majoring in criminal justice.
>
> Some people who are Miami Hurricane fans are also Miami Dolphin fans.
>
> There is at least one professor at this school who does not have brown eyes.

The first two statements use universal quantifiers, while the third and fourth use existential quantifiers. Note that the statements using existential quantifiers are not "all inclusive" (or all exclusive) as the other two are.

Negation

The *negation* of a statement is a corresponding statement with the opposite truth value. This means that if a statement is true its negation is false, and if a statement is false its negation is

Sidelight A Brief History of Logic

The basic concepts of logic can be attributed to Aristotle, who lived in the fourth century BCE. He used words, sentences, and deduction to prove arguments using techniques we'll study in this chapter. He was also remarkably hairy, as you can see from the accompanying picture. About a hundred years later, Euclid formalized geometry using deductive proofs. Both subjects were considered to be the "inevitable truths" of the universe revealed to rational people.

In the 19th century, people began to reject the idea of inevitable truths and realized that a deductive system like Euclidean geometry is only true based on the original assumptions. When the original assumptions are changed, a new deductive system can be created. This is why there are different types of geometry. (Bet you didn't know that! See Section 9-7 on non-Euclidean geometry.)

Eventually, several people developed the use of symbols rather than words and sentences in logic. One such person was George Boole (1815–1864). Boole created the symbols used in this chapter and developed the theory of symbolic logic. He also used symbolic logic in mathematics. His manuscript, entitled "An Investigation into the Laws of Thought, on Which Are Founded the Mathematical Theories of Logic and Probabilities," was published in 1854.

Boole was a friend of Augustus De Morgan, who formulated the laws bearing his name that we studied in Chapter 2. As we'll see later in this chapter, there are logic versions as

Source: National Library of Medicine

well. Much earlier, Leonhard Euler (1707–1783) used circles to represent logical statements and proofs: we'll study these in Section 3-5. The idea was refined by John Venn (1834–1923) into the Venn diagrams that proved so useful in our study of sets. In fact, the term *Venn diagram* wasn't used by Venn (which is good, because that would have been a tad pompous, don't you think?). What did he call his diagrams? Eulerian circles.

Math Note

The words *each*, *every*, and *all* mean the same thing, so what we say about *all* in this section applies to the others as well. Likewise, *some*, *there exists*, and *at least one* are considered to be the same and are treated that way as well.

true. For example, for the statement "My dorm room is blue," the negation is "My dorm room is not blue." It's important to note that the truth values of these two are completely opposite: one is true, and the other is false—period. You can't negate "My dorm room is blue" by saying "My dorm room is yellow," because it's completely possible that *both* statements are false. To make sure that you have a correct negation, check that if one of the statements is true, the other must be false, and vice versa. The typical way to negate a simple statement is by adding the word *not*, as in these examples:

Statement	Negation
Auburn will win Saturday.	Auburn will not win Saturday.
I took a shower today.	I did not take a shower today.
My car is clean.	My car is not clean.

You have to be especially careful when negating quantified statements. Consider the example statement "All dogs are fuzzy." It's not quite right to say that the negation is "All dogs are not fuzzy," because if some dogs are fuzzy and others aren't, then both statements are false. All we need for the statement "All dogs are fuzzy" to be false is to find at least one dog that is not fuzzy, so the negation of the statement "All dogs are fuzzy" is "Some dogs are not fuzzy." (In this setting, we define the word *some* to mean *at least one*.)

The negation of quantified statements is summarized in Table 3-1.

Table 3-1	Negations of Quantified Statements			
Statement contains . . .	**Example**	**Negation**	**Example**	
All do	All of my meals are low in fat.	Some do not, or not all do	Some of my meals are not low in fat.	
Some do	Some majors require 5 years of study.	None do, or all do not	There are no majors that require 5 years of study.	
Some do not	Some people don't go to football games.	All do	Everyone goes to football games.	
None do	No airlines include checked bags for free.	Some do	At least one airline allows a checked bag for free.	

Example 3   Writing Negations

Write the negation of each of the following quantified statements.

(a) Every student taking philosophy this semester will pass.
(b) Some people who are Miami Hurricane fans are also Miami Dolphin fans.
(c) There is at least one professor in this school who does not have brown eyes.
(d) No nursing student is also majoring in criminal justice.

SOLUTION

(a) Some student taking philosophy this semester will not pass (or, not every student taking philosophy this semester will pass).
(b) No people who are Miami Hurricane fans are also Miami Dolphin fans.
(c) All professors in this school have brown eyes.
(d) At least one nursing student is also majoring in criminal justice.

CAUTION

Be especially careful when negating statements. It's tempting to say that the negation of a statement like "Every student will pass" is "Every student will fail," but it isn't. Not only is it *possible* that both statements are false, in most classes it's *likely* that both are false.

Try This One   3

3. Write the negation of a statement.

Write the negation of each of the following quantified statements.

(a) All smartphones have cameras.
(b) No inmate can win the lottery.
(c) Some professors have Ph.Ds.
(d) Someone in this class will get a B.

Symbolic Notation

Our main goal in the study of formal logic is to be able to evaluate logical arguments objectively. In order to do that, we'll need to write statements in symbolic form. Now we'll introduce the symbols and methods that will be used. The symbols for the connectives *and*, *or*, *if . . . then*, and *if and only if* are shown in Table 3-2.

Table 3-2	Symbols for the Connectives		
	Connective	**Symbol**	**Name**
	and	∧	Conjunction
	or	∨	Disjunction
	if . . . then	→	Conditional
	if and only if	↔	Biconditional

Math Note

For three of the four connectives in Table 3-2, the order of the simple statements doesn't matter: for example, $p \wedge q$ and $q \wedge p$ represent the same compound statement.

The same is true for the connectives ∨ (disjunction) and ↔ (biconditional). The one exception is the conditional (→), where order is crucial.

Simple statements in logic are usually denoted with lowercase letters like p, q, and r. For example, we could use p to represent the statement "I get paid Friday" and q to represent the statement "I will go out this weekend." Then the conditional statement "If I get paid Friday, then I will go out this weekend" can be written in symbols as $p \rightarrow q$.

The symbol ~ (tilde) represents a negation. If p still represents "I get paid Friday," then ~p represents "I do not get paid Friday."

We often use parentheses in logical statements when more than one connective is involved in order to specify an order. (We'll deal with this in greater detail in the next section.) Let's see if there is a difference between the compound statements ~$p \wedge q$ and ~$(p \wedge q)$.

Example 4 | Studying Order in Logical Connectives

David Sobecki/McGraw Hill

Let's use the symbol p to represent the statement "Large Coney is a dog," and q to represent the statement "Guinness is a cat."

(a) What does the statement (~p) ∧ q mean?
(b) What does the statement ~$(p \wedge q)$ mean?
(c) Do these two statements say the same thing?

SOLUTION

(a) The parentheses tell us that the first thing we should do is negate statement p. Since p = Large Coney is a dog, ~p = Large Coney is not a dog. (Incidentally, as the photo clearly shows, this negation is false.) So (~p) ∧ q would be translated into words as "Large Coney is not a dog and Guinness is a cat."

(b) This time, the parentheses tell us to first form the conjunction of p and q. This is "Large Coney is a dog and Guinness is a cat." Then we need to negate this statement, giving us ~$(p \wedge q)$ = It is not the case that Large Coney is a dog and Guinness is a cat.

(c) These two statements don't say the same thing. The first clearly says that Large Coney is not a dog. But the second one says it's not true that BOTH Large Coney is a dog and Guinness is a cat. So that statement could be true if Large Coney is a dog, but Guinness is not a cat.

Try This One | 4

Use the statements p and q to decide if ~$q \rightarrow p$ means the same as ~$(q \rightarrow p)$.

If you found Example 4 hard to follow, hang in there: we'll study a systematic way to decide whether or not two statements really say the same thing later in this chapter. In Example 5, we'll practice writing verbal statements in symbols.

Example 5 Writing Statements Symbolically

Let p represent the statement "It is cloudy" and q represent the statement "I will go to the beach." Write each statement in symbols.

(a) I will not go to the beach.
(b) It is cloudy, and I will go to the beach.
(c) If it is cloudy, then I will not go to the beach.
(d) I will go to the beach if and only if it is not cloudy.

SOLUTION

(a) This is the negation of statement q, which we write as $\sim q$.
(b) This is the conjunction of p and q, written as $p \wedge q$.
(c) This is the conditional of p and the negation of q: $p \rightarrow \sim q$.
(d) This is the biconditional of q and not p: $q \leftrightarrow \sim p$.

Medioimages/Photodisc/
Getty Images

Try This One   **5**

Let p represent the statement "I will buy a Coke" and q represent the statement "I will buy some popcorn." Write each statement in symbols.

(a) I will buy a Coke, and I will buy some popcorn.
(b) I will not buy a Coke.
(c) If I buy some popcorn, then I will buy a Coke.
(d) I will not buy a Coke, and I will buy some popcorn.

You probably noticed that some of the compound statements we've written sound a little awkward. It isn't always necessary to repeat the subject and verb in a compound statement using *and* or *or*. For example, the statement "It is cold, and it is snowing" can be written "It is cold and snowing." The statement "I will go to a movie, or I will go to a play" can be written "I will go to a movie or a play." Also the words *but* and *although* can be used in place of *and*. For example, the statement "I will not buy a television set, and I will buy a CD player" can also be written as "I will not buy a television set, but I will buy a CD player."

Statements written in symbols can also be written in words, as shown in Example 5.

Example 6 Translating Statements from Symbols to Words

Write each statement in words. Let $p =$ "My dog is a golden retriever" and $q =$ "My dog is fuzzy."

(a) $\sim p$ (b) $p \vee q$ (c) $\sim p \rightarrow q$ (d) $q \leftrightarrow p$ (e) $q \wedge p$

SOLUTION

(a) My dog is not a golden retriever.
(b) My dog is a golden retriever or my dog is fuzzy.
(c) If my dog is not a golden retriever, then my dog is fuzzy.
(d) My dog is fuzzy if and only if my dog is a golden retriever.
(e) My dog is fuzzy, and my dog is a golden retriever.

David Sobecki/McGraw Hill

If this is your dog (which it's not, because it's mine), statement (e) describes it pretty well.

Try This One   **6**

Write each statement in words. Let $p =$ "The suspect is guilty" and $q =$ "The witness is lying like a rug."

(a) $\sim p$ (b) $p \vee q$ (c) $\sim p \rightarrow q$ (d) $p \leftrightarrow q$ (e) $p \wedge q$

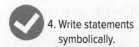

4. Write statements symbolically.

In this section, we defined the basic terms of symbolic logic and practiced writing statements using symbols. These skills will be crucial in our objective study of logical arguments, so we're off to a good start.

Answers to Try This One

1 (b) and (d) are statements.

2 (a) (conjunction), (c) (conditional), and (d) (disjunction) are compound; (b) and (e) are simple.

3 (a) Some smartphones don't have cameras.
(b) Some inmates can win the lottery.
(c) No professors have Ph.Ds.
(d) No one in this class will get a B.

4 $\sim q \to p$ means "If Guinness is not a cat, then Large Coney is a dog." $\sim(q \to p)$ means "It is not the case that if Guinness is a cat then Large Coney is a dog." These two statements don't mean the same thing.

5 (a) $p \wedge q$ (b) $\sim p$ (c) $q \to p$ (d) $\sim p \wedge q$

6 (a) The suspect is not guilty.
(b) The suspect is guilty or the witness is lying like a rug.
(c) If the suspect is not guilty, then the witness is lying like a rug.
(d) The suspect is guilty if and only if the witness is lying like a rug.
(e) The suspect is guilty, and the witness is lying like a rug.

Exercise Set   3-1

Writing Exercises

1. Define the term *statement* in your own words.
2. Is the sentence "This sentence is a statement" a statement? Explain.
3. Explain the difference between a simple and a compound statement.
4. Describe the terms and symbols used for the four connectives.
5. Write an example of each type of compound statement we studied: conjunction, disjunction, conditional, and biconditional. The topics should be things you find interesting.
6. What is the negation of a statement?
7. Explain why the negation of "All spring breaks are fun" is not "All spring breaks are not fun."
8. Explain why we're interested in writing statements in symbols.

Applications in Our World

For Exercises 9–18, state whether the sentence is a statement or not.

9. Please don't use your phone in class.
10. $5 + 9 = 14$
11. $9 - 3 = 2$
12. Nicki is a student in vet school.
13. Who will win the student government presidency?
14. Neither Sam nor Mary arrives to the exam on time.
15. You should carry a phone with you.
16. Bill Gates is the founder of Microsoft.
17. Go with the flow.
18. Math is not hard.

For Exercises 19–28, decide if each statement is simple or compound.

19. He goes to parties and hangs out at the coffee shop.
20. Sara got her hair highlighted.
21. Raj will buy an iMac or a Dell computer.
22. Euchre is fun if and only if you win.
23. February is when Valentine's Day occurs.
24. Diane is a chemistry major.
25. If you win the Megabucks multistate lottery, you'll be rich.
26. The patient is bleeding and going into shock.
27. $\sqrt{9} = 3$ or -3

28. Malcolm and Alisha will both miss the spring break trip.

For Exercises 29–36, identify each statement as a conjunction, disjunction, conditional, or biconditional.

29. Bob and Tom like stand-up comedians.
30. Either he passes the test, or he fails the course.
31. A number is even if and only if it is divisible by 2.
32. Politicians are dishonest and evasive.
33. I haven't decided yet if I'm going to the game or to the library to study.
34. If I don't get something to eat soon, I'm going to pass out.
35. I'm going to pass bio lab if and only if my experiment works out the way I hope it will.
36. When your battery dies, you need to charge your phone overnight.

For Exercises 37–42, write the negation of the statement.

37. The shirt I'm wearing to my interview is white.
38. Don't worry, your computer doesn't have a virus.
39. The hospital isn't full.
40. My name is not Richard Smoker.
41. Come on, you're not going to flunk this class.
42. Wow, that dude has some big biceps.

For Exercises 43–54, identify the quantifier in the statement as either universal or existential.

43. All fish swim in water.
44. Everyone that buys this hat gets a free bowl of soup.
45. Some people who live in glass houses throw stones.
46. There is at least one person in this class who won't pass.
47. Every happy dog wags its tail.
48. No men can join a sorority.
49. I've seen a four-leaf clover.
50. Each student that participates in this study gets a hundred bucks.
51. As far as I know, nobody's ever survived a fall from Mt. Catherine.
52. Everyone in the class was bored by the professor's lecture.
53. At least one of my friends has a Mac laptop.
54. No one here gets out alive.

For Exercises 55–66, write the negation of the statements in Exercises 43–54.

For Exercises 67–76, write each statement in symbols. Let p = "Depression is a public health issue" and let q = "Not enough people seek treatment."

67. Depression is a public health issue and not enough people seek treatment.
68. Depression is not a public health issue.
69. If not enough people seek treatment, then depression is a public health issue.

70. It is not the case that not enough people seek treatment or depression is a public health issue.
71. It is false that not enough people seek treatment.
72. It is not true that depression is a public health issue.
73. Not enough people seek treatment, or depression is not a public health issue.
74. Enough people seek treatment, or depression is a public health issue.
75. Not enough people seek treatment if and only if depression is a public health issue.
76. If depression is a public health issue, then not enough people seek treatment.

For Exercises 77–86, write each statement in symbols. Let p = "Sophie has been arrested" and q = "Bubba's never been arrested."

77. Bubba has been arrested at least once.
78. Sophie and Bubba have both been arrested.
79. If Bubba's been arrested, then Sophie has been arrested too.
80. You're saying that Sophie's been arrested? That is totally not true.
81. Either Sophie or Bubba has been arrested.
82. When the newspaper reported that Sophie and Bubba had both been arrested, that was inaccurate.
83. Sophie has been arrested if and only if Bubba has not.
84. Neither Bubba nor Sophie has ever been arrested.
85. If Sophie has not been arrested, then Bubba has not.
86. Bubba has never been arrested if and only if Sophie hasn't been arrested either.

For Exercises 87–96, write each statement in words. Let p = "The plane is on time." Let q = "The sky is clear."

87. $p \wedge q$
88. $\sim p \vee q$
89. $q \rightarrow p$
90. $q \rightarrow \sim p$
91. $\sim p \wedge \sim q$
92. $q \leftrightarrow p$
93. $p \vee \sim q$
94. $\sim p \leftrightarrow \sim q$
95. $q \rightarrow (p \vee \sim p)$
96. $(p \rightarrow q) \vee \sim p$

For Exercises 97–106, write each statement in words. Let p = "This resort is all-inclusive." Let q = "Water sports cost extra."

97. $\sim q$
98. $p \rightarrow q$
99. $p \vee \sim q$
100. $q \leftrightarrow p$
101. $\sim p \rightarrow \sim q$
102. $\sim p$
103. $p \vee q$
104. $(\sim p \vee q) \vee \sim q$
105. $q \vee p$
106. $(p \vee q) \rightarrow \sim(\sim q)$

Critical Thinking

107. Explain why the sentence "This statement is false" is not a statement.

108. Explain why each of the alleged statements listed at the end of the Sidelight on page 102 is actually a paradox.

109. (a) Write a verbal translation of the statement $a < 20$.
     (b) Write the negation of the statement you wrote in (a).
     (c) Write your statement from (b) in inequality form.

110. (a) Write a verbal translation of the statement $b > 10$.
     (b) Write the negation of the statement you wrote in (a).
     (c) Write your statement from (b) in inequality form.

111. Each of the sentences below is not a statement because none can be objectively evaluated as true or false. In each case, write a similar sentence that qualifies as a statement. Doing a little bit of online research might help on some questions.

     (a) That movie was really good.
     (b) Too many people in our society are obese.
     (c) Violent crime is less of a problem than it used to be.
     (d) Basketball players are really tall.
     (e) It's way too cold in Chicago during the winter.
     (f) Holy cow, that car is fast!

112. Think about the way you answered Exercise 109, then look back at the definition of what makes a set well-defined from Section 2-1. What is the relationship between a declarative sentence not being a statement and a set not being well-defined?

Statements involving negations and quantifiers can be confusing—sometimes intentionally. Evaluate each statement in Exercises 113–116 and try to write exactly what it actually says in simpler language.

113. All of our fans will not be attending all of our games.

114. You can't fool some of the people all of the time.

115. I wouldn't say that everybody doesn't like my history professor.

116. Everyone in this class doesn't have time to do their homework.

Section 3-2	**Truth Tables**

LEARNING OBJECTIVES

1. Construct truth tables for the negation, disjunction, and conjunction.

2. Construct truth tables for the conditional and biconditional.

3. Construct truth tables for compound statements.

4. Use the hierarchy of logical connectives.

"You can't believe everything you hear." Chances are you were taught this when you were younger, and it's pretty good advice. In an ideal world, everyone would tell the truth all the time, but in the real world, it is extremely important to be able to separate fact from fiction. When someone is trying to convince you of some point of view, the ability to logically evaluate the validity of an argument can be the difference between being informed and being deceived—and maybe between keeping and being separated from your hard-earned money!

This section is all about deciding when a compound statement is or is not true, based not on the topic itself, but simply on the structure of the statement and the truth of the underlying components. We learned about logical connectives in Section 3-1. In this section, we'll analyze these connectives using *truth tables*. A **truth table** is a diagram in table form that is used to show when a compound statement is true or false based on the truth values of the simple statements that make up that compound statement. This will allow us to analyze arguments objectively.

Negation

According to our definition of *statement*, a statement is either true or false, but never both. Consider the simple statement p = "Today is Tuesday." If it is in fact Tuesday, then p is true, and its negation $(\sim p)$ "Today is not Tuesday" is false. If it's not Tuesday, then p is false and $\sim p$ is true. The truth table for the negation of p looks like this.

p	$\sim p$	
T	F	*If statement p is true, its negation is false.*
F	T	*If statement p is false, its negation is true.*

There are two possible conditions for the statement p—true or false—and the table tells us that in each case, the negation $\sim p$ has the opposite truth value.

Conjunction

If we have a compound statement with two component statements p and q, there are four possible combinations of truth values for these two statements:

Possibilities		Symbolic value of each	
		p	q
1. p and q are both true.		T	T
2. p is true and q is false.		T	F
3. p is false and q is true.		F	T
4. p and q are both false.		F	F

So when setting up a truth table for a compound statement with two component statements, we'll need a row for each of the four possibilities.

Now we're ready to analyze conjunctions. Recall that a conjunction is a compound statement involving the word *and*. Suppose a friend who's prone to exaggeration tells you, "I bought a new laptop and a new iPad." This compound statement can be symbolically represented by $p \wedge q$, where p = "I bought a new laptop" and q = "I bought a new iPad." When would this conjunctive statement be true? If your friend actually had made both purchases, then the statement "I bought a new laptop and a new iPad" would be true, right? In terms of a truth table, that tells us that if p and q are both true, then the conjunction $p \wedge q$ is true as well, as shown below.

p	q	$p \wedge q$
T	T	T

On the other hand, suppose your friend bought only a new laptop or only a new iPad, or maybe neither of those things. Then the statement "I bought a new laptop and a new iPad" would be false. In other words, if either or both of p and q are false, then the compound statement $p \wedge q$ is false as well. With this information, we complete the truth table for a basic conjunction:

Mark Dierker/McGraw Hill

	p	q	$p \wedge q$
Bought laptop and iPad	T	T	T
Bought laptop, not iPad	T	F	F
Bought iPad, not laptop	F	T	F
Bought neither	F	F	F

Truth Values for a Conjunction

The conjunction $p \wedge q$ is true only when both p and q are true.

Disjunction

Next, we'll look at truth tables for *or* statements. Suppose your friend from the previous example made the statement, "I bought a new laptop *or* a new iPad" (as opposed to *and*). If your friend actually did buy one or the other, then this statement would be true. And if they had bought neither, then the statement would be false. So a partial truth table looks like this:

	p	q	$p \vee q$
Bought laptop and iPad	T	T	
Bought laptop, not iPad	T	F	T
Bought iPad, not laptop	F	T	T
Bought neither	F	F	F

Sidelight **Logical Gates and Computer Design**

Logic is used in electrical engineering in designing circuits, which are the heart of computers. The truth tables for *and*, *or*, and *not* are used for computer gates. These gates determine whether electricity flows through a circuit. When a switch is closed, the current has an uninterrupted path and will flow through the circuit. This is designated by a 1. When a switch is open, the path of the current is broken, and it will not flow. This is designated by a 0. The logical gates are illustrated here—notice that they correspond exactly with our truth tables.

This simple little structure is responsible for the operation of almost every computer in the world—at least until quantum computers become a reality. If you're interested, do a Google search for *quantum computer* to read about the future of computing.

But what if the person actually bought both items? You might lean toward the statement "I bought a new laptop or a new iPad" being false. Believe it or not, it depends on what we mean by the word *or*. There are two interpretations of that word, known as the *inclusive or* and the *exclusive or*. The inclusive or has the possibility of both statements being true; but the exclusive or does not allow for this. That is, exactly one of the two simple statements must be true.

Let's look at an example of each. If I said "Tomorrow, I'm going to class or I'm going to the beach," you would interpret that to mean one or the other but not both will occur: that's the exclusive or. But if an admissions counselor says "You'll get a scholarship if you scored over 30 on the ACT or were in the top 5% of your class," it would be silly to interpret that as one or the other but not both: no school would say "Well, we were going to give you a scholarship if you scored over 30 or were at the top of your class, but since you did both, you'll get nothing and like it." That's the inclusive or.

In the study of logic, we'll use the inclusive or, so a disjunctive statement like "I bought a new laptop or an iPad" is considered true if both things occur. To summarize that in simple terms, we'll interpret a statement like "Jiang walks or takes the bus to campus" to mean "Jiang walks, or takes the bus, or both."

Going back to the laptop and iPad example, we will say that if your friend had bought both, the disjunction "bought a laptop or bought an iPad" is true. This completes our truth table for a disjunction:

		p	q	$p \lor q$
Bought laptop and iPad		T	T	T
Bought laptop, not iPad		T	F	T
Bought iPad, not laptop		F	T	T
Bought neither		F	F	F

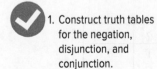

1. Construct truth tables for the negation, disjunction, and conjunction.

Truth Values for a Disjunction

The disjunction $p \lor q$ is true when either p or q or both are true. It is false only when both p and q are false.

Conditional Statement

A conditional statement, which is sometimes called an *implication*, consists of two simple statements using the connective if . . . then. For example, the statement "If I bought a ticket, then I can go to the concert" is a conditional statement. The first component, in this case "I bought a ticket," is called the **antecedent**. The second component, in this case "I can go to the concert," is called the **consequent**.

Conditional statements are used all the time in math, not just in logic. "If an element is in both set A and set B, then it's in the intersection of A and B" is an example from earlier in the book.

To illustrate the truth table for the conditional statement, we'll think about the following example:

> If the Cubs win tomorrow, they make the playoffs.

We'll use p = "the Cubs win tomorrow" and q = "they make the playoffs." This makes our conditional statement $p \rightarrow q$. We'll fill out the truth table by considering four cases.

Case 1: The Cubs win tomorrow, and they make the playoffs (both p and q are true). The statement was that if the Cubs won, they would make the playoffs, so if they win and make the playoffs, the statement was definitely true. So the first line of the truth table is

Cubs win, make playoffs

p	q	$p \rightarrow q$
T	T	T

Case 2: The Cubs win tomorrow, but don't make the playoffs (p is true, but q is false). I told you that if the Cubs won, they'd make the playoffs; if they won and didn't make the playoffs, I'm a liar liar pants on fire, and the conditional statement is false. The second line of the truth table is

Cubs win, don't make playoffs

p	q	$p \rightarrow q$
T	T	T
T	F	F

Case 3: The Cubs lose tomorrow and still make the playoffs (p is false and q is true). This requires some serious thought. My claim was that if the Cubs won, they'd make the playoffs. In order for that claim to be false, the Cubs would have to win and not make the playoffs. That's not the case if they didn't win, so the statement is not false. And we know that if a statement isn't false, it's true! That makes the next line of the truth table

Cubs lose, make playoffs

p	q	$p \rightarrow q$
T	T	T
T	F	F
F	T	T

Case 4: The Cubs lose tomorrow and don't make the playoffs (p and q are both false). This is pretty much the same as Case 3: the statement is only false if the Cubs win and don't make the playoffs, so again if the Cubs lose, the statement isn't false, making it true. This completes the truth table for a conditional.

Math Note

If you're totally unconvinced by the discussions in Cases 3 and 4, you should take a look at Exercises 65 and 66. In those problems, we'll develop a way to rewrite a conditional statement as a disjunction, in which case we can use what we already know about disjunctions to study truth values.

Cubs lose, don't make playoffs

p	q	$p \rightarrow q$
T	T	T
T	F	F
F	T	T
F	F	T

For Cases 3 and 4, it might help to think of it this way: we'll be optimists and consider a statement to be true unless we have absolute proof that it's false. And the only time that happens for a conditional statement is when the antecedent is true and the consequent is false.

Truth Values for a Conditional Statement

The conditional statement $p \rightarrow q$ is false only when the antecedent p is true and the consequent q is false.

Biconditional Statement

A biconditional statement is really two statements; it's the conjunction of two conditional statements. For example, the statement "I will stay in and study Friday if and only if I don't have any money" is the same as "If I don't have any money, then I will stay in and study Friday *and* if I stay in and study Friday, then I don't have any money." In symbols, we can write either $p \leftrightarrow q$ or $(p \rightarrow q) \wedge (q \rightarrow p)$. Since the biconditional is a conjunction, for it to be true, both of the statements $p \rightarrow q$ and $q \rightarrow p$ must be true. We will once again look at cases to build the truth table.

Case 1: Both p and q are true. Then both $p \rightarrow q$ and $q \rightarrow p$ are true, and the conjunction $(p \rightarrow q) \wedge (q \rightarrow p)$, which is also $p \leftrightarrow q$, is true as well.

p	q	$p \leftrightarrow q$
T	T	T

Case 2: p is true and q is false. In this case, the implication $p \rightarrow q$ is false, so it doesn't even matter whether $q \rightarrow p$ is true or false—the conjunction has to be false.

p	q	$p \leftrightarrow q$
T	T	T
T	F	F

Case 3: p is false and q is true. This is case 2 in reverse. The implication $q \rightarrow p$ is false, so the conjunction must be as well.

p	q	$p \leftrightarrow q$
T	T	T
T	F	F
F	T	F

Case 4: p is false and q is false. According to the truth table for a conditional statement, both $p \rightarrow q$ and $q \rightarrow p$ are true in this case, so the conjunction is as well. This completes the truth table.

p	q	$p \leftrightarrow q$
T	T	T
T	F	F
F	T	F
F	F	T

Comstock Images/Alamy Stock Photo

A technician who designs an automated irrigation system needs to decide whether the system should turn on *if* the water in the soil falls below a certain level or *if and only if* the water in the soil falls below a certain level. In the first instance, other inputs could also turn on the system.

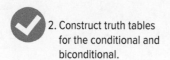

2. Construct truth tables for the conditional and biconditional.

Truth Values for a Biconditional Statement

The biconditional statement $p \leftrightarrow q$ is true when p and q have the same truth value and false when they have opposite truth values.

Table 3-3 provides a summary of the truth tables for the basic compound statements and the negation. *The last thing you should do is to try and memorize these tables!* If you understand how we built them, you can rebuild them on your own when you need them.

Table 3-3	Truth Tables for the Connectives and Negation

Conjunction (and)

p	q	$p \wedge q$
T	T	T
T	F	F
F	T	F
F	F	F

Disjunction (or)

p	q	$p \vee q$
T	T	T
T	F	T
F	T	T
F	F	F

Conditional (if . . . then)

p	q	$p \rightarrow q$
T	T	T
T	F	F
F	T	T
F	F	T

Biconditional (if and only if)

p	q	$p \leftrightarrow q$
T	T	T
T	F	F
F	T	F
F	F	T

Negation (not)

p	$\sim p$
T	F
F	T

Truth Tables for Compound Statements

Once we know truth values for the basic connectives, we can use truth tables to find the truth values for any logical statement. The key to the procedure is to take it step by step, so that in every case, you're deciding on truth values based on one of the truth tables in Table 3-3. The procedure is illustrated in Example 1.

Example 1	Constructing a Truth Table

Construct a truth table for the statement $\sim p \vee q$.

SOLUTION

Step 1 Set up a table as shown.

p	q
T	T
T	F
F	T
F	F

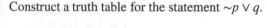

The order in which you list the Ts and Fs doesn't matter as long as you cover all the possible combinations. For consistency in this book, we'll always use the order TTFF for p and TFTF for q when these are the only two letters in the logical statement.

Step 2 Find the truth values for $\sim p$ by negating the values for p, and put them into a new column marked $\sim p$.

p	q	$\sim p$
T	T	F
T	F	F
F	T	T
F	F	T

Truth values in the $\sim p$ column are opposite those in the p column.

Chris Ryan/age fotostock

"My leg isn't better, or I'm taking a break" is an example of a statement that can be written as $\sim p \vee q$.

Step 3 Find the truth values for the disjunction ~$p \vee q$. Use the T and F values for ~p and q in the second and third columns, and use the disjunction truth table from earlier in the section. To make this simpler, it's a GREAT idea to use thin strips of paper to cover up all but the two columns we're interested in. In this case, we're covering up the p column. This allows us to focus on the only two columns that are relevant here.

p	q	~p	~$p \vee q$
	T	F	T
	F	F	F
	T	T	T
	F	T	T

The disjunction is true unless ~p and q are both false.

The truth values for the statement ~$p \vee q$ are found in the last column. The statement is true unless p is true and q is false.

Try This One   1

Construct a truth table for the statement $p \vee$ ~q.

When a compound statement has multiple connectives, it will sometimes have parentheses to indicate the order in which we should work with those connectives. We'll find the truth value of statements in parentheses first, as shown in Example 2. (If this doesn't remind you of the order of operations for arithmetic and algebra, you're just not paying attention.)

Example 2   Constructing a Truth Table

Creatas/Getty Images

"It is not true that if it rains, then we can't go out" is an example of a statement that can be written as ~$(p \rightarrow$ ~$q)$.

Construct a truth table for the statement ~$(p \rightarrow$ ~$q)$.

SOLUTION

Step 1 Set up the table as in Example 1.

p	q
T	T
T	F
F	T
F	F

Step 2 Find the truth values for ~q by negating the values for q, and put them into a new column.

p	q	~q
T	T	F
T	F	T
F	T	F
F	F	T

Truth values in the ~q column are opposite those in the q column.

Step 3   Find the truth values for the implication $p \rightarrow \sim q$, using the values in the first and third columns and the implication truth table from earlier in the section. To make it easier to focus on the first and third columns, we'll cover up the second.

p	q	$\sim q$	$p \rightarrow \sim q$
T		F	F
T		T	T
F		F	T
F		T	T

The conditional is true unless p is true and ~q is false.

Step 4   Find the truth values for the negation $\sim(p \rightarrow \sim q)$ by negating the values we just found for $p \rightarrow \sim q$.

p	q	$\sim q$	$p \rightarrow \sim q$	$\sim(p \rightarrow \sim q)$
T	T	F	F	T
T	F	T	T	F
F	T	F	T	F
F	F	T	T	F

The negation has opposite values from the p → ~q column.

The truth values for $\sim(p \rightarrow \sim q)$ are in the last column. The statement is true only when p and q are both true.

Try This One   2

Construct a truth table for the statement $p \leftrightarrow (\sim p \wedge q)$.

We can also construct truth tables for compound statements that involve three or more components. For a compound statement with three simple statements p, q, and r, there are eight possible combinations of Ts and Fs to consider. The truth table is set up as shown in Step 1 of Example 3.

Example 3   Constructing a Truth Table with Three Components

Construct a truth table for the statement $p \vee (q \rightarrow r)$.

SOLUTION

Step 1   Set up the table as shown.

p	q	r
T	T	T
T	T	F
T	F	T
T	F	F
F	T	T
F	T	F
F	F	T
F	F	F

Rubberball/Getty Images

"I'll do my math assignment, or if I think of a good topic, then I'll start my English essay" is an example of a statement that can be written as $p \vee (q \rightarrow r)$.

Again, the order of the Ts and Fs doesn't matter as long as all the possible combinations are listed. Whenever there are three letters in the statement, we'll use the order shown in Step 1 for consistency, so you probably should get in the habit of doing the same when you practice.

Step 2 Find the truth value for the statement in parentheses, $q \rightarrow r$. Use the values in the q and r columns and the conditional truth table from earlier in the section. Put those values in a new column labeled $q \rightarrow r$.

p	q	r	$q \rightarrow r$
	T	T	T
	T	F	F
	F	T	T
	F	F	T
	T	T	T
	T	F	F
	F	T	T
	F	F	T

The conditional is true unless q is true and r is false.

Step 3 Find the truth values for the disjunction $p \vee (q \rightarrow r)$, using the values for p from the first column and those we just put in the $q \rightarrow r$ column. Use the truth table for disjunction from earlier in the section, and put the results in a new column.

p	q	r	$q \rightarrow r$	$p \vee (q \rightarrow r)$
T			T	T
T			F	T
T			T	T
T			T	T
F			T	T
F			F	F
F			T	T
F			T	T

The disjunction is true unless both p and q → r are false.

The truth values for the statement $p \vee (q \rightarrow r)$ are found in the last column. The statement is true unless p and r are false while q is true.

Try This One   3

Construct a truth table for the statement $(p \wedge q) \vee \sim r$.

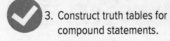

3. Construct truth tables for compound statements.

Here's a summary of the method we've developed for building a truth table to analyze a compound statement: we set up a table with all possible combinations of truth values for the simple statements that make up the compound statement. Then we build new columns, one at a time, by finding truth values for parts of the compound statement using the basic truth tables we developed earlier in this section. If there are parentheses in the compound statement, we find the truth value of the statement or statements in the parentheses first. So one way to ensure that the order in which we evaluate compound statements isn't ambiguous is to always use parentheses.

Once again, this sounds an awful lot like the order of operations for arithmetic: in order to avoid *always* having to use parentheses, we agree on a standard order for multiple operations, using parentheses only when the intent is to violate that order. To accomplish the same thing in dealing with statements in logic, a **hierarchy of connectives** has been agreed upon somewhere along the line. This hierarchy tells us which connectives should be done first when there are no parentheses to guide us:

1. Negation ($\sim$)
2. Conjunction ($\wedge$) or disjunction ($\vee$)
3. Conditional ($\rightarrow$)
4. Biconditional ($\leftrightarrow$)

Math Note
When parentheses are used to emphasize order, the statement $p \vee q \rightarrow r$ is written as $(p \vee q) \rightarrow r$. The statement $p \leftrightarrow q \wedge r$ is written as $p \leftrightarrow (q \wedge r)$.

The connectives higher on the list are done first. The connective that's lowest on the list is done last, so it describes the type of statement overall. For example, $p \vee q \rightarrow r$ is a conditional statement; the hierarchy tells us to first find the disjunction, which means we could write the compound statement as $(p \vee q) \rightarrow r$. That shows that the statement is a conditional: if p or q, then r.

When a compound statement has both a conjunction and a disjunction, we'll need to use parentheses to indicate which should be considered first. A statement like $p \wedge q \vee r$ is ambiguous without parentheses: $(p \wedge q) \vee r$ is a conjunction, and $p \wedge (q \vee r)$ is a disjunction. (In Exercise 61, you'll prove that these two statements are in fact different.)

Example 4	**Using the Hierarchy of Connectives**

For each, identify the type of statement using the hierarchy of connectives, and rewrite using parentheses to indicate order.

(a) $\sim p \vee \sim q$
(b) $p \rightarrow \sim q \wedge r$
(c) $p \vee q \leftrightarrow q \vee r$
(d) $p \rightarrow q \leftrightarrow r$

SOLUTION

(a) The negations are higher on the list than the disjunction, so if we add parentheses, we get $(\sim p) \vee (\sim q)$. The statement is a disjunction.
(b) The lowest connective in this statement is conditional, so this is a conditional statement. With parentheses, it looks like $p \rightarrow (\sim q \wedge r)$.
(c) The biconditional is lowest on the list, so the overall statement is a biconditional: $(p \vee q) \leftrightarrow (q \vee r)$.
(d) Again, the biconditional comes last, so the statement is a biconditional: $(p \rightarrow q) \leftrightarrow r$.

Try This One	4

4. Use the hierarchy of logical connectives.

For each, identify the type of statement using the hierarchy of connectives, and rewrite using parentheses to indicate order.

(a) $\sim p \vee q$
(b) $p \vee \sim q \rightarrow r$
(c) $p \vee q \leftrightarrow \sim p \vee \sim q$
(d) $p \wedge \sim q$
(e) $p \leftrightarrow q \rightarrow r$

Example 5	An Application of Truth Tables

Use the truth value of each simple statement to determine the truth value of the compound statement.

> *p:* Streaming music accounts for most of music sales.
> *q:* Vinyl LPs are no longer widely available.
> *r:* Digital music downloads are available through Apple and Google.

> Statement: $p \vee q \rightarrow r$

SOLUTION

Revenue from streaming music bypassed digital downloads and physical media in 2016 and now accounts for over 80% of revenue, so statement *p* is true. Statement *q* is false—vinyl has made an unexpected comeback in recent years and can be found in many stores. You can, in fact, download music from Apple or Google, so statement *r* is true.

Now we'll analyze the compound statement. First, according to the hierarchy of connectives, the disjunction should be evaluated: $p \vee q$ is true when either *p* or *q* is true, so in this case, $p \vee q$ is true. Next, the implication $(p \vee q) \rightarrow r$ is true if both *r* and $p \vee q$ is true, which is the case here. So the compound statement $p \vee q \rightarrow r$ is true.

Try This One 5

Using the simple statements in Example 5, find the truth value of the compound statement $(\sim p \wedge \sim q) \rightarrow r$.

We have seen that truth tables are an effective way to organize truth values for statements, allowing us to determine the truth values of some very complicated statements in a systematic way. In the next section, we'll see why we want to be able to do this when we use truth tables to decide when an argument is logically valid.

Answers to Try This One

1

p	q	$\sim q$	$p \vee \sim q$
T	T	F	T
T	F	T	T
F	T	F	F
F	F	T	T

2

p	q	$\sim p$	$\sim p \wedge q$	$p \leftrightarrow (\sim p \wedge q)$
T	T	F	F	F
T	F	F	F	F
F	T	T	T	F
F	F	T	F	T

3

p	q	r	$p \wedge q$	$\sim r$	$(p \wedge q) \vee \sim r$
T	T	T	T	F	T
T	T	F	T	T	T
T	F	T	F	F	F
T	F	F	F	T	T
F	T	T	F	F	F
F	T	F	F	T	T
F	F	T	F	F	F
F	F	F	F	T	T

4
  (a) Disjunction; $(\sim p) \vee q$
  (b) Conditional; $(p \vee \sim q) \rightarrow r$
  (c) Biconditional; $(p \vee q) \leftrightarrow (\sim p \vee \sim q)$
  (d) Conjunction; $p \wedge (\sim q)$
  (e) Biconditional; $p \leftrightarrow (q \rightarrow r)$

5 True

Exercise Set   **3-2**

Writing Exercises

1. A friend randomly opens your math book and ends up seeing a truth table in Section 3-2. How would you explain to them what the point of a truth table is?
2. Explain the difference between the inclusive and exclusive disjunctions. Write an example of each in plain English, and explain why each is the type of disjunction described.

3. I claim that a biconditional statement is really a conjunction of two conditional statements. Explain why that makes sense.
4. Describe the hierarchy of connectives. What's the point of having one? How do you use it?
5. How do you find the truth value of a negation? Why?
6. Explain how we developed the truth values for a conjunction, and for a disjunction.

Computational Exercises

For Exercises 7–36, construct a truth table for each.

7. $\sim(p \vee q)$
8. $q \rightarrow p$
9. $\sim p \wedge q$
10. $\sim q \rightarrow \sim p$
11. $\sim p \leftrightarrow q$
12. $(p \vee q) \rightarrow \sim p$
13. $\sim(p \wedge q) \rightarrow p$
14. $(p \vee q) \wedge (q \vee p)$
15. $(\sim q \wedge p) \rightarrow \sim p$
16. $q \wedge \sim p$

17. $(p \wedge q) \leftrightarrow (q \vee \sim p)$
18. $p \rightarrow (q \vee \sim p)$
19. $(p \wedge q) \vee p$
20. $(q \rightarrow p) \vee \sim r$
21. $(r \wedge q) \vee (p \wedge q)$
22. $(r \rightarrow q) \vee (p \rightarrow r)$
23. $\sim(p \vee q) \rightarrow \sim(p \wedge r)$
24. $(\sim p \vee \sim q)~ \sim r$
25. $(\sim p \vee q) \wedge r$
26. $p \wedge (q \vee \sim r)$

27. $(p \wedge q) \leftrightarrow (\sim r \vee q)$
28. $\sim(p \wedge r) \rightarrow (q \wedge r)$
29. $r \rightarrow \sim(p \vee q)$
30. $(p \vee q) \vee (\sim p \vee \sim r)$
31. $p \rightarrow (\sim q \wedge \sim r)$

32. $(q \vee \sim r) \leftrightarrow (p \wedge \sim q)$
33. $\sim(q \rightarrow p) \wedge r$
34. $q \rightarrow (p \wedge r)$
35. $(r \vee q) \wedge (r \wedge p)$
36. $(p \wedge q) \leftrightarrow \sim r$

If p and r are false statements, and q is a true statement, find the truth value of each compound statement in Exercises 37–42.

37. $q \vee (p \wedge \sim r)$
38. $(p \wedge q) \vee (q \wedge r)$
39. $r \rightarrow \sim(p \vee q)$

40. $\sim(p \wedge q) \vee \sim r$
41. $\sim p \wedge \sim(r \vee \sim q)$
42. $(p \rightarrow r) \rightarrow (\sim q \wedge p)$

Applications in Our World

For Exercises 43–48, use the truth value of each simple statement to determine the truth value of the compound statement. Use the Internet if you need help determining the truth value of a simple statement.

43. *p:* Japan bombs Pearl Harbor.
    q: The United States stays out of World War II.
    Statement: $p \rightarrow q$
44. *p:* The Cleveland Cavaliers win the NBA championship in 2016.
    q: The world ends.
    Statement: $p \wedge q$
45. *p:* NASA sends a spacecraft to the Moon with a crew aboard.
    q: NASA sends a spacecraft to Mars with a crew aboard.
    Statement: $p \vee q$
46. *p:* The war in Afghanistan continues into 2021.
    q: The United States commits more troops to Afghanistan in August 2021.
    Statement: $p \rightarrow q$
47. *p:* Apple releases the iPad.
    q: Apple stops making desktop computers.
    r: Samsung and Amazon release tablet computers.
    Statement: $(p \vee q) \wedge r$
48. *p:* A world pandemic causes millions of deaths across the globe.
    q: Several vaccines are developed to combat the pandemic.
    r: All eligible Americans get vaccinated by the end of 2021.
    Statement: $p \wedge q \rightarrow \sim r$

Exercises 49–54 are based on the compound statement below.

A new weight loss supplement claims that if you take the product daily and cut your calorie intake by 10%, you will lose at least 10 pounds in the next 4 months.

49. This compound statement is made up of three simple statements. Identify them and assign a letter to each.
50. Write the compound statement in symbolic form, using conjunctions and the conditional.
51. Construct a truth table for the compound statement you wrote in Exercise 50.
52. If you take this product daily and don't cut your calorie intake by 10%, and then don't lose 10 pounds, is the claim made by the advertiser true or false?
53. If you take the product daily, don't cut your calorie intake by 10%, and do lose 10 pounds, is the claim true or false?
54. If you don't take the product daily, cut your calorie intake by 10%, and do lose 10 pounds, is the claim true or false?

Exercises 55–60 are based on the compound statement below.

The owner of a professional baseball team publishes an open letter to fans after another losing season. The letter states that if attendance for the following season is over 2 million, then ownership will add $20 million to the payroll and the team will make the playoffs the following year.

55. This compound statement is made up of three simple statements. Identify them and assign a letter to each.

56. Write the compound statement in symbolic form, using conjunction and the conditional.
57. Construct a truth table for the compound statement you wrote in Exercise 56.
58. If attendance goes over 2 million the next year and the owner raises payroll by $20 million, but the team fails to make the playoffs, is the owner's claim true or false?
59. If attendance is less than 2 million but the owner still raises the payroll by $20 million and the team makes the playoffs, is the owner's claim true or false?
60. If attendance is over 2 million, the owner doesn't raise the payroll, but the team still makes the playoffs, is the owner's claim true or false?

Critical Thinking

61. Construct two truth tables to show that the statement $p \wedge q \vee r$ is ambiguous. (*Hint:* Look back at our discussion of the hierarchy of connectives.)
62. Let's look a little deeper at the statement $p \wedge q \vee r$. Write three simple statements for p, q, and r so that $(p \wedge q) \vee r$ and $p \wedge (q \vee r)$ have different meanings. (*Hint:* The truth tables from Exercise 61 will probably help.)
63. Using the hierarchy for connectives, write the statement $p \rightarrow q \vee r$ by using parentheses to indicate the proper order. Then construct truth tables for $(p \rightarrow q) \vee r$ and $p \rightarrow (q \vee r)$. Are the resulting truth values the same? Are you surprised? Why or why not?
64. In 2003, the New York City Council was considering banning indoor smoking in bars and restaurants. Opponents of the ban claimed that it would have a negligible effect on indoor pollution, but a huge negative effect on the economic success of these businesses. Eventually, the ban was enacted, and a 2004 study by the city department of health found that there was a sixfold decrease in indoor air pollution in bars and restaurants, but jobs, liquor licenses, and tax revenues all increased. Assign truth values to all the premises of the opponents'

claim; then write the claim as a compound statement and determine its validity.
65. Consider the following two statements:

   "If we scored more points, we won!"
   "We didn't score more points, or we won."

   Explain why these statements say exactly the same thing from an English standpoint. (We'll deal with them logically in the next question.)
66. Write the two statements in Exercise 65 as compound statements using letters p and q, then construct a truth table for each compound statement. How can you use this to help convince you that the truth table we built for the conditional is correct?
67. Here's a cool question to ponder: when looking at truth values for conditional statements, a lot of folks feel like calling a conditional true if the antecedent is false doesn't make sense. So let's pretend for a second that we didn't do that. Instead, pretend that the only time we consider a conditional statement $p \rightarrow q$ to be true is if p and q are both true. From a standpoint of truth tables and formal logic, why would that be silly?

Section 3-3	**Types of Statements**

LEARNING OBJECTIVES

1. Classify a statement as a tautology, a self-contradiction, or neither.
2. Identify logically equivalent statements.
3. Write negations of compound statements.
4. Write the converse, inverse, and contrapositive of a statement.

It's no secret that weight loss has become big business in the United States. It seems like almost every week, a new company pops into existence with the latest miracle pill to turn you into a supermodel.

A typical advertisement will say something like "Use of our product may result in significant weight loss." That sounds great, but think about what that statement really means. If use of the product "may" result in significant weight loss, it also may not result in any weight loss at all, it may result in

Jill Braaten/McGraw Hill

weight gain, or it may result in turning you into a pumpkin. In fact, the statement could be translated into "You will lose weight or you will not lose weight." Of course, this statement is always true. In this section, we'll study statements of this type (and others).

Tautologies and Self-Contradictions

In our study of truth tables in Section 3-2, we saw that most compound statements are true in some cases and false in others. What we haven't done is think about whether that's true for *every* compound statement. Let's look at a couple of examples to decide if statements have to be true sometimes and false others.

Consider the simple statement "I'm going to look for a new job this year." Its negation is "I'm not going to look for a new job this year." Now think about these two compound statements:

"I'm going to look for a new job this year, or I'm not going to look for a new job this year."
"I'm going to look for a new job this year, and I'm not going to look for a new job this year."

Hopefully, it's pretty clear to you that the first statement is always true, while the second statement is always false (whether you look for a new job or not). The first is an example of a *tautology*, while the second is an example of a *self-contradiction*.

> A **tautology** is a compound statement that's always true, regardless of the truth values of the simple statements that make it up. A **self-contradiction** is a compound statement that is always false.

CAUTION

Don't make the mistake of thinking that every statement is either a tautology or a self-contradiction. We've seen many examples of statements that are sometimes true and other times false.

The sample statements above are simple enough that it's easy to tell that they are always true or always false based on common sense. But what if the statements are more complicated? It might not be at all obvious if a statement is always true, never true, or sometimes true. In that case, we can build a truth table to help us decide.

Example 1   Using a Truth Table to Classify a Statement

Let p = "I'm going to a metal concert," and q = "I'll wear black leather." For each statement below, first write the statement in words using the given statements for p and q, and try to predict if the statement is a tautology, a self-contradiction, or neither. Then build a truth table to decide for sure.

(a) $(p \wedge q) \rightarrow p$ (b) $(p \wedge q) \wedge (\sim p \wedge \sim q)$ (c) $(p \vee q) \rightarrow q$

SOLUTION

(a) The statement $(p \wedge q) \rightarrow p$ translates as "If I'm going to a metal concert and I'll wear black leather, then I'm going to a metal concert." This certainly sounds like it's always true. Notice that it's a LOT easier to make this judgment when using specific statements for p and q, as opposed to just looking at the statement in symbolic form.

Now let's use a truth table to see if we're right.

p	q	$p \wedge q$	$(p \wedge q) \rightarrow p$
T	T	T	T
T	F	F	T
F	T	F	T
F	F	F	T

David Sobecki/McGraw Hill

Since the truth table value consists of all Ts, the statement is always true, making it a tautology.

(b) The statement $(p \wedge q) \wedge (\sim p \wedge \sim q)$ translates as "I'm going to a metal concert and I'll wear black leather, and I'm not going to a metal concert and I won't wear black leather." Sounds like a self-contradiction to me! Let's check it out.

p	q	$\sim p$	$\sim q$	$p \wedge q$	$\sim p \wedge \sim q$	$(p \wedge q) \wedge (\sim p \wedge \sim q)$
T	T	F	F	T	F	F
T	F	F	T	F	F	F
F	T	T	F	F	F	F
F	F	T	T	F	T	F

Since the truth value consists of all Fs, the statement is always false, so it is a self-contradiction.

(c) The statement $(p \vee q) \rightarrow q$ translates as "If I'm going to a metal concert or I'll wear black leather, then I'll wear black leather." This one's a little tougher, right? I'm thinking it's probably true sometimes and false others, but I wouldn't bet a limb on it, so using a truth table should be particularly helpful here.

p	q	$p \vee q$	$(p \vee q) \rightarrow q$
T	T	T	T
T	F	T	F
F	T	T	T
F	F	F	T

Since the statement can be true in some cases and false in others, it is neither a tautology nor a self-contradiction.

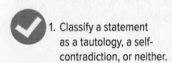

1. Classify a statement as a tautology, a self-contradiction, or neither.

Try This One   1

Write each statement verbally using the statements p and q from Example 1, then use a truth table to decide if each statement is a tautology, a self-contradiction, or neither.

(a) $(p \vee q) \wedge (\sim p \rightarrow q)$
(b) $(p \wedge \sim q) \wedge \sim p$
(c) $(p \rightarrow q) \vee \sim q$

Logically Equivalent Statements

In Exercises 65 and 66 of Section 3-2, we studied the two logical statements $p \rightarrow q$ and $\sim p \vee q$. The truth tables for the two statements are combined into one here:

p	q	$\sim p$	$p \rightarrow q$	$\sim p \vee q$
T	T	F	T	T
T	F	F	F	F
F	T	T	T	T
F	F	T	T	T

Notice that the truth values for both statements are *identical*: TFTT. When this occurs, the statements are said to be *logically equivalent*; that is, both compositions of the same simple statements have the same meaning. For example, the statement "If it snows, I will go skiing" is logically equivalent to saying "It won't snow or I will go skiing." In plain English, they say the same thing, just in different ways. Formally defined,

Math Note

The symbol ⇔ is often used interchangeably with ≡.

Two compound statements are **logically equivalent** if and only if they have the same truth values for all possible combinations of truth values for the simple statements that compose them. The symbol for logically equivalent statements is ≡.

Example 2   Identifying Logically Equivalent Statements

Decide if the two statements $p \rightarrow q$ and $\sim q \rightarrow \sim p$ are logically equivalent. Then write examples of simple statements p and q and write each compound statement verbally.

SOLUTION

The truth table for the statements is

p	q	$\sim p$	$\sim q$	$p \rightarrow q$	$\sim q \rightarrow \sim p$
T	T	F	F	T	T
T	F	F	T	F	F
F	T	T	F	T	T
F	F	T	T	T	T

Math Note

It would be a really good exercise for you to think about WHY the two verbal statements at the end of this example say the same thing.

Since both statements have the same truth values, they are logically equivalent. Suppose we let p = John's salary level is below the poverty line and q = John qualifies for food stamps. Then $p \rightarrow q$ reads as "If John's salary is below the poverty line, then he qualifies for food stamps," and $\sim q \rightarrow \sim p$ reads as "If John does not qualify for food stamps, then his salary is not below the poverty line."

✓ 2. Identify logically equivalent statements.

Try This One   2

Decide which two statements are logically equivalent. Then write examples of simple statements p and q and write each compound statement verbally.

(a) $\sim(p \wedge \sim q)$ (b) $\sim p \wedge q$ (c) $\sim p \vee q$

Think about these two compound statements: "Today is Thursday and it's snowing," and "Today is not Thursday or it's not snowing." The only way the first could be true is if it is Thursday, and if it is snowing. If either (or both) of those statements is false, then the conjunction is false. If you think that makes the second statement sound a lot like the negation of the first, then you're becoming a master of logic before our very eyes. This fact is one of De Morgan's laws for logic, which give us some examples of equivalent statements.

De Morgan's Laws for Logic

For any statements p and q,

- The statement $\sim(p \vee q)$ is logically equivalent to $\sim p \wedge \sim q$.

- The statement $\sim(p \wedge q)$ is logically equivalent to $\sim p \vee \sim q$.

Do you see the similarities between De Morgan's laws for logic and the ones for sets? We're exchanging sets like A and B for statements p and q, exchanging negation for complement, and exchanging intersection and union for "and" and "or." We can prove De Morgan's laws using truth tables, but it's more fun to make you do it in the exercises, so that's what we did.

De Morgan's laws are most often used to write the negation of conjunctions and disjunctions. For example, the negation of the statement "I will go to work or I will go to the beach" is "I will not go to work and I will not go to the beach." Notice that when you negate a conjunction, it becomes a disjunction; and when you negate a disjunction, it becomes a conjunction—that is, the *and* becomes an *or*, and the *or* becomes an *and*.

| Example 3 | Using De Morgan's Laws to Write Negations |

Write the negations of the following statements, using De Morgan's laws.

(a) Studying is necessary and I am a hard worker.
(b) Shoplifting is a felony or a misdemeanor.
(c) I will pass this test or I will drop this class.
(d) The patient needs an RN or an LPN, and is very sick.

Math Note

Notice that De Morgan's laws were used twice in part (d). Clever!

SOLUTION

(a) Studying is not necessary or I am not a hard worker.
(b) Shoplifting is not a felony and is not a misdemeanor.
(c) I will not pass this test and I will not drop this class.
(d) The patient doesn't need an RN and doesn't need an LPN, or she's not very sick.

Try This One   3

Write the negations of the following statements, using De Morgan's laws.

(a) I will study for this class or I will fail.
(b) I will go to the dance club and the restaurant.
(c) It is not silly or I have no sense of humor.
(d) The movie is a comedy or a thriller, and it is awesome.

Think about this for a second: what's the negation of the statement "If I pass this class, I won't have to take any classes this summer"? Not so easy, is it?

Earlier in this section, we saw that the two statements $p \rightarrow q$ and $\sim p \lor q$ are logically equivalent. Now that we know De Morgan's laws, we can use this fact to find the negation of a conditional statement $p \rightarrow q$.

$$\sim(p \rightarrow q) \equiv \sim(\sim p \lor q) \quad \textit{Use De Morgan's Law}$$
$$\equiv \sim(\sim p) \land \sim q \quad \sim(\sim p) \textit{ is the same as } p$$
$$\equiv p \land \sim q$$

This can be checked by using a truth table as shown. In the fifth column we have the negation of $p \rightarrow q$. In the sixth, we have $p \land \sim q$.

p	q	$\sim q$	$p \rightarrow q$	$\sim(p \rightarrow q)$	$p \land \sim q$
T	T	F	T	F	F
T	F	T	F	T	T
F	T	F	T	F	F
F	F	T	T	F	F

$p \rightarrow q$ has truth values that are opposite $p \land \sim q$.

The fifth and sixth columns have the same truth values, so the negation of $p \rightarrow q$ is $p \land \sim q$.

Looking back at our example, we can now see that the negation of "If I pass this class I won't have to take any classes this summer" is "I passed this class and I do have to take a class this summer."

| **Example 4** | Writing the Negation of a Conditional Statement |

Write the negation of the statement "If you go to the movies with me, I'll buy a 2-gallon bucket of popcorn for $46."

SOLUTION

We can write the statement as $p \rightarrow q$, where $p =$ "You go to the movies with me" and $q =$ "I buy a 2-gallon bucket of popcorn for $46." We just found that the negation of $p \rightarrow q$ is $p \wedge \sim q$. This translates to "You go to the movies with me and I do not buy a 2-gallon bucket of popcorn for $46."

| Try This One | 4 |

Write the negation of the statement "If the video is popular, then it can be found on YouTube."

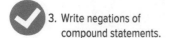
3. Write negations of compound statements.

Table 3-4 summarizes the negations of the basic compound statements.

| **Table 3-4** | Negation of Compound Statements |

Statement	Negation	Equivalent Negation
$p \wedge q$	$\sim(p \wedge q)$	$\sim p \vee \sim q$
$p \vee q$	$\sim(p \vee q)$	$\sim p \wedge \sim q$
$p \rightarrow q$	$\sim(p \rightarrow q)$	$p \wedge \sim q$

Variations of the Conditional Statement

Conditional statements play a very big role in logic (as well as in math in general), and one of the ways we can learn more about them is to study three related statements, the *converse*, the *inverse*, and the *contrapositive*. These are defined in Table 3-5.

| **Table 3-5** | Variations of a Conditional Statement |

Name	In Symbols	In Words
Conditional	$p \rightarrow q$	"If p, then q."
Converse	$q \rightarrow p$	"If q, then p."
Inverse	$\sim p \rightarrow \sim q$	"If not p, then not q."
Contrapositive	$\sim q \rightarrow \sim p$	"If not q, then not p."

We'll use the example "If Tessa is a chocolate Lab, then Tessa is brown" to illustrate the variations of a conditional, and to decide if each is logically equivalent to the original conditional statement. Note that the original conditional statement is true—the thing that makes a Labrador retriever a chocolate Lab is brown fur.

Converse: if Tessa is brown, then Tessa is a chocolate Lab.

Unlike the original statement, this is not true: there are plenty of brown dogs that are not chocolate Labs.

Inverse: if Tessa is not a chocolate Lab, then Tessa is not brown.

David Sobecki/McGraw Hill

In case you're wondering, this puppy dog is mine, too.

Again, not true: Tessa could be a breed other than chocolate Lab and still be brown. (For the record, she isn't, which I can tell because she's sitting on my feet at the moment. But she *could* be.)

Contrapositive: if Tessa is not brown, then Tessa is not a chocolate Lab.

Now this one is true: all chocolate Labs are brown, so if Tessa weren't brown, she couldn't possibly be a chocolate Lab. What these examples show is that we know for sure that the converse and inverse are not logically equivalent to an original conditional statement. This is because we've found a specific example where the converse and inverse have different truth values than the original.

On the other hand, we don't yet know for sure that any original statement and its contrapositive are equivalent: we just showed that in this particular case they happen to have the same truth value. To see if they're actually equivalent in general, we can use truth tables.

p	q	$p \to q$	$\sim p$	$\sim q$	$\sim q \to \sim p$
T	T	T	F	F	T
T	F	F	F	T	F
F	T	T	T	F	T
F	F	T	T	T	T

This shows that a conditional and its contrapositive are in fact logically equivalent.

Example 5 Writing the Converse, Inverse, and Contrapositive

Write the converse, the inverse, and the contrapositive for the statement "If you earned a bachelor's degree, then you got a high-paying job."

SOLUTION

It's helpful to write the original implication in symbols: $p \to q$, where $p =$ "You earned a bachelor's degree" and $q =$ "You got a high-paying job."

Converse: $q \to p$. "If you got a high-paying job, then you earned a bachelor's degree."
Inverse: $\sim p \to \sim q$. "If you did not earn a bachelor's degree, then you did not get a high-paying job."
Contrapositive: $\sim q \to \sim p$. "If you did not get a high-paying job, then you did not earn a bachelor's degree."

Try This One **5**

Write the converse, the inverse, and the contrapositive for the statement "If you do well in math classes, then you are intelligent." Then explain why the contrapositive says the same thing as the original statement.

Next, we'll study a very common issue in logic that causes a lot of people to draw faulty conclusions in their daily lives.

Example 6 Studying the Converse of a Statement

Consider the statement "If Sofia makes at least $400,000 a year, she can choose to buy a new Mercedes." I think we can all agree that this statement is true. Write the converse of this statement, and discuss whether or not it's also true.

SOLUTION

The converse of the statement is "If Sofia can choose to buy a new Mercedes, then Sofia makes at least $400,000 a year." This is not necessarily a true statement. Maybe Sofia makes far less than that, but is living in a refrigerator box so she can afford the payments on a $90,000 car. Or maybe Sofia is scraping by on the paltry sum of $300,000 per year and can still choose to buy a new Mercedes.

The conclusion we can draw from this example is that if a conditional statement is true, its converse may or may not be true.

Try This One   6

Write the converse of the true statement below, and explain why the converse might not be true.

"If you paid income tax last year, then you must have earned some money."

Since we run into conditional statements so often, it's useful to be able to recognize them when words other than "if . . . then" are used. Let's do a quick review: a conditional statement $p \to q$ is sometimes called an implication, and consists of two simple statements: the first (p) is called the antecedent, and the second (q) is called the consequent. For example, the statement "If you jump into the Arctic Ocean you will freeze your patootie off" consists of the antecedent "You jump into the Arctic Ocean" and the consequent "You will freeze your patootie off" connected by the "if . . . then" connective.

Here are some other ways a conditional can be stated:

p implies q
q if p
p only if q
p is sufficient for q
q is necessary for p
All p are q

Doug Menuez/Photodisc Green/
Getty Images

In four of these six forms, the antecedent comes first, but for "q if p" and "q is necessary for p," the consequent comes first. So identifying the antecedent and consequent is important.

For example, think about the statement "If you drink and drive, you get arrested." Writing it in the different possible forms, we get:

Drinking and driving implies you get arrested.
You get arrested if you drink and drive.
You drink and drive only if you get arrested.
Drinking and driving is sufficient for getting arrested.
Getting arrested is necessary for drinking and driving.
All those who drink and drive get arrested.

These all say the same thing. To illustrate the importance of getting the antecedent and consequent in the correct order, look back at Example 6. Switching the order results in the converse, which isn't the same as the original statement.

Example 7   Writing Variations of a Conditional Statement

Write each statement in symbols. Let p = "A building uses solar heat" and q = "The owner will pay less for electricity."

(a) If a building uses solar heat, the owner will pay less for electricity.
(b) Using less electricity is necessary for a building using solar heat.

(c) A building uses solar heat only if the owner pays less for electricity.

(d) Using solar heat is sufficient for paying less on your electric bill.

(e) The owner pays less for electricity if a building uses solar heat.

SOLUTION

(a) If p, then q; $p \rightarrow q$

(b) q is necessary for p; $p \rightarrow q$

(c) p only if q; $p \rightarrow q$

(d) p is sufficient for q; $p \rightarrow q$

(e) q if p; $p \rightarrow q$

Actually, these statements all say exactly the same thing!

Try This One   7

4. Write the converse, inverse, and contrapositive of a statement.

Write each statement in symbols. Let $p =$ "A student comes to class every day" and $q =$ "A student gets a good grade."

(a) A student gets a good grade if a student comes to class every day.

(b) Coming to class every day is necessary for getting a good grade.

(c) A student gets a good grade only if a student comes to class every day.

(d) Coming to class every day is sufficient for getting a good grade.

In this section, we saw that some statements are always true (tautologies) and others are always false (self-contradictions). We also defined what it means for two statements to be logically equivalent—they have the same truth values. Now we're ready to tie it all together to accomplish our original goal: analyzing logical arguments to decide if they make sense or not.

Answers to Try This One

1 (a) I'm going to a metal concert or I'll wear black leather, and if I'm not going to a metal concert then I'll wear black leather. This is neither.

(b) I'm going to a metal concert and I won't wear black leather, and I'm not going to a metal concert. This is a self-contradiction.

(c) If I go to a metal concert I'll wear black leather or I won't wear black leather. This is a tautology.

2 (a) and (c); Answers will vary.

3 (a) I will not study for this class and I will not fail.

(b) I will not go to the dance club or the restaurant.

(c) It is silly and I have a sense of humor.

(d) The movie is not a comedy and it is not a thriller, or it is not awesome.

4 The video is popular and it cannot be found on YouTube.

5 *Converse*: if you are intelligent, then you do well in math classes.

Inverse: if you do not do well in math classes, then you are not intelligent.

Contrapositive: if you are not intelligent, then you do not do well in math classes.

The original statement claimed that if you do well in math classes, you must be intelligent. If you weren't intelligent and still did well in math classes, that would contradict the original statement.

6 *Converse*: if you earned money last year, then you paid income taxes. This isn't necessarily true for a handful of reasons. For one, no income tax is due for folks making below a certain level. Also, there are many deductions that in some cases allow even wealthy people to avoid paying income tax.

7 (a) $p \rightarrow q$ (b) $q \rightarrow p$

(c) $q \rightarrow p$ (d) $p \rightarrow q$

Exercise Set 3-3

Writing Exercises

1. Explain the difference between a tautology and a self-contradiction.
2. Is every statement either a tautology or a self-contradiction? Why or why not?
3. Describe how to find the converse, inverse, and contrapositive of a conditional statement.
4. How can you decide if two statements are logically equivalent?
5. How can you decide if one statement is the negation of another?
6. Is a statement always logically equivalent to its converse? Explain.
7. Describe De Morgan's laws for logic. What are they used for?
8. How are De Morgan's laws for logic similar to De Morgan's laws for sets?

Computational Exercises

For Exercises 9–18, classify each statement as a tautology, a self-contradiction, or neither.

9. $(p \vee q) \vee (\sim p \wedge \sim q)$
10. $(p \to q) \wedge (p \vee q)$
11. $(p \wedge q) \wedge (\sim p \vee \sim q)$
12. $(p \vee q) \wedge (\sim p \wedge \sim q)$
13. $(p \leftrightarrow q) \vee \sim(q \to p)$

14. $(p \wedge q) \leftrightarrow (p \to \sim q)$
15. $(\sim p \to q) \leftrightarrow (p \vee q)$
16. $(p \to q) \wedge (q \to p)$
17. $(p \leftrightarrow q) \wedge (\sim p \leftrightarrow \sim q)$
18. $(p \to q) \wedge (\sim p \vee q)$

For Exercises 19–28, decide if the two statements are logically equivalent statements, negations, or neither.

19. $\sim q \to p;\ \sim p \to q$
20. $p \wedge q \sim q \vee \sim p$
21. $\sim(p \vee q);\ p \to \sim q$
22. $\sim(p \to q);\ \sim p \wedge q$

23. $q \to p;\ \sim(p \to q)$
24. $p \vee (\sim q \wedge r);\ (p \wedge \sim q) \vee (p \wedge r)$
25. $\sim(p \vee q);\ \sim(\sim p \wedge \sim q)$
26. $(p \vee q) \to r;\ \sim r \to \sim(p \vee q)$
27. $(p \wedge q) \vee r;\ p \wedge (q \vee r)$
28. $p \leftrightarrow \sim q;\ (p \wedge \sim q) \vee (\sim p \wedge q)$

For Exercises 29–34, write the converse, inverse, and contrapositive of each.

29. $p \to q$
30. $\sim p \to \sim q$
31. $\sim p \to \sim(q \wedge p)$
32. $(q \vee \sim r) \to (p \vee r)$
33. $p \to (q \vee r)$
34. $(p \vee \sim q) \to r$

Applications in Our World

In Exercises 35–44, use De Morgan's laws to write the negation of the statement.

35. The patient is septic or is in shock.
36. The experimental seedlings are growing quickly or they are not diseased.
37. It is not cold and I am soaked.
38. I will walk in the Race for the Cure walkathon and I will be tired.
39. I will go to the beach and I will not get sunburned.
40. The coffee is a latte or an espresso.
41. The suspect is a white male or the witness is not correct.
42. I will go to college and I will get a degree.
43. It is right or it is wrong.
44. The hotel custodial staff is not on strike or it is not at a union meeting.

In Exercises 45–50, use De Morgan's laws to write an equivalent statement.

45. It is not the case that my grade is an A or a B.
46. It's totally false to say that the student has special needs and doesn't belong in this classroom.

47. The prosecuting attorney for this case is not experienced and is not prepared.
48. This firm's managing partner isn't white or isn't male.
49. It's not true that my friends are not serious about school or are not prepared to work hard.
50. No way is that patient unresponsive and not able to breathe on his own.

For Exercises 51–57, let p = "I need to talk to my friend" and q = "I will send her a text message." Write each of the following in symbols (see Example 7).

51. If I need to talk to my friend, I will send her a text message.
52. If I will not send her a text message, I do not need to talk to my friend.
53. Sending a text message is necessary for needing to talk to my friend.
54. I will send her a text message if I need to talk to my friend.
55. Needing to talk to my friend is sufficient for sending her a text message.

56. I need to talk to my friend only if I will send her a text message.

57. I do not need to talk to my friend only if I will not send her a text message.

58. Are any of the statements in Exercises 51–57 logically equivalent?

For Exercises 59–64, write the converse, inverse, and contrapositive of the conditional statement. Then explain why the contrapositive says the same thing as the original statement, and why the converse does not.

59. If Mariel graduated with a Bachelor's degree in Management Information Systems, then she will get a good job.

60. If I don't earn $5,000 this summer as a barista at the coffeehouse, then I can't buy the green Ford Focus.

61. If we make a mistake on the dosage, that patient may not survive.

62. If my cell phone will not charge, then I will replace the battery.

63. I will go to Nassau for spring break if I lose 10 pounds by March 1.

64. The politician will go to jail if he gets caught taking kickbacks.

For Exercises 65–70, write the negation of each statement in Exercises 59–64.

65. Negation of Exercise 59.
66. Negation of Exercise 60.
67. Negation of Exercise 61.
68. Negation of Exercise 62.
69. Negation of Exercise 63.
70. Negation of Exercise 64.

Critical Thinking

In Exercises 71–76, decide if the given compound statement is true or false. Be careful, and explain your reasoning.

71. The moon is made of green cheese if and only if Justin Bieber is the king of Siam.

72. The average student takes less than 6 years to complete an undergraduate degree if the University of Southern California is in Tahiti.

73. Drinking and driving increases the risk of an accident if and only if driving on the left side of the road is standard in the United States.

74. If my parents were born in outer space, then Venus orbits around the Earth every 28 days.

75. The current year is before 1970 if there are more than a million people in the United States.

76. Fewer than 20 people watched the finale of *The Voice* last year if and only if the winner was 14 inches tall.

77. In this section, we wrote the negation of $p \rightarrow q$ by using a disjunction. See if you can write the negation of $p \rightarrow q$ by using a conjunction.

78. Try to write the negation of the biconditional $p \leftrightarrow q$ by using only conjunctions, disjunctions, and negations.

79. Can you think of a true conditional statement about someone you know so that the converse is true as well? How about so that the converse is false?

80. Can you think of a true conditional statement about someone you know so that the inverse is true as well? How about so that the inverse is false?

81. Use truth tables to prove both of De Morgan's laws for logic (see page 126).

82. We defined converse, inverse, and contrapositive for conditional statements. Using these as models, define the converse, inverse, and contrapositive for a biconditional statement $p \leftrightarrow q$. Which, if any, are logically equivalent to the original biconditional?

Section 3-4 | Logical Arguments

LEARNING OBJECTIVES

1. Define *valid argument* and *fallacy.*

2. Use truth tables to decide if an argument is valid.

3. Identify common argument forms.

4. Use common argument forms to decide if arguments are valid.

Common sense is a funny thing in our society: we all think we have it, and we also think that most other people don't. Isn't that awesome? This thing that we call common sense is really the ability to think logically, to evaluate an argument or situation and decide what is and is not reasonable. It doesn't take a lot of imagination to picture how valuable it is to be able to think logically. We're pretty well protected by caregivers for our first few years of life, but after that the main tool we have to guide us through the perils of life is our brain. The more effectively that brain can analyze and evaluate

Ian McKinnell/Alamy Stock Photo

the mass of information we're all exposed to every day, the more successful we're likely to be.

The work we've done in building the basics of symbolic logic in the first three sections in this chapter has prepared us for the real point: analyzing logical arguments objectively. That's the topic of this important section. Remember why we're using letters to represent statements: your goal is to ignore the TOPIC of an argument, and simply focus on whether or not a conclusion can be reasonably drawn from a preliminary set of statements. Check your emotions and opinions at the door, and get ready to be argumentative.

Valid Arguments and Fallacies

A logical argument consists of two parts: a set of premises and a conclusion based on those premises. Premises are statements that are offered as supporting evidence for the conclusion. Our goal is to decide whether an argument is *valid* or *invalid*.

> An argument is **valid** if the conclusion necessarily follows from the premises, and **invalid** if it's not valid. An error in reasoning that leads to an invalid argument is known as a **fallacy**.

Notice that our definition of valid doesn't use the word *true*, because logic is not about deciding if a claim is true: it's about deciding if the claim can be deduced from the premises.

Let's look at an example.

Premise 1:   All students in this class will pass.
Premise 2:   Rachel is a student in this class.
Conclusion: Rachel will pass this class.

Since we are told that ALL students in the class will pass, we can be sure that if Rachel is a student in the class, she will pass. This is an example of a valid argument because the conclusion logically follows from the premises.

It's very important at this point to understand the difference between a true statement and a conclusion to a valid argument. A statement that is known to be false can still be a valid conclusion if it follows logically from the given premises. For example, consider this argument:

Los Angeles is in California or Mexico.
Los Angeles is not in California.
Therefore, Los Angeles is in Mexico.

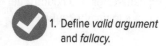

1. Define *valid argument* and *fallacy*.

This is a valid argument: if we accept the two premises, then Los Angeles would in fact be in Mexico. We know, however, that Los Angeles is NOT in Mexico, and there's the tricky part. *To be valid, the conclusion of an argument has to follow from the premises whether they're true or not.* In this case, we're accepting the premise "Los Angeles is not in California" even though we know that it's actually false. We can then conclude that the conclusion (Los Angeles is in Mexico) follows logically from that premise. Again, this emphasizes that the validity of an argument is not about whether or not the conclusion is a true statement.

Truth Table Method

One method for determining the validity of an argument is by using truth tables. We'll use the following procedure.

> **Procedure for Determining the Validity of Arguments**
>
> **Step 1**   Write the argument in symbols.
>
> **Step 2**   Write the argument as a conditional statement; use a conjunction between all premises and the implication ($\Rightarrow$) for the conclusion. (*Note:* the $\Rightarrow$ is the same as $\rightarrow$ but will be used to designate an argument.)

Sidelight   Logic and The Art of Forensics

Many students find it troubling that an argument can be considered valid even if the conclusion is clearly false. But arguing in favor of something that you don't necessarily believe to be true isn't a new idea by any means—lawyers do it all the time, and it's commonly practiced in the area of formal debate, a style of intellectual competition that has its roots in ancient times.

In formal debate (also known as forensics), speakers are given a topic and asked to argue one side of a related issue. Judges determine which speakers make the most effective arguments and declare the winners accordingly. One of the most interesting aspects is that in many cases, the contestants don't know which side of the issue they will be arguing until right before the competition begins. While that aspect is intended to test the debater's flexibility and preparation, a major consequence is that opinion, and sometimes truth, is taken out of the mix, and contestants and judges must focus on the validity of arguments.

A variety of organizations sponsor national competitions in formal debate for colleges. The largest is an annual

Nikolay Mamluke/iStock/Getty Images

championship organized by the National Forensics Association. Students from well over 100 schools participate in a wide variety of categories. In 2021, the overall team champion was Western Kentucky University.

Math Note

There can be more than two premises in an argument; in that case, put the conjunction sign ∧ between all premises.

Step 3 Set up and construct a truth table as follows:

Symbols	Premise ∧ Premise ⇒ Conclusion

Step 4 If all truth values under ⇒ are Ts (that is, the statement in the last column is a tautology), then the argument is valid; otherwise, it is invalid.

Notice that we're not looking for all Ts under the conclusion, because we don't care if the conclusion is true or not. We're looking for all Ts under the *implication of the conclusion from the premises* because we're interested in the conclusion following logically from the premises.

Example 1   Deciding If an Argument Is Valid

Decide if the following argument is valid.

> If the ad campaign is a success, the marketing manager won't lose her job.
> The marketing manager lost her job.
> _____
> Therefore, the ad campaign wasn't a success.

SOLUTION

Step 1 *Write the argument in symbols.* Let p = "The ad campaign is a success," and let q = "The marketing manager loses her job."

Translated into symbols:

$p \rightarrow \sim q$ (Premise)
q (Premise)
_____
$\therefore \sim p$ (Conclusion)

A line is used to separate the premises from the conclusion and the three triangular dots ∴ mean "therefore."

Step 2 *Write the argument as an implication* by connecting the premises with a conjunction and implying the conclusion as shown.

Premise 1		Premise 2		Conclusion
$(p \to \sim q)$	$\wedge$	q	$\Rightarrow$	$\sim p$

Step 3 *Construct a truth table* as shown. This is one of the main skills we learned in the last section. Notice how the final column in the truth table is the whole argument written in symbols from Step 2.

p	q	$\sim p$	$\sim q$	$p \to \sim q$	$(p \to \sim q) \wedge q$	$[(p \to \sim q) \wedge q] \Rightarrow \sim p$
T	T	F	F	F	F	T
T	F	F	T	T	F	T
F	T	T	F	T	T	T
F	F	T	T	T	F	T

Step 4 *Determine the validity of the argument.* Since all the values under the $\Rightarrow$ are true, the argument is valid.

Try This One   1

Decide if the argument is valid or invalid.

> I will run for student government or I will join the athletic boosters.
>
> I did not join the athletic boosters.
> ───────────────────────────────
> Therefore, I will run for student government.

Example 2   **Deciding If an Argument Is Valid**

Decide if this argument is valid: emerging research shows that if cancer patients undergo aromatherapy, depression is lessened. One of my cancer patients is reporting a decrease in depression, so she must have undergone aromatherapy.

SOLUTION

Step 1 *Write the argument in symbols.* Let p = "A cancer patient undergoes aromatherapy" and let q = "The patient suffers less from depression." The argument can then be written as

$p \to q$ *If aromatherapy, then less depression*

q *Less depression*

$\therefore p$ *Therefore, aromatherapy*

Step 2 *Write the argument as an implication.*

$(p \to q) \wedge q \Rightarrow p$

Step 3 *Construct a truth table* for the argument.

p	q	$p \to q$	$(p \to q) \wedge q$	$[(p \to q) \wedge q] \Rightarrow p$
T	T	T	T	T
T	F	F	F	T
F	T	T	T	F
F	F	T	F	T

Step 4 *Determine the validity of the argument.* This argument is invalid since it is not a tautology. (Remember, when the values are not all Ts, the argument is invalid.) In this case, we can't conclude that the patient underwent aromatherapy.

Math Note

Notice that we're putting our Ts and Fs under the connective symbol in the later columns, not centered in the column. I've found that this really helps me to focus on the truth values for the given connective without getting distracted by all the other stuff.

Math Note

While our goal in this chapter is to evaluate arguments using formal logic and truth tables, if you want some extra brain exercise, you might consider trying to decide if arguments are valid from the verbal descriptions, then seeing how you did using truth tables.

Tetra Images/Shutterstock

Try This One 2

Decide if this argument is valid: John's boss warned him that if he blew off work to go to the playoff game, he'd get fired. I heard John got fired, so I guess he must have gone to that playoff game. Cool!

CAUTION

Remember that in symbolic logic, whether or not the conclusion is true is not important. The main concern is whether the conclusion follows from the premises.

Consider the following two arguments.

1. Either $2 + 2 \neq 4$ or $2 + 2 = 5$

   $\dfrac{2 + 2 = 4}{\text{So } 2 + 2 = 5.}$

2. If $2 + 2 \neq 5$ then my feet hurt

   $\dfrac{\text{My feet don't hurt.}}{\text{So } 2 + 2 \neq 5.}$

In Exercises 63 and 64, you'll construct truth tables for these arguments and find that the one with the false conclusion is valid, while the one with the true conclusion is not. Did we mention that the validity of an argument is not about whether or not the conclusion is true?

The validity of arguments that have three premises can also be tested using truth tables, as shown in Example 3. In this case, the last column will contain a conjunction of three premises. You might want to write small on these babies, as we'll see in Example 3. In this example, we'll use an abstract argument so we can just focus on the procedure.

Example 3 | Deciding If an Argument Is Valid

Decide if the argument is valid.

$$p \rightarrow r$$
$$q \wedge r$$
$$\dfrac{p}{\therefore \sim q \rightarrow p}$$

SOLUTION

Step 1 *Write the argument in symbols.* This has been done already.

Step 2 *Write the argument as an implication.* Make a conjunction of all three premises and imply the conclusion:

$$(p \rightarrow r) \wedge (q \wedge r) \wedge p \Rightarrow (\sim q \rightarrow p)$$

Step 3 *Construct a truth table.* When there are three premises, we will begin by finding the truth values for each premise and then we'll work the conjunction from left to right as shown.

p	q	r	$\sim q$	$p \to r$	$q \wedge r$	$\sim q \to p$	$(p \to r) \wedge (q \wedge r) \wedge p$	$[(p \to r) \wedge (q \wedge r) \wedge p] \Rightarrow (\sim q \to p)$
T	T	T	F	T	T	T	T	T
T	T	F	F	F	F	T	F	T
T	F	T	T	T	F	T	F	T
T	F	F	T	F	F	T	F	T
F	T	T	F	T	T	T	F	T
F	T	F	F	T	F	T	F	T
F	F	T	T	T	F	F	F	T
F	F	F	T	T	F	F	F	T

Since the truth value for $\Rightarrow$ is all Ts, the argument is valid.

Try This One   3

Decide if the argument is valid.

$$p \vee q$$
$$\underline{q \vee \sim r}$$
$$\therefore q$$

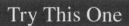

2. Use truth tables to decide if an argument is valid.

Common Valid Argument Forms

We have seen that truth tables can be used to test an argument for validity. But because we're focusing on the structure of an argument and not on the actual statements involved, once we've tested a certain argument, we wouldn't need to do the same one again just because the topic changes. In particular, some argument forms are common enough that they are recognized by special names. When an argument fits one of these forms, we can decide if it is valid or not just by knowing the general form, rather than constructing a truth table. This can be a big time-saver.

We'll start with a description of some commonly used valid arguments. It's a really good exercise in logical thinking to think about why these arguments are valid, rather than just trying to memorize them.

1. **Law of detachment**

$$p \to q$$
$$\underline{p}$$
$$\therefore q$$

Example:

If our team wins Saturday, then they go to a bowl game.

Our team won Saturday.

Therefore, our team goes to a bowl game.

2. **Law of contraposition**

$$p \to q$$
$$\underline{\sim q}$$
$$\therefore \sim p$$

Example:

If I try hard, I'll get an A.

I didn't get an A.

Therefore, I didn't try hard.

3. **Law of syllogism**, also known as **law of transitivity**:

$$p \to q$$
$$\underline{q \to r}$$
$$\therefore p \to r$$

Math Note

The law of contraposition isn't too hard to remember if you think of why it's named that way: in essence, it says that the contrapositive of a conditional statement says the same thing as the original statement. If p implies q, then not q implies not p.

Jason Pack/FEMA

Example:

If I make an illegal U-turn, I'll get a ticket.

If I get a ticket, I'll get points on my driving record.

Therefore, if I make an illegal U-turn, I'll get points on my driving record.

4. **Law of disjunctive syllogism:**

$$p \vee q$$
$$\sim p$$
$$\therefore q$$

Example:

My top client demands the penthouse or an executive suite.

He couldn't stay in the penthouse.

Therefore, he stayed in an executive suite.

Common Fallacies

Next, we'll list some commonly used arguments that are invalid.

1. **Fallacy of the converse:**

$$p \rightarrow q$$
$$q$$
$$\therefore p$$

Example:

If it's Friday, then I will go to happy hour.

I am at happy hour.

Therefore, it is Friday.

This is not valid! You can go to happy hour other days, too. I've done it.

2. **Fallacy of the inverse:**

$$p \rightarrow q$$
$$\sim p$$
$$\therefore \sim q$$

Example:

If I exercise every day, then I will lose weight.

I don't exercise every day.

Therefore, I won't lose weight.

This is also not valid. You could still lose weight without exercising every day.

3. **Fallacy of the inclusive or:**

$$p \vee q$$
$$p$$
$$\therefore \sim q$$

Example:

I'm going to take chemistry or physics.

I'm taking chemistry.

Therefore, I'm not taking physics.

Math Note

If you look closely at the fallacy of the converse and the fallacy of the inverse, they should look familiar. Calling each argument invalid is basically the same as saying the converse and the inverse, respectively, are not logically equivalent to an original conditional statement.

Comstock Images/Getty Images

Sidelight   **Circular Reasoning**

Circular reasoning (sometimes called "begging the question") is a sneaky type of fallacy in which the premises of an argument contain a claim that the conclusion is true, so naturally if the premises are true, so is the conclusion. But this doesn't constitute evidence that a conclusion is true. Consider the following example: a suspect in a criminal investigation tells the police detective that his statements can be trusted because his friend Sue can vouch for him. The detective asks the suspect how he knows that Sue can be trusted, and he says, "I can assure you of her honesty." Ultimately, the suspect becomes even more suspect because of circular reasoning: his argument boils down to "I am honest because I am honest."

While this example might seem blatantly silly, you'd be surprised how often people try to get away with this fallacy. A Google search for the string *circular reasoning* brings up hundreds of arguments that are thought to be circular.

Remember, we've agreed that by *or* we mean *one or the other, or both*. So you could be taking both classes.

You will be asked to prove that some of these argument forms are valid or invalid by using truth tables in the exercises.

Example 4   Recognizing Common Argument Forms

Decide if the following arguments are valid, using the given forms of valid arguments and fallacies.

(a) $p \to q$
p
$\therefore q$

(b) $\sim p \to q$
$\sim q$
$\therefore p$

(c) $\sim p \to \sim q$
$\sim q$
$\therefore \sim p$

(d) $\sim r \to s$
$s \to t$
$\therefore \sim r \to t$

Math Note

Parts (b) through (d) of Example 4 show that you can use known argument forms in some cases even if the given argument isn't completely identical to the basic form if negations are used.

SOLUTION

(a) This is the law of detachment, therefore a *valid* argument.
(b) This fits the law of contraposition with the statement $\sim p$ substituted in place of p, so it is valid.
(c) This fits the fallacy of the converse, using statement $\sim p$ and $\sim q$ rather than p and q, so it is an invalid argument.
(d) This is the law of syllogism, with statements $\sim r$, s, and t, so the argument is valid.

Try This One   4

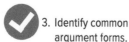

3. Identify common argument forms.

Decide if the arguments are valid, using the commonly used valid arguments and fallacies.

(a) $\sim p \lor q$
p
$\therefore q$

(b) $r \lor s$
s
$\therefore \sim r$

(c) $\sim p \to q$
q
$\therefore \sim p$

(d) $(p \land q) \to \sim r$
r
$\therefore \sim(p \land q)$

Example 5   Using Common Argument Forms to Decide If an Argument Is Valid

Use the given forms of valid arguments and fallacies to decide if each argument is valid.

(a) If you eat a healthy diet, you'll live past 70.
You've really made an effort to eat healthy foods.

Therefore, you'll live past 70.

(b) You can access the Web at this hotel if you pay for WiFi.
You're too broke to pay for WiFi.

So you won't be able to access the Web.

(c) If you watch *Big Brother*, you watch reality shows.
If you watch reality shows, you have time to kill.

Therefore, if you have time to kill, you watch *Big Brother*.

(d) You planned to major in criminal justice or prelaw. Since you're majoring in criminal justice, I guess that means you're not majoring in prelaw.

(e) Botany grads work either in parks or at research labs. Linh got her degree in botany last year and doesn't work at a park, so she must work at a research lab.

SOLUTION

I'm not going to say it's *necessary* to write the arguments in symbols first, but it sure makes it a lot easier to recognize the known forms.

Math Note

The first premise of part (b) is a little tricky because the consequent is written first. It could be restated as "If you pay for WiFi, you can access the Web."

(a) In symbolic form this argument is $(p \rightarrow q) \wedge p \Rightarrow q$. We can see that this is the law of detachment, so the argument is *valid*.
(b) In symbolic form this argument is $(p \rightarrow q) \wedge \sim p \Rightarrow \sim q$. This is the fallacy of the inverse, so the argument is *invalid*.
(c) In symbolic form this argument is $(p \rightarrow q) \wedge (q \rightarrow r) \Rightarrow (r \rightarrow p)$. We know by the law of transitivity that if $(p \rightarrow q) \wedge (q \rightarrow r)$, then $p \rightarrow r$. The given conclusion, $r \rightarrow p$, is the converse of $p \rightarrow r$, so is not equivalent to $p \rightarrow r$ (the valid conclusion). So the argument is *invalid*.
(d) In symbolic form the argument is $(p \vee q) \wedge p \Rightarrow \sim q$. This is the fallacy of the inclusive or, so the argument is *invalid*.
(e) In symbolic form the argument is $(p \vee q) \wedge \sim p \Rightarrow q$. This is the law of disjunctive syllogism, so the argument is *valid*.

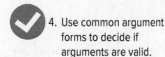

4. Use common argument forms to decide if arguments are valid.

Try This One 5

Determine whether the following arguments are valid using the given forms of valid arguments and fallacies.

(a) If you do well in hospitality management, you'll get a job at a great resort.
You graduated with a 3.8 GPA in hospitality management.

Therefore, you should get a job at a great resort.

(b) If you work hard, you will be a success.
You are not a success.

Therefore, you do not work hard.

(c) Jon is either really cheap, or flat broke. I got a look at his checking account statement and he's not broke, so he must just be cheap.

(d) If my lab rats don't die, I'll get a passing lab grade. Since I passed bio lab, that means the rats didn't die.

Many books define a valid argument this way: an argument is valid if the conclusion must be true when all the premises are true. We chose not to use that terminology because it makes it seem like there's no point in setting up an entire truth table to test an argument—just the row where all premises are true. But that's not accurate. The last column of the truth tables we've used does NOT show truth or falsehood of the *conclusion*: it shows truth or falseness of the *implication* premises → conclusion. Only when this implication is always true can we conclude that an argument is valid.

Answers to Try This One

1	Valid	**4**	(a) Valid (b) Invalid (c) Invalid (d) Valid
2	Invalid	**5**	(a) Valid (b) Valid (c) Valid (d) Invalid
3	Invalid		

Exercise Set   **3-4**

Writing Exercises

1. Describe the structure of an argument.
2. Describe in your own words what it means for an argument to be valid.
3. When setting up a truth table to test the validity of an argument, why don't we only look at the case where all premises are true?
4. Is it possible for an argument to be valid, yet have a false conclusion? Explain your answer.
5. Is it possible for an argument to be invalid, yet have a true conclusion? Explain your answer.

6. When you are setting up a truth table to determine the validity of an argument, what connective is used between the premises of an argument? What connective is used between the premises and the conclusion? Why do these make sense?
7. Describe what the law of syllogism says, in your own words.
8. Describe why the fallacy of the inclusive or is a fallacy.

Computational Exercises

For Exercises 9–18, using truth tables, decide if each argument is valid.

9. $p \to q$
$p \wedge q$
$\therefore p$

10. $p \vee \sim q$
q
$\therefore p \wedge q$

11. $\sim p \vee q$
p
$\therefore p \wedge \sim q$

12. $p \leftrightarrow \sim q$
$p \wedge \sim q$
$\therefore p \vee q$

13. $\sim p \vee q$
$q \to \sim p$
$\therefore p$

14. $\sim p \leftrightarrow q$
$\sim p \wedge q$
$\therefore \sim p \vee q$

15. $p \leftrightarrow q$
$\sim q$
$\therefore \sim p$

16. $p \vee \sim q$
$\sim q \to p$
$\therefore p$

17. $p \wedge \sim q$
$\sim r \to q$
$\therefore q$

18. $p \leftrightarrow q$
$q \leftrightarrow r$
$\therefore p \wedge q$

In Exercises 19–24, write the given common argument form in symbols, then use a truth table to prove that it either is or is not a valid argument.

19. Law of detachment
20. Law of contraposition

21. Law of syllogism
22. Law of disjunctive syllogism
23. Fallacy of the converse
24. Fallacy of the inclusive or

For Exercises 25–34, decide if the following arguments are valid, using the given forms of valid arguments and fallacies. Make sure you list the form that you used.

25. $p \to q$
$\sim q$
$\therefore \sim p$

26. $p \vee q$
q
$\therefore \sim p$

27. $\sim p \to q$
$\sim q$
$\therefore \sim p$

28. $p \vee \sim q$
q
$\therefore p$

29. $p \to q$
$r \to \sim q$
$\therefore p \to \sim r$

30. $p \to \sim q$
$\sim q$
$\therefore p$

31. $\sim p \vee q$
$\sim q$
$\therefore \sim p$

32. $\sim p \to q$
$\sim q$
$\therefore p$

33. $p \vee \sim q$
q
$\therefore \sim p$

34. $p \to \sim q$
$\sim r \to q$
$\therefore p \to r$

Applications in Our World

For Exercises 35–48, identify p, q, and r if necessary. Then translate each argument to symbols and use a truth table to decide if the argument is valid or invalid.

35. If I don't have to go to summer school, I'll get an internship.

    I have to go to summer school.

    ∴ I won't get an internship.

36. I need to take a grad class in evidence gathering or civil rights.

    I couldn't get into the class on civil rights.

    ∴ I'm taking a class in evidence gathering.

37. If Julia uses monster.com to send out her resume, she will get an interview.

    Julia got an interview.

    ∴ Julia used monster.com to send out her resume.

38. If it snows, I can go snowboarding.

    It did not snow.

    ∴ I cannot go snowboarding.

39. I will go to the party if and only if my ex is not going.

    My ex is not going to the party.

    ∴ I will go to the party.

40. If the gallery opening is Friday, you should finish the piece you're working on.

    If you finish the piece you're working on, then the gallery opening is Friday.

    Therefore, the gallery opening is Friday and you'll finish the piece you're working on.

41. Either I did not study or I passed the exam.

    I did not study.

    ∴ I failed the exam.

42. I will run the marathon if and only if I can run 30 miles by January 1.

    I can run 30 miles by January 1 or I will not run the marathon.

    ∴ If I ran the marathon, then I was able to run 30 miles by January 1.

43. If that post-op patient has a high fever, she must have an infection of some sort. If she does have an infection, we'll have to keep her overnight for observation. So if she has a high fever, we'll have to keep her overnight for observation.

44. If you back up your hard drive, then you're protected from data loss. Either you're protected from data loss, or you're some sort of daredevil. That means that if you are a daredevil, you won't back up your hard drive.

45. Mayor Dryer will get elected if and only if he spends the most money on his campaign. An inside source told me last week that either Mayor Dryer will spend the most money or his opponent will run a smear campaign to influence voters. Just today I found out that Mayor Dryer's opponent decided not to run a smear campaign, so I conclude that Dryer won't win the election.

46. If Allie is a fine arts major, then she lives on campus. Allie's Facebook page says that she lives on campus and her best friend's name is Moose. So if Allie's best friend's name isn't Moose, she's majoring in something other than fine arts.

47. My company will get a huge contract if and only if I convince the client that our company will best represent their interests. If I manage to convince the client of this, I'll get a big bonus this year. So if I don't get a big bonus this year, you'll know that my company didn't get the huge contract.

48. Defendants who are convicted of manslaughter always serve time in prison. The defendant in the case Elena tried was represented by a public defender and was convicted of manslaughter. So if the defendant was represented by a public defender, the defendant will not serve time in prison.

For Exercises 49–56, write the argument in symbols; then decide whether the argument is valid by using the common forms of valid arguments and fallacies.

49. I studied or I failed the class.

    I did not fail the class.

    ∴ I studied.

50. If I go to the student symposium on environmental issues, I will fall asleep.

    If the speaker is interesting, I will not fall asleep.

    ∴ If I go to the student symposium on environmental issues, the speaker will not be interesting.

51. If it is sunny, I will wear SPF 50 sun block.

    It is not sunny.

    ∴ I will not wear SPF 50 sun block.

52. I will backpack through Europe if I get at least a 3.5 grade point average.

    I do not get at least a 3.5 grade point average.

    ∴ I will not backpack through Europe.

53. If we don't lobby the senator's office relentlessly, bills we oppose will pass. Since the Senate just passed a bill we opposed, we must not have lobbied the senator's office relentlessly.

54. Jason told me that if he got an A in anthropology, he'd run across the main campus quad wearing nothing but a

sombrero. Since he did that yesterday (yuck), he must have gotten an A in anthropology.

55. I will absolutely not wear a Speedo at the beach or I'll be embarrassed. I'm not embarrassed at the beach today, so you can be sure I didn't wear a Speedo.

56. If an internship requires collecting specimens at sea, I'll get seasick. I just found out that I got an internship at Woods Hole Institute that involves collecting mussel specimens off the coast of Massachusetts. Seasickness here I come!

Critical Thinking

57. Oscar Wilde once said, "Few parents nowadays pay any regard to what their children say to them. The old-fashioned respect for the young is fast dying out." This statement can be translated to an argument, as shown next.

If parents respected their children, then parents would listen to them.

Parents do not listen to their children.

∴ Parents do not respect their children.

Using a truth table, determine whether the argument is valid or invalid.

58. Winston Churchill once said, "If you have an important point to make, don't try to be subtle or clever. Use a pile driver. Hit the point once. Then come back and hit it again. Then a third time—a tremendous whack!" This statement can be translated to an argument as shown.

If you have an important point to make, then you should not be subtle and you should not be clever.

You are not being subtle and you are not being clever.

∴ You will make your point.

Using a truth table, determine whether the argument is valid or invalid.

59. Make up your own example for each of the four common valid argument forms discussed in Section 3-4. Use topics that have some relevance to your life.

60. Make up your own example for each of the four common fallacies discussed in Section 3-4 (including the one in the Sidelight on page 140). Use topics that have some relevance to your life.

61. The law of detachment is sometimes given the Latin name *modus ponens*. Look up the literal translation of that phrase on the Internet and describe how it applies.

62. The law of contraposition is also known by a Latin name, *modus tollens*. Look up the literal translation of that phrase and explain how it applies.

63. Write the argument labeled 1 on page 137 in symbols and make a truth table for it. Is the argument valid? Is the conclusion a true or false statement? What can you conclude?

64. Repeat Exercise 63 for the argument labeled 2 on page 137.

Section 3-5	Euler Circles

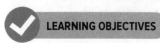 **LEARNING OBJECTIVES**

1. Define *syllogism*.
2. Use Euler circles to decide if an argument is valid.

Abraham Lincoln once said, "You can fool some of the people all of the time, and all of the people some of the time, but you cannot fool all of the people all of the time." Lincoln was a really smart guy—he understood the power of logical arguments and the fact that cleverly crafted phrases could be an effective tool in the art of persuasion. What's interesting about this quote from our perspective is the liberal use of the quantifiers *some* and *all*. In this section, we will study a particular type of argument that uses these quantifiers, along with *no* or *none*. A technique developed by Leonhard Euler way back in the 1700s is a useful method for analyzing these arguments and testing their validity. And it also gives us a historical look at the genesis of the Venn diagrams that we found so useful in Chapter 2.

Euler circles are diagrams similar to Venn diagrams. We will use them to study arguments using four

Source: Library of Congress Prints & Photographs Division [LC-USZ62-13016]

Table 3-6	Types of Statements Illustrated by Euler Circles		
Type		**General Form**	**Example**
Universal affirmative		All *A* is *B*	All chickens have wings.
Universal negative		No *A* is *B*	No horses have wings.
Particular affirmative		Some *A* is *B*	Some horses are black.
Particular negative		Some *A* is not *B*	Some horses are not black.

types of statements. The statement types are listed in Table 3-6, and the Euler circle that illustrates each is shown in Figure 3-1.

Each statement can be represented by a specific diagram. The universal affirmative "All *A* is *B*" means that every member of set *A* is also a member of set *B*. For example, the statement "All chickens have wings" means that the set of all chickens is a subset of the set of animals that have wings. This is illustrated in Figure 3-1(a).

The universal negative "No *A* is *B*" means that no member of set *A* is a member of set *B*. In other words, set *A* and set *B* are *disjoint sets*. For example, "No horses have wings" means that the set of all horses and the set of all animals with wings are disjoint (nonintersecting): see Figure 3-1(b).

The particular affirmative "Some *A* is *B*" means that there is at least one member of set *A* that is also a member of set *B*. For example, the statement "Some horses are black" means that there is at least one horse that is a member of the set of black animals. The × in Figure 3-1(c) means that there is at least one black horse.

The particular negative "Some *A* is not *B*" means that there is at least one member of set *A* that is not a member of set *B*. For example, the statement "Some horses are not black" means that there is at least one horse that does not belong to the set of black animals. The diagram for the particular negative is shown in Figure 3-1(d). The × is placed in circle *A* but not in circle *B*. The × in this example means that there exists at least one horse that is some color other than black.

CORBIS

Math Note

The "some" quantifier doesn't necessarily mean that "all" is not a more accurate description of a situation. For example, if I say "Some nursing jobs pay over $40,000," it's entirely possible that ALL of them actually do. When you read a statement like "Some horses are black," you can't assume that there are some that are not black.

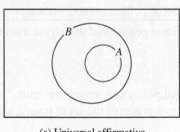

(a) Universal affirmative
"All *A* is *B*"

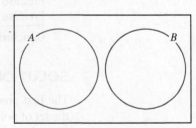

(b) Universal negative
"No *A* is *B*"

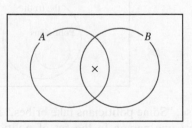

(c) Particular affirmative
"Some *A* is *B*"

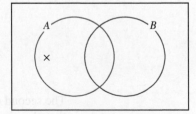

(d) Particular negative
"Some *A* is not *B*"

Figure 3-1

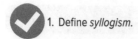

1. Define *syllogism*.

Many of the arguments we studied in Section 3-4 consisted of two premises and a conclusion. This type of argument is called a **syllogism**. We will use Euler circles to test the validity of syllogisms involving the statement types in Table 3-6. Here's a simple example:

Premise	All politicians stretch the truth.
Premise	Some politicians take bribes.
Conclusion	Some people that stretch the truth take bribes.

Remember that we are not concerned with whether the conclusion is true or false, but only whether the conclusion logically follows from the premises. If it does, the argument is valid. If not, the argument is invalid.

Euler Circle Method for Testing the Validity of an Argument

To decide if an argument is valid, diagram both premises in the same figure. If the conclusion is shown in the figure, the argument is valid. But if the premises can be diagrammed so that a different conclusion can be shown, the argument is invalid.

In many cases, the premises can be diagrammed more than one way. Our job will be to find every way the premises can possibly be diagrammed, and see if they match the conclusion. If even one diagram contradicts the conclusion, it's possible to get a different conclusion from the premises, and we've proved that the argument is invalid.

The examples in this section illustrate how we can decide if an argument is valid using Euler circles.

Example 1   Using Euler Circles to Decide If an Argument Is Valid

Use Euler circles to decide if the argument is valid.

Premise	All politicians stretch the truth.
Premise	Some politicians take bribes.
Conclusion	Some people that stretch the truth take bribes.

SOLUTION

The first premise, "All politicians stretch the truth," is the universal affirmative; it says that the set of politicians is a subset of the set of people who stretch the truth. This is diagrammed like this:

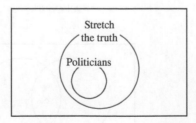

The second premise, "Some politicians take bribes," is the particular affirmative; it tells us that there's at least one person in the set of politicians that's also in the set of people who take bribes. So we need to put an ✕ in the intersection of the "politicians" circle and the "take bribes" circle; this can be done by either putting the entire "take bribes" circle inside the "stretch the truth" circle, or by having the "take bribes" circle go outside the "stretch the truth" circle.

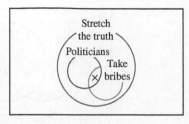 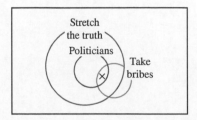

The conclusion of the argument is that some people that stretch the truth take bribes. For this to be valid, every possible diagram of the premises has to have at least one object in the set of people that stretch the truth that's also in the set of people that take bribes. The × in each diagram shows that this is the case, so the argument is valid.

Try This One   1

Use Euler circles to decide if the argument is valid.

All college students buy textbooks.

Some book dealers buy textbooks.

Therefore, some college students are book dealers.

It isn't necessary to use actual subjects such as crooked politicians and truth-stretchers in syllogisms. Arguments can use letters to represent the various sets, as shown in Example 2.

Example 2   **Using Euler Circles to Decide If an Argument Is Valid**

Use Euler circles to decide if the argument is valid.

Some *A* is not *B*.

All *C* is *B*.

∴ Some *A* is *C*.

SOLUTION

The first premise, "Some *A* is not *B*," is diagrammed by drawing circles for *A* and *B* with at least one element in *A* that is not in *B*.

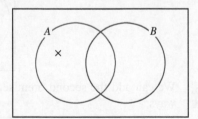

The second premise, "All *C* is *B*," is diagrammed by placing circle *C* inside circle *B*. This can be done in three different ways: the circle for *C* has to live entirely inside the circle for *B*. It can either intersect partially with the circle for *A*, live completely inside it, or miss it entirely.

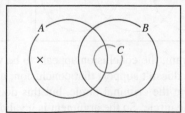 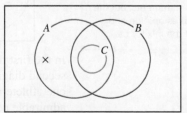

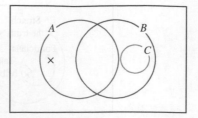

The third diagram shows that the argument is invalid. It matches both premises, but there are no members of *A* that are also in *C,* so it contradicts the conclusion "Some *A* is *C.*"

Try This One   2

Use Euler circles to decide if the argument is valid.
 Some *A* is *B*.
 Some *A* is not *C*.
 ∴ Some *B* is not *C*.

Let's try one more specific example.

Example 3   Using Euler Circles to Decide If an Argument Is Valid

Use Euler circles to decide if the argument is valid.

 No criminal is admirable.
 Some athletes are not criminals.

 ∴ Some admirable people are athletes.

SOLUTION

Diagram the first premise, "No criminal is admirable."

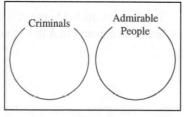

We can add the second premise, "Some athletes are not criminals," in at least two different ways:

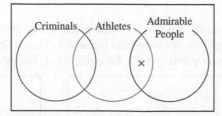   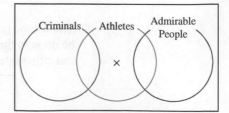

Math Note

One of the nice things about using Euler circles to evaluate arguments is that we don't have to translate the statements to symbols: we can use the verbal statements, but the diagram takes opinion and bias out of the mix for us.

In the first diagram, the conclusion appears to be valid: some athletes are admirable. But the second diagram doesn't support that conclusion; all we know for sure is that there is at least one athlete not in the criminal circle, but this doesn't guarantee that they're also inside the admirable people circle. So the argument is invalid.

Try This One   3

Use Euler circles to decide if the argument is valid.

All cruise directors have degrees in hospitality management.
No one with a degree in hospitality management services cabins.

∴ No cruise directors service cabins.

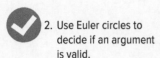

2. Use Euler circles to decide if an argument is valid.

Remember that Euler circles can be useful for certain types of arguments, but they're definitely not a substitute for truth tables. Euler circles are a convenient way to evaluate syllogisms made up of special types of statements involving quantifiers; truth tables work for arguments in general. In Exercises 47 and 48, we'll see how truth tables can be used in place of Euler circles.

Answers to Try This One

1 Invalid
2 Invalid

3 Valid

Exercise Set   3-5

Writing Exercises

1. Name and give an example of each of the four types of statements that can be diagrammed with Euler circles.
2. Explain how to decide whether an argument is valid or invalid after drawing Euler circles.
3. What is a syllogism?
4. How do Euler circles differ from Venn diagrams?

Computational Exercises

For Exercises 5–14, draw an Euler circle diagram for each statement.

5. Some environmentalists ride motorcycles.
6. No hybrid cars cost less than $26,000.
7. Some people do not go to college.
8. No textbooks are perfect. (Not even this one.)
9. Some law enforcement officers are women.
10. Some fad diets do not result in weight loss.
11. Some laws in the United States are laws in Mexico.
12. Some lobbyists do not offer bribes to politicians.
13. No cheeseburgers are low in fat.
14. Some politicians are crooks.

For Exercises 15–24, use Euler circles to decide if each argument is valid or invalid.

15. All X is Y.
    Some Y is Z.
    ∴ Some X is Z.

16. Some A is not B.
    No B is C.
    ∴ Some A is not C.

17. Some P is Q.
    No Q is R.
    ∴ Some P is not R.

18. All S is T.
    No S is R.
    ∴ Some T is R.

19. No M is N.
    No N is O.
    ∴ Some M is not O.

20. Some U is V.
    Some U is not W.
    ∴ No W is U.

21. Some A is not B.
     No A is C.
     ∴ Some A is not C.

22. All P is Q.
     All Q is R.
     ∴ All P is R.

23. No S is T.
     No T is R.
     ∴ No S is R.

24. Some M is N.
     Some N is O.
     ∴ Some M is O.

Applications in Our World

For Exercises 25–42, use Euler circles to decide if the argument is valid.

25. All phones are communication devices.
     Some communication devices are inexpensive.
     ∴ Some phones are inexpensive.

26. Some students are overachievers.
     No overachiever is lazy.
     ∴ Some students are not lazy.

27. Some dietitians are overweight.
     No personal trainers are overweight.
     ∴ No personal trainers are dietitians.

28. Some protesters are angry.
     Some protesters are not civil.
     ∴ Some civil people are not angry.

29. Some teachers are underpaid.
     Nobody that's underpaid reports complete satisfaction with their job.
     ∴ Some teachers are not completely satisfied with their jobs.

30. Some Democrats vote for Independents.
     Some people that vote for Independents are not politically active.
     ∴ Some Democrats are not politically active.

31. Some biologists are park rangers.
     No biologist is a rodeo clown.
     ∴ No park rangers are rodeo clowns.

32. Some CEOs are women.
     Some women are tech-savvy.
     ∴ Some CEOs are not tech-savvy.

33. Some juices have antioxidants.
     Some fruits have antioxidants.
     ∴ No juices are fruits.

34. Some theater majors perform in musicals.
     No art appreciation majors perform in musicals.
     ∴ No theater majors are art appreciation majors.

35. Some nurses have RN degrees.
     All nurses went to college.
     All people that went to college have high school diplomas.
     ∴ Some people with high school diplomas went to college.

36. All students text message during class.
     Some students in class take notes.
     All students who take notes in class pass the test.
     ∴ Some students pass the test.

37. Some birds can talk.
     Some animals that can talk can also moo.
     All cows can moo.
     ∴ Some cows can talk.

38. All cars use gasoline.
     All things that use gasoline emit carbon dioxide.
     Some cars have four doors.
     ∴ Some things with four doors emit carbon dioxide.

39. Every police officer carries a gun, and some people that carry guns cannot be trusted. So it's clear to me that some police officers can't be trusted.

40. I've seen bosses that don't treat their employees well. Everyone I know has a boss, so some of them don't get treated well.

41. All of the teams in the NCAA tournament won more games than they lost this year, and some of them won at least two-thirds of their games. So all of the teams that won more than two-thirds of their games are playing in the NCAA tournament.

42. Some of the people that got flu shots this year got the flu anyhow. Nobody in my family got a flu shot this year, so I'm sure that none of us will get the flu.

Critical Thinking

For Exercises 43–46, write a conclusion so that the argument is valid. Use Euler circles.

43. All *A* is *B*.
    All *B* is *C*.
    ∴

44. No *M* is *P*.
    All *S* is *M*.
    ∴

45. All calculators can add.
    No adding machines can make breakfast.
    ∴

46. Some people are prejudiced.
    All people have brains.
    ∴

47. Here's a guide for translating the types of statements studied with Euler circles into statements using connectives so that we can use truth tables to evaluate the same arguments. For two sets *A* and *B*, let *p* = "the object belongs to set *A*," and *q* = "the object belongs to set *B*." Then "All *A* is *B*" can be translated as $p \rightarrow \sim q$; "No *A* is *B*" can be translated as $p \rightarrow \sim q$; "Some *A* is *B*" translates to $p \wedge q$; and "Some *A* is not *B*" translates to $p \wedge \sim q$. Use these to translate the arguments in Examples 1 and 3 in this section into symbols, like we did in Section 3-4.

48. Use your answers from Problem 47 to decide if each argument is valid. Did your results match the results of Examples 1 and 3? If not, try again!

CHAPTER 3 Summary

Section	Important Terms	Important Ideas
3-1	Statement Simple statement Compound statement Connective Conjunction Disjunction Conditional Biconditional Negation	**Formal symbolic logic** is based on statements. A statement is a sentence that is either true or false but not both. A simple statement contains only one idea. A compound statement is formed by joining two or more simple statements with connectives. The four basic connectives are the conjunction (which uses the word *and* and the symbol ∧), the disjunction (which uses the word *or* and the symbol ∨), the conditional (which uses the words *if . . . then* and the symbol →), and the biconditional (which uses the words *if and only if* and the symbol ↔). The symbol for negation is ~. Statements are usually written using logical symbols and letters of the alphabet to represent simple statements. This removes the actual topic from our study of the argument and makes it less likely that we will be influenced by emotion or personal opinions.
3-2	Truth table Antecedent Consequent	**Truth tables** can be used to decide if compound statements are true or false based on the truth value of the simple statements that make them up. This is largely based on knowing how to evaluate the truth values associated with connectives. A conjunction $p \wedge q$ is true only when both p and q are true; a disjunction $p \vee q$ is true as long as at least one of p or q is true; a conditional $p \rightarrow q$ is true unless p is true and q is false; a biconditional $p \leftrightarrow q$ is true when p and q have the same truth value.
3-3	Tautology Self-contradiction Logically equivalent statements Converse Inverse Contrapositive De Morgan's laws	**A statement** that is always true is called a tautology. A statement that is always false is called a self-contradiction. Two statements that have the same truth values are said to be logically equivalent. De Morgan's laws are used to find the negation of a conjunction or disjunction. From the conditional statement, three other statements can be made: the converse, the inverse, and the contrapositive. Of these three, only the contrapositive is logically equivalent to the original conditional statement.
3-4	Argument Fallacy Premise Conclusion	**A logical argument** consists of two parts: a set of premises, and a conclusion based on those premises. Premises are statements that are offered as supporting evidence for the conclusion. An argument is called valid when the conclusion follows necessarily from the premises regardless of the truth value of the individual statements. If not, it is called invalid. Truth tables can be used to decide if a given argument is valid. We write a conditional statement where the antecedent is the conjunction of all premises and the consequent is the conclusion of the argument. If that conditional statement is a tautology, then the argument is valid. Otherwise, it is invalid.
3-5	Syllogism Euler circles Universal affirmative Universal negative Particular affirmative Particular negative	**Euler circles** are diagrams similar to Venn diagrams that can be used to decide if certain types of arguments are valid. This type of argument involves statements of four forms: all *A* is *B*, no *A* is *B*, some *A* is *B*, and some *A* is not *B*.

Math in Mind Control REVISITED

Colin Anderson/Blend Images

All of the arguments listed are invalid except the last one. If you were to write that argument using connectives, it would be "Woman of color elected vice president → world ends; woman of color elected vice president; therefore, world ends." Of course this conclusion is false (at least as of this writing, but I guess if it happened after that, you wouldn't be reading this), but this is because the first premise is false, not because the argument isn't valid.

The first argument is basically "Fire → smoke, so smoke → fire." This is the fallacy of the converse. So is the second: money → happiness, so happiness → money. The third argument looks like this: being in SEC → good enough for bowl, so not being in SEC → not good enough for bowl. This is the fallacy of the inverse. The fourth argument is classical circular reasoning (see the Sidelight on page 140). The fifth is another fallacy of the inverse: win popular vote → win election, so not winning popular vote → not win election.

Review Exercises

Section 3-1

For Exercises 1–5, decide whether the sentence is a statement.

1. Let's go with the flow.
2. Medical assistants report a high level of job satisfaction.
3. That politician's ideas are crazy.
4. Ignorance is always a choice.
5. Are we there yet?

In Exercises 6–12, decide if the statement is simple or compound. If it's compound, classify it as a conjunction, disjunction, conditional, or biconditional.

6. I'm majoring in communications and public relations.
7. The fine arts are widely supported by public grants.
8. If we hadn't found the obstruction in time, the patient probably would have died.
9. Defendants are convicted if and only if the jury agrees on their guilt unanimously.
10. Anthropology is the study of an obscure animal known as the anthrop.
11. I'm going to pass this final or change my major.
12. We should call a tree surgeon if our prized buckeye tree loses any more leaves.

For Exercises 13–18, write the negation of the statement.

13. The cell phone is out of juice.
14. No people who live in glass houses throw stones.
15. Some failing students can learn new study methods.
16. There is a printer that has no ink.
17. All SUVs are gas guzzlers.
18. The tires on that hybrid are black.

For Exercises 19–28, let p = "It is ambitious" and let q = "It is worthwhile." Write each statement in symbols.

19. It is ambitious and worthwhile.
20. If it is worthwhile, then it is ambitious.
21. It is worthwhile if and only if it is ambitious.
22. It is worthwhile and not ambitious.
23. If it is not ambitious, then it is not worthwhile.
24. It is not true that it is worthwhile and ambitious.
25. It is not true that if it is ambitious, then it is worthwhile.
26. It is not worthwhile if and only if it is not ambitious.
27. It is not true that it is not worthwhile.
28. It is neither ambitious nor worthwhile.

For Exercises 29–33, let p = "It is cool." Let q = "It is cloudy." Write each statement in words.

29. $p \vee \sim q$
30. $q \rightarrow p$
31. $p \leftrightarrow q$
32. $(p \vee q) \rightarrow p$
33. $\sim(\sim p \vee q)$

Section 3-2

34. True or false: a conditional statement is true unless the consequent is false.
35. True or false: a disjunction is true only if one of the two simple statements involved is true.

For Exercises 36–43, construct a truth table for each statement.

36. $p \leftrightarrow \sim q$
37. $\sim p \rightarrow (\sim q \vee p)$
38. $(p \rightarrow q) \wedge \sim q$
39. $\sim p \vee (\sim q \rightarrow p)$
40. $\sim q \leftrightarrow (p \rightarrow q)$
41. $(p \rightarrow \sim q) \vee r$
42. $(p \vee \sim q) \wedge r$
43. $r \rightarrow (\sim p \vee q)$

For Exercises 44–47, use the truth value of each simple statement to determine the truth value of the compound statement. You may need to use the Internet as a resource.

44. *p:* the population of the United States surpassed 400 million in 2021.
    q: the United States was the third most populous country in the world in 2021.
    Statement: $p \rightarrow q$

45. *p:* arctic glaciers continued to melt at an increasing rate in 2021.
    q: climate change was a top priority of the U.S. government in 2021.
    Statement: $p \wedge q$

46. *p*: Joe Biden was a presidential candidate in 2020.
    q: Snoop Dogg was a presidential candidate in 2020.
    r: Joe Biden won the Democratic nomination in 2020.
    Statement: $(p \vee q) \rightarrow r$
47. *p*: attending college costs thousands of dollars.
    q: lack of education does not lead to lower salaries.
    r: the average college graduate will make back more than
        they paid for school.
    Statement: $(p \wedge \sim q) \leftrightarrow r$

Section 3-3

48. Write an example of a conditional statement $p \rightarrow q$ that has
    a different truth value than its converse. Then explain why
    you can't do that for the contrapositive.

For Exercises 49–53, decide if the statement is a tautology, self-contradiction, or neither.

49. $p \rightarrow (p \vee q)$
50. $(p \rightarrow q) \rightarrow (p \vee q)$
51. $(p \wedge \sim q) \leftrightarrow q \wedge \sim p$
52. $q \rightarrow p \vee \sim p$
53. $(\sim q \vee p) \wedge q$

For Exercises 54–56, decide if the two statements are logically equivalent.

54. $\sim(p \rightarrow q)$; $\sim p \wedge \sim q$
55. $\sim p \vee q$; $\sim(p \leftrightarrow q)$
56. $(\sim p \wedge q) \vee r$; $(\sim p \vee r) \wedge (q \vee r)$

For Exercises 57–60, use De Morgan's laws to write the negation of each statement.

57. Social work is lucrative or fulfilling.
58. We will increase sales or our profit margin will go down.
59. The signature is not authentic and the check is not valid.
60. It is not strenuous and I am tired.

For Exercises 61 and 62, assign a letter to each simple statement and write the compound statement in symbols.

61. I will be happy only if I get rich.
62. Having a good career is sufficient for a fulfilling life.

For Exercises 63–65, write the converse, inverse, and contrapositive of the statement.

63. If gas prices go any higher, I will start riding my bike to work.
64. If I don't pass this class, my parents will kill me.
65. The patient will get an MRI only if the X-rays are inconclusive.

Section 3-4

For Exercises 66–69, use truth tables to decide if each argument is valid or invalid.

66. $p \rightarrow \sim q$
    $\sim q \leftrightarrow \sim p$
    $\therefore p$
67. $\sim q \vee p$
    $p \wedge q$
    $\therefore \sim q \leftrightarrow p$

68. $\sim p \vee q$
    $q \vee \sim r$
    $\therefore q \rightarrow (\sim p \wedge \sim r)$
69. $\sim r \rightarrow \sim p$
    $\sim q \vee \sim r$
    $\therefore p \leftrightarrow q$

For Exercises 70–73, write the argument in symbols; then use a truth table to decide if the argument is valid.

70. I'm going to Wal-Mart and McDonald's.
    If I go to McDonald's, I will get the sweet tea.
    $\therefore$ I did not get sweet tea and go to Wal-Mart.
71. If we don't hire two more workers, the union will strike.
    If the union strikes, our profit will not increase.
    We hired two more workers.
    $\therefore$ Our profit will not increase.
72. Whenever I show my clients a new workout routine, they
    end up getting in better shape. T.J. is chunkier than ever, so
    obviously I didn't show him a new routine.
73. I won't go check on my bacteria experiment today if I'm
    not finished with my paper. I finished my paper, so either
    I didn't check on my bacteria experiment or I went
    bowling.

For Exercises 74 and 75, use the commonly used forms of arguments from Section 3-4 to decide if the argument is valid.

74. If it is early, I will get tickets to the comedy club.
    It is not early.
    $\therefore$ I will not get tickets to the comedy club.
75. If pigs fly, then I'm a monkey's uncle.
    Pigs fly or birds don't sing.
    Birds do sing.
    $\therefore$ I'm a monkey's uncle.

Section 3-5

For Exercises 76–80, use Euler circles to determine whether the argument is valid or invalid.

76. No *A* is *B*.
    Some *B* is *C*.
    $\therefore$ No *A* is *C*
77. Some *A* is not *C*.
    Some *B* is not *C*.
    $\therefore$ Some *A* is not *B*.
78. All money is green.
    All grass is green.
    $\therefore$ Grass is money.
79. No psychologists write prescriptions.
    Some psychiatrists write prescriptions.
    $\therefore$ No psychiatrists are psychologists.
80. Some policemen are college grads, and all college grads
    expect to make at least $35,000 per year, so some policemen
    expect to make $35,000 or greater.

Chapter Test

1. True or false: an argument can only be valid if its conclusion is true.
2. True or false: the last column in a truth table used to decide if an argument is valid represents a conditional statement.
3. Explain why we use letters to represent statements when studying arguments in logic.
4. Decide if each sentence is a statement.
   (a) My degree is in microbiology.
   (b) The capital of California is Paris.
   (c) Hang in there, buddy.
   (d) A career in law enforcement is WAY better than a career in teaching.

For Exercises 5–8, write the negation of the statement.

5. The image is uploading to my online bio.
6. All men have goatees.
7. Some students ride a bike to school.
8. No nursing majors have trouble finding jobs.

For Exercises 9–14, let p = "It is warm." Let q = "It is sunny." Write each statement in symbols.

9. It is warm and sunny.
10. If it is sunny, then it is warm.
11. It is warm if and only if it is sunny.
12. It is warm or sunny.
13. It is false that it is not warm and sunny.
14. It is not sunny, and it is not warm.

In Exercises 15–18, let p = "Congress is in session" and q = "My representative is in Aruba." Write each statement in words.

15. $p \wedge \sim q$
16. $q \rightarrow p$
17. $(p \vee q) \rightarrow p$
18. $\sim(\sim p \vee q)$

For Exercises 19–22, construct a truth table for each statement.

19. $p \rightarrow \sim q$
20. $(p \rightarrow \sim q) \wedge r$
21. $(\sim q \vee p) \wedge p$
22. $p \rightarrow (\sim q \vee r)$

For Exercises 23–25, decide if each statement is a tautology, self-contradiction, or neither.

23. $(p \wedge q) \wedge \sim p$
24. $(p \vee q) \rightarrow (p \rightarrow q)$
25. $\sim(p \wedge q) \vee p$
26. Are the two statements logically equivalent? $(p \vee q) \wedge r; (p \wedge r) \vee (q \wedge r)$
27. Write the converse, inverse, and contrapositive for the statement "If I exercise regularly, then I will be healthy." Which are equivalent to the original statement?
28. Use De Morgan's laws to write the negation of the compound statement.
   (a) It is not cold and it is snowing.
   (b) I am hungry or thirsty.

For Exercises 29 and 30, use truth tables to determine the validity of each argument.

29. $\sim q \vee p$
    $\underline{\quad p \vee q \quad}$
    $\therefore \sim q \rightarrow p$

30. $p \rightarrow q$
    $\underline{\sim q \vee \sim r}$
    $\therefore q \leftrightarrow (\sim p \wedge \sim r)$

For Exercises 31 and 32, decide if the argument is valid or invalid by using the given forms of valid arguments and fallacies.

31. If I finish my paper early, I will have my professor proofread it.
    $\underline{\text{I have my professor proofread my paper.}}$
    $\therefore$ I finish my paper early.

32. If Starbucks isn't too busy, I'll study there, but if Starbucks is too busy, I'll study at the library. If I don't end up studying at Starbucks, you can find me at the library.

For Exercises 33 and 34, use Euler circles to determine whether the argument is valid or invalid.

33. No B is A.
    $\underline{\text{Some } A \text{ is } C.}$
    $\therefore$ No B is C.

34. Some of the arts are inaccessible to the general public.
    $\underline{\text{All pursuits that are inaccessible need public funding.}}$
    $\therefore$ Some of the arts need public funding.

Projects

1. Politicians argue in favor of positions all the time. An informed voter doesn't vote for a candidate because of the candidate's party, gender, race, or how good they look on TV—an informed voter listens to the candidates' positions and evaluates them.

   Do a Google search for the text of a speech by each of the main candidates in the most recent presidential election. Then find at least three logical arguments within the text, write the arguments in symbols, and use truth tables or commonly used argument forms to analyze the arguments, and see if they are valid.

2. Electric circuits are designed using truth tables. A circuit consists of switches. Two switches wired in *series* can be represented as $p \wedge q$. Two switches wired in *parallel* can be represented as $p \vee q$.

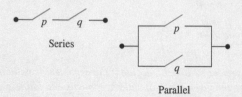

   Series

   Parallel

   In a series, circuit electricity will flow only when both switches p and q are closed. In a parallel circuit, electricity will flow when one or the other or both switches are closed. In a truth table, T represents a closed switch and F represents an open switch. So the truth table for $p \wedge q$ shows electricity flowing only when both switches are closed.

| **Truth table** | | | | **Circuit** | | |
p	q	$p \wedge q$		p	q	$p \wedge q$
T	T	T		closed	closed	current
T	F	F		closed	open	no current
F	T	F		open	closed	no current
F	F	F		open	open	no current

   Also, when switch p is closed, switch $\sim p$ will be open and vice versa, and p and $\sim p$ are different switches. Using this knowledge, design a circuit for a hall light that has switches at both ends of the hall so that the light be turned on or off from either switch.

3. One of the biggest benefits to studying formal logic is generally improving your ability to think logically, even if you're not using the specific methods we studied in this chapter. The ability to think logically is an important part of almost all standardized tests. In particular, the Law School Admissions Test (LSAT) is well known for featuring logic puzzles like the following. How logically can you think now that you've completed this chapter?

   A therapist is scheduling sessions for an upcoming day with plans to see seven patients, plus take an hour to grab a salad and run some errands. All of the patients are being treated for either depression or posttraumatic stress. All of the patients will be scheduled for a separate session. In addition:

   - The therapist likes to separate out conditions, so none of the patients suffering from posttraumatic stress will be scheduled consecutively.
   - Jyoti will be scheduled before both Henry and Paul.
   - Juan, who suffers from posttraumatic stress, absolutely needs the third appointment of the day.
   - Henry is also being treated for posttraumatic stress.
   - Jyoti is not being treated for posttraumatic stress only if Paul is not being treated for depression.
   - Paul can't make the seventh and final appointment.

   (a) Which of the following could be a complete list of the appointment order and respective conditions from earliest to latest?
      (i) Jyoti (PTSD), Mariska (Depression), Juan (PTSD), Paul (Depression), Henry (PTSD), Chanel (Depression), and Jake (Depression)
      (ii) Jyoti (PTSD), Paul (Depression), Henry (PTSD), Jake (Depression), Juan (PTSD), Chanel (Depression), and Mariska (PTSD)
      (iii) Henry (PTSD), Jyoti (Depression), Juan (PTSD), Mariska (Depression), Jake (Depression), Paul (PTSD), and Chanel (Depression)
      (iv) Chanel (Depression), Jyoti (Depression), Juan (PTSD), Mariska (Depression), Henry (PTSD), Jake (Depression), and Paul (PTSD)
      (v) Mariska (PTSD), Jyoti (Depression), Juan (PTSD), Jake (Depression), Paul (Depression), Henry (PTSD), Chanel (Depression)

   (b) What's the smallest possible number of patients with appointments before Henry?

   (c) If Jyoti has the second appointment, which of the following has to be true?
      (i) Jake's and Juan's appointments are not consecutive.
      (ii) Mariska is scheduled before Paul.
      (iii) Henry has the last appointment of the day.
      (iv) The first patient of the day suffers from PTSD.
      (v) Exactly four of the patients suffer from PTSD.

   (d) Each one of the following statements must be false EXCEPT:
      (i) Exactly five of the patients suffer from depression.
      (ii) Exactly three of the patients with appointments after Juan have PTSD.
      (iii) Jyoti arrives second, and Henry arrives fifth.
      (iv) Jyoti suffers from depression, and Jake from PTSD.
      (v) There are three appointments between Juan and Paul, regardless of which has the earlier appointment.

(e) If Henry doesn't have the last appointment, then which one of the following CANNOT be true?
  (i) Chanel arrives before Mariska.
  (ii) Exactly four of the patients have PTSD.
  (iii) Paul arrives before Juan.
  (iv) Jake is being treated for depression.
  (v) Jyoti is being treated for depression.

(f) If Chanel is scheduled after Jake but before Mariska, then which one of the following statements, if true, would provide enough information to determine the patient's exact order of arrival and condition each is being treated for?

  (i) Chanel is scheduled fourth, and exactly three of the patients have PTSD.
  (ii) Chanel is scheduled sixth, and exactly four of the patients have PTSD.
  (iii) Jake is scheduled second, and exactly four of the patients have depression.
  (iv) Jyoti is scheduled fourth, and no more than three of the patients have depression.
  (v) Jyoti is scheduled first, and exactly three of the patients have depression.

Design elements: Front matter, Chapter Opener, Summary and End Matter header design (random numbers background illustration): ©pixeldreams.eu/Shutterstock RF

Photo: mirtmirt/Shutterstock

Outline

Math in | Computer Graphics

If you're like most college students, having endured 10 to 12 years of math class as you grew up, you probably feel like you know an awful lot about numbers. That may well be true if you're talking about the base 10 number system that you started learning even before you went to school. But you may be surprised to learn that there are many other number systems that can be used. Some of them are historical artifacts that give us a glimpse into the developmental stages of humankind's study of mathematics. But others are used commonly today. In fact, you're using some of them every single time you use a computer, even if you don't realize it.

At first, it seems odd to many students to study different types of numbers, especially ancient number systems that are no longer in common use. But just as studying other languages gives you a deeper appreciation for the nuances of language itself, the study of other number systems provides deeper insight into the one we commonly use. The history of human thought is intimately tied to the development of language and mathematics, so studying early number systems gives us an interesting look at our species' intellectual past.

The first systems we will study use symbols to represent numbers that are completely different from the ones you're accustomed to. But we'll also study different systems that use familiar numerals like 0 and 1, and even letters from A through F. The binary system, which uses only 0 and 1, is in a very real sense responsible for the operation of every computer system. At its core, a computer is a series of millions and millions of switches that are either on or off. If 1 represents on and 0 represents off, you've got a numeration system with those two symbols—the binary system. But computer architecture goes deeper than that: the switches are organized into groups of 8 and 16, and programmers use numeration systems with 8 and 16 numerals as well.

In particular, all of the colors displayed by computer screens are created by a code in the hexadecimal system, which you'll learn about here in Chapter 4. The colors are a combination of red, green, and blue, with the amount of each determining the result. So the code that defines a color consists of six characters: the first two are a numeral in the hexadecimal system that determines the amount of red, the next two the amount of green, and the last two the amount of blue. When you've completed this chapter, you should be able to answer the questions below. For now, we'll just say that 00 represents none of that color (0%), and FF represents the maximum amount of that color (100%). Based on that you can make some educated guesses, and you can check your answers with a Google search for "color code ######," where the last six characters are the six-digit hexadecimal code.

1. What color do you think is displayed by the code 000000? What about FFFFFF? Why?
2. What color should be displayed by each of the following codes? Why?
   a. FF0000   b. 00FF00   c. 0000FF
3. Convert the hexadecimal number FF into base 10. What does that tell you about the number of different amounts of each of red, green, and blue that can be coded using 00 to FF in hexadecimal?
4. Convert the hexadecimal numbers A1, 23, and FF into base 10 numbers. Based on your answer to Question 3, what percentage of each of red, green, and blue does the color code A123FF represent? Can you predict what common color that shade is likely to be similar to?
5. Use what you learned in Question 4 to write a color code that you think might correspond to a shade of orange, then look up your code in Google and write a description of how you did.

For answers, see Math in Computer Graphics Revisited on page 207

Section 4-1 | **Early and Modern Numeration Systems**

✓ **LEARNING OBJECTIVES**

1. Use a tally system.
2. Define and use simple grouping systems.
3. Define and use multiplicative grouping systems.
4. Define a positional system and identify place values.
5. Use numeration systems that combine aspects of other types of systems.

Are the words "number" and "numeral" synonymous? Most people answer yes to that question, but the answer is actually no. A **number** is a *concept*, or an idea, used to represent some quantity. A **numeral**, on the other hand, is a *symbol* used to represent a number. For example, there's only one concept of the number "five," but there are many different numerals (symbols) that can be used to represent the number five; 5, V, cinco, and ||||| are just a few.

Just as there are a wide variety of languages that developed in various parts of the world, there are a wide variety of systems for representing numbers that developed at different times and in different places. In this section, we'll study a few of them. In doing so, we're looking at a microcosm of the history of human intellectual development.

The Studio Dog/Photodisc/ Getty Images

> A **numeration system** consists of a set of symbols (numerals) to represent numbers, and a set of rules for combining those symbols.

The numeration system that you're familiar with is called the *Hindu-Arabic numeration system*, also known as the *base 10 numeration system*. It originated in India and was brought to Europe during the 15th century by Arab traders.

There are four main types of numeration systems we'll study: tally, simple grouping, multiplicative grouping, and positional.

Tally Systems

A **tally system** is the simplest kind of numeration system, and certainly the oldest. In fact, it's possible that tally systems are the oldest form of writing period. In a tally system there is only one symbol needed and a number is represented by repeating that symbol. For example, an ancient cave dweller might have drawn three stick-figure children on the wall of his cave to indicate that he had three children living there.

In modern times, tally systems are still used as a crude method of counting items. Most often, they are used to keep track of the number of occurrences of some event. For example, umpires in baseball often keep track of the runs scored by each team by making a mark for each run as it scores.

The most common symbol used in tally systems is |, which we call a stroke. Tallies are usually grouped by fives, with the fifth stroke crossing the first four, as in ||||.

Math Note

The oldest known examples of tally systems date to almost 45,000 years ago. They're notches carved into animal bones, and quite literally represent the very beginning of formal math.

Example 1 | Using a Tally System

An emergency room nurse is interested in keeping track of the number of patients she treats during her shift. There were six patients before her first break, eight more between break and lunch, and only four during a slow afternoon. If she used a tally system to keep track of the patients, write out what that may have looked like, and use it to total the number of patients.

SOLUTION

If we go with the tally system described before this example, we'd have this:

|||| |||| |||| |||

The nice thing about the groups of five is that it makes it easy to find the total: 3 fives with 3 ones, which is 18 patients.

Try This One

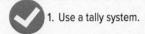

1. Use a tally system.

The pharmacist at the hospital filled four prescriptions before 10 A.M., six more between 10 A.M. and lunch, and then 12 in the afternoon. Represent this with a tally system and find the total number of prescriptions filled.

Tally systems are easy to work with, but have a pretty obvious disadvantage: representing large numbers is cumbersome (to put it politely). This probably wasn't a big deal to prehistoric people, but as society developed, larger numbers started to be useful, and grouping systems developed.

Simple Grouping Systems

In a **simple grouping system** there are symbols that represent select numbers. Often, these numbers are powers of 10. To write a number in a simple grouping system, repeat the symbol representing the appropriate value(s) until the desired quantity is reached. For example, suppose in a simple grouping system the symbol Δ represents the quantity "ten" and the symbol Γ represents the quantity "one." Then to write the numeral representing the quantity "fifty-three" in this system we would use five Δ's and three Γ's as follows: ΔΔΔΔΔΓΓΓ.

The Egyptian Numeration System

One of the earliest formal numeration systems was developed by the Egyptians over 5,000 years ago. It used a system of pictures (known as hieroglyphs) to represent certain numbers. These symbols are shown in Figure 4-1.

Symbol	Number	Description	Symbol	Number	Description
|	1	Vertical staff	⌒	10,000	Pointing finger
∩	10	Heel bone	⋊	100,000	Burbot fish (or tadpole)
◉	100	Scroll	⚥	1,000,000	Astonished person
⨏	1,000	Lotus flower			

Figure 4-1

Sidelight   Math in the Ancient World

It's pretty reasonable to hypothesize that the earliest uses of math were based on counting objects, and evidence indicates that the idea of "number" evolved gradually. In fact, many early languages had words for "one," "two," and "many," but not for any number greater than two. But the world moved on and numeration systems began to develop.

The oldest known object that is thought to be evidence of mathematical thinking is the Lebombo bone, discovered in the Lebombo Mountains of Swaziland in 1973. It's a 43,000-year-old fibula from a baboon with 29 distinct notches carved into it. This leads us to two possible conclusions: an early human had nothing better to do other than carve notches into a bone, or the bone was used to tally something. In fact, archaeologists believe it was used to tally the days of a cycle, most likely either lunar or menstrual.

Another bone found near the headwaters of the Nile River in what is now the Congo, called the Ishango bone, shows a more sophisticated tally approach, with notches carved in distinct columns. This bone is about 15,000 years newer than the Lebombo bone, so as you might expect mathematical progress was relatively slow for prehistoric people. (Before you criticize, let's see you try solving math problems while in constant danger of being eaten by a tiger.)

Of course, we're just making educated guesses as to the true uses of these artifacts; the oldest undisputed examples of mathematics are systems we'll study in this section, the Babylonian and Egyptian systems. Each of these may be more than 5,000 years old. But in any case it's interesting to note that language is thought to have existed for something like 45,000 years before any real formal development of mathematics. Compared to that long period of relative inactivity, math has developed at breakneck speed since Egyptian times, resulting in technological wonders that would have been considered witchcraft just a couple centuries ago.

The Egyptian system is a simple grouping system: the value of any numeral is determined by counting up the number of each symbol and multiplying the number of occurrences by the corresponding value from Figure 4-1. Then the numbers for each symbol are added, as we see in Example 2.

Example 2   Using the Egyptian Numeration System

Find the numerical value of each Egyptian numeral.

(a) ⳥⳥⳥ᐱᐱᐱ⳥⳥⳥⳥⳥⳥⳥⳥⳥⳥⳥⳥⳥⳥⳥

(b) ⳥⳥⳥⳥⳥⳥

(c) Compare the number of symbols needed for each number in the Egyptian system and in our system. What can you conclude?

SOLUTION

(a) This number is made up of 3 fish (which means 3 hundred thousands), 3 pointing fingers (3 ten thousands), 2 scrolls (2 hundreds), 3 heel bones (3 tens), and 6 vertical staffs (6 ones), which makes the number $3 \times 100{,}000 + 3 \times 10{,}000 + 2 \times 100 + 3 \times 10 + 6 \times 1 = 300{,}000 + 30{,}000 + 200 + 30 + 6 = 330{,}236$.

(b) The number consists of 1 million, 2 thousands, 1 hundred, 1 ten, and 1 one; the number is $1{,}000{,}000 + 2{,}000 + 100 + 10 + 1 = 1{,}002{,}111$.

(c) The first number requires 17 symbols in the Egyptian system and just 6 in ours (if you don't count the comma). The second needs 6 in the Egyptian system but 7 in ours. Conclusion? In many cases the Egyptian system requires more symbols to be written, but not always.

Try This One   2

Find the numerical value of each Egyptian numeral.

(a) ⳥⳥⳥⳥⳥⳥

(b) ⳥⳥⳥⳥⳥⳥⳥⳥⳥⳥⳥⳥⳥

In order to write numbers using the Egyptian system, simple groupings of ones, tens, hundreds, etc. are used. For example, 28 is equal to $10 + 10 + 8$, or 2 tens and 8 ones, and is written as

Example 3   Writing Numbers in the Egyptian System

Write each number as an Egyptian numeral.

(a) 237   (b) 3,202,419

SOLUTION

(a) We can write 237 as $2 \times 100 + 3 \times 10 + 7 \times 1$. So we need 2 of the hundreds symbol (scroll), 3 of the tens symbol (heel bone), and 7 of the ones symbol (vertical staff).

⳥⳥⳥⳥⳥⳥⳥⳥⳥⳥⳥⳥

(b) Since 3,202,419 consists of 3 millions, 2 hundred thousands, 2 thousands, 4 hundreds, 1 ten, and 9 ones, it is written as

⳥⳥⳥⳥⳥⳥⳥⳥⳥⳥⳥⳥⳥⳥⳥⳥⳥⳥⳥⳥⳥⳥⳥

Math Note

The choice of symbol to represent 1,000,000 in the Egyptian system says a lot about how common numbers of that size were in the ancient world.

Try This One 3

Write each number as an Egyptian numeral.

(a) 627   (b) 511,120

Addition and subtraction in the Egyptian system can be performed by grouping symbols, as shown in Example 4.

Example 4 | Adding and Subtracting in the Egyptian System

Perform each operation in the Egyptian numeration system.

(a) 𝓢𝓢⌒⌒⌒|||||| + 𝓢⌒⌒⌒⌒⌒⌒|||||

(b) 𝓢𝓢𝓢𝓢𝓢⌒⌒ || − 𝓢𝓢⌒⌒⌒||||

SOLUTION

(a) First we'll find the total number of each symbol. But if any symbol appears more than 10 times, 10 of them can be replaced with 1 of the symbol representing the next larger number. The total number of each symbol is

𝓢𝓢𝓢 ⌒⌒⌒⌒⌒⌒⌒⌒⌒⌒⌒⌒ |||||||||||

Ten heel bones represents 10 tens, which is the same as 1 hundred. So we can replace 10 heel bones with 1 scroll, giving us four of them. We can also replace 10 vertical staffs with 1 heel bone, giving us three of them when combined with the two left.

𝓢𝓢𝓢𝓢⌒⌒⌒|

(b) In this case, we're going to have to do some rewriting before we subtract since there are more heel bones and vertical staffs in the number being subtracted. In the top number, we can convert one heel bone (10) into 10 vertical staffs, and one scroll (100) into 10 heel bones. (If this reminds you of the way we "borrow" from the next left digit when subtracting in our system, then you get it!) Once this is done, the number of symbols can be subtracted as shown below, with the answer on the bottom line. You might find it helpful to cross out matching symbols in both lines.

𝓢𝓢𝓢𝓢⌒⌒⌒⌒⌒⌒⌒⌒⌒⌒⌒⌒ ||||||||||||
− 𝓢𝓢⌒⌒⌒ |||||
_____
𝓢 ⌒⌒⌒⌒⌒⌒⌒⌒ ||||||||

Math Note

One of the objectives of this section is to expand your numerical ability by performing calculations using numerical systems other than our own. So in problems like Example 4, you should be using these systems in your solution.

But don't forget that you can check your answer by converting the original problem into our system, performing the calculation, and seeing if it matches your result.

Try This One 4

Perform each operation in the Egyptian numeration system.

(a) 𝓢𝓢⌒⌒⌒⌒⌒|||||||| + 𝓢𝓢𝓢𝓢⌒⌒⌒||||

(b) 𝘧𝓢𝓢𝓢⌒⌒|| − 𝓢𝓢𝓢𝓢𝓢⌒⌒⌒⌒⌒|||

If you're familiar with Roman numerals, the Egyptian system might remind you of them a bit. Before we study that comparison, we need to become familiar with the symbols the Romans used.

Symbol	Number
I	1
V	5
X	10
L	50
C	100
D	500
M	1,000

In fact, the Roman numeration system is like a simple grouping system, with a clever twist that reduces the number of symbols necessary. In order to keep from having to repeat any symbol more than three times, subtraction is used. For example, 8 is written as VIII (one five plus three ones). But 9 is not written as VIIII; instead, we write IX. The fact that the ones symbol is written before the tens symbol tells us to *subtract* 1 from 10, leaving 9. Here are the rules you need to write numbers in Roman numerals:

1. When a letter is repeated in sequence, its numerical value is added. For example, XXX represents $10 + 10 + 10$, or 30.

2. When smaller-value letters follow larger-value letters, the numerical values of each are added. For example, LXVI represents $50 + 10 + 5 + 1$, or 66.

3. When a smaller-value letter precedes a larger-value letter, the smaller value is subtracted from the larger value. For example, IV represents $5 - 1$, or 4, and XC represents $100 - 10$, or 90. (You might find it easier to think of this as "one taken away from five" and "ten taken away from 100.")

In addition, I can only precede V or X, X can only precede L or C, and C can only precede D or M. So 4 is written as IV, 9 is written as IX, 40 is written as XL, 90 is written XC, 400 is written as CD, and 900 is written as CM. But you wouldn't write 99 as IC because I can only precede V or X; 99 is written as XCIX.

Example 5 shows how to convert Roman numerals to Hindu-Arabic numerals.

Example 5	**Finding the Value of Roman Numerals**

Find the value of each Roman numeral.

(a) LXVIII   (b) XCIV   (c) MCML   (d) CCCXLVI   (e) DCCCLV

SOLUTION

(a) L = 50, X = 10, V = 5, and III = 3; so LXVIII = $50 + 10 + 5 + 3 = 68$.
(b) XC = 90 and IV = 4; so XCIV = 94.
(c) M = 1,000, CM = 900, L = 50; so MCML = 1,950.
(d) CCC = 300, XL = 40, V = 5, and I = 1; so CCCXLVI = 346.
(e) D = 500, CCC = 300, L = 50, V = 5; so DCCCLV = 855.

Jupiterimages/Stockbyte/Getty Images

Roman numerals are still in use today. For example, many clocks and watches contain Roman numerals. Can you think of other places Roman numerals are still used?

Try This One	**5**

Find the value of each Roman numeral.

(a) XXXIX   (b) MCLXIV   (c) CCCXXXIII

Going back to studying the number of symbols needed to represent a given number, if we were to write the five numbers from Example 5 in the Egyptian system, it would have taken 14, 13, 15, 13, and 18 symbols respectively. Hmm . . . maybe the Romans were on to something. . . .

| Example 6 | Writing Numbers Using Roman Numerals |

Math Note

For larger numbers, the Romans placed a bar over their symbols. The bar means to multiply the numerical value of the number under the bar by 1,000. For example, $\overline{\text{VII}}$ means $7 \times 1,000$ or 7,000, and $\overline{\text{XL}}$ means 40,000.

Write each number using Roman numerals.

(a) 19   (b) 238   (c) 1,999   (d) 840

SOLUTION

(a) 19 is written as $10 + 9$ or XIX.
(b) 238 is written as $200 + 30 + 8$ or CCXXXVIII.
(c) 1,999 is written as $1,000 + 900 + 90 + 9$ or MCMXCIX.
(d) 840 is written as $500 + 300 + 40$ or DCCCXL.

Try This One   6

Write each number using Roman numerals.

(a) 67   (b) 192   (c) 202   (d) 960

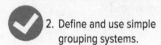

2. Define and use simple grouping systems.

Multiplicative Grouping Systems

The most cumbersome thing about simple grouping systems is that you often have to write a given symbol many times. The Romans addressed this by using the idea of subtraction: unbeknownst to them, the Chinese attacked the same issue using multiplication over a thousand years earlier. In a **multiplicative grouping system,** there's a symbol for each value 1 through 9 (we call these the multipliers), and also for select other numbers (often powers of 10). If we wanted to represent 40, which is four 10s, rather than writing the symbol for 10 four times, we'd write the multiplier for four followed by the symbol for 10. Example 7 illustrates a multiplicative grouping system that we made up; after that, we'll study the Chinese system that is thought to be the first multiplicative grouping system.

| Example 7 | Using a Multiplicative Grouping System |

The symbols used in a multiplicative grouping system are as follows:

one	α	seven	γ
two	β	eight	η
three	χ	nine	ι
four	δ	ten	φ
five	ε	one hundred	ρ
six	θ	one thousand	ω

Write the symbols that would be used to represent the numbers 68 and 1,950. How does the number of symbols needed compare to the number needed in the Roman system? (See Example 5.)

Symbol	Value
零 or ◯	0
一	1
二	2
三	3
四	4
五	5
六	6
七	7
八	8
九	9
十	10
百	100
千	1,000

Figure 4-2

SOLUTION

Sixty-eight is made up of 6 tens and 8 ones. To represent 6 tens, we write the multiplier 6 (θ) next to the base value 10 (φ), or $\theta\varphi$. To represent 8 ones, we can either write $\eta\alpha$ (the multiplier 8 times the base value 1), or we could just write η. So, the number 68 is written as either $\theta\varphi\,\eta\alpha$ or $\theta\varphi\,\eta$.

We can write out 1,950 as

$$1 \times 1,000 + 9 \times 100 + 5 \times 10$$
$$\quad\ \ \alpha\omega \qquad\quad \iota\rho \qquad\quad \varepsilon\varphi$$

In parts (a) and (c) of Example 5, we needed six and four symbols, respectively, so this doesn't appear to be reducing the number needed significantly.

Try This One   7

Using the symbols in Example 7, write the symbols that would be used to represent the numbers 96 and 2,870.

The symbols used for the Chinese numeration system are shown in Figure 4-2. Because Chinese is written vertically rather than horizontally, their numbers are also represented vertically. Fifty-three would be written:

五 five
十 tens
and
三 three ones

Example 8 — Using the Chinese Numeration System

Math Note

If you think some of the numerals in the Chinese system are hard to write, you should see the financial versions.

Since some of the basic symbols can be easily changed to symbols representing larger numbers with a couple of strokes, a system with more complicated symbols is often used for important financial records to discourage forgery.

Wikipedia describes this distinction pretty well, and displays the more complicated symbols.

Find the value of each Chinese numeral.

(a) 六
百
五
十
四

(b) 三
千
七
百
二
十
六

(c) 五
千
六
十
五

SOLUTION

Reading from the top down, we can calculate each value as below. Remember that in each group of symbols, the multiplier comes first, followed by the power of 10.

(a)
六
百 } $6 \times 100 = 600$
$+$
五
十 } $5 \times 10 = 50$
$+$
四 } 4 $= 4$
654

(b)
三
千 } $3 \times 1,000 = 3,000$
$+$
七
百 } $7 \times 100 = 700$
$+$
二
十 } $2 \times 10 = 20$
$+$
六 } 6 $= 6$
3,726

(c)
五
千 } $5 \times 1,000 = 5,000$
$+$
六
十 } $6 \times 10 = 60$
$+$
五 } 5 $= 5$
5,065

Try This One 8

Find the value of each Chinese numeral.

(a) 四
百
二
十
七

(b) 六
千
七
十
五

(c) 二
十
六

Example 9 Writing Numbers in the Chinese Numeration System

Math Note

By now, you're probably thinking that these other numeration systems are ridiculously complicated to use, but that's most likely because you're unfamiliar with the basic symbols. If the Chinese system used the familiar digits from 0 through 9 along with 10, 100, etc., you'd probably pick it up in no time.

Try not to let the unfamiliar symbols keep you from understanding the structure of the systems, because it's the structure that we're really interested in studying.

3. Define and use multiplicative grouping systems.

Write each number as a Chinese numeral.

(a) 65   (b) 183   (c) 8,749

SOLUTION

(a) 六
十 } 6×10

五 } 5

(b) 一
百 } 1×100

八
十 } 8×10

三 } 3

(c) 八
千 } $8 \times 1{,}000$

七
百 } 7×100

四
十 } 4×10

九 } 9

Try This One 9

Write each number as a Chinese numeral.

(a) 45   (b) 256   (c) 6,321

Positional Systems

The next big leap in the evolution of numeration systems was the realization that numerals written in a multiplicative grouping system have twice as many symbols as they need. In a **positional system**, instead of writing the multiplier with another number to multiply it by (like a power of 10), we just write the multiplier alone; the number to multiply it by is understood from that multiplier's position. If this sounds a lot like the number system you're familiar with, it should.

The numeration system you grew up with is a positional system that requires 10 symbols—the digits from 0 through 9—to represent any number, no matter how big or small. A fundamental understanding of how that system is designed requires a clear understanding of exponents, so a quick review seems like a good idea.

For any number b and natural number n, we define the **exponential expression** b^n as

$$b^n = b \cdot b \cdot b \cdots b$$

where b appears as a factor n times. The number b is called the **base**, and n is called the **exponent**. We also define $b^1 = b$ for any base b, and $b^0 = 1$ for any nonzero base b.

Sidelight   **Roman and Hindu-Arabic Numerals**

The Romans spread their system of numerals throughout the world as they conquered other nations. The system was well entrenched in Europe until the 1500s, when our present system, called the Hindu-Arabic system, became widely accepted.

   The present system is thought to have been invented by the Hindus before 200 BCE. It was spread throughout Europe by the Arabs, who traded with the Europeans and traveled throughout the Mediterranean region. It is interesting to note that for about 400 years, the mathematicians of early Europe were divided into two groups—those favoring the use of the Roman system and those favoring the use of the Hindu-Arabic system. The Hindu-Arabic system eventually won out, although Roman numerals are still in use today.

John Wang/Royalty Free/Getty Images

As we mentioned earlier, the numeration system we use today is called the Hindu-Arabic system. (See the Sidelight above for some perspective on this name.) It is a positional system since the position of each digit indicates a specific value. The **place value** of each number is given as

	hundred	ten		hundred	ten				
billion	million	million	million	thousand	thousand	thousand	hundred	ten	one
10^9	10^8	10^7	10^6	10^5	10^4	10^3	10^2	10^1	1

The number 82,653 means there are 8 ten thousands, 2 thousands, 6 hundreds, 5 tens, and 3 ones. We say that the place value of the 6 in this numeral is hundreds.

Example 10   Hindu-Arabic Numerals as a Multiplicative Grouping System

Suppose that we build a multiplicative grouping system similar to the one in Example 7, but instead of making up symbols to represent certain numbers, we just use the digits from 0 through 9, 10, 100, 1,000, and 10,000. Write each of the numbers below in this system.

(a) 84   (b) 4,912   (c) 28,506

SOLUTION

(a) Eighty-four is 8 tens and 4 oncs, which would look like 8 10 4 1.
(b) 4 thousands, 9 hundreds, 1 ten, and 2 ones: 4 1,000 9 100 1 10 2 1.
(c) 2 ten thousands, 8 thousands, 5 hundreds, and 6 ones: 2 10,000 8 1,000 5 100 6 1.

Try This One   10

Write 911 and 34,062 in the system described in Example 10.

Do you see the issue with using Hindu-Arabic numerals in a multiplicative grouping system? The lack of symbols between the numerals makes it awfully hard to read. Enter arithmetic! We know that 8 tens and 4 ones really means the same thing as 8 times 10 plus 4 times 1. So, it would be clearer to write our answer to part (a) of Example 10 as

$$8 \times 10 + 4 \times 1$$

This is an example of **expanded notation** for a Hindu-Arabic numeral. Here's another example:

$$32{,}569 = 30{,}000 \quad + 2{,}000 \quad + 500 \quad + 60 \quad + 9$$
$$= 3 \times 10{,}000 + 2 \times 1{,}000 + 5 \times 100 + 6 \times 10 + 9$$
$$= 3 \times 10^4 \quad + 2 \times 10^3 \quad + 5 \times 10^2 + 6 \times 10^1 + 9 \times 10^0$$

Working with expanded notation serves two purposes. First, it helps to clarify the place values in the system. Second, it helps us to see why positional systems are so brilliantly efficient: we don't have to write out all of the multiplications and additions. Instead, all that information is encoded in where each digit lives in the numeral. And that's pretty darn smart. And that was one of the main objectives of this section: to give you an appreciation for our own number system by comparing it to others.

Since all of the place values in the Hindu-Arabic system correspond to powers of 10, the system is known as a **base 10 system**. We'll study other base number systems in Section 4-3.

| **Example 11** | Writing a Base 10 Number in Expanded Notation |

Write 9,034,761 in expanded notation.

SOLUTION
9,034,761 can be written as

$$9{,}000{,}000 + 30{,}000 \quad + 4{,}000 \quad + 700 \quad + 60 \quad + 1$$
$$= 9 \times 1{,}000{,}000 + 3 \times 10{,}000 + 4 \times 1{,}000 + 7 \times 100 + 6 \times 10 + 1$$
$$= 9 \times 10^6 \quad + 3 \times 10^4 \quad + 4 \times 10^3 \quad + 7 \times 10^2 + 6 \times 10^1 + 1 \times 10^0$$

Math Note

Because you're so familiar with the Hindu-Arabic system, you might not think that practicing writing numbers in expanded form is useful. But I PROMISE you'll change your mind when we study base number systems in Section 4-3.

| Try This One | 11 |

Write each number in expanded notation.

(a) 573   (b) 86,471   (c) 2,201,567

4. Define a positional system and identify place values.

The four types of numeration systems we've studied (tally, grouping, multiplicative grouping, and positional) in some sense trace the history of numerals from simplest to most efficient. In studying that history, there are some interesting systems to consider that combine elements from two or three types of systems. Of course, the list of systems we're studying is not comprehensive—there are many other systems that we haven't discussed, and in the exercises you'll be asked to make up your own.

Other Numeration Systems

The Babylonians had a numerical system consisting of just two symbols, ❮ and ❯. (These wedge-shaped symbols are known as "cuneiform.") The ❮ represents the number of 10s, and ❯ represents the number of 1s.

| **Example 12** | Using the Babylonian Numeration System |

What number does ❮❮❮❯❯❯❯❯❯ represent?

SOLUTION
Since there are 3 tens and 6 ones, the number represents 36.

Try This One   **12**

What number does **<<<<<‖** represent?

If this were the whole story on the Babylonian system, (a) it would be quite the pain to write large numbers, and (b) it would be a simple grouping system. But the Babylonian system was also positional (and in fact was the earliest known example of a positional system). Numbers from 1 to 59 were written using the two symbols as shown in Example 12, but after the number 60, a space was left between the groups of numbers. For example, the number 2,538 was written as

<<<<‖   <‖‖‖‖‖‖

which means that there are 42 sixties and 18 ones. The space separates the 60s from the ones. The value is found as follows:

$$\begin{aligned} 42 \times 60 &= 2,520 \\ +18 \times\ 1 &=\ \ \ \underline{18} \\ &\ \ \ 2,538 \end{aligned}$$

When there are three groupings of numbers, the symbols to the left of the first space represent the number of 3,600s (note that $3,600 = 60 \times 60$). The next group of symbols represents the number of 60s, and the final group represents the number of 1s.

Math Note

The Babylonian system dates back to at least 3100 BCE, and is essentially a base 60 positional system. There are theories on why base 60 was used, but nobody knows for sure. (One prominent theory is that 360 was used because the Babylonians believed the year had 360 days.) It's interesting to note that the influence of the system can still be seen in everyday life. If you've ever wondered why hours are divided into 60 minutes, and minutes into 60 seconds, you can stop wondering.

We know a lot about the Babylonian system because it was recorded on clay tablets rather than paper. Over 400 such tablets have been found, giving archaeologists a lot of information to study.

David Lees/Getty Images
A Babylonian clay tablet with numeric symbols.

Example 13   Using the Babylonian Numeration System

Write the numbers represented.

(a) **<<<<<<‖   <<<‖‖‖‖**

(b) **<‖‖   <<<<<<‖   <<‖‖‖**

SOLUTION

(a) There are 52 sixties and 34 ones; so the number represents

$$\begin{array}{r} 52 \times 60 = 3{,}120 \\ +\,34 \times \;\;1 = \quad 34 \\ \hline 3{,}154 \end{array}$$

(b) There are twelve 3,600s, fifty-one 60s, and twenty-three 1s. The numeral represents

$$\begin{array}{r} 12 \times 3{,}600 = 43{,}200 \\ 51 \times \quad 60 = \;\;3{,}060 \\ 23 \times \quad\;\; 1 = \qquad 23 \\ \hline 46{,}283 \end{array}$$

Try This One 13

Write the number represented.

(a)  (b) 

Example 14 Writing a Number in the Babylonian System

Write 5,217 using the Babylonian numeration system.

SOLUTION

Since the number is greater than 3,600, we need to divide by 3,600 to see how many 3,600s are contained in the number.

$$5{,}217 \div 3{,}600 = 1 \text{ remainder } 1{,}617$$

The remainder, 1,617, is then divided by 60 to see how many 60s are in 1,617.

$$1{,}617 \div 60 = 26 \text{ remainder } 57$$

So the number 5,217 consists of

$$\begin{array}{r} 1 \times 3{,}600 = 3{,}600 \\ 26 \times \quad 60 = 1{,}560 \\ 57 \times \quad\;\; 1 = \quad\;\; 57 \\ \hline \text{Total} = 5{,}217 \end{array}$$

It can be written as

Math Note

The Babylonians didn't have a symbol for zero. This complicated their writings. For example, how is the number 7,200 distinguished from the number 72?

Try This One 14

Write each number using the Babylonian numeration system.

(a) 42 (b) 384 (c) 4,278

Moving to the Western Hemisphere, the last system we'll study was used by the Mayans, an advanced civilization that occupied southeastern Mexico and parts of Central America for over 3,000 years beginning around 2000 BCE. The Mayan system is similar to the Babylonian system in a number of ways; it uses a small number of symbols (three), and it combines the elements of grouping and positional systems.

The three symbols in the Mayan system are a dot (representing one), a horizontal line (representing five), and the symbol ⬭ (representing zero). Numbers are written vertically, except for ones. So the number 3 is written as • • •, and the number 12 is written as ⬭ (two fives and two ones). For numbers greater than 19, the system becomes positional, with the top grouping describing the number of twenties, and the bottom one describing the number of ones. For example, the numeral

represents 7 twenties plus 14 ones, making it $7 \times 20 + 14 \times 1$, or 154.

The next place value after 20 is not 20^2 as you might expect, but 360 (which is 18×20). This was used because the Mayan calendar was made up of 18 months of 20 days each, with 5 "nameless days" to match the 365-day solar year. The next higher place value is 18×20^2, or 7,200; then 18×20^3 and so on. The symbol for zero is used to indicate that there are no digits in a given place value. For example, the numeral

Three hundred sixties

Twenties

Ones

represents $18 \times 360 + 0 \times 20 + 11 \times 1$, which is 6,491.

| Example 15 | Using the Mayan Numeration System |

Math Note

Mayan mathematicians got a lot of press near the end of 2012 when some people interpreted the Mayan calendar to foretell the end of the world on December 21, 2012.

But this was a misinterpretation: that particular date signalled the end of a calendar cycle and the beginning of a new one, not the "end of the world" (whatever that means).

(a) The Mayan calendar lists the creation date as 3114 BCE, which was 5,126 years before I wrote this question. Write that number of years in Mayan numerals.

(b) Convert the Mayan numeral below into a Hindu-Arabic numeral.

SOLUTION

(a) First, we divide 5,126 by 360 to get 14 with a remainder of 86. So we need 14 three hundred sixties, 4 twenties, and 6 ones.

(b) With four distinct groupings, the top tells us the number of 7,200s; the second, the number of 360s; the third, the number of 20s, and the last tells us the number of 1s. In this case, we have sixteen 7,200s, no 360s, nine 20s, and no 1s. This is $16 \times 7,200 + 9 \times 20 = 115,380$.

Try This One   15

(a) Write the year of Mayan creation in Mayan numerals.
(b) Convert the Mayan numeral below into a Hindu-Arabic numeral.

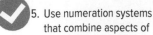

5. Use numeration systems that combine aspects of other types of systems.

It's easy to think "why would we study number systems that haven't been used in thousands of years?", but try not to think of this section in those terms. Learning about the actual development of numeration systems gives us a much deeper understanding and appreciation of the Hindu-Arabic system, which is fundamental to EVERY way that math is used in our world.

Answers to Try This One

1 ⫴⫴ ⫴⫴ ⫴⫴ ⫴⫴ ‖; 22 prescriptions

2 (a) 1,211   (b) 1,102,041

3 (a) ◖◖◖◖◖◖◖∩∩‖‖‖‖‖‖

   (b) ⊲⊲⊲⊲⊲𝄐ϙ∩∩

4 (a) ◖◖◖◖◖◖◖∩‖‖

   (b) ◖◖◖◖◖◖◖∩∩∩∩∩∩‖‖‖‖‖‖‖‖

5 (a) 39   (b) 1,164   (c) 333

6 (a) LXVII (b) CXCII (c) CCII (d) CMLX

7 ιφ θ βω ηρ γφ

8 (a) 427   (b) 6,075   (c) 26

9 (a) 四   (b) 二   (c) 六
       十   百   千
       五   五   三
                   十   百
                   六   二
                               十
                               一

10 9 100 1 10 1 1; 3 10,000 4 1,000 6 10 2 1

11 (a) $5 \times 10^2 + 7 \times 10^1 + 3$

   (b) $8 \times 10^4 + 6 \times 10^3 + 4 \times 10^2 + 7 \times 10^1 + 1$

   (c) $2 \times 10^6 + 2 \times 10^5 + 1 \times 10^3 + 5 \times 10^2 + 6 \times 10^1 + 7$

12 52

13 (a) 1,375   (b) 40,873

14 (a) ⟨⟨⟨⟨▼▼

   (b) ▼▼▼▼▼▼   ⟨⟨▼▼▼▼

   (c) ▼   ⟨▼   ⟨▼▼▼▼▼▼▼▼▼

15 (a) •••
           —
           •
           —
           ••••

   (b) 7,292

Exercise Set   4-1

Writing Exercises

1. Describe the difference between a number and a numeral.
2. Briefly describe how a grouping system works.
3. Briefly describe how a multiplicative grouping system works.
4. Describe what place values are and what they represent in the Hindu-Arabic numeration system.
5. Both the Roman and Egyptian systems use symbols to represent certain numbers. Explain how they differ (aside from the fact that they use different symbols).

6. Describe how to write a number in the Hindu-Arabic system in expanded notation.
7. Which of the systems we studied do you think is most efficient? Why?
8. The tally system only needs one symbol, the Babylonian needs just two, and the Mayan three. Why aren't these simpler to work with than the Hindu-Arabic system?

Computational Exercises

For Exercises 9–16, write each number using Hindu-Arabic numerals.

9. 𓍢𓏏𓏏𓏺𓏺𓏺

10. 𓏭𓍢𓇾𓇾𓏏𓏏𓏏𓏺𓏺𓏺𓏺

11. 𓏏𓏏𓇾𓇾𓏏𓏏𓏺𓏺𓏺𓏺𓏺

12. (Egyptian/hieratic numerals)

13. (Egyptian/hieratic numerals)

14. (Egyptian/hieratic numerals)

15. (Egyptian/hieratic numerals)

16. (Egyptian/hieratic numerals)

For Exercises 17–26, write each number using Egyptian numerals.

17. 37	21. 1,256	25. 1,000,030
18. 52	22. 8,261	26. 4,313,911
19. 801	23. 362,430	
20. 955	24. 151,040	

For Exercises 27–32, perform the indicated operations. Write your answers as Egyptian numerals.

27. (Egyptian numerals) ∩∩∩∩∩||| + ∩∩∩∩∩∩|||

28. (Egyptian numerals) 𓂋𓂋∩∩|| + 𓂋∩∩∩||

29. (Egyptian numerals) +

30. (Egyptian numerals) ∩∩∩|| − ∩||||

31. (Egyptian numerals)

32. (Egyptian numerals)

For Exercises 33–38, write each number using Hindu-Arabic numerals.

33.
一
百
八
十
九

34.
三
千
四
百
七

35.
五
十
二

36.
九
千
八
百
三
十
四

37.
七
百
一
十
三

38.
八
十
九

For Exercises 39–44, write each number using Chinese numerals.

39. 89	41. 284	43. 2,356
40. 567	42. 9,857	44. 21

In Example 10, we used the familiar numerals from the Hindu-Arabic system in a multiplicative grouping system. For Exercises 45–50, write each number in that system.

45. 47	47. 917	49. 23,450
46. 93	48. 358	50. 98,076

For Exercises 51–54, write each number in expanded notation.

51. (a) 1,805	(b) 32,714
52. (a) 6,002	(b) 29,300
53. (a) 162,873	(b) 200,321,416
54. (a) 742,311	(b) 17,531,801

For Exercises 55–58, write the number in the Hindu-Arabic system.

55. (a) $7 \times 1,000 + 3 \times 100 + 9 \times 1$
    (b) $4 \times 10,000 + 8 \times 100 + 5 \times 10$
56. (a) $3 \times 1,000 + 6 \times 10 + 4 \times 1$
    (b) $2 \times 10,000 + 3 \times 1,000 + 4 \times 100 + 5 \times 1$
57. (a) $6 \times 100,000 + 4 \times 10,000 + 6 \times 1,000 + 9 \times 10$
    (b) $8 \times 10,000,000 + 6 \times 100,000 + 4 \times 1,000 + 2 \times 10$
58. (a) $7 \times 100,000 + 3 \times 1,000 + 4 \times 100 + 6 \times 1$
    (b) $5 \times 1,000,000 + 3 \times 100,000 + 9 \times 1,000 + 2 \times 10$

For Exercises 59–66, write each number using Hindu-Arabic numerals.

59. ≪≪≪⟨𝖳

60. ≪≪≪𝖳𝖳𝖳𝖳𝖳

61. ⟨𝖳   ≪⟨𝖳   ⟨𝖳

62. ≪⟨𝖳𝖳   ≪≪≪   ⟨𝖳𝖳𝖳

63. ≪⟨𝖳𝖳   ⟨𝖳   ≪≪⟨𝖳𝖳𝖳

64. ≪≪   𝖳𝖳𝖳   ⟨

65. 𝖳   ⟨𝖳   ≪≪≪⟨𝖳𝖳𝖳

66. ≪⟨𝖳𝖳   ⟨𝖳   ≪≪⟨𝖳𝖳𝖳

For Exercises 67–74, write each number using Babylonian numerals.

67. 78	71. 1,023
68. 156	72. 1,776
69. 292	73. 12,583
70. 514	74. 7,360

For Exercises 75–82, write each number in the Hindu-Arabic system. (For 81 and 82, refer to the Math Note on page 165.)

75. (a) XVIII   (b) XCIX
76. (a) CXV   (b) XLIII
77. (a) CCXVI   (b) MCX
78. (a) LXXXVI   (b) CCXXXIII
79. (a) CDXVIII   (b) MMCMLXVII
80. (a) CMXCIV   (b) MCMXCIV
81. (a) $\overline{\text{XC}}$MMD   (b) $\overline{\text{MMMCC}}$
82. (a) $\overline{\text{LXML}}$ (b) $\overline{\text{MMVIII}}$

For Exercises 83–90, write each number in Roman numerals. (For 89 and 90, refer to the Math Note on page 165.)

83. (a) 58   (b) 147
84. (a) 36   (b) 209
85. (a) 567   (b) 1,258
86. (a) 893   (b) 3,720
87. (a) 1,462   (b) 3,909
88. (a) 2,222   (b) 3,449
89. (a) 70,311   (b) 2,060,000
90. (a) 42,104   (b) 3,000,400

In Exercises 91 and 92, convert each number from the multiplicative grouping system in Example 7 on page 165 to the Hindu-Arabic system. In Exercises 93 and 94, convert each number from the Hindu-Arabic system to the system in Example 7.

91. (a) $\iota\varphi\,\eta\alpha$ (b) $\delta\varphi\,\theta\alpha$ (c) $\varepsilon\omega\,\chi\rho\,\beta\varphi\,\delta\alpha$
92. (a) $\alpha\varphi\,\chi\alpha$ (b) $\alpha\rho\,\alpha\alpha$ (c) $\alpha\omega\,\gamma\rho\,\beta\varphi\,\varepsilon\alpha$
93. (a) 85   (b) 39   (c) 4,107
94. (a) 44   (b) 21   (c) 8,403

In Exercises 95–98, perform the operation using the multiplicative grouping system in Example 7 on page 165. First, try to perform the operations without changing to Hindu-Arabic numerals, then check your answers by converting to Hindu-Arabic numerals, performing the operation, and converting back.

95. $\theta\varphi\,\gamma\alpha + \chi\varphi\,\chi\alpha$
96. $\varepsilon\varphi\,\iota\alpha - \delta\varphi\,\beta\alpha$
97. $\beta\alpha\,(\varphi + \chi\varphi\,\eta\alpha)$
98. $\delta\varphi\,\alpha(\chi\varphi - \beta\varphi\,\eta)$

In Exercises 99–102, write each Mayan numeral in the Hindu-Arabic system.

99. (a)   (b)

100. (a)   (b)

101. (a)   (b)

102. (a)   (b)

In Exercises 103–106, write the given number in the Mayan numeration system.

103. (a) 119   (b) 380
104. (a) 94   (b) 400
105. (a) 4,901   (b) 96,030
106. (a) 6,010   (b) 104,311

Applications in Our World

Most movies use Roman numerals in the credits to indicate the date the film was made. Here are some movies and their dates. Find the year the movie was made.

Movie	Year
107. *Gone With the Wind*	MCMXXXIX
108. *Casablanca*	MCMXLII
109. *Animal House*	MCMLXXVIII
110. *Raiders of the Lost Ark*	MCMLXXXI
111. *The Hunger Games*	MMXII
112. *Rogue One*	MMXVI

In the world of international finance, account numbers need to be protected, so they are often encoded. Each of the account numbers in Exercises 113–118 has been encoded using a multiplicative grouping system with the symbol representations shown in the table. Decode each account number.

A	B	C	D	E	F	G	H	I	J	S	X	Z	Q
0	1	2	3	4	5	6	7	8	9	10	100	1,000	10,000

113. EX CS DB – JZ DX BS FB
114. HX BS JB – CZ CX CS CB
115. GX AS F – BQ CZ JX AS I
116. X ES D – JQ DZ JX S B
117. IZ EX DB – EZ JS GB
118. HZ ES B – FZ FS G

Critical Thinking

119. Look at some of the numbers we wrote in the Egyptian system, and count the number of symbols needed. Then compare that to the number written in the Hindu-Arabic system. If you're clever, find a method for finding the number of symbols needed in the Egyptian system without having to actually write the number in that system.

120. In the Math Note on page 167, I pointed out that if the Chinese system used the numerals from 0 to 10, 100, etc. that you're familiar with, you'd pick it up in no time. So prove me right! Write the numbers from Try This One 9 in the style of the Chinese system, but use Hindu-Arabic numerals rather than Chinese characters. Then

121. Make up your own numeration system using your own symbols. Indicate whether it is a simple grouping, multiplicative grouping, or positional system. Explain how to add and subtract in your numeration system.

122. Most clocks that use Roman numerals have four written incorrectly, as IIII rather than IV. Think of as many potential reasons for this as you can, then do an Internet search to see if you can find the reason.

123. A colleague of mine once gave a quiz with the question "Why is it useful to learn about the Babylonian numeration system?" and one student answered "If you have any Babylonian friends, you could communicate with them about numbers." Explain why we laughed hysterically at that answer.

124. Which of the ancient numeration systems we studied (Egyptian, Babylonian, Roman) do you think is the most efficient? Why?

125. Using information from within this section, and supplementing with some Internet research, draw a timeline of numerical development that includes all of the systems we studied.

126. In the Hindu-Arabic system, if two numbers are positive and have a different number of digits, the one with the smaller amount of numerals is always the smaller number. For each of the other systems discussed in this section, decide whether or not a number can be smaller than a different number with fewer numerals. Include examples.

127. Find the largest number you can write with exactly four symbols in each of the numeration systems studied in this section.

128. Find the largest number you can write with three groupings in the Mayan numeration system, and with four groupings.

129. Explain why you would never write a Roman numeral with two IIs before a V, or two XXs before a C.

130. Why would it be silly to write the strings DM or LC in Roman numerals?

131. Look at the solution to Example 1, then the paragraph just after Try This One 2. What can you conclude about a tally system that uses groups of five tallies?

Section 4-2   Tools and Algorithms in Arithmetic

LEARNING OBJECTIVES

1. Multiply using the Egyptian algorithm.

2. Multiply using the Russian peasant method.

3. Multiply using the lattice method.

4. Multiply using Napier's bones.

Handheld calculators have been widely available since the early 1970s, which means that for most college students, they've always existed. Compared to your lifespan, 45 years may seem like a long time, but compared to the history of mathematics, it's practically the blink of an eye. In order to have a greater appreciation of the modern tools we have, it's useful to look back at what people had to do before calculators and computers were invented. In this section, we'll look at a handful of extremely clever methods that were developed to perform multiplications.

Howard Bartrop/Image Source. All rights reserved.

While you'll probably never use them in real life, studying them can give you insight into the types of innovations that helped advance human thought from the simple to the abstract, paving the way for all of the modern advances that we too often take for granted.

The Egyptian Algorithm

An **algorithm** is a method or process for solving a mathematical problem.

The Egyptian algorithm is an ancient method of multiplication that can be done by hand because it requires only doubling numbers and addition. We'll illustrate it with an example, then summarize.

Example 1   Using the Egyptian Algorithm

Use the Egyptian algorithm to multiply 13×24.

SOLUTION

Step 1 Form two columns with 1 at the top of the first column and 24 at the top of the second column:

1   24

Math Note

The Egyptian algorithm is described on the Rhind papyrus, a priceless document that was written around 1650 BCE; much of it was transcribed from a lost document about 150 years older than that.

The Rhind papyrus is undoubtedly one of the oldest math papers in the world. There are older mathematical carvings on stone or clay tablets, but documents of that age written on the paper-like material papyrus are extremely rare, to say the least.

Step 2 Double the numbers in each column, and continue to do so until the first column contains numbers that can be added to get the other number in the product, 13:

1	24
2	48
4	96
8	192

We stop here because we can get 13 from adding 1, 4, and 8.

Step 3 Add the numbers in the second column that are next to 1, 4, and 8: $24 + 96 + 192 = 312$. This is the product of 13 and 24.

Try This One 1

Use the Egyptian algorithm to multiply 22×15.

We could have put either original number at the top of the second column, but it's usually a little quicker if we put the larger number there, as we did in Example 1.

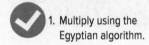

1. Multiply using the Egyptian algorithm.

The Egyptian Algorithm

To multiply two numbers A and B:

1. Form two columns with the numeral 1 at the top of the first, and one of the numbers (we'll say B) to be multiplied at the top of the second.

2. Double the numbers in each column repeatedly until the first column contains numbers that can be added to A.

3. Add the numbers in the second column that are next to the numbers in the first column that add to A. This sum is the product of A and B.

The Russian Peasant Method

Another method for multiplying by hand is known as the Russian peasant method. As you will see from Example 2, it's similar to the Egyptian algorithm, but maybe a bit simpler in that you don't have to keep searching for numbers that add to one of the factors.

Sidelight   Dividing with the Egyptian Algorithm

The Egyptian algorithm can be used for dividing as well. Suppose, for example, we want to divide 1,584 by 24. Make two columns headed with 1 and 24 and double as before:

1	24
2	48
4	96
8	192
16	384
32	768
64	1,536

We can stop here since the next entry would be larger than 1,584. Now find the numbers in the right column that add up to 1,584 (in this case 48 and 1,536), then add the corresponding numbers in the left column. The sum $2 + 64 = 66$ is the quotient of 1,584 and 24.

This method will not work unless the numbers divide evenly, but a modified Egyptian method will work in such cases (see Exercise 52).

| Example 2 | Using the Russian Peasant Method |

Use the Russian peasant method to multiply 26×15.

SOLUTION

Step 1 Form two columns with 26 and 15 at the top.

26   15

Step 2 Divide the numbers in the first column by 2 (ignoring remainders), and double the numbers in the second column, until you reach one in the first column.

26	15	
13	30	
6	60	*13/2 is 6 with a remainder: write 6*
3	120	
1	240	*3/2 is 1 with a remainder: write 1*

Step 3 Add the numbers in the second column that are next to odd numbers in the first column: $30 + 120 + 240 = 390$. This is the product of 26 and 15.

Try This One   2

Use the Russian peasant method to multiply 18×12.

Math Note

The Russian peasant method is often called the method of *mediation and duplation*. "Mediation" refers to halving objects, while "duplation" refers to doubling them.

The Russian Peasant Method

To multiply two numbers A and B:

1. Form two columns with A at the top of one column and B at the top of the other.

2. Divide the numbers in the first column by two repeatedly, ignoring remainders, until you reach one. Double the numbers in the second column, with the last result next to the one in the first column.

3. Add the numbers in the second column that are next to odd numbers. The result is the product of A and B.

The Russian peasant method can be used to multiply numbers with more digits, as in Example 3.

| Example 3 | Using the Russian Peasant Method |

Use the Russian peasant method to multiply 103×19.

SOLUTION

Form the columns as described in the colored box above:

103	19
51	38
25	76
12	152
6	304
3	608
1	1,216

Now add the numbers in the second column that are next to odd numbers in the first:

$$19 + 38 + 76 + 608 + 1,216 = 1,957$$

So $103 \times 19 = 1,957$.

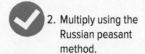

2. Multiply using the Russian peasant method.

Try This One   3

Use the Russian peasant method to multiply 210×21.

The Lattice Method

The lattice method for multiplication was used in both India and Persia as early as the year 1010. It was later introduced in Europe in 1202 by Leonardo of Pisa (more commonly known as Fibonacci) in his work entitled *Liber Abacii* (Book of the Abacus). The lattice method reduces multiplying large numbers into multiplying single digit numbers, as illustrated in the next two examples.

Example 4   Using the Lattice Method

Math Note

The original definition of the term *lattice* refers to a framework or structure made of crossed wood or metal strips. The boxes used to arrange the calculations in the lattice method match that description pretty well.

Find the product 36×568 using the lattice method.

SOLUTION

Step 1 Form a lattice as illustrated with one of the numbers to be multiplied across the top, and the other written vertically along the right side.

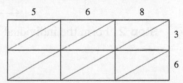

Step 2 Within each box, write the product of the numbers from the top and side that are above and next to that box. Write the first digit above the diagonal and the second below it, using zero as first digit if necessary.

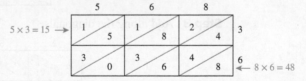

Step 3 Starting at the bottom right of the lattice, add the numbers along successive diagonals, working toward the left. If the sum along a diagonal is more than 9, write the last digit of the sum and carry the first digit to the addition along the next diagonal.

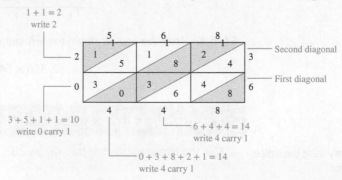

Step 4   Read the answer, starting down the left side then across the bottom:

$$36 \times 568 = 20,448$$

Try This One   4

Find the product 53×844 using the lattice method.

In theory, you can perform multiplication with any number of digits using the lattice method: you just need more boxes in your lattice.

Example 5 | **Using the Lattice Method**

Find the product $2,356 \times 547$ using the lattice method.

SOLUTION

Step 1   Form a lattice with one of the numbers to be multiplied across the top and the other one down the right side.

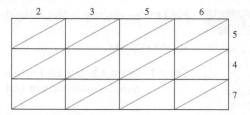

Step 2   Form the individual products in each box.

$4 \times 2 = 8$; write zero for first digit →

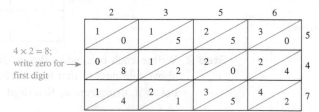

Step 3   Add along the diagonals.

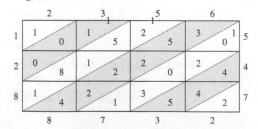

Step 4   Read the answer down the left and across the bottom.

$$2,356 \times 547 = 1,288,732$$

Try This One   5

Use the lattice method to find the product 568×478.

✓ 3. Multiply using the lattice method.

Napier's Bones

ManuelVelasco/Getty Images

John Napier (1550–1617), a Scottish mathematician, introduced Napier's bones as a calculating tool based on the lattice method of multiplication. Napier's bones consist of a set of 11 rods: the first rod (called the index) and 1 rod for each digit 0–9, with multiples of each digit written on the rod in a lattice column as illustrated in Figure 4-3.

The next example illustrates how Napier's bones are used to multiply by a single-digit number.

Figure 4-3

Example 6 | **Using Napier's Bones**

Use Napier's bones to find the product $2,745 \times 8$.

SOLUTION

Choose the rods labeled 2, 7, 4, and 5 and place them side by side; also, place the index to the left. Then locate the level for the multiplier 8, as shown in Figure 4-4.

Figure 4-4

Add the numbers diagonally as in the lattice method (Figure 4-5).

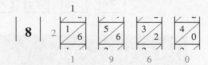

Figure 4-5

The product is 21,960.

Sidelight   **John Napier (1550–1617)**

John Napier was born to a wealthy family in Merchiston, Scotland, in 1550. He lived in the time of Copernicus and Kepler, an age of great discoveries (and great fame) for those involved in astronomy. Napier was a young genius, entering St. Salvator's College in St. Andrews at the age of 13, but he left after two years, spending time traveling throughout Europe and studying a variety of subjects in different countries. A man of varied talents and interests, Napier found himself especially interested in simplifying the calculations with large numbers that were so crucial to the astronomical observations of the day. He is widely credited with inventing logarithms, a computational tool still in widespread use today. Napier worked with the Englishman Henry Briggs on his logarithms, and Briggs published the first logarithm table after Napier's death in 1617. Logarithm tables maintained their status as vitally important to calculations in a variety of sciences well into the 20th century, when the advent of computers reduced them to an interesting historical artifact.

Hulton Archive/Getty Images

Try This One   6

Use Napier's bones to multiply 6 × 8,973.

| Example 7 | Using Napier's Bones |

Use Napier's bones to multiply 234 × 36.

SOLUTION

Choose the 2, 3, and 4 rods and place them side by side, with the index to the left. Locate the multipliers 3 and 6.

Add the diagonals as in the lattice method.

The product is 8,424.

Try This One   7

4. Multiply using Napier's bones.

Use Napier's bones to multiply 126 × 73.

Answers to Try This One

1 330   **3** 4,410   **5** 271,504   **7** 9,198

2 216   **4** 44,732   **6** 53,838

Exercise Set   4-2

Writing Exercises

1. Describe the connection between the Egyptian algorithm and the Russian peasant method. How are they different?

2. Describe the connection between the lattice method and Napier's bones. How are they different?

Computational Exercises

For Exercises 3–12, use the Egyptian algorithm to multiply.

3. 12×21
4. 15×30
5. 23×17
6. 29×15
7. 34×110

8. 17×45
9. 56×8
10. 11×13
11. 18×12
12. 7×35

For Exercises 13–22, use the Russian peasant method to multiply.

13. 11×23
14. 14×32
15. 23×16
16. 29×17
17. 34×11

18. 17×42
19. 56×7
20. 11×17
21. 18×15
22. 7×45

For Exercises 23–32, use the lattice method to multiply.

23. 23×456
24. 453×938

25. $89 \times 1,874$
26. $287 \times 7,643$

27. 876×903
28. 45×583
29. 67×875

30. 359×83
31. 568×359
32. $2,348 \times 83,145$

Construct a set of Napier's bones out of poster board or heavy paper and use them to do the multiplications in Exercises 33–40.

33. 9×523
34. 8×731
35. 23×45
36. 71×52

37. 47×123
38. 69×328
39. 154×236
40. 211×416

In Exercises 41–46, use the method in the Sidelight on page 177 to perform each division.

41. $372 \div 12$
42. $624 \div 16$
43. $1,387 \div 19$

44. $2,200 \div 25$
45. $8,853 \div 227$
46. $7,503 \div 183$

Critical Thinking

47. Perform the multiplication 38×147 three different ways: using the traditional method taught in grade school, the Egyptian algorithm, and the Russian peasant method. Did you get the same answer all three times? Which method was the fastest? If you got more than one answer, use a calculator to check. Which do you think you're most likely to get the correct answer with most often?

48. Repeat Exercise 47, this time using the multiplication 140×276, and substituting the lattice method for the Russian peasant method.

49. If you had to perform a multiplication of two 3-digit numbers without a calculator for a million dollars on a game show, and you could only do it once, what method would you use? Why?

50. Perform the multiplication in Example 4 using the setup below (like you learned in elementary school), multiplying down the digits and carrying the first digit of products over 10. First, make sure that you got the same answer as in Example 4. Then describe how the two methods compare. How are they similar? How are they different?

$$\begin{array}{r} 568 \\ \times\ 36 \\ \hline \end{array}$$

51. So far, we've used historic methods of multiplication only on whole numbers. Now let's take a look at decimals. Figure out a way to modify the lattice method to multiply 47.3 by 12.9. (*Hint:* Ignore the decimals at first, then decide how they fit into your answer.)

52. (a) Perform the division 230 ÷ 18 using regular long division.
    (b) To divide 230 ÷ 18 using a modified Egyptian algorithm, set up the division as in Exercises 41–46. After you finish doubling the two columns (stopping when the next row would have a number greater than 230), find the combination of numbers in the right column whose sum is closest to 230 without going over. Then add the corresponding numbers in the left column, and find the difference between the sum of numbers in the right column and 230. How do these results compare with numbers you came up with in the long division? Use the result to develop a method for division using the Egyptian algorithm.

53. Refer to Exercise 52. Use the method developed in part (b) to perform the following divisions.
    (a) 427 ÷ 30   (b) 900 ÷ 21   (c) 1,324 ÷ 18

Section 4-3   Base Number Systems

✓ **LEARNING OBJECTIVES**

1. Convert between base 10 and other bases.
2. Convert between binary, octal, and hexadecimal.

In Section 4-1, we studied a variety of numeration systems other than our own. The thing that they all have in common is that they use numerals different from the ones we're all so familiar with in the Hindu-Arabic system. In this section, you'll find out that there are numeration systems that use the numerals you're familiar with, but are still different from the Hindu-Arabic system. The key difference is that all of the digits in our system are based on powers of 10. You can also define systems based on powers of other numbers. If a system uses some of our "regular" numerals, but is based on powers other than 10, we will call it a **base number system**.

The best way to get some perspective on base number systems is to review the base 10 positional system that we use, and to be completely clear on what the significance of every digit is. A number like 453 can be expanded out as

$$453 = 4 \times 100 + 5 \times 10 + 3 \times 1$$
$$= 4 \times 10^2 + 5 \times 10^1 + 3 \times 10^0$$

and we understand from experience that a 5 in the second digit from the right means five 10s. We can expand numbers in positional systems with bases other than 10 in the same way. The only difference is that the digits represent powers of some number other than 10.

Base Five System

Math Note

Recall that we have defined 10^0 to be 1, $10^1 = 10$, $10^2 = 10 \cdot 10$, $10^3 = 10 \cdot 10 \cdot 10$, etc. We'll study exponents in depth in Section 5-6.

In a base five system it isn't necessary to have 10 numerals as in the Hindu-Arabic system; only five numerals (symbols) are needed; we use only the numerals 0, 1, 2, 3, and 4. Just as each digit in the Hindu-Arabic system represents a power of 10, each digit in a base five system represents a power of five. The place values for the digits in base five are:

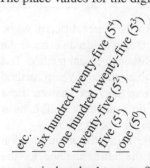

etc. | six hundred twenty-five (5^4) | one hundred twenty-five (5^3) | twenty-five (5^2) | five (5^1) | one (5^0)

You might wonder why we wrote out six hundred twenty-five and the other numbers above, rather than using 625, 125, etc. This is to remind you that there's only one number that we call six hundred twenty-five, but there are many numerals to represent it. In the base 10 system,

Sidelight   Math and the Search for Life Out There

One of humanity's greatest questions is "Are we alone in the universe?" The search for life beyond our solar system is going on every hour of every day, with giant antennas listening for radio signals from outer space. In the hit movie *Contact*, a scientist recognizes that a signal she's receiving must be from an intelligent source because she recognizes that it's based on mathematics (prime numbers). In fact, most scholars feel like our best chance of communicating with an alien civilization is through the universal nature of mathematics! While almost every aspect of languages is dependent on the nature and experiences of the speaker or listener, numbers may be the one true universal constant.

If you think of it, this is intimately tied to the difference between numbers and numerals: while there may be many different ways to represent numbers, the concept of a number is always the same in any language, culture,

time, or place. You may then wonder how we would decipher messages from another culture based on mathematics. A good guess is in the same way historians have been able to learn about the numeration systems of ancient cultures. The concrete nature of numbers gives us a big advantage in decoding messages, providing our best chance at finding life out there.

Mike Norton/Hemera/Getty Images

that numeral is 625. As we'll see, in the base five system, it's 1000. Sound crazy? Read on. You might think this topic is kind of cool.

When writing numbers in base five, we use the subscript "five" to distinguish them from base 10 numbers, because a numeral like 423 in base five corresponds to a number different from the numeral 423 in base 10. (This is where understanding the difference between a number and a numeral is crucial—remember that a number is an unchanging concept, while a numeral is a symbol used to represent a given number.) Table 4-1 shows some base 10 numbers also written in base five.

This might look confusing, but it can be clarified using the reasoning in the comments next to the table. The numbers 1 through 4 are written the same in both systems. The number 5 can't

Table 4-1	Base Five Numbers	
Base 10 number	**Corresponding base five number**	
1	1_{five}	
2	2_{five}	
3	3_{five}	
4	4_{five}	
5	10_{five}	1 five and no ones
6	11_{five}	
7	12_{five}	
8	13_{five}	1 five and 3 ones
9	14_{five}	
10	20_{five}	2 fives and 0 ones
11	21_{five}	
25	100_{five}	1 twenty-five, no fives or ones
30	110_{five}	1 twenty-five, 1 five, no ones
50	200_{five}	2 twenty-fives
125	1000_{five}	1×5^3
625	10000_{five}	1×5^4

be written in base five using the numeral 5 because the base five system only uses the digits 0, 1, 2, 3, and 4. Instead, we write it as 10_{five}, meaning $1 \times 5 + 0 \times 1$, or 1 five and no ones. (Actually, this shows why base five doesn't need any numerals beyond 4: we can write the numbers 5 through 9 using the numerals from 0 through 4.) In the same way, the number 8 would be written in base five as 13_{five}, meaning 1 five and 3 ones.

Before we move on, we should talk about pronunciation. We have seen that the numeral 10_{five} represents the number 5. There's a pretty good chance that you want to pronounce that numeral "ten base five," but that's not a great idea: When you do that, you're telling your brain that it represents the number 10 when it doesn't. It's a much better idea to pronounce it "one zero base five."

Converting Base Five Numbers to Base 10 Numbers

Base five numbers can be converted to base 10 numbers using the place values of the base five numbers and expanded notation. For example, the number 242_{five} can be expanded as

$$
\begin{aligned}
242_{\text{five}} &= 2 \times 5^2 + 4 \times 5^1 + 2 \times 5^0 \\
&= 2 \times 25 + 4 \times 5 + 2 \times 1 \\
&= 50 \quad + 20 \quad + 2 \\
&= 72
\end{aligned}
$$

Example 1	**Converting Numbers to Base 10**

Write each number in base 10.

(a) 42_{five}
(b) 134_{five}
(c) 4213_{five}
(d) What number would the numeral 42 represent in a base seven system?

SOLUTION

The place value chart for base five is used in each case.

(a) $42_{\text{five}} \quad = 4 \times 5 + 2 \times 1 = 20 + 2 = 22$
(b) $134_{\text{five}} \quad = 1 \times 5^2 + 3 \times 5 + 4 \times 1$
$\qquad\qquad = 1 \times 25 + 3 \times 5 + 4 \times 1$
$\qquad\qquad = 25 + 15 + 4 = 44$
(c) $4213_{\text{five}} = 4 \times 5^3 + 2 \times 5^2 + 1 \times 5 + 3 \times 1$
$\qquad\qquad = 4 \times 125 + 2 \times 25 + 1 \times 5 + 3 \times 1$
$\qquad\qquad = 500 + 50 + 5 + 3 = 558$
(d) In base seven, the rightmost digit is the number of ones, and the next moving left is the number of sevens. So 42 in base seven is $4 \cdot 7 + 2 \cdot 1 = 30$ in base 10.

Math Note

To find the place value for each digit of a base five number, it's helpful to read from right to left: the rightmost digit is in the 5^0 (or 1) place, the next digit to the left is in the 5^1 place, then 5^2, and so forth.

Try This One	**1**

Write each number in the base 10 system.

(a) 302_{five} (b) 1324_{five} (c) 40000_{five} (d) 2112_{three}

Converting Base 10 Numbers to Base Five Numbers

Base 10 numbers can be written in the base five system using the place values of the base five system and successive division. This method is illustrated in Examples 2 and 3.

| Example 2 | Converting Numbers from Base 10 to Base Five |

Write 84 in the base five system.

Math Note

The answer to Example 2 can be checked using multiplication and addition:

$3 \times 25 + 1 \times 5 + 4 \times 1$
$= 75 + 5 + 4 = 84$

SOLUTION

Step 1 Identify the largest place value number (1, 5, 25, 125, etc.) that will divide into the base 10 number. In this case, it's 25.

Step 2 Divide 25 into 84, as shown.

$$\begin{array}{r} 3 \\ 25\overline{)84} \\ -75 \\ \hline 9 \end{array}$$ ← *Three 25s in 84*

← *9 left over*

This tells us that there are three 25s in 84, with 9 left over after the 25s are accounted for.

Step 3 Divide the remainder by the next lower place value. In this case, that's 5 because we're in the base five system and already found the digit for 5^2.

$$\begin{array}{r} 1 \\ 5\overline{)9} \\ -5 \\ \hline 4 \end{array}$$ ← *One 5 in 9*

← *4 left over*

Step 4 Continue dividing until the remainder is less than 5. In this case, the remainder is 4, so we're done dividing, and there are four 1s left over. The answer, then, is 314_{five}. This tells us that in 84, there are three 25s, one 5, and four 1s.

Try This One   2

Write 73 in the base five system.

| Example 3 | Converting Numbers from Base 10 to Base Five |

Write 653 in the base five system.

SOLUTION

Step 1 Since 625 is the largest place value that will divide into 653, we divide by 625.

$$\begin{array}{r} 1 \\ 625\overline{)653} \\ -625 \\ \hline 28 \end{array}$$ *There is one 625 in 653.*

Step 2 Divide the remainder by 125.

$$\begin{array}{r} 0 \\ 125\overline{)28} \\ -0 \\ \hline 28 \end{array}$$ *There are no 125s in 28.*

Even though 125 doesn't divide into the 28, we still need to write zero to account for its place value in the base five number system.

Step 3 Divide by 25.

$$\begin{array}{r} 1 \\ 25\overline{)28} \\ -25 \\ \hline 3 \end{array}$$ *There is one 25 in 28.*

Step 4 Divide the remainder by 5.

$$\begin{array}{r} 0 \\ 5\overline{)3} \\ -0 \\ \hline 3 \end{array}$$

There are no 5s in 3, and 3 left over.

The solution is 10103_{five}. It's useful to note that after you've done all the successive divisions, the answer is all of the quotients from the divisions, with the final remainder in the last digit.

Check: $1 \times 625 + 0 \times 125 + 1 \times 25 + 0 \times 5 + 3 \times 1 = 653$.

Try This One   3

Write each number in the base five system.

(a) 52   (b) 486   (c) 1,000

Photodisc/Getty Images
You probably recognize this as the power button from an electronic device, but do you know what the symbol really represents? It's designed to look like a one and a zero, a tribute to the importance of the binary number system in computing.

Other Number Bases

Once we understand the idea of alternative bases, we can define new number systems with as few as two symbols, or digits. (Remember, we only needed digits zero through four for base five numbers.) For example, a base two, or **binary system**, uses only two digits, 0 and 1. The place values of the digits in the base two numeration system are powers of two:

etc. | sixteen (2^4) | eight (2^3) | four (2^2) | two (2^1) | one (2^0)

The base eight or **octal system** consists of eight digits, 0, 1, 2, 3, 4, 5, 6, and 7. The place values of the digits in the base eight system are powers of eight:

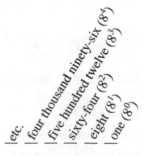

etc. | four thousand ninety-six (8^4) | five hundred twelve (8^3) | sixty-four (8^2) | eight (8^1) | one (8^0)

Math Note

The hexadecimal system is also used quite a lot in computer programming. If you didn't know that, shame on you for not reading the chapter opener on color codes in computer graphics. We'll explore the connection in Problems 103–108.

For a base number greater than 10, we have a slight problem: we need more than 10 digits, so we can't use only the numerals we typically associate with numbers. For example, in base 16 (called the hexadecimal system), we need 15 digits. So we use the digits from 0 through 9 as you would expect, then go to letters: A is used to represent 10, B represents 11, C represents 12, and so on. (We can't actually use 10 through 15 because they have two digits!) The place values of the digits in base 16 are powers of 16:

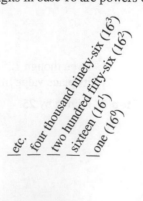

etc. | four thousand ninety-six (16^3) | two hundred fifty-six (16^2) | sixteen (16^1) | one (16^0)

Table 4-2	Base Number Systems

Base two (binary system)
Digits used: 0, 1
Place values: 2^6 2^5 2^4 2^3 2^2 2^1 2^0
Numbers: 0, 1, 10, 11, 100, 101, 110, 111, 1000, 1001, 1010, …

Math Note

To save space, we're omitting the base subscript from the list of numbers in Table 4-2. This is NOT something you should do in general!

Base three
Digits used: 0, 1, 2
Place values: 3^6 3^5 3^4 3^3 3^2 3^1 3^0
Numbers: 0, 1, 2, 10, 11, 12, 20, 21, 22, 100, 101, 102, 110, …

Base five
Digits used: 0, 1, 2, 3, 4
Place values: 5^6 5^5 5^4 5^3 5^2 5^1 5^0
Numbers: 0, 1, 2, 3, 4, 10, 11, 12, 13, 14, 20, 21, 22, …

Base eight (octal system)
Digits used: 0, 1, 2, 3, 4, 5, 6, 7
Place values: 8^6 8^5 8^4 8^3 8^2 8^1 8^0
Numbers: 0, 1, 2, 3, 4, 5, 6, 7, 10, 11, 12, 13, 14, 15, 16, 17, 20, …

Base 10
Digits used: 0, 1, 2, 3, 4, 5, 6, 7, 8, 9
Place values: 10^6 10^5 10^4 10^3 10^2 10^1 10^0
Numbers: 0, 1, 2, 3, 4, 5, 6, 7, 8, 9, 10, 11, 12, 13, 14, 15, …

Base 16 (hexadecimal system)
Digits used: 0, 1, 2, 3, 4, 5, 6, 7, 8, 9, A, B, C, D, E, F
Place values: 16^6 16^5 16^4 16^3 16^2 16^1 16^0
Numbers: 0, 1, 2, 3, 4, 5, 6, 7, 8, 9, A, B, C, D, E, F, 10, 11, …

Table 4-2 summarizes the digits for some of the base number systems and the place values of the digits in the system. It's kind of a one-stop reference for the systems we discussed. Of course, we can't list ALL of the place values for any system: they can continue indefinitely with higher and higher powers of the base number.

Looking at Table 4-2, we can make some key observations. First, the number of symbols is equal to the base. Second, the place values for any base are

$$\cdots \underline{b^6}\ \underline{b^5}\ \underline{b^4}\ \underline{b^3}\ \underline{b^2}\ \underline{b^1}\ \underline{b^0}$$

where b is the base. For example, the place values for base six are

$$\cdots \frac{46{,}656}{6^6}\ \frac{7{,}776}{6^5}\ \frac{1{,}296}{6^4}\ \frac{216}{6^3}\ \frac{36}{6^2}\ \frac{6}{6^1}\ \frac{1}{6^0}$$

Math Note

$b^0 = 1$ for any $b \neq 0$

In order to convert from numbers written in bases other than 10 to base 10, we'll use expanded notation. This is the same procedure we used in Example 1 for base five. Example 4 shows how this works for other bases.

Example 4	Converting Numbers to Base 10

Write each number in base 10.

(a) 132_{six} (b) 10110_{two} (c) $5BD8_{\text{sixteen}}$ (d) 2102_{three} (e) $A582_{\text{twelve}}$

Kevin Winter/Getty Images

A microphone converts sound to a voltage signal, which in turn is converted into a binary number. Each measurement is recorded as a 16-digit binary number and then interpreted by an amplifier.

SOLUTION

(a) The place values of the digits in base six are powers of 6:

$$132_{\text{six}} = 1 \times 6^2 + 3 \times 6^1 + 2 \times 1$$
$$= 1 \times 36 + 3 \times 6 + 2 \times 1$$
$$= 36 + 18 + 2 = 56$$

(b) The place values for base two are powers of 2:

$$10110_{\text{two}} = 1 \times 2^4 + 0 \times 2^3 + 1 \times 2^2 + 1 \times 2^1 + 0 \times 1$$
$$= 1 \times 16 + 0 \times 8 + 1 \times 4 + 1 \times 2 + 0 \times 1$$
$$= 16 + 0 + 4 + 2 + 0 = 22$$

(c) The place values for base 16 are powers of 16; B represents 11 and D represents 13.

$$5BD8_{\text{sixteen}} = 5 \times 16^3 + 11 \times 16^2 + 13 \times 16 + 8 \times 1$$
$$= 5 \times 4,096 + 11 \times 256 + 13 \times 16 + 8 \times 1$$
$$= 20,480 + 2,816 + 208 + 8 = 23,512$$

(d) The place values for base three are powers of 3:

$$2102_{\text{three}} = 2 \times 3^3 + 1 \times 3^2 + 0 \times 3^1 + 2 \times 1$$
$$= 2 \times 27 + 1 \times 9 + 0 \times 3 + 2 \times 1$$
$$= 54 + 9 + 0 + 2 = 65$$

(e) The place values for base 12 are powers of 12; A represents 10 and B represents 11.

$$A5B2_{\text{twelve}} = 10 \times 12^3 + 5 \times 12^2 + 11 \times 12 + 2 \times 1$$
$$= 10 \times 1,728 + 5 \times 144 + 11 \times 12 + 2 \times 1$$
$$= 17,280 + 720 + 132 + 2 = 18,134$$

Try This One   4

Write each number in base 10.

(a) 5320_{seven} (b) 110110_{two} (c) $42AE_{\text{sixteen}}$ (d) $9B1C_{\text{thirteen}}$

Converting Base 10 Numbers to Other Base Numbers

In Examples 2 and 3, we used division to convert base 10 numbers to base five. The same procedure can be used to convert to other base number systems as well.

Example 5   Converting Numbers to Bases Other Than 10

(a) Write 48 in base three.
(b) Write 51 in base two.
(c) Write 19,443 in base 16.

SOLUTION

(a) **Step 1** The place values for base three are powers of three. The largest power of three less than 48 is 3^3, or 27, so we divide 48 by 27.

$$\begin{array}{r} 1 \\ 27\overline{)48} \\ -27 \\ \hline 21 \end{array}$$ *There's one 27 in 48.*

Step 2 Divide the remainder by 3^2 or 9.

$$9 \overline{)21} \quad \begin{array}{l} 2 \\ \end{array}$$
$$\underline{-18} \quad \textit{There are two 9s in 21.}$$
$$3$$

Step 3 Divide the remainder by 3^1 or 3.

$$3 \overline{)3} \quad \begin{array}{l} 1 \\ \end{array} \quad \textit{There's one 3 in 3.}$$
$$\underline{-3}$$
$$0 \quad \textit{Zero 1s left over.}$$

So 48 is $1 \times 3^3 + 2 \times 3^2 + 1 \times 3^1 + 0 \times 3^0$, which makes it 1210_{three}.

(b) The place values for base two are 1, 2, 4, 8, 16, 32, etc. Use successive division, as shown.

$$\begin{array}{ccccc}
1 & 1 & 0 & 0 & 1 \\
32\overline{)51} & 16\overline{)19} & 8\overline{)3} & 4\overline{)3} & 2\overline{)3} \\
\underline{32} & \underline{16} & \underline{0} & \underline{0} & \underline{2} \\
19 & 3 & 3 & 3 & 1
\end{array}$$

So $51 = 110011_{two}$.

(c) The place values in base 16 are 1, 16, 256 (16^2), 4,096 (16^3), etc. Use successive division as shown. (Remember, in base 16, B plays the role of 11 and F plays the role of 15.)

$$\begin{array}{ccc}
4 & B & F \\
4,096\overline{)19,443} & 256\overline{)3,059} & 16\overline{)243} \\
\underline{16,384} & \underline{2,816} & \underline{240} \\
3,059 & 243 & 3
\end{array}$$

So $19,443 = 4BF3_{sixteen}$.

Try This One   5

1. Convert between base 10 and other bases.

(a) Write 84 in base two.
(b) Write 1,258 in base 12.
(c) Write 122 in base three.
(d) Write 874 in base 16.

Base Numbers and Computers

Computers use three bases to perform operations: base two (binary), base eight (octal), and base 16 (hexadecimal).

Base two is used because it has just two characters, 0 and 1. The processor of a computer is like a series of tiny switches, with programs that tell each switch whether it should be on or off. This works perfectly with the binary system.

In computer programming, binary characters (bits) are often grouped into groups of eight (a byte). This is where the base eight system is useful. See the Sidelight on page 194 for more information.

The base 16 system is used for several reasons. First of all, 16 characters consist of 2 bytes. Also, 16 is 2^4, which means one hexadecimal character can replace four binary characters. This increases the speed at which the computer is able to perform numerical applications since there are fewer characters for the computer to read and fewer operations to perform. Finally, large numbers can be written in base 16 with fewer characters than in base two or base 10, saving much needed space in the computer's memory.

There's a shortcut for converting between binary and octal or binary and hexadecimal. To use this shortcut, first notice that every octal digit can be written as a three-digit binary number, and every hexadecimal digit can be written as a four-digit binary number, as illustrated in Table 4-3.

Table 4-3		Binary Equivalents for Octal and Hexadecimal Digits			

Octal Digit	Binary Equivalent	Hex Digit	Binary Equivalent	Hex Digit	Binary Equivalent
0	000	0	0000	8	1000
1	001	1	0001	9	1001
2	010	2	0010	A	1010
3	011	3	0011	B	1011
4	100	4	0100	C	1100
5	101	5	0101	D	1101
6	110	6	0110	E	1110
7	111	7	0111	F	1111

Sidelight   Bar Codes

Bar codes are a series of black and white stripes that vary in width. The width of each stripe determines a binary digit that the scanner then decodes. The most familiar bar codes are the UPC codes that appear on almost every product you buy. Most UPCs have a left and right margin that tells the reader where to begin and end, a five-digit manufacturer's code, a check digit in the middle, then a five-digit product code. Once a register reads the code and knows what the product is, it computes the price that was programmed in for that product.

Nick Koudis/Photodisc/Getty Images

Left margin   Check digit   Right margin

1 8 2 0 0   5 3 0 4 7

Manufacturer's code   Product code

To read a UPC, you need the information in the table below. Every digit in the base 10 system is represented by a seven-digit binary number.

Digit	Manufacturer's Number	Product Number
0	0001101	1110010
1	0011001	1100110
2	0010011	1101100
3	0111101	1000010
4	0100011	1011100
5	0110001	1001110
6	0101111	1010000
7	0111011	1000100
8	0110111	1001000
9	0001011	1110100

Notice that all of the binary numbers in the manufacturer's code begin with zero, while those in the product code begin with one. This is done so that the scan can be done from left to right, or right to left. The computer can tell the correct direction by recognizing the difference in the digits. As an example of a digit, the digit five is represented by 1001110 in the product code. This number is represented with black and white bars, with a single-width white bar representing a zero and a single-width black bar representing a one.

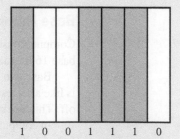

1   0   0   1   1   1   0

Notice how this matches the first four stripes to the right of the check digit in the UPC to the left (without the borders around the individual bars). Every digit in the left half of a UPC is represented by four stripes: white, then black, then white, then black. On the right half, it's black, then white, then black, then white. The key is to recognize the width of each stripe to decide how many bars it represents.

It takes a little bit of practice to distinguish bars that represent one, two, three, and four consecutive digits, but once you can do that, you can decode UPCs—just not as fast as a computer can.

The following examples illustrate the shortcut for converting between these bases.

| Example 6 | Converting between Binary and Octal |

(a) Convert the binary number 1001110110_{two} to octal.

(b) Convert the octal number 7643_{eight} to binary.

SOLUTION

(a) Starting at the rightmost digit, group the digits of the binary number into groups of three (if there are not three digits that remain at the left of the number, fill them in with zeros), then use Table 4-3 to change each group to an octal digit as follows.

$$001 \quad 001 \quad 110 \quad 110$$
$$1 \quad\quad 1 \quad\quad 6 \quad\quad 6$$

So $1001110110_{two} = 1166_{eight}$.

(b) First, convert each octal digit into a three-digit binary digit using Table 4-3, and then string them together to form a binary number.

$$7 \quad\quad 6 \quad\quad 4 \quad\quad 3$$
$$111 \quad 110 \quad 100 \quad 011$$

So $7643_{eight} = 111110100011_{two}$.

| Try This One | 6 |

(a) Convert 1100101_{two} to octal.   (b) Convert 6147_{eight} to binary.

| Example 7 | Converting between Binary and Hexadecimal |

(a) Convert the binary number 1110011001111_{two} to hexadecimal.

(b) Convert the hexadecimal number $9D7A3_{sixteen}$ to binary.

SOLUTION

(a) Starting at the rightmost digit, group the binary number into groups of four (adding zeros in front as needed) and then convert each group of four to a hexadecimal digit using Table 4-3.

$$0001 \quad 1100 \quad 1100 \quad 1111$$
$$1 \quad\quad C \quad\quad C \quad\quad F$$

So $1110011001111_{two} = 1CCF_{sixteen}$.

(b) First, convert each hexadecimal number to a four-digit binary number, and then string them together to form a binary number.

$$9 \quad\quad D \quad\quad 7 \quad\quad A \quad\quad 3$$
$$1001 \quad 1101 \quad 0111 \quad 1010 \quad 0011$$

So $9D7A3_{sixteen} = 10011101011110100011_{two}$.

| Try This One | 7 |

(a) Convert 100011001_{two} to hexadecimal.   (b) Convert $ABCD_{sixteen}$ to binary.

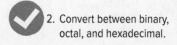

2. Convert between binary, octal, and hexadecimal.

Sidelight   The Binary System and Computer Storage

Most computer users are familiar with the term megabytes, or MB—it's one of the standard measurements of file size. Our study of octal numbers can give us some insight into just how much information goes into a typical computer file. One character is called a bit (short for "binary digit"), and 8 bits make up 1 byte of storage. A kilobyte (kB) is 1,000 bytes, and a megabyte (MB) is 1,000 kilobytes, or 1 million bytes. (There is a slight difference between processor memory, where 1 MB is 1,024 kB, and disk storage, where 1 MB is 1,000 kB.)

This raises the question: exactly how many characters are in certain computer files? A typical MP3 song file is in the neighborhood of 4 MB. Doing some simple multiplying, this means there are 4 million bytes, each of which contains 8 characters, so there are 32 million characters in a typical MP3. To get some perspective on how much information that is, think of it this way: if you were to try to write out those characters by hand, writing at a reasonable pace of one per second, you better pack a lunch. In fact, you better pack a little heavier than

that—it would take a little bit over 370 *days* to write that many characters, and that's writing constantly with no sleep breaks! An iPhone with 64 gigabytes of storage (a gigabyte is 1,000 megabytes) contains 5.12×10^{11}, or 512 billion characters. It's not very easy to get some perspective on how big a number that is, but if every single person on the face of the Earth got together to write out those characters, they'd have to write 73 characters each. In fact, the number of characters is just about half of the total number of human beings that have ever lived! And all on a phone that fits in your pocket.

Screen capture by Dave Sobecki, artwork by Mircea Gabriel Eftemie, ©Benedictum

Answers to Try This One

1 (a) 77   (b) 214   (c) 2,500   (d) 68

2 243_{five}

3 (a) 202_{five} (b) 3421_{five} (c) 13000_{five}

4 (a) 1,876   (b) 54   (c) 17,070   (d) 21,657

5 (a) 1010100_{two} (b) $88A_{twelve}$ (c) 11112_{three}
(d) $36A_{sixteen}$

6 (a) 145_{eight} (b) 110001100111_{two}

7 (a) $119_{sixteen}$ (b) 1010101111001101_{two}

Exercise Set   4-3

Writing Exercises

1. Explain why the base two system requires only two digits.
2. Explain the significance of place values in base number systems. What does each place value represent relative to the base number?
3. In computer terms, what is a bit? What about a byte? What do these terms have to do with base number systems?
4. How can you tell how many characters are needed for a given base number system?
5. Explain why letters are needed in base number systems with a base larger than 10.
6. What does the first digit in a four-digit base eight number represent? How did you come up with your answer?

Computational Exercises

For Exercises 7–26, convert each number to base 10.

7. 1011_{two}
8. 372_{twelve}
9. 53_{six}
10. $8A21_{twelve}$

11. 99_{eleven}
12. 3451_{seven}
13. 10221_{three}
14. 110011_{two}

15. 2221_{five}
16. 5320_{eight}
17. 153_{six}
18. 11001_{two}
19. 438_{nine}
20. 2561_{eight}

21. 352_{seven}
22. 11112_{five}
23. $921E_{sixteen}$
24. $7CBA_{thirteen}$
25. $9E52_{sixteen}$
26. 10000001_{two}

For Exercises 27–46, write each base 10 number in the given base.

27. 31 in base two
28. 186 in base eight
29. 345 in base six
30. 266 in base three
31. 16 in base seven
32. 3,050 in base 12
33. 745 in base nine
34. 10,455 in base 14
35. 22 in base two
36. 5,621 in base 11
37. 18 in base five
38. 97 in base four
39. 2,361 in base 16
40. 96 in base two
41. 18,432 in base five
42. 25,000 in base 16
43. 24,189 in base 13
44. 88 in base three
45. 256 in base two
46. 497 in base four

For Exercises 47–56, convert each number to base 10, then change to the specified base.

47. 134_{six} to base two
48. 1011010_{two} to base eight
49. 342_{five} to base 12
50. 4711_{nine} to base three
51. 1221_{three} to base four
52. 1521_{seven} to base three
53. 432_{eight} to base 12
54. $3AB_{twelve}$ to base six
55. $1782_{fourteen}$ to base seven
56. 3000_{four} to base 11

For Exercises 57–64, convert each binary number (a) to octal then (b) to hexadecimal. (The subscript two is omitted since it's understood that all numbers are binary.)

57. 110011111
58. 1011101
59. 10111011001
60. 101100001
61. 11111111101
62. 11110110111
63. 100011000111
64. 11110100110111

For Exercises 65–72, convert each octal number to binary. (The subscript eight is omitted since it's understood that all numbers are octal.)

65. 1570
66. 354
67. 6504
68. 76241
69. 620
70. 2742
71. 72165
72. 4731

For Exercises 73–80, convert each hexadecimal number to binary. (The subscript 16 is omitted since it's understood that all numbers are hexadecimal.)

73. 6A2B
74. 657F1
75. 362
76. 9E73
77. A5D
78. 95C1
79. 654B
80. AF3

Applications in Our World

Problems similar to base number problems occur often in everyday life. Using a procedure like the one used in Examples 2, 3, and 5, solve each.

81. Change 87 ounces to pounds and ounces (1 lb = 16 oz).
82. Change 237 ounces to quarts, pints, and ounces (1 qt = 2 pt = 32 oz).
83. Change 1,256 inches to yards, feet, and inches (1 yd = 3 ft = 36 in.).
84. Change $5.88 to quarters, dimes, nickels, and pennies using the smallest number of coins.

A process of encoding data using two symbols is called binary coding. One of the earliest codes was the Morse code. Words were encoded using dots and dashes and transmitted by sound. If we consider a dot as a zero and a dash as a one, the various letters of the alphabet can be transformed into a binary code, as follows.

A	01	J	0111	S	000
B	1000	K	101	T	1
C	1010	L	0100	U	001
D	100	M	11	V	0001
E	0	N	10	W	011
F	0010	O	111	X	1001
G	110	P	0110	Y	1011
H	0000	Q	1101	Z	1100
I	00	R	010		

For Exercises 85–90, decode the following messages.

85. 000 1 111 0110
86. 10 111 0000 111 11 0 011 111 010 101
87. 1010 0100 01 000 000 00 000 111 0001 0 010

88. 010 0 000 1 0010 111 010 0100 001 10 1010 0000
89. 00 1 00 000 010 01 00 10 00 10 110
90. 1 001 010 10 01 010 111 001 10 100

The U.S. Postal Service uses a Postnet code on business reply forms. This code consists of a five-digit Zip code plus a four-digit extension and a check digit. This code consists of long and short vertical bars.

Digit	Bar code	Digit	Bar code
1	ⅠⅠⅠⅠⅠ	6	Ⅰ Ⅰ ⅠⅠ
2	ⅠⅠⅠⅠⅠ	7	Ⅰ ⅠⅠⅠ
3	ⅠⅠ ⅠⅠ	8	Ⅰ Ⅰ ⅠⅠ
4	Ⅰ ⅠⅠⅠ	9	Ⅰ Ⅰ ⅠⅠ
5	Ⅰ Ⅰ ⅠⅠ	0	ⅠⅠⅠⅠⅠ

For example, the Zip code 15131 would be encoded as follows:

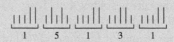

Using the Postnet code, find the following five-digit Zip codes for Exercises 91–96.

91.
92.

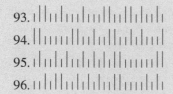

93.

94.

95.

96.

For Exercises 97–102, write each Zip code using the Postnet code.

97. 26135
98. 14157
99. 18423
100. 30214
101. 11672
102. 54901

Colors on Web pages are determined by a six-digit hexadecimal code number that describes the amount of red, green, and blue to mix. The first pair of digits determines the amount of red, the second pair the amount of green, and the final pair the amount of blue. The number 00 represents none of that particular color, while the number FF represents the maximum amount. In Exercises 103–108, you're given six codes for common colors. Determine the percentage of the maximum amount used for each of red, green, and blue.

103. Light blue: ADD8E6
104. Brown: A52A2A
105. Steel blue: 4863A0
106. Violet: 8D38C9
107. Magenta: FF00FF
108. Orange: FFA500

Critical Thinking

For Exercises 109–114, use the information found in the Sidelight found on page 192. Identify the digit shown by the bar code. State whether the digit belongs to the manufacturer's number or the product number, then describe how the corresponding bars would be drawn.

109. 0111101
110. 0101111
111. 1100110

112. 1110010
113. 0111011
114. 1001000

115. Think about how you would write numbers in a base one system. The result is actually a numeration system we discussed in Section 4-1. Which system? Explain.

116. What base number system could be defined using just the symbols from the Babylonian numeration system? What about the Egyptian and Mayan systems?

117. (a)   Fill in the table below with binary equivalents for base 10 numbers.

Base 10	1	2	3	4	5	6	7	8	9
Binary									

Base 10	10	11	12	13	14	15	16	17	18
Binary									

(b)   Which base 10 numbers have two digits in their binary form? Three digits? Four digits?

(c)   Make a conjecture as to which base 10 numbers have five digits in their binary form. How can you check your answer by finding the binary form of just one base 10 number?

118. Based on the results of Exercise 117, find a formula for the number of nonzero base 10 numbers that have n digits in their binary form. Use your formula and the results of Exercise 117 to find the base 10 number that has binary form 1000000 without directly converting

that binary number to base 10, then check your answer by converting directly.

119. (a)   Build a table similar to the one in Exercise 117, but for base three rather than binary forms. Continue the table until the first base 10 number that has four digits in its base three form.

(b)   How many nonzero base 10 numbers have one digit in their base three form? Two digits? Three digits?

(c)   Make a conjecture as to which base 10 numbers have four digits in their base three form, then write a formula for the number of nonzero base 10 numbers that have n digits in their base three form.

120. (a)   Let's change our approach from Exercises 117–119. We have seen that for any base n, the first two-digit number is 10, the first three-digit number is 100, and so on. Find the base 10 equivalent of 10_{four}, 100_{four}, 1000_{four}, and 10000_{four}. How many numbers in base four have one digit? Two digits? Three digits? Four digits?

(b)   Make a conjecture as to how many base four numbers have five digits.

(c)   (This one is challenging! Demand extra credit.) Based on all of the information in Exercises 117–120, write a general formula that tells you how many nonzero numbers have n digits in their base b expansion.

121. When working in base number systems, it's really important to know that $b^0 = 1$ for any positive base b. But why is this the case? You certainly can't multiply a number by itself zero times and get 1. Is there a reason, or is it just made up? Let's see. Throughout, we'll assume that b is some positive number.

(a)   What is $b^2 \cdot b^5$? Write a verbal description of how you got your answer.

(b)   Using the same concept from part (a), what is $b^2 \cdot b^0$?

(c)   For *any* exponent n, what is $b^n \cdot b^0$?

(d)   Why does this tell you what b^0 has to equal?

| Section 4-4 | **Operations in Base Number Systems** |

LEARNING OBJECTIVES

1. Add in bases other than 10.
2. Subtract in bases other than 10.
3. Multiply in bases other than 10.
4. Divide in bases other than 10.

Now that we know how to write numbers in other base systems, it seems reasonable to take a look at working with numbers in those systems using the basic operations of addition, subtraction, multiplication, and division. To be honest you could convert the numbers to base 10, perform the operations the way you always have, then convert the answer back. But we're going to advise you against doing that. The real value of this section is the mental exercise of adapting arithmetic skills you learned a long time ago to a new situation. This mental flexibility has a lot of benefits: it helps give a deeper understanding of the familiar operations in arithmetic and it's also good practice for problem solving, which almost always involves using some basic skills you've acquired and adapting them to a certain situation.

For the most part, arithmetic operations can be performed in other base number systems just as they are performed in base 10. For example, $4_{\text{five}} + 3_{\text{five}} = 12_{\text{five}}$. Why? For the same reason that $9 + 3 = 12$ in base 10. When adding 3 to 9, we get past the highest digit in base 10 (which is nine), so we "wrap around" and get one 10, with two 1s left over: in other words, 12. In $4_{\text{five}} + 3_{\text{five}} = 12_{\text{five}}$, adding three gets us past the highest digit in base five (which is four), and we get one 5, with two 1s left over.

You're probably able to add very well in base 10 because you learned the sums of all of the digits back when you first started school. When working in a different base, you can build an addition table that displays the sums of all the digits in that system to stand in for those years of experience. A table for base five is shown on the next page. (The subscript five has been omitted.) All of the values were obtained using the same reasoning we used to add 4 and 3 in base five. It would be *great* practice for you to write out a table like this by yourself.

Math Note

Computations like $4_{\text{five}} + 3_{\text{five}} = 12_{\text{five}}$ probably look bizarre to you, since your past experience tells you that 4 plus 3 is not 12. This is where it really helps a lot to read 12_{five} not as "twelve base five," but rather "one two base five."

+	0	1	2	3	4
0	0	1	2	3	4
1	1	2	3	4	10
2	2	3	4	10	11
3	3	4	10	11	12
4	4	10	11	12	13

In order to add using the table, find one number in the left column and the other number in the top row. Then draw a horizontal and a vertical line. The intersection of the lines is the sum. For example, $2_{\text{five}} + 4_{\text{five}} = 11_{\text{five}}$, as shown below.

+	0	1	2	3	④
0	0	1	2	3	4
1	1	2	3	4	10
②	2	3	4	10	⑪
3	3	4	10	11	12
4	4	10	11	12	13

We perform addition of numbers with more than one digit in base five the same way we do in base 10: by lining up the numbers vertically and adding one digit at a time. When a result is two digits, we will apply the first digit to the next column on the left. (You probably called this "carrying" a digit in elementary school.)

| **Example 1** | Adding in Base Five |

Add in base five: $324_{five} + 24_{five}$.

SOLUTION

Step 1 Add the digits in the ones column, 4 and 4. $4_{five} + 4_{five} = 13_{five}$ (see the table). Write the 3 in the ones place; the digit 1 represents one 5, so we apply it to the fives column.

$$\begin{array}{r} \overset{1}{324}_{five} \\ + \ 24_{five} \\ \hline 3_{five} \end{array}$$

Step 2 Add $1_{five} + 2_{five} + 2_{five} = 10_{five}$ (see the table). Write the 0 and apply the 1 to the 25s column as shown.

$$\begin{array}{r} \overset{1}{324}_{five} \\ + \ 24_{five} \\ \hline 03_{five} \end{array}$$

Step 3 Add $1_{five} + 3_{five} = 4_{five}$ (see the table). Write the 4 as shown.

$$\begin{array}{r} 324_{five} \\ + \ 24_{five} \\ \hline 403_{five} \end{array}$$

The sum in base five is $324_{five} + 24_{five} = 403_{five}$.

Math Note

The answer can be checked by converting the numbers to base 10 and seeing if the answers are equal:

$$\begin{array}{r} 324_{five} = \ 89_{ten} \\ +24_{five} = \ 14_{ten} \\ \hline 403_{five} = 103_{ten} \end{array}$$

Try This One   1

Add in base five: $324_{five} + 203_{five}$.

In the remainder of this section, to save a bit of space, we're going to use the term *carrying* to refer to applying a second digit in a sum to the next place column to the left. But please don't interpret that as an invitation to memorize that routine as a "trick" of some sort. Think about what we're actually doing by using Example 1 as a reference: in that case, the extra digit in a sum represents 1 of the next higher power of base five, so adding that 1 into the other numbers with that power makes perfect sense.

| **Example 2** | Adding in Base Five |

Add in base five: $1244_{five} + 333_{five}$.

SOLUTION

$$\begin{array}{r} \overset{111}{1244}_{five} \\ + \ 333_{five} \\ \hline 2132_{five} \end{array}$$

$4_{five} + 3_{five} = 12_{five}$ Write the 2 and carry the 1.
$1_{five} + 4_{five} + 3_{five} = 13_{five}$ Write the 3 and carry the 1.
$1_{five} + 2_{five} + 3_{five} = 11_{five}$ Write the 1 and carry the 1.
$1_{five} + 1_{five} = 2_{five}$

The sum is 2132_{five}.

Try This One   2

Add in base five: $4301_{five} + 2024_{five}$.

Of course, addition can be performed in other bases as well as in base five. It might help to construct an addition table for the given base.

Example 3	Adding in Base Two

Add in base two: $10111_{two} + 110_{two}$.

SOLUTION

The addition table for base two is

+	0	1
0	0	1
1	1	10

Then

$$\begin{array}{r} {}^{11} \\ 10111_{two} \\ + \quad 110_{two} \\ \hline 11101_{two} \end{array}$$

$1_{two} + 0_{two} = 1_{two}$

$1_{two} + 1_{two} = 10_{two}$ Write the 0 and carry 1.

$1_{two} + 1_{two} + 1_{two} = 11_{two}$ Write the 1 and carry 1.

$1_{two} + 0_{two} = 1_{two}$

bring down the 1

The sum is 11101_{two}.

Try This One 3

Add in base two: $11001_{two} + 1001_{two}$.

The addition table for the hexadecimal system is shown next. Remember that in base 16, the digit A corresponds to 10, B to 11, C to 12, D to 13, E to 14, and F to 15.

+	0	1	2	3	4	5	6	7	8	9	A	B	C	D	E	F
0	0	1	2	3	4	5	6	7	8	9	A	B	C	D	E	F
1	1	2	3	4	5	6	7	8	9	A	B	C	D	E	F	10
2	2	3	4	5	6	7	8	9	A	B	C	D	E	F	10	11
3	3	4	5	6	7	8	9	A	B	C	D	E	F	10	11	12
4	4	5	6	7	8	9	A	B	C	D	E	F	10	11	12	13
5	5	6	7	8	9	A	B	C	D	E	F	10	11	12	13	14
6	6	7	8	9	A	B	C	D	E	F	10	11	12	13	14	15
7	7	8	9	A	B	C	D	E	F	10	11	12	13	14	15	16
8	8	9	A	B	C	D	E	F	10	11	12	13	14	15	16	17
9	9	A	B	C	D	E	F	10	11	12	13	14	15	16	17	18
A	A	B	C	D	E	F	10	11	12	13	14	15	16	17	18	19
B	B	C	D	E	F	10	11	12	13	14	15	16	17	18	19	1A
C	C	D	E	F	10	11	12	13	14	15	16	17	18	19	1A	1B
D	D	E	F	10	11	12	13	14	15	16	17	18	19	1A	1B	1C
E	E	F	10	11	12	13	14	15	16	17	18	19	1A	1B	1C	1D
F	F	10	11	12	13	14	15	16	17	18	19	1A	1B	1C	1D	1E

| Example 4 | Adding in Base 16 |

Add in base 16: $135E_{sixteen} + 21C_{sixteen}$.

SOLUTION

Math Note

If I were a betting man, I'd wager upwards of 10 dollars that you never thought you'd add E and C to get 1A. It's amazing what you can do when you open your mind to thinking outside the base 10 box!

$$\overset{1}{135E_{sixteen}}$$
$$\underline{+\ 21C_{sixteen}}$$
$$157A_{sixteen}$$

— $E_{sixteen} + C_{sixteen} = 1A_{sixteen}$ Write the A and carry 1.

— $1_{sixteen} + 5_{sixteen} + 1_{sixteen} = 7_{sixteen}$

— $3_{sixteen} + 2_{sixteen} = 5_{sixteen}$

— Bring down the 1

The sum is $157A_{sixteen}$.

| Try This One | 4 |

Add in base 16: $8D51_{sixteen} + 947A_{sixteen}$

1. Add in bases other than 10.

Now that we know how to perform addition in other bases, we should be able to subtract as well. Addition tables can help with subtraction; the addition table for base five follows. To perform a subtraction like $12_{five} - 4_{five}$, we find 4 in the far left column of the table, then move across that row until we find 12. The number at the top of the column is the difference: $12_{five} - 4_{five} = 3_{five}$.

+	0	1	2	③	4
0	0	1	2	3	4
1	1	2	3	4	10
2	2	3	4	10	11
3	3	4	10	11	12
④	4	10	11	⑫	13

We will subtract numbers with more than one digit using a method similar to addition. When the digit to be subtracted is larger, as in the first step of Example 5, we'll need to "borrow" a one from the next larger place value.

| Example 5 | Subtracting in Base Five |

Subtract in base five:

$$\begin{array}{r} 321_{five} \\ -\ 123_{five} \end{array}$$

SOLUTION

Step 1   Since 3 is larger than 1, we need to borrow a one from the next column; this makes the subtraction $11_{five} - 3_{five} = 3_{five}$. Change 2 in the fives column to a 1.

$$\begin{array}{r} \overset{1}{}\ \overset{11}{2}\ \ \\ 3\ \ \cancel{2}\ \ \cancel{1}_{five} \\ -1\ \ 2\ \ 3_{five} \\ \hline 3_{five} \end{array}$$

Math Note

The answer can be checked by adding $143_{\text{five}} + 123_{\text{five}}$ and seeing if the answer is 321_{five} (which it is).

Step 2 In the second column, $1_{\text{five}} - 2_{\text{five}}$ requires borrowing; change 3 in the third column to 2 and take $11_{\text{five}} - 2_{\text{five}}$ to get 4_{five}.

$$\begin{array}{ccc} 2 & 11 & \\ \cancel{3} & \cancel{2} & 1_{\text{five}} \\ -1 & 2 & 3_{\text{five}} \\ \hline & 4 & 3_{\text{five}} \end{array}$$

Step 3 Subtract $2_{\text{five}} - 1_{\text{five}}$ to get 1_{five} (no borrowing necessary).

$$\begin{array}{ccc} 2 & & \\ \cancel{3} & 2 & 1_{\text{five}} \\ -1 & 2 & 3_{\text{five}} \\ \hline 1 & 4 & 3_{\text{five}} \end{array}$$

The difference is $321_{\text{five}} - 123_{\text{five}} = 143_{\text{five}}$.

Try This One 5

Perform the indicated operation: $7316_{\text{eight}} - 1257_{\text{eight}}$.

2. Subtract in bases other than 10.

The multiplication table for base five is shown next.

×	0	1	2	3	4
0	0	0	0	0	0
1	0	1	2	3	4
2	0	2	4	11	13
3	0	3	11	14	22
4	0	4	13	22	31

For example, $3_{\text{five}} \times 4_{\text{five}} = 22_{\text{five}}$ ($3 \times 4 = 12$ in base 10, which is 22_{five}).
Multiplication is done in base five using the same basic procedure you learned in base 10.

Example 6 **Multiplying in Bases Five and Two**

Math Note

As always, you can check your answer by converting to base 10.

Perform each multiplication in the given base.

(a) $\begin{array}{r} 314_{\text{five}} \\ \times\ 23_{\text{five}} \end{array}$ (b) $\begin{array}{r} 1011_{\text{two}} \\ \times\ 11_{\text{two}} \end{array}$

SOLUTION

(a) First, we multiply each digit in 314_{five} by the last digit in 23_{five}.

Step 1 Multiply $4_{\text{five}} \times 3_{\text{five}} = 22_{\text{five}}$. Write the second digit and carry the first to the fives column.

$$\begin{array}{r} 2 \\ 314_{\text{five}} \\ \times\ 23_{\text{five}} \\ \hline 2_{\text{five}} \end{array}$$

Step 2 Multiply $1_{\text{five}} \times 3_{\text{five}}$ to get 3_{five} and then add the carried 2_{five} to get 10_{five}. Write the 0 and carry the 1 to the 25s column.

$$\begin{array}{r} 1 \\ 314_{\text{five}} \\ \times\ 23_{\text{five}} \\ \hline 02_{\text{five}} \end{array}$$

Step 3 Multiply $3_{five} \times 3_{five} = 14_{five}$ and add the carried 1_{five} to get 20_{five}.

$$
\begin{array}{r}
1 \\
314_{five} \\
\times\ 23_{five} \\
\hline
2002_{five}
\end{array}
$$

Now, repeat, multiplying each digit in 314_{five} by the first digit in 23_{five}.

Step 4 Multiply $4_{five} \times 2_{five} = 13_{five}$. Write the 3 in the fives column and carry the 1.

$$
\begin{array}{r}
1 \\
314_{five} \\
\times\ 23_{five} \\
\hline
2002_{five} \\
30
\end{array}
$$ *Put zero in the ones column.*

Step 5 Multiply $1_{five} \times 2_{five} = 2_{five}$ and add the carried 1 to get 3_{five}.

$$
\begin{array}{r}
314_{five} \\
\times\ 23_{five} \\
\hline
2002_{five} \\
330
\end{array}
$$

Step 6 Multiply $3_{five} \times 2_{five}$ to get 11_{five}.

$$
\begin{array}{r}
314_{five} \\
\times\ 23_{five} \\
\hline
2002_{five} \\
11330_{five}
\end{array}
$$

Step 7 Add the partial products in base five.

$$
\begin{array}{r}
314_{five} \\
\times\ 23_{five} \\
\hline
2002_{five} \\
11330_{five} \\
\hline
13332_{five}
\end{array}
$$

The product is $314_{five} \times 23_{five} = 13332_{five}$.

(b) The multiplication table for base two is

$$
\begin{array}{c|cc}
\times & 0 & 1 \\
\hline
0 & 0 & 0 \\
1 & 0 & 1
\end{array}
$$

Using the procedure from part (a),

$$
\begin{array}{r}
1011_{two} \\
\times\ 11_{two} \\
\hline
1011 \\
10110 \\
\hline
100001_{two}
\end{array}
$$

(Remember to add the partial products in base two!) The product is 100001_{two}.

Try This One   6

Perform each multiplication in the given base.

(a) $\begin{array}{r} 321_{four} \\ \times\ 12_{four} \end{array}$

(b) $\begin{array}{r} 621_{eight} \\ \times\ 45_{eight} \end{array}$

✓ 3. Multiply in bases other than 10.

Division is performed in the same way that long division is performed in base 10. The basic procedure for long division is illustrated in Example 7.

It will be helpful to look at the multiplication table for the given base in order to find the quotients.

| Example 7 | Dividing in Base Five |

Divide in base five:

$$3_{\text{five}}\overline{)2032_{\text{five}}}$$

SOLUTION

Step 1 Using the multiplication table for base five, we need to find a product less than or equal to 20_{five} that is divisible by 3_{five}. This is done as follows:

×	0	1	2	3	4
0	0	0	0	0	0
1	0	1	2	3	4
2	0	2	4	11	13
3	0	3	11	14	22
4	0	4	13	22	31

The number we need is 14_{five} and $3_{\text{five}} \times 3_{\text{five}} = 14_{\text{five}}$. The first digit in the quotient is 3_{five}.

$$3_{\text{five}}\overline{)\overset{3}{2032_{\text{five}}}}$$

Step 2 Then multiply $3_{\text{five}} \times 3_{\text{five}} = 14_{\text{five}}$ and write the quotient under 20. Subtract $20_{\text{five}} - 14_{\text{five}}$ to get 1_{five}, then bring down the next digit.

$$3_{\text{five}}\overline{)\overset{3}{2032_{\text{five}}}}$$
$$\underline{14}\downarrow$$
$$13$$

Step 3 Next find a product smaller than or equal to 13_{five} in the table. It is 11_{five}. Since $3_{\text{five}} \times 2_{\text{five}} = 11_{\text{five}}$, write the 2 in the quotient. Then multiply $2_{\text{five}} \times 3_{\text{five}} = 11_{\text{five}}$. Write the 11 below the 13. Subtract $13_{\text{five}} - 11_{\text{five}}$, which is 2_{five}, and then bring down the 2.

$$3_{\text{five}}\overline{)\overset{32}{2032_{\text{five}}}}$$
$$\underline{14}$$
$$13$$
$$\underline{11}\downarrow$$
$$22$$

Step 4 Find a product in the multiplication table divisible by 3_{five} that is less than or equal to 22_{five}. It is 22_{five} since $3_{\text{five}} \times 4_{\text{five}} = 22_{\text{five}}$. Write the 4 in the quotient and the 22 below the 22 in the problem. Subtract.

$$3_{\text{five}}\overline{)\overset{324}{2032_{\text{five}}}}$$
$$\underline{14}$$
$$13$$
$$\underline{11}$$
$$22$$
$$\underline{22}$$
$$0$$

The remainder is 0. So $2032_{\text{five}} \div 3_{\text{five}} = 324_{\text{five}}$.

This answer can be checked by multiplication, as in Example 6.

$$324_{\text{five}}$$
$$\times\ \ 3_{\text{five}}$$
$$\overline{2032_{\text{five}}}$$

Try This One 7

Divide in base five:

$$4_{\text{five}}\overline{)112_{\text{five}}}$$

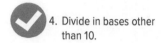

4. Divide in bases other than 10.

Hopefully, studying a variety of numeration systems in this chapter has given you some perspective on the awesome scope of the study of math by human beings, from scratching tallies on a cave wall to maybe one day finding life among the stars. Sometimes, you need to take a step back to appreciate something that you take for granted. In this case, maybe you'll feel a little bit better about the system of numbers and operations you learned as a child by comparing it to other systems that are unfamiliar to you.

Sidelight Grace Murray Hopper (1906–1992)

Computer Genius

Grace Murray Hopper received a doctorate in mathematics from Yale in 1934, and taught at Vassar College until she felt the call of duty in 1943. She was granted a leave of absence from Vassar and volunteered for the U.S. Naval Reserve. Grace had to receive a special exemption to join, since at 105 pounds soaking wet, she was 15 pounds below the minimum weight. After graduating first in her class from the Naval Reserve Shipman's School in Massachusetts in 1944, she was commissioned a lieutenant, junior grade, and was assigned to work with Dr. Howard Aiken at the Harvard Computer Laboratory. Drs. Hopper and Aiken collaborated with others on the MARK I, which is considered by some to be the first modern computer.

During her time at Harvard, she developed a computer known as "FLOWMATIC," which enabled programmers to write programs in simple languages such as COBOL and have them translated into the complicated language that the computer uses.

During her lifetime, she received honorary degrees from more than 40 colleges and universities in the United States. She retired from the U.S. Navy in 1966 at the age of 60 only to return to active duty a year later. She stayed on until the mid-1980s, when she retired for the last time as a Rear Admiral. Her retirement ceremony was held aboard the *USS Constitution,* and she was recognized as "the greatest living female authority in the computer field" at that time.

Naval Historical Center

The girl who was too small to join the Navy ended up being awarded the Defense Distinguished Service Medal, the Legion of Merit, the Meritorious Service Medal, the American Campaign Medal, the World War II Victory Medal, the National Defense Service Medal, the Armed Forces Reserve Medal, and the Naval Reserve Medal. Three months after her passing in 1992, the U.S. Navy destroyer USS *Hopper* was commissioned in her honor, and continues to carry on the service that Grace dedicated her life to.

Answers to Try This One

1 1032_{five}

2 11330_{five}

3 100010_{two}

4 $121CB_{\text{sixteen}}$

5 6037_{eight}

6 (a) 11112_{four} (b) 34765_{eight}

7 13_{five}

Exercise Set   **4-4**

Writing Exercises

1. While hanging around one evening, your roommate is paging through your math book, and sees the calculation $5_{six} + 4_{six} = 13_{six}$, then says "This is totally stupid . . . five plus four isn't thirteen." How would you explain why that calculation is correct?

2. When asked to do arithmetic in a base number system, you could convert all numbers to base 10, perform the calculation, then convert the answer back. Explain why this is a bad idea.

3. Explain how you can use an addition table for a given base to perform subtraction.

4. When adding in a base number system, whenever the result of an addition in a column has two digits, we "carry" the first digit to the next column to the left. Explain what we're really doing when we do this, and why it makes sense.

Computational Exercises

5. Make an addition table for base three.
6. Make an addition table for base four.
7. Make a multiplication table for base four.
8. Make a multiplication table for base three.

For Exercises 9–40, perform the indicated operations.

9. 33_{four}
   $+22_{four}$

10. 44_{five}
    $+33_{five}$

11. $7BB4_{fourteen}$
    $+3A6D_{fourteen}$

12. $8A2B_{twelve}$
    $+191A_{twelve}$

13. 321_{six}
    $+1255_{six}$

14. 143_{five}
    $+432_{five}$

15. 76874_{nine}
    $+33137_{nine}$

16. 64605_{seven}
    $+23456_{seven}$

17. 1432_{five}
    $- 413_{five}$

18. 143_{five}
    $- 34_{five}$

19. 212_{seven}
    -136_{seven}

20. 9327_{eleven}
    -7318_{eleven}

21. 42831_{nine}
    $- 2781_{nine}$

22. 6323_{seven}
    $- 415_{seven}$

23. $12AB_{twelve}$
    $- 93A_{twelve}$

24. $7A4C_{fourteen}$
    $-2CCD_{fourteen}$

25. 525_{six}
    $\times \ \ 4_{six}$

26. 241_{seven}
    $\times \ \ 6_{seven}$

27. 818_{nine}
    $\times \ 62_{nine}$

28. 423_{five}
    $\times 332_{five}$

29. $AB5_{twelve}$
    $\times \ 42_{twelve}$

30. 5186_{nine}
    $\times \ 23_{nine}$

31. 5432_{six}
    $\times 153_{six}$

32. $C429_{thirteen}$
    $\times \ B24_{thirteen}$

33. $3_{nine}\overline{)1568_{nine}}$

34. $2_{three}\overline{)1202_{three}}$

35. $4_{five}\overline{)2023_{five}}$

36. $6_{seven}\overline{)1425_{seven}}$

37. $2434_{five} \div 14_{five}$

38. $3222_{four} \div 21_{four}$

39. $ABC2_{fourteen} \div 5C_{fourteen}$

40. $92A0_{twelve} \div 2B_{twelve}$

Applications in Our World

The binary, octal, and hexidecimal systems are used extensively in computer programming; arithmetic in these systems has very real applications. For Exercises 41–56, perform the indicated operations.

41. 1001_{two}
    $+ \ 111_{two}$

42. 62_{eight}
    $+145_{eight}$

43. $3BA_{sixteen}$
    $+ \ \ 49_{sixteen}$

44. 10111_{two}
    $+ 1101_{two}$

45. 1100_{two}
    $- \ \ 11_{two}$

46. 732_{eight}
    $- 45_{eight}$

47. $526B_{sixteen}$
    $- 4A1_{sixteen}$

48. 1000_{two}
    $- 101_{two}$

49. 1010_{two}
    $\times 101_{two}$

50. 54_{eight}
    $\times \ 2_{eight}$

51. $A25_{sixteen}$
    $\times \ \ 4_{sixteen}$

52. 326_{eight}
    $\times \ 21_{eight}$

53. $11_{two}\overline{)1011_{two}}$

54. $6_{eight}\overline{)437_{eight}}$

55. $5_{sixteen}\overline{)37B1_{sixteen}}$

56. $10_{two}\overline{)11111_{two}}$

In Exercises 57–60, a coded message is provided, along with a binary number used to encrypt the message. Each letter of the alphabet is assigned the letter corresponding to its position: A = 1, B = 2, C = 3, and so on. The numbers are converted to their binary equivalents, then a binary coding number is added to each character. To decode the message, the recipient subtracts the given binary number from each character, then converts to the equivalent letter. Decode each message.

57. 11110   10110   11001   10010   1111   1000
    11001   1011;   encryption number: 111
58. 1110   11001   11000   11110   1101   10010   1111
    1011   11110;   encryption number: 1010
59. 11010   10110   10000   10101   1110   10010
    11001   10110   100000   1110   100000   11101
    100110;   encryption number: 1101
60. 10010   1010   1010   11001   1110   10011   1100
    110   11001   10010   1110   1001   10011   1110
    1100   1101   11001;   encryption number: 101

The American Standard Code for Information Interchange (ASCII) is used to encode characters of the alphabet as binary numbers. Each character is assigned an eight-digit binary number written in two groups of four digits as follows:

A–O are prefixed by 0100, and the second grouping starts with A = 0001, B = 0010, C = 0011, etc.

P–Z are prefixed by 0101, and the second grouping starts with P = 0000, Q = 0001, R = 0010, etc. For example, C = 0100 0011 and Q = 0101 0001.

For Exercises 61–64, find the letter of the alphabet corresponding to the binary code.

61. 0100 1100   63. 0100 0111
62. 0101 0101   64. 0101 1010

For Exercises 65–68, write each word in ASCII code.

65. DORM
66. PARTY
67. UNION
68. QUAD

Critical Thinking

69. In a certain base number system, $5 + 6 = 13$. What is the base?
70. In a certain base number system, $15 - 6 = 6$. What is the base?
71. When adding in a base number system, will you ever have to carry a digit other than one to the next column to the left? Why or why not?
72. Convert each fraction in the given base number system to a mixed numeral:

    (a) $\dfrac{14_{\text{five}}}{2_{\text{five}}}$ (b) $\dfrac{46_{\text{eight}}}{3_{\text{eight}}}$ (c) $\dfrac{65_{\text{sixteen}}}{8_{\text{sixteen}}}$

73. Each of the symbols ♥, ♠, and ♦ represents one of the numerals in a base three system. Use the addition problem below to figure out which numerals correspond to each symbol.

74. Refer to Exercise 73. Use the subtraction problem to figure out which numerals correspond to each

symbol. The answer may or may not be the same as Exercise 73.

75. Each of the symbols ♣, ♥, ♠, and ♦ represents one of the numerals in a base four system. Use the addition problem below to figure out which numerals correspond to each symbol.

76. Each of the symbols A, E, I, O, and U represents one of the numerals in a base five system. Use the addition problem below to figure out which numerals correspond to each symbol.

$$\begin{array}{r} \text{E\ \ O\ \ I\ \ U}_{\text{five}} \\ + \quad \text{E\ \ U\ \ A\ \ U}_{\text{five}} \\ \hline \text{O\ \ O\ \ E\ \ A\ \ A}_{\text{five}} \end{array}$$

Section	Important Terms	Important Ideas
4-1	Number Numeral Numeration system Tally system Simple grouping system Multiplicative grouping system Digit Place value Positional system Expanded notation	**Throughout history**, different cultures have used different numeration systems; that is, different ways to represent numbers. The simplest numeration system is the tally system, where a number is represented by repeating a single symbol multiple times. In simple grouping systems, there are different symbols for different size numbers, and groups of those symbols represent a number. Multiplicative grouping systems use a multiplier to indicate how many times to include a given symbol, rather than writing the symbol multiple times. Positional systems are like multiplicative grouping systems, but the number to apply a multiplier to is understood from a position in the numeral. The system used in our world today is called the Hindu-Arabic system. It's a positional system based on powers of 10, and uses 10 symbols called digits to represent numbers.
4-2	Algorithm	**Many methods** were developed for performing calculations before the development of electronic calculators. The Egyptian algorithm, the Russian peasant method, the lattice method, and Napier's bones are all procedures that can be used to multiply by hand.
4-3	Base number system Base Binary system Octal system Hexadecimal system	**The base five** system has only five digits: 0, 1, 2, 3, and 4. The place values of the numbers written in the base five system are $5^0 = 1$, $5^1 = 5$, $5^2 = 25$, etc. Base number systems can be defined for other whole number bases in the same way, using powers of that base rather than powers of five. The binary (base two), octal (base eight), and hexadecimal (base 16) systems are especially important in computer programming.
4-4		**Operations** such as addition, subtraction, multiplication, and division can be performed in other number bases using procedures analogous to the procedures you learned for base 10 in elementary school.

Math in Computer Graphics REVISITED

mirtmirt/Shutterstock

1. 000000 is black, because black is the absence of all color. FFFFFF is white, because if you have the full amount of each color of light being emitted, that's as bright as you can get—white.

2. a. FF0000 has to be red: it's full red and none of the other colors.
   b. 00FF00 is full green and none of the other colors: that's green.
   c. 0000FF is full blue and none of the other colors: that's blue.

3. In hexadecimal, F plays the role of 15, and the place values for a two-digit hexadecimal number are 16^1, or 16, and 16^0, or 1. So $FF_{sixteen} = 15 \cdot 16 + 15 \cdot 1 = 255$. That means there are 256 different possible amounts of each of red, green, and blue, from 0 to 255.

4. A1$_{sixteen}$ = 176, 23$_{sixteen}$ = 35, and FF$_{sixteen}$ = 255. This is a whole lot of blue, a decent amount of red, and just a bit of green. Based on a lot of red and blue, a reasonable guess would be something in the purple range.

5. Orange is a combination of red and yellow. Yellow isn't one of the colors in the red/green/blue color scheme, but green has a lot of yellow in it. So my guess would be something that has a lot of red, maybe half green, and very little blue. To get half green, we'd want a base 10 number around 125 or so, and in hexadecimal that would be 7D, so I'd try something like FF7D11, which I swear I typed in before checking. And I rocked it! I put that code, and nothing else, into a Google search, and the color that came up isn't just orange—it's super orange. Cool.

Review Exercises

Section 4-1

For Exercises 1–10, write each number using Hindu-Arabic numerals.

1. ☥◎◎∩∩⏐⏐
2. 𓂧𓏲𓏲𓈖∩∩⏐⏐⏐⏐⏐
3. ◀⏐   ◀◀⏐
4. MCXLVII
5. CDXIX
6. ◀◀◀⏐   ◀◀⏐⏐
7. 二千六百四
8. 九百五十七
9. ⎯ / ••• / ⎯ (three-line symbol with dots)
10. ⎯⎯ •• / shell symbol / ••••

For Exercises 11–20, write each number in the system given.

11. 49 in the Egyptian system
12. 896 in the Roman system
13. 88 in the Babylonian system
14. 125 in the Egyptian system
15. 503 in the Roman system
16. 8,325 in the Chinese system
17. 165 in the Babylonian system
18. 74 in the Chinese system
19. 97 in the Mayan system
20. 7,901 in the Mayan system

For Exercises 21 and 22, perform the indicated operation. Leave answers in the Egyptian system.

21. ◎◎◎∩∩∩∩⏐⏐⏐⏐⏐ + 𓂧◎◎∩∩⏐⏐⏐⏐⏐⏐
22. 𓂧𓂧◎◎◎◎∩∩⏐⏐⏐ + 𓂧𓈖∩∩⏐⏐⏐⏐⏐

Super Bowls (played every year) and Olympic Games (every four years) are numbered using Roman numerals. Super Bowl LV was played in Tampa on February 7, 2021, and the games of the XXXII Olympiad took place in Tokyo in the summer of 2021.

Note that these games were supposed to be played in 2020 but were delayed due to the COVID-19 pandemic. Use these facts for Exercises 23–26.

23. In what year was Super Bowl XXIV played?
24. What Super Bowl is scheduled to take place in 2030?
25. What Olympiad will occur in 2028?
26. In what year did the games of the XVI Olympiad take place?

Section 4-2

For Exercises 27–30, use the Egyptian algorithm to multiply.

27. 23×12
28. 13×18
29. 36×21
30. 15×16

For Exercises 31–34, multiply using the Russian peasant method.

31. 15×22
32. 12×17
33. 13×12
34. 22×45

For Exercises 35–38, use the lattice method of multiplication to find each product.

35. 23×85
36. $45 \times 3,981$
37. 439×833
38. 548×505

For Exercises 39–42, use Napier's bones (constructed for the exercises in Section 4-2) to find each product.

39. 31×82
40. 74×53
41. 147×95
42. 88×796

Section 4-3

For Exercises 43–50, write each number in base 10.

43. 1110111_{two}
44. 672_{eight}
45. $A03B_{twelve}$
46. 14441_{five}
47. 2012_{three}
48. 6000_{seven}
49. 28645_{nine}
50. $1A214_{eleven}$

For Exercises 51–58, write each number in the specified base.

51. 232 in base six
52. 905 in base 12
53. 2,001 in base nine
54. 43 in base two
55. 119 in base four
56. 51 in base two
57. 3,343 in base seven
58. 2,460 in base 14

In Exercises 59–66, subscripts are omitted since the base is provided in the instructions.

For Exercises 59–62, convert each binary number to (a) octal and (b) hexadecimal.

59. 111011011
60. 10001110111
61. 1101100111
62. 111000111101

For Exercises 63 and 64, convert each octal number to binary.

63. 7324
64. 643

For Exercises 65 and 66, convert each hexadecimal number to binary.

65. A5B3
66. 9F87

For Exercises 67 and 68, refer to Exercises 103 to 108 in Section 4-3. Find the percentage of maximum color for each of red, blue, and green in the given colors.

67. Royal Blue: 2B60DE
68. Plum: B93B8F

Section 4-4

For Exercises 69–80, perform the indicated operation.

69. 156_{nine}
    $+\ 84_{nine}$

70. 101110_{two}
    $+\ 1101_{two}$

71. $6A20_{twelve}$
    $+B096_{twelve}$

72. 7267_{nine}
    $-\ 354_{nine}$

73. 3312_{four}
    -2321_{four}

74. 65602_{eleven}
    -46031_{eleven}

75. 371_{nine}
    $\times 51_{nine}$

76. 1101_{two}
    $\times 111_{two}$

77. $6A5_{sixteen}$
    $\times B8_{sixteen}$

78. $3_{five}\overline{)1242_{five}}$

79. $10_{eight}\overline{)3426_{eight}}$

80. $5_{sixteen}\overline{)324_{sixteen}}$

Chapter Test

For Exercises 1–5, write each number using Hindu-Arabic numerals.

1.
2. ◁◁˥ ◁˥
3. MCMLXVI
4. 三
   百
   六
   十
   八
5. ☰
   ◉
   • • •

For Exercises 6–10, write each number in the system given.

6. 567 in the Roman system
7. 55 in the Babylonian system
8. 521 in the Egyptian system
9. 873 in the Chinese system
10. 2,400 in the Mayan system

For Exercises 11–13, multiply using the given method.

11. Multiply 17×13 by the Egyptian algorithm.
12. Multiply 23×15 by the Russian peasant method.
13. Use the lattice method to multiply 364×736.

For Exercises 14–19, write each number in base 10.

14. 341_{five}
15. $A07B_{twelve}$
16. 21101_{three}
17. 1100111_{two}
18. 463_{seven}
19. $1A436_{eleven}$

For Exercises 20–27, write each number in the specified base.

20. 183 in base 12
21. 4,673 in base nine
22. 65 in base three
23. 48 in base two
24. 434 in base seven
25. 889 in base 12
26. CCXLI in base eight
27. ◁⌒ϯ☉∩ in base eight
28. Convert 111011001_{two} to octal and hexadecimal.
29. Convert 7324_{eight} and $A6D92_{sixteen}$ to binary.

For Exercises 30–37, perform the indicated operation.

30. 263_{nine}
    $+\ 18_{nine}$

31. $5A79_{twelve}$
    $+B068_{twelve}$

32. 6772_{eight}
    $-\ 735_{eight}$

33. 11001010_{two}
    $-\ 110011_{two}$

34. 254_{six}
    $\times\ 3_{six}$

35. 413_{five}
    $\times 21_{five}$

36. $7_{eight}\overline{)1342_{eight}}$

37. $2_{three}\overline{)1012_{three}}$

38. In Chapter 4, we studied four main types of numeration systems: tally, simple grouping, multiplicative grouping, and positional. Explain how each system on that list builds on the previous.

Projects

1. In this project, we'll test various methods of multiplication, with a goal of making an individual decision on which method is most effective. We'll consider both ease of use, amount of time needed, and likelihood of getting the correct product. You will need a calculator to check the products you compute, as well as a watch to time each calculation.

   (a) **Two-digit multiplication** Multiply the first two digits of your social security number and the last two digits using (i) the Egyptian algorithm, (ii) the Russian peasant method, (iii) the lattice method, (iv) Napier's bones, and (v) the method you learned in grade school. In each case, time how long it takes the method from start to finish, and make note of whether you got the right answer for each.

   If you are working in a group, each group member should repeat the above steps. If you are working on your own, repeat the process, multiplying the first two digits of your phone number by the last two, then again multiplying the first two digits of your zip code by the last two.

   (b) **Three-digit multiplication** Repeat question (a), but this time choose numbers this way: turn to a random page past this one in the book and note the page number, then to another page at least 100 pages further on and note that page number. Use each method of multiplication again. If you are working in a group, each group member should choose different page numbers. If you are working on your own, choose three sets of random pages.

   (c) Calculate the average amount of time it took to perform all calculations with each of the five methods, and rank them in order from fastest to slowest.

   (d) Calculate the percentage of problems in which the correct answer was obtained for each method, and rank them from highest to lowest.

   (e) Based on your experience in using the methods, and your rankings in Questions 3 and 4, write a short essay identifying the method that you think is most effective, and justify that choice.

2. The two oldest numeration systems that we know a lot about are the Babylonian and Egyptian systems. In this project, you'll research both systems, comparing and contrasting them. Use either the Internet or a good old-fashioned library to find sources.

   (a) Find as much information as you can about how historians and archaeologists know the details of these numerical systems. How long have we known about them? What do we know about how long ago they were developed? By how much time do the systems predate the oldest existing documents? After answering these questions, and including any other relevant information you find, draw a timeline with key dates for both systems.

   (b) Based on your research and your timeline, do you think the systems were developed completely independently, or was one developed with prior knowledge of the other? Back up your opinion.

   (c) Critique the two systems by comparing and contrasting them. Which do you think is more efficient? (Make sure you define what "more efficient" means to you.) Which is easier to learn? Which would be easier to perform arithmetic operations in?

   (d) Try to come up with a method for representing fractional quantities, like one-half, in each system. Is it easier in one than the other? Why do you feel that way?

Design elements: Front matter, Chapter Opener, Summary and End Matter header design (random numbers background illustration): ©pixeldreams.eu/Shutterstock RF

The Real Number System

Photo: James P. Blair/Photodisc/Getty Images

Outline

Math in | Government Spending

Most Americans have some vague idea that the government spends way more than it has, and that our nation is badly in debt. But very few people have any idea just how much money we're talking about. Maybe it's because ignorance is bliss, or maybe the numbers are just so staggering that not just anyone can appreciate their size.

One of the really interesting things about being a math professor is that when you tell people what you do, they almost always respond with something like "Wow, I'm not very good at math." Why are people so eager to brag about not being good at math? Nobody has ever met an English professor and said "Wow, I stink at English . . . I'm practically illiterate!" Of course, literacy is important for success in life, but so is *numeracy*—the ability to understand and work with numbers.

This chapter is all about studying different categories of numbers, and the way we combine them using operations. It's unlikely that you'll encounter operations that you have never seen before in this chapter. Instead, we'll focus on a deeper understanding of numbers and operations, much the same way you focus on *understanding* sentences rather than just reading individual words in English class.

That brings us back to our good friends in Washington. Here are some frightening numbers about the financial state of our government, followed by some questions. By the time you finish this chapter, you should be able to answer all of the questions. But more importantly, you should have a real understanding of the significance of these numbers. All of the dollar amounts are written in scientific notation, which we will study

in Section 5-6. Write your answers in decimal notation, and then write how those numbers would be read aloud.

Fact: The government budget deficit for 2020 was 3.13×10^{12}. (This means that the government spent over 3 trillion dollars more than it took in.)

Fact: In 2020, the government spent $\$3.45 \times 10^{11}$ just paying *interest* on the national debt, making no progress toward paying off the actual amount. By contrast, total federal spending for 2020 was $\$6.55 \times 10^{12}$.

Fact: The national debt, which is the amount of money that is owed by the government to various creditors, was about $\$2.876 \times 10^{13}$ in September 2020.

Question 1: The population of the United States in 2020 was about 331 million. How many dollars per person did the government spend total in 2020, and how many dollars per person did it spend above the amount it took in?

Question 2: What percentage of the total amount spent for 2020 went to simply pay interest on the national debt?

Question 3: If the national debt were divided evenly among all citizens, how much would each person owe?

Question 4: Most congressional representatives are very well off. There are 435 members of the House of Representatives, and 100 senators. If they all got together and decided to pay off the national debt (yeah, right), how much would each owe?

Question 5: If you round the deficit for 2020 provided above to $\$3.1 \times 10^{12}$ (which seems like no big deal, right?), how much did you just save the American taxpayers?

For answers, see Math in Government Spending Revisited on page 292

Section 5-1   **The Natural Numbers**

LEARNING OBJECTIVES

1. Find the factors of a natural number.
2. Identify prime and composite numbers.
3. Find the prime factorization of a number.
4. Find the greatest common factor of two or more numbers.
5. Find the least common multiple of two or more numbers.

Your first exposure to numbers had nothing to do with symbols written on a page—you learned about the concept of numbers by counting the things around you. Maybe you had one dog, two parents, three sisters, four stuffed animals, and five pairs of those little footy pajamas. At some point you could count to 10 (using your fingers), and maybe even 20 if you were barefoot. Eventually, you headed off to school and learned about other types of numbers—negatives, fractions, decimals— but you can still get pretty far in the material world with the counting numbers you started with.

JGI/Jamie Grill/Blend Images LLC

In this section, we'll study the counting numbers in more depth than you did when counting with your fingers and toes. You might think this is kind of silly in a college math course, but our focus will be on a deep, fundamental understanding of rules and operations that you probably just memorized when you were a child.

Prime and Composite Numbers

The counting numbers we talked about above are often called the *natural numbers* because we use them to enumerate objects in the natural world.

> The set of **natural numbers**, also known as the **counting numbers**, consists of the numbers 1, 2, 3, 4, 5, ... The letter N is often used to represent the set of natural numbers: $N = \{1, 2, 3, 4, 5, \ldots\}$.

Multiplication is used extensively in studying the natural numbers, largely because of the following fact: every natural number can be written as the product of two or more natural numbers. For example,

$$12 = 3 \times 4 \qquad 16 = 4 \times 4 \qquad 19 = 1 \times 19 \qquad 30 = 2 \times 3 \times 5$$

The natural numbers that are multiplied to get a product are called the **factors** of that product. So the statement $12 = 3 \times 4$ tells us that 3 and 4 are factors of 12. The example $16 = 4 \times 4$ shows that the factors of a number don't have to be distinct, and the example $19 = 1 \times 19$ shows that 1 is a perfectly acceptable factor.

We know that 3 and 4 are factors of 12, but they're not the only factors. We can also write 12 as 1×12 or 2×6. Now we have a list of *all* factors of 12: 1, 2, 3, 4, 6, and 12.

Example 1   Finding Factors

(a) Find all factors of 24.
(b) Can you think of a number that has only two factors?
(c) Can you think of a way to write 24 as a product of more than two natural numbers without using 1?

 1. Find the factors of a natural number.

SOLUTION

(a) Think of all the ways you can write 24 as a product of two natural numbers, starting with 1 as one of the factors, and working upward.

$$1 \times 24 \qquad 2 \times 12 \qquad 3 \times 8 \qquad 4 \times 6$$

These are all of the possibilities, so the factors of 24 are 1, 2, 3, 4, 6, 8, 12, and 24.

(b) I don't know about you, but I sure can: 2 is an easy example. The only way to write 2 as the product of two natural numbers is 1×2.

(c) I can do this one, too! $2 \times 2 \times 6$ is one of several possibilities.

Try This One 1

(a) Find all factors of 50.
(b) Write 50 as a product of more than two natural numbers without using 1.

Notice that if you divide 24 by any of its factors, the remainder is zero. In fact, this is a good way to test to see if one number is a factor of another. Six is a factor of 24 because $24 \div 6 = 4$ with no remainder, but 5 is not a factor of 24 because $24 \div 5$ has remainder 4. For this reason, we will sometimes call the factors of a number its **divisors**.

A natural number a is **divisible** by another natural number b if dividing a by b results in a remainder of zero. We might also say "b divides a."

We know from above that 6 is a factor of 24, which means that 24 is divisible by 6, or 6 divides 24.

Using Technology

Deciding if a Number Is a Factor

There are many clever ways to decide if one number divides another (these are discussed in a group project at the end of this chapter). But in the 21st century, where a calculator is as close as the phone in your pocket, technology provides a quick solution. We have seen that a is a factor of b if dividing b by a results in a whole number. And calculators are fantastic at calculations like division because they're, you know, calculators.

Are 8 and 12 factors of 156?

Typical Scientific or Online Calculator:
156 ÷ 8 = and 156 ÷ 12 =

Typical Graphing Calculator:
156 ÷ 8 ENTER and 156 ÷ 12 ENTER

Spreadsheet:
Enter in any cell: = 156/8 and = 156/12

In each case, the results are 19.5 and 13. Since 19.5 isn't a whole number, 8 isn't a factor of 156. But 12 is, because the result of the division is 13, which is a whole number.

As we saw in the second part of Example 1, most numbers have more than two factors, but not all: for example, 7 can only be written as 1×7. We give these numbers a special name.

A natural number greater than 1 is called **prime** if it has exactly two factors, 1 and itself.

Not surprisingly, we also have a name for numbers with more than two factors.

A natural number greater than 1 is called **composite** if it isn't prime; that is, if it has at least one factor other than itself and 1.

| **Example 2** | Deciding if a Number Is Prime |

Decide whether each number is prime or composite.

(a) 25   (b) 17   (c) 12   (d) 31

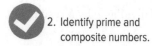

Math Note

Is the number 1 prime? See Exercise 112 at the end of the section for a discussion. (Today, 1 is considered neither prime nor composite.)

SOLUTION

If we can find even one factor other than 1 and the number itself, then it is composite.

(a) Five is a factor of 25, since $5 \times 5 = 25$, so 25 is composite.
(b) The only factors of 17 are 1 and 17, so it is prime.
(c) Two is a factor of 12 (as are several other numbers), so 12 is composite.
(d) There are no factors of 31 other than 1 and 31, so it is prime.

✓ 2. Identify prime and composite numbers.

| Try This One | 2 |

Decide whether each number is prime or composite.

(a) 34   (b) 29   (c) 27   (d) 10

You might wonder if there's a formula for finding all of the prime numbers. If you find one, we can guarantee you'll be famous, because people have been looking for one for well over 2,000 years. But here's a clever way to generate a list of the prime numbers up to 50. (See Figure 5-1). We write a list of all natural numbers up to 50 and cross out the ones that can't be prime. Begin by crossing out 1, since it's not considered to be prime. Any number that has 2 as a factor can't be prime (other than 2 itself), so we cross out 4, 6, 8, 10, and all the even numbers. Any number other than 3 that is divisible by 3 also can't be prime, so we cross out 3, 6, 9, 12, ... (The ones that 2 also divides are already crossed out.) We continue this process for numbers divisible by 5 and 7. At this point, all remaining numbers are prime. (In Exercise 115, we'll look at why stopping at 7 is sufficient.)

All of the prime numbers less than 50 are circled in Figure 5-1.

Math Note

The process in Figure 5-1 is known as the Sieve of Eratosthenes, named after the Greek scholar from the third century BCE who invented it. He was also the first human to calculate the circumference of Earth to a reasonably accurate value—almost 2,500 years ago!

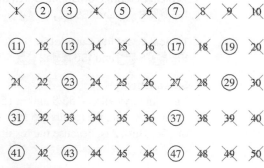

Figure 5-1

Prime Factorization

We know that there's more than one way to write 12 as a product: for example, we could write it as $12 = 2 \times 6$, or $12 = 3 \times 4$. But notice that some of those factors can also be written as products: $6 = 2 \times 3$ and $4 = 2 \times 2$. Now let's rewrite each multiplication statement:

$$12 = 2 \times 6 = 2 \times 2 \times 3$$
$$12 = 3 \times 4 = 3 \times 2 \times 2$$

Sidelight Twin Primes

There are six pairs of prime numbers that differ by 2 in Figure 5-1: 3 and 5, 5 and 7, 11 and 13, 17 and 19, 29 and 31, and 41 and 43. Pairs of prime numbers that differ by 2 are called twin primes, and mathematicians have been trying for a very long time to answer the question, "Are there infinitely many pairs of twin primes?" It was proved over 2,000 years ago by Euclid that there are infinitely many prime numbers, but nobody has been able to prove that there is an infinite number of twin primes. Computer evidence makes it likely that there are, but this little conjecture has escaped proof since before the time of the Roman empire.

You might think that twin primes are relatively rare as you get into the larger numbers, and you would be wrong. There are 14,871 pairs of twin primes between 1 and 2,000,000, and well over 800 trillion known pairs. The largest known twin primes, at least in early 2012, are so huge they have over 200,000 digits, and would take more than 55 hours just to write.

We ended up with the same result (aside from the order, which doesn't matter), and all of the factors are prime numbers. This result indicates a very important property of natural numbers known as the fundamental theorem of arithmetic. (A **theorem** is a statement that has been proved to be true.)

The Fundamental Theorem of Arithmetic

Every composite number can be written as the product of prime numbers in exactly one way. (The order of the factors is unimportant.)

The next two examples illustrate two methods for finding the prime factorization of a number. In the *tree method*, a diagram is built by finding factorizations successively.

Example 3 Finding Prime Factorization Using the Tree Method

Find the prime factorization of 100 using the tree method.

Math Note

When you use the tree method, it doesn't matter which pair of factors you start with. If you redo Example 3 starting with 4 × 25 or 10 × 10, you'll end up with the same result (give it a try). I know this because the fundamental theorem of arithmetic guarantees that 100 (and every other composite number) has a unique prime factorization.

SOLUTION

Start with any factorization of 100, say 2×50, then factor 50 as 5×10. Finally factor 10 as 2×5. This is shown using a tree.

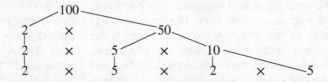

Rearrange the factors in order: $2 \times 2 \times 5 \times 5$ or $2^2 \times 5^2$.

Try This One   3

Use the tree method to find the prime factorization of 360.

In the *division method*, we start with the original number and keep dividing by prime numbers until the result of the division is prime.

| **Example 4** | Finding a Prime Factorization Using the Division Method |

Math Note

To avoid mistakes, always start with the smallest prime factor and keep using it until you can't divide any further, and then move to the next prime number. Don't skip around.

Use the division method to find the prime factorization of 140.

SOLUTION

We begin by dividing by 2 as many times as we can with no remainder.

$$140 \div 2 = 70; \ 70 \div 2 = 35$$

This tells us that there are two factors of 2. Since the next prime (3) does not divide 35 evenly, there are no factors of 3. We move on to 5:

$$35 \div 5 = 7$$

At this point the result of the division is a prime number, so we're done. We found that 140 has two factors of 2, one of 5, and one of 7: $140 = 2 \times 2 \times 5 \times 7$ or $2^2 \times 5 \times 7$.

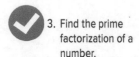

3. Find the prime factorization of a number.

| Try This One | 4 |

Use the division method to find the prime factorization of 240.

Greatest Common Factors

For a variety of reasons that will become apparent later, finding factors that are common to two or more numbers is useful. Let's look at the factors of 18 and 24:

18: 1, 2, 3, 6, 9, 18
24: 1, 2, 3, 4, 6, 8, 12, 24

These two numbers have four factors in common: 1, 2, 3, and 6. Obviously, the largest of these is 6, so we will call it the *greatest common factor* of 18 and 24.

Sidelight   **Perfect Numbers**

The factors of 6 are 1, 2, 3, and 6. If you add these factors, leaving out 6, something unusual happens: $1 + 2 + 3 = 6$, and you get the original number back. Numbers with this property—the sum of the factors (other than itself) is the original number—are called *perfect numbers*. (Personally, I don't think they're better than any other number, but evidently someone thought they were pretty awesome based on the name.) The next perfect number is 28: $1 + 2 + 4 + 7 + 14 = 28$.

After 28, you'd have to search for a while: the third perfect number is 496, and the fourth is 8,128. The first three were known to the Pythagoreans, a group of scholars from the fifth century BCE. The fourth was believed to have been discovered around the year 100. A student waiting for the next perfect number after that would have had a looooong wait—almost 1400 years, when 33,550,336 was found to be perfect in 1461. (If you have any idea how someone figured that out in the Middle Ages, I'd love to hear it.)

Perfect numbers six and seven were discovered around 1600, and the eighth in 1732. (It has 19 digits, by the way. Even today it's extremely difficult to verify that it is in fact perfect without a computer.) Perfect numbers 9 through 12 were discovered in the 1800s; the twelfth has 77 digits and is the last to be discovered without the aid of a computer. Things took off in the 1950s, and as of this writing, there are 47 known perfect numbers, the largest having just a bit less than 26 million digits.

Here's an oddity: every known perfect number ends in either 6 or 8. It's unknown whether there are any with a last digit different from 6 or 8, but it's strongly suspected that there are no odd perfect numbers. It also hasn't been proven whether there are infinitely many perfect numbers, so if you aspire to be a famous mathematician, get to work on that.

<table>
<tr><td>

Math Note

The GCF is also sometimes called the greatest common divisor.

</td><td>

The **greatest common factor** of two or more numbers is the largest number that is a factor of all of the original numbers. We will use the abbreviation GCF to represent the greatest common factor.

</td></tr>
</table>

For relatively small numbers like 18 and 24, we can list all of the factors and find the largest. But this is really inconvenient for large numbers, so we'll use a different strategy, based on prime factorization.

Procedure for Finding the Greatest Common Factor of Two or More Numbers

Step 1 Write the prime factorization of each number.

Step 2 Make a list of each prime factor that appears in all prime factorizations. For prime factors with exponents, choose the smallest power that appears in each.

Step 3 The GCF is the product of the numbers you listed in Step 2.

Example 5 illustrates the procedure.

Example 5   Finding the Greatest Common Factor of Two Numbers

Math Note

When the GCF of two numbers is 1, we call the numbers **relatively prime**. For example, 15 and 17 have no factors in common other than 1, so they are relatively prime.

Find the GCF of 72 and 180.

SOLUTION

Step 1 Write the prime factorizations of 72 and 180:
$$72 = 2 \times 2 \times 2 \times 3 \times 3 = 2^3 \times 3^2$$
$$180 = 2 \times 2 \times 3 \times 3 \times 5 = 2^2 \times 3^2 \times 5$$

Step 2 List the common factors: 2 (with exponent 2) and 3 (with exponent 2).

Step 3 The GCF is $2^2 \times 3^2 = 36$.

Try This One   5

Find the GCF of 54 and 144.

Now let's try finding the GCF of three numbers.

Example 6   Finding the Greatest Common Factor of Three Numbers

Find the GCF of 40, 60, and 100.

SOLUTION

Step 1 Write the prime factorization of each number:
$$40 = 2 \times 2 \times 2 \times 5 = 2^3 \times 5$$
$$60 = 2 \times 2 \times 3 \times 5 = 2^2 \times 3 \times 5$$
$$100 = 2 \times 2 \times 5 \times 5 = 2^2 \times 5^2$$

Step 2 List the common factors: 2 (with exponent 2), and 5 (with exponent 1).

Step 3 The GCF is $2^2 \times 5 = 20$.

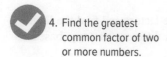

4. Find the greatest common factor of two or more numbers.

Try This One   6

Find the GCF of 45, 75, and 150.

We now know a whole lot about the natural numbers. In the next section, we'll include zero and negative numbers to form a larger set of numbers known as the *integers*.

Finding the greatest common factor will come in handy when reducing fractions in Section 5-3, but it also has applications in our world.

Example 7   **Applying the GCF to Packaging Goods for Sale**

An enterprising college student gets a great deal on slightly past-their-prime packets of instant coffee from a chain of coffee stores. He acquires 200 packets of decaf and 280 packets of regular coffee. The plan is to package them for resale at a college fair so that there's only one type of coffee in each box, and every box has the same number of packets. How can he do this so that each box contains the largest number of packets possible? (Of course, he wants to package all of the coffee he has.)

SOLUTION

The number of packets per box has to be a factor of both 200 and 280 so that there won't be any left over. If the boxes are to contain the largest number of packets possible, our entrepreneur will need to find the greatest common factor.

$$200 = 2 \times 2 \times 2 \times 5 \times 5 = 2^3 \times 5^2$$
$$280 = 2 \times 2 \times 2 \times 5 \times 7 = 2^3 \times 5 \times 7$$

The factors common to both numbers are 2 (with exponent 3) and 5 (with exponent 1). So the GCF is $2^3 \times 5 = 40$. That tells us that he should put 40 packets in each box. He'll have 5 boxes of decaf and 7 boxes of regular to sell.

Try This One   7

The enterprising student from Example 7 later buys unsold Halloween candy for resale. If he has 360 "fun size" bags of peanut M&M's and 420 bags of plain M&M's, how should he package them to follow the same guidelines as in Example 7?

Least Common Multiples

If you multiply any natural number by 1, then 2, then 3, and so on, you will generate a list of numbers. We call this list the **multiples** of a number. For example, starting with 6, we would get $6 \times 1 = 6$, $6 \times 2 = 12$, $6 \times 3 = 18$, etc. So the multiples of 6 are 6, 12, 18, 24, 30, ...

Take a look at the multiples of 4 and 6:

Multiples of 4: 4, 8, 12, 16, 20, 24, . . .
Multiples of 6: 6, 12, 18, 24, 30, 36, . . .

The smallest number on both lists is 12, so we will call it the *least common multiple* of 4 and 6.

> The **least common multiple (LCM)** of two or more numbers is the smallest number that is a multiple of each. If you like, you can think of it as the smallest number that is divisible by all of the numbers.

Example 8 | **Studying a Least Common Multiple**

For the numbers 18 and 24:

(a) Write the first five multiples of each number.
(b) Look at your two lists and find the least common multiple.
(c) Write the prime factorization of each of 18, 24, and the least common multiple.
(d) What can you observe about the LCM?

SOLUTION

(a) For 18: 18, 36, 54, 72, 90
    For 24: 24, 48, 72, 96, 120
(b) The first number that's on both lists is 72, so that's the least common multiple.
(c) $18 = 2 \times 9 = 2 \times 3 \times 3 = 2 \times 3^2$
    $24 = 2 \times 12 = 2 \times 2 \times 6 = 2 \times 2 \times 2 \times 3 = 2^3 \times 3$
    $72 = 8 \times 9 = 2 \times 2 \times 2 \times 3 \times 3 = 2^3 \times 3^2$
(d) The LCM has the biggest number of factors of 2 that appear in either of 18 or 24 (that's the 2^3), and also has the biggest number of factors of 3 that appear (the 3^2).

Try This One 8

Repeat Example 8 for the numbers 12 and 20.

As we saw in Example 8, you can always find the least common multiple by listing out all multiples, but if the numbers are big or have few factors in common, this can be really time-consuming. What we observed in part (d) of Example 8 leads to a better procedure: it's a lot like the procedure we used to find the GCF, and it also relies on prime factorization.

Procedure for Finding the Least Common Multiple of Two or More Numbers

Step 1 Write the prime factorization of each number.

Step 2 Make a list of every prime factor that appears in *any* of the prime factorizations. For prime factors with exponents, choose the largest that appears in any factorization.

Step 3 The LCM is the product of the numbers you listed in Step 2.

We'll illustrate this procedure with an example.

| **Example 9** | Finding the Least Common Multiple of Three Numbers |

Find the LCM of 24, 30, and 42.

SOLUTION

Step 1 Write the prime factorization of each number:

$$24 = 2 \times 2 \times 2 \times 3 = 2^3 \times 3$$
$$30 = 2 \times 3 \times 5$$
$$42 = 2 \times 3 \times 7$$

Step 2 List all the factors that appear: 2 (with largest exponent 3), 3, 5, and 7.

Step 3 The LCM is $2^3 \times 3 \times 5 \times 7 = 840$.

| **Try This One** | **9** |

5. Find the least common multiple of two or more numbers.

Find the LCM of each.

(a) 40, 50   (b) 28, 35, 49   (c) 16, 24, 32

The least common multiple is very useful in adding and subtracting fractions, but it also has applications in our world.

| **Example 10** | Applying the LCM to Grocery Shopping |

Math Note

In Problems 107–110, we'll study a really interesting example of using the LCM in the world of HD video.

Have you ever noticed that many hot dogs come in packages of 10, but most hot dog buns come in packages of eight? (Who thought *that* was a good idea?) What's the smallest number of packages you can buy of each so that you end up with the same number of hot dogs and buns?

SOLUTION

The total number of hot dogs we buy will be a multiple of 10, while the number of buns will be a multiple of 8. What we need to find is the least common multiple of 8 and 10. So we write the prime factorization of each:

$$8 = 2 \times 2 \times 2 = 2^3$$
$$10 = 2 \times 5$$

The factors that appear in either list are 2 (with largest exponent 3) and 5, so the LCM is $2^3 \times 5 = 40$. That means we would need four packages of hot dogs and five packages of buns. Hope you're hungry!

| **Try This One** | **10** |

After getting her first job out of college, Aliyah vows to buy herself a new item of clothing every 15 days, and a new pair of shoes every 18 days just because she can. How long will it be until she buys both items on the same day?

Sidelight · **Fermat Numbers**

In 1650, a French lawyer who studied mathematics in his spare time made a strange observation: the first five Integers of the form $2^{2^n} + 1$ are prime. For $n = 0, 1, 2, 3,$ and 4, the numbers are $2^1 + 1 = 3$, $2^2 + 1 = 5$, $2^4 + 1 = 17$, $2^8 + 1 = 257$, and $2^{16} + 1 = 65{,}537$. At that time, the lawyer, named Pierre de Fermat, speculated that all such numbers were prime. Needless to say, it was not easy to decide that a number like 65,537 was prime in 1650; the next such number, for $n = 5$, is $2^{32} + 1$, which has 10 digits. It turns out that Fermat was incorrect, but it took until the advent of computers to attack the problem in real depth.

The numbers of the form $2^{2^n} + 1$ are called Fermat numbers in his honor, and mathematicians have now been able to show that the Fermat numbers for values of n from 5 to 32 are composite. Beyond that, it gets a little spotty. But we can forgive mathematicians for this limitation: the Fermat number for $n = 33$ has almost 26 billion digits!

Lebrecht Music and Arts Photo Library/Alamy Stock Photo

Answers to Try This One

1 (a) 1, 2, 5, 10, 25, 50   (b) $2 \times 5 \times 5$

2 (a) Composite   (c) Composite
(b) Prime   (d) Composite

3 $360 = 2^3 \times 3^2 \times 5$

4 $240 = 2^4 \times 3 \times 5$

5 18

6 15

7 60 bags per box, for 6 boxes of peanut and 7 boxes of plain

8 (a) For 12: 12, 24, 36, 48, 60; for 20: 20, 40, 60, 80, 100
(b) The LCM is 60.
(c) $12 = 2^2 \times 3$; $20 = 2^2 \times 5$; $60 = 2^2 \times 3 \times 5$
(d) Again, the LCM has the biggest power of each factor that appears.

9 (a) 200   (b) 980   (c) 96

10 90 days

Exercise Set   5-1

Writing Exercises

1. Describe the set of natural numbers in your own words.
2. What's the difference between a prime number and a composite number?
3. Explain why every natural number other than 1 has at least two factors.
4. What is the prime factorization of a number? Can you find the prime factorization for every natural number?
5. Explain what the greatest common factor of a list of numbers is.
6. Explain what the least common multiple of a list of numbers is.
7. What does it mean to say that a number a divides another number b?
8. How can you use a calculator or spreadsheet to decide if one number divides another?

Computational Exercises

For Exercises 9–28, find all factors of each number.

9. 16
10. 225
11. 126
12. 54
13. 32
14. 48
15. 70
16. 66
17. 96
18. 100
19. 17
20. 19
21. 64
22. 120
23. 105
24. 365
25. 98
26. 36
27. 71
28. 47

For Exercises 29–38, find the first five multiples of each.

29. 3	33. 15	37. 1
30. 7	34. 20	38. 25
31. 10	35. 17	
32. 12	36. 19	

For Exercises 39–58, find the prime factorization of each.

39. 20	46. 1,800	53. 2,175
40. 32	47. 378	54. 1,008
41. 48	48. 432	55. 440
42. 31	49. 825	56. 990
43. 67	50. 576	57. 515
44. 150	51. 320	58. 1,296
45. 200	52. 1,250	

For Exercises 59–76, find the greatest common factor of the given numbers.

59. 6, 9	61. 7, 10
60. 10, 35	62. 6, 11

63. 30, 36	70. 5, 15, 25
64. 75, 105	71. 12, 18, 30
65. 105, 126	72. 42, 60, 18
66. 210, 140	73. 36, 60, 108
67. 440, 660	74. 60, 90, 84
68. 85, 102	75. 210, 336, 546
69. 12, 24, 48	76. 57, 114, 171

For Exercises 77–94, find the least common multiple of the given numbers.

77. 5, 10	86. 195, 390
78. 12, 24	87. 4, 7, 11
79. 7, 5	88. 5, 6, 13
80. 6, 10	89. 12, 15, 24
81. 18, 21	90. 30, 18, 42
82. 25, 35	91. 18, 24, 36
83. 50, 75	92. 42, 48, 56
84. 60, 90	93. 16, 20, 104
85. 70, 90	94. 65, 78, 104

Applications in Our World

95. In one college class, there are 28 students taking the class for a grade and 20 taking it pass-fail. The instructor wants to assign groups for a project, and has two requirements: every group should contain only students with the same grading option, and all the groups should have the same number of students. What's the greatest number of students that the instructor can put in each group?

96. The manager of a cable company that offers both digital TV and high-speed Internet is trying to arrange service calls efficiently. The plan is to divide the people awaiting service into groups based on whether they need TV or Internet service so that the company's techs can concentrate on one type of service each day. If there are 40 people waiting for TV service and 32 waiting for Internet service, and the scheduled number of visits should be the same each day, what's the greatest number of visits that can be scheduled each day?

97. Two people bike on a circular trail. One person can ride around the trail in 24 minutes, and another person can ride around the trail in 36 minutes. If both start at the same place at the same time, when will they be at the starting place at the same time again?

98. Two clubs offer a "free day" (no admission charge) to recruit new members. The health club has a free day every 45 days. A nearby swimming club has a free day every 30 days. If today is a free day for both clubs, how long will it be until a person can again use both clubs for free on the same day?

99. There are 30 women and 36 men in a bowling league. The president wants to divide the members into all-male and all-female teams, each of the same size. Find the number of members and the number of teams for each sex.

100. In a study designed to test the effectiveness of an experimental medication for ADHD, 96 boys and 72 girls are divided into groups. Each group will have only boys or only girls, and all groups should have the same number of subjects. How many groups of boys and groups of girls will there be? How many children in each group?

101. An amusement park has two shuttle buses. Shuttle bus A makes six stops and shuttle bus B makes eight stops. The buses take 5 minutes to go from one stop to the next. Each bus takes a different route. If they start at 10:00 A.M. from station one, at what time will they both return to station one? Station one is at the beginning and the end of the loop and is not counted twice.

102. Four lighthouses can be seen from a boat offshore. One light blinks every 10 seconds. The second light blinks every 15 seconds. The third light blinks every 20 seconds, and the last light blinks every 30 seconds. If all four were started at the same time, how often will all the lights be on at the same time?

103. In a political science class, the teacher wants to assign students with the same political affiliation to groups in order to prepare for a debate. If there are 12 Republicans, 15 Democrats, and 21 Independents, and the teacher wants the same number of people in each group, how many groups will there be for each affiliation?

104. A bakery makes three types of muffins to be distributed to local grocery stores: chocolate chip, blueberry, and banana nut. Each morning, the bakers produce 120 chocolate chip muffins, 150 blueberry, and 90 banana nut. They want to package the muffins efficiently for shipping. The plan is to pack them in boxes so that every box has just one kind of muffin, and every box has the same number of muffins. What's the smallest number of boxes they can ship in total?

105. There are two classes of periodic cicadas that are native to the Midwest: one group emerges every 13 years and eats every tree they can find, while the other emerges every 17 years. If they both emerge in a given year, how long will it be until they both emerge together again?

106. Suppose that in one city, the baseball team has a winning season every 5 years, while the football team has a winning season every 3 years. If they both had winning records this year, how long will it be until that happens again?

The video that you see on TV or a computer screen might appear to be "moving," but in fact it's a series of still pictures that are displayed so quickly they trick your brain into seeing motion. The frame rate (frames per second, or fps) of video is a description of how many individual pictures are shown each second. Film is shot at a frame rate of 24 fps, but that's not the rate that is displayed on TV screens. Use this information in Exercises 107–110.

107. All of North America and some parts of South America and Asia use the NTSC format, which displays at 30 frames per second. It's difficult to convert 24 fps directly to 30 fps, so instead the frame rate is converted to a multiple of 24 that can then be easily converted back down to any frame rate that's a factor of 24. Find the smallest frame rate that 24 fps can be converted to in order to make this work.

108. Refer to Problem 107. The PAL video standard is used in most of Europe, Asia, and Africa. It uses a frame rate of 25 fps. Find the smallest frame rate that standard 24 fps video can be upconverted to in order to ultimately convert to PAL.

109. What's the smallest frame rate that can be used as an intermediate conversion if you want to convert PAL video to NTSC?

110. A relatively new frame rate used in high-def television is 60 fps. Find the smallest frame rate that both PAL and NTSC video needs to be upconverted to in order to convert back to this new standard.

Critical Thinking

111. How many prime numbers are even? Explain how you decided.

112. Until the 19th century, most mathematicians considered 1 to be a prime number. First, make a case for why it would be reasonable to consider 1 to be prime. Then look at the definition of prime number on page 215, but remove the phrase "greater than 1" from it. Now make a case for why 1 shouldn't be considered prime.

113. German mathematician Christian Goldbach (1690–1764) made the following conjecture: every even number greater than 2 can be expressed as the sum of two prime numbers. For example, $6 = 3 + 3$, $8 = 5 + 3$, etc. Express every even number from 4 through 20 as a sum of two prime numbers.

114. Another conjecture attributed to Goldbach is that any odd number greater than 7 can be expressed as the sum of three odd prime numbers. For example, $11 = 3 + 3 + 5$. Express every odd number from 9 through 25 as the sum of three odd prime numbers.

115. In using the sieve of Eratosthenes to generate a list of prime numbers less than 50, we were able to stop after crossing off the multiples of the prime number 7. Why was this sufficient? (*Hint:* Think about the numbers less than 50 that are multiples of 11—why are they already crossed out?) If we wanted to generate a list of all prime numbers less than 200, what's the largest prime we'd need to cross out the multiples of?

116. Fill in each blank with "less than or equal to" or "greater than or equal to," then explain your responses.
    (a) The GCF of a list of numbers is always _____ all of the numbers.
    (b) The LCM of a list of numbers is always _____ all of the numbers.

Using Technology

We've seen that we can use technology to perform divisions, helping us to decide if one number is a factor of another. Because spreadsheets are so good at doing repeated calculations, we can set up a spreadsheet to help us find the factors of any number. We'll help you do that in Questions 117–121.

117. Open a new spreadsheet. Enter "Potential factors" in cell Al, and enter the number 2 in cell A2. Then enter "=A2+1" in cell A3. This will add 1 to the value in cell A2, which in this case gives you 3. Now select cell A3 and copy, then select all of the cells from A4 to A25 and paste. This should give you a list of potential factors from 2 through 25.

118. Now put 540 in cell Bl. This is the number we'll be finding the factors of. In the cell right under that (B2), enter "=B1/A2." This asks the computer to divide the number you put in B2 (540) by the number in A2 (2). The dollar signs in front of the B and the 1 tell the software to divide every number by what's in that exact cell without adjusting when you copy this formula down. So now copy and paste that formula all the way down to cell B25. This will show you the result of dividing 540 by all of the numbers from 2 through 25. The ones that divide evenly, and the results of those divisions, are the factors. List the factors of 540.

119. The spreadsheet we've built will find all factors of every number up to 625. (See Exercise 113 for a discussion of why.) In general, if you divide by all natural numbers up to n, that will find all of the factors of a number that is n^2 or smaller. How high would our list of potential factors have to go in order to find all factors of 3,780?

120. Copy and paste the formula in cell A25 down as far as you need to in order to find the factors of 3,780 (see previous question). Then copy the cells in column B down as well, and list all factors of 3,780.

121. Is there any limit on the size of numbers that you can find the factors of using your shiny new spreadsheet? Discuss.

Section 5-2 | **The Integers**

 LEARNING OBJECTIVES

1. Define whole numbers and integers.

2. Find the opposite and absolute value of a number.

3. Compare numbers using >, <, and =.

4. Add and subtract integers.

5. Multiply and divide integers.

6. Perform calculations using the order of operations.

At the beginning of Section 5-1, we talked about how the natural numbers have that name because they're naturally used to count objects. That makes them about as old as human thought itself. With the idea of counting in mind, you might wonder why zero isn't included in the natural numbers—it's possible to have zero cows, zero children, and zero paintings in your cave, after all. Surprisingly, the number zero wasn't used until the year 200 or so, which is thousands of years after the natural numbers came into use.

The story for negative numbers starts even later—they entered the picture at least a hundred years after zero. It's interesting that numbers that we take for granted were in some sense unknown to some of the greatest mathematical minds in human history. But that's the nature of progress.

Steve Bronstein/Getty Images

In this section, we'll extend our study of numbers to include zero and the negatives of natural numbers, which will allow us to study the basic operations of arithmetic in depth. It will also allow us to apply numbers to more situations in our world.

Definition of Integers

When we add the number zero to the natural numbers, we get a new set called the *whole numbers*.

> The set of **whole numbers** is defined as {0, 1, 2, 3, 4, . . .}.

Math Note

Zero is neither positive nor negative, and the opposite of zero is zero.

Every natural number has an **opposite**, or **additive inverse**. The opposite of 2 is −2, for example. For that reason, the opposite of a number is commonly called its **negative**. When we expand the set of whole numbers by including the negatives of the natural numbers (−1, −2, −3, . . .) the resulting set is called the *integers*.

> The set of **integers** is defined as {. . . , −3, −2, −1, 0, 1, 2, 3, . . .}.

 1. Define whole numbers and integers.

One good way to study the integers is to picture them on a number line, as in Figure 5-2.

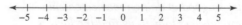

Figure 5-2

Calculator Guide

You can't use the subtraction sign to enter negative numbers into most calculators. To enter the negative integer −20:

Typical Scientific or Online Calculator

20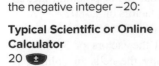

Typical Graphing Calculator
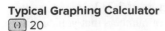

The arrows on each end indicate that the line extends infinitely far in each direction. Zero is called the *origin*. The numbers to the right of zero are the *positive integers*, and those to the left are the *negative integers*. Notice that when looking at a number line, two numbers are opposites when they are the same distance away from zero, but in opposite directions, as seen in Figure 5-3.

Opposites

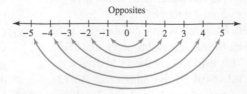

Figure 5-3

Math Note

You can write a positive sign (+) in front of a positive number if you like, but we'll assume that any number except zero with no sign in front of it is positive.

If we use the symbol x to represent any arbitrary number, its opposite would look like $-x$, or if you like, $-(x)$. For example, the opposite of 8 is $-(8)$. But this should be true for negative numbers also: the opposite of -8 (which we know is 8) should be written as $-(-8)$. This tells us that $-(-8) = 8$! (Two wrongs don't make a right, but two negatives do make a positive, at least in this case.)

Earlier, we pointed out that a number and its negative are always the same distance away from zero. The term **absolute value** is used to describe how far away a number is from zero on a number line. Since distance is a physical quantity, negative distances don't make sense; this tells us that the absolute value of any number has to be at least zero. The symbol for absolute value is a pair of vertical bars: the expression $|8|$ represents the absolute value of the number 8.

It's easy to decide that the absolute value of any positive number is the number itself—for example, 8 is most definitely 8 units away from zero. But we know that any negative number is the same distance away from zero as its opposite; this would mean, for example, that $|-8|$ is also 8. In other words, the absolute value of a negative number is the opposite of that number. Now we have rules:

Rules for Absolute Value

The absolute value of any positive number is the number itself.
The absolute value of any negative number is the opposite of that number.
The absolute value of zero is zero.

Here's the simple way that most people remember the rules for absolute value: if the number you're finding the absolute value of isn't negative, do nothing. If it is, drop the negative sign. In Problem 110, we'll see how to write these rules in symbols.

Example 1 — Finding Opposites and Absolute Values

Furnace Creek
ELEVATION -190 FEET

John Henshall/Alamy Stock Photo

Integers are often used for applications where it makes sense to set a zero and then talk about distances to either side. For example, sea level corresponds to 0; the distance above sea level corresponds to the positive integers; and depth below sea level corresponds to negative integers.

2. Find the opposite and absolute value of a number.

Find each number.

(a) The opposite of 100.
(b) The opposite of -50.
(c) The opposite of 0.
(d) $|26|$
(e) $|-99|$
(f) $-(-10)$
(g) $-|-10|$

SOLUTION

(a) The opposite of 100 is -100.
(b) The opposite of -50 is 50 (or $-(-50)$).
(c) Zero is the only number that is its own opposite.
(d) The absolute value of any positive number is the number itself, so $|26| = 26$.
(e) The absolute value of a negative number is its opposite, so $|-99| = 99$.
(f) The opposite of -10 is 10.
(g) Careful! The absolute value of -10 is 10, and the opposite of 10 is -10.

Try This One   1

Find each number.

(a) The opposite of -9.
(b) The opposite of 24.
(c) $|-15|$
(d) $|18|$
(e) $|0|$
(f) $-(-6)$
(g) $-|-30|$

Number lines are useful in deciding which of two integers is larger. Of course, you know that 10 is bigger than 5: on a number line, this is seen from the fact that 10 lives farther to the right. This idea is helpful for comparing sizes when negative numbers are involved. For example, your first instinct might be that −10 is larger than −5, but it isn't because −5 is farther to the right on a number line. We would say that −5 is **greater than** −10, and write −5 > −10. The integer −1 is to the left of 3 on a number line, so we say that −1 is **less than** 3, and write −1 < 3.

| **Example 2** | Using the Symbols >, <, and = |

Fill in the space between the two numbers with >, <, or =.

(a) 4   −2   (b) −20   0   (c) −10   −7   (d) |−4| 4

SOLUTION

(a) Since 4 is to the right of −2 on a number line, 4 > −2.
(b) Since −20 is to the left of 0 on a number line, −20 < 0.
(c) Since −10 is to the left of −7 on a number line, −10 < −7.
(d) Since |−4| is 4, the two numbers are the same; |−4| = 4.

> **Math Note**
>
> The symbols > and < are known as *inequality signs.* There are three others: ≥ means "is greater than or equal to," ≤ means "is less than or equal to," and ≠ means "is not equal to."

✓ 3. Compare numbers using >, <, and =.

Try This One   2

Fill in the space between the two numbers with >, <, or =.

(a) −5   −15   (b) −3   2   (c) |6| |−6| (d) 4   −2

Addition and Subtraction of Integers

You learned how to add and subtract integers pretty early in life, but there's a chance you memorized a lot of the results. In studying these operations as an adult, we will focus on really understanding them, and we'll use a number line as a tool. To add 4 and 1, we'll start at 4, then move one unit to the right; this shows that 4 + 1 = 5, as in Figure 5-4 (a).

A little more thought is required when negative numbers are involved. To add 1 + (−4), we start at 1, then move 4 units to the left (since we're adding *negative* 4). We end up at −3, so 1 + (−4) = −3. See Figure 5-4 (b).

To add −3 + (−4), start at −3 on the number line, then count 4 units to the left to get −7. This shows that −3 + (−4) = −7. See Figure 5-4 (c).

> **Math Note**
>
> The result of adding two numbers is called their *sum.*

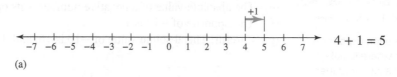

(a)

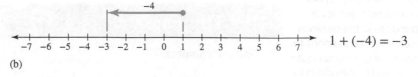

(b)

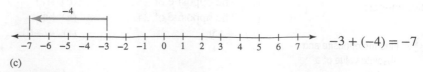

(c)

Figure 5-4

We can use these examples to summarize rules for adding integers.

Rules for Addition of Integers

Rule 1   To add two integers with the same sign, add the absolute values of the numbers and give the answer the common sign.

Rule 2   To add two integers with different signs, find the difference between the two absolute values and give the answer the sign of the number with the larger absolute value.

| Example 3 | Adding Integers |

Find each sum:

(a) $-2 + (-8)$ (b) $-6 + 8$ (c) $6 + (-12)$

SOLUTION

(a) The numbers have the same sign, so use Rule 1. The sum of the absolute values is $2 + 8 = 10$, so $-2 + (-8) = -10$.

(b) This time, the numbers have opposite signs, so we use Rule 2. The difference between the absolute values is 2, and the number with the larger absolute value is positive, so $-6 + 8 = 2$.

(c) Again, the numbers have opposite signs. The difference between the absolute values is 6, and the negative number has the larger absolute value, so $6 + (-12) = -6$.

Try This One   3

Find each sum.

(a) $(-3) + (-8)$ (b) $(-14) + 22$ (c) $19 + (-32)$

When you first learned how to subtract, it was probably taught to you in terms of taking away objects. If you had 8 cookies and that little monster that sat in front of you took 3, you had $8 - 3$ or 5 left. This works fine for natural numbers, but loses its effectiveness when negative numbers are involved. (I challenge you to take away negative 3 cookies.) So a better definition of subtraction uses what we already know about addition.

Math Note

The rule for subtracting integers illustrates a good method of problem solving: we turned a new problem (subtraction) into one we already know how to solve (addition).

Rule for Subtracting Integers

To subtract two integers $a - b$, add the opposite of b to a. That is, $a - b = a + (-b)$.

| Example 4 | Subtracting Integers |

Find each difference:

(a) $-2 - (-8)$ (b) $3 - (-7)$ (c) $-2 - 6$

SOLUTION

(a) Rewrite as $-2 + (8)$ since 8 is the opposite of -8. Now use Rule 2 for addition: the difference between absolute values is 6, and the positive number has the larger absolute value, so $-2 - (-8) = 6$.

(b) Rewrite as $3 + (7)$ since 7 is the opposite of -7. Now add to get $3 - (-7) = 10$.

(c) Rewrite as $-2 + (-6)$ since -6 is the opposite of 6. Now use Rule 1 for addition. The sum of the absolute values is 8, and we give the result a negative sign, so $-2 - 6 = -8$.

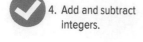

Try This One   4

4. Add and subtract integers.

Find each difference.

(a) $5 - 12$ (d) $25 - (-6)$
(b) $-3 - 6$ (e) $-14 - 8$
(c) $-8 - (-2)$

Example 5   **Applying Addition and Subtraction to Account Balances**

If you have a Paypal account, you can add money to your account, giving you a positive balance, or you can spend more than you currently have in your account, giving you a negative balance that must be paid off. A 5-day history for an active Internet shopper is given below:

Monday: Balance of +$12
Tuesday: Balance of −$5
Wednesday: Balance of −$16
Thursday: Balance of −$25
Friday: Balance of $20

(a) Find the sum of the balances for Monday and Tuesday.
(b) Find the difference of the balances for Thursday and Friday. What does this tell you about the account activity?
(c) What does this result say about subtracting a negative?

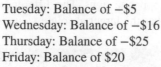

Antonio M. Rosario/Getty Images

SOLUTION

(a) The sum of the balances is $12 + (-\$5) = \$7$.
(b) The difference is $20 - (-\$25) = \$20 + \$25 = \$45$. This tells us that the shopper deposited $45 into her Paypal account on Friday.
(c) In subtracting the balance from Thursday, we had to subtract a negative, and we decided that what must have happened was that $45 was added to the account. This is another way to see why it makes sense that subtracting a negative results in addition.

Try This One   5

Find the sum of the balances for Wednesday and Thursday, and the difference of the balances for Monday and Friday.

Multiplication and Division of Integers

Many college students are creatures of habit. Let's suppose that every Friday afternoon, you stop by the ATM in the union to withdraw $40 for the weekend. Six weeks into the semester,

Sidelight **Of Borrowing, Lending, and Arithmetic with Integers**

Shakespeare once wrote, "Neither a borrower nor a lender be." William would have had to change his tune had he survived to the 21st century, I think. Borrowing money is a necessity for almost everyone that needs to pay for a car, tuition, or a home, and when you deposit money into a bank, you are in essence lending the bank your money. So borrowing and lending are at least somewhat familiar to just about everyone. It turns out that we can use this familiarity to help understand arithmetic with negative numbers easily.

The key is to think of positive integers as amounts of money you have, and negative integers as amounts that you owe. If you have 50 bucks, then incur a debt of $20, you're left with $30: this shows that 50 + (−20) is the same thing as 50 − 20, or 30. This idea is most useful in understanding why subtracting a negative results in addition. Suppose I'm the guy you owe that $20 to, leaving you with $30. In a wholly uncharacteristic display of kindness, I decide to wipe out your debt. So you are removing (subtracting) the debt (−$20), which puts you back to the $50 you had originally. In essence, subtracting negative $20 is the same thing as adding $20! If you think of it that way, it makes perfect sense that subtracting a negative results in addition. And of course, understanding an operation or procedure ALWAYS makes it easier to learn how to do it efficiently, especially in a clutch situation like a test.

you would have withdrawn $40 six times. You could find the total amount withdrawn in dollars using addition:

$$\underbrace{40 + 40 + 40 + 40 + 40 + 40}_{40 \text{ added 6 times}} = 240$$

Maybe you looked at the calculation above and thought "Wouldn't it make more sense to just multiply 40 by 6?" If so, that's great! It means you completely understand the fact that multiplication is nothing other than a shortcut for repeated addition. (That's why we use the word "times" when describing multiplication—the phrase "six times forty" is an indication of the fact that the multiplication means to add 40 six *times*.) So we could represent the repeated addition as $6 \times 40 = 240$. We would read the left side as "six times forty" or "the **product** of six and forty."

If we multiply a positive number times a negative, the same idea works just as well. In fact, you can think of your weekly withdrawals as a negative balance on your account, in which case the overall effect on your account would be

$$-40 - 40 - 40 - 40 - 40 - 40 = -240 \text{ dollars}$$

This could also be written as $6 \times (-40) = -240$. This result suggests that when we multiply a positive number and a negative number, the result is negative. But what if we switch the order, and multiply -40×6? We already know that we can think of 40×6 as adding 6 forty times. So maybe we can think of -40×6 as *subtracting* 6 forty times. This would result in -240 as well, which is the same result we got for $6 \times (-40)$. Again, the product of two integers with opposite signs is negative.

What about multiplying two negatives, like $-3 \times (-2)$? We could think of this as subtracting -2 three times:

$$-(-2) - (-2) - (-2) = 2 + 2 + 2 = 6$$

So $-3 \times (-2) = 6$. Conclusion? The product of two negative integers is positive. Here's a summary of the rules for multiplication:

Rules for Multiplying Integers

Rule 1   The product of two integers with the same sign is positive.

Rule 2   The product of two integers with opposite signs is negative.

Example 6   Multiplying Integers

Find each product:

(a) $(-6) \times (-4)$ (c) -5×16

(b) $3 \times (-9)$ (d) $(-6) \times (-3) \times 8 \times (-5)$

SOLUTION

(a) The signs are the same, so the product is positive: $(-6) \times (-4) = 24$.
(b) The signs are opposite, so the product is negative: $3 \times (-9) = -27$.
(c) The signs are again opposite, so the product is negative: $-5 \times 16 = -80$.
(d) Whenever we repeat the same operation, we work from left to right: first, $(-6) \times (-3) = 18$. Next, $18 \times 8 = 144$. Finally, $144 \times (-5) = -720$.

Try This One   6

Find each product:

(a) $(-16) \times 3$ (c) $12 \times (-8)$

(b) $(-10) \times (-8)$ (d) $(-2) \times (-16) \times 7 \times (-3)$

M G Therin Weise/Photographer's Choice/Getty Images

One application of negative integers is temperature. For example, if over 5 days the low temperatures in degrees Fahrenheit were $-25°, -21°, -15°, -23°, -21°$, find the average temperature by adding the integers and dividing the result by 5: $-105°/5 = -21°$.

The final operation we'll cover in this section is division. Division is the opposite of multiplication in the same way that subtraction is the opposite of addition. For example, $6 - 4 = 2$ because $2 + 4 = 6$. In the same way, $12 \div 2 = 6$ because $6 \times 2 = 12$. We read the operation $12 \div 2$ as "twelve divided by two" or "the **quotient** of twelve and two." Since division and multiplication are so closely related, it comes as no surprise that the rules for dividing signed numbers are the same as the rules for multiplying:

Rules for Dividing Integers

Rule 1   The quotient of two integers with the same sign is positive.

Rule 2   The quotient of two integers with opposite signs is negative.

Example 7   Dividing Integers

Find each quotient:

(a) $32 \div 8$ (b) $-27 \div (-3)$ (c) $-42 \div 7$ (d) $20 \div (-5)$ (e) $8 \div 0$

SOLUTION

(a) The signs are the same, so the quotient is positive: $32 \div 8 = 4$.
(b) Again, the signs are the same, so the quotient is positive: $-27 \div (-3) = 9$.
(c) The signs are opposite, so the quotient is negative: $-42 \div 7 = -6$.
(d) Again, the signs are the opposite, so the quotient is negative: $20 \div (-5) = -4$.
(e) This operation can't be done: we say that the result is undefined. Here's why: think of division as cutting something into a certain number of pieces. If you cut a pizza into eight slices, you've divided the total size by 8. Now . . . I challenge you to cut a pizza into zero pieces! Unless you can make it wink out of existence, you will fail. Conclusion: you can't divide ANYTHING by zero.

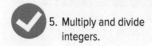

5. Multiply and divide integers.

Try This One 7

Find each quotient.

(a) $96 \div 6$ (b) $-48 \div 3$ (c) $-84 \div (-2)$ (d) $100 \div (-25)$

Order of Operations

How would you interpret the following sentence?

> My girlfriend said Charlie stole my heart and my cat.

Exactly who is the cat thief here—Charlie or the girlfriend? That depends on how the sentence is punctuated. Here are two different interpretations:

> My girlfriend said, "Charlie stole my heart and my cat."
> "My girlfriend," said Charlie, "stole my heart and my cat."

Isn't that cool? (The fact that the same words in the same order can have totally different meanings, not stealing someone's cat, which is totally NOT cool.)

The point is that the exact same string of words in the same order can have two completely different meanings, depending on how the words are grouped. The same is true for mathematical operations. The expression $3 + 5 \times 7$ could be interpreted as:

$$(3 + 5) \times 7 \quad \text{or} \quad 3 + (5 \times 7)$$

In the first case, the result is 8×7, or 56. In the second case, the result is $3 + 35$, or 38.

Whenever different operations are combined in a calculation, there's always the potential for the result to be ambiguous. One way to avoid this is to use parentheses, as we did in the example above. But this requires using parentheses in practically *every* calculation with multiple operations. To avoid this, people long ago agreed on a set of rules to clarify the meaning of such expressions. These rules are known as the **order of operations**.

Steps for Using Order of Operations to Perform Calculations

Step 1 Perform all calculations inside grouping symbols first. Grouping symbols commonly used are parentheses ( ), brackets [ ], braces { }, and absolute values | |.

Step 2 Evaluate all exponents.

Step 3 Perform all multiplication and division in order from left to right.

Step 4 Perform all addition and subtraction in order from left to right.

Note: If an expression contains absolute values, after performing the calculation inside the absolute value, find the absolute value of the result before moving to Step 2.

In the next four examples, we'll illustrate how to use the order of operations.

> **Math Note**
>
> The order of operations is commonly remembered as Parentheses—Exponents—Multiplication/Division—Addition/Subtraction, or PEMDAS for short. There are a wide variety of sayings to help remember PEMDAS, ranging from the silly to the . . . well, let's just say colorful. You can probably find one you like with a Google search.

Example 8   Using the Order of Operations

Perform the calculation: $9 \cdot 3 - (15 \div 5)$

> **Math Note**
>
> A single dot is often used in place of a multiplication sign. In Example 8, $9 \cdot 3$ means the same thing as 9×3.

SOLUTION

$$9 \cdot 3 - (15 \div 5) \quad \textit{Divide inside parentheses}$$
$$= 9 \cdot 3 - 3 \quad \textit{Multiply before subtracting}$$
$$= 27 - 3 \quad \textit{Subtract}$$
$$= 24$$

Try This One   8

Perform the calculation: $(2 + 12) - 15 \div 3$

Example 9 | Using the Order of Operations

Math Note

Remember that exponents represent repeated multiplication: 2^3 means $2 \times 2 \times 2$.

Perform the calculation: $5 \cdot (8 - 10) + 2^3 \div 4$

SOLUTION

$$5 \cdot (8 - 10) + 2^3 \div 4 \qquad \textit{Subtract inside parentheses}$$
$$= 5 \cdot (-2) + 2^3 \div 4 \qquad \textit{Apply the exponent:} \ 2 \times 2 \times 2 = 8$$
$$= 5 \cdot (-2) + 8 \div 4 \qquad \textit{Multiply and divide}$$
$$= -10 + 2 \qquad \textit{Add}$$
$$= -8$$

Try This One   9

Perform the calculation: $216 + (4 \times 5)^2 - 13 \cdot 2$

When an expression contains more than one set of grouping symbols, start by performing the operations in the innermost set and work your way outward.

Example 10 | Using the Order of Operations

Perform the calculation: $84 \div 4 - \{3 \times [10 + (15 - 2)]\}$

SOLUTION

$$84 \div 4 - \{3 \times [10 + (15 - 2)]\} \qquad \textit{Subtract inside parentheses}$$
$$= 84 \div 4 - \{3 \times [10 + 13]\} \qquad \textit{Add inside brackets}$$
$$= 84 \div 4 - \{3 \times 23\} \qquad \textit{Multiply inside braces}$$
$$= 84 \div 4 - 69 \qquad \textit{Divide before subtracting}$$
$$= 21 - 69 \qquad \textit{Subtract}$$
$$= -48$$

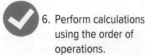

6. Perform calculations using the order of operations.

Try This One   10

Perform the calculation: $9 \times 5 + \{[32 - (6 \times 4)] - 5\}$

The world is a pretty complicated place, so it's not a big surprise that calculations in our world often involve more than one operation. This is why it's important to remember the order of operations.

Example 11 | Applying Order of Operations to a Checking Account

Suppose you have a checking account set up just to pay your share of the rent each month. At the start of the school year, there's $1,100 in the account. You write a check for $240 on the first of every month from August to May, and you deposit a student loan check for $600 at the

C Squared Studios/Photodisc/Getty Images

beginning of each semester (fall and spring). Write an order of operations problem based on this situation, then find the ending balance in your account.

SOLUTION

This problem involves repeated additions and subtractions, so multiplication will come in handy. There are 10 months from August to May, so 10 payments of $240. There are also two deposits of $600. So the amount in dollars is given by

$$1,100 - 10 \times 240 + 2 \times 600 \quad \textit{Multiply first}$$
$$= 1,100 - 2,400 + 1,200$$
$$= -100$$

Uh oh—you're overdrawn! Time to get a summer job. By the way, if we were to simply perform that calculation from left to right without regard for the order of operations, it would show an ending balance of $156,961,200. Nice, but unrealistic.

Try This One   11

A new company makes a profit of $22,340 in the first year. The owner wants to pocket $10,000 of that, then divide the remainder among three salespeople and two associates. Write an order of operations problem based on this situation. How much will each salesperson and associate get?

Answers to Try This One

1   (a) 9   (b) −24   (c) 15   (d) 18   (e) 0
        (f) 6   (g) −30

2   (a) >   (b) <   (c) =   (d) >

3   (a) −11   (b) 8   (c) −13

4   (a) −7   (b) −9   (c) −6
        (d) 31   (e) −22

5   Sum: −$41; difference: $8

6   (a) −48   (b) 80   (c) −96   (d) −672

7   (a) 16   (b) −16   (c) 42   (d) −4

8   9

9   590

10 48

11 (22,340 − 10,000) ÷ (3 + 2); $2,468

Exercise Set   5-2

Writing Exercises

1. What is the difference between the natural numbers and the whole numbers?
2. What is the difference between the whole numbers and the integers?
3. How can we use rules for addition to help us subtract?
4. What is the connection between multiplication and addition?
5. How do you find the absolute value of a number?
6. Describe the order of operations.
7. Explain why subtracting a negative results in addition, using either a number line or financial transactions.
8. Explain why multiplying two negative integers results in a positive product.

Computational Exercises

For Exercises 9–18, find each.

9. $|-8|$
10. $|-12|$
11. $|+10|$
12. $|+14|$
13. The opposite of -8
14. The opposite of $+27$
15. The opposite of $+10$
16. The opposite of -16
17. The opposite of 0
18. The opposite of -9

For Exercises 19–28, insert $>$, $<$, or $=$.

19. $16 \quad 22$
20. $8 \quad 14$
21. $-5 \quad -10$
22. $-6 \quad -22$
23. $0 \quad -3$
24. $-5 \quad 0$
25. $-9 \quad +8$
26. $16 \quad -32$
27. $-10 \quad -7$
28. $-14 \quad +3$

For Exercises 29–90, perform the indicated operation(s).

29. $-6 + 5$
30. $-8 + 4$
31. $16 + (-7)$
32. $(-5) + (-7)$
33. $(-8) + (-3)$
34. $(-4) + 9$
35. $-31 + (-92)$
36. $-26 + 49$
37. $-3 + (-14) + (-16)$
38. $-5 + (-16) + (-28)$
39. $58 - (-20)$
40. $111 - (-51)$
41. $60 - 125$
42. $114 - 86$
43. $-13 - (-41)$
44. $-28 - (-10)$
45. $-210 - (-142)$
46. $-96 - (-403)$
47. $-240 - (-(-138))$
48. $106 - (-(-40))$
49. $(-3)(8)$
50. $(-12)(6)$
51. $4(-9)$
52. $6(-14)$
53. $(-3)(-14)$
54. $(-7)(-14)$
55. $-20(-17)$
56. $-8(-250)$
57. $(-9)(0)$
58. $0(6)$
59. $64 \div 8$
60. $72 \div 9$
61. $-25 \div 5$
62. $-42 \div 7$
63. $32 \div (-8)$
64. $49 \div (-7)$
65. $-14 \div (-2)$
66. $-15 \div (-3)$
67. $-90 \div (-90)$
68. $-56 \div 4$
69. $0 \div 16$
70. $0 \div (-10)$
71. $197 \div 0$
72. $-18 \div 0$
73. $-42 \div 6 + 7$
74. $32 \div (8 \times 2)$
75. $5^3 - 2 \cdot 7$
76. $4 \cdot 3^2 - 2 \cdot 4$
77. $9 \cdot 9 - 5 \cdot 6$
78. $32 - (-6)(4)$
79. $3^3 + 5^2 - 2^4$
80. $14^2 - 5^3 + 8^2$

81. $-3[6 + (-10) - (-2)]$
82. $-5 \cdot 4 - [-3 + 8 - (-5)]$
83. $376 - 14 \cdot 3^4$
84. $82 - 9 \cdot 6 - (-2)^2$
85. $256 - 4^3 \cdot 5 + (8 \cdot 4 - 6 \cdot 4)$
86. $6^2 + 5 \cdot 9 - (-27 + 3 \cdot 2)$
87. $-56 \div 8 - \{3 \times [-10 - (4 \times 3)]\}$
88. $(96 - 70) + [(-4 \times 9) - 32 \div 8]$
89. $32 - \{-16 + 5[25 + 9^2 + (8 - 6)]\}$
90. $2\{-5 - 6[3^2 - 7 \cdot (4 + 1)]\}$

Applications in Our World

91. A student's bank balance at the beginning of the month was $867. During the month, she made deposits of $83, $562, $37, and $43, and withdrawals of $74, $86, and $252. What was her bank balance at the end of the month?
92. Pike's Peak in Colorado is 14,110 feet high, while Death Valley is 282 feet below sea level. Find the vertical distance from the top of Pike's Peak to the bottom of Death Valley.
93. The manager of a biological supply company runs a breeding facility for baby rats. At the beginning of a week, there were 1,286 baby rats. The table shows the number of new rats born each day of the week and the number that were sold. How many rats were left at the end of the week?

	Mon.	Tue.	Wed.	Thur.	Fri.	Sat.
Born	382	494	327	778	256	641
Sold	105	850	416	237	192	965

94. A large grocery store has 354 cases of canned vegetables in the storeroom. During the past month, the store removed 87 cases, 53 cases, 42 cases, and 67 cases to put on the shelves. Also, the store received two lots of 80 cases each. How many cases are in the storeroom now?
95. The 30-year average snowfall in Bismarck, North Dakota, is approximately 18 inches per year. As of January 10, Bismarck received approximately 10 inches. How much more snow will Bismarck receive this year if this is an average year?
96. The average annual snowfall in Salt Lake City is about 59 inches. As of January 10, the recorded snowfall was 34 inches. How much more snow will Salt Lake City receive this year if this is an average year?
97. The table shows the change in population for the states/territories with the biggest percentage of population gains and losses from 2010 to 2020. Find the change in population for each state or territory, with positive numbers representing gain and negative representing loss.

State/Territory	Pop. 2020	Pop. 2010
Texas	29,145,505	25,145,561
Idaho	1,839,106	1,567,582
Utah	3,271,616	2,763,885
Puerto Rico	3,285,874	3,725,789

State	Pop. 2020	Pop. 2010
West Virginia	1,793,716	1,852,994
Mississippi	2,961,279	2,967,297

98. The average high temperatures in my hometown, beautiful Cleveland, Ohio, for the months from October to April are shown in the table. Find the change in average high for each 1-month period, with a raised temperature represented by a positive change, and a lowered temperature represented by a negative change.

Month	Oct	Nov	Dec	Jan	Feb	Mar	Apr
Temp (°F)	60.8	48.7	37.4	32.6	35.8	46.1	57.3

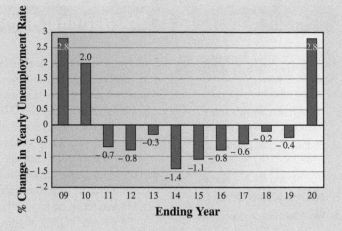

99. A food bank sends a truck with 400 pounds of food to each of five local food pantries every week, and receives a government surplus shipment of 3,000 pounds every other week. If it has 10,000 pounds of food on hand at the beginning of July, how much will it have six weeks later?

100. A small college admits a first-year class of 330 students at the beginning of each year, and graduates 145 students twice a year. What will be the net change in student population over a 5-year span?

Problems 101–104 are based on the following bar graph, which shows the yearly change in the U.S. unemployment rate from the previous year. For example, the percent change from 2015 to 2016 was a decrease of 0.8%.

101. Find the difference between the change from 2009 to 2010 and the change from 2013 to 2014.

102. Find the sum of the changes for all of the years in which unemployment went down.

103. Find the difference between the change in the year with the biggest increase, and the change in the year with the biggest decrease.

104. (a) Find the average of the change for all years in which unemployment went up.

    (b) Find the average of the change for all years in which unemployment went down.

    (c) Based on your answers to parts (a) and (b), make an educated guess on the overall average for all years shown in the chart. Then compute the actual average. How accurate was your guess?

Critical Thinking

105. If x represents an arbitrary integer, can you make any general statement about the sign of $-x$? What about $-(-x)$?

106. Write a sentence in English that can have two different meanings depending on the punctuation, then write a mathematical expression whose value can be different depending on the order in which the operations are performed.

107. When you first learned to subtract in grade school, you were probably taught to think of it as "taking away," as in "Suzie has four cookies, and Johnny takes away two of them." Explain why this isn't a reliable way to think of subtraction when working with integers.

108. Example 11 illustrates a practical situation that leads to a calculation requiring the order of operations. Write a word problem that does the same, but doesn't use the same situation as any of the problems in the book. Extra points for creativity!

109. In the order of operations, multiplication and division live at the same priority level, and if there is more than one occurrence of these operations, we're instructed to work from left to right. But does it really matter? Write some examples of calculations that involve multiple occurrences of multiplication and division, and check to see if the order matters. Summarize your results. Repeat for addition and subtraction. Do the signs of the numbers involved matter?

110. Suppose that a is any integer, either positive, negative, or zero. Using the rules for absolute value on page 227, fill in the blanks in the following formula for computing absolute value:

$$|a| = \begin{cases} \underline{} & \text{if } a \geq 0 \\ \underline{} & \text{if } a < 0 \end{cases}$$

Section 5-3 — The Rational Numbers

LEARNING OBJECTIVES

1. Define rational numbers.
2. Convert between improper fractions and mixed numbers.
3. Reduce fractions to lowest terms.
4. Multiply and divide fractions.
5. Add and subtract fractions.
6. Write fractions in decimal form.
7. Write terminating and repeating decimals in fraction form.

So far in this chapter, we've seen many realistic situations that involve integers. But everything in the world doesn't come in whole pieces. For example, Authentic was given a $\frac{1}{9}$ chance of winning the 2020 Kentucky Derby, making him the third betting choice of the 15 horses in the field. The big guy wasn't expected to win, but he came through, covering the $1\frac{1}{4}$ mile distance in 2 minutes, $\frac{3}{5}$ seconds to win by $1\frac{1}{4}$ lengths over Tiz the Law. Had you bet two bucks on Authentic to win, the good folks at Churchill Downs would have returned to you the princely sum of $18.80.

Garry Jones/ASSOCIATED PRESS

There are TONS of things that come in fractional parts, so many that an inability to understand and work with fractions will have a real impact on your world. This section is all about studying things that come in fractional parts by studying the numbers that can be used to represent them.

Definition of Rational Numbers

The word "ratio" in math refers to a comparison of the sizes of two different quantities. For example, if there is one junior for every three sophomores in a class, we would say that the ratio of juniors to sophomores is 1 to 3. Ratios are often written as fractions: in this case $\frac{1}{3}$. For that reason, numbers that can be written as fractions are called *rational numbers*.

> A **rational number** is any number that can be written as a fraction in the form $\frac{a}{b}$, where a and b are both integers (and b is not zero). The integer a is called the **numerator** of the fraction, and b is called the **denominator**.

 1. Define rational numbers.

Math Note

Rational numbers are the result of division: if you divide 11 by 2, the result is not an integer, but instead the rational number $\frac{11}{2}$.

We can locate rational numbers on a number line; most of them will be between integers. For example, $\frac{2}{3}$ is between 0 and 1, and $-\frac{11}{2}$ is between -5 and -6. See Figure 5-5.

All of the integers we've studied are also rational numbers because they can be written as fractions with denominator 1. For example, $3 = \frac{3}{1}$, which fits the definition of rational number. This means that a number like 3 is a natural number, a whole number, an integer, and a rational number. Of course, there are many rational numbers that are not integers, like $\frac{2}{3}$. Every rational number can also be written in decimal form. We'll study decimals later in this section.

Chocolate Chip Cookies

1	cup softened butter (2 sticks)
$\frac{1}{2}$	cup granulated sugar
$1\frac{1}{2}$	cups packed brown sugar
2	eggs
$2\frac{1}{2}$	cups all-purpose flour
$\frac{3}{4}$	teaspoon salt
1	teaspoon baking powder
1	teaspoon baking soda
18	ounces chocolate chips

McGraw Hill

Fractions are very commonly used in recipes.

When the numerator of a fraction has a smaller absolute value than the denominator, we call the fraction a **proper fraction**. For example, $\frac{2}{3}, \frac{1}{2}$, and $-\frac{7}{8}$ are all proper fractions. When the absolute value of the numerator is larger than or equal to that of the denominator, we call the fraction an **improper fraction**. For example, $\frac{4}{3}, \frac{10}{4}$, and $-\frac{11}{2}$ are all improper fractions.

A **mixed number** consists of a whole number and a fraction. For example, $3\frac{1}{2}$ and $-10\frac{3}{5}$ are mixed numbers. It's important to understand that a mixed number is really an addition without the $+$ sign : $3\frac{1}{2}$ means the same thing as $3 + \frac{1}{2}$. (The mixed number

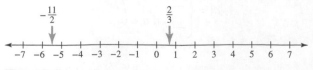

$$-\frac{11}{2} \qquad\qquad \frac{2}{3}$$

Figure 5-5

is read as "three and one half.") We can use division to write improper fractions as integers or mixed numbers. This can be helpful in interpreting their size.

| Example 1 | Writing an Improper Fraction as a Mixed Number |

Math Note

Notice that every proper fraction has absolute value less than 1, and every improper fraction has absolute value greater than or equal to 1.

Write $\frac{17}{6}$ as a mixed number.

SOLUTION

First, divide 17 by 6 using long division:

$$6\overline{)17} \\ \underline{-12} \\ 5$$

with quotient 2.

The quotient is 2, with remainder 5. That tells us that $17 = 2 + \frac{5}{6}$. In mixed number form, we write $\frac{17}{6} = 2\frac{5}{6}$.

Math Note

Dividing 17 by 6 on a calculator won't result in a mixed number, but rather a decimal approximation: 2.833333333. In many cases, in exchanging a fraction for a decimal approximation, we lose accuracy.

Try This One 1

Write $\frac{23}{7}$ as a mixed number.

Since we use division to rewrite improper fractions as mixed numbers, it shouldn't come as a surprise that we can use multiplication to do the opposite—rewrite mixed numbers as improper fractions. This will be useful in performing operations on mixed numbers.

| Example 2 | Writing a Mixed Number as an Improper Fraction |

Calculator Guide

Graphing calculators have a feature that allows you to find the improper fraction form for a mixed number. For the number in Example 2:

7  2 ÷ 5 [ENTER], then hit

[MATH] 1 [ENTER] to convert to fractional form.

Write $7\frac{2}{5}$ as an improper fraction.

SOLUTION

Step 1 Multiply the whole number part (7 in this case) by the denominator (5), then add the numerator (2):

$$7 \cdot 5 + 2 = 35 + 2 = 37$$

Step 2 Write as a fraction with the result of step 1 as numerator, and the original denominator of the fractional part.

$$7\frac{2}{5} = \frac{37}{5}$$

In Problem 114, we'll think about exactly why this two-step procedure works.

2. Convert between improper fractions and mixed numbers.

Try This One 2

Write $10\frac{1}{4}$ as an improper fraction.

Since fractions can be thought of as the division of two integers, we can take advantage of rules for division of signed numbers to address the sign of a fraction. If either the numerator or denominator is negative while the other is positive, it's like dividing two numbers with opposite signs. In this case, the result is negative. So rather than write fractions with a negative in the numerator or denominator, we will write a negative in front of the fraction to indicate that it represents a negative number. For example, all of $\frac{-3}{5}, \frac{3}{-5}$, and $-\frac{3}{5}$ represent the same number.

On the other hand, if both the numerator and denominator are negative, the value of the fraction is positive, so we wouldn't bother to write the negatives at all. For example, $\frac{-3}{-5} = \frac{3}{5}$.

Reducing Fractions

Math Note

Over 3 billion pizzas are sold in the United States every year; on average, every human being in America eats 23 pounds of pizza annually.

Suppose that you and your roommate order a pizza that you want to split evenly. When it gets delivered, you find that the pizza place didn't cut it. If you cut it into two equal pieces, each of you is getting $\frac{1}{2}$ of the pizza. But if the pizza place had cut it into 8 equal slices, you would each take 4 slices, and get $\frac{4}{8}$ of the pizza. In either case, you got half of the pizza, so it must be true that $\frac{4}{8} = \frac{1}{2}$. We could recognize this mathematically by dividing the numerator and denominator of $\frac{4}{8}$ by 4, resulting in $\frac{1}{2}$. This is known as **reducing a fraction to lowest terms**. In this case, we say that $\frac{4}{8}$ and $\frac{1}{2}$ are **equivalent fractions**, which means they represent the same number.

Example 3   Reducing a Fraction to Lowest Terms

Reduce each fraction to lowest terms.

(a) $\dfrac{18}{24}$ (b) $\dfrac{8}{21}$

SOLUTION

(a) Both the numerator and denominator can be divided by 2 with no remainder:

$$\frac{18 \div 2}{24 \div 2} = \frac{9}{12}$$

Before we congratulate ourselves, notice that there's still more that can be done: each of 9 and 12 is divisible by 3!

$$\frac{9 \div 3}{12 \div 3} = \frac{3}{4}$$

Now the fraction is in lowest terms because 3 and 4 have no common divisors. We could have accomplished this in one step by finding the greatest common factor (GCF) of 18 and 24, which is 6, then dividing the numerator and denominator by it:

$$\frac{18 \div 6}{24 \div 6} = \frac{3}{4}$$

Either method is acceptable.

(b) Nothing jumps out at me that both numerator and denominator can be divided by, so let's try writing the prime factorization of each.

$$8 = 2 \cdot 2 \cdot 2$$
$$21 = 3 \cdot 7$$

Aha! I couldn't find anything to divide by because there isn't a common factor. This fraction is already in lowest terms and can't be reduced.

Try This One   3

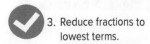
3. Reduce fractions to lowest terms.

Reduce each fraction to lowest terms.

(a) $\dfrac{56}{64}$ (b) $\dfrac{15}{77}$

Sidelight   A Word to Cancel from Your Math Vocabulary

When reducing fractions, it's common to not write out the division as we did in Example 3. Often, we represent dividing the numerator and denominator of a fraction by some number (in this case 6) like this:

$$\frac{\overset{3}{\cancel{18}}}{\underset{4}{\cancel{24}}}$$

There is nothing wrong with this approach, unless you forget that what you did was to divide numerator and denominator by the same number. Here's another way to look at it:

$$\frac{18}{24} = \frac{\cancel{6} \cdot 3}{\cancel{6} \cdot 4}$$

It's common to use the word "canceling" to refer to this process. The problem with that term is that it makes it really easy to forget the math of what you've done, and just think of it as "crossing out" the same symbol in the numerator and denominator. And here's why that's dangerous:

$$\frac{\cancel{3}+1}{\cancel{3}+3} = \frac{1}{3} \quad \textit{This is wrong!}$$

The original fraction, before "canceling," is $\frac{4}{6}$, or $\frac{2}{3}$. Hopefully, it's clear that the statement $\frac{2}{3} = \frac{1}{3}$ is nonsense. What went wrong here is that we didn't *divide* the numerator and denominator by 3—we *subtracted* 3 from the numerator and denominator. This is a useful example because it shows that adding or subtracting something to the numerator and denominator in this manner completely changes the value of the fraction.

We can guarantee from past experience that if you get in the habit of making this mistake, it will be *extremely* difficult to break that habit. One good way to avoid it is to resist the temptation to use the word "canceling." I like to explain it to my students this way: **there is no operation called "canceling."** If you call the process of reducing fractions what it is—dividing the numerator and denominator by the same number—you'll be far less likely to fall into this common trap.

If we can divide both sides of a fraction by the same nonzero number, it seems reasonable that we can also multiply both sides by the same nonzero number. This is basically the opposite of reducing fractions, and it will come in handy later in the section.

| Example 4 | Rewriting a Fraction with a Larger Denominator |

Change each fraction to an equivalent fraction with the indicated denominator.

(a) $\dfrac{3}{8} = \dfrac{?}{32}$ (b) $\dfrac{5}{4} = \dfrac{?}{56}$

SOLUTION

(a) To change the denominator from 8 to 32, we have to multiply by 4. So to make the fractions equivalent, we also need to multiply the numerator by 4. The equivalent fraction is

$$\frac{3 \cdot 4}{8 \cdot 4} = \frac{12}{32}$$

(b) It may be a bit harder to notice what number we need to multiply 4 by to get 56, so we divide $56 \div 4 = 14$. The number we need to multiply the numerator and denominator by is 14.

$$\frac{5 \cdot 14}{4 \cdot 14} = \frac{70}{56}$$

Math Note

Some people call the process of "unreducing" fractions in Example 4 "building up" fractions. Notice that all we're really doing is multiplying by 1!

Try This One   4

Change each fraction to an equivalent fraction with the indicated denominator.

(a) $\dfrac{3}{8} = \dfrac{?}{24}$ (b) $\dfrac{7}{13} = \dfrac{?}{65}$ (c) $\dfrac{5}{9} = \dfrac{?}{99}$ (d) $\dfrac{4}{5} = \dfrac{?}{50}$

Multiplying and Dividing Fractions

Multiplication is the simplest operation when working with fractions.

Multiplying Fractions

To multiply two fractions, multiply the numerators and the denominators separately. That is,

$$\frac{a}{b} \cdot \frac{c}{d} = \frac{a \cdot c}{b \cdot d}$$

Example 5 | **Multiplying Fractions**

Math Note

The rules for multiplying positive and negative integers from Section 5-2 also apply to rational numbers. We'll study exactly why in Problems 117 and 118.

Find each product, and write the answer in lowest terms.

(a) $\dfrac{5}{8} \cdot \dfrac{3}{5}$ (b) $-\dfrac{5}{9} \cdot \dfrac{3}{11}$ (c) $1\dfrac{3}{4} \times 2\dfrac{2}{5}$

SOLUTION

(a) Rather than actually multiplying out the numerator and denominator, we'll write as $5 \cdot 3$ and $8 \cdot 5$, which allows us to reduce easily.

$$\frac{5}{8} \cdot \frac{3}{5} = \frac{^1\cancel{5} \cdot 3}{8 \cdot \cancel{5}_1} = \frac{3}{8}$$

(b) The product will be negative since the two fractions have opposite signs.

$$-\frac{5}{9} \cdot \frac{3}{11} = -\frac{5 \cdot \cancel{3}^1}{_3\cancel{9} \cdot 11} = -\frac{5}{33}$$

(c) Our multiplication rule doesn't apply to mixed numbers, so we should first rewrite each as an improper fraction, then multiply.

$$1\frac{3}{4} \times 2\frac{2}{5} = \frac{7}{4} \times \frac{12}{5} = \frac{7 \cdot \cancel{12}^3}{_1\cancel{4} \cdot 5} = \frac{21}{5} \quad \text{or} \quad 4\frac{1}{5}$$

Try This One   5

Find each product, and write the answer in lowest terms.

(a) $\dfrac{2}{7} \cdot \left(-\dfrac{21}{8}\right)$ (b) $\dfrac{11}{9} \times \dfrac{12}{55}$ (c) $3\dfrac{1}{2} \cdot 2\dfrac{2}{3}$

If we switch the numerator and denominator of a fraction, then multiply by the original fraction, something interesting happens:

$$\frac{3}{5} \cdot \frac{5}{3} = \frac{15}{15} = 1$$

For any nonzero fraction $\frac{a}{b}$, the fraction $\frac{b}{a}$ is called the **multiplicative inverse**, or **reciprocal**, of $\frac{a}{b}$. This term is useful in dividing fractions.

Dividing Fractions

To divide two fractions, multiply the first by the reciprocal of the second. That is,

$$\frac{a}{b} \div \frac{c}{d} = \frac{a}{b} \cdot \frac{d}{c}$$

| Example 6 | Dividing Fractions |

Math Note

A convenient way to think of dividing by a fraction is the phrase "invert and multiply." "Invert" refers to flipping the fraction you're dividing by upside down.

Find each quotient, and write the answer in lowest terms.

(a) $\dfrac{3}{4} \div \left(-\dfrac{5}{8}\right)$

(b) $\dfrac{11}{2} \div \dfrac{4}{9}$

SOLUTION

(a) Multiply $\frac{3}{4}$ by the reciprocal of $-\frac{5}{8}$. The two fractions have opposite signs, so the quotient will be negative.

$$\frac{3}{4} \div \left(-\frac{5}{8}\right) = \frac{3}{4} \times \left(-\frac{8}{5}\right) = -\frac{3 \cdot \overset{2}{\cancel{8}}}{\underset{1}{\cancel{4}} \cdot 5} = -\frac{6}{5}$$

(b) Again, multiply the first fraction by the reciprocal of the second.

$$\frac{11}{2} \div \frac{4}{9} = \frac{11}{2} \times \frac{9}{4} = \frac{99}{8}$$

This time, the quotient can't be reduced.

| Try This One | 6 |

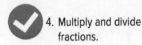

4. Multiply and divide fractions.

Find each quotient, and write the answer in lowest terms.

(a) $-\dfrac{5}{3} \div \left(-\dfrac{10}{9}\right)$

(b) $\dfrac{14}{5} \div \dfrac{3}{8}$

Adding and Subtracting Fractions

Based on the method we use for multiplying fractions, it's tempting to assume we do the same for addition: add the numerators and denominators separately. But there's a small problem with this approach—it doesn't work!

   Think back to our earlier example of a pizza cut into eight equal pieces. Suppose you were sharing the pizza with two friends, and you got three slices. Then you'd have $\frac{3}{8}$ of the pizza. If one of your friends decided she wasn't that hungry and gave you one more piece, you'd now have $\frac{3}{8} + \frac{1}{8}$ of the pizza. But if we add numerators and denominators separately, we get

$$\frac{3}{8} + \frac{1}{8} \overset{?}{=} \frac{3+1}{8+8} = \frac{4}{16}$$

This is $\frac{1}{4}$ of the pizza, which is a silly answer; it's less than the $\frac{3}{8}$ you started with!

   This example shows that addition of fractions is different from multiplication. If two fractions happen to have the same denominator, there's a simple rule for adding (or subtracting).

Adding and Subtracting Fractions with a Common Denominator

To add or subtract two fractions with the same denominator, add or subtract the numerators, and keep the common denominator the same in your answer.

| **Example 7** | **Adding and Subtracting Fractions with a Common Denominator** |

Math Note

It will help you to understand the importance of common denominators if you think of "twelfths" as objects in Example 7(a). Five twelfths plus eleven twelfths is sixteen twelfths, just like five tacos plus eleven tacos is sixteen tacos. That's a lot of tacos.

Find each sum or difference.

(a) $\dfrac{5}{12} + \dfrac{11}{12}$ (b) $\dfrac{7}{3} - \dfrac{4}{3}$

SOLUTION

(a) $\dfrac{5}{12} + \dfrac{11}{12}$ *Add numerators, keep denominator 12*

$= \dfrac{16}{12}$ *Reduce*

$= \dfrac{4}{3}$

(b) $\dfrac{7}{3} - \dfrac{4}{3}$ *Subtract numerators, keep denominator 3*

$= \dfrac{3}{3}$ *Reduce*

$= 1$

Math Note

From this point on, we'll assume that any answer that's a fraction should be reduced to lowest terms.

Try This One   7

Find each sum or difference:

(a) $\dfrac{17}{9} + \dfrac{4}{9}$ (b) $\dfrac{2}{15} - \dfrac{11}{15}$

To add or subtract fractions with different denominators, we'll need to rewrite the fractions so that they have the same denominator.

Steps for Adding or Subtracting Fractions with Different Denominators

Step 1   Find the least common multiple of the denominators. (This is usually called the **least common denominator**, or **LCD**.)

Step 2   Rewrite each fraction as an equivalent fraction with denominator equal to the LCD.

Step 3   Add or subtract as in Example 7.

| **Example 8** | **Adding and Subtracting Fractions** |

Find each sum or difference.

(a) $\dfrac{1}{4} + \dfrac{5}{6}$ (b) $\dfrac{4}{9} - \dfrac{2}{5}$ (c) $2\dfrac{1}{2} + 3\dfrac{1}{4}$

SOLUTION

(a) The LCD of 4 and 6 is 12, so we rewrite each fraction with denominator 12, then add.

$$\frac{1}{4} + \frac{5}{6} = \frac{1 \cdot 3}{4 \cdot 3} + \frac{5 \cdot 2}{6 \cdot 2} = \frac{3}{12} + \frac{10}{12} = \frac{13}{12} \quad \text{or} \quad 1\frac{1}{12}$$

(b) The LCD of 9 and 5 is 45. Rewrite each fraction with denominator 45 and subtract:

$$\frac{4}{9} - \frac{2}{5} = \frac{4 \cdot 5}{9 \cdot 5} - \frac{2 \cdot 9}{5 \cdot 9} = \frac{20}{45} - \frac{18}{45} = \frac{2}{45}$$

(c) First, we need to rewrite the mixed numbers as improper fractions.

$$2\frac{1}{2} = \frac{5}{2} \quad \text{and} \quad 3\frac{1}{4} = \frac{13}{4}$$

The LCD of 2 and 4 is 4, so we rewrite $\frac{5}{2}$ as $\frac{10}{4}$, then add.

$$\frac{10}{4} + \frac{13}{4} = \frac{23}{4} \quad \text{or} \quad 5\frac{3}{4}$$

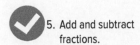 5. Add and subtract fractions.

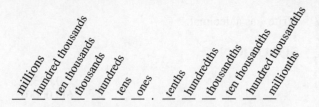

Try This One   8

Find each sum or difference.

(a) $-\frac{3}{8} + \frac{5}{6}$ (b) $\frac{9}{10} - \frac{1}{2}$ (c) $5\frac{3}{4} - 4\frac{1}{3}$

Roberto Westbrook/Spaces Images/ Blend Images LLC

Gas prices are often given in thousandths of a dollar.

Fractions and Decimals

You might be familiar with the decimal form of fractions, especially when using a calculator. Any fraction can be written in decimal form. (The opposite is not true—many decimals can be written as fractions, but not all.) In order to work with numbers in decimal form, you need to be familiar with **place value**, which is described by Table 5-1.

Table 5-1	Place Value Chart for Decimals

millions | hundred thousands | ten thousands | thousands | hundreds | tens | ones . tenths | hundredths | thousandths | ten thousandths | hundred thousandths | millionths

---

Sidelight   **Music and Fractions**

Music is based on math, and fractions play a big role. The timing of music is described by fractions: there are half notes, quarter notes, eighth notes, and so forth. The tones of the notes themselves illustrate fractions as well. When the string of a guitar is plucked, a note is sounded, and if that string is shortened to make it $\frac{1}{2}$ as long, the same note will be heard, but a full octave higher. By increasing the length of the string, other notes can be sounded. For example, if a string $\frac{16}{15}$ as long as the string that plays note C is used, it will play the next lower note, B. If it is stretched to $\frac{6}{5}$ as long, it will sound an A note. That's pretty much how a guitar works: by pressing the strings against the fret

board, the player is in essence changing their length. This might sound like a pretty modern idea, but the famous Greek mathematician Pythagoras discovered the mathematical relationship between the notes C, F, and G over 2,500 years ago.

Lawrence Manning/ Digital Stock/ CORBIS

To change a fraction to decimal form, we use division.

Example 9   **Writing a Fraction in Decimal Form**

Math Note

If your instructor is okay with you using a calculator, then the long division we're demonstrating here isn't necessary. But you still need to be aware of the difference between decimals that end, like this one, and those that don't, like In Example 10.

Write $\frac{5}{8}$ as a decimal.

SOLUTION

Divide 5 by 8, as shown.

$$
\begin{array}{r}
0.625 \\
8\overline{)5.000} \\
-48 \\
\hline
20 \\
-16 \\
\hline
40 \\
-40 \\
\hline
0
\end{array}
$$

Once we get a remainder of zero, we're done: $\frac{5}{8} = 0.625$.

Try This One   **9**

Write each fraction as a decimal.

(a) $\dfrac{3}{8}$ (b) $\dfrac{7}{20}$

Because we eventually got a remainder of zero, the decimal equivalent for $\frac{5}{8}$ ends after three digits. We call this a **terminating decimal**. Not all fractions can be converted to terminating decimals, as we'll see in Example 10.

Example 10   **Writing a Fraction in Decimal Form**

Math Note

While it's not at all obvious, it can be shown that the decimal expansion for every rational number is either terminating or repeating. In fact, the converse of that fact is true as well—if a decimal either terminates or repeats, it's the decimal expansion for a rational number.

Write $\frac{5}{6}$ as a decimal.

SOLUTION

$$
\begin{array}{r}
0.8333\ldots \\
6\overline{)5.0000} \\
-48 \\
\hline
20 \\
-18 \\
\hline
20 \\
-18 \\
\hline
20 \\
-18 \\
\hline
2
\end{array}
$$

Notice that the pattern will keep repeating, so $\frac{5}{6} = 0.8333\ldots$.

Try This One   **10**

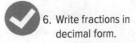
6. Write fractions in decimal form.

Write each fraction as a decimal.

(a) $\dfrac{5}{12}$ (b) $\dfrac{19}{33}$

Math Note

The decimal $0.5\overline{6}$ is not the same as $0.\overline{56}$. In the first case, only the 6 repeats $(0.5666\ldots)$. In the second case, the 56 repeats $(0.565656\ldots)$.

The decimal for $\frac{5}{6}$, $0.8333\ldots$, is called a **repeating decimal**. Repeating decimals can be written by placing a line over the digits that repeat.

$0.8333\ldots$ is written as $0.8\overline{3}$

$0.626262\ldots$ is written as $0.\overline{62}$

Any terminating decimal can be written in fraction form, using the following procedure.

Procedure for Writing a Terminating Decimal as a Fraction

Step 1 Drop the decimal point and place the resulting number in the numerator of a fraction.

Step 2 Use a denominator of 10 if there was one digit to the right of the decimal point, a denominator of 100 if there were two digits to the right of the decimal point, a denominator of 1,000 if there were three digits to the right of the decimal point, and so on.

Step 3 Reduce the fraction if possible.

Example 11 **Writing a Terminating Decimal as a Fraction**

Math Note

The MATH—FRAC command described in the Calculator Guide next to Example 2 can be used to find the rational number corresponding to any terminating decimal.

Write each decimal as a fraction.

(a) 0.8   (b) 0.65   (c) 0.024   (d) 2.65

SOLUTION

(a) $0.8 = \dfrac{8}{10} = \dfrac{4}{5}$ *One digit to the right of the decimal point*

(b) $0.65 = \dfrac{65}{100} = \dfrac{13}{20}$ *Two digits to the right of the decimal point*

(c) $0.024 = \dfrac{24}{1,000} = \dfrac{3}{125}$ *Three digits to the right of the decimal point*

(d) Even with a digit to the left of the decimal point, the procedure in the colored box still works:

$2.65 = \dfrac{265}{100} = \dfrac{53}{20}$ *Two digits to the right of the decimal point*

Try This One   **11**

Write each decimal as a fraction.

(a) 0.4   (b) 0.48   (c) 0.325   (d) 5.12

A different procedure is used to find the fractional equivalent of a repeating decimal.

Procedure for Writing a Repeating Decimal as a Fraction

Step 1   Write the equation n = the repeating decimal, then write a second equation by multiplying both sides of the first equation by 10 if one digit repeats, 100 if two digits, repeat, etc.

Step 2   Now you will have two equations: subtract the first equation from the second. The repeating part of the decimal will subtract away.

Step 3   Divide both sides of the resulting equation by the number in front of n. The result will be the fractional equivalent of the repeating decimal. (Reduce if necessary.)

Example 12   **Writing a Repeating Decimal as a Fraction**

Change $0.\overline{8}$ to a fraction.

SOLUTION

Step 1   Write $n = 0.\overline{8}$ and multiply both sides of the equation by 10 to get $10n = 8.\overline{8}$.

Step 2   Subtract the first equation from the second one as shown.

$$\begin{array}{r} 10n = 8.\overline{8} \\ -n = 0.\overline{8} \\ \hline 9n = 8 \end{array}$$

Step 3   Divide both sides by 9.

$$\frac{9n}{9} = \frac{8}{9}$$

$$n = \frac{8}{9}$$

> **Math Note**
>
> The result can be checked by changing $\frac{8}{9}$ to a decimal.

Try This One   **12**

Change $0.\overline{4}$ to a fraction.

Example 13   **Writing a Repeating Decimal as a Fraction**

Change $0.\overline{63}$ to a fraction.

SOLUTION

Step 1   Write $n = 0.\overline{63}$ *Multiply both sides by 100*
             $100n = 63.\overline{63}$

Step 2 $\begin{array}{r} 100n = 63.\overline{63} \\ -n = \ \ \ 0.\overline{63} \\ \hline 99n = \ \ \ \ \ 63 \end{array}$ *Subtract first equation from second*

   Divide both sides by 99

Step 3 $\dfrac{99n}{99} = \dfrac{63}{99}$ *Reduce*

$$n = \frac{63}{99} = \frac{21}{33} = \frac{7}{11}, \quad \text{so} \quad 0.\overline{63} = \frac{7}{11}.$$

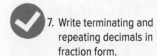

7. Write terminating and repeating decimals in fraction form.

Try This One 13

Change $0.\overline{56}$ to a fraction.

We'll close the section with an application of operations involving rational numbers.

Example 14 · Applying Rational Numbers to Fitness Training

In her final 2 days of preparation for a triathlon, Cat hopes to swim, run, and bike a total of 50 miles. She works out at a nearby state park with a swimming quarry and running/biking trail. One lap in the quarry is $\frac{1}{4}$ mile and the trail is $3\frac{1}{4}$ miles. The first day, Cat swims six laps, runs the trail twice, and bikes it five times. How many more miles does she need to cover the second day?

SOLUTION

The total distance she covers on the first day is

$$6 \cdot \frac{1}{4} + 2\left(3\frac{1}{4}\right) + 5\left(3\frac{1}{4}\right)$$

The mixed number $3\frac{1}{4}$ is $\frac{13}{4}$ as an improper fraction. Using order of operations, we perform the multiplications first, then add.

$$\frac{6}{1} \cdot \frac{1}{4} + \frac{2}{1} \cdot \frac{13}{4} + \frac{5}{1} \cdot \frac{13}{4} = \frac{6}{4} + \frac{26}{4} + \frac{65}{4} = \frac{97}{4} \text{ miles}$$

Now we subtract from 50 to get the distance Cat needs to cover the second day.

$$50 - \frac{97}{4} = \frac{200}{4} - \frac{97}{4} = \frac{103}{4} \quad \text{or} \quad 25\frac{3}{4} \text{ miles}$$

Try This One 14

If Cat swims three laps and runs and bikes the trail four times each on the second day, will she reach her goal? If she is over or under her goal, by how much?

Answers to Try This One

1 $3\frac{2}{7}$

2 $\frac{41}{4}$

3 (a) $\frac{7}{8}$ (b) already in lowest terms

4 (a) $\frac{9}{24}$ (b) $\frac{35}{65}$ (c) $\frac{55}{99}$ (d) $\frac{40}{50}$

5 (a) $-\frac{3}{4}$ (b) $\frac{4}{15}$ (c) $\frac{28}{3}$ or $9\frac{1}{3}$

6 (a) $\frac{3}{2}$ (b) $\frac{112}{15}$

7 (a) $\frac{7}{3}$ (b) $-\frac{3}{5}$

8 (a) $\frac{11}{24}$ (b) $\frac{2}{5}$ (c) $\frac{17}{12}$ or $1\frac{5}{12}$

9 (a) 0.375   (b) 0.35

10 (a) 0.416666 . . .   (b) 0.57575757 . . .

11 (a) $\frac{2}{5}$ (b) $\frac{12}{25}$ (c) $\frac{13}{40}$ (d) $\frac{128}{25}$

12 $\frac{4}{9}$

13 $\frac{56}{99}$

14 Yes, with 1 mile to spare

Exercise Set   5-3

Writing Exercises

1. Define *rational number* in your own words.
2. Explain why every integer is also a rational number, but the reverse is not true.
3. How can you tell if a fraction is in lowest terms?
4. Explain how to reduce a fraction to lowest terms.
5. Describe how to multiply two fractions.
6. Describe how to divide two fractions.
7. Describe how to add or subtract two fractions.

8. What is a repeating decimal? What about a terminating decimal?
9. Use a specific example to show that you can't add two fractions by just adding the numerators and denominators separately. Explain what you're doing.
10. Explain why it makes sense that if two fractions have the same denominator, you can add them by just adding numerators and keeping the common denominator.

Computational Exercises

For Exercises 11–26, reduce each fraction to lowest terms.

11. $\dfrac{7}{42}$

12. $\dfrac{8}{24}$

13. $\dfrac{42}{60}$

14. $\dfrac{16}{20}$

15. $\dfrac{10}{33}$

16. $\dfrac{12}{25}$

17. $\dfrac{30}{36}$

18. $\dfrac{25}{75}$

19. $\dfrac{91}{104}$

20. $\dfrac{68}{119}$

21. $\dfrac{27}{224}$

22. $\dfrac{105}{128}$

23. $\dfrac{420}{756}$

24. $\dfrac{950}{2,400}$

25. $\dfrac{427}{305}$

26. $\dfrac{261}{609}$

For Exercises 27–36, change each fraction to an equivalent fraction with the given denominator.

27. $\dfrac{5}{16} = \dfrac{?}{48}$

28. $\dfrac{15}{32} = \dfrac{?}{96}$

29. $\dfrac{19}{24} = \dfrac{?}{48}$

30. $\dfrac{5}{8} = \dfrac{?}{40}$

31. $\dfrac{21}{16} = \dfrac{?}{144}$

32. $\dfrac{30}{11} = \dfrac{?}{220}$

33. $\dfrac{9}{55} = \dfrac{?}{330}$

34. $\dfrac{48}{57} = \dfrac{?}{741}$

35. $\dfrac{71}{335} = \dfrac{?}{15,075}$

36. $\dfrac{111}{112} = \dfrac{?}{23,520}$

For Exercises 37–58, perform the indicated operations and reduce the answer to lowest terms.

37. $-\dfrac{5}{6} + \dfrac{2}{3}$

38. $\dfrac{3}{4} + \dfrac{7}{10}$

39. $-\dfrac{11}{12} - \dfrac{5}{8}$

40. $\dfrac{19}{24} - \dfrac{7}{18}$

41. $-\dfrac{5}{12} \times -\dfrac{7}{10}$

42. $\dfrac{5}{18} \times \dfrac{9}{25}$

43. $\dfrac{7}{9} \div \dfrac{2}{3}$

44. $-\dfrac{7}{24} \div \dfrac{23}{30}$

45. $\left(\dfrac{7}{16} \div \dfrac{3}{8}\right) \times \dfrac{3}{5}$

46. $-\dfrac{7}{8} \div \left(\dfrac{2}{3} \div \dfrac{15}{16}\right)$

47. $-\dfrac{11}{22} \times \left(\dfrac{1}{6} \times \dfrac{3}{4}\right)$

48. $\left(\dfrac{9}{10} - \dfrac{2}{3}\right) \times \dfrac{1}{2}$

49. $\left(\dfrac{5}{8} + \dfrac{3}{4}\right) \times \dfrac{2}{3}$

50. $\left(\dfrac{3}{7} + \dfrac{5}{9}\right) \times \dfrac{7}{12}$

51. $\left(\dfrac{9}{14} \div \dfrac{3}{7}\right) \times \dfrac{1}{2}$

52. $\left(\dfrac{4}{5} + \dfrac{7}{8}\right) \div \dfrac{1}{9}$

53. $\left(\dfrac{9}{10} - \dfrac{2}{3}\right) \times \dfrac{5}{6}$

54. $\dfrac{3}{4} \div \left(\dfrac{5}{8} + \dfrac{1}{2}\right)$

55. $\left(\dfrac{1}{8} \div \dfrac{1}{4}\right)^2 + \left(\dfrac{3}{4}\right)^3$

56. $\left(\dfrac{5}{12} - \dfrac{3}{8}\right)\left(\dfrac{4}{3} + \dfrac{5}{12}\right)$

57. $\dfrac{1}{3}\left(\dfrac{1}{6} - \dfrac{5}{8}\right)^2$

58. $-\dfrac{1}{2}\left(\dfrac{2}{3} + \dfrac{3}{4}\right)^3 + \dfrac{7}{8}\left(\dfrac{9}{16} - \dfrac{1}{8}\right)$

For Exercises 59–70, change each fraction to a decimal.

59. $\dfrac{1}{5}$

60. $\dfrac{3}{10}$

61. $\dfrac{2}{3}$

62. $\dfrac{7}{5}$

63. $\dfrac{9}{4}$

64. $\dfrac{11}{9}$

65. $\dfrac{11}{36}$

66. $\dfrac{12}{7}$

67. $\dfrac{3}{4}$

68. $\dfrac{15}{8}$

69. $\dfrac{48}{51}$

70. $\dfrac{17}{24}$

For Exercises 71–84, change each decimal to a reduced fraction.

71. 0.12

72. 0.36

73. 0.375

74. 0.925

75. 4.315

76. 9.002

77. $0.\overline{7}$

78. $0.\overline{2}$

79. $0.\overline{54}$

80. $0.\overline{62}$

81. $2.\overline{12}$

82. $8.\overline{15}$

83. $0.45\overline{3}$ (*Hint:* If $x = 0.45\overline{3}$, what are $1,000x$ and $100x$?)

84. $0.27\overline{4}$ (See hint for Exercise 83.)

In Exercises 85–88, perform the calculation. (Hint: Think about order of operations.)

85. $\dfrac{\dfrac{5}{3}+\dfrac{11}{6}}{\dfrac{3}{4}-\dfrac{21}{2}}$

86. $\dfrac{\dfrac{9}{4}-\dfrac{11}{3}}{\dfrac{2}{5}+\dfrac{1}{6}}$

87. $\dfrac{-\dfrac{2}{3}+1-\dfrac{1}{2}}{7-\dfrac{3}{8}+\dfrac{23}{6}}$

88. $\dfrac{\dfrac{2}{3}+\dfrac{3}{4}+\dfrac{4}{5}}{\dfrac{5}{6}+\dfrac{6}{7}+\dfrac{7}{8}}$

Applications in Our World

89. Lamont and Marge are driving to a business conference. They stop for gas, and Lamont estimates that they've driven $\frac{3}{5}$ of the distance. If the total trip is 285 miles, how many miles have they driven?

90. The Daytona International Speedway is $\frac{5}{2}$ miles long, and the Iowa Speedway is $\frac{7}{8}$ miles long. How much longer is a 250-lap race in Daytona?

91. A company uses $\frac{2}{7}$ of its budget for advertising. Of that, $\frac{1}{2}$ is spent on television advertisement. What part of its budget is spent on television advertisement?

92. According to the U.S. Census Bureau, $\frac{67}{300}$ of Americans without a high school diploma didn't have health insurance in 2019. In a group of 100 high school dropouts, how many would be likely to have health insurance?

93. According to the U.S. Census Bureau, $\frac{57}{500}$ of the population of the United States lived in poverty in 2020. Out of 200 randomly selected people, how many would be likely to live in poverty?

94. According to the U.S. Census Bureau, in 2020, about $\frac{3}{8}$ of the citizens of the United States lived in one of the five most populated states (California, Texas, New York, Florida, and Pennsylvania). If the total population was about 331 million in 2020, how many people lived in one of those five states?

95. According to the Bureau of Labor Statistics, about $\frac{13}{250}$ adults in the workforce in the United States were unemployed in August 2021. How many adults in a representative group of 5,000 were employed at that time?

96. On a map, 1 inch represents 80 miles. If two cities are $2\frac{3}{8}$ inches apart, how far apart in miles are they?

97. An architect's rendering of a house plan shows that $\frac{1}{4}$ inch represents 1 foot. If the family room plan is 3 inches long, how long will the actual family room be?

98. A ¾-mile stretch of beach just north of the Everglades is being developed into an exclusive oceanfront community. If there will be 12 lots of equal sizes, how much beach will each lot have?

99. An estate was divided among five people. The first person received $\frac{1}{8}$ of the estate. The next two people each received $\frac{1}{3}$ of the estate. The fourth person received $\frac{1}{10}$ of the estate. What fractional part of the estate did the last person receive?

100. For a certain municipality, $\frac{2}{3}$ of the waste generated consisted of paper products, $\frac{1}{10}$ consisted of glass products, and $\frac{1}{5}$ consisted of plastic products. Eight thousand tons of waste were hauled. How many tons of it consisted of paper, glass, and plastic products?

101. A recipe calls for $2\frac{1}{2}$ cups of flour and $\frac{2}{3}$ cup of sugar. If a person wanted to cut the recipe in half, how many cups of flour and sugar are needed?

102. The recipe for original Toll House chocolate chip cookies calls for, among other things, $2\frac{1}{4}$ cups of flour, $\frac{3}{4}$ cup of granulated sugar, and $\frac{3}{4}$ cup of brown sugar. It makes 60 cookies. A student group wants to bake 300 cookies for a fundraiser. How much flour, sugar, and brown sugar will they need?

Exercises 103–108 refer to the 2020 presidential election.

103. Of all votes cast, $\frac{64}{125}$ were cast for Joe Biden and $\frac{117}{250}$ were cast for Donald Trump. What fraction of ballots was cast for someone other than Trump or Biden?

104. Among voters, $\frac{18}{50}$ identified themselves as Republicans and $\frac{37}{100}$ as Democrats. What fraction identified themselves as neither?

105. Among voters in the election, $\frac{67}{100}$ listed white as their race and $\frac{13}{100}$ listed Black. What fraction listed some other race?

106. Of voters with a postgraduate degree, $\frac{31}{50}$ voted for Biden and $\frac{1}{100}$ voted for someone other than the two major-party candidates. What fraction voted for Trump?

107. The table below lists the fraction of the total electorate made up by different groups based on education level.

High school or less	$\frac{9}{50}$
Some college	$\frac{23}{100}$
Associate degree	$\frac{4}{25}$

What fraction of voters had at least an undergraduate degree?

108. Refer to the table in Exercise 107. In addition, $\frac{27}{100}$ of the voters finished their schooling with an undergraduate degree. What fraction went on to graduate study?

Critical Thinking

109. When adding fractions, we usually find the least common denominator and rewrite both fractions with that denominator. But do you really need the *least* common denominator? In the sum $\frac{3}{8}+\frac{5}{12}$, first add by using the least common denominator. Then add by using a common denominator that is the product of the two original

denominators. Do you get the same answer? Try again for the sum $\frac{5}{6} + \frac{5}{9}$. What can you conclude? What is the advantage of finding the least common denominator?

110. We added mixed numbers by first rewriting them as improper fractions, then adding as usual. For the sum in Example 8, part *c*, instead of rewriting as improper fractions, add the whole number parts, then the fractional parts. What would you do next to find the overall answer? Did you get the same answer as in the solution to the example problem? Which method do you prefer?

111. We have seen how to find the fractional equivalent for repeating decimals where one, two, or three digits repeat. It seems reasonable, then, to conclude that we can find a fractional equivalent for any repeating decimal. Is there a decimal that has the largest possible number of repeating digits? Why or why not?

112. We know that we can find a fractional equivalent for any decimal that is repeating or terminating. Does that mean that all decimals are rational numbers? Why or why not?

113. One property of the rational numbers is that they are **dense**. This means that between any two rational numbers, you can always find another rational number. Given two arbitrary rational numbers, how would you find another rational number in between them?

114. Review the procedure for converting a mixed number to an improper fraction illustrated in Example 2. Explain why that procedure works using addition of fractions.

115. (a) Fill in each circle with greater than, less than, or equal to:

0.99999 . . . ◯ 1   0.4999... ◯ $\frac{1}{2}$

0.24999... ◯ $\frac{1}{4}$

(b) Use the technique presented in this section to find the fractional form for each repeating decimal. Does this change your answers to part (a)?

116. Use the results of Exercise 115 to write the rational equivalent for each repeating decimal without doing any calculations:
   (a) 3.9999 . . .   (b) 0.749999 . . .   (c) 0.39999 . . .

117. From Section 5-2, we know that the product of two integers with opposite signs is negative. Using the rule for multiplying fractions in the colored box on page 242, prove that the same is true for rational numbers. (*Hint:* It's helpful to examine four different cases; one case is the first numerator is negative, while all remaining parts are positive.)

118. Mimic the result of Problem 117 to prove that the product of two rational numbers with the same sign is positive.

119. Joe Einstein thinks he's found a genius new shortcut way to reduce fractions, demonstrated below. Explain why Joe isn't as clever as he thinks, even though he ended up with four correct answers.

$$\frac{49}{98} = \frac{4\cancel{9}}{\cancel{9}8} = \frac{4}{8} = \frac{1}{2}$$

$$\frac{26}{65} = \frac{2\cancel{6}}{\cancel{6}5} = \frac{2}{5}$$

$$\frac{16}{64} = \frac{1\cancel{6}}{\cancel{6}4} = \frac{1}{4}$$

$$\frac{19}{95} = \frac{1\cancel{9}}{\cancel{9}5} = \frac{1}{5}$$

120. Find the decimal equivalents for $\frac{1}{7}, \frac{2}{7}, \frac{3}{7}, \dots, \frac{6}{7}$. Do you see any pattern? Explain.

Section 5-4   The Irrational Numbers

LEARNING OBJECTIVES

1. Define irrational numbers.
2. Simplify radicals.
3. Multiply and divide square roots.
4. Add and subtract square roots.
5. Rationalize denominators.
6. Approximate irrational numbers with a calculator.

If you're paying close attention, you may have noticed that in each of the last two sections, we've built upon the set of numbers defined in the previous section. We started with natural numbers, then included zero and the negatives of whole numbers to get the integers. Then we expanded that further, defining the rational numbers, a set that contains all of the natural numbers, whole numbers, and integers. In this section, we break that pattern, defining a new set of numbers distinct from all of those we've studied so far.

To accomplish that, we will think about numbers that are not rational—that is, cannot be written as fractions with integers in the numerator and denominator. The problem is that it's not so easy to find such numbers when using that approach. Instead, we'll focus on a key fact about rational numbers: any decimal that is either terminating or repeating can be written as a fraction. But there are decimals that neither terminate nor repeat. Here's an example: 0.0103050709011013. . . . Can you see the pattern? A zero is followed by an odd number, with that odd number increasing regularly. This pattern continues forever, so the decimal is not terminating. But there's no definite string that repeats, so it isn't repeating either. Numbers of this form are called *irrational numbers* because it turns out that they can't be written as the ratio of integers. In other words, they're not rational.

A number is **irrational** if it can be written as a decimal that neither terminates nor repeats.

Even though this definition may be new to you, there's one big class of irrational numbers that you probably have some experience working with: certain square roots. For example, the number $\sqrt{2} \approx 1.41421356237\ldots$ is irrational.

The symbol $\sqrt{2}$ is read as "the square root of two," and the symbol around 2 is called a *radical* sign.

An expression with a radical sign is called a **radical**, and the number under the radical sign is called the **radicand**.

Here's a definition of square roots:

The **square root** of a number a, symbolized $\sqrt{a}$, is the nonnegative number you have to multiply by itself (or square) to get a.

So $\sqrt{2}$ is the number you have to square to get 2.

It's not particularly easy to prove conclusively that $\sqrt{2}$ is an irrational number, but here's a way of understanding why that's likely to be the case.

You can approximate $\sqrt{2}$ by finding a number that when squared gives an answer close to 2. For example, $(1.4)^2 = 1.96$ and $(1.5)^2 = 2.25$. So the square root of 2 is a number between 1.4 and 1.5. You could guess 1.41, but $(1.41)^2 = 1.9881$. Although $(1.41)^2$ is closer to 2 than $(1.4)^2$, it's still too small. Now try 1.42; $(1.42)^2 = 2.0164$. This number is too large. So $\sqrt{2}$ is between 1.41 and 1.42.

You can continue the process as shown.

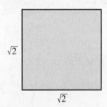

A square with an area of 2 square units would have sides $\sqrt{2}$ units long.

$$1 < \sqrt{2} < 2$$
$$1.4 < \sqrt{2} < 1.5$$
$$1.41 < \sqrt{2} < 1.42$$
$$1.414 < \sqrt{2} < 1.415$$
$$1.4142 < \sqrt{2} < 1.4143$$
etc.

Notice that squaring the number on the left side of each inequality gives a number slightly smaller than 2 and squaring the number on the right side of each inequality gives a number slightly larger than 2. You could continue this process forever, but you would never get a number that when squared would give an answer that is exactly equal to 2. (This fact was proved by Euclid over 3,000 years ago.)

It turns out that $\sqrt{2}$ is not alone when it comes to square roots that are irrational numbers. One thing that is easy to determine is that not *every* square root is irrational. Since $2^2 = 4$, $\sqrt{4} = 2$, so $\sqrt{4}$ is an integer, and consequently a rational number. The same is true for $\sqrt{9}$, $\sqrt{16}$, $\sqrt{25}$, $\sqrt{36}$, and so on. In each case, the number under the radical is the square of an integer. We'll call these numbers **perfect squares**, and observe that the square root of a perfect square is always an integer. But for any other natural number, the square root happens to be irrational.

In case you're wondering, there are plenty of irrational numbers that are not square roots. The most famous of them is probably π (pi), which is defined to be the distance around a circle (the circumference) divided by the diameter.

Simplifying Radicals

There is a simple property of square roots, commonly known as the product rule, that will help us in working with irrational numbers that are defined by square roots.

Math Note

Notice that we used the "approximately equal to" sign ($\approx$) rather than the "equal to" sign ($=$). This is because we can't write the exact value of an irrational number in decimal form, because there are infinitely many digits.

Calculator Guide

You can find decimal approximations of irrational numbers using a calculator. To approximate $\sqrt{2}$:

Typical Scientific or Online Calculator

Typical Graphing Calculator

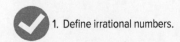
1. Define irrational numbers.

Sidelight   A Slice of pi

You would more than likely be amazed, if not appalled, at the tremendous amount of effort that has gone into the study of the number pi. The earliest mentions of pi come from about 1900 BCE, when it was approximated as 25/8 in Babylon and 256/81 in Egypt. As hard as it may be to believe, almost 4,000 years later mathematicians are still working on better and better approximations! The value of pi is about 3.14, but that's a tremendously crude approximation by modern standards. In fact, the number of digits computed as of the end of 2016 was over 22 trillion. It's almost hard to imagine how many digits that is. If you recited two digits each and every second of every day, it would take over 300 *years* to read that many digits.

One of the more interesting aspects of pi is that the digits appear to be totally random. Remember, it's been calculated out to 22 trillion digits, and nobody has found any pattern whatsoever. In fact, it's been speculated that the digits are completely randomly distributed, which would mean that *every possible string of numbers* can be found somewhere in the decimal expansion of pi! Your birthdate? It's in there. Your cell phone number? It's in there. Social security number? In there. All assuming, that is, that the digits really are randomly distributed. There is a Web page at http://www.angio.net/pi/piquery.html where you can search for any string of numbers in the first 200 million digits of pi.

But if you don't find the string you're looking for, remember: you're looking in a very, very small portion of the actual number—a measly 200 million out of *infinitely many* digits!

The Product Rule For Square Roots

For any two positive numbers a and b, $\sqrt{ab} = \sqrt{a} \cdot \sqrt{b}$.

One way to use the product rule is to simplify square roots. We will call a square root *simplified* if the number inside the radical has no factors that are perfect squares. The procedure for simplifying square roots is illustrated in Example 1.

Example 1   **Simplifying Radicals**

Math Note

The product rule $(\sqrt{ab} = \sqrt{a} \cdot \sqrt{b})$ is usually stated for a and b positive. It's also true if a and/or b are equal to zero, but it's not particularly useful in that case. No negatives though! Because the square of any number is at least zero, square roots of negative numbers are undefined.

Simplify each radical.

(a) $\sqrt{40}$ (b) $\sqrt{200}$ (c) $\sqrt{26}$

SOLUTION

(a) Notice that 40 can be written as $4 \cdot 10$, and that 4 is a perfect square. Also, 4 is the largest factor of 40 that is a perfect square. Using the product rule, we can write

$$\sqrt{40} = \sqrt{4 \cdot 10} \qquad \textit{Use the product rule}$$
$$= \sqrt{4} \cdot \sqrt{10} \qquad \sqrt{4} = 2$$
$$= 2\sqrt{10}$$

(b) The largest perfect square factor of 200 is 100.

$$\sqrt{200} = \sqrt{100 \cdot 2} \qquad \textit{Use the product rule}$$
$$= \sqrt{100} \cdot \sqrt{2} \qquad \sqrt{100} = 10$$
$$= 10\sqrt{2}$$

(c) The prime factorization of 26 is $2 \cdot 13$. It has no perfect square factors, so $\sqrt{26}$ is already simplified.

Try This One   **1**

Simplify each radical.

(a) $\sqrt{75}$ (b) $\sqrt{56}$ (c) $\sqrt{74}$

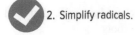
2. Simplify radicals.

| Example 2 | Is There a Sum Rule for Square Roots? |

We know that $\sqrt{ab} = \sqrt{a} \cdot \sqrt{b}$. Is it also true that $\sqrt{a+b} = \sqrt{a} + \sqrt{b}$? Try some sample numbers and see what you can conclude.

SOLUTION

I'm going to be clever and choose a and b so that all of the numbers I need to compute the square root of are perfect squares. Then I won't need a calculator. Let's go with $a = 9$ and $b = 16$:

$$\sqrt{a+b} = \sqrt{9+16} = \sqrt{25} = 5$$
$$\sqrt{a} + \sqrt{b} = \sqrt{9} + \sqrt{16} = 3 + 4 = 7$$

With just one choice, we were able to show that the product rule does NOT carry over to sums: for this choice of a and b, $\sqrt{a+b}$ and $\sqrt{a} + \sqrt{b}$ are NOT equal.

Try This One   2

Try to find numbers for which $\sqrt{a+b}$ and $\sqrt{a} + \sqrt{b}$ ARE equal. What has to happen?

| CAUTION | Splitting up the sum of a root into two separate roots is almost certainly the most common mistake students make when working with roots. Keep Example 2 in mind! There is no "sum rule" for roots. |

The product rule can also be used to multiply two square roots.

| Example 3 | Multiplying Square Roots |

Find each product.

(a) $\sqrt{6} \cdot \sqrt{2}$ (b) $\sqrt{5} \cdot \sqrt{20}$ (c) $\sqrt{2} \cdot \sqrt{3} \cdot \sqrt{3}$

SOLUTION

(a) Using the product rule (reading it from right to left), we can write the product as a single square root by multiplying the numbers under the radical.

$$\sqrt{6} \cdot \sqrt{2} = \sqrt{6 \cdot 2} = \sqrt{12}$$

Now we can simplify like we did in Example 1.

$$\sqrt{12} = \sqrt{4 \cdot 3} = \sqrt{4} \cdot \sqrt{3} = 2\sqrt{3}$$

(b) Again, we can multiply the two numbers under the radical.

$$\sqrt{5} \cdot \sqrt{20} = \sqrt{5 \cdot 20} = \sqrt{100}$$

Math Note

In the same way that we assume any fractional answer should be reduced, from now on we'll assume that any square root answer should be simplified, like we did in Example 1.

This time, it's easier to simplify: $\sqrt{100}$ is 10 because $10^2 = 100$. This shows that the product of two irrational numbers is not always irrational.

(c) The product rule doesn't specifically tell us how to multiply three radicals, but we can use it in stages (multiply the first two, then multiply the result by the third) to show that we can multiply the three numbers inside the radical.

$$\sqrt{2} \cdot \sqrt{3} \cdot \sqrt{3} = \sqrt{2 \cdot 3 \cdot 3} = \sqrt{18}$$

Now we can simplify.

$$\sqrt{18} = \sqrt{9 \cdot 2} = \sqrt{9} \cdot \sqrt{2} = 3\sqrt{2}$$

In this case, it would have been simpler to notice that the product of the second and third factors, $\sqrt{3} \cdot \sqrt{3}$, is 3, in which case we would have obtained the same answer, but much more quickly.

Try This One   3

Find each product.

(a) $\sqrt{35} \cdot \sqrt{5}$ (b) $\sqrt{18} \cdot \sqrt{6}$ (c) $\sqrt{42} \cdot \sqrt{15} \cdot \sqrt{15}$

Since multiplication and division are strongly related, it's not a big surprise that there is a quotient rule for square roots, similar to the product rule. We use it to divide two square roots, and to simplify square roots of fractions.

The Quotient Rule for Square Roots

For any two positive numbers a and b, $\sqrt{\dfrac{a}{b}} = \dfrac{\sqrt{a}}{\sqrt{b}}$.

Note: The quotient rule is also true if $a = 0$, but not if $b = 0$, as this would make the denominator zero.

Sidelight   Wow, That is One Hot . . . Rectangle?

Can a rectangle be attractive? Maybe "aesthetically pleasing" would be a better description—and many folks that study architecture, art, and even human beauty will answer with an emphatic "yes." And the rectangle in question is not only based on math, it's based on an irrational number.

If you divide a line into two parts so that the ratio of the longer part to the smaller is the same as the ratio of the total length to the longer part (see Figure 5-6), that ratio is called the **golden ratio**. The golden ratio has been popping up in art and architecture, often unintentionally, for literally thousands of years. For whatever reason, when a rectangle is drawn so that its sides match the golden ratio, it just looks "right" to most people, like the shape is just exactly what a rectangle should look like. In fact, there's a pretty good chance that the room you're sitting in right now has a golden rectangle in it someplace: windows, furniture, and picture frames feature it regularly.

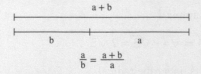

$$\frac{a}{b} = \frac{a+b}{a}$$

Figure 5-6

One of the most often-cited examples of the golden ratio in architecture is the Parthenon, probably the most famous and important surviving structure from the ancient Greek empire. There's no evidence that the Parthenon was designed with the golden ratio consciously in mind, but to me that makes it even cooler—there are a variety of examples of golden rectangles throughout the structure, almost certainly just because they looked right.

DAJ/Getty Images

In Section 6-9, we'll see how to use the equation that describes the golden ratio to calculate its value: it works out to be the irrational number $\frac{1+\sqrt{5}}{2}$. In much the same way that the Greek letter π (pi) is used to represent an irrational number that's about 3.14, the Greek letter φ (phi) is used to represent the golden ratio, which is about 1.618.

Although there's not much hard science behind it, there is some evidence that faces that most people perceive to be beautiful have the golden ratio in them—both the ratio of face length to width, and the ratio of the length of the mouth to the width of the nose are two examples.

So is all of this perception of beauty math, science, or mysticism? We may never know for sure, but it's kind of fun to debate.

| Example 4 | Using the Quotient Rule |

Find each quotient.

(a) $\dfrac{\sqrt{27}}{\sqrt{3}}$ (b) $\dfrac{\sqrt{60}}{\sqrt{5}}$

SOLUTION

In each case, we'll use the quotient rule to write the quotient as a single square root, dividing the numbers underneath the radical, then simplify.

(a) $\dfrac{\sqrt{27}}{\sqrt{3}} = \sqrt{\dfrac{27}{3}} = \sqrt{9} = 3$ (b) $\dfrac{\sqrt{60}}{\sqrt{5}} = \sqrt{\dfrac{60}{5}} = \sqrt{12} = \sqrt{4 \cdot 3} = \sqrt{4} \cdot \sqrt{3} = 2\sqrt{3}$

Try This One   4

Find each quotient.

(a) $\dfrac{\sqrt{80}}{\sqrt{10}}$ (b) $\dfrac{\sqrt{48}}{\sqrt{3}}$

3. Multiply and divide square roots.

As we saw in Example 2, there is no "sum" or "difference" rule for radicals: $\sqrt{a+b}$ isn't equal to $\sqrt{a} + \sqrt{b}$. So adding and subtracting square roots is very different from multiplying and dividing. In an expression like $10\sqrt{5}$, the 10 is called a **coefficient**. If two radicals are identical other than the coefficient, we call them **like radicals**. (You might prefer to think of like radicals as radicals that have the exact same radicand.) For example, $10\sqrt{5}, 7\sqrt{5}$, and $-\sqrt{5}$ are like radicals, but $10\sqrt{5}$ and $10\sqrt{3}$ are not because they have different radicands. Only like radicals can be added or subtracted.

Addition and Subtraction of Like Radicals

To add or subtract like radicals, add or subtract their coefficients and keep the radical the same. In symbols,

$$a\sqrt{c} + b\sqrt{c} = (a+b)\sqrt{c} \quad \text{and} \quad a\sqrt{c} - b\sqrt{c} = (a-b)\sqrt{c}.$$

| Example 5 | Adding Square Roots |

Find the sum: $6\sqrt{3} + 8\sqrt{3} + 5\sqrt{3}$

SOLUTION

Since all the radicals are like radicals, the sum can be found by adding the coefficients of the radicals.

$$6\sqrt{3} + 8\sqrt{3} + 5\sqrt{3} = (6+8+5)\sqrt{3} = 19\sqrt{3}$$

Try This One   5

Find the sum: $7\sqrt{5} + 8\sqrt{5} + 9\sqrt{5}$

A familiar use of radicals is finding sides of right triangles with the Pythagorean theorem, $a^2 + b^2 = c^2$. If the sides of the garden a and b are 10 ft and 24 ft, respectively, then $10^2 + 24^2 = c^2$, and $c = \sqrt{10^2 + 24^2}$, or 26 ft.

If you have a hard time visualizing the addition of coefficients in Example 4, it might help to mentally replace the number $\sqrt{3}$ with some objects, like cats. In that case, we'd have 6 cats + 8 cats + 5 cats, which would be 19 cats. (We'd also need to have our heads examined.) This will work for subtraction as well.

| **Example 6** | Subtracting Square Roots |

Find the difference: $3\sqrt{10} - 7\sqrt{10}$

SOLUTION

$$3\sqrt{10} - 7\sqrt{10} = (3 - 7)\sqrt{10} = -4\sqrt{10}$$

Try This One   6

Find the difference: $12\sqrt{11} - 20\sqrt{11}$

In life, things are not always as they appear. Based on our definition of like radicals, it seems reasonable to conclude that $\sqrt{2}$ and $\sqrt{8}$ are not like, since their radicands are different. But as we'll see in Example 7, simplifying can expose like radicals in disguise.

| **Example 7** | Adding and Subtracting Square Roots |

Perform the indicated operations.

$$4\sqrt{12} - 3\sqrt{8} + 5\sqrt{32}$$

SOLUTION

Simplify each radical first (if possible):

$$4\sqrt{12} = 4\sqrt{4}\sqrt{3} = 4 \cdot 2\sqrt{3} = 8\sqrt{3}$$
$$3\sqrt{8} = 3\sqrt{4}\sqrt{2} = 3 \cdot 2\sqrt{2} = 6\sqrt{2}$$
$$5\sqrt{32} = 5\sqrt{16}\sqrt{2} = 5 \cdot 4\sqrt{2} = 20\sqrt{2}$$

Two of the radicals are like, but the third is not.

$$8\sqrt{3} - 6\sqrt{2} + 20\sqrt{2} = 8\sqrt{3} + 14\sqrt{2}$$

Since $8\sqrt{3}$ and $14\sqrt{2}$ aren't like radicals, we can't simplify any further.

Try This One   7

4. Add and subtract square roots.

Perform the indicated operations.

(a) $2\sqrt{8} + 5\sqrt{50}$

(b) $6\sqrt{28} + 4\sqrt{112} - 2\sqrt{12}$

Another method used to simplify radical expressions is called **rationalizing the denominator**. When a radical expression contains a square root sign in the denominator of a fraction, it can be simplified by multiplying the numerator and denominator by a radical expression that will make the radicand in the denominator a perfect square. This is called rationalizing the denominator. Examples 8 and 9 illustrate the process.

| **Example 8** | Rationalizing Denominators |

Simplify each radical expression.

(a) $\dfrac{18}{\sqrt{3}}$ (b) $\dfrac{6}{\sqrt{18}}$

Math Note

When rationalizing the denominator, make sure you multiply BOTH the numerator and denominator by the same number. (In that case, you're not changing the actual number because you're just multiplying by 1.)

SOLUTION

(a) If we multiply the numerator and denominator by $\sqrt{3}$, the denominator will become $\sqrt{9}$, which is, of course, 3.

$$\frac{18}{\sqrt{3}} = \frac{18}{\sqrt{3}} \cdot \frac{\sqrt{3}}{\sqrt{3}} = \frac{18\sqrt{3}}{\sqrt{9}} = \frac{18\sqrt{3}}{3} = 6\sqrt{3}$$

(b) We could mimic what happened in part (a) and multiply the numerator and denominator by $\sqrt{18}$, but it will be easier to simplify if we multiply instead by $\sqrt{2}$. This will make the denominator $\sqrt{36}$, which is 6.

$$\frac{6}{\sqrt{18}} = \frac{6}{\sqrt{18}} \cdot \frac{\sqrt{2}}{\sqrt{2}} = \frac{6\sqrt{2}}{\sqrt{36}} = \frac{6\sqrt{2}}{6} = \sqrt{2}$$

Try This One   8

Simplify each radical expression.

(a) $\dfrac{20}{\sqrt{5}}$ (b) $\dfrac{3}{\sqrt{12}}$

When the radicand of a square root is a fraction, we can use the quotient rule to apply the root to the numerator and denominator separately, then simplify by rationalizing the denominator.

Example 9 — Simplifying the Square Root of a Fraction

Math Note

Radical expressions such as $\frac{\sqrt{30}}{6}$ can also be written as $\left(\frac{1}{6}\right)\sqrt{30}$. A radical expression like $\frac{2\sqrt{3}}{5}$ can also be written as $\frac{2}{5}\sqrt{3}$.

Simplify $\sqrt{\dfrac{5}{6}}$.

SOLUTION

Apply the quotient rule to split into two separate roots:

$$\frac{\sqrt{5}}{\sqrt{6}}$$

Now multiply the numerator and denominator by $\sqrt{6}$ to rationalize the denominator:

$$\frac{\sqrt{5}}{\sqrt{6}} \cdot \frac{\sqrt{6}}{\sqrt{6}} = \frac{\sqrt{30}}{\sqrt{36}} = \frac{\sqrt{30}}{6}$$

Try This One   9

Simplify each radical expression.

(a) $\sqrt{\dfrac{5}{18}}$ (b) $\sqrt{\dfrac{75}{8}}$

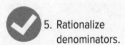
5. Rationalize denominators.

In many cases, we prefer to write square roots in radical form because that represents the exact value of that particular number. But sometimes a decimal approximation is useful for getting some perspective on the exact size of a number. In this case, a calculator comes in handy.

Example 10 | Approximating Square Roots with a Calculator

Approximate each irrational number to three decimal places.

(a) $\sqrt{311}$ (b) $\sqrt{1,416}$

SOLUTION

(a) Scientific or online calculator: 311 then ⬚, SQRT, or ⬚ ⬚
     Graphing calculator: ⬚ ⬚ 311 ⬚ ENTER
     In either case, the result is 17.63519209. . . . To three decimal places, this rounds to 17.635.

(b) $\sqrt{1,416} \approx 37.630$

Try This One   10

Approximate each irrational number to four decimal places.

(a) $\sqrt{23}$ (b) $\sqrt{2,915}$

6. Approximate irrational numbers with a calculator.

In this section, we focused on operations on certain types of irrational numbers—square roots. But don't forget that there are many other irrational numbers, among them the multiples of π. When working with irrational numbers, one of the key features you should keep in mind is that decimal forms are always approximations because by definition, the irrational numbers have decimal forms with infinitely many digits.

Answers to Try This One

1 (a) $5\sqrt{3}$ (b) $2\sqrt{14}$ (c) $\sqrt{74}$

2 If one of the numbers is zero, it works.

3 (a) $5\sqrt{7}$ (b) $6\sqrt{3}$ (c) $15\sqrt{42}$

4 (a) $2\sqrt{2}$ (b) 4

5 $24\sqrt{5}$

6 $-8\sqrt{11}$

7 (a) $29\sqrt{2}$ (b) $28\sqrt{7} - 4\sqrt{3}$

8 (a) $4\sqrt{5}$ (b) $\dfrac{\sqrt{3}}{2}$

9 (a) $\dfrac{\sqrt{10}}{6}$ (b) $\dfrac{5\sqrt{6}}{4}$

10 (a) 4.7958   (b) 53.9907

Exercise Set   5-4

Writing Exercises

1. How can you tell the difference between an irrational number and a rational number?
2. Is every square root an irrational number? Explain.
3. Are there irrational numbers other than square roots? Explain.
4. Describe two ways that we use the product rule for square roots.
5. How can you tell if a square root is simplified or not?
6. Explain how to rationalize the denominator of a square root.
7. What is the golden ratio? What is a golden rectangle?
8. Describe how the golden ratio plays a part in architecture.

Computational Exercises

For Exercises 9–14, state whether each number is rational or irrational.

9. $\sqrt{49}$

10. $\sqrt{37}$

11. $0.232332333\ldots$

12. $\dfrac{5}{6}$

13. π

14. 0

For Exercises 15–20, without using a calculator, name two integers that the given square root is between on a number line.

15. $\sqrt{11}$

16. $\sqrt{28}$

17. $\sqrt{100}$

18. $\sqrt{75}$

19. $\sqrt{200}$

20. $\sqrt{160}$

In Exercises 21–26, use a calculator to approximate each irrational number to three decimal places.

21. $\sqrt{30}$

22. $\sqrt{72}$

23. $\sqrt{1,431}$

24. $\sqrt{993}$

25. $-\sqrt{\dfrac{126}{3}}$

26. $-\sqrt{\dfrac{57}{2}}$

For Exercises 27–40, simplify the radical.

27. $\sqrt{24}$

28. $\sqrt{27}$

29. $\sqrt{80}$

30. $\sqrt{175}$

31. $\sqrt{30}$

32. $\sqrt{42}$

33. $10\sqrt{20}$

34. $4\sqrt{8}$

35. $3\sqrt{700}$

36. $2\sqrt{162}$

37. $\dfrac{3}{4}\sqrt{800}$

38. $\dfrac{8}{9}\sqrt{405}$

39. $-\dfrac{5}{6}\sqrt{1,008}$

40. $-\dfrac{3}{22}\sqrt{1,210}$

For Exercises 41–72, perform the operations and simplify the answer.

41. $\sqrt{2}\cdot\sqrt{10}$

42. $\sqrt{15}\cdot\sqrt{6}$

43. $\sqrt{18}\cdot\sqrt{15}$

44. $\sqrt{5}\cdot\sqrt{25}$

45. $2\sqrt{6}\cdot3\sqrt{8}$

46. $6\sqrt{15}\cdot2\sqrt{5}$

47. $\dfrac{\sqrt{60}}{\sqrt{2}}$

48. $\dfrac{\sqrt{42}}{\sqrt{6}}$

49. $\dfrac{\sqrt{64}}{\sqrt{8}}$

50. $\dfrac{\sqrt{15}}{\sqrt{3}}$

51. $2\sqrt{7}+10\sqrt{7}$

52. $50\sqrt{11}+11\sqrt{11}$

53. $8\sqrt{3}-15\sqrt{3}$

54. $5\sqrt{7}-11\sqrt{7}$

55. $2\sqrt{3}+5\sqrt{3}-9\sqrt{3}$

56. $8\sqrt{5}-6\sqrt{5}-7\sqrt{5}$

57. $\sqrt{320}-\sqrt{80}$

58. $\sqrt{125}+\sqrt{20}$

59. $6\sqrt{5}-3\sqrt{80}$

60. $13\sqrt{90}+5\sqrt{40}$

61. $6\sqrt{72}-9\sqrt{8}$

62. $5\sqrt{10}+2\sqrt{40}$

63. $3\sqrt{2}-\sqrt{8}+4\sqrt{12}$

64. $10\sqrt{20}-20\sqrt{10}+6\sqrt{5}$

65. $5\sqrt{40}+4\sqrt{50}-6\sqrt{32}$

66. $8\sqrt{12}-9\sqrt{20}+\sqrt{75}$

67. $\sqrt{5}(\sqrt{75}+\sqrt{12})$

68. $\sqrt{3}(\sqrt{48}-\sqrt{27})$

69. $-\sqrt{6}(\sqrt{108}-4\sqrt{27})^2$

70. $\sqrt{3}(\sqrt{28}-3\sqrt{252})^2$

71. $\sqrt{2}(\sqrt{192}+\sqrt{108})^2$

72. $\sqrt{11}(-3\sqrt{44}+2\sqrt{99})^2$

For Exercises 73–82, rationalize the denominator and simplify.

73. $\dfrac{1}{\sqrt{5}}$

74. $\dfrac{3}{\sqrt{8}}$

75. $\dfrac{3}{\sqrt{6}}$

76. $\dfrac{10}{\sqrt{20}}$

77. $\sqrt{\dfrac{3}{28}}$

78. $\sqrt{\dfrac{7}{8}}$

79. $\dfrac{42}{\sqrt{118}}$

80. $\dfrac{7}{\sqrt{168}}$

81. $\sqrt{\dfrac{18}{243}}$

82. $\sqrt{\dfrac{27}{343}}$

Applications in Our World

If an object is dropped from a height of h feet, the time in seconds it takes to reach the ground is given by the formula

$$t=\sqrt{\dfrac{2h}{32}}$$

(This formula ignores air resistance.) Use this formula for Exercises 83–88.

83. How long will it take an object to reach the ground when dropped from a 144-foot building?

84. How long will it take an object to reach the ground when dropped from a 256-foot bridge?

85. The tallest observation deck in the United States is the Willis Tower in Chicago, at a height of 1,353 feet. Use a calculator to find how long it will take an object to reach the ground when dropped from that height. Round to the nearest tenth of a second.

86. The tallest observation deck in the world is the Shanghai Tower in China, at a height of 1,844 feet. Use a calculator to find how long it will take an object to reach the ground

when dropped from that height. Round to the nearest tenth of a second.

87. How much longer will it take an object dropped from 200 feet to reach the ground than one dropped from 100 feet? Is the answer surprising? Why?

88. How much longer will it take an object dropped from 400 feet to reach the ground than one dropped from 100 feet? Is the answer surprising? Why?

Use this information for Exercises 89–92: the voltage of an electric circuit can be found by the formula

$$V = \sqrt{P \cdot r}$$

where V = volts, P = power in watts, and r = resistance in ohms.

89. Find the voltage when $P = 80$ watts and $r = 5$ ohms.

90. Find the voltage when $P = 360$ watts and $r = 10$ ohms.

91. Use a calculator to approximate to one decimal place the voltage in a circuit with 1,000 watts of power and 20 ohms of resistance.

92. Use a calculator to approximate to one decimal place the voltage in a circuit with 1,200 watts of power and 25 ohms of resistance.

From the time of their invention in 1656 until the 1930s, pendulum clocks were the most accurate in the world, and they are still in use today. The time it takes for a pendulum to complete one swing and return to its starting point is called its **period**. *The period of a pendulum is important in using one as a timekeeping device, and it can be calculated using the formula*

$$t = 2\pi\sqrt{\frac{l}{32}}$$

where t is the period in seconds and l is the length in feet of the pendulum arm.

93. Find the period of a pendulum whose arm length is 128 feet.

94. Find the period of a pendulum whose arm length is 64 feet.

95. Using a calculator and trial-and-error, find the approximate arm length of a pendulum with a period of 1 second.

96. Using a calculator and trial-and-error, find the approximate arm length of a pendulum with a period of 5 seconds.

Critical Thinking

97. When two integers are multiplied, the result is always another integer. We say that the integers are **closed under multiplication**. Are the irrational numbers closed under multiplication? What about the rational numbers? Explain.

98. Use several examples to decide whether $\sqrt{a - b} = \sqrt{a} - \sqrt{b}$ for positive numbers a and b.

99. Based on the way we defined square roots in this section, why can't you compute the square root of a negative number?

100. In the same way that we define square roots in terms of squares, we can define **cube roots** in terms of cubes. So the cube root of a number a, denoted $\sqrt[3]{a}$, is the number whose cube, or third power, is a. Do you think that cube roots of integers are irrational numbers? Always? Sometimes? Never? Discuss your answer.

As mentioned in Exercise 100, we can define roots other than square roots in terms of powers other than 2. The cube root of a number a $(\sqrt[3]{a})$ is the number you have to raise to the third power to obtain a. For example, $\sqrt[3]{8} = 2$ because $2^3 = 8$. Similarly, the fourth root of a $(\sqrt[4]{a})$ is the number you have to raise to the fourth power to obtain a, and so on.

101. Find each requested root without using a calculator:
    (a) $\sqrt[3]{27}$ (b) $\sqrt[3]{216}$
    (c) $\sqrt[4]{16}$ (d) $\sqrt[4]{625}$

102. Find each requested root without using a calculator, if possible:
    (a) $\sqrt[3]{-8}$ (b) $\sqrt[3]{-64}$
    (c) $\sqrt[4]{-16}$ (d) $\sqrt[5]{-32}$

103. Without using a calculator, name two integers that the given root is between on a number line:
    (a) $\sqrt[3]{20}$ (b) $\sqrt[4]{100}$
    (c) $\sqrt[3]{-100}$ (d) $\sqrt[4]{300}$

104. Explain why you can find the cube root of a negative number, but not the square root of a negative number.

105. For an expression of the form $a + b$, the **conjugate** is the expression of the form $a - b$. Multiply each irrational number by its conjugate, then use inductive reasoning to make a conjecture based on the results.
    (a) $3 + \sqrt{5}$ (b) $1 + \sqrt{12}$
    (c) $-2 - \sqrt{20}$

Exercises 106 and 107 are based on the golden ratio and golden rectangle, discussed in the Sidelight on page 256.

106. The dimensions of some common rectangles are provided: find the ratio of longer side to shorter side, and determine which are closest to a golden rectangle. Then evaluate whether or not you think aesthetics are involved in the design of each.
    (a) 32″ HDTV: 27.5″ by 15.5″
    (b) Legal size paper: 8.5″ by 14″
    (c) Traditional photo print size: 3″ by 5″
    (d) 27″ iMac screen: $23\frac{3}{8}$″ by $13\frac{1}{8}$″
    (e) Paperback book: 4″ × 7″

107. Find pictures of three celebrities you think are very good-looking in a magazine or on a website. Then see how the ratios of measurements on their face listed at the end of the Sidelight compare to the golden ratio. What can you conclude?

Section 5-5

The Real Numbers

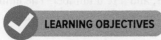

LEARNING OBJECTIVES

1. Define the real numbers.
2. Identify properties of the real numbers.

The Russian art of nesting dolls, or *matryoshka*, dates back to the 1890s. A series of hollow carved dolls with similar proportions fits one inside the other, with all of them fitting inside the largest. This is a pretty good (but not perfect) model for the sets of numbers that we've built while working through this chapter. With the exception of the irrational numbers, we have a group of sets that live inside one another: natural numbers, whole numbers, integers, rational numbers.

In this section, we'll study a larger set of numbers that contains all of the other sets we've studied so far, including the irrationals. This set is called the *real numbers,* and they make up the set of numbers that are used throughout the remainder of the book.

Stockdisc/Getty Images

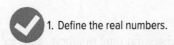

1. Define the real numbers.

The set of **real numbers** consists of the union of the set of rational numbers and the set of irrational numbers.

Using set notation, {real numbers} = {rational numbers} ∪ {irrational numbers}.

Figure 5-7 traces the way we built the real numbers step-by-step through the first four sections of this chapter, always including a new group of numbers to build a larger set.

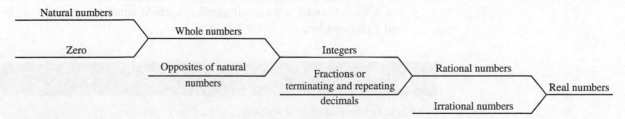

Figure 5-7

Example 1

Drawing a Venn Diagram for the Real Number System

Draw a Venn diagram that represents the relationships between these sets: real numbers, rational numbers, irrational numbers, integers, whole numbers, and natural numbers.

SOLUTION

Since all the numbers we will work with are real numbers, that will be our universal set. The rational numbers and irrational numbers are disjoint, so they need two separate circles that don't intersect. All of the other sets live inside the rational numbers circle, with natural numbers inside whole numbers, which are in turn inside integers.

Try This One   1

Draw a Venn diagram for these sets: rational numbers, positive integers, negative integers, and natural numbers.

One good way to make sure you understand the definitions of the various sets of numbers is to classify numbers according to the sets that they're in. For example, the number 2 is a natural number, a whole number, an integer, a rational number, and a real number. The number -5 is an integer, a rational number, and a real number. The number $\sqrt{3}$ is an irrational number and a real number.

Example 2	**Classifying Numbers**

Classify each number according to type.

(a) 0

(b) $\sqrt{15}$

(c) $-\dfrac{3}{4}$

(d) $0.8\overline{6}$

(e) $\sqrt{25}$

SOLUTION

(a) Zero is a whole number, and all whole numbers are integers, rational numbers, and real numbers.

(b) Since 15 is not a perfect square, $\sqrt{15}$ is an irrational number. Since all irrational numbers are real numbers, $\sqrt{15}$ is also a real number.

(c) The number $-\frac{3}{4}$ is a rational number. Since all rational numbers are real numbers, it is also a real number.

(d) The number $0.8\overline{6}$ is a repeating decimal; so it is a rational number and a real number.

(e) At first glance, you might think $\sqrt{25}$ is an irrational number because of the radical sign, but $\sqrt{25} = 5$, and 5 is a natural number, a whole number, an integer, a rational number, and a real number.

Try This One   2

Classify each number according to type.

(a) π

(b) $-9.\overline{2}$

(c) $\sqrt{6}$

(d) -10

(e) $\sqrt{36}$

Properties of Real Numbers

You might recall that when you multiply two numbers, the order in which they are multiplied doesn't matter. For example, 3×2 and 2×3 are both equal to 6. In other words, $3 \times 2 = 2 \times 3$. But when you subtract two numbers, order is very important. For example $5 - 3 = 2$, but $3 - 5 = -2$, which tells us that $5 - 3 \neq 3 - 5$.

The fact that two numbers can be multiplied in any order is called the *commutative property of multiplication*. Subtraction is *not* commutative. This is just one example of a property of real numbers. By **property**, we mean a fact that is always true, regardless of the specific numbers chosen. In this section, we'll study 11 basic properties of the real numbers.

Sidelight   Keeping It Real

Why are the real numbers called the real numbers? Probably because they can all represent the size or measure of some real object. Integers represent tons of real quantities—anything that can be counted, sizes in whole-number units like 10 feet, and things like temperatures, elevations, and debts, which can be negative. Rational numbers very commonly describe measurements and sizes: a standard piece of paper is $8\frac{1}{2} \times 11$ inches, for example. We can even think of physical quantities that are represented by irrational numbers. The floor of a room that measures 10 feet by 10 feet is $\sqrt{200}$ feet from one corner to the opposite corner, and a soccer ball, which has diameter 10 inches, has a volume of $\frac{500}{3}\pi$ cubic inches.

But what about numbers that are not real? Think about a number like $\sqrt{-4}$. Based on our definition of square root, we would need to find a number whose square is -4. But whenever you multiply a number by itself, the result is positive because (obviously) in this case the two factors have the same sign. So there is no real number whose square is -4, and we define $\sqrt{-4}$ to be an imaginary number. Surprisingly, imaginary numbers actually do have applications in our world! They play a role in electrical engineering and quantum mechanics, among other areas.

For any set of numbers, the *closure property* for a given operation says that if two numbers from the set are combined using that operation, the result is another number in that set. For example, the set of natural numbers is closed under addition because the sum of two natural numbers is always a natural number. But the natural numbers are not closed under subtraction. Both 5 and 10 are natural numbers, but $5 - 10$ is not, since it's negative.

The set of real numbers is closed under addition and multiplication since the sum or product of any two real numbers will always be a real number. This provides our first two properties of real numbers.

> **The closure property of addition:** For any two real numbers a and b, the sum $a + b$ is also a real number.

> **The closure property of multiplication:** For any two real numbers a and b, the product $a \cdot b$ is also a real number.

The *commutative property* for an operation states that for any two numbers, the order in which the operation is performed on the numbers doesn't matter. Addition of real numbers is commutative since $a + b = b + a$. For example, $6 + 7 = 7 + 6$, and $5 + 10 = 10 + 5$.

Multiplication of real numbers is also commutative. For example, $8 \times 6 = 6 \times 8$, and $15 \times 3 = 3 \times 15$.

> **The commutative property of addition:** For any real numbers a and b, $a + b = b + a$.

> **The commutative property of multiplication:** For any real numbers a and b, $a \times b = b \times a$.

Remember, to show that a conjecture is not true, all that's needed is one counterexample. So we can easily see that subtraction and division are not commutative—choosing any two nonzero numbers will work. For example, $6 - 3$ and $3 - 6$ are not equal; $6 \div 3$ and $3 \div 6$ are also not equal.

Our next property involves operations with three numbers. Think about the sum $4 + 5 + 8$. If we use parentheses to group the 4 and the 5, we get $(4 + 5) + 8 = 9 + 8 = 17$. If instead we group the 5 and the 8, we get $4 + (5 + 8) = 4 + 13 = 17$. This is different than the commutative property—we didn't change the order, but rather the grouping. When grouping numbers differently in this manner doesn't affect the result, we say that an operation is *associative*. Both addition and multiplication are associative.

Sidelight   Math and Body Composition

As obesity becomes more and more of a health issue in the United States, an index that compares height and weight has become widely used to evaluate body composition. The body mass index, or BMI, was actually developed in the mid-1800s by a Belgian scholar who dabbled in astronomy, mathematics, statistics, and sociology, but it's probably used today more than ever.

To determine your BMI, (1) square your height in inches, (2) divide your weight in pounds by that square, then (3) multiply the result by 703. The following guidelines have been created by the World Health Organization:

Under 18.5	Underweight
18.5 – 24.9	Normal
25 – 29.9	Overweight
30 – 39.9	Obese
40 and over	Severely obese

Rick Gomez/Getty Images

Here's a sample calculation for a person who is 5 foot 7 inches tall and weighs 135 pounds.

$$5'7'' = 67''$$
$$67^2 = 4{,}489$$
$$135 \div 4{,}489 \approx 0.0301$$
$$0.0301 \times 703 \approx 21.1$$

You will need this formula to determine the BMIs for some well-known athletes in Exercises 49–56.

The associative property of addition: For any real numbers a, b, and c, $(a + b) + c = a + (b + c)$.

The associative property of multiplication: For any real numbers a, b, and c, $(a \times b) \times c = a \times (b \times c)$.

Subtraction and division are not associative, as we will see in Problems 65 and 66.

You surely know that any number multiplied by 1 is the original number. We could say that 1 identifies any number when multiplying; in math terms, we call 1 the *identity* for multiplication. The identity for addition is zero because any number added to zero is the original number.

The identity property of addition: The sum of any real number a and zero is the original number a. The number zero is called the **identity for addition**.

The identity property of multiplication: The product of any real number a and 1 is the original number a. One is called the **identity for multiplication**.

Numbers in a set can have *inverses* for a specific operation. If an operation is performed on a number and its inverse, the answer will be the identity for that operation.

For addition, 2 and -2 are *additive inverses* since $2 + (-2) = 0$. (Note that 0 is the identity for addition.) In algebra, the inverses for addition are called *opposites*. Every real number has an additive inverse (or opposite). The additive inverse of 0 is 0. The additive inverse for a number a is designated by $-a$.

Math Note

Recall that the expression $-a$ doesn't mean that the expression is negative; it depends on the value of a. If a is positive, then $-a$ is negative. But if a is negative, then the expression $-a$ is positive.

Inverse property of addition: For any real number a, there exists a real number $-a$ such that $a + (-a) = 0$ and $-a + a = 0$. The number $-a$ is called the **additive inverse** or opposite of a.

For multiplication, 6 and $\frac{1}{6}$ are *multiplicative inverses* since $6 \times \frac{1}{6} = 1$. (Recall that 1 is the identity for multiplication.) The multiplicative inverse of a number is also called its *reciprocal*. Every real number a except zero has a multiplicative inverse, denoted $\frac{1}{a}$.

Inverse property of multiplication: For any real number a except zero, there exists a real number $\frac{1}{a}$ such that $a \times \frac{1}{a} = 1$ and $\frac{1}{a} \times a = 1$. The number $\frac{1}{a}$ is called the **multiplicative inverse,** or reciprocal of a.

Math Note

Remember that the reciprocal of a fraction is obtained by "flipping the fraction upside down." For example, the reciprocal of $\frac{2}{3}$ is $\frac{3}{2}$.

Make sure you don't confuse inverses for addition with inverses for multiplication. To find an additive inverse for a number, change its sign. To find the multiplicative inverse for a number, use its reciprocal. The additive inverse of $-\frac{2}{3}$ is $+\frac{2}{3}$. The multiplicative inverse of $-\frac{2}{3}$ is $-\frac{3}{2}$. The sign does not change.

The *distributive property of multiplication over addition* states that when a number is multiplied by a sum, the number can first be multiplied by each number in the sum, and then the addition can be performed. For example,

$$5(6 + 3) = 5 \cdot 6 + 5 \cdot 3$$
$$5(9) = 30 + 15$$
$$45 = 45$$

Stated formally,

> **The distributive property of multiplication over addition:** For any real numbers a, b, and c, $a \cdot (b + c) = a \cdot b + a \cdot c$.

This property will be referred to simply as the distributive property since addition cannot be distributed over multiplication, as we'll see in Problem 67.

Example 3   Studying the Distributive Property and Exponents

Does the distributive property apply to exponents? Check to see if it works for squares by deciding if $(a + b)^2 = a^2 + b^2$ for any real numbers a and b.

SOLUTION

Let's begin by just picking some specific numbers for a and b and seeing what happens. If $a = 3$ and $b = 7$, we get

$$(3 + 7)^2 = 10^2 = 100$$
$$3^2 + 7^2 = 9 + 49 = 58$$

In this one specific example, $(a + b)^2$ is NOT equal to $a^2 + b^2$, so it certainly can't be true for EVERY a and b. This means that the distributive property does NOT apply to exponents.

Try This One   3

Try to find numbers a and b so that $(a + b)^2$ IS equal to $a^2 + b^2$. What has to happen?

Example 4   Identifying Properties of Real Numbers

Math Note

The distributive property specifically applies when a number is multiplied by a sum, but it can be used for differences also. (We'll study why in Problem 69.) But it's important to note that *these are the only two situations where the distributive property applies!*

Identify the property illustrated by each calculation.

(a) $6 + 5 = 5 + 6$

(b) $0 + 3 = 3$

(c) $4(3 + 9) = 12 + 36$

(d) $\frac{2}{5} \times \frac{5}{2} = 1$

(e) $(6 \cdot 7) \cdot 3 = 6 \cdot (7 \cdot 3)$

SOLUTION

(a) Commutative property of addition

(b) Identity property of addition

(c) Distributive property

(d) Inverse property of multiplication

(e) Associative property of multiplication

Try This One   4

Identify the property illustrated by each calculation.

(a) $-2(3 + 5) = -6 - 10$ (d) $(-5 \cdot 1 + 3 \cdot 2) + 4 \cdot 7 = -5 \cdot 1 + (3 \cdot 2 + 4 \cdot 7)$
(b) $127 \times 1 = 127$
(c) $18 + (-18) = 0$ (e) $\dfrac{2}{3} \cdot \dfrac{4}{5} = \dfrac{4}{5} \cdot \dfrac{2}{3}$

The properties illustrated in Example 4 are straightforward. Identifying the properties in Example 5 is a little more challenging.

Example 5   Identifying Properties of Real Numbers

Identify the property illustrated by each calculation.

(a) $(6 + 5) + 3 = (5 + 6) + 3$ (b) $1 \times 7 = 7 \times 1$ (c) $5 \cdot (6 + 2) = (6 + 2) \cdot 5$

SOLUTION

(a) This is not the associative property of addition since different numbers would have to be in parentheses. Notice that only the order of the 6 and 5 was changed. This is an example of the commutative property of addition, $6 + 5 = 5 + 6$.
(b) This is not an example of the identity property of multiplication. That property states that $1 \times 7 = 7$. This is an example of the commutative property of multiplication since $a \cdot b = b \cdot a$.
(c) This is not an example of the distributive property. The distributive property states that $5 \cdot (6 + 2) = 5 \cdot 6 + 5 \cdot 2$. Here the order of the multiplication was changed. So this is an example of the commutative property of multiplication.

Try This One   5

Identify the property illustrated by each calculation.

(a) $4 + (2 + 3) = 4 + (3 + 2)$ (b) $5 \cdot (6 + 1) = 5 \cdot (1 + 6)$ (c) $3 \cdot \dfrac{1}{3} = \dfrac{1}{3} \cdot 3$

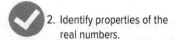
2. Identify properties of the
   real numbers.

The 11 properties of real numbers that we studied in this section are summarized in Table 5-2.

Table 5-2   Properties of the Real Number System

Name	Property	Example
(For any real numbers *a*, *b*, and *c*)		
Closure property of addition	$a + b$ is a real number	$8 + (-3)$ is a real number
Closure property of multiplication	$a \times b$ is a real number	-5×8 is a real number
Commutative property of addition	$a + b = b + a$	$9 + 8 = 8 + 9$
Commutative property of multiplication	$a \cdot b = b \cdot a$	$6 \cdot 8 = 8 \cdot 6$
Associative property of addition	$(a + b) + c = a + (b + c)$	$(12 + 7) + 3 = 12 + (7 + 3)$
Associative property of multiplication	$(a \cdot b) \cdot c = a \cdot (b \cdot c)$	$(5 \cdot 3) \cdot 2 = 5 \cdot (3 \cdot 2)$
Identity property of addition	$0 + a = a$	$0 + 14 = 14$
Identity property of multiplication	$1 \times a = a$	$1 \times (-3) = -3$
Inverse property of addition	$a + (-a) = 0$	$6 + (-6) = 0$
Inverse property of multiplication	$a \cdot \frac{1}{a} = 1, a \neq 0$	$4 \cdot \frac{1}{4} = 1$
Distributive property	$a(b + c) = a \cdot b + a \cdot c$	$6(2 + 5) = 6 \cdot 2 + 6 \cdot 5$

Answers to Try This One

1

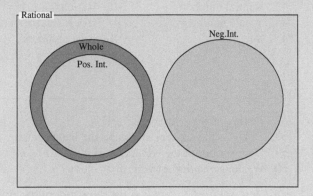

(c) Irrational and real
(d) Integer, rational, and real
(e) Natural number, whole number, integer, rational, and real

3 This only happens if at least one of a or b is zero.

4 (a) Distributive property
(b) Identity property of multiplication
(c) Inverse property of addition
(d) Associative property of addition
(e) Commutative property of multiplication

5 (a) Commutative property of addition
(b) Commutative property of addition
(c) Commutative property of multiplication

2 (a) Irrational and real
(b) Rational and real

Exercise Set   5-5

Writing Exercises

1. Describe the set of real numbers in your own words.
2. Explain what is meant by the term "property of real numbers."
3. Explain what commutative properties refer to. Which operations are commutative?
4. Explain what associative properties refer to. Which operations are associative?

5. What does it mean for a set to be closed under a certain operation?
6. Explain why zero is the identity element for addition, while 1 is the identity element for multiplication.
7. There are exactly two situations that the distributive property applies to. Describe them.
8. Discuss the truth value of this statement: every real number is either rational or irrational.

Computational Exercises

For Exercises 9–24, classify each number by using one or more of the categories—natural, whole, integer, rational, irrational, real.

9. -5
10. 18
11. $\dfrac{3}{4}$
12. $-\dfrac{2}{3}$
13. 6.25
14. -18.376
15. $-\sqrt{6}$
16. $-\sqrt{18}$
17. 0.03030030003 . . .
18. $-\pi$
19. 2.8
20. 13.6
21. 33
22. -17
23. $\sqrt{9}$
24. $\sqrt{100}$

For Exercises 25–40, name the property illustrated.

25. $4 + 8$ is a real number
26. $6 \cdot 1 = 6$
27. $17 + 6 = 6 + 17$
28. $-5 \times (3 + 4) = -5(3) + (-5)(4)$
29. $4 \times 8 = 8 \times 4$
30. $6 + (-6) = 0$
31. $4 \cdot \left(\sqrt{5} + \sqrt{11}\right) = 4\sqrt{5} + 4\sqrt{11}$

32. $\sqrt{3} + \sqrt{4}$ is a real number
33. $\dfrac{5}{8} \cdot \dfrac{8}{5} = 1$
34. $-5 + (+5) = 0$
35. $-6 \cdot (2 + 3) = -6 \cdot (3 + 2)$
36. $(16 + 3) + 5 = 16 + (3 + 5)$
37. $5 + 0 = 5$
38. $-6 + (+6) = (+6) + (-6)$
39. $\dfrac{3}{4} \times \dfrac{4}{3} = \dfrac{4}{3} \times \dfrac{3}{4}$
40. $(8 \times 4) \times 2 = (4 \times 8) \times 2$

The process of simplifying expressions in algebra actually makes extensive use of the properties of real numbers. In each simplification, name the property that has been used in obtaining each new form of the expression. In some cases, there may be more than one.

41. $4(3y - 12) + 6y - 5$
$= 12y - 48 + 6y - 5$ _____
$= 12y + 6y - 48 - 5$ _____
$= (12 + 6)y - 48 - 5$ _____
$= 18y - 53$

42. $(-5 \cdot 2x)(4x) + x(x-3)$
$= (-5)(2x \cdot 4x) + x(x-3)$ _____
$= (-5)(8x^2) + x(x-3)$ _____
$= -40x^2 + x^2 - 3x$ _____
$= (-40+1)x^2 - 3x$ _____
$= -39x^2 - 3x$

For Exercises 43–48, determine under which operations (addition, subtraction, multiplication, division) the system is closed.

43. Natural numbers   46. Rational numbers
44. Whole numbers   47. Irrational numbers
45. Integers   48. Real numbers

Applications in Our World

For Exercises 49–56, use the body mass index formula found in the Sidelight on page 266 to find the BMI for the given athlete, then use the table in the Sidelight to classify each according to World Health Organization standards.

49. LeBron James (basketball): 6′8″, 250 lb
50. Kristaps Porzingis (basketball): 7′3″, 240 lb
51. Novak Djokovic (tennis): 6′2″, 170 lb
52. Connor McGregor (MMA): 5′9″, 154 lb
53. Bryson DeChambeau (golf): 6′1″, 240 lb
54. Serena Williams (tennis): 5′9″, 159 lb
55. J.J. Watt (football): 6′5″, 295 lb
56. Patrick Mahomes (football): 6′3″, 229 lb

For Exercises 57–60, decide whether the two procedures described are commutative or not, and explain your answer.

57. Putting on your shoes and socks.
58. Going to the grocery store and the laundromat.
59. Doing your homework and working out.
60. Washing and drying your dirty clothes.

For Exercises 61–64, decide if the three procedures described are associative or not, and explain your answer.

61. Driving to the bank and McDonalds, and walking your dog.
62. Riding an exercise bike, talking on the phone, and watching TV.
63. Graduating from college, buying a house, and getting married.
64. Getting on a plane, flying to Mexico, and renting a car.

Critical Thinking

For Exercises 65–68, complete the two operations and compare the results. Then describe the result in terms of a property from this section.

65. $(12 - 8) - 15$ $12 - (8 - 15)$
66. $100 \div (25 \div 5)$ $(100 \div 25) \div 5$
67. $10 + (3 \cdot 5)$ $(10 + 3) \cdot (10 + 5)$
68. $8(20 - 4)$ $8 \cdot 20 - 8 \cdot 4$

69. The distributive property specifically applies to a situation where a number is being multiplied by a sum. Using a generic computation $a(b - c)$ and the distributive property we know, explain why the distributive property also applies to a number multiplied by a difference.
70. How do we know that the distributive property also applies to a sum or difference in parentheses with more than two terms, like $5(3 + 7 - 4)$? (*Hint:* One of the properties of addition we studied in this section will help!)
71. The distributive property applies to a situation where a number is being *multiplied* by a sum or difference. Compute $\frac{25-10}{5}$ using the order of operations, then "distribute" the denominator: $\frac{25}{5} - \frac{10}{5}$. What does the result tell you about "distributing" division? Why does that make sense?
72. Write a description of all of the properties of real numbers that are either used in, or illustrated by, the following computation, including where and how they are used.

$$\frac{11}{4} + \frac{19}{6} = \frac{3}{3} \cdot \frac{11}{4} + \frac{19}{6} \cdot \frac{2}{2} = \frac{33}{12} + \frac{38}{12} = \frac{33+38}{12} = \frac{71}{12}$$

73. We know that the real numbers are closed under addition. Explain how we can use this fact to prove that they're also closed under subtraction.
74. We know that the real numbers are closed under multiplication. Does this mean that they're also closed under division? (*Hint:* Be careful!)
75. This diagram is an attempt to draw the relationships between sets of numbers in a different way than the Venn diagram in Example 1. But this diagram is wrong. Why?

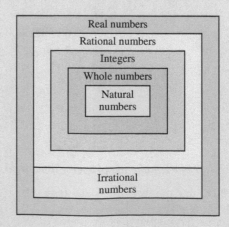

| **Section 5-6** | **Exponents and Scientific Notation** |

REUTERS/Alamy Stock Photo

LEARNING OBJECTIVES

1. Define integer exponents.
2. Use rules for exponents.
3. Convert between scientific and decimal notation.
4. Perform operations with numbers in scientific notation.
5. Use scientific notation in applied problems.

It wasn't that long ago that a million dollars was an almost unimaginable amount of money. The first million dollar lottery was held in New York in 1970, and at that time, winning a million dollars instantly made someone ultra-rich.

But as they always do, times have changed. In the 21st century, a lot of lottery players don't even bother to buy a ticket when the jackpot is "only" a million dollars! The amount of money made by CEOs, entertainers, and athletes continues to grow at an amazing clip, and even the average Joe can make a salary that would have been considered a king's ransom 50 years ago. As the numbers that describe our monetary system get larger and larger, the ability to work with and interpret really large numbers will become more and more important. In this section, we'll study a way to work with really large (and really small) numbers. It starts with an understanding of exponents.

In 2020, Patrick Mahomes signed a 10-year contract worth $450 million. His average annual salary is over 800 times that of the average American household.

Exponents are a concise method of representing repeated multiplications. For example, $5 \cdot 5 \cdot 5 \cdot 5$ can be written as 5^4, the exponent 4 describing how many times to multiply 5 by itself. In general:

For any positive integer n,

$$a^n = \underbrace{a \cdot a \cdot a \cdot a \cdot \cdots \cdot a}_{n \text{ factors}}$$

where a is called the **base** and n is called the **exponent**.

The expression a^n is read as "a to the nth power." When the exponent is 2, such as 5^2, it can be read as "five squared" or "five to second power." When the exponent is 1, it is usually not written; that is, $a^1 = a$.

Math Note

It's hard to make sense of zero and negative exponents in terms of repeated multiplication—you can't multiply a number by itself −3 times. In Problem 105, we'll see why the way we've defined them makes sense.

When an exponent is negative, it is defined as follows: for any positive integer n,

$$a^{-n} = \frac{1}{a^n}$$

For example, $5^{-3} = \frac{1}{5^3}$ or $\frac{1}{125}$. Finally, when the exponent is zero, it is defined as follows:

For any nonzero number a, $a^0 = 1$.

That is, any nonzero number to the 0 power is equal to 1. For example, $6^0 = 1$.

| **Example 1** | **Evaluating Expressions with Exponents** |

Evaluate each expression.

(a) 6^3 (b) 3^{-4} (c) 9^0

SOLUTION

(a) $6^3 = 6 \cdot 6 \cdot 6 = 216$

(b) $3^{-4} = \dfrac{1}{3^4} = \dfrac{1}{3 \cdot 3 \cdot 3 \cdot 3} = \dfrac{1}{81}$

(c) $9^0 = 1$

Try This One   1

1. Define integer exponents.

Evaluate each expression.

(a) 3^6 (b) 2^{-6} (c) 7^0

There are three rules for working with exponents that help us greatly in performing calculations. Think about the expression $4^2 \cdot 4^3$. Using the definition of exponents, we can write this as $(4 \cdot 4) \cdot (4 \cdot 4 \cdot 4)$, which is the same as 4^5. The net result is that we added the exponents: there were two factors from 4^2 and three factors from 4^3, for a total of five factors. This is an example of an exponent rule.

Rules for Exponents

For any nonzero real number a and any integers m and n:

1. (The product rule) $a^m \cdot a^n = a^{m+n}$

2. (The quotient rule) $\dfrac{a^m}{a^n} = a^{m-n}$

3. (The power rule) $(a^m)^n = a^{m \cdot n}$

Most students find it easier to understand and use these rules when they're described in words rather than symbols.

Math Note

Neither the product rule nor the quotient rule says *anything* about multiplying or dividing exponential expressions with *different bases*!

1. **Product rule.** When two expressions with the same base are multiplied, keep the base unchanged and add the exponents.

2. **Quotient rule.** When two expressions with the same base are divided, keep the base unchanged and subtract the exponents.

3. **Power rule.** When an expression with an exponent is raised to a power, keep the base unchanged and multiply the two exponents.

Example 2 illustrates the use of these rules in simplifying and performing calculations.

Example 2   Using Rules For Exponents

Simplify the expression using rules for exponents, and then evaluate the resulting expression.

(a) $3^3 \cdot 3^5$ (b) $\dfrac{5^4}{5^6}$ (c) $(2^3)^4$

SOLUTION

(a)   The bases are the same and we're multiplying, so we add exponents (product rule).

$$3^3 \cdot 3^5 = 3^{3+5} = 3^8 \text{ which is } 6{,}561$$

Calculator Guide

To find 3^8 for Example 2 using a calculator:

Typical Scientific or Online Calculator

Typical Graphing Calculator

 2. Use rules for exponents.

(b) The bases are the same and we're dividing, so we subtract the exponents (quotient rule)

$$\frac{5^4}{5^6} = 5^{4-6} = 5^{-2} \qquad \text{which is} \qquad \frac{1}{5^2} \qquad \text{or} \qquad \frac{1}{25}$$

(c) When raising an exponential expression to a power, multiply the exponents (power rule).

$$(2^3)^4 = 2^{3\cdot4} = 2^{12} \qquad \text{which is} \qquad 4{,}096$$

Try This One   2

Simplify the expression using rules for exponents, and then evaluate the resulting expression.

(a) $6^2 \cdot 6^4$ (b) $\dfrac{10^8}{10^{11}}$ (c) $(4^3)^2$

Scientific Notation

Now we turn our attention to working with really large and really small numbers. We will express numbers in terms of powers of 10, because this allows us to write a lot of digits with just a few symbols:

Powers of 10

	$10^0 = 1$	$10^{-1} = 0.1$
	$10^1 = 10$	$10^{-2} = 0.01$
	$10^2 = 100$	$10^{-3} = 0.001$ *Thousandth*
Thousand	$10^3 = 1{,}000$	$10^{-4} = 0.0001$
	$10^4 = 10{,}000$	$10^{-5} = 0.00001$
	$10^5 = 100{,}000$	$10^{-6} = 0.000001$ *Millionth*
Million	$10^6 = 1{,}000{,}000$	$10^{-7} = 0.0000001$
	$10^7 = 10{,}000{,}000$	$10^{-8} = 0.00000001$
	$10^8 = 100{,}000{,}000$	
Billion	$10^9 = 1{,}000{,}000{,}000$	

Math Note

Because it's very common to work with really large or really small numbers in science, the system for writing them concisely is called scientific notation.

For example, Earth is about 93,000,000 (93 million) miles from the sun. Another way to write this number is 9.3 times 10 million, or 9.3×10^7. A number is in **scientific notation** when it's written as a decimal with one nonzero digit to the left of the decimal point times some power of 10. The rules for writing a number in scientific notation are below.

Procedure for Writing a Number in Scientific Notation

Step 1   Move the decimal point either right or left so that there is exactly one nonzero digit to the left of the decimal point.

Step 2   Write the resulting number times 10 to some power:

      (a) If the decimal point was moved to the left, the exponent of 10 is the number of places the decimal point was moved.

      (b) If the decimal point was moved right, the exponent of 10 is the negative of the number of places the decimal point was moved.

Example 3   **Writing Numbers in Scientific Notation**

Write each number in scientific notation.

(a) 3,572,000,000   (b) 0.000087

SOLUTION

(a) Move the decimal point 9 places to the left so that it falls between the 3 and 5.

In scientific notation, $3{,}572{,}000{,}000 = 3.572 \times 10^9$.

(b) Move the point to the right 5 places so that it will fall between the 8 and 7.

In scientific notation, $0.000087 = 8.7 \times 10^{-5}$.

> **Math Note**
>
> If you have trouble remembering whether to use a positive or negative exponent, think of it this way: a power of 10 with a positive exponent, like 10^8, is a really large number, so large numbers get positive exponents in scientific notation, and small numbers get negative exponents.

Try This One   3

Write each number in scientific notation.

(a) 516,000,000   (b) 0.000162

As we'll see shortly, writing numbers in scientific notation is very useful in performing calculations, but you should also be able to convert from scientific to decimal notation. This sometimes makes it easier to get perspective on the size of the numbers involved. To do so, we just reverse the procedure above.

Rules for Converting a Number in Scientific Notation to Decimal Notation

(a) If the exponent of the power of 10 is positive, move the decimal point to the right the same number of places as the exponent. You may need to put in zeros as placeholders when you run out of digits.

(b) If the exponent is negative, move the decimal point to the left the same number of places as the exponent, again putting in zeros as needed.

Example 4   **Converting from Scientific to Decimal Notation**

Write each number in decimal notation.

(a) 4.192×10^8 (b) 6.37×10^{-8}

SOLUTION

(a) Since the exponent is positive, move the decimal point 8 places to the right.

$$4.\underset{\underset{1\,2\,3\,4\,5\,6\,7\,8}{|\,|\,|\,|\,|\,|\,|\,|}}{19200000}$$

In decimal notation, $4.192 \times 10^8 = 419{,}200{,}000$.

Calculator Guide

Here's how to enter
4.192×10^8 into a calculator:

Typical Scientific or Online Calculator

4.192 [EE] 8 [=]

Typical Graphing Calculator

4.192 [EE] or [2nd] [,] 8 [ENTER]

3. Convert between scientific and decimal notation.

(b) Since the exponent is negative, move the decimal point 8 places to the left.

$$\underbrace{00000006.37}_{8\,7\,6\,5\,4\,3\,2\,1}$$

In decimal notation, $6.37 \times 10^{-8} = 0.0000000637$.

Try This One   4

Write each number in decimal notation.

(a) 9.61×10^{10} (b) 2.77×10^{-6}

Operations with Numbers in Scientific Notation

Suppose you're asked to multiply 20,000,000 by 42,000,000,000. In decimal form, this would be challenging, to say the very least. (Try it if you have an hour or so to kill.) But if we convert the numbers to scientific notation and write out the multiplication. . .

$$(2 \times 10^7) \times (4.2 \times 10^{10}) = (2 \times 4.2) \times (10^7 \times 10^{10}) = 8.4 \times 10^{17}$$

Commutative Property of Multiplication

Wow! That was much easier. And that, friends, is one of the best things about scientific notation.

> **Multiplying Numbers in Scientific Notation**
>
> To multiply two numbers in scientific notation, multiply the numbers to the left of the powers of 10 and add the exponents of the 10s. (You may have to rewrite to put the result in proper scientific notation.)

Example 5   Multiplying Numbers in Scientific Notation

Find each product. Write your answer in scientific notation.

(a) $(3 \times 10^5)(2.2 \times 10^3)$ (b) $(5 \times 10^2)(7 \times 10^3)$

SOLUTION

(a) Multiply 3×2.2 to get 6.6, and add exponents $5 + 3$ to get 8.

$$(3 \times 10^5)(2.2 \times 10^3) = 6.6 \times 10^8$$

(b) This time, when we multiply 5×7, we get a two-digit number, so our answer won't be in proper scientific notation:

$$(5 \times 10^2)(7 \times 10^3) = 35 \times 10^5$$

We can remedy this by moving the decimal point one place left, which raises the exponent by 1:

$$35 \times 10^5 = 3.5 \times 10^6$$

To make sure we chose the right exponent, notice that in changing 35 to 3.5, we made it smaller, so we needed to make the exponent larger.

Calculator Guide

Many calculators will display very small or large answers in scientific notation, even if the original numbers were entered in decimal notation. For example, when multiplying 3 million and 5 million entered in decimal form, the calculator displays:

Typical Scientific or Online Calculator

1.5e + 13

Typical Graphing Calculator

1.5E13

This represents 1.5×10^{13}.

Try This One   5

Find each product. Write your answer in scientific notation.

(a) $(6 \times 10^5)(1 \times 10^4)$ (b) $(5 \times 10^3)(3.1 \times 10^6)$

The same rule for multiplication works just fine when the exponents are negative.

Example 6   Multiplying Numbers in Scientific Notation

Math Note

The part of a number in scientific notation that comes before the power of 10 is often called the *decimal part* of the number, as we did in Example 6.

Find the product. Round the decimal part of your answer to two decimal places.

(a) $(3.25 \times 10^{-4})(5.1 \times 10^{-3})$ (b) $(8.6 \times 10^3)(9.7 \times 10^{-6})$

SOLUTION

(a) $(3.25 \times 10^{-4})(5.1 \times 10^{-3}) = 16.575 \times 10^{-7} \approx 1.66 \times 10^{-6}$

   Be careful when moving the decimal point! Adding 1 to -7 makes it -6.

(b) $(8.6 \times 10^3)(9.7 \times 10^{-6}) = 83.42 \times 10^{-3} \approx 8.34 \times 10^{-2}$

Try This One   6

Find the product. Write each answer in scientific notation, rounding the decimal part of the answer to two decimal places.

(a) $(4.2 \times 10^{-4})(2 \times 10^{-2})$ (c) $(8.32 \times 10^{-3})(6.48 \times 10^5)$
(b) $(7.3 \times 10^{-9})(5.1 \times 10^6)$

The rule for dividing numbers in scientific notation is similar to the one for multiplying. In Problem 106, we'll see why it makes perfect sense.

Dividing Numbers in Scientific Notation

To divide two numbers in scientific notation, divide the numbers to the left of the power of 10 and subtract the exponents of the 10s. (You may have to rewrite to put the result in proper scientific notation.)

Example 7   Dividing Numbers in Scientific Notation

Find each quotient. Write your answer in scientific notation.

(a) $\dfrac{9 \times 10^6}{3 \times 10^4}$ (b) $\dfrac{9 \times 10^{-3}}{2 \times 10^5}$ (c) $\dfrac{1.2 \times 10^{-5}}{4.8 \times 10^{-4}}$

SOLUTION

In each case, we'll divide the decimal parts, then subtract the exponents to find the appropriate power of 10.

(a) $\dfrac{9 \times 10^6}{3 \times 10^4} = \dfrac{9}{3} \times 10^{6-4} = 3 \times 10^2$

(b) $\dfrac{9 \times 10^{-3}}{2 \times 10^{5}} = \dfrac{9}{2} \times 10^{-3-5} = 4.5 \times 10^{-8}$

(c) $\dfrac{1.2 \times 10^{-5}}{4.8 \times 10^{-4}} = \dfrac{1.2}{4.8} \times 10^{-5-(-4)} = 0.25 \times 10^{-5+4} = 0.25 \times 10^{-1}$

This time, the answer needs to be rewritten to put it in proper scientific notation. The decimal point will be moved one place to the right, so we need to subtract one from the exponent (we made the decimal part larger).

$$0.25 \times 10^{-1} = 2.5 \times 10^{-2}$$

Try This One 7

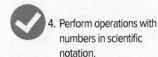

4. Perform operations with numbers in scientific notation.

Find each quotient. Round the decimal part of the answer to two decimal places.

(a) $\dfrac{3.2 \times 10^{7}}{8 \times 10^{4}}$

(b) $\dfrac{9.6 \times 10^{-5}}{1.3 \times 10^{-8}}$

(c) $\dfrac{2.2 \times 10^{-6}}{4.8 \times 10^{5}}$

Applications of Scientific Notation

As we pointed out earlier, many areas in science and economics require the use of very large or very small numbers. This is when scientific notation really comes in handy. Many calculators display only 8 to 12 digits, so numbers with more digits can only be entered in scientific notation.

Example 8 | Applying Scientific Notation to Astronomy

Digital Vision/Getty Images

Earth's orbit around the sun is not circular, but it's reasonably close to a circle with radius 9.3×10^{7} miles. How far does Earth travel in 1 year? (The formula for circumference of a circle is $C = 2\pi r$.)

SOLUTION

The distance Earth travels in 1 year is one full turn around a circle with radius 9.3×10^{7} miles. Using the formula for circumference, and approximating π with 3.14,

$$\begin{aligned} C &\approx 2(3.14)(9.3 \times 10^{7}) \\ &\approx 58.4 \times 10^{7} \\ &= 5.84 \times 10^{8} \text{ miles} \end{aligned}$$

In decimal form, this is 584,000,000, or 584 million miles!

Try This One 8

Lawrence Lawry/Photodisc/Getty Images

5. Use scientific notation in applied problems.

The speed of light in glass is about 6.56×10^{8} feet per second, and the circumference of Earth is about 1.31×10^{8} feet. Using the formula

$$\text{Time} = \dfrac{\text{distance}}{\text{speed}}$$

find how long it takes light to circle the globe in a glass fiber-optic cable.

Answers to Try This One

1   (a) 729

(b) $\dfrac{1}{64}$

(c) 1

2   (a) $6^6 = 46{,}656$

(b) $10^{-3} = \dfrac{1}{1{,}000}$

(c) $4^6 = 4{,}096$

3   (a) 5.16×10^8

(b) 1.62×10^{-4}

4   (a) 96,100,000,000

(b) 0.00000277

5   (a) 6×10^9

(b) 1.55×10^{10}

6   (a) 8.4×10^{-6}

(b) 3.72×10^{-2}

(c) 5.39×10^3

7   (a) 4×10^2

(b) 7.38×10^3

(c) 4.58×10^{-12}

8   About 0.2 seconds

Exercise Set   5-6

Writing Exercises

1. Describe how exponents are defined using multiplication.
2. What is scientific notation? How do you convert a number from decimal to scientific notation?
3. Explain how to convert a number from scientific to decimal notation.
4. What are some of the advantages of using scientific notation?
5. Describe in your own words how to multiply two numbers in scientific notation.
6. Describe in your own words how to divide two numbers in scientific notation.
7. When a number written in scientific notation has a positive exponent on 10, what can we say for sure about the number? Why?
8. If a number between 0 and 1 is written in scientific notation, what can we say about the exponent on 10? Why?

Computational Exercises

For Exercises 9–18, evaluate each without a calculator.

9. 3^5
10. 6^4
11. 8^0
12. 9^0
13. $(-5)^0$

14. $(-4)^0$
15. 3^{-5}
16. 6^{-4}
17. 2^{-6}
18. 7^{-2}

For Exercises 19–38, simplify the expression using rules for exponents, then evaluate the resulting expression.

19. $3^4 \cdot 3^2$
20. $5^3 \cdot 5^3$
21. $4^4 \cdot 4^3$
22. $2^6 \cdot 2^4$
23. $\dfrac{3^4}{3^2}$
24. $\dfrac{6^5}{6^3}$
25. $\dfrac{12^5}{12^4}$

26. $\dfrac{25^3}{25}$
27. $(5^2)^3$
28. $(4^4)^2$
29. $3^2 \cdot 3^{-4}$
30. $4^{-3} \cdot 4^5$
31. $35^{-3} \cdot 35^{-2}$
32. $16^3 \cdot 16^{-3}$
33. $\dfrac{2^5}{2^7}$

34. $\dfrac{3^4}{3^6}$
35. $\dfrac{100^4}{100^7}$
36. $\dfrac{13^2}{13^5}$
37. $\dfrac{9^2}{3^6}$
38. $\dfrac{5^9}{25^5}$

In Exercises 39–46, simplify the expression using rules for exponents.

39. $(a^4)^5$
40. $(k^{-3})^2$
41. $(m^{-4})^2$
42. $(b^5)^6$

43. $\dfrac{a^{10} \cdot a^{12}}{a^8}$
44. $\dfrac{b^7 \cdot b^5}{b^6}$

45. $\dfrac{x^{-5} \cdot x^3}{x^{-9}}$
46. $\dfrac{y^4 \cdot y^{-8}}{y^{-5}}$

For Exercises 47–56, write each number in scientific notation.

47. 625,000,000
48. 9,910,000
49. 0.0073
50. 0.261
51. 528,000,000,000

52. 2,220,000
53. 0.00000618
54. 0.0000000077
55. 43,200
56. 56,000

For Exercises 57–66, write each number in decimal notation.

57. 5.9×10^4
58. 6.28×10^6
59. 3.75×10^{-5}

60. 9×10^{-10}
61. 2.4×10^3
62. 7.72×10^5

63. 3×10^{-6}

64. 4×10^{-9}

65. -1.46×10^{-8}

66. -2.26×10^{-7}

For Exercises 67–80, perform the indicated operations. Write the answers in scientific notation, rounding to two decimal places. Then write the rounded answer in decimal notation as well.

67. $(3 \times 10^4)(2 \times 10^6)$

68. $(5 \times 10^3)(8 \times 10^5)$

69. $(6.2 \times 10^{-2})(4.3 \times 10^{-6})$

70. $(1.7 \times 10^{-5})(3.8 \times 10^{-6})$

71. $(4 \times 10^4)(2.2 \times 10^{-7})$

72. $(2.2 \times 10^5)(3.6 \times 10^{-4})$

73. $(5 \times 10^{-2})(3 \times 10^{-8})$

74. $(4.3 \times 10^5)(2.2 \times 10^{-6})$

75. $\dfrac{5 \times 10^4}{2.5 \times 10^2}$

76. $\dfrac{9 \times 10^6}{3 \times 10^2}$

77. $\dfrac{4.2 \times 10^{-2}}{7 \times 10^{-3}}$

78. $\dfrac{6.4 \times 10^8}{8 \times 10^{-2}}$

79. $\dfrac{6.6 \times 10^3}{1.1 \times 10^5}$

80. $\dfrac{3 \times 10^7}{1.5 \times 10^{-5}}$

For Exercises 81–86, write each number in scientific notation and perform the indicated operations. Leave answers in scientific notation. Round to two decimal places.

81. $(63,000,000)(41,000,000)$

82. $(52,000)(3,000,000)$

83. $\dfrac{600,000,000}{25,000,000}$

84. $\dfrac{32,000,000}{64,000,000}$

85. $(0.00000025)(0.000004)$

86. $\dfrac{0.0000036}{0.0009}$

Applications in Our World

87. When a new rumor about a team's star running back starts to spread across a large college campus, the number of students that have heard the rumor d days after it starts can be described by the formula $N = 50d^4$. The number of students that have heard the rumor but don't believe it can be described by the formula $D = 40d^2$. Find a formula that describes the percentage of students that have heard the rumor that don't believe it in terms of days after it starts to spread. What happens to that percentage for days 1, 2, 3, and 4 after the rumor starts?

88. The total box office revenue for a new movie can be described by the formula $M = 200w^4$, where M is in dollars, and w is the number of weeks since the movie debuts. The total number of people that have seen the movie in theaters can be described by the formula $T = 300w^3$. Find a formula that describes the average revenue per person in terms of weeks since the movie debuted. Then find the average revenue for weeks 4, 6, 8, and 10. What trend do you notice?

89. Light travels through air at 1.86×10^5 miles per second. How many miles does light travel in 10 minutes? Write your answer in decimal notation.

90. There are about 1×10^{14} cells in the human body. About how many body cells are there in the United States? (Use 331 million as the population of the United States.)

91. The mass of a proton is about 1.67×10^{-24} grams. There are 1.27×10^{26} molecules in a gallon of water, and 10 protons in 1 molecule of water. Also, 1 gallon of water weighs 3,778 grams. Based on this information, what percentage of the weight of water comes from protons?

92. It has been estimated that there are 1×10^{20} grains of sand on the beach at Coney Island, New York. By one estimate, the beach has an area of 3.3 million square feet. How many grains of sand are there on average per square foot of beach?

For Exercises 93–96, use the following information. The number of miles that light travels in one year is called a light-year. One light-year is equal to 5.88×10^{12} miles.

93. The nearest star (other than the sun) to Earth is Proxima Centauri, at 4.2 light-years away. How far is this in miles?

94. The star Gruis is 280 light-years from Earth. How far is this in miles?

95. The (former) planet Pluto is 4,681 million miles from Earth. Write this number in scientific notation. How far is that in light-years?

96. The average distance from Earth to the sun is 93 million miles. Write this number in scientific notation. Use the number of miles in a light-year to find how many minutes it takes light to reach Earth from the sun.

97. One atom is 1×10^{-8} centimeters in length. How many atoms could be laid end-to-end on a meter stick (which is 100 cm in length)?

98. One grain of spruce pollen weighs about 7×10^{-5} grams. How many grains are there in a pound of spruce pollen? (One pound is about 454 grams.)

99. The planet Venus is about 67 million miles from the sun. Assuming that its orbit is approximately circular, how far does it travel in 1 year? (One Venus year, not one Earth year.)

100. Each red blood cell contains 250 million molecules of hemoglobin. There are about 25 trillion red blood cells in the average human. How many molecules of hemoglobin are there in an average human?

101. The largest bank in the United States, JP Morgan Chase, announced a profit of $\$2.91 \times 10^{10}$ in 2020. How much on average did the company earn every day in that year?

102. Refer to Exercise 101. If the CEO of JP Morgan Chase had gone stark raving mad and decided to equally distribute the company's 2020 profit to the 331 million

people in the United States, how much would each person have received?

103. The total of all player salaries in 2021 for the Los Angeles Dodgers was of 2.67×10^8. The team was projected to have an average ticket price of $49. How many tickets would the team have to sell to cover the cost of those salaries?

104. At the end of the college football season, there are three bowl games that are part of the college football playoff. ESPN signed a contract to televise these three games from 2014 through 2025 at a cost of 4.7×10^8 per year. In 2020, an estimated 56.7 million viewers watched the games. If viewership continues at that rate, how much is ESPN spending per viewer?

Critical Thinking

105. Read the second column of the table below from top to bottom. Using inductive reasoning, complete the missing entries in the table. Then explain how that justifies our definitions of zero and negative exponents.

Exponent (n)	2^n
5	32
4	16
3	8
2	4
1	2
0	?
−1	?
−2	?

106. The quotient rule for exponents is $a^m/a^n = a^{m-n}$, where a is any nonzero number, and m and n are integers. Use the definition of exponents from the beginning of the section and reducing fractions to justify this rule. Does your justification require any restrictions on m or n? Explain.

107. The power rule for exponents is $(a^m)^n = a^{m \cdot n}$, where a is any nonzero number, and m and n are integers. Use the

definition of exponents from the beginning of the section and multiplication to justify this rule. Does your justification require any restrictions on m or n? Explain.

108. We have already developed rules for multiplying and dividing numbers in scientific notation. In this exercise, we'll develop a power rule for scientific notation.
    (a) Using the definition of exponent from the beginning of this section and our rule for multiplying numbers in scientific notation, compute $(2 \times 10^4)^3$.
    (b) Repeat for $(a \times 10^n)^3$.
    (c) Use the results of (a) and (b) to verbally describe a rule for raising a number in scientific notation to a power. Does your rule work for zero and negative powers?

109. Refer to the result of Example 8 on page 277. How fast is Earth traveling around the sun in miles per hour? If you could fly that quickly, how long would it take to get from New York to Los Angeles (2,462 miles)?

110. Is there a largest number that you can write in decimal form using fewer symbols than it would take to write it in scientific notation? If there is, find it. If not, explain why.

Section 5-7   Arithmetic and Geometric Sequences

LEARNING OBJECTIVES

1. Define arithmetic sequence.
2. Find a particular term of an arithmetic sequence.
3. Find the sum of n terms of an arithmetic sequence.
4. Define geometric sequence.
5. Find a particular term of a geometric sequence.
6. Find the sum of n terms of a geometric sequence.

After a steady rise in home prices from 2010 to 2020, something really strange happened in 2021: there was an absolute buying frenzy, with many houses being sold in hours after being put on the market. Prices increased dramatically, which was good news if you were selling a house but bad news for millions of folks who could no longer afford one. This forced a lot of people to rent housing and try to save money for a home.

Suppose that this is the predicament a young couple find themselves in. They determine that they can afford to sock away $400 each month. They don't trust banks, so they plan to put $400 in cash in their cookie jar each month, which, by the way, is not the brightest idea I've ever heard. The amount they save would grow like this as the months pass: $400, $800, $1,200, $1,600, $2,000. . . . This is a special type of *sequence* that we will study in this section.

wragg/iStock/Getty Images

A **sequence** is a list of related numbers in a definite order. Each number in the sequence is called a **term** of the sequence.

Note: In many cases, there is a rule, or formula, that determines the terms of a sequence. But not always.

Arithmetic Sequences

In the sequence defined by saving money for a down payment above, every term after the first can be obtained by adding 400 to the previous term. This is an example of an **arithmetic sequence**. A sequence is arithmetic when every term after the first is obtained by adding the same fixed number to the previous term. This amount is called the **common difference**.

Examples of Arithmetic Sequences	Common Difference
$1, 3, 5, 7, 9, 11, \ldots$	2
$2, 7, 12, 17, 22, 27, \ldots$	5
$100, 97, 94, 91, 88, 85, \ldots$	-3
$\frac{1}{2}, \frac{2}{2}, \frac{3}{2}, \frac{4}{2}, \frac{5}{2}, \frac{6}{2}, \ldots$	$\frac{1}{2}$

Example 1 | Finding the Terms of an Arithmetic Sequence

Write the first five terms of an arithmetic sequence as described.

(a) The value of an account that starts with $700 and has $80 deposited each month
(b) First term 9 and common difference 7
(c) First term $\frac{1}{16}$ and common difference $-\frac{1}{8}$

SOLUTION

(a) We begin at $700, and add $80 to get each of the next four terms:

$$\$700, \$780, \$860, \$940, \$1{,}020$$

(b) We begin at 9, and add 7 to get the next four terms:

$$9, 16, 23, 30, 37$$

(c) This time we will start at $\frac{1}{16}$ and *subtract* $\frac{1}{8}$ to get the next four terms: it will be helpful to write $\frac{1}{8}$ as $\frac{2}{16}$ so we can perform the subtractions easily.

$$\frac{1}{16}, -\frac{1}{16}, -\frac{3}{16}, -\frac{5}{16}, -\frac{7}{16}$$

Try This One   1

Write the first five terms of an arithmetic sequence as described.

(a) The weight of an infant that was 7 lb at birth and has been gaining 1.3 lb per week
(b) First term 6 and common difference 10
(c) First term $\frac{3}{8}$ and common difference $-\frac{1}{2}$

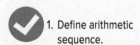

1. Define arithmetic sequence.

We often use the letter a with a subscript to represent the terms of a generic sequence. The symbol a_1 represents the first term, a_2 the second term, a_3 the third, and so on. So an arbitrary sequence looks like $a_1, a_2, a_3, a_4, a_5, \ldots$. We call a_n the nth term of the sequence.

Math Note

A **finite sequence** has a finite number of terms; an **infinite sequence** has infinitely many terms, like 2, 4, 6, 8, 10, . . .

When a sequence is arithmetic with first term a_1 and common difference d, we can find a formula to quickly calculate any term of the sequence without having to find all of the terms that come before it. In this case, the sequence looks like

$$a_1, \quad a_1 + d, \quad a_1 + d + d, \quad a_1 + d + d + d, \ldots$$

Doing some simplifying, we find that

$$a_1 = a_1 \quad a_2 = a_1 + d \quad a_3 = a_1 + 2d \quad a_4 = a_1 + 3d$$

Every term a_n looks like a_1 plus some coefficient times d, and the coefficient is one less than the subscript of that term in the sequence. This gives us a formula:

Finding the *n*th Term of an Arithmetic Sequence

The *n*th term of an arithmetic sequence can be found using the formula

$$a_n = a_1 + (n-1)d$$

where n is the number of the term you want to find, a_1 is the first term, and d is the common difference.

Example 2	Finding a Particular Term of an Arithmetic Sequence

Find the value of the account in Example 1 after 14 months.

Math Note

You *could* check your answer by writing out the first 14 terms of the sequence, but that kind of defeats the purpose of developing a formula for the *n*th term in the first place.

SOLUTION

The account values were an arithmetic sequence with first term $700 and common difference $80. So we substitute into the *n*th term formula using $a_1 = \$700$, $d = \$80$, and $n = 14$.

$$a_n = \$700 + (14 - 1)(\$80)$$
$$a_{14} = \$700 + 13(\$80)$$
$$= \$700 + \$1{,}040$$
$$= \$1{,}740$$

The account is worth $1,740 after 14 months.

Try This One   2

Use the *n*th term formula to find the weight of the baby in Try This One 1 after 20 weeks.

Example 3	Finding a Particular Term of an Arithmetic Sequence

Find the 10th term of an arithmetic sequence with first term $\frac{1}{7}$ and common difference $\frac{2}{5}$.

Thinkstock/JupiterImages

The trombone plays a different note when the musician changes the length of the instrument while setting up a vibrating column of air through the mouthpiece. For a given length of trombone *L*, it will resonate for notes with wavelengths

$$\frac{2L}{1}, \frac{2L}{2}, \frac{2L}{3}, \frac{2L}{4}, \frac{2L}{5}, \ldots$$

(This is true of other brass and woodwind instruments as well.) What is the *n*th term of this sequence?

SOLUTION

Substitute in the formula using $a_1 = \frac{1}{7}$, $n = 10$, and $d = \frac{2}{5}$.

$$a_n = a_1 + (n - 1)d$$
$$a_{10} = \frac{1}{7} + (10 - 1)\left(\frac{2}{5}\right)$$
$$= \frac{1}{7} + (9)\left(\frac{2}{5}\right)$$
$$= \frac{5}{35} + \frac{126}{35}$$
$$= \frac{131}{35}$$

Try This One   3

Find the requested term of the arithmetic sequence with the given first term and common difference.

(a) First term $\frac{1}{9}$, common difference $\frac{1}{3}$: find the seventh term.
(b) First term $-\frac{5}{6}$, common difference $-\frac{1}{10}$: find the sixth term.

Now let's change our savings account scenario a bit. Suppose that you put aside $10 this month, then $15 next, $20 the following, and so on. This time it's the amount you're saving that forms an arithmetic sequence (because it's going up by $5 each month) rather than the value of the account. In this case, it would be of interest to find how much you'd put aside in, say, the 14th month, but it would be a whole lot more useful to find the SUM of the contributions for the first 14 months.

The point is that sometimes it's more useful to find the *sum* of terms in an arithmetic sequence rather than the terms themselves. Our next formula accomplishes just that. (If you're interested, we'll work out a proof of this formula in Project 3 at the end of this chapter.)

2. Find a particular term of an arithmetic sequence.

Finding the Sum of an Arithmetic Sequence

The sum of the first *n* terms of an arithmetic sequence is given by the formula

$$S_n = \frac{n(a_1 + a_n)}{2}$$

where a_1 is the first term of the sequence and a_n is the *n*th term.

| Example 4 | Finding the Sum of an Arithmetic Sequence |

Find the sum of the first 14 terms in the sequence of deposits described in the text above the sum formula.

SOLUTION

First, we need to find the 14th term (that's the a_n in the sum formula). Using the formula shown previously,

$$a_n = a_1 + (n - 1)d$$

where $a_1 = \$10$, $n = 14$, and $d = \$5$.

$$a_{14} = \$10 + (14 - 1)\$5$$
$$= \$10 + 13(\$5)$$
$$= \$75$$

Next, substitute into the formula for finding the sum.

$$S_n = \frac{n(a_1 + a_n)}{2}$$
$$= \frac{14(\$10 + \$75)}{2}$$
$$= \$595$$

The total of all deposits over the first 14 months is $595.

Try This One   4

3. Find the sum of n terms of an arithmetic sequence.

Find the sum of the first 12 terms of each sequence.

(a) 5, 12, 19, 26, 33, . . .   (b) −1, −3, −5, −7, −9, . . .   (c) $\frac{1}{5}, \frac{2}{5}, \frac{3}{5}, \frac{4}{5}, \ldots$

Geometric Sequences

Remember the geniuses from the beginning of the lesson that decided to open a savings account in their sock drawer? Suppose that a month after starting to save $400 a month, the wife gets discovered by a major modeling agency, and the cash starts rolling in. Instead of adding $400 to their savings every month, now they're able to double the total each month. The amount would now look like $400, $800, $1,600, $3,200, $6,400, ... This is an example of a **geometric sequence**. A sequence is geometric when every term after the first is obtained by multiplying the previous term by the same fixed nonzero number. This multiplier is called the **common ratio**.

Examples of Geometric Sequences	Common Ratio
1, 3, 9, 27, 81, 243, . . .	$r = 3$
2, 10, 50, 250, 1,250, . . .	$r = 5$
5, −10, 20, −40, 80, . . .	$r = -2$
$1, \frac{1}{4}, \frac{1}{16}, \frac{1}{64}, \frac{1}{256}, \ldots$	$r = \frac{1}{4}$

Sidelight   **The Fibonacci Sequence**

Sometime around the year 1200, an Italian mathematician named Leonardo Pisano, who went by the name Fibonacci, discovered the sequence 1, 1, 2, 3, 5, 8, 13, 21, 34, ... Do you see the pattern? After the first two terms, each successive term comes from adding the two previous terms. Not exactly earth-shattering, right? The truly amazing thing about this innocent little list of numbers is that numbers from this sequence are found all over the place in nature. The number of petals that many flowers have correspond to Fibonacci numbers. Lilies and irises have 3 petals, buttercups have 5, delphiniums have 8, and marigolds have 13. Different varieties of daisies have 21, 34, 55, or 89 petals. The seeds on the head of a sunflower spiral out from the center in clockwise and counterclockwise directions: some sunflowers have 21 spirals in one direction and 34 in the other. A giant sunflower has 89 in one direction and 144 in the other, and these are Fibonacci numbers as well.

Speaking of spirals, the beautiful shell of the chambered nautilus, pictured here, is a smooth spiral that can be built using a series of squares with areas matching—you guessed it—the Fibonacci sequence.

Interestingly, Fibonacci numbers have a very strong connection to the golden ratio we studied in the Sidelight on page 256. This connection will be explored in Project 2 at the end of this chapter. Even more surprising, ratios of Fibonacci numbers are used by some stock analysts to find ideal times to buy or sell a certain stock. The theory is based on the idea that the golden ratio is not just naturally attractive to the human eye, but to the human subconscious as well. (Now that's *deep*.) A Google search for "Fibonacci" and "stock market" yields a veritable cornucopia of pages that explain the technique (as well as a bunch that claim it's bunk . . . you be the judge).

Kaz Chiba/Photodisc/Getty Images

Example 5 | Finding the Terms of a Geometric Sequence

Write the first five terms of a geometric sequence with first term 4 and common ratio −3.

SOLUTION

Start with first term 4, then multiply by −3 to obtain each successive term:

$$4, -12, 36, -108, 324$$

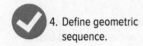

4. Define geometric sequence.

Try This One   5

Write the first five terms of a geometric sequence with first term $\frac{1}{2}$ and common ratio 4.

We can also find a formula to quickly calculate any term of a geometric sequence without having to find all of the terms that come before it. A geometric sequence with first term a_1 and common ratio r looks like

$$a_1, \quad a_1 \cdot r, \quad a_1 \cdot r \cdot r, \quad a_1 \cdot r \cdot r \cdot r, \ldots$$

Doing some simplifying, we find that

$$a_1 = a_1 \quad a_2 = a_1 \cdot r \quad a_3 = a_1 \cdot r^2 \quad a_4 = a_1 \cdot r^3$$

Every term a_n looks like a_1 times some power of r, and the power is one less than the subscript of that term in the sequence. This gives us a formula:

Finding the nth Term of a Geometric Sequence

The nth term of a geometric sequence can be found using the formula

$$a_n = a_1 \cdot r^{n-1}$$

where n is the number of the term you want to find, a_1 is the first term, and r is the common ratio.

Example 6 | Finding a Particular Term of a Geometric Sequence

Find the 12th term of a geometric sequence with first term $\frac{1}{2}$ and common ratio 6.

SOLUTION

Substitute in the formula using $a_1 = \frac{1}{2}, r = 6$ and $n = 12$.

$$a_n = a_1 r^{n-1}$$

$$a_{12} = \frac{1}{2}(6)^{12-1}$$

$$= \frac{1}{2}(6^{11}) \quad \textit{Don't forget order of operations!}$$

$$= 181,398,528$$

The 12th term is 181,398,528.

Example 6 illustrates the distinguishing feature of geometric sequences: they can grow VERY quickly.

Sidelight   **Sequences and the Planets**

In the late 1700s, the world of scientists became very excited when two German astronomers discovered a mathematical sequence that actually predicted the average distance the then-known planets were from the sun. This distance is measured in what are called astronomical units. One astronomical unit (AU) is equal to the average distance Earth is located from the sun (about 93 million miles).

The sequence, called the Titius-Bode law (named for its discoverers in 1777), is 0, 3, 6, 12, 24, 48, 96, 192, ... When 4 is added to each number and the sum is divided by 10, the result gives the approximate distance in AUs each planet is from the sun as shown.

Between Mars and Jupiter, no planet exists, and it looks like the sequence breaks down. However, an asteroid belt is located between Mars and Jupiter, and some astronomers thought that this was once a planet.

More amazing was the fact that in 1781 William Herchel discovered the planet Uranus, which is located at 19.2 AU from the sun!

Unfortunately, the next two planets discovered didn't fit the pattern at all. The predicted locations are 38 and 76.4 AU: Neptune is just 30.1 AU from the sun, and Pluto is 39.5. And now poor Pluto isn't even considered a planet anymore. So maybe the whole thing was just a big coincidence to begin with. This is a cautionary tale about thinking you've found a definite pattern from looking at a small handful of numbers. Sometimes the pattern continues, and sometimes it does not.

Planet	Sequence	AU
Mercury	$(0 + 4) \div 10$	0.4
Venus	$(3 + 4) \div 10$	0.7
Earth	$(6 + 4) \div 10$	1.0
Mars	$(12 + 4) \div 10$	1.6
_____	$(24 + 4) \div 10$	2.8
Jupiter	$(48 + 4) \div 10$	5.2
Saturn	$(96 + 4) \div 10$	10.0
_____	$(192 + 4) \div 10$	19.6

Try This One   6

 5. Find a particular term of a geometric sequence.

Find the ninth term of a geometric sequence with first term 3 and common ratio (-2).

There is also a formula for adding terms of a geometric sequence, which we'll prove in Project 3 at the end of this chapter.

Finding the Sum of a Geometric Sequence

The sum of the first n terms of a geometric sequence is $S_n = \dfrac{a_1(1 - r^n)}{1 - r}$, where a_1 is the first term and r is the common ratio.

Example 7   Finding the Sum of a Geometric Sequence

Find the sum of the first 15 terms of a geometric sequence with first term 8 and common ratio $-\frac{1}{2}$.

SOLUTION

Using $a_1 = 8$, $r = -\frac{1}{2}$, and $n = 15$, substitute into the formula

$$S_n = \frac{a_1(1 - r^n)}{1 - r}$$

$$S_{15} = \frac{8[1 - (-\frac{1}{2})^{15}]}{[1 - (-\frac{1}{2})]}$$ *A calculator is a big help here.*

$$\approx 5.33$$

The sum is about 5.33.

Try This One   **7**

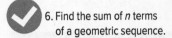
6. Find the sum of n terms of a geometric sequence.

Find a decimal approximation for the sum of the first 20 terms of a geometric sequence with first term 6 and common ratio $\frac{1}{4}$.

Applications of Sequences

Example 8   **Applying Arithmetic Sequences to Saving for a Home**

We began this section with an example in which a couple is saving money for a down payment on a house. Suppose that your plan is similar, but you decide to put away $150 the first month, $175 the second, $200 the third, and so on. Would you save more or less in the first 2 years than if you put away $400 each month?

SOLUTION

If we list the amount saved each month, we get an arithmetic sequence with first term 150 and common difference 25. To find the amount after 2 years, we need to find the sum of the first 24 terms, or S_{24}, for this sequence. But first we need to know a_{24} since it's part of the formula for S_{24}.

$$a_n = a_1 + (n - 1)d$$
$$a_{24} = 150 + (24 - 1)25 = 150 + 23 \cdot 25 = 725$$

Now we use the formula for the sum, S_n.

$$S_n = \frac{n(a_1 + a_n)}{2}$$

$$S_{24} = \frac{24(150 + 725)}{2} = 12(875) = \$10,500$$

If you had put away $400 each month, the total after 2 years would simply be $24 \times 400 = \$9,600$. The first plan saves more money in 2 years.

Try This One   **8**

If you're offered a 1-year job where you get paid just $20 the first week, but $20 is added to your pay each week, how much money would you earn in that year?

Example 9   **Applying Geometric Sequences to Salary**

When a worker earns a 4% raise each year, to find his or her salary in any given year, you multiply the salary from the year before by 1.04. If you take a job with a starting annual salary

Math Note

In calculus, we study infinite series, which is what results when you add ALL of the terms of an infinite sequence. Believe it or not, often the result is a finite number! The sum of the infinitely many terms of the sequence described in Example 7 is 16/3, for example. See Problems 81 and 82 for more on this topic.

of $40,000 and earn a 4% raise each year, how much money would you earn in your first 10 years?

SOLUTION

Since the salary gets multiplied by 1.04 each year, the annual salaries form a geometric sequence with first term 40,000 and common ratio 1.04. We can use the formula for the sum of a geometric sequence with $n = 10$:

$$S_n = \frac{a_1(1 - r^n)}{1 - r}$$

$$S_{10} = \frac{40,000(1 - 1.04^{10})}{1 - 1.04}$$

$$= \$480,244.28$$

Try This One   9

For the job in Example 9, if you're offered the option of starting at $30,000 and taking an 8% raise each year, would that earn you more or less in the first 10 years? By how much?

Answers to Try This One

1 (a) 7, 8.3, 9.6, 10.9, 12.2

(b) 6, 16, 26, 36, 46

(c) $\frac{3}{8}, -\frac{1}{8}, -\frac{5}{8}, -\frac{9}{8}, -\frac{13}{8}$

2 31.7 lb

3 (a) $\frac{19}{9}$ (b) $-\frac{4}{3}$

4 (a) 522 (b) −144 (c) $\frac{78}{5}$

5 $\frac{1}{2}, 2, 8, 32, 128$

6 768

7 About 8

8 $27,560

9 $45,647.41 less

Exercise Set   5-7

Writing Exercises

1. If you're given a list of numbers, explain how to tell if the list is an arithmetic sequence.
2. If you're given a list of numbers, explain how to tell if the list is a geometric sequence.
3. For an arbitrary sequence, explain what the symbol a_n represents.
4. For an arbitrary sequence, explain what the symbol S_n represents.
5. Give an example of a quantity in our world that might be described by an arithmetic sequence, and justify your answer.
6. Give an example of a quantity in our world that might be described by a geometric sequence, and justify your answer.
7. What's the difference between a sequence and a set of numbers?
8. A sequence is called **alternating** if its signs alternate from positive to negative. Is it possible for an arithmetic sequence to be alternating? What about a geometric sequence?

Computational Exercises

For the arithmetic sequence in Exercises 9–24, find each:

(a) the first five terms
(b) the common difference

(c) the 12th term
(d) the sum of the first 12 terms

9. $a_1 = 1, d = 6$

10. $a_1 = 10, d = 5$

11. $a_1 = -9, d = -3$

12. $a_1 = -15, d = -2$

13. $a_1 = \frac{1}{4}, d = \frac{3}{8}$

14. $a_1 = \frac{3}{7}, d = \frac{1}{7}$

15. $a_1 = 4, d = -\frac{1}{3}$

16. $a_1 = \frac{5}{2}, d = -\frac{1}{4}$

17. 5, 13, 21, 29, 37, . . .

18. 2, 12, 22, 32, 42, . . .

19. 50, 48, 46, 44, 42, . . .

20. 12, 7, 2, −3, −8, . . .

21. $\frac{1}{8}, \frac{19}{24}, \frac{35}{24}, \frac{17}{8}, \frac{67}{24}, \ldots$

22. $\frac{1}{2}, \frac{9}{10}, \frac{13}{10}, \frac{17}{10}, \frac{21}{10}, \ldots$

23. 0.6, 1.6, 2.6, 3.6, 4.6, . . .

24. 0.3, 0.7, 1.1, 1.5, 1.9, . . .

For the geometric sequence in Exercises 25–40, find each:

(a) the first five terms

(b) the common ratio

(c) the 12th term

(d) the sum of the first 12 terms

25. $a_1 = 12, r = 2$

26. $a_1 = 8, r = 3$

27. $a_1 = -5, r = \frac{1}{4}$

28. $a_1 = -9, r = \frac{2}{3}$

29. $a_1 = \frac{1}{6}, r = -6$

30. $a_1 = \frac{3}{7}, r = -3$

31. $a_1 = 100, r = -\frac{1}{4}$

32. $a_1 = 10, r = -\frac{1}{10}$

33. 4, 12, 36, 108, 324, . . .

34. 6, 12, 24, 48, 96, . . .

35. $\frac{1}{2}, \frac{1}{4}, \frac{1}{8}, \frac{1}{16}, \frac{1}{32}, \ldots$

36. $\frac{2}{3}, \frac{2}{9}, \frac{2}{27}, \frac{2}{81}, \frac{2}{243}, \ldots$

37. −3, 15, −75, 375, −1,875, . . .

38. −3, 12, −48, 192, −768, . . .

39. 1, 3, 9, 27, 81, . . .

40. 8, 2, $\frac{1}{2}, \frac{1}{8}, \frac{1}{32}, \ldots$

For Exercises 41–48, determine whether each sequence is an arithmetic sequence, a geometric sequence, or neither.

41. 5, −15, 45, −135, 405, . . .

42. 42, 35, 28, 21, 14, . . .

43. 2, 4, 8, 14, 22, . . .

44. 2, 4, 12, 48, 240, . . .

45. 6, 2, −2, −6, −10, . . .

46. $\frac{1}{10}, \frac{3}{40}, \frac{9}{160}, \frac{27}{640}, \frac{81}{2,560}, \ldots$

47. $\frac{5}{8}, \frac{1}{8}, -\frac{3}{8}, -\frac{7}{8}, -\frac{11}{8}, \ldots$

48. 4, −3, $\frac{9}{16}, -\frac{27}{64}, \frac{81}{256}, \ldots$

In Exercises 49–60, find and simplify a formula for a_n, the nth term of the given sequence. (Hint: Decide what kind of sequence it is, then use information from within the section.)

49. 3, 12, 48, 192, 768, . . .

50. −1, 2, −4, 8, −16, . . .

51. 11, 19, 27, 35, 43, . . .

52. −8, −5, −2, 1, 4, 7, . . .

53. 40, 30, 20, 10, 0, −10, . . .

54. 8, 2, −4, −10, −16, −22, . . .

55. 0.5, 0.1, 0.02, 0.004, 0.0008, . . .

56. 243, 81, 27, 9, 3, 1, . . .

57. $5, \frac{15}{4}, \frac{45}{16}, \frac{135}{64}, \frac{405}{256}, \ldots$

58. $-2, -\frac{1}{3}, -\frac{1}{18}, -\frac{1}{108}, -\frac{1}{648}, \ldots$

59. 4, 2.4, 1.44, 0.864, 0.5184, . . .

60. −2, 0.8, −0.32, 0.128, −0.0512, . . .

61. If the fifth term of a geometric sequence is 81, and the common ratio is $\frac{3}{4}$, find the sum of the first 20 terms to two decimal places.

62. If the fifth term of a geometric sequence is $\frac{64}{27}$ and the sixth term is $\frac{256}{81}$, find the sum of the first 50 terms to the nearest whole number.

63. If the tenth term of an arithmetic sequence is 10 and the common difference is 3, find the sum of the first 30 terms.

64. If the eighth term of an arithmetic sequence is $\frac{5}{2}$ and the common difference is $-\frac{3}{2}$, find the sum of the first 100 terms.

Applications in Our World

65. A new car that cost $27,000 originally depreciates in value by $3,500 in the first year, $3,000 in the second year, $2,500 in the third year, and so on.
   (a) How much value does it lose in the seventh year?
   (b) How much is the car worth at the end of the seventh year?

66. A large piece of machinery at a factory originally cost $50,000. It depreciates by $1,800 the first year, $1,750 the second year, $1,700 the third year, and so on.
   (a) What is the amount of depreciation in the fifth year?
   (b) What is the value of the machinery after the fifth year?

67. At one university, the first parking violation in the union lot results in a $5 fine. The second violation is an $8 fine, the third $11, and so on. One particular student has the fines sent home to her parents, who inform her that if she racks up over $100 in fines, they're selling her car on

eBay. If she has been given eight tickets, will she be able to keep her car?

68. A company decided to fine its workers for parking violations on its property. The first offense carries a fine of $25, the second offense is $30, the third offense is $35, and so on. What is the fine for the eighth offense?

69. A contractor is hired to build a cell phone tower, and its price depends on the height of the tower. The contractor charges $1,000 for the first 10 feet, and after that the price for each successive 10-foot section is $250 more than the previous section. What would be the cost of a 90-foot tower?

70. At one plant that produces auto parts, management had to lay off 25 workers in the first quarter of 2020, 30 workers in the second quarter, 35 in the third, and continued this pattern through 2021. If the plant originally had 700 workers, how many remained at the end of 2021?

71. A bungee jumper reaches the bottom of a jump and rebounds 80 feet upward on the first bounce. Each successive bounce is $\frac{1}{2}$ as much as the previous. If the jumper bounces a total of 10 times before coming to a stop, what is the total height of all bounces?

72. A ball that rebounds $\frac{7}{8}$ as high as it bounced on the previous bounce is dropped from a height of 8 feet. How high does it bounce on the fourth bounce and how far has it traveled after the fourth bounce? The fourth bounce is over when the fifth bounce begins.

73. A desperate gambler named Phil borrows $1,200 from Vito the loan shark. Vito informs him that he will be charging 10% interest each month, and Phil must pay off his debt in 1 year or his thumbs will be broken. How much cash does Phil have to cough up at the end of 1 year to save his thumbs?

74. A student deposited $500 in a savings account that pays 5% annual interest. At the end of 10 years, how much money will be in the savings account?

75. A contestant on a quiz show gets $1,000 for answering the first question correctly, $2,000 more for answering the second question correctly, $4,000 more for the third question, $8,000 for the fourth, and so on. The contestants can stop at any time, but if they get a question wrong, they lose it all. Xavier's goal is to make enough money to pay cash for a $225,000 house. How many questions does he have to answer correctly?

76. An eccentric business owner gives a new employee a choice: work for $4,000 per month, or get 1 cent the first day, 2 cents the second, 4 cents the third, 8 cents the

fourth, and so on. The employee, completely offended at the thought of working all day for a measly penny, chooses the $4,000 without a second thought. If there are 22 workdays in an average month, how much money per month did haste cost this employee?

77. Suppose that you're offered two jobs: both have a starting salary of $30,000 per year. The first includes a raise of $2,000 each year, and the second a raise of 4% each year. Which job will pay more in the 10th year, and by how much?

78. Your financial advisor calls, offering you two can't-miss propositions if you can scrape together an initial investment of $5,000. The first investment will increase in value by $1,000 per year, every year. The second will increase at the rate of 10% every year. Which is the better investment if you plan to keep the money invested for 12 years?

79. Rhonda and her coworkers receive a 4% raise every year. Rhonda's starting salary was $38,000 per year. One day, she sneaks a peek at the pay stub of Keiko (one of her coworkers) and finds that Keiko is in her sixth year at the company, and her current pay is $4,100 per month. Did the coworker start at a higher salary than Rhonda? If so, by how much?

80. A utility company claims that it has been cutting its emissions by 9% each year for the last 4 years. This year, the EPA takes measurements and finds that its carbon dioxide emissions are at 84% of the acceptable maximum. Assuming that the company's claim is accurate, how far above or below the acceptable maximum was it 4 years ago?

Critical Thinking

When a geometric sequence has infinitely many terms, if the common ratio is between −1 and 1, you can still find the sum of all the terms. The formula for the sum in this case is $S = \frac{a_1}{1-r}$. Use this formula for Exercises 81 and 82.

81. A repeating decimal between −1 and 1 can be written as the sum of an infinite geometric sequence as shown below.

$$0.333\ldots = \frac{3}{10} + \frac{3}{100} + \frac{3}{1,000} + \ldots$$

Find a_1 and r, then find the sum of all of the terms using the sum formula above.

82. Refer to Problem 81. Write the repeating decimal $0.151515\ldots$ as the sum of an infinite geometric sequence and find the sum using the given sum formula.

83. Consider the following situation: in a mythical post-apocalyptic world, there are just two people left to repopulate. Fortunately, one is a man and one is a woman,

and they're both of reproductive age. Do you think that a sequence describing the population is more likely to be arithmetic or geometric? Explain your answer.

84. In most cases, reducing fractions is a good idea, but not always. When trying to find a pattern in a sequence of numbers to determine what kind of sequence it is, reducing fractions is notorious for obscuring the pattern. For each sequence below, if you "unreduce" some of the fractions, you can recognize the type of sequence. Do so, then find a formula for a_n, the nth term of the sequence.

(a) $\frac{1}{4}, 1, \frac{7}{4}, \frac{5}{2}, \frac{13}{4}, 4, \ldots$

(b) $2, \frac{1}{3}, -\frac{4}{3}, -3, -\frac{14}{3}, \ldots$

(c) $\frac{25}{2}, \frac{93}{8}, \frac{86}{8}, \frac{79}{8}, \frac{72}{8}, \ldots$

Section	Important Terms	Important Ideas
5-1	Natural number Factor Divisor Divisible Multiples Prime number Prime factorization Composite number Fundamental theorem of arithmetic Greatest common factor Relatively prime Least common multiple	**The set** of natural numbers is {1, 2, 3, 4, 5, . . .}. If we can write a natural number as the product of two numbers, we call those numbers factors of the original number. A natural number is prime if its only factors are 1 and itself and composite if it has other factors. Every composite number can be factored uniquely as the product of prime numbers—we call this the prime factorization for a number. This fact is known as the fundamental theorem of arithmetic. For a set of two or more numbers, we can find the greatest number that is a factor of each (the greatest common factor, or GCF). We can also find the smallest number that is a multiple of each (the least common multiple, or LCM).
5-2	Whole number Integer Opposite Absolute value Order of operations	**The set** of whole numbers is {0, 1, 2, 3, . . .}. The set of integers is {. . . − 3, −2, −1, 0, 1, 2, 3, . . .}. Every integer has an opposite, or negative. The absolute value of a negative integer is the opposite of that integer (making it positive). The absolute value of zero or any positive integer is itself. When we perform calculations that involve more than one operation, the order in which we perform them affects the outcome, so there is an order of operations that must be followed.
5-3	Rational number Proper fraction Improper fraction Mixed number Lowest terms Reciprocal Least common denominator Place value Terminating decimal Repeating decimal	**Rational** numbers can be written either as fractions or decimals. The decimals are either terminating or repeating; the fractions have integers in the numerator and denominator. A fraction is in lowest terms when the numerator and denominator have no common factors. The rules for performing operations on rational numbers are given in this section.
5-4	Irrational number Radical Perfect square Like radicals Rationalizing the denominator	**Irrational** numbers are nonterminating nonrepeating decimals. Square roots of numbers that are not perfect squares are irrational numbers. Rules for performing operations on square roots are given in this section.
5-5	Real number Property Identity	**A real number** is either rational or irrational. We studied 11 properties of the real numbers: the closure properties for addition and multiplication, the commutative properties for addition and multiplication, the associative properties for addition and multiplication, the identity properties for addition and multiplication, the inverse properties for addition and multiplication, and the distributive property.

Section	Important Terms	Important Ideas
5-6	Exponential notation Base Exponent Scientific notation	**In order** to write very large or very small real numbers without a string of zeros, mathematicians and scientists use scientific notation. Using powers of 10, scientific notation simplifies operations such as multiplication and division of large or small numbers. When the power of 10 is positive, the number written in scientific notation is greater than 1. When the power is negative, the number is less than 1.
5-7	Sequence Arithmetic sequence Common difference Geometric sequence Common ratio	**A sequence** of numbers is a list of related numbers in a definite order. We studied two basic types of sequences: arithmetic sequences and geometric sequences. Many problems in our world can be solved using sequences. A sequence is arithmetic if every member is obtained from adding the same number to the previous member. A sequence is geometric if every member is obtained from multiplying the previous member by the same number.

Math in | Government Spending REVISITED

James P. Blair/Photodisc/Getty Images

Question 1: Divide the total spending, 6.55×10^{12}, by the population, 3.31×10^8, to get 1.9789×10^4. This means the government spent \$19,789 per person. Next divide the deficit, 3.13×10^{12}, by the population to get $\$9.456 \times 10^3$. The government spent \$9,456 per person more than it took in.

Question 2: Divide the amount spent on interest, 3.45×10^{11}, by the total spending, 6.55×10^{12}, to get 0.0527. This means that 5.27% of every dollar spent went just to pay interest on the national debt.

Question 3: Divide the national debt, 2.876×10^{13}, by the population to get 8.6888×10^4. This tells us that it would

take \$86,888 dollars from every human being in the United States to pay off the national debt.

Question 4: Divide the national debt by 535 to get 5.376×10^{10}. This means that every representative in Congress would have to contribute almost \$54 billion to pay off the national debt. Almost nobody is that well off!

Question 5: Subtract the actual deficit, 3.13×10^{12}, minus the rounded deficit, 3.1×10^{12}. The result is 0.03×10^{12}, or 3×10^{10}. You just saved the American people 30 BILLION dollars. Well done.

Review Exercises

Section 5-1

For Exercises 1–4, find all factors of each.

1. 60
2. 45
3. 380
4. 650

For Exercises 5–8, find the first five multiples of each.

5. 4
6. 32
7. 9
8. 60

For Exercises 9–12, find the prime factorization for each.

9. 96
10. 44
11. 250
12. 720

For Exercises 13–16, find the GCF and LCM.

13. 6, 10
14. 35, 40
15. 60, 80, 100
16. 27, 54, 72

17. An investor takes out two certificates of deposit. One matures every 18 months, the other every 22 months. They decide to continue rolling them over until they both mature at the same time. How long will it be until that happens?

Section 5-2

For Exercises 18–27, perform the indicated operations.

18. $-6 + 24$
19. $18 - 32$
20. $5(-9)$
21. $32 \div (-8)$
22. $6 + (-2) - (-3)$
23. $6 \cdot 8 - (-2)^2$
24. $4 \cdot 3 \div (-3) + (-2)$
25. $100 - \{[6 + (2 \cdot 3) - 5] + 4\}$

26. $\{8 \cdot 7^3 - 55[(3 + 4) - 6]\} + 20$

27. $(-5)^3 + (-7)^2 - 3^4$

28. Tyron has a separate checking account set aside for housing, utilities, and car payment. For fall semester, the account starts with $2,400. During the 5-month semester, Tyron writes checks each month for rent ($340), utilities ($45), and car payment ($170). His parents make three deposits of $350 during the semester. How much money is left at the end of the semester?

Section 5-3

For Exercises 29–31, reduce each fraction to lowest terms.

29. $\dfrac{75}{95}$

30. $\dfrac{56}{64}$

31. $\dfrac{265}{30}$

For Exercises 32–43, perform the indicated operations. Write your answer in lowest terms.

32. $\dfrac{1}{8} + \dfrac{5}{6}$

33. $\dfrac{3}{10} - \dfrac{2}{5} + \dfrac{1}{4}$

34. $\dfrac{5}{9} \times \dfrac{3}{7}$

35. $\dfrac{15}{16} \div \left(-\dfrac{21}{40}\right)$

36. $\dfrac{1}{2} \div \left(\dfrac{2}{3} + \dfrac{3}{4}\right)$

37. $\dfrac{9}{10} \times \left(\dfrac{5}{6} - \dfrac{1}{8}\right)$

38. $\dfrac{2}{3}\left(\dfrac{3}{4} + \dfrac{1}{2} - \dfrac{1}{6}\right)$

39. $1\dfrac{7}{8} - \left(\dfrac{3^2}{4}\right)$

40. $-\dfrac{6}{7}\left(\dfrac{1}{2} + 2\dfrac{1}{3}\right)$

41. $\dfrac{9}{10} + \left(\dfrac{2}{5} - \dfrac{1}{4}\right)$

42. $\dfrac{5}{8} - \dfrac{2}{3}\left(-1 + \dfrac{2}{5}\right)$

43. $\dfrac{1}{2} - \dfrac{3}{4} - \dfrac{7}{8} \cdot \dfrac{1}{6}$

For Exercises 44–45, change each fraction to a decimal.

44. $\dfrac{5}{16}$

45. $\dfrac{6}{7}$

46. Change $\frac{23}{6}$ into a mixed number, then into a decimal.

For Exercises 47–50, change each decimal to a reduced fraction.

47. 0.6875

48. 0.22

49. $0.2\overline{5}$

50. $0.4\overline{5}$

51. In the National Football League, $\frac{3}{8}$ of the teams make the playoffs; in Major League Baseball, $\frac{4}{15}$ make the playoffs, and in the National Basketball Association and the National Hockey League, $\frac{8}{15}$ do. There are 32 teams in the NFL, and 30 teams in MLB, NBA, and NHL. How many teams make the playoffs in each sport?

Section 5-4

For Exercises 52–55, simplify each.

52. $\sqrt{48}$

53. $\sqrt{112}$

54. $\dfrac{5}{\sqrt{20}}$

55. $\sqrt{\dfrac{5}{12}}$

For Exercises 56–61, perform the indicated operations.

56. $\sqrt{20} + 2\sqrt{75} - 3\sqrt{5}$

57. $\sqrt{18} - 5\sqrt{2} + 4\sqrt{72}$

58. $\sqrt{27} \cdot \sqrt{63}$

59. $\dfrac{\sqrt{20}}{\sqrt{5}}$

60. $\sqrt{6}(\sqrt{2} + \sqrt{5})$

61. $(\sqrt{14} + \sqrt{6})(\sqrt{14} - \sqrt{6})$

Section 5-5

For Exercises 62–67, classify each number as natural, whole, integer, rational, irrational, and/or real.

62. $-\frac{5}{16}$

63. 0.86

64. $0.3\overline{7}$

65. $\sqrt{15}$

66. 0

67. 16

For Exercises 68–71, state which property of the real numbers is being illustrated.

68. $8 \cdot \frac{1}{8} = 1$

69. $3 + 5 = 5 + 3$

70. $6 + 5$ is a real number

71. $2(3 + 8) = 2 \cdot 3 + 2 \cdot 8$

Section 5-6

For Exercises 72–79, evaluate each.

72. 4^5

73. $(-3)^0$

74. 3^{-4}

75. $7^2 \cdot 7^4$

76. $\frac{5^6}{5^2}$

77. $(3^4)^2$

78. $2^3 \cdot 2^{-5}$

79. $6^{22} \cdot 6^{-3}$

For Exercises 80–83, write each number in scientific notation. Round to two decimal places.

80. 3,826

81. 25,946,000,000

82. 0.00000327

83. 0.00048

For Exercises 84–87, write each number in decimal notation.

84. 5.8×10^{11}

85. 2.33×10^9

86. 6.27×10^{-4}

87. 8.8×10^{-6}

For Exercises 88–89, perform the indicated operations and write the answers in scientific notation.

88. $(3.2 \times 10^{-5})(8.9 \times 10^{-7})$

89. $\dfrac{1.8 \times 10^{-5}}{3 \times 10^2}$

90. According to numerous websites, the average American drinks the equivalent of 5.97×10^2 cans of soda per year. According to the U.S. Census Bureau, the population of the United States at the end of 2020 was about 3.31×10^8. How many cans of soda were consumed in the United States in 2020? Write your answer in both scientific and decimal notation, then write how the decimal answer would be read aloud.

91. The speed of sound in air is about 1.126×10^3 feet per second. When the volcano Krakatoa erupted on August 26, 1883, the blast could be heard 3,000 miles away. How long would it have taken for the sound to reach that far?

Section 5-7

For Exercises 92–93, write the first six terms of the arithmetic sequence. Find the ninth term and the sum of the first nine terms.

92. $a_1 = 8, d = 10$

93. $a_1 = -\frac{1}{5}, d = \frac{1}{2}$

For Exercises 94–95, write the first six terms of the geometric sequence. Find the ninth term and the sum of the first nine terms.

94. $a_1 = -3, r = 3$

95. $a_1 = -\frac{2}{5}, r = -\frac{1}{2}$

96. The population of the United States is increasing by about 2.2 million people per year. Given that the population was 331 million in 2020, find the expected population in 2030.

97. The net profit of a small company is increasing by 5% each year. If the net profit for this year is $20,000, find the projected profit for the sixth year of operation and the total amount of money the company can be expected to make in the next 6 years.

Chapter Test

For Exercises 1–6, classify each number as natural, whole, integer, rational, irrational, and/or real.

1. -27
2. 8.6
3. $\frac{5}{9}$
4. $0.6\overline{2}$
5. $\sqrt{50}$
6. $-\sqrt{25}$

For Exercises 7–8, find the GCF and LCM of each group of numbers.

7. $42, 56$
8. $150, 175, 200$

For Exercises 9–11, reduce each fraction to lowest terms.

9. $\dfrac{15}{35}$
10. $\dfrac{81}{108}$
11. $\dfrac{112}{175}$

For Exercises 12–13, simplify the radical.

12. $\sqrt{48}$
13. $\sqrt{243}$

For Exercises 14–22, perform the indicated operations.

14. $-5 \cdot (-6) + 3 \cdot 2$
15. $18 - 3^2 - 4^2 + 6 \div 3$
16. $\left(\dfrac{5}{6} \cdot \dfrac{3}{4}\right) \div \dfrac{2}{3}$
17. $\left(\dfrac{1}{7} + \dfrac{1}{9}\right) - \dfrac{2}{3} \cdot \dfrac{3}{4}$
18. $-6 + \dfrac{1}{4} \div \dfrac{2}{3} + \sqrt{81}$
19. $[4 + (2 \times 3) - 6^2] + 18$
20. $\sqrt{27} + \sqrt{3}(2\sqrt{2} - 1)$
21. $\dfrac{16}{\sqrt{32}}$
22. $2\sqrt{50} - 3\sqrt{32}$

For Exercises 23–24, change each decimal into a reduced fraction.

23. 0.875 24. $0.\overline{2}$

For Exercises 25–30, state the property illustrated.

25. $0 + 15 = 15 + 0$
26. 6×7 is a real number.
27. $0 + (-2) = -2$
28. $\frac{1}{5} \cdot 5 = 1$
29. $(4 \times 6) \times 10 = 4 \times (6 \times 10)$
30. $6(5 + 7) = 6 \cdot 5 + 6 \cdot 7$

For Exercises 31–35, evaluate each.

31. 8^4
32. 7^{-3}
33. 6^0
34. $5^{-3} \cdot 5^{-2}$
35. $\dfrac{2^{-3}}{2^2}$

36. Write 52,000,000 in scientific notation.
37. Write 0.00236 in scientific notation.
38. Write 9.77×10^3 in decimal notation.
39. Write -6×10^{-5} in decimal notation.
40. $(5.2 \times 10^8)(3 \times 10^{-5}) = $ _____ in scientific notation.
41. Divide $\dfrac{2.1 \times 10^9}{7 \times 10^5}$

42. Write the first seven terms, the 20th term, and the sum of the first 20 terms for the arithmetic sequence where $a_1 = 1$ and $d = 2.5$.
43. Write the first seven terms, the 15th term, and the sum of the first 15 terms for the geometric sequence where $a_1 = \frac{3}{4}$ and $r = -\frac{1}{6}$.
44. A runner decides to train for a marathon by increasing the distance she runs by $\frac{1}{2}$ mile each week. If she can run 15 miles now, how long will it take her to run 26 miles?
45. A gambler decides to double his bet each time he wins. If his first bet is $20 and he wins five times in a row, how much did he bet on the fifth game? Find the total amount he bet.

Projects

1. To add together the first 100 natural numbers, you can think of them as an arithmetic sequence with first term and common difference both equal to 1. In this project, we'll develop a different method.
   (a) Find the sum using the formula for the sum of an arithmetic sequence.
   (b) Write out the sum of the first 100 natural numbers. You don't have to write out all 100 numbers, but write at least the first five and the last five with an ellipsis (. . .) between.
   (c) Write the same sum underneath the first one, but in the opposite order. What property of real numbers can you use to conclude that the two sums are equal?
   (d) Add the two sums together one term at a time. You should now have a sum that's easy to compute using the fact that multiplication is repeated addition by the same number.

   (e) The sum of those two lists is twice the number we're looking for, so divide by 2. Did you get the same answer as in part (a)?
   (f) Repeat steps (b) through (e), but this time find the sum of the first n natural numbers, $1 + 2 + 3 + \cdots + n$. The result should be a formula with variable n.
   (g) Use your answer from part (f) with $n = 100$. Do you again get the same answer as in part (a)?
   (h) Now compare your answer from part (f) to the formula for the sum of an arithmetic sequence with first term and common difference both equal to 1. Do you get the same formula?

2. In the Sidelights on pages 256 and 284, we learned about the golden ratio and the Fibonacci sequence. Based on the way they're defined, there appears to be absolutely no connection between them. But appearances can be deceiving, can't they?

Recall that the Fibonacci sequence starts out 1, 1, then each term after that is obtained by adding the previous two.

(a) Write the first 12 terms of the Fibonacci sequence.

(b) Now we're going to define a new sequence R_n, with R standing for ratio. R_1 is the ratio of the second term in the Fibonacci sequence to the first: that is $\frac{1}{1}$. R_2 is the ratio of the third term to the second: $\frac{2}{1}$. In general, R_n is the ratio of the $n + 1$st term in the Fibonacci sequence to the nth term. Using your answer to (a), find the first 11 terms of the sequence R_n.

(c) The terms of the sequence R_n should be closing in on a certain number. What is the significance of that number?

(d) Now let's define a variation on the Fibonacci sequence. Rather than starting out 1, 1, start with 2, 2. Then find the other terms as usual, by adding the two previous. Write the first 12 terms of this sequence.

(e) Repeat part (b), using the sequence you wrote in part (d). Do you get the same result?

(f) Repeat parts (d) and (e), this time starting out your new Fibonacci sequence with any natural number you like (other than 1 or 2, obviously). Conclusions?

(g) What if we define a Fibonacci sequence starting with a noninteger rational number? Try it with $\frac{1}{2}, \frac{1}{2}$, then $\frac{5}{3}, \frac{5}{3}$. What do you find?

3. (a) In this part, we'll prove the formula for finding the sum of the first n terms of an arithmetic sequence:
$$S_n = \frac{n(a_1 + a_n)}{2}$$

   (i) The first term is a_1; the second is $a_1 + d$ (where d is the common difference); the third is $a_1 + 2d$; and in general, $a_n = a_1 + (n - 1)d$. Using these expressions, write the sum of a_1 through a_n.

   (ii) Use the commutative property to regroup, putting all of the a_1's next to each other. How many are there? How can you simplify that sum?

   (iii) All of the remaining terms have a d in them. Factor out the d; what remains is the sum of a

group of consecutive natural numbers. Using the Internet as a resource, find a formula for the sum of the first $(n - 1)$ natural numbers and use it to replace that sum. (This formula was developed in Project 1 if you did it, so you won't need the Internet.)

   (iv) You should now have a two-term expression, with a fraction in the second term. Get a common denominator and perform the addition.

   (v) Factor n out of the two terms in the numerator, then write $2a_1$ in the first term as $a_1 + a_1$; the sum of the second and third term can now be simplified using the formula listed above for the nth term of an arithmetic sequence, and if all went well, you've proved the formula for the sum.

(b) Now we'll prove the formula for the sum of the first n terms in a geometric sequence: $S_n = \frac{a(1 - r^n)}{1 - r}$

   (i) The first term is a_1; the second is $a_1 r$ (where r is the common ratio); the third is $a_1 r^2$; and in general, $a_n = a_1 r^{n-1}$. Using these expressions write the sum of a_1 through a_n, and call this S_n.

   (ii) Find rS_n by multiplying both sides of your equation from (i) by r. Make sure to use the distributive property on the right side!

   (iii) Subtract the equation from part (ii) from the equation from part (i). (*Hint:* Most of the stuff on the right side will subtract away to zero!)

   (iv) Solve the resulting equation for S_n using factoring and dividing, and you have proved the formula. Good job.

   (v) Extra credit: if the common ratio is a positive number less than 1, what happens to the expression r^n in the long run (as n approaches positive infinity)? What does this say about the sum of infinitely many terms in a geometric series with common ratio between zero and 1?

Design elements: Front Matter, Chapter Opener, Summary and End Matter header design (random numbers background illustration): ©pixeldreams.eu/Shutterstock RF

Topics in Algebra

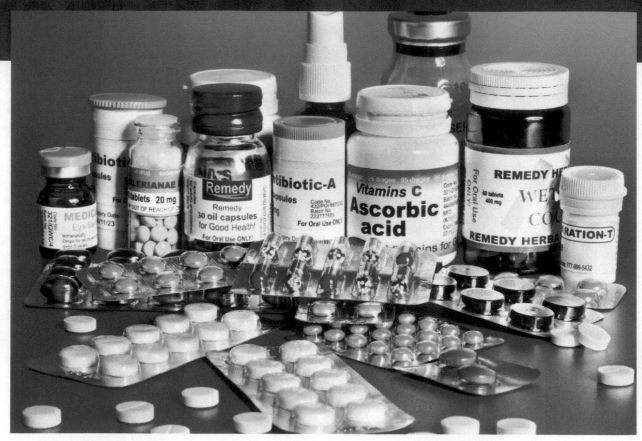

Photo: Gennadiy Poznyakov/123RF

Outline

Math in | Drug Administration

Have you ever looked at the dosage information on a bottle of aspirin and thought, "It just doesn't seem reasonable to recommend the same dosage for all adults"? People come in all shapes and sizes, and the effect of a certain dosage is in large part dependent on the size of the individual. If a 105-pound woman and her 230-pound husband both take two aspirin the morning after their wedding reception, she is in effect getting more than twice as much medicine as he is.

Because there's a lot of variation in the world, the study of algebra was created based on a brilliantly simple idea: using a symbol, rather than a number, to represent a quantity that can change. Since the word "vary" is a synonym for change, we call such a symbol a variable. The use of variables is the thing that distinguishes algebra from arithmetic and makes it extremely useful to describe phenomena in a world in which very few things stay the same for very long. An understanding of the basics of algebra is, in a very real sense, the gateway to higher mathematics and its applications. Simply put, you can only go so far with just arithmetic. To model real situations, expressions involving variables are almost always required.

In this unit, we'll review some key ideas from algebra, but one thing we won't do is find x just for the heck of it. A lot of students, maybe even most, believe that algebra was conceived as an abstract study, with early nerds sitting around simplifying expressions and solving equations on scrolls. Then a few centuries later, some genius said "Holy cow, we can use this stuff to find when two trains left San Francisco!"

Nothing could be farther from the truth. Algebra, and really all of classical math, was developed 100% as a tool for solving very real problems. So our study of algebra will focus on modeling, which is using algebra to represent situations in our world, like dosage calculations. You're accustomed to losing points when you make a mistake in a calculation, but outside the classroom, the stakes can be a LOT higher. The wrong dose of aspirin might upset your stomach, but more serious drugs carry with them more serious consequences. In many cases, an incorrect dosage could lead to death. Using the

skills you learn in the chapter, you should be able to answer the following questions.

Suppose that a new pain-killing drug is being tested for safety and effectiveness in a variety of people. The recommended dosage for an average 5′10″, 170-pound man is 400 mg. The manufacturer claims that the minimum effective dose for a person of that size is 250 mg, and that anything over 1,500 mg could be lethal. Two of the patients in the test are a 6′5″, 275-pound college football player recovering from a major injury, and a 4′2″, 65-pound girl being treated for sickle-cell disease. According to the manufacturer, dosages for this drug are proportional to body weight.

1. What would the recommended dosage be for each patient?

2. If the medications are mixed up and the recommended dosage for the football player is given to the child, is she in danger of dying?

3. Would the child's recommended dosage be at all effective for the football player?

For many medications, the effective dosage depends not just on body weight but also on body surface area (BSA), which accounts for the overall size of the patient. One model for body surface area is

$$BSA = \frac{\sqrt{1.15\, wh}}{60}$$

where w is weight in pounds and h is height in inches.

4. Find the ratio of body surface area for the football player compared to the girl. How does it compare to the ratio of their weights?

5. Find the body surface area for an average 5′10″, 170-pound man.

6. If the dosages for the medication in the drug trial are based on body surface area rather than weight, rework Questions 2 and 3 to see if the child would be in danger, or if the medication would be at all effective for the football player.

7. Write any conclusions about using body surface area rather than weight when doing dosage calculations.

For answers, see Math in Drug Administration Revisited on page 357

Applications of Linear Equations

LEARNING OBJECTIVES

1. Translate verbal expressions into mathematical symbols.
2. Solve problems using linear equations.

Two of the most common questions that students ask when learning about solving equations are "When would I ever *use* this?" and "Who cares what *x* is?" (If you want to drive your professor up the wall, ask these questions at least once a day.) Nobody will claim that you will solve multiple equations every day once you leave the hallowed halls of college. But the fact is that solving equations is a topic that is widely applied to almost every area of study. And even if it weren't, the problem-solving skills that you learn and hone while solving applied problems are among the best "brain exercise" you can get. And what could possibly be more useful to your education than training your brain to work better?

So in this section, we will solve problems that relate to issues in our world. In most cases, the plan is to write an equation that describes a situation, then solve that equation to find some quantity of interest. And even if you don't find the situations applicable to your life, keep the bigger picture in mind: you'll be practicing useful problem-solving skills with every question.

Before we get into problem solving, let's take a quick look at what makes algebra algebra. Think about this question for a second: what is a variable? If you answered something along the lines of "It's a letter, like *x*," then rest assured that (1) you're not alone, and (2) your answer is (to put it politely) not very good. (To put it impolitely, it kind of stinks.) A lot of people think of algebra in those terms, but if you do, you'll probably never *get it*.

Look at the word itself: variable. Able to vary. THAT'S what a variable is: it's a quantity that is able to vary. We typically use letters to *represent* variables, but the letter itself isn't a variable: the quantity that it represents is. For example, if you work a part-time job, the number of hours you work each week probably changes. So the number of hours is a variable, and if you wanted, you could represent that variable with *h*, *W*, a smiley face, a picture of your mom, or any symbol that isn't a numeral. I'll stick to using letters because I don't know what your mom looks like.

In the box below, we've outlined a general strategy for attacking word problems using algebra. If you look close enough, you'll notice that our strategy is based on Polya's problem-solving strategy that we studied in Chapter 1, with a few elaborations matching this specific type of problem.

A General Procedure for Solving Word Problems Using Equations

Step 1   Read the problem carefully, but *don't read it all at once without doing anything!* As you're reading, write down any information provided by the problem that seems relevant. This will at least get you started. Make sure you carefully note what it is you're being asked to find. Draw a diagram if the situation calls for one.

Step 2   Assign a variable to an unknown quantity in the problem. Most of the time, the variable should represent the quantity you're being asked to find.

Step 3   Write an equation based on the information given in the problem. Remember, an equation is a statement that two quantities are equal, so keep an eye out for statements in the problem indicating two different ways to express the same quantity.

Step 4   Solve the equation.

Step 5   Make sure that you answer the question! The best approach is to reread the question, then write your answer in sentence form.

Step 6   Check to see if your solution makes sense based on the original wording of the problem.

The step that almost everyone finds most challenging is Step 3. In order to write an equation that describes a situation, you have to translate verbal statements into mathematical symbols. For example, the verbal statement "six more than three times some number" can be written in symbols as "$6 + 3x$" or "$3x + 6$." A careful read of Table 6-1 will help get you started on these types of translations. The table is also very useful to refer back to as you work on problems.

Table 6-1	**Common Phrases That Represent Operations**

Phrases that represent addition

6 more than a number	$x + 6$
A number increased by 8	$x + 8$
5 added to a number	$x + 5$
The sum of a number and 17	$x + 17$

Phrases that represent subtraction

18 decreased by a number	$18 - x$
6.5 less than a number	$x - 6.5$
3 subtracted from a number	$x - 3$
The difference between a number and 5	$x - 5$

Phrases that represent multiplication

8 times a number	$8x$
Twice a number	$2x$
A number multiplied by 4	$4x$
The product of a number and 19	$19x$
$\frac{2}{3}$ of a number	$\frac{2}{3}x$

Phrases that represent division

A number divided by 5	$x \div 5$
35 divided by a number	$35 \div x$
The quotient of a number and 6	$x \div 6$

Example 1	**Translating Verbal Statements into Symbols**

Translate each verbal statement into symbols.

(a) 14 times a number
(b) A number divided by 7
(c) 10 more than the product of 8 and a number
(d) 3 less than 4 times a number
(e) 6 times the sum of a number and 18

Math Note

Remember, the actual letter you choose for a variable is unimportant. You can choose any letter (or other symbol) you like.

SOLUTION

(a) Using variable x to represent the unspecified number, we can write this as $14x$.
(b) $\dfrac{x}{7}$
(c) $8x + 10$
(d) It might help to reword this as 3 subtracted from 4 times a number: $4x - 3$
(e) Parentheses are required here because the multiplication is 6 times the sum: $6(x + 18)$

Try This One	1

1. Translate verbal expressions into mathematical symbols.

Translate each verbal statement into symbols.

(a) 100 divided by a number
(b) 5 more than the product of a number and 7
(c) The difference between 25 and a number
(d) The product of 8 and the difference of a number and 4

Solving Word Problems

We'll begin our study of solving word problems with a basic translation problem. It's not terribly realistic (to say the least), but it gives you a start on the basic steps for solving.

Example 2   Solving a Basic Translation Problem

If 8 times a number plus 3 is 27, find the number.

SOLUTION

Math Note

When translating a statement into an equation, the word "is" usually indicates where the equal sign should go.

Step 1 Write the relevant information:

$$8 \text{ times a number plus } 3 \text{ is } 27$$

Identify what we're asked to find: that unknown number.

Step 2 Use variable x to represent the unknown number.

Step 3 Translate the relevant information into an equation:

$$8 \text{ times a number plus } 3 \text{ is } 27$$
$$8x \qquad\qquad + \quad 3 = 27$$

Step 4 Solve the equation:

$$8x + 3 = 27 \qquad \textit{Subtract 3 from both sides.}$$
$$8x + 3 - 3 = 27 - 3 \qquad \textit{Simplify.}$$
$$8x = 24 \qquad \textit{Divide both sides by 8.}$$
$$\frac{8x}{8} = \frac{24}{8} \qquad \textit{Simplify.}$$
$$x = 3$$

Math Note

When checking your answer to a word problem, don't just substitute it into the equation you wrote—if you wrote the wrong equation, you won't know. Instead, check that it matches the verbal description of the problem.

Step 5 Answer the question: the requested number is 3.

Step 6 Check: 8 times 3 is 24, and when you add 3, you get 27. This matches the description of the problem.

Try This One   2

Ten less than twice a number is 42. Find the number.

In the next example, see if you can recognize the similarity to the abstract problem in Example 2.

Example 3   A Problem Involving Contract Negotiations

Two baseball teams are interested in signing a free-agent pitcher. An inside source informs the general manager of one team that the other has made an offer, and the player's agent said "Double that and add an extra million per year, and you're in our league." According to a published report, the player is seeking a contract of $18 million per year. What was the rival team's offer?

SOLUTION

Step 1 Relevant information: twice the offer plus 1 million is 18 million. We're asked to find the offer.

Step 2 Use the variable x to represent the offer. Since the numbers are in millions, we'll let x stand for the offer in million dollar units—that will keep the arithmetic simpler.

Mike Flippo/Shutterstock

Math Note

Remember, in Example 3 we're using million dollar units, so 1 million and 18 million are represented by 1 and 18.

Step 3 Translate the relevant information into an equation:

Twice the offer plus one is eighteen.

$$2x + 1 = 18$$

Step 4 Solve the equation:

$2x + 1 = 18$ *Subtract 1 from both sides.*

$2x + 1 - 1 = 18 - 1$ *Simplify.*

$2x = 17$ *Divide both sides by 2.*

$$x = \frac{17}{2} \text{ or } 8.5$$

Step 5 Answer the question: the team's offer was $8.5 million. Only $8.5 million? That's an insult!

Step 6 Check: doubling $8.5 million gives $17 million, and adding 1 million more makes it $18 million as required.

Try This One   3

A teacher with a fondness for joking around with students tells one that if he doubled his score on the last test and subtracted 12, he would have just barely gotten an A. The syllabus says that the minimum cutoff for A is 92. What was the student's score?

Sometimes when there are two unknowns in a word problem, one unknown can be represented in terms of the other. For example, if I know that one number is 5 more than another number, then the first number can be represented by x and the second number can be represented by $x + 5$. Example 4 uses this idea.

Example 4   An Application to Home Improvement

Ryan McVay/Photodisc/Getty Images

Pat and Ron are planning to build a deck off the back of their house, and they buy some plans from the Internet. The plans can be customized to the required deck height, which in this case will be 92 inches. They call for support posts of two different heights. The taller ones are 8 inches longer than the shorter ones, and the plans say that the sum of the lengths should be the height of the deck. How long should the support posts be cut?

SOLUTION

Step 1 Relevant information: the posts are 8 inches different in length, and the lengths should add to 92 inches.

Step 2 We'll call the length of the shorter posts x. The other posts are 8 inches longer, so they must be $x + 8$.

Step 3 Translate the relevant information into an equation:

Length of shorter post + length of longer post is 92.

$$x + x + 8 = 92$$

Step 4 Solve the equation:
$$x + x + 8 = 92$$
$$2x + 8 = 92$$
$$2x = 84$$
$$x = 42$$

Math Note

We could also call the length of the longer posts x; in that case, the shorter ones would have length $x - 8$.

Step 5 Answer the question: this is where it becomes really important to reread the original question. We were asked to find *two* lengths, so $x = 42$ isn't a valid answer. We used x to represent the length of the shorter posts, and found that it's 42 in. The longer posts are supposed to be 8 in. longer, so the two lengths are 42 in. and 50 in.

Step 6 Check: the two lengths are definitely separated by 8 inches and 42 inches + 50 inches = 92 inches, as required.

Try This One   4

The railings for the deck in Example 4 have three different lengths of board. The shortest is 10 inches less than the next shortest, which is 14 inches shorter than the longest. Their combined length is supposed to match the overall length of the deck, which in this case is 20 feet. How long should each piece be?

Example 5   An Application Involving Money

After a busy Friday evening, the tip jar at an off-campus bar is stuffed full of dollar bills and quarters. The tradition is that the bartenders split the dollars, while the barbacks split the quarters. There's $245 in the jar, with three times as many quarters as dollar bills (college students aren't known to be the best tippers in the universe). How much money goes to the bartenders, and how much to the barbacks?

Tom Grill /Royalty-Free/ Corbis/Getty Images

SOLUTION

Step 1 Relevant information: three times as many quarters as dollar bills, and the total value is $245.

Step 2 Use variable d to represent the number of dollar bills. Then $3d$ is the number of quarters (because there are three times as many).

Math Note

In this case, we could have used a variable like q to represent the number of quarters, but then the number of dollar bills would be $\frac{q}{3}$, and we'd be introducing fractions. So it's simpler to use d = the number of dollar bills.

Step 3 Translate the relevant information into an equation: the value in dollars of the quarters is the number of quarters ($3d$) times $0.25. The value of the dollar bills is the number of them (d) since, of course, each is worth a dollar.

$$\text{Total value is } \$245.$$
$$0.25(3d) + d = 245$$

Step 4 Solve the equation:

$$0.25(3d) + d = 245 \qquad \textit{Multiply.}$$
$$0.75d + d = 245 \qquad \textit{Combine like terms.}$$
$$1.75d = 245 \qquad \textit{Divide both sides by 1.75.}$$
$$d = \frac{245}{1.75} = 140$$

Step 5 Answer the question: there are 140 dollar bills, so the bartenders split $140. Three times as many quarters is 420 quarters; multiply by $0.25 to get $105 to be split by the barbacks.

Step 6 Check: 420 quarters is three times as many as 140 dollar bills, and $140 + $105 = $245. Sounds like a winner to me.

Try This One   5

Renata collected $137 in tips during a Friday evening shift waiting tables, split among one- and five-dollar bills. She got 53 more singles than fives. How many of each did she get?

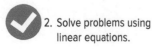

2. Solve problems using linear equations.

As you work the exercises, keep in mind that one of the main goals is to practice organized thinking and problem-solving skills. You shouldn't worry too much about how realistic or interesting you think the situations are—focus on the process of setting up and solving the equations. But if you find the situations interesting and relevant, feel free to send me some cash. I prefer hundreds.

Answers to Try This One

1 (a) $\dfrac{100}{x}$ (b) $7x + 5$ (c) $25 - x$ (d) $8(x - 4)$

2 26

3 52 (ouch!)

4 $68\dfrac{2}{3}$ in., $78\dfrac{2}{3}$ in., $92\dfrac{2}{3}$ in.

5 67 singles and 14 fives

Exercise Set 6-1

Writing Exercises

1. Describe in your own words what a variable in algebra is.
2. Write a real situation where you could represent a variable quantity with a symbol.
3. Write some reasons why it's a bad idea to read an entire word problem without writing down any information.
4. Write some reasons why it's a great idea to write your answer to a word problem in the form of a sentence.
5. What's the difference between checking your answer when solving an equation, and checking your answer when solving a word problem?
6. How do you choose what the variable in a word problem should represent?

Computational Exercises

For Exercises 7–26, write each phrase in symbols.

7. 3 less than a number
8. A number decreased by 17
9. A number increased by 9
10. 6 increased by a number
11. 11 decreased by a number
12. 8 more than a number
13. 6 subtracted from a number
14. 7 times a number
15. One-half a number added to that number
16. 5 more than 3 times a number
17. The quotient of 3 times a number and 6
18. 4 less than 6 times a number
19. The quotient of a number and 14
20. The product of 7 and a number, all subtracted from 10.
21. Triple the sum of a number and pi.
22. One-fourth the difference of 12 and a number.
23. Three times a number subtracted from the quotient of twice that number and the sum of that number and 8.
24. The sum of 146 and the product of a number raised to the third power and 18.
25. The square of the sum formed from adding five times a number to three times a different number.
26. The square root of the difference between half a number and two-thirds of a different number.

For Exercises 27–36, solve each.

27. Six times a certain number plus the number is equal to 56. Find the number.
28. The sum of a number and the number plus 2 is equal to 20. Find the number.
29. Twice a number is 32 less than 4 times the number. Find the number.
30. The larger of two numbers is 10 more than the smaller number. The sum of the numbers is 42. Find the numbers.
31. The difference of two numbers is 6. The sum of the numbers is 28. Find the numbers.
32. Five times a number is equal to the number increased by 12. Find the number.
33. Twice a number is 24 less than 4 times the number. Find the number.
34. The difference between one-half a number and the number is 8. Find the number.
35. Twelve more than a number is divided by 2. The result is 20. Find the number.
36. Eighteen less than a number is tripled, and the result is 10 more than the number. Find the number.

Applications in Our World

37. A math class containing 57 students was divided into two sections. One section has three more students than the other. How many students were in each section?
38. In 2020 the Coca-Cola company and PepsiCo had combined revenues of $103.4 billion, with PepsiCo revenues $37.4 billion higher. Find the revenue for each company.

39. The cost, including sales tax, of a Ford Focus SE is $19,392.70. If the sales tax is 6%, find the cost of the car before the tax was added.

40. During the first day of a heat wave, an emergency room had three times as many patients as the day before. There were 48 patients seen total over the 2-day period. How many patients visited the E.R. on each day?

41. Three students that share a townhouse find that their electric bill for October is $2.32 less than the September bill. The total of both bills is $119.48, and each bill is split evenly among the roommates. How much did each owe in September?

42. If Marita invested half of an insurance settlement at 8% and half at 6% and received $210 simple interest, find the total amount of money invested.

43. A basketball team played 32 games and won 4 more games than it lost. Find the number of games the team won.

44. The enrollment of students in evening classes at a local university decreased by 6% between the years 2020 and 2021. If the total number of students attending evening classes in both years was 16,983, find how many students enrolled in evening classes in each of those years.

45. If a television set is marked $\frac{1}{3}$ off and sells for $180, what was the original price?

46. A carpenter wanted to cut a 6-foot board into three pieces so that each piece is 6 inches longer than the preceding one. Find the length of each piece.

47. The Halloween Association reported that last year, Americans spent $0.68 billion more on candy than on costumes for Halloween. If the total spent by Americans for both items was $3.18 billion, how much did Americans spend on each item?

48. In a charity triathlon, Mark ran half the distance and swam a quarter of the distance. When taking a quick break to get a drink of Gatorade, Mark was just starting to bike the remaining 15 miles. What was the total distance of the race?

49. A nurse is told to give a patient recovering from surgery a total of 21 units of a potent antibiotic over 3 days. The dosage should be cut in half the second day, then in half again for the third day. How many units should be administered on the first day?

50. An investor flips a house, selling it for $82,000. If the profit was 20%, what price did the investor pay for the house?

51. A father left $\frac{1}{2}$ of his estate to his son, $\frac{1}{3}$ of his estate to his granddaughter, and the remaining $6,000 to charity. What was his total estate?

52. In 2021 there were 167 female officials in Congress, and there were 119 more female members of the House of Representatives than female senators. Find the number of females in each house of Congress.

53. While shopping on BlueFly.com, Amina notices a special where if she buys two items, the third will be half off. She buys one item, then another item that is half of that amount, and then a third item that is a quarter of the original item amount. The discount she is given is half off the cheapest item. She ends up spending $65 on the order (neglect taxes). What is the price of the first item she bought?

54. There were five winning lottery tickets for a total jackpot of $24 million. Three of the winners won twice as much as the other two. How much did each of the two that won the least get?

55. In Mary's purse, there are $3.15 worth of nickels and dimes. There are 5 times as many nickels as dimes. The vending machine is only taking dimes, and Mary needs 10 dimes for a purchase. Does Mary have enough dimes?

56. If the perimeter of a triangular flower bed is 15 feet with two sides the same length and the third side 3 feet longer, what are the measures of the three sides of the flower bed?

57. Last semester, Marcus's tuition bill was 12% cheaper than this semester's tuition bill of $640. How much did Marcus pay for tuition last semester?

58. Kalini and her two friends will rent an apartment for $875 a month, but Kalini will pay double what each friend does because she will have her own bedroom. How much will Kalini pay a month?

59. In 2021 NBA all-star voting, 16,970,693 votes were cast for the top three vote-getters. LeBron James got 355,448 more votes than Kevin Durant, who got 86,073 votes more than Steph Curry. What percentage of the ballots cast for the top three vote-getters did each receive?

60. Three sisters inherited $100,000 from a rich uncle. The uncle's favorite niece got twice as much as his second favorite niece and the second favorite niece got twice as much as the least favorite niece. How much did the favorite niece get?

61. A bounty hunter makes a base monthly salary of $1,700, plus $900 for every bail-jumper brought in. Write an algebraic expression that describes total monthly earnings, using a variable that stands for the number of fugitives captured. Then use your equation to find how many fugitives they would need to average per month in order to make $60,000 per year.

62. At one telemarketing firm, workers are paid $3.70 per hour, plus $0.30 for every caller they keep on the line for at least a minute. Write an algebraic expression describing hourly earnings, using a variable that stands for the number of callers kept on the line for at least a minute. Then use your equation to find how many callers they'd have to keep on the line for at least a minute over an 8-hour shift to make $104.

Critical Thinking

63. The temperature and the wind combine to cause body surfaces to lose heat. Meteorologists call this effect "the windchill factor." For example, if the actual temperature outside is 10°F and the wind speed is 20 miles per hour, it will feel like it is −20° F outside, so −20° F is called the windchill. When it is 25°F outside and the wind speed is 40 miles per hour, it will feel like it is −35° F outside. From the information given, write a linear equation for determining the windchill using the actual temperature and the wind speed. Then use your equation to find the windchill factor on a day with temperature 30°F and 18 mile per hour winds.

64. Suppose your roommate brags about making $250 in singles and fives one night waiting tables, with four times as many one-dollar bills as five-dollar bills. How can you tell that they're not telling the exact truth?

65. The prevailing winds for air travel in the United States typically blow from west to east, slowing down travel to the west considerably. Last year, I flew from Cincinnati to Salt Lake City, a distance of 1,447 miles from east to west. The return flight took three-fourths as long as the flight out west. If the average speed of the plane in still air is 422 mph, what was the wind speed? (Assume it was the same for both flights.)

(*Hint:* Use a variable to represent what you're asked to find, and fill in the following chart. The information will help you to write an equation.)

	Distance	Speed	Time
Cincinnati → Salt Lake			
Salt Lake → Cincinnati			

66. In an attempt to conserve energy (and, let's be honest, save some cash), Bob decides to ride his bike to work every day. He starts out by riding a half mile uphill, which slows him down by 4 miles per hour. The rest of the ride is a mile downhill, which speeds him up by 5 miles per hour. After several days, he notices that when he reaches the top of the hill, he's exactly halfway there if you measure in terms of time. How fast would Bob be riding if the trip were on level ground? (*Hint:* Make a table similar to the one in Problem 65.)

67. Winona has two jobs: one pays $11.25 per hour, and the other pays $9.50 per hour plus an average of 60% of the hourly pay in tips and bonuses. Each pay period, 22% of her total pay goes to taxes.
    (a) Write and simplify an equation that describes Winona's total take-home pay in terms of the number of hours worked at the first job and the number of hours worked at the second.
    (b) If Winona is committed to work 30 hours per week at the first job, how many hours per week would she need to work at the second job if she needs her biweekly take-home pay to be $800?

68. Refer to Problem 67. If Winona decides that she'll work exactly 45 hours per week total and is not committed to a certain number of hours at either job, how many hours should she work at each job to earn $1,000 in take-home pay every two weeks? What about $1,200?

Section 6-2	**Ratio, Proportion, and Variation**

 LEARNING OBJECTIVES

1. Write ratios in fraction form.
2. Solve proportions.
3. Solve problems using proportions.
4. Solve problems using direct variation.
5. Solve problems using inverse variation.

The most obvious way to compare the sizes of two numbers is to subtract them. But is that the *best* way? Suppose you're comparing the cost of an item at two different stores, and you find that the item is a dollar more at Target than at Wal-Mart. If that item is a bottle of Coke, and the prices are $1 and $2, that dollar difference is significant. But if the item is a 55-inch curved-screen TV and the prices are $1,201 and $1,200, would you really care? It's essentially the same price. If you *divide* the prices rather than subtracting them, however, something interesting happens:

$$\text{Coke: } \frac{\$2}{\$1} = 2$$

$$\text{TV: } \frac{\$1,201}{\$1,200} = 1.0008$$

Do you see the point? For the Coke, a dollar more is twice as much—kind of a big deal. For the TV, when you divide the prices, you essentially get 1, meaning the two prices are pretty much the same. The most meaningful way to compare the sizes of two numbers is to divide them, forming what we called a *ratio* in Section 5-3.

Ratios

A **ratio** is a comparison of two quantities using division.

For example, in 2021 about 52% of all smartphones in the United States were iPhones, and about 48% were some other model, so we would say that the ratio of iPhones to other smartphones was 52 to 48.

For two nonzero numbers, a and b, the **ratio of a to b** is written as $a{:}b$ (read a to b) or $\frac{a}{b}$.

Ratios can be written using either a colon or a fraction as shown in the definition, but in math we'll typically use the fraction so that we can do arithmetic with ratios.

Example 1   Writing Ratios

Math Note

To set up a correct ratio, whatever number comes first in the ratio statement should be placed in the numerator of the fraction and whatever number comes second in the ratio statement should be placed in the denominator of the fraction.

According to the Sporting Goods Manufacturers' Association, 95.1 million Americans participate in recreational swimming, 56.2 million Americans participate in recreational biking, 52.6 million Americans participate in bowling, and 44.5 million Americans participate in freshwater fishing. Find each:

(a) The ratio of recreational swimmers to recreational bikers
(b) The ratio of people who fish to people who bowl

SOLUTION

(a) $\dfrac{\text{Number of swimmers}}{\text{Number of bikers}} = \dfrac{95.1}{56.2}$ *We omitted millions from each number because they would divide out.*

(b) $\dfrac{\text{Number of people who fish}}{\text{Number of people who bowl}} = \dfrac{44.5}{52.6}$

Try This One   1

From 1969 through 1977, there were 24 teams in Major League Baseball, and only 4 made the playoffs each year. Now, there are 30 teams, and 10 make the playoffs. Find the ratio of teams making the playoffs to those not making the playoffs in 1969 and today.

Since ratios can be expressed as fractions, they can be simplified by reducing the fraction. For example, the ratio of 10 to 15 is written 10:15 or $\frac{10}{15}$, and the fraction $\frac{10}{15}$ can be reduced to $\frac{2}{3}$. So the ratio 10:15 is the same as 2:3.

Some of the ratios we'll be interested in compare different measurements, like lengths, weights, and others. In that case, we have to make sure that the units match, or we'll get some really deceiving comparisons. A ratio is supposed to compare the sizes of two quantities, but you can't always get the story from just numbers.

Example 2   Writing a Ratio Involving Units

Suppose that we want to compare the lengths of a 2-foot board, and one that is 18 inches long. First, explain why the ratio below is deceiving. Then write a ratio that provides an accurate comparison.

$$\frac{2}{18}$$

Math Note

We could have written 18 inches as $\frac{3}{2}$ feet in Example 2, and the ratio would have been $\frac{2 \text{ feet}}{3/2 \text{ feet}}$, which also simplifies to $\frac{4}{3}$.

SOLUTION

This is deceiving because 18 is 9 times as big as 2, so this looks like the 18-inch board is a lot longer. Of course, it isn't: 18 inches is less than 2 feet. To make the comparison accurate, we need to rewrite so that both measurements have the same units. We could convert both to feet, or both to inches: either is fine. But it's easier to convert 2 feet to 24 inches, giving us the ratio

$$\frac{24}{18} = \frac{4}{3}$$

So the ratio of 2 feet to 18 inches is $\frac{4}{3}$.

Try This One   2

1. Write ratios in fraction form.

Find the ratio of 40 ounces to 2 pounds. (There are 16 ounces in 1 pound.)

Proportions

When two ratios are equal, they can be written as a *proportion*.

> A **proportion** is an equation in which two ratios are stated to be equal.

For example, the ratios 4:7 and 8:14 are equal; this fact can be expressed as a proportion:

$$\frac{4}{7} = \frac{8}{14}$$

Two fractions, $\frac{a}{b}$ and $\frac{c}{d}$, are equal if $ad = bc$. (This will be shown in Exercise 59.) The product of the numerator of one fraction and the denominator of the other fraction is called a cross product. For example, $\frac{3}{4} = \frac{6}{8}$ since $3 \cdot 8 = 4 \cdot 6$, or $24 = 24$.

This tells us that two ratios form a proportion if the cross products of their numerators and denominators are equal. For example, the two ratios $\frac{5}{6}$ and $\frac{15}{18}$ can be written as a proportion since

$$\frac{5}{6} \bowtie \frac{15}{18} \qquad \textit{This is called cross multiplying.}$$

$$5 \cdot 18 = 6 \cdot 15$$
$$90 = 90$$

So we can write 5:6 = 15:18, or $\frac{5}{6} = \frac{15}{18}$.

Example 3   Deciding if a Proportion Is True

Math Note

Cross multiplying is a really convenient procedure for working with proportions, but there's a catch: it ONLY works when the equation in question looks like single fraction = single fraction. Don't try to apply cross multiplying to any equation that's not of that form.

Decide if each proportion is true or false.

(a) $\dfrac{3}{5} = \dfrac{9}{15}$ (b) $\dfrac{5}{3} = \dfrac{7}{2}$ (c) $\dfrac{14}{16} = \dfrac{7}{8}$

SOLUTION

In each case, we will cross multiply and see if the two products are equal.

(a) $3 \cdot 15 = 45$; $5 \cdot 9 = 45$ The proportion is true.
(b) $5 \cdot 2 = 10$; $3 \cdot 7 = 21$ The proportion is false.
(c) $14 \cdot 8 = 112$; $16 \cdot 7 = 112$ The proportion is true.

Try This One   3

Decide if each proportion is true or false.

(a) $\dfrac{2}{9} = \dfrac{6}{25}$ (b) $\dfrac{5}{2} = \dfrac{25}{4}$ (c) $\dfrac{11}{2} = \dfrac{55}{10}$

If there is an unknown value in a proportion, we can solve for the unknown value by cross multiplying as shown in Examples 4 and 5.

Example 4   Solving a Proportion

Solve the proportion for x.

$$\frac{12}{48} = \frac{3}{x}$$

SOLUTION

We begin by cross multiplying.

$$\frac{12}{48} \bowtie \frac{3}{x} \qquad \text{Cross multiply.}$$

$$12x = 3 \cdot 48$$
$$12x = 144 \qquad \text{Divide both sides by 12.}$$
$$\frac{12x}{12} = \frac{144}{12} \qquad \text{Simplify.}$$
$$x = 12$$

Check: $\dfrac{12}{48} \overset{?}{=} \dfrac{3}{12}$

$\dfrac{1}{4} \overset{?}{=} \dfrac{1}{4} \checkmark$

Arterra Picture Library/Alamy
Stock Photo

The height of the people and the statue are in proportion. If we know the ratio of the heights and the height of the people, we can find the height of the statue.

Try This One   4

Solve the proportion: $\dfrac{x}{7} = \dfrac{22}{25}$

Example 5   Solving a Proportion

Solve the proportion. $\dfrac{x-5}{10} = \dfrac{x+2}{20}$

SOLUTION

$$\frac{x-5}{10} \bowtie \frac{x+2}{20} \qquad \text{Cross multiply.}$$

$$20(x-5) = 10(x+2) \qquad \text{Multiply out parentheses.}$$
$$20x - 100 = 10x + 20 \qquad \text{Subtract 10x from both sides.}$$
$$10x - 100 = 20 \qquad \text{Add 100 to both sides.}$$
$$10x = 120 \qquad \text{Divide both sides by 10.}$$
$$x = 12$$

Check: $\dfrac{x-5}{10} = \dfrac{x+2}{20}$

$$\dfrac{12-5}{10} \overset{?}{=} \dfrac{12+2}{20}$$

$$\dfrac{7}{10} \overset{?}{=} \dfrac{14}{20}$$

$$\dfrac{7}{10} = \dfrac{7}{10} \checkmark$$

Try This One   5

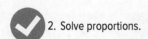 2. Solve proportions.

Solve the proportion: $\dfrac{x+6}{15} = \dfrac{x-2}{5}$

Applications of Proportions

Proportions have been around for a really long time in one form or another—a written record of their use goes back at least to 400 BCE or so, but most math historians feel that the idea is almost as old as formal numeric thought. This is because they're very useful in solving problems in our world. Example 6 illustrates a procedure that works well for problems where ratios are provided in some way.

| **Example 6** | **Applying Proportions to Fuel Consumption** |

JG Photography/Alamy Stock Photo

While on a spring break trip, a group of friends burns 12 gallons of gas in the first 228 miles, then stops to refuel. If they have 380 miles yet to drive, and the SUV has a 21-gallon tank, can they make it without refueling again?

SOLUTION

Step 1 *Identify the ratio statement.* The ratio the problem gives us is 12 gallons of gas to drive 228 miles.

Step 2 *Write the ratio as a fraction.* The ratio is $\dfrac{12 \text{ gallons}}{228 \text{ miles}}$.

Step 3 *Set up the proportion.* We need to find the number of gallons of gas needed to drive 380 miles, so we'll call that x. The ratio we already have is gallons compared to miles, so the second ratio in our proportion should be as well. We have x gallons, and 380 miles, so the proportion is

$$\dfrac{12 \text{ gallons}}{228 \text{ miles}} = \dfrac{x \text{ gallons}}{380 \text{ miles}}$$

Step 4 Solve the proportion.

Math Note
When setting up a proportion, be sure to put like quantities in the numerators and like quantities in the denominators. In Example 6, gallons were placed in the numerators and miles in the denominators.

$$\dfrac{12}{228} = \dfrac{x}{380} \qquad \textit{Cross multiply.}$$

$$\dfrac{12}{228} \diagdown\!\!\!\!\diagup \dfrac{x}{380}$$

$$228x = 12 \cdot 380 \qquad \textit{Simplify.}$$
$$228x = 4{,}560 \qquad \textit{Divide both sides by 228.}$$
$$\dfrac{228x}{228} = \dfrac{4{,}560}{228}$$
$$x = 20$$

Step 5 *Answer the question.* The SUV will burn 20 gallons of gas to cover the last 380 miles, so they can make it without stopping.

> ## Try This One 6
>
> In 2021 roughly 13 of every 100 people in the United States were African-American. A marketing company wants to select a group of 250 people that accurately reflects the racial makeup of the country. How many African-Americans should be included?

In order to decide that a certain species is endangered, biologists have to know how many individuals are in a population. But how do they do that? It's actually an interesting application of proportions illustrated in Example 7.

| **Example 7** | **Applying Proportions to Wildlife Population** |

Photo by Scott Bauer, USDA-ARS

As part of a research project, a biology class plans to estimate the number of fish living in a lake thought to be polluted. They catch a sample of 35 fish, tag them, and release them back into the lake. A week later, they catch 80 fish and find that 5 of them are tagged. About how many fish live in the lake?

SOLUTION

Step 1 *Identify the ratio statement.* Five of 80 fish caught were tagged.

Step 2 *Write the ratio as a fraction.* $\dfrac{5 \text{ tagged}}{80 \text{ total}}$

Step 3 *Set up the proportion.* We want to know the number of fish in the lake, so call that x. The comparison in the lake overall is $\dfrac{35 \text{ tagged}}{x \text{ total}}$, so the proportion is

$$\frac{5 \text{ tagged}}{80 \text{ total}} = \frac{35 \text{ tagged}}{x \text{ total}}$$

Step 4 *Solve the proportion.*

$$\frac{5}{80} = \frac{35}{x}$$
$$5x = 35 \cdot 80$$
$$5x = 2{,}800$$
$$x = \frac{2{,}800}{5} = 560$$

Step 5 *Answer the question.* There are about 560 fish in the lake.

> ## Try This One 7
>
> The staff biologists at a wildlife preserve tagged 74 protected woodpeckers shortly after hatching season. In midsummer, 30 of the birds were caught and 16 were tagged. Estimate the total population of these woodpeckers in the preserve.

✓ 3. Solve problems using proportions

Variation

Two quantities are often related in such a way that if one goes up, the other does too, and if one goes down, the other goes down as well. For example, if you have a job that pays $95 a day, the amount you make goes up or down depending on how many days you work. This is an example of what is called **direct variation**. In this case, we can write a ratio statement based on the pay: $\frac{\$95}{1 \text{ day}}$. We could then use this to write an equation that describes your total pay depending on how many days you work: $y = \frac{\$95}{1 \text{ day}} \cdot x$ days, or just $y = 95x$.

Sidelight   Proportions in My World

Proportions are a topic that it's easy to write application problems about because they're useful in a wide variety of everyday settings. In fact, I just used a proportion yesterday in my car. I have one of those cool touch screen systems with GPS, entertainment options and so forth, and you can

David Maxwell/Stringer

upload a picture to be the background for the home screen. The catch is that the screen resolution is 800 × 384 pixels, and to make a picture display right, it should be cropped to make it fit.

But it doesn't have to be exactly those dimensions—it just needs to have the same ratio of width to length. So the picture I wanted to use had a width of 1,164 pixels, and I had to figure out the height it should be cropped to. How did I do it? Using a proportion. The ratio of width to height needed to be the same as it is for the 800 × 384 screen, so I got

$$\frac{800}{384} = \frac{1,164}{x}$$

The solution, 559 (rounded), told me the number of pixels to crop to, and I think the result looks pretty darn good. Yay proportions.

A quantity y is said to **vary directly** with x if there is some nonzero constant k so that $y = kx$. The constant k is called the **constant of proportionality**.

| **Example 8** | Using Direct Variation to Find Wages |

REB Images/Blend Images LLC

Suppose you earn $95 per day. Write a variation equation that describes total pay in terms of days worked, and use it to find your total pay if you work 6 days and if you work 15 days.

SOLUTION

Let y = the total amount earned
      x = the number of days you work
      k = $95 per day (as we saw above)
Then $y = 95x$ is the variation equation.
For $x = 6$ days: $y = 95 \cdot 6 = \$570$
For $x = 15$ days: $y = 95 \cdot 15 = \$1,425$

Try This One   8

A 6-month-old Labrador puppy gets $4\frac{1}{2}$ cups of food per day. Write a variation equation that describes how many cups of food she eats in terms of days, then use it to find how much she eats in 6 days and in 2 weeks.

When two quantities vary directly, if we know the size of each for some specific case, we can use that information to find the constant of proportionality. That gives us the equation of variation, which we can then use to solve problems.

| **Example 9** | Using Direct Variation to Find a Weight |

When utility cables are strung above ground, the weight is an important consideration. The weight of a certain type of cable varies directly with its length. If 20 feet of cable weighs 4 pounds, find k and determine the weight of 75 feet of cable.

SOLUTION

Step 1 Write the equation of variation.

$$y = kx \quad \text{where } y = \text{the weight}$$
$$x = \text{length of cable in feet}$$
$$k = \text{the constant}$$

Step 2 Find k.

$$4\text{ lb} = k \cdot 20\text{ ft} \quad \textit{Substitute } y = 4 \textit{ and } x = 20$$
$$4\text{ lb} = k \cdot 20\text{ ft} \quad \textit{Divide both sides by 20 ft.}$$
$$\frac{4\text{ lb}}{20\text{ ft}} = \frac{k \cdot \cancel{20\text{ ft}}}{\cancel{20\text{ ft}}}$$
$$k = 0.2\text{ lb/ft}$$

Now we know that the equation of variation can be written as $y = 0.2x$.

Step 3 Solve the problem for the new values of x and y using $k = 0.2$.

$$y = 0.2x \qquad \textit{Substitute } x = 75$$
$$y = 0.2 \cdot 75$$
$$y = 15\text{ pounds}$$

So 75 feet of cable will weigh 15 pounds.

Try This One   9

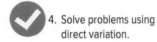
4. Solve problems using direct variation.

The weight (in pounds) of a hollow statue varies directly with the square of its height (in feet); i.e., $y = kx^2$, where $y =$ the weight and $x =$ the height. If a statue that's $4\frac{1}{2}$ feet tall weighs 12 pounds, find the weight of a statue that's 10 feet tall.

Sometimes quantities that are connected vary so that if one goes up, the other goes down. Think about driving a certain distance. If your speed goes up, the amount of time it takes goes down, right? If you average 60 miles per hour, you'll get where you're going in half the time it would take if you average 30 miles per hour. This is an example of **inverse variation**.

A quantity y is said to **vary inversely** with x if there is some nonzero constant k such that $y = \frac{k}{x}$.

Example 10   Using Inverse Variation to Find Driving Time

Rob Melnychuk/Photodisc/Getty Images

The time it takes to drive a certain distance varies inversely with the speed, and the constant of proportionality is the distance. A family has a vacation cabin that is 378 miles from their residence. Write a variation equation describing driving time in terms of speed. Then use it to find the time it takes to drive that distance if they take the freeway and average 60 miles per hour, and if they take the scenic route and average 35 miles per hour.

SOLUTION

Let $y =$ the time it takes to drive the distance
$\quad x =$ the average speed
$\quad k = 378$ miles (the distance)

Then the variation equation is $y = \dfrac{k}{x}$ or $y = \dfrac{378}{x}$.

If they average 60 miles per hour:

$$y = \frac{378}{60} = 6.3\text{ hours}$$

If they average 35 miles per hour:

$$y = \frac{378}{35} = 10.8 \text{ hours}$$

I vote for the freeway.

Try This One   10

A student attends a college 120 miles from home. Write a variation equation describing the time it takes to drive home for a weekend in terms of speed. Then use it to find driving time if the student averages 72 miles per hour and manages to not get pulled over by the state patrol.

Example 11 | Applying Inverse Variation to Construction

In construction, the strength of a support beam varies inversely with the cube of its length. If a 12-foot beam can support 1,800 pounds, how many pounds can a 15-foot beam support?

SOLUTION

Step 1 *Write the variation equation.* Let

y = strength of the beam in pounds it can support
x = length of the beam
k = the constant of proportionality

The variation equation is $y = \frac{k}{x^3}$, since y varies inversely with the cube of x.

Step 2 Find k.

$$y = \frac{k}{x^3} \qquad \textit{Substitute } y = 1,800 \textit{ and } x = 12$$

$$1,800 = \frac{k}{12^3} \qquad \textit{Solve for } k.$$

$$1,800 = \frac{k}{1,728}$$

$$k = 1,800 \cdot 1,728 = 3,110,400$$

Step 3 *Substitute in the given value for x.* In this case, the given length is 15.

$$y = \frac{3,110,400}{x^3}$$

$$y = \frac{3,110,400}{15^3} = 921.6$$

A 15-foot beam can support 921.6 pounds.

Try This One   11

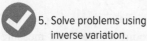

5. Solve problems using inverse variation.

If the temperature of a gas is held constant, the pressure the gas exerts on a container varies inversely with its volume. If a gas has a volume of 38 cubic inches and exerts a pressure of 8 pounds per square inch, find the volume when the pressure is 64 pounds per square inch.

Answers to Try This One

1 In 1969: $\dfrac{4}{20}$; Today: $\dfrac{10}{20}$

2 $\dfrac{5}{4}$

3 (a) False   (b) False   (c) True

4 $x = \dfrac{154}{25}$

5 $x = 6$

6 Either 32 or 33

7 139 woodpeckers

8 $y = \dfrac{9}{2}x$; In 6 days: 27 cups; In 2 weeks: 63 cups

9 $59\dfrac{7}{27}$ lb

10 $y = \dfrac{120}{x}$; about 1.7 hours

11 4.75 cubic inches

Exercise Set   6-2

Writing Exercises

1. Write an example of a ratio in an applied situation.
2. Why are units important when writing ratios?
3. What's the difference between a ratio and a proportion?
4. How are proportions used in solving problems?
5. Write an example of two quantities that vary directly in an applied situation. No using the examples in this section!
6. Write an example of two quantities that vary inversely in an applied situation. No using the examples in this section!
7. Describe the procedure for solving a proportion.
8. What is the constant of proportionality in a variation problem?

Computational Exercises

For Exercises 9–18, write each ratio statement as a fraction and reduce to lowest terms if possible.

9. 18 to 28
10. 5 to 12
11. 14:32
12. 40:75
13. 12 cents to 15 cents
14. 18 inches to 42 inches
15. 3 weeks to 8 days
16. 2 pounds to 12 ounces
17. 5 feet to 30 inches
18. 12 years to 2 decades

For Exercises 19–28, solve each proportion.

19. $\dfrac{3}{x} = \dfrac{14}{45}$

20. $\dfrac{x}{2} = \dfrac{18}{6}$

21. $\dfrac{5}{6} = \dfrac{x}{42}$

22. $\dfrac{9}{8} = \dfrac{45}{x}$

23. $\dfrac{x-6}{12} = \dfrac{1}{3}$

24. $\dfrac{x+3}{5} = \dfrac{35}{25}$

25. $\dfrac{2}{x-3} = \dfrac{5}{x+8}$

26. $\dfrac{4}{x-3} = \dfrac{16}{x-2}$

27. $\dfrac{x-3}{4} = \dfrac{x+6}{20}$

28. $\dfrac{x}{10} = \dfrac{x-2}{20}$

Applications in Our World

29. The Information Resources Institute reports that one out of every five people who buy ice cream buys vanilla ice cream. If a store sells 75 ice cream cones in one day, about how many will be vanilla?
30. The U.S. Department of Agriculture reported that 57 out of every 100 milk drinkers drink skim milk. If a storeowner orders 25 gallons of milk, how many should be skim?
31. Under normal conditions, 1.5 feet of snow will melt into 2 inches of water. After a monster snowstorm, there were 3.5 feet of snow. How many inches of water will there be when the snow melts?
32. The Travel Industry Association of America reports that 4 out of every 35 people who travel do so by air. If there are 180 students who are traveling for spring break, how many of them will fly?
33. A gallon of paint will cover 640 square feet of wall space. If I plan to paint a room whose walls measure 2,560 square feet, how many gallons of paint will I need?

34. The American Dietetic Association reported that 31 out of every 100 people want to lose weight. If 384 students were surveyed at random in the student union, how many would want to lose weight?

35. The U.S. Census Bureau reported that 9 out of every 20 joggers are female. On a trail, there were 230 joggers on July 4. Approximately how many were female?

36. Angel took a 2-year lease on a new car and after 8 months of driving, he'd put 6,600 miles on it. The lease allows 10,000 miles per year. If his driving habits stay consistent, will he stay under the allotted mileage for 2 years? By how much?

37. Out of every 80 phone chargers sold by a discount website, 3 were returned as defective. If the website sold 1,000 phone chargers this holiday season, how many should they expect will be returned as defective?

38. The American Dietetic Association states that 11 out of every 25 people do not eat breakfast. If there are 175 students in a large lecture hall, about how many of them did not eat breakfast?

39. Malik is interested in measuring the height of a tree in his yard. He measures the length of the tree's shadow at 18 feet at the same time his own shadow is 3 feet 6 inches. If Mark is 5′10″, how tall is the tree?

40. An online photo printing service will put any photo you like on a commemorative baseball. The preferred dimensions for photo uploads are 640 × 480 pixels. Quan wants to have her son's Little League photo put on a ball. The file she has is 1,100 pixels wide. To what height should she crop it so that the proportions fit the requirements?

41. A small college has 1,200 students and 80 professors. The college is planning to increase enrollment to 1,500 students next year. How many new professors should be hired, assuming they want to maintain the same ratio?

42. The taxes on a house assessed at $128,000 are $3,200 a year. If the assessment is raised to $160,000 and the tax rate did not change, how much would the taxes be now?

43. According to a poll conducted by the Pew Research Center in early 2021, for every 71 Americans in the 18–29 age group who used Instagram, there were 29 who did not. At that time there were about 53.3 million Americans in that age group. How many were not Instagram users?

44. The Gallup corporation conducts a world poll every few years in which people are asked if they would like to move to another country if they could. In the most recent poll, for every 15 people who would like to move, 85 are content to stay in their country. Among those expressing a desire to move, 43 out of 200 would like to move to the United States. The next highest choices were Canada and Germany, each at 6 out of 99.

(a) Based on these numbers, if a random sample of 1 million people was chosen from around the globe, how many would like to move to the United States?

(b) How many more would like to move to the United States than Canada and Germany combined?

45. According to a study reported by WalletHub in 2021, the metropolitan areas with the highest and lowest proportions of adults suffering from obesity were McAllen, TX, and Asheville, NC, respectively. In McAllen, for every 56 people with obesity, there were 69 without. In Asheville, for every 37 people with obesity, there were 163 without. At that time, the estimated populations were 141,968 for McAllen and 91,560 for Asheville. How many more people were suffering from obesity in McAllen than in Asheville?

46. Many people worked from home during 2020 and found that they liked it. As a result, people thought about moving further away from their workplaces. In a March 2021 survey reported by pwc.com, 11 of every 50 workers were considering moving more than 50 miles from their workplace, 3 of 25 had already done so, and 33 of 50 planned to stay put or move less than 50 miles from their workplace.

(a) If a company had 1,250 workers, how many more would be considering moving more than 50 miles from their workplace than actually did move?

(b) In a city with a workforce of 48,000 workers, how many would be expected to definitely stay in town, or within 50 miles of their workplace?

47. The amount of simple interest on a specific amount of money varies directly with the time the money is kept in a savings account when the interest rate is constant. Find the amount of interest on a $5,000 savings account, if the interest rate is 6%, and the money has been invested for 4 years.

48. The number of tickets purchased for a prize varies directly with the amount of the prize. For a prize of $1,000, 250 tickets are purchased. Find the approximate number of tickets that will be purchased on a prize worth $5,000.

Use the following information for Exercises 49–52. If everyone had the same body proportions, your weight in pounds would vary directly with the cube of your height in feet. According to the Centers for Disease Control, in 2021 the average height and weight for an adult male in the United States was 5 feet 9.4 inches and 196 lb. Use this information to write a variation equation, then use it to find the weight that each of the following famous athletes would be if they had the same body type as the average male.

49. Basketball player Lebron James: 6′8″ (Actual weight is 250 lb.)

50. Basketball player Kristaps Porzingas: 7′3″ (Actual weight is 240 lb.)

51. Jockey Pat Day: 4′11″ (Actual weight is 105 lb.)

52. Football player Chase Young: 6'5" (Actual weight is 265 lb.)

53. Under certain conditions, the pressure of a gas varies inversely with its volume. If a gas with a volume of 20 cubic inches has 36 pounds of pressure, find the amount of pressure 30 cubic inches of gas is under.

54. The strength of a particular beam varies inversely with the square of its length. If a 10-foot beam can support 500 pounds, how many pounds can a 12-foot beam support?

55. In karate, the force needed to break a board varies inversely with the length of the board. If it takes 5 lb of force to break a board that is 3 feet long, how many pounds of force will it take to break a board that is 5 feet long?

56. The weight of a body varies inversely with the square of the distance from the center of the earth. If the radius of the earth is 4,000 miles, how much would a 150-lb woman weigh 500 miles above the surface of the earth?

57. The time to complete a project is inversely proportional to the number of people who are working on the project. A class project can be completed by 3 students in 20 days. In order to finish the project in 5 days, how many more students should the group add?

58. The intensity of sound varies inversely as the square of the distance from the source. A sound with an intensity of 300 watts/m^2 is heard from 20 feet away from a speaker. What is the intensity of the sound 50 feet away from the same speaker?

Critical Thinking

59. Starting with the equation $\frac{a}{b} = \frac{c}{d}$, find the LCD and multiply both sides by it. Make sure you simplify fractions. This proves something that's very important in working with proportions. What is it?

60. Write the generic variation equations for direct and inverse proportionality. Now fill in the blanks in the next two sentences, and explain how you can deduce your answer from the equations.

    When a quantity goes up in proportion to another going up, this is _____ variation. When a quantity goes down in proportion to another going up, this is _____ variation.

Stores are required by law to display unit prices. A unit price is the ratio of the total price to the number of units. For Exercises 61–66, decide which is a better buy. These prices were obtained from actual foods.

61. Flour: 10 pounds for $3.39 or 25 pounds for $7.49

62. Candy: 20 ounces for $1.50 or 24 ounces for $1.75

63. Potato sticks: 7 ounces for $1.99 or 1.5 ounces for $0.50

64. Cookies: 7 ounces for $0.99 or 14 ounces for $1.50

65. Coffee: 11.5 ounces for $2.75 or 34.5 ounces for $7.49

66. Beggin' Strips dog treats: $11.20 for 1.4 pounds, $2.75 for 8 ounces.

67. If a varies directly with the square of b, and b in turn varies inversely with d:
    (a) Give a verbal description of the variation of a with respect to d.
    (b) If a is 90 when d is 5, find a when d is 9.

    (c) Given the information in (b), can you find b? Why or why not?

68. If x varies inversely with the cube of y, and y in turn varies inversely with the square root of z:
    (a) Give a verbal description of the variation of x with respect to z.
    (b) If x is 14 when z is 4, find x when z is 16.
    (c) If x is 20 when y is 2 and y is 16 when z is 4, find x when z is 100.

In Exercises 69–72, decide if the two quantities are likely to vary directly or inversely, and explain your answer.

69. (a) The amount of time you spend on Facebook and your GPA
    (b) The square footage of an apartment and the monthly rent

70. (a) The outdoor temperature and the total weight of clothing you wear
    (b) The number of miles you run per week and your weight

71. (a) The age of your car and the total cost of maintaining it
    (b) The crime rate in a given area and the rate of unemployment in that area

72. (a) The distance you drive in an hour and a half, and your average speed
    (b) The magnitude of an earthquake and the cost of property damage that it causes

| **Section 6-3** | **The Rectangular Coordinate System and Linear Equations in Two Variables** |

LEARNING OBJECTIVES

1. Plot points in a rectangular coordinate system.

2. Graph linear equations.

3. Find the intercepts of a linear equation.

4. Find the slope of a line.

5. Graph linear equations in slope-intercept form.

6. Graph horizontal and vertical lines.

7. Find linear equations that describe situations in our world.

As the pandemic of 2020 ravaged the U.S. economy, the unemployment rate became a big story. But what did unemployment look like before the crisis? Here's a look at unemployment rates from 2000 to 2020, according to the U.S. Bureau of Labor Statistics.

Year	'00	'01	'02	'03	'04	'05	'06	'07	'08	'09	'10
Rate (%)	4.0	4.7	5.8	6.0	5.5	5.1	4.6	4.6	5.8	9.3	9.6

Year	'11	'12	'13	'14	'15	'16	'17	'18	'19	'20
Rate (%)	8.9	8.1	7.4	6.2	5.3	4.9	4.4	3.9	3.7	8.1

There's quite a lot of data here, and in order to make sense of the data, we'll have to follow along the table one number at a time. The rate increased quite a bit from 2000 to 2003. It then went down steadily from 2003 to 2006 before increasing dramatically from 2007 to 2010. From 2011 to 2019 it fell a lot. Then in 2020, it had a huge increase.

Now let's look at the same information in graphic form:

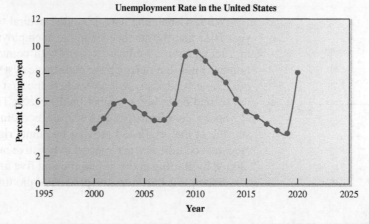

When I look at this graph, two things occur to me immediately. One, it's much, much easier to see the trends in unemployment compared to when the information was in table form. Two, the graph looks a lot like a super-awesome roller coaster, which is totally irrelevant but still pretty cool. The first observation is the important one because it illustrates the big advantage of being able to draw graphs: graphing data almost always makes it easier to understand their significance than simply looking at raw numbers.

In this section, we'll discuss the basics of graphing and study a particular kind of graph that we can use to explore many situations in our world.

The Rectangular Coordinate System

The foundation of graphing in math is a system for locating data points using a pair of perpendicular number lines. We call each one an **axis**. The horizontal line is called the *x* **axis**, and the vertical line is called the *y* **axis**. The point where the two intersect is called the **origin**. Collectively, they form what is known as a **rectangular coordinate system**, sometimes called the **Cartesian plane**. The two axes divide the plane into four regions called **quadrants**, which we number using Roman numerals I, II, III, and IV as shown in Figure 6-1.

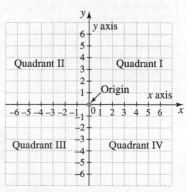

Figure 6-1

Sidelight   **The Controversial Descartes**

The rectangular coordinate system we use in graphing is called the Cartesian plane because the whole study of graphing was developed by the 17th-century philosopher René Descartes (pronounced "day-cart"). His work revolutionized math because he brought together the ancient fields of arithmetic, geometry, and algebra into a single subject that we now call *analytic geometry*. This alone makes Descartes one of the most important figures in the history of human thought. In 1637, he published a book with the lengthy and somewhat grandiose title *Discourse on the Method of Rightly Conducting the Reason and Seeking for Truth in the Sciences*. (Descartes was not the most humble guy in town.) In the book, he advocated the position that all knowledge should be devised using mathematical reasoning. Terrific for math, not so good for compassion.

Because of his extreme feelings on the importance of reasoning in every area, Descartes felt that any being incapable of logical reasoning was merely a biological machine, also incapable of having feelings, emotions, or even feeling physical pain. In particular, his mistreatment of animals, some of which was far too graphic for a family-friendly textbook, is used to this day as an argument in opposition to animal cruelty. Anyone that's ever owned a pet will find it hard to believe Descartes was smart enough to develop an entire branch of mathematics, but dense enough to not recognize that animals can feel pain.

In popular culture, Descartes is best known for his famous statement "I think, therefore I am," which roughly means that thinking about existence is proof that one exists. And how did Descartes meet his end? One day, he walked into a bar and ordered an ale, and the bartender asked him "do you want some peanuts with that?" When Descartes replied "I think not," he ceased to exist. (Actually, he died of pneumonia at age 53 in 1650, but my story is much more fun.)

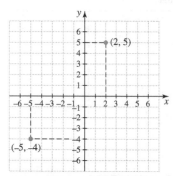

Figure 6-2

When visualizing data, it's often natural to have two quantities paired together, like the year 2015 and the number 5.3 in the unemployment table. We can associate pairs of numbers in a rectangular coordinate system with points by locating each number on one of the two number lines that make up the coordinate system. We call the two numbers **coordinates** of a point, and write them as (x, y), where the first number describes a number on the x axis and the second describes a number on the y axis. The coordinates of the origin are $(0, 0)$. A point P whose x coordinate is 2 and whose y coordinate is 5 is written as $P = (2, 5)$. It is plotted by starting at the origin and moving two units right and five units up, as shown in Figure 6-2. Negative coordinates correspond to negative numbers on the axes, so a point like $(-5, -4)$ is plotted by starting at the origin, moving five units left and four units down.

Example 1 illustrates the process of plotting points.

Example 1   **Plotting Points**

Math Note

One common way to go wrong when plotting points is to mix up the order of the coordinates, so think alphabetical. If you're thinking in terms of x and y axis, x comes first alphabetically, and the x coordinate is first. If you think horizontal and vertical axis, it still works.

Plot the points $(5, -3)$, $(0, 4)$, $(-3, -2)$, $(-2, 0)$, and $(2, 6)$.

SOLUTION

To plot each point, start at the origin and move left or right according to the x value, and then up or down according to the y value.

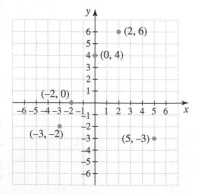

Try This One   1

Plot the points whose coordinates are $(5, 3)$, $(-1, 5)$, $(0, -5)$, $(-3, 0)$, and $(-4, -2)$.

Aaron Roeth Photography

Archaeological digs use a rectangular coordinate system to track where objects are found.

Given a point on the plane, its coordinates can be found by drawing a vertical line back to the x axis and a horizontal line back to the y axis. For example, the coordinates of point C shown in Figure 6-3 are $(-3, 4)$. This skill comes in VERY handy in interpreting information from graphs displaying real data, like the one that begins this section.

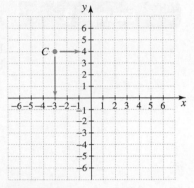

Figure 6-3

| Example 2 | Finding the Coordinates of Points |

Find the coordinates of each point shown on the plane below.

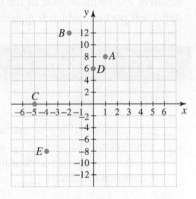

SOLUTION

The easiest way to make a mistake in graphing is to not be conscious of the scale marked on each axis. In this case, notice that every tick mark on the y axis represents TWO units, not one.

 1. Plot points in a rectangular coordinate system.

$A = (1, 8)$ $B = (-2, 12)$ $C = (-5, 0)$ $D = (0, 6)$ $E = (-4, -8)$

Try This One 2

Find the coordinates of the points shown.

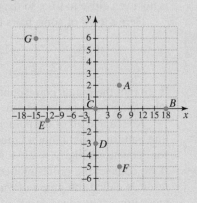

Blend Images/Getty Images

The interface between a mouse and a computer uses a coordinate system to match the cursor's motion to the mouse's motion.

Example 3	Identifying the Significance of the Signs of Coordinates

Earlier we saw that the two axes divide a coordinate plane into four quadrants. Fill in each blank.

- If both coordinates of a point are positive, the point is in quadrant _____.
- If both coordinates are negative, the point is in quadrant _____.
- If the x coordinate is positive and the y coordinate is negative, the point is in quadrant _____.
- If the y coordinate is positive and the x coordinate is negative, the point is in quadrant _____.

SOLUTION

Two positive coordinates puts a point in the upper right portion of the graph, which is quadrant I. Two negative coordinates is the bottom left portion, which is quadrant III. If the x coordinate is positive, a point is to the right of the origin; a y coordinate that's negative means the point is below the y axis. So positive x and negative y is quadrant IV (lower right portion). A negative x coordinate (left of the origin) and a positive y coordinate (above the y axis) is the upper left portion, which is quadrant II.

Try This One   3

Fill in each blank.

- A point with first coordinate zero and positive y coordinate is on the _____ axis, between quadrants _____ and _____.
- A point with second coordinate zero and negative x coordinate is on the _____ axis, between quadrants _____ and _____.

Linear Equations in Two Variables

Suppose you get offered six bucks to stand on a street corner and hand out fliers for a restaurant, plus an extra \$2 for every customer that brings your flier into the restaurant and orders a meal. If we use variable x to represent the number of customers ordering a meal with your flier, and variable y to represent the amount of money you make, we could write an equation to describe y: $y = 2x + 6$ (amount you make is \$2 times the number of customers plus \$6). If we choose a pair of numbers to substitute into this equation for x and y, the resulting equation is either true or false. For example, for $x = 4$ and $y = 14$, the equation is $14 = 2(4) + 6$, which is a true statement. We call the pair $(4, 14)$ a **solution** to the equation, and say that the pair of numbers **satisfies** the equation. But for $x = 14$ and $y = 4$, the equation is $4 = 2(14) + 6$, which is not even close to true, so the pair $(14, 4)$ is not a solution to the equation.

Here's a list of some pairs that satisfy the equation $y = 2x + 6$:

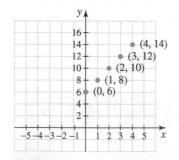

Figure 6-4

Math Note

Notice that in Figures 6-4 and 6-5, each box along the x axis represents one unit, but each box along the y axis represents two units.

It's very common to label the two axes with a different scale, especially in applied situations where the quantities represented by x and y could be totally different in size.

$$(0, 6) \qquad (1, 8) \qquad (2, 10)$$
$$(3, 12) \qquad (4, 14)$$

If we plot the points corresponding to these pairs, something interesting happens (see Figure 6-4).

Notice that all of the points appear to line up in a straight line pattern. This is not a coincidence. In fact, for this reason, equations like $y = 2x + 6$ are called *linear equations*. If we connect the points plotted with a line (Figure 6-5), the result is called the *graph of the equation*. This is one of the most important ideas in all of math, so it deserves a cool definition box:

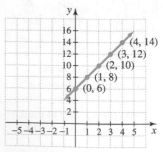

Figure 6-5

The **graph of an equation** is a way to geometrically represent *every* pair of numbers that is a solution to the equation. Each of those pairs corresponds to a point on the graph.

When an equation can be written in the form $ax + by = c$, where a, b, and c are real numbers, it's called a **linear equation in two variables**. (Our example can be rearranged to look like $-2x + y = 6$, so it fits that definition with $a = -2$, $b = 1$, and $c = 6$.)

In Example 4, we illustrate the process of drawing the graph of a linear equation in two variables.

| Example 4 | Graphing a Linear Equation in Two Variables |

Graph $x + 2y = 5$.

SOLUTION

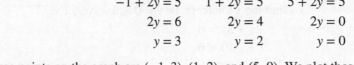

Only two points are *necessary* to find the graph of a line, but it's a good idea to find three to make sure that you haven't made a mistake. To find pairs of numbers that make the equation true, we will choose some numbers to substitute in for x, then solve the resulting equation to find the associated y. I decided to choose $x = -1$, $x = 1$, and $x = 5$, but any three will do.

$$x + 2y = 5 \qquad x + 2y = 5 \qquad x + 2y = 5$$
$$-1 + 2y = 5 \qquad 1 + 2y = 5 \qquad 5 + 2y = 5$$
$$2y = 6 \qquad\quad 2y = 4 \qquad\quad 2y = 0$$
$$y = 3 \qquad\qquad y = 2 \qquad\qquad y = 0$$

Three points on the graph are $(-1, 3)$, $(1, 2)$, and $(5, 0)$. We plot those three points and draw a straight line through them.

The arrows at each end of the graph are important! They indicate that the line continues indefinitely in each direction. The graph doesn't suddenly end because we ran out of space on our drawing.

Try This One 4

Graph $2x - y = 10$.

Using Technology: Making a Table of Values

Both graphing calculators and spreadsheets can be used to make a table of values for graphing an equation. In each case you first have to solve the equation for y. In Example 3, $x + 2y = 5$ becomes $y = \frac{5 - x}{2}$.

Graphing Calculator

Step 1: Press [Y=], then enter the right side of this equation next to "$y_1 =$." Make sure you enclose the entire numerator in parentheses!

Step 2: Press [2nd] [WINDOW] to get to the table setup window.

Step 3: Use the arrow keys and enter to select "Ask" next to "Indpnt."

Step 4: Press [2nd] [GRAPH] to display the table. Key in any x value and press [ENTER] to calculate the associated y value.

Spreadsheet

Step 1: Enter x values you'd like to input in column A.

Step 2: In cell B1, enter "=" followed by the formula found for y, with A1 in place of x. In this case, we would enter "=(5–A1)/2." This will put the y value in cell B1.

Step 3: Copy the formula in B1 down to as many rows as you have x values for.

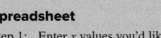

Microsoft Corporation

Intercepts

When drawing the graph of an equation, the first thing you need to decide is how to scale the two axes. A good way to help decide is to find the key points where the graph crosses each axis. The point where a graph crosses the x axis is called the **x intercept**, and the point where a graph crosses the y axis is called the **y intercept**. Every point on the x axis has y coordinate zero, and every point on the y axis has x coordinate zero, so we get the following rules.

2. Graph linear equations.

Finding Intercepts

To find the x intercept, substitute zero for y and solve the equation for x.
To find the y intercept, substitute zero for x and solve the equation for y.

Example 5

Math Note

Intercepts are *points on the graph*, not numbers. Make sure you give both coordinates, not just the result of solving the equation in the procedure.

Finding Intercepts

Find the intercepts of $2x - 3y = 6$, and use them to draw the graph.

SOLUTION

To find the x intercept, let $y = 0$ and solve for x.

$$2x - 3y = 6$$
$$2x - 3(0) = 6$$
$$2x = 6$$
$$x = 3$$

The x intercept has the coordinates $(3, 0)$.

To find the y intercept, let $x = 0$ and solve for y.

$$2x - 3y = 6$$
$$2(0) - 3y = 6$$
$$-3y = 6$$
$$y = -2$$

The y intercept has the coordinates $(0, -2)$.

Now we plot the points $(3, 0)$ and $(0, -2)$, and draw a straight line through them. (It would still be a good idea to find one additional point to check your work. If the three points don't line up, there must be a mistake.)

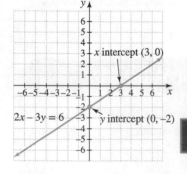

Try This One   5

For each equation, find the intercepts and use them to draw the graph.

(a) $x + 5y = 10$ (b) $4x - 3y = 12$

3. Find the intercepts of a linear equation.

Slope

One of the key features of any line is its steepness. To describe the steepness of a line, we'll use the term *slope*. The slope of a line on a graph is very much like the slope (or steepness) of a road. Look at the two roads in Figure 6-6.

The "slope" of a road can be defined as the "rise" (vertical height) divided by the "run" (horizontal distance) or as the change in y compared to the change in x. In road A, we have

$$\frac{30 \text{ ft}}{50 \text{ ft}} = 0.6$$

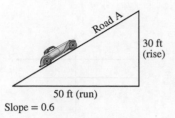

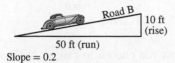

30 ft (rise)

50 ft (run)

Slope = 0.6

10 ft (rise)

50 ft (run)

Slope = 0.2

Figure 6-6

That is, for every 50 feet horizontally the road rises a height of 30 feet. Road B has a slope of

$$\frac{10 \text{ ft}}{50 \text{ ft}} = 0.2$$

Since the slope of road A is larger than the slope of road B, we say that road A is steeper than road B.

On the Cartesian plane, slope is defined as follows:

The **slope** of a line (designated by m) is

$$m = \frac{y_2 - y_1}{x_2 - x_1} \qquad \begin{array}{l} \textit{Change in y coordinate} \\ \hline \textit{Change in x coordinate} \end{array}$$

where (x_1, y_1) and (x_2, y_2) are two points on the line.

Arthur S. Aubry/Photodisc/Getty Images

The slope of this road is 0, because the rise is 0.

In words, the slope of a line can be found by subtracting the y coordinates (the vertical height) of two points and dividing that difference by the difference obtained from subtracting the x coordinates (the horizontal distance) of the same two points. See Figure 6-7.

Example 6 shows the procedure for finding the slope of a line when we know two points on that line.

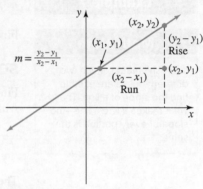

Figure 6-7

Example 6	Finding the Slope of a Line

Math Note

If the line goes "uphill" from left to right, the slope will be positive. If a line goes "downhill" from left to right, the slope will be negative. The slope of a vertical line is *undefined*. The slope of a horizontal line is 0 (see sample graphs at the top of page 324).

Find the slope of a line passing through the points (2, 3) and (5, 8).

SOLUTION

Designate the two points as follows:

$$(2, 3) \quad \text{and} \quad (5, 8)$$
$$\downarrow \downarrow \qquad\qquad \downarrow \downarrow$$
$$(x_1, y_1) \qquad\quad (x_2, y_2)$$

Substitute into the formula:

$$m = \frac{y_2 - y_1}{x_2 - x_1} = \frac{8 - 3}{5 - 2} = \frac{5}{3}$$

The slope of the line is $\frac{5}{3}$. A line with that slope would rise 5 feet vertically for every 3 feet horizontally.

Try This One 6

Find the slope of a line passing through the points $(-1, 4)$ and $(2, -8)$.

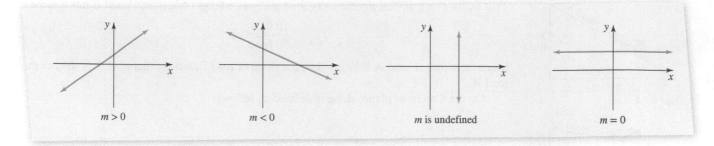

When finding slope, it doesn't matter which of the two points you choose to call (x_1, y_1) and which you call (x_2, y_2). But the order of the subtraction in the numerator and denominator has to be consistent—if you subtract $y_2 - y_1$ in the numerator, you have to subtract $x_2 - x_1$ in the denominator.

If you know the equation of a line, you can find the slope by first finding two points like we did in graphing, then using the slope formula.

Example 7	Finding Slope Given the Equation of a Line

Math Note

It doesn't matter which two different points are used to find the slope of a line. That's what makes a line a line—the slope of every portion is the same.

Find the slope of the line $5x - 3y = 15$.

SOLUTION

Find the coordinates of any two points on the line. In this case, we choose the intercepts, which are $(3, 0)$ and $(0, -5)$. Then substitute into the slope formula.

$$m = \frac{y_2 - y_1}{x_2 - x_1} = \frac{-5 - 0}{0 - 3} = \frac{-5}{-3} = \frac{5}{3}$$

The slope of the line $5x - 3y = 15$ is $\frac{5}{3}$.

Try This One   7

4. Find the slope of a line.

Find the slope of the line $-2x + 4y = 8$.

If we start with the equation $5x - 3y = 15$ from Example 7 and solve the equation for y, the result suggests a useful fact, which we'll examine further in Problem 93.

$$5x - 3y = 15$$
$$-3y = -5x + 15$$
$$y = \frac{5}{3}x - 5$$

Notice that the coefficient of x is 5/3, which is the same as the slope of the line, as found in Example 7. This is not a coincidence. Also, notice that in the final equation, the y intercept is easy to find: substituting in $x = 0$ gives us the value $y = -5$. So solving the equation for y was useful because we can get important information about the line very easily when the equation is in that form.

> The **slope-intercept form** for an equation in two variables is $y = mx + b$, where m is the slope and $(0, b)$ is the y intercept.

The graph of a line in slope-intercept form can be drawn by using the y intercept as a point, then plotting the "rise," which is the numerator of the slope in fraction form, and then the run, which is the denominator, as shown in Example 8.

| **Example 8** | **Using Slope-Intercept Form to Draw a Graph** |

Math Note

When the slope of a line is negative, like $-\frac{2}{3}$, start at the y intercept and move 2 units *down* and 3 units to the right to get the second point.

Graph the line $y = \dfrac{5}{3}x - 6$.

SOLUTION

The slope is 5/3 and the y intercept is $(0, -6)$. Starting at the point $(0, -6)$, we move vertically upward 5 units for the rise, and move horizontally 3 units right for the run. That gives us the second point $(3, -1)$. Then draw a line through these points. To check, notice that $(3, -1)$ satisfies the equation.

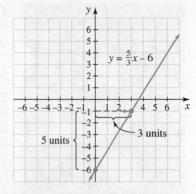

Try This One   8

Graph the line $y = \dfrac{2}{5}x - 2$.

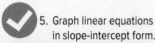

5. Graph linear equations in slope-intercept form.

Math Note

The graph of $x = 0$ is the y axis, and the graph of $y = 0$ is the x axis.

Horizontal and Vertical Lines

Think about what the equation $y = 3$ says in words: that the y coordinate is always 3. This is a line whose height is always 3, which is a horizontal line. Similarly, an equation like $x = -6$ is a vertical line with every point having x coordinate -6.

| **Example 9** | **Graphing Vertical and Horizontal Lines** |

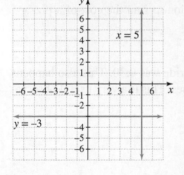

Graph each line, and describe why the graph makes sense.

(a) $x = 5$ (b) $y = -3$

SOLUTION

(a) The graph of $x = 5$ is a vertical line with every point having x coordinate 5 because the equation literally says "the x coordinate of every point is 5." We draw it so that it passes through 5 on the x axis.

(b) The graph of $y = -3$ is a horizontal line with every point having y coordinate -3 because the equation literally says "the y coordinate of every point is -3." We draw it so that it passes through -3 on the y axis.

Try This One   9

Graph each line.

(a) $x = -7$ (b) $y = 1$

6. Graph horizontal and vertical lines.

Applications of Linear Equations in Two Variables

A wide variety of situations in our world can be modeled using linear equations in two variables. These equations are usually written in slope-intercept form, $y = mx + b$. We will close the section with two examples.

| Example 10 | Finding a Linear Equation Describing Cab Fare |

The standard fare for a taxi in one city is $5.50, plus $0.30 per mile. Write a linear equation that describes the cost of a cab ride in terms of the length of the ride in miles. Then use your equation to find the cost of a 6-mile ride, an 8.5-mile ride, and a 12-mile ride.

SOLUTION

The first quantity that varies in this situation is the length of the trip, so we can assign variable x to the number of miles. The corresponding quantity that changes is the cost, so we will let $y =$ the cost of the ride. Since each mile costs $0.30, the total mileage cost is 0.30 times the number of miles, or $0.30x$. Adding the upfront cost of $5.50, the total cost is given by $y = 0.30x + 5.50$.

For $x = 6$ miles,

$$y = 0.30(6) + 5.50$$
$$y = \$7.30$$

For $x = 8.5$ miles,

$$y = 0.30(8.5) + 5.50$$
$$y = \$8.05$$

For $x = 12$ miles,

$$y = 0.30(12) + 5.50$$
$$y = \$9.10$$

Steve Mason/Photodisc/Getty Images

> **Math Note**
>
> Remember, when assigning variables in a word problem, think about what "variable" really means: a quantity in the problem that can change, or vary.

Try This One   10

The cost of a medium cheese pizza at Mario's Campus Pizzeria is $6.75, and each additional topping costs $0.35. Write a linear equation that describes the cost of a pizza in terms of the number of toppings. Then use your equation to find the cost of a pizza with three toppings and one with five toppings.

In Example 10, the slope of the equation describing the cab ride was 0.30. This also happens to be the rate at which the cost changes as the mileage increases. This is a very useful observation in applying linear equations to situations in our world.

> **Slope and Rate of Change**
>
> The slope of any line is the rate of change of y with respect to x.

In Example 11, we'll see how interpreting rate of change as slope can help to model a situation.

| Example 11 | Using Rate of Change to Model with a Linear Equation |

The United States Census Bureau uses demographic information to set a poverty threshold that is used to determine how many Americans are living in poverty based on annual income. For an individual on their own, the poverty threshold was $4,190 in 1980, and it has increased at the rate of about $220 per year since then. Write a linear equation that describes the poverty threshold in dollars in terms of years after 1980. Then use your equation to estimate the poverty threshold in 2020, and the year that it will pass $15,000 per year.

Math Note

The way we've set the time scale here is very common and very useful. Instead of using calendar years, like 1980, we call the first time we're interested in time zero. Then every later time is years after 1980.

SOLUTION

The problem gives us two key pieces of information: the poverty threshold was $4,190 at time zero (we're told to use years after 1980, and 1980 is zero years after 1980), and the threshold is changing at the rate of +$220 per year (positive because the threshold is increasing). The rate of change is the slope of a line describing the poverty threshold, and the threshold at time zero is the y intercept.

When we're asked to write an equation describing some quantity, we represent that quantity with y so that when we have an equation of the form "$y =$ some expression," it will express the quantity we were asked for. So in this case, $y =$ the poverty threshold, and $x =$ number of years after 1980. Our equation is

$$y = 220x + 4,190$$

To estimate the threshold in 2020 we need to substitute in $x = 40$ because 2020 is 40 years after 1980.

$$y = 220(40) + 4,190 = \$12,990$$

To find the year when the threshold is expected to pass $15,000 per year, we need to substitute in 15,000 for y and solve for x.

$$15,000 = 220x + 4,190$$
$$220x = 10,810$$
$$x = \frac{10,810}{220} \approx 49.1$$

So the threshold should pass $15,000 about 49 years after 1980, which is 2029.

Try This One   11

After posting some racially insensitive comments, an influencer begins to lose followers. At the beginning of the summer they had 211,000 followers, but that number went down by 6,200 a week. Write a linear equation that describes the number of followers in terms of weeks, and use it to find how long it will take before 100,000 followers remain.

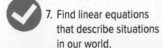

7. Find linear equations that describe situations in our world.

Answers to Try This One

1

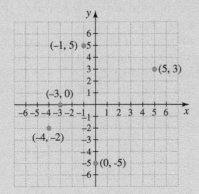

2 $A = (6, 2); B = (18, 0); C = (0, 0);$
$D = (0, -3); E = (-12, -1); F = (6, -5);$
$G = (-15, 6)$

3 y, I, II; x II, III

4

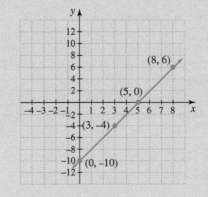

5   (a) and (b)

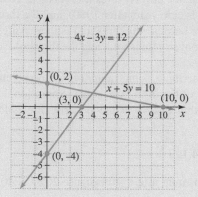

6 $m = -4$

7 $m = \dfrac{1}{2}$

8

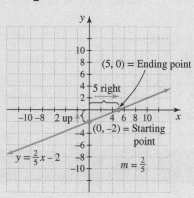

9

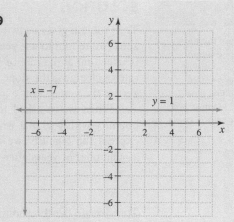

10 $y = 6.75 + 0.35x$; 3 toppings; \$7.80;
        5 toppings: \$8.50

11 $y = -6{,}200x + 211{,}000$; about 17.9 weeks

Exercise Set   6-3

Writing Exercises

1. How can you tell what quadrant a point is in by looking at its coordinates?
2. What does the slope of a line really mean?
3. Describe the two different types of intercepts for a graph. How do you find them from an equation?
4. How can you tell if a line is vertical from looking at its equation? What about horizontal?
5. If a line is neither horizontal nor vertical, how can you find the slope without finding two points on the line?

6. Describe the formula used to find slope when you know two points on a line.
7. What's the advantage of having a graph describing some quantity rather than an equation or a table of data?
8. When some quantity is described by a linear equation, what does the slope of the graph tell us about that quantity?

Computational Exercises

For Exercises 9–22, plot each point on the Cartesian plane.

9. $(-2, -5)$

10. $(-3, -8)$

11. $(-6, 4)$

12. $(3, -7)$

13. $(6, 0)$

14. $(-4, 0)$

15. $(0, 3)$
16. $(0, -2)$
17. $(5.6, -3.2)$
18. $(-4.8, 7.3)$

19. $(3\frac{1}{2}, 6\frac{2}{3})$
20. $(4\frac{1}{4}, 5\frac{1}{8})$
21. $(-45, -2)$
22. $(-4, -200)$

In Problems 23–30, write the coordinates of the point on one of the two following coordinate planes:

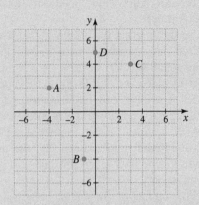

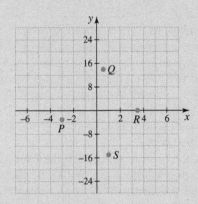

23. A
24. B
25. C
26. D
27. P
28. Q
29. R
30. S

For Exercises 31–40, draw the graph for each equation by first finding at least two points on the line.

31. $5x + y = 20$
32. $x + 4y = 24$
33. $3x - y = 15$
34. $2x - y = 10$
35. $4x + 7y = 140$
36. $3x - 8y = 120$
37. $2x + 7y = -120$
38. $5x - 3y = -180$
39. $\frac{3}{4}x - \frac{5}{6}y = 12$
40. $\frac{2}{3}x - \frac{5}{8}y = 6$

For Exercises 41–48, find the slope of the line passing through the two points.

41. $(-3, -2), (6, 7)$
42. $(4, 0), (3, -5)$
43. $(4, 0), (0, 7)$
44. $(2, 10), (4, 9)$
45. $(4\frac{5}{6}, 3\frac{2}{3}), (8, -2)$
46. $(-2\frac{3}{4}, 11\frac{1}{2}), (-1, 6)$
47. $(3.8, -1.2), (2.2, 3.1)$
48. $(2, -6.1), (3.4, -2.8)$

For Exercises 49–56, find the coordinates of both intercepts for each line.

49. $3x + 4y = 24$
50. $-2x + 7y = -28$
51. $-5x - 6y = 30$
52. $x + 6y = 10$
53. $\frac{1}{4}x - \frac{2}{3}y = 18$
54. $-\frac{9}{10}x + \frac{1}{3}y = -3$
55. $-0.7x - 0.5y = 28$
56. $0.04x + 0.07y = 28$

For Exercises 57–64, write the equation in slope-intercept form, then find the slope and the y intercept. Finally, draw the graph of the line.

57. $7x + 5y = 35$
58. $-2x + 7y = 14$
59. $x - 4y = 16$
60. $4x - 8y = 15$
61. $3x - 8y = 100$
62. $-12y + 5x = 160$
63. $y - \frac{5}{6} = -\frac{2}{3}\left(x - \frac{3}{4}\right)$
64. $y + \frac{7}{2} = \frac{7}{4}\left(2x + \frac{11}{2}\right)$

For Exercises 65–68, draw the graph for each equation.

65. $x = -3$
66. $y = 2$
67. $-3y = 8$
68. $5x = 42$

Applications in Our World

69. A newspaper's website charges $6.50 per week to run an ad plus a one-time setup fee of $50. Write an equation that describes the cost of an ad in terms of the number of weeks it runs, then use your equation to find the cost of running an ad for
    (a) 3 weeks. (b) 5 weeks. (c) 10 weeks.

70. A house painter charges $60 for a home visit and $210 per room painted. Write an equation that describes the cost of hiring the painter in terms of the number of rooms painted, then use your equation to find the cost of painting
    (a) a 5–room house. (b) a 7–room house.
    (c) a 20–room mansion.

71. You can rent a late-model SUV at Nifty Rent-a-Car for $26 a day, plus 20 cents for each mile driven. Write an

equation describing the daily rental charge in terms of miles driven, then use your equation to find the cost if you drive
(a) 30 miles. (b) 55 miles. (c) 220 miles.

72. The cost of reserving a particular domain name is $25 up front plus $4.99 per month. Find the cost of reserving that domain name for
    (a) 3 months. (b) 6 months. (c) 2 years.

73. According to the technology research company the Radicati Group, in 2021 the average user of a corporate email account received 121 messages per day, and that number increases at the rate of 5 daily messages per year. Write a linear equation describing the average daily number of emails received. Then use your equation to predict how many emails will be received

by an average corporate account in 2025, and the year in which the average number of emails will pass 175.

74. In 2005, the Veteran's Administration reported 10.5 million visits to mental health care professionals. That number has since increased steadily by about 1.02 million visits per year. Write a linear equation describing the number of mental health care visits (in millions). Then use your equation to estimate the number of visits in 2020, and how long it would take for the number of visits to reach 30 million.

75. In 2012, 28% of 10th graders reported drinking alcohol in the last month. That percentage went down by 1.42% per year for the next several years. Write a linear equation that describes the percentage in terms of years after 2012. The percentage in 2019 was 17.8%. How does that compare to the number predicted by your equation?

76. From 1970 to 2000, the percentage of high school dropouts in America declined at the rate of about 0.17% per year. The percentage was 15.0 in 1970. Write a linear equation describing the percentage in terms of years after 1970. The percentage was at 5.1 in 2019. When did your equation predict this percentage would be reached?

77. In 2004, the U.S. Census Bureau announced that there were 36.3 million Americans aged 65 or over, and they projected that number to grow to 86.7 million by 2050. If this projection is accurate and the population of older Americans grows at a constant rate, write an equation describing the number of Americans aged 65 or over (in millions) in terms of years after 2004. Then use your equation to find when this population would be expected to reach 100 million.

78. The amount of $10 in 2000 had the same buying power as $15.90 in 2021. If the cost of an item that was valued at $10 in 2000 increased at a constant rate, write an equation describing that cost in terms of years after 2000. Then use your equation to decide how long it will take the cost of that item to double.

79. Total government expenditures for Medicare from 2010 to 2019 can be modeled by the equation $y = 29.835x + 507.32$, where x is years after 2010, and y is expenditures in billions of dollars. (*Source:* CMS.gov)
    (a) Find Medicare expenditures for 2010, 2015, and 2019.
    (b) If expenditures continued increasing at the same average rate, in what year would they pass $1 trillion? (If you don't know how much a trillion is in billions, look it up!)
    (c) Use the Internet to find the actual Medicare expenditures for 2020, and compare them to the amount predicted by the equation.

80. Total U.S. tax revenues from 2010 to 2019 can be modeled by the equation $y = 147.48x + 2{,}263.8$, where x is years after 2010 and y is revenue in billions of dollars. (*Source:* OMB)
    (a) Find tax revenue for 2011, 2013, and 2015.
    (b) If revenue continued increasing at the same average rate, in what year would it get above $4 trillion? (If you don't know how much a trillion is in billions, look it up!)
    (c) Use the Internet to find the actual tax revenues for 2020, and compare them to the amount predicted by the equation.

81. It costs Damage Inc. $1.73 for each super slingshot it produces. The initial cost to the company for production, marketing, and paying workers is $12,000.
    (a) Write a linear equation that describes the total cost of producing x slingshots, and use your equation to find how much it would cost to make 1,000 slingshots.
    (b) If the company can sell each slingshot for $12.99, will it make a profit if it makes 1,000 slingshots and manages to sell them all? If so, how much will it make? If not, how much will it lose?

82. After paying a recording studio a fee of $4,000 for studio time, an independent band self-produces a demo CD. It costs them $1.29 for production and printing of each copy.
    (a) Write a linear equation that describes the total cost of producing x copies of the demo CD, and use your equation to find how much it would cost to make 2,000 copies.
    (b) If the band is able to sell the CDs for $7.99 at its shows, will it make a profit if it makes 2,000 copies and sells them all? If so, how much will it make? If not, now much will it lose?

83. Sandi weighs 160 lb, but has started a new diet plan that promises a loss of 3 lb per month every month until reaching a goal weight. Write a linear equation that describes how much Sandi weighs each month, and then use your equation to find out how many months it will take to get to 130 lb.

84. A state lottery starts at $2 million, and then for each $1 ticket purchased, it adds $0.50 to the pot. Write a linear equation that describes how much money is in the pot in terms of number of tickets sold, and then use your equation to find out how much will be in the pot if 250,000 tickets are sold.

85. There are 350 known diseases within a community, and doctors can eliminate 5 of them per year. Assuming no new diseases are introduced to the community, write a linear equation that represents the number of diseases in terms of years from now. Use your equation to predict how many years it will take for the community to be free of all diseases.

86. It's estimated that at the beginning of 2021, there were 960 million computer viruses circulating on the Internet, but only about 100,000 were widely circulated. If malicious losers were producing and widely circulating 3,000 new viruses weekly, find a formula that describes the number of widely circulating viruses in terms of weeks from the beginning of 2021. Then use your equation to predict when that number would top 200,000.

87. The table displays the percentage of the voting age population that voted in federal elections in presidential election years from 1976 to 2020. (*Source:* Wikipedia)

Year	1976	1980	1984	1988	1992	1996
% voting	53.6	52.6	53.1	50.1	55.1	49.1

Year	2000	2004	2008	2012	2016	2020
% voting	51.3	55.3	56.8	54.9	54.6	61.9

(a) Plot points corresponding to all of the data in the table. Put the year on the *x* axis and the percentage on the *y* axis. Then connect the points with line segments.

(b) Use your graph to describe the trends in percentage over that 40-year span.

88. (a) Based on your graph in Exercise 87, in what year did the largest percentage of eligible voters cast a vote? What about the smallest percentage?

(b) In what year did the largest decrease from the previous election take place? The largest increase?

(c) Do some research on possible reasons for the variations in voter turnout, and report what you find.

89. The number of Chinese children adopted by American families for the years from 2010 to 2019 is shown in the table. (*Source:* Johnstonsarchive.net)

Year	2010	2011	2012	2013	2014
Adoptions	3,401	2,587	2,697	2,306	2,040

Year	2015	2016	2017	2018	2019
Adoptions	2,354	2,231	1,905	1,475	819

(a) Plot points corresponding to all of the data in the table. Put the year on the *x* axis and the number of adoptions on the *y* axis. Then connect the points with line segments.

(b) Use your graph to describe trends in the number of adoptions over that 10-year period.

90. (a) Based on your graph in Exercise 89, in what year did the largest number of adoptions occur? What about the smallest number?

(b) In what year did the largest increase from the previous year take place? The largest decrease?

(c) Do some research on possible reasons for the changes in adoption numbers, and report what you find.

Critical Thinking

91. Show why the slope of a vertical line is said to be undefined. (*Hint:* Pick two points on a vertical line and calculate the slope.)

92. Why is the slope of a horizontal line zero?

93. Start with a generic linear equation $ax + by = c$, and show that if you solve the equation for y, the coefficient of x is the slope of the line, and the constant term is the second coordinate of the y intercept. (*Hint:* Find the slope of the original equation by finding two points on the line.)

94. Suppose that two quantities can each be modeled by a linear equation, and that the graphs of those two lines happen to be parallel. What does that tell you about the relationship between the quantities?

95. (a) The data in the table show the total military spending, in billions, for the U.S. government between 2011 and 2015. Find the rate of change for each time period in the table, and use that information to discuss whether or not you think the data could be modeled accurately using a linear equation.

(b) Military spending for 2019 was $731.8. How would that affect your answer to part (a)?

Year	2011	2012	2013	2014	2015
Spending (billion $)	752.3	725.2	679.2	647.8	633.8

96. (a) The table provides historical poverty thresholds for families of six people according to the U.S. Census Bureau. Find the rate of change for each time period in the table, and use that information to discuss whether or not you think the data could be modeled accurately using a linear equation.

Year	1980	1990	2000	2010	2015
Poverty Threshold	$11,269	$17,839	$23,533	$29,887	$32,570

(b) The poverty threshold for 2020 was $35,499. Does this affect your answer to part (a)?

Section 6-4

Functions

LEARNING OBJECTIVES

1. Identify functions.
2. Write functions in function notation.
3. Evaluate functions.
4. Find the domain and range of functions.
5. Graph linear functions.
6. Interpret the graph of a function.

Each of the following statements came up in a Google search for the string "is a function of":

> "Intelligence is a function of experience."
> "Health is a function of proper nutrition."
> "Freedom is a function of economics."
> "What we wear each day is a function of atmospheric conditions."

These statements illustrate the simple idea behind one of the most important concepts in all of math, the function. In every case, the statement is telling us that one thing depends on another, and this is the essence of functions.

But first, let's take a look at a related idea, the relation.

Rainer Elstermann/zefa/Corbis/Getty Images

A **relation** is a rule matching up two sets of objects. Relations are often represented by sets of ordered pairs.

The following are examples of relations:

$$A = \{(\text{Apple, AAPL}), (\text{Ford, F}), (\text{Google, GOOG}), (\text{Kraft Foods, KFT})\}$$
$$B = \{(9, 0), (-3, 1), (3, 9), (-1, 5), (3, 3)\}$$
$$C = \{(x, y) \mid 3x + 5y = 7\}$$

Often, equations are used to represent sets of ordered pairs, as in relation C above. Usually we will simplify the notation by simply writing the equation; it is understood that the equation represents a relation between two sets, with x representing elements from one set, and y representing elements from another.

A function is a special type of relation.

A **function** is a set of ordered pairs in which no two ordered pairs have the same first coordinate and different second coordinates.

Relation A above is a function because no first coordinate is repeated (every company corresponds to its unique stock symbol). Relation B isn't a function because the ordered pairs (3, 9) and (3, 3) have the same first coordinate. Relation C is also a function (although it's not as clear as the other two) because for each value of x that you substitute into the equation, there is only one possible value of y that corresponds to it.

Example 1

Identifying Functions

Which of the following relations represent functions? (Assume that x represents the first coordinate.)

(a) $\{(5, 10), (-3, 10), (11, 10), (-10, 10), (\sqrt{3}, 10)\}$ (c) $3x^2 + y - 2x = 5$
(b) $y = x^2$ (d) $x = |y|$

SOLUTION

(a) Careful! Even though all ordered pairs have the same second coordinate, this is still a function. It's repeats in the FIRST coordinate that are bad.

(b) This is a function. Every number has only one square, so every value of x has only one associated y.

(c) This is also a function. Every x will again have only one associated y.

(d) This is not a function. Positive values of x will correspond to two possible values of y. For example, if $x = 2$, y can be 2 or -2.

Try This One   1

✓ 1. Identify functions.

Which of the following equations represent functions? (Assume that x represents the first coordinate.)

(a) $\{(11, -2), (3, 2), (-11, 2), (-3, -2)\}$ (c) $y = |x|$

(b) $x = y^2$ (d) $x^2 + y^2 = 4$

Function Notation

The equation $y = x^2$ represents a function that relates variables x and y. We call x the **independent variable** and y the **dependent variable** because its value depends on the choice of x. Another way to write the same function is $f(x) = x^2$. This is known as **function notation**, and is read aloud as "f of x equals x squared." This is the notation most commonly used to describe functions.

Math Note

The function symbol $f(x)$ is NOT a product, and should NEVER be read as "f times x." It designates that f is a function with variable x. If you read it as "f times x," you'll never understand functions.

The independent variable is sometimes called the **input** of a function, and the dependent variable is called the **output**. This is a really useful way to think about functions—sort of like a machine that inputs the first coordinate of an ordered pair, then outputs the matching second coordinate. Thinking of a function this way makes it pretty easy to understand why we pronounce $f(x)$ as "f OF x"; $f(x)$ represents the output of function f, and it's the result of f doing something to input x.

Functions can also be called by names other than f. Letters like f, g, h, and k are commonly used to represent functions, but a letter that is more representative can be used. For example, the circumference of a circle is a function of the radius (meaning that it depends on the value of the radius), so we could use the letter C to represent this function:

$$C(r) = 2\pi r$$

To write an equation in function notation, solve for y in terms of x, then change the letter y to the symbol $f(x)$.

Example 2   Writing a Function in Function Notation

Write $3x - 2y = 6$ in function notation.

SOLUTION

Math Note

Notice that the equation $3x - 2y = 6$ does in fact define a function: when we solve for y, there's only one possible output for any input x.

We need to solve the equation for y, then replace y with $f(x)$.

$$3x - 2y = 6 \qquad \textit{Subtract 3x from both sides.}$$
$$-2y = -3x + 6 \qquad \textit{Divide both sides by } -2.$$
$$y = \frac{3}{2}x - 3 \qquad \textit{Replace y with f(x).}$$
$$f(x) = \frac{3}{2}x - 3$$

Note that both $y = \frac{3}{2}x - 3$ and $f(x) = \frac{3}{2}x - 3$ define the same function, but in some ways function notation is a better choice. Not only does it indicate that the relation defined is indeed a function, but it also tells us immediately that the letter x is used to represent the input of that function: that's the x part in "$f(x)$."

Try This One   2

Write $10y + 30 = 5x$ in function notation. How can you tell that this equation actually defines a function?

When a function is written as $f(x)$, $f(2)$ means to find the value (another name for output) of the function when $x = 2$. This is known as **evaluating** a function.

Example 3   Evaluating a Function

Math Note

When finding $f(-2)$, make sure you replace *all* occurrences of the variable x with (-2). The parentheses are almost always a good idea.

Let $f(x) = x^2 + 3x - 5$ and $g(x) = 4 + \sqrt{x + 1}$. Find $f(3)$, $f(-2)$, $g(0)$, and $g(-3)$.

SOLUTION

$$f(3) = (3)^2 + 3(3) - 5 = 9 + 9 - 5 = 13$$
$$f(-2) = (-2)^2 + 3(-2) - 5 = 4 - 6 - 5 = -7$$
$$g(0) = 4 + \sqrt{0 + 1} = 4 + \sqrt{1} = 4 + 1 = 5$$
$$g(-3) = 4 + \sqrt{(-3) + 1} = 4 + \sqrt{-2}$$

Since $\sqrt{-2}$ isn't a real number, we say that $g(-3)$ is **undefined**. A lot of students want to say "no solution" here: don't be one of them! We're not solving an equation; we're just looking for the output of a function. If there is no output, we just say that the function is not defined for that input.

Try This One   3

3. Evaluate functions.

Let $f(x) = 3x^2 - 2x + 5$ and $g(x) = \sqrt{1 - 4x}$. Find $f(-1)$, $f(2)$, $g(-2)$, and $g(2)$.

In the last part of Example 3, we found an input (-3) for which the function g had no output. Sometimes that happens. We give a special name to all of the inputs that DO have an associated output.

The **domain** of a function is the set of all values of the independent variable x that result in real number values for y. The **range** of a function is the set of all possible y values.

Example 4   Finding the Domain and Range of a Function

Find the domain and range of each function:

(a) $f(x) = x^2$ (b) $f(x) = \sqrt{x}$ (c) $f(x) = \dfrac{3x - 2}{x + 1}$

Math Note

Remember that the symbol $\sqrt{}$ is defined to be the *positive* square root of the number inside the radical (as long as that number isn't zero).

SOLUTION

(a) There are no restrictions on what values x can be; therefore, the domain is all real numbers. Since x^2 is never negative, the range is $\{y \,|\, y \geq 0\}$.
(b) Since the square root of a negative number is undefined, x can't be negative. Therefore, the domain is $\{x \,|\, x \geq 0\}$. Since $\sqrt{x}$ is never negative, the range is $\{y \,|\, y \geq 0\}$.
(c) Since the denominator of a fraction can't be zero, we have to exclude $x = -1$. Every other x value will result in a real number output, so the domain is all real numbers except -1, which we write as $\{x \,|\, x \neq -1\}$. The range is not obvious, but notice that an output of 3 would make the equation

$$3 = \frac{3x - 2}{x + 1}$$

Multiplying both sides by $x + 1$, we get the contradiction $3x + 3 = 3x - 2$, so the range is $\{y \,|\, y \neq 3\}$.

Try This One   **4**

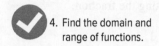 4. Find the domain and range of functions.

Find the domain and range of the following functions:

(a) $f(x) = |x|$ (b) $f(x) = \sqrt{x - 2}$ (c) $f(x) = \dfrac{2x - 1}{x - 3}$

Linear Functions

It's no secret that the makeup of families in the United States has changed considerably in the last 50 years. In 1975, only 14.3% of all children were born to a single mother. That number was 40.0% in 2019. The function $f(x) = 0.584x + 11.38$ can be used to model this percentage, where x is years after 1970 and the output f is the percentage. For example, $f(20) \approx 23.1$; this tells us that about 23.1% of babies born in the United States in 1990 (which is 20 years after 1970) had single mothers.

This example illustrates the main reason why functions are so important in math: they can be used to model a practically limitless variety of phenomena in our world. Anything that can be quantified can be modeled with a function, from the number of wins by your favorite football team to important economic data.

When we defined functions in terms of ordered pairs, it probably occurred to you that those ordered pairs look just like the ones we used to graph equations. This makes it simple to define what the graph of a function is:

> The **graph** of a function is a diagram of all ordered pairs matched up by that function.

Studying graphs of functions is useful for the same reason that studying graphs of equations is: when a function represents some data of interest, the graph is a great way to get an overview of what the data mean. In the remainder of this chapter, we'll study certain types of functions that are really useful in modeling situations in our world. The function representing the percentage of babies born to single mothers is an example of our first type, the *linear function*.

> A **linear function** is a function of the form $f(x) = ax + b$, where a and b are real numbers.

We learned a lot about linear equations in two variables in Section 6-3, and most of what we learned carries over to working with linear functions. The graph of every linear function is a straight line. For the function $f(x) = ax + b$, a is the slope of the graph and $(0, b)$ is the y intercept. In Example 5, we'll look at two different quick methods for graphing a linear function. You should decide which method you understand better and use that one.

Math Note

This practical example of a function also illustrates another key idea: between 2010 and 2016, the percentage of children born to single mothers actually decreased a bit. Even when a quantity can be modeled well by a function, that doesn't mean it can predict what will happen in the future.

Example 5   Graphing Linear Functions

Graph each linear function.

(a) $f(x) = 3x - 2$

(b) $f(x) = \dfrac{2}{3}x + 3$

SOLUTION

(a) In this method, we'll take advantage of knowing the slope, which in this case is 3, and the y intercept, which is $(0, -2)$. Plot the point $(0, -2)$, and then use a rise of 3 and a run of 1 to find a second point. Then we draw the line connecting those points.

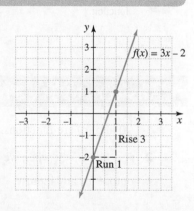

Math Note

Remember that if the slope is negative, move downward from left to right.

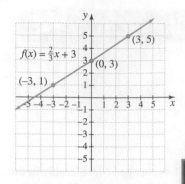

$f(x) = \frac{2}{3}x + 3$

(b) In the second method, we'll evaluate the function for three x values then plot the associated points, and again draw a line connecting the points. Notice that we cleverly chose inputs that make it easy to evaluate the function by eliminating the fraction.

$$f(-3) = \frac{2}{3}(-3) + 3 = -2 + 3 = 1$$

$$f(0) = 3$$

$$f(3) = \frac{2}{3}(3) + 3 = 2 + 3 = 5$$

Try This One   5

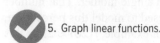

5. Graph linear functions.

Graph each linear function.

(a) $f(x) = 2x - 4$

(b) $f(x) = -\frac{3}{4}x = 1$

Being able to draw graphs is a useful skill: being able to *interpret* them is even more useful.

Example 6   Interpreting the Graph of a Function

The graph of the function described earlier that represents the percentage of babies in the United States born to single mothers is shown in Figure 6-8. Recall that the input x represents years after 1970. Use it to answer these questions:

(a) About what percentage of babies born in 2015 had single mothers?

(b) About when did the percentage go above 25%?

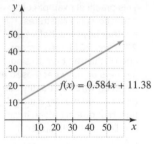

$f(x) = 0.584x + 11.38$

Figure 6-8

SOLUTION

(a) The year 2015 was 45 years after 1970, so we need the output of the function, which corresponds to the height of the graph, at $x = 45$. The point with that x coordinate looks to be about $(45, 38)$, so we conclude that the percentage was about 38% in 2015.

(b) The percentage went above 25% when the height of the graph reached heights higher than 25. It looks like the point $(24, 25)$ is on the graph, so the percentage rose above 25% about 24 years after 1970, or in 1994.

Try This One   6

6. Interpret the graph of a function.

Using Figure 6-8:

(a) How much did the percentage go up from 1970 to 2000?

(b) In which years were more than 70% of the babies born to mothers that were married?

Answers to Try This One

1 (a) and (c) are functions.

2 $f(x) = \frac{1}{2}x - 3$; when we solve for y, there's only one output for any input x.

3 10; 13; 3; undefined

4 (a) Domain: all real numbers. Range: $\{y \mid y \geq 0\}$.

(b) Domain: $\{x \mid x \geq 2\}$. Range: $\{y \mid y \geq 0\}$.

(c) Domain: $\{x \mid x \neq 3\}$. Range: $\{y \mid y \neq 2\}$.

5 (a)

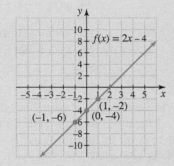

(b)

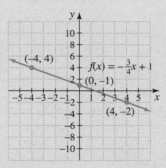

6 (a) About 18%   (b) 1970–2001

Exercise Set   6-4

Writing Exercises

1. What is the difference between a function and a relation?
2. What is the domain of a function?
3. What is the range of a function?
4. Explain how the terms "input" and "output" are used in the study of functions.
5. Why do we read the symbol $f(x)$ as "f of x" rather than "f times x"?

6. What does it mean to evaluate a function?
7. Write how the expression $f(x) = 3x - 7$ would be read aloud.
8. In the linear function $f(x) = ax + b$, what is the significance of the constants a and b?

Computational Exercises

For Exercises 9–22, state whether or not the relation is a function. Assume that x represents the first coordinate.

9. $\{(5, 8), (6, 9), (7, 10), (8, 11)\}$
10. $\{(0, 1), (0, 2), (0, 3), (0, 4), (0, 5)\}$
11. $\{(6, 11), (7, 11), (8, 11), (9, 11)\}$
12. $\{(-2, 4), (-1, 5), (0, 3), (1, 8), (-2, 2)\}$
13. $3x + 2y = 6$
14. $-6x = y + 5$
15. $y = 5x$
16. $x = 2y^2$
17. $|x - 3| = 2y$
18. $3x = |2y|$
19. $y^2 = \sqrt{x^2 + 6}$
20. $|2y^2 = 8x^4|$
21. A relation that has input all students at the University of District Columbia, and output the professors teaching them this semester
22. A relation that has input all babies born in a 6-month period at Mercy Hospital, and output their biological mothers

For Exercises 23–34, evaluate the function for the indicated values.

23. $f(x) = 3x + 8; f(3), f(-2)$
24. $f(x) = -2x + 5; f(-5), f(2)$

25. $f(x) = 4x - 8; f(10), f(-10)$
26. $f(x) = 6x^2 - 2x + 5; f(2), f(-4)$
27. $f(x) = 8x^2 + 3x; f(0), f(6)$
28. $f(x) = -3x^2 + 5; f(1.5), f(-2.1)$
29. $f(x) = x^2 + 4x + 7; f(-3.6), f(4.5)$
30. $f(x) = -x^2 + 6x - 3; f(0), f(-20)$
31. $f(x) = \sqrt{4x + 3};\ f\left(\dfrac{1}{2}\right),\ f\left(-2\dfrac{1}{4}\right)$
32. $f(x) = |2x - 7|;\ f\left(\dfrac{11}{2}\right),\ f\left(-6\dfrac{1}{2}\right)$
33. $f(x) = |2x| + 6;\ f\left(-\dfrac{1}{2}\right),\ f\left(9\dfrac{1}{2}\right)$
34. $f(x) = 11 + \sqrt{4 - 3x};\ f\left(-\dfrac{7}{3}\right),\ f(2)$

For Exercises 35–46, find the domain and range of each function.

35. $f(x) = (x - 1)^2$
36. $f(x) = \sqrt{3 - x}$
37. $f(x) = \dfrac{x}{x - 2}$
38. $y = \dfrac{2}{3}x + 3$
39. $y = 3 + \sqrt{x}$
40. $f(x) = -\sqrt{x}$
41. $2x - y = 7$
42. $y = \dfrac{2x + 1}{x}$
43. $f(x) = -\sqrt{x + 1} - 1$
44. $f(x) = 2(x + 3)^2$
45. $g(x) = 7 + \sqrt{10 - 2x}$
46. $h(x) = \dfrac{4|4 - x|}{9}$

In Exercises 47–56, graph each linear function.

47. $f(x) = 7x - 8$

48. $f(x) = 6x$

49. $f(x) = -3x + 2$

50. $f(x) = -5x + 1$

51. $g(x) = \dfrac{2}{3}x - 4$

52. $g(x) = -\dfrac{1}{4}x + 2$

53. $g(x) = -\dfrac{3}{5}x + 1$

54. $g(x) = \dfrac{3}{4}x + 2$

55. $k(x) = -100x + 750$

56. $k(x) = 90x - 300$

Applications in Our World

The percentage of the adult population that smokes can be modeled by the function $P(x)$, whose graph is shown below. The input x is years after 1965. Use the graph for Problems 57–66.

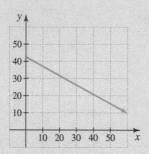

57. Estimate the percentage of adults that smoked in 1965.

58. Estimate the percentage of adults that smoked in 2015.

59. In what year did the percentage of smokers drop below 30%?

60. In what year did the percentage of smokers drop below 20%?

61. Estimate the change in the percentage of smokers from 1975 to 1985.

62. Estimate the change in the percentage of smokers from 1995 to 2015.

63. In what years was the percentage of nonsmokers between 75 and 85?

64. In what years was the percentage of nonsmokers less than 70?

65. Use your results from Problems 57 and 58 to find a formula for the function $P(x)$.

66. Use your answer from Problem 65 to calculate answers to Problems 61 and 62. How accurate were your estimates?

67. A restaurant server made $2.52 per hour plus $78 in tips on a Friday evening shift. The amount of money made for the night is given by the function $P(t) = 2.52t + 78$, where t is the number of hours worked. How much would a server make for a 4-hour shift if the pay rate is the same? An 8-hour shift?

68. A realtor makes a commission of 6% of the purchase price of a house plus a third of the $3,000 closing costs. The amount made on a house is given by the function $A(s) = 1,000 + 0.06s$, where s is the selling price of the house. How much more would be made on a $300,000 dollar house than a $120,000 house?

69. A company started out spending $578 to get up and running, and then it planned on charging $25 for its product. The total amount of profit (which is money

brought in minus cost) is given by the function $p(x) = 25x - 578$. If the company sells 18 units, will it make enough to cover its start-up costs? How much will it make if it sells 120 units?

70. On a history test, Miley got a quarter of the true/false questions wrong, two multiple choice questions wrong, and every other question right. The function $w(q) = \frac{1}{4}q + 2$ describes the number of questions she got wrong in terms of the number of true/false questions on the test. If there were 20 true/false questions on the test, how many total questions did Miley get wrong?

71. The Wertz car rental company charges $45 to rent a Mustang convertible, plus $42 per day. Write the cost of renting the car as a function of the number of days rented. Use your function to find how much it would cost to keep the car for a week.

72. A salesperson earns a base income of $250 a week plus 10% commission on the dollar value of each sale. Write a function to represent the amount of weekly income based on dollar value of sales. If sales for the week totaled $5,000, what is the income for the week?

73. Mr. Trent and his son Gary are both hospitalized after being exposed to a tropical disease while on vacation. Based on their relative sizes, Mr. Trent's recommended dosage for the treatment medication is three times Gary's minus 10 milligrams. Write a function describing Mr. Trent's dosage in terms of Gary's. If the attending physician orders a dose of 25 mg for Gary, use your function to find the size of his dad's dose.

74. Ichiko spent $234.56 at the bookstore this semester on textbooks. The average price per book is found by taking the total amount and dividing by the number of books purchased. Write the average price per book as a function of the total number of books purchased. If Ichiko bought six books, what was the average price per book?

75. During a road trip, Frank and Mandy are 80 miles into the trip when their GPS informs them that they'll reach their destination in time for a party if they average 65 miles per hour the rest of the way. Write the total distance traveled as a function of hours after that point. If their total trip is 275 miles, how much longer do they have to drive?

76. A company has endorsed a team of runners in the citywide marathon and says it will pay the runners $0.40 per mile they run plus donate a fixed amount of $500 to the charity the runners are sponsoring. Write the total amount the company will pay as a function of the total number of miles the runners cover during the

marathon. If the company pays $600 and there were 10 runners who all ran the same distance, how far did each runner cover?

77. While Kiara is studying for class, she realizes that reading her math book and taking notes as her professor suggested takes her triple the time it used to take her to just read without taking notes. However, she also notices her grades rise dramatically, so she decides this is a good idea. Write the time it takes Kiara to read and take notes as a function of the time it takes her to only read her book. If it took Kiara 45 minutes to read a section of her book without taking notes, how long would it take her to read the same section while taking notes?

78. When Jose is walking on a treadmill, he notes that he is only burning 80% of the calories he would be if he were jogging. Write the amount of calories Jose burns walking as a function of the amount of calories he would burn by jogging. If he burns 120 calories jogging, how much would he burn by walking?

79. A shoe store was having a sale. Customers could get 20% off their entire purchase if they bought at least two pairs of shoes. Write the sale price that a customer who bought at least two pairs of shoes would pay as a function of the price before the discount. If Cindy bought three pairs of shoes totaling $119.78, what would the sale price be?

80. A company has determined its price per unit is $15.40 divided by the product of 0.07 and the number of units it sells. Write the price per unit as a function of the number of units it sells. How much will the price per unit be if it sells 2,000 units?

81. In the first 2 hours after opening for dinner, a chef used a third of the chicken breasts on hand. Over the next 2 hours the chef used half of what remained, then used four more in the final hour before closing. Write the number of chicken breasts used as a function of the number left over. If there were 30 chicken breasts on hand at the start of the night, how many were used during the evening?

82. At a grocery store, the cashier rang up a total of $140.28 for groceries before taxes. The customer told her child to get the sodas she forgot, and they cost $1.59 each. Write the total amount of the grocery bill as a function of the number of sodas the customer buys. If the customer buys four sodas, how much will her new total be after a 6% sales tax has been applied?

83. At a poker tournament, the dealer makes 75% of the amount the players must contribute to play plus $48 in tips for the night. Write the amount the dealer makes if there are 10 players at the table as a function of the amount the players must contribute to play. If the dealer made $198 for the night, how much did each player have to pay to play the game?

84. At a dieter's meeting, DeAndre weighed 90% of his original weight less the 5 pounds he'd lost that month.

Write DeAndre's current weight as a function of his original weight. If DeAndre started at 220 lb, how much does he weigh now?

85. The manager of a winery found that when the price charged for a wine-tasting tour was $25, 500 customers were willing to pay that price for the tour. When the price was changed to $20, the number of customers willing to pay that price increased to 600.

(a) Write the number of customers as a linear function of the price. Use independent variable p and dependent variable C.

(b) Graph your linear function, labeling the axes with correct variables. Does it make sense to extend the graph into the second and fourth quadrants? Why or why not?

(c) If the price was $45, how many customers would pay for the tour?

(d) According to your model, if 300 customers paid for the tour, what price would they pay?

(e) What is the p intercept? What does it mean?

(f) What is the C intercept? What does it mean?

86. In 2013 a house was appraised at a value of $256,000. In 2018, a new freeway was built right next to the neighborhood, and by 2021 the same house was valued at $200,000. Let $t = 0$ correspond to the year 2013.

(a) Write the value of the house as a linear function of the year.

(b) Graph your function, labeling axes with correct variables. Start your graph with the year 2013.

(c) According to this trend, in what year would the house have a value of $116,000?

(d) According to this trend, what will the house's value be in the year 2041?

(e) What can you conclude about using a function based on past data to predict the future?

87. While traveling across state lines to retrieve an escaped convict, a sheriff drove 140 miles in 2 hours. He drove at the same speed the entire way and didn't take a break, so in 5 hours, he had gone 350 miles.

(a) Write the distance he traveled as a linear function of the time in hours.

(b) Graph your function, labeling the axes with correct variables. Start your graph at $t = 0$, which is when he started the trip.

(c) If he drove for 3 more hours at the same speed without stopping, how far would he have traveled?

(d) How long had he been driving when he was 210 miles into the trip?

88. In 2016, sales for Stacy's Super Slingshots were $32,000. In 2020, the novelty of the slingshots wore off, and sales had dwindled to $20,000. Let $t = 0$ represent the year 2012.

(a) Write the sales (S) as a linear function of time (t) in years.

(b) Graph your function, labeling the axes with correct variables. Start your graph at $t = 0$.

(c) What is the S intercept? What does this intercept mean?

(d) What is the t intercept? What does this intercept mean?

(e) According to this trend, how much did Stacy's Super Slingshots have in sales in 2007?

(f) How much will sales be in 2027?

(g) What can you conclude about using a function based on a certain time period to make predictions outside that period?

Critical Thinking

89. For the function $f(x) = x^2 + 3x + 1$, find $f(3)$, $f(-3)$, $f(x + 3)$, and $f(x - 3)$.

90. Think of four situations you've encountered today outside of math class that could be modeled by a function.

In Exercises 91–94, a function representing some quantity is described, and you're asked to think about what the function might look like. There is no single answer; think about the situation and use some creativity. (Thanks to my friend Bob Davis for inspiring these problems.)

91. An oven is set to 450 degrees when it's turned on. It's left on for 1 hour, then turned off. The function D describes the temperature inside the oven, and the variable t is minutes after the oven was turned on.

(a) Write a plausible domain for $D(t)$, and explain why you chose that domain.

(b) Write a plausible range for $D(t)$, and explain why you chose that range.

(c) Draw a plausible graph of the function $D(t)$, and make sure it matches your answers to parts (a) and (b).

92. The independent variable of a function is n, the number of times during the semester that you do your homework. The output is P, your average score at the end of the semester written as a percent.

(a) Write a plausible domain for $P(n)$, and explain why you chose that domain.

(b) Write a plausible range for $P(n)$, and explain why you chose that range.

(c) Draw a plausible graph of the function $P(n)$, and make sure it matches your answers to parts (a) and (b).

93. A variable y describes the age of a man in years. The function $m(y)$ gives the number of miles an average man of age y can cover on foot in 1 hour.

(a) Write a plausible domain for $m(y)$, and explain why you chose that domain.

(b) Write a plausible range for $m(y)$, and explain why you chose that range.

(c) Draw a plausible graph of the function $m(y)$, and make sure it matches your answers to parts (a) and (b).

94. The variable s describes the amount of sugar a company uses in each can of a new soft drink being developed. The function $t(s)$ gives the resulting taste rating (on a

scale of 0 to 10, with 0 being awful and 10 being awesome) of the drink if s teaspoons of sugar are used. (A survey of soda consumers is being performed to determine taste rating.)

(a) Write a plausible domain for $t(s)$, and explain why you chose that domain.

(b) Write a plausible range for $t(s)$, and explain why you chose that range.

(c) Draw a plausible graph of the function $t(s)$, and make sure it matches your answers to parts (a) and (b).

It's possible to decide if a relation defines a function by just looking at its graph. Decide if each relation in Problems 95–98 is a function. (Hint: Look carefully at the definition of function on page 332 and think about how the ordered pairs correspond to the graph.)

95.

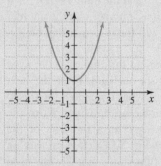

97.

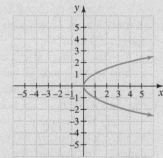

96.

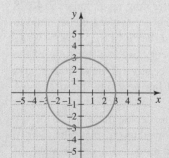

98.

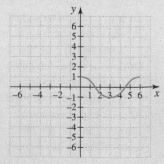

99. Explain why the graphical method for deciding if a relation is a function is called the vertical line test.

Section 6-5	**Quadratic, Exponential, and Logarithmic Functions**

LEARNING OBJECTIVES

1. Graph quadratic functions.
2. Apply quadratic functions to problems in our world.
3. Graph exponential functions.
4. Apply exponential and logarithmic functions to problems in our world.

We saw in Section 6-4 that linear functions can be used to model a lot of things in our world. But there's an obvious problem that limits the modeling capabilities of linear functions: they either always increase or always decrease, and they do so at a constant rate. So they can't accurately model quantities that change direction or have variable rates of change. In this section, we'll study different types of functions that prove useful in modeling other types of data.

Have you ever wondered how much your driving speed affects fuel efficiency? I know I have, and I decided to do some online research to see what I could learn. As you might expect, going either really slow or really fast isn't great for gas mileage: somewhere in between is the sweet spot. I found the numbers in the table, which come from an experiment involving over 2,000 driving hours for a group of 2009–2013 Subaru Outbacks, reported on a site called Automatic.com.

Speed	MPG		Speed	MPG
30	26.6		55	33.0
35	30.2		60	32.3
40	33.1		65	31.5
45	34.2		70	28.6
50	33.9		75	27.3

A quick glance shows that mileage increases until you get to about 45 miles per hour, then starts to decrease at higher speeds. As usual, if we look at a graph of the data, the pattern is even more clear:

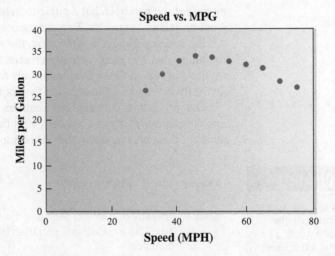

Quadratic Functions

Linear functions are relatively easy to work with for two big reasons: algebraically, the formulas are simple because there are only two types of terms: constants and first powers of the input variable. Graphically, all of the graphs are the same shape: a line. You just need to find a couple of points to decide *which* line.

The next type of function we'll study is a bit more complicated algebraically—there will be one new type of term, with second powers of the variable—and it's also a bit more complicated graphically. But we'll see that all of the graphs have the same basic shape (similar to the points describing speed vs. gas mileage), so we can still draw graphs without too much effort.

A function of the form $f(x) = ax^2 + bx + c$, where a, b, and c are real numbers and $a \neq 0$, is called a **quadratic function**. The graph of a quadratic function is called a **parabola**.

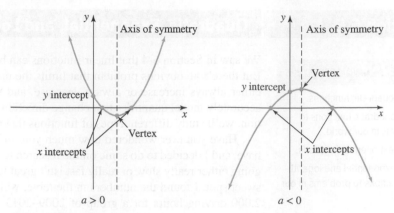

Figure 6-9

Every parabola is shaped kind of like a capital U, or a bowl, like the graphs in Figure 6-9. There are three key things that distinguish a parabola: whether it opens up or down, where the point where it changes direction lives, and how wide it is.

When a parabola decreases in height, reaches a low point, and then increases, we say that it *opens upward*. This occurs when a, the coefficient of x^2, is positive. When a parabola increases in height, reaches a high point, then decreases, we say that it *opens downward*. This occurs when a is negative.

The point where a parabola changes direction is called the **vertex**. For parabolas that open upward, the vertex is the lowest point. For parabolas that open downward, the vertex is the highest point. Every parabola has two halves that are mirror images of each other, with the divider being a vertical line through the vertex. We call this line the **axis of symmetry**. It's not part of the parabola, but a guide to help us draw one. We say that a parabola is *symmetric* about its axis of symmetry. Two representative parabolas are shown in Figure 6-9.

Because a parabola is symmetric, the vertex occurs halfway between the two x intercepts (assuming that the graph actually crosses the x axis). In Exercise 82, we'll see how we can use this fun fact to find a simple formula for finding the vertex of a parabola, which is pretty obviously the key point in drawing the graph.

Our procedure for graphing requires solving quadratic equations by factoring or the quadratic formula. If you need help with those topics, they're covered both in the corequisite guide for this chapter, and in the online supplements.

Math Note	**Procedure for Graphing a Parabola**
You can check to make sure your parabola opens in the correct direction by noting the sign of a. If $a > 0$, it should open upward, and if $a < 0$, it should open downward.	**Step 1** Identify a, b, and c; then use the formula $x = \frac{-b}{2a}$ to find the x coordinate of the vertex. The y coordinate can then be found by substituting the x coordinate into the function.
	Step 2 Find the y intercept by evaluating the function for $x = 0$.
	Step 3 Find the x intercepts, if any, by substituting 0 for $f(x)$ and solving for x, using either factoring or the quadratic formula.
	Step 4 If there are no x intercepts, find at least one other point on each side of the vertex to help determine the shape.
	Step 5 Plot all the points, then connect them with a smooth curve.

Example 1   **Graphing a Quadratic Function**

Graph the function

$$f(x) = x^2 - 6x + 5$$

Math Note

Since the axis of symmetry is a vertical line and passes through the vertex, its equation is $x = 3$.

SOLUTION

Step 1 In this case, $a = 1$, $b = -6$, and $c = 5$. We can begin by finding the x coordinate of the vertex:

$$x = \frac{-b}{2a} = \frac{-(-6)}{2(1)} = \frac{6}{2} = 3$$

Evaluate $f(3)$ to find the y coordinate:

$$f(3) = 3^2 - 6(3) + 5 = 9 - 18 + 5 = -4$$

The vertex is $(3, -4)$.

Step 2 Find the y intercept by evaluating $f(0)$.

$$f(0) = 0^2 - 6(0) + 5 = 5$$

The y intercept is $(0, 5)$.

Step 3 Find the x intercepts by substituting 0 for $f(x)$ and solving.

Math Note

If the equation can't be solved by factoring, use the quadratic formula. If $b^2 - 4ac$ is negative, the parabola has no x intercepts since the square root of a negative number isn't a real number.

$$0 = x^2 - 6x + 5 \qquad \textit{Factor.}$$
$$0 = (x - 5)(x - 1) \qquad \textit{Set each factor} = 0.$$

$0 = x - 5$	$0 = x - 1$
$5 = x$ or	$1 = x$ or
$x = 5$	$x = 1$

The x intercepts are $(5, 0)$ and $(1, 0)$.

Step 4 We already have at least one point on each side of the vertex, so this should be enough to draw the graph.

Step 5 Plot all of the points we found: the vertex $(3, -4)$, and intercepts $(0, 5)$, $(5, 0)$, and $(1, 0)$. Then connect them with a smooth curve, making sure to change direction at the vertex.

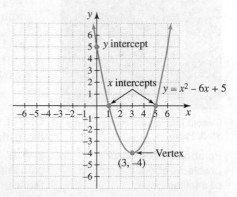

Try This One   1

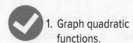

1. Graph quadratic functions.

Graph $f(x) = x^2 - 3x - 10$.

Applications of Quadratic Functions

The next two examples illustrate some of the ways that quadratic functions can be used to solve problems in our world.

Example 2   Modeling Demographic Data with a Quadratic Function

Comstock/Comstock Images/PunchStock

The percentage of the population in the United States that was foreign-born can be modeled by the function $P(x) = 0.0025x^2 - 0.3194x + 16.637$, where x is the number of years after 1900.

(a) When did the percentage reach its low point?

(b) What was the percentage in 2018?

(c) The Pew Research Center has projected that the foreign-born population will reach 18% by 2065. Is that projection in line with this model? Explain.

Math Note
How in the world would anyone find a function that describes some data, like the percentage of the population that's foreign born? By using **regression analysis**, which you'll learn a bit about in an upcoming Sidelight, and a whole lot about in Chapter 11.

SOLUTION

(a) The graph of this function is a parabola that opens upward ($a = 0.0025$ is positive), so the lowest point occurs at the vertex.

$$x = \frac{-b}{2a} = \frac{-(-0.3194)}{2(0.0025)} = \frac{0.3194}{0.005} \approx 63.88$$

The vertex occurs pretty close to $x = 64$. Since x represents years after 1900, the lowest percentage was in 1964.

(b) The year 2018 is 118 years after 1900, so we need to find the output for input $x = 118$.

$$P(118) = 0.0025(118)^2 - 0.3194(118) + 16.637 \approx 13.8$$

In 2018, about 13.8% of the population was foreign-born.

(c) For 2065, we'll need to use $x = 165$ in our model:

$$P(165) = 0.0025(165)^2 - 0.3194(165) + 16.637 \approx 32.0$$

The model predicts a much higher percentage in 2065 than the Pew Center. This is another example of the fact that projecting too far into the future with a model based on past data is dangerous.

Try This One   2

The percentage of cars sold in the United States that are imported can be modeled by the formula $P(x) = 0.082x^2 - 1.44x + 27.2$, where x is years after 1988.

(a) Find the year when the percentage was lowest.
(b) Use the model to estimate the percentage in 2020.
(c) Why is this model unlikely to be useful in the long term?

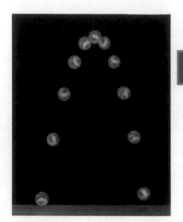

Custom Life Science Images/Alamy
Stock Photo

Parabolas often occur in problems about the motion of a falling object.

One of the most common applications of quadratic functions is *projectile motion*: studying the path of an object that has been propelled with some initial velocity and is subject to gravity. As you can see from the time-lapse photo of a ball in flight shown in the margin, the path appears to be parabolic. This fact was proved by Galileo in the early 17th century, using the world's first high-speed digital camera. (Just kidding. NO idea how he pulled it off.)

Example 3   Studying Projectile Motion

When an object is launched or thrown from some height h_0 (in feet) with an initial upward velocity v_0 (in feet per second), the height in feet of that object t seconds later is given by the function $h(t) = -16t^2 + v_0 t + h_0$. (This ignores the effects of friction and air resistance, which can be negligible for an object like a ball bearing and significant for an object like a paper airplane.) If a bowling ball is launched upward from the top of a 230-foot building with initial velocity 60 ft/sec, find its maximum height and how long it takes to reach the ground. (Legal disclaimer: if you try this, you're a jerk.)

SOLUTION

We're given an initial height of 230 feet, so $h_0 = 230$. We're also given initial velocity $v_0 = 60$, so the height function is

$$h(t) = -16t^2 + 60t + 230$$

The highest point is the vertex of this parabola (notice that the parabola opens downward because $a = -16$ is less than zero).

$$t = -\frac{b}{2a} = -\frac{60}{2(-16)} = 1.875 \quad \textit{High point reached after 1.875 seconds.}$$

$$h(1.875) = -16(1.875)^2 + 60(1.875) + 230 = 286.25$$

The maximum height is 286.25 feet above ground.

To find how long it takes to reach the ground, we substitute $h = 0$ (it hits the ground when the height is zero) and solve for t (find *time*—get it?).

$$0 = -16t^2 + 60t + 230$$

$$t = \frac{-60 \pm \sqrt{60^2 - 4(-16)(230)}}{2(-16)} \approx -2.35, 6.10$$

If you find a way to go back in time, let me know, but until then the negative answer is silly. It takes 6.1 seconds for the ball to reach the ground.

Try This One   3

A decent Major League pitcher can throw a ball at 90 miles per hour, which is 132 feet per second. How high could one of these guys throw a ball straight up, and how long do they have to get out of the way to avoid getting hit in the melon when it comes back down? (Let's say the pitcher is 6 feet tall, and he releases the ball from a point a foot above his head.)

In some situations, a quadratic function can be written based on a description.

Example 4 | Fencing a Yard Efficiently

A family adopts a new puppy and plans to fence in a rectangular portion of their backyard for an exercise area. They buy 60 feet of fencing and plan to put the fence against the house so that fencing is needed on only three sides. Find a quadratic function that describes the area enclosed, and use it to find the dimensions that will enclose the largest area.

Comstock Images/Alamy Stock Photo

SOLUTION

In this situation, a diagram will be very helpful. If we let $x =$ the length of the sides touching the house, then the side parallel to the house will have length $60 - 2x$ (60 feet total minus two sides of length x).

The area of a rectangle is length times width, so in this case, we get

$$A(x) = x(60 - 2x)$$

To find a, b, and c, we should multiply out the parentheses:

$$A(x) = 60x - 2x^2 \quad \text{or} \quad A(x) = -2x^2 + 60x \qquad a = -2, b = 60, c = 0$$

The maximum area will occur at the vertex of the parabola:

$$x = \frac{-b}{2a} = \frac{-60}{2(-2)} = \frac{-60}{-4} = 15$$

The sides touching the house should be 15 feet long. The side parallel to the house should be $60 - 2(15) = 30$ feet long.

2. Apply quadratic functions to problems in our world.

Try This One
4

Suppose the family in Example 4 decides to move the dog pen away from the house, so that fence is needed on all four sides. What dimensions will provide the largest area?

Exponential Functions

On March 1, 2020, there had been a total of 29 confirmed cases of COVID-19 in the United States. Some very prominent people proclaimed that would be down to zero in short order. As we know, they were tragically wrong. The cumulative number of cases for the next 11 days is shown in the graph, and the graph has one striking feature: it starts out low, but grows in a BIG hurry.

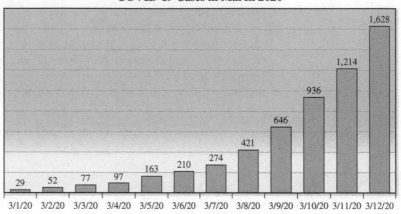

COVID-19 Cases in March 2020

The types of functions we've studied so far aren't particularly effective at modeling such rapid growth. The next type of function on our list is really great for this type of quantity.

Some of the functions we've worked with so far have exponents, but in every case the exponent has been a number, while the base has been a variable. This next batch of functions reverses that—the variable appears in an exponent. These functions are very useful in modeling situations like investments and loans, depreciation, disease spread, growth and decay, and population changes.

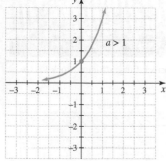

(a)

An **exponential function** has the form $f(x) = c \cdot a^x$, where a is a positive real number, but not 1, and c is any nonzero real number.

Examples of exponential functions are

$$f(x) = 2^x \qquad f(x) = 200 \cdot 10^x$$

$$f(x) = 5\left(\frac{1}{3}\right)^x \qquad f(x) = (0.5)^x$$

The graph of an exponential function has two forms.

1. When $a > 1$, the function increases as x increases, as in Figure 6-10a.

2. When $0 < a < 1$, the function decreases as x increases, as in Figure 6-10b.

For any acceptable value of a, the expression $a^0 = 1$, so every exponential function of the form $f(x) = a^x$ has y intercept $(0, 1)$. Also, the graph approaches the x axis in one direction but never touches it. When this happens, we say that the x axis is a **horizontal asymptote** of the graph.

Since all of the exponential functions have the same basic shape, we can draw the graph by plotting a handful of points and drawing a curve similar to the ones in Figure 6-10.

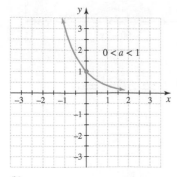

(b)

Figure 6-10

| **Example 5** | Graphing Exponential Functions |

Draw the graph of each function.

(a) $f(x) = 2^x$ (b) $g(x) = \left(\dfrac{1}{2}\right)^x$

SOLUTION

(a) Pick several numbers for x and find $f(x)$:

$$x = -2: f(-2) = 2^{-2} = \frac{1}{2^2} = \frac{1}{4}$$

$$x = -1: f(-1) = 2^{-1} = \frac{1}{2}$$

$$x = 0: f(0) = 2^0 = 1$$

$$x = 1: f(1) = 2^1 = 2$$

$$x = 2: f(2) = 2^2 = 4$$

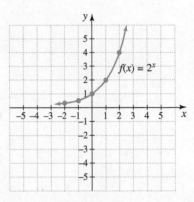

Plot the points, then connect them with a smooth curve (first graph at right). Note the graph approaching the x axis to the left.

(b) Pick several numbers for x and find $f(x)$.

$$\text{For } x = -2, \ f(-2) = \left(\frac{1}{2}\right)^{-2} = 4$$

$$\text{For } x = -1, \ f(-1) = \left(\frac{1}{2}\right)^{-1} = 2$$

$$\text{For } x = 0, \ f(0) = \left(\frac{1}{2}\right)^{0} = 1$$

$$\text{For } x = 1, \ f(1) = \left(\frac{1}{2}\right)^{1} = 0.5$$

$$\text{For } x = 2, \ f(2) = \left(\frac{1}{2}\right)^{2} = 0.25$$

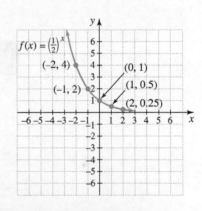

Plot the points and draw the graph. Note the graph approaching the x axis to the right.

Calculator Guide

The keys needed to compute exponential expressions are different on scientific and graphing calculators. To find 2^{-2} on a calculator:

Typical Scientific or Online Calculator:

2 [y^x] or [x^y] 2 [±] [=]

Result: 0.25

Typical Graphing Calculator:

2 [^] [(-)] 2 [ENTER] Result: 0.25

3. Graph exponential functions.

| Try This One | 5 |

Draw the graph of each function.

(a) $f(x) = 2^{x-2}$ (b) $g(x) = \left(\dfrac{1}{3}\right)^x$

We mentioned earlier that exponential functions are useful for modeling a variety of situations. The next three examples illustrate some of these applications; the third will require defining a new function first.

| **Example 6** | Exponential Population Growth |

Williams County, ND, was one of the fastest-growing counties in America in 2015. One estimate of that growth uses the function $A(t) = A_0(1.095)^t$ to model the population, where A_0 is the population at time $t = 0$, and t is time in years. According to the U.S. Census Bureau, the population was 22,398 in 2010.

(a) Use the model to find the population in 2015.
(b) Use the model to predict the population in 2050, and discuss whether this seems like a reasonable prediction.

David Buffington/Photodisc/Getty Images

SOLUTION

(a) In this case, we'll use $A_0 = 22{,}398$ and use 2010 as $t = 0$. That makes 2015 $t = 5$:

$$A(5) = 22{,}398(1.095)^5 \approx 35{,}259.8$$

Of course, we can't have 0.8 of a person living in North Dakota (or anywhere else, for that matter), so the predicted population in 2015 is 35,259 or 35,260 depending on whether you choose to round up, or decide that 0.8 is still less than a full person and round down.

(b) The year 2050 corresponds to $t = 40$:

$$A(40) = 22{,}398(1.095)^{40} \approx 844{,}839.1$$

The 2050 prediction is 844,839 people. Whether this is reasonable is to some extent an opinion, but . . . from what I can tell, Williams County is a very nice place to live. But I have a hard time believing that an area that small will have almost a million people in less than 35 years, so I'm going to say that our model loses effectiveness that far out.

Try This One 6

The actual population of Williams County in 2020 was 40,950, which is quite a bit below what the original model predicted. A less optimistic projection at that point can be modeled by $A(t) = A_0(1.062)^t$, where A_0 is again the population at time $t = 0$ and t is time in years. Using 2015 as beginning time and the answer to Example 6(a) as beginning population, what does this model predict the population will be in 2050? How realistic do you think this is?

Sidelight   Where Do Modeling Functions Come From?

In some of the examples In this section, a function that models some real quantity was provided. This leads to an obvious question: where do these functions come from? How do you find a formula that models some actual data? This was a much harder question 20 or so years ago, before the rise of graphing calculators and spreadsheets.

The process of finding a certain type of function to model a data set is known as **regression**. It originated with a method to find a straight line that best fits certain data, which was discovered in the early 1800s. The advent of computers changed everything, though, as other types of models can be found using techniques that would be at best extremely tedious by hand. Today's graphing calculators come preprogrammed with a variety of options that can be used to model a wide variety of data, not just data that happen to resemble a straight line when you plot the data

points. The screen shot to the right, from a TI-84, shows some of the options available. Of particular interest to us are LinReg, which finds a linear function; QuadReg, which finds a quadratic function; LnReg, which finds a logarithmic function; and ExpReg, which finds an exponential function. Each of those choices was used to find at least one function mentioned in this section.

```
EDIT CALC TESTS
4:LinReg(ax+b)
5:QuadReg
6:CubicReg
7:QuartReg
8:LinReg(a+bx)
9:LnReg
0↓ExpReg
```

There are a number of excellent resources available on the Internet to help you learn to use this extremely valuable feature of a graphing calculator, and there is also a wide variety of online regression calculators that will find models based on data that you input.

Example 7	**Computing the Value of an Investment**

Rolf Bruderer/Blend Images LLC

It's often said that people would save more if they understood how compound interest worked. The calculation in Example 7 shows that if you invest just $10,000 at age 25 and receive 8% interest (a reasonable long-term expectation) until you retire at age 65, that amount will have grown to almost a quarter of a million dollars.

Interest on a savings account can be compounded annually, semiannually, quarterly, or daily. The formula for compound interest is given by

$$A = P\left(1 + \frac{r}{n}\right)^{nt}$$

where A is the amount of money, which includes the principal plus the earned interest, P is the principal (amount initially invested), r is the yearly interest rate in decimal form, n is the number of times the interest is compounded a year, and t is the time in years that the principal has been invested. If $10,000 is invested at 8% per year compounded quarterly, find the value of the investment (amount) after 40 years.

SOLUTION

$P = \$10,000$
$r = 8\% = 0.08$
$t = 40$ years
$n = 4$, since the interest is compounded quarterly, or four times a year.

$$A = P\left(1 + \frac{r}{n}\right)^{nt}$$
$$= \$10,000\left(1 + \frac{0.08}{4}\right)^{4\cdot40}$$
$$= \$237,699.07$$

At the end of 40 years, the $10,000 will grow to $237,699.07!

Try This One   7

Find the value of a $5,000 investment compounded semiannually (i.e., twice a year) for 30 years at 6% interest.

Here's a reasonable question you might ask when investing in an account like the one in Example 7: how long would you have to wait for the investment to grow to a certain amount? Being able to answer questions like this requires a new function that allows us to solve equations where the variable (time in this case) is in an exponent. In much the same way that we use roots to "undo" powers when solving equations, we can use a group of functions called **logarithms** to "undo" exponentials.

Roots are actually a really good analogy for understanding logarithms, or *logs* for short. There are many different powers: square, cube, fourth power, etc., so there are many different roots: square root, cube root, fourth root, etc. Exponential functions have many potential bases: we graphed exponentials with bases 2, 1/2, and 1/3 in Example 5 and Try This One 5. So there are many different logarithms, depending on the base. For example, the logarithm function with base 2, denoted $\log_2 x$, is the function that "undoes" an exponential with base 2. That is,

$$\log_2(2^x) = x \qquad \text{The log "undoes" the exponential,}$$
leaving the x unchanged.

Math Note

The definition we've given of logarithm is a little bit limited. In a college algebra book, there are multiple sections devoted to studying log functions in depth.

In this section, we're restricting our study to using logs to solve equations with exponentials in them. We'll study a couple of applications of log functions in the exercises.

For any base b that is positive but not 1, the **logarithm function with base b**, denoted $\log_b x$, is a function with the following property: for any real number x, $\log_b(b^x) = x$.

The fact that there are many different logarithms (infinitely many, in fact) presents a problem when programming a calculator. You can't put infinitely many log buttons on a calculator. You could try, but you would not be successful. Most calculators have a button for log base 10, which is simply denoted log (if there's no base listed, it means the base is 10). Fortunately, there's a formula that allows us to change the base of a logarithm. Shockingly, it's called the **change of base formula**. We'll see how to use it in Example 8.

> **Math Note**
>
> In words, the change of base formula says that to calculate the log to any base of some number, you find log base 10 of that number, then divide by log base 10 of the original base.

The Change of Base Formula

For any acceptable base b and any positive real number a,

$$\log_b a = \frac{\log a}{\log b}$$

Example 8 | Finding the Amount of Time Needed to Reach an Investment Goal

After being injured in an accident with a driver who was texting while driving, Maureen is awarded an $80,000 settlement. She wants to invest the money so that she has $100,000 to put toward a home. If the investment draws 6% interest compounded monthly, how long will it take for her to reach that goal?

SOLUTION

First, we need the interest formula with $r = 0.06$ (6% interest), $n = 12$ (compounded monthly), $P = 80,000$ (initial investment), and $A = 100,000$ (final amount we're hoping for).

$$A = P\left(1 + \frac{r}{n}\right)^{nt}$$

$$100,000 = 80,000\left(1 + \frac{0.06}{12}\right)^{12t}$$

$$100,000 = 80,000(1.005)^{12t}$$

> **Calculator Guide**
>
> Computing log base 10 of 1.25 with a calculator is really simple:
>
> **Typical Scientific or Online Calculator:**
>
> 1.25 [LOG]
>
> **Typical Graphing Calculator:**
>
> [LOG] 1.25 [ENTER]

We want to solve this equation for t. To do so, we'll isolate the exponential expression and "undo" the exponential with a logarithm.

$100,000 = 80,000(1.005)^{12t}$ *Divide both sides by 80,000.*

$1.25 = (1.005)^{12t}$ *Apply log base 1.005 to both sides.*

$\log_{1.005} 1.25 = \log_{1.005}(1.005)^{12t}$ $log_{1.005}(1.005)^{12t} = 12t$

$\log_{1.005} 1.25 = 12t$ *Divide both sides by 12.*

$$t = \frac{\log_{1.005} 1.25}{12}$$

To see what number of years this actually is, we need to find $\log_{1.005} 1.25$ using the change of base formula.

$$\log_{1.005} 1.25 = \frac{\log 1.25}{\log 1.005} \approx 44.7$$

$$t = \frac{44.7}{12} \approx 3.7$$

It will take about 3.7 years for the investment to grow to $100,000.

Try This One   **8**

If Maureen finds an investment that offers 7.5% interest compounded daily, how much sooner will she have the $100,000 she's hoping for?

| **Example 9** | **Carbon Dating** |

age fotostock/Alamy Stock Photo

Carbon-14 is a radioactive isotope found in all living things. It begins to decay when an organism dies, and scientists can use the proportion remaining to estimate the age of objects derived from living matter, like bones, wooden tools, or textiles. The function $f(x) = A_0 2^{-0.0175x}$ describes the amount of carbon-14 in a sample, where A_0 is the original amount and x is the number of centuries (100 years) since the carbon-14 began decaying.

A wooden bowl is found at an archaeological dig in Peru, and when analyzed, it's found to have 62% of the amount of carbon-14 found in a living sample of that type of wood. About how old is the bowl?

SOLUTION

We're given that the amount of carbon-14 is 62% of the original amount, which makes $f(x) = 0.62A_0$. We need to set up and solve an equation using the function provided.

$$0.62A_0 = A_0 2^{-0.0175x}$$ *Divide both sides by A_0.*

$$0.62 = 2^{-0.0175x}$$ *Apply log base 2 to both sides.*

$$\log_2 0.62 = \log_2 2^{-0.0175x}$$ $\log_2 2^{-0.0175x} = -0.0175x$.

$$\log_2 0.62 = -0.0175x$$ *Divide both sides by -0.0175.*

$$x = \frac{\log_2 0.62}{-0.0175} \approx 39.4$$

The bowl is 39.4 centuries, or 3,940 years, old.

Try This One 9

4. Apply exponential and logarithmic functions to problems in our world.

It's very common for archaeological digs to have several layers of artifacts from different eras. An ax handle found above the wooden bowl in Example 9 has 72% of the original amount of carbon-14 remaining. How much later was it made than the bowl?

Answers to Try This One

1

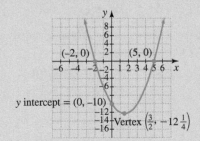

$(-2, 0)$ $(5, 0)$
y intercept = $(0, -10)$
Vertex $\left(\frac{3}{2}, -12\frac{1}{4}\right)$

2 (a) 1997 (b) 65.1%
(c) After reaching a low point, a parabola then goes upward indefinitely. Since this is a percentage, it can't go higher than 100%.

3 279.25 feet; 8.26 sec.

4 15 feet by 15 feet

5 (a)

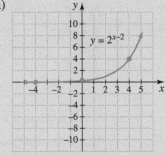

$y = 2^{x-2}$

(b)

6   The prediction is now 453,352 people. This is quite a lot lower than the other projection, but still seems too high to be realistic.

7 $29,458.02

8   About 0.7 years, or a bit less than 8½ months.

9   About 1,230 years

Exercise Set   6-5

Writing Exercises

1. Describe in your own words what a parabola is.
2. How can you tell whether a parabola opens up or down from its equation?
3. Explain how to find the maximum or minimum of a quadratic function.
4. What is the major difference between exponential functions and quadratic functions?
5. What are logarithms used for?
6. Why are there many different logarithms?
7. Explain what the change of base formula is good for, and how to use it.

8. What does it mean to say that a function models some quantity?
9. What type of quantity is a quadratic function likely to be a reasonable model for?
10. Name some quantities that exponential functions are used to model.
11. What are some differences between quantities modeled by quadratic functions, and quantities modeled by linear functions?
12. Describe what you've learned about using functions based on data to make predictions about future quantities.

Computational Exercises

For Exercises 13–24, graph each parabola.

13. $f(x) = x^2$
14. $f(x) = x^2 - 6x$
15. $f(x) = x^2 + 6x + 9$
16. $f(x) = 4x^2 - 4x + 1$

17. $f(x) = -x^2 + 12x - 36$
18. $f(x) = -2x^2 + 3x + 4$
19. $f(x) = -10x^2 + 20x$
20. $f(x) = -3x^2 + 5x + 2$

21. $f(x) = \frac{3}{4}(x - 5)^2 + 6$
22. $f(x) = \frac{1}{2}(x + 3)^2 + 10$
23. $f(x) = -\frac{2}{3}(x + \frac{1}{2})^2 + 10$
24. $f(x) = -\frac{11}{4}(x + \frac{17}{2})^2 - \frac{21}{4}$

For Exercises 25–34, graph each exponential function.

25. $f(x) = 5^x$
26. $f(x) = 3^{x+1}$
27. $f(x) = (\frac{1}{2})^{x-2}$

28. $f(x) = (\frac{1}{4})^{x+1}$
29. $f(x) = 2^{-x}$
30. $f(x) = 3^{-x}$

31. $f(x) = -\frac{2}{3} \cdot 5^{-x}$
32. $f(x) = 0.8 \cdot 5^{x+1}$

33. $f(x) = -3^{|x|}$
34. $f(x) = 2^{-|x|}$

In Exercises 35–38, use the change of base formula to calculate each value. Round your answer to one decimal place.

35. $\log_2 10$
36. $\log_{10} 2$

37. $\log_{1.7} 25.4$
38. $\log_{2.36} 1{,}000$

In Exercises 39–46, solve the equation. Round your answer to one decimal place if necessary.

39. $3^x = 243$
40. $4^x = 4{,}096$
41. $2^{11t} = 90$
42. $3^{4t} = 55$

43. $5(1.1)^{2x} = 10$
44. $9(2.3)^{3x} = 18$
45. $13 - 8(4)^{y+3} = 2$
46. $6 + 3(7)^{y-4} = 21$

Applications in Our World

Problems 47–50 use the projectile motion formula from Example 3, $h(t) = -16t^2 + v_0 t + h_0$.

47. How high does a ball go if it's thrown upward from 6 feet above ground level at 60 feet per second? How long does it take to reach the ground?

48. A model rocket engine is designed to provide an initial upward velocity of 425 feet per second, and is designed so that the second stage pops the recovery parachute

15 seconds after launch. How far away will the rocket be from its highest point when the parachute deploys? Will it be on the way up or the way back down?

49. A spy needs to climb the outside of a castle in Latvia (don't ask why—if I told you I'd have to kill you). So they throw one of those cool hooks they use in movies to climb up walls. Our spy's top throwing velocity for the 4-pound hook is 74 feet per second, and the hook is released from a height of 3 feet. Will the hook reach over the castle wall, which is 80 feet high?

50. A bottlenose dolphin has a top swimming speed of about 32 feet per second. If a dolphin in an aquarium show jumps out of the water at that speed, can it jump through a circular hoop with diameter 4 feet that's centered 17 feet above the water?

51. The profit for a company that produces discount sneakers can be modeled by the quadratic function $P(x) = 4x^2 - 32x + 210$, where x is the number of pairs of sneakers sold.
    (a) For how many pairs of sneakers did the profit reach its lowest point?
    (b) What was the lowest profit made by the company?
    (c) What is the profit for selling 18 pairs of sneakers?
    (d) How many pairs of sneakers must be sold to have a profit of $162?

52. The cost (in hundreds of dollars) for Jake's Snowplowing business can be modeled by the quadratic function $C(t) = -t^2 + 12t$, where t is the time in months. Let $t = 0$ correspond to the start of a new year.
    (a) In what month does the snowplowing business have the most cost? Why do you think that is?
    (b) What is the most the company will have to spend to stay in business?
    (c) In March, what will the costs be to run the business?
    (d) In which month(s) will the cost be $3,200? (*Hint:* Let $C = 32$.)

53. The revenue for jet ski rentals can be modeled by $R(x) = -x^2 - 6x + 432$, where x represents the number of $0.50 price hikes.
    (a) What is the most revenue and how many price hikes should occur to obtain the most revenue? What is the meaning of your answer?
    (b) If the rental company charges $2.00 more, how much revenue will they make?
    (c) If the rental company charges $5.00 less, how much revenue will they make?

54. An online vendor that sells boxes of computer paper directly to businesses has weekly revenue function $f(x) = 2,400x - 60x^2$, where x is the number of boxes sold in hundreds.
    (a) How many boxes sold will maximize the revenue?
    (b) What's the highest possible weekly revenue?
    (c) What will the revenue be from selling 1,000 boxes?
    (d) How many boxes would need to be sold for the revenue to be $20,000?

55. The length of a picture is $2\frac{1}{2}$ times its width.
    (a) Find an equation for the area as a function of the width.
    (b) If the width is 10 inches, what is the length and what is the area?
    (c) If the area is 40 square inches, what is the length and width of the picture?
    (d) If the width was doubled and the relationship between length and width stayed the same, what would be the new area function?

56. Find the length of the sides of a gutter consisting of three sides that can be made from a piece of aluminum that is 16 inches wide in order for it to carry the maximum capacity of water. (The shape of the gutter is ⊔.)

57. The population growth of an emerging African nation is defined by the function $f(t) = A_0(1.4)^t$, where A_0 is the present population and t is the time in decades. If the present population is 4,200,000, find the population in 5 years.

58. In a biology experiment, there were 300,000 cells present initially and then the number started decreasing every second. The number of cells present can be modeled by the function $f(t) = 300,000(5)^{-0.4307t}$, where t is the time in seconds. Find how many cells remain after 10 seconds.

59. The monthly mortgage payment on a house that was financed for $100,000 at a 3.5% interest rate can be calculated using the function

$$P(t) = \frac{(0.035)(100,000)}{1 - \left(1 + \dfrac{0.035}{12}\right)^{-12t}} \div 12$$

where t is the time the mortgage was financed in years. What is the monthly payment if the house was financed for 30 years?

60. A baby that weighs 7 lb at birth may increase in weight by 11% per month. Use the function $f(t) = K(1 + r)^t$, where K is the initial value, r is the decimal value of the percentage of increase ($r = 0.11$), and t is the time in months. How much would the baby weigh 6 months from birth?

Problems 61–64 use the interest formula from Example 7,
$$A = P\left(1 + \frac{r}{n}\right)^{nt}.$$

61. An investment offers 5.5% interest compounded weekly if the investor keeps the money untouched for a 3-year period. (This is known as a *certificate of deposit.*)
    (a) How much would an initial investment of $8,000 be worth at the end of the initial 3-year period?
    (b) If the investor decides to roll that new amount into another certificate of deposit with the same terms, find the value at the end of the second 3-year period.

62. Facing some stiff gambling debts, Lucky Louie turns to a quick loan from a local loan shark. Louie borrows $2,500 at 70% interest compounded daily.
    (a) The friendly neighborhood loan shark informs Louie that his thumbs will be broken unless he pays the loan off in full in exactly 1 year. How much will he owe at that point?

(b) Louie decides that in the interest of avoiding broken thumbs, he'd just as soon stay well away from the 1-year deadline. Instead, after hitting a winning streak, he decides to pay the loan back when the balance reaches $3,500. When should he pay?

63. If you invest $20,000 at 4% interest for 5 years, make a table displaying the ending value of the investment if interest is compounded quarterly (four times a year), monthly, weekly, and daily. What can you conclude?

64. Repeat Problem 63, but for compounding hourly, every minute, and every second. Now what can you conclude?

Problems 65–68 use the carbon dating equation from Example 9, $f(x) = A_0 2^{-0.0175x}$.

65. In 1998, the Shroud of Turin was examined by researchers, who found that plant fibers in the fabric had 92.1% of the amount of carbon-14 in a living sample. If this is accurate, when was the fabric made?

66. In 2003, Japanese scientists announced the beginning of an effort to bring the long-extinct woolly mammoth back to life using modern cloning techniques, a goal that I think we can all agree is pretty cool. Their efforts were focused on an especially well-preserved specimen discovered frozen in the Siberian ice. Nearby samples of plant material were found to have 28.9% of the amount of carbon-14 in a living sample. What was the approximate age of these samples?

67. Fossilized dental remains found in England in 1964 that were previously thought to be Neanderthal were found in 2011 to actually be human. That startled the scientific community, as they showed that modern humans lived in Europe over 7,000 years earlier than it was previously thought. The testing indicated that 0.42% of the carbon-14 remained. How old were these remains?

68. There is considerable debate over how long humans have lived in the Americas, fueled in part by a discovery made in 2004 by archaeologist Al Goodyear from the University of South Carolina. He discovered a site in 2004 that purportedly contains evidence of a human settlement that predates other estimates by 10,000 years. Carbon dating of burned plant material indicated 0.2% of the amount of carbon-14 in a live sample. How old was that sample if this was accurate?

You probably know that when an earthquake is reported, how powerful it was (known as the magnitude) is given as a decimal number somewhere between 0 and 10. The scale used to measure the magnitude is known as the Richter scale, named in honor of a seismologist in California who devised the first version of the scale in 1935. Magnitude is given by the function $M(E) = \frac{2}{3}\log\frac{E}{10^{4.4}}$, where E is the amount of energy released by the quake. Use this function in Problems 69–72. (If you need help with scientific notation, it's in Section 5-6.)

69. The deadliest earthquake of 2020 struck Turkey on October 30, resulting in 119 deaths. The quake released 7.943×10^{14} joules of energy. What did it measure on the Richter scale?

70. In 2021 there was a much more powerful earthquake off the coast of Alaska that fortunately did not result in any fatalities. It released 5.012×10^{16} joules of energy. This is roughly 60 times as much energy as the Turkish quake of 2020. How much higher did it measure on the Richter scale? (See Problem 69.)

71. The average daily energy consumption for the entire United States for 2015 was 2.62×10^{17} joules. What would be the magnitude of an earthquake that released that much energy?

72. The largest and most powerful nuclear weapon ever detonated was tested by the Soviet Union on October 30, 1961, on an island in the Arctic Sea. The blast was so powerful that there were reports of windows breaking in Finland over 700 miles away. The detonation released about 2.1×10^{17} joules of energy. What would be the magnitude of an earthquake that released that much energy?

*The **decibel** scale for measuring sound intensity describes how powerful a sound is. Zero decibels is the quietest sound an average human can hear, and 120 decibels is where a sound starts to cause pain. The function $D(I) = 10\log\left(\frac{I}{10^{-12}}\right)$ calculates the decibel level, where I is the intensity of the sound measured in watts of power per square meter. Use this function to find the decibel level of some common sounds with the given intensities in Problems 73–76. (If you need help with scientific notation, it's in Section 5-6.)*

73. Whisper: Intensity $= 5.2 \times 10^{-10}$
74. Heavy traffic: Intensity $= 8.5 \times 10^{-4}$
75. Normal conversation: 3.2×10^{-6}
76. Jet plane: 8.3×10^{2}

Critical Thinking

77. What is the difference between the function $f(x) = 2^{-x}$ and the function $f(x) = \left(\frac{1}{2}\right)^x$?

78. Suppose that you know the following about the graph of some quadratic function $f(x)$: the vertex is $(4, -10)$ and one of the x intercepts is $(-1, 0)$. What is the other one?

79. If an equation of the form $ax^2 + bx + c = 0$ has no real solutions and $a > 0$, what does that tell you about the graph of the function $f(x) = ax^2 + bx + c$?

80. For any acceptable base, the function $f(x) = \log_b x$ is undefined if x is zero or negative. Explain why. (*Hint:* Think about why negative numbers don't have square roots.)

81. Dallas Cowboys Stadium has the world's largest high-def video screen, which is cool. What's not quite so cool is that it's only 90 feet above the field, and in the first exhibition game at the stadium, a punted football hit the video screen. Using the projectile motion formula from Example 3, find the initial velocity that would be necessary for a ball kicked upward from 2 feet off the ground to just reach a height of 90 feet. (This is NOT easy. . . .)

82. Because a parabola is symmetric, its vertex has to be halfway between any two points at the same height. In particular, it's halfway between the two intercepts (if there are any intercepts). Use the quadratic formula to show how this leads to the formula $x = -\frac{b}{2a}$ for finding the first coordinate of a vertex.

83. Suppose that some quantity doubles in the span of 1 year, starting at 2 in year 1 and finishing at 4 in year 2. A graph of this quantity would contain the points $(1, 2)$ and $(2, 4)$. We'll use this example to compare linear growth, quadratic growth, and exponential growth.

    (a) Verify that each of the functions $f(x) = 2x$, $g(x) = 2x^2 - 4x + 4$, and $h(x) = 2^x$ contains those two points. Then fill in the table of values:

x	3	4	5	6	7
$2x$					
$2x^2 - 4x + 4$					
2^x					

    (b) Draw the graphs of all three functions for $x > 1$ on the same set of axes.

    (c) What can you conclude about the differences between linear, quadratic, and exponential growth?

84. Now let's compare linear, quadratic, and exponential decline. Suppose that some quantity is halved in 1 year, starting at 20 in year 1 and finishing at 10 in year 2. A graph of this quantity would contain the points $(1, 20)$ and $(2, 10)$.

    (a) Verify that each of the functions $f(x) = -10x + 30$, $g(x) = -10x^2 + 20x + 10$, and $h(x) = 20(\frac{1}{2})^{x-1}$ contains those two points. Then fill in the table of values:

x	3	4	5	6	7
$-10x + 30$					
$-10x^2 + 20x + 10$					
$20\left(\frac{1}{2}\right)^{x-1}$					

    (b) Draw the graphs of all three functions for $x > 1$ on the same set of axes.

    (c) What can you conclude about the differences between linear, quadratic, and exponential decline?

85. The richest family in town hires you to do 2 weeks of work landscaping their estate, but they're a little bit eccentric. Instead of just making an offer, they give you three choices of payment. With choice 1, you start out at $10 the first day, then $10 is added to the payment for each additional day until the 14th day. For choice 2, you get one dollar the first day, and your pay in dollars for any day is the square of the number of days into the job. For choice 3, you get a dime the first day, 20 cents the second day, 40 cents the third, and your pay doubles every day until the 14th.

    (a) You have 10 seconds to decide or the job goes to someone else. Which do you choose?

    (b) Calculate how much money you would make in those 14 days with each option. How well did you choose?

    (c) Which type of growth is illustrated by each option?

86. A new resort in Costa Rica is managed by a retired math professor with a strange sense of humor. Patrons can choose one of three payment options. In each case, the cost decreases each day, and you have to leave when it reaches zero. Option 1: per-person cost is $200 the first day, then decreases by $25 every day. Option 2: per-person cost is $300 the first day, and in each successive day, you subtract 10 times the square of the number of days you've been there from the previous day's cost. Option 3: per-person cost is $500 the first day and $250 the second day; in general, every day is half as much as the previous day.

    (a) Which option is the best if your primary goal is to spend the least amount of money? How much would you spend and for how many days?

    (b) Which option is best if your primary goal is to stay the longest? How long would you end up staying?

    (c) Which type of decline is illustrated by each option?

Section	Important Terms	Important Ideas
6-1	Variable	**Many problems** in our world can be solved by writing an equation that describes the situation in the problem and then evaluating or solving that equation to find desired quantities. The first step in doing so is usually identifying quantities that can change (or vary) and represent them with letters. In many cases, a statement within the problem can be translated into an equation.
6-2	Ratio Proportion Cross multiply Direct variation Inverse variation	**The most efficient** way to compare the sizes of two quantities is to divide them, forming a ratio. A proportion is a statement that two ratios are equal; proportions come in handy in solving a ton of problems in our world. When two quantities vary directly, if one goes up the other does proportionally. When two quantities vary inversely, if one goes up the other goes down proportionally.
6-3	Rectangular coordinate system Cartesian plane x axis y axis Origin Quadrants Coordinates Linear equation in two variables Graph of an equation x intercept y intercept Slope Slope-intercept form	**The rectangular** coordinate system (also called the Cartesian plane) consists of two number lines: one vertical axis called the y axis and one horizontal axis called the x axis. The axes divide the plane into four quadrants. A point is located on the plane by its coordinates, which consist of an ordered pair of numbers (x, y). The graph of an equation is the set of all points in the plane whose coordinates make the equation true when substituted in for the variable. The graph of a linear equation in two variables, which is an equation that can be written in the form $ax + by = c$, is a straight line. The slope of a line is a measure of how steep it is, and is defined to be the change in y coordinate divided by the change in x coordinate. The slope describes the rate at which the y coordinate is changing with respect to the x coordinate. A horizontal line has slope zero, while a vertical line has undefined slope. When an equation is written in the form $y = mx + b$, m is the slope, and the point $(0, b)$ is the y intercept. Linear equations can be used to effectively model situations where the rate of change of some quantity is either constant, or very close to constant.
6-4	Relation Function Function notation Evaluate Domain Range Vertical line test Linear function	**A relation** is a rule matching up two sets of objects. Relations can be represented by sets of ordered pairs, graphs, or equations relating two variables. A relation is called a function if every element in the domain (the set of first coordinates) corresponds to exactly one element in the range (the set of second coordinates). Function notation is used to name functions: $f(x)$ represents a function named f whose variable is x, and is read "f of x." A linear function has the form $f(x) = ax + b$. The graph is a straight line with slope a and y intercept $(0, b)$. Linear functions are used to model data with a rate of change that is constant, or close to constant.
6-5	Quadratic function Parabola Vertex Axis of symmetry Exponential function Asymptote Logarithmic Function	**A quadratic** function has the form $f(x) = ax^2 + bx + c$. The graph is a parabola that opens up if $a > 0$ and down if $a < 0$. Quadratic functions are useful in modeling data that show a change in direction. Exponential functions have the variable in an exponent. They have many applications to finance, population studies, archaeology, and other areas. Logarithms are functions defined to "undo" exponential functions. They are used to solve equations in which the variable is in an exponent. Logs are very useful in analyzing applications of exponential functions.

Math in Drug Administration REVISITED

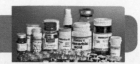

Gennadiy Poznyakov/123RF

1. Football player: 647 mg; girl: 153 mg
2. The potentially lethal dose for the girl is 574 mg, so his dose could kill her.
3. The minimum effective dose for the player is 404 mg, so her dose will do him no good.
4. The BSA ratio is $\frac{2.60}{1.02} \approx 2.54$. The weight ratio is $\frac{275}{65} \approx 4.23$. The ratio is much bigger for weight than for BSA.
5. 1.95
6. The recommended dosages are now 533 mg for the player and 209 mg for the girl. The lethal dose for the girl is 785 mg, so she'll be okay with his dose. The minimum effective dose for the player is now 333 mg, so her dosage will still do him no good.
7. Conclusions can vary, but it certainly seems like if the dosage is based on weight rather than BSA, it exaggerates the differences in dosage.

Review Exercises

Section 6-1

For Exercises 1–4, write each phrase in symbols.

1. 8 times a number decreased by 4
2. 3 added to four times a number
3. The product of 6 and the sum of one-half and a number
4. Three-fourths subtracted from the quotient of 18 and five times a number
5. A student is making the drive home for Thanksgiving, which takes 8 hours round trip. With the holiday traffic, this student averages 40 miles per hour on the way home and 50 miles per hour back to school. Find the time it took to get home and the time it took to get back to school. (Use the formula distance = rate × time).
6. The health food store sells mixed soy nuts for $1.20 a pound and Asian trail mix for $1.80 a pound. If the store sold 2 more pounds of Asian trail mix than soy nuts and the total sales were $21.60 for these items for the day, how many pounds of each did the store sell?
7. The manager of a concert venue is adding up the proceeds from a Friday night show. Tickets sold for $8, $10, and $12 depending on where the seats were. Twice as many $8 seats were sold as $10, and 10 more $12 seats were sold than $10. The total proceeds were $3,122. How many of each type of ticket were sold?
8. A patient that returned from Jamaica with a great tan and a tropical foodborne illness is given an aggressive course of antibiotics. A total of 5,000 milligrams is to be taken over 5 days, with the amount for each of the third, fourth, and fifth days half as much as the amount for each of the first 2 days. How much should be taken on the first day? *Note:* That patient was me.
9. In a department store, Juaquin notices a jacket that he wanted to buy on the "60% off" rack. If the original price of the jacket was $75 before sales tax, what is the sale price of the jacket before sales tax?
10. At one county detention facility, inmates convicted of petty larceny receive a 30-day sentence, those convicted of DUI get 6 days, and simple assault results in 45 days. At one point, there were twice as many larceny inmates as assault, and three times as many DUI inmates as larceny. Collectively, the prisoners at that point were serving 423 days in jail. How many DUI inmates were there at that time?
11. At a New Year's Eve party in a certain restaurant, the cost was $60 per couple. If more than 50 couples attended, the restaurant agreed to drop the price by 50 cents a couple for each couple in excess of 50. If 76 couples attended the party, what was the cost per couple?

Section 6-2

For Exercises 12–15, write each ratio as a fraction.

12. 82 miles to 15 gallons of gasoline
13. 16 ounces cost $2.37
14. 4 months to 2 years
15. 18 minutes to 2 hours

For Exercises 16–18, solve each proportion.

16. $\dfrac{2}{x} = \dfrac{14}{63}$

17. $\dfrac{y - 3}{8} = \dfrac{40}{23}$

18. $\dfrac{120}{x + 30} = \dfrac{51}{2x - 3}$

19. If you burn 90 calories when exercising for 12 minutes, how many calories will you burn when exercising for 30 minutes?
20. The U.S. Center for Disease Control reported that 4 out of 10 people with incomes between $15,000 and $24,999 exercise regularly. About how many people exercise regularly in a group of 85 people who are in that income bracket?
21. In his will, a man's estate was divided according to a ratio of three parts for his wife and two parts for his son. If his estate amounted to $280,000, how much did each receive?
22. A professor states that if a student misses a unit test (worth 30 points), he will use the score on the final exam (100 points), proportionally reduced, for the score on the unit test. If a

student scored 85 on the final exam, what would be the student's score on the unit test?

23. The cost of building a deck varies directly with the area of the deck. If a 6-foot by 9-foot deck costs $2,160, find the cost of building a 9-foot by 12-foot deck.

24. The amount of paint needed to paint a spherical object varies directly with the square of the diameter. If 3 pints of paint are needed to paint a model of Mercury with a diameter of 36 inches, how much paint will be needed to paint a model of Earth with a diameter of 60 inches?

25. The amount of amperage in amps of electricity passing through a wire varies inversely with the resistance in ohms of the wire when the potential remains the same. If the resistance is 20 ohms when the amperage is 10 amps, find the amperage when the resistance is 45 ohms.

26. The cost of producing an item varies inversely with the square root of the number of items produced. Find the cost of producing 1,600 items if the cost of producing 900 items is $600.

Section 6-3

27. What is a rectangular coordinate system? How is it used?
28. What exactly is the graph of an equation?

For Exercises 29–34, draw the graph for each line.

29. $4x - y = 8$
30. $y = 2x + 6$
31. $x = -5$
32. $y = 8$
33. $y = -3x + 12$
34. $x - 6y = 11$

For Exercises 35–38, find the slope of the line containing the two given points.

35. $(3, 8), (-2, 6)$
36. $(-5, -6), (-3, 8)$
37. $(5, 9), (-3, 9)$
38. $(-3, -8), (6, -2)$

For Exercises 39–42, write the equation in slope-intercept form and find the slope, x intercept, and y intercept, and graph the line.

39. $3x + y = 12$
40. $-2x + 8y = 15$
41. $4x - 7y = 28$
42. $x - 3y = 9$

43. Detroit was one of the hardest-hit cities when housing prices declined starting in 2008. According to Zillow.com, the average value of a house in the city limits was $81,000 at the beginning of 2008. The values declined steadily after that, bottoming out at $32,000 at the beginning of 2011.
    (a) Write a linear equation describing the home values in Detroit in terms of years after the beginning of 2008.
    (b) Use your equation to find the average home value in Detroit at the beginning of September 2009.
    (c) At what rate was the average home value changing between January 2008 and January 2011?
    (d) Use your equation to estimate the average value in January 2014, then explain why your answer shows that home values couldn't keep declining at that rate.

44. Marcus invested a sum of money at 5% annual interest and twice as much at 8%. At the end of the year, the total interest earned was $89. How much was put into each investment?

Section 6-4

45. Explain in your own words what a function is and how the words "input" and "output" apply.
46. Describe how to read the symbol "$g(t)$" out loud, and describe what the significance is.

For Exercises 47–49, state whether or not the relation is a function.

47. $\{(2, 5), (5, -7), (6, -10)\}$
48. $\{(-1, 5), (2, 6), (-1, 3)\}$
49. A relation that corresponds dogs with their owners.

For Exercises 50–52, find the domain and range for each relation and state whether or not the relation is a function.

50. $y = \sqrt{3 - x}$
51. $3x + 2y = 6$
52. $y^2 = x$

For Exercises 53–55, evaluate each function for the specific value.

53. $f(x) = -2x + 10$ $\qquad f(10), f(-3)$
54. $f(x) = x^2 + 7x + 10$ $\qquad f(-10), f\left(\dfrac{3}{4}\right)$
55. $f(x) = 2x^2 - 3x$ $\qquad f(8), f(-5)$

For Exercises 56 and 57, graph each linear function.

56. $f(x) = -2x + 5$
57. $f(x) = -\frac{3}{4}x + 4$

58. The graph shown here is of a function describing the number of rounds of golf played in the United States annually (in millions) in terms of years after 2000. Use the graph to answer these questions.

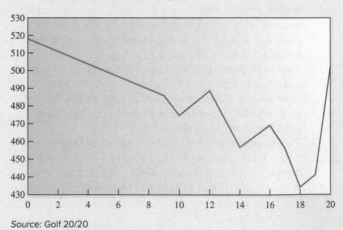

Source: Golf 20/20

(a) About how many more rounds were played in 2000 than in 2020?
(b) In what year did the number of rounds fall below 500 million?
(c) Find a formula for the function describing the number of rounds played, in millions, for the years from 2000 to 2009, and use it to estimate how many rounds were likely played in 1995.
(d) Why can't your formula from part (c) give an accurate projection of the number of rounds in 2025?

59. Margarita focused on her social life a whole lot more than her classes in her first semester of college, and got a 1.8 GPA for her efforts. Uh oh. Since then, her cumulative GPA has increased at the rate of 0.15 per semester. Write a function describing Margarita's cumulative GPA in terms of the number of semesters she's been in college, and use it to predict the GPA she'll graduate with if she finishes in 5 years without going to summer school.

Section 6-5

60. Describe the types of quantities that are likely to be modeled effectively by quadratic functions.
61. What is the reason we defined logarithms in Section 6-5?

For Exercises 62–64, graph each parabola.

62. $f(x) = x^2 + 10x + 25$
63. $f(x) = 3x^2 - 4x - 4$
64. $f(x) = -6x^2 + 12x$

For Exercises 65–68, graph each exponential function.

65. $f(x) = -3^x$ 66. $f(x) = 6^x$

67. $f(x) = \left(\dfrac{1}{3}\right)^x$ 68. $f(x) = -\left(\dfrac{1}{2}\right)^x$

69. Use a calculator to evaluate $\log_{3.2} 119$. Round to two decimal places.
70. Solve the equation: $3(2)^{y-7} = 99$
71. A ball is thrown upward from a height of 4 feet with a velocity of 80 feet per second. The function $f(x) = -16x^2 + v_0x + h_0$ describes the height after x seconds, where v_0 is the initial velocity and h_0 is the initial height. Find the maximum height reached by the ball, and how long it takes to reach the ground.
72. A radioactive isotope decays according to the function $F(x) = A_0 2^{20.5x}$ where A_0 is the initial amount and x is the number of days since it started decaying.
    (a) If there were 200 pounds initially, how much remains after 10 days?
    (b) How long will it take for 80% of the isotope to decay?

Chapter Test

For Exercises 1 and 2, solve the proportion.

1. $\dfrac{x}{9} = \dfrac{16}{36}$

2. $\dfrac{3}{7} = \dfrac{20}{x-4}$

3. A person has invested part of $5,000 in stocks paying a 4% dividend and the rest in stocks paying a 6% dividend. If the total of the dividends was $270, how much did the person invest in each stock?
4. A marketing company bought high-capacity flash drives for six new employees at a cost of $400. How much would it cost them to buy the same drives for the entire existing sales force of 37 people?
5. If you can bike 2 miles in $12\frac{1}{2}$ minutes, how many hours will it take to bike 210 miles without counting rest stops?
6. The number of vibrations per second of a metal string varies directly with the square root of the tension when all other factors remain unchanged. Find the number of vibrations per second of a string under a tension of 64 pounds when a string under a tension of 25 pounds makes 125 vibrations per second.
7. The number of hours it takes to do a certain job varies inversely with the number of people working on the job. If it takes 12 people 8 hours to build a front porch on a Habitat for Humanity house, how many hours will it take 8 people to build a porch?

For Exercises 8–11, draw the graph for each.

8. $3x - 5y = -15$
9. $x = -10$
10. $y = \dfrac{11}{2}$
11. $2x + 3y = -8$

12. Find the slope of the line containing the points $(-5, -10)$ and $(2, 7)$. Does the line slope upward or downward?

For Exercises 13 and 14, write each equation in slope-intercept form and find the slope, x intercept, and y intercept and graph the line.

13. $x + 5y = 20$
14. $2x - 11y = 22$

For Exercises 15 and 16, state whether or not each relation is a function.

15. $\{(-4, 6), (-10, 18), (12, 5), (12, 6), (-3, 5)\}$
16. A relation with input a day in 2013 and output the closing value of one share of Exxon Mobil stock on that day.

For Exercises 17 and 18, evaluate each function for the specific value.

17. $f(x) = -3x + 10$; find $f(12)$ and $f(-15)$
18. $f(x) = 2x^2 + 6x - 5$; find $f(3)$ and $f\left(\dfrac{5}{2}\right)$

For Exercises 19–22, graph each.

19. $f(x) = x^2 - 9x + 14$
20. $f(x) = -2x^2 + x - 3$
21. $f(x) = 2^{1.5x}$
22. $f(x) = -3^{-0.8x}$

23. A contractor needs to make a rectangular drain consisting of three sides from a piece of aluminum that is 24 inches wide. Find the length of each side if it needs to carry the maximum capacity of water. The drain is ⊔ shaped.
24. Find the amount of interest earned on a $4,000 investment held for 5 years, if the rate is 8.5% compounded semiannually. Use $A = P\left(1 + \frac{r}{n}\right)^{nt}$.
25. How long would it take the investment in Problem 24 to triple in value?

Projects

1. We started Chapter 6 by talking about drug dosage as it relates to the size of a patient. We looked at two different methods for calculating dosages: body weight and body surface area (BSA). It turns out that the latter is used more commonly. BSA is the area of all of your body's skin if you could take it off and stretch it flat. While doing so would result in a very accurate reading for BSA, it would also render the patient deceased, which makes any dosage kind of pointless (and that doesn't even consider the legal and moral ramifications).

   The point is that it's not terribly easy to measure a living person's BSA, and a wide variety of formulas has been developed to estimate this important number. In this project, your task is to rework the questions in the chapter opener using several different ways of measuring the BSA for each patient. Here's some extra information you need:

   - The football player is 6′4″, while the child is 4′5″.
   - The recommended dose for an average man with BSA of 1.9 square meters is 400 mg.
   - The minimum effective dose for that size is 250 mg, and anything over 1,500 mg could be lethal.

   Use each of the BSA formulas listed, and compare the amount of medication that would be prescribed based on each. How widely do the amounts vary for different formulas? Write a short report discussing this variation and decide if there's one formula that you think is most useful based on ease of use and calculated result. Finally, discuss whether or not you think that it's really possible to get a good measure of body surface area using only height and weight.

   (*Note:* to convert height in inches to centimeters (cm), multiply by 2.54; to convert weight in pounds to kilograms (kg), divide by 2.2.) All formulas provide BSA in square meters; all heights are in cm, and all weights are in kg.

   (a) The Mosteller formula: $\text{BSA} = \sqrt{\dfrac{\text{Height} \times \text{Weight}}{3,600}}$

   (b) The DuBois formula:
   $\text{BSA} = 0.007184 \times \text{Height}^{0.725} \times \text{Weight}^{0.425}$

   (c) The Haycock formula:
   $\text{BSA} = 0.024265 \times \text{Height}^{0.3964} \times \text{Weight}^{0.5378}$

   (d) The Gehan/George formula:
   $\text{BSA} = 0.0235 \times \text{Height}^{0.42246} \times \text{Weight}^{0.51456}$

   (e) The Boyd formula:
   $\text{BSA} = \dfrac{\text{Height}^{0.3} \times \text{Weight}^{\{0.6157-[0.00816474\,\ln\,(\text{Weight})]\}}}{30.03316}$

   (*Note:* "ln" in the Boyd formula represents the natural log formula, which is log to a particular base. You can calculate it using the "LN" key on a calculator.)

2. Using the Internet, find population statistics for your state for the beginning of every decade from 1950 to 2020. Then either use the regression feature on your calculator or do an Internet search for a website that computes regression equations. Find a linear regression model for the population, a quadratic regression model, and an exponential regression model. Then use each of those models to calculate the current population of your state, and find the most recent estimate you can. Which of the models gave the best estimate? Based on the graphs we studied in this chapter, explain why the best model did so based on what the graph of that type of function tends to look like.

3. A function is linear when it increases or decreases at a steady rate. Make a list of at least five real quantities that you suspect might be modeled well by a linear function. Then do some research, finding values for the quantities on your list and plotting them on a graph. Rank your list from best to worst in terms of how linear the graphs appear to be.

4. One of the interesting aspects of quantities that can be modeled by exponential functions is that the rate of growth accelerates rapidly as the value of the variable (often time) increases. Let's study just how significant that change in growth rate is.

   (a) Write a function that describes the value of a modest $500 investment if it earns 5% interest compounded monthly.

   (b) Fill in the table below with the value of the investment at the end of each 5-year period.

Years Passed	5	10	15	20	25	30	35	40	45	50
Value										

   (c) Use the results in the table to find the average rate of change of the value for each 5-year period, and put those rates of change in the next table. (Recall that rate of change is the slope of a line.)

End of 5-year period	5	10	15	20	25	30	35	40	45	50
Rate of change										

   (d) Write a short essay discussing how significant you feel the growth of the rate of change is for an exponential growth function.

   (e) Now let's look at the effect of changing the growth rate variable, r, in an exponential function. If an initial population P_0 grows at $r\%$ per year, the population in t years is given by the function $P(t) = P_0(2)^{1.443rt}$. Suppose that a city has 400,000 people currently, and the population is projected to grow at 1% annually for the next 30 years. Find the population in 30 years.

(Remember that the growth rate has to be written as a decimal, not a percentage.)

(f) Now fill in the table for the new projected population if the growth rate is changed to the given percentage.

Growth rate	1%	1.1%	1.2%	1.3%	1.4%	1.5%
Population						

Growth rate	1.6%	1.7%	1.8%	1.9%	2.0%
Population					

(g) What can you conclude about the long-term effects of changing the growth rate by just a little bit?

(h) Many home loans have a 30-year term. How can your conclusion from part (g) be applied to securing a loan with a term that long?

Design elements: Front matter, Chapter Opener, Summary and End Matter header design (random numbers background illustration): ©pixeldreams.eu/Shutterstock RF

Consumer Math

Photo: Digital Vision/SuperStock

Outline

362

Math in College Budgeting

"If you think education is expensive, try the cost of ignorance." This famous quote comes from Dr. Derek Bok, the former president of Harvard University. There are many nuanced interpretations of exactly what Dr. Bok was trying to say, but if you think of it strictly in terms of financial costs, it says that your earning power increases dramatically as you attain more education. This isn't a cliché or a myth: this table is based on cold, hard facts.

Education level	Median salary in 2020
Less than high school diploma	$30,784
High school diploma	$38,792
Some college/no degree	$43,316
Associate degree	$46,124
Bachelor's degree	$64,896
Master's degree	$77,844
Professional degree	$96,772
Doctorate degree	$97,916

Source: Bureau of Labor Statistics

And that's assuming you can even *get* a job without a degree. Here are some stunning statistics, from a Georgetown University study: of 11.6 million jobs created between 2008 and 2016, 8.4 million went to workers with at least a bachelor's degree, and another 3.1 million went to folks with some college or an associate degree. Think about that for a second: 11.6 million jobs, and only 80,000 went to workers without a degree.

The downside is that college is expensive, and costs continue to rise at a rate far greater than inflation. According to the Bureau of Labor Statistics, between 1980 and 2020, the price of average consumer goods in the United States went up by 236%. Ouch. But during the same period, college costs went up by nearly 1,200%! The truth is that a pretty small percentage of families have the means to pay all of the costs associated with college, meaning that student loans are a reality for an increasing segment of the population. Even the most optimistic students worry about whether or not they'll be able to afford college, and how much debt they'll run up while earning a degree.

This might lead you to a larger realization: long ago, survival depended on overcoming physical challenges that modern humans can only imagine, but today, survival is more about navigating the waters of the modern financial system. There are over 300 million people in the United States, and at any given time, it seems like half of them are trying to figure out a way to separate you from your hard-earned money. That's why this chapter is all about different aspects of our financial system that are particularly important to the average adult. Success in this chapter will help you to become a more well-informed consumer, which makes it less likely that those 150 million people will succeed in getting your cash.

We begin with a thorough look at percentages, which play a big role in almost all areas of consumer math. Then we'll study budgeting, loans, and investments, three topics of particular interest in the current financial climate. The goal is to help you take the guesswork out of financial planning, so that rather than hoping you're doing the right things for your future, you can be sure that you are.

Now to return to the original issue: how much debt can you incur while paying for an education? In a 2020 article, the College Board reported that a "moderate" college budget for an in-state public 4-year college for the 2020–2021 academic year averaged $26,821. (This takes into account tuition, fees, and other living expenses.) Let's study the ramifications of this frightening statistic.

1. How much would it cost to attain a 4-year degree with this budget?

2. If you are able to obtain grants and scholarships to pay for 25% of these costs and can also pay $7,000 per year out of pocket, how much money would you have to borrow to cover the rest?

3. If you obtain a federal student loan with an interest rate of 6.8% and a term of 10 years for the amount in Question 2, what will your monthly payment be upon graduation if you pay interest on the loan while still in school? How much total will you pay in interest?

4. Repeat Question 3 if you capitalize the interest on your loan.

5. Based on the median salaries in the table, how long would you have to work after obtaining a bachelor's degree to make back the total amount spent on student loan payments and the amount paid out of pocket? (Use the loan with interest capitalized, and don't forget to factor in how much you would have been able to make without going to college at all!)

For answers, see Math in College Budgeting Revisited on page 441

Section 7-1   **Percents**

LEARNING OBJECTIVES

1. Convert between percent, decimal, and fraction form.

2. Solve problems involving percents.

3. Find percent increase or decrease.

4. Evaluate the validity of claims based on percents.

Have you ever been shopping, come across a clearance rack that said something like "40% off lowest ticketed price," and had to ask someone what the discounted price of a certain item would be? If so, you're certainly not alone. Any math teacher will tell you that people ask them questions like that all of the time.

Percents are a math topic with a huge number of applications to everyday life, so they should be high on our priority list if we're trying to learn about math in our world. In this section, we'll review key ideas about percents: what they really mean, how to use them in calculations, and even how they can be deceptively misused.

The word "percent" can be translated literally as "per hundred."

hxdbzxy/iStock/Getty Images

Percent means hundredths, or per hundred. That is, $1\% = \dfrac{1}{100}$.

For example, *Variety* magazine reported the results of a poll indicating that 70% of Americans prefer to watch movies at home. Just 13% said they prefer watching movies in a theater, while 17% were undecided. This tells us that 70 out of every 100 people prefer to watch at home, 13 of every 100 prefer the theater, and 17 of every 100 are undecided.

Percent Conversions

To work with percents in calculations, we will need to convert them to either decimal or fractional form.

Converting from Percents

- To change a percent to a decimal, drop the % sign and move the decimal point two places to the left.

- To convert a percent to a fraction, write a fraction with the percent in the numerator and 100 in the denominator, then simplify or reduce if necessary.

Example 1   **Converting from Percent Form**

Convert each of 80%, 37.5%, and 6% to (a) decimal and (b) fraction form.

SOLUTION

(a) For each of these, our job is to move the decimal two places left and drop the percent sign. Of course, if there is no decimal point to begin with, put it in after the last digit.

$$80\% = 0.80 \qquad 37.5\% = 0.375$$

For 6%, we'd run out of digits when moving the decimal point two places left, so we need to put in a zero placeholder.

$$6\% = 0.06$$

(b) Now we build a fraction by putting the percent in the numerator and 100 in the denominator, then simplify.

$$80\% = \frac{80}{100} = \frac{8}{10} = \frac{4}{5} \quad \textit{Divide numerator and denominator by 10, then 2}$$

For 37.5%, we'll need to work a little harder to simplify.

$$37.5\% = \frac{37.5}{100}$$

First we'll multiply the numerator and denominator by some number that eliminates the decimal in the numerator. Multiplying by 10 works reliably, but in this case multiplying by 2 is a little quicker.

$$37.5\% = \frac{37.5}{100} \cdot \frac{2}{2} = \frac{75}{200} = \frac{3}{8} \quad \textit{Divide numerator and denominator by 25}$$

The last one is straightforward.

$$6\% = \frac{6}{100} = \frac{3}{50}$$

Try This One 1

Convert each of 62.5%, 3%, and 150% to (a) decimal and (b) fraction form.

Most of the time when we're calculating a percentage, we'll get our answer in decimal or fractional form. So we'll want to convert back to percent form to interpret our result.

Converting to Percents

- To change a decimal to a percent, move the decimal point two places to the right and add a percent sign.
- To change a fraction to a percent, first change the fraction to a decimal by dividing, then change the decimal to a percent.

Example 2 Converting to Percents

Convert each to a percent:
(a) 0.472   (b) 1.03   (c) $\frac{7}{8}$ (d) $\frac{5}{6}$

SOLUTION

(a) $0.472 = 47.2\%$ *Decimal point moved two places right*

(b) $1.03 = 103\%$

(c) $\frac{7}{8} = 7 \div 8 = 0.875 = 87.5\%$

(d) $\frac{5}{6} = 5 \div 6 = 0.83\overline{3} = 83.\overline{3}\%$ *Note the repeating decimal*

Try This One 2

Convert each to a percent:
(a) 0.974   (b) 0.04   (c) $\frac{5}{16}$ (d) $\frac{7}{9}$

1. Convert between percent, decimal, and fraction form.

Problems Involving Percents

The most common calculations involving percents involve finding a percentage of some quantity. To understand how to do so, consider the following example. You probably know that 50% of 10 is 5. Let's rewrite that statement, then turn it into a calculation:

50% of 10 is 5
$0.5 \times 10 = 5$

When writing a percentage statement in symbols, the word "of" becomes multiplication, and the word "is" becomes an equal sign. Also, we change the percent into decimal or fractional form. The next three examples show how to use this procedure to set up calculations.

| **Example 3** | **Finding a Certain Percentage of a Whole** |

According to nursingcenter.com, about 9.6% of all registered nurses in the United States are men. The average hospital in 2020 had 284 registered nurses. How many would likely be male?

SOLUTION

First, write 9.6% in decimal form, as 0.096. The question is "what is 9.6% of 284?" which we translate into symbols:

9.6% of 284 is _____
$0.096 \times 284 = 27.264$

You can't have 0.264 nurses, so it would make sense to round down and conclude that there are 27 male nurses at an average hospital.

XiXinXing/Shutterstock

Try This One   3

According to Google, 13.2% of registered nurses have at least a master's degree. If 430 nurses attend a professional conference, how many would be likely to have at least a master's degree?

| **CAUTION** | In calculations with percents, it's very common for the result to contain digits after a decimal point, like 27.264 in Example 3. You should always think about what quantity your answer represents to decide if it's appropriate to round to the nearest whole number. |

| **Example 4** | **Finding a Percentage from a Discount** |

My wife is an absolute wizard with Macy's coupons: as a high-volume shopper there (sigh) she gets quite a few coupons in the mail. On one recent mall outing, she found a lovely red strapless number with a selling price of $79. She had two coupons to choose from: one offers $15 off any purchase of $50 or more, the other offers 20% off any item.

(a) Which was the better choice?
(b) How much will the dress cost?
(c) The sales tax in our county is 6.5%. What's the total cost including tax?

David Sobecki/McGraw Hill

SOLUTION

(a) The question really asks if the $15 discount is more or less than 20%, so we need to answer this question:

$$15 \text{ is what percent of } 79?$$
$$15 = \quad x \quad \times 79 \qquad x \text{ is the percent in decimal form.}$$

This is the equation $79x = 15$, which we can solve for x.

$$79x = 15 \qquad \textit{Divide both sides by 79.}$$
$$x = \frac{15}{79} \approx 0.19$$

The decimal 0.19 corresponds to 19%, so the 20% off coupon will be a little better.

(b) The price was 20% less than the original price of $79, so we need to find 20% of $79: $0.20 \times \$79 = \$15.80$. So the cost of the dress is $\$79 - \$15.80 = \$63.20$. (You could also note that if the dress is 20% off, then the actual cost is 80% of the original price. That allows you to find the discount price in one step.)

(c) Next, find 6.5% of the sale price, $63.20: $0.065 \times \$63.20 = \$4.11$. Finally, add tax to the selling price: $\$63.20 + \$4.11 = \$67.31$. (Again, this could be done in one step: adding 6.5% sales tax makes the total price 106.5% of the sale price, so you could just do 1.065 times $63.20.)

Try This One   4

A refrigerator has a regular price of $1,149 at Home Depot and $1,219 at Lowe's. Home Depot is offering 15% off, but Lowe's offers $200 off if you put the purchase on store credit.

(a) Which is the better option?
(b) Find the sale price of the best option.
(c) If you buy at a store in a county with 7% sales tax, find the total cost including tax.

Example 5 | Finding a Whole Amount Based on a Percentage

A medium-sized company reported that it had to cut its workforce back to 70% of what it was last year. If it has 63 workers now, how many did it have a year ago?

SOLUTION

Convert 70% to a decimal: $70\% = 0.70$. Now write as a question and translate to symbols:

$$70\% \text{ of what number is } 63?$$
$$0.70 \quad \times \quad x \quad = 63$$

This gives us the equation $0.70x = 63$, which we solve for x.

$$0.70x = 63 \qquad \textit{Divide both sides by 0.70.}$$
$$\frac{0.70x}{0.70} = \frac{63}{0.70}$$
$$x = 90$$

The company had 90 workers a year ago.

Try This One   5

The 2020 Cleveland Browns won 11 games, which was 183% of their 2019 win total. How many did they win in 2019?

| **Example 6** | Calculating Cost of Sale from Commission |

A real estate agent receives a 7% commission on all home sales. How expensive was a home if the commission was $5,775.00?

SOLUTION

In this case, the problem can be written as $5,775.00 is 7% of what number?

$$\$5{,}775 \text{ is } 7\% \text{ of } x$$
$$5{,}775 = 0.07 \times x$$ *Divide both sides by 0.07.*
$$\frac{5{,}775}{0.07} = \frac{0.07 \times x}{0.07}$$
$$82{,}500 = x$$

The home was purchased for $82,500.00.

Trinette Reed/Blend Images LLC

 2. Solve problems involving percents.

Try This One   6

A sales clerk receives a 9% commission on all sales. Find the total sales the clerk made if the commission was $486.00.

Sometimes it's useful to find the percent increase or the percent decrease in a specific situation. In this case, we can use the following method.

> **Procedure for Finding Percent Increase or Decrease**
>
> **Step 1**   Find the amount of the increase or the decrease.
> **Step 2**   Make a fraction as shown:
>
> $$\frac{\text{Amount of increase}}{\text{Original amount}} \quad \text{or} \quad \frac{\text{Amount of decrease}}{\text{Original amount}}$$
>
> **Step 3**   Change the fraction to a percent.

| **Example 7** | Evaluating a Percent Decrease |

A municipal government promises voters that there will be a 15% cut in spending for the next fiscal year. The city's budget was $72.5 million this past year, and is projected to be $64.3 million for this year. Is the government keeping its promise? Explain.

SOLUTION

Let's find the percent change in the budget. The decrease is $8.2 million, which certainly sounds like an awful lot of money. This is exactly why percent change is so useful to compute: it's not about the raw size of the change—it's about how the change compares to the original amount.

$$\frac{\text{Change}}{\text{Original amount}} = \frac{-\$8.2 \text{ million}}{72.5 \text{ million}} \approx -0.113$$

This is only an 11.3% cut, so the government will not be keeping its promise.

Math Note

When finding percent increase or decrease, make sure you divide by the *original* amount, not the ending amount.

Try This One 7

The city budget manager justifies the smaller budget cut by claiming that the cut is in line with the decrease in tax revenues for the year. In fact, tax revenue went from $43.4 million to $39.6 million. Evaluate this claim.

3. Find percent increase or decrease.

Percent increase or decrease is often misused, sometimes intentionally, sometimes not. In Example 8, we'll look at a common deceptive use of percents in advertising.

Example 8 Recognizing Misuse of Percents in Advertising

Math Note

This example brings up a really useful point: when you're not sure how to tackle a problem that sounds abstract, turn it into a concrete problem by choosing some specific numbers. That can be really helpful.

A department store advertised that certain merchandise was reduced 25%. Also, an additional 10% discount card would be given to the first 200 people who entered the store on a specific day. The advertisement then stated that this amounted to a 35% reduction in the price of an item. Is the advertiser being honest?

SOLUTION

Let's say that an item was originally priced at $50.00. (We could actually use any price here: I just picked fifty bucks randomly.) First find the discount amount.

$$\begin{aligned} \text{Discount} &= \text{rate} \times \text{selling price} \\ &= 25\% \times \$50.00 \\ &= 0.25 \times \$50.00 \\ &= \$12.50 \end{aligned}$$

Then find the reduced price.

$$\begin{aligned} \text{Reduced price} &= \text{original price} - \text{discount} \\ &= \$50.00 - \$12.50 \\ &= \$37.50 \end{aligned}$$

Next find 10% of the reduced price.

$$\begin{aligned} \text{Discount} &= \text{rate} \times \text{reduced price} \\ &= 10\% \times \$37.50 \\ &= \$3.75 \end{aligned}$$

Find the second reduced price.

$$\begin{aligned} \text{Reduced price} &= \$37.50 - \$3.75 \\ &= \$33.75 \end{aligned}$$

The final amount is $50 - $33.75 = $16.25 less than the original price. Now find the percent of the total reduction.

$$\frac{\text{Reduction}}{\text{Original price}} = \frac{16.25}{50.00} = 0.325 = 32.5\%$$

The total percent of the reduction was 32.5%, and not 35% as advertised. Liars!

Try This One   8

4. Evaluate the validity of claims based on percents.

A department store offered a 20% discount on all television sets. They also stated that the first 50 customers would receive an additional 5% discount. Find the total percent discount. You can use any selling price for the televisions.

Answers to Try This One

1   (a) 0.625, 0.03, 1.5   (b) $\dfrac{5}{8}, \dfrac{3}{100}, \dfrac{3}{2}$

2   (a) 97.4%   (b) 4%   (c) 31.25%   (d) 77.$\overline{7}$%

3   57

4   (a) Home Depot   (b) $976.65   (c) $1,045.02

5   6 games

6 $5,400

7   This claim is not legitimate. Tax revenue only decreased by about 8.76%.

8   24%

Exercise Set   **7-1**

Writing Exercises

1. What exactly does the word "percent" mean?
2. Explain how to change percents into decimal and fraction form.
3. Explain how to change decimals and fractions into percent form.
4. Explain how the word "of" plays an important role in calculations involving percents.
5. How do you find the percent increase or decrease of a quantity?
6. Is it possible to have more than 100% of a quantity? Explain.

Computational Exercises

For Exercises 7–18, express each as a percent.

7. 0.63
8. 0.87
9. 0.025
10. 0.0872
11. 1.56
12. 3.875
13. $\frac{1}{5}$
14. $\frac{5}{8}$
15. $\frac{2}{3}$
16. $\frac{1}{6}$
17. $1\frac{1}{4}$
18. $2\frac{3}{8}$

For Exercises 19–28, express each as a decimal.

19. 18%
20. 23%
21. 6%
22. 2%
23. 62.5%
24. 75.6%
25. 320%
26. 275%
27. $66\frac{2}{3}$%
28. $20\frac{1}{3}$%

For Exercises 29–36, express each as a fraction or mixed number.

29. 24%
30. 36%
31. 236%
32. 520%
33. $\frac{1}{2}$%
34. $12\frac{1}{2}$%
35. $16\frac{2}{3}$%
36. $4\frac{1}{6}$%

Applications in Our World

37. Find the sales tax and total cost of a laser printer that costs $299.99. The tax rate is 5%.
38. Find the sales tax and total cost of an espresso machine that costs $59.95. The tax rate is 7%.
39. Find the sales tax and total cost of a Sony Playstation that costs $249.99. The tax rate is 6%.
40. Find the sales tax and total cost of a wireless mouse that costs $19.99. The tax rate is 4.5%.
41. A diamond ring was reduced from $999.99 to $399.99. Find the percent reduction in the price.
42. A pair of headphones was reduced from $109.99 to $99.99. Find the percent reduction in price.
43. A 27-inch flat panel computer monitor is on sale for $249.99. It was reduced $80.00 from the original price. Find the percent reduction in price.

44. The sale price of a spring break vacation package was $179.99, and the travel agent said by booking early, you saved $20. Find the percent reduction in price.

45. You've had your eye on a luggage set that sells for $159.99, waiting for a sale. Finally, success! To make room for next year's model, the price has been reduced by 40%. You also have a 20% off coupon you can use on top of that. How much will you pay, not including tax?

46. Some really nice beach towels that usually sell for $24.99 each are on sale over Memorial Day weekend at 25% off. In addition, the store has distributed a coupon good for an additional 10% off your entire purchase. How much will it cost to buy four towels, not including tax?

47. If a sales clerk receives a 7% commission on all sales, find the commission the clerk receives on the sale of a computer system costing $1,799.99.

48. If the commission for selling a 70-inch high-definition television set is 12%, find the commission on a television set that costs $2,499.99.

49. Milo receives a commission of 6% on all sales. If his commission on a sale was $75.36, find the cost of the item he sold.

50. The sales tax in Pennsylvania is 6%. If the tax on an item is $96, find the cost of the item.

51. In the 1995–1996 school year, there were 86,087 bachelor's degrees awarded nationwide in health care fields. That number grew to 129,623 for 2009–2010, and reached 251,355 for 2018–2019. First, find the actual change in number of degrees awarded for the time period from 1996 to 2010, and then from 2010 to 2019. Then find the percent change for each time period. Finally, find the average rate of change in degrees per year for each time period.

52. According to the U.S. Bureau of Labor Statistics, there were 110,700 chefs/head cooks employed in the United States in 2010 and 309,800 food service managers. Those numbers were projected to increase to 138,700 and 356,000 by 2030. Which job was expected to have the larger percent increase? By how much?

53. You saved $200 on your new laptop because you bought it online. If this was a 25% savings from the original price, find the original cost of the laptop.

54. The average teachers' and superintendents' salaries in a school district in western Pennsylvania was $50,480. Five years later, the new average was $54,747. Find the percent increase.

55. According to CTIA—The Wireless Association, in 2001, there were 127,540 cellular tower sites in the United States. That number rose to 253,086 for 2010 and 395,562 for 2019. Find the actual change in number of cell tower sites and the percent change for each 9-year period. What can you conclude?

56. Based on statistics from UNAIDS.org, in 2005 there were 31.8 million people worldwide living with HIV and 2 million AIDS-related deaths. By 2020, the number of people living with the virus had grown to 37.7 million, but 680,000 deaths were reported. Find the percent

change for each statistic, and write any conclusions you can draw.

57. According to information found on statista.com, there were 501,000 new homes sold in the United States in 2015, which was an increase of 178,000 homes from five years earlier. That number increased to 820,000 homes in 2020. Find the percent change for each period, as well as the average percent change per year.

58. According to the College Board, the average cost of tuition, room and board, and fees at public 4-year universities was $14,212 in the 2008–2009 academic year, $18,100 for 2013–2014, and $20,598 for 2018–2019. Find the percent increase from 2009 to 2014 and from 2014 to 2019. Then find the average percent change per year for each period.

59. You buy $129 in clothes at a department store. You can choose between coupons that offer $20 off your entire purchase or 25% off. Which will save you more money? By how much?

60. While shopping at Ross you pick out two pairs of jeans at $19 each and a pair of shoes on sale for $17. Then you splurge on a $46 sweater. You have two coupons but can only use one. The first offers 40% off any single item. The second offers 15% off your entire purchase. Which is the better choice?

61. The U.S. Department of Housing and Urban Development issued a report in early 2011 reporting some grim numbers on the housing industry. There were 523,000 building permits issued for single-family homes in 2010, but that number was projected to decrease by 27.0% in 2011. Construction was expected to start on 375,000 new homes in 2011, which would represent a 28.8% decrease from 2010. Also, in the 12-month period through January 2011, 298,000 new homes were sold, but this represented just 66.0% of new homes completed.
    (a) How many fewer permits were expected to be issued for single-family homes in 2011 compared to 2010?
    (b) How many fewer new homes were started in 2011 compared to 2010?
    (c) How many new homes were completed in the 12-month period through January 2011, and how did that compare to the number of new homes started in 2010?

62. Refer to Exercise 61. Fast forward 5 years, and things were looking much rosier indeed for the housing industry. There were 745,000 building permits issued for single-family homes in 2016, which was a 7.1% increase from 2015. Construction started on 781,000 new homes in 2016, which represents a 9.3% increase from 2015. Also, in the 12-month period through December 2016, 563,000 new homes were sold; this represented 76.3% of new homes completed.
    (a) How did the increase in the number of permits for single-family homes from 2015–2016 compare to the expected decrease from 2010–2011?

(b) How many more new homes were started in 2016 compared to 2015?

(c) How many new homes were completed in the 12-month period through December 2016, and how did that compare to the number of new homes started in 2016? How did it compare to the number started in 2010?

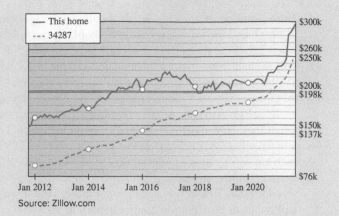

Source: Zillow.com

The graph shown here, based on information from Zillow.com, displays the estimated value of my father-in-law's house (orange) and the average for all houses in his town (green) for the period from 2012 to late 2021. The tick marks along the axis occur on January 1 of each year. Use the graph to answer Exercises 63–66.

63. Find the percent increase in the value of my father-in-law's house from January 1, 2012, to January 1, 2020. Then find the percent increase in the average price of all houses in his town over the same period.

64. Find the percent change for both my father-in-law's home and the average home in his town for each 2-year period starting on January 1, 2012.

65. Find the percent increase in the value of my father-in-law's home from January 2020 to the end of the graph, which is 21 months later. Then find the percent increase per month. If that trend continues, how much will the house be worth in January 2024?

66. Find the sum of the 2-year percent changes for both my father-in-law's home and all homes in his town. How do the results compare to the answers for Exercise 63? What can you conclude?

Critical Thinking

67. A store has a sale with 30% off every item. When you enter the store, you receive a coupon that states that you receive an additional 20% off. Is this equal to a 50% discount? Explain your answer.

68. You purchase a stock at $100 per share. It drops 30% the next day; however, a week later, it increases in value by 30%. If you sell it, will you break even? Explain your answer.

69. Suppose a friend planning a shopping spree on the day after Thanksgiving tells you he plans to buy a 65-inch plasma TV, and you say "There's no way you can afford that!" He then tells you that the store is offering 50% off any one item, and he has an Internet coupon good for 50% off any price, even a discounted one. So that's 100% off, and he'll get it for free! Explain why your friend will come home very disappointed.

70. A store that used to sell a grill for $90 now offers it at $60, and advertises "33% off our best-selling grill!" An amusement park used to have 60 rides and now boasts 90 rides, claiming "50% more rides this year!" Which one of them is lying?

71. While grocery shopping last week, I saw this claim on a package of some product: "NOW 20% MORE!"

Sounds nice, right? Explain why this claim is completely meaningless.

Another common gimmick is for companies to make a big deal about lowering the price on a product, but fail to mention that they also lowered the size of the package. A famous example is ice cream. Large name brands for many years came in half-gallons, which contained 64 ounces of creamy goodness. At some point, some brands kept the shape of the cartons the same, and also the price, but reduced the size to 56 ounces, and then again a couple of years later to 48 ounces. Let's say that the original half-gallon was priced at $4.

72. By what percent was the size reduced the first time?

73. At $4 for 56 ounces, how much would you have to pay to get 64 ounces of ice cream?

74. Use your answer from Problem 73 to find what the effective percent increase in price was.

75. Repeat Problems 72–74 for the reduction in size from 56 to 48 ounces.

76. Repeat Problems 72–74 for the overall reduction in size from 64 to 48 ounces. How do the results compare to a combination of the results for the two smaller reductions?

Section 7-2	**Personal Budgeting**

LEARNING OBJECTIVES

1. Calculate monthly take-home pay.
2. Calculate the amount spent on regular expenditures.
3. Prepare a budget.
4. Prorate long-term expenses to save in advance for them.

How much money did you spend last month? If you're like most folks, you probably can't answer that question exactly, and if you're like a lot of folks, it would take you a while to even come up with a ballpark guess. By some estimates, American consumers have a combined debt of nearly 15 trillion dollars. Trillion! That's an almost unimaginable amount of money. Imagine spending a million dollars a day, every day. That would be pretty much impossible, wouldn't it? But if you could, it would still take you over 2,700 YEARS to spend a trillion dollars. This debt problem happened because people spend more money than they make. And that happens because so few people put forth the effort to track exactly what they make and spend. Learning how to budget your money is a good first step in avoiding massive debt; the hard part is sticking to that budget!

Glow Images

Budgeting is the process of deciding how much money you can spend on various expenses based on your **income**: that is, the amount of money you're earning. In this section, we'll study budgeting and hopefully help you to make good decisions in regard to your expenses.

Income

The first step in building a budget has to be identifying exactly how much money you're making. Most people can tell you their hourly wage or salary, but because of deductions for taxes, insurance, and retirement, a lot of people don't know how much money is actually left for them to spend. This amount is known as **take-home pay**. Since many common expenses occur on monthly cycles, it's useful to know exactly how much money you bring home per month.

Example 1	Calculating Monthly Take-Home Pay

Math Note

The percentage of your pay that is withheld for taxes depends on a large variety of factors including income level, where you live, and family status. Nationally, it's hard to pin down an exact average, but standard estimates are that the average person takes home about 70–75% of their pay.

Camila works full-time as an executive assistant and has a check for $1,023.07 direct-deposited into a checking account every other Friday. What is Camila's monthly take-home pay?

SOLUTION

For a quick estimate, we can double the biweekly pay, since four weeks is close to a month. That gives us about $2,046.14 per month. A more accurate calculation would be:

$$\frac{\$1,023.07}{14 \text{ days}} = \$73.08 \text{ per day}$$

$$\$73.08 \times 365 \text{ days} = \$26,674.20 \text{ per year}$$

$$\frac{\$26,674.20}{12 \text{ months}} = \$2,222.85 \text{ per month}$$

Try This One   1

Eddie works 25 hours a week at a campus bookstore and gets a weekly paycheck in the amount of $195.30. What is the monthly take-home pay?

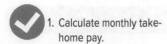

1. Calculate monthly take-home pay.

Of course, the total amount of money you have available to spend each month depends on all sources of income, including paychecks, student loan disbursements, gifts or allowances from relatives, and investment income.

Expenses

While there are a lot of people who don't know how much money they bring home each month, it probably wouldn't be a stretch to say that few people know exactly how much they spend each month. Expenses that you think of as minor can add up very quickly.

For example, my friend Charles noticed that one of his students came to class every day with two cups of coffee from a well-known coffee chain that isn't exactly famous for their low prices. He asked her if she'd ever thought about how much she spent on that coffee over the course of a year; not surprisingly, she had not. Let's help her out.

Example 2   **Calculating the Sum of Regular Expenditures**

(a) If you buy two cups of name-brand coffee every day at $3.25 per cup, how much would you spend in one year?

(b) If you spend $4.79 for lunch at Burger King three times per week, how much would you spend during a 16-week semester?

SOLUTION

(a) Two cups a day at $3.25 each is $6.50 per day. Multiplying by the number of days in a year, we get

$$\$6.50 \times 365 = \$2,372.50 \qquad Wow$$

(b) $3 \times \$4.79 \times 16 = \$229.92$

John Flournoy/McGraw Hill

Try This One   2

(a) Refer to Example 2(a). How much would you save over the course of a year if you instead bought coffee from a convenience store at 79 cents per cup?

(b) The average price of a pack of cigarettes in the United States in 2021 was about $6.65. How much money would a pack-a-day smoker set fire to in a year?

---

2. Calculate the amount spent on regular expenditures.

We'll call expenses like daily gourmet coffee **luxuries**, which are things you can live without. Everyone wants to be able to spend money on luxuries, but you won't know how much you can comfortably spend until you calculate your **necessary expenses**. These are the basic things you have to pay for: food, shelter, insurance, transportation, clothing, utilities, student loan payments. Some of those expenses are **fixed**: the amount you spend each month is consistent. Housing, insurance, utilities, loans, and transportation usually fall into this category. Other necessary expenses are **variable**, meaning they can change from month to month, like food and clothing.

The key to maintaining a sensible budget is to add up your necessary expenses and subtract them from your income. The result will let you know how much you have to spend on entertainment and luxuries, and will also let you know if you're able to save any money (which should be one of your goals every month).

Example 3   **Preparing a Budget**

Tremaine brings home $2,146.79 per month, and has the fixed expenses shown below. How much can he afford to spend on food, clothing, and luxuries?

Rent:	$625	Insurance:	$97.50
Car Payment:	$199.23	Utilities:	$175
Cell Phone:	$79.50	Student Loan:	$211.53
Gas:	$110		

SOLUTION

Adding all of the fixed expenses, we get a total of $1,497.76. This leaves $2,146.79 − $1,497.76 = $649.03 to spend on food, clothing, and luxuries.

Try This One 3

Leslie shares an apartment with two friends, and the rent and utilities are split equally by all three tenants. Find the amount left for variable expenses and luxuries if her monthly income is $1,556.27 and her fixed expenses are the following:

Total rent:	$1,100	Total utilities:	$327
Public transportation:	$88	Insurance:	$48.20
Cell phone:	$91.50	Gym membership:	$38.95

Example 4 — Finding the Effect of Eating Out on a Budget

Tremaine, our friend from Example 3, goes out for happy hour drinks and dinner with friends every Friday, and usually spends about $50. On average, what percentage of his budget after fixed expenses is eaten up by these happy hours?

SOLUTION

A typical month has four Fridays (although not all of them, of course), so Tremaine is spending on average $200 per month at happy hour, out of $649.03 left after fixed expenses. This represents

$$\frac{\$200}{\$649.03} = 30.8\%$$

of his budget.

Try This One 4

Leslie (Try This One 3) treats her roommates to salads and pizzas from Papa Antonio's every Monday and Thursday, at a cost of $29 each day, including delivery and tip. What percentage of her budget after fixed expenses goes toward these semi-weekly gatherings on average?

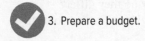
3. Prepare a budget.

Prorating Long-Term Expenses

So far, we've focused on monthly expenses. But some regular expenses occur on an annual or semiannual basis: car insurance, tuition payments, and property taxes are three examples that come to mind. If you manage your money by spending every penny you have each month, when those occasional large expenses pop up, it can be very hard to find the money to pay for them. The result is usually mounting debt that can take years to overcome.

One way to avoid the negative effects of these long-term expenses is to budget money every month in advance of the due dates, maybe putting it into a separate savings account. Then when the big expense is due, the money is there waiting.

Example 5 — Prorating Long-Term Expenses

Jin pays $337.24 every 6 months for car insurance. He also pays $2,623 for tuition at the beginning of both fall and spring semesters, and budgets $650 for books and supplies for each semester. If he wants to plan ahead, how much money should he put into savings every month?

Math Note

We neglected to consider interest earned in Example 5, but that's an additional benefit of saving ahead for long-term expenses: you'll actually have a little more money than you need when the expense comes due.

SOLUTION

Each of the expenses listed is paid every 6 months, so the total amount will need to be saved in a 6-month period. The sum of the expenses is $337.24 + $2,623 + $650 = $3,610.24. Dividing by 6 we get

$$\frac{\$3,610.24}{6} = \$601.71$$

Jin should put aside $601.71 each month.

4. Prorate long-term expenses to save in advance for them.

Try This One   5

Dharma and Greg have a combined monthly income of $5,280.94. They want to save ahead for semiannual property taxes ($1,146), semiannual car insurance ($493.21), and annual homeowners insurance ($893.20). What percentage of their income should be saved each month?

Answers to Try This One

1 $848.63   **3** $813.95   **5**   6.6%

2   (a) $1,795.80   **4**   28.5%
        (b) $2,427.25

Exercise Set   **7-2**

Writing Exercises

1. Describe what budgeting is in your own words.
2. What is meant by the term "take-home pay"? How does it differ from income?
3. Describe some typical sources of income for college students.
4. We discussed two types of regular expenses in this section. What are they, and how do they differ?
5. What do we mean by luxury items in the context of budgeting?
6. How do you go about prorating long-term expenses?

Applications in Our World

In Exercises 7–16, calculate the monthly take-home pay for each worker.

7. $1,246.32 every other Friday
8. $911.13 every two weeks
9. $423.35 weekly
10. $1,784.21 each Monday
11. $1,412.35 the first and third Friday of every month
12. $937.89 the second and fourth Monday of every month
13. 40 hours per week at $13.25 per hour with 27% deducted for taxes and insurance
14. 25 hours per week at $9.80 per hour with 22% deducted for taxes and insurance

15. 46 hours per week; regular hourly rate is $15.90, but for all hours over 40, the pay is 1.5 times as much; 32% is deducted for taxes and insurance
16. 40 hours from Monday to Friday and 8 hours two Saturdays per month; regular hourly rate is $18.40; pay is double for weekends; 30% is deducted for taxes and insurance

In Exercises 17–22, find the annual cost of each regular expense.

17. A bag of chips ($1.25) from a vending machine twice a day
18. Two cans of Diet Coke ($0.90 each) every day
19. A steak, egg, and cheese bagel ($3.69) for breakfast every weekday
20. A medium three-item pizza ($7.25) four times a week

21. Three pitchers of beer ($9.50 each) every weekend
22. Two new video games ($29.99) every month

In Exercises 23–28, a monthly income and fixed expenses are provided. Find the amount that would be left in a budget for variable expenses and luxuries.

23. Income: $1,150; rent: $250; car payment: $173.25; cell phone: $49.99; utilities: $95; insurance: $49.70
24. Income: $1,475; rent: $450; car payment: $205.50; cell phone: $71.99; utilities: $152.70; insurance: $98.25; student loan payment: $194.30
25. Income: $2,775.20; house payment: $975.23; utilities: $295; car payment: $195.52; insurance: $110.25; cell phone and Internet: $120.95; furniture loan payment: $87.50
26. Income: $5,888.89; house payment: $1,950.11; utilities: $412.50; car payments: $198.23 and $304.29; cell phones: $149.80; insurance: $284.12; property tax: $295.18
27. Income: $975 from work, $800 from parents; rent: $695 split among three roommates; utilities: $198 split among three roommates; on-campus dining plan: $130; cell phone and iPad data plan: $98; renter's insurance: $27.95; gas: $79.40
28. Income: $595 from work, $1,143 from student loan disbursement; rent: $940 split among four roommates; utilities: $198 split among four roommates; sorority dues: $119; car payment: $112.45; cell phone: $79.50; insurance: $79.45

29. If the person in Problem 23 eats at Chili's twice a week, spending an average of $21.50 each time, what percentage of their budget after fixed expenses goes to Chili's?
30. If the couple in Problem 26 saves $1,100 per month for their kids' education, what percentage of their budget after fixed expenses is saved?
31. If the student in Problem 27 spends $9.99 per week on new iPad apps, what percentage of their budget after fixed expenses and iPad app spending is left?
32. If the student in Problem 28 goes to an average of five concerts and sporting events per month, spending an average of $19.75 on tickets, what percentage of their budget is left after fixed expenses and tickets?

In Exercises 33–36, find the amount that would need to be saved monthly to prorate the given long-term expenses.

33. Car insurance: $295.40 every 6 months; tuition and fees: $3,804 annually; books and supplies: $530 twice a year
34. Homeowner's insurance: $894 annually; property tax: $2,150.40 every 6 months
35. Business liability insurance: $2,140 every 6 months; property tax: $3,461.45 every 6 months; worker's compensation coverage: $1,280 every 3 months
36. Tuition: $1,450 at the beginning of August and the beginning of January; on-campus housing: $2,150 twice per year; campus meal plan: $1,475 twice per year

Critical Thinking

The point of Exercises 37–42 is for you to build a personal budget, so of course answers will vary from individual to individual.

37. Keep a record of income for two weeks and make a list of all regular income. List any source of income individually, and whether or not the source is regular or variable.
38. Make a list of regular fixed monthly expenses that you have. These should be expenses that are the same amount, or very close to it, every month.
39. Over a two-week span, keep a list of all expenses, and categorize them as regular variable expenses or luxury spending.

40. Based on your answers to Exercises 37 and 38, draw up a personal budget that allows you to figure out how much you have to spend each month on variable expenses and luxuries.
41. Write an analysis of how the spending you kept track of in Exercise 39 fits within your budget from Exercise 40. Do you feel that you need to make any changes in your personal spending habits?
42. What percentage of your income do you spend on things that you could do without? How does that percentage compare to the percentage spent on things that you must have?

Section 7-3 | Simple Interest

LEARNING OBJECTIVES

1. Compute simple interest and future value.
2. Compute principal, rate, or time.
3. Compute interest using the Banker's rule.
4. Compute the true rate for a discounted loan.

The topic of the next two sections is of interest to anyone who plans to buy a house or a car, have a credit card, invest money, have a savings account—in short, pretty much everyone. This interesting topic is interest—a description of how fees are calculated when money is borrowed, and how your money grows when you save. Unless you don't mind being separated from your hard-earned money, this is a topic you should be eager to understand well.

Interest is a fee paid for the use of money. For example, if you borrow money from a bank to buy a car, of course you have to pay back the money that you borrowed. But you also have to pay an additional amount (the interest) for the privilege of using the bank's money.

Otherwise, they wouldn't lend it to you. On the other hand, if you deposit money in a savings account, the bank will pay you interest for saving money since it will be using your money to provide loans, mortgages, etc. to people who are borrowing money. The stated rate of interest is generally given as a yearly percentage of the amount borrowed or deposited.

There are two kinds of interest. *Simple interest* is a one-time percent of an amount of money. *Compound interest* is a percentage of an original amount, as well as a percentage of the new amount including previously calculated interest. We'll study simple interest in this section, and compound interest in the next.

Stockbyte/Getty Images

Simple Interest

In order to compute simple interest, we will need three pieces of information: the *principal*, the *rate*, and the *time*.

Math Note

Remember: *P* (principal) is the beginning amount borrowed or invested, and *A* (future value) is the final amount repaid or accumulated.

Interest (I) is the fee charged for the use of money.

Principal (P) is the amount of money borrowed or placed into a savings account.

Rate (r) is the percent of the principal paid for having money loaned, or earned for investing money. Unless indicated otherwise, rates are given as a percent for 1 year. When doing calculations, we'll always write r in decimal form.

Time (t) or **term** is the length of time that the money is being borrowed or invested. When the rate is given as a percent per year, time has to be written in years.

Future value (A) is the amount of the loan or investment plus the interest paid or earned.

We can use what we know about percents to develop a formula for simple interest.

Example 1   Developing a Formula for Simple Interest

(a) An account is opened with a deposit of $500 and earns 4% per year in simple interest. That means that after the end of the first year, 4% of the current value will be added to the account. How much interest will be earned?

(b) If that much interest is earned in 1 year, how much would be earned in 5 years? What about in t years?

(c) Use parts (a) and (b) to write a formula for finding the amount of simple interest earned on an account with principal value P, interest rate r, and time t.

SOLUTION

(a) All we need to do is compute 4% of $500—easy!

$$0.04(500) = \$20$$

(b) This is kind of common sense: if $20 is earned in 1 year, then $5 \times \$20 = \$100$ is earned in 5 years. Using that line of reasoning, it doesn't matter what the number of years is: you always multiply $20 by the number of years, so in t years, $20t$ dollars is earned.

(c) In part (a), we multiplied the interest rate (in decimal form) by the principal amount. In part (b), we multiplied that amount by the number of years. So overall, we get the product of principal, interest rate, and number of years:

$$I = Prt$$

where I is the interest earned.

Try This One 1

If an account has a beginning balance of $1,200 and earns 2.5% simple interest per year, write a formula that describes the amount of interest earned in t years.

What did we learn? The amount of interest is the interest rate r times the original amount P times the number of years passed t. And the future value A is the principal amount plus the interest.

Math Note

The formula $A = P(1 + rt)$ can be used to find future value without explicitly computing the interest first.

Formulas for Computing Simple Interest and Future Value

1. Interest = principal × rate × time:

$$I = Prt$$

2. Future value = principal + interest:

$$A = P + I \quad \text{or} \quad A = P(1 + rt)$$

Example 2 | Computing Simple Interest and Future Value

payphoto/123RF

Find the simple interest and future value for a loan of $3,600 for 3 years at a rate of 8% per year.

SOLUTION

First, change the rate to a decimal and substitute into the formula $I = Prt$:

$$8\% = 0.08$$
$$I = Prt$$
$$= (\$3,600)(0.08)(3)$$
$$= \$864$$

The interest on the loan is $864. To find the future value, substitute into the formula $A = P + I$

$$A = P + I$$
$$= \$3,600 + \$864$$
$$= \$4,464$$

The total amount of money to be paid back is $4,464.

Note that if we hadn't specifically been asked to find the interest, we could have found the future value in one step.

Substitute into the formula

$$A = P(1 + rt)$$
$$= \$3,600(1 + 0.08 \cdot 3)$$
$$= \$4,464$$

Try This One 2

Find the simple interest and future value for a $12,000 loan for 5 years at 7%.

Since rates are typically given in terms of percent per year, when the time of a loan or investment is given in months, we need to divide the time by 12 to convert to years.

Example 3 | **Computing Simple Interest for a Term in Months**

To meet payroll during a down period, United Ceramics Inc. needed to borrow $2,000 at 4% simple interest for 3 months. Find the interest.

SOLUTION

Change 3 months to years by dividing by 12, and change the rate to a decimal. Substitute in the formula $I = Prt$.

$$I = (\$2,000)(0.04)\left(\frac{3}{12}\right) \qquad 4\% = 0.04$$

$$= \$20$$

The interest is $20.

Try This One 3

1. Compute simple interest and future value.

Needing some quick cash for books at the beginning of the semester, Marta borrows $600 at 11% simple interest for 2 months. How much interest will Marta pay?

Often, a simple interest loan is paid off in monthly installments. To find the monthly payment, divide the future value of the loan by the number of months in the term of the loan.

Example 4 | **Computing Monthly Payments**

Admiral Chauffeur Services borrowed $600 at 9% simple interest for $1\frac{1}{2}$ years to repair a limousine. Find the interest, future value, and the monthly payment.

SOLUTION

Step 1 Find the interest.

$$I = Prt$$

$$= (\$600)(0.09)\left(1\frac{1}{2}\right) \qquad 9\% = 0.09$$

$$= \$81$$

The interest is $81.

Step 2 Find the future value of the loan.

$$A = P + I$$

$$= \$600 + \$81$$

$$= \$681$$

Step 3 Divide the future value of the loan by the number of months. Since $1\frac{1}{2}$ years = 18 months, divide $681 by 18 to get $37.83. The monthly payment is $37.83.

DreamPictures/Blend Images LLC

Try This One 4

The Lookout Restaurant took out a loan for $5,000. The simple interest rate was 6.5%, and the term of the loan was 3 years. Find the interest, future value, and monthly payment.

Finding the Principal, Rate, or Time

In addition to finding the interest and future value for a loan or investment, we can find the principal, the rate, and the time period by substituting into the formula $I = Prt$ and solving for the unknown.

Examples 5–7 show how to find the principal, rate, or time.

| Example 5 | Computing Principal |

Calculator Guide

In the calculation for Example 5, order of operations is very important. The parentheses are, too!

Typical Scientific or Online Calculator

 93.5 ÷ (.055 × 2) =

Typical Graphing Calculator

93.5 ÷ (.055 × 2)
 ENTER

Phillips Health and Beauty Spa is replacing one of its workstations. The interest on a loan secured by the spa was $93.50. The money was borrowed at 5.5% simple interest for 2 years. Find the principal.

SOLUTION

$$I = \$93.50, r = 5.5\% = 0.055, \text{ and } t = 2$$
$$I = Prt$$
$$\$93.50 = P(0.055)(2) \quad \text{\textit{Divide both sides}}$$
$$\frac{\$93.50}{(0.055)(2)} = \frac{P(0.055)(2)}{(0.055)(2)} \quad \text{\textit{by (0.055)(2).}}$$
$$P = \$850$$

The amount of the loan was $850.

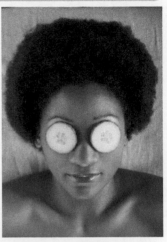

Trinette Reed/Blend Images LLC

Try This One   5

Find the principal on a savings account that paid $76.50 in simple interest at 6% over 3 years.

The same formulas can be used for investments as well. Example 6 shows this.

| Example 6 | Computing Interest Rate |

R & S Furnace Company invested $15,250 for 10 years and received $9,150 in simple interest. What was the rate that the investment paid?

SOLUTION

$$P = \$15,250, t = 10, \text{ and } I = \$9,150$$
$$I = Prt$$
$$\$9,150 = (\$15,250)(r)(10) \quad \text{\textit{Divide both sides by ($15,250)(10).}}$$
$$\frac{\$9,150}{(\$15,250)(10)} = \frac{(\$15,250)(r)(10)}{(\$15,250)(10)}$$
$$0.06 = r$$
$$r = 0.06 \text{ or } 6\%$$

The interest paid on the investment was 6%.

Try This One   6

If you invest $8,000 for 30 months and receive $1,000 in simple interest, what was the rate?

CAUTION	Be sure to change the decimal to a percent since rates are given in percents.

Example 7	Computing the Term of a Loan

Judi and Laura borrowed $4,500 at $8\frac{3}{4}\%$ to put in a hot tub. They had to pay $2,756.25 interest. Find the term of the loan and the monthly payment.

SOLUTION

$$P = \$4,500, \ r = 8\tfrac{3}{4}\% = 0.0875, \text{ and } I = \$2,756.25$$

$$I = Prt$$

$$\$2,756.25 = (\$4,500)(0.0875)t \qquad \textit{Divide both sides}$$

$$\frac{\$2,756.25}{(\$4,500)(0.0875)} = \frac{(\$4,500)(0.0875)t}{(\$4,500)(0.0875)} \qquad \textit{by (\$4,500)(0.0875).}$$

$$7 = t$$

The term of the loan was 7 years, which is 84 months. The total amount paid is the principal plus the interest:

$$\$4,500 + \$2,756.25 = \$7,256.25$$

Divide by 84 months to find the monthly payment:

$$\frac{\$7,256.25}{84} = \$86.38$$

Not bad for hours and hours of steamy relaxation, although the total amount that needs to be paid back should give you some pause about financing expenses over that long a period.

Try This One   7

2. Compute principal, rate, or time.

A pawn shop offers to finance a guitar costing $750 at 4% simple interest. The total interest charged will be $150. What is the term of the loan and the monthly payment?

The Banker's Rule

Simple interest for short-term loans is sometimes computed in days. For example, the term of a loan may be 90 days. In this case, the time would be $\frac{90 \text{ days}}{365 \text{ days}} = \frac{90}{365}$ since there are 365 days in a year. However, many lending institutions use what is called the *Banker's rule*. The Banker's rule treats every month like it has 30 days, so it uses 360 days in a year. They claim that the computations are easier to do. When a lending institution uses 360 days instead of 365, how does that affect the amount of interest? For example, on a $5,000 loan at 8% for 90 days, the interest would be

$$I = Prt$$

$$= (\$5,000)(0.08)\left(\frac{90}{365}\right) \qquad \textit{8\% = 0.08; 90 days is } \frac{90}{365} \textit{ year.}$$

$$= \$98.63$$

Using the Banker's rule, the interest is

$$I = Prt$$

$$= (\$5,000)(0.08)\left(\frac{90}{360}\right) \qquad \textit{We used a 360-day year here.}$$

$$= \$100.00$$

We can see why this is called the Banker's rule and not the customer's rule!

Sidelight   An Outdated Rule

The Banker's rule is very old, and there was a time when it made sense. Originally, interest on savings and loans had to be calculated by hand, and standardizing every month to 30 days did in fact make the calculations a lot simpler. But those types of calculations haven't been done by hand for over 50 years, and in the age of computers, it's just plain silly to worry about how difficult it is to compute the interest. That's exactly why computers are called computers—they're really good at computing things! So why do some lenders still use the Banker's rule? Because they can, probably, and it makes them more money.

Example 8   Using the Banker's Rule

Find the simple interest on a $1,800 loan at 6% for 120 days. Use the Banker's rule.

SOLUTION

$$P = \$1,800, \ r = 6\% = 0.06, \ t = \frac{120}{360}$$

$$I = Prt$$

$$= (\$1,800)(0.06)\left(\frac{120}{360}\right) = \$36$$

The interest using the Banker's rule is $36.

Try This One   8

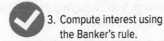
3. Compute interest using the Banker's rule.

Find the simple interest on a $2,200 loan at 7% interest for 100 days. Use the Banker's rule, then compare to the amount of interest using a 365-day year.

Discounted Loans

Sometimes the interest on a loan is paid up front by deducting the amount of the interest from the amount the bank gives you. This type of loan is called a **discounted loan**. The interest that is deducted from the amount you receive is called the **discount**. Example 9 illustrates how it works.

Example 9   Finding the True Rate of a Discounted Loan

A student obtained a 2-year $4,000 loan for college tuition. The rate was 9% simple interest and the loan was a discounted loan.

(a) Find the discount.
(b) Find the amount of money the student received.
(c) Find the true interest rate.
(d) Discuss whether this seems like a deceptive practice.

SOLUTION

(a) The discount is the total interest for the loan.

$$P = \$4,000, \ r = 9\%, \ t = 2 \text{ years}$$

$$I = Prt$$

$$= (\$4,000)(0.09)(2) \qquad 9\% = 0.09$$

$$= \$720$$

The discount is $720.

Sidelight   Who Gets the Discount?

Everyone likes getting discounts, so "discounted loan" sounds great, right? This is just another way that lenders can take advantage of customers. Since you're paying the interest up front, out of the amount you're borrowing, the effect is that you're borrowing less money and paying the same amount of interest. As Example 9 shows, the actual interest rate you end up paying is quite a bit higher than the rate you're quoted. The only thing they're really "discounting" is your ability to make smart financial decisions. If you're ever offered a discounted loan, walk away immediately. Or run if you're wearing sensible shoes.

Beathan, Fuse/Corbis RF/Getty Images

(b) The student received $4,000 − $720 = $3,280.

(c) The true interest rate is calculated by finding the rate on a $3,280 loan with $720 interest.

$$I = Prt$$
$$\$720 = (\$3,280)\, r(2) \qquad \textit{Multiply.}$$
$$\$720 = \$6,560r \qquad \textit{Divide both sides by } \$6,560.$$
$$r = \frac{\$720}{\$6,560} = 0.1098 \text{ (rounded)}$$

The true interest rate is approximately 10.98%.

(d) If you're being quoted a loan at 9%, and the actual percentage you're paying is almost 11%, that surely qualifies as a deceptive lending practice. Buyer beware!

Try This One   9

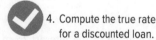

4. Compute the true rate for a discounted loan.

Mary Dixon obtained a $5,000 discounted loan for 3 years at 6% simple interest.

(a) Find the discount.
(b) Find the amount of money Mary received.
(c) Find the true interest rate.

In the next lesson, we'll learn about a different type of interest, compound interest. A lot of people (including instructors!) think that simple interest isn't really used much in the modern world, but that's not true. Student loans use simple interest, as do payday loans, and many short-term loans for things like furniture or other goods. Simple interest is also used in certificates of deposit.

All of these financial situations, and more, will come up in the remainder of this chapter.

Answers to Try This One

1 $I = 1,200(0.025)t$

2 Interest = $4,200, future value = $16,200

3 $11

4 $975, $5,975, $165.97

5 $425

6 5%

7 5 years; $15

8 $42.78; $42.19

9 (a) $900   (b) $4,100   (c) 7.32%

Exercise Set   **7-3**

Writing Exercises

1. Describe what interest is, and how simple interest is calculated.
2. What do the terms "principal" and "future value" refer to?
3. What is meant by the term of a loan?
4. How is the interest rate for a loan or a savings account typically described?

5. Describe how we developed a formula for finding the simple interest on an account.
6. What is the point of interest? Think of it from both the standpoint of borrowing money and from saving money.
7. What is the Banker's rule? Does it help borrowers or lenders?
8. What does it mean for a loan to be discounted?

Computational Exercises

For Exercises 9–24, find the missing value.

Principal	Rate	Time	Simple Interest
9. $12,000	6%	2 years	_____
10. $6,150	0.25%	_____	$61.50
11. $154,625	4.75%	30 years	_____
12. $600	4%	_____	$72
13. $7,300	_____	6 years	$1,927.20
14. $200	_____	3 years	$45
15. _____	9%	4 years	$354.60
16. _____	15%	7 years	$65,625
17. _____	1.16%	5 years	$464
18. $1,250	5%	_____	$375
19. $15,600	1.8%	4.75 years	_____
20. $420	_____	30 months	$31
21. $1,975	7.2%	$3\frac{1}{2}$ years	_____
22. $325	_____	8 years	$156
23. $700	$6\frac{3}{4}$%	_____	$141.75
24. _____	5.6%	$3\frac{1}{2}$ years	$1,372

For Exercises 25–28, find the future value of the loan

25. $P = \$800$, $r = 4\%$, $t = 5$ years
26. $P = \$15,000$, $r = 9\%$, $t = 7$ years
27. $P = \$960$, $r = 2.28\%$, $t = 6\frac{1}{2}$ years
28. $P = \$1,350$, $r = 3.38\%$, $t = 4$ years

For Exercises 29–32, find the interest on each loan using the Banker's rule.

29. $P = \$2,200$, $r = 4.3\%$, $t = 30$ days
30. $P = \$550$, $r = 1.75\%$, $t = 135$ days
31. $P = \$1,750$, $r = 2.3\%$, $t = 45$ days
32. $P = \$660$, $r = 5.6\%$, $t = 225$ days

For Exercises 33–36, the loans are discounted. For each exercise, find (a) the discount, (b) the amount of money received, and (c) the true interest rate.

33. $P = \$3,000$, $r = 8\%$, $t = 3$ years
34. $P = \$1,750$, $r = 4\frac{1}{2}\%$, $t = 6$ years
35. $P = \$6,000$, $r = 7\frac{1}{2}\%$, $t = 4.5$ years
36. $P = \$33,000$, $r = 3.6\%$, $t = 7$ years

Applications in Our World

For Exercises 37–70, all interest is simple interest.

37. In addition to working and a family contribution, Jane had to borrow $8,000 over the course of 6 years to complete a degree. The interest is $4,046.40. Find the rate.
38. Fareed started a new e-commerce business and borrowed $15,000 for 12 years to get the business up and running. The interest is $18,000. Find the rate.
39. To take advantage of a going-out-of-business sale, the College Corner Furniture Store had to borrow some money. It paid back a total of $150,000 on a 6-month loan at 12%. Find the principal.
40. To purchase two new copiers, a campus bookstore paid $1,350 interest on a 9% loan for 3 years. Find the principal.
41. To train employees to use new equipment, Williams Muffler Repair had to borrow $4,500.00 at $9\frac{1}{2}\%$. The

company paid $1,282.50 in interest. Find the term of the loan.
42. Berger Car Rental borrowed $8,650.00 at 6.8% interest to cover the increasing cost of auto insurance. Find the term of the loan if the interest is $441.15.
43. To pay for new supplies, Jiffy Photo Company borrowed $9,325.00 at 8% and paid $3,170.50 in interest. Find the term of the loan.
44. Mary Beck earned $216 interest on a savings account at 8% over 2 years. Find the principal.
45. John White has savings of $4,300, which earned $9\frac{3}{4}\%$ interest for 5 years. Find the interest.
46. Marco Rivera had savings of $816 invested at $4\frac{1}{2}\%$ for 3 years. Find the interest.
47. Matt's Appliance Store borrowed $6,200 for 3 years for repairs. The rate was 6%. Find the future value of the loan.

48. Gecko's Pub borrowed $12,000 for 6 years at 9%. Find the future value of the loan.

49. Martin wants to buy a new bedroom set that costs $1,829 including tax. Unfortunately, he doesn't have $1,829, so he secures a 2-year loan from the furniture store at 11% interest to be repaid in 24 equal monthly installments. Find the monthly payment.

50. When Cecilia's car needs a new transmission, she's stuck with a bill for $1,800. The dealership offers her an 18-month loan at 12.9% interest to be repaid in 18 equal installments. Find the monthly payment.

51. To pay for a new roof on the family home, Jiro is offered a 10-year loan at 6% and a 5-year loan at 10%. If the roof cost $6,300:
    (a) Without doing any calculations, which loan sounds like the better deal to you? Why?
    (b) Find the monthly payment for each loan.
    (c) Which loan results in less total interest being paid? By how much?

52. Elena needs $530 to cover the cost of books for this semester and decides to borrow an extra 25% above that in case of unanticipated expenses. Find the monthly payment if she borrows that amount at 6.29% interest for $1\frac{1}{2}$ years.

Businesses that offer repayment plans for purchases are required by law to disclose the interest rate. But that doesn't mean they go out of their way to let you know what it is. You have to read all the paperwork. In Problems 53–56, find the interest rate for each purchase.

53. To finance a new laptop, Emilie is offered a 3-year payment plan with low monthly payments of $34.20. The cost of the laptop was $884.29 including tax.

54. For a new set of tires ($412.23 including tax), you're offered 18 payments of $28.80.

55. For a 312-square-foot family room, you choose carpet that costs $1.49 per square foot. The tax rate is 6.5%, and you're offered 24 payments of $27.41.

56. To fence their backyard, Juan and Cheryl choose fence at $8.49 per foot. They need 194 feet of fence, and the tax rate is 7.6%. The financing is for $2\frac{1}{2}$ years with monthly payments of $72.12.

57. Sally borrowed $600 for 90 days at 4%. Find the interest using the Banker's rule.

58. Asma borrowed $950 for 120 days at $6\frac{3}{4}$%. Find the interest using the Banker's rule.

59. If you borrow $2,200 for college expenses at 7.2% interest for 2 years, and the lender uses the Banker's rule, how much more interest are you paying than if the lender used a 365-day year? (Ignore leap years.)

60. Repeat Problem 59 if the loan is for 5 years, then again for 10 years. What can you conclude?

61. The West Penn Finance group secured an $18,000 discounted loan to remodel its offices. The rate was 5% and the term was 6 years. Find
    (a) the discount.
    (b) the amount of money the group received.
    (c) the true interest rate.

62. The University Center obtained a $20,000 discounted loan for 3 years to remodel the student game room. The rate was 9%. Find
    (a) the discount.
    (b) the amount of money the center received.
    (c) the true interest rate.

63. Susan would like to buy a new car. Which loan would have the higher interest amount: a personal loan of $10,000 at 9% for 6 years or an auto loan of $10,000 at 8% for 60 months? Why?

64. Sea Drift Motel is converting its rooms into privately owned condominiums. The interest on a $1,000,000, 20-year construction loan is $98,000. What is the rate of interest? Does the rate seem unreasonable?

65. The Laurel Township Fire Department is deciding whether to purchase a new tanker truck or repair the one they now use. For a new truck loan, the interest rate on $25,000 is 18% for a 10-year period; to repair the existing truck, the department must borrow $18,000 at $12\frac{1}{2}$% for 8 years. Which loan is less expensive?

66. A local miniature golf course owner has to recarpet the greens and replace the golf clubs. A bank will lend the owner the necessary $7,800 at 9.5% interest over 48 months. A savings and loan company will lend the owner $7,800 at 8.5% interest for 54 months. Which loan will be less expensive for the golf course owner to assume?

67. Suppose that you borrow $10,000 for school expenses at 6% interest for 5 years.
    (a) How much simple interest would you pay?
    (b) Suppose the bank splits the 5-year loan into five 1-year loans, so that the future value of the loan would be recalculated at the end of each 1-year period, with interest charged on the new amount for the next year. Fill in the table below, which will show the future value of the loan at the end of each 1-year period. Round to the nearest dollar.

End of year	1	2	3	4	5
Future value					

    (c) How much more interest would you end up paying with the loan being split this way?
    (d) Repeat parts (b) and (c) for the situation where the loan is split into two loans, each with length $2\frac{1}{2}$ years.

68. A local bank offers two choices for a certificate of deposit with a term of 4 years. In option 1, you get 4% simple interest for the entire term. In option 2, you get just 3.5% simple interest per year, but the CD is split into four separate 1-year CDs, similar to the loan in Exercise 65.
    (a) If you plan to invest $5,000, which CD is the better option? How much more will the better option be worth?
    (b) Would your answers to part (a) change if the amount you invest is $50,000? Why or why not?

Critical Thinking

69. As you know, when computing interest it's important to convert the interest rate, given as a percentage, to decimal form. But everyone makes mistakes, so occasionally you might forget to do that. If so, how can you tell? Try to repeat some of the problems from this section without converting the rate to decimal form, then explain how you could catch this mistake.

70. Suppose that when making a major purchase, you're offered a 10-year loan at 5% and a 5-year loan at 10%. You decide on what amount would constitute a "major purchase" for you.
    (a) Without doing any calculations, which sounds like the better deal to you? Why?
    (b) Find the total amount of interest paid on each loan.
    (c) Find the monthly payment for each.
    (d) Now discuss which loan you would prefer, considering as many aspects of the loan as you can think of.

71. Suppose that you have a choice of two loans: one at 5% simple interest for 6 years, and one at 6% simple interest for 5 years. Which will result in the smaller future value? Does it depend on the principal?

72. When a loan is discounted, is it better or worse for the borrower if the term is longer? Try some specific examples and make a conjecture.

73. Let's trace the growth of an account that draws simple interest. Suppose the account starts with a value of $750 and earns 4% simple interest.
    (a) Find the amount of simple interest earned in one year.
    (b) Use your answer to part (a) to fill in the table.

Years	Interest Earned in that Year	New Balance
1		
2		
3		
4		
5		

    (c) Now the big question: let's say that a friend of our investor starts a new account after year 3, with an initial deposit of the same amount that's in this account, which also earns 4% simple interest. Explain why this is totally unfair to the person who started the account 3 years earlier.

74. Let's try to fix what makes the account in Question 73 unfair. Fill in the table again, but this time, compute the interest each year as 4% of the NEW BALANCE, rather than 4% of the original investment. How much more does the investor end up earning?

| **Section 7-4** | **Compound Interest** |

LEARNING OBJECTIVES

1. Develop and use compound interest formulas.

2. Find the time needed to reach an investment goal.

3. Compute the effective interest rate of an investment.

4. Compare the effective rate of two investments.

5. Find the future value of an annuity.

6. Compute the periodic payment needed to meet an investment goal.

If you think about simple interest over a long period of time, it doesn't sound like a great deal for the investor. Suppose you put $1,000 into an account that pays 5% simple interest, and you keep it untouched for 30 years. Each year, you're getting 5% of $1,000 in interest. But for all that time, the bank could have been increasing your money through loans and investments, and they're still paying interest only on the original amount.

This is where compound interest comes into play. It seems more fair for the bank to pay interest on the actual value of the account each year, not just the original amount. When interest is computed on the principal *and* any previously earned interest, it's called **compound interest**. Let's look at an example that compares simple and compound interest.

| **Example 1** | Comparing Simple and Compound Interest |

Suppose that $5,000 is invested for 3 years at 8%.

(a) Find the amount of simple interest.
(b) Find the compound interest if interest is calculated once per year.

SOLUTION

(a) Using the formula $I = Prt$ with $P = \$5,000$, $r = 0.08$, and $t = 3$, we get

$$I = \$5,000 \times 0.08 \times 3 = \$1,200 \quad \textit{Simple interest over 3 years}$$

(b) **First year**   For the first year, we have $P = \$5,000$, $r = 0.08$ and $t = 1$:

$$I = Prt = \$5,000 \times 0.08 \times 1 = \$400$$

The interest for the first year is $400.

Second year   At the beginning of the second year, the account now contains $5,400, so we use this as principal for the second year. The rate and time remain the same.

$$I = Prt = \$5,400 \times 0.08 \times 1 = \$432$$

The interest for the second year is $432.

Third year   The principal is now $5,400 + $432 = $5,832.

$$I = Prt = \$5,832 \times 0.08 \times 1 = \$466.56$$

The interest for the third year is $466.56, and the total interest for 3 years is $400 + $432 + $466.56 = $1,298.56. This is almost a hundred dollars more than with simple interest.

Try This One   1

For an investment of $100,000 at 6% interest for 4 years, find (a) the simple interest, and (b) the compound interest if interest is calculated once per year.

Example 2   A Further Look at Compound Interest

In Section 7-3, we saw that the future value of an account earning simple interest can be computed using the formula $A = P(1 + rt)$.

(a) Use this formula to find the future value of the second account in Example 1 after each of the first 3 years. (This is the compound interest account.)
(b) Based on your results, write a formula involving an exponent that would compute the future value after 3 years in one calculation.

SOLUTION

(a) Once again, we have $P = \$5,000$ and $r = 0.08$.

   First year: $A = \$5,000(1 + 0.08) = \$5,400$

   Second year: $A = \$5,400(1 + 0.08) = \$5,832$

   Third year: $A = \$5,832(1 + 0.08) = \$6,298.56$

   Notice that this matches our result from Example 1: we found $1,298.56 in interest, which makes the future value $6,298.56.

(b) Now we have to be clever, and the exponent thing is a hint. In the first year, we took the principal amount and multiplied it by $(1 + 0.08)$. We then multiplied by that same factor again for the second year: this could be written as $\$5,000(1 + 0.08)(1 + 0.08)$. AHA! There's where we get an exponent: this is the same as $\$5,000(1 + 0.08)^2$. For the third year, we'd just multiply by $(1 + 0.08)$ again, giving us $\$5,000(1 + 0.08)^3$. For the record, this does provide the same future value that we got previously.

Try This One   2

Use what you learned in Example 2 to find the future value of the compound interest account after 10 years.

The result of Example 2 allows us to write a formula for calculating the future value of an account that earns compound interest once a year: that formula is $A = P(1 + r)^t$. But here's the problem with interest only being calculated once per year: suppose you deposit a certain amount, and then withdraw the amount 364 days later. If interest is calculated just once per year, you would get absolutely nothing for letting the bank use your money for over 99.7% of the year. That hardly seems fair!

For this reason, compound interest is typically calculated more than once per year. It can be computed **semiannually** (twice a year), **quarterly** (four times a year), **monthly** (12 times a year), or even **daily** (365 times a year).

When the interest is calculated more than once per year, you don't get the entire percentage each time. If our $5,000 investment at 8% were compounded twice a year, we'd get $\frac{1}{2}$ of the interest, or 4%, each time. So when interest is compounded n times per year, the r in our calculation above becomes r/n, and the exponent t becomes nt (t years at n times per year). This gives us the most important formula in this section:

> ### Math Note
>
> A savings account where the principal and interest can't be withdrawn for some fixed period of time without penalty is called a *certificate of deposit.*

> ### Math Note
>
> When interest is compounded yearly, $n = 1$; semiannually, $n = 2$; quarterly, $n = 4$; and daily, $n = 365$.

Formula for Computing Compound Interest

$$A = P\left(1 + \frac{r}{n}\right)^{nt}$$

where A is the future value (principal + interest)

r is the yearly interest rate in decimal form

n is the number of times per year the interest is compounded

t is the term of the investment in years

| Example 3 | Computing Compound Interest |

> ### Calculator Guide
>
> To compute the future value in Example 3:
>
> **Typical Scientific or Online Calculator**
>
> 7000 ⊠ 〔 1.0075 〕 y^x
>
> 20 ▱
>
> **Typical Graphing Calculator**
>
> 7000 〔 1.0075 〕 ⌃
>
> 20 〔ENTER〕
>
> The parentheses are critical.

Find the interest on $7,000 compounded quarterly at 3% for 5 years.

SOLUTION

Quarterly means 4 times a year, so $n = 4$.

$$P = \$7,000, r = 3\% = 0.03, t = 5$$
$$A = P\left(1 + \frac{r}{n}\right)^{nt}$$
$$= \$7,000\left(1 + \frac{0.03}{4}\right)^{4(5)}$$
$$= \$7,000(1.0075)^{20} \quad \textit{See Calculator Guide.}$$
$$= \$8,128.29$$

To find the interest, subtract the principal from the future value.

$$I = \$8,128.29 - \$7,000$$
$$= \$1,128.29$$

The interest is $1,128.29.

Try This One 3

Find the interest on $600 compounded semiannually at 4.5% for 6 years.

| **Example 4** | Computing Compound Interest |

Find the interest on $11,000 compounded daily at 5% for 6 years. Assume a 365-day year.

SOLUTION

$$P = \$11,000, r = 5\% = 0.05, n = 365, t = 6$$

$$A = P\left(1 + \frac{r}{n}\right)^{nt}$$

$$= \$11,000\left(1 + \frac{0.05}{365}\right)^{365(6)} \qquad \textit{See Using Technology box.}$$

$$= \$14,848.14$$

To find the interest, subtract the principal from the future value.

$$I = \$14,848.14 - \$11,000$$
$$= \$3,848.14$$

The interest is $3,848.14.

1. Develop and use compound interest formulas.

| Try This One | 4 |

Find the interest on $50,000 compounded weekly at 7% for 20 years. Assume a 52-week year.

| **CAUTION** |

In Example 3, we simplified $1 + \frac{0.03}{4}$ to 1.0075 before calculating, but didn't do the same for $1 + \frac{0.05}{365}$ in Example 4. This is because the second number has a much longer decimal expansion, and would likely have required rounding. This affects the accuracy considerably. If you rework the calculation in Example 4, but round $1 + \frac{0.05}{365}$ to four decimal places, the result is $2,692.99, which is off by over a thousand dollars!

We'll study the potential effects of rounding in Exercises 71 and 72.

Using Technology: Technology and Compound Interest

As we get further into our study of financial math, the formulas are going to get more complicated, so appropriate use of technology becomes more important. Here are some thoughts on using scientific or online calculators, graphing calculators, and spreadsheets to do compound interest calculations.

Let's use the calculation in Example 4 as a sample: $11,000\left(1 + \dfrac{0.05}{365}\right)^{365(6)}$

Typical Scientific or Online Calculator

These types of calculations are most challenging with a scientific calculator because we need to work in stages from inside the parentheses, rather than working from left to right. Note also that we had to calculate $365 \times 6 = 2,190$ first.

1 [+] .05 [÷] 365 [=] [y^x] 2190 [=] [×] 11000

Typical Graphing Calculator

The use of parentheses will allow us to do this all in one calculation.

11000 [(] 1 [+] .05 [÷] 365 [)] [^] [(] 365 [×] 6 [)] [ENTER]

Spreadsheet

To set up a spreadsheet to do these calculations, we'll want a separate column for each of principal P, rate r, number of times calculated per year n, and time t. The order doesn't matter, but for illustration we'll put P in column A, r in column B, n in column C, and t in column D. Then we'll put the compound interest formula in column E, using cell references to substitute in values for the variables. The formula will be very similar to the way we entered the calculation into a graphing calculator.

$$=A2*(1+B2/C2)^\wedge(C2*D2)$$

Then when values are entered for P, r, n, and t for a particular account, the formula can be copied down to calculate the future value. A template that allows you to perform these calculations is provided in the instructor resources for this section: ask if your instructor wants to provide the template for your use.

E2	f_x =A2*(1+B2/C2)^(C2*D2)			
	A	B	C	D	E
1	P	r	n	t	A
2	11000	0.05	365	6	14848.1418

Microsoft Corporation

One of the most useful aspects of the compound interest formula is using it to find the amount of time needed to meet a certain investment goal. Since t lives in the exponent in our formula, we're going to need to use the logarithms that we studied in Section 6-5 to solve for t when setting up an equation, as in Example 5.

Example 5	**Finding the Time Needed to Reach an Investment Goal**

If you want to save $5,000 before buying your first new car, and you have $3,000 right now to invest at 3% interest compounded monthly, how long will you have to wait?

SOLUTION

$$A = \$5,000, P = \$3,000, r = 3\% = 0.03, n = 12$$

$$A = P\left(1 + \frac{r}{n}\right)^{nt}$$

$$5,000 = 3,000\left(1 + \frac{0.03}{12}\right)^{12t} \qquad \textit{Divide both sides by 3,000.}$$

$$\frac{5,000}{3,000} = \left(1 + \frac{0.03}{12}\right)^{12t} \qquad \textit{Simplify.}$$

$$\frac{5}{3} = (1.0025)^{12t} \qquad \textit{Apply } \log_{1.0025} \textit{ to both sides.}$$

$$\log_{1.0025}\frac{5}{3} = \log_{1.0025}(1.0025)^{12t} \qquad \textit{Simplify right side.}$$

$$\log_{1.0025}\frac{5}{3} = 12t \qquad \textit{Divide both sides by 12.}$$

$$t = \frac{\log_{1.0025}\frac{5}{3}}{12} = \frac{\frac{\log\left(\frac{5}{3}\right)}{\log 1.0025}}{12} \approx 17 \qquad \textit{Use change of base formula.}$$

Uh oh . . . better figure out a way to save more money unless you're okay with waiting for 17 years.

Try This One	5

2. Find the time needed to reach an investment goal.

If you owe $1,400 on a high-interest credit card at 21% compounded monthly, how long will it take the amount you owe to double if you don't make any payments?

Effective Rate

As the number of times per year that interest is compounded goes up, the amount of interest does too, but not by as much as you might think. For a $1,000 investment at 4% interest for 10 years, the difference between compounding yearly and compounding daily is only about $11. (We'll study the effects of more compounding periods in critical thinking Exercises 61–64.) Still, because of this relatively small difference, when interest is compounded more than once per year, the interest earned on a savings account is actually a bit higher than the stated rate.

For example, consider a savings account with a principal of $5,000 and an interest rate of 4% compounded semiannually. The interest is compounded twice a year at 2%. Using the formula shown in Example 3, the actual interest for 1 year is $202. The stated rate is 4%, but the actual rate can be found by dividing $202 by $5,000; it is 4.04%. This rate is called the *effective rate* or *annual yield*.

> The **effective rate** (also known as the **annual yield**) is the simple interest rate which would yield the same future value over 1 year as the compound interest rate.

The next formula can be used to calculate the effective interest rate.

Formula for Effective Interest Rate

$$E = \left(1 + \frac{r}{n}\right)^n - 1$$

where

E = effective rate
n = number of periods per year the interest is calculated
r = interest rate per year (i.e., stated rate)

The stated rate is also called the **nominal rate**.
Example 6 illustrates the use of this formula.

| **Example 6** | **Finding Effective Interest Rate** |

Calculator Guide

To compute the effective interest rate:

Typical Scientific or Online Calculator

1 ⊞ .04 ⊟ 52 ⊟ ⊙ʸ 52 ⊟

1 ⊟

Typical Graphing Calculator

◉ 1 ⊞ .04 ⊟ 52 ⊙ ⌃

52 ⊟ 1 ᴇɴᴛᴇʀ

The template we mentioned earlier for computing compound interest also has a tab for computing the effective interest rate. Ask if your instructor would like to provide it.

Find the effective interest rate when the stated rate is 4% and the interest is compounded weekly, then describe what your result means.

SOLUTION

Let $r = 0.04$ (rate is 4%) and $n = 52$ (compounded weekly) and then substitute into the formula.

$$E = \left(1 + \frac{r}{n}\right)^n - 1$$

$$= \left(1 + \frac{0.04}{52}\right)^{52} - 1 \quad \textit{See Calculator Guide.}$$

$$\approx 0.0408 = 4.08\%$$

The effective rate is 4.08%. This tells us that an account at 4% compounded weekly will earn the same amount of interest in 1 year as a simple interest account at 4.08%.

Try This One   6

Find the effective rate for a stated interest rate of 8% compounded quarterly. What does your answer mean?

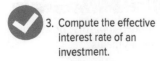

3. Compute the effective interest rate of an investment.

The effective rates of two savings accounts can be used to determine which account would be a better investment. The next example shows how to do this.

| Example 7 | Comparing the Effective Rate of Two Investments |

Which savings account is a better investment: 6.2% compounded daily or 6.25% compounded semiannually?

SOLUTION

Find the effective rates of both accounts and compare them.

6.2% daily

$r = 0.062, n = 365$

$$E = \left(1 + \frac{r}{n}\right)^n - 1$$

$$= \left(1 + \frac{0.062}{365}\right)^{365} - 1$$

$$\approx 0.0640 = 6.40\%$$

6.25% semiannually

$r = 0.0625, n = 2$

$$E = \left(1 + \frac{r}{n}\right)^n - 1$$

$$= \left(1 + \frac{0.0625}{2}\right)^2 - 1$$

$$\approx 0.0635 = 6.35\%$$

The 6.2% daily investment has a slightly better effective rate than 6.25% semiannually.

Try This One 7

Which is a better investment, 3% compounded monthly or 3.25% compounded semiannually?

4. Compare the effective rate of two investments.

Math Note

A scary fact: in 2016, the Retirement Confidence Survey reported that almost 35% of all working Americans at that time were currently not saving ANY money for retirement.

Annuities

An **annuity** is a savings investment plan in which the investor makes a regular, fixed payment into a compound-interest account where the interest rate doesn't change during the term of the investment. This differs from a regular compound-interest account in that you don't invest an entire amount at the beginning of the investment period, but rather break the investment up into smaller payments. For example, you might pay $500 annually for 3 years into an account that yields 6% interest compounded annually. The total amount accumulated (payments plus interest) is called the **future value** of the annuity.

Annuities are often used by individuals or families to save money to pay for things like college expenses, vacations, home improvements, or (most commonly) retirement. Businesses and governments set up annuities to pay future expenses like business expansion, new equipment, and health care costs. The payments in a regular annuity are made at the end of each period.

Example 8 shows how an annuity works.

| Example 8 | Finding the Future Value of an Annuity |

Find the future value of an annuity where a $500 payment is made annually for 3 years at 6%.

SOLUTION

The interest rate is 6% and the payment is $500 each year for 3 years.

  I. End of the first year $500 (payment)
 II. End of the second year

The $500 collected 6% interest and a $500 payment is made; the value of the annuity at the end of the second year is

$$\begin{aligned}
\$500(0.06) = \ \$\ 30 &\quad \text{Interest} \\
\$500 &\quad \text{Principal paid at end of first year} \\
\underline{+\$500} &\quad \text{Payment at the end of the second year} \\
\$1,030 &
\end{aligned}$$

III. End of the third year

During the third year, the $1,030 earns 6% interest and a payment of $500 is made at the end of the third year. The annuity is worth

$1,030(0.06) = $ 61.80   Interest
$1,030.00   Principal at end of second year
+$500.00   Payment at end of third year
—————
$1,591.80

The future value of the annuity at the end of the 3 years is $1,591.80.

Try This One   8

Find the future value of an annuity where a $300 payment is made annually for 4 years at 9%.

Example 8 shows that an annuity really is just a series of compound-interest problems. But calculating each of those compound-interest problems one at a time can be pretty time-consuming for a long-term annuity. In many cases, annuities are used to save for retirement, and regular payments might be made for 30 years or more. Mix in the fact that the payments might be made monthly, and in that case you'd need over 360 separate compound interest calculations to find the future value! Who has time for that? (Well, a computer. But who else?) To avoid this mess, the following formula can be used to find the future value of an annuity. Memorize this formula and then eat it. Just kidding. Obviously this formula is very complicated and requires great care to use. There's a spreadsheet template available in the instructor online resources in case your instructor decides to make it available.

Formula for Finding the Future Value of an Annuity

$$A = \frac{R\left[\left(1 + \frac{r}{n}\right)^{nt} - 1\right]}{\frac{r}{n}}$$

where A is the future value of the annuity
R is the regular periodic payment
r is the annual interest rate
n is the number of payments made per year
t is the term of the annuity in years

CAUTION

You should be especially careful in using the annuity formula because it uses both capital R and lowercase r. The lowercase represents what it always does—annual interest rate. Think of the capital R as standing for "Regular" or "Recurring" payment.

Example 9   Finding the Future Value of an Annuity

If you open an annuity with semiannual payments of $800 at 5% compounded semiannually for 4 years:

(a) Find the future value of the annuity.
(b) How much interest will you earn?
(c) How much money would you have to invest in a regular savings account at 5% compounded semiannually to get the same future value after 3 years? (Note that your first payment on the annuity is at the end of the first year, so that would be when you'd make the lump-sum investment, giving you just 3 years of interest.)

Sidelight **The State of Retirement Savings in the United States**

I have no doubt that there are some people that would be perfectly happy to work for a living until they die. I don't know any of them, mind you, but they probably exist. Most of us, though, want to retire at some point. There are two things required to retire: you have to live long enough, and you need money. The amount you need depends on way too many factors to give reasonable estimates. One thing I can tell you, though, is that for MANY Americans, the picture is pretty grim.

The following information comes from the 2020 Retirement Confidence Survey.

- 32% of Americans had saved not one single penny for retirement as of 2020.

- Among those that have some retirement savings, 18% have less than $1,000, 27% have less than $10,000, and 41% have less than $25,000.

- Less than half of workers have even TRIED to calculate how much they might need for retirement.

- In spite of these numbers, 40% report believing that they'll need over a million dollars to retire.

The best time to begin saving for retirement is like 20 years ago. The second best time is now!

SOLUTION

$$R = \$800,\ r = 5\% = 0.05,\ n = 2 \text{ (semiannual)},\ t = 4$$

(a) The calculations in the annuity formula are a little complicated, so we'll be especially careful to work in stages.

$$A = \frac{R\left[\left(1 + \dfrac{r}{n}\right)^{nt} - 1\right]}{\dfrac{r}{n}}$$

$$A = \frac{\$800\left[\left(1 + \dfrac{0.05}{2}\right)^{2(4)} - 1\right]}{\dfrac{0.05}{2}} = \frac{\$800\left[(1.025)^8 - 1\right]}{0.025} = \$6,988.89$$

The future value of the annuity at the end of 4 years is $6,988.89.

(b) You would have made eight payments of $800 (two each year for 4 years), so your total investment would be $6,400. The interest earned is $6,988.89 − $6,400 = $588.89.

(c) This is a compound interest problem with $r = 0.05$, $A = \$6,988.89$, $n = 2$, and $t = 3$; we need to solve for the principal.

$$A = P\left(1 + \frac{r}{n}\right)^{nt}$$

$$6,988.89 = P\left(1 + \frac{0.05}{2}\right)^{2(3)}$$

$$P = \frac{6,988.89}{(1.025)^6} = 6,026.50$$

You'd have to invest $6,026.50 at the end of the first year to get the same final amount.

Math Note

An IRA is an annuity where the regular payments are taken directly from your paycheck before the government calculates taxes.

This decreases the amount of tax you pay in the short term, but you have to pay tax on the amount when you withdraw the money from the IRA at a later date.

Try This One 9

✓ 5. Find the future value of an annuity.

(a) Find the future value of an annuity when the payment is $275 quarterly, the interest is 6.5% compounded quarterly, and the term is 6 years.

(b) Find the interest earned.

(c) How much would you have to invest at the end of the first year to get the same return over 5 years in a regular savings account at 6.5% compounded quarterly?

One way that annuities are commonly used is to save money for some specific future expense. In that case, you'd more than likely know what future value you're shooting for, and you'd be interested in finding the regular payment necessary to reach that goal. In that case, we can use the annuity formula and solve for the regular payment, as in Example 10.

| **Example 10** | **Finding the Monthly Payment for an Annuity** |

Photodisc/Getty Images

Suppose you've always dreamed of opening your own tattoo parlor, and decide it's time to do something about it. A financial planner estimates that you would need a $35,000 initial investment to start the business, and you plan to save that amount over the course of 5 years by investing in an annuity that pays 7.5% compounded weekly. How much would you need to invest each week?

SOLUTION

We know the following values: $A = 35,000$, $r = 7.5\% = 0.075$, $t = 5$, and $n = 52$.

$$A = \frac{R\left[\left(1 + \frac{r}{n}\right)^{nt} - 1\right]}{\frac{r}{n}}$$

$$35,000 = \frac{R\left[\left(1 + \frac{0.075}{52}\right)^{52(5)} - 1\right]}{\frac{0.075}{52}} \qquad \textit{Multiply both sides by } \frac{0.075}{52}.$$

$$35,000 \cdot \frac{0.075}{52} = R\left[\left(1 + \frac{0.075}{52}\right)^{52(5)} - 1\right] \qquad \textit{Divide to isolate R.}$$

$$R = \frac{35,000\left(\frac{0.075}{52}\right)}{\left(1 + \frac{0.075}{52}\right)^{52(5)} - 1} \qquad \textit{Reduce error by not rounding } \frac{0.075}{52}.$$

$$= \frac{\frac{2,625}{52}}{\left(1 + \frac{0.075}{52}\right)^{260} - 1} = 111.04$$

A payment of $111.04 per week would be necessary to save the required $35,000.

6. Compute the periodic payment needed to meet an investment goal.

| **Try This One** | **10** |

Find the monthly payment needed to save a $6,000 down payment for a car in an annuity that pays 5.5% interest compounded monthly over 2 years.

Answers to Try This One

1 (a) $24,000 (b) $26,247.70

2 $10,794.62 **3** $183.63

4 $152,569.20 **5** About 3.3 years

6 8.24%; a compound interest account at 8% compounded quarterly earns the same amount of interest in 1 year as a simple interest account at 8.24%.

7 3.25% semiannually   **8** $1,371.94

9 (a) $7,993.75 (b) $1,393.75 (c) $5,790.81

10 $237.07

Exercise Set   **7-4**

Writing Exercises

1. Describe the difference between simple interest and compound interest.
2. What does it mean to say that interest is compounded quarterly? What about compounded monthly?
3. What is the effective rate of an investment?
4. Describe how an annuity works.

5. What's the difference between an annuity and an investment using compound interest? What are the advantages of each?
6. Why does it make sense that the interest rate r is divided by n in the compound interest formula?

Computational Exercises

For Exercises 7–12, find the compound interest and future value for each.

Principal	Rate	Compounded	Time
7. $825	4%	Annually	10 years
8. $1,495	0.6%	Semiannually	11 years
9. $560	4.25%	Quarterly	3 years
10. $750	9%	Daily	1 year
11. $320	2.5%	Daily	$4\frac{1}{2}$ years
12. $115,000	4.19%	Weekly	32 years

For Exercises 13–16, find the future value of each annuity.

Payment	Rate	Compounded	Time
13. $430	7%	Annually	15 years
14. $8,000	1.61%	Semiannually	3 years
15. $2,750	0.34%	Quarterly	6 years
16. $3,500	1.73%	Quarterly	$8\frac{1}{2}$ years

For Exercises 17–20, find the periodic payment needed to attain the future value of each annuity.

Future Value	Rate	Compounded	Time
17. $7,200	3.1%	Quarterly	3 years
18. $27,500	5.3%	Monthly	8 years
19. $1.1 million	7.2%	Monthly	28 years
20. $2.4 million	9.25%	Weekly	32 years

For Exercises 21–24, find the effective interest rate.

21. Rate: 6%   Compounded: Quarterly
22. Rate: 10%   Compounded: Semiannually
23. Rate: 6.5%   Compounded: Quarterly
24. Rate: 9.55%   Compounded: Semiannually

For Exercises 25–28, determine which is the better investment.

25. 4.5% compounded semiannually or 4.25% compounded quarterly.
26. 7% compounded monthly or 7.2% compounded semiannually.
27. 3% compounded daily or 3.1% compounded quarterly.
28. 5.74% compounded semiannually or 5.6% compounded daily.

Applications in Our World

29. A couple decides to set aside $5,000 in a savings account for a second honeymoon trip. It is compounded quarterly for 10 years at 9%. Find the amount of money they will have in 10 years.
30. In order to pay for college, the parents of a child invest $20,000 in a bond that pays 8% interest compounded semiannually. How much money will there be in 18 years?
31. A 25-year-old plans to retire at age 50. She decided to invest an inheritance of $60,000 at 7% interest compounded semiannually. How much will she have at age 50?
32. To pay for new machinery in 5 years, a company owner invests $10,000 at $7\frac{1}{2}$% compounded quarterly. How much money will be available in 5 years?

33. The Elders hope to accumulate $12,000 for a sunroom over the next 6 years. How much would they need to invest right now at 3.9% compounded quarterly to reach that goal?
34. A recent college grad accepts a job but plans to save money to go back to grad school in 6 years. She wants to have $22,000 saved at that point to help pay for school. How much would she have to invest right now at $5\frac{1}{4}$% compounded quarterly to reach that goal?
35. How much would you have to invest into a 5-year certificate of deposit paying 2.3% compounded weekly to make it worth $4,500 at the end of the term?

36. After the birth of their first child, the Bartons plan to set up an account to pay for a college education. The goal is to save $80,000 over the next 18 years, and their financial planner suggests a bond fund that historically pays 6.2% interest compounded monthly. How much should they put into the fund now?

37. After winning $73,000 on a game show, Jasmine invests the money in a fixed-rate account offering 7.2% interest compounded semiannually. How long will it take that amount to grow to $100,000?

38. Jamaal is offered a new job with a signing bonus of $17,000. After accepting the job, rather than spending the bonus, Jamaal decides to invest it with a long-term goal of turning it into $40,000 so he can take 6 months off and travel. If he finds an investment that pays 7.1% compounded daily, how long will it take to reach that goal?

39. How many years does it take an investment to double if the interest rate is 6.8% compounded annually? What if it's compounded daily?

40. If you have a $700 balance on a credit card that charges 18% interest compounded monthly, how long will it take that amount to reach $2,000 if you don't make any payments?

41. A married couple plan to save money for their daughter's college education in 4 years. They decide to purchase an annuity with a semiannual payment earning 7.5% compounded semiannually. Find the future value of the annuity in 4 years if the semiannual payment is $2,250.

42. A business owner decided to purchase an annuity to pay for new copy machines in 3 years. The payment is $600 quarterly at 8% interest compounded quarterly. Find the future value of the annuity in 3 years.

43. Find the future value of an annuity if you invest $200 quarterly for 20 years at 5% interest compounded quarterly.

44. The Washingtons decide to save money for a vacation in 2 years. They purchase an annuity with semiannual payments of $200 at 9% interest compounded semiannually. Find the amount of money they can spend for the vacation.

45. The owner of the Campus Café plans to open a second location on a satellite campus in 5 years, so the owner buys an annuity that pays 10.5% interest compounded annually.
    (a) If the payment is $4,000 a year, find the future value of the annuity in 5 years.
    (b) How much more interest would be earned if the owner could invest the full amount paid into the annuity in a regular compound-interest account with the same terms for 4 years?

46. Titan Thigh, the owner of Work-it-out Fitness, wants to buy an annuity that pays 4% interest compounded quarterly for 4 years.
    (a) If the quarterly payment is $160, find the future value of the annuity.
    (b) How much more interest would be earned if the owner could invest the full amount paid into the annuity in a regular compound-interest account with the same terms for 3 years?

47. In order to plan for their retirement, a married couple decides to buy an annuity that pays 8% interest compounded semiannually.
    (a) If the semiannual payment is $2,000, how much will they have saved in 10 years?
    (b) Find the total interest earned.
    (c) How much would this couple have to invest at the end of the first year to get the same return over 9 years in a regular savings account?

48. The Wash-n-Surf, an Internet café and laundromat, plans to replace two wash and surf hubs in 3 years.
    (a) If the owner buys an annuity at 6% interest compounded annually with an annual payment of $800, find the value of the annuity in 3 years.
    (b) Find the total interest earned.
    (c) How much would the owner have to invest at the end of the first year to get the same return over 2 years in a regular savings account?

49. Petty Marine Co. has a long-term plan to expand to a second location that's actually near some water, so they want to start a monthly annuity to save $150,000 in capital over 5 years. The best rate they can find is 7%. Find the monthly payment.

50. Suppose you plan to work right after you graduate, but still save money for grad school. You decide to save $10,000 before starting, and find a weekly annuity that pays 6.5% interest for 4 years. How much will you need to pay each week?

51. The Massive Chemical Corporation starts an annuity to pay for the huge government penalties they expect in 10 years when a pending case finally gets litigated. Their lead attorney informs them that they can expect a $4,000,000 fine. An investment house offers 11% interest on annuities of that size, compounded semiannually. What will the semiannual payment be on this annuity?

52. A 25-year-old decides that their goal is to retire at age 50 with at least $2,000,000 in savings. The company investment annuity offers 7.4% annual returns, compounded monthly. What amount will they need to invest each month?

53. If the Elders (Exercise 33) are also offered an annuity at 3.9% compounded monthly to save the $12,000 they need over 6 years, what would the monthly payments be? How much more would they end up contributing to the final amount?

54. If the college grad in Exercise 34 is also offered an annuity at $5\frac{1}{4}$% compounded quarterly to save the $22,000 she needs over 6 years, what would the quarterly payments be? How much more would she end up contributing to the final amount?

55. If you have $5,000 to invest and you're offered a 6-year investment at 4.5% paying simple interest, and the same rate and time with compound interest compounded daily, and you make the ridiculously bad choice of simple interest, how much money would you cost yourself?

56. Suppose you're offered the following three accounts to invest $10,000 for 10 years: 12% simple interest, 6% interest compounded monthly, and an annuity with quarterly payments of $250 at 8% interest compounded quarterly. Which is the best choice?

57. Which is a better way to invest $4,000 if the concern is simply the future value: a 3-year certificate of deposit paying 4.1% compounded quarterly, or a 3-year annuity that divides that $4,000 into 12 quarterly payments and pays 5.4% compounded quarterly?

58. Which investment results in a greater future value: a $30,000 investment into an account paying 3.8% interest compounded monthly for 4 years, or an annuity that divides the $30,000 into monthly payments for those 4 years and pays 5.9% interest compounded monthly?

59. Bill and Ted open an annuity with $80 monthly payments that pays 4.9% interest compounded monthly. How long will it take for the future value to reach $15,000?

60. How long will it take to accumulate $300,000 in an annuity that pays 7.2% compounded monthly if the monthly payment is $520?

Critical Thinking

In Problems 61–64, we'll study what happens to the compound interest formula when we keep compounding more and more often.

61. The compound interest formula is $A = P(1 + \frac{r}{n})^{nt}$, where n is the number of times per year that interest is compounded.

    (a) If we perform the algebraic substitution $u = \frac{n}{r}$, show that the compound interest formula becomes $A = P(1 + \frac{1}{u})^{urt}$, and then can be rewritten as $A = P[(1 + \frac{1}{u})^{u}]^{rt}$.

    (b) When the number of times per year that interest is compounded grows larger and larger, what happens to the expression we called u? (In this case, we say that n is *tending to infinity*.)

    (c) Using a calculator, fill in the table below. Round to three decimal places.

u	50	100	500	1,000	1,500	2,000	2,500
$\left(1+\dfrac{1}{u}\right)^{u}$							

    What can you conclude about the value of $(1 + \frac{1}{u})^{u}$ as the number of times interest is compounded gets very large?

62. The number that you should have found in Exercise 61(c) is about 2.718, and is denoted e. (Actually, e is irrational, so its decimal equivalent neither terminates nor repeats: $e = 2.718...$) What is the result of the compound interest formula from Exercise 61(a) when the number of times interest is compounded tends to infinity? (In this case, we say that interest is **compounded continuously**. You can think of this as the interest being compounded every instant of every day.)

63. Use the formula $A = Pe^{rt}$ to find the future value after 20 years of a $5,000 account that earns 4% interest compounded continuously, then compare that to the future value of the same account if interest is compounded annually.

64. Now let's compare interest compounded continuously to simple interest. Suppose that $50,000 is invested in two accounts: one earns 6% simple interest, the other earns 6% compounded continuously. Fill in the following table with the future value of each account after each term.

Years	5	10	15	20	25	30
6% simple interest						
6% compounded continuously						

65. An investor deposits $5,000 into an account paying 4% compounded semiannually. Two years later the investor deposited $2,000 into the same account. How much money was there at the end of 5 years?

66. Sam deposited $3,500 into a savings account paying 3% compounded quarterly. Two years later he withdrew $800. How much money was in the account at the end of 6 years?

67. Akish deposited $900 into a savings account paying 2% compounded quarterly. Two years later she deposited $400 into the same account. One year after that, she withdrew $200. How much money was in the account at the end of 6 years?

68. Which account would draw more interest over 5 years: 12% simple interest or 7.5% interest compounded hourly? What about over 20 years?

69. What percent simple interest would be needed on a 10-year investment to have the same future value as one that pays 5% compounded monthly?

70. As you know, some years have 366 days. Does this have any significant effect on interest compounded daily? Do some sample calculations to compare a 365-day year to a 366-day year to decide.

71. As we pointed out in the Caution on page 390, rounding can cause significant error in the compound interest formula. For $30,000 invested at 6% interest compounded weekly, the formula is $A = 30,000(1 + \frac{0.06}{52})^{52t}$, and the number in parentheses is 1.001153846 to nine decimal places.

    (a) Calculate the future value after 5 years without rounding at all: enter the calculation just as it looks above.

    (b) Now calculate the future value if you round the number in the parentheses to three decimal places. How inaccurate is the result?

72. Repeat Problem 71 with $t = 20$. Is the effect of rounding exaggerated as the term increases?

73. Using Technology. If your instructor decided not to supply the spreadsheet that calculates compound interest, why not build one yourself? The variables in that formula are the principal P, the interest rate r, the number of times interest is compounded per year n, and the time in years t. Put headers for these four variables at the top of columns A through D, and a header for future value at the top of column E. Then, in cell E2, input the compound interest formula, using cell references A1, B1, C1, and D1 in place of the variables P, r, n, and t. When you input values for those variables in cells A1 through D1, the future value should appear in cell E1. Begin using values from a problem you've already solved so that you can check to see that the formula is working properly. Then find the difference in future value for a $20,000 initial investment at 6% annual interest for 5 years if interest is compounded annually, monthly, weekly, daily, hourly, and every minute. Then report on your results.

74. Using Technology. Repeat Exercise 73 for the annuity formula on page 394. This time, you'll need columns for recurring payment R, interest rate r, number of payments made per year n, and time in years t. After building the spreadsheet, do an Internet search for the average current interest rate on annuities, and use your spreadsheet with that rate and trial-and-error to find how much you would need to put aside monthly to accumulate $100,000 over 10 years.

Section 7-5   Installment Buying

LEARNING OBJECTIVES

1. Find amount financed, total installment price, and finance charge for a fixed installment loan.

2. Use a table to find APR for a loan.

3. Compute unearned interest and payoff amount for a loan paid off early.

4. Compute credit card finance charges using the unpaid balance and average daily balance methods.

5. Study the effects of making minimum payments.

A lot of people celebrate their college graduation by getting a new car. Unless you're very wealthy, you won't be plunking down a stack of crisp hundred dollar bills to make that happen—you'll need to get a loan. In this case, you'll be doing what is called **installment buying**. This is when an item is purchased and the buyer pays for it by making periodic partial payments, or installments.

There are natural advantages and disadvantages to installment buying. The most obvious advantage is that it allows you to buy an item that you don't have enough money to pay for, and use it while you're raising that money. The most obvious disadvantage is that you pay interest on the amount borrowed, so you end up paying more for the item—in some cases a *lot* more.

SW Productions/Brand X Pictures/ Getty Images

Fixed Installment Loans

A **fixed installment loan** is a loan that is repaid in equal payments. Sometimes the buyer will pay part of the cost at the time of purchase. This is known as a **down payment**. The box contains the other key terms we'll use to describe the wonderful world of installment loans.

The **amount financed** is the amount a borrower will pay interest on.

Amount financed = Price of item − Down payment

The **total installment price** is the total amount of money the buyer will ultimately pay.

Total installment price = Sum of all payments + Down payment

The **finance charge** is the interest charged for borrowing the amount financed.

Finance charge = Total installment price − Price of item

Example 1	Calculating Information About a Car Loan

David Sobecki/McGraw Hill

Cat bought a 2-year old Genesis for $24,650. Her down payment was $6,000, and she will have to pay $414 for 48 months. Find the amount financed, the total installment price, and the finance charge.

SOLUTION

Using the formulas on page 400,

$$\text{Amount financed} = \text{Cash price} - \text{Down payment}$$
$$= \$24,650 - \$6,000$$
$$= \$18,650$$

Since she paid $414 for 48 months and her down payment was $6,000,

$$\text{Total installment price} = \text{Total of monthly payments} + \text{Down payment}$$
$$= 48 \times \$414 + \$6,000 \quad \textit{48 payments at \$414.}$$
$$= \$25,872$$

Now we can find the finance charge:

$$\text{Finance charge} = \text{Total installment price} - \text{Cash price}$$
$$= \$25,872 - \$24,650$$
$$= \$1,222$$

The amount financed was $18,650; the total installment price was $25,872, and the finance charge was $1,222.

Try This One   1

If you buy a used car for $8,200 with a down payment of $1,000 and 36 monthly payments of $270, find the amount financed, the total installment price, and the finance charge.

In some sense, an installment loan is the opposite of an annuity. You make equal regular payments, but instead of saving money for future use, you're paying off money that you borrowed to use at the beginning of the loan.

Sidelight   To Buy or to Lease? That is the Question.

Obvious statement of the day: new cars are expensive. (And used ones aren't particularly cheap, either.) If you don't have a lot of money to make a down payment but you want that new-car smell, leasing may be the way to go. Like an auto loan, with a lease you're making regular equal payments to a financial institution. The difference is simple: after making payments for a set number of months, you don't own the vehicle and you have to give it back. In essence, you're renting the car rather than buying it. So why would you want to do that? There are several reasons.

For one, many leases require little or even no down payment, so you can get a new car without having to save a lot of money in advance. Second, lease payments are usually quite a bit less than those you'd have on a loan (as they should be, since you don't get to keep the car!). Third, most leases last for 2 or 3 years, allowing regular leasers to get a brand new car

every few years. And you do have the option of buying the car at a pre-arranged price after the term of the lease expires.

The biggest downside to leasing is obvious—you're paying for the car but you don't get to keep it. Also, most leases have mileage limits, and if you go over the number of miles you're allotted, you can end up paying a substantial penalty at the end of the lease. You may pay more for insurance with a leased car as well. In addition, you're committed to making payments for the life of the lease, so if you decide you don't like the car that much or your financial situation changes, you're pretty much out of luck.

Depending on what you're looking for and your financial situation, leasing might be the right thing for you. But make sure you understand all of the provisions before signing anything because leases are commitments that are *very* difficult to get out of.

| **Example 2** | Computing a Monthly Payment |

Ryan McVay/Photodisc/Getty
Images

After a big promotion, a young couple bought $9,000 worth of furniture. The down payment was $1,000. The balance was financed for 3 years at 8% simple interest per year.

(a) Find the amount financed.
(b) Find the finance charge (interest).
(c) Find the total installment price.
(d) Find the monthly payment.
(e) The furniture store was offering a promotion that would have allowed them to get the furniture for no money down if they financed at 8.7%, and they'd have 5 years to pay off the balance in equal installments. Would they pay more or less in the long run? By how much? What would be the difference in the monthly payment? What are some factors that would go into deciding which offer to take?

SOLUTION

(a) Amount financed = Price of item − Down payment
$$= \$9{,}000 - \$1{,}000 = \$8{,}000$$

Math Note

The monthly payment is the total installment price minus the down payment, divided by the number of payments.

(b) To find the finance charge, we use the simple interest formula:

$$I = Prt$$
$$= \$8{,}000 \times 0.08 \times 3 \qquad \textit{8\% = 0.08}$$
$$= \$1{,}920$$

(c) In this case, the total installment price is simply the cost of the furniture plus the finance charge:

$$\text{Total installment price} = \$9{,}000 + \$1{,}920$$
$$= \$10{,}920$$

(d) To calculate the monthly payment, divide the amount financed plus the finance charge ($8,000 + $1,920) by the number of payments:

$$\text{Monthly payment} = \$9{,}920 \div 36$$
$$= \$275.56$$

In summary, the amount financed is $8,000, the finance charge is $1,920, the total installment price is $10,920, and the monthly payment is $275.56.

(e) In this case, the amount financed would be the full $9,000. The finance charge would be $I = Prt = \$9{,}000(0.087)(5) = \$3{,}915$. The total installment price is now $9,000 + $3,915 = $12,915. Since there was no money down, the monthly payment is this full amount divided by 60 payments, which gives us $215.25. Overall, they'd end up paying $1,995 more, but the monthly payment would be $60.31 less. The advantages of this are that they don't have to come up with any money up front and the monthly payment is less. But they end up spending a LOT more overall. So they'd have to decide which is more important to them: saving money in the long run, or saving some money up front.

Try This One   2

1. Find amount financed, total installment price, and finance charge for a fixed installment loan.

A graphic design pro buys a new iMac for $1,499 with a $200 down payment, and gets manufacturer financing for 5 years at 12% simple interest. Find (a) the amount financed, (b) the finance charge, (c) the total installment price, and (d) the monthly payment. (e) Is this a better offer than no money down and 11.5% simple interest for 4 years? Calculate the total installment price and monthly payment, then discuss.

Annual Percentage Rate

Justin Sullivan/Getty Images

APR on a loan is similar to the effective rate on an investment, but may also factor in up-front fees.

Many lenders add up-front fees to a loan and then spread them over the life of the loan. This has the effect of making the actual interest rate that a borrower pays higher than the quoted rate. Because this can get confusing, lenders are required by law to disclose an **annual percentage rate**, or APR, that reflects the true interest charged. This allows consumers to compare loans with different terms. The mathematical procedures for computing APR are extremely complicated, so tables have been compiled that help you to estimate APR for a loan. There is also a wide variety of APR calculators available online. A partial APR table is shown in Table 7-1. There are three steps required to find an APR from the table; they are listed below.

Using the APR Table

Step 1 Find the finance charge per $100 borrowed using the formula

$$\frac{\text{Finance charge}}{\text{Amount financed}} \times \$100$$

Step 2 Find the row in the table marked with the number of payments and move to the right until you find the amount closest to the number from Step 1.

Step 3 The APR (to the nearest half percent) is at the top of the corresponding column.

Table 7-1	APR Table

Annual Percentage Rate

Numbers of payments	4.00%	4.50%	5.00%	5.50%	6.00%	6.50%	7.00%	7.50%	8.00%	8.50%	9.00%	9.50%	10.00%
	Finance charge per $100 of amount financed												
6	1.17	1.32	1.46	1.61	1.76	1.9	2.05	2.2	2.35	2.49	2.64	2.79	2.93
12	2.18	2.45	2.73	3	3.28	3.56	3.83	4.11	4.39	4.66	4.94	5.22	5.5
18	3.2	3.6	4	4.41	4.82	5.22	5.63	6.04	6.45	6.86	7.28	7.69	8.1
24	4.22	4.75	5.29	5.83	6.37	6.91	7.45	8	8.54	9.09	9.64	10.19	10.75
30	5.25	5.92	6.59	7.26	7.94	8.61	9.3	9.98	10.66	11.35	12.04	12.74	13.43
36	6.29	7.09	7.9	8.71	9.52	10.34	11.16	11.98	12.81	13.64	14.48	15.32	16.16
48	8.38	9.46	10.54	11.63	12.73	13.83	14.94	16.06	17.18	18.31	19.45	20.59	21.74
60	10.5	11.86	13.23	14.61	16	17.4	18.81	20.23	21.66	23.1	24.55	26.01	27.4

Example 3	Finding APR

Ever had your transmission go out? Not fun. One couple ends up with a $1,900 repair bill, and can handle paying $400 of it up front. They finance the rest for 2 years with a monthly payment of $67.34. Find the APR.

SOLUTION

Step 1 Find the finance charge per $100. The total amount they will pay is $67.34 per month × 24 payments, or $1,616.16. Since they financed $1,500, the finance charge is $1,616.16 − $1,500 = $116.16. Now we use our formula for finding the finance charge per $100.

Math Note

You're not likely to get a finance charge per $100 that's EXACTLY in the table, so you'll typically have to estimate by figuring how far the number you get is away from the two closest entries in the table, like we did in Example 3.

$$\text{Finance charge per } \$100 = \frac{\text{Finance charge}}{\text{Amount financed}} \times \$100$$

$$= \frac{\$116.16}{\$1,500} \times \$100$$

$$\approx \$7.74$$

Step 2 Find the row for 24 payments and move across the row until you find the number closest to $7.74. The two closest entries in the table are $7.45 and $8.00; $7.74 is pretty close to halfway in between those two values.

Step 3 Move to the top of the column to get the APR; $7.45 corresponds to 7.0% and $8.00 corresponds to 7.5%, so we can estimate the APR at just about 7.25%.

✓ 2. Use a table to find APR for a loan.

Try This One   3

Suppose that you buy a like-new Corolla for $19,900 with a down payment of $3,000, and finance the balance at $623 per month for 30 months. Find the APR.

One way to save money on a fixed installment loan is to pay it off early. This allows you to avoid paying the entire finance charge. The amount of the finance charge that is saved when a loan is paid off early is called **unearned interest**. There are two methods for calculating unearned interest, the **actuarial method** and the **rule of 78**. The actuarial method uses the APR table, and the following formula:

The Actuarial Method

$$u = \frac{kRh}{100 + h}$$

where
- u = unearned interest
- k = number of payments remaining, excluding the current one
- R = monthly payment
- h = finance charge per $100 for a loan with the same APR and k monthly payments

Math Note

The value h is found using Table 7-1: it's the entry in the row with the number of remaining payments and the column matching the loan's APR.

Example 4   Using the Actuarial Method

Our friends from Example 3 decide to use their tax refund to pay off the full amount of the repair with their 12th payment. Find the unearned interest and the payoff amount. Do you think it was worthwhile for them to pay the loan off early?

SOLUTION

To use the formula for the actuarial method, we'll need values for k, R, and h. Half of the original 24 payments will remain, so $k = 12$. From Example 3, the monthly payment is $67.34 and the APR is 7.25%. Using Table 7-1, we find the row for 12 payments and the columns for 7% and 7.5%. The values in these columns and the row for 12 payments are $3.83 and $4.11. These are separated by $0.28, and half of that is 14 cents. So our best estimate on h would be $3.83 + $0.14 = $3.97. Now we substitute those values into the formula:

$$u = \frac{kRh}{100 + h} \qquad k = 12, R = \$67.34, h = \$3.97$$

$$= \frac{(12)(67.34)(3.97)}{100 + 3.97} \qquad \textit{Multiply in the numerator, add in the denominator.}$$

$$= \frac{3,208.08}{103.97} \approx 30.86$$

The unearned interest is $30.86.

The payoff amount is the amount remaining on the loan minus unearned interest. At this point, 11 payments have been made, so there would be 13 remaining if the plan wasn't to pay the loan off early.

$$\text{Payoff amount} = 13 \times \$67.34 - 30.86$$
$$= \$844.56$$

The couple would have to come up with $799.41 to pay off the loan. They'll be paying almost $850 all at once to save about thirty bucks in interest. I don't think it's worth it, but of course you may disagree.

Try This One   4

Suppose that after buying the car in Try This One 3, you decide to pay the loan off in 24 months rather than 30. Use the actuarial method to find the unearned interest and the payoff amount. Do you think it would be worth paying the loan off early?

The next example will show how to find the unearned interest and payoff amount of a fixed installment loan using the rule of 78. This method has the advantage of not requiring the APR table.

The Rule of 78

$$u = \frac{fk(k + 1)}{n(n + 1)}$$

where u = unearned interest
f = finance charge
k = number of remaining monthly payments
n = original number of payments

Notice that the formula has the number of payments remaining in the numerator and the original number of payments in the denominator. That means as the number of remaining payments gets smaller, the unearned interest does as well. The reason it's not simply the proportion $u = \frac{fk}{n}$ is that the lenders collect more interest and less principal in the beginning of a loan. We'll discuss that in a very revealing Sidelight in Section 7-6. Stay tuned.

Example 5   Using the Rule of 78

Use the rule of 78 to find the unearned interest on the auto repair loan from Examples 3 and 4.

SOLUTION

We've already seen that the finance charge was $116.16. The number of remaining payments is 12, and there were originally 24 payments.

Substitute into the formula using f = $116.16, n = 24, and k = 12.

$$u = \frac{fk(k + 1)}{n(n + 1)}$$
$$= \frac{116.16(12)(12 + 1)}{24(24 + 1)} = \frac{18,120.96}{600} \approx 30.20$$

Interesting. This time we got $30.20, which is 66 cents less than using the actuarial method.

Try This One   5

A $16,500 truck loan is to be paid off in 48 monthly installments of $386.50. The borrower decides to pay off the loan after 40 payments have been made. Find the amount of interest saved. Use the rule of 78.

3. Compute unearned interest and payoff amount for a loan paid off early.

So far, we've covered examples of closed-ended credit; this is credit with a fixed number of payments and a specific payoff date. We turn our attention now to open-ended credit, where there is no fixed number of payments or payoff date. This is also called **revolving credit**. By far the most common example of this is credit cards. In theory, credit cards are awesome: you don't have to worry about carrying cash, or getting a pocketful of coins you don't want in change. But of course, if you're not vigilant about paying your cards off every month, you can end up paying a small fortune in interest. Since more and more people use credit cards for convenience, an understanding of open-ended credit is more important than ever. The next thing we'll do is study how the interest on a credit card is calculated.

The Unpaid Balance Method

With the **unpaid balance method**, interest is charged only on the balance from the previous month. Example 6 shows how to find the interest on the unpaid balance.

Example 6   Computing a Credit Card Finance Charge

Math Note

The billing cycle for credit cards and loans can begin and end on any day of the month; however, for the examples given in this section, we will assume the cycle ends on the last day of the month.

For the month of April, Elliot had an unpaid balance of $356.75 at the beginning of the month and made purchases of $436.50. A payment of $200.00 was made during the month. The interest on the unpaid balance is 1.8% per month. Find the finance charge and the balance on May 1.

SOLUTION

Step 1 Find the finance charge on the unpaid balance using the simple interest formula with rate 1.8%.

$$I = Prt$$
$$= \$356.75 \times 0.018 \times 1 \quad \textit{1.8\% = 0.018}$$
$$= \$6.42 \text{ (rounded)}$$

The finance charge is $6.42.
(*Note:* since the interest rate is given per month, the time is always equal to 1 month.)

Step 2 To the unpaid balance, add the finance charge and the purchases for the month; then subtract the payment to get the new balance.

$$\text{New balance} = \$356.75 + \$6.42 + \$436.50 - \$200$$
$$= \$599.67$$

The new balance as of May 1 is $599.67.

Try This One   6

For the month of January, Christina has an unpaid balance of $846.50 from December. Purchases for the month were $532.86, and a payment of $350.00 was made during the month. If the interest on the unpaid balance is 2% per month, find the finance charge and the balance on February 1.

Sidelight   Money for Nothing?

Since most credit card companies either use the unpaid balance method, or don't charge interest on new purchases, if you pay off the full balance each month, you never pay any interest. In essence, you're getting to use the bank's money for nothing! But don't shed any tears for the poor banks. They charge the merchant a fee for each transaction. Also, the freedom that comes from buying stuff and walking out the door without handing over any money makes it extremely difficult for people to spend only what they can afford, especially young people using credit cards for the first time. Consider the following statistics:

- The average college senior will graduate with $1,183 in credit card debt in addition to any student loan debt.
- 54% of all credit card users pay interest fees by not paying off the card every month.
- Total credit card debt in the United States in 2021 was almost 807 *billion* dollars.
- The average credit card debt per household in 2021 was $6,270. This was up significantly from 2020, when it was $5,315.
- The average household that was carrying a credit card balance in 2020 paid almost $1,400 in credit card interest that year.
- In 2010, over 90% of college students had a credit card, but new regulations on credit card companies targeting

Fuse/Corbis/Getty Images

students decreased that number by quite a bit. Debit cards are now far more common—debit cards draw from an account, so no money, no purchases. Still, in 2020, 57% of college students had at least one credit card.

If you find yourself spending more than you can pay off, and having to keep a balance on your card, the best approach is probably the one pictured above.

Average Daily Balance Method

If your credit card company uses the unpaid balance method (most do these days), then you'll never pay any interest if you pay off your balance in full each month. Needless to say, the companies that issue credit cards are not fond of this approach, so they sometimes charge an annual fee for use of the card. Another way for them to make money from all users is to use the **average daily balance method** for computing the finance charge. In this method, the balance for each day of the month is used to compute an average daily balance, and interest is computed on that average. So in effect, consumers start paying interest on the day they make a purchase. Example 7 shows how to compute interest using this method.

Example 7	Computing a Credit Card Finance Charge

Ryan McVay/Photodisc/Getty Images

Betty's credit card statement showed the following transactions during the month of August.

August 1	Previous balance	$165.50
August 7	Purchases	59.95
August 12	Purchases	23.75
August 18	Payment	75.00
August 24	Purchases	107.43

Find the average daily balance, the finance charge for the month, and the new balance on September 1. The interest rate is 1.5% per month on the average daily balance.

SOLUTION

Step 1 Find the balance as of each transaction.

August 1	$165.50
August 7	$165.50 + $59.95 = $225.45
August 12	$225.45 + $23.75 = $249.20
August 18	$249.20 − $75.00 = $174.20
August 24	$174.20 + $107.43 = $281.63

Step 2 Find the number of days for each balance.

Date	Balance	Days	Calculations
August 1	$165.50	6	(7 − 1 = 6)
August 7	$225.45	5	(12 − 7 = 5)
August 12	$249.20	6	(18 − 12 = 6)
August 18	$174.20	6	(24 − 18 = 6)
August 24	$281.63	8	(31 − 24 + 1 = 8)

> **Math Note**
>
> Since the transaction period starts on August 1 and ends on the last day of the month, which must be included, we had to add 1 to the last period of days in Step 2. The total number of days has to equal the number of days in the given month.

Step 3 Multiply each balance by the number of days, and add these products.

Date	Balance	Days	Calculations
August 1	$165.50	6	$165.50(6) = $993.00
August 7	$225.45	5	$225.45(5) = $1,127.25
August 12	$249.20	6	$249.20(6) = $1,495.20
August 18	$174.20	6	$174.20(6) = $1,045.20
August 24	$281.63	8	$281.63(8) = $2,253.04
		31	$6,913.69

Step 4 Divide the total by the number of days in the month to get the average daily balance.

$$\text{Average daily balance} = \frac{\$6,913.69}{31} \approx \$223.02$$

The average daily balance is $223.02.

Step 5 Find the finance charge. Multiply the average daily balance by the rate, which is 1.5%, or 0.015.

$$\text{Finance charge} = \$223.02 \times 0.015 \approx \$3.35$$

Step 6 Find the new balance. Add the finance charge to the balance as of the last transaction.

$$\text{New balance:} \quad \$281.63 + \$3.35 = \$284.98$$

The average daily balance is $223.02. The finance charge is $3.35, and the new balance is $284.98.

Try This One   7

4. Compute credit card finance charges using the unpaid balance and average daily balance methods.

A credit card statement for the month of November showed the following transactions.

November 1	Previous balance	$937.25
November 4	Purchases	$531.62
November 13	Payment	$400.00
November 20	Purchases	$89.95
November 28	Payment	$100.00

(a) Find the average daily balance.
(b) Find the finance charge. The interest rate is 1.9% per month on the average daily balance.
(c) Find the new balance on December 1.

The procedure for finding the average daily balance is summarized next.

Procedure for the Average Daily Balance Method

Step 1   Find the balance as of each transaction.

Step 2   Find the number of days for each balance.

Step 3   Multiply the balances by the number of days and find the sum.

Step 4   Divide the sum by the number of days in the month.

Step 5   Find the finance charge (multiply the average daily balance by the monthly rate).

Step 6   Find the new balance (add the finance charge to the balance as of the last transaction).

It's definitely a good idea to check with your credit card company to see what method it uses for computing finance charges. If it uses average daily balance, you will end up paying interest even if you pay off the full amount each month. There's a lot of competition out there, so if that's the case, you can probably find another card with a better deal.

Each month when you get a credit card statement, there's a minimum payment that must be paid. Before 2003, these minimum payments had shrunk to the point that they didn't even cover the interest for the month, so if a cardholder paid just the minimum payment, the balance would continue to rise. In response to a public outcry, the federal government set new regulations in 2003 requiring credit card companies to set the minimum payment high enough that it covers the interest for the month, as well as part of the principal balance. (The exact minimum payment formula varies some from company to company.)

This is not to say that making minimum payments only is a good idea. If at all possible, you should always pay off the full amount each month. This is the only foolproof way to make sure that you don't get in the habit of spending more money than you can afford to. In Example 8, we'll look at how a credit card balance changes if only minimum payments are made.

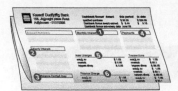

Microsoft Corporation

Example 8 | **Studying the Effect of Making Minimum Payments**

Suppose you have a $2,300 balance on a credit card with an interest rate of 1.35% per month, and the minimum payment for any month is the amount of interest plus 1% of the principal balance. If you don't make any more purchases on that card and make the minimum payment for 6 months, how much will you pay down the balance?

SOLUTION

This sounds like we'll have to compute interest, but actually we won't since the minimum payment will cover the interest each month. In short, we'll be paying the interest each month plus 1% of the principal. So our balance will decrease by 1% each month.

Month 1: 1% of $2,300 = $23; new balance:

$$\$2,300 - \$23 = \$2,277$$

Month 2: 1% of $2,277 = $22.77; new balance:

$$\$2,277 - \$22.77 = \$2,254.23$$

Month 3: 1% of $2,254.23 = $22.54; new balance:

$$\$2,254.23 - \$22.54 = \$2,231.69$$

Math Note

The terms of the credit card in Example 8 (1.35% interest per month, and a minimum payment equal to interest plus 1% of principal) represent industry averages in September 2021.

Month 4: 1% of $2,231.69 = $22.32; new balance:

$$\$2,231.69 - \$22.32 = \$2,209.37$$

Month 5: 1% of $2,209.37 = $22.09; new balance:

$$\$2,209.37 - \$22.09 = \$2,187.28$$

Month 6: 1% of $2,187.28 = $21.87; new balance:

$$\$2,187.28 - \$21.87 = \$2,165.41$$

So after 6 months, you would have paid off $2,300 − $2,165.41, or $134.59. If it seems like it would take you forever to pay off the full amount at that rate, you're pretty close.

In Problems 65 and 66 we'll study some methods to simplify these calculations. Better still, in Problem 71 we'll use a spreadsheet to see how long it would actually take to pay off the card when only making minimum payments.

Try This One   8

6. Study the effects of making minimum payments.

If the credit card company in Example 8 instead sets its minimum payment at interest plus 3% of the principal balance, how much more progress would you make after making minimum payments for 6 months?

Answers to Try This One

1 Amount financed, $7,200; total installment price, $10,720; finance charge, $2,520

2 (a) $1,299   (d) $34.64
    (b) $779.40   (e) The total installment price
    (c) $2,278.40   is less ($2,188.54), but the
                        monthly payment is higher
                        ($4,559). Each loan has its
                        advantages.

3 8.0%

4 Unearned interest, $85.83; Payoff amount, $4,275.17
    Whether it's worth paying off early is an opinion.

5 $62.82

6 Finance charge = $16.93; balance on Feb.
    1 = $1,046.29

7 (a) $1,198.69
    (b) $22.78
    (c) $1,081.60

8 $249.57

Exercise Set   7-5

Writing Exercises

1. Explain what is meant by the term "installment loan."
2. What is a finance charge?
3. What is the difference between the purchase price of an item and the amount financed?
4. What is the difference between closed-ended credit and open-ended credit?
5. What is the annual percentage rate (APR) for a loan? Why is it typically different than the stated interest rate?

6. What is the difference between the unpaid balance method and the average daily balance method for computing interest? Which is better for the consumer?
7. How are minimum payments on credit cards calculated?
8. What's the difference between buying and leasing a car?
9. What are the advantages and disadvantages of paying a fixed-installment loan off early?
10. Describe the only way to avoid paying interest charges on most credit cards.

Applications in Our World

11. Mary Lee bought a stove for $460 with a down payment of $60, then paid $42 a month for 10 months. Find the total installment price of the stove.

12. Martin Dennis bought a wide-screen television for $1,720, making a down payment of 15% and then paying the balance over 18 months. The finance charge was 4% of the amount financed. Find the down payment and the installment price of the television and the monthly payment.

13. Joy Lansung bought a microwave oven for $375, making a down payment of 15% and financing the rest for 12 months with payments of $27.25. Find the down payment and the total installment price of the oven.

14. Mary Scherer bought a wristwatch for $845. She made a down payment of $95 and financed the rest with six monthly payments of $128.75. Find the total installment price.

15. Hang Yo bought a treadmill for $925. He made a 20% down payment and financed the rest over 18 months. Find the monthly payment if the finance charge was 5% of the amount financed.

16. Stacie Howard bought a water softener for $550, making a down payment of $75 and paying the balance off in 12 monthly payments. Find the monthly payment if the interest rate was 11%.

In Exercises 17–22, for the automotive loan, find the amount financed, the total installment price, and the finance charge.

17. Purchase price including taxes and fees: $14,295.41; down payment: $2,600; payments: $279.40 for 48 months.

18. Purchase price including taxes and fees: $22,152.37; down payment: $6,300; payments: $312.15 for 60 months.

19. Purchase price: $19,500; sales tax: 6.5%; license and title fees: $375; down payment: $3,500; payments: $390.25 for 60 months.

20. Purchase price: $6,052; sales tax: 8%; license and title fees: $260; down payment: $1,500; payments: $180.10 for 36 months.

21. Purchase price: $31,600; sales tax 5.5%; license and title fees: $410; money credited for trade-in (subtracted from price of vehicle before taxes): $12,400; payments: $471.22 for 48 months.

22. Purchase price: $20,600; sales tax 4.5%; license and title fees: $304; money credited for trade-in (subtracted from price of vehicle before taxes): $3,600; payments: $371.48 for 60 months.

23. Richard Johnston bought a GMC Terrain for $29,725. He made a down payment of $10,000 and paid $454 monthly for 4 years. Find the APR.

24. Jennifer Siegel bought a new Ford Focus for $17,400. She made a down payment of $4,000 and made monthly payments of $396 for 3 years. Find the APR.

25. Erin LaRochelle bought four comforter sets for $900, making a down payment of $100 and paying off the balance in 12 monthly payments of $71.05. Find the APR.

26. Hector Rondón bought his wife a turquoise bracelet, earrings, and pendant for her birthday. He paid $1,125 for the set and had a down payment of $175. He paid the balance with 12 monthly payments of $84. Find the APR.

27. Matt Brawn bought a diamond engagement ring for $2,560. The down payment was $600, and Matt made 24 monthly payments of $90. Find the APR.

28. Mary Sinclair bought a preowned Corvette for $27,900.00. She made a down payment of $8,000.00 and financed the rest over 5 years with monthly payments of $389.50. Find the APR.

29. In Exercise 23, Richard was able to pay off the loan at the end of 30 months. Using the actuarial method, find the unearned interest and payoff amount.

30. In Exercise 24, Jennifer decided to pay off the loan at the end of 2 years. Using the actuarial method, find the unearned interest and payoff amount.

31. In Exercise 25, Erin decided to pay off the loan at the end of 6 months. Using the actuarial method, find the unearned interest and the payoff amount.

32. In Exercise 26, Hector was able to pay off the loan at the end of 6 months. Using the actuarial method, find the unearned interest and the payoff amount.

33. In Exercise 27, Matt was able to pay off the loan 6 months early. Using the actuarial method, find the unearned interest and the payoff amount.

34. In Exercise 28, Mary decided to pay off the loan 6 months early. Using the actuarial method, find the unearned interest and the payoff amount.

For Exercises 35–40, use the rule of 78.

35. A $4,200.00 loan is to be paid off in 36 monthly payments of $141.17. The borrower decides to pay off the loan after 20 payments have been made. Find the amount of interest saved.

36. Malika borrowed $150.00 for 1 year. Her payments are $13.75 per month. If she decides to pay off the loan after 6 months, find the amount of interest that she will save.

37. Greentree Limousine Service borrowed $200.00 to repair a limousine. The loan was to be paid off in 18 monthly installments of $13.28. After a good season, they decide to pay off the loan early. If they pay off the loan after 10 payments, how much interest do they save?

38. Household Lighting Company borrowed $600 to purchase items from another store that was going out of business. The loan required 24 monthly payments of $29.50. After 18 payments were made, the company decided to pay off the loan. How much interest was saved?

39. Lydia needed to have her roof repaired. She borrowed $950 for 10 months. The monthly payments were $99.75

each. After seven payments, she decided to pay off the balance of the loan. How much interest did she save?

40. The owners of Scottdale Village Inn decided to remodel the dining room at a cost of $3,250. They borrowed the money for 1 year and repaid it in monthly payments of $292.50. After eight payments were made, the owners decided to pay off the loan. Find the interest saved.

41. Ticha is interested in a new Chevy Cruze and is given the option to lease or buy. If she buys, with a down payment of $4,000 the monthly payments will be $340.27 for 48 months. If she leases, with $1,000 up front, the monthly lease payment will be $199.
    (a) How much will Ticha pay total if she chooses the lease option?
    (b) How much more will she pay in the first 2 years if she chooses to buy instead?
    (c) Which option would you choose? Explain your answer.

42. Bui is checking out a Ford Explorer, which has a sticker price of $32,415. She's able to negotiate $3,000 off sticker price, and the tax rate is 6.5%. License and title fees are $395. If she trades in her old minivan for $9,000 (deducted from the negotiated price before taxes), the monthly payment will be $500.95 for 60 months. She's also offered a lease with $2,900 down and payments of $429 for 36 months.
    (a) How much will Bui pay total if she chooses the lease option?
    (b) How much more will she pay in the first 3 years if she chooses to buy instead?
    (c) How much interest will she pay if she buys?
    (d) Which option would you choose? Explain your answer.

43. If the lender uses the rule of 78 rather than the actuarial method in Problem 29, does this benefit the borrower or the lender? What's the difference in unearned interest?

44. (a) If the lender uses the rule of 78 rather than the actuarial method in Problem 34, does this benefit the borrower or the lender? What's the difference in unearned interest?
    (b) Based on Problems 43 and 44, which method for computing unearned interest appears to be better for the borrower? Does it matter how far from the end of the loan the payoff is made?

45. When the engine falls out of Rhonda's old car, it's time to shop for something newer. The plan is to keep the monthly payment at $150, and the loan will be 6.2% simple interest.
    (a) If Rhonda plans to make a down payment of $2,000 and finance the car for 48 months, what's the price of the most expensive car that will fit those parameters? (*Hint:* Let x = the price of the car and use the formulas on p. 400 and the simple interest formula to set up an equation.)

(b) How much does the highest affordable price go up if the down payment is $4,000?

46. Refer to Exercise 45. A salesperson tries to convince Rhonda that raising the payment by just $30 per month would allow for a much better car.
    (a) With the down payment of $2,000, how much higher is the maximum affordable price if $30 is added to the monthly payment?
    (b) How much more will Rhonda end up paying total with the $180 payment?

47. For the month of January, Juan had an unpaid balance on a credit card statement of $832.50 at the beginning of the month and made purchases of $675.00. A payment of $400.00 was made during the month. If the interest rate was 2% per month on the unpaid balance, find the finance charge and the new balance on February 1.

48. For the month of July, the unpaid balance on Sue's credit card statement was $1,131.63 at the beginning of the billing cycle. She made purchases of $512.58. She also made a payment of $750.00 during the month. If the interest rate was 1.75% per month on the unpaid balance, determine the finance charge and the new balance on the first day of the August billing cycle.

49. Sam and Hector are redecorating their apartment. On the first day of their credit card billing cycle, the balance was $2,364.79. They recently made purchases totaling $1,964.32. They were able to make a payment of $1,000.00 during this billing cycle. If the interest rate is 1.67% per month on the unpaid balance, what is the finance charge and what will their new balance be on the first day of the next billing cycle?

50. Janine has recently accepted a position with an upscale clothing store. On the first day of Janine's March credit card billing cycle, the unpaid balance was $678.34. Clothing purchases were made totaling $3,479.03, as was a payment of $525.00 during the billing cycle. If the interest rate is 2.25% per month on the unpaid balance, find the finance charge and the new balance on the first day of the April billing cycle.

51. Joe's credit card statement on the first day of the May billing cycle shows a balance of $986.53. During this billing cycle, he charged $186.50 to his account and made a payment of $775.00. At 1.35% interest per month on the unpaid balance, what is the finance charge? Also, find the balance on the first day of the next billing cycle.

52. K'Vaughn's credit card statement shows a balance of $638.19 on the first day of the billing cycle. If he makes a payment of $475.00 and charges $317.98 during this billing period, what will his finance charge be (the interest rate is 1.50% of the unpaid balance per month)? What will his beginning balance be at the start of the next billing cycle?

53. Mary's credit card statement showed these transactions during September:

September 1	Previous balance	$627.75
September 10	Purchase	$87.95
September 15	Payment	$200.00
September 27	Purchases	$146.22

(a) Find the average daily balance.
(b) Find the finance charge for the month. The interest rate is 1.2% per month on the average daily balance.
(c) Find the new balance on October 1.

54. Pablo's credit card statement showed these transactions during March:

March 1	Previous balance	$2,162.56
March 3	Payment	$800.00
March 10	Purchases	$329.27
March 21	Payment	$500.00
March 29	Purchases	$197.26

(a) Find the average daily balance.
(b) Find the finance charge for the month. The interest rate is 2% per month on the average daily balance.
(c) Find the new balance on April 1.

55. Mike's credit card statement showed these transactions during the month of June:

June 1	Previous balance	$157.95
June 5	Purchases	$287.62
June 20	Payment	$100.00

(a) Find the average daily balance.
(b) Find the finance charge for the month. The interest rate is 1.4% per month on the average daily balance.
(c) Find the new balance on July 1.

56. Charmaine's credit card statement showed these transactions during the month of December:

December 1	Previous balance	$1,325.65
December 15	Purchases	$287.62
December 16	Purchases	$439.16
December 22	Payment	$700.00

(a) Find the average daily balance.
(b) Find the finance charge for the month. The interest rate is 2% per month on the average daily balance.
(c) Find the new balance on January 1.

57. Ruth's credit card statement showed these transactions for the month of July:

July 1	Previous balance	$65.00
July 2	Purchases	$720.25
July 8	Payment	$500.00
July 17	Payment	$100.00
July 28	Purchases	$343.97

(a) Find the average daily balance.
(b) Find the finance charge for the month. The interest rate is 1.1% per month.
(c) Find the new balance on August 1.

58. Tamera's credit card statement showed these transactions for the month of September:

September 1	Previous balance	$50.00
September 13	Purchases	$260.88
September 17	Payment	$100.00
September 19	Purchases	$324.15

(a) Find the average daily balance.
(b) Find the finance charge for the month. The interest rate is 1.9% per month on the average daily balance.
(c) Find the new balance on October 1.

59. Ellen has maxed out a credit card at $11,500 and vows not to make any other credit card purchases. The credit card company charges 1.2% interest per month, and the minimum monthly payment is all interest due plus 2% of the principal balance. How much of the balance can Ellen pay down if she pays the minimum payment only for 4 months?

60. If you have $500 on a credit card at 1.4% per month, and the minimum payment is interest due plus 1% of the principal balance, what will the balance be after 6 months of minimum payments?

61. (a) For the credit cards in Exercises 53, 55, and 57, find the new balance on the first of the month following the given purchases if the credit card company uses the unpaid balance method, rather than the average daily balance method. Assume that the monthly interest rate remains the same.
(b) Calculate the difference in the amount each customer would owe with the two methods.
(c) Under what circumstances does the unpaid balance method work out better for the consumer?

62. Repeat Exercise 61 parts (a) and (b) for the accounts in Exercises 54, 56, and 58. Does this help you draw any conclusions about when each method is better for the consumer?

Critical Thinking

We'll study student loans in depth in Section 7–6. Problems 63 and 64 provide a brief glimpse at some aspects of student loans.

63. In most cases, payments on student loans are deferred until the borrower graduates, although interest does get added to the principal while the student is in school. If Jaime takes out a student loan for $40,000 at 4.5% simple interest and graduates in 4 years, find the monthly payment that would be required when they graduate if they plan to pay off the loan in 8 years after graduation.

64. Refer to Exercise 63. Rather than borrowing a lump sum at the beginning of college, Jaime's friend Cheryl decides to borrow $10,000 each year at the beginning of the school year. If these are treated as four individual loans at 4.5% simple interest, how much more or less would Cheryl owe at the end of 4 years compared to Jaime?

65. Find the principal on a loan at 8% for 4 years when the monthly payments are $100 per month.

66. A couple borrowed $800 for 1 year at 12% interest. Payments were made monthly. After eight payments were made, they decided to pay it off. Find the interest that was saved if it was computed equally over 12 months. Then find the interest saved using the rule of 78. Explain which is a better deal for the borrower.

67. In Example 8, we calculated the amount of progress that would be made in paying down a credit card balance if only minimum payments are made. Since 1% of the balance will be paid off each month, 99% of the balance will remain. Use this fact to rework the calculations. How does it help?

68. Here's another approach to the calculation in Example 8. It's actually like a savings account with *negative* interest: instead of 1% of the amount being added each compounding period (month in this case), 1% is being *subtracted* from the amount. Use the compound interest formula with a principal balance of $2,300 and interest of −1% per month compounded monthly for 6 months.

How does the result compare to the calculations in Example 8?

69. Use the result of Problem 68 to find how many months it would take to reduce the balance to $500.

70. Use the result of Problem 68 to show that if the minimum monthly payments are always calculated the same way and the borrower never pays more than the minimum, the loan will technically never reach a zero balance. Then explain why that doesn't mean that realistically it won't ever be paid off.

71. Using Technology. In Example 8, we began to study the effects of making minimum payments on credit card debt. We can set up a spreadsheet to help calculate how long it would take to pay down the $2,300 debt. *The key is to recognize that if 1% of the debt is paid every month, then the new balance is 99% of the previous balance.* This allows you to set up a formula that calculates 99% of the previous balance. So begin by putting $2,300 in cell A1. Then use a formula in A2 to calculate 99% of the amount in A1. You can then copy that down as far as you like to watch how the balance changes.

Also, the bank (like most) requires that the minimum payment each month is not less than $20, so once the balance gets below $1,000, the monthly payment will be $20 per month. At that point, change your formula so that it subtracts $20 from the previous balance. Use your spreadsheet to find how long it takes to pay off the balance, in months and in years.

Section 7-6   Student Loans and Home Buying

LEARNING OBJECTIVES

1. Compute the interest on a student loan.

2. Compute payments on a student loan.

3. Find a monthly mortgage payment.

4. Find the total interest on a home loan.

5. Compare two mortgages with different lengths.

6. Make an amortization schedule for a home loan.

What do paying for college and paying for a home have in common? Simple: they both cost a LOT of money, which means the loans needed to obtain them tend to have much longer terms than loans for cars and other items. That alone makes them different from other types of loans, but there are other issues that make student loans and home loans worth studying separately.

We'll begin with student loans, a topic of interest to more students today than ever before. Many economists worry that student loan debt could be the root cause of

FAFSA

the next great financial crisis. In August 2021, total student loan debt in America was almost $1.6 trillion dollars. It's hard to even imagine how much money that is, but put it this way: that's over $4,800 for every human being in the United States, regardless of whether they went to college or not. The average college graduate in the class of 2020 left school with just about $30,000 in student loan debt. When it comes to credit card debt, the best advice is often "Don't." But that's not a viable option for college costs: as we pointed out in the chapter opener, the cost of NOT going to college is far more in the long run than the cost of attending.

That makes student loans a necessary evil for most folks, and if there's one common theme in this chapter, it's that understanding the nuances of the financial transactions you're likely to deal with makes you far less likely to end up in a bad situation.

There are three basic types of loans that most undergraduates might consider: federal loans made by the government, federal loans made by banks or other lenders but guaranteed by the government, and private loans. Every student's first choice should be the federal loans due to stability: the interest rates are fixed for the life of the loan and are regulated by Congress, so you're unlikely to end up with a nasty surprise somewhere down the line. There are, however, limits on the amount that can be borrowed for one student through federal loan programs, and private loans might be the only alternative for an especially expensive education.

Other than the length of the loan, the big difference between student loans and other installment loans is that the payments are typically deferred until after graduation. This makes perfect sense, as in many cases students aren't working enough to make the money required to keep up with payments. So that's a pretty good deal. But it comes with a catch: in some cases, interest will accumulate on your loan while you're in school even though you're not making any payments. This can increase the principal balance considerably. The loan term can vary depending on the amount borrowed, but the standard term is 10 years, with payments deferred until 6 months after graduation.

Interest on student loans is simple interest: the interest accrues only on the principal balance, not on previously accrued interest. So while the student is still in school, the amount of interest that accrues each month remains constant. The interest is calculated using the simplified daily interest formula:

The Simplified Daily Interest Formula

Annual interest = Principal balance × interest rate

$$\text{Daily interest amount} = \frac{\text{Principal balance} \times \text{interest rate}}{365.25}$$

Monthly interest amount = Daily interest amount × days in month

---

Example 1 | Computing Interest on a Student Loan

If Sonia borrows $6,500 to cover one-time college expenses in a federal student loan program at 6.8% interest:

(a) Find the monthly interest for 30- and 31-day months.
(b) How much interest will accrue while Sonia is still in school? Assume school starts in August, graduation is in May (3 years and 9 months later), and payments begin 6 months after graduation.

SOLUTION

(a) The daily interest is

$$\text{Daily interest} = \frac{\text{Balance} \times \text{interest}}{365.25} = \frac{\$6,500 \times 0.068}{365.25}$$

$$= \$1.21$$

In a 30-day month, the interest will be

$$30 \times \$1.21 = \$36.30$$

and in a 31-day month, it will be

$$31 \times \$1.21 = 37.51$$

(b) Interest will accrue for 4 years and 3 months. The 3 months are September, October, and November: two 30-day months, and one 31-day month. So the interest accrued is

$$4 \times \text{Balance} \times \text{rate} + 2(36.30) + 37.51$$
$$= 4(6{,}500)(0.068) + 72.60 + 37.51$$
$$= \$1{,}878.11$$

Try This One   1

1. Compute the interest on a student loan.

Max borrows \$11,500 for the last 2 years of college, acquiring a federal student loan at 6.2% interest.

(a) Find the monthly interest for 30- and 31-day months.
(b) Find the amount of interest that accrues if the loan was acquired in August, graduation is 2 years later, and payments begin 3 months later.

When interest is covered by the government while the borrower is still in school, a student loan is called **subsidized**. If interest is added to the principal while the borrower is in school the loan is called **unsubsidized**. For the loan in Example 1, the student would have saved \$1,878.11 by acquiring a subsidized loan, which makes it pretty obvious that this type of loan is always the best choice if possible.

When monthly payments begin on a student loan, they're based on the principal balance at that point in time. These payments can be computed using the following formula:

Math Note

The payment formula is very similar to the formula you would use to find the regular payments for an annuity. This makes sense because a loan of this nature is like a reverse annuity, where money is being paid down rather than accumulating. If the term isn't 10 years, the exponent is −12 times the number of years.

Monthly Payments on a Student Loan

For a 10-year student loan with principal P and interest rate r (written as a decimal), the monthly payment R is given by

$$R = \frac{P \cdot \dfrac{r}{12}}{1 - \left(1 + \dfrac{r}{12}\right)^{-120}}$$

CAUTION

Obviously, the payment formula is pretty complicated. We will have to be extra careful in doing the calculations and take it in stages. This is another formula that we put together a spreadsheet calculator for. Ask your instructor if they would like to make it available to you.

Example 2   Finding the Monthly Payment on a Student Loan

Find the monthly payment on the loan in Example 1 if the term is 10 years.

SOLUTION

In Example 1, we found that the interest accrued before payments start is \$1,878.11. Since the original loan amount was \$6,500, the principal at the time payments begin is

$$\$6{,}500 + \$1{,}878.11 = \$8{,}378.11 = P$$

Calculator Guide

The interest rate is $r = 0.068$. Now we can find the payment:

The keystrokes for the final calculation in Example 2 are:

Typical Scientific or Online Calculator

47.48 ÷ (1 − 1.00566667

y^x 120 ±) =

Typical Graphing Calculator

47.48 ÷ (1 − 1.00566667

∧ (-) 120) ENTER

Like other interest formulas, the payment formula can be very sensitive to rounding.

$$R = \frac{P \cdot \frac{r}{12}}{1 - \left(1 + \frac{r}{12}\right)^{-120}}$$

$$R = \frac{8{,}378.11 \cdot \frac{0.068}{12}}{1 - \left(1 + \frac{0.068}{12}\right)^{-120}} \qquad \textit{Multiply in numerator}$$

$$\qquad \qquad \qquad \qquad \qquad 1 + \frac{0.068}{12} \approx 1.00566667$$

$$= \frac{47.48}{1 - (1.00566667)^{-120}} \qquad \textit{See calculator guide.}$$

$$= \$96.42$$

Try This One   2

Find the monthly payment on the loan in Try This One 1 if the term is 10 years.

When interest on a student loan is not paid during college, we say that the interest is **capitalized**. As we've seen, when interest is capitalized, it adds to the principal balance, resulting in larger payments. When interest is not capitalized, payments must be made during college but only covering the amount of interest.

These terms are easy to get mixed up, so you can use this summary as a reference.

Capitalized Interest vs. Not Capitalized Interest

When interest is capitalized:

- No interest is paid while the borrower is still in college.
- The principal balance continues to grow while in college.
- The resulting payments will be larger.

When interest is not capitalized:

- Interest only is paid each month while the borrower is still in college.
- The principal balance doesn't grow while the borrower is still in college.
- The resulting payments will be smaller.

Note that these apply only to unsubsidized loans: for subsidized loans, there IS no interest while the borrower is in college, so there's nothing to capitalize.

The monthly payment calculation we did in Example 2 included accrued interest, which means that interest was capitalized on that loan. Let's see how not capitalizing interest would affect the monthly payment.

Example 3   Studying the Effects of Capitalizing Interest

If interest on the loan in Examples 1 and 2 is not capitalized:

(a) Find the interest payment while Sonia is still in school.
(b) Find the payment on the loan when full payments begin.
(c) How much less would Sonia pay in total if interest was not capitalized?

SOLUTION

(a) We already did this in Example 1! The interest due each month will be the interest that accrues on the loan at the rate of $1.21 per day. So the student will owe that amount times the number of days in any given month.

(b) The principal remains $P = \$6,500$ since the interest is being paid while Sonia is still in school. The interest rate is $r = 0.068$.

Math Note

When interest isn't capitalized, the payments can vary depending on what day of the month the payment is processed due to weekends or holidays, but it will always add up to the correct yearly amount.

$$R = \frac{P \cdot \dfrac{r}{12}}{1 - \left(1 + \dfrac{r}{12}\right)^{-120}}$$

$$R = \frac{\$6,500 \cdot \dfrac{0.068}{12}}{1 - \left(1 + \dfrac{0.068}{12}\right)^{-120}} = \$74.80$$

(c) In Example 2, we found that when all interest is deferred, Sonia would make 120 payments of $96.42 for a total of $120 \times \$96.42 = \$11,570.40$. If the interest that accrued while Sonia was in school has already been paid, there would be 120 payments of $74.80, which totals $120 \times \$74.80 = \$8,976$. Adding the interest paid up front ($1,878.11), we get $10,854.11. By not capitalizing the interest, Sonia saved $\$11,570.40 - \$10,854.11 = \$716.29$ over the life of the loan. This is exactly the kind of calculation that needs to be done when acquiring a student loan so that you can decide if making those payments while still in school is worth the overall savings that will result.

Try This One   3

If interest on the loan in Try This Ones 1 and 2 is not capitalized:

(a) Find the interest payment while Max is still in school.
(b) Find the payment on the loan when full payments begin.
(c) How much less would Max pay in total if interest was not capitalized?

2. Compute payments on a student loan.

Home Buying

For many people, the day they buy their first home is one of the proudest days of their life—and one of the scariest. There is nothing that compares to the feeling of looking at a house and knowing that it's all yours. But the buying process is tremendously intimidating. There are dozens of documents to sign, and the sheer numbers involved are enough to make almost everyone wonder if they're making a colossal mistake.

The most common home loans are paid over a 30-year span. That's a major commitment, and one that nobody should enter into without an understanding of the mathematics that go into the process. In the remainder of this section, we will study that math, hopefully helping you to become a well-informed home buyer.

SW Productions/Brand X Pictures/ Getty Images

Mortgages

A **mortgage** is a long-term loan where the lender has the right to seize the property purchased if the payments are not made. Homes are the most common items bought using mortgages. The most common mortgage term is 30 years, but they are widely available in terms from 15 to as many as 50 years.

There are several types of mortgages. A **fixed-rate mortgage** means that the rate of interest remains the same for the entire term of the loan. The payments (usually monthly) stay the same. An **adjustable-rate mortgage** means that the rate of interest may fluctuate (i.e., increase and decrease) during the period of the loan. Some lending institutions will allow you to make **graduated payments**. This means that even though the interest doesn't change for the period of the loan, you can make smaller payments in the first few years and larger payments at the end of the loan period.

Finding Monthly Payments and Total Interest

At one time, it was common to use tables to compute the monthly payment on a mortgage. Of course, at one time it was common to compute square roots using a table also. But then calculators and computers came into widespread use, and the need for tables was reduced. Not only can we compute monthly payments on a home loan with a formula, it's a formula that we've already used!

The monthly payment formula that we used for student loans is actually a special case of the formula we'll use for home loans. The only difference is that we adapted that formula for a 10-year term, which is typical for most student loans. Home loans come in a wide variety of terms, so we'll want to be able to adapt to different lengths. This general formula also allows for payments at some regular interval other than monthly. We've provided a spreadsheet with this formula preprogrammed if your instructor would like you to have it. It can't hurt to ask, right?

> **Formula for Computing Monthly Payments on a Mortgage**
>
> $$R = \frac{P\left(\frac{r}{n}\right)}{1 - \left(1 + \frac{r}{n}\right)^{-nt}}$$
>
> R = regular monthly payment
> P = amount financed, or principal
> r = rate written as a decimal
> n = number of payments per year
> t = number of years

Math Note

The word "mortgage" comes from a combination of Old French words "mort" (dead) and "gage" (pledge). It is believed the intent was that the debtor pledged the property to secure the loan, and if they failed to pay, the property was taken, and was therefore "dead" to the debtor.

Example 4 | **Finding a Monthly Payment Using the Formula**

Martin Barraud/Caia Image/Glow Images

After one hit single, a young singer unwisely decides that she needs a $2.2 million mansion. With some of the proceeds from her album she puts down $500,000, leaving $1,700,000 to finance at 6% for 20 years. Find the monthly payment.

SOLUTION

In the formula above, use $P = 1,700,000$, $r = 0.06$, $n = 12$, and $t = 20$.

$$R = \frac{P\left(\frac{r}{n}\right)}{1 - \left(1 + \frac{r}{n}\right)^{-nt}}$$

$$R = \frac{1,700,000\left(\frac{0.06}{12}\right)}{1 - \left(1 + \frac{0.06}{12}\right)^{-20(12)}}$$

Multiply in numerator. $1 + \frac{0.06}{12} = 1.005$ (in some cases, rounding might be necessary here): $-20(12) = -240$.

$$= \frac{8,500}{1 - (1.005)^{-240}}$$

$$\approx \$12,179.33$$

The monthly payment is $12,179.33, and the singer better hope her next album does well too.

Sidelight   **The Great Mortgage Crisis of 2008**

Throughout much of the first decade of the 21st century, homeowners were very happy people. Housing prices were rising at an almost unprecedented rate, and people watched the value of their investment soar. New homes were being built everywhere you looked, and lending institutions were practically climbing over each other to hand out home loans.

Then a funny thing happened—the housing market got oversaturated and prices started to fall. At the same time, people who took adjustable-rate mortgages to buy larger houses had their rates go up and couldn't make their payments anymore. Foreclosures (when the lending institution takes back a home) started to rise, causing even more houses to go on the market. Soon, the whole house of cards came crashing down, taking the U.S. economy with it.

There is plenty of blame to go around, but put in its simplest terms, the blame is to be shared equally between home buyers and lenders. Millions of people bought homes they couldn't afford with adjustable-rate mortgages, and the

Stockbyte/Getty Images

lenders gave out loans to millions of people who couldn't afford them. The result was an economic bust that cost 8.8 million American jobs before the job market bottomed out in February 2010. One lesson has been learned from this mess—lenders will be far less likely to put people in homes that they can't afford from now on.

Try This One   4

Use the payment formula to find the monthly payment on the one-hit-wonder pop star's second house, a $120,000 mortgage at 5.2% for 15 years.

Well that was fun. But not many of us will ever buy a home with a monthly payment over ten grand. In the next three examples, we'll study the numbers involved in a more realistic transaction for most folks.

Example 5   Finding Monthly Mortgage Payments

Math Note

A Google search for the phrase "mortgage calculator" results in over 50 million hits! There are thousands of pages available that can be used to quickly calculate monthly payments on a mortgage.

But doing the calculations here will help you to become very familiar with the terms involved in mortgages.

The Petteys family plans to buy a home for $174,900, and has been offered a 30-year mortgage with a rate of 4.5% if they make a 20% down payment. What will the monthly payment be with this loan?

SOLUTION

We know that $r = 0.045$, $n = 12$, and $t = 30$. What we don't know is the amount borrowed: we were given the price of the home. Let's be clever: if the down payment is 20%, then the amount financed will be the remaining 80% of the selling price.

$$0.80 \times \$174,900 = \$139,920$$

This is our P and we're ready to use the formula.

$$R = \frac{P\left(\dfrac{r}{n}\right)}{1 - \left(1 + \dfrac{r}{n}\right)^{-nt}}$$

$$R = \frac{139{,}920\left(\dfrac{0.045}{12}\right)}{1 - \left(1 + \dfrac{0.045}{12}\right)^{-30(12)}}$$

$$= \frac{524.7}{1 - (1.00375)^{-360}}$$

$$\approx 708.95$$

Now that's more like it! $708.95 is a reasonable monthly payment.

3. Find a monthly mortgage payment.

Try This One 5

The Trissel family agreed on a price of $229,500 for a home. Their company credit union offers a 5.0% 20-year loan with 15% down. Calculate the monthly payment.

It's an eye-opening experience to calculate the total interest on a mortgage. To do so, multiply the monthly payments by the total number of payments and then subtract the principal.

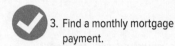

Example 6 — Finding Total Interest on a Mortgage

Math Note

The long term of a home loan means you pay a LOT higher percentage of interest than on a shorter-term loan.

In September of 2021, the average loan on a new car in the United States was 5.27% for 5 years: with these terms, the interest on the loan amounts to about 14% of the amount borrowed.

For the 30-year mortgage in Example 6, the interest is 82.4% of the amount borrowed.

Find the total amount of interest the Petteys family would pay if they take the loan in Example 5.

SOLUTION

On a 30-year mortgage, there are $30 \times 12 = 360$ payments. We found that the monthly payment would be $708.95.

$$\$708.95 \times 360 = \$255{,}222$$

This is the total of payments. We subtract the amount financed from Example 5:

$$\$255{,}222 - \$139{,}920 = \$115{,}302 \quad \textit{Interest on the loan.}$$

That looks like a typo, doesn't it? The interest on the loan is over $100,000. Wow.

Try This One 6

Find the total interest paid on the loan in Try This One 5.

Not surprisingly, the length of a loan has a profound effect on how much interest is paid. In the next example, we'll weigh the amount of extra monthly payment required versus the amount of interest saved.

4. Find the total interest on a home loan.

Example 7 — Comparing Mortgages with Different Terms

Suppose that the Petteys family from Examples 5 and 6 is also offered a 15-year mortgage with the same rate and down payment. Find the difference in monthly payment and interest paid between the 15- and 30-year mortgages.

SOLUTION

We pretty much need to rework Examples 5 and 6 with a 15-year mortgage, then compare the results. So we have $P = \$139{,}920$, $r = 0.045$, $n = 12$, and $t = 15$.

DreamPictures/Blend Images/CORBIS

Math Note

There is a series of costs associated with initiating a mortgage that add to the principal value of the loan. Known as **closing costs**, they averaged about $3,500 nationally in 2020 but can vary dramatically from state to state.

These costs are added after the down payment is calculated as a percentage of selling price. When deciding on how much you can afford to pay for a house, it's important to find what the closing costs are.

$$R = \frac{P\left(\frac{r}{n}\right)}{1 - \left(1 + \frac{r}{n}\right)^{-nt}}$$

$$R = \frac{139,920\left(\frac{0.045}{12}\right)}{1 - \left(1 + \frac{0.045}{12}\right)^{-15(12)}}$$

$$= \frac{524.7}{1 - (1.00375)^{-180}}$$

$$\approx 1,070.38$$

The difference in monthly payments is

$$\$1,070.38 - \$708.95 = \$361.43 \quad \textit{\$708.95 was payment for 30 years.}$$

With a monthly payment of $1,070.38 for 15 years (which is 180 months) the total payments are

$$\$1,070.38 \times 180 = \$192,668.40$$

and the interest paid is

$$\$192,668.40 - \$139,920 = \$52,748.40$$

The interest paid on the 30-year mortgage was $115,302:

$$\$115,302 - \$52,748.40 = \$62,553.60$$

If the Petteys family can manage an extra $361.43 a month, they will save over $60,000 in interest!

Sidelight   **Where Are Your Payments Actually Going?**

When studying fixed-installment loans in Section 7-5, we saw that paying off a loan early doesn't save you nearly as much money in interest as you might think. For example, in paying off the $1,500, 3-year loan in Example 4 of that section in 24 months, the couple saved a whopping 30 bucks. Don't spend it all in one place! Why does this happen? You would think shaving off a full year from a 3-year loan would save a whole lot more than that.

The answer, as it turns out, is that the interest you pay each month is calculated as a percentage of the total amount you still owe. So when a loan is brand new, you're paying a LOT in interest every month because the balance is high. As you begin to pay down that balance, though, the amount of interest you have to pay gets lower. Keep in mind that we're talking about fixed-installment loans: that means you pay the exact same amount every month. What changes is how much of that payment goes into paying interest, and how much subtracts from the balance of the loan.

Let's take a look at a specific example. I looked up some numbers online last night and found that in 2020 the average home buyer secured a 30-year fixed-interest loan for $370,200 at an interest rate of 3.72%, leading to monthly payments of about $1,708. This means that when the very first

payment is made, the interest charge is one-twelfth of 3.72% (because it's for 1 month, and 3.72% is an annual rate) times the balance of $370,200. This works out to be about $1,148.

What does that tell us? That of the $1,708 the average home buyer paid in their first month of ownership, $1,148 of that money went to pay interest, and only $560 was subtracted from the balance they owed. The next month, the interest would be calculated on a slightly smaller balance, meaning slightly less interest, and a bit more toward paying down the balance.

Fast-forward 15 years, or halfway through the loan period. Using an online calculation tool, I found that for the first payment in the 16th year, only $730 would go to interest, with $978 paying down the balance. The net result is that early on, most of your payment goes to interest, while closer to the end of the loan term, most of the total interest has been paid, and the majority of your payment is going toward the balance.

So that's the story of why paying off a loan early usually doesn't save much in interest. By the time you get to a point where it's manageable to pay off the entire balance at once, most of the interest has already been paid. We'll study this idea in a bit more depth in the last example in this section.

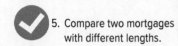

5. Compare two mortgages with different lengths.

Try This One   7

If the Trissel family from Try This One 5 chooses a 15-year mortgage instead of 20, find the increase in monthly payment and total interest saved.

Computing an Amortization Schedule

After securing a mortgage, the lending institution will prepare an **amortization schedule**. This schedule shows what part of the monthly payment is paid on the principal and what part of the monthly payment is paid in interest.

In order to prepare an amortization schedule, the next procedure can be used.

Math Note

Be sure to subtract any down payment from the cost of the home before beginning the amortization schedule.

Procedure for Computing an Amortization Schedule

Step 1 Find the interest for the first month. Use $I = Prt$, where $t = \frac{1}{12}$. Enter this value in a column labeled Interest.

Step 2 Subtract the interest from the monthly payment to get the amount paid on the principal. Enter this amount in a column labeled Payment on Principal.

Step 3 Subtract the amount of the payment on principal found in Step 2 from the principal to get the balance of the loan. Enter this in a column labeled Balance of Loan.

Step 4 Repeat the steps using the amount of the balance found in Step 3 for the new principal.

Example 8   Preparing an Amortization Schedule

Math Note

Preparing an amortization schedule makes it easy to see why more interest is paid earlier in a fixed installment loan, and more principal is paid later.

Compute the first two months of an amortization schedule for the loan in Example 5.

SOLUTION

The value of the mortgage is $139,920, the interest rate is 4.5%, and the monthly payment is $708.95.

Step 1 Find the interest for month 1.

$$I = Prt$$
$$= \$139,920 \times 0.045 \times \frac{1}{12}$$
$$= \$524.70$$

Enter this in a column labeled Interest.

Step 2 Subtract the interest from the monthly payment.

$$708.95 - \$524.70 = \$184.25$$

This goes into the Payment on Principal column.

Step 3 Subtract principal payment from principal.

$$\$139,920 - \$184.25 = \$139,735.75$$

This goes into the Balance of Loan column. Now we repeat Steps 1–3 using the balance of $139,735.75.

Step 4 $I = Prt$

$$= \$139{,}735.75 \times 0.045 \times \frac{1}{12}$$

$$= \$524.01$$

Step 5 $\$708.95 - \$524.01 = \$184.94$

Step 6 $\$139{,}735.75 - \$184.94 = \$139{,}550.81$

The first 2 months of the amortization schedule are:

Payment Number	Interest	Payment on Principal	Balance of Loan
1	$524.70	$184.25	$139,735.75
2	$524.01	$184.94	$139,550.81

Try This One   8

6. Make an amortization schedule for a home loan.

Make an amortization schedule for the first 3 months for the loan in Try This One 5.

Answers to Try This One

1 (a) 30-day: $58.50; 31-day: $60.45
   (b) $1,603.45

2 $146.79

3 (a) $1.95 times the number of days in the month
   (b) $128.83   (c) $551.55

4 $961.50

5 $1,287.41

6 $113,903.40

7 Increase in payment = $255.54; interest saved = $31,252.80

8 Payment Number	Interest	Payment on Principal	Balance of Loan
1	$812.81	$474.60	$194,600.40
2	$810.84	$476.57	$194,123.83
3	$808.85	$478.56	$193,645.27

Exercise Set   7-6

Writing Exercises

1. What's the difference between a subsidized and unsubsidized student loan?
2. What is the advantage of a federal student loan over private loans?
3. What does it mean to capitalize interest on a student loan?
4. What's the biggest difference between a student loan and a regular installment loan?
5. What specifically makes a loan a mortgage?
6. Explain how to find the total interest paid on a mortgage.
7. What are the advantages and disadvantages of getting a mortgage with a shorter term?
8. What is an amortization schedule?

Applications in Our World

For Problems 9–12, for the given student loan, find the interest that accrues in a 30-day month, then find the total amount of interest that will accrue before regular payments begin, again using 30-day months.

9. $7,200 at 6.8% interest; student graduates 3 years and 9 months after loan is acquired; payments deferred for 6 months after graduation

10. $11,500 at 6.8% interest; student graduates 2 years after loan is acquired; payments deferred for 3 months after graduation

11. $21,500 at 6.2% interest; student graduates 4 years after loan is acquired; payments deferred for 3 months after graduation

12. $16,500 at 6.2% interest; student graduates 1 year and 9 months after loan is acquired; payments deferred for 6 months after graduation

In Problems 13–16, find the monthly payment on the loans in Problems 9–12. Assume that the term of each loan is 10 years.

13. Monthly payment for Problem 9.
14. Monthly payment for Problem 10.
15. Monthly payment for Problem 11.
16. Monthly payment for Problem 12.

17. Mona takes out a $12,100 student loan to help pay for the first 2 years of college, then 2 years later needs another loan of $14,000 to get through to graduation. Mona graduates 4 years and 9 months after acquiring the first loan, and payments are deferred until 6 months after graduation. The interest rate on both loans is 6.8%. Find the total amount of interest that will accrue until payments begin.

18. Huai takes out a $3,200 student loan at 6.8% to help with 2 years of community college. After finishing the 2 years, Huai transfers to a state university and borrows another $12,400 to defray expenses for the 5 semesters needed to graduate. Graduation is 4 years and 4 months after acquiring the first loan, and payments are deferred for 3 months after graduation. The second loan was acquired 2 years after the first and had an interest rate of 7.2%. Find the total amount of interest that will accrue until payments begin.

19. Repeat Problem 17 if the first loan is subsidized and the second is not.

20. Repeat Problem 18 if the first loan is subsidized and the second is not.

21. Refer to Problem 17; find Mona's monthly payment when regular payments begin.

22. Refer to Problem 18; find Huai's monthly payment when regular payments begin.

23. Juanita takes out a federal unsubsidized loan for $13,100 with an interest rate of 6.8%, a term of 10 years, and payments deferred until 6 months after graduation. She decides not to capitalize the interest.
    (a) Find Juanita's monthly interest payment while in school. Assume a 30-day month.

(b) What will her monthly payment be when payments begin? She graduates 5 years after acquiring the loan.

(c) How much less will she pay over the life of the loan by not capitalizing the interest?

24. Ben takes out a federal unsubsidized loan for $9,275 with an interest rate of 6.2%, a term of 10 years, and payments deferred until 3 months after graduation. He decides not to capitalize the interest.
    (a) Find Ben's monthly interest payment while in school. Assume a 30-day month.
    (b) What will his monthly payment be when payments begin? He graduates 4 years after acquiring the loan.
    (c) How much less will he pay over the life of the loan by not capitalizing the interest?

In Problems 25 and 26, the interest is capitalized.

25. If you need to take out a $20,000 student loan 2 years before graduating, which loan option will result in the lowest overall cost to you: a subsidized loan with 6.8% interest for 10 years, a federal unsubsidized loan with 6.2% interest for 10 years, or a private loan with 6.0% interest and a term of 15 years? How much would you save over the other options? All payments are deferred for 6 months after graduation.

26. If you need to take out a $40,000 student loan 5 years before graduating, which loan option will result in the lowest overall cost to you: a subsidized loan with 7.0% interest for 10 years, a federal unsubsidized loan with 6.8% interest for 10 years, or a private loan with 6.5% interest and a term of 14 years? How much would you save over the other options? All payments are deferred for 6 months after graduation.

27. You receive a series of four annual federally subsidized loans to pay for 4 years of college, each for $7,100 at 6.8%. If you graduate 4 years and 6 months after acquiring the first loan, payments are deferred for 6 months after graduation, and the term is 10 years, find your monthly payment.

28. Pat receives a series of four annual federally subsidized student loans, each for $5,300 at 6.8%. To defray rising costs for her senior year, 3 years after acquiring the first loan she takes out a private student loan for $4,000 at 7.5% interest with a term of 10 years and capitalizes the interest for her last year of college. She graduates 9 months after getting the private loan. Payments on all loans are deferred until 6 months after graduation. Find her monthly payment.

29. A house sells for $145,000 and a 15% down payment is made. A mortgage was secured at 7% for 25 years.
    (a) Find the down payment.
    (b) Find the amount of the mortgage.
    (c) Find the monthly payment.
    (d) Find the total interest paid.

30. A house sells for $182,500 and a 5% down payment is made. A mortgage is secured at 7% for 15 years.
    (a) Find the down payment.
    (b) Find the amount of the mortgage.
    (c) Find the monthly payment.
    (d) Find the total interest paid.

31. A building sells for $200,000 and a 40% down payment is made. A 30-year mortgage at 6% is obtained.
    (a) Find the down payment.
    (b) Find the amount of the mortgage.
    (c) Find the monthly payment.
    (d) Find the total interest paid.

32. An ice cream store sells for $125,000 and a 12% down payment is made. A 25-year mortgage at 7.5% is obtained.
    (a) Find the down payment.
    (b) Find the amount of the mortgage.
    (c) Find the monthly payment.
    (d) Find the total interest paid.

33. An auto parts store sells for $325,000 and a 10% down payment is made. A 40-year mortgage at 7.5% is obtained, and closing costs are $3,200.
    (a) Find the down payment.
    (b) Find the amount of the mortgage.
    (c) Find the monthly payment.
    (d) Find the total interest paid.

34. A beauty shop sells for $175,000 and a 22% down payment is made. A 20-year mortgage at 6.5% is obtained, and closing costs are $2,700.
    (a) Find the down payment.
    (b) Find the amount of the mortgage.
    (c) Find the monthly payment.
    (d) Find the total interest paid.

35. A computer store sells for $1,200,000. The buyer made a 30% down payment and secured a 20-year mortgage on the balance at 5.5%.
    (a) Find the down payment.
    (b) Find the amount of the mortgage.
    (c) Find the monthly payment.
    (d) Find the total interest paid.

36. A grocery store sells for $550,000 and a 25% down payment is made. A 40-year mortgage at 6% is obtained.
    (a) Find the down payment.
    (b) Find the amount of the mortgage.
    (c) Find the monthly payment.
    (d) Find the total interest paid.

37. Find the monthly payment on the loan in Problem 29 if the term is reduced to 15 years. How much will be saved in total interest?

38. Find the monthly payment on the loan in Problem 30 if the term is increased to 30 years. How much extra interest will be paid?

39. The buyer of the store in Problem 33 is also offered a 30-year mortgage at 6.5%. How much would this option save over the life of the loan?

40. The buyer of the beauty shop in Problem 34 is also offered a 15-year mortgage at 6.0%. How much would this save over the life of the loan?

41. A house is bought for $232,000 with a 15% down payment. A mortgage is secured at 6.75% for 20 years. Find the monthly payment.

42. A building sells for $330,000 with a down payment of 25%. Find the monthly payment on a 25-year mortgage at 5.3%.

43. A home is bought for $163,000 with a 12% down payment. Find the monthly payment if the mortgage is 6.75% for 18 years.

44. A house sells for $289,000. The buyer made a 20% down payment and financed the balance with a 5.75% loan for 15 years. Find the monthly payment.

45. A store was bought for $725,000 and the buyer made a 10% down payment. The balance was financed with a 6.35% loan for 27 years. Find the monthly payment.

46. A house was bought for $162,000.00 with a 5% down payment. The balance was financed at 5.6% for 18 years. Find the monthly payment.

47. A pizza parlor was bought for $327,000 with no down payment and a 6.7% loan for 10 years. Find the monthly payment.

48. A supermarket building was bought for $375,000. The down payment was 10%. The balance was financed at 7.2% for 15 years. Find the monthly payment.

49. Compute an amortization schedule for the first 3 months for the loan in Exercise 29 of this section.

50. Compute an amortization schedule for the first 3 months for the loan in Exercise 30 of this section.

51. Compute an amortization schedule for the first 3 months for the loan in Exercise 35 of this section.

52. Compute an amortization schedule for the first 3 months for the loan in Exercise 36 of this section.

A young couple has saved up $14,000 for a down payment on a home. They are currently paying $1,300 per month to rent a condo. Use this information in Exercises 53–56.

53. The couple is preapproved for a 30-year mortgage at 4.9%, and their realtor estimates that they will need to set aside $3,000 for closing costs at the time of sale. What is the price of the most expensive home they can buy without raising their monthly housing payment?

54. If the couple instead opts for a 15-year mortgage at 4.2%, how much is the most expensive home they can buy?

55. If the couple were to borrow the maximum amount you found in Exercises 53 and 54, how much more do you think they would pay in interest with the 30-year mortgage? Make a guess, then calculate the exact amount.

56. If the couple decides that they can afford to go up to $1,500 per month, how much more can they afford to spend on a home for each loan option?

Critical Thinking

57. You decide to buy a $180,000 home. If you make a 25% down payment, you can get a 20-year mortgage at 9%, but if you make a 10% down payment, you can get a 25-year mortgage at 7%. Which is the better option for you?

58. Which mortgage would cost you less, a 30-year mortgage at 6.5% or a 15-year mortgage at 10%?

59. Find the total amount of interest paid on the student loan in Problem 23(b), and the total amount of interest paid on the mortgage in Problem 29. What percentage of the amount borrowed is the interest in each case?

60. Find the total amount of interest paid on the loan in Problem 37, then the percentage this represents of the amount borrowed. How does this compare to the percentages you found in Problem 59? What can you conclude?

61. Using Technology. As you probably noticed, making an amortization table is a bit time-intensive. There sure are a lot of repetitive calculations. Sounds like another job for technology! Your job here is to use a spreadsheet to make an amortization table for the Petteys loan we studied in the lesson. We already created the first two months in Example 8, so you can use that to double-check your spreadsheet results. A template to help you get started can be found in the online resources for this lesson.

62. Using Technology. In my humble opinion, the most valuable thing you can get out of this lesson is the ability to compare different loans. The second template in the class resources is built so that you can input the principal, percent down payment, interest rate, and term of a loan, then calculate the monthly payment, total amount of interest, and total cost over the life of the loan. Use this spreadsheet to explore different options for a home that you think you might pay for some day, and report on what you found.

Section 7-7 — Investing in Stocks and Bonds

LEARNING OBJECTIVES

1. Find information from a stock listing.
2. Compute the P/E ratio for a stock.
3. Compute the total cost of a stock purchase.
4. Compute the profit or loss from a stock sale.
5. Compute profit from a bond sale.

Comstock Images/Getty Images

We have seen that the magic of compound interest allows your money to grow considerably over long periods of time. However, interest rates on basic savings accounts are usually quite low, so if you rely on savings alone to build a nest egg, you better hope that you live a very, very long life indeed. Most successful investors grow their money much more quickly using the stock market. In this section, we'll learn about the basics of stocks and bonds, and how to get information about the performance of stocks that you might be interested in.

When a company files legal papers to become a corporation, it is able to issue **stock**. When investors buy shares of stock, they become part owners of the company; for example, if a company issues 1,000 shares of stock and an investor buys 250 shares, then the investor owns one-quarter of the company. The investor is called a **shareholder**.

When a company makes money, it can choose to distribute part of the profit to its shareholders. This money is called a **dividend**. Stockholders receive a sum of money based on the number of shares of the stock that they own. Sometimes if a company doesn't make a profit or its owners or managers decide to reinvest the money into the company, no dividends are paid.

Besides issuing stock, a company can also issue **bonds**. Usually bonds are issued to raise money for the company for start-up costs or special projects. A person who purchases a bond is really lending money to the company. The company, in turn, repays the owner of the bond its **face value** plus interest. As a general rule, bonds are a safer investment than stocks, but stocks have greater growth potential.

Stocks can be bought and sold on a **stock exchange**. The price of a stock varies from day to day (even from minute to minute) depending on the amount that investors are willing to pay for it. This can be affected by the profitability of the company, the economy, scandals, even global political concerns. Investors buy or sell stock through a **stockbroker**. Traditionally, this was an individual working for a brokerage firm, but it has become common for people to use online brokers, in which the investor initiates the buying and selling of stocks. In either case, the brokerage charges a fee, called a **commission**, for the service of having their representatives buy or sell the stock at an exchange. Bonds can also be bought and sold like stocks.

Investors often own a combination of stocks and bonds. The set of all stocks and bonds owned is called an investor's **portfolio**. Sometimes a group of investors hire a manager to handle their investments. The manager invests in stocks and bonds, follows the activities of companies, and buys and sells in an attempt to achieve maximum profit for the group. This type of investment is called a **mutual fund**.

Stocks

In order to get information about a certain stock, you can refer to a stock table. These tables can be found in newspapers and online financial sites. The listings vary somewhat depending on the source. In this case, a stock listing for a company called Computer Programming and Systems, Inc. will be used as an example.

Computer Programs & Systems Inc. (CPSI) - NasdaqGS

45.49 ▲ **0.05 (0.11%)** 10:37AM EDT - Nasdaq Real Time Price

Prev Close:	45.44	Day's Range:	45.02–45.60
Open:	45.02	52wk Range:	43.24–64.86
Bid:	45.36 × 100	Volume:	5,119
Ask:	45.52 × 100	Avg Vol (3m):	96,911
1y Target Est:	46.20	Market Cap:	514.16M
Beta:	0.780949	P/E (ttm):	18.61
Earnings Date:	Oct 28–Nov 2 (Est.)	EPS (ttm):	2.44
		Div & Yield:	2.56 (5.64%)

Let's take a look at what all this stuff means.

- At the top, we see the name of the company and the **ticker symbol** (CPSI in this case). This is a shorthand symbol assigned by the exchange that represents the stock. Every company has a short string of letters to represent it for trading purposes. For example, Apple Computer is AAPL, McGraw-Hill Education is MHED, and Ford Motor Company is just F. (Ford's been around for a LONG time.)

- The "NasdaqGS" indicates that this particular stock is traded on the Nasdaq, one of the American stock exchanges.

- The big bold number at the top is the current price of the stock. The current price isn't like the price of milk at the grocery store: it reflects the most recent price at which the stock has been bought from someone trying to sell. This is followed by how much the price has changed during the current day's trading session.

- **Prev Close**: If you can add and subtract, this is redundant: note that this stock has gone up by $0.05 in this session and is currently worth $45.49, so of course the closing price on the previous day was $45.44.

- **Open**: This is the price at the time today's session began. This is often different than the closing price because many factors affect what people are willing to pay from day to day.

- **Bid** and **Ask**: These are a snapshot of what people have been bidding to buy the stock, and what price current shareholders have been asking for to sell. In this case, $45.36 × 100 indicates that an offer of $45.36 for up to 100 shares has been made. $45.52 × 100 means that someone has offered up to 100 shares for sale at $45.52.

- **1y Target Est** is what analysts are predicting the stock price will be in one year. While this is at best an educated guess, it can at least give you some guidance as to whether the stock might be a good buy.

- **Beta** refers to the **volatility** of the stock, which is a measure of how much the price fluctuates. If the beta is close to 1, then the stock's stability is similar to the rest of the market. Lower numbers indicate that the price is more stable than average.

- The **Earnings Date** reflects the next time the company is expected to release data on how well the company is doing financially.

- **Day's Range** and **52wk Range**: These ranges indicate the highest and lowest price during the current session, as well as the last full year. The latter is particularly useful, because it gives you a quick snapshot of how the stock has fluctuated and how its current price compares.

- **Volume** describes how many shares have been traded in the current session, while **Avg Vol (3m)** provides a 3-month average of shares traded. Lower-volume stocks can be more difficult to buy and sell due to availability.

- **Market Cap** is the total value of the company: it's calculated by multiplying the number of shares that have been issued by the current price, so a little math allows you to calculate how many shares there are. This in turn makes the volume of sales more meaningful.

- **P/E (ttm)** refers to the P/E (price to earnings) ratio. This is a particularly important number, so we'll study that in detail a bit later.

- **EPS (ttm)** represents earnings per share, or the amount of money the company made for each share of stock that has been issued. Clearly, bigger is better here.

- **Div & Yield** is the dividend that the company pays for each share, and what percent of the current price that represents. If nothing else, this tells you whether or not the company pays a dividend on shares, but you can also use it for comparison purposes with other stocks. A dividend is essentially a company distributing profit to the shareholders, so if you own 100 shares of a stock that pays a dividend of $1.50, the company pays you $150.

Now let's put our new knowledge into practice by studying shares of Facebook stock.

| Example 1 | Reading a Stock Listing |

Use the Facebook stock listing to answer each question.

Facebook, Inc. (FB) - NasdaqGS

87.27 ↓ 0.88 (**1.00%**) 1:29PM EDT - Nasdaq Real Time Price

Prev Close:	88.15	Day's Range:	86.70–87.79
Open:	87.15	52wk Range:	70.32–99.24
Bid:	87.06 × 400	Volume:	16,715,420
Ask:	87.07 × 300	Avg Vol (3m):	31,684,600
1y Target Est:	111.75	Market Cap:	245.89B
Beta:	0.89812	P/E (ttm):	88.69
Earnings Date:	Oct 26–Oct 30 (Est.)	EPS (ttm):	0.98
		Div & Yield:	N/A (N/A)

(a) What is the current price of the stock? What are shareholders trying to sell it currently asking for?

(b) What do analysts predict the price will be in one year?

(c) Is this stock more or less volatile than the market as a whole? How do you know?

(d) What's the highest stock price in the last 52 weeks?

(e) Does Facebook pay a dividend to shareholders? If so, how much?

(f) Use the market cap to find how many shares the company issued.

SOLUTION

(a) The current price is the big number near the top of the listing: $87.27 per share. Shareholders are asking for $87.07 (as seen in the "Ask" row).

(b) According to the 1y Target Est row, the prediction is $111.75 in one year. (For what it's worth, the actual price one year after this listing was published was $127.31, so the stock did even better than expected.)

(c) The beta is less than 1, so the stock has been more stable than the market in general.

(d) The 52wk Range row shows 70.32 − 99.24, so the highest price in the last 52 weeks is $99.24.

(e) The N/A in the dividend row shows that they do not pay a dividend. (It stands for "not applicable".)

(f) The market cap is the total value of all shares, and it's listed as $245.89 billion. Divide this by the current price of $87.27 to get about 2.82 billion. About 2.82 billion shares have been issued.

Try This One   1

1. Find information from a stock listing.

Use this stock listing to answer each question.

Starbucks Corporation (SBUX)
NasdaqGS - NasdaqGS Real Time Price. Currecncy in USD

57.35 +0.62 (+1.09%)
At close: February 17 4:00PM EST

Previous Close	56.73	Market Cap	83.58B
Open	56.80	Beta	0.62
Bid	57.20 × 500	PE Ratio (TTM)	29.41
Ask	57.28 × 2200	EPS (TTM)	1.95
Day's Range	56.71 – 57.57	Earnings Date	Apr 27, 2017
52 Week Range	50.84 – 61.64	Dividend & Yield	0.90 (1.59%)
Volume	7,879,844	Ex-Dividend Date	N/A
Avg. Volume	9,200,479	1y Target Est	64.56

(a) What's the current price of the stock? What was it selling for at the opening of trading?

(b) What are the highest and lowest prices over the last year?

(c) What are potential buyers willing to pay per share at this point?

(d) Does Starbucks pay a dividend to shareholders? How much per share?

(e) How many shares have been issued?

(f) What is the ticker symbol used to represent Starbucks on the stock exchange?

P/E Ratio

The P/E ratio of a stock is a comparison of the current selling price to the company's earnings per share.

Math Note

The abbreviation P/E is used to remind you that this is a ratio of Price to Earnings.

Formula for the P/E ratio

$$P/E \text{ ratio} = \frac{\text{Current price}}{\text{Annual earnings per share}}$$

The annual earnings per share is found by dividing a company's total earnings by the number of shares that are owned by the stockholders for the last year. The annual earnings per share for a stock is found by subtracting expenses, taxes, losses, etc. from the gross revenues. These figures can be found in a company's annual reports.

Example 2	Computing a P/E Ratio

Use information in the Facebook stock listing from Example 1 to calculate the P/E ratio. Does it match what's listed in the table? Why or why not?

SOLUTION

The P/E ratio is the ratio of current price to earnings per share. According to the listing, the current price is $87.27, and earnings per share are $0.98.

$$\frac{\text{Price}}{\text{Earnings}} = \frac{\$87.27}{\$0.98} \approx 89.05$$

The P/E ratio listed in the table is 88.69, which is close but not exact. The most likely cause of this is rounding in the earnings per share. That number is rounded to the nearest cent, but it could be as much as 0.9849 and still round to 0.98. In fact, if you recalculate the P/E ratio using 0.984, you get pretty much exactly the value in the table.

Try This One   2

2. Compute the P/E ratio for a stock.

On one recent day, the price of Facebook stock was $352.96, and earnings per share were listed as $13.47. Find the P/E ratio.

Our answer to Example 2 means that the price of a share of Facebook stock is about 89 times the company's annual earnings per share. If you divide $1.00 by the P/E ratio 89, you get about 0.011, which means that for every dollar you invest in the company by purchasing its stock, the company makes about 1.1¢. This doesn't necessarily mean that they're going to give you that 1.1¢ in the form of a dividend, though. The dividends paid are determined by the board of directors of the company, and they might choose to use that profit for other purposes, like expansion, lobbying, investing in new equipment, or any number of things.

Another way of looking at this P/E ratio is that you are paying the company $1.00 so it can earn 1.1¢. The P/E ratio for Starbucks, on the other hand, was 29.41, and $1/29.41 is about 0.034. This means that shareholders were paying the company $1 so that it could earn 3.4¢. Based strictly on price and earnings, the investment in Starbucks is better than in Facebook. So, in general, the lower the P/E ratio is, the better the investment, but there are many other factors to consider. Ultimately the price of a stock is determined by what investors are willing to pay for it, and investors are human beings, so you can't make definite judgments based on raw numbers. Also remember that since the price of a company's stock is constantly changing, the P/E ratio also changes.

Knowing the price per share of stock and the P/E ratio, you can find the annual earnings per share for the last 12 months by using the following formula:

Math Note

This formula is obtained from solving the P/E ratio formula for earnings.

Formula for Annual Earnings per Share

$$\text{Annual earnings per share} = \frac{\text{Yesterday's closing price}}{\text{P/E ratio}}$$

| **Example 3** | Computing Earnings per Share |

Math Note

If you're a bargain shopper, you'll want to keep an eye on the P/E ratio when looking to buy stocks. A low P/E ratio typically indicates that a stock is currently selling at a bargain price.

On a recent day T-Mobile stock closed at \$63.92 per share, and the P/E ratio was listed as 38.05. What were the earnings per share for the last year?

SOLUTION

$$\text{Annual earnings per share} = \frac{\text{Selling price}}{\text{P/E ratio}} = \frac{\$63.92}{38.05} \approx \$1.68$$

The company earned \$1.68 per share over the last year.

| Try This One | 3 |

On the same day, Verizon was trading at \$49.19, with a P/E ratio of 15.28. How did the company's earnings per share compare to their rival T-Mobile?

The table below summarizes some basic rules of thumb when it comes to interpreting the P/E ratio for a stock. Remember that the best approach is to compare a stock's P/E ratio to other companies in the same industry, because current market conditions can have a pretty profound effect on P/E ratio.

P/E Ratio	Common Interpretations
N/A	Company has no earnings; it's also standard for a company with negative earnings (that is, a loss) to have an undefined P/E ratio, even though mathematically a negative P/E ratio does make sense.
0–10	The stock may be undervalued, making it a good buy; alternatively investors might feel like the earnings are likely to decline. The company may have shown short-term profit from selling assets that will hurt earnings in the long term.
10–17	P/E ratios in this range are historically considered to represent stocks being sold at fair market value.
17–25	The stock may be overvalued, so shop carefully. On the other hand, earnings may have increased since the last official earnings report was released. The company might be up and coming, with earnings expected to increase dramatically at some point.
>25	The stock might be very overpriced at the moment, or the company may be expected to have a big jump in earnings coming up. It could also be the case that the company's most recent earnings are unusually low for that company, with an expectation of recovery.

There are two ways to make money from stocks: buy shares of a stock that pays dividends, or buy stock at a low price and sell it at a higher price. But of course, you can't just go buy stock at the corner store—you need to use a brokerage firm, placing an order which is then carried out by representatives at the stock exchange. In exchange for that service, the broker charges a commission, which varies among brokers. Brokers can also make recommendations concerning what stocks to buy and sell, which further justifies their commissions.

The amount that an investor receives from the sale of a stock is called the **proceeds**. The proceeds are equal to the amount of the sale minus the broker's commission. The next two examples illustrate the buying and selling of stocks.

Example 4	Finding the Total Cost of Buying Stock

Jill Braaten/McGraw Hill

Shares of Apple Computer (AAPL) closed at $12.89 on April 1, 2004. Suppose that an investor bought 600 shares at that price using a broker that charged a 2% commission. Find the amount of commission and the total cost to the investor.

SOLUTION

Step 1 Find the purchase price.

$$600 \text{ shares} \times \$12.89 = \$7,734.00$$

Step 2 Find the broker's commission.

$$2\% \text{ of purchase price} = 0.02 \times \$7,734.00$$
$$= \$154.68$$

Step 3 Add the commission to the purchase price.

$$\$7,734.00 + \$154.68 = \$7,888.68$$

The investor paid a total of $7,888.68 for the transaction.

Try This One	4

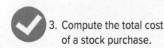

3. Compute the total cost of a stock purchase.

Apple closed at $38.45 on January 3, 2005. If 250 shares were bought at that price through a broker with a 1.6% commission, find the commission and total cost to the investor.

A **stock split** occurs when a company increases the number of outstanding shares. Most commonly, this is a **2-for-1 split**, in which the number of shares doubles. This can be done for a variety of reasons, but the simplest explanation is that when the price of one share gets particularly high, it tends to scare away smaller investors, and the split effectively halves the price of one share. In any case, the net result is that in an instant, the number of shares owned by shareholders doubles. If you owned 100 shares of a stock selling at $50, you'd have $5,000 in stock. If the company did a 2-for-1 split, you'd now have 200 shares, but each would be worth only $25.

Example 5	Studying a Stock Split

(a) Apple did a 2-for-1 split on February 28, 2005, then a 7-for-1 split on June 9, 2014. How many shares would our investor from Example 4 have owned after the second split?
(b) The trading price on 2/27/05 was $82.98. What was the trading price at opening the next day? What was the value of our investor's stock before and after the split?

SOLUTION

(a) The investor had 600 shares prior to the 2-for-1 split, so had $2 \times 600 = 1,200$ afterward. Then the big 7-for-1 split resulted in $7 \times 1,200 = 8,400$ shares.
(b) The point of splitting the stock is to lower the price per share, so the price would have been cut in half after a 2-for-1 split. The price was $41.49 at the beginning of trading the day after the split. The original value was $600 \times \$82.98 = \$49,788$. Of course, it was the same after the split because a split doesn't affect the total value for current shareholders. If the number of shares doubles, the price is cut in half to account for that.

> ## Try This One   5
>
> CF Industries manufactures and distributes fertilizer products worldwide. On June 18, 2015, their stock split 5-for-1, and the adjusted price after the split was $63.37.
>
> (a) What was the price before the split?
> (b) If a shareholder had 220 shares before the split, how many would they have had after? What was the total value of their stock?

Example 6	Finding the Amount Made from Selling Stock

Our clever friend the Apple investor from Examples 4 and 5 could have sold the shares for $130.28 on June 1, 2015. If the sale had been made using a brokerage with a 1.5% commission, find the amount of the commission, the proceeds, and the amount of profit made.

SOLUTION

First, we need the amount of the sale. From Example 5, this investor had 8,400 shares.

$$8,400 \times \$130.28 = \$1,094,352.00$$

The commission is 1.5% of this amount:

$$0.015 \times \$1,094,352 = \$16,415.28$$

The total proceeds are the amount of sale minus the commission:

$$\$1,094,352 - \$16,415.28 = \$1,077,936.72$$

From Example 4, the initial investment was $7,888.68, so the profit is

$$\$1,077,936.72 - \$7,888.68 = \$1,070,048.04$$

I could live with that. What are your chances of ever making over a million dollars on an $8,000 investment? Not good.

> ### Math Note
>
> Stockbrokers charge commission for both buying and selling your stock, so you end up paying on both ends when investing in stock.

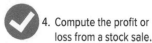

4. Compute the profit or loss from a stock sale.

> ## Try This One   6
>
> When Apple founder Steve Jobs died in October 2011, the company's stock saw a temporary dip, falling to $369.80 on October 7. If the investor in Try This One 4 sold all of their Apple stock that day, and the broker charged 1.6% for sales, find the commission, proceeds, and profit made. Don't forget to factor in the February 2005 stock split.

Bonds

When an investor buys bonds, the investor is actually loaning money to the company or government entity that issues the bonds. In exchange for lending that money, the investor will receive a fixed return on their investment for a given period of time. The ending value of the bond, which represents the original cost plus interest, is called the **face value**; this is typically $1,000, but can vary. Bonds also have a **maturity date**, which is the date that the interest is paid. In some cases, no interest is paid until the maturity date, at which time the full amount of interest is paid. **Coupon bonds**, on the other hand, pay an annual or semiannual interest payment, known as a **coupon**.

You can think of it this way: if someone offers to sell you a $20 bill for $15, that's a pretty darn good deal. But the catch is that you don't get the $20 until some predetermined time later down the road. That's what a bond is; it's actually pretty simple.

But here's where it can get more interesting. Suppose that you get tired of waiting for the time when you get that $20 bill, and you sell the bond to your neighbor for $18. You've made $3 on your investment and you're done. Your neighbor now waits until the maturity date, cashes in the bond, and makes the other $2 of the original $5 in profit. This is how bonds are bought and sold on the open market. Of course, your neighbor wouldn't pay more than $20 for a bond with a maturity value of $20. But if you were desperate for cash, you might be compelled to unload the bond for less than the $15 you originally paid. You've taken a loss, but at least you got some money out at the time you needed it.

Investors buy bonds for two main reasons. First, they're much safer investments than stocks, even though they have much less opportunity for a big profit. The interest is guaranteed, as long as the company that issued the bond stays in business and has the money to pay its debts. This makes bonds a safe, but not guaranteed, investment. Second, bonds are used to offset losses on stocks when the economy is in a down cycle, because the value of bonds tends to go up when the value of stocks goes down.

Here's a short version of why that makes sense: in our earlier hypothetical example where you're making a $5 profit for waiting to cash in your bond, if the economy is booming in the meantime, prices in general will tend to go up, so the $5 you're getting at some point in the future has less buying power than the $5 you were expecting when you made the investment. This makes the bond in effect worth less. The opposite is also true: if the economy fizzles after you buy that bond, the $5 you eventually get will in effect be worth more than the $5 you thought you were getting, so the bond is worth more. Stocks tend to rise when the economy is doing well and fall when it's not, so the value of bonds usually goes in the opposite direction of the stock market. We'll study bond trading in Example 7.

Example 7	Finding the Profit on a Bond Trade

Monique buys a bond with face value $1,000 that was originally issued 30 months ago. The maturity date is 4 years from the time it was issued, and the interest rate is 4% simple interest per year. If she pays $820 for the bond and keeps it until the maturity date, what is her profit? What percent return does she get per year?

SOLUTION

With 4% simple interest, we can use the simple interest formula to find the value of the bond at maturity.

$$I = Prt = 1,000(0.04)(4) = \$160$$

After drawing $160 in interest, the value of the bond will be $1,160, so if Monique paid $820, her profit is $1,160 − $820 = $340.

Her overall percent return is

$$\frac{\$340}{\$820} = 41.5\%$$

and she'll need to wait 18 months, or 1.5 years to cash in the bond, for a return of

$$\frac{41.5\%}{1.5} = 27.7\% \text{ per year.}$$

Monique is pretty smart.

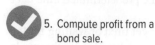

5. Compute profit from a
   bond sale.

Try This One   7

Rashard bought four bonds with face values of $1,000 each, simple interest rate of 5.8% per year, and a maturity date 10 years after they were issued. He paid $4,725 3 years after the bonds were issued. If he keeps the bonds until maturity, find his total profit, and his percent return per year.

Mutual Funds

Many times investors purchase a group of stocks and bonds called a **mutual fund**. Mutual funds are managed by professional managers and include money from other investors. The manager follows the markets and makes the decisions of when to buy or sell the stocks and bonds. Mutual funds usually consist of a large number of relatively small investments in companies. This way, if a single stock doesn't perform well, only a small amount of money is lost. Sometimes mutual funds can be high return but also high risk.

Ratings for mutual funds can be found on most financial websites, and in business publications like the *Wall Street Journal*. They are rated either from A to F, or from 5 to 1, with 5 being the best. Often, two separate ratings are given. An overall rating compares the fund to all other stock funds. A category rating compares a fund to other funds that have similar holdings. For example, there are funds that invest strictly in smaller businesses, and it makes sense to compare those funds to others like them, as well as to the market as a whole.

Answers to Try This One

1   (a) $57.35; $56.80
       (b) Highest; $61.64; lowest; $50.84
       (c) $57.20
       (d) It does; 90 cents per share
       (e) About 1.457 billion
       (f) SBUX

2   26.2 (rounded)

3   Verizon's earnings per share ($3.22) was $1.54 higher.

4   Commission = $153.80, total cost = $9,766.30

5   (a) $316.85
       (b) 1,100 shares valued at $69,707

6   Commission = $2,958.40,
       proceeds = $181,941.60, profit = $172,175.30

7 $1,595; 4.8% per year

Exercise Set   7-7

Writing Exercises

1. Explain in your own words what stock is.
2. What's the difference between stocks and bonds?
3. What is a mutual fund?
4. What is meant by the term P/E ratio?
5. What does a stockbroker do?
6. When selling stock, what's the difference between sale price, proceeds, and profit or loss?

7. Explain why the price of bonds tends to go up when the price of stocks goes down.
8. Why would an investor sell a bond for less than he paid for it?
9. If a bond is issued by a private company, is the investor guaranteed to make a profit? Why or why not?
10. If bonds are safer investments than stocks, why do more people invest in the stock market?

Applications in Our World

Use the following stock listing for Sunoco for Exercises 11–20.

Sunoco LP (SUN)
NYSE - NYSE Delayed Price. Currency in USD

28.04 –0.47 (–1.65%)
At close: February 17 4:00PM EST

Previous Close	28.51	Market Cap	2.68B
Open	28.44	Beta	0.70
Bid	0.00 ×	PE Ratio (TTM)	19.61
Ask	0.00 ×	EPS (TTM)	1.43
Day's Range	27.91–28.66	Earnings Date	Feb 22, 2017
52 Week Range	21.01–37.25	Dividend & Yield	3.30 (11.58%)
Volume	1,022,112	Ex-Dividend Date	N/A
Avg. Volume	1,277,000	1y Target Est	30.06

11. What is the highest price that the stock sold for during the last 52 weeks?
12. What was the lowest price that the stock sold for during the last 52 weeks?
13. What was the amount of the dividend per share that the company paid last year?
14. If you own 175 shares, how much in dividends did you make last year?
15. How many shares were traded yesterday?
16. What was the closing price of the stock yesterday?
17. Find the annual earnings per share.
18. If you buy 480 shares of Sunoco stock at the current listed price and the broker's commission is 1.5%, find the total cost of the purchase.
19. If the 1 year target estimate is accurate, how much will those 480 shares be worth a year from now?
20. Is the price of this stock more or less stable than the average stock? What about the Facebook stock from Example 1?

Use the following stock listing for WABTEC for Exercises 21–30.

Westinghouse Air Brake Technologies Corporation (WAB)
NYSE - NYSE Delayed Price. Currency in USD

87.91 +0.38 (+0.28%)
At close: February 17 4:02PM EST

Previous Close	88.05	Market Cap	7.82B
Open	87.53	Beta	1.19
Bid	0.00 ×	PE Ratio (TTM)	22.09
Ask	0.00 ×	EPS (TTM)	3.97
Day's Range	86.22–88.70	Earnings Date	Feb 21, 2017
52 Week Range	65.54–89.18	Dividend & Yield	0.40 (0.45%)
Volume	447,720	Ex–Dividend Date	N/A
Avg. Volume	783,674	1y Target Est	90.10

21. What was the highest price that the stock sold for during the last 52 weeks?
22. What was the lowest price that the stock sold for during the last 52 weeks?
23. What was the amount of the dividend per share that the company paid last year?
24. If you own 357 shares, how much in dividends did you make last year?
25. How many shares were traded yesterday?
26. What was the closing price of the stock yesterday?
27. Find the annual earnings per share.
28. If you purchase 1,247 shares of stock at the closing price and the broker's commission is 2.6%, find the total cost of the purchase.
29. What is the current total value of the company?
30. Is the price of this stock more or less stable than the average stock? What about the Facebook stock from Example 1?

Use the following stock listing for Wal-Mart for Exercises 31–40.

Wal-Mart Stores Inc. (WMT) - NYSE

60.75 ↓ 0.08 (0.13%) 4:01PM EST
After Hours : **60.75** 0.00 (0.00%) 4:12PM EST

Prev Close:	60.83	Day's Range:	60.50–60.97
Open:	60.58	52wk Range:	56.30–90.97
Bid:	60.73 × 4300	Volume:	5,296,196
Ask:	60.74 × 2900	Avg Vol (3m):	12,578,300
1y Target Est:	63.63	Market Cap:	194.51B
Beta:	0.444396	P/E (ttm):	13.01
Next Earnings Date:	N/A	EPS (ttm):	4.67
		Div & Yield:	1.96 (3.24%)

31. What was the highest price that the stock sold for today?
32. What was the lowest price that the stock sold for today?
33. What was the amount of the dividend per share that the company paid last year?
34. If you own 682 shares, how much in dividends did you make last year?
35. How many shares were traded yesterday?
36. What was the closing price of the stock yesterday?
37. Find the annual earnings per share.
38. What would be the total cost of buying 842 shares of Wal-Mart stock at the current ask price if the broker charges a 2% commission?
39. How much more would the total cost have been if the shares had been bought at the 52-week high price?

40. Is the price of this stock more or less stable than the average stock? What about the Facebook stock from Example 1?

41. If the closing price of a stock is $21.92 and the annual earnings per share is $0.88, find the P/E ratio.

42. If the closing price of a stock is $6.65 and the annual earnings per share is $0.35, find the P/E ratio.

43. If the closing price of a stock is $24.19 and the annual earnings per share is $1.61, find the P/E ratio.

44. If the closing price of DirecTV stock is $20.18 and the annual earnings per share is $1.06, find the P/E ratio.

45. If the closing price of Gaither stock is $18.53 and the P/E ratio is 55, find the annual earnings per share.

46. If the closing price of Jacob Energy is $75.66 and the P/E ratio is 25, find the annual earnings per share.

47. If the closing price of Marine Max is $25.76 and the P/E ratio is 13, find the annual earnings per share.

48. If the closing price of Omnicare is $43.73 and the P/E ratio is 30, find the annual earnings per share.

49. An investor purchased 800 shares of stock for $63.25 per share and sold them later for $65.28 per share. The broker's commission was 2% of the purchase price and 2% of the selling price. Find the amount the investor made or lost on the stock.

50. An investor purchased 200 shares of a stock at $93.75 per share and sold it later at $89.50 per share. The broker's commission on the purchase and sale of the stock is 2.5%. Find the amount of money the investor made or lost on the sale.

51. An investor purchased 550 shares of stock at $51.60 per share. The shares were later sold at $49.70. The broker's commission on the purchase was 2% and 1.5% on the sale. Find the amount of money the investor made or lost on the stock.

52. An investor purchased 670 shares of a stock at $73.20 per share, then sold the stock at $82.35. If the broker's commission was 2.5% on the purchase and sale of the stocks, how much money did the investor make or lose on the transaction?

53. Some companies pay an annual dividend to stockholders, while others choose to instead invest that money back into the company. Suppose that you buy 500 shares of stock at $22 in a company that pays an annual dividend of $1.70 per share, then sell all of your shares at $38 three years later. Your best friend buys 500 shares of stock at $20 in a company that doesn't pay dividends, and sells it at $38 3 years later. Which of you will make more money? By how much? (You can ignore commissions on the sales.)

54. Suppose that the investor in Examples 5 and 6 of Section 7-7 had used an online brokerage that charges a flat fee of $39.95 for all trades, rather than a percentage of the sale. How much greater would the investor's profit have been?

55. Shares of stock in the Ford Motor Company reached a low of $1.43 on November 10, 2008. One year later, the stock closed at $8.41. If you had invested $20,000 in Ford stock on 11/10/08, buying as many shares as possible while paying an online brokerage fee of $19.99 for the purchase, then sold the stock a year later with the same fee, how much profit would you have made? How much money would you have made per day for holding the stock for a year?

56. Refer to Exercise 55.

    (a) If you'd been wise and patient enough to wait until January 10, 2011, to sell the stock at $18.65, how much greater would your profit have been?

    (b) Using the Internet as a resource, investigate why Ford stock was so low in November 2008, and why it went up so much, then write a short essay summarizing your findings.

    (c) In September 2021, Ford stock was trading at $12.58. How much money would you have cost yourself if you had kept it until that point?

In Exercises 57–60, find the value at maturity for the bond described. All interest is simple interest.

57. Face value: $1,000; term: 5 years; rate: 4.9%

58. Face value: $5,000; term: 10 years; rate: 5.7%

59. Face value: $2,500; term: 90 months; rate: 6.19%

60. Face value: $3,750; term: 42 months; rate: 3.79%

61. If the bond in Exercise 57 is sold for $925 and is kept by the buyer until it matures, find the buyer's profit.

62. If the bond in Exercise 58 is sold for $4,000 and is kept by the buyer until it matures, find the buyer's profit.

63. If the bond in Exercise 59 is sold for $2,800 and is kept by the buyer until it matures, find the buyer's profit.

64. If the bond in Exercise 60 is sold for $4,190 and is kept by the buyer until it matures, find the buyer's profit.

65. Find the buyer's percent return per year for the bond purchase in Exercise 61 if it were bought 2 years after being issued.

66. Find the buyer's percent return per year for the bond purchase in Exercise 62 if it were bought $6\frac{1}{2}$ years after being issued.

67. Find the buyer's percent return per year for the bond purchase in Exercise 63 if it were bought 28 months after being issued.

68. Find the buyer's percent return per year for the bond purchase in Exercise 64 if it were bought 32 months after being issued.

Critical Thinking

69. Compare the two investments below and decide which would have been the better choice.
    Investment 1: $10,000 was invested in a 24-month CD that earned 5.1% annual interest compounded daily.
    Investment 2: 1,400 shares of stock in the Lybarger Aviation Company were bought at $7.11 per share using a brokerage with a 0.75% commision rate on both buying and selling stock. Over the 2 years the stock was held, it paid a dividend of $0.48 per share in the first year and $0.36 per share in the second year. The stock was sold through the same brokerage for $7.95 per share.

70. Compare the two investments and write a paragraph or two describing which you think is the better choice, and why.
    Investment 1: four 10-year bonds with face value $1,000 that draw 3.9% interest with 7 years remaining until maturity at a cost of $3,750 plus a 0.9% commission.
    Investment 2: putting $3,750 into a 4-year CD at 5.85% interest compounded monthly with no commissions or fees.

71. A bond with face value $10,000, simple interest 6.45%, and term 12 years is originally bought by Larry. After 33 months, he sells it to Curly for $11,400. Curly then holds onto it for 7 years, eventually selling it to Moe for $14,950. Moe keeps the bond until it matures and cashes it in.
    (a) Which investor made the greatest profit?
    (b) Which got the greatest percent return on his investment?
    (c) Which got the greatest percent return per year?

72. When workers invest for retirement, a general rule of thumb is to invest more in stocks when far from retirement, and convert more to bonds as retirement nears. Explain why this rule makes sense.

73. Using the Internet or a newspaper as a resource, look up a current stock quote for Facebook corporation (see Example 1).
    (a) If you had bought 400 shares of Facebook at the price shown in Example 1 at an online brokerage with a $29.95 commission, then sold all shares at the current price (same commission), what would your profit/loss be? What percent is that of your original investment?
    (b) Based on the P/E ratio and the current closing price, find Facebook's earnings per share for last year.

74. Repeat Problem 73 for Wal-Mart stock (see Problems 31–40).

75. Stock in Buckeye Brewers, Inc., is currently selling for $42.29 and earnings per share were $2.35 last year. One of their competitors, Nittany Beverage Distributors, is selling for $24.36 and earnings per share last year were $1.42. Which stock is the better buy? Why do you feel that way?

76. Some investors prefer stocks that pay a dividend, while others favor stocks that do not. Give some possible reasons for each perspective.

77. Now let's look a little deeper at that awesome investment in Apple that we studied during the lesson.
    (a) We found the profit for the investor was $1,070,048.04. Look up the dates of the original purchase and sale of the stock, and calculate both the percent return and percent per year.
    (b) As of this writing (9/26/21) the price of Apple stock is $146.92. If this investor had held onto the investment until that point and then sold, find the profit, percent return, and percent per year return. Then describe whether or not you think hanging onto the stock for that extra amount of time would have been worthwhile.

78. It seems hard to believe now, but at one point in the late 1990s, Apple was struggling terribly. In fact, many analysts expected the company to go bankrupt before too long, and in March of 1999, the stock reached a low price of $8.05 per share. A true story: at that point, one of my students, a real estate broker who returned to college because he wanted to become a computer programmer, told me that if I could scrape together any money to invest in Apple, I should. (It turns out they had this revolutionary new product called the iPod in secret development at the time.) Sadly, I had just bought my first house and could barely afford to pay attention, let alone buy stock in a company that 75% of America thought would be out of business in 5 years. Let's say that I had managed to scrape together $2,500 to invest in Apple at that time. Ignoring commissions, but factoring in stock splits (2-for-1 in June 2000, 2-for-1 in February 2005, and 7-for-1 in June 2014) and the current price as of this writing (see Exercise 77), how much did I cost myself by ignoring his advice? And just how stupid do you think I am? Explain at length.

Section	Important Terms	Important Ideas
7-1	Percent Percent increase Percent decrease	**Percent** means "per hundred," or "hundredths." So 45% means 45 per hundred. In order to do calculations with percents, they have to be changed to fractions or decimals. The word "of" is important in calculations with percents: the phrase "40% of 80 is 32" translates to the equation $0.40 \times 80 = 32$. This allows us to set up many percent calculations.
7-2	Budget Take-home pay Necessary expenses Luxuries Fixed expenses Variable expenses Prorating	**A budget** is a plan that lays out income and expenses, allowing you to keep track of what you spend and how much you can afford to spend. Since your take-home pay is largely eaten up by regular expenses like food, shelter, transportation, and insurance, it's important to realize how much you can spend on other items. One way to plan for regular expenses is to save a prorated amount every so often so that you have the full amount available to you at the time it's due.
7-3	Interest Simple interest Principal Rate Term Future value Banker's rule Discounted loan	**When you** borrow money, you pay a fee for its use. This fee is called interest. Likewise, when you put money into a savings account, the bank pays interest for the use of your money. Simple interest is interest computed only as a percentage of the principal. The formula $I = Prt$ is used to compute simple interest. Future value is the sum of the principal and any interest earned.
7-4	Compound interest Effective rate Annual yield Annuity Future value of an annuity	**Compound interest** is interest calculated on both the principal and any interest previously earned. Compound interest investments earn more interest than simple interest investments at the same rate. Compound interest can be calculated any number of times per year; typical compounding frequencies are annually (once per year), semiannually (twice per year), quarterly (four times per year), monthly, or daily. We can use the compound interest formula to find the future value of an account; using logarithms we can find the time needed to reach an investment goal. Since the actual rate is higher when interest is compounded more than once per year, the true rate is called the effective rate or annual yield. An annuity is a savings plan where an individual or business makes the same payment each period into a compound interest account where the rate remains the same for the term of the annuity.
7-5	Fixed installment loan Finance charge Down payment Total installment price Lease Annual percentage rate (APR) Payoff amount Actuarial method Rule of 78 Closed-ended credit	**A fixed** installment loan is a loan that is repaid in equal (usually monthly) payments. A down payment is a cash payment made on the purchase. Many times a finance charge is added to the amount financed. The total installment price is found by summing the monthly payments and adding the down payment. Because you pay back some of the principal each month, you do not have the full use of the money for the term of the loan. This means that the actual interest is higher than the stated interest rate. This actual interest rate is called the annual percentage rate and can be computed approximately using an APR table. When an installment loan is paid off early, the amount of interest saved can be determined by the rule of 78, or the actuarial method.

Section	Important Terms	Important Ideas
	Open-ended credit Revolving credit Unpaid balance method Average daily balance method	Credit card companies also charge interest. There are two ways the companies compute interest. One method is computing interest on the unpaid balance. In this case you are charged interest only on last month's balance. The other method is called the average daily balance. Here the interest is computed on the average balance on all of the days of the month. This includes any purchases and payments made during the month.
7-6	Federal student loan Private student loan Subsidized Unsubsidized Capitalized Mortgage Fixed-rate mortgage Adjustable-rate mortgage Graduated payments Amortization schedule	**There are** three basic types of student loans available: federal loans made by the government, federal loans made by other lenders but guaranteed by the government, and private. In addition, a loan can be subsidized (no interest accrues while in school) or unsubsidized (interest accrues while in school). Interest on student loans is simple interest and can be paid while in school or capitalized (deferred until after graduation). Monthly payments can be calculated using a formula based on the principal at the time payments begin and the interest rate. When a loan is acquired to pay for the purchase of property, the loan is called a mortgage. The lender has the right to take ownership of the property if payments aren't made. Monthly payments can be calculated using a table, a formula, or an online calculator. A table listing the amount of each payment going to pay interest, the amount toward principal, and the remaining balance of the loan is called an amortization schedule.
7-7	Stock Dividend Bond Face value Maturity date Stock exchange Stockbroker Commission Yield P/E ratio Proceeds Mutual fund Ticker symbol Stock split 2-for-1 split	**Investors** can purchase stocks and bonds. A stock is a share of ownership in a company. A bond is actually a loan to a company. A mutual fund is a combination of stocks and bonds that is managed by a professional investor. Newspapers and websites show information about stocks and bonds by using tables. The tables show the 52-week high price and low price of the stocks and bonds. The table also shows the yield, the P/E ratio, the dividend, the volume of sales, the closing price, and the net change of a stock. P/E ratio is a comparison between the share price of a stock and the company's earnings per share. After being issued, but before their maturity date, bonds can be bought and sold like stocks. Bonds tend to be safer investments than stocks but have less growth potential. In addition, bonds tend to go up in value when stocks are declining, and vice versa.

Math in College Budgeting REVISITED

Digital Vision/SuperStock

1. Multiplying the yearly amount by 4, we get $107,284 needed for 4 years.

2. With 25% paid for, that leaves 75% we're responsible for, minus $7,000 a year we're contributing up front: 0.75 (26,821) − 7,000 = $13,115.75 per year. Multiply this by 4 and we get $52,463 we'll need to borrow.

3. If we pay interest while in school, the principal at the time payments start will remain $52,463. Use the student loan payment formula with $P = 52,463$, $r = 0.068$, and $n = 12$: the result gives us monthly payments of $603.75. The total of all payments is 120 × $603.75 = $72,450. Now

calculate the simple interest paid while in school: $P = 52{,}463$, $r = 0.068$, $t = 4$, so $I = \$52{,}463 *$ $0.068 * 4 = \$14{,}269.94$. (Note: This assumes that you borrowed the entire amount up front, which is not the smartest idea in the world, but simplifies the calculation. More likely, you'd borrow one year at a time, which would lower the total interest.) Adding the two amounts, we get a total of $\$72{,}450 +$ $\$14{,}269.94 = \$86{,}719.94$ in total payments. Subtracting the amount borrowed, we find that total interest is $\$86{,}719.94 - \$52{,}463 = \$34{,}256.94$.

4. If we capitalize the interest, upon graduation our principal will be $\$52{,}463 + \$14{,}269.94 = \$66{,}732.94$.

This time the monthly payment formula yields $\$767.96$. Multiply this by 120 payments to get total payments of $\$92{,}155.20$. Subtract the principal of $\$52{,}463$ to get total interest of $\$39{,}692.20$.

5. According to the table, the median salary with a bachelor's degree is $\$26{,}104$ higher than with a high school diploma. In Question 4, we found that the total amount paid on loans is $\$92{,}155.20$; add the $\$28{,}000$ contributed while in school, and the total cost of the degree is $\$120{,}155.20$. Finally, divide this amount spent by the extra $\$26{,}104$ we'll make each year to find that it will take 4.6 years to make back the amount spent.

Review Exercises

Section 7-1

For Exercises 1–6, find the missing value.

	Fraction	Decimal	Percent
1.	$\frac{7}{8}$	_____	_____
2.	_____	0.54	_____
3.	_____	_____	185%
4.	_____	0.06	_____
5.	$5\frac{3}{4}$	_____	_____
6.	_____	_____	45.5%

7. Find 72% of 96.
8. 18 is what percent of 60?
9. 25% of what number is 275?
10. If the sales tax is 5% on a calculator, find the tax and the total cost if the calculator is $19.95.
11. If the sales tax on a coffee table is $3.60, find the cost of the table if the tax rate is 6%.
12. Marcia received a commission of $2,275 for selling a small home. If she receives a 7% commission, find the price of the home.
13. In 2000, there were 385 million credit card accounts in the U.S. In 2020, that had grown to 511 million. Find the percent increase.
14. In 2001, there were 3,147 adolescents under 18 being held in state prisons. In 2018, the number was 699. Find the percent decrease.
15. A pair of jeans that usually sells for $71.50 is on sale at 30% off, and you also have a coupon for 10% off any purchase, including discounted merchandise.
    (a) Explain why you're not going to get 40% off.
    (b) Find the sale price, including the coupon.
16. Annual in-state tuition for undergrads at Enormous State U. was $7,326 in 2020 and was projected to increase

by 12% by 2022. If that happened, what was tuition in 2022?

Section 7-2

For Exercises 17–19, calculate the monthly take-home pay.

17. $1,463 every other Friday
18. $712 once a week
19. 40 hours per week at $17.75 per hour with 23% deducted for taxes

In Exercises 20 and 21, find the annual cost of each regular expense.

20. A $1.25 candy bar from a vending machine every Monday through Friday
21. A twice-weekly visit to a Chinese buffet ($11.95)

In Exercises 22 and 23, a monthly income and fixed expenses are provided. Find the amount that would be left in a budget for variable expenses and luxuries.

22. Income: $3,200; house payment: $1,142; car payment: $304; cell phone and Internet: $94.30; utilities: $148; insurance: $154
23. Income: $945 from work, $400 from parents; rent of $850 split evenly among three residents; utilities: $48; cell phone: $60; gas: $24; renters insurance: $12
24. If the person in Exercise 22 spends an average of $160 on clothes at a high-end mall each week, what percentage of their budget are they spending after fixed expenses on these clothes?

In Exercises 25 and 26, find the amount that would need to be saved monthly to prorate the given long-term expenses.

25. Homeowner's insurance: $955 annually; property tax: $1,723 every 6 months
26. Tuition: $2,190 at the beginning of August and again at the beginning of January; school books: $400 at the beginning of each semester

Section 7-3

For Exercises 27–34, find the missing value.

	Principal	Rate	Time	Simple Interest
27.	$4,300	9%	6 years	_____
28.	$16,000	_____	3 years	$1,920
29.	$875	12%	_____	$262.50
30.	$50	6%	18 months	_____
31.	$230	_____	6.5 years	$104.65
32.	_____	3%	5 years	$63.75
33.	_____	14%	2 years	$385
34.	$785.00	12%	_____	$1,130.40

35. Ace Auto Parts borrowed $6,000 at 6% for 5 years to enlarge its display area. Find the simple interest and future value of the loan.
36. Sam's Sound Shack borrowed $13,450 at 8% for 15 years to remodel its existing store. Find the simple interest and future value of the loan.
37. Julie earned $60.48 in simple interest on a savings account balance of $4,320.00 over a 12-month period. Find the rate of interest.
38. John has an opportunity to buy a new boat. He has to borrow $5,300 at 11% simple interest for 36 months. Find the monthly payment.
39. Find the simple interest on a $2,300 loan at 5% for 80 days. Use the Banker's rule.
40. Find the simple interest on an $8,750 loan at 8.5% for 100 days. Use the Banker's rule.
41. Raul obtained a 3-year, $6,000 discounted loan at 6%. Find the discount and the amount of money Raul received.
42. Marla obtained a 4-year $9,250 discounted loan at 12%. Find the discount and the amount of money Marla received.

Section 7-4

For Exercises 43–46, find the compound interest and future value.

	Principal	Rate	Compounded	Time
43.	$1,775	5%	annually	6 years
44.	$200	4.2%	semiannually	10 years
45.	$45	3.04%	quarterly	42 months
46.	$21,000	5.19%	monthly	74 months

47. Find the effective rate when the stated rate is 12% and the interest is computed quarterly.
48. Which is the better investment: 4.3% compounded semiannually or 4.27% compounded daily?
49. How much money would you need to invest now in a 3-year certificate of deposit that pays 3.75% interest compounded monthly in order to have a future value of $10,000?
50. Charles inherits $12,500 from a favorite uncle and decides to invest it in an account that pays 4.6% compounded quarterly. The plan is to leave it in until it reaches $20,000. How long will Charles have to wait?
51. How many years does it take an investment to double if it draws 8.4% interest compounded semiannually? How much sooner will it double if interest is compounded daily?

52. The Evergreen Landscaping Company will need to purchase a new backhoe in 7 years. The owner purchases an annuity that pays 8.3% interest compounded semiannually. If the semiannual payment is $4,000, find the future value of the annuity in 7 years.
53. Mike and Marie plan to take an African vacation in 3 years. In order to save money for the trip, they purchase an annuity that pays 3% interest compounded quarterly. Find their monthly payment if they need $9,000 for the trip.

Section 7-5

In Exercises 54 and 55, for the automotive loan, find the amount financed, the total installment price, and the finance charge.

54. Purchase price including taxes and fees: $12,942.49; down payment: $4,300; payments: $261.34 for 36 months.
55. Purchase price: $22,400; sales tax 6.25%; license and title fees: $325; money credited for trade-in (subtracted from price of vehicle before taxes): $8,100; payments: $290.40 for 60 months.
56. Alex purchased a four-piece luggage set for $750. The down payment was 15% and the interest rate was 6%. Find the total installment price and the monthly payment if the loan was paid off in 8 months.
57. Judy purchased a Chevy Cobalt for $10,900. Her down payment was $1,000. She paid the balance with monthly payments of $310 for 3 years. Find the APR.
58. Max Dunbar bought a used BMW for $20,500 on Autotrader.com. The down payment was $6,000 and the balance was paid off with monthly payments of $311 for 5 years. Find the APR.
59. Levon bought a mobile home for $149,500. He made a down payment of $8,000 and financed the remainder with an 8.5% simple interest loan for 25 years. Find his monthly payment.
60. In Exercise 57, Judy decided to pay off her loan at the end of 24 months. Use the actuarial method and find the unearned interest and the payoff amount.
61. In Exercise 58, Max was able to pay off the loan at the end of 3 years. Use the actuarial method and find the unearned interest and the payoff amount.
62. A loan for $1,500 is to be paid back in 30 monthly installments of $61.25. The borrower decides to pay off the balance after 24 payments have been made. Find the amount of interest saved. Use the rule of 78.
63. For the month of February, Pedro had an unpaid balance on his credit card of $563.25 at the beginning of the month. He had purchases of $563.25 and made a payment of $350.00 during the month. Find the finance charge if the interest rate is 1.75% per month on the unpaid balance, and find the new balance on March 1.
64. Sid's Used Cars had these transactions on its credit card statement:

April 1	Unpaid balance	$5,628.00
April 10	Purchases	$2,134.60
April 22	Payment	$ 900.00
April 28	Purchases	$ 437.80

Find the finance charge if the interest rate is 1.8% on the average daily balance and find the new balance on May 1.

65. If you have $845.32 on a credit card that charges 1.19% interest per month, and the minimum payment is interest due plus 2% of the principal balance, what will the balance be after 6 months of minimum payments?

Section 7-6

66. Latwan borrowed $4,700 to cover one-time college expenses in an unsubsidized federal student loan at 6.8% interest.
    (a) Find the amount of interest that accrues in a 30- and 31-day month.
    (b) The loan was acquired in August of 2019, and Latwan graduated in May of 2021. If payments began 6 months after graduation, how much interest will accrue while Latwan is in school?

67. Find the monthly payment on the loan in Exercise 66. The term is 10 years.

68. Find the monthly payment on the loan in Exercise 66 if it's a federally subsidized loan.

69. How much less interest would Latwan (Exercise 66) pay over the life of the student loan if interest on the loan is not capitalized?

70. A home was purchased for $145,000 with a 20% down payment. The mortgage rate was 8.5% and the term of the mortgage was 25 years.
    (a) Find the amount of the down payment.
    (b) Find the amount of the mortgage.
    (c) Find the monthly payment.
    (d) Compute an amortization schedule for the first 2 months.

71. A business sold for $252,000. The down payment was 8%. The buyer financed the balance at 8.25% for 25 years. Find the monthly payment on the mortgage.

72. How much would the home buyers in Exercise 70 have saved over the life of the loan if they'd taken a 15-year mortgage?

Section 7-7

Use the table shown for Exercises 73–77.

Wells Fargo & Company (WFC)
NYSE - NYSE Delayed Price. Currency in USD

58.09 –0.03 (–0.05%)
At close: February 17 4:00PM EST

Previous Close	58.12	Market Cap	291.38B
Open	57.58	Beta	1.01
Bid	0.00 ×	PE Ratio (TTM)	14.56
Ask	0.00 ×	EPS (TTM)	3.99
Day's Range	57.45–58.10	Earnings Date	Apr13, 2017
52 Week Range	43.55–59.01	Dividend & Yield	1.52 (2.62%)
Volume	15,783,230	Ex-Dividend Date	N/A
Avg. Volume	21,623,522	1y Target Est	58.35

73. What was the high price and low price of the stock for the last 52 weeks?

74. If you own 475 shares of this stock, how much was the dividend you received?

75. How many shares of the stock were sold yesterday?

76. What do analysts predict the stock price will be in a year?

77. Find the annual earnings per share of the stock.

78. An investor purchased 90 shares of stock for $86.43 per share and later sold it for $92.27 per share. How much profit was made if the broker's fee was 2% on the purchase and the sale of the stock? Ignore the dividends.

79. Explain the difference between stocks and bonds.

80. Find the value at maturity for a bond with face value $10,000 with simple interest rate 4.2% and term 4 years.

81. A bond with face value $1,000, simple interest 5.3%, and term 6 years is bought by an investor for $1,145 with a commission of 1%. Who makes a greater profit on the bond, the original owner or the buyer?

82. If the buyer in Exercise 81 bought the bond with 40 months left until the maturity date and kept it until maturity, find the percent return on investment, and percent return per year.

Chapter Test

1. Change $\frac{5}{16}$ to a percent.
2. Write 0.63 as a percent.
3. Write 28% as a fraction in lowest terms.
4. Change 16.7% to a decimal.
5. Thirty-two of 40 people surveyed in a shopping mall said that they had used a credit card to make at least one purchase that day. What percent is that?
6. Of the 48 states in the continental United States, 89.6% have some form of state lottery. How many states is that?
7. Sixty-eight teams made the NCAA basketball tournament in 2021, which represented 19.3% of all Division 1 teams. How many Division 1 teams were there in 2012?
8. Find the sales tax and total price on a toaster oven that sells for $29.95. The tax rate is 8%.
9. If a salesperson receives a 15% commission on all merchandise sold, find the amount sold if their commission is $385.20.

10. On the first day of math class, 28 students were present. The next day, 7 more students enrolled in the class because the other section was canceled. Find the percent increase in enrollment.

11. If you get a paycheck every other Friday for $936.75, what is your monthly take-home pay?

12. Elaine works 40 hours a week at a job that pays $16.90 per hour. Each month, 21% of her pay is deducted for taxes, and another 4% for insurance. Find her monthly take-home pay.

13. Elaine from Exercise 12 has the following monthly fixed expenses: rent: $825; car payment: $211; utilities: $125; insurance: $96; student loan payment: $194. Find the amount that she has left for variable expenses and luxuries.

14. Elaine (Exercise 12) has a weakness for frozen yogurt. In fact, she visits the local froyo shop three times a week, at an

average of $4.75 per visit. How much is she spending per year on those treats? What percentage of her budget after fixed expenses is she spending? (Legal disclaimer: I'm eating froyo right now, so I'm not really in a position to point fingers.)

15. Montez is tired of having to come up with $340 every 6 months for car insurance, and $812 every year for medical insurance, so he decides to start saving every month to prorate these expenses. How much should he save monthly?

16. Find the simple interest on $1,350 at 12% for 3 years.

17. Find the rate for a principal of $200 invested for 15 years if the simple interest earned is $150.

18. Ron's Detailing Service borrowed $435 at 3.75% for 6 months to purchase new equipment. Find the simple interest and future value of the loan, and the monthly payment.

19. Explain the difference between simple and compound interest.

20. Which account will draw more interest on a $6,000 investment: 7.2% simple interest for 4 years or 6.4% interest compounded monthly for 3 years?

21. Find the simple interest on a $5,000 loan at 4% for 60 days. Use the Banker's rule.

22. Latoya obtained a 6-year $12,650 discounted loan at 7.5%. Find the discount and the amount of money Latoya received.

23. Find the interest and future value for a principal of $500 invested at 6.5% compounded semiannually for 4 years.

24. Find the interest and future value on a principal of $9,750 invested at 10% compounded quarterly for 6 years.

25. In order to purchase a motorcycle, Jayden borrowed $12,000 at 9.5% for 4 years. Find the monthly payment.

26. Find the effective rate when the stated interest rate is 8% and the interest is compounded semiannually.

27. In order to open a new branch in 3 years, the owner of Quick Fit Fitness Center purchases an annuity that pays 4.5% interest compounded semiannually. If the semiannual payment is $3,000, find the future value of the annuity in 3 years.

28. Sara bought furniture for her first apartment at a price of $935. She made a down payment of 30% and financed the rest for 6 months at 10% interest. Find the total installment charge and the monthly payment.

29. Bart Johnston purchased a Mazda for $15,000 and had a down payment of $2,000. The balance was financed at $305 per month for 48 months. Find the APR.

30. In Exercise 29, Bart was able to pay off the loan at the end of 24 months. Using the actuarial method, find the unearned interest and the payoff amount.

31. A loan for $2,200 is to be paid off in 24 monthly installments of $111.85. The borrower decides to pay off the loan after 20 payments have been made. Find the amount of interest saved, using the rule of 78.

32. For the month of November, Harry had an unpaid balance of $1,250 on his credit card. During the month, he made purchases of $560 and a payment of $800. Find the finance charge if the interest rate is 1.6% per month on the unpaid balance, and find the new balance on December 1.

33. Rhonda's credit card statement for the month of May shows these transactions.

May 1	Unpaid balance	$474.00
May 11	Payment	$300.00
May 20	Purchases	$ 86.50
May 25	Purchases	$120.00

Find the finance charge if the interest rate is 2% on the average daily balance, and find the new balance on June 1.

34. Yasmil took out an unsubsidized student loan when starting college. The amount was $11,200, the rate was 6.8%, and the term was 10 years.
    (a) Find the amount of interest that will accrue while Yasmil is in school during a 30-day month.
    (b) If interest is capitalized, how much will Yasmil owe when payments begin (54 months after acquiring the loan)?
    (c) Find Yasmil's monthly payment.

35. (a) How much less would payments be on the loan in Exercise 34 if interest had not been capitalized?
    (b) How much would not capitalizing the interest save over the life of the loan?

36. A home is purchased for $180,000 with a 5% down payment. The mortgage rate is 6% and the term is 30 years.
    (a) Find the amount of the down payment.
    (b) Find the amount of the mortgage.
    (c) Find the monthly payment.
    (d) Compute an amortization schedule for the first 2 months.

37. How much more would the monthly payment be on the loan in Exercise 36 if the term were 15 years instead of 30? How much would be saved over the life of the loan?

Use the following table for Exercises 38–41.

Twitter, Inc. (TWTR)
NYSE - NYSE Delayed Price. Currency in USD

16.62 +0.27 (+1.65%)
At close: February 17 4:00PM EST

Previous Close	16.35	Market Cap	11.79B
Open	16.36	Beta	0.99
Bid	0.00 ×	PE Ratio (TTM)	−25.53
Ask	0.00 ×	EPS (TTM)	−0.65
Day's Range	16.36–16.62	Earnings Date	Apr 24, 2017–Apr 28, 2017
52 Week Range	13.73–25.25	Dividend & Yield	N/A (N/A)
Volume	14,378,181	Ex-Dividend Date	N/A
Avg. Volume	19,303,982	1y Target Est	14.32

38. What were the 52-week high and low prices of the stock?

39. Is this stock more or less volatile than the market as a whole? How can you tell? What does that mean?

40. What do analysts predict the stock price will be in a year?

41. Find the annual earnings per share of the stock.
42. A bond with face value $1,000, simple interest 5.5%, and term 8 years is sold for $1,250 with $5\frac{1}{2}$ years remaining until the maturity date. Find the amount of profit for the

original owner and the buyer (assuming that the buyer keeps the bond until its maturity date). Who do you think made the better investment? Why?

Projects

1. Compare the investments below to decide which you think is the best. Consider such things as total profit, length of time, and amount of money needed up front.
   (a) $20,000 placed into a savings account at 3.8% compounded monthly for 10 years.
   (b) A 10-year annuity that pays 4.5% interest with monthly payments of $170.
   (c) Buying 700 shares of stock at $14.30 per share; selling 200 shares 4 years later at $25.10, and the rest 3 years after that at $28.05. The brokerage charges 1% commission on both buying and selling.
   (d) Buying a $150,000 house with $20,000 down and financing the rest with a 15-year mortgage at 5%. Then selling the house and paying off the balance of the loan in 8 years at a selling price, after commission, of $192,000. (*Hint:* You will need to calculate the monthly payment, then compute unearned interest and payoff amount.)
2. You have $1,000 to invest. Investigate the advantages and disadvantages of each type of investment.
   (a) Checking account
   (b) Money market account
   (c) Passbook savings account
   (d) Certificate of deposit

   Write a paper indicating which type of account you have chosen and why you chose that account.
3. Time to play fantasy stock market. Everyone in your group gets $10,000 to invest in whatever stocks you like. Here are the parameters:
   • You can invest as much or as little of that $10,000 as you choose.
   • Assume that you'd be trading through an online broker with a flat commission fee of $19.95 on all trades.
   • You can choose the time period for the game, but all remaining stocks have to be sold at the closing value on the last day of the time period you set.
   • If any companies you choose paid a dividend in the last year, divide that dividend by 365 and multiply by the number of days you owned the stock to prorate the dividend. (Don't forget to multiply by the number of shares!)
   • If you feel the market conditions are bad, you're perfectly welcome to not buy into the market. In that case, you'll need to compute your final amount by finding the best rate you could get locally on a savings account and compute the amount of interest you'd make during the chosen time period.
   • To find the final value of your account at the end of the time period, calculate the proceeds from any sales on the last day, add any dividends or cash value left in your account, and subtract the commission fees.

• Ask your instructor to decide on what an appropriate reward would be for the winner.
4. In Section 7-5 we used simple interest on the amount borrowed to find the monthly payment on an auto loan, but in real life the amount you owe decreases as you pay down the loan, so calculating the monthly payment is more involved than that. Fortunately, there are a ton of online calculators out there to help.
   (a) Find an online calculator and use it to calculate payments for each car listed.

Car	Loan	Payments
Toyota Matrix	$12,000, 1.9%, 60 months	
Toyota Camry	$14,000, 4.75%, 36 months	
Toyota Sienna	$14,000, 1.9%, 2 years	
Mazda 929	$58,000, 3.9%, 6 years	
Mercedes S-class sedan	$92,000, 3.9%, 5 years	

   (b) Now let's explore a quick way to estimate a monthly payment for a situation where the Internet might not be available. At the beginning of your loan, you owe interest on the whole amount. At the end you owe nothing. So it's reasonable to guess that the average amount you owe interest on during the life of the loan is half of the principal balance. So we can estimate the payments with these steps:
   • Find the simple interest on half of the loan balance for the length of the loan.
   • Add that estimated interest to the principal balance.
   • Divide by the number of payments.
   Recalculate the payments for each loan using this estimation method.
   (c) Which loans did the method work best and worst for? What can you conclude?
   (d) Based on the work you've done, make a guess as to how accurately this estimation method would calculate the monthly payment for a 30-year home loan at 4.75% interest using a loan amount you choose. Explain how you made that guess. Then calculate the payment using one of our methods from Section 7-6, and this estimation method. How well did you guess?
5. Obviously, new cars cost more than used ones. But how much more? That question isn't as easy as you might think because used cars are more likely to require expensive maintenance and repairs. Pick a brand new car model that you might be interested in buying. (Make sure that this

model has been available for at least the last 4 years.) The website Edmunds.com is an excellent source of information on pricing and maintenance costs for new and used vehicles. Find what the price would be for the new model. Then do some research to find what the interest rate is in your area on a standard new car loan with term 5 years. Finally, use an online loan calculator to find the monthly payments for 60 months. Then repeat for the same model car, but one that's 4 years old and has average mileage for that age. (Note that the interest rates on used cars are almost always different than on new.) For both the new and used car purchase, use a 10% down payment. Finally, add in the maintenance and repair costs for each vehicle. Including the total amount paid over 5 years (based on the monthly payment), and total repair and maintenance costs, how do the prices compare?

Design elements: Front matter, Chapter Opener, Summary and End Matter header design (random numbers background illustration): ©pixeldreams.eu/Shutterstock RF

Photo: clintscholz/iStock/Getty Images

Outline

Math in | Travel

I love my house, I love my yard, I love my neighborhood, and I REALLY love my dogs. But I still get the urge to travel—a lot. Traveling to new cities, basking in new surroundings, experiences, cultures—there's just something magical about travel that makes you want to do it more and more once you catch the bug. When you start to travel outside of the States, you realize that a lot of things are different, among them the way that many things are measured.

To measure something means to assign a number that represents its size. In fact, measurement might be the most common use of numbers in everyday life. Numbers are used to measure heights, weights, distances, grades, weather, sizes of homes, capacities of bottles and cans, and much more. Even in our monetary system, we're using numbers to measure sizes: little ones, like the cost of a candy bar, and big ones, like an annual salary.

The one thing that every measurement has in common is that it's good for absolutely nothing unless there are units attached to the number. If a friend asks you how far you live from campus and you answer "three," at best they'll look at you funny and cautiously back away. An answer like that is meaningless because without units, there's no context to describe what the number represents. And that's entirely because there are many different units that can be used to describe similar measurements. That number 3 could mean 3 miles, kilometers, blocks, houses, even parsecs if you come from a galaxy far, far away.

In this unit, we'll study the different ways that things are measured in our world, with a special focus on the different units that are used. You probably have at least a passing familiarity with the metric system, since measurements in units like liters and kilometers are becoming more common. When you gather information of interest to you, you don't always get to choose the units that the information is provided in, so the ability to convert measurements from one unit to another is a particularly useful skill in the information age.

One obvious application of this skill comes into play when traveling outside the United States. Most countries use the metric system almost exclusively, meaning that to understand various bits of information, you need to be able to interpret measurements in an unfamiliar system. Suppose that you and your sweetie are planning a dream vacation to Australia. Who wouldn't want to go to a place where you might see a kangaroo hop across the road in front of you? Seriously, it happens all the time. The little guys are like squirrels over there. The following questions might be of interest to you on the trip. By the time you finish this unit, you'll be able to answer these and other questions, making you a more well-informed traveler.

1. A check of a tourist website informs you that the temperature is expected to range from 24° to 32° Celsius. How should you pack?

2. The airline that will fly you down under allows a maximum of 30 kilograms for each piece of luggage. If your empty suitcase weighs 4 pounds, how many pounds can you pack into it?

3. The resort you will be staying at is 4.2 kilometers from the airport and 0.6 kilometers from downtown. Can you walk downtown? And how long will it take to drive your rental car from the airport at 35 miles per hour?

4. The gas station closest to the airport caters to many tourists, so it has prices listed in both Australian and American dollars. Gas costs $1.43 per liter (in U.S. dollars). If you use 3/4 of the car's 16-gallon tank, how much will it cost to refill?

5. At the time of this writing, one Australian dollar is worth 0.72 U.S. dollars. If the resort quotes you a price of 289 dollars per night Australian, how much are you paying in American dollars?

For answers, see Math in Travel Revisited on page 477

| **Section 8-1** | **Measures of Length: Converting Units and the Metric System** |

✓ **LEARNING OBJECTIVES**

1. Convert measurements of length in the English system.
2. Convert measurements in the metric system.
3. Convert between English and metric units of length.

As I reflect back on the history of numeracy in my family (the human family, that is), counting up objects certainly came first. But I bet that second place goes to measuring the lengths of objects. In studying length, we're looking back into our distant past—the earliest units of measure. The fact that lengths can vary so widely describes why different units of measurement are necessary: while inches are perfectly good for measuring your jacket size, the Earth is about 5,892,000,000,000 inches from the sun: clearly, a larger unit of measure would be a better idea for such a huge distance. In learning how to convert distances from one unit to another, you'll learn a basic approach that can be used to convert any units of measure, and you'll study a variety of situations in our world where this skill comes in handy.

Pressmaster/Shutterstock

The units of length that are commonly used in the United States are the inch, foot, yard, and mile. These are part of the **English system of measurement**, which is very old. (See the Sidelight on page 452 for some historical perspective.)

A list of equivalences between units of measure is shown below.

Units of Length in the English System

12 inches (in.) = 1 foot (ft)

3 ft = 1 yard (yd)

5,280 ft = 1 mile (mi)

One way to use these conversions is multiplication or division. For example, there are 3 feet in 1 yard, so to convert 8 yards to feet, you multiply 8 by 3 to get 24. There are 12 inches in a foot, so to convert 564 inches to feet, you divide by 12 to get 47 feet. This works okay, but requires some careful thought to make sure you're not multiplying when you should divide or vice versa.

A better approach is provided by **dimensional analysis**, which is based on a fiendishly simple idea: if measurements in two different units represent the same actual length, like 1 yard and 3 feet, then a fraction formed from dividing those units is just a fancy way to say 1. For example,

$$\frac{1 \text{ yd}}{3 \text{ ft}} = 1 \quad \text{and} \quad \frac{3 \text{ ft}}{1 \text{ yd}} = 1$$

We call a fraction of this type a **conversion factor**, and multiplying any measurement by a conversion factor won't change the size of the measurement because we're just multiplying by 1.

| **Example 1** | **Writing and Using Conversion Factors** |

(a) There are 5,280 feet in 1 mile. Use this fact to write two conversion factors.
(b) Multiply each of the conversion factors you wrote by 2.3 miles. What are the units of the product in each case? What can you conclude?

SOLUTION

(a) Since 5,280 feet and 1 mile represent the exact same length, a fraction with one of them in the numerator and the other in the denominator is just another way to write the number 1. There are two ways to do this:

$$\frac{5{,}280 \text{ ft}}{1 \text{ mi}} \text{ and } \frac{1 \text{ mi}}{5{,}280 \text{ ft}}$$

Those are the two conversion factors.

(b) Let's look at the first requested multiplication: $2.3 \text{ mi} \times \frac{5{,}280 \text{ ft}}{1 \text{ mi}}$. Notice that we have the units of miles in the numerator, and then also in the denominator. Those units will divide out, and the remaining units will be feet. So we get

$$2.3 \text{ mi} \times \frac{5{,}280 \text{ ft}}{1 \text{ mi}} = 12{,}144 \text{ ft}$$

This tells us that there are 12,144 feet in 2.3 miles.

Now the second multiplication: $2.3 \text{ mi} \times \frac{1 \text{ mi}}{5{,}280 \text{ ft}}$. This time, the units of miles are both in the numerator, and nothing divides out. We get

$$2.3 \text{ mi} \times \frac{1 \text{ mi}}{5{,}280 \text{ ft}} \approx 0.000436 \text{ mi}^2/\text{ft}$$

This is a perfectly legitimate product: it just doesn't *mean* anything useful. Square miles per foot is not a length. So what can we conclude? That you can use multiplication by a conversion factor to convert units in some cases, but you have to be careful about which conversion factor you choose.

Try This One   1

(a) As you know, there are 12 inches in one foot. Write two conversion factors based on this fact.
(b) Multiply each of your conversion factors by 140 inches and describe which one of them provides a meaningful result.

Now here comes the most important sentence in this chapter, so don't blink: *Our clever procedure for converting units will be to multiply by one or more conversion factors in a way that the units we don't want will divide out and leave behind the units we do want.* Dimensional analysis is illustrated in Examples 2 through 5 but is used in most of the examples in this chapter.

Example 2   Converting Yards to Feet

Convert 6 yards to feet.

SOLUTION

The goal is to eliminate the units of yards and leave behind feet. So we would like to use a conversion factor that has yards in the denominator (to divide out yards) and feet in the numerator. This is the conversion factor $\frac{3 \text{ ft}}{1 \text{ yd}}$:

$$\frac{6 \text{ yd}}{1} \cdot \frac{3 \text{ ft}}{1 \text{ yd}} = 18 \text{ ft}$$

The yards divided out, leaving behind feet in the numerator.

Math Note

You might find it helpful to write the original measurement as a fraction with denominator 1 as in Example 2, but it isn't necessary.

Try This One   2

Convert 81 feet to yards.

Example 3	Converting Feet to Miles

The highest point on Earth is Mount Everest, at 29,035 feet. How many miles is that?

SOLUTION

This time, we want a unit fraction with feet in the denominator and miles in the numerator. We know that 1 mile = 5,280 feet, so the unit fraction we need is $\frac{1\,mi}{5,280\,ft}$.

$$\frac{29,035\ \cancel{ft}}{1} \cdot \frac{1\ mi}{5,280\ \cancel{ft}} = \frac{29,035\ mi}{5,280} \approx 5.5\ mi$$

Pixtal/age fotostock

Try This One   3

The deepest point in the oceans of Earth is the Mariana Trench, located south of Japan in the Pacific Ocean. Its depth is about 6.83 miles. How many feet is that?

Sometimes we will need to multiply by more than one unit fraction to make a conversion.

Example 4	Converting Measurements Using Two Unit Fractions

Math Note

Notice that we put the units we wanted to eliminate in the denominator, and the units we wanted to keep in the numerator. If you understand this simple idea, dimensional analysis will be a snap for you no matter how complicated the conversions might look.

The distance by air from JFK airport to LaGuardia airport in New York is about 11 miles. How many inches is that?

SOLUTION

Looking at the colored box at the beginning of this section, we don't have an equivalence between inches and miles. But we can go from miles to feet (using $\frac{5,280\,ft}{1\,mi}$) then from feet to inches (using $\frac{12\,in.}{1\,ft}$):

$$\frac{11\ \cancel{mi}}{1} \cdot \frac{5,280\ \cancel{ft}}{1\ \cancel{mi}} \cdot \frac{12\ in.}{1\ \cancel{ft}} = 11 \times 5,280 \times 12\ in. = 696,960\ in.$$

Sidelight   **A Brief History of Measuring Lengths**

Nobody knows how and when human beings started measuring physical objects, but historians seem to be in general agreement that the earliest recorded measurements were lengths. In particular, they were based on parts of the human body. If you search the Internet for early units of measurement, the most common one you will find is the **cubit**, first used by the Egyptians around 5,000 years ago. The cubit was equal to the distance from a person's elbow to the outstretched middle finger. (If you try to measure something that way, make sure to stretch out your other fingers as well. That's funny if you think about it.) Other early units were the **palm** (the distance across the base of a person's four fingers) and the **digit** (the thickness of a person's middle finger).

This leads us to believe that one of two things was true: either everyone was exactly the same size in ancient Egypt, or their units of measurement were pretty imprecise. Things got a little more specific sometime around the year 1100 CE, when King Henry I of England declared that the distance

from the tip of his nose to the end of his thumb would be known as a **yard**. (At least this distance was based on one specific person's physical size.) The oldest known yardstick still in existence, believed to have been made in 1445, is accurate to a modern yard within about $\frac{1}{300}$th of an inch.

A couple hundred years later, King Edward I declared that one-third of a yard should be called a **foot**. In 1595, Queen Elizabeth changed the length of a mile from the Roman tradition of 5,000 feet to the current 5,280 feet so that it was exactly 8 furlongs. (Perhaps she was a big fan of horse racing, which still uses furlongs.) Eventually, the English system of measurement became standard throughout the world. Of course, these units are still used in the United States today. The real irony? England has been officially using the metric system since 1965, which would be sort of like England deciding not to speak English anymore. Go figure. In fact, only three countries in the world have not officially adopted the metric system, which we'll study later in this section: the United States, Liberia, and Myanmar.

Try This One   4

A football field is 100 yards long. How many inches is that?

Dimensional analysis can also be used to convert more complicated units, like those for speed, as shown in Example 5.

Example 5 | **Converting Units for Speed**

Richard Wear/Getty Images

A decent Major League pitcher can throw a baseball 90 miles per hour. How fast is that in feet per second?

SOLUTION

Notice that we can write 90 miles per hour as a fraction: $\frac{90 \text{ mi}}{1 \text{ h}}$. Now we can multiply by conversion factors that convert miles to feet $\left(\frac{5,280 \text{ ft}}{1 \text{ mi}}\right)$, hours to minutes $\left(\frac{1 \text{ h}}{60 \text{ min}}\right)$, and minutes to seconds $\left(\frac{1 \text{ min}}{60 \text{ sec}}\right)$.

$$\frac{90 \text{ mi}}{1 \text{ h}} \cdot \frac{5,280 \text{ ft}}{1 \text{ mi}} \cdot \frac{1 \text{ h}}{60 \text{ min}} \cdot \frac{1 \text{ min}}{60 \text{ sec}} = \frac{90 \times 5,280 \text{ ft}}{60 \times 60 \text{ sec}} = 132 \frac{\text{ft}}{\text{sec}}$$

CAUTION

In this case, we wanted seconds in the denominator, so we had to be extra careful to put the units we wanted to eliminate in the numerator and the ones we wanted to keep in the denominator.

Try This One   5

1. Convert measurements of length in the English system.

The speed of sound at sea level and room temperature is about 1,126 feet per second. How fast is that in miles per hour?

The Metric System

If you grew up in the United States, you're probably perfectly happy with the English system of measurement. You've been using it your whole life, after all. But the truth is that it's not that great of a system because there's no rhyme or reason to it. Why is 12 inches a foot? Why is 5,280 feet a mile? They just are. Somebody decided, and that was that. The metric system, on the other hand, makes a bit more sense because the different units of measure are all related by powers of 10. We'll see exactly what that means after introducing the base units. The metric system uses three basic units of measure: the *meter,* the *liter,* and the *gram.*

Length in the metric system is measured using **meters**. One meter is a bit more than a yard, about 39.4 inches. The symbol for meter is "m."

The basic unit for capacity in the metric system is the **liter**, which is a little bit more than a quart. (You're probably familiar with 2-liter bottles of soda—that's close to a half gallon.) The symbol for liter is "L."

The basic unit for measuring weight in the metric system is the **gram**. This is a very small unit—a nickel weighs about 5 grams. The symbol for gram is "g."

Now back to other units in the metric system. All units other than the three basic units are based on multiples of the basic units, with those multiples all being powers of 10. For example, a standard unit of weight for people in the metric system is the kilogram: this is equal to 1,000 grams. A commonly used unit of length for smaller measurements is the centimeter, which is $\frac{1}{100}$ meter. It's the prefix in front of the basic unit that determines the size. The most common metric prefixes are listed in Table 8-1.

Math Note

Technically speaking, the gram is a unit of mass, not weight, and the two are slightly different. The weight of an object is the measure of the gravitational pull on it, so your weight can vary depending on where you are. (See Sidelight on the next page.) Mass is a constant measure of the physical size of an object regardless of location.

Since we all live on the Earth, in this book we'll consider mass and weight to be synonymous. If you don't live on Earth, my apologies, and take me to your leader.

Table 8-1	Metric Prefixes

Math Note

Notice that the prefixes for each of the three most common metric units that are *smaller* than the basic unit all end in the letter *i*: deci-, centi-, and milli-.

Prefix	Symbol	Meaning	
kilo	k	1,000 units	
hecto	h	100 units	*Bigger than basic unit*
deka	da	10 units	
	m, L, g	1 unit	
deci	d	$\frac{1}{10}$ of a unit	
centi	c	$\frac{1}{100}$ of a unit	*Smaller than basic unit*
milli	m	$\frac{1}{1,000}$ of a unit	

The relationship between units in the metric system is a lot like the units used in our monetary system. Table 8-2 demonstrates this comparison for the base unit of meter.

Table 8-2	Relationship Between Metric Units and U.S. Currency

Metric Unit	Meaning	Monetary Comparison
kilometer	1,000 meters	$1,000
hectometer	100 meters	100
dekameter	10 meters	10
meter (base unit)	1 meter	dollar (base unit)
decimeter	0.1 meter	0.10 (10 cents)
centimeter	0.01 meter	0.01 (1 cent)
millimeter	0.001 meter	0.001 (1/10 cent)

Sidelight   The Lunar Weight-Loss Plan

Since weight is a measure of gravitational pull, your weight is less in locations where gravity is not as strong. The effect is negligible pretty much anywhere on Earth, although you would weigh about 0.3% less at the top of Mt. Everest than you would at sea level. But once you leave Earth, weight can vary widely. The force of gravity on the moon, for example, is about one-sixth of what it is on Earth. So by heading to the moon, you would instantly shed five-sixths of your weight! Of course, you would still be the same size (i.e., your mass won't change), but the scale would seem like a much friendlier place to most of us.

NASA

There are several ways to do conversions in the metric system. The easiest way is to multiply or divide by powers of 10. Here are the basic rules:

Conversion When You Use the Metric System

To change a larger unit to a smaller unit in the metric system, multiply by 10^n, where n is the number of steps that you move down in Table 8-3.

To change a smaller unit to a larger unit in the metric system, divide by 10^n, where n is the number of steps that you move up in Table 8-3.

Table 8-3	Metric Unit Conversions

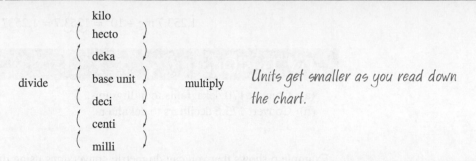

divide $\left\{\begin{array}{l} \text{kilo} \\ \text{hecto} \\ \text{deka} \\ \text{base unit} \\ \text{deci} \\ \text{centi} \\ \text{milli} \end{array}\right\}$ multiply

Units get smaller as you read down the chart.

Let's see why these rules for converting within the metric system make perfect sense.

Example 6	Using Dimensional Analysis to Develop Metric Conversions

Based on Table 8-2, 1 meter is the same as 100 centimeters. Use dimensional analysis to convert 455 cm to meters. Then describe the arithmetic operation that you did.

SOLUTION

If we start with 455 cm, we'll want cm in the denominator and m in the numerator:

$$455 \text{ cm} \times \frac{1 \text{ m}}{100 \text{ cm}} = 4.55 \text{ m}$$

Arithmetically what we did was to divide by 100. This matches the procedure in the colored box: centi- is two steps below the base unit, so the box tells us to divide by 10^2.

Try This One 6

Use dimensional analysis to convert 3.8 km to meters, then describe what you did arithmetically.

Example 7	Converting Metric Units

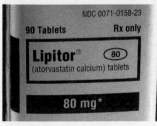

Jill Braaten/McGraw Hill

Almost all solid medications are measured in milligrams.

(a) Convert 42.5 kilometers to centimeters.
(b) Convert 1,253.7 milligrams to grams.

SOLUTION

(a) The prefix "kilo" is higher in Table 8-3 than "centi," so we're converting to smaller units, which means we need to multiply. There are five steps from "kilo" to "centi" in the table, so the power of 10 we multiply by is 10^5. This can be quickly accomplished by moving the decimal point five places to the right:

$$42.5 \text{ km} \times 10^5 = 42.50000 = 4,250,000 \text{ cm}$$

(b) The gram is the base unit of weight, and the prefix "milli" is below the base unit in the table, so we are converting to a larger unit and must divide. There are three steps from "milli" to the base unit in the table, so we divide by 10^3. This can be quickly done by moving the decimal point three places to the left:

$$1,253.7 \text{ mg} \div 10^3 = 1253.7 = 1.2537 \text{ grams}$$

Try This One   7

(a) Convert 170 dekagrams to milligrams.
(b) Convert 173.6 deciliters to dekaliters.

Example 6 shows that you can do metric conversions using dimensional analysis if you want to, but it's quite a bit quicker to learn how to move the decimal point, like we did in Example 7.

Because most of the world uses the metric system while the United States uses the English system, it's often necessary to convert between the two systems. We'll use dimensional analysis, along with the table below, which provides unit equivalences for units of length.

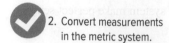

2. Convert measurements in the metric system.

David Sobecki/McGraw Hill

Many road signs now list mileages in both miles and kilometers.

English and Metric Equivalents for Length

1 in. = 2.54 cm	1 mm ≈ 0.03937 in.
1 ft = 30.48 cm = 0.3048 m	1 cm ≈ 0.3937 in.
1 yd = 91.44 cm = 0.9144 m	1 m ≈ 1.0936 yd = 3.2808 ft
1 mi ≈ 1.61 km	1 km ≈ 0.6214 mi

Looking at a ruler that has both inches and centimeters on it is a good way to get a feel for the connection between them. Notice that 2 inches is just about 5 centimeters.

Example 8   Converting Between English and Metric Lengths

Convert each.

(a) 135 feet to centimeters
(b) 87 centimeters to inches
(c) 213.36 millimeters to feet

SOLUTION

(a) To convert feet to centimeters use the unit fraction $\frac{30.48 \text{ cm}}{1 \text{ ft}}$.

$$135 \text{ ft} \cdot \frac{30.48 \text{ cm}}{1 \text{ ft}} = 135 \cdot 30.48 \text{ cm} = 4,114.8 \text{ cm}$$

(b) To convert centimeters to inches use the unit fraction $\frac{1 \text{ in.}}{2.54 \text{ cm}}$.

$$87 \text{ cm} \cdot \frac{1 \text{ in.}}{2.54 \text{ cm}} = \frac{87 \text{ in.}}{2.54} = 34.25 \text{ in. (rounded)}$$

Math Note

Because some of the English-metric conversions are approximations due to rounding, the converted measurements are often approximations as well. This accounts for the fact that the two methods shown for part (c) lead to slightly different results.

(c) To convert from millimeters to feet, we first convert millimeters to centimeters using the unit fraction $\frac{1\,cm}{10\,mm}$, then convert centimeters to feet using the unit fraction $\frac{1\,ft}{30.48\,cm}$.

$$213.36\ \text{mm} \cdot \frac{1\ \text{cm}}{10\ \text{mm}} \cdot \frac{1\ \text{ft}}{30.48\ \text{cm}} = \frac{213.36\ \text{ft}}{(10 \cdot 30.48)} = 0.7\ \text{ft}$$

Alternatively, we could convert millimeters to inches, then inches to feet:

$$213.36\ \text{mm} \cdot \frac{0.03937\ \text{in.}}{1\ \text{mm}} \cdot \frac{1\ \text{ft}}{12\ \text{in.}} \approx \frac{213.36 \cdot 0.0397}{12}\ \text{ft} \approx 0.7\ \text{ft}$$

The results are close, but not exactly the same. The first worked out to be exactly 0.7, while the second came out to 0.6999986. I think we can all agree that's pretty darn close to 0.7.

Try This One   8

3. Convert between English and metric units of length.

Convert each of the following. Round to two decimal places.

(a) 237 feet to meters   (b) 128 cm to inches   (c) 23,342 mm to feet

---

Sidelight   The (Really) Long and Short of It

Table 8-1 introduces the most common metric unit prefixes, but there are some others that are used to describe measurements that are either really big or really small. In biology and physics, for example, microscopic measurements are common for things like parts of cells, molecules, and atoms. The average cell in a human body is about a hundred-thousandth of a meter; a convenient measurement for this is 10 micrometers. The prefix "micro" represents 10^{-6} or one millionth. A nanometer is a thousand times smaller: "nano" represents 10^{-9} or one billionth. A single molecule of glucose is about one nanometer in width. Divide by a thousand again, and you get a picometer, with "pico" representing 10^{-12}, or a trillionth. The radius of a hydrogen atom is about 120 picometers.

On the larger side, we have "mega" (one million), "giga" (one billion), and "tera" (one trillion), prefixes you're probably familiar with from the computer world. A byte is a string of eight digits used to transmit information. A megabyte (MB) of data is a million bytes: a typical MP3 file is around 4 MB. A gigabyte (GB) is a thousand megabytes, making it a billion bytes. The DVDs that most movies come on have between 4 and 8 GB of storage. A terabyte (TB) is a thousand gigabytes, making it a trillion bytes. Hard drives with capacities in the terabytes are now pretty common, although by the time you read this, we might be measuring in petabytes, which is the next leap up by a factor of a thousand.

An interesting fact: anyone familiar with computers or smartphones is comfortable with the term gigabytes. But when *Back to the Future* was made (1985), giga- was so uncommon as a prefix that it's consistently mispronounced in the movie (jiga, rather than giga) and NOT ONE SINGLE PERSON involved in production of the movie knew to correct the cast.

Answers to Try This One

1 (a) $\frac{12\,\text{in.}}{1\,\text{ft}}$ and $\frac{1\,\text{ft}}{12\,\text{in.}}$.

    (b) $\frac{1,680\,\text{in.}^2}{\text{ft}}$ and 11.7 ft; only 11.7 ft has meaningful units

2 27 yd

3 36,062.4 ft

4 3,600 in.

5 About 767.7 mi/h

6 3,800 m; multiply by 1,000

7 (a) 1,700,000 mg
    (b) 1.736 daL

8 (a) 72.24 m
    (b) 50.39 in.
    (c) 76.58 ft

Exercise Set   8-1

Writing Exercises

1. What are the basic units of length in the English measurement system, and how are they related?
2. What are the basic units of length, capacity, and weight in the metric system?
3. What importance do prefixes play in different units of measure in the metric system?
4. What is a conversion factor in measurement? What are they used for?
5. Describe how to use dimensional analysis to convert units of measure.
6. Describe a good rule of thumb for relating inches and centimeters.
7. Why is using dimensional analysis better for converting units than just multiplying or dividing by a number?
8. Describe the rules for converting units within the metric system.

Computational Exercises

For Exercises 9–20, convert each using dimensional analysis. Round to 2 decimal places if necessary.

9. 12 feet to yards
10. 36 inches to feet
11. 17 feet to yards
12. 10,345 feet to miles
13. $13\frac{1}{4}$ yards to feet
14. 0.35 mile to yards

15. 21 inches to feet
16. $47\frac{1}{2}$ yards to feet
17. 18 inches to yards
18. 5,237,834 inches to miles
19. 876 inches to yards
20. 182 feet to yards

For Exercises 21–60, convert each measurement to the specified unit.

21. 8 m = _____ cm
22. 0.25 m = _____ dm
23. 12 dam = _____ m
24. 24 hm = _____ dam
25. 0.6 km = _____ hm
26. 30 cm = _____ dm
27. 90 m = _____ dm
28. 18,426 mm = _____ m
29. 375.6 cm = _____ m
30. 63 m = _____ km
31. 405.3 m = _____ km
32. 0.0034 dam = _____ hm
33. 0.0237 km = _____ cm
34. 50,000 cm = _____ km
35. 0.00056 km = _____ mm

36. 650,000 mm = _____ km
37. $12\frac{3}{4}$ km = _____ dm
38. $39\frac{2}{5}$ m = _____ cm
39. 8 hm = _____ km
40. 3,256.4 dm = _____ cm
41. 5 meters = _____ inches
42. 14 yards = _____ meters
43. 16 inches = _____ millimeters
44. 50 meters = _____ yards
45. 235 feet = _____ dekameter
46. 563 decimeters = _____ inches
47. 1,350 meters = _____ feet
48. 4,375 dekameters = _____ feet
49. $19\frac{2}{3}$ inches = _____ millimeters
50. 256 kilometers = _____ miles
51. 0.06 hectometer = _____ feet
52. $143\frac{3}{4}$ inches = _____ centimeters
53. $56\frac{1}{2}$ feet = _____ decimeters
54. 44,000 millimeters = _____ yards
55. 2.35 kilometers = _____ miles
56. 837 miles = _____ kilometers
57. 42 decimeters = _____ yards
58. $254\frac{3}{8}$ centimeters = _____ inches
59. 333 inches = _____ meters
60. 1,256 kilometers = _____ inches

Applications in Our World

61. Steve sprinted the last $25\frac{1}{2}$ yards of a race. How many feet did he sprint? How many meters?
62. A professional basketball court is 94 feet long. How many inches is that? How many meters?
63. On a baseball diamond, the bases are 90 feet apart. How many yards does a batter run in reaching second base? How many meters?
64. A football player completed a 45-yard pass. How many inches did he throw the ball? How many centimeters?
65. A traffic reporter announced a backup on I-95 of 1,500 yards. How far is that in miles? In kilometers?
66. A couple wants to buy a new queen-sized mattress. They only have an area that measures 5 ft × 6 ft to put the mattress. The mattress size is 60 in. × 80 in. Will the mattress fit? Convert the measurements to feet to determine this.
67. Celia used a tape measure to section off areas for a vegetable garden. The section for tomatoes measures 126 inches long. How many yards long is the section for Celia's tomatoes?
68. What are the dimensions of a 6 dm by 5 dm piece of poster board in centimeters? In millimeters?
69. To make a batch of an energy drink, Jantel used 2.3 grams of citric acid. How many milligrams of citric acid is that?

70. If a millipede has 1,000 legs, how many legs would a decipede have?

71. Sanford's gym coach made him run 20 laps on a 700-m track. How many kilometers did he run? How many miles?

72. How many miles will a runner cover in a 10-km race?

73. Raheem drives 65 miles per hour on a freeway in Detroit. When he crosses into Windsor, Ontario, Raheem resumes the same speed. How fast is that in kilometers per hour?

74. Mary brings a 20-inch by 24-inch picture to the frame shop, but the frame she wants only lists measurements in terms of centimeters. What are the measurements of her picture in centimeters?

75. A bowling lane from the foul line to the pins is 62 feet $10\frac{3}{16}$ inches long. How many meters is a bowling lane?

76. A 5-carat round-cut diamond measures half an inch in diameter. How many millimeters is the diameter of the diamond?

77. The new Mars Pathfinder postage stamp measures 1.5×3 inches. What are the stamp's measurements in millimeters?

78. Light travels at 186,283 miles every second. How many feet per hour does light travel?

79. In April of 2015, a magnetic levitation train in Japan broke the speed record for a train, reaching a top speed of 603 kilometers per hour. How many miles can that baby go in an hour? How many feet in a minute?

80. Here's a fact that can blow your mind if you're not careful. As you read these words, you're living on a projectile (we call it Earth) that is traveling through space at close to 30 kilometers per second. Okay, so maybe that didn't blow your mind. But it will if you convert to miles per hour, so go for it.

81. The Earth is on average 92,900,000 miles from the sun. In July, the distance is 94,400,000 miles, the farthest of any month. What is the distance in feet of the difference of the distance in July from the average?

82. To build a loft in his dorm room, Jay got 8-foot planks from the lumber yard. To get the right size, he had to cut off a quarter of each plank. How many inches did he cut off each?

83. A college is building a new basketball court that is 94 feet long and 50 feet wide, but the wood for the court comes only in 10-meter planks that are 8 centimeters wide. How many planks are needed to surface the court? (*Hint:* Round each measurement up to estimate the number of planks needed.)

84. If three servings of ice cream have 20 grams of fat, how many milligrams of fat does one serving have?

85. A popular pain-relieving drug has 200 mg of its active ingredient in one capsule. If a bottle has 100 capsules, how many grams of the active ingredient are in the bottle?

86. The campus fitness center is installing an Olympic-sized swimming pool that measures 164 ft × 82 ft. The builders want to put 6-inch tiles around the border. How many tiles will they need? (*Hint:* Don't forget the corners!)

87. If one U.S. dollar was worth 1.22 Canadian dollars when Sal went to a debate tournament in Toronto, and he had $400 in U.S. currency as spending money, how much was his money worth in Canadian dollars?

88. While in Chile, where $1 in U.S. currency was worth 610.045 in Chilean pesos, you won 10,000,000 Chilean pesos. Did you plan to eat a nice dinner out, buy a new car, or buy a new house? Explain your answer.

89. On her summer abroad in France, Jane bought a pair of shoes for 54.23 euros. The store owner only had francs to give her as change. She gave him 60 euros. How much did he give her back in francs? At the time, 1 euro was worth 6.55957 francs.

90. In January of 2017, one dollar could buy 21.96 Mexican pesos. In January of 2021, that same dollar could buy 19.66 Mexican pesos. If you paid 1,200 pesos per night at a hotel in Cancun in January 2017, and your friend paid the same rate 4 years later, how much less per night in dollars did you spend?

Critical Thinking

In Problems 91–98, decide which of millimeters, centimeters, meters, or kilometers would be most appropriate for measuring the given length or distance.

91. The height of a polar bear

92. The altitude of a commercial plane in flight

93. The font size in a Word document

94. The length from your bedroom to your first class

95. The outside length of a size 8 shoe

96. The length of a marathon

97. The size of a bacterial colony in a medical sample

98. The size of a prison cell

In Problems 99–108, decide what units in both the English and metric systems would be a reasonable choice for describing each quantity.

99. The height of a tree

100. A cross-country race

101. The amount of snowfall during a big storm in Denver

102. The length of a city block

103. The width of a brain tumor

104. The distance from your seat in a classroom to the door

105. The height of an office building in Manhattan

106. The diameter of a contact lens

107. The height of the tallest player on your school's basketball team

108. The width of a swimming pool

109. Suppose that four world-class sprinters compare times after running different distances. The first ran the 100-yard dash in 9.12 seconds; the second ran the 100-meter dash in 9.72 seconds; the third ran the 220-yard dash

in 19.75 seconds; the fourth ran the 200-meter dash in 19.71 seconds. Which is the most impressive feat? Explain your reasoning.

110. What units would 7,000 cm have to be converted to so that the result has no zeros?

111. What units would 0.006 m have to be converted to so that the result has no zeros?

112. Using Technology. We developed a method for converting units within the metric system by multiplying or dividing by an appropriate power of 10. These

divisions can be programmed into a spreadsheet that will use formulas to perform the conversions. You should be able to enter any length in meters and have the sheet automatically convert it to mm, cm, dm, dam, hm, and km. A template to help you get started can be found in the online resources for this lesson.

113. Using Technology. Actually, why quit there? You can also program in formulas that will convert that length in meters to inches, feet, yards, and miles as well. Same template, more units!

Section 8-2 | Measures of Area, Volume, and Capacity

LEARNING OBJECTIVES

1. Convert units of area in the English system.

2. Convert measurements between English and metric units of area.

3. Convert measurements of volume in the English system.

4. Convert units of volume in the metric system.

5. Convert measurements between English and metric units of volume.

The units of length we studied in Section 8-1 are useful for measuring many things, but they have a major limitation: length is a one-dimensional measurement, and we live in a three-dimensional world. It's equally important to be able to measure in two dimensions, like the size of a lot for a new home, and in three dimensions, like the capacity of a swimming pool. So in this section, we will study area (two dimensions) and volume and capacity (both three dimensions).

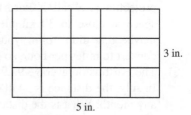

jpgfactory/Getty Images

Conversions of Area

Units of length are also called **linear units**, because they are used to measure along a line. The units used to measure the size of an object in two dimensions are called **square units**. They are based on a very simple idea: one square unit is defined to be the area of a square that is one unit long on each side. For example, a square inch is the area of a square that is 1 inch on each side, as shown in Figure 8-1.

The area of other figures is then defined to be the number of square units that will fit inside the figure. For example, the rectangle in Figure 8-2 is 3 inches by 5 inches, and we can see that 15 one-inch by one-inch squares will fit inside. So its area is 15 square inches, which we abbreviate as 15 in².

Finding the area of various geometric shapes will be covered in Chapter 9. In this section we'll focus on conversions between different units of area.

Figure 8-1 One square inch

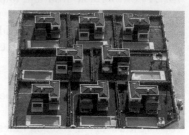

Figure 8-2 15 in²

Example 1 | Developing Conversions for Square Units

The goal is to convert 1 square foot into square inches. If you guessed that 1 ft² = 12 in², you're not alone. You're also not correct. Convert 1 ft² to square inches by multiplying by an appropriate conversion factor twice. What can you conclude?

SOLUTION

$$1 \text{ ft}^2 \cdot \frac{12 \text{ in.}}{1 \text{ ft}} \cdot \frac{12 \text{ in.}}{1 \text{ ft}} = 144 \text{ in}^2$$

Notice that there are two factors of feet in the denominator, which is exactly what we need to divide out the square feet we started with. The resulting units are square inches. And the conversion we get between square feet and square inches (1 square foot = 144 square inches) is the square of the conversion factor between linear feet and inches.

Math Note

The multiplication in the numerator of Example 1 shows why in² is a great symbol for representing square inches.

The conclusion we can draw is that if we know a linear conversion factor, we can square the number that isn't 1 to get a related conversion factor for square units.

Try This One   1

Make a conjecture about a conversion factor between square yards and square feet. Then use dimensional analysis to prove that your conjecture is correct.

Example 1 shows that the conversion factors for square units will be the squares of the conversion factors for linear units. In case you're more of a visual learner, Figure 8-3 illustrates graphically why 1 square foot is 144 square inches. The conversions for area are summarized in Table 8-4.

Table 8-4   English and Metric Equivalents for Area

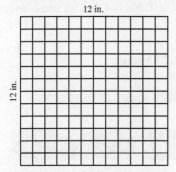

English Conversions for Area	English/Metric Conversions for Area
$1 \text{ ft}^2 = 144 \text{ in}^2$	$1 \text{ in}^2 \approx 6.45 \text{ cm}^2$
$1 \text{ yd}^2 = 9 \text{ ft}^2 = 1{,}296 \text{ in}^2$	$1 \text{ ft}^2 \approx 0.093 \text{ m}^2$
$1 \text{ mi}^2 = 27{,}878{,}400 \text{ ft}^2 = 3{,}097{,}600 \text{ yd}^2$	$1 \text{ yd}^2 \approx 0.836 \text{ m}^2$
$1 \text{ acre} = 43{,}560 \text{ ft}^2$	$1 \text{ mi}^2 \approx 2.59 \text{ km}^2$
$1 \text{ mi}^2 = 640 \text{ acres}$	$1 \text{ acre} \approx 4{,}047 \text{ m}^2$

Figure 8-3 1 square foot is made up of 144 1-square-inch boxes.

Notice that some of the conversions in Table 8-4 are exact, while others are approximate. Just as we did with lengths in Section 8-1, we'll use dimensional analysis in Examples 2 through 6 to convert units of area.

Example 2   Converting Square Inches to Square Feet

Jeff Henry/Corbis

A 52-in. wide-screen TV has a screen area of 1,064 in². How many square feet is that?

SOLUTION

We want to replace in² with ft², so we multiply by the conversion factor $\frac{1 \text{ ft}^2}{144 \text{ in}^2}$.

$$\frac{1{,}064 \text{ in}^2}{1} \cdot \frac{1 \text{ ft}^2}{144 \text{ in}^2} \approx 7.39 \text{ ft}^2$$

Try This One   2

A 32-in. wide-screen TV has screen area 3.05 ft². How many square inches is that?

Example 3   Converting Acres to Square Miles

Yellowstone National Park has an area of 2,219,789 acres. How large is it in square miles?

SOLUTION

From Table 8-4, the conversion factor we need is $\frac{1 \text{ mi}^2}{640 \text{ acres}}$.

$$\frac{2,219,789 \text{ acres}}{1} \cdot \frac{1 \text{ mi}^2}{640 \text{ acres}} \approx 3,468.42 \text{ mi}^2$$

Try This One   3

The state of Delaware has an area of 2,490 square miles. How many acres is that?

Example 4   Converting Square Yards to Acres

Math Note

The acre is a very old measure of area, traditionally thought to be the area that could be plowed by one person behind one ox in one day.

An average of 2,191,781 square yards of carpet is discarded into landfills every day in the United States. How many acres would that carpet cover?

SOLUTION

There is no direct conversion between square yards and acres in Table 8-4, but we can convert to square feet (using $\frac{9 \text{ ft}^2}{1 \text{ yd}^2}$) and then to acres (using $\frac{1 \text{ acre}}{43,560 \text{ ft}^2}$).

$$\frac{2,191,781 \text{ yd}^2}{1} \cdot \frac{9 \text{ ft}^2}{1 \text{ yd}^2} \cdot \frac{1 \text{ acre}}{43,560 \text{ ft}^2} \approx 452.85 \text{ acres}$$

We could have also converted square yards to square miles, then square miles to acres.

Try This One   4

An American football field is about 1.32 acres, counting the end zones. Promoters are planning to unfurl a giant flag covering the entire field at a playoff game. How many square yards of fabric will be needed?

1. Convert units of area in the English system.

Because both the metric and English systems are widely used today, it's useful to be able to convert between English and metric units of area.

Example 5   Converting Area from English to Metric Units

Ryan McVay/Getty Images

The plans for building a deck call for 28 square meters of deck boards. John has already bought 300 square feet. Does he have enough?

SOLUTION

John needs to convert 28 square meters to square feet to see how much wood is required. This is done using the unit fraction $\frac{1 \text{ ft}^2}{0.093 \text{ m}^2}$.

$$28 \text{ m}^2 \cdot \frac{1 \text{ ft}^2}{0.093 \text{ m}^2} \approx 301.1 \text{ ft}^2$$

The 300 ft^2 that John bought is not quite enough.

Try This One   5

Fussy University prohibits any displays on bulletin boards that have an area greater than 300 square inches. Does a poster with an area of 1,700 square centimeters meet this restriction?

Example 6 | **Converting Area from Metric to English Units**

Prince Edward County in Ontario, Canada, has an area of 1,050.1 square kilometers. Is it larger or smaller than Butler County, Ohio, which is 470 square miles?

SOLUTION

To find how many square miles in 1,050.1 square kilometers, multiply by the unit fraction $\frac{1 \text{ mi}^2}{2.59 \text{ km}^2}$.

$$1{,}050.1 \text{ km}^2 \cdot \frac{1 \text{ mi}^2}{2.59 \text{ km}^2} \approx 405.4 \text{ mi}^2$$

Butler County is about 64.6 square miles larger.

Try This One   6

The campus of the University of Guelph in Ontario has an area of 4.1 km^2. How many square miles is that?

The next example shows a very common and very useful application of area conversions for homeowners.

Example 7 | **Finding the Cost of Carpeting a Room**

Pete and Chasten plan to have new carpet installed in their living room. They measure and find that the room is 300 ft^2. The carpet they want sells for $22 per square yard installed. How much will it cost to have the room carpeted?

SOLUTION

We first convert square feet to square yards, then use the unit fraction $\frac{\$22}{1 \text{ yd}^2}$ to find the cost.

$$300 \text{ ft}^2 \cdot \frac{1 \text{ yd}^2}{9 \text{ ft}^2} \cdot \frac{\$22}{1 \text{ yd}^2} \approx \$733.33$$

Try This One   7

2. Convert measurements between English and metric units of area.

Find the cost of carpeting Pete and Chasten's basement, which is 520 square feet, with cheaper carpet costing $14.50 per square yard.

Conversions of Volume and Capacity

Just as the measurements of area are based on the two-dimensional square, measurements of volume are based on the three-dimensional analog, the cube. So volume is measured in cubic units.

In the English system, this is typically cubic inches (in^3), cubic feet (ft^3), and cubic yards (yd^3). For example, 1 cubic foot consists of a cube whose measure is 12 inches on each side. Recall that to find the volume of a cube, we multiply the length times the width times the height. Since all these measures are the same for a cube, 1 cubic foot = 12 inches × 12 inches × 12 inches or 1,728 cubic inches. Also, 1 cubic yard = 3 feet × 3 feet × 3 feet or 27 cubic feet, as shown in Figure 8-4.

Measures of volume also include measures of capacity. The capacity of a container is equal to the amount of fluid the container can hold. In the English system, capacity is measured in fluid ounces, pints, quarts, and gallons. The unit equivalences for capacity are shown in Table 8-5.

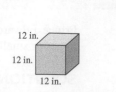

1 cubic foot = 1,728 cubic inches

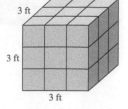

1 cubic yard = 27 cubic feet

Figure 8-4

Table 8-5	**Units of Capacity in the English System**

1 pint (pt) = 16 fluid ounces (oz)

1 quart (qt) = 2 pints (pt)

1 gallon (gal) = 4 quarts (qt)

Since volume and capacity are both measures of size in three dimensions, we can compare the units.

A cubic foot (ft^3) of water is about 7.48 gallons, and a cubic yard of water is about 202 gallons. If a gallon of water is poured into a container, it would take up about 231 cubic inches. Finally, a cubic foot of fresh water weighs about 62.5 pounds and a cubic foot of salt water weighs about 64 pounds. These measures are summarized in Table 8-6.

Table 8-6	**Unit Equivalences for Capacity in the English System**

1 cubic foot ≈ 7.48 gal

1 cubic yard ≈ 202 gal

1 gal. ≈ 231 in^3

1 ft^3 fresh water ≈ 62.5 lb

1 ft^3 salt water ≈ 64 lb

Example 8	**Converting Cubic Feet to Gallons**

Erica Simone Leeds

If a water tank at an aquarium has a volume of 3,000 cubic feet, how many gallons of water will the tank hold? If the tank contains salt water, how much does the water weigh?

SOLUTION

From Table 8-6, 1 cubic foot = 7.48 gallons.
  Use dimensional analysis as shown.

$$3{,}000 \ \text{ft}^3 \cdot \frac{7.48 \ \text{gal}}{1 \ \text{ft}^3} = 22{,}440 \ \text{gallons}$$

Now we use the fact that 1 ft^3 of salt water weighs 64 pounds. This gives us a conversion factor of $\frac{64 \ \text{lb}}{1 \ \text{ft}^3}$.

$$3{,}000 \ \text{ft}^3 \cdot \frac{64 \ \text{lb}}{1 \ \text{ft}^3} = 192{,}000 \ \text{lb}$$

Try This One   8

How many cubic feet of water are in a 100-gallon fresh water fish tank? How much does the water weigh?

Example 9	Computing the Weight of Water in Bottles

The team managers for a college football team carry quart bottles of water for the players to drink during breaks at practice. The bottles are stored in carrying racks that weigh 6 pounds and hold 16 bottles. How heavy (in pounds) is a rack when full? The bottles weigh 1/4 pound each when empty.

SOLUTION

Unless the team managers want to get fired, beaten up, and possibly arrested, they'll serve fresh water, so first we will need to know how much a quart of fresh water weighs.

$$1\,\cancel{qt} \cdot \frac{1\,\cancel{gal}}{4\,\cancel{qt}} \cdot \frac{1\,\cancel{ft^3}}{7.48\,\cancel{gal}} \cdot \frac{62.5\,lb}{1\,\cancel{ft^3}} = 2.1\ lb$$

So the water in 16 bottles weighs $16 \times 2.1 = 33.6$ lb. Each empty bottle weighs 1/4 pound, so 16 of them will weigh 4 pounds. The rack weighs another 6 pounds, so that's a total of 10 extra pounds, and the total weight is 43.6 pounds.

Try This One   9

A pallet of 1-pint water bottles delivered to a grocery store contains 288 bottles. Twenty-four of the empty bottles weigh a pound, and the pallet itself weighs 18 pounds. What's the total weight?

In the metric system, a cubic centimeter consists of a cube whose measure is 1 centimeter on each side. Since there are 100 centimeters in 1 meter, 1 cubic meter is equal to 100 cm × 100 cm × 100 cm or 1,000,000 cubic centimeters. See Figure 8-5.

The base unit for volume in the metric system is the liter, defined to be 1,000 cubic centimeters (see Figure 8-6). Cubic centimeters are abbreviated cm^3 or cc.

This provides a simple and interesting connection between two of the base units in the metric system, meters and liters: 1 cubic centimeter is the same as 1 milliliter. Comparisons of other volume units are shown in Table 8-7.

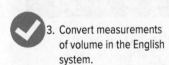

3. Convert measurements of volume in the English system.

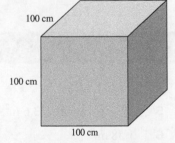

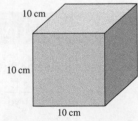

1 cubic centimeter   1 cubic meter = 1,000,000 cubic centimeters   1 liter = 1,000 cubic centimeters

Figure 8-5   **Figure 8-6**

Table 8-7	Unit Equivalences for Capacity in the Metric System

$1\ cm^3 = 1\ mL$

$1\ dm^3 = 1,000\ cm^3 = 1\ L$

$1\ m^3 = 1,000\ dm^3 = 1\ kL$

$1\ L \approx 1.06\ quarts$

| **Example 10** | **Converting Metric Units of Volume** |

A typical backyard in-ground pool has a volume of about 82.5 cubic meters. How many milliliters of water are needed to fill such a pool?

SOLUTION

There's a conversion in Table 8-7 from mL to cm³, so we first convert to cm³, then to mL. Since 100 cm = 1 m, (100 cm)³ = (1 m)³; that is, 1 m³ = 1,000,000 cm³.

$$82.5 \text{ m}^3 \cdot \frac{1,000,000 \text{ cm}^3}{1 \text{ m}^3} \cdot \frac{1 \text{ mL}}{1 \text{ cm}^3} = 82,500,000 \text{ mL}$$

| **Try This One** | **10** |

4. Convert units of volume in the metric system.

On average, 2,476,000 cubic centimeters of Coke are consumed worldwide every second. How many kiloliters is that?

The last equation in Table 8-7 provides a conversion between English and metric units of volume.

| **Example 11** | **Converting Liters to Gallons** |

How many gallons of Mountain Dew are in one 2-liter bottle?

SOLUTION

We know that 1 liter is about 1.06 quarts, and that 1 gallon is 4 quarts:

$$2 \text{ L} \cdot \frac{1.06 \text{ qt}}{1 \text{ L}} \cdot \frac{1 \text{ gal}}{4 \text{ qt}} \approx 0.53 \text{ gal}$$

| **Try This One** | **11** |

5. Convert measurements between English and metric units of volume.

How many liters of gas does an SUV with a 21-gallon tank hold?

Answers to Try This One

1 1 yd² should be 9 ft²;

$1 \text{ yd}^2 \cdot \dfrac{3 \text{ ft}}{1 \text{ yd}} \cdot \dfrac{3 \text{ ft}}{1 \text{ yd}} = 9 \text{ ft}^2$

2 439.2 in²

3 1,593,600 acres

4 6,388.8 yd²

5 Yes, it is about 263.6 in².

6 1.58 mi²

7 $837.78

8 13.37 ft³; 835.6 lb

9 330.8 lb

10 2.476 kL

11 79.25 L

| **Exercise Set** | **8-2** |

Writing Exercises

1. Explain what "square units" mean. Use examples like "square inches" and "square miles."

2. If 1 yd = 3 ft, how come 1 yd² ≠ 3 ft²?

3. What is the connection between the base units of length and capacity in the metric system?
4. Describe three things in your house, dorm, or apartment that could be measured using linear units, and explain why those units would be appropriate.

5. Repeat Exercise 4 for square units.
6. Repeat Exercise 4 for cubic units.

Computational Exercises

For Exercises 7–68, convert each measurement to the specified equivalent unit.

7. $3,456 \text{ in}^2 = $ _____ ft^2
8. $245,678 \text{ ft}^2 = $ _____ acres
9. $1,642\frac{3}{4}$ acres $= $ _____ mi^2
10. $45,678,231 \text{ yd}^2 = $ _____ mi^2
11. $18\frac{1}{2} \text{ yd}^2 = $ _____ ft^2
12. $7 \text{ mi}^2 = $ _____ acres
13. $345 \text{ yd}^2 = $ _____ acres
14. $77,234,587 \text{ in}^2 = $ _____ acres
15. $654,543 \text{ in}^2 = $ _____ yd^2
16. $10 \text{ yd}^2 = $ _____ in^2
17. $52 \text{ ft}^2 = $ _____ m^2
18. $40\frac{5}{6} \text{ m}^2 = $ _____ yd^2
19. $16.6 \text{ mi}^2 = $ _____ km^2
20. 32 acres $= $ _____ km^2
21. $19.3 \text{ in}^2 = $ _____ cm^2
22. $18 \text{ ft}^2 = $ _____ dm^2
23. $5\frac{2}{3} \text{ yd}^2 = $ _____ dm^2
24. $15.6 \text{ m}^2 = $ _____ ft^2
25. $156\frac{7}{8} \text{ km}^2 = $ __ acres
26. $80,000 \text{ cm}^2 = $ _____ in^2
27. $19.4 \text{ cm}^2 = $ _____ ft^2
28. $256\frac{2}{3} \text{ in}^2 = $ _____ dm^2
29. $35 \text{ yd}^2 = $ _____ m^2
30. 152 acres $= $ _____ km^2
31. $1,327\frac{1}{2} \text{ dm}^2 = $ _____ ft^2
32. $1,000,000 \text{ m}^2 = $ _____ yd^2
33. 3 cubic feet $= $ _____ fluid ounces
34. 400 gallons $= $ _____ cubic feet
35. 3,724 gallons $= $ _____ cubic yards
36. 12,561 fluid ounces $= $ _____ gallons
37. 12,000 cubic inches $= $ _____ cubic feet

38. 22,000 cubic feet $= $ _____ gallons
39. $54\frac{3}{8}$ pints $= $ _____ cubic inches
40. 4,532 cubic inches $= $ _____ gallons
41. 97 milliliters $= $ _____ cubic meters
42. 28.5 cubic centimeters $= $ _____ milliliters
43. 38.251 cubic decimeters $= $ _____ deciliters
44. 433 milliliters $= $ _____ cubic centimeters
45. 87,250 cubic centimeters $= $ _____ liters
46. $67\frac{5}{12}$ liters $= $ _____ cubic centimeters
47. 437 centiliters $= $ _____ cubic meters
48. 1.6 liters $= $ _____ cubic centimeters
49. 500 mL $= $ _____ L
50. 96.7 hl $= $ _____ L
51. 92 L $= $ _____ dL
52. 84.5 daL $= $ _____ hL
53. 42 L $= $ _____ mL
54. 92 L $= $ _____ hL
55. 12 gallons $= $ _____ L
56. 1.9 L $= $ _____ gallons
57. $378\frac{3}{5}$ quarts $= $ _____ mL
58. 1,500 mL $= $ _____ quarts
59. 21,050 mL $= $ _____ quarts
60. $27\frac{1}{2}$ quarts $= $ _____ mL
61. 175.2 kL $= $ _____ gallons
62. 940 pints $= $ _____ kL
63. 11,460 ounces $= $ _____ kL
64. 84.1 kL $= $ _____ gallons
65. Find the weight of $733\frac{1}{3}$ cubic feet of fresh water.
66. Find the weight of 23 cubic feet of salt water.
67. Find the volume in cubic feet of $205\frac{3}{4}$ pounds of fresh water.
68. Find the volume in cubic feet of 742.4 pounds of salt water.

Applications in Our World

69. I need to paint the walls in a spare bedroom. A gallon of paint will cover 400 ft^2. The surface to be painted measures $69,120 \text{ in}^2$. How many gallons of paint should I buy?
70. Jose has a 15-m^2 garden and plans to buy some mulch to cover his garden. On the bag of mulch it says that one bag covers 20 ft^2. If each bag costs $2, how much will it cost to mulch his garden?
71. Dan and Kareem want to pour a concrete patio in their backyard. They measure the length, width, and thickness of the patio and find that the amount of concrete they need is 200 ft^3. When they called the concrete company, the company rep asked how many cubic yards of concrete they wanted. Help them answer the question.

72. Chloe drank two 20-oz bottles of Diet Pepsi while she was studying. She poured the soda into the only clean glass she could find, which was a 1-pint beer glass. How many glasses did she fill?
73. A new campus was built on a 1,000-acre plot of land. How many square miles does it cover?
74. A new extension center is being built on 5 acres of campus north of the engineering building. How many square yards is the extension center being built on?
75. A pump added 5 gallons of water to a pool to level it after a diver displaced too much water during a cannonball jump. How many cubic feet of water were added?

76. A bottle containing a quarter of a cubic meter of a potent drug is half empty. Nurses give doses of 100 cc at a time. How many doses are left?

77. An installer is hired to tile the backsplash in a kitchen. The area needing tile covers 32 square feet, and the homeowner wants custom tiles that measure 8 square inches. How many tiles should they order, allowing 10% extra for breakage?

78. A store has 122 1-quart containers of milk, but the manager wants to pour the milk into gallon containers to sell at a higher price. How many gallon containers can the manager fill, allowing 5% extra for spillage?

79. Five cubic feet of salt water have flooded into a boat through a leak in the side. If the boat itself weighs 200 lb, how much does it weigh with the salt water?

80. Gianna determines she needs 10,000 square inches of material to make a slipcover for her sofa. At the fabric store, the material is sold by the square yard. How much of the fabric she chose should she tell the clerk to cut if she allows 10% extra for cutting waste?

81. A bicycle club captain buys eight 20-oz containers of Gatorade for a race, but the bikers' water bottles are quart-sized. If they fill each biker's bottle completely, how many bikers will have Gatorade for the race?

82. A dump truck with 1,500 gallons of soil arrives on campus to fill in some new planters on the quad. Each planter needs 2 cubic yards of soil. How many planters can be filled?

83. A search party covered 6 square miles of territory looking for a prized statue that fell from a cargo plane. How many square kilometers did they search?

84. A Jacuzzi was filled with 200 gallons of water. The Jacuzzi itself weighed 50 lb. If a person who weighs 60 kg and another who weighs 105 kg get into the Jacuzzi, how much is the total weight in pounds including the two people, the tub, and the water? (*Note:* 1 kg ≈ 2.2 lb.)

85. In 1970, the largest engine available on a Ford Mustang was a 429-cubic inch model that produced 375 horsepower. In 2021, you could top out at a 5.2-liter engine that produced 760 horsepower. Which of the two cars had the bigger engine? How much bigger was that engine by percentage?

86. Rate each engine in Exercise 85 in terms of horsepower per cubic inch of volume.

Critical Thinking

In Problems 87–92, which unit would be most appropriate: acres, square miles, or square feet?

87. The size of a lot in a new subdivision
88. The amount of planking needed to install a hardwood floor
89. Nassau County, NY
90. The area in a suburban yard that needs to be covered with mulch
91. Lake Erie
92. The total desk space in a small office

In Problems 93–96, which unit would be the most appropriate: liters, milliliters, or kiloliters?

93. The amount of coffee that one McDonald's location serves between 7 and 8 A.M.
94. The amount of coffee served by all Starbucks locations in an hour
95. The dosage of an experimental cancer drug
96. A backyard swimming pool
97. The capacity of a dump truck is listed by the manufacturer as 10 yd³. A construction site requires the removal of 200 m³ of dirt, and the contractor has three identical trucks. How many trips will each need to make?
98. After the excavation at the construction site in Exercise 97, a period of bad weather delays the construction. When the weather clears up, the crew chief finds that the excavated dirt has been replaced with water that now needs to be pumped out of the hole dug for the foundation. The contractor has two industrial pumps that can each pump water out at the rate of 750 gallons per minute. How long will the pumps need to empty the site?

99. Traditionally, carpet and tile have been priced by the square yard, but in recent years, more and more stores are listing prices by the square foot. Why do you suppose they made this change?

100. Without doing any calculating, which sounds like the better deal: gas for $3.89 per gallon, or $1.27 per liter? Which is actually the better price?

Exercises 101–106 may require doing some research, either online or maybe even making some good old-fashioned phone calls, if you can imagine that.

101. (a) Find the total area of your campus, and write that area in acres, square miles, square yards, and square feet.
    (b) Find the approximate enrollment at your campus. How many square feet of area are there for every student?
    (c) Berry College in Rome, GA, has the country's largest college campus, at 26,000 acres. How many of your campuses would fit on Berry's campus? (This is a really easy question if you go to Berry.)

102. About 600,000 gallons of fresh water flow over Horseshoe Falls in Niagara Falls, Canada, every second.
    (a) Find the total weight in tons of water that flows over the falls in an hour.
    (b) The nearest domed stadium to the Horseshoe Falls is the Rogers Centre in Toronto, home of the Toronto Blue Jays. Using the Internet, find the volume of the Rogers Centre with the roof closed. If the water flowing over Horseshoe Falls were somehow diverted into the Rogers Centre, how long

would it take to completely fill the building? (Legal disclaimer: this would be a REALLY bad idea.)

103. (a) The floor area of a rectangular room is found by multiplying the length and width. Find the floor area of your classroom, and write it in square feet, square inches, square meters, and square centimeters. (If the room is not rectangular, take some measurements and estimate the area as best you can.)

(b) The Pentagon in suburban Washington, DC, has total floor area of just over 6,500,000 square feet. If a campus building consisted of 100 identical copies of your classroom, 8 restrooms, each with an area of 320 square feet, and 150 square meters of hallway, how would it compare in size to the Pentagon?

104. Refer to Exercise 103. The volume of a rectangular room in cubic feet is found by multiplying the floor area in square feet and the height of the ceiling in feet.

(a) Find the volume of your classroom in cubic feet. How many gallons of water would the classroom hold? (If it were watertight, obviously.)

(b) How much would the water needed to fill your classroom weigh in pounds? (Assume it's fresh water.)

(c) The Vehicle Assembly Building at the Kennedy Space Center in Florida is the fourth-largest structure in the world by volume. Find the volume of the building online. How many of your classrooms would it hold? How many gallons of water? How many tons would that water weigh? (Again, fresh water.)

105. As a general rule, there are about 20 drops of water in 1 mL.

(a) How many drops would it take to fill a 16-ounce plastic water bottle?

(b) How much does 1,000,000 drops of fresh water weigh?

(c) If you could put in one drop per second, how long would it take you to fill the swimming pool in Example 10 of Section 8-2 with an eyedropper?

106. A standard teaspoon has a capacity of $\frac{1}{6}$ fluid ounce. Assuming you could fill a teaspoon to capacity on each scoop, and you can manage a scoop every 5 seconds, how long would it take, using only a teaspoon, to empty:

(a) A 20-gallon aquarium

(b) The swimming pool in Example 10 of Section 8-2

(c) Lake Okechobee (1.25 cu mi) (*Note:* There are about 1.48×10^{11} cubic feet in one cubic mile.)

(d) The Atlantic Ocean (77,640,000 cu mi)

107. Using Technology. Find a unit conversion website or app, and use it to check at least six of your homework problems. In each case, describe how closely your answer matched what the online converter gave you, and draw any conclusions you can about such converters.

Section 8-3   Measures of Weight and Temperature

Mirelle/Shutterstock

LEARNING OBJECTIVES

1. Convert weights in the English system.

2. Convert weights in the metric system.

3. Convert weights between English and metric units.

4. Convert temperatures between the Fahrenheit and Celsius scales.

According to their website, the maximum weight for a parasailing company in Cancun is 180 kg. If you're like the average American, you have no idea exactly how heavy you can be and still parasail with that company. This, of course, is because you are so accustomed to weights being measured in pounds. But like any other form of measurement, there are a variety of units that can be used for weight. We will study them in this section, then conclude with a quick look at measuring temperatures.

Conversions for Weight

In Section 8-1, we learned that weight and mass are not quite the same thing. But in this section, we'll assume that all the objects we're weighing are earthbound, so that mass and weight are equivalent.

In the English system, weight is most often measured in ounces, pounds, and tons. In the metric system, the base unit for weight is the gram, but this is a very small unit—a dollar bill weighs about a gram, for example, so weighing anything bigger than a bread box in grams is kind of silly. So a more common unit is the kilogram, which is 1,000 grams. There is also a metric equivalent of the ton: a metric ton, sometimes written "tonne," is 1,000 kg. The basic conversions for weight, along with the symbols used to represent units, are shown in Table 8-8.

Table 8-8	Units of Weight		
	English Units	**Metric Units**	**English/Metric Conversions**
	16 ounces (oz) = 1 pound (lb)	1 kilogram (kg) = 1,000 grams (g)	1 kg ≈ 2.2 lb
	2,000 lb = 1 ton (T)	1,000 kg = 1 metric ton (t)	1 oz ≈ 28 g

Conversions within the metric system follow the same procedures we learned for converting lengths in Section 8-1. In the English system, our old friend dimensional analysis will do the work for us. We love that guy.

Example 1   Converting Weights in the English System

Let's be clear on one thing: M&Ms rule. You can buy a $2\frac{5}{8}$ lb party bag of regular M&Ms for $9.98 at Wal-Mart. Or, if you're a theater buff, you can buy a case of 12 movie-sized boxes, each containing 3.4 ounces of chocolatey bliss, for $18.28 at Sam's Club. Which has more candy? Which is the better buy?

SOLUTION

To decide which has more, we can either convert $2\frac{5}{8}$ lb to ounces, or convert the number of ounces in the boxes to pounds. Either is fine. First, with the case of movie boxes, you're getting a total of $12 \times 3.4 = 40.8$ ounces. Let's convert that to pounds:

$$40.8 \text{ oz} \cdot \frac{1 \text{ lb}}{16 \text{ oz}} = 2.55 \text{ lb}$$

Notice that $2\frac{5}{8}$ lb is 2.625, so the party bag has more candy, but just by a little.

The better buy, on the other hand, is clear. The party bag has more candy and costs a LOT less, so that's the way to go.

Try This One   1

Skittles aren't in the same league as M&Ms, but they'll work in a pinch. You can get a $3\frac{3}{8}$ pound bag for $14.99 at Staples (because when I think of office supplies, I think Skittles). Or you can get a box of 36 2.17 ounce bags on Amazon for $21.99. Which purchase has more Skittles? Which costs more per pound?

Example 2   Converting Weights in the English System

An empty Boeing 747 weighs about 175 tons. How much does it weigh in ounces?

SOLUTION

We will need to first convert tons to pounds, then pounds to ounces.

$$175 \text{ T} \cdot \frac{2,000 \text{ lb}}{1 \text{ T}} \cdot \frac{16 \text{ oz}}{1 \text{ lb}} = 5,600,000 \text{ oz}$$

David Frazier/Corbis Getty Images

Try This One   2

An average male African elephant weighs about 176,000 ounces. How many tons does he weigh?

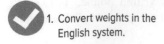

1. Convert weights in the English system.

Admittedly, it's silly to measure an airplane or an elephant in ounces. We chose examples like these as a reminder of exactly why it's so valuable to have different units of measure for any given quantity.

Example 3	Converting Metric Units of Weight

Convert each of the following:
(a) 150 grams of protein powder to milligrams
(b) 23 grams of salt to kilograms
(c) 3 kilograms of chicken to milligrams

SOLUTION

For reference, here is the order of the metric prefixes from Table 8-3 from largest to smallest: kilo, hecto, deka, base unit, deci, centi, milli.

Math Note

Recall that when converting metric units, we multiply or divide by powers of 10. When moving three steps in Table 8-3, we multiply or divide by 10^3, or 1,000. When moving six steps, we multiply or divide by 10^6, or 1,000,000.

(a) The prefix milli is three steps below the base unit, so multiply by 1,000 (or move the decimal three places right).

$$150 \text{ g} = 150.000 \text{ g} = 150,000 \text{ mg}$$

(b) The prefix kilo is three units above the base unit, so divide by 1,000 (or move the decimal three places left).

$$23 \text{ g} = 0023 \text{ g} = 0.023 \text{ kg}$$

(c) The prefix milli is six steps below the prefix kilo, so multiply by 1,000,000 (which is easy in this case, without moving the decimal).

$$3 \text{ kg} \times 1,000,000 = 3,000,000 \text{ mg}$$

Try This One	3

Convert each of the following:

(a) A 10-kilogram dumbbell to grams
(b) A 250-milligram aspirin to grams
(c) 400 milligrams of a steroid to kilograms

2. Convert weights in the metric system.

One of the most common uses of weight conversion is converting between English and metric units, as in Example 4.

Example 4	Converting between English and Metric Units of Weight

(a) The heaviest amount ever successfully lifted in Olympic competition was in the clean and jerk competition in Athens in 2004: 263.5 kg. How many pounds is that?
(b) An average adult hummingbird weighs one-eighth of an ounce. How many grams does it weigh?

Paul Kitagaki Jr./ZUMA Wire/Alamy Stock Photo

SOLUTION

(a) We can use the unit fraction $\frac{2.2 \text{ lb}}{1 \text{ kg}}$:

$$263.5 \text{ kg} \cdot \frac{2.2 \text{ lb}}{1 \text{ kg}} \approx 579.7 \text{ lb}$$

(b) This time, the unit fraction $\frac{28 \text{ g}}{1 \text{ oz}}$ is helpful:

$$\frac{1}{8} \text{ oz} \cdot \frac{28 \text{ g}}{1 \text{ oz}} \approx = \frac{28}{8} \text{ g} \approx 3.5 \text{ g}$$

Math Note
Since 1 kg ≈ 2.2 lb, for a very quick (and very imprecise) idea of the equivalent in pounds for a weight in kilograms, double the number of kilograms and add a bit more.

Try This One   4

(a) One of the weight classes in Olympic boxing is 81 kg. How heavy is that in pounds?

(b) An average golden retriever puppy weighs $12\frac{1}{2}$ ounces at birth. How many grams does it weigh?

Example 5   Assessing Weight Loss

A mixed martial arts fighter has a big bout coming up in 3 months and needs to lose 15 pounds in order to make weight for the fight. She finds an advertisement online for a British weight loss system that guarantees a loss of 3 kg per month. If that claim is true, will she reach her goal?

SOLUTION

We can do this in one calculation by multiplying by the unit fraction $\frac{2.2\,lb}{1\,kg}$ to eliminate kg, then multiplying by 3 months:

$$\frac{3\ \cancel{kg}}{1\ \cancel{month}} \cdot \frac{2.2\ lb}{1\ \cancel{kg}} \cdot 3\ \cancel{months} \approx 19.8\ lb$$

If the claim is accurate (a *very* big if), the fighter will make weight easily.

Try This One   5

Bart's football coach tells him that he'll start next year if he gains 10 pounds in the 14 weeks between spring practice and summer drills. He hires a personal trainer who promises him a gain of 1 kg every 3 weeks. Will he make his goal?

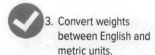

3. Convert weights between English and metric units.

Conversions for Temperature

Temperature is measured in the English system using degrees **Fahrenheit**. On this scale, water freezes at 32° and boils at 212°. The average temperature of the human body is 98.6°. In the metric system, temperatures are measured in degrees **Celsius** (also known as **centigrade**). Like all metric measurements, the Celsius scale is based on powers of 10 in some form: water freezes at 0° and boils at 100°. The average temperature of the human body is 37°.

One degree Celsius corresponds to $\frac{9}{5}$ degrees Fahrenheit, and there are relatively simple formulas for converting between the two.

Fahrenheit–Celsius Conversions
To convert Celsius to Fahrenheit:
$$F = \frac{9}{5}C + 32$$
To convert Fahrenheit to Celsius:
$$C = \frac{5}{9}(F - 32)$$

Ken Cavanagh/McGraw Hill

Example 6   Converting Celsius to Fahrenheit

In preparing for vacation, Randi and Catalina check the Internet and find that the average temperature over the next week at their destination is predicted to be 28° Celsius. Should they pack coats or bathing suits?

Sidelight   It's Absolutely Freezing!

You're probably perfectly satisfied with the Fahrenheit scale, and when you mix in Celsius, that seems more than sufficient for measuring temperatures. Physicists and chemists would tend to disagree with you, however. In their view, temperature is a measure of how much heat is contained in some medium, so negative temperatures just don't seem right. In 1948, a third temperature scale was developed to address this issue. It is known as the **Kelvin** scale, in honor of the British physicist who introduced it. Zero degrees Celsius might be a bit warm for our friend the polar bear, and zero degrees Fahrenheit is more to his liking. But zero kelvins (the terminology used, rather than "zero degrees Kelvin") would be far too cold even for him.

On the Kelvin scale, one degree corresponds to the same amount of temperature difference as the Celsius scale. The difference is the temperature that corresponds to zero degrees. On the Kelvin scale, this temperature is known as

mysticbengal/Shutterstock

absolute zero, which is the coldest possible temperature. Heat comes from an interaction of molecules, and at absolute zero, there is no heat because molecules stop moving. This temperature corresponds to $-459.67°F$ and $-273.15°C$, and because one kelvin is the same amount of heat as one degree Celsius, you can convert temperatures between those two scales by simply adding or subtracting $273.15°$.

It is theoretically impossible to cool anything to absolute zero, but in 2000, Finnish scientists reported reaching a low of 1 ten-billionth of a kelvin in the lab.

SOLUTION

Substitute 28 for C in the first formula above:

$$F = \frac{9}{5}C + 32$$

$$= \frac{9}{5} \cdot 28 + 32 = 82.4°$$

Woo hoo! Swimsuits and sunblock it is!

Try This One   6

The average high temperature for March in Moose Jaw, Saskatchewan, is $-1°$ Celsius. Find the temperature on the Fahrenheit scale.

Example 7   Converting Fahrenheit to Celsius

A frozen pizza needs a 450° Fahrenheit oven to cook. Suppose you bought a cheap oven on eBay, and found that it had been scavenged from Romania, so the temperatures on the dial are Celsius. At what temperature should you set the oven?

SOLUTION

Substitute $F = 450$ into the second conversion formula:

$$C = \frac{5}{9}(F - 32)$$

$$C = \frac{5}{9}(450 - 32) \approx 232.2°$$

You should set it close to 230°.

Try This One   7

The pizza box recommends 225° Fahrenheit for rewarming. Where should the dial be set for rewarming?

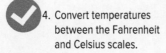

4. Convert temperatures between the Fahrenheit and Celsius scales.

Answers to Try This One

1   The box of 36 is 4.8825 lb, so it has more Skittles. It's also more per pound: $4.50/lb compared to $4.44/lb.

2 $5\frac{1}{2}$ T

3   (a) 10,000 g
      (b) 0.250 g
      (c) 0.000400 kg

4   (a) $\approx$ 178.2 lb
      (b) $\approx$ 350 g

5   Yes, by about 0.3 lb

6   30.2°F

7   About 110°C (225°F $\approx$ 107.2°C)

Exercise Set   **8-3**

Writing Exercises

1. Explain the difference between mass and weight.
2. In most applications, weight and mass are treated as if they mean the same thing. Why is it reasonable to do that?
3. Even though the base unit of mass/weight in the metric system is the gram, kilograms are used far more commonly. Why is that?
4. Measuring temperatures in metric units is different from the other measurements we studied. How?
5. How do powers of 10 fit into the metric system of measuring temperatures?
6. What is a kelvin? (*Hint:* See Sidelight on page 473.)

Computational Exercises

For Exercises 7–16, convert to the indicated unit.

7. 33 oz to pounds
8. 92.6 oz to pounds
9. 2.3 lb to ounces
10. $3\frac{5}{16}$ lb to ounces
11. 4.2 T to pounds
12. $7\frac{9}{20}$ T to pounds
13. 3,500 lb to tons
14. 2,350 lb to tons
15. 34,567 oz to tons
16. $5\frac{7}{8}$ T to ounces

For Exercises 17–30, convert each metric weight to the indicated unit.

17. 9 dg = _____ cg
18. 6 mg = _____ cg
19. 44 kg = _____ hg
20. 16.34 hg = _____ g
21. $51\frac{1}{2}$ cg = _____ dg
22. 181.6 g = _____ dag
23. 7.26 g = _____ dg
24. 0.325 g = _____ mg
25. 3,217 cg = _____ g
26. $19\frac{1}{4}$ dag = _____ kg
27. $5\frac{3}{7}$ mg = _____ g
28. 400 g = _____ kg
29. $22\frac{5}{6}$ kg = _____ g
30. 5,632 g = _____ hg

For Exercises 31–46, convert each amount to the specified unit.

31. 120 grams = _____ ounces
32. $15\frac{3}{4}$ ounces = _____ grams
33. 4,823 centigrams = _____ pounds
34. 27 metric tons = _____ pounds
35. 3 tons = _____ hectograms
36. 14 decigrams = _____ ounces
37. 357,201 pounds = _____ metric tons
38. 13 pounds = _____ decigrams
39. 5.75 tons = _____ metric tons
40. 16.3 metric tons = _____ tons
41. 213 ounces = _____ decigrams
42. 64 pounds = _____ dekagrams
43. 815 dekagrams = _____ ounces
44. 550 hectograms = _____ ounces
45. 14,625 milligrams = _____ pounds
46. 41 pounds = _____ grams

For Exercises 47–52, convert each Celsius temperature to an equivalent Fahrenheit temperature.

47. 14°C
48. 27°C
49. $33\frac{1}{3}$°C
50. −5°C
51. −18°C
52. $-30\frac{1}{2}$°C

For Exercises 53–58, convert each Fahrenheit temperature to an equivalent Celsius temperature.

53. 5°F
54. 27°F
55. 96.2°F
56. −3°F
57. −10°F
58. $-5\frac{1}{4}$°F

In Exercises 59–64, convert each temperature in kelvins to the units given. See the Sidelight on page 473 for guidance.

59. 219.3 K to Celsius
60. −5 K to Celsius
61. 251 K to Fahrenheit
62. −256 K to Fahrenheit
63. 100.32 K to Fahrenheit
64. $462\frac{1}{4}$ K to Fahrenheit

Applications in Our World

65. After giving birth to her first child, Tricia was informed that he weighed 111 ounces. Her first thought was, "Holy smokes—no wonder it felt like I was carrying around a bowling ball." The average newborn boy in the United States weighs about 7 pounds, 10 ounces, and an average bowling ball weighs 12 pounds. Find the baby's weight in pounds and rate Tricia's first instinct for accuracy.

66. Billy's pickup truck is capable of carrying $\frac{3}{4}$ ton. Billy is picking up a load of rock that weighs 1,700 pounds. Will the pickup truck be able to handle the load?

67. The maximum weight a certain elevator can withstand is 4,400 kilograms. Ten big dudes, weighing about 210 pounds each, step into the elevator. Will the elevator hold them?

68. On the scale at the doctor's office Susan weighed 68 kilograms. The healthy weight for a person her height is 130 lb. Does Susan need to reduce her weight?

69. Olivia Sanchez is going to visit Europe on vacation; she checked the local weather of the country she's visiting and found that the average temperature this time of year is about 35°C. This temperature sounds pretty cold to the American-born Olivia, but she decides before packing she should convert this to °F. After converting, will Olivia pack a coat?

70. For a chicken Marsala recipe, Janine needed 24 oz of boneless, skinless chicken breasts. At the supermarket, the chicken was measured in pounds. How many pounds did Janine need to buy?

71. Five cars weighing 2,540 lb each are put onto a ferry. The ferry has a maximum weight capacity of 6 tons. Will the ferry be able to hold all the cars without sinking into the bay?

72. A chemistry teacher has 750 grams of a substance and wants to separate the substance into 3-oz jars. How many 3-oz jars can be filled?

73. A thermometer in the window of your car reads 20 degrees Celsius, but the other side where the Fahrenheit scale is located has faded in the sun. What is the Fahrenheit temperature?

74. On a summer trip to France, you catch a virus and feel like you have a fever. The only thermometer they sell at the store shows you have a 38.8 degrees Celsius temperature. What is your Fahrenheit temperature?

75. The border patrol seizes 5 lb of an illegal substance at a check point. How many grams of the substance did they seize?

76. While building a new playground at the city recreation center, a truck pours 3,000 kg of sand into the foundation. The specs require 2.5 metric tons of sand. Did the truck pour the right amount into the foundation?

77. Shirl lost 3.6 lb of fat on a diet. How many ounces was that?

78. During a dinner at Outback Steakhouse, a family of six eats four 12-oz filet mignons and two 20-oz porterhouse steaks. How many pounds of steak did the family consume during the dinner?

79. An elevator says the maximum capacity is 1,000 kg. Twelve people are already on the elevator, their weight totaling 2,000 lb. Another person who weighs 260 lb wants to get on. Should the other passengers let them onto the elevator?

80. To mold silver into a ring, Aaliyah has to heat it up to its boiling point, which is 3,924 degrees Fahrenheit. The heating element can only heat to 2,000 degrees Celsius before it will cool down for safety reasons. Can Aaliyah make the ring?

81. If the temperature goes down by 7 degrees Celsius when a cold front comes in, how much did it drop on the Fahrenheit scale?

82. Between 8 A.M. and noon, the temperature rose by 22 degrees Fahrenheit. How much did it go up on the Celsius scale?

83. The cement Yuan bought at the hardware store weighs 52 lb per cubic foot. To pour a driveway, the specifications require at least a weight of 25 kg per cubic foot. Did Yuan buy the right type of cement?

84. A 1.7-ton truck crosses the border into Canada. The scale measures in kilograms. How much does the scale say the truck weighs?

85. The sugar fructose melts at between 103 and 105 degrees Celsius. Julia heats up fructose on a burner to 222.8 degrees Fahrenheit. Is the sugar melted yet?

86. A lab tech had a fifth of a kilogram left of a substance and wanted to divide it into 50 bags for the next day's experiment. Each bag needed 4,500 mg of the substance. Would the lab tech have enough?

87. On a see-saw, three Americans weighing a total of 450 lb are sitting on one side. Two Europeans weighing a total of 137 kg sit on the other side. What would a person have to weigh in kilograms to balance out the see-saw?

88. (a) According to the Guinness Book of World Records, the largest healthy baby ever delivered weighed 360 ounces. How much did he weigh in pounds? How does this weight compare to the average weight of a newborn boy, which is 7.6 lb?

    (b) Suppose the child was a little older and weighed 50 lb when his parents took him to the doctor, who read his temperature as 40°C. Does he have a fever? If so, how many degrees Fahrenheit above normal body temperature is the poor little guy? (Little being a relative term in this case.) Recall that 98.6°F is considered the normal body temperature for humans.

    (c) If he has a fever, the recommended dose is 15 mg of acetaminophen per kg of body weight. If there are 160 mg of acetaminophen in 5 mL of children's Tylenol, how many mL of children's Tylenol should the child receive?

    (d) If the entire bottle of liquid children's Tylenol contains 4 fluid ounces, how many mg of acetaminophen are in the bottle? How many grams? How many kilograms?

89. (a) An empty Airbus A320 aircraft weighs about 47 tons. How much does it weigh in ounces?

    (b) If this same plane burns about 1.2 tons of fuel per hour when at cruising altitude, determine the number of gallons per minute this jet consumes (assuming jet fuel weighs about 6.8 lb per gallon).

    (c) At $4.55 per gallon for jet fuel, how much will fuel cost for a 4-hour flight?

    (d) An A320 can carry up to 180 passengers. How much would each have to pay just to cover fuel costs for the flight?

In Exercises 90–93, which would be the most appropriate units of weight: gram, kilogram, or metric ton?

90. A cruise ship
91. The eight-person rowing crew at Harvard
92. A crew-neck sweater
93. Tom Cruise

Critical Thinking Questions

94. We know that 100°C corresponds to 212°F, and we can see that you multiply 100 by 2.12 to get 212. Without using the temperature conversion formulas, does this mean that you can get from Celsius to Fahrenheit by multiplying by 2.12? Why or why not?

95. (a) The bar graph shows the average monthly high temperatures in degrees Celsius for Las Vegas. Convert each temperature to the Fahrenheit scale (round to the nearest degree), then draw a similar graph using those temperatures. Discuss how the graphs compare visually. Which one appears to show a greater variation in temperature?

    (b) Redraw your graph of Fahrenheit temperatures with a scale on the vertical axis that begins at 50° rather than 0°. How does this affect your perception of the graph?

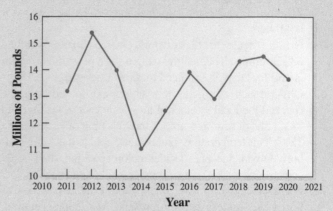

Source: World Nuclear Association

97. If the temperature is 26°C, are you more likely to be in Seattle or San Antonio?

98. If the temperature is 9°C, are you more likely to be in Louisville or Anchorage?

99. Is Blaine, at 112 kg, more likely to be a football player or a golfer?

100. Is Damon, at 50 kg, more likely to be a school bus driver or one of the kids riding the bus?

101. If you step into a bath at 90°F, what will be the result?

102. If you step into a bath at 90°C, what will be the result?

103. If you step into a bath at 90 K, what will be the result?

104. Climate change is a very real, and very serious, problem. But you can't believe everything you read. This sentence appeared in *The Time Before History: Five Million Years of Human Impact* by Colin Tudge: "Twenty years could see a 7 degree celsius, or 44.6 degree fahrenheit, rise in temperature, the difference between a frozen landscape and a temperate one." Explain why this claim doesn't hold up mathematically.

105. Using Technology. The bar graph in Problem 95 shows the average monthly high temperatures in degrees Celsius (rounded to the nearest degree) for Las Vegas. Read the graph and record the temperatures in a spreadsheet. Use the correct formula to convert each temperature to the Fahrenheit scale, then create a bar graph using those temperatures. Discuss how the graphs compare visually. Which one appears to show a greater variation in temperature?

Average Monthly High Temperatures for Las Vegas

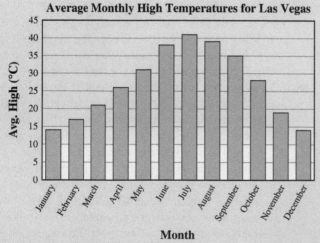

Source: National Weather Service

96. Australia is one of the world's leading producers of uranium for use in nuclear power production. The graph shows annual uranium production in Australia from 2011 to 2020 in millions of pounds.

    (a) Convert the weights into metric tons, then draw a line graph using your data. Scale the vertical axis so that the values range from zero to 8,000 metric tons.

    (b) Describe the apparent change in uranium production for each graph based on a quick look, and discuss possible reasons for the discrepancy.

Section	Important Terms	Important Ideas
8-1	English system of measurement Dimensional analysis Conversion factor The metric system Base unit Mass Weight Meter Liter Gram	**Dimensional analysis** is a process that can be used to convert from one unit of measurement to another. It's based on writing conversion factors: since 1 foot and 12 inches represent the same quantity, multiplying a measurement by a fraction like $\frac{1 \text{ foot}}{12 \text{ inches}}$ can divide out unwanted units without changing the size of the measurement. Conversions within the metric system can be done by moving the decimal the appropriate number of places, based on the prefix of the units. Dimensional analysis can also be used. Conversions between metric and English units of measure can also be done using dimensional analysis.
8-2	Linear units Square units Cubic units	**Area** is a measure of the size of a two-dimensional object. Area is measured in square units. Volume is a measure of the size of a three-dimensional object and is measured in cubic units. If we know a conversion factor for a linear measure (length), we can square or cube that factor to find a corresponding conversion factor for area or volume. Liquid capacity, measured in units like ounces or gallons, is also a measure of volume.
8-3	Pounds Ounces Tons Metric tons Fahrenheit Celsius Kelvin	**Pounds,** ounces, and tons are units commonly used to measure weight in the English system. The base unit for weight in the metric system is the gram. Dimensional analysis is used to convert between units of weight. Degrees Fahrenheit are the units for temperature in the English system, and degrees Celsius are the units in the metric system. There are simple formulas to convert from one to the other. The Kelvin scale is used commonly in the sciences to measure temperatures.

Math in Travel REVISITED

clintscholz/iStock/Getty Images

1. The Fahrenheit equivalents are 75.2° and 89.6°, so beach wear is the way to go.

2. Thirty kilograms is about 66 pounds, so you can pack 62 pounds of clothes and personal items.

3. Walking to downtown shouldn't be a problem—0.6 kilometers is about 0.37 miles. The distance from the airport is about 2.6 miles, which will take just about 4½ minutes at 35 miles per hour.

4. You used 12 gallons, which is about 45.3 liters. At $1.43 per liter, the refill costs $64.78.

5. It will cost $208.08.

Review Exercises

Section 8-1

For Exercises 1–16, convert each of the following using dimensional analysis.

1. $4\frac{1}{2}$ yards to feet
2. 23,564 feet to miles
3. 21 inches to feet
4. 356 inches to yards
5. 8,245,264 inches to miles
6. 10.2 m to cm
7. 0.36 m to dm
8. 13.54 km to dm

9. 34,500 mm to m
10. 0.56 dam to hm
11. $7\frac{3}{8}$ m to inches
12. 35.21 m to yards
13. 85 cm to inches
14. $23\frac{1}{3}$ yd to dm
15. 345 feet to dm
16. 1,235,543 in. to km

17. How many meters are run in a 100-yard dash?
18. Traffic is backed up for 10 km on a major highway. How many miles is the backup?
19. The 2021 Chevrolet Corvette Stingray has a top speed of 195 miles per hour. How fast is that in kilometers per hour? In feet per second?
20. It is about 3,600 meters from the Lincoln Memorial to the United States Capitol. How long would it take to walk that distance at a speed of 3 miles per hour?
21. The Eiffel Tower stands 1,986 feet tall. The Chrysler Building in New York City is 319 meters. Which is taller?
22. Which units would be more appropriate for measuring the length of a cab ride, centimeters or kilometers?

Section 8-2

For Exercises 23–41, convert each of the following to the indicated units.

23. 5,698 in^2 to square feet
24. 543 yd^2 to acres
25. 2.45 mi^2 to square feet
26. 456,321 in^2 to square yards
27. 5,643 acres to square miles
28. $23\frac{1}{3}$ in^2 to square centimeters
29. 14.1 ft^2 to square decimeters
30. $123\frac{1}{2}$ acres to square kilometers

31. 32.5 cm^2 to square feet
32. 324 km^2 to square feet
33. 7 ft^3 to fluid ounces
34. 42,000 ft^3 to cubic inches
35. 6 yd^3 to gallons
36. 3.5 L to cubic centimeters
37. 45 pints to cubic inches
38. 312 L to mL
39. 673 hL to liters
40. 54,457 mL to kL
41. 231 cL to L

42. A nurse is instructed to give a patient 500 cubic centimeters of a drug. The only thing available to measure with is marked off in milliliters, so the nurse measures out 5,000 mL and gives it to the patient (the reasoning, be it right or not, is that milli is 10 times centi). Is the patient in danger?
43. The instructions on a dechlorinator for a fish tank say to put in five drops per gallon. Geoff has a fish tank that holds 2.27 cubic feet of water. How many drops should Geoff use?
44. The football field and surrounding sideline area at Enormous State U. covers an area of 79,000 square feet. The athletic department decides to replace the natural grass with Field Turf, a popular synthetic alternative. The cost of installing Field Turf is $119 per square yard. How much will the project cost?

45. Juan is considering two properties to buy for investment purposes. One is 2 acres and will cost $41,000. The other is 9,500 square meters and will cost $46,000. Which costs less per square foot?
46. Ron's Lawnscaping Artists charge $45 for one fertilizer treatment on a 5,000-square-foot lawn. How much should they charge for a lawn that is 5,000 square meters?
47. Which units would be more appropriate for measuring the area of a shopping mall, acres or square meters?
48. Which units would be a better choice for measuring the volume of a mouse's heart, cubic centimeters or liters?

Section 8-3

For Exercises 49–62, convert each of the following to the indicated units.

49. 65 oz to pounds
50. 2.4 T to pounds
51. 34,500 lb to tons
52. 1.3 T to ounces
53. 567,376 oz to tons
54. $9\frac{3}{4}$ dg to cg
55. 0.457 g to mg

56. 5.6 kg to g
57. 34,345 g to hg
58. 130 g to ounces
59. 45.5 oz to grams
60. $192\frac{1}{8}$ oz to decigrams
61. 23,456 mg to pounds
62. 456.6 hg to ounces

For Exercises 63 and 64, convert each Celsius temperature to Fahrenheit.

63. 13°C

64. $-2\frac{1}{2}$°C

For Exercises 65 and 66, convert each Fahrenheit temperature to Celsius.

65. 92.3°F

66. −21° F

67. Convert 241.7 kelvins to degrees Celsius.
68. Convert −8.2 degrees Celsius to kelvins.
69. How many pounds is a baby who weighs 3,200 g?
70. Heinrich's temperature is 38.5°C. A person is running a fever if their temperature is above 100°F. Does Heinrich have a fever?
71. Welcome to the new normal: I'm at a conference in Falls of Rough, KY, and the high temperature yesterday was 22° Celsius higher than it is right now. It's a chilly 37°F now. What was the high yesterday in degrees Fahrenheit?
72. A contestant on a weight loss show started out weighing 346 pounds and lost 58 kg on the show. What was the contestant's ending weight?
73. On the American side of Niagara Falls, a home improvement center sells landscaping rock for $65 per ton. On the Canadian side, a different business sells a similar stone for $58 (American) per metric ton. Which is the better buy?

Chapter Test

For Exercises 1–21, convert each of the following:

1. 132 in. to yards
2. 8,665 ft to miles

3. 0.123 m to centimeters
4. 2 hm to decimeters

5. 34 cm to inches
6. $27\frac{1}{2}$ ft to meters
7. 3,475 in^2 to square feet

8. 3,235 acres to square miles

9. 21.32 cm^2 to square inches
10. $19\frac{1}{3} \text{ ft}^2$ to square meters
11. 100 cm^3 to cubic inches
12. $15,154 \text{ in}^3$ to cubic meters
13. 4.19 yd^3 to gallons
14. $34\frac{1}{4}$ pints to cubic inches

15. 365 oz to pounds
16. 34,675 oz to tons
17. 5.91 dg to cg
18. 250 g to ounces
19. 2,854 cg to pounds
20. 13°C to °F
21. 85.2°F to °C

22. A marathon is 26 miles, 385 yards. How many kilometers is that?
23. If a runner averages 5.9 feet per second, how long would it take them to complete a marathon (see Problem 22)?
24. According to WebMD, a healthy weight for a 6-foot male is between 140 and 180 pounds. Bobby is 6′0″ and 88 kilograms. How does his weight rank according to WebMD?
25. Tyron likes two carpets equally well and has to choose one. One costs $12.50 per square yard; the other costs $1.70 per square foot. Which is the better buy?
26. How many feet is a 5-km race?
27. It is about 353 km from Toronto to Ottawa. How long would it take to make that drive averaging 65 miles per hour?

28. The Ironman Triathlon is a race in which participants swim 2.4 miles, bike 112 miles, and run 26.2 miles. Is the total distance covered more than a 200-km car race?
29. A carpet installer brags that he can install industrial carpet and pad at the rate of 20 square yards per hour. If he's not exaggerating, how long would it take him to carpet an office building with 3,500 square feet of carpet space?
30. The famous rotunda of the Capitol building has a floor area of about 7,238 square feet. A marble restoration company charges $40 per square yard for its services. How much would it cost to restore the entire floor in the rotunda?
31. A young family is looking to buy a house and decides they need at least a ¾-acre lot so there's room for the kids and their dog Moose to play. Their agent plans to show them a house with a 30,000-sq-ft lot. Is this a waste of the agent's time?
32. The ill-fated ship *Titanic* weighed 46,000 tons. How many kilograms did it weigh?
33. The warmest month in Aruba is September, with an average high of 32.8° Celsius. The coolest month is January, with an average low of 24.4°C. What is the difference between these temperatures in degrees Fahrenheit?

Projects

1. Do you remember where most of the English units of measurement came from? They were just flat-out made up by someone. So why can't you make up your own system of measure? You can, and that's what you'll do in this project. You can use the length of body parts, I guess, but keep it clean and keep in mind that body parts vary from person to person so that might not be the best choice. A better choice might be some common objects that are completely uniform, like maybe coins, dollar bills, iPhones . . . be creative. So here we go:
   (a) Define three units in your system: something small enough to measure small objects (analogous to inches, maybe), something maybe 10 times or so bigger (analogous to feet), and something much larger to measure large distances. You can either start with one object for the base measure and define the two larger lengths in terms of a certain number of that object, or you can use objects for the small and medium-sized units then define the big one in terms of one of the smaller.
   (b) Using a ruler, measure the objects you're using as accurately as you possibly can, then write conversion factors between your measurement system and the English system, as well as the metric system.
   (c) Use your conversion factors to find the lengths of each of the following in your system:
      (1) Your height
      (2) The width of one of your thumbs
      (3) The distance between where you live and the nearest airport
      (4) The distance from Earth to the moon

   (d) Now let's talk about area: use the conversion factors from part (b) to write conversions between square units in your system and square units in the English and metric systems. Then find the area in your system of each rectangular object below, using the fact that the area of a rectangle is the product of the length and width.
      (1) A postage stamp
      (2) The front cover of a textbook
      (3) The room that you sleep in. If it's not rectangular, the room someone else sleeps in, but make sure it's someone you know so you don't get arrested.

2. While the metric system has never been officially adopted by the United States government, at various times the government and other entities have made attempts to make use of the metric system more common. Using a library or the Internet as a resource:
   (a) Make a list of the top 20 countries in the world in terms of population, and find when each of those countries officially adopted the metric system.
   (b) Research the metric system in Great Britain. (While it has been officially adopted there, many people still cling to the English system. See if you can find any information on governmental attempts to encourage more people to abandon the English system.)
   (c) Write about any initiatives you can find in the United States to make the metric system either official, or at least more commonly used, both today and historically.
   (d) Make a list of difficulties that you think our society would encounter if the government decided to switch to the metric system for all measurements overnight.

Design elements: Front matter, Chapter Opener, Summary and End Matter header design (random numbers background illustration): ©pixeldreams.eu/Shutterstock RF

Geometry

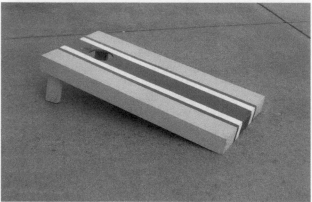

Photos: David Sobecki/McGraw Hill

Outline

It probably won't surprise you to learn that one of the most common questions that math students ask is, "How can I actually use this stuff?" Of course, how math is used in our world is the main theme of this book, and this chapter fits that framework especially well. The ideas presented in this chapter are commonly used in everyday things like working around the home, so we'll present some actual projects that geometry was used for in my home.

Geometry is one of those topics that most people recognize when they see it, but might have a hard time writing an actual definition for. If you look up a mathematical definition on the Web, it will refer to specific figures, like points, lines, triangles, and so forth. A more casual definition might say that geometry is the study of physical shapes and objects. Since we live in a physical world, in a very real sense, geometry is the study of the world around us.

If you really look for simple geometric figures, you can see them almost everywhere. The roads or sidewalks you took to get where you are right now are kind of like lines. The desk or table you're working at is probably rectangular, and the pen or pencil you're writing with is basically a cylinder. One way to think of the value of geometry in studying our world is that most of the things we encounter are in some way made up of basic geometric figures. This makes geometry one of the easiest areas of math to apply in our world.

In each of the projects described below, one or more of the techniques you will learn in this chapter was used to solve a problem. By the time you finish the chapter, you should be able to solve the problems yourself.

1. The diagram shows the measurements of all counter-tops in our kitchen. We originally decided on solid surface countertops at $56.62 per square foot including installation. We hated those countertops, and replaced them with granite 4 years later at $97.46 per square foot installed. How much did we originally save by going with the cheaper material?

How much more did we end up spending total than if we'd gone with granite in the first place?

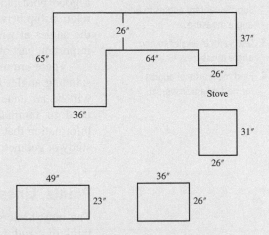

2. When finishing the basement, we needed trim along the new stairwell. From the landing at the top of the stairs in the top right photo on the facing page, the stairs extend 57 inches horizontally and drop 42 inches. I wanted to cut the trim running along the stairs so that the bottom was parallel to the floor. At what angle did it need to be cut?

3. When you start to worry about family members falling through the cracked boards, it's time for a new floor on the old deck. The deck is 10 feet wide and 26 feet long, and the deck boards will run in the longer direction. Standard decking is $5\frac{1}{2}$ inches wide, and I'll be buying 16-foot lengths. To allow water to run through, it's common to leave a $\frac{1}{8}''$ gap between the boards. How many boards should I buy?

4. The boards for a cornhole game are 4 feet long. The bottom edge has a height of 4 inches, and the top edge has a height of 12 inches. The legs are attached so that they are perpendicular to the top. What angle did the bottom of the legs need to be cut at so that they sit flat on the ground?

For answers, see Math in Home Improvement Revisited on page 550

✓ **LEARNING OBJECTIVES**

1. Write names for angles.
2. Use complementary and supplementary angles to find angle measure.
3. Use vertical angles to find angle measure.
4. Find measures of angles formed by a transversal.

If you've ever watched really good billiards players, you probably noticed that they make bank shots look very easy. But if you try them yourself, you find they're not so easy at all. The secret to bank shots, and really to being a good pool player at all, is angles. Understanding the relationship between the angle at which two balls hit and the angles at which they move after impact is the most important part of the game.

There are many, many applications in which understanding angles is key, so our main focus in this section will be an understanding of angles. To begin, we will need to familiarize ourselves with some background information that will form the fundamentals of our entire study of geometry.

David Sobecki/McGraw Hill

Points, Lines, and Planes

The most basic geometric figures we will study are points, lines, and planes. It's easiest to think of a **point** as a location, like a particular spot on this page. We represent points with dots, but in actuality a point has no length, width, or thickness. (We call it *dimensionless.*) A **line** is a set of connected points that has an infinite length, but no width. We draw representations of lines, but again, in actuality a line cannot be seen because it has no thickness. We will assume that lines are straight, meaning that they follow the shortest path between any two points on the line. This means that only two points are needed to describe an entire line. A **plane** is a two-dimensional flat surface that is infinite in length and width, but has no thickness. You might find it helpful to think of a plane as an infinitely thin piece of paper that extends infinitely far in each direction. Examples of the way we represent these basic figures are shown in Figure 9-1.

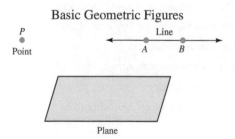

Basic Geometric Figures

Figure 9-1

Guo Xiang Chia/Moment/ Getty Images

Rays can be represented by rays of light starting at a source and continuing outward.

We typically use capital letters to represent points, and we symbolize them as dots. To name lines, we usually identify two points on the line, like A and B in Figure 9-1, and write $\overleftrightarrow{AB}$. We will sometimes assign a lowercase letter to represent a line, like l.

Points and lines can be used to make other geometric figures. A **line segment** is a finite portion of a line consisting of two distinct points, called **endpoints**, and all of the points on a line between them. The line segment connecting two points A and B is written as $\overline{AB}$.

Any point on a line separates the line into two halves, which we call **half lines**. A half-line beginning at point A and continuing through point B is written as $\overset{\circ}{\longrightarrow}{AB}$. The open circle over A indicates that the point A is not included. When the endpoint of a half line is included, we call the resulting figure a **ray**. A ray with endpoint A that continues through point B is written as $\overrightarrow{AB}$. The figures described in the last few paragraphs are summarized in Table 9-1.

Table 9-1   Lines and Portions of Lines

Name	Figure	Symbol
Line	A B	$\overleftrightarrow{AB}$
Line segment	A B	$\overline{AB}$
Half line	A B	$\overset{\circ\longrightarrow}{AB}$
Ray	A B	$\overrightarrow{AB}$

Rays are used to define angles, which are the most important figures in this section.

> An **angle** is a figure formed by two rays with a common endpoint. The rays are called the **sides** of the angle, and the endpoint is called the **vertex**.

Math Note

When we use three letters to name an angle, the vertex of the angle is always represented by the letter in the middle. The order of the other two letters doesn't matter.

Some angles are represented in Figure 9-2. The symbol for angle is $\measuredangle$, and there are a number of ways to name angles. The angle in Figure 9-2(a) could be called $\measuredangle ABC$, $\measuredangle CBA$, $\measuredangle B$, or $\measuredangle 1$.

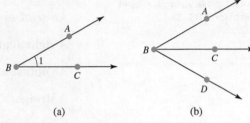

(a)   (b)

Figure 9-2

CAUTION

You should only use a single letter to denote an angle if there's no question as to the angle represented. In Figure 9-2(b), $\measuredangle B$ is ambiguous, because there are three different angles with vertex at point B. (Can you name all three of them?)

Example 1   Naming Angles

(a) Name the angle shown here in four different ways.

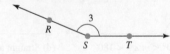

(b) Write three different angles that have point B as their vertex.

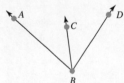

SOLUTION

(a) $\measuredangle RST$, $\measuredangle TSR$, $\measuredangle S$, and $\measuredangle 3$.

(b) The two that everyone sees at first are $\measuredangle ABC$ and $\measuredangle CBD$. To see the third, you have to kind of ignore ray $\overrightarrow{BC}$; $\measuredangle ABD$ also has vertex B.

Try This One   1

1. Write names for angles.

(a) Write three different ways to represent the bottom angle in the diagram.

(b) Write three different angles that have vertex A.

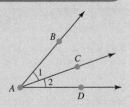

Math Note

Why is a full rotation divided into 360 degrees? That question is actually open to some debate. The practice is most commonly attributed to the ancient Babylonians, whose number system was base 60, meaning that multiples of 60 would be particularly easy to work with.

Some historians believe, however, that the fact that 360 is so close to the number of days in a year is too coincidental to ignore.
    We'll study a different unit for measuring angles in Problems 85–88.

One way to measure an angle is in **degrees**, symbolized by °. One degree is defined to be $\frac{1}{360}$ of a complete rotation.

The instrument that is used to measure an angle is called a **protractor**. Figure 9-3 shows how to measure an angle using a protractor. The center of the base of the protractor is placed at the vertex of the angle, and the bottom of the protractor is placed on one side of the angle. The angle measure is marked where the other side falls on the scale. The angle shown in Figure 9-3 has a measure of 40°. The symbol for the measure of an angle is $m\angle$; so we would write $m\angle ABC = 40°$.

Angles can be classified by their measures.

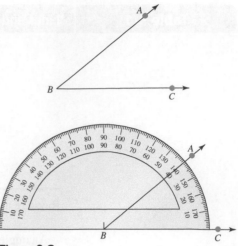

Figure 9-3

An **acute angle** has a measure between 0° and 90°.

A **right angle** has a measure of 90°.

An **obtuse angle** has a measure between 90° and 180°.

A **straight angle** has a measure of 180°.

See Figure 9-4.

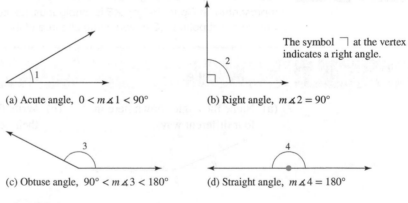

(a) Acute angle, $0 < m\angle 1 < 90°$

(b) Right angle, $m\angle 2 = 90°$

The symbol ⌐ at the vertex indicates a right angle.

(c) Obtuse angle, $90° < m\angle 3 < 180°$

(d) Straight angle, $m\angle 4 = 180°$

Figure 9-4

Pairs of Angles

Pairs of angles have various names depending on how they are related.

Math Note

If two complementary angles are adjacent, together they form a right angle. If two supplementary angles are adjacent, together they form a straight angle.

Two angles are called **adjacent angles** if they have a common vertex and a common side. Figure 9-5(a) shows a pair of adjacent angles, $\angle ABC$ and $\angle CBD$. The common vertex is B and the common side is $\overrightarrow{BC}$.

Two angles are said to be **complementary** if the sum of their measures is 90°. Figure 9-5(b) shows two complementary angles. The sum of the measures of $\angle GHI$ and $\angle IHJ$ is 90°.

Two angles are said to be **supplementary** if the sum of their measures is equal to 180°. Figure 9-5(c) shows two supplementary angles. The sum of the measures of $\angle WXY$ and $\angle YXZ$ is 180°.

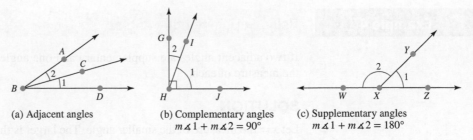

(a) Adjacent angles   (b) Complementary angles   (c) Supplementary angles
$m\angle 1 + m\angle 2 = 90°$ $m\angle 1 + m\angle 2 = 180°$

Figure 9-5

Example 2   **Using Complementary Angles**

If $\angle FEG$ and $\angle GED$ are complementary and $m\angle FEG$ is 28°, find $m\angle GED$.

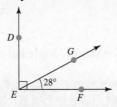

SOLUTION

Since the two angles are complementary, the sum of their measures is 90°, so $m\angle GED + m\angle FEG = 90°$, and solving we get:

$$m\angle GED = 90° - m\angle FEG$$
$$= 90° - 28°$$
$$= 62°$$

We say that the **complement** of an angle with measure 28° has measure 62°.

Math Note

It's not always *necessary* to draw a diagram for a geometry problem, but it is always *helpful*.

Try This One   **2**

Find the measure of the complement of an angle with measure 41°.

Example 3   **Using Supplementary Angles**

If $\angle RQS$ and $\angle SQP$ are supplementary and $m\angle RQS = 135°$, find the measure of $\angle SQP$.

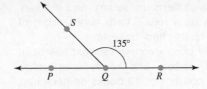

SOLUTION

Since $\angle RQS$ and $\angle SQP$ are supplementary, the sum of their measures is 180°, so $m\angle SQP + m\angle RQS = 180°$, and solving for the measure of $\angle SQP$ we get

$$m\angle SQP = 180° - m\angle RQS$$
$$= 180° - 135°$$
$$= 45°$$

We say that the **supplement** of an angle with measure 135° has measure 45°.

Try This One   **3**

Find the measure of the supplement of an angle with measure 74°.

| **Example 4** | Using Supplementary Angles |

If two adjacent angles are supplementary and one angle is 3 times as large as the other, find the measure of each.

SOLUTION

Let x = the measure of the smaller angle. The larger is three times as big, so $3x$ = the measure of the larger. The angles are supplementary, so their measures add to 180°. This gives us an equation:

$$x + 3x = 180°$$

Smaller angle plus larger angle is 180°.

$$4x = 180°$$

$$\frac{4x}{4} = \frac{180°}{4}$$

$$x = 45°$$

The smaller angle has measure 45°, and the larger has measure 3 × 45°, or 135°.

2. Use complementary and supplementary angles to find angle measure.

| Try This One | 4 |

Two angles are complementary, and the smaller has measure 17° less than the larger. Find the measure of each.

Just as two intersecting streets have four corners at which you can cross, when two lines intersect, four angles are formed, as we see in Figure 9-6. The angles opposite each other are called **vertical angles**. Angles 1 and 3 are vertical angles, as are angles 2 and 4.

Sidelight   **The Father of Geometry**

The geometry that we're studying in most of this chapter is known as Euclidean geometry, in honor of Euclid, who was a professor of mathematics at the University of Alexandria sometime around 300 BCE. That particular university was formed to serve as the center of all of the knowledge that had been accumulated to that point. By all indications, it wasn't terribly different from a modern university—it had lecture rooms, laboratories, gardens, museums, and a library with over 600,000 papyrus scrolls. There is no record of off-campus bars, but it probably had those, too.

The Bible is the only book ever written that has been more widely translated, edited, and read than Euclid's famous *Elements,* which consists of 13 books on geometry, number theory, and algebra. The book remained for the most part unchanged for over 2,000 years, but in the 1800s some logical flaws were corrected by mathematicians of the time.

All of the geometry we study in the first six sections of this chapter has its roots in the works of Euclid. He didn't discover all of these results himself, but he was the first scholar to build the field of geometry into a comprehensive study.

For about 2,100 years, the term *Euclidean geometry* was redundant, because it was believed that Euclid's geometry was the only true way to represent our physical reality, making it the only possible geometry. That began to change in the early 1800s, when mathematicians discovered that alternate

Hulton Archive/Getty Images

geometries that violate some of Euclid's principles could be developed into perfectly consistent systems. A century or so later, Einstein's theory of general relativity showed that Euclidean geometry describes our physical world well only where the gravitational field is relatively weak—like say here on Earth. We'll study some non-Euclidean geometries in Section 9-7.

If you study Figure 9-6 for a few seconds, a useful fact might occur to you: *two vertical angles have the same measure.* (You'll prove this in Problems 89–92.) Combining this fact with what we know about supplementary angles will allow us to find the measure of all four angles in a diagram like Figure 9-6 if we know just one of the angles. This is illustrated in Example 5.

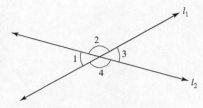

∡1 and ∡3 are vertical angles; $m∡1 = m∡3$
∡2 and ∡4 are vertical angles; $m∡2 = m∡4$

Figure 9-6

Example 5	**Using Vertical Angles**

Math Note

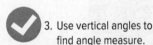

When two angles are supplementary, you can always find the measure of one by subtracting the measure of the other from 180°.

Find $m∡2$, $m∡3$, and $m∡4$ when $m∡1 = 40°$.

SOLUTION

Since ∡1 and ∡3 are vertical angles and $m∡1 = 40°$, $m∡3 = 40°$. Since ∡1 and ∡2 form a straight angle (180°), $m∡1 + m∡2 = 180°$, and

$$m∡2 = 180° - m∡1$$
$$= 180° - 40°$$
$$= 140°$$

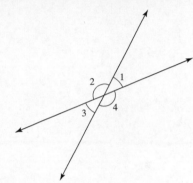

Finally, since ∡2 and ∡4 are vertical angles, $m∡4 = 140°$.

Try This One 5

Find $m∡1$, $m∡2$, and $m∡4$ when $m∡3 = 75°$.

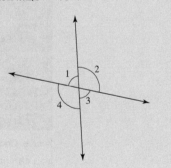

3. Use vertical angles to find angle measure.

Two lines in the same plane are called **parallel** if they never intersect. You might find it helpful to think of them as lines that go in the exact same direction. If two lines l_1 and l_2 are parallel, we write $l_1 \parallel l_2$. When two parallel lines are intersected by a third line, we call the third line a **transversal**. As you can see in Figure 9-7, eight angles are formed. The angles between the parallel lines (angles 3–6) are called **interior angles**, and the ones outside the parallel lines (angles 1, 2, 7, and 8) are called **exterior angles**. The following box defines special relationships among these angles that allow us to find the measure of all of the angles if we know just one.

When two parallel lines are intersected by a transversal:

• **Alternate interior angles** are the angles formed between the parallel lines on opposite sides of the transversal. Alternate interior angles have equal measures.

- **Alternate exterior angles** are the angles formed outside of the parallel lines on opposite sides of the transversal. Alternate exterior angles have equal measure.

- **Corresponding angles** consist of one exterior and one interior angle with no common vertex on the same side of the transversal. Corresponding angles have equal measures.

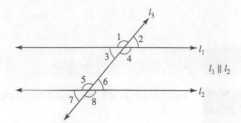

$l_1 \parallel l_2$

Pairs of alternate interior angles   Pairs of alternate exterior angles
∡3 and ∡6 $m\angle 3 = m\angle 6$ ∡1 and ∡8 $m\angle 1 = m\angle 8$
∡4 and ∡5 $m\angle 4 = m\angle 5$ ∡2 and ∡7 $m\angle 2 = m\angle 7$

Pairs of corresponding angles
∡1 and ∡5 $m\angle 1 = m\angle 5$
∡2 and ∡6 $m\angle 2 = m\angle 6$
∡3 and ∡7 $m\angle 3 = m\angle 7$
∡4 and ∡8 $m\angle 4 = m\angle 8$

Figure 9-7

Source: Google
You can find parallel roads and transversals on almost any city map.

| Example 6 | Finding Angles Formed by a Transversal |

Find the measures of all the angles shown when the measure of ∡2 is 50°.

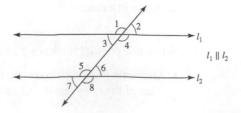

$l_1 \parallel l_2$

SOLUTION

First, let's identify the angles that have the same measure as ∡2. They are ∡3 (vertical angles), ∡6 (corresponding angles), and ∡7 (alternate exterior angles). Mark all of these as 50° on the diagram (Figure 9-8).

Since ∡1 and ∡2 are supplementary, $m\angle 1 = 180° - 50° = 130°$. This allows us to find the remaining angles: ∡4 is a vertical angle with ∡1, so it has measure 130° as well. Now ∡8 is a corresponding angle with ∡4, and ∡5 is an alternate interior angle with ∡4, which means they both have measure 130° as well.

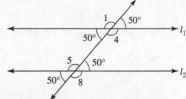

Figure 9-8

Try This One   6

4. Find measures of angles formed by a transversal.

Find the measures of all the angles shown when the measure of ∡1 is 165°.

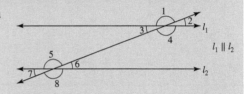

Answers to Try This One

1   (a) ∡2, ∡CAD, ∡DAC
      (b) ∡BAC, ∡CAD, ∡BAD

2   49°

3   106°

4 $36\frac{1}{2}°$ and $53\frac{1}{2}°$

5 $m\angle 1 = 75°,\ m\angle 2 = 105°,\ m\angle 4 = 105°$

6 $m\angle 2, \angle 3, \angle 6, \angle 7 = 15°;\ m\angle 4, \angle 5, \angle 8 = 165°$

Exercise Set   9-1

Writing Exercises

1. Explain why you can't actually draw a point or a line, just figures that represent them.
2. What is the difference between a half line and a ray?
3. Describe the four different ways that we name angles.
4. Explain why the two lines below are not parallel even though they don't meet in the diagram.

5. Describe how to find the complement and supplement of an angle.
6. When parallel lines are intersected by a transversal, there are four types of pairs of angles with equal measures formed. Describe each, using a diagram to illustrate.

Computational Exercises

For Exercises 7–12, identify and name each figure.

7.   A •——→ B

8.   ←——•——•——→ R   S

9.   ←————→ *l*

10.   • P

11.   •——• T   U

12.   ○——•——→ E   F

For Exercises 13 and 14, name each angle in four different ways.

13.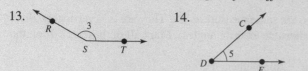

14.

For Exercises 15–18, name every angle in the figure.

15.

16.

17.

18.

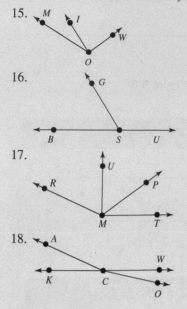

For Exercises 19–22, classify each angle as acute, right, obtuse, or straight.

19.   21.

20.   22.

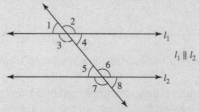

For Exercises 23–30, identify each pair of angles as alternate interior, alternate exterior, corresponding, or vertical.

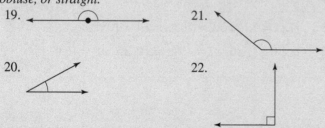

23. ∡1 and ∡4   27. ∡1 and ∡5
24. ∡3 and ∡6   28. ∡2 and ∡7
25. ∡2 and ∡6   29. ∡1 and ∡8
26. ∡5 and ∡8   30. ∡4 and ∡8

For Exercises 31–38, find the measure of the complement of each angle.

31. 8°   35. $18\frac{1}{4}°$
32. 24°   36. $81\frac{5}{8}°$
33. 32.4°   37. $(x + 10)°$
34. 56.8°   38. $(y - 5)°$

For Exercises 39–46, find the measure of the supplement for each angle.

39. 156°   43. $111\frac{5}{6}°$
40. 90°   44. $5\frac{3}{4}°$
41. 62.5°   45. $(y - 15)°$
42. 143.1°   46. $(x + 20)°$

In Problems 47–50, find the value of x if ∡A and ∡B are complementary angles.

47. $∡A = (x + 5)°$ $∡B = (2x - 15)°$
48. $∡A = (x - 30)°$ $∡B = (x + 45)°$
49. $∡A = (x - 8)°$ $∡B = (3x + 12)°$
50. $∡A = (5x + 1)°$ $∡B = (3x - 11)°$

In Problems 51–54, find the value of x if ∡A and ∡B are supplementary angles.

51. $∡C = (x + 27)°$ $∡D = (x + 18)°$
52. $∡C = (3x - 44)°$ $∡D = (5x - 20)°$
53. $∡C = (2x + 10)°$ $∡D = (4x - 28)°$
54. $∡C = (10x + 20)°$ $∡D = (20x + 10)°$

For Exercises 55–58, find the measures of ∡1, ∡2, and ∡3.

55.   56.

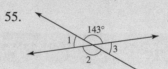

57.   58.

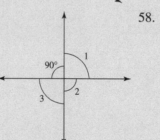

For Exercises 59–62, find the measure of ∡1 through ∡7.

59.

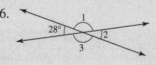
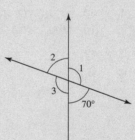

60.

61.   62.

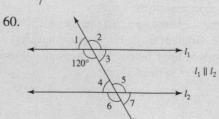

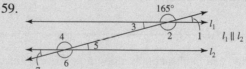

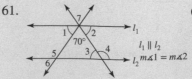

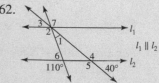

In Problems 63–66, find the measure of each marked angle.

63.

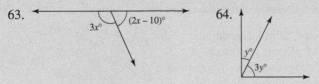

64.

65.

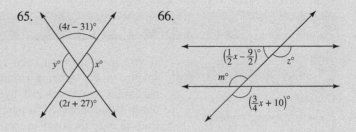

66.

Applications in Our World

For Exercises 67–72, identify the measure of the angle made by the hands of a clock at these times:

67. 3 o'clock
68. 6 o'clock
69. 2 o'clock
70. 4 o'clock
71. 5:30
72. 9:15

Exercises 73–80 use the following description of streets in downtown Cleveland: Euclid Avenue and Prospect Avenue are parallel, both running west to east, with Euclid north of Prospect. East Ninth Street is a transversal running northwest to southeast. The angle made with Euclid on the northwest corner is 76°. Find the angle made by the path of the car for each of the following turns.

73. Driving west on Euclid and turning northwest on E. Ninth

74. Driving east on Prospect and turning northwest on E. Ninth
75. Driving southeast on E. Ninth and turning east on Euclid
76. Driving northwest on E. Ninth and turning east on Euclid
77. Driving east on Euclid and turning southeast on E. Ninth
78. Driving west on Prospect and turning northwest on E. Ninth
79. Driving northwest on E. Ninth and turning west on Prospect
80. Driving southeast on E. Ninth and turning west on Prospect

In Problems 81–84, think about both the interior and exterior of a house. Write all the examples you can of:

81. A pair of parallel lines
82. A pair of lines that meet in a right angle
83. A pair of parallel planes
84. A pair of planes that meet in a right angle

Critical Thinking

Degrees are a perfectly good way to measure angles, but they're made up: somebody decided that a full rotation is 360°, and that's that. Another measure used for angles, radian measure, is based on a physical size. We'll use radian measure in Problems 85–88.

85. Draw a circle with center (0, 0) and radius 1, then using a piece of string, mark off the exact length of the radius on that string. The circle has to be perfect, so trace around a glass, or something round. Next, attach one mark on the string to the circle at the point (1, 0), and align the string along the curve of the circle, going counterclockwise. Put a mark on the circle where the other mark on the string is, then draw a ray from the center of the circle to the point you just marked. Congratulations! You've drawn an angle that measures **one radian**.

86. The length all the way around a circle is called the **circumference**; it can be calculated using the formula Circumference = $2\pi r$, where r is the radius. Based on this formula, what is the radian measure of one complete rotation?

87. Use your answer to Question 86 to find a conversion factor between degree and radian measure. You'll need to remember the number of degrees in one full rotation.

88. Use your conversion factor from Question 87 to find the radian measure of an angle that measures:
    (a) 90°   (b) 45°   (c) 120°   (d) 225°

In Problems 89–92, you'll prove one of the results from this section.

89. Write an equation describing the sum of the measures of angles 1 and 2 in the diagram below.

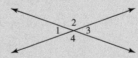

90. Repeat Question 89, but this time for angles 2 and 3.
91. Subtract the equation you wrote in Question 90 from the equation you wrote in Question 89.
92. Solve the result of Question 91 for the measure of angle 1. What result did you prove?

93. Fill in each blank. It might be helpful to refer to Exercises 15–18.
    When two rays meet at a common vertex, 1 angle is formed.
    When three rays meet at a common vertex, _____ angles are formed.
    When four rays meet at a common vertex, _____ angles are formed.
    Use your results to make a conjecture on the number of angles formed when five rays meet at a common vertex.

94. Draw a diagram with five rays meeting at a common vertex, and use it to decide if your answer to Question 93 is correct.

| **Triangles**

LEARNING OBJECTIVES

1. Identify types of triangles.
2. Find one missing angle in a triangle.
3. Use the Pythagorean theorem to find side lengths.
4. Use similar triangles to find side lengths.

If you look for them, you can find familiar geometric shapes in a surprising number of places. Architects in particular are fascinated by creating complex designs out of basic shapes. The pyramid structure at the famous Louvre in Paris is one of the most well-known examples, blending an assortment of simple shapes into an architectural masterpiece.

Photov.com/AGE Fotostock

As we continue our study of geometry in our world, we use points, lines, rays, and angles to form more complex figures, beginning with the triangle. Although they are very simple figures, triangles have many more applications to applied situations than you might think.

A geometric figure is said to be **closed** when you can start at one point, trace the entire figure, and finish at the point you started without lifting your pen or pencil off the paper. One such figure is the triangle:

A **triangle** is a closed geometric figure that has three sides and three angles. The three sides of the triangle are line segments, and the points where the sides intersect are called the **vertices** (plural of "vertex").

One way we can name a triangle is according to its vertices, using the symbol Δ. For example, we could call the triangle in Figure 9-9 Δ*ABC*. (In this case, the order of the vertices doesn't matter.)

Types of Triangles

Special names are given to certain types of triangles based on either the lengths of their sides or the measures of their angles.

For sides:

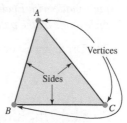

Figure 9-9

An **isosceles triangle** has at least two sides with the same length.

An **equilateral triangle** has three sides with the same length.

A **scalene triangle** has sides that are three different lengths.

Examples are shown in Figure 9-10(a).

Math Note

Because an isosceles triangle has two equal sides, it also has two equal angles. The equilateral triangle has three equal angles. The dash marks on the sides of the triangles indicate the sides have the same length.

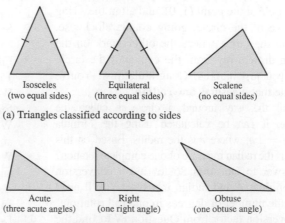

Isosceles
(two equal sides)

Equilateral
(three equal sides)

Scalene
(no equal sides)

(a) Triangles classified according to sides

Acute
(three acute angles)

Right
(one right angle)

Obtuse
(one obtuse angle)

(b) Triangles classified according to angles

Figure 9-10

For angles:

> An **acute triangle** has three acute angles (less than 90°).
>
> A **right triangle** has one right angle (90°).
>
> An **obtuse triangle** has one obtuse angle (greater than 90° and less than 180°).
>
> Examples are shown in Figure 9-10(b).

Example 1   Identifying Types of Triangles

Identify the type of triangle.

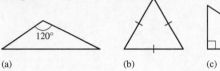

(a)   (b)   (c)

SOLUTION

(a) Obtuse triangle since one angle is greater than 90°.
(b) Equilateral triangle since all sides are equal, or acute triangle since all angles are acute. Also isosceles because it has at least two sides equal.
(c) Right triangle since one angle is 90°.

Try This One   1

Identify the type of triangle.

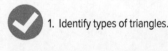

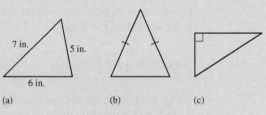

(a)   (b)   (c)

1. Identify types of triangles.

For a variety of reasons, it's very useful to know the following fact:

> **The Sum of Angle Measures in a Triangle**
>
> In any triangle, the measures of the three angles add to 180°.

We will outline a proof of this fact in Exercises 57–61, but you can see that it's true with a simple demonstration. Cut any triangle you like out of a piece of paper, then tear it into three pieces, one containing each angle. Now arrange the three angles next to each other, as shown in Figure 9-11: you will find that when the vertices of the three angles are placed together, they always form a straight line (or a 180° angle). Try it!

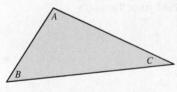

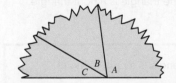

Figure 9-11

Example 2   Finding an Angle in a Triangle

Find the measure of angle C in the triangle.

SOLUTION

Since the sum of the measures of the angles of a triangle is 180°,

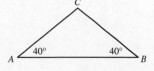

$$m\angle A + m\angle B + m\angle C = 180°$$
$$40° + 40° + m\angle C = 180°$$
$$80° + m\angle C = 180°$$
$$m\angle C = 180° - 80°$$
$$= 100°$$

$m\angle A$ and $m\angle B = 40°$.
Add on left side.
Subtract $80°$ from both sides.

The measure of angle C is $100°$.

Try This One   2

Find the measure of angle B.

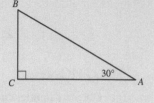

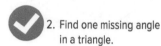
2. Find one missing angle in a triangle.

The Pythagorean Theorem

Right triangles are very special because there is a simple relationship among their sides. This is the famous Pythagorean theorem, which is attributed to the Greek mathematician Pythagoras around 500 BCE, although it was actually known much earlier in some form (see Sidelight on page 496). In any case, a **theorem** is a fact that has been proven true using deductive reasoning. This particular theorem allows us to find the third side of a right triangle if we know two sides. The side across from the right angle is called the **hypotenuse**, and the other two sides are called the **legs**.

The Pythagorean Theorem

The sum of the squares of the lengths of the two legs in a right triangle always equals the square of the length of the hypotenuse. If we use a and b to represent the lengths of the legs and c to represent the length of the hypotenuse, as in the figure, then

$$a^2 + b^2 = c^2$$

In addition, if the sides of any triangle satisfy this equation, the triangle is a right triangle.

CAUTION

Remember, the Pythagorean theorem only holds true if the triangle is a right triangle!

Example 3   Using the Pythagorean Theorem

(a) For the right triangle shown, find the length of side a.

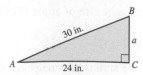

(b) Why does the hypotenuse have to be the longest side in any right triangle?

(c) Which of the triangles is right?

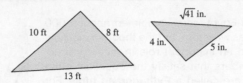

SOLUTION

(a) $a^2 + b^2 = c^2$ *Side b is 24 in. and side c is 30 in.*

    $a^2 + 24^2 = 30^2$ *Subtract 24^2 from both sides.*

        $a^2 = 30^2 - 24^2 = 324$ *Apply square root to both sides.*

        $a = \sqrt{324} = 18$ *Ignore the negative root; a is a length.*

The missing side is 18 inches long.

(b) The squares of the two legs have to add up to the square of the hypotenuse. Each of those squares has to be positive (both because they're squares, and because they come from physical lengths), so when you add them up, you get a result larger than either of the two original numbers. This shows that the square of the hypotenuse has to be bigger than the square of either two legs. That in turn guarantees that the hypotenuse is the longest side.

(c) If the triangle is right, the side lengths have to satisfy the Pythagorean theorem, and the longest side would need to be the hypotenuse. Let's try the first triangle:

$$10^2 + 8^2 \overset{?}{=} 13^2$$
$$100 + 64 \overset{?}{=} 169$$

This isn't true, so the triangle isn't right. It's close, and kind of looks like a right triangle, but not good enough.

For the second triangle:

$$4^2 + 5^2 \overset{?}{=} \sqrt{41}^2$$
$$16 + 25 \overset{?}{=} 41$$

This is a true statement, so the triangle is right.

Calculator Guide

To find the length in Example 3:

Typical Scientific Calculator:

Typical Graphing Calculator:

2nd x^2 30 x^2 − 24 x^2)
ENTER

Online Calculator:

√ 30 x^2 − 24 x^2 =

Most online calculators will either insert parentheses after you choose the square root operator or will insert the calculation after it inside a radical.

Try This One   3

(a) If the hypotenuse of a right triangle is 13 feet long and one leg is 5 feet long, find the length of the other leg.

(b) Explain why the right angle has to be the largest angle in any right triangle.

(c) Which of the triangles is right?

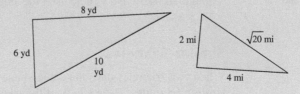

Notice that we ignored the negative root when solving for the length of side a in Example 3. Since we're finding a physical length when using the Pythagorean theorem, we'll keep doing that as we move forward.

Sidelight   Pythagoras and the Pyramids

The great pyramids in Egypt were completed about 2,000 years before Pythagoras was born, but the theorem that now bears his name played an important role in their construction. Three whole numbers that satisfy the equation in the Pythagorean theorem are called a **Pythagorean triple**. For example, $3^2 + 4^2 = 5^2$, so 3, 4, and 5 form a Pythagorean triple. So if a triangle has legs of length 3 units and 4 units, and a hypotenuse of length 5 units, it must be a right triangle.

This brings us back to the pyramid builders. Right angles are tremendously important in building structures, especially massive structures like the pyramids. If all of the angles are off by just a little bit, the result is a crooked pyramid. To ensure that angles

Ablestock/Alamy Stock Photo

were right angles, the Egyptians used a rope divided into even sections of 3, 4, and 5 units long with knots. If the rope wouldn't stretch into a taut triangle with sides of length 3 and 4 against an edge that was supposed to be a right angle, then they would know that the angle was off.

Example 4   Applying the Pythagorean Theorem

SW Productions/Brand X Pictures/ Getty Images

Triangles play a very important role in roof construction.

To build the frame for the roof of a shed, a carpenter has to cut a 2 × 4 to fit on the diagonal. If the length of the horizontal beam is 12 feet and the height is 3 feet, find the length of the diagonal beam.

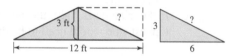

SOLUTION

The frame forms two right triangles with legs of 3 feet and 6 feet. The length of the diagonal beam can be found using the Pythagorean theorem.

$$c^2 = a^2 + b^2 \quad \textit{Use } a = 3, b = 6.$$
$$= 3^2 + 6^2 \quad \textit{Square each.}$$
$$= 9 + 36 \quad \textit{Add.}$$
$$= 45 \qquad \textit{Apply square root to both sides.}$$
$$c = \sqrt{45}$$
$$\approx 6.7 \text{ feet}$$

The length of the beam should be about 6.7 feet, or about 6 feet $8\frac{1}{2}$ inches.

Try This One   4

The rectangular frame for a large sign is 10 feet long and 8 feet high. Find the length of a diagonal beam that will be used for bracing. (*Hint:* Draw a diagram.)

 3. Use the Pythagorean theorem to find side lengths.

Similar Triangles

When two triangles have the same shape but not necessarily the same size, they are called **similar triangles**. Have a look at the triangles in Figure 9-12. Since the measure of angle A and the measure of angle A' are equal, they are called *corresponding angles*. Likewise, angle B and angle B' are corresponding angles since they have the same measure. Finally, angle C and angle C' are corresponding angles. When two triangles are similar, then their corresponding angles will have the same measure. It works both ways, actually: if all of the corresponding angles have equal measures, then the triangles have to be similar.

Math Note

Actually, since the sum of all three angle measures in a triangle is 180°, if two triangles have two pairs of corresponding angles equal, the third pair of angles must be equal as well, and they are similar triangles.

The sides that are opposite the corresponding angles are called *corresponding sides*. When two triangles are similar, the ratios of the corresponding sides are equal. For the two triangles shown in Figure 9-12,

$$\frac{\text{Length of side } AB}{\text{Length of side } A'B'} = \frac{10}{30} = \frac{1}{3}$$

$$\frac{\text{Length of side } AC}{\text{Length of side } A'C'} = \frac{4}{12} = \frac{1}{3}$$

$$\frac{\text{Length of side } BC}{\text{Length of side } B'C'} = \frac{6}{18} = \frac{1}{3}$$

Since these three ratios are all equal, we can describe the situation by saying that the lengths of the corresponding sides are in proportion. What does it really tell us? That the bigger triangle is exactly

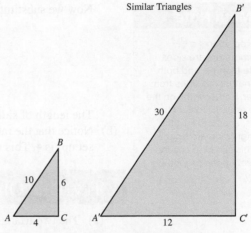

Similar Triangles

$m\angle A = m\angle A', m\angle B = m\angle B', m\angle C = m\angle C'$

$$\frac{\text{Length of side } AB}{\text{Length of side } A'B'} = \frac{\text{Length of side } BC}{\text{Length of side } B'C'} = \frac{\text{Length of side } AC}{\text{Length of side } A'C'}$$

Figure 9-12

three times the size of the smaller one. Think of it as a computer graphic that's been blown up to 300%. When you look at it that way, it makes perfect sense that when two triangles are similar, the ratio of corresponding sides is constant: if one side is three times as big, the others have to be also, or the triangles wouldn't be the same shape. In summary:

Math Note

The similar triangle relationships can be written more concisely as a single equation with two equal signs in it, but we chose to write it as three equations since, in practice, we choose two sets of corresponding sides and set up an equation with just one equal sign and two sides.

Similar Triangle Relationships

If triangle *ABC* is similar to triangle *A'B'C'* then

$$\frac{\text{Length of side } AB}{\text{Length of side } A'B'} = \frac{\text{Length of side } AC}{\text{Length of side } A'C'}$$

$$\frac{\text{Length of side } AB}{\text{Length of side } A'B'} = \frac{\text{Length of side } BC}{\text{Length of side } B'C'}$$

$$\frac{\text{Length of side } AC}{\text{Length of side } A'C'} = \frac{\text{Length of side } BC}{\text{Length of side } B'C'}$$

Example 5   **Using Similar Triangles**

(a) If the two triangles below are similar, find the length of side $B'C'$.

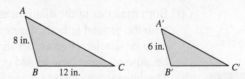

(b) How would you describe the relationship between the sizes of the two triangles?

SOLUTION

(a) To begin, we will use the variable x to represent the length we're asked to find. Now we want to write ratios of corresponding sides. Sides AB and $A'B'$ are corresponding, and sides BC and $B'C'$ are as well. So the ratios formed from these corresponding sides will be equal:

$$\frac{\text{Length of side } AB}{\text{Length of side } A'B'} = \frac{\text{Length of side } BC}{\text{Length of side } B'C'}$$

Math Note

In the proportion $\frac{8}{6} = \frac{12}{x}$, both numerators are sides from one triangle, and both denominators are sides from the other. Both lengths on the left side are left sides of the triangles, and both lengths on the right side are bottom sides. This can help ensure you've set up the proportion correctly.

Now we substitute in the lengths and solve for x.

$$\frac{8}{6} = \frac{12}{x} \quad \textit{Cross multiply.}$$

$$8x = 72 \quad \textit{Divide both sides by 8.}$$

$$x = 9$$

The length of side $B'C'$ is 9 inches.

(b) Notice that the ratio of the corresponding sides (seen on the left side of the proportion we set up) is $\frac{4}{3}$. This tells us that the first triangle is $1\frac{1}{3}$ times as big as the second.

Try This One   **5**

(a) The two triangles shown are similar. Find the length of side AC.

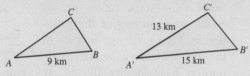

(b) How would you describe the relationship between the sizes of the two triangles?

One clever use of similar triangles is to measure the height of objects that are difficult to measure directly. Example 6 shows how to use this form of indirect measurement.

Example 6   **Using Similar Triangles in Measurement**

Morey Milbradt/Alamy Stock Photo

If a tree casts a shadow 12 feet long and at the same time a person who is 5 feet 10 inches tall casts a shadow of 5 feet:

(a) Explain why the two triangles in the diagram are similar.
(b) Find the height of the tree.

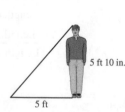

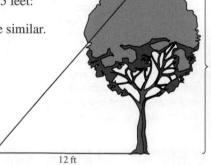

SOLUTION

(a) Both triangles in the diagram are right triangles, and since the sun makes the same angle with the ground in both diagrams, the triangles have two corresponding angles equal, and must be similar. (Remember that if two of the angles are equal, the third has to be as well because all three have to add up to 180°.)

Math Note

In this solution, note that we had to write the height of the person (5′10″) as the mixed number $5\frac{10}{12}$.

(b) That means we can set up and solve a proportion:

$$\frac{\text{Height of person}}{\text{Length of person's shadow}} = \frac{\text{Height of tree}}{\text{Length of tree's shadow}}$$

$$\frac{5\frac{10}{12}}{5} = \frac{x}{12} \quad \textit{Cross multiply.}$$

$$5x = 70 \quad \textit{Divide both sides by 5.}$$

$$x = 14 \text{ feet}$$

The height of the tree is 14 feet.

4. Use similar triangles to find side lengths.

Try This One 6

Find the length of a pole if it casts a 20-foot shadow at the same time that a 5′3″ woman casts a 15-foot shadow.

Answers to Try This One

1 (a) Scalene and acute
(b) Isosceles and acute
(c) Right

2 60°

3 (a) 12 feet
(b) The angles add up to 180°. If one is 90°, the other two are 90° combined, so both have to be less than 90°.
(c) Both

4 12.8 feet

5 (a) 7.8 km
(b) The second triangle is $1\frac{2}{3}$ times as big as the first.

6 7 feet

Exercise Set 9-2

Writing Exercises

1. Describe the three ways that triangles can be classified based on lengths of sides.
2. Describe the three ways that triangles can be classified based on measures of angles.
3. Explain why the following statement is incorrect: the sum of the squares of the lengths of two sides in a triangle is equal to the square of the length of the third side.
4. What does it mean for two triangles to be similar?
5. Explain how to find the measure of the third angle in a triangle if you know the measures of the other two.

6. Explain how to find the length of the third side of a right triangle when you know the lengths of the other two sides.
7. What does it mean for a geometric figure to be closed? Is a plane a closed figure?
8. Explain why two triangles have to be similar if they have two pairs of corresponding angles that are equal.
9. What did right triangles have to do with the construction of the great pyramids in Egypt?
10. Explain how you can use the length of a shadow to measure a tall object that you can't measure directly.

Computational Exercises

For Exercises 11–16, classify each triangle.

11.
3 ft 3 ft
2 ft

12.

13.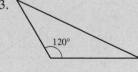
120°

14.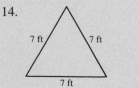
7 ft 7 ft
7 ft

15.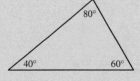
80°
40° 60°

16.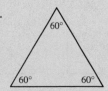
60°
60° 60°

For Exercises 17–22, find the measure of angle C.

17.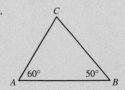
C
60° 50°
A B

18.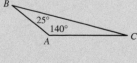
B
25°
140°
A
C

19.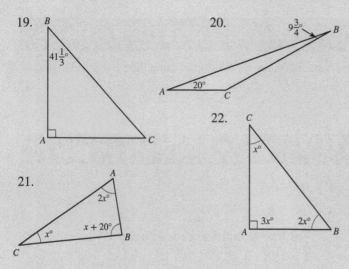

20.

21.

22.

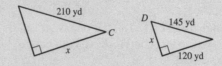

For Exercises 35–40, explain why the two triangles are similar, then find the length x.

35. $m \angle A = 35°$; $m \angle B = 35°$

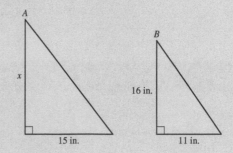

36. $m \angle C = 49°$; $m \angle D = 41°$

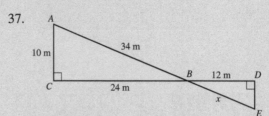

In Exercises 23–30, a and b represent the lengths of the legs of a right triangle with hypotenuse length c. Find the length of the missing side.

23. $a = 16$ ft, $b = 30$ ft
24. $a = 272$ in., $b = 510$ in.
25. $a = 8.4$ yd, $b = 11$ yd
26. $a = 5$ cm, $b = 9.3$ cm

27. $b = 9$ mi, $c = 21$ mi
28. $b = 26$ ft, $c = 29$ ft
29. $a = 4\frac{1}{2}$ ft, $c = 12\frac{1}{8}$ yd
30. $b = 27\frac{3}{4}$ in., $c = 4\frac{1}{2}$ ft

37.

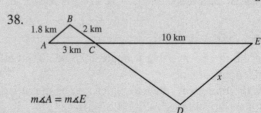

For Exercises 31–34, the two triangles drawn are similar. Find the value of x. Then describe the relationship between the sizes of the two triangles.

31.

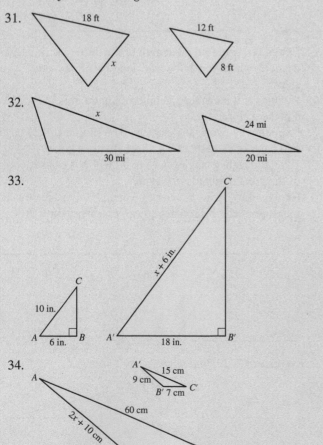

32.

33.

34.

38.

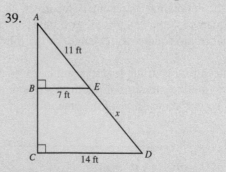

$m \angle A = m \angle E$

39.

40.

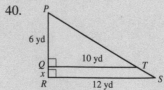

Applications in Our World

41. A baseball diamond is really a square with the bases at the corners. It's 90 feet from home plate to first base. How far is it from home plate to second base?

42. Television screens are sized according to the length of a diagonal across the screen, from one corner to another. The screen of a 52 in. wide-screen TV is 24.5 in. high. How wide is it?

43. Kendall needs to fix a loose screen in a second-story bedroom. The bad spot is 17′4″ off the ground. The manufacturer of a 14-foot ladder recommends for safety reasons that the base be placed at least 5 feet from a wall it's leaning against. Will Kendall be able to safely reach the screen if she can reach 4′3″ above the spot where the ladder hits the wall?

44. My swimming pool is an 18 × 36 foot rectangle. The vacuum hose has to be long enough to reach from one corner to the one diagonally opposite. What's the length of the shortest hose that will work?

45. A carpenter needs to build a stairway down to a basement in a house under construction. The vertical drop is 8′9″, and in order to keep the staircase from being too steep it will extend 14 feet horizontally. The main supports will go diagonally from the top of the stairs to the basement floor. How long do they need to be?

46. What's the tallest piece of plywood sheeting that you can take through a door that's 6′8″ tall and 32″ wide? (You can disregard the thickness of the plywood.)

47. For a triathlon, the athletes start at point *A*, swim directly across Siegel Lake to point *B*, then cycle to point *C*, a distance of 7 miles. They then discard their bikes and run 2 miles to point *D*, then turn to run 4 miles to point *E*, then go under the tunnel another 2 miles back to point *C*. They then finish the race back on their bicycles for 7 miles to the finish line at point *A*. What is the distance they swam across Siegel Lake if the measure of angle *A* is the same as the measure of angle *E*?

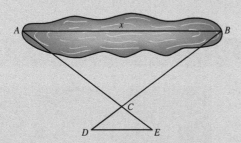

48. At the local mall, a group of friends ate lunch at the food court, located at point *A*. Afterward, they walked 60 yd to the Apple store (point *C*) to look at new iPads. From there, they went 25 yd to Macy's at point *B* to try on some shoes. After, they continued walking 150 yd to point *D*, where they saw the latest rom-com. After the show, they found their car at point *E*. How far away from the movie theater was the car?

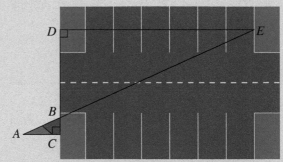

49. Find the height of the tree.

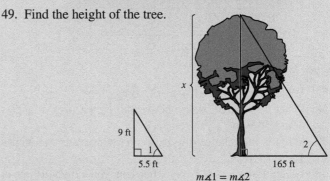

$m\angle 1 = m\angle 2$

50. Find the height of the tower.

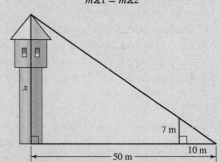

51. To build a loft in their dorm room, Kevin and Neil had to figure out how long to cut a beam so that the beam would run diagonally from a point on the floor 6 feet away from the wall to the top of the 8-foot-high wall. How long should they cut the beam?

52. How high up on a wall is the top of a 20-foot ladder if its bottom is 6 feet from the base of the wall?

53. A plane flies 175 miles north then 120 miles due east. How far diagonally is the plane from its starting point?

54. Two cell phone towers are 20 feet tall and 38 feet tall and they sit 30 feet apart. How far is it from the top of one tower to the other?

55. Two cars are stopped next to each other at the corner of Fifth and Elm, two streets that are perpendicular.
    (a) If Juliette goes straight on Fifth Street at 30 miles per hour, and Omar stalls at the light, how far apart will they be in 5 minutes?
    (b) If Juliette turns right onto Elm and Omar turns left onto Elm, and they both drive at 30 miles per hour, how far apart will they be in 5 minutes?
    (c) If Juliette goes straight on Fifth and Omar turns left onto Elm and they both drive 30 miles per hour, how far apart will they be in 5 minutes?

56. (a) Refer to Exercise 55. If Juliette turns right onto Elm and Omar makes a U-turn, heading back down Fifth, and they're 1.7 miles apart after 2 minutes, how fast were they going (in miles per hour) if their average speeds were the same?

(b) Under the same circumstances, what was the average speed for each (in miles per hour) if Juliette was driving twice as fast as Omar?

Critical Thinking

In Exercises 57–61, we will prove that the measures of the angles in a triangle sum to 180°, using the diagram below.

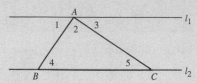

Lines l_1 and l_2 are parallel.

57. What is $m\angle 1 + m\angle 2 + m\angle 3$? Why?
58. What is the relationship between the measures of angles 3 and 5? Why?
59. What is the relationship between the measures of angles 1 and 4? Why?
60. Using Exercises 47–49, what can you conclude about $m\angle 2 + m\angle 4 + m\angle 5$?
61. Discuss whether this proves the result for every triangle, or if there is anything special about this diagram that makes it apply only to certain triangles.
62. Explain how the Pythagorean theorem guarantees that the hypotenuse is the longest side in any right triangle.
63. The length of a side in a triangle depends on the measure of the angle across from it—larger angles means longer sides. This observation, combined with Exercise 62, shows that in a right triangle, the right angle is the largest angle. How else can you show this is true?
64. In the right triangle below, it can be shown that the lengths of the segments have these properties:

$$\frac{\text{Length of side } AD}{\text{Length of side } CD} = \frac{\text{Length of side } CD}{\text{Length of side } DB}$$

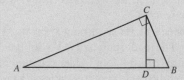

Show the proportion is true by using similar triangles.

65. Michelle and Brad hire a local contractor to build a stone retaining wall in their yard. When he's finished, Michelle thinks the left edge of the wall looks like it's not quite perpendicular to the ground. The wall is 12′7″ long and 2′8″ tall at the left edge. Describe how Michelle and Brad could use just a tape measure and their knowledge of triangles to see if the left side is in fact perpendicular to the ground. Then find what the measurement your method describes will be if the contractor did a good job.

66. Mika hired a landscaper to build three triangular flower beds in her back yard. She insisted that all three have the exact same shape but differ in size depending on the type of flowers to be planted (kind of a picky request, but the customer's always right). When the work was done, Mika measured the sides of all three beds and got the following measurements:

Bed 1: 10′3″ × 7′9″ × 14′
Bed 2: 17′5″ × 13′2″ × 23′9½″
Bed 3: 7′8¼″ × 5′9¾″ × 10′6″

Did Mika ask for her money back? How did you decide?

Section 9-3　　Polygons and Perimeter

 LEARNING OBJECTIVES

1. Find the sum of angle measures of a polygon.
2. Find the angle measures of a regular polygon.
3. Find the perimeter of a polygon.

High-rise buildings capture our imagination and get most of the attention, but did you know that as recently as 2008, the largest building in the United States had only five floors? The Pentagon in suburban Washington, D.C., has a floor area of over $6\frac{1}{2}$ million square feet, and is one of the world's finest examples of the importance of geometry in architecture.

　　Triangles, as we've seen, are very useful in many settings. But of course they are limited to three sides. In this section, we will study closed figures with more than three sides, including the five-sided figure known as the pentagon.

Digital Vision/Getty Images

Polygons

Closed geometric figures whose sides are line segments are classified according to the number of sides. These figures are called *polygons.* Table 9-2 shows the number of sides and some of the shapes of these polygons: **triangle, quadrilateral, pentagon, hexagon, heptagon, octagon, nonagon, decagon, dodecagon,** and **icosagon.**

Table 9-2	Basic Polygons	
Name	**Number of sides**	
Triangle	3	
Quadrilateral	4	
Pentagon	5	
Hexagon	6	
Heptagon	7	
Octagon	8	
Nonagon	9	
Decagon	10	
Dodecagon	12	
Icosagon	20	

Math Note

The word "polygon" comes from the Greek word "polygonos," meaning "many-angled."

We know from Section 9-2 that the measures of the angles in a triangle always add up to 180°. Are triangles special in that regard? Or is there something similar we can say about polygons that have more than three sides? Let's check it out.

Example 1	Studying Angles in a Polygon

(a) Notice that the diagram at right shows that any quadrilateral (four-sided polygon) can be divided into two nonoverlapping triangles. Use the diagram to find the sum of the measures of the interior angles.

(b) Try to draw a similar diagram for a pentagon (five-sided polygon), and use it to find the sum of the measures of the interior angles.

SOLUTION

(a) The angles in each triangle add up to 180°, and all of the angles in BOTH triangles combine to form all of the angles in the quadrilateral. So, those angle measures must add up to $2 \times 180° = 360°$.

(b) At right is what the diagram looks like:
Since the figure is completely filled in by three triangles, the sum of the angle measures will be $3 \times 180° = 540°$.

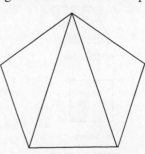

Try This One   1

(a) Draw a diagram that shows a six-sided polygon filled in with a certain number of triangles and use it to find the sum of the angle measures in a hexagon.

(b) Come up with a general procedure or formula for finding the sum of angle measures in a polygon with *n* sides.

Ingram Publishing/Alamy Stock Photo

Viktor_Gladkov/Image Club/ Stockphoto/Getty Images

Stockbyte/Getty Images

Many shapes found in nature are polygons.

If all went well in Example 1 and Try This One 1, you noticed that in every case, you can fill in a polygon with *n* sides by using *n* − 2 triangles. Each of those has angles that sum to 180°, which gives us the following general formula for the sum of angle measures in a polygon.

> **The Sum of Angle Measures of a Polygon**
>
> The sum of the measures of the interior angles of a polygon with *n* sides is $(n - 2)180°$.

Example 2 | **Finding the Sum of Angle Measures of a Polygon**

1. Find the sum of angle measures of a polygon.

Find the sum of the measures of the angles of a heptagon.

SOLUTION

According to Table 9-2, a heptagon has seven sides, so the sum of the measures of the angles of the heptagon is

Use n = 7.

$$(n - 2)180° = (7 - 2)180°$$
$$= 5 \cdot 180°$$
$$= 900°$$

The sum of the measures of the angles of a heptagon is 900°.

Try This One   2

Find the sum of the measures of the angles of an icosagon.

Quadrilaterals

Just as there are special names for certain types of triangles, there are names for certain types of quadrilaterals as well. (Recall that a quadrilateral is a polygon with four sides.)

A **trapezoid** is a quadrilateral that has exactly two parallel sides. See Figure 9-13(a).

A **parallelogram** is a quadrilateral in which opposite sides are parallel and equal in measure. See Figure 9-13(b).

A **rectangle** is a parallelogram with four right angles. See Figure 9-13(c).

A **rhombus** is a parallelogram in which all sides are equal in length. See Figure 9-13(d).

A **square** is a rhombus with four right angles. See Figure 9-13(e).

Types of Quadrilaterals

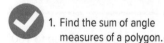

(a) Trapezoid

(b) Parallelogram

(c) Rectangle

(d) Rhombus

(e) Square

Figure 9-13

The next diagram explains the relationships of the quadrilaterals. Looking at the relationships, you can see that a square is also a rectangle and a rhombus. A rhombus and a rectangle are also parallelograms.

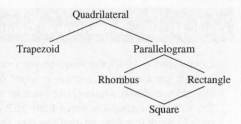

Most of the polygons drawn in Table 9-2 are actually special polygons: all of the sides have the same length, and all of the angles are equal in measure. We call such a figure a **regular polygon**. The most common examples of regular polygons are squares and equilateral triangles.

| Example 3 | Finding Angle Measure for a Regular Polygon |

2. Find the angle measures of a regular polygon.

Find the measure of each angle of a regular hexagon.

SOLUTION

First, find the sum of the measures of the angles for a hexagon. The formula is $(n - 2) \cdot 180°$, where n is the number of sides. Since a hexagon has six sides, the sum of the measures of the angles is $(6 - 2) \cdot 180° = 720°$. Next, divide the sum by 6 since a hexagon has six angles: $720 \div 6 = 120°$.

Each angle of a regular hexagon has a measure of 120°.

Try This One   3

Find the measure of each angle of a regular pentagon.

Perimeter

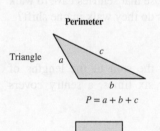

Triangle

$P = a + b + c$

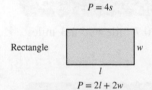

Square

$P = 4s$

Rectangle

$P = 2l + 2w$

Figure 9-14

Perimeter

The **perimeter** of a polygon is the sum of the lengths of its sides. The perimeter of a triangle with sides of length a, b, and c is simply $P = a + b + c$. For a square with side length s, the perimeter is $P = s + s + s + s$, or $P = 4s$. For a rectangle with length l and width w, there are two sides that are l units long, and two that are w units long, so the perimeter is $P = l + l + w + w$, or $P = 2l + 2w$. See Figure 9-14.

| Example 4 | Finding the Perimeter of a Rectangle |

The Houser family finds their dream home perfect in every way except one: the backyard is not fenced in, and their dog Bunch needs room to roam. The rectangular portion they plan to enclose is 95 feet wide and 70 feet long. How much fence will they need to enclose the yard on all four sides?

SOLUTION

The amount of fence needed is the perimeter of the rectangle.

$$P = 2l + 2w$$
$$= 2(70) + 2(95)$$
$$= 140 + 190 = 330$$

The Housers need 330 feet of fence.

Try This One   4

The intramural field at Fiesta University is a large square measuring 600 yards on a side. Find the perimeter.

Sidelight   **A Triumph of Geometry**

Completed in 1943, the Pentagon is a marvel of architectural design, packing an incredible amount of floor space into a five-story building. Although it covers an area of just 29 acres, the total floor space is more than 152 acres. Over 25,000 people work there on an average day; this would make it the fourth-largest city in Alaska. There are 17.5 *miles* of corridors in the building; at an average walking pace, it would take almost 6 hours to walk all of them. And yet because of the geometric design, it takes at most 7 minutes to walk from any point in the building to any other.

Getty Images/Digital Vision/Getty Images

Example 5	Finding the Perimeter of a Polygon

The length of each outside wall of the Pentagon is 921 feet. Suppose that sentries have to walk the outside wall six times during a 4-hour shift. How many miles do they walk in one shift?

SOLUTION

A pentagon has five sides, and each has length 921 feet, so the sum of the lengths of the sides is $5 \times 921 = 4{,}605$ feet. In walking the perimeter six times, a sentry covers $6 \times 4{,}605 = 27{,}630$ feet. Now we convert to miles:

$$\frac{27{,}630 \text{ feet}}{1} \times \frac{1 \text{ mi}}{5{,}280 \text{ feet}} \approx 5.23 \text{ miles}$$

Try This One   5

The running path at a state park is a right triangle with legs 0.7 mile and 1.3 miles. What's the total length around the path?

 3. Find the perimeter of a polygon.

Answers to Try This One

1   (a) 720°

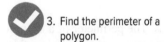

(b) The sum of the angles for an *n*-sided polygon is $(n - 2) \cdot 180°$.

2   3,240°

3   108°

4   2,400 yards

5   About 3.5 miles

Exercise Set	9-3

Writing Exercises

1. Is a circle a polygon? Why or why not?
2. How can you find the sum of the angle measures for a polygon?
3. What makes a polygon regular?
4. How can you find the measure of the angles for a regular polygon?

5. What is the perimeter of a polygon?
6. Is the perimeter of a pentagon five times the length of one side? Why or why not?

Computational Exercises

For Exercises 7–12, identify each polygon and find the sum of the measures of the angles.

7.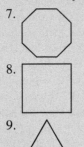

10.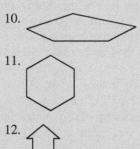

11.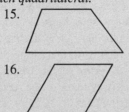

8.

9.

12.

For Exercises 13–16, identify each quadrilateral.

13.

15.

14.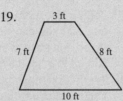

16.

For Exercises 17–32, find the perimeter of the pictured or described polygon. Dotted lines are for measurement reference only and don't affect the perimeter. If a polygon appears to be regular, you can assume that it is.

17.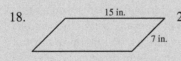
22 yd
16 yd

19.
3 ft
7 ft   8 ft
10 ft

18.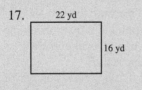
15 in.
7 in.

20.
9 cm

21.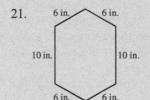
6 in.   6 in.
10 in.   10 in.
6 in.   6 in.

22.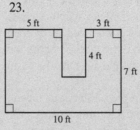
8 yd
10 yd
5 yd
20 yd

23.
5 ft   3 ft
4 ft
7 ft
10 ft

24.
3 in.   4.3 in.
2 in.   1.6 in.
3.4 in.

25.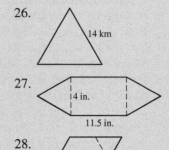
7 mi

26.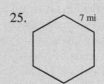
14 km

27.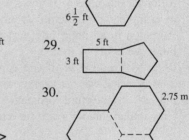
4 in.
11.5 in.

28.
$6\frac{1}{2}$ ft

29.
5 ft
3 ft

30.
2.75 m

31. A regular icosagon with sides of length 11.2 cm
32. A regular dodecagon with sides of length 4′3″
33. A heptagon has perimeter 110 feet. Four of the sides are the same length, and the remaining sides are half as long. How long are the shorter sides?
34. The four sides of an octagon that are vertical or horizontal are all one length, while the four slanted sides are 4 inches less. If the perimeter is 124 inches, how long are the shorter sides?

Applications in Our World

35. At least how far does a Major League player run when he hits a home run? The baseball diamond is a square with sides 90 feet.
36. A rectangular plot of land where outdoor concerts are held is about to be enclosed by a fence. How many feet of fence will be needed if the plot is 110 yards on one side and 270 feet on the other, and requires a 4-foot opening for entrances on two of the four sides?
37. How many feet of hedges will be needed to enclose a triangular display at an amusement park if the sides measure 62 feet, 85 feet, and 94 feet?
38. Jim and Joe have built a rectangular stage for their band that measures 40 feet by 56 feet. They want to put fiber optic lighting around the border that comes in 8-foot sections for $24.00 each. Find the cost of the lighting for their stage.
39. How much molding in length will be needed to frame an 11 × 14-inch picture if there is to be a 2-inch mat around the picture?
40. A carpenter needs to put baseboard around the room shown. How many feet are needed? Each door is 30 inches wide.

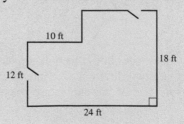
10 ft
12 ft
18 ft
24 ft

41. How many times would you have to walk around a football field in order to walk a mile? The dimensions of a football field are 360 feet by 160 feet. One mile is 5,280 feet.

42. How many times would you have to walk around a soccer field in order to walk a mile? The dimensions of a soccer field are 345 feet by 223 feet.

43. Ultimate Fighting matches take place in a steel cage that is a regular octagon with sides 12 feet 3 inches. Jed and Bubba decide to build an Ultimate Fighting cage in Jed's backyard, using fence they can buy at Home Depot for $4 per foot. How much will it cost to build the cage?

44. For a LiveStrong walkathon, Cat is walking her dogs Macleod and Tessa around the perimeter of a rectangular park that is 0.4 miles by 0.7 miles. Her sponsors have pledged a total of $22.50 per mile. If Cat and the girls do four laps around the park, how much money will they raise?

45. Refer to Example 5. If the sentry was told he needed to complete two circuits around the perimeter of the Pentagon in an hour, how fast would he have to go in miles per hour?

46. Refer to Exercise 44. If Cat completed her four laps in 2 hours, what was her average speed in miles per hour?

47. For an architecture project, Lauren is designing a rectangular outdoor space for a town center. She wants a perimeter of 620 yards, and because she's a fan of the golden ratio (see Sidelight on page 256 if you're interested), she wants the longer sides to be 1.618 times as long as the shorter. Find the lengths of each side.

48. An A-frame cabin is being designed so that the front of the building is an isosceles triangle with base 6 feet less than the lengths of the identical sides forming the roof line. The perimeter of the front is 67′6″. How long is the base?

49. A triangular plot of land is being surveyed as a possible building site for a company planning to relocate its factory. The company's architect believes that the site will need at least 1,000 feet of frontage to fit the company's needs. The longest side is 400 feet longer than the next shorter side, which is 175 feet longer than the shortest. If the perimeter is 3,000 feet, which (if any) of the sides will work for frontage?

50. A two-story house is designed so that the front of the house looks like an isosceles triangle resting on a rectangle, which is 46 feet along the bottom and 19 feet high. The top vertex of the triangle is 12 feet directly above the halfway mark of the rectangle's top. The owner wants to string Christmas lights around the entire perimeter of the front. How many feet of lights will be needed?

Critical Thinking

51. For the triangle shown, ∡*BCD* is called an exterior angle. If you know the measures of ∡*A* and ∡*B*, explain how the measure of ∡*BCD* can be found.

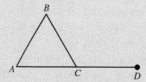

52. What is the measure of each exterior angle of a regular pentagon? (See Problem 51.)

53. Make a table with the number of sides in a regular polygon in one column and the measure of each angle in the other (see Example 2). Start with 8 sides and increment by 2 until you reach 20 sides. Do you see any pattern? Where do you think that pattern is headed as the number of sides gets larger and larger?

54. When asked to draw a polygon with a certain number of sides, almost everyone draws a regular polygon, or at least something close to it. What about funkier polygons? Do they still follow the same rules? Consider the polygon below:

(a) How many sides does this polygon have?
(b) There are two different size interior angles in this polygon: five of the angles are acute and five are obtuse. Use a protractor to measure the two sizes of angles. (You can find a printable protractor online if you don't have access to one.) Do all of the angles add up to what they're supposed to?

(c) Try to divide the polygon into a certain number of nonoverlapping triangles. If you can do so, do you get the right number of triangles to give you a sum of angles that matches the formula within the section?

(d) Based on your observations, do you think the work we did on finding the sum of interior angles in Example 1 can be generalized to polygons that contain obtuse angles?

55. A city park is being planned with a rectangular walking path around the outside. The original plans called for the path to be 2,400 feet on one side and 1,300 on the other. Due to a dispute with local landowners, the plans need to be altered so that the path is moved 50 feet inward on all sides.

(a) Find the length of the original path in miles.
(b) How many feet less is the new path?
(c) What percentage of the length was lost by moving the sides in by 50 feet?

56. Each of the following figures is a regular polygon. Measure the side length for each, then the distance *d* from any angle to the one directly across from it (making sure to go through the center of the figure), and use that information to fill in the table. What do you notice? (It would be a BIG help to make a copy of the figures and enlarge them with a copier or scanner. Big, I tell you.)

# sides	
Perimeter	
Distance	
Perimeter	
Distance	

Section 9-4

Areas of Polygons and Circles

✓ LEARNING OBJECTIVES

1. Find areas of rectangles and parallelograms.

2. Find areas of triangles and trapezoids.

3. Find circumferences and areas of circles.

In the 21st century, more and more people are taking on home improvement projects that would have been done only by professionals 20 years ago. Stores like Lowe's and Home Depot have become regular stops for homeowners looking to personalize their little corner of the world.

Suppose that you plan to install ceramic tile in your kitchen to cover up that ridiculous yellow linoleum—what were the previous owners thinking? Tiles come in many different sizes, and while planning the job, you want to make sure you buy enough tile. But you also don't want to buy TOO much. A calculation of area is just the ticket to make sure the job is well planned and ultimately successful. In this section we will learn how to find the area of various geometric figures.

Areas of Polygons

From our study of measurement in Chapter 8, we already know that the area of a geometric figure is a measure of the region bounded by its sides. We also know that area is measured in square units, like square feet or square meters. In Example 1, we'll think about computing the area of a rectangle.

Comstock/PunchStock/Getty Images

Example 1 — Finding the Area of a Rectangular Room

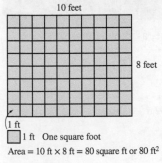

10 feet

8 feet

1 ft

☐ 1 ft   One square foot

Area = 10 ft × 8 ft = 80 square ft or 80 ft²

Figure 9-15

Let's say that your kitchen is a rectangle measuring 8 feet by 10 feet, and that the tiles you've picked out are 12 inches on a side. Then each tile covers exactly one square foot of space, because that's the definition of square foot: a square that is one foot on each side. Figure 9-15 shows a nice, efficient way to lay out the tiles in your kitchen.

(a) How many tiles would be needed?
(b) How many square feet is your kitchen?
(c) Use the result of this example to write a formula for finding the area of a rectangle when you know the length and width.

SOLUTION

(a) If you look at the tiles in the diagram, you can see that there are 8 rows, each of which has 10 tiles. So there must be a total of $8 \times 10 = 80$ tiles.

(b) Each tile covers 1 square foot, so the area of the kitchen is 80 square feet.

(c) There's nothing special about 10 feet and 8 feet here: no matter what the length and width are, you'd have a number of rows corresponding to one dimension, each of which has the number of tiles corresponding to the other dimension. Conclusion: the area of any rectangle is found from multiplying the length and width: $A = lw$.

Try This One   1

A homeowner plans to lay a patio in the backyard using paver stones that are 12 in. by 6 in. rectangles. The patio will be 16 ft by 12 ft.

(a) What will be the total area of the patio?
(b) How many paver stones will be needed?

Area Formulas for Rectangles and Squares

The area of a rectangle is the product of the length and width. If l is the length and w is the width, then

$$A = lw$$

In a square, the length and width are equal, so if s is the length of the sides,

$$A = s^2$$

Example 2   Finding the Cost of Installing Carpet

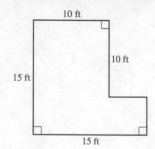

A couple plans to carpet an L-shaped living room, as shown to the left. Find the total cost of the carpet if it is priced at $25.00 per square yard.

SOLUTION

With a little bit of ingenuity, we can divide the room into two figures we know the area of: a rectangle and a square.

$$A = lw \qquad\qquad A = s^2$$
$$= 15 \cdot 10 \qquad\quad = 5^2$$
$$= 150 \text{ square feet} \quad = 25 \text{ square feet}$$

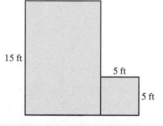

The total area is 150 square feet + 25 square feet = 175 square feet.
    Now we can use dimensional analysis to finish the calculation:

$$175 \text{ ft}^2 \times \frac{1 \text{ yd}^2}{9 \text{ ft}^2} \times \frac{\$25}{1 \text{ yd}^2} = \$486.11$$

It will cost $486.11 to carpet the room.

Try This One   2

Two homeowners plan to install sod around their new house, as shown.

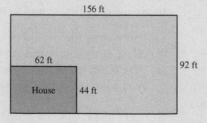

If sod costs $3.98 per square yard, find the total cost.

Figure 9-16

Once we know the formula for the area of a rectangle, we can use it to develop formulas for other polygons. We'll start with the parallelogram, shown in Figure 9-16.
    The key dimensions here are the base (length of the bottom side) and height (vertical distance from the bottom side to the top side). The trick is to "cut" the bottom left corner piece off and attach it to the right side. This turns the parallelogram into a rectangle, and we can find the area using length times width.

Area Formula for Parallelograms

The area of a parallelogram is the product of the base and the height. If b is the length of the base and h is the height,

$$A = bh$$

CAUTION

Be careful when working with parallelograms! The base is the length of the bottom (or top) side, but the height is *not* the length of the left or right side. It is the vertical distance from the bottom side to the top side.

Example 3   **Finding the Area of a Parallelogram**

Find the area of the parallelogram:

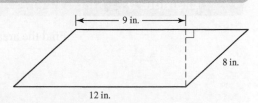

SOLUTION

We're given the base but not the height, so we'll need to find that first. The dashed line forms a right triangle with hypotenuse 8 inches and one leg 3 inches (we know the top side of the parallelogram is the same length as the bottom side (12 inches) by the definition of parallelogram). So we can use the Pythagorean theorem to find the height. See how this stuff is all fitting together?

$$h^2 + 3^2 = 8^2 \Rightarrow h^2 = 55 \Rightarrow h = \sqrt{55}$$

Now we use the formula for the area of a parallelogram.

$$A = bh = 12 \text{ in.} \times \sqrt{55} \text{ in.} = 12\sqrt{55} \text{ in}^2 \approx 89 \text{ in}^2$$

Try This One   **3**

Find the area of the parallelogram:

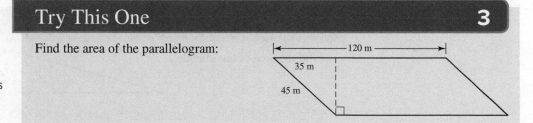

 1. Find areas of rectangles and parallelograms.

Now that we're on a roll, let's keep it going.

Example 4   **Finding the Area of a Triangle**

Figure 9-17

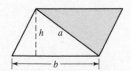

(a) Start with the diagram in Figure 9-17. If you make an identical copy of that triangle, flip it upside down, and place it on top of the original triangle so that the sides labeled a match up, what figure will you get?
(b) Find the area of that figure in terms of b and h.
(c) This new figure is made up of two copies of the original triangle. Use that fact to develop a formula for the area of a triangle.

SOLUTION

(a) Following the provided directions, the diagram will look as shown at left.
    The resulting figure is a parallelogram.
(b) Not only is the figure a parallelogram, it's a parallelogram with base b and height h, which means it has area $A = bh$.
(c) To turn this into the area of the original triangle, we just need to divide by 2, because the parallelogram has two copies of the triangle. So the area of the triangle is $A = \frac{1}{2}bh$.

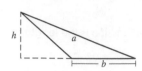

Try This One 4

Does the procedure in Example 4 still work for the obtuse triangle shown here? Draw the diagram and discuss.

Area Formula for Triangles

The area of a triangle is half the base times the height. If b is the length of the base and h is the height, then

$$A = \frac{1}{2}bh$$

Example 5 | Finding the Area of a Triangle

Find the area of the triangle shown.

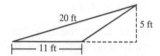

SOLUTION

There are two things we need to be careful of here. First, notice that the height of the triangle is still the perpendicular height from the base to the highest vertex, even if that height (seen as a dotted line here) lives outside the triangle. Second, there's an extraneous piece of information in the diagram: all we need to know is the base (11 ft) and the height (5 ft). The 20-foot length isn't needed. So the area is

$$A = \frac{1}{2}(11)(5) = \frac{55}{2}\,\text{ft}^2$$

Try This One 5

Find the area of the triangle:

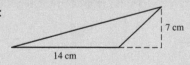

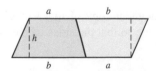

Figure 9-18

We can also use parallelograms to find the area of a trapezoid. In the first drawing in Figure 9-18, we see the relevant dimensions. It's important to recall that the top and bottom sides are parallel. In the second figure, we have an identical copy of the original trapezoid flipped both horizontally and vertically: it fits together with the original to form a parallelogram with area $h(a + b)$. Since this parallelogram consists of two copies of the trapezoid, the area of the trapezoid is half as much.

Elliot, Elliot/Getty Images

How would you find the area of the front of this house to estimate the amount of paint needed to paint it?

Math Note

When finding the area of a trapezoid, make sure that a and b are the lengths of the parallel sides.

Area Formula for Trapezoids

The area of a trapezoid with parallel sides a and b and height h is

$$A = \frac{1}{2}h(a + b)$$

Example 6	**Finding the Area of a Trapezoid**

Find the area of the trapezoid:

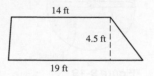

SOLUTION

The two parallel sides have length 14 and 19 feet, and the height is 4.5 feet:

$$A = \frac{1}{2}h(a + b) = \frac{1}{2}(4.5)(14 + 19) = 74.25 \text{ square feet.}$$

Try This One   6

Find the area of the trapezoid shown.

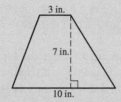

 2. Find areas of triangles and trapezoids.

Sidelight **Calculus and Area**

We now have a variety of formulas for finding area, but how would you find the area of a figure that's curved or oddly shaped, like the front of the building shown here? This is one of the major topics of calculus. In calculus, we learn how to approximate areas, and in many cases find exact areas of figures when we don't have basic formulas for them.

It's all based on a really simple idea: we're good at computing areas of rectangles, so to approximate areas of other shapes, we'll fit little rectangles inside the shape, and sum the areas of the rectangles. This is not terribly different than the way we started with the simple rectangle formula in this section and used it to build other area formulas.

Tony Weller/Getty Images

Math Note
The center of a circle is inside the circle, but is not actually part of the circle.

Circles

By definition, polygons have sides that are line segments. The most common geometric figure that doesn't fit that criterion is the circle—no part of a circle is straight.

> A **circle** is the set of all points in a plane that are the same distance from a fixed point, which we call the **center** of the circle.

Math Note
An interesting fact: manhole covers are circular because there are just a small handful of shapes that make it completely impossible for the lid to fall into the hole, and the circle is by far the simplest of them.

Based on the definition of circle, a line segment from any point on a circle to the center is always the same length. We call this length the **radius** of the circle. A line segment starting at a point on a circle, going through the center, and ending at a point on the opposite side is called a **diameter** of the circle, and its length is twice the radius. This simply represents the distance all the way across the circle. If we use r to represent radius and d to represent diameter,

$$d = 2r \quad \text{and} \quad r = \frac{d}{2}$$

The distance around the outside of a circle is called the **circumference** (C) of the circle. (This is analogous to the perimeter of a polygon.) The key parts of a circle are illustrated in Figure 9-19.

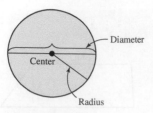

- Diameter
- Center
- Radius

Figure 9-19

Thousands of years ago, people started to realize that if you divide the circumference of any circle by its diameter, the result is always the same number. Eventually, this number was given a special name, the Greek letter π (pronounced **pi**). We can't write an exact value of pi, because it's irrational, so its decimal expansion is infinitely long. But 3.14 is commonly used as an approximation. (See the Sidelight on page 254 for a discussion of pi.)

The fact that circumference divided by diameter always equals pi gives us a formula for the circumference of a circle.

Circumference Formula for Circles

The circumference of a circle is π times the diameter, or 2π times the radius:

$$C = \pi d \quad \text{or} \quad C = 2\pi r$$

Example 7   Finding the Distance Around a Track

Calculator Guide

As a substitute for using the approximation $\pi \approx 3.14$, most calculators have π built in. To find $\pi \cdot 300$ in Example 4:

Typical Scientific or Online Calculator

 × 300 =

Typical Graphing Calculator

 × 300 ENTER

A dirt track is set up for amateur auto racing. It consists of a rectangle with half-circles on the ends, as shown in Figure 9-20. The track is 300 yards wide, and 700 yards from end to end. What is the distance around the track?

SOLUTION

From the diagram, we can see that the diameter of the circular ends is 300 yards, so the circumference is $C = \pi d = \pi(300) \approx 942$ yards. (This is the total length of the curved portion, since the two half-circles make one full circle.) The length of each straightaway is the total length of the track (700 yards) minus twice the radius of the circular ends, which is 150 yards (see Figure 9-21). So each straightaway is $700 - 2(150) = 400$ yards. Now we can find the total length:

$$L = 942 + 400 + 400 = 1{,}742 \text{ yards}$$

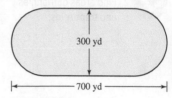

Figure 9-20

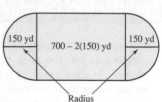

- Radius

Figure 9-21

Try This One   **7**

The entrance to Antonio's living room is an arch consisting of a rectangle that is 7 feet high and 4 feet wide with a half-circle on top. How many feet of trim would be needed to go around the outside of the arch, not including the bottom?

Math Note

It's kind of cool to note that what took hundreds, if not thousands of years, for humanity's great thinkers to identify—that pi is the ratio of a circle's area to the square of its radius—became a very simple computation with the invention of calculus.

Through exhaustive experiments involving comparing areas of circles to other known figures, the Greeks found that the ratio of a circle's area to the square of its radius is π. This provides our next area formula:

Area Formula for Circles

The area of a circle is pi times the square of the radius: $A = \pi r^2$

Example 8   Finding the Area Enclosed by a Track

Find the area enclosed by the track in Example 7.

SOLUTION

The area is the sum of the area of a rectangle and the area of a circle (again because the two half-circles form one full circle).

Rectangle

$A = lw$
$= 400 \cdot 300$
$= 120{,}000$

Circle

$A = \pi r^2$
$= \pi \cdot 150^2$
$\approx 70{,}650$

The combined area is about $120{,}000 + 70{,}650 = 190{,}650$ square yards.

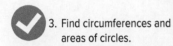
3. Find circumferences and areas of circles.

Try This One   8

Find the area of the doorway in Try This One 7.

Sidelight   Math and the Orbits of Comets

For most of our history, humans have been fascinated and mystified by the appearance of comets. But starting in the 16th century, astronomers were able to use math to unlock the secrets of comets and were able to predict their appearances.

The most famous such comet is Halley's Comet. It has been appearing for millions of years, but its occasional appearances remained a mystery until British astronomer Sir Edmond Halley began studying it in 1704. He noticed that earlier records showed that a comet had appeared in the same region of the sky in 1456, 1531, 1607, and 1682. He was the first person to suggest that it was the same comet appearing every 75 or 76 years, and correctly predicted that

it would reappear in 1758. (Sadly, he didn't live to see his prediction come true.)

When a circle is stretched in one direction, the resulting figure is called an ellipse. Scientists have determined that the paths comets take through the solar system are elliptical in shape. Armed with this knowledge and some data, they can use math to calculate the path, speed, and appearance dates of some comets.

solarseven/iStock/Getty Images

Answers to Try This One

1   (a) 192 ft^2 (b) 384   **2** $5,140.39

3 $\approx 3{,}394$ square meters

4   Yes, it still results in a parallelogram with area bh.

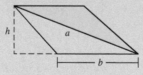

5   49 square centimeters

6   45.5 square inches

7   20.3 feet

8   34.3 square feet

Exercise Set   9-4

Writing Exercises

1. How can we use the formula for the area of a rectangle to find the area of a parallelogram?
2. How can we use the formula for the area of a parallelogram to find the area of a triangle?
3. Explain the difference between perimeter and circumference.
4. What is the connection between the number π and the dimensions of a circle?

5. Explain the difference between area and perimeter.
6. Which polygon measure is of interest to you in each situation? Explain.
   (a)   The amount of tile needed to refloor a room
   (b)   Installing a fence around the outside of your back yard

Computational Exercises

For Exercises 7–26, find the area of each figure.

7.
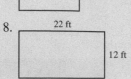
17 in.

8.
22 ft
12 ft

9.
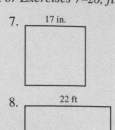
15 yd
30 yd

10.
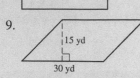
105 cm
150 cm

11.
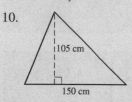
20 m
20 m

12.
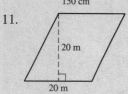
10 ft
14 ft

13.
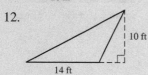
26 mi
24 mi

14.
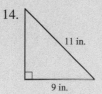
11 in.
9 in.

15.
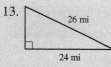
20 in.
17 in.
35 in.

16.
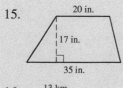
13 km
19 km
24 km

17.
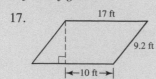
17 ft
9.2 ft
←10 ft→

18.
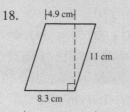
|4.9 cm|
11 cm
8.3 cm

19.
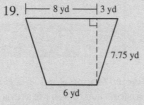
|— 8 yd —|3 yd|
7.75 yd
6 yd

20.
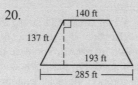
140 ft
137 ft
193 ft
285 ft

21.
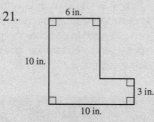
6 in.
10 in.
3 in.
10 in.

22.
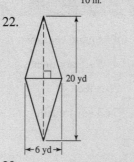
20 yd
←6 yd→

23.
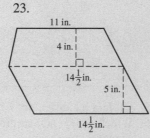
11 in.
4 in.
$14\frac{1}{2}$ in.
5 in.
$14\frac{1}{2}$ in.

24.
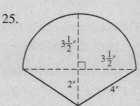
$6\frac{1}{2}$ m
2 m
$8\frac{3}{4}$ m
1.7 m

25.
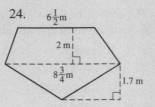
$3\frac{1}{2}''$
$3\frac{1}{2}''$
2"
4"

26.
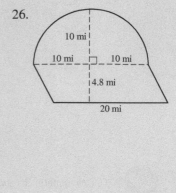
10 mi
10 mi   10 mi
4.8 mi
20 mi

For all calculations involving π, answers may vary slightly depending on whether you use π = 3.14 or the π key on a calculator.

For Exercises 27–32, find the circumference and the area of each circle.

27.
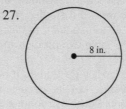
8 in.

28.
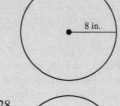
10 ft

29.
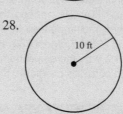
16 m

30.
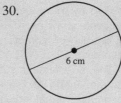
6 cm

31.
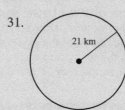
21 km

32.
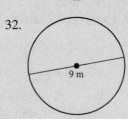
9 m

In Problems 33–38, find the area of the shaded region.

33.
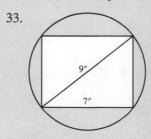
9"
7"

34.
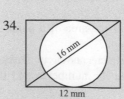
16 mm
12 mm

35.
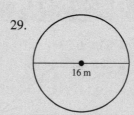
10 ft
10 ft
15 ft

36.
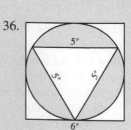
5"
5"   5"
6"

37. 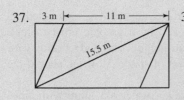 38.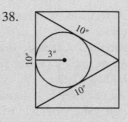

39. A circle has area 32 square feet. Find the radius.
40. A circle has area 9 square inches. Find the diameter.
41. A parallelogram has area 100 square miles and two of the parallel sides have length 12 miles. How far apart are those parallel sides?
42. A trapezoid has bases 9.2 feet and 11.3 feet and an area of 48 square feet. How far apart are the bases?

Applications in Our World

43. How many square yards of carpeting are needed to cover a square dorm room that measures 10 feet on a side?
44. Find the cost of coating a rectangular driveway that measures 11 feet by 21 feet, at $2.50 per square foot.
45. Jewell printed off several 3-inch by 5-inch pictures from her digital camera onto a sheet of glossy photo paper measuring 24 by 25 inches. How many can she paste onto a 600-square-inch poster board if the images are aligned side by side so no white space is left?
46. Find the amount and the cost of artificial turf needed to cover a football field that measures 360 feet by 160 feet. The cost of the turf is $20.00 per square foot.
47. A stage floor shaped like a trapezoid with bases of 60 feet and 75 feet and a height of 40 feet is to be covered with plywood for a play. What would be the cost for the plywood if it sells for $0.60 a square foot?
48. Derek and Amir have started a lawn care service, and their first big job is installing new sod on a trapezoid-shaped lawn. The parallel sides are 41 and 62 feet, and the vertical distance between the parallel sides is 80 feet.
    (a) If the sod costs them $7.00 per square yard, find the cost of sod.
    (b) If Derek and Amir can put down sod at the rate of 900 square feet per hour and charge $65 per hour for their services plus a 20% markup on the sod, how much should they write the invoice for when finishing the job?
49. The front of an A-frame cabin is an isosceles triangle with base 16 feet. The two other sides of the triangle are each 18 feet long. The front has a rectangular doorway measuring 88″ by 34″ and two windows that are 3-foot by 3-foot squares. The rest is covered with wood siding.
    (a) Find the area of the sided part of the house front.
    (b) The owner of the cabin wants to repaint the front of the cabin because it faces the sun all day and has faded. The paint chosen covers 350 square feet per gallon, and the owner plans on two coats. If the paint comes in gallon cans, how many should the owner buy? How much paint will be left over?
50. The fabric for a triangular team banner with base 6 feet and height twice as much costs $6.98 per square yard. How much will the fabric cost if 10% extra is bought to allow for waste?
51. Janelle and Sandra are planting various colors of flowers in a circular shape on the front lawn of their university to display the school's new logo. The circle has a diameter

of 12 feet and the flowers cost $15.00 per square yard. Find the cost of the flowers.
52. How much more pizza do you get in a large pizza that has a diameter of 15 inches than you do in a small pizza that has a diameter of 10 inches?
53. The broadcast antenna at a college radio station sends out a signal with a range of 18 miles. Advertisers are charged 45 cents per week for each square mile of area the signal covers. How much is a week's worth of advertising?
54. Find the distance around the inside lane of a track that has the dimensions shown below.

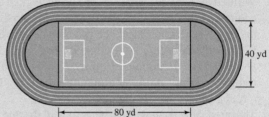

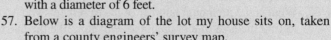

55. For a Homecoming Game Gala, the cheerleading squad wants to sell ice cream to earn money for new equipment. They plan to make a sign in the shape of an ice cream cone with the dimensions shown here. Find the area of the sign so they know how much plywood to buy.

56. Find the area of the walkway (24 inches wide) that will be in-stalled around a circular jacuzzi with a diameter of 6 feet.
57. Below is a diagram of the lot my house sits on, taken from a county engineers' survey map.
    (a) What assumptions do you have to make in order to find the area using information from this section?
    (b) Using your assumptions from part (a), find the area in square feet and acres. (One acre is 43,560 square feet.)

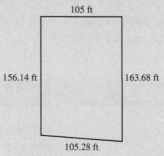

58. The next diagram is the lot owned by our friends across the street, the Pettys.
    (a) What assumptions do you have to make in order to find the area using information from this section?
    (b) Using your assumptions from part (a): Is their lot bigger or smaller than ours? By how many square feet? How many acres?

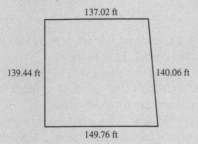

137.02 ft

139.44 ft   140.06 ft

149.76 ft

59. A rectangular billboard to advertise a new medical center opening is designed by a graphic artist to have length 12 feet more than height. The budget allows for an area of 240 square feet of space. How tall should the billboard be?

60. An engineer is designing a water ski jump with sides in the shape of right triangles. To get the desired slope, the length of the base should be 3.4 times the height of the triangles, and in order to fit in the allotted space at the ski park, the area of each triangle should be no more than 30 square feet. Find the dimensions of the largest triangle that fits these criteria.

Critical Thinking

61. Under what conditions will the area of a circle be the same as its circumference?

62. We used the formula for the area of a parallelogram to develop a formula for the area of a trapezoid, but it can also be done by dividing a generic trapezoid into two triangles. Give it a try!

63. Review the definition of a circle on page 513. How can you use that definition to draw a perfect circle using things you can find lying around the house? (Tracing a circular object doesn't count!)

64. The area of a triangle can be found if the measures of the three sides are known. This formula was discovered about 100 BCE by a Greek mathematician known as Heron. Heron's formula is

$$A = \sqrt{s(s-a)(s-b)(s-c)}$$

    where $s = \frac{1}{2}(a+b+c)$ and a, b, and c are the measures of the lengths of the sides of the triangle. Using the formula, find the area of a triangle if the sides are 5 in., 12 in., and 13 in.

65. The triangle in Problem 64 is also a right triangle. Find the area using the formula $A = \frac{1}{2}bh$ and see if you get the same answer.

66. What is the minimum number of measurements you need to find the area of a nonregular pentagon, like the one in the diagram? Explain how you came up with your answer.

67. Suppose that you want to fence in a rectangular plot of land, and you buy 80 feet of fence.

    (a) How much area will you enclose if you make one side 5 feet long?
    (b) Find the area enclosed if you make one side 10 feet, 15 feet, and 20 feet.
    (c) What do you think is the largest area you can enclose? What is the smallest?

68. Suppose that you plan to enclose 400 square feet of space in the shape of a rectangle with a fence.
    (a) Find the amount of fence you will need if you make one side 40 feet.
    (b) Find the amount of fence you will need if you make one side 35 feet, 30 feet, 25 feet, and 20 feet.
    (c) What do you think is the least amount of fence you can use? What about the most?

69. Suppose that the track in Exercise 54 has 8 lanes, each of which is 48 inches wide.
    (a) How much longer is one lap around the outside lane than the inside lane? (Assume that a runner has the sense to run as far inside their lane as possible.)
    (b) If two runners both run 10 laps at a pace of 6 minutes per mile, one in the inside lane and one in the outside lane, how much sooner would the inside runner finish?

70. A regular hexagon can be divided into six equilateral triangles. Use this fact to find the area of a regular hexagon that has 2-foot-long sides.

71. Look back at Figure 9-16. How do we know the cut-off piece of parallelogram fits perfectly on the other side? Use theorems about angles from earlier in this chapter to prove that it fits.

| Section 9-5 | Volume and Surface Area |

Owning a swimming pool sounds pretty great, especially during those hot summer weekends. But a lot of work goes into maintaining a pool, and it takes a while to learn everything you need to know. A surprising amount of math goes into it, too. Once you fill up your pool with

LEARNING OBJECTIVES

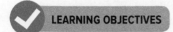

1. Find the volumes of three-dimensional figures.
2. Find the surface areas of three-dimensional figures.

sparkling clear water, it won't stay that way very long if you don't add chemicals. But how much should you add? The amount of chemicals you need depends on the capacity of the pool, and that's where the math comes in.

In Section 8-2, we learned that capacity is a measure of volume, which is the amount of space enclosed by a three-dimensional object. We know that volume is measured in cubic units, like cubic inches or cubic meters, and we also have conversions to write volume in terms of familiar capacity units like gallons or liters.

Comstock Images/Jupiter Images/ Alamy Stock Photo

A **polyhedron** is a three-dimensional figure bounded on all sides by polygons. The simplest polyhedra are rectangular solids and cubes, which are bounded on all sides by rectangles or squares.

We learned that the definition of cubic units in measuring volume is based on the capacity of a cube with the same unit dimension on all sides. For example, 1 cubic inch is defined to be the volume of a cube that is 1 inch on all sides. We can use this to develop a formula for the volume of a rectangular solid.

Example 1   Developing the Formula for the Volume of a Rectangular Solid

(a) A rectangular solid is shown in Figure 9-22(a). We could divide it into a number of smaller cubes, as shown in Figure 9-22(b). Use this to find the volume of the solid.
(b) Adapt the procedure you did in part (a) to find a formula for the volume of any rectangular solid.

SOLUTION

(a) With a little bit of visualization, we can literally count the number of cubes the solid is divided into. There are two layers, each of which has six cubes, so there are 12 cubes total. Each of those cubes is 1 in. on all sides, so each has a volume of 1 cubic inch. That means the 12 cubes combined (which forms the rectangular solid) have volume 12 cubic inches.
(b) Here's how we found the number of cubes: the width and length of the solid (3 in. and 2 in., in this case) were multiplied to tell us how many cubes there are in each layer. We then recognize that there are two layers because the height is 2 in. Ultimately, we multiply length, width, and height to get the volume, so there's our formula: if a rectangular solid has length l, width w, and height h, the volume is $V = lwh$.

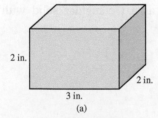

2 in.

2 in.

3 in.

(a)

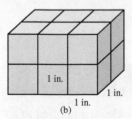

1 in.

1 in.

1 in.

(b)

Figure 9-22

Try This One   **1**

(a) A cube is a special case of a rectangular solid: one that has the same measurement for all of length, width, and height. Use that fact to find a formula for the volume of a cube.
(b) How many cubes that are 1 ft on all sides will fit inside a larger cube that is 3 ft by 3 ft by 3 ft?

Here's a summary of the first two volume formulas that we've discovered:

Volume Formulas for Rectangular Solids and Cubes

The volume of a rectangular solid is the product of the length, width, and height. For a cube, all three dimensions are equal, so the volume is the length of a side raised to the third power (cubed).

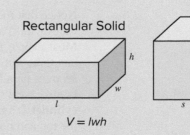

Cube

Rectangular Solid

h

w

l

$V = lwh$

s

$V = s^3$

| **Example 2** | Finding the Volume of a Rectangular Solid |

11 ft

30 in.

4 ft

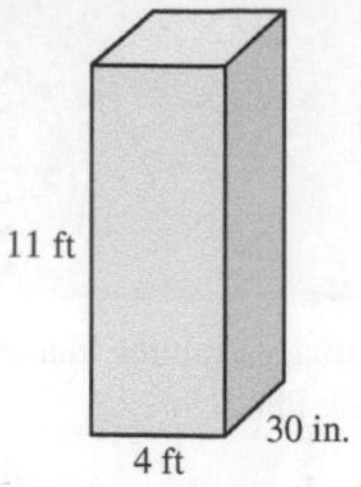

Figure 9-23

Find the volume of the rectangular solid pictured in Figure 9-23.

SOLUTION

The length is 4 feet, the width is 30 inches, and the height is 11 feet. First, we need to rewrite 30 inches in terms of feet:

$$30 \text{ in.} \times \frac{1 \text{ ft}}{12 \text{ in.}} = \frac{30}{12} \text{ ft} = 2.5 \text{ ft}$$

Now we use the volume formula with $l = 4$, $w = 2.5$, and $h = 11$:

$$V = lwh = 4 \times 2.5 \times 11 = 110 \text{ ft}^3$$

Try This One   2

Find the volume of a rectangular solid with length 10 meters, width 12 meters, and height 50 centimeters.

| **Example 3** | Finding the Volume of a Swimming Pool |

In order to figure out how much chlorine to use, Judi needs to know the capacity of her pool. The pool is a rectangle 18 feet wide and 36 feet long, and has an average depth of 4 feet. How many gallons does it hold? (There are 7.48 gallons in 1 cubic foot.)

SOLUTION

With an average depth of 4 feet, we can think of the pool as a rectangular solid with $l = 36$ ft, $w = 18$ ft, and $h = 4$ ft.

$$V = lwh = 36 \times 18 \times 4 = 2,592 \text{ ft}^3$$

Now we use dimensional analysis to convert to gallons:

$$2,592 \text{ ft}^3 \times \frac{7.48 \text{ gal}}{1 \text{ ft}^3} \approx 19,388 \text{ gal}$$

The capacity of the pool is about 19,388 gallons.

Try This One   3

A rectangular dunking tank at a carnival is 4 feet long and 5 feet wide, and is filled to a depth of 4 feet. How many gallons of water were needed to fill it?

Notice that the volume of a rectangular solid can be thought of as the area of the base (length times width) multiplied by the height. The same is true for our next solid figure, the cylinder. Most people think "round" when they hear the word "cylinder," but actually a cylinder is any three-dimensional figure with two parallel bases that are identical shapes, and all cross-sections parallel to those bases are that same shape as well. So technically a rectangular solid is also a cylinder. In this section, we'll study a special type of cylinder: a **right circular cylinder** is a figure with a circular top and bottom, and with sides that make a right angle with the top and bottom. (Think of a soda can.)

Math Note

Since the area of a circle with radius r is πr^2, the volume of a right circular cylinder is the area of the base times the height, just like a rectangular solid.

Volume Formula for a Right Circular Cylinder

Cylinder

The volume of a right circular cylinder is given by the formula

$$V = \pi r^2 h$$

where r is the radius of the circular ends and h is the height.

$$V = \pi r^2 h$$

Example 4   Finding the Volume of a Cylinder

How many cubic inches does a soup can hold if it is a right circular cylinder with height 4 inches and radius 1.5 inches? How many ounces? (One ounce is about 1.8 cubic inches.)

SOLUTION

Using $r = 1.5$ in. and $h = 4$ in., we get

$$V = \pi r^2 h = \pi \cdot 1.5^2 \cdot 4 = 9\pi \approx 28.3 \text{ in}^3$$

Back to our old friend dimensional analysis to convert to ounces:

$$28.3 \text{ in}^3 \times \frac{1 \text{ oz}}{1.8 \text{ in}^3} = 15.7 \text{ oz}$$

Try This One   4

Find the volume of an oil storage tank that is a right circular cylinder with height 40 feet and radius 15 feet. How many quarts of oil will it hold? (One cubic foot is about 29.9 quarts.)

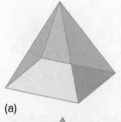

(a)

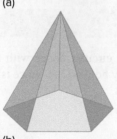

(b)

Figure 9-24

A right circular cylinder is not a polyhedron because the edges aren't polygons, but our next figure is. A **pyramid** is a polyhedron whose base is a polygon and whose sides are triangles. When you think of pyramids, you probably picture pyramids with a square base, like the one in Figure 9-24(a). But the base can be any polygon. A pyramid with pentagonal base is shown in Figure 9-24(b).

Volume Formula for a Pyramid

The volume of a pyramid is given by the formula

$$V = \frac{1}{3} Bh$$

where B is the area of the base and h is the height.

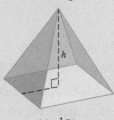

$$V = \tfrac{1}{3} Bh$$

Math Note

Notice that the formula for the volume of a pyramid also multiplies the area of the base by the height. But the factor of $\frac{1}{3}$ is needed because with the pyramid tapering to a point, the volume has to be a lot less than a cylinder with the same base.

Example 5 | Finding the Volume of a Pyramid

The Great Pyramid at Giza was built by the Egyptians roughly 4,570 years ago. It has a square base measuring 230.6 meters on a side and is 138.8 meters high. Find its volume.

SOLUTION

Since the base is a square that is 230.6 meters on a side, its area is 230.6^2, or 53,176.36 square meters. Using the volume formula with $B = 53,176.36$ and $h = 138.8$, we get

$$V = \frac{1}{3} \cdot 53{,}176.36 \cdot 138.8 \approx 2{,}460{,}293 \text{ m}^3$$

Math Note

For some perspective, the volume of the Great Pyramid found in Example 5 is almost two and a half times the volume of the Empire State Building, which was the tallest building in New York City until the new World Trade Center surpassed it in 2012.

Try This One   5

The smallest of the three great pyramids is the Pyramid of Menkaure. Its base measures 339 feet on each side, and it is 215 feet high. Find the volume.

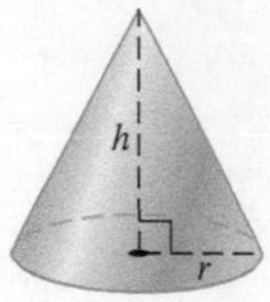

Figure 9-25

The next figure we will discuss is the **right circular cone**. This is similar to a pyramid with a circular base, as shown in Figure 9-25.

Despite the clear differences—a round base and sides that aren't triangles—it's kind of interesting that the volume formula for a right circular cone matches the pyramid formula: one-third times the area of the base times the height. Since the base is a circle with area πr^2, we get the following:

Volume Formula for a Right Circular Cone

The volume of a right circular cone is given by

$$V = \frac{1}{3}\pi r^2 h$$

where r is the radius of the circular bottom and h is the height.

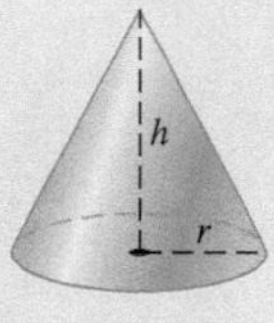

$V = \frac{1}{3}\pi r^2 h$

Example 6 | Finding the Volume of a Cone

The cups attached to a water cooler on a golf course are right circular cones with radius 1.5 inches and height 3 inches. How many ounces of water do they hold? (One ounce is about 1.8 cubic inches.)

SOLUTION

Using $r = 1.5$ and $h = 3$, we get

$$V = \frac{1}{3}\pi r^2 h = \frac{1}{3}\pi(1.5 \text{ in.})^2(3 \text{ in.}) \approx 7.1 \text{ in}^3$$

Now we convert to ounces:

$$7.1 \text{ in}^3 \times \frac{1 \text{ oz}}{1.8 \text{ in}^3} \approx 3.9 \text{ oz}$$

The cups hold about 3.9 ounces of water.

Try This One   6

How many ounces of melted ice cream does an ice cream cone hold if it has a radius of 1.1 inches and a height of 5 inches?

Many of the games we play use solid objects that are circular in three dimensions—basketballs, baseballs, golf balls, pool balls, etc. The geometric name for such an object is a **sphere**. Just as a circle is the set of all points in a plane that are the same distance from a fixed point, a sphere is the set of all points in space that are the same distance away from a fixed point. And we use the same terminology: the fixed point is called the **center** of the sphere, and the distance is called the **radius**. (Twice the radius is called the **diameter**.)

Math Note

The formula for the volume of a sphere was originally developed experimentally, and its original proof is considered to be one of the greatest works of ancient Greek mathematics. But calculus turned the proof into a very simple exercise.

Volume Formula for a Sphere

The volume of a sphere is given by the formula

$$V = \frac{4}{3}\pi r^3$$

where r is the radius of the sphere.

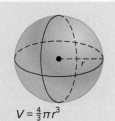
$V = \frac{4}{3}\pi r^3$

Example 7   Finding the Volume of a Sphere

The famous ball at Epcot center in Orlando has a diameter of 164 feet. Find its volume.

SOLUTION

To find the volume, we need the radius. The diameter is twice the radius, so the radius in this case is half of 164 feet, or 82 feet.

$$V = \frac{4}{3}\pi r^3 = \frac{4}{3}\pi(82)^3 \approx 2,309,565 \text{ ft}^3$$

The volume of the ball is about 2,309,565 cubic feet. That's a lot of cubic feet.

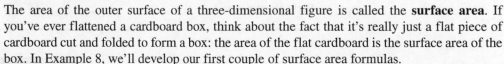

Try This One   7

An official NBA basketball is a sphere with diameter 9.4 inches. What is the volume?

David Sobecki/McGraw Hill

Can you find the author in this picture?

1. Find the volumes of three-dimensional figures.

Surface Area

The area of the outer surface of a three-dimensional figure is called the **surface area**. If you've ever flattened a cardboard box, think about the fact that it's really just a flat piece of cardboard cut and folded to form a box: the area of the flat cardboard is the surface area of the box. In Example 8, we'll develop our first couple of surface area formulas.

Example 8   Developing Surface Area Formulas

(a) The rectangular box in the diagram has six sides: top, bottom, front, back, left, and right. Each of those sides is a rectangle. Find the surface area by finding the area of each side and adding the results together.

(b) Use what you learned in part (a) to write a general formula for the surface area of a rectangular solid with length l, width w, and height h.

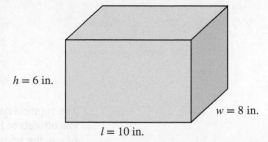
$h = 6$ in.
$w = 8$ in.
$l = 10$ in.

SOLUTION

(a) Every side is a rectangle, so we'll be repeatedly using the formula for the area of a rectangle to find the areas of the sides. The top is 10 in. by 8 in., so the area is 80 in^2. Now a key observation: the bottom is identical, so that's two sides with area 80 in^2. Next, the front side has area (6 in.)(10 in.) = 60 in^2, as does the back side. Finally, the right and left sides both have area (6 in.)(8 in.) = 48 in^2. Now we can add up the surface area:

$$2(80 \text{ in}^2) + 2(60 \text{ in}^2) + 2(48 \text{ in}^2) = 376 \text{ in}^2$$

(b) The last line in the computation is our guide for a general formula. The top and bottom sides each have area *lw*, the front and back each have area *lh*, and the right and left sides each have area *wh*, so the formula is

$$A = 2lw + 2lh + 2wh$$

Try This One   8

Find the surface area of a cube that is 4 ft. on each side, then use what you learned to write a formula for the surface area of a cube with side length *s*.

The formulas we just developed, along with the surface areas of some of the other nine figures we've studied, are listed in Table 9-3.

Table 9-3	Surface Area Formulas for Three-Dimensional Figures

All variables represent the same dimensions as in the volume formulas.
SA **represents surface area.**

Cube	$SA = 6s^2$
Rectangular	$SA = 2lw + 2lh + 2wh$
Right circular cylinder	$SA = 2\pi r^2 + 2\pi rh$
Sphere	$SA = 4\pi r^2$
Pyramid (Square base)	$SA = l^2 + 2l\sqrt{\dfrac{l^2}{4} + h^2}$ (l = length of base)
Right circular cone	$SA = \pi r^2 + \pi r\sqrt{r^2 + h^2}$

Ken Cavanagh/McGraw Hill

John A. Rizzo/Photodisc/Getty Images

The relation between surface area and volume is important when determining how something will absorb or lose heat. Ice absorbs heat through its surface. When you break up a large piece of ice, the volume stays the same, but the surface area is greatly increased. So small ice cubes melt quickly, and large blocks of ice last much longer.

| Example 9 | Finding the Surface Area of a Cylinder |

How many square inches of sheet metal are needed to form the soup can in Example 4?

SOLUTION

The radius of the can is 1.5 inches, and the height is 4 inches. Using the formula from Table 9-3:

$$SA = 2\pi r^2 + 2\pi rh$$
$$= 2\pi(1.5)^2 + 2\pi(1.5)(4)$$
$$= 4.5\pi + 12\pi$$
$$= 16.5\pi \approx 51.8 \text{ in}^2$$

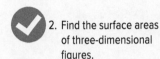

2. Find the surface areas of three-dimensional figures.

Try This One   9

How many square inches of leather are needed to cover an NBA basketball? (See Try This One 7 for dimensions, and ignore the seams on the ball.)

| Example 10 | Finding the Amount of Building Material Needed |

One important application of surface area comes from constructing containers. The surface area represents the amount of material that's needed, which in turn has a big effect on the cost of construction.

An aluminum silo is being made from a right circular cylinder with a right circular cone on top. The radius of the cylinder is $5\frac{1}{2}$ feet and the height is 14 feet; the cone has height 4 feet. How many square feet of aluminum will be needed for construction?

SOLUTION

The most important thing to note is that the amount of material in square feet makes this a surface area problem. A diagram will be helpful: see Figure 9-26.

We'll have to be careful with the cylinder part: we'll want a bottom but not a top, so instead of $2\pi r^2$ in the formula, we use just πr^2.

Cylinder with no top: $SA = \pi r^2 + 2\pi rh$
$$= \pi(5.5)^2 + 2\pi(5.5)(14)$$
$$\approx 578.8 \text{ ft}^3$$

For the cone, we'll also have to remove the bottom, which is the πr^2 part of the formula.

Cone with no bottom: $SA = \pi r \sqrt{r^2 + h^2}$
$$= \pi(5.5)\sqrt{(5.5)^2 + 4^2}$$
$$\approx 117.5 \text{ ft}^3$$

Total surface area: $578.8 + 117.5 = 696.3 \text{ ft}^3$.

4 ft

14 ft

5.5 ft

Figure 9-26

Try This One   10

A sheet metal storage shed will be a rectangular solid with square base (8 feet on a side) and height 7 feet. The roof will be a pyramid with base matching the rectangular solid and height 2 feet. Find the number of square feet of sheet metal needed.

Answers to Try This One

1 a. $V = s^3$, where s is the length of each side.
b. 27

2 60 cubic meters

3 598.4 gallons

4 About 28,274 cubic feet; about 845,392.6 quarts

5 8,236,005 cubic feet

6 About 3.5 ounces

7 About 435 cubic inches

8 96 ft³; $SA = 6s^2$

9 About 277.6 square inches

10 About 295.6 square feet

Exercise Set   **9-5**

Writing Exercises

1. Explain why the volume of a rectangular solid is the product of length, width, and height.
2. Explain why the volume of a cube is the length of a side raised to the third power.
3. What is the difference between volume and surface area?
4. Describe the connection between the volume formulas $V = lwh$ and $V = \pi r^2 h$.
5. Describe what the surface area of a three-dimensional figure means.
6. What is the significance of surface area in constructing product packaging?
7. Explain why the surface area of a cube is $6s^2$, where s is the length of a side.
8. Is the height of a pyramid the length of one of the sides? Why or why not?

Computational Exercises

For Exercises 9–20, find the volume of each figure.

9.

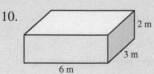

5 in.

10.

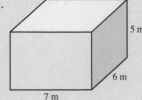

2 m
3 m
6 m

11.

5 m
6 m
7 m

12.

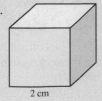

2 cm

13.

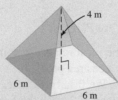

4 m
6 m
6 m

14.

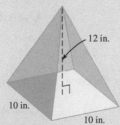

12 in.
10 in.
10 in.

15.

32 cm
28 cm

16.

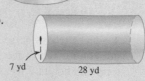

7 yd
28 yd

17.

18.

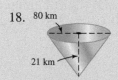

19.

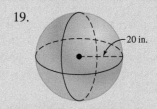

20.

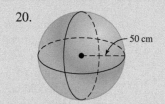

For Exercises 21 and 22, find the volume of the solid figure, not including the hole cutout.

21.

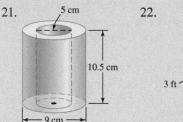

22.

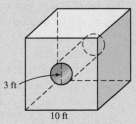

In Exercises 23–34, find the surface area of the figure from the given exercise.

23. Exercise 9
24. Exercise 10
25. Exercise 11
26. Exercise 12
27. Exercise 13
28. Exercise 14

29. Exercise 15
30. Exercise 16
31. Exercise 17
32. Exercise 18
33. Exercise 19
34. Exercise 20

Applications in Our World

35. How many cubic feet of dirt will be removed to build an in-ground swimming pool whose dimensions are 18 feet in length, 12 feet in width, and 3 feet in depth?

36. At a college tech expo, Jamal wants to build a simple display table in the shape of a cube with each side measuring 6 feet. How many square feet of plywood will be needed?

37. A pyramid found in Central America has a square base measuring 932 feet on each side and a height of 657 feet. Find the volume of the pyramid.

38. For a class in entrepreneurship, Jenny designed a pop-up tent made for dogs in the shape of a pyramid with a square base measuring 6 feet on a side with a height of 4 feet. Find the volume of the tent.

39. A cylindrical-shaped gasoline tank has a 22-inch diameter and is 36 inches long. Find its volume.

40. Find the surface area of a can of Zoom Energy Drink with a diameter of 8 cm and a height of 8.5 cm.

41. Find the volume of a cone-shaped funnel whose base diameter is 7 inches and whose height is 11 inches.

42. Find the volume of a cone-shaped Christmas tree that is 6 feet tall and has a base diameter of 4 feet.

43. The diameter of Mars is 4,200 miles. How many square miles is the surface of Mars?

44. The diameter of the moon is 3,474 kilometers. If the moon were made of green cheese, how many cubic kilometers of cheese would there be?

45. How many cubic inches of packing peanuts can be held by a rectangular box that is 1 foot long, 8 inches wide, and 2 feet tall?

46. A storage shed maker charges $2.50 per cubic foot. Find the price of a custom shed measuring 4 yards long, 5 yards wide, and 8 feet tall.

47. What is the minimum number of square inches of cardboard required to make a rectangular box that is 1 foot long, 2 feet wide, and 7 inches high?

48. What is the minimum number of square centimeters of wrapping paper needed to wrap a rectangular box that is 40 cm wide, 60 cm long, and 85 mm high?

49. A cereal box is required to have length 9 inches and height 13 inches to fit in the space allotted by major grocery chains. To hold the weight of this particular cereal required, it needs to have a volume of 400 cubic inches. How wide should it be?

50. A motor oil manufacturer is designing a new cylindrical can for its economy size. They want it to be 23 cm tall so it fits store shelves, and they want it to hold 2 liters of oil. What should the radius be? (Recall that 1 cubic centimeter = 1 milliliter.)

For Exercises 51–54, suppose that a particular brand of paint claims that under normal conditions, each gallon will cover 350 square feet. The paint is sold in 5-gallon and 1-gallon cans, with the 5-gallon costing the same as 4 1-gallon cans. Find how many cans of each should be bought for each situation to keep the cost as low as possible.

51. To paint the walls and flat roof of a rectangular building that is 40 feet long, 30 feet wide, and 12 feet high.

52. To paint forty 55-gallon drums that are being reconditioned for sale, including the top and bottom. Each drum has a height of 35 inches and a diameter of 24 inches.

53. To paint the outside walls of a round above-ground swimming pool with a diameter of 25 feet and a height of 4 feet.

54. To paint a giant papier-mâché model of the moon, which is a sphere with a diameter of 26 feet.

55. Sarah and Jack got an air purifier as a wedding gift. The maker of this particular model claims that it processes 10,000 cubic feet per hour and that all air in a room is processed five times every hour.
    (a) Find the volume of a room for which those two claims match.
    (b) If Sarah and Jack's master bedroom measures 22 feet by 18 feet and has 10-foot ceilings, how many times per hour can the purifier process the air in the room?

56. Now that they have a shiny new air purifier, Sarah and Jack (Problem 55) decide to repaint and recarpet the master bedroom as well, so it's off to Home Depot. Paint for the walls is $27 per gallon, and each gallon covers 350 square feet. The ceiling paint is $16 per gallon with the same coverage, and the carpet is $26 per square yard installed. How much will they spend? (Don't forget that they have to buy paint by the full gallon. And you can ignore doors and windows.)

Critical Thinking

57. Twelve rubber balls with 3-inch diameters are placed in a box with dimensions 12 inches × 9 inches × 3 inches. Find the volume of the space that is left over.

58. A rain gutter (cross section shown) is 24 feet long. How many gallons of water will it hold when it's full? One cubic foot of water is equal to 7.48 gallons.

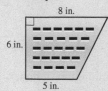

59. Find the surface area of the pyramid in Exercise 13 without using the formula provided in Table 9-3 by finding areas of the sides first. (*Hint:* You'll need the Pythagorean theorem.)

60. Use the technique from Question 59 to prove the formula for the surface area of a pyramid provided in Table 9-3.

61. By finding the area of the top, bottom, and sides separately, prove the formula for the surface area of a cylinder provided in Table 9-3. (*Hint:* What would the sides look like if you unrolled the cylinder?)

62. Find the surface area of a silo that is a cylinder with radius 9 feet and height 22 feet, with a dome-shaped top that is half of a sphere.

63. A bowling ball is a sphere (duh) with diameter $8\frac{1}{2}$ inches.
    (a) How many bowling balls can you fit into a rectangular room with a square base ($8\frac{1}{2}$ feet on a side) that is 10 feet tall if the balls are stacked directly on top of each other?
    (b) How much empty space will there be in the room in cubic feet?
    (c) What percentage of the space in the room is filled?
    (d) The bowling balls as stacked in part (a) don't quite reach the ceiling. Suppose that the ceiling is lowered so that the number of bowling balls found in (a) exactly touches all the walls and the ceiling. What percentage of the space in the room is filled now?

64. A golf ball is a sphere (again, duh) with radius 2.13 cm.
    (a) How many golf balls can you fit inside the room in Exercise 63 if the balls are stacked directly on top of each other?
    (b) How much empty space will there be in the room in cubic feet?

    (c) What percentage of the space in the room is filled?
    (d) If the walls and ceiling were moved in a bit so that the number of golf balls from part (a) fit exactly inside [as in 63(d)], what percentage of the space in the room would be filled?

65. A pyramid with a square base is placed flat side down into a rectangular box with a square base. The dimensions of the base of the pyramid are the same as those for the base of the box, and the pyramid reaches just to the top of the box. What percentage of the box is filled by the pyramid?

66. (a) A right circular cone is placed inside a right circular cylinder with the same radius as the base of the cone. The cone just reaches to the top of the cylinder. What percentage of the cylinder is filled by the cone?
    (b) Another cone is sitting on a flat surface covered by a dome that is made from half of a sphere. The base of the cone is the same size as the bottom of the dome, and the cone just reaches the top of the dome. What percentage of the dome is filled by the cone?

67. When reading the expression x^2, instead of saying "x to the second power," we say "x squared." For x^3, instead of saying "x to the third power," we say "x cubed." Why doesn't x^4 have a cool nickname like x^2 and x^3 do?

68. The volume of a package measures how much it can hold. The surface area measures how much material is required, which in turn affects the cost of the packaging. So the ratio of volume to surface area is a measurement of efficiency: a larger ratio indicates that the package can hold a lot of stuff for a smaller cost. The table provides the measurements of several food packages, three rectangular and three cylindrical. Find the ratio of volume to surface area for each, and decide which is the most efficient.

Dimensions	Shape
$l = 12.5, w = 5.4, h = 20.5$	Rectangular
$l = 9.5, w = 2.5, h = 16$	Rectangular
$l = 7.5, w = 4, h = 13.7$	Rectangular
$r = 3.7, h = 11$	Cylindrical
$r = 3.3, h = 6.8$	Cylindrical
$r = 3.9, h = 12.5$	Cylindrical

| **Section 9-6** | **Right Triangle Trigonometry** |

✓ **LEARNING OBJECTIVES**

1. Find basic trigonometric ratios.
2. Use trigonometric ratios to find sides of a right triangle.
3. Use trigonometric ratios to find angles of a right triangle.
4. Solve problems using trigonometric ratios.

We have seen that triangles can be used to solve many practical problems, but there are still many problems involving triangles that we can't solve using what we've learned so far. Here's a simple example. Like pretty much everything you buy today, ladders come with a variety of safety warnings designed to keep the manufacturer from getting sued. Suppose you buy a 12-foot ladder to do some work around the house, and the safety warning tells you it should not be placed at an angle steeper than 65° with the ground. Assuming that you don't have a protractor handy, how can you decide if it's safe to place the bottom 4 feet from the wall?

SW Productions/Brand X Pictures/ Getty Images

In this section we will study the basics of **trigonometry**, a very old subject whose name literally means "angle measurement." In trigonometry, we use relationships among the sides and angles of triangles to solve problems. This is in some ways like the work we did earlier with similar triangles, but is more widely applicable.

The big restriction we work with when studying the basics of trigonometry is that all triangles have to be right triangles. In this section, we'll use capital letters A, B, and C to represent the angles, and lowercase letters a, b, and c to represent the lengths of the sides. **We will always use C to represent the right angle**. In each case, the letter of an angle matches the letter of the side across from it. So, for example, a will represent the length of the side across from angle A. (See Figure 9-27.)

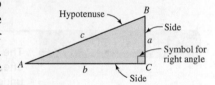

Figure 9-27

> **Math Note**
>
> Remember, in a right triangle, the side across from the right angle is called the hypotenuse. It's always the longest side.

The Trigonometric Ratios

There are three basic trigonometric ratios. They are called the **sine** (abbreviated sin), the **cosine** (abbreviated cos), and the **tangent** (abbreviated tan). The trigonometric ratios are defined for a given angle A as follows:

> **Math Note**
>
> Generations of students have remembered the trigonometric ratios using this simple mnemonic device: SOHCAHTOA (which most students pronounce "sow-cah-tow-ah").
>
> This stands for Sine is Opposite over Hypotenuse, Cosine is Adjacent over Hypotenuse, Tangent is Opposite over Adjacent.

The Trigonometric Ratios

$$\sin A = \frac{\text{Length of side opposite angle } A}{\text{Length of hypotenuse}} = \frac{a}{c}$$

$$\cos A = \frac{\text{Length of side adjacent to angle } A}{\text{Length of hypotenuse}} = \frac{b}{c}$$

$$\tan A = \frac{\text{Length of side opposite angle } A}{\text{Length of side adjacent to angle } A} = \frac{a}{b}$$

| **Example 1** | **Finding Basic Trigonometric Ratios** |

For the triangle shown, find sin B, cos B, and tan B.

SOLUTION

Since two of the three ratios involve the length of the hypotenuse, we need to find that first, using the Pythagorean theorem. (The hypotenuse is across from angle C, so we label its length c.)

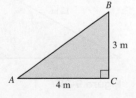

Math Note

Notice that the units divide out when you compute the trigonometric ratios, so the units of length are unimportant in defining the ratios.

But the units have to be consistent! If the hypotenuse is given in feet and one of the legs in inches, you have to convert one or the other to make the units match before finding the trig ratios.

$$c^2 = a^2 + b^2$$
$$= 4^2 + 3^2 = 16 + 9 = 25 \qquad \textit{Apply the square root to both sides.}$$
$$c = \sqrt{25} = 5 \text{ m} \qquad \textit{Discard the negative answer: c is a length.}$$

Now we can use the trigonometric ratios from the definitions above; the hypotenuse is 5 m, the side opposite B is 4 m, and the side adjacent to B is 3 m.

$$\sin B = \frac{\text{Opposite}}{\text{Hypotenuse}} = \frac{4 \text{ m}}{5 \text{ m}} = \frac{4}{5}$$

$$\cos B = \frac{\text{Adjacent}}{\text{Hypotenuse}} = \frac{3 \text{ m}}{5 \text{ m}} = \frac{3}{5}$$

$$\tan B = \frac{\text{Opposite}}{\text{Adjacent}} = \frac{4 \text{ m}}{3 \text{ m}} = \frac{4}{3}$$

Try This One   1

Find sin A, cos A, and tan A for the triangle below.

1. Find basic trigonometric ratios.

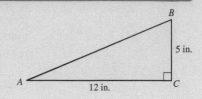

CAUTION

The symbols "sin," "cos," and "tan" are instructions to do something to the angle that comes after them, much like the symbol $\sqrt{}$ is an instruction to do something to the number underneath. So you should NEVER write one of those symbols with nothing after it: cos is just as meaningless as a square root symbol with nothing under it.

Example 2   Trig Ratios and Similar Triangles

(a) How can you tell that the two triangles in Figure 9-28 are similar?
(b) Find sin 30° using the side lengths in both triangles. Do you get the same result?
(c) Will you always get the same result for sin 30° in any right triangle with a 30° angle? Why or why not?

SOLUTION

(a) Since two of the angles are equal (the right angle and the 30° angle), all of the angles must be equal (since all have to add to 180°), so the triangles have the same shape and are similar.
(b) For the two triangles we get

$$\sin 30° = \frac{\text{opp}}{\text{hyp}} = \frac{3}{6} = \frac{1}{2} \qquad \sin 30° = \frac{\text{opp}}{\text{hyp}} = \frac{6}{12} = \frac{1}{2}$$

We did get the same result.

(c) Fortunately, the same thing will happen for any right triangle with a 30° angle. All such triangles are similar, so their corresponding sides are in proportion. The ratio of opposite to hypotenuse will be identical in all of them.

Try This One   2

Find cos 30° using each of the triangles in Figure 9-28. Do you get the same result for each triangle? Why is this a good thing in terms of trigonometry?

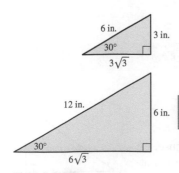

Figure 9-28

The importance of Example 2 is that it shows that the trigonometric ratios will be the same for a given angle regardless of the *size* of the triangle it lives in. This is a good thing because sin 30° wouldn't mean a whole lot if it were different depending on the situation. How useful would square roots be if $\sqrt{2}$ changed from problem to problem?

One situation that trigonometry is often used for is to find the lengths of sides of a triangle. If we know one side and one of the acute angles, we can find either of the remaining sides, as in Example 3.

Example 3	Finding a Side of a Triangle Using Tangent

In the right triangle *ABC*, find the length of side *a* when $m\angle A = 30°$ and $b = 200$ feet.

SOLUTION

Step 1 Draw and label a figure that matches the given information.

Step 2 Choose an appropriate trigonometric ratio and substitute values into it. In this case, we know the side adjacent to angle *A*, and we are asked to find the side opposite, so tangent is a good choice because it's the trig ratio that has opposite and adjacent in it.

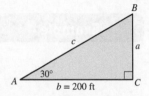

$$\tan A = \frac{\text{Opposite}}{\text{Adjacent}} = \frac{a}{b}$$

$$\tan 30° = \frac{a}{200}$$

Step 3 Solve the resulting equation for *a*.

$$\tan 30° = \frac{a}{200} \qquad \textit{Multiply both sides by 200.}$$

$$a = 200 \tan 30°$$

To evaluate this answer, we can use either a calculator or a spreadsheet (see the Using Technology box following Try This One 3). The result, rounded to two decimal places, is 115.47, so the length of side *a* is 115.47 feet.

Math Note

There's usually more than one way to find the length of a side using the trigonometric ratios. In Example 2, we could have noticed that $m\angle B = 60°$ and used the tangent of *B* to set up an equation.

Try This One   3

In the right triangle *ABC*, find the length of side *b* if $m\angle B = 58°$ and $a = 12$ inches.

Using Technology

Computing Values for Trig Ratios

If using a calculator, we'll have to first make sure that it's in degree mode. There are other units that degrees are measured in, but we'll be using degrees exclusively. In this case, we'll illustrate computing tan 30°.

Typical Scientific Calculator:
1. Press **MODE**, **DRG**, or **RAD** until you see DEG in the display window. This puts the calculator in degree mode.
2. Press 30 **TAN**.

Typical Graphing Calculator:
1. Press **MODE**, use arrow keys and **ENTER** to select "Degree," then press **2nd** **MODE** to exit.
2. Press **TAN**, then 30, then **ENTER**.

Online Calculator:
Most online calculators will either have a DEG button or a toggle switch between DEG and RAD. In either case, make sure you have DEG selected. **TAN** 30 **=**

Spreadsheet:
Pick a blank cell and enter "=TAN(RADIANS(30))." Radians are a different unit that angles are measured in: this command converts the angle to radians so that the spreadsheet can evaluate properly.
   For sine or cosine, substitute SIN or COS for TAN.

| **Example 4** | **Finding a Side of a Triangle Using Cosine** |

In the right triangle ABC, find the measure of side c when $m\angle B = 72°$ and the length of side a is 24 feet.

SOLUTION

Step 1   Draw and label a figure that matches the given information.

Step 2   We know the side adjacent to B and want to find the hypotenuse, so cosine is a good choice because it's the trig ratio that contains both adjacent and hypotenuse.

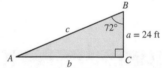

$$\cos B = \frac{a}{c}$$

$$\cos 72° = \frac{24}{c}$$

Step 3   Solve for c.

$$\cos 72° = \frac{24}{c} \qquad \textit{Multiply both sides by } c.$$

$$c \cdot \cos 72° = 24 \qquad \textit{Divide both sides by } \cos 72°.$$

$$c = \frac{24}{\cos 72°} \approx 77.67 \text{ ft (to two decimal places)}$$

Try This One   4

In the right triangle ABC, find the measure of side b when $m\angle A = 53°$ and the hypotenuse (side c) is 18 cm.

2. Use trigonometric ratios to find sides of a right triangle.

Trigonometric ratios can be used to find angles as well, provided that we know at least two sides of a right triangle. To accomplish this, we'll need to use special routines programmed into a calculator known as **inverse trigonometric functions**.

| **Example 5** | **Understanding Inverse Trig Functions** |

(a) What would you need to do in order to solve the equation $x^2 = 20$? More importantly, WHY would that work?
(b) What would you need in order to solve an equation that looks like $\sin B = 0.96$?

SOLUTION

(a) I can't tell you what YOU would do, but I can tell you what I would do: I'd apply the square root to both sides. This would work because the square root function is specifically designed to "undo" squaring.

(b) You would need a function that is designed to "undo" the sine function: that's what would leave the variable B isolated.

Try This One 5

(a) Explain what you would do to solve the equation $\sqrt[3]{x} = -4$, and why it would work.
(b) What would you need in order to solve an equation like $\tan A = 1.3$?

What we learned from Example 5 is that in order to solve many equations involving trig functions, we'll need functions that undo them: that's the point of the inverse trig functions. In Example 6, we'll see how to use one.

Example 6 Finding an Angle Using Trigonometric Ratios

Math Note

The symbol $\sin^{-1}$ is used to represent the inverse of sine. Likewise, $\cos^{-1}$ and $\tan^{-1}$ represent the inverses of cosine and tangent, respectively. When working with right triangles, you can think of these symbols as instructions that "undo" sine, cosine, or tangent, leaving the angle behind.

In the right triangle ABC, side c (the hypotenuse) measures 25 inches and side b measures 24 inches. Find the measure of angle B.

SOLUTION

Step 1 Draw and label a figure that matches the given information.

Step 2 We know the hypotenuse and the side opposite angle B, so sine is a good choice because it's the trig ratio that contains both opposite and hypotenuse.

$$\sin B = \frac{\text{Opposite}}{\text{Hypotenuse}} = \frac{24}{25} = 0.96$$

Step 3 Solve for B. This is where we need the inverse sine function, either on a calculator or spreadsheet. For instructions, see the Using Technology box following Try This One 6: regardless of the technology you choose, the result, rounded to two decimal places, is 73.74, so $m\angle B \approx 73.74°$.

 3. Use trigonometric ratios to find angles of a right triangle.

Try This One 6

Find the measure of angle A for the right triangle ABC if the measure of side a is 42 inches and the measure of side b is 18 inches.

Using Technology

Computing Values for Inverse Trig Functions
To solve the equation $\sin B = 0.96$, we need to compute $\sin^{-1} 0.96$.

Typical Scientific Calculator .96 [2nd] [SIN]

Typical Graphing Calculator [2nd] [SIN] .96 [ENTER]

Online Calculator
Some online calculators have an INV button that will turn the trig buttons into their inverse trig counterparts. For others there's a separate FUNC keyboard that contains the inverse trig functions. In either case, most commonly you'll use the $\sin^{-1}$ button first, then enter .96.

Spreadsheet
=ASIN(.96) provides an answer in radians, which then needs to be converted to degree measure. So, the command you need is =DEGREES(ASIN(.96))

In the remainder of the section, we will demonstrate just a few of the many, many problems in our world that can be solved using trigonometry. To begin, we return to the ladder question that began the section.

| **Example 7** | An Application of Trigonometry to Home Improvement |

The safety label on a 12-foot ladder says that the ladder should not be placed at an angle steeper than 65° with the ground. What is the closest safe distance between the base of the ladder and the wall?

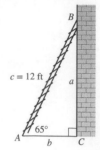

SOLUTION

Step 1   Draw and label a figure that matches the description of the situation. You can just draw the triangle without the cool bricks and ladder if you want.

Step 2   Choose the appropriate formula and substitute the values for the variables. Since we need to find the length of side b (which is adjacent to the given angle A) and we are given the measure of side c (which is the hypotenuse), cosine is a good choice.

$$\cos A = \frac{\text{Adjacent}}{\text{Hypotenuse}} = \frac{b}{c}$$

$$\cos 65° = \frac{b}{12}$$

Step 3   Solve for b. $\cos 65° = \dfrac{b}{12}$

$$b = 12 \cos 65° \approx 5.07 \text{ feet}$$

The bottom of the ladder can't be any closer than 5.07 feet from the wall.

Try This One   7

Being afraid of heights, a homeowner decides that he's unwilling to place the ladder in Example 7 any steeper than a 50° angle with the ground. How far up the wall will the ladder reach at that angle?

Angle of Elevation and Angle of Depression

A lot of applications of right triangle trigonometry in our world use what we call the *angle of elevation* or the *angle of depression*.

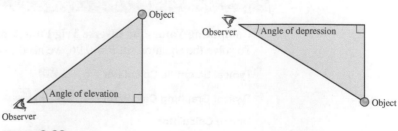

Figure 9-29

Comstock Images/Getty Images
Trigonometry is used extensively in surveying.

The **angle of elevation** of an object is the measure of the angle from a horizontal line at the point of an observer upward to the line of sight to the object. The angle of **depression** is the measure of an angle from a horizontal line at the point of an observer downward to the line of sight to the object. See Figure 9-29.

| Example 8 | Finding the Height of a Tall Object Using Trigonometry |

In order to find the height of a building, an observer who is 6 feet tall measures the angle of elevation from their head to the top of the building to be 32° when they are standing at a point 200 feet from the building. How tall is the building?

SOLUTION

Step 1 Draw and label a figure that matches the description of the situation.

Step 2 Choose the appropriate formula and substitute the values for the variables. We are given the measure of angle A and the measure of side b, which is adjacent to the angle. It would be helpful to find side a of the triangle because that's part of the building's height. That's the side opposite the angle we know, so tangent is a good choice.

$$\tan A = \frac{\text{Opposite}}{\text{Adjacent}} = \frac{a}{b}$$

$$\tan 32° = \frac{a}{200}$$

Step 3 Solve for a.

$$\tan 32° = \frac{a}{200} \quad \textit{Multiply both sides by 200.}$$

$$a = 200 \tan 32° \approx 125 \text{ feet (to the nearest foot)}$$

Looking back at the diagram, we see that the height of the building is side a (125 feet) plus 6 feet, so the building is 131 feet tall.

Try This One   8

A hiker standing on top of a 150-foot cliff sights a boat at an angle of depression of 24°. How far is the boat from the base of the cliff? (The hiker's eye level is 5.6 feet above the top of the cliff.)

| Example 9 | Finding an Angle of Depression |

A local photographer gets a hot tip that a famous starlet is sunbathing on a boat cruising down a river in a state of undress that her parents would not be particularly happy about. The photographer sets up a camera on a bridge that is 22 feet above the water hoping to get a shot as the boat rounds a nearby bend in the river. A laser range finder indicates that eye level is 72 feet from the point where the boat will come into view. At what angle of depression should the camera's tripod be aimed?

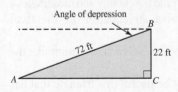

Javier Perini CM/Image Source

SOLUTION

Step 1 Draw and label a figure that matches the description of the situation (in the margin).

Step 2 The angle of depression is not part of the triangle, but it does form a 90° angle with angle B. We know the side adjacent to B and the hypotenuse, so we can use cosine to find angle B, then subtract from 90° to get the angle of depression.

$$\cos B = \frac{\text{Adjacent}}{\text{Hypotenuse}} = \frac{22}{72}$$

Step 3 Solve for B using the inverse cosine feature on a calculator (see Calculator Guide).

$$B \approx 72.2°$$

Step 4 Subtract angle B from 90°: $90° - 72.2° = 17.8°$. The angle of depression should be a bit less than 18°.

Try This One   9

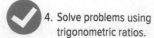
4. Solve problems using trigonometric ratios.

Lanae hopes to photograph the rising sun as it just peeks over the top of a 20-foot-tall house across the street, 120 feet from where the camera is set up. At what angle of elevation should the camera be aimed?

CAUTION   If you're planning to use trig ratios to solve an applied problem, the most important step is to make sure that there's actually a right triangle in the diagram. If not, the trig ratios simply don't apply.

Answers to Try This One

1 $\sin A = \frac{5}{13}$; $\cos A = \frac{12}{13}$; $\tan A = \frac{5}{12}$

2 $\cos 30° = \frac{\sqrt{3}}{2}$ in each case; this is good because we want $\cos 30°$ to be the same for every 30° angle.

3   19.20 inches

4   10.83 cm

5   a. Cube each side, because cubing is the opposite of cube root.
        b. We'd need a function that's the opposite of tangent.

6   66.80°   **7** 9.19 feet   **8** 349.48 feet

9   9.46°

Exercise Set   **9-6**

Writing Exercises

1. Describe each of the trigonometric ratios in terms of sides of a right triangle.
2. Given a right triangle ABC with right angle C, explain how we would label the lengths of the sides using the letters we used in this section.
3. Why can we be sure that $\sin 30°$ always has the same value regardless of the size of the right triangle containing the 30° angle?

4. Describe what is meant by an angle of elevation and an angle of depression.
5. Explain why the question "Find tan" is silly and meaningless.
6. What do we need to solve an equation of the form $\sin B = 0.4$?
7. How can we use technology to solve an equation like the one in Question 6?
8. Why would trig ratios not be useful in a triangle that has two 70° angles?

Computational Exercises

For Exercises 9–16, use the given right triangle to find cos A, sin A, and tan A.

9.

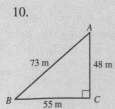

10.

11.

12.

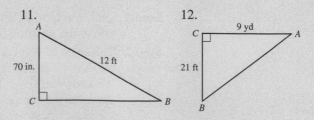

13.

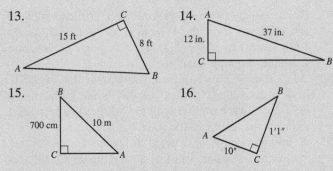

14.

15.

16.

28. Side c if $m \angle A = 15°$ and side $b = 1,250$ ft
29. $m \angle A$ if side $b = 183$ ft and side $c = 275$ ft
30. $m \angle B$ if side $b = 104$ yd and side $c = 132$ yd
31. $m \angle A$ if side $b = 18$ mi and side $c = 36$ mi
32. $m \angle B$ if side $a = 529$ mi and side $c = 1,000$ mi
33. $m \angle A$ if side $a = 306$ in. and side $b = 560$ in.
34. $m \angle B$ if side $a = 1,428$ ft and side $c = 1,800$ ft
35. $m \angle B$ if side $a = 62\frac{1}{2}$ ft and side $b = 18$ yd
36. $m \angle A$ if side $a = 192.6$ yd and side $c = 701$ ft

Exercises 17–36 refer to right triangles labeled like the one below. Use trigonometric ratios to find each measure.

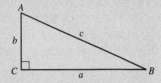

Every triangle has three sides and three angles. "Solving" a triangle means finding the lengths of all three sides and the measures of all three angles. Solve each triangle in Exercises 37–40.

17. Side a if $m \angle B = 72°$ and side $c = 300$ cm
18. Side b if $m \angle A = 41°$ and side $c = 200$ yd
19. Side c if $m \angle A = 76°$ and side $b = 18.6$ in.
20. Side a if $m \angle A = 30°$ and side $c = 40$ km
21. Side b if $m \angle B = 8°$ and side $c = 10$ mm
22. Side c if $m \angle B = 62°$ and side $a = 313$ mi
23. Side a if $m \angle A = 32°$ and side $b = 86\frac{1}{8}$ in.
24. Side b if $m \angle B = 82.4°$ and side $a = 634.8$ ft
25. Side c if $m \angle A = 28°$ and side $a = 872$ ft
26. Side a if $m \angle A = 22°$ and side $b = 27$ ft
27. Side b if $m \angle B = 53°$ and side $c = 97$ mi

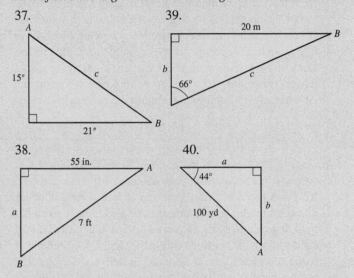

37.

38.

39.

40.

Applications in Our World

41. An airplane is flying at an altitude of 6,780 feet and sights the angle of depression to a control tower at the next airport to be 16°. Find the horizontal distance the plane is from the control tower. (Disregard the height of the tower.)
42. If the angle of elevation from ground level to the top of a building is 38° and the observer is 500 feet from the building, find the height of the building.
43. How high is a satellite radio tower if the angle of elevation from the ground is 82 degrees and the observation point is 700 meters from the base of the tower?
44. Trey locked his keys in his house again and desperately needs to get his term paper from his desk on the second floor of his house. He grabs a 25-foot ladder and leans it 63 degrees against the house to the bottom of the second floor window. How high is the window from the ground?
45. One common use of trigonometry is to measure inaccessible objects. Based on the measurements in the diagram, how wide is the lake?

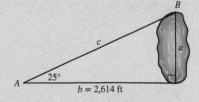

46. The flag in front of a police station is brought in each night by tradition. One day the mechanism jams and an officer has to climb the pole to bring the flag in. When standing 105 feet from the pole, the angle of elevation to the top of the pole is 40°. How high will this brave defender of public safety have to climb?

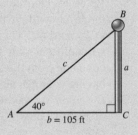

47. A treadmill that's 6 feet long has an adjustable back leg so that the back edge can be anywhere from 0 to 8 inches lower than the front. Find the range of angles at which this model can be inclined.
48. A two-story building that's 20 feet tall casts an 8-foot shadow, and 2 hours later the shadow is $12\frac{1}{2}$ feet long. How much did the sun's angle of elevation change in those 2 hours?
49. In a moment of frustration, Kelly tosses a math book out of the dorm room window. (Legal disclaimer: don't try this at home.) The window is 138 feet from the ground, and the book was thrown at a 21-degree angle of depression. Find the horizontal distance from where the book landed to Kelly's dorm. (Assume that the book traveled a straight-line path. Kelly has a really strong throwing arm.)

50. Archaeologists find a lost city with a map room leading them to the treasure of a long-dead civilization. When the instructions carved into a stone tablet are translated, they find that that location of the treasure room is found by following the shadow of a 10-foot staff to a location on the map at 2 P.M. on the day of the winter solstice. They check with meteorologists at a nearby university and find that the angle of elevation of the sun at that location on that day and time is 44°. How far away from the staff is the location of the treasure on the map?

51. Patrice sits on a bench with a laptop 2,456 feet from the base of the WiFi antenna, and the angle of elevation from the laptop, which is 2.5 feet from the ground, to the top of the antenna is 58 degrees. How tall is the antenna?

52. A hotel security camera is installed at a point 9 feet above the ground. According to the manufacturer, the widest field of view will result if the camera is aimed at a spot 22 feet down the hall from the point where it's mounted. At what angle of depression should the camera be aimed?

53. On a military training exercise, an artillery crew is instructed to train their weapon on a target that is $\frac{3}{4}$ mile away and 80 feet high. At what angle of elevation should they set the weapon?

54. While finishing his basement, Dave finds that the stairs have a horizontal span of 12 feet while going down 8.5 feet. At what angle should the hand railing be placed to match the incline of the stairs?

55. A business jet is instructed to make a straight-line approach and landing, starting at an altitude of 4,500 feet and a distance of 7 miles from the runway. At what angle of depression should the pilot fly?

56. A sniper sets up on a roof across the street from a hostage situation. The roof is 36 feet high, and a laser distance finder indicates that the straight-line distance from the target is 90 yards. At what angle of depression should the rifle be aimed to train in on the target?

Critical Thinking

57. Look at the instructions in the Using Technology box on page 531 for putting your calculator in degree mode. You can use a similar approach to put your calculator in radian mode, which measures angles in radians rather than degrees. Rework Example 7 with your calculator in radian mode or without using the degrees command if using a spreadsheet. How can you tell that you get the wrong answer without actually looking at the correct answer?

58. Refer to Question 57. Rework Example 8 in radian mode or without using the degrees command if using a spreadsheet. Can you tell that the answer is wrong without checking the correct answer? What can you conclude from the last two questions about the danger of having your calculator in the wrong mode?

59. Based on the way we've defined sine and cosine in terms of right triangles, explain why the values of these trig ratios always have to be between zero and one.

60. Refer to Question 59. Can we make any analogous statements about the value of tangent? Why or why not?

61. From the top of a building 300 feet high, the angle of elevation to a plane is 33° and the angle of depression to an automobile directly below the plane is 24°. Find the height of the plane and the distance the automobile is from the base of the building.

62. The angle of elevation to the top of a tree sighted from ground level is 18°. If the observer moves 80 feet closer, the angle of elevation from ground level to the top of the tree is 38°. Find the height of the tree.

63. Suppose that there is no trigonometric ratio known as tangent. Can you solve the problem in Example 3 without it? Describe the procedure you would use.

64. How can you solve the problem in Example 4 without using the cosine ratio?

65. When you're asked to solve a right triangle and you're given two of the sides, you've really been given all three. When you're given one of the nonright angles, you've actually been given both of them. Explain these statements.

66. In Exercises 37 through 40, you were given two pieces of information about the triangle in addition to the fact that it's a right triangle. Explain why being given only the other two angles will not allow you to solve a triangle.

67. Write and solve your own problem that uses either angle of elevation or angle of depression. Make sure you explain the context, and why you chose that context.

68. There are three other trigonometric ratios that are defined in a trig class: each of them is the reciprocal of one of the three trig ratios we already know. Here are the definitions:

Secant: $\sec x = \dfrac{1}{\cos x}$ Cosecant: $\csc x = \dfrac{1}{\sin x}$

Cotangent: $\dfrac{1}{\tan x}$

(a) Write each of these new trig ratios in terms of adjacent side, opposite side, and hypotenuse of a right triangle.

(b) Why are there SIN, COS, and TAN keys on your calculator, but not SEC, CSC, and COT keys? Discuss.

Section 9-7 A Brief Survey of Non-Euclidean and Other Geometries

LEARNING OBJECTIVES

1. Identify the basics of elliptic geometry.
2. Identify the basics of hyperbolic geometry.
3. Create fractal figures.
4. Create tessellations.

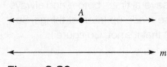

Figure 9-30

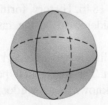

Figure 9-31 Great circles on a sphere.

Figure 9-32 A triangle in elliptic geometry.

The Euclidean geometry that we have studied so far in this chapter helps us to solve many problems in our world, and because of this, many people would say that Euclidean geometry is the geometry that describes our physical world. But does it really? Suppose you hop in your car and drive 500 miles due south. In your perception, you have driven a straight-line path. But in fact, you were driving along a curved path, matching the surface of Earth. This is why when you look out over the ocean, you can't see the other side—the water literally curves out of your sight at some point, which we call the horizon.

Stockbyte/Getty Images

After thousands of years of accepting Euclid's geometry as the only one, in the 1800s mathematicians and scientists started to question whether the principles of Euclidean geometry really were universal, or if maybe there were other systems in which Euclid's principles didn't apply. And so the field of non-Euclidean geometry was born. There are entire courses taught on non-Euclidean geometries, so in this section we will simply take a quick look at some of them.

A **postulate** is a fundamental assumption underlying an area of study that cannot be proved, but rather is assumed to be true. It turns out that Euclidean geometry has at its foundation the **parallel postulate**, which says that given any line m and a point A not on that line, there is exactly one line parallel to m through A. See Figure 9-30.

Mathematicians spent about 2,000 years trying to prove the parallel postulate without success. You'd think that they would have given up after about 1,500 years of failure, but we're a pretty patient bunch. Eventually, however, it started to occur to some of the world's great thinkers that maybe nobody could prove the parallel postulate because it couldn't be proven. Many of the attempts used a technique called proof by contradiction. Here's a quick overview of how that works: if you're trying to prove that something is true, you say "What if it's NOT?" If by assuming that the result is not true you end up with something happening as a consequence that you KNOW FOR SURE is false, then your assumption that the original result was false must be wrong. In other words, the original result is true.

So math folks started with the assumption that the parallel postulate was false, and tried to find some consequence of that assumption that contradicted the known laws of geometry. Not only could they not do it, but in some cases entirely new systems of geometry were discovered that don't follow the rules set out by Euclid around 300 BCE. One such system is elliptic geometry.

Elliptic Geometry

Elliptic geometry was developed by the German mathematician Bernhard Riemann (1826–1866). While Euclidean geometry is based on planes, elliptic geometry is based on spheres. In elliptic geometry, the lines are finite in length and are defined to be *great circles* of a sphere. A **great circle** is a circle on the sphere that has the same center as the sphere. See Figure 9-31.

This explains why elliptic geometry doesn't satisfy the parallel postulate: since all lines are great circles, there ARE no parallel lines. Think first of lines of longitude on a globe: all of them meet at the poles. Any other great circle that doesn't meet those at the poles (think of the equator) have to intersect them in two other points. In any case, if there are no parallel lines period, given a line and a point not on it, you certainly can't find a unique line through that point parallel to the line. So the parallel postulate is out the window.

Another interesting feature of elliptic geometry is that the sum of the angles in a triangle is more than 180° (see Figure 9-32). Not only that, but the sum of the angles isn't fixed—it depends on the area of the triangle. As the area gets smaller, the sum approaches but never reaches 180°. The Pythagorean theorem also fails in elliptic geometry: it's actually possible to draw a triangle with three 90° angles, in which case all the sides have the same length. But it still qualifies as a right triangle, and $a^2 + b^2 = c^2$ can only hold if side c is longer than the other two sides.

Sidelight   Cool Facts About Living on a Sphere

Finding an example of an elliptic geometry is as easy as a flea finding an example of a dog. Here are some interesting things about living on the surface of a very large sphere.

- We mentioned that if you drive 500 miles in one direction, you're not actually driving on a straight line because of the curvature of the earth. In fact, if you were able to drive on a perfectly straight line in some *Star Wars*-inspired hovercraft, when you got to your destination you'd be about 35 feet above the ground. (Although, to be fair, this assumes that the altitude is the same as where you began.)

- My wife and I take ocean cruises a lot, and when I'm standing on the top deck of a ship, I often wonder how far I can see before the horizon curves away from me. Eventually I realized that I'm a professional mathematician, so I figured it out. From the top deck of a large cruise ship, which puts you at 236 feet above the water line, you can see for a little bit less than 19 miles on a clear day.

- What about if you're NOT on a cruise ship? If you're standing on the beach looking out over the ocean, and your eyes are 5 feet off the ground, you have just 2.7 miles

NAN/Alamy Stock Photo

of sight line before the world curves away, not to be seen again until you turn around and look in the other direction.

- The height of the highest point on Earth (Mt. Everest, at 29,029 feet) is just 0.13887% of the Earth's radius. To put that in perspective, a bump on a billiard ball to the same scale as Mt. Everest would be a bit less than 0.0017 inches in height. In order to be legal for use in world-class billiard events, a ball can't have any bumps more than 0.005 inches in height. This means (as hard as it is to believe) that for all the mountains and valleys on the elliptic geometry we call home, it's literally smoother than a billiard ball. (Legal disclaimer: I had heard this several times before and always assumed it was one of those made-up Internet things. So I did that math and found out that it's not. Go figure.)

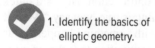

1. Identify the basics of elliptic geometry.

If you find it hard to imagine an elliptic geometry, consider the fact that you live on one! That's why that 500-mile path you drove felt like a line to you even though it was curved—when you live on a sphere, lines follow the contour of the sphere.

Hyperbolic Geometry

Hyperbolic geometry was developed independently by the Russian mathematician Nikolay Lobachevsky (1792–1856) and the Hungarian mathematician János Bolyai (1802–1860). Lobachevsky was actually the first person to publish an account of a non-Euclidean geometry, in 1829. His work, originally written in Russian, was mostly ignored until it was translated into German in 1840. Eventually, however, he may have longed for the days of being ignored, as his controversial ideas got him fired from his university post in 1846. History, fortunately, has viewed him much more favorably, and he is generally considered to be the father of non-Euclidean geometry.

Hyperbolic geometry is based on the assumption that, given a line m and a point A not on the line, there are an infinite number of lines through A parallel to m. The shapes in hyperbolic geometry are drawn not on planes, but on a funnel-shaped surface known as a **pseudosphere**. A line on a pseudosphere is pictured in Figure 9-33.

Here are some facts about hyperbolic geometry:

- The shortest distance between two points is not a straight line but rather a curve.

- Like elliptic geometry, the role of lines is played by curves, so the sides of triangles are arcs along the pseudosphere (as shown in Figure 9-34). Notice that the curve of the pseudosphere arcs inward rather than outward like a sphere, so the sum of the angles in a triangle is always *less* than 180°. Just like in elliptical geometry, as the area of the triangle gets smaller, the sum of the angles approaches 180°.

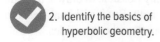

2. Identify the basics of hyperbolic geometry.

The remaining types of geometries we look at take place in Euclidean space, so they're not non-Euclidean geometries. They do, however, go beyond the basic shapes and

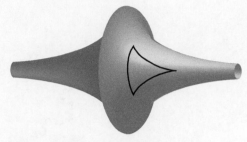

Figure 9-33 A line in hyperbolic geometry.

Figure 9-34 A triangle in hyperbolic geometry.

formulas we've used so far. We'll have a look at a couple of procedures that lead to fascinating patterns, and in some cases have a strange way of appearing in nature as well, starting with fractals.

Fractal Geometry

Let's start the next topic with some geography. The first photo, courtesy of Google Earth, is a satellite view of the southern Maine coast, near Kennebunkport. Next to it, we've traced a portion of the coastline in light blue. The third photo is a zoomed-in view of the same portion that was traced, and the fourth traces the portion of the coastline on the zoomed-in view.

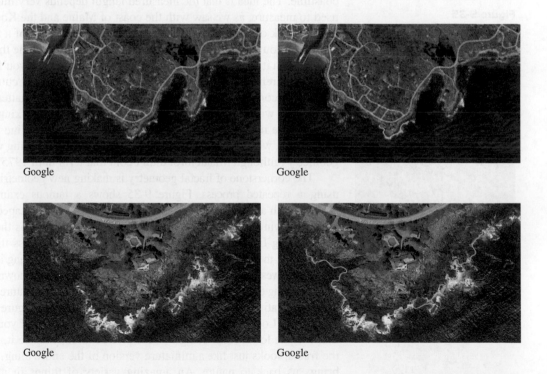

Google

Google

Google

Google

If you were to use the photos to find the length of that portion of coastline, you would most likely conclude that the coastline is a LOT longer using the fourth photo than the second. And if you were to zoom in further, you'd find it to be longer still.

Now let's get back to geometry. The next graphic shows a figure that we'll learn about in a bit known as the Koch snowflake. Try to picture tracing along the very top portion to estimate a length. Then look at the second graphic, which is a zoomed-in look at that top portion.

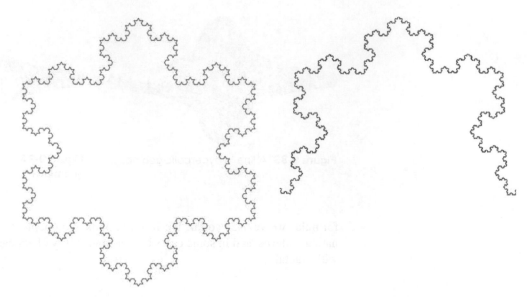

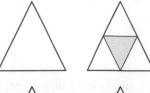

Figure 9-35

Steve Allen/Getty Images

Figure 9-36

Viktar Malyshchyts/Shutterstock

Figure 9-37

Once again, we can see that there's actually a lot more length than we could see from a far-away view. Which brings us to our next topic.

It's pretty unusual when an entire area in math can be traced to a single paper, but such is the case with fractal geometry. In 1965, the French mathematician Benoit Mandelbrot (1924–2010) published a fascinating paper conjecturing that Great Britain has an infinitely long coastline. The idea is that the measured length depends very much on the scale of the device used to measure, as we saw with the coast of Maine and the Koch snowflake. Here's another way to think of it: if you measure around the outside rails of a round swimming pool with a yardstick, you will be missing a lot of the curve and conclude the circumference is quite a bit smaller than it actually is. If you switch to a foot-long ruler, you'll get a larger (and more accurate) measurement but still too small because you'll still be "cutting corners."

If you continue this process, you'll conclude that no matter how short your ruler is, the same thing will happen, just on a much smaller scale. Looking at it this way, the only way to get a true measurement of a detailed object like a coastline would be to use a ruler with length zero, which of course is impossible to measure anything with! And so was born fractal geometry (although Mandelbrot didn't coin that term until 1975).

The cornerstone of fractal geometry is making new geometric figures from simpler shapes using a repeated process. Figure 9-35 shows a famous example, the Sierpinski triangle. Starting with an equilateral triangle, an equilateral triangle-shaped portion is removed from the middle, leaving behind three smaller triangles. This process is then continued indefinitely, and the resulting figure is called a **fractal**. The repeating process that leads to a fractal is called **iteration**. If there's anything computers are good at, it's repeating iterations of a process, so computers are awesome at drawing fractals—one cool example is shown in Figure 9-36. Oddly enough for something discovered in the late 1960s, fractals appear in nature as well, as we will see shortly.

So iteration is one feature that makes a geometric figure a fractal. A second is **self-similarity**. Look back at the fourth picture in Figure 9-35. If you look carefully at the portion that was the lower right triangle in the picture directly above it, you'll see that this portion of the fractal looks just like a miniature version of the entire thing. That's self-similarity. Which brings us back to nature. An amazing variety of things in nature exhibit either exact or approximate self-similarity. One striking example is a particular variety of broccoli known as Romanesco broccoli, pictured in Figure 9-37. As you look more and more closely, you see that the individual portions look amazingly like the entire head no matter how far you zoom in.

Other examples of self-similarity in nature include mountain ranges, coastlines, snowflakes, fern leafs, clouds, DNA strands, galaxies . . . should I go on?

The third feature defining a fractal figure is called the **fractal dimension**. This is a ratio that describes how much the measurement of an object changes based on the scale at which it is measured. A straight line has fractal dimension 1 because you'll get the same length no matter

Sidelight The Mobius Strip

An unusual mathematical shape can be created from a flat strip of paper. After cutting a strip, give it a half twist and tape the ends together to make a closed ring as shown.

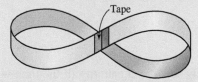

The shape you have made is called a Mobius strip, named for the German mathematician Augustus Ferdinand Mobius (1790–1868) who created it. What is unusual about the Mobius strip is that it has only one side. You can draw a line through the center of the strip, starting anywhere and ending where you started without crossing over an edge. You can color the entire strip with one color by starting anywhere and ending where you started without crossing over an edge.

The Mobius strip has more unusual properties. Take a scissors and cut through the center of the strip following the line you drew as shown.

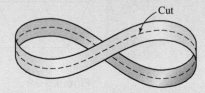

You won't get two strips as you might suspect, but you will have one long continuous two-sided strip.

Finally, make a new strip and cut it one-third of the way from the edge. You will find that you can make two complete trips around the strip using one continuous cut. The result of this cut will be two distinct strips: one will be a two-sided strip, and the other will be another one-sided Mobius strip.

The Mobius strip is a shape that belongs to a recently developed branch of mathematics called *topology.* Topology is a kind of geometry in which solid shapes are bent or stretched into different shapes. The properties of the shape that remain unchanged in the new shape are then studied.

how small your measuring device is. Almost all fractals have dimension that is not an integer, unlike regular Euclidean figures, which have dimension zero (points), 1 (lines), 2 (planes), etc. As you can picture from the "coast of Britain" example, the more complex a fractal is, the more error you're likely to have measuring with a large scale, and the higher the fractal dimension.

Let's have a look at a method for creating a different fractal figure.

Step 1 Start with an equilateral triangle.
Step 2 Replace each – side with ⌃.
Step 3 Repeat Step 2 in the new figure.
Step 4 Continue the process as long as you like. See Figure 9-38 for several iterations of the process. The figure is called the Koch snowflake.

Figure 9-38 exhibits the three classic characteristics of a fractal: it comes from iterations of a process; it exhibits perfect self-similarity (look carefully at the "arms" of the snowflake, and if you're really interested there's a very cool animation on Wikipedia illustrating this); and it turns out to have fractal dimension about 1.26.

Fractals play a much more important role in modern society than you would think. For example, because of them, your cell phone works much better than cell phones did 15 or 20 years ago. The effectiveness of an antenna is largely determined by how much perimeter it has compared to area. With a fractal figure like the Koch snowflake, with each iteration, the ratio of the perimeter to the area increases, so most modern cell phone antennas have a fractal design.

Math Note

Fun facts about some of the fractals we looked at:

1. The total area of the Sierpinski triangle is zero.

2. The area of the Koch snowflake is 8/5 times the area of the triangle you start with, but the perimeter is infinite.

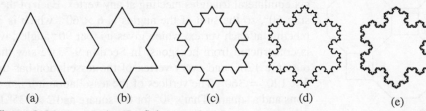

(a)   (b)   (c)   (d)   (e)

Figure 9-38 The Koch snowflake.

Radius Images/CORBIS

bu7shimeji/Shutterstock

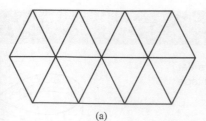

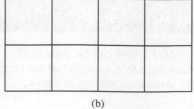

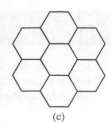

(a)   (b)   (c)

Figure 9-39

If you're a movie fan, you most likely see fractal patterns all the time without realizing it. Remember when we pointed out that fractal patterns appear often in nature? That makes the mathematical structures that generate fractal patterns incredibly useful in modeling nature in computer graphics, and this is used extensively in filmmaking. It's used even more in creating worlds other than our own in fantasy and science fiction movies. If you find this interesting, a quick Internet search for "fractals in movies" will provide a treasure trove of further information.

3. Create fractal figures.

Denise Dolan/Shutterstock

Tessellations

Yet another method for designing interesting geometric shapes is known as *tessellation*. The formal mathematical study of tessellations dates to the 1600s, but the concept itself is almost surely one of the oldest in all of math. For thousands and thousands of years, civilizations have used tessellations in art and architecture.

> A **tessellation**, also called a **tiling**, is a pattern that uses the same geometric shape to cover a plane without any gaps.

Figure 9-39 shows several tessellations. These figures are made by using regular polygons. (Recall that a regular polygon is a polygon in which all sides have the same length and all interior angles have the same measure.) The tessellation shown in Figure 9-39(a) is made from equilateral triangles. The one shown in Figure 9-39(b) is made from squares, and the one shown in Figure 9-39(c) is made from regular hexagons.

Tessellations can also be formed using more than one regular polygon. Figure 9-40 shows a tessellation created by a square and a regular octagon.

You may have noticed that there's not a single formula in this section so far: let's fix that. Recall that the endpoints of the segments that make up a polygon are called its vertices.

Math Note

The angle measures for regular polygons we use here come from the formula for the sum of the measures of the angles in a regular polygon back in Section 9-3.

> **Polygons That Can Form Tessellations**
>
> A tessellation can be formed by polygons exactly when the sum of the angles around the vertices is 360°.

Let's see how this applies to the four tessellations in Figures 9-39 and 9-40. The first has six equilateral triangles meeting at any vertex. Each of the angles in a right triangle has measure 60°, so the sum of the angles is 6 × 60°, which is 360°. The second has four squares meeting at each vertex, which gives us four 90° angles, which again adds to 360°. The third is constructed from hexagons. In Section 9-3, we saw that the angles in a regular hexagon measure 120°, and every vertex of the tessellation has three hexagons around it, so we get 3 × 120° = 360°. The vertices of the tessellation in Figure 9-40 are surrounded by two octagons and a square. That's 90° for the square and 2 × 135° for the octagons (again using something we learned in Section 9-3). The total of course is 360°.

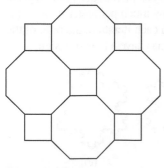

Figure 9-40

If you're not into repeated copies of regular polygons, you can make your own tessellation using these steps, which are illustrated in Figure 9-41(e).

Step 1   Draw a regular geometric polygon. In this case, we'll use a square.
Step 2   Cut the square from top to bottom using any design. See Figure 9-41(a).
Step 3   Move the left piece to the right side. Tape the pieces together. See Figure 9-41(b).
Step 4   Cut the piece from left to right using any design. See Figure 9-41(c).
Step 5   Move the top piece to the bottom piece and tape the pieces together. See Figure 9-41(d).
Step 6   Trace the figure on another sheet of paper and then move it so that the sides line up like puzzle pieces.
Step 7   Continue the process as many times as you like until a tessellation figure is complete. See Figure 9-41(e).

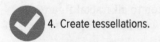

Math Note

Not every regular polygon can be used to create a tessellation. For example, it can be shown that regular pentagons alone cannot create a tessellation.

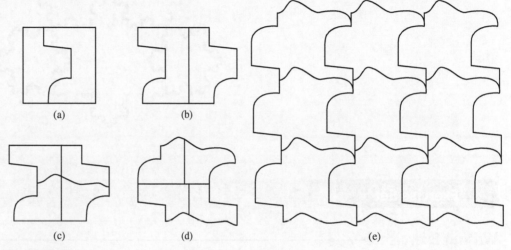

(a)   (b)

(c)   (d)   (e)

Figure 9-41

4. Create tessellations.

Tessellations are very common in architectural design, but are most commonly used in mosaic designs of floors, walls, and windows. (There's a really good chance that the next restroom you use will have a tessellation in it somewhere.) The most famous user of tessellations is without question the Dutch artist M. C. Escher, profiled in the next sidelight.

Sidelight M. C. Escher

A Dutch graphic artist, Maurits Cornelius Escher (1898–1972), used tessellations in his artwork and lithography. Escher had very limited formal training in math; he studied architecture briefly but switched to graphic arts, not a big stretch since he had excelled at drawing from a young age. At the age of 24 he traveled through Spain and Italy, drawing a lot of his inspiration from the intricate designs of decorations at the Alhambra Palace in Granada, Spain. These designs were largely based on mathematical formulas and featured interlocking repetitive patterns. Sound familiar?

Escher then read articles on symmetry, non-Euclidean geometry, and other areas of mathematics associated with tessellations. He is recognized for his works containing infinite repeating geometric patterns and created paintings of objects that could be drawn in two dimensions but were impossible to create in three dimensions. The illustration here is a computer graphic inspired by a famous Escher work, and a Google search for Escher will allow you to see many of his most famous tessellations.

vexworldwide/Shutterstock

To close the lesson, here's a look at a cool combination of the last two topics: a tessellation formed by the fractal figure we looked at called the Koch snowflake. Mind blown.

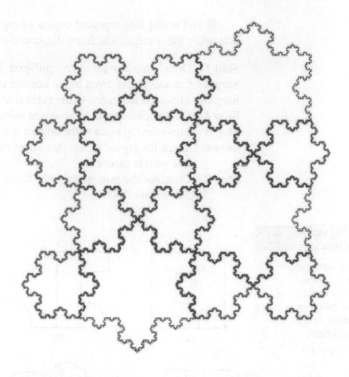

Exercise Set   9-7

Writing Exercises

1. Describe what Euclid's parallel postulate says about Euclidean geometry.
2. What is the surface on which shapes are drawn in Euclidean geometry?
3. What is the surface on which shapes are drawn in elliptic geometry?
4. How are lines represented in elliptic geometry?
5. Explain why the parallel postulate fails in elliptic geometry.
6. Why are there no parallel lines in elliptic geometry?
7. Explain how triangles in elliptic geometry differ from triangles in Euclidean geometry.
8. Explain why the parallel postulate fails in hyperbolic geometry.
9. Describe the shortest path between two points in hyperbolic geometry.
10. What is a pseudosphere?
11. Explain how triangles in hyperbolic geometry differ from triangles in Euclidean geometry.
12. Name three features common to all fractal figures.
13. What does it mean for a geometric figure to be self-similar?
14. Describe how fractals are drawn using iterations.
15. List at least three things in nature that are self-similar and describe how they exhibit that feature.
16. What is a tessellation? How are tessellations used?
17. What has to happen at vertices in order for polygons to tessellate?
18. True or false, and explain: you can make a tessellation by having three regular hexagons meet at a common vertex.

Computational Exercises

19. Using the letter H and putting an H at each corner, create a fractal-like figure.
20. Develop a fractal by replacing each _____ side in a square with ⌐_⌐⌐.
21. Do you think that you can build a Sierpinski triangle with a nonequilateral triangle? Try it with a triangle that is isosceles but not equilateral.
22. Try to build a Sierpinski triangle starting with a right triangle with legs that are the same length.
23. Try to build a Sierpinski triangle starting with a triangle that has three different side lengths.
24. What can you conclude from the results of Exercises 21–23?

Tessellations are also called "tilings" for a reason: the most common application for literally thousands of years has been designing decorative tile for floors, walls, etc.

25. Design a tile pattern using a tessellation where the basic shape is a square and an equilateral triangle with side length the same as the square.
26. Design a tile pattern using a tessellation where the basic shape is a square and two equilateral triangles with side

lengths the same as the square. How does your pattern compare to the one in Exercise 25?

27. In Figure 9-39, we saw how to tessellate equilateral triangles. Can the same be done with isosceles triangles that are not equilateral? See if you can design a pattern.

28. What is the result of tessellating any right triangle? What is the result of tessellating any scalene triangle that is not a right triangle?

29. Use a square and the procedure described on page 545 to create a tessellation.

30. Use a regular hexagon and the procedure described on page 545 to create a tessellation.

Critical Thinking

31. Try to form a tessellation using only regular pentagons, then explain why it's not possible.

32. Repeat Exercise 31 using regular octagons.

33. Use the rule for tessellating polygons to develop a rule for when a regular polygon can be tessellated with itself *based on the number of sides it has.*

34. Many of the sports we commonly play use spheres (or as we call them, balls). Make a list of as many sports you can think of that use a ball, and using the Internet as a resource where necessary, decide if the pattern on the ball for each does or does not represent a tessellation.

35. Use Google Maps or another Internet site that shows satellite images of landmasses and look at a satellite view of the coast of California. Use the zoom button to zoom in one level at a time, then describe how the changing view applies to self-similarity.

36. Use the Internet to find and download a high-resolution image of a fern leaf, then zoom in repeatedly on the leaf. Describe how close you think the image is to exact self-similarity.

37. When building the tessellation shown in Figure 9-40, I decided to start with a square and find a regular polygon that I could use two copies of to tessellate with the square. I used the formula $(n - 2)180°$ for finding the sum of angles in an n-sided regular polygon. How did I come up with octagon? (You can do this numerically, but your instructor should give you extra credit if you do it algebraically, which is both trickier and nearly 70% cooler.)

38. In the Sidelight on page 540, I mentioned figuring out how far it is to the horizon when you're on the top deck of a large cruise ship. How did I do that? You tell me. The diagram is a hint. You'll need to look up the radius of Earth, by the way.

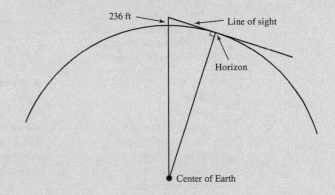

Section	Important Terms	Important Ideas
9-1	Point Line Plane Line segment Half line Ray Angle Vertex Degree Acute angle Right angle Obtuse angle Straight angle Complementary angles Supplementary angles Vertical angles Parallel lines Transversal Alternate interior angles Alternate exterior angles Corresponding angles	**The principles** of geometry have long been useful in helping people understand the physical world. The basic geometric figures are the point, the line, and the plane. From these figures, other figures such as segments, rays, and half lines can be made. When two rays have a common endpoint, they form an angle. Angles can be measured in degrees using a protractor. When lines intersect, several different types of angles are formed; vertical, corresponding, alternate interior, and alternate exterior. Two angles are called complementary if the sum of their measures is equal to 90°. Two angles are called supplementary if the sum of their measures is equal to 180°. Pairs of corresponding angles have the same measure, as do pairs of alternate interior angles, and pairs of alternate exterior angles.
9-2	Triangle Sides Vertices Isosceles triangle Equilateral triangle Scalene triangle Acute triangle Obtuse triangle Right triangle Hypotenuse Legs Pythagorean theorem Similar triangles	**A closed** geometric figure with three sides is called a triangle. A triangle can be classified according to the lengths of its sides or according to the measures of its angles. For a right triangle, the Pythagorean theorem states that $c^2 = a^2 + b^2$, where c is the length of the hypotenuse and a and b are the lengths of the legs. Two triangles with the same shape are called similar triangles. In this case, all of the angles are the same measure. Corresponding sides of similar triangles are in proportion. This allows us to solve problems involving finding the lengths of sides.
9-3	Polygon Quadrilateral Pentagon Hexagon Heptagon Octagon Nonagon Decagon Dodecagon Icosagon Trapezoid Parallelogram Rectangle Rhombus Square Regular polygon Perimeter	**A polygon** is classified according to the number of sides. A quadrilateral has four sides; a pentagon has five sides, etc. The sum of the measures of the angles of a polygon with n sides is $(n - 2)180°$. Special quadrilaterals such as the trapezoid, parallelogram, rhombus, and square are used in this chapter. A regular polygon is a polygon in which all sides are the same length and all angles have the same measure. The distance around the outside of a polygon is called the perimeter.

Section	Important Terms	Important Ideas
9-4	Area Circle Center Radius Diameter Circumference π (pi)	**The measure** of the portion of the plane enclosed by a geometric figure is called the area of the geometric figure. Area is measured in square units: if lengths are measured in inches, area is measured in square inches. One square inch is the area of a square that measures 1 inch on each side. Area formulas can be developed for many familiar geometric figures. A circle is a closed geometric figure in which all the points are the same distance from a fixed point called the center. A segment connecting the center with any point on the circle is called a radius. A segment connecting two points on the circle and passing through the center of the circle is called a diameter. The circumference of a circle is the distance around the outside of the circle.
9-5	Volume Rectangular solid Cube Right circular cylinder Pyramid Cone Sphere Surface area	**In geometry**, we also study three-dimensional figures. Some of the familiar three-dimensional figures are the cube, the rectangular solid, the pyramid, the cone, the cylinder, and the sphere. The volume of a three-dimensional geometric figure is the amount of space that is enclosed by the surfaces of the figure. Volume is measured in cubic units: if lengths are measured in inches, volume is measured in cubic inches. One cubic inch is the volume of a cube that measures 1 inch on each side. The surface area of a three-dimensional figure is the area of the faces or surfaces of the figure.
9-6	Trigonometry Sine Cosine Tangent Inverse trigonometric functions Angle of elevation Angle of depression	**Trigonometry** is the study of the relationship between the angles and the sides of a triangle. The trigonometric ratios of sine, cosine, and tangent are defined with a right triangle. Using the trig ratios, we can find side lengths and angle measures in a right triangle when we are given some of those measurements. The concepts of right triangle trigonometry can be used to solve many problems in navigation, engineering, physics, and many other areas.
9-7	Parallel postulate Elliptic geometry Great circle Hyperbolic geometry Pseudosphere Fractal Iteration Self-similarity Fractal dimension Tessellation	**Euclidean geometry** is based on the parallel postulate, which says that given any line m and a point not on it, there is exactly one line parallel to m through that point. In elliptic geometry, there are no parallel lines, and shapes are drawn on a sphere. In hyperbolic geometry, there are infinitely many lines parallel to any given line through a point, and shapes are drawn on a pseudosphere. Fractal figures are formed by starting with a basic shape and repeating the shape indefinitely in a series of steps called iterations. Fractal figures exhibit self-similarity and typically have a non-integer fractal dimension. Tessellations are formed when the same geometric shapes are used repeatedly to cover a plane without any gaps.

Math in Home Improvement REVISITED

David Sobecki/McGraw Hill

1. The U-shaped section can be divided into three rectangles, giving a total of six rectangles we can find the area of using length times width. The combined area is 7,835 square inches, which is about 54.4 square feet. The cost of the solid surface countertops was 54.4 × $56.62 = $3,080.13. The cost of the granite was 54.4 × $97.46 = $5,301.82. So we initially saved the difference, which is $2,221.69, but in the long run spent $3,080.13 more than we needed to. Sigh.

2. Draw a right triangle diagram with the given measurements:

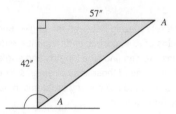

57″

42″

A

A

Angle A of the triangle is found using the equation $\tan A = \frac{42}{57}$. Solving for A, we get 36.4°. The angle marked A along the bottom is also 36.4° because of alternate interior angles, and the angle marked with the arc, which the trim should be cut at, is the supplement of 36.4°, or 143.6°.

3. First, let's find how many rows of boards we need to cover the 10-foot width. Each board is $5\frac{1}{2}″$ wide, and we add the $\frac{1}{8}″$ gap between boards, so the

width of each row is $5\frac{1}{2}″ + \frac{1}{8}″ = 5\frac{5}{8}″$. Dividing by 12, we find that each row is 0.46875 ft. wide. So there will be 10/0.46875 rows; this is $21\frac{1}{3}$ rows. Each row is 26 feet long, so we multiply the number of rows by 26 feet to find that we need $554\frac{2}{3}$ total feet of boards. Each is 16 feet long, so we divide that total length by 16 to get $34\frac{2}{3}$. This means I need 35 boards, and will have a bit left over.

4. This is another triangle diagram:

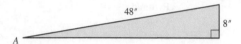

48″

8″

A

The 8″ side comes from the difference between the front edge and back edge heights. Solving the equation $\sin A = \frac{8}{48}$, we find that angle A measures 9.6°. If the legs were perpendicular to the hypotenuse and cut at a 90° angle, the bottom would make the same 9.6° angle with the ground (see below, left drawing). So to make it sit flat, it needs to be cut 9.6° short of 90°, or 80.4°.

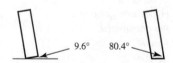

9.6°   80.4°

None of these were made up—they are the actual calculations done by the author while doing the projects.

Review Exercises

Mixed Review

For Exercises 1–10, identify each figure.

1.
   A   B

2.
   R   S

3.
   C   D

4.
   5

5.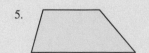

6.

7.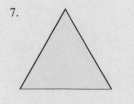

8.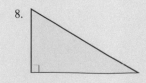

9.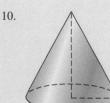

10.

Section 9-1

For Exercises 11–16, identify each type of angle or pairs of angles.

11.

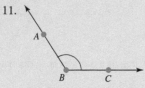

12.

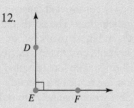

13.

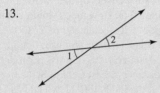

∡1 and ∡2

14.

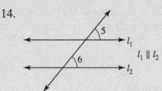

$l_1 \parallel l_2$

∡5 and ∡6

15.

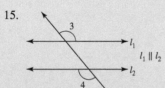

$l_1 \parallel l_2$

∡3 and ∡4

16.

∡7 and ∡8

17. Find the complement of each angle.
  (a) 27° (b) 88°
18. Find the supplement of each angle.
  (a) 172° (b) 13°
19. Find the measures of ∡1, ∡2, and ∡3.

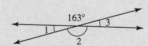

20. Find the measures of ∡1 through ∡7.

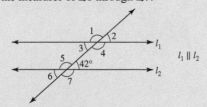

$l_1 \parallel l_2$

Section 9-2

21. Find the measure of the third angle of a triangle if the measures of the other two angles are 95° and 42°.
22. Find the measure of the hypotenuse of a right triangle if the lengths of the two legs are 8 inches and 15 inches.
23. Find the height of a water tower if its shadow is 24 feet when a 6-foot fence pole casts a shadow of 2 feet.

In Exercises 24–27, classify each triangle by angles (acute, obtuse, or right) and by sides (equilateral, isosceles, or scalene).

24. 26.

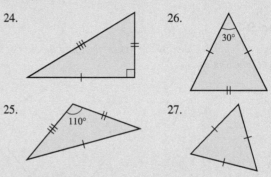

25. 27.

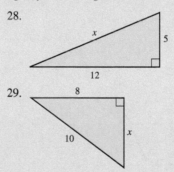

In Exercises 28–29, use the Pythagorean theorem to find the length of the missing side.

28.

29.

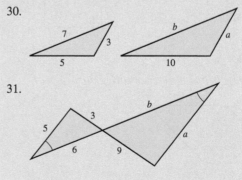

In Exercises 30–31, use the proportional property of similar triangles to find the missing sides (assume the two triangles in each exercise are similar).

30.

31.

32. If a 6-foot-tall person casts a 10-foot-long shadow, find the height of a building that casts a 150-foot-long shadow at the same time.
33. If a 12-foot ladder reaches 9 feet, 6 inches up a wall, how far up would a 20-foot ladder reach when placed at the same angle?

34. If the 20-foot ladder from Question 33 is placed with its base 9 feet from the wall, how high up does it reach?

35. A series of guy wires hold a radio antenna tower in place. If the tower is 140 feet high and the wires are attached to a point on the ground 90 feet from the base of the tower, how long are the wires?

36. While doing conditioning drills for spring training, a baseball player is supposed to run the length and width of a rectangular field that is 500 yards long and 800 feet wide. If he cheats and runs straight across on a diagonal, how much less distance does he run?

Section 9-3

In Exercises 37 and 38, identify the figure, then find the sum of the measures of the angles.

37.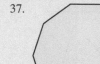

38.

In Exercises 39–42, find the perimeter.

39.

40.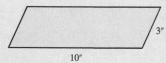

41.

42.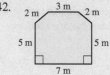

Section 9-4

For Exercises 43–46, find the area of each figure.

43.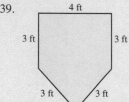

44.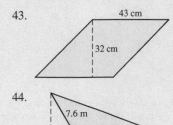

45.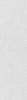

46.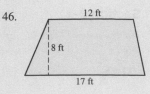

47. Find the circumference and area of a circle whose diameter is 16.5 yards.

48. The main gathering area of a newly constructed museum of modern art is shaped like a parallelogram. The north and south sides are 40 feet in length, and those two sides are 65 feet apart. Find how much it would cost to have the room tiled with marble that costs $12 per square foot installed.

49. Solar covers for swimming pools are made of a special plastic material, similar to bubble wrap, that helps to hold in heat. One supplier sells solar covers by the square foot, charging 22 cents per square foot. How much would a solar cover cost for a circular pool with a diameter of 17 feet?

50. The manufacturer of a new lawn tractor promises that in average use, it can cut 130 square yards of grass per minute. It currently takes the landscaper at East Central State an hour and 20 minutes to cut the football field, which is a rectangle with length 120 yards and width 53 yards. How much time would the new tractor save?

Section 9-5

For Exercises 51–54, find the volume.

51.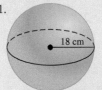

52.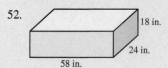

53.

54.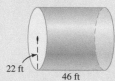

55. Find the surface area of the figure in Exercise 52.
56. Find the surface area of the figure in Exercise 54.
57. If the diameter of a bicycle wheel is 26 inches, how many revolutions will the wheel make if the rider rides 1 mile? (1 mile = 5,280 feet.)
58. Find how many square inches of fabric are needed to make the kite shown.

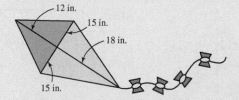

59. Find the volume of a ball if the diameter is 4.2 cm.
60. Find the surface area for the walls of the building shown. The door is a 3-foot by 6-foot rectangle with a half-circle on top of it, and there are two windows that measure 4 feet wide and 3 feet high.

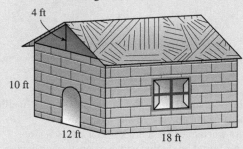

Section 9-6

61. For the triangle shown, find sin B, cos B, and tan B.

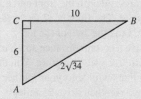

In Exercises 62–65, find the measure of the requested side or angle in triangle ABC. Note that C is the right angle.

62. Find b if $c = 27$ inches and $m\angle B = 49°$.
63. Find a if $b = 12$ yards and $m\angle A = 11°$.

64. Find $m\angle A$ if $c = 11$ meters and $b = 7.5$ meters.
65. Find $m\angle B$ if $a = 100$ miles and $b = 127$ miles.
66. If a tree 32 feet tall casts a shadow of 40 feet, find the angle of elevation of the sun.
67. A pole is leaning against a wall. The base is 15 feet from the wall, and from that point the angle of elevation to the top of the pole is 63°. How long is the pole?
68. The ramp for a motorcycle jump is a straight, flat surface 130 feet long. The beginning is at ground level, and the end is 27 feet high. What angle does the ramp make with the ground?
69. A plane takes off in a straight line and flies for a half-hour at 160 miles per hour. At that point, the pilots are informed by air traffic control that they have flown 17° off course. They are instructed to change direction and fly at the same speed along a perpendicular path back to the line they should have flown on, then turn and continue to the destination. How long will it take them to get back on course?

Section 9-7

70. Starting with an equilateral triangle, create a tessellation using the procedure described in Section 9-7.
71. Make a fractal figure by starting with a square piece of paper 8 inches on a side, then drawing a square that is 3 inches on a side in each corner. Continue the process for at least four iterations, each time drawing squares in each corner that are $\frac{3}{8}$ the length of the square they are inside.
72. Describe lines in elliptic geometry in terms of a globe.
73. The figure shows the first iteration of a fractal starting with a square. Draw the next two iterations.

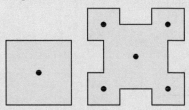

74. Discuss the self-similarity of the fractal in Exercise 73.

Chapter Test

1. Find the complement and supplement of an angle whose measure is 73°.
2. Find the measures of $\angle 1$, $\angle 2$, and $\angle 3$.

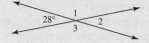

3. Find the measures of $\angle 1$ through $\angle 7$.

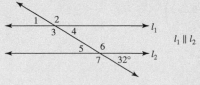

4. If two angles of a triangle have a measure of 85° and 47°, respectively, find the measure of the third angle.

5. Find the measure of side x. The two triangles are similar.

6. One way to measure the height of a tree is to cut it down and use a tape measure. But that's pretty drastic . . . another is to measure the length of its shadow, then measure the length of your shadow at the same time. Suppose the shadow of the tree is 13′11″ long at a time when a person who is 5′10″ tall casts a shadow 2′8″ long. How tall is the tree?

7. Find the sum of the angles of an octagon.

8. Isobel is building a flowerbed in her backyard in the shape below. She wants to put three layers of landscape timbers along the border. If the timbers come in 8-foot lengths, how many will she need to buy?

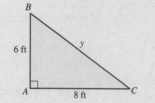

For Exercises 9–12, find the area.

9. The figure in Problem 8.

10.

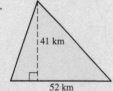

11.

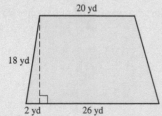

12.

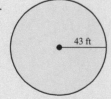

For Exercises 13–17, find the volume and surface area of each.

13.

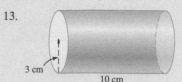

14.   15.

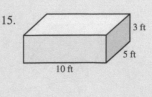

16.   17.

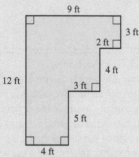

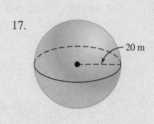

18. Are lines of latitude on a globe examples of lines in elliptic geometry? Explain why or why not.

19. Discuss whether or not a brick wall is a tessellation.

20. Design a fractal by starting with a square then drawing a smaller square whose vertices are the midpoints of the sides of the original square. For the next iteration, draw a square inside the second one using the same technique. Continue for at least five iterations, then discuss the self-similarity of your fractal.

21. A city lot is shaped like a trapezoid with the front and back sides parallel and 157 feet apart. The front is 32 feet wide, and the back is 48 feet wide. Find the area of the lot.

22. Starting from home, you drive a certain distance south to Best Buy to get a new monitor for your computer. You then drive 5 miles west to return that awful shirt your mother bought you for your birthday. You then drive 25 miles home. How far is it from your house to Best Buy?

23. You create a masterpiece in your art studio and it is selected for an art exhibition. Your painting is 70 inches by 100 inches, and you choose framing that costs $4.35 per yard. Find the cost of the framing.

24. During your chemistry lab, you fill a cone-shaped beaker that has a 1-inch diameter and a 2-inch height with lead, which weighs 0.41 pounds per cubic inch. Find the weight of the lead in the beaker.

25. To install a satellite television transmitter, the technician digs a cylindrical hole that is 3 feet in diameter with a depth of 5 feet. How much dirt in pounds must be removed if 1 cubic foot of dirt weighs 98 pounds?

26. A bank sets up a security camera on top of a building across the street from their entrance. The camera is controlled by a computer, and the head of security needs to input an angle of depression for proper aim. If the camera is at a height of 46 feet and the base of the building across the street is 70 feet from the bank entrance, what angle should the manager input?

Projects

1. The number pi played a big role in our study of area and volume. In a Sidelight feature back in Section 5-4, we referred to a website (http://www.angio.net/pi/piquery.html) where you can search for strings of numbers within the first 200 million digits of pi. Go to the site and search for the following:
   (a) Your birthday, in the form MMDDYYYY
   (b) Your phone number, without area code
   (c) Your phone number with the area code
   (d) The first 8 natural numbers consecutively
   (e) The first 9 natural numbers consecutively
   (f) The first 6 even natural numbers consecutively
   (g) The first 7 odd natural numbers consecutively
   Based on these results, what can you say about the likelihood of finding a given string based on the number of digits in that string? Try to make some general conjectures about how likely you are to find a given string based on the number of digits, and try several numbers to test out your conjectures.

2. The numbers 3, 4, and 5 are called a Pythagorean triple, because a right triangle with sides 3, 4, and 5 units satisfies the Pythagorean theorem. That is, $3^2 + 4^2 = 5^2$. The numbers 5, 12, and 13 form another Pythagorean triple.
   (a) Do 6, 8, and 10 form a Pythagorean triple?
   (b) Do 9, 12, and 15 form a Pythagorean triple?
   (c) Do 10, 24, and 26 form a Pythagorean triple?
   (d) Based on your answers for parts (a) through (c), can you make any definite statements about how many Pythagorean triples there are?
   (e) Based on parts (a) through (c), write a general formula or formulas for finding Pythagorean triples.
   (f) Can you find a Pythagorean triple where the numbers are not multiples of either 3, 4, 5 or 5, 12, 13?
   (g) Search the Internet for the string "Pythagorean triple" and find how many Pythagorean triples there are where all three numbers are less than 100.

   (h) Research the question "are Pythagorean triples more common, or more rare, among three-digit numbers than among two-digit numbers?" In general, are they harder or easier to find as the lengths get larger?

3. Gather a variety of objects that you could pour a liquid into: glasses, buckets, pitchers, coolers, aquariums, sinks, or anything else that you could calculate (or estimate) the volume of, using the volume formulas we learned in Section 9-5.
   (a) Calculate the volume of each in both cubic inches and cubic centimeters by taking measurements and using volume formulas. If the containers are not exactly regular, like a bucket that is wider at the top than the bottom, make your best estimate of the volume.
   (b) Convert the measurements in cubic inches to gallons using the conversions from Chapter 8. Convert the measurements in cubic centimeters to liters also.
   (c) Calculate the volume of each container in gallons and liters by pouring measured amounts of water into it using a measuring cup, pitcher, or beaker.
   (d) Compare the measurements done with water to your calculations. How well did you do?

4. Math class . . . photography class . . . whatever. In this project, your job is to search the world around you for as many examples as you can find of tessellations and fractals, then photograph them and submit a report on each. For tessellations, discuss the basic shape or shapes that are tessellated. If they happen to be polygons, show that the sum of the measures of the angles around the vertices is 360°. If not, discuss how the shapes fit together and how you think they may have been designed. For fractals, describe why you think your picture shows a fractal, and discuss the self-similarity.

Design elements: Front matter, Chapter Opener, Summary and End Matter header design (random numbers background illustration): ©pixeldreams.eu/Shutterstock RF

Probability and Counting Techniques

Photo: Steve Allen/Stockbyte/Getty Images

Outline

Math in Gambling

The fact that you're reading this sentence means that you're probably taking a math class right now. But maybe not . . . you could be an instructor evaluating the book, or maybe an editor looking for mistakes (unsuccessfully, no doubt). Still, I would be willing to bet that you're taking a math class. The word "probably" indicates a certain likelihood of something happening, and that basic idea is the topic of this unit. We call the study of the likelihood of events occurring *probability*.

Probability is one of the most useful concepts in math because being able to anticipate the likelihood of events can be useful in so many different areas. Games of chance, business and investing, sports, and weather forecasting are just a few samples from an essentially limitless list of applications. What are the chances of your team winning the championship? Should you take an umbrella to the golf course today? Will stock in a company you're keeping an eye on go up or down? Is that new job offer a good opportunity, or a disaster waiting to happen? Every day you make decisions regarding possible events that are governed at least in part by chance. The more you know about the likelihood of events, the more informed your decisions are likely to be.

We titled this chapter opener "Math in Gambling," not because everything we deal with will involve traditional gambling games, but to encourage you to think about the fact that almost everything we do is a gamble to some extent. And in that regard, almost everything we do relates to probability in some way. Have you ever gone outside during a thunderstorm? Not worn a seatbelt? Texted while driving? Smoked? Eaten a fatty diet? Ridden a motorcycle? Flown on a plane? Sped up to beat a red light?

In each of those instances, you were gambling, and not with something silly like casino chips. You were literally gambling with your life as the stakes. To be fair, the odds were DRAMATICALLY in your favor, and the fact that you're not currently deceased proves that you're on a heck of a winning streak. But the more you learn about probability, the more likely you'll be to give some deeper thought to the consequences of actions. And that's what we in academia call "getting an education."

So let's talk about gambling, both traditional and otherwise. You can think about these questions now, make some educated guesses, and maybe discuss them with your classmates. Eventually, you can come back and answer all of them, and see how accurate your guesses were.

1. Just last week the Powerball lottery reached a jackpot of $1.5 billion. Billion. With a "b." If you bought 10 tickets, were you more likely to win the jackpot or get struck by lightning? Make a guess, then look up the odds of winning the jackpot, and your probability of getting struck by lightning, on the Internet. Which is more likely?

2. What does texting while driving do to the probability that you'll be in a fatal accident? Again, make an educated guess, then do some research to see how you did.

3. A handful of gambling scenarios is provided below. In each case, find the expected value (that is, the average amount a person would win or lose) if placing the bet 100 times. Then rank the scenarios from best to worst in terms of your likelihood of winning or losing money. It's a GREAT idea to rank them before you do the calculations. You may be surprised at the results.

   a. At a church fair, you bet $1 and roll two dice. If the sum is 2, 3, 11, or 12, you get back your dollar plus four more. On any other roll, you lose.

   b. In a casino, you bet $1 on 33 at a roulette table. There are 38 possible numbers that can come up. If you win, you get your dollar back, plus 35 more.

   c. You buy a one-dollar ticket to a multistate lottery. If you match all six numbers, including the Mega Ball, you win the $20 million jackpot. If not, you lose. There are 175,711,536 possible combinations, and only one of them will be a winner.

   d. You bet $1 on flipping a coin with your roommate. Heads, you win; tails, your roommate wins.

For answers, see Math in Gambling Revisited on page 637

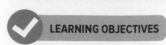

LEARNING OBJECTIVES

1. Compute theoretical probabilities.
2. Compute empirical probabilities.

Walking into a casino without knowing anything about probability is kind of like going to a stick fight without a stick—you're likely to take a beating. Casinos aren't in the business of losing money, and the games are designed so that most people lose more than they win. But an understanding of what is likely to happen in a given situation can give you an advantage over other players, giving you a better chance of walking out the door with some cash in your pockets. (Although if you learn enough about probability, you might decide that staying away from casinos is your best approach.)

Pe Dra/Shutterstock

The study of probability originated in an effort to understand games of chance, like those that use coins, dice, and playing cards. We'll start out using games of chance to illustrate the ideas, but will eventually see that probability has many applications beyond simple games. In this section, we'll study the basic concepts involved in studying probability. In short, the probability of something occurring is a number that represents how likely it is to occur. You can think of it as the percent chance of something happening. The major difference between percent chance and probability is that we write probability as a number between zero and one. So, in that regard, it's like percent chance with the percentage written in decimal or fractional form.

Sample Spaces

Processes such as flipping a coin, rolling a die, or drawing a card from a deck are called *probability experiments*.

Example 1   Finding Probability from Percent Chance

When you flip a coin, what's the percent chance that it will land heads up? What's the probability?

SOLUTION

There are only two possible results when you flip a coin: heads or tails. This is a classic example of a 50-50 chance: there's a 50% chance that it lands heads up, and a 50% chance that it lands tails up. We were told that probability is like percent chance, but it's a number between zero and one, and we know that the decimal form of 50% is 0.5. So the probability of getting heads on one flip of a coin is 0.5, or 1/2.

Try This One   1

If you play the game of rock, paper, scissors, there are three possible results: you win, you lose, or you tie. If neither of the people playing can read minds, you'd expect that each option would happen as often as the others. Write the percent chance of losing one game, and the probability of losing one game.

A **probability experiment** is a process that leads to well-defined results called outcomes. An **outcome** is the result of a single trial of a probability experiment.

Some examples of a trial are flipping a coin once, rolling a single die, and drawing one card from a deck. When a coin is tossed, there are two possible outcomes: heads or tails. When rolling a single die, there are six possible outcomes: 1, 2, 3, 4, 5, or 6.

In a probability experiment, we can predict what outcomes are possible, but we can't predict with certainty which one will occur. We say that the outcomes occur at **random**. In any experiment, the set of all possible outcomes is called the *sample space*.

A **sample space** is the set of all possible outcomes of a probability experiment.

Some sample spaces for various probability experiments are shown here.

Hocus Focus Studio/E+/Getty Images

Brandon Laufenberg/E+/Getty Images

Experiment	Sample Space
Flip one coin	{head, tail}
Roll a die	{1, 2, 3, 4, 5, 6}
Answer a true/false question	{true, false}
Flip two coins	{head/head, tail/tail, head/tail, tail/head}

It's a pretty good idea for you to take a moment to think about why there are *four* possible outcomes when two coins are flipped. Think about flipping a quarter and a dime at the same time. Both coins could land heads up; both coins could land tails up; the quarter could land heads up and the dime could land tails up; or the quarter could land tails up and the dime could land heads up. The situation is the same even if the coins are both quarters, dimes, whatever.

When working with probability, we'll often want to think of sets of outcomes. If you roll a single six-sided die, the clear outcomes are 1, 2, 3, 4, 5, and 6. But what if you bet on getting an even number? Then you don't care so much about what number comes up, as long as it's 2, 4, or 6. Getting an even number when rolling a die is an example of an *event*.

An **event** is any subset of the sample space of a probability experiment.

There's a subtle distinction between an outcome and an event: an outcome is a single occurrence, while an event can contain a number of outcomes. So rolling 2 with a single die is an outcome, but can also be considered an event. Rolling an odd number, on the other hand, is an event made up of three outcomes, but is not itself an outcome.

Theoretical Probability

Now we're ready to specifically define what is meant by probability. The first type we'll study is called **theoretical probability**. The goal is to determine all of the possible outcomes in a sample space and determine the probability, or likelihood, of an event occurring without actually performing experiments. There is one key assumption we make in theoretical probability: that every outcome in a sample space is equally likely. For example, when a single die is rolled, we assume that each number is equally likely to come up. When a card is chosen from a deck of 52 cards, we assume that each card has the same probability of being drawn.

Photodisc/Getty Images

Experiment: draw a card.
Sample space: 52 cards.
Event: drawing an ace.

Math Note

Theoretical probability is also called classical probability because it was the first type studied in the 17th and 18th centuries.

Formula for Theoretical Probability

Let E be an event in the sample space S, $n(E)$ be the number of outcomes in E, and $n(S)$ the number of outcomes in S. The probability of E is

$$P(E) = \frac{n(E)}{n(S)} = \frac{\text{Number of outcomes in } E}{\text{Number of outcomes in } S}$$

This probably looks a lot fancier than it actually is. What the formula really says is just what you might guess based on the fact that the probability of flipping a coin and having it land heads up is 1/2: you divide the number of ways that the event you're interested in can happen by the total number of outcomes.

In Example 2, we'll compute some theoretical probabilities.

Example 2   Computing Theoretical Probabilities

A single die is rolled. For the three events listed below, without calculating probabilities, put them in order from least to most likely based on an educated guess. Then compute each probability.

(a) A 5
(b) A number less than 5
(c) An odd number

SOLUTION

The formula for probability has the number of outcomes in the numerator, so we should expect that an event with more outcomes is more likely. A reasonable guess for order would be a 5, an odd number, and a number less than 5.

In this case the sample space is 1, 2, 3, 4, 5, and 6, and there are six outcomes: $n(S) = 6$.

(a) There is one possible outcome that gives a 5, so $P(5) = \frac{1}{6}$.
(b) There are four possible outcomes for the event of getting a number less than 5: 1, 2, 3, or 4. So $n(E) = 4$, and

$$P(\text{a number less than 5}) = \frac{n(E)}{n(S)} = \frac{4}{6} = \frac{2}{3}$$

(c) There are three possible outcomes for the event of getting an odd number: 1, 3, or 5. So $n(E) = 3$, and

$$P(\text{odd number}) = \frac{n(E)}{n(S)} = \frac{3}{6} = \frac{1}{2}$$

Jules Frazier/Photodisc/Getty Images

A die roll has six outcomes. If E = roll a 2, then $P(E) = \frac{1}{6}$. If E = roll an even number, then $P(E) = \frac{3}{6} = \frac{1}{2}$.

Try This One   2

Each number from 1 to 12 is written on a card, and the 12 cards are placed in a box. One card is picked at random. For the four events listed below, without calculating probabilities, put them in order from least to most likely based on an educated guess. Then compute each probability.

(a) A 7
(b) An odd number
(c) A number less than 4
(d) A number greater than 7

Example 3	Computing Theoretical Probabilities

Two coins are flipped. Find the probability of getting

(a) Two heads.   (b) At least one head.   (c) At most one head.

SOLUTION

The sample space is {HH, HT, TH, TT}; so $n(S) = 4$.

Math Note

A good problem-solving strategy to use is to make a list of all possible outcomes in the sample space before computing the probabilities of events.

(a) There's only one way to get two heads: HH. So

$$P(\text{two heads}) = \frac{n(E)}{n(S)} = \frac{1}{4}$$

(b) "At least one head" means one or more heads; i.e., one head or two heads. There are three ways to get at least one head: HT, TH, and HH. So $n(E) = 3$, and

$$P(\text{at least one head}) = \frac{n(E)}{n(S)} = \frac{3}{4}$$

(c) "At most one head" means no heads or one head: TT, TH, HT. So $n(E) = 3$, and

$$P(\text{at most one head}) = \frac{n(E)}{n(S)} = \frac{3}{4}$$

Try This One   3

Suppose that in a certain game, it's equally likely that you will win, lose, or tie. Find the probability of

(a) Losing twice in a row.   (b) Winning at least once in two tries.
(c) Having the same outcome twice in a row.

1. Compute theoretical probabilities.

Now that we know a little bit about probability, we can make a series of simple but important observations. Here's a really good idea: after reading each observation in boldface, close your eyes and *think* about why it should be true before reading on. If you can figure out half or more of them, it means you're off to a great start on understanding probability.

George Doyle/Stockbyte/Alamy Stock Photo

A 10% chance of rain doesn't mean it won't rain. You would expect it to rain on roughly 1 in 10 days when a 10% chance of rain was forecast.

1. **Probability is never negative.** Both $n(E)$ and $n(S)$ have to be zero or positive, so we can't get a negative number by dividing them.

2. **Probability is never greater than 1.** An event is a subset of the sample space, so there can't be more outcomes in any event than in the entire sample space; that means the numerator is less than or equal to the denominator in the probability formula.

3. **When an event can't possibly occur, its probability is zero. When an event is certain to occur, the probability is 1.** If an event can't occur, then none of the outcomes in the sample space satisfy it, and $n(E) = 0$. If an event has to occur, then every outcome in the sample space satisfies it, and $n(E) = n(S)$, so $n(E)/n(S) = 1$.

4. **If you add the probabilities for every outcome in the sample space, the result is always 1.** For example, when a die is rolled, each of the six outcomes has probability $\frac{1}{6}$, so the sum of the probabilities for those six outcomes is 1.

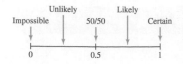

Figure 10-1

Probabilities are usually expressed as fractions or decimals between (and including) zero and one. But occasionally we'll express probabilities as percents. As we saw in Example 1, when the probability of an event is $\frac{1}{2}$, there's a 50% chance that it'll occur. If an event has probability close to zero, it's very unlikely to occur, and if the probability is close to 1, it's very likely to occur. See Figure 10-1.

In addition to finding the probability that an event will occur, it can be useful to find the probability that the event will *not* occur. For example, when rolling a die, if E is the event of rolling 2, then the **complement of E,** written E', is the event of not rolling 2. We know that $P(E) = \frac{1}{6}$; there are five ways for E not to occur: 1, 3, 4, 5, and 6, so $P(E') = \frac{5}{6}$. Notice that these probabilities add to 1, which is no coincidence.

> **The Probability of a Complement**
>
> For any event E, if E' is the event "E does not occur," then
> $$P(E') = 1 - P(E)$$

Example 4	**Finding a Probability Using Complements**

Of the next 32 trials on the docket in a county court, 5 are homicides, 12 are drug offenses, 6 are assaults, and 9 are property crimes. If jurors are assigned to trials randomly, what's the probability that a given juror won't get a homicide case? Find the probability using both the formula for theoretical probability on page 560 and the formula for complements. Then compare the two methods.

SOLUTION

To use the theoretical probability formula, we need to count up the number of trials that aren't homicides. There were 12 drug offenses, 6 assaults, and 9 property crimes. That's a total of 27 trials not for homicide, so the probability of not getting a homicide case is

$$\frac{\text{Number of cases not homicides}}{\text{Total number of cases}} = \frac{27}{32}$$

To use the complement formula, all we need to do is recognize that 5 cases are homicides, so the probability of getting a homicide case is $\frac{5}{32}$. That makes the probability of not getting a homicide

$$1 - \frac{5}{32} = \frac{27}{32}$$

This shows why knowing the formula for probability of a complement is useful: it was quicker to find the probability of getting a homicide case and subtract from 1 than it was to count up all the other types of cases.

Try This One   4

For the docket in Example 4, what's the probability that a juror gets assigned to a case that isn't a drug offense?

Probability and Sets

The theory of probability is related to the theory of sets discussed in Chapter 2. For a given probability experiment, the sample space can be considered the universal set, and an event E can be considered as a subset of the universal set.

Sidelight   You Bet Your Life!

You probably think of gambling as betting money at a casino or on a sporting event, but people gamble all the time in many ways. In fact, people bet their lives every day by engaging in unhealthy activities like smoking, using drugs, eating a high-fat diet, and texting while driving. Maybe people don't care about the risks involved in these activities because they don't understand the concept of probability. On the other hand, people tend to fear things that are far less likely to harm them, like flying, because the occasional negative consequence is sensationalized in the press.

In his book *Probabilities in Everyday Life* (Ivy Books, 1986), author John D. McGrevey states:

When people have been asked to estimate the frequency of death from various causes, the most overestimated causes are those involving pregnancy, tornados, floods, fire, and homicide. The most underestimated categories include death from diseases such as diabetes, stroke, tuberculosis, asthma, and stomach cancer (although cancer in general is overestimated).

Nick Rowe/Getty Images

Which do you think is safer: flying across the United States on a commercial airline, or driving cross country? According to our friend McGrevey, the probability of being killed on any given airline flight is about 1/1,000,000, while the probability of being killed on a transcontinental automobile trip is just 1/8,000. That means that driving across the country is 125 times more dangerous than flying!

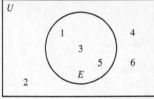

$E \cup E' = U$
$P(E) + P(E') = 1$

Figure 10-2

For example, when rolling a die, the sample space is $\{1, 2, 3, 4, 5, 6\}$; so the universal set is $U = \{1, 2, 3, 4, 5, 6\}$. Let the event E be getting an odd number; i.e., 1, 3, or 5. In sets, $E = \{1, 3, 5\}$. A Venn diagram can now be drawn illustrating this example. See Figure 10-2.

Notice that set E contains 1, 3, and 5 while the numbers 2, 4, and 6 are in the universal set but not in E. So $E' = \{2, 4, 6\}$. Now recall from set theory that $E \cup E' = U$. We know from earlier in the section that the sum of the probabilities of all outcomes in a sample space is 1. So if U is the set of all outcomes, then $P(U) = 1$. Also, E' represents the outcomes in U but not in E, so $P(E) + P(E') = 1$. Now we just need to subtract $P(E)$ from both sides, and we've proved the formula for the probability of a complement using things from set theory that we already know to be true.

There are several other times in this chapter where we'll take advantage of the set theory connection to develop facts about probability.

Empirical Probability

The second approach to probability we will study is to compute it using experimental data, rather than counting equally likely outcomes. For example, suppose 100 games into the season, your favorite baseball team has won 60 games and lost 40. You might reasonably guess that since they've won 60 of their 100 games so far, the probability of them winning any given game is about 60/100, or 0.6. This type of probability is called **empirical probability**, and is based on *observed frequencies*—that is, the number of times a particular event has occurred out of a certain number of trials. In this case, the observed frequency of wins is 60, the observed frequency of losses is 40, and the total number of trials is $60 + 40 = 100$.

Math Note

The total number of trials is the sum of all observed frequencies.

Formula for Empirical Probability

$$P(E) = \frac{\text{Observed frequency of the specific event } (f)}{\text{Total number of trials } (n)} = \frac{f}{n}$$

In Example 4, we based probability calculations on observed data (the number of trials of each type on the court docket). But that wasn't an empirical probability because we were computing probabilities for ONLY THOSE CASES. This next point is key: in empirical probability, we use observed frequencies to make predictions on probability for similar situations that we don't

Brandon Laufenberg/E+/Getty Images

In this coin toss, the empirical probability of heads was $\frac{6}{10}$, or $\frac{3}{5}$.
With more tosses, you would expect P (heads) to approach $\frac{1}{2}$.

have complete data for. In the baseball game example, you'd be predicting that the probability the team wins any given game during the rest of the season is 0.6. You can't KNOW that's true like we can in theoretical probability: you're using available data to make an educated guess.

The information in the baseball problem can be written in the form of a **frequency distribution** that consists of classes and frequencies for the classes, as shown below:

Result (Class)	Observed Frequency
Win	60
Lose	40
Total	100

This technique is often helpful in working out empirical probabilities.

Example 5 | **Computing Empirical Probabilities**

Everyone has a blood type that falls into one of four categories: O, A, B, or AB. When a blood transfusion is needed in an emergency medical situation, you might think that you'd have to get the same type as your blood, but that's not actually the case. The chart below describes which types can donate, and receive, various blood types.

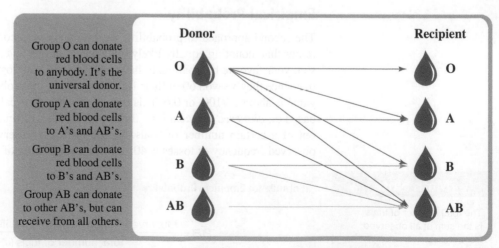

Group O can donate red blood cells to anybody. It's the universal donor.

Group A can donate red blood cells to A's and AB's.

Group B can donate red blood cells to B's and AB's.

Group AB can donate to other AB's, but can receive from all others.

In a random sample of 500 people, 210 had type O blood, 223 had type A, 51 had type B, and 16 had type AB. First, set up a frequency distribution. Then use it to answer the following questions.

(a) What's the probability that a randomly selected patient can receive only one type of blood?
(b) If an accident victim has type B blood, and an EMT on site is the only person willing to donate blood, what's the probability that it will be compatible with the victim?

(c) If a person with blood type A gets a transfusion, what's the probability that the donor had type O?

(d) Find the probability that a randomly selected donor's blood can be given to patients of more than just one blood type.

SOLUTION

Type (class)	Observed Frequency
A	223
B	51
AB	16
O	210
Total	500

(a) The only group that can receive just one blood type is folks with type O, so we need to find the probability of a random patient having type O. $P(O) = \frac{f}{n} = \frac{210}{500} = 0.42$

(b) A patient with type B can receive from either type O or type B, so we need to find the probability that the EMT is one of those two types. The frequency for O or B is $210 + 51 = 261$, so $P(O \text{ or } B) = \frac{261}{500} = 0.522$.

(c) In order for the blood to be compatible with a type A patient, the donor had to be either A or O. So $223 + 210 = 433$ donors would be eligible. If the donor wasn't A, they must have been O, which has a frequency of 210. So the probability is $\frac{210}{433} \approx 0.485$.

(d) There's only one group whose blood can be donated to just one type: AB. So the question is really asking us to find the probability of a random donor NOT belonging to blood type AB. We can find the probability of not AB by subtracting the probability of AB from 1.

$$P(\text{not AB}) = 1 - \frac{16}{500} = \frac{484}{500} = 0.968$$

Try This One 5

Use the information from Example 5 to answer these questions:

(a) What's the probability that a randomly selected patient donor's blood can be given only to a patient of the same blood type?

(b) Find the probability that a randomly selected donor's blood can be given to exactly two different blood types.

(c) What's the probability that a randomly selected patient cannot receive blood from a donor of the same type?

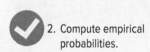

2. Compute empirical probabilities.

It's important to understand the relationship between theoretical probability and empirical probability in certain situations. In theoretical probability, the probability of rolling a 3 when a die is thrown is found by looking at the sample space, and is $\frac{1}{6}$. To find the probability of getting a 3 when a die is thrown using empirical probability, you would actually roll a die a certain number of times and count the number of times a 3 was obtained, then divide that number by the number of times the die was rolled. For example, suppose a die was rolled 60 times, and a 3 occurred 12 times. Then the empirical probability of getting a 3 would be $\frac{12}{60} = \frac{1}{5}$. Most of the time, the probability obtained from empirical methods will differ from that obtained using theoretical probability. The question is, then, "How many times should I roll the die when using empirical probability?" There is no specific answer except to say that the more times the die is rolled, the closer the results obtained from empirical probability will be to those of theoretical probability.

We should also point out that other than helping to understand probability, there's no good reason to try to find an empirical probability for a reasonably simple probability experiment like rolling a die. It's far more sensible to just compute the theoretical probability. Empirical probability is useful for situations where it's either really difficult or impossible to gather enough data to calculate a theoretical probability.

Sidelight   Probability and Your Fears

All of us at one time or another have thought about dying. Some people have fears of dying in a plane crash or dying from a shark attack. In the Sidelight on page 563, we saw that it's safer to fly across the United States than to drive. Statisticians who work for insurance companies (called actuaries) also calculate probabilities for dying from other causes. For example, based on deaths in the United States, the risks of dying from various other causes are shown.

Motor vehicle accident	1 in 7,000
Shot by a gun	1 in 10,000
Accident while walking across the street	1 in 60,000
Lightning strike	1 in 3 million
Shark attack	1 in 100 million

The death risk for various diseases is much higher:

Heart attack	1 in 400
Cancer	1 in 600
Stroke	1 in 2,000

As you can see, the chances of dying from diseases are much higher than the chances of dying from accidents. (For the record, this is intended to ease some of your fears, not encourage you to act recklessly. Your chances of dying in a motor vehicle accident go up by an awful lot if you're sending texts and speeding while you drive.)

In summary, then, theoretical probability uses sample spaces and assumes the outcomes are equally likely. Empirical probability uses observed frequencies and the total number of trials.

Answers to Try This One

1   33 1/3%; 1/3

2   (a) 1/12   (b) 1/2   (c) 1/4   (d) 5/12

3   (a) 1/9   (b) 5/9   (c) 1/3

4   5/8

5   (a) 0.032   (b) 0.548   (c) 0

Exercise Set   10-1

Writing Exercises

1. Define in your own words what the probability of an event means.
2. What is a sample space?
3. What's the difference between an outcome and an event?
4. What is the range of numbers that can represent probabilities? Why?
5. What is the probability of an event that can't occur? Explain why.
6. What is the probability of an event that is certain to occur? Explain why.
7. Explain the difference between theoretical and empirical probability.
8. Describe how to find the empirical probability of an event after conducting an experiment.
9. How does probability apply to percentages?
10. If you know the probability of an event occurring, how can you find the probability that it doesn't occur?

Computational Exercises

For Exercises 11–18, decide whether or not the given number could represent a probability.

11. $\frac{3}{4}$
12. 0.75
13. $-\frac{1}{2}$
14. 0
15. $\frac{41}{40}$
16. $\frac{40}{41}$
17. 72%
18. 111%

For Exercises 19–22, decide whether the probability described is theoretical or empirical.

19. At one school, 59% of the students having lunch in the union are female, so the probability of a randomly selected student from the campus phone directory being male is 0.41.

20. A pool table has 15 balls labeled 1 through 15. The probability of a ball made on the break having a number less than 6 is $\frac{1}{3}$.

21. The probability of a randomly selected state beginning with the letter A is $\frac{2}{25}$.

22. While at a casino, Catalina won 10 of the first 15 hands of blackjack she played, so she has a $\frac{2}{3}$ chance of winning the next hand.

Applications in Our World

23. If a die is rolled one time, find the probability of
    (a) Getting a 4.
    (b) Getting an even number.
    (c) Getting a number greater than 4.
    (d) Getting a number less than 7.
    (e) Getting a number greater than 0.
    (f) Getting a number greater than 3 or an odd number.
    (g) Getting a number greater than 3 and an odd number.

24. A couple has two children. Find the probability that
    (a) Both children are girls.
    (b) At least one child is a girl.
    (c) Both children are of the same sex.

25. On the *Price Is Right* game show, a contestant spins a wheel with numbers 1 through 7, with equally sized regions for each of these numbers. If the contestant spins once, find the probability that the number is
    (a) A 6.
    (b) An even number.
    (c) A number greater than 4.
    (d) A number less than 8.
    (e) A number greater than 7.

26. A list contains the names of five anthropology students, two sociology students, and three psychology students. If one name is selected at random to assist in the professor's new study, find the probability that the chosen student is
    (a) An anthropology student.
    (b) A psychology student.
    (c) An anthropology student or a sociology student.
    (d) Not a psychology student.
    (e) Not an anthropology student.

27. On a shelf at a gaming store, there are five Sony Playstations and four Nintendo Wii consoles left. If one gaming system is selected at random, find the probability that the system is a Wii console.

28. If there are only 50 lottery tickets for the Big Game, one of which is a winning ticket, and you buy 7 of those tickets at random, what is the probability that you'll win the super jackpot?

29. In a math class of seven women and nine men, if one person is selected at random to come to the board to show the solution to a problem, what is the probability that the student is a man?

30. A recent survey reported that 67% of Americans approve of human embryonic stem cell research. If an American is selected at random, find the probability that they will disapprove or have no opinion on the issue.

31. The Federal Bureau of Investigation reported that in 2019 there were 3,963 single-bias hate crimes that were racially motivated. Of those, 1,930 were motivated by anti-Black bias, 666 by anti-white bias, 134 by bias against a multiracial person, 158 by anti-Asian bias, and 527 by anti-Hispanic bias. Make a frequency distribution and use it to find the probability that a racially motivated hate crime picked at random was
    (a) Motivated by bias against Asians.
    (b) Motivated by bias against Blacks or Hispanics.
    (c) Not motivated by bias against whites.

32. Of 8,302 single-bias hate crime offenses reported in 2019, 4,784 were based on racial or ethnic bias, 1,650 on religious bias, 1,395 on sexual orientation bias, 224 on gender identity bias, and 169 on bias against those with disabilities. Make a frequency distribution and use it to find the probability that a randomly selected hate crime was
    (a) Motivated by racial or ethnic bias.
    (b) Motivated by bias against sexual orientation or gender identity.
    (c) Not motivated by bias against religion or those with disabilities.

33. In a survey, 16% of male college students said they lie sometimes to get a woman to go out on a date with them. If a male college student is chosen at random, find the probability that he does not lie to get a date with a woman.

34. While conducting a survey on smartphone use in a shopping mall, a marketing consultant found that 24 people surveyed had an iPhone, 16 had a Samsung running Android, 8 had another brand phone running Android, and 4 didn't have a smartphone. Find the probability that a randomly selected person in the mall has
    (a) An iPhone.
    (b) A smartphone.
    (c) An Android phone.

Exercises 35–38 refer to a standard deck of playing cards. If you are unfamiliar with playing cards, see the description on page 588.

35. During a game of Texas Hold'em poker, each of four players is dealt two cards, then the dealer "burns" a card (puts it face down), then deals the "flop" (three cards face up). The dealer then burns another card, then flips over the "turn" card (one card face up). One player needs a spade on the "turn" to make a flush. No one else has a spade, that player has two in his hand, and there are two on the flop. If neither of the burn cards are spades, what's the probability the turn card will be a spade?

36. During a game of Gin Rummy, Sven needs the eight of diamonds to make a straight in his hand. He and the other player have been dealt 10 cards each, the other

player does not have the card he wants, and all other cards are in the deck. What is the probability that the next card picked from the deck is the eight of diamonds?

37. During a game of Blackjack, three players are dealt two cards each, and the dealer has two cards. No one has a card that is worth 10 or 11, which would be a 10, a face card, or an ace. What is the probability that the next player dealt a card would get a card worth 10 or 11?

38. (a) Before the cards are dealt for a game of poker, what's the probability that the 31st card in the deck is a spade?
    (b) After five cards have been dealt to each of six players, with eight spades among those dealt, what's the probability that the 31st card in the deck is a spade?

39. A survey on campus revealed that 68% of the students felt that a new attendance policy was unfair. If a student is randomly asked to give an opinion of the new attendance policy, find the probability that the student will either think it's fair or have no opinion.

40. A survey of 25 students in line during registration revealed that 3 were math majors, 10 were history majors, 2 were psychology majors, 7 were biology majors, and the rest were undecided. If the clerk calls a name from the same line of students at random, find the probability the student would be either a history or biology major.

41. On one college campus with 5,300 students, 31% are first-year students, 28% are sophomores, 26% are juniors, and the rest are seniors. If a student is randomly chosen from the campus phone book to win a $500 gift card to the bookstore, find the probability that
    (a) The student isn't a first-year student.
    (b) The student is a senior.
    (c) The student is a sophomore or junior.

42. There are 248 students in a large-lecture history course. On the first exam, 12% got an A, 28% a B, 41% a C, 11% a D, and the rest failed. Based on these results, what's the probability that a randomly selected student will
    (a) Get better than a C on the second test?
    (b) Fail the second test?
    (c) Not get an A on the second test?

43. On a 10-question true/false test, there are seven false questions and the rest are true. If Marcus answered the first eight questions correctly, and five of them were false, find the probability that when he answers true for the next question, his answer will be correct.

44. According to namecensus.com, the five most common male names and their percentages are as follows:

Name	Percentage
James	3.318%
John	3.271%
Robert	3.143%
Michael	2.629%
William	2.451%

(a) If Mary meets a man at a party, find the probability his name is one of the most popular five male names in the country.

(b) If Sonia goes to the grocery store and the clerk is a man, find the probability his name would be John or Robert.
(c) If you rent a new apartment and the landlord is a man, find the probability his name would be in the top three most popular male names.

45. In March 2017, Congress began debating a new health care bill. A poll of 1,000 Americans at the time indicated that 451 opposed the bill, 239 favored the bill, and 310 were not sure. Based on these results, if you ask a randomly selected American for their opinion on this issue,
    (a) What is the probability that a randomly selected American was in favor of the bill?
    (b) What is the probability that they were not in support of the bill?

46. The Harris Poll surveyed 2,494 Americans who were employed full time about the gender gap in pay. Half were men, and half were women. One question asked for agreement on the statement "I believe men and women at my company are paid equally for equal work." The number giving each response is shown below.

Response	Men	Women
Agree	973	748
Disagree	224	387
Not sure	50	112

(a) If a person who participated in the survey is selected at random, what is the probability that they disagreed?
(b) What is the probability that the person is a man who either agreed or disagreed?
(c) Based on the data from the survey, if you had stopped a random woman on the street what is the probability that the woman would have said that women were paid less for equal work at their company?

47. In a survey conducted by Bank of America, college graduates were asked how much money they typically donate to their alma mater each year. The responses are summarized below:

Nothing	58%
Something, but less than $500	32%
$500 or more	10%

Based on these results:
(a) What is the probability that a randomly selected college graduate gives at least something in a typical year?
(b) What is the probability that a randomly selected college graduate gives less than $500 in a typical year?

48. Jockey International surveyed men to find out how old their oldest pair of underwear is. The results are both a little bit gross, and summarized below:

Less than 1 year	17%
1–4 years	59%
5–9 years	15%
10–19 years	7%
20 or more years	2%

Based on these results:

(a) What is the probability that a randomly selected man has a pair of underwear that is older than 4 years? 9 years?

(b) What is the probability that a randomly selected man has no pair of underwear more than a year old?

(c) What is the probability that a randomly selected man has no underwear more than 9 years old?

49. The students at a university are classified by a 0 for first-year students, a 1 for sophomores, a 2 for juniors, a 3 for seniors, and a 4 for graduate students. There are two extra scholarships to assign, so an administrator randomly selects from a box with only the numbers 0, 1, 2, 3, and 4 to choose the class of the first recipient. The numbers are placed back into the box and one is randomly selected for the class of the second recipient. Find the sample space, and then find the probability of the following events:

(a) An odd number is chosen first and an even number is chosen second. (*Note:* 0 is considered an even number.)

(b) The sum of the two numbers selected is greater than 4.

(c) For both selections, an even number is drawn.

(d) The sum of the two numbers selected is odd.

(e) The same number is drawn twice.

50. Only six students attended a school charity event, so each of their names was placed into three boxes for the three raffles for the event. What is the probability that the same student's name will be drawn from each box?

Critical Thinking

51. (a) Compute the empirical probability of a randomly selected person in your math class being left-handed. (You'll need to find out which hand everyone writes with.)

(b) Do some research on the Internet to compare that probability to the probability that a person in the general population is left-handed. How do they compare?

(c) The website mlb.com lists rosters for all 30 Major League Baseball teams, along with which hand each player throws with. Look up your favorite team (or the one closest to your campus if you don't have one) and see how the empirical probability of a player being left-handed compares to your class and the general population. Try to explain any apparent discrepancies.

52. (a) Compute the empirical probability of a randomly selected person in your math class having brown eyes.

(b) Do some research on the Internet to compare that probability to the analogous probability for the general public in the United States. If the results for (a) and (b) are significantly different, do some more research to make up a list of possible reasons.

53. (a) Find the theoretical probability of rolling a number less than 3 with one roll of a single die.

(b) Roll a die each of the number of times in the following table, record the results, and fill in the empirical probabilities. Describe any trends or anomalies that you observe, with possible explanations. (If you don't have any dice, do a Google search for "roll dice online.")

Number of rolls	10	20	30	40	50
Probability of rolling less than 3					

54. (a) Find the theoretical probability of rolling either 6, 7, or 8 with two dice.

(b) Roll two dice each of the number of times in the next table, record the results, and fill in the empirical probabilities. Describe any trends or anomalies that you observe, with possible explanations. (If you don't have any dice, do a Google search for "roll dice online.")

Number of rolls	10	20	30	40	50
Probability of rolling 6, 7, or 8					

55. Mike and Jose flip the same coin every day at work to decide who pays for the morning coffee. After looking over credit card receipts, Jose realizes that he's paid 73 of the last 100 days. Can Jose conclude that there's something funny about the coin they're flipping? Discuss.

56. When entering a chemistry class one semester, a student decided that they had a 50% chance of passing the class. They justified this by saying "Only two things can happen: I'll either pass or I won't." Critique the student's reasoning.

57. USING TECHNOLOGY. Have you ever seen a 15-sided die? I know I haven't. But you can roll one virtually in a spreadsheet. Actually, you can pick as many sides as you want using the **randbetween** command. Entering "=randbetween(1,15)" will generate a randomly selected number between 1 and 15, which is exactly the same thing as rolling a 15-sided die. The goal of this question is to set up a die-rolling experiment, and we've done some of the work for you. In the online resources for this lesson, there's an Excel template that you can download. You can enter the number of sides you want, then copy down the formula in cell B2 as many times as you want. Each represents a single roll of the die. You decide on a number of trials, then compute empirical probabilities for each outcome. Finally, double the number of trials and repeat, comparing the second result to the theoretical probability of each number. What should the probability work out to be for each number in the long run?

✓ LEARNING OBJECTIVES

1. Use the fundamental counting principle.
2. Calculate the value of factorial expressions.
3. Find the number of permutations of n objects.
4. Find the number of permutations of n objects taken r at a time.
5. Find the number of permutations when some objects are alike.

In the first section of this chapter, we saw that the key to being able to compute theoretical probabilities is being able to count up the number of outcomes in a sample space and the number of outcomes that make up a certain event. When a sample space is small, like the outcomes for rolling a single die, you can literally list and count them. But as probability experiments get more complicated, we're going to need to learn some techniques that go beyond simply listing and counting. So we're going to take a break from probability for the next two sections,

Corbis/age fotostock

and get really good at counting outcomes. Then we'll use our superior abilities to compute probability in a wide range of situations in the remainder of the chapter.

Let's start with an example of a counting problem:

Suppose that as part of an exciting new job, you're responsible for designing a new license plate for your state. You want it to look cool, of course. But if you want to keep your job, you better make sure that the sequence of letters and numbers you choose guarantees that there are enough different combinations so that every registered vehicle has a different plate number.

In the next two sections, we'll study three basic rules for counting the number of outcomes for a sequence of events. The first is the *fundamental counting principle*.

The Fundamental Counting Principle

After getting that new job, you naturally want a new apartment, and furniture to go along with it. The hip furniture boutique around the corner has the couch you want in either leather or microsuede, and each comes in your choice of four colors. How many different couches do you have to choose from?

We'll illustrate the situation with a *tree diagram*, which displays all the possible combinations.

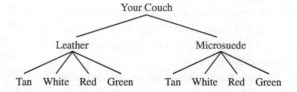

Each of the colors at the bottom of the diagram actually represents a different couch. For example, if you follow the left branch at each stage, you get a tan leather couch. The eight branches at the bottom shows that there are eight possible couches. Notice that this is the same number you'll get if you multiply the number of choices at each stage.

Now what if each couch can also come with or without an end recliner? Each of the eight choices in our diagram would have two more possibilities beneath it.

This would give us a total of 16 couches. This is $2 \cdot 4 \cdot 2$, which is again the product of the number of choices at each stage. This illustrates our first key counting principle.

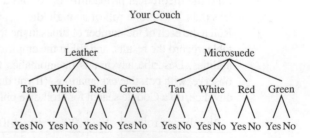

Math Note

The occurrence of the first event in no way affects the occurrence of the second event, which in turn does not affect the occurrence of the third event, etc.

The Fundamental Counting Principle

In a sequence of n procedures, if the first can occur in k_1 ways, the second in k_2 ways, the third in k_3 ways, and so on, the total number of ways the sequence of procedures can occur is

$$k_1 \cdot k_2 \cdot k_3 \cdot \cdots \cdot k_n$$

| Example 1 | Using the Fundamental Counting Principle |

There are four basic blood types, represented by letters: A, B, AB, and O. There is also an Rh factor represented by either + or −. If a local blood bank labels donations according to type, Rh factor, and sex of the donor, how many different ways can a blood sample be labeled?

SOLUTION

There are four possibilities for blood type, two for Rh factor, and two for sex of the donor. Using the fundamental counting principle, there are

$$4 \cdot 2 \cdot 2 = 16$$

different ways that blood could be labeled.

Try This One 1

A discount paint manufacturer plans to make several different paints. The categories include

Color	Red, blue, white, black, green, brown, yellow
Type	Latex, oil
Texture	Flat, semigloss, high gloss
Use	Outdoor, indoor

How many different kinds of paint can be made?

Royalty-free/Getty Images

When determining the number of different ways a sequence of procedures can occur, we'll need to know whether or not repetitions are permitted. The next example shows the difference between the two situations.

| Example 2 | Using the Fundamental Counting Principle with Repetition |

(a) The letters A, B, C, D, and E are to be used in a four-letter ID card. How many different cards are possible if letters are allowed to be repeated?
(b) How many cards are possible if each letter can only be used once?

SOLUTION

(a) There are four spaces to fill and five choices for each. The fundamental counting principle gives us

$$5 \cdot 5 \cdot 5 \cdot 5 = 5^4 = 625$$

(b) The first letter can still be chosen in five ways. But with no repetition allowed, there are only four choices for the second letter, three for the third, and two for the last. The number of potential cards is

$$5 \cdot 4 \cdot 3 \cdot 2 = 120$$

Hopefully, this ID card is for a pretty small organization.

How many three-digit codes are possible if repetition is not permitted?

Try This One 2

The lock on a storage facility is controlled by a keypad containing digits 1 through 5.

(a) How many three-digit codes are possible if digits can be repeated?
(b) How many three-digit codes are possible if digits cannot be repeated?

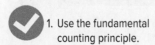

1. Use the fundamental counting principle.

Factorial Notation

The next couple of counting techniques we'll learn use **factorial notation**. The symbol for a factorial is the exclamation mark (!). In general, $n!$ means to multiply the whole numbers from n down to 1. For example,

$$1! = 1 = 1$$
$$2! = 2 \cdot 1 = 2$$
$$3! = 3 \cdot 2 \cdot 1 = 6$$
$$4! = 4 \cdot 3 \cdot 2 \cdot 1 = 24$$
$$5! = 5 \cdot 4 \cdot 3 \cdot 2 \cdot 1 = 120$$

The formal definition of factorial notation is given next.

> For any natural number n
>
> $$n! = n(n-1)(n-2)(n-3) \cdots 3 \cdot 2 \cdot 1$$
>
> $n!$ is read as "**n factorial**."
> 0! is defined as 1. (This might seem strange, but it will be explained later.)

Some of the formulas we'll be working with require division of factorials. This will be simple if we make two key observations:

- $\dfrac{n!}{n!}$ is always 1. For example, $\dfrac{3!}{3!} = \dfrac{3 \cdot 2 \cdot 1}{3 \cdot 2 \cdot 1} = \dfrac{6}{6} = 1$
- You can write factorials without including all of the factors down to 1. Let's look at an example. We know that $5! = 5 \cdot 4 \cdot 3 \cdot 2 \cdot 1$. But notice that if you ignore the 5 for a moment, what's left is $4 \cdot 3 \cdot 2 \cdot 1$, which we recognize as 4!. So we can write 5! As $5 \cdot 4!$. Starting over with $5! = 5 \cdot 4 \cdot 3 \cdot 2 \cdot 1$, if you ignore the 5 and the 4 for a moment, what's left is $3 \cdot 2 \cdot 1$, which is 3!. So we can also write 5! as $5 \cdot 4 \cdot 3!$. We'll see in the next example why this technique is useful.

Calculator Guide

Typical Scientific Calculator
5 `x!` or `SHIFT` `x⁻¹`

Typical Graphing Calculator
5 `MATH`; use right arrow to choose PRB, then press `4` `ENTER`.

Online Calculator
Most online calculators have an `x!` button, which makes them just like scientific calculators, although you may have to hit the `=` button to display the answer. Some will require changing modes to get to the factorial key, usually by hitting a button labeled `FUNC`.

Example 3   Evaluating Factorial Expressions

Evaluate each expression:

(a) 8!   (b) $\dfrac{12!}{10!}$

(c) We know that $\dfrac{12}{10} = \dfrac{6}{5}$. Is $\dfrac{12!}{10!} = \dfrac{6!}{5!}$? What can you conclude?

SOLUTION

(a) $8! = 8 \cdot 7 \cdot 6 \cdot 5 \cdot 4 \cdot 3 \cdot 2 \cdot 1 = 40{,}320$

(b) This is where the technique we just learned for rewriting factorials really comes in handy. First, write 12! as $12 \cdot 11 \cdot 10!$, then note that $\dfrac{10!}{10!} = 1$.

$$\frac{12!}{10!} = \frac{12 \cdot 11 \cdot 10!}{10!} = 12 \cdot 11 = 132$$

(c) We've already found that $\frac{12!}{10!} = 132$. On the other hand, $\frac{6!}{5!} = \frac{6 \cdot 5 \cdot 4 \cdot 3 \cdot 2 \cdot 1}{5 \cdot 4 \cdot 3 \cdot 2 \cdot 1} = 6$, so the two expressions are not equal. We can conclude that when dividing two factorials, you can't just ignore the factorials and reduce the fractions as usual. Well, I guess we can't stop you, but you'll get the wrong answer if you do.

Try This One   3

Evaluate each expression:

(a) 6!   (b) $\dfrac{12!}{8!}$ (c) $\dfrac{6!}{4!}$

✓ 2. Calculate the value of factorial expressions.

Permutations

The second rule that we can use to find the total number of outcomes for a sequence of events is the *permutation rule*.

> An arrangement of *n* distinct objects in a specific order is called a **permutation** of the objects.

| **Example 4** | **Counting the Number of Permutations** |

MedioImages/Photodisc/Getty Images

Suppose that a photographer wants to arrange three people, Carmen, Juan, and Christina, in a specific order for a portrait.

(a) List the possible orders.
(b) Use the fundamental counting principle to find the number of possible orders, and make sure that it matches the answer to part (a).
(c) How many orders do you think there would be for eight people?

SOLUTION

(a) Here are all possible orders listed out horizontally:

Carmen	Juan	Christina	Carmen	Christina	Juan
Juan	Carmen	Christina	Juan	Christina	Carmen
Christina	Juan	Carmen	Christina	Carmen	Juan

Note that there are six possibilities.

(b) There are three ways we could choose the person in the first position. After that, there are two left to choose from for the second position, and just one for the last position. So the fundamental counting principle gives us

$$\underset{\substack{\text{Choices for} \\ \text{1st position}}}{3} \cdot \underset{\substack{\text{Choices for} \\ \text{2nd position}}}{2} \cdot \underset{\substack{\text{Choices for} \\ \text{3rd position}}}{1} = 3!$$

This matches the number we got by listing them out.

(c) If there are eight people, listing all the orders would be, to say the least, not fun. But we could use the same idea with the fundamental counting principle: eight choices for the first position, seven for the second, and so on. So we'd get $8! = 40{,}320$.

Try This One   4

(a) List all the orders in which you could line up four photos in a frame.
(b) Use the fundamental counting principle to find the number of possible orders, and make sure that it matches the answer to part (a).

What really happened in Example 4? We discovered a general formula for finding a number of permutations.

Permutations of *n* Objects

The number of permutations of *n* distinct objects using all of the objects is *n*!.

Now we can just use our formula to count permutations, and not have to list them or use the fundamental counting principle.

| Example 5 | Calculating the Number of Permutations |

There are seven horses in a race and a large jackpot for anyone who can pick the finishing order of all seven exactly right. How difficult is that? Find the number of possible orders they can finish in.

SOLUTION

This is a permutation problem because we're interested in the number of ways to order seven objects (not that I consider race horses to be merely objects). There are

$$7! = 5,040$$

possible finishing orders.

Try This One 5

3. Find the number of permutations of n objects.

In how many different orders can the 14 basketball teams in the Big Ten conference finish? (Seriously, there are 14 teams in the Big Ten, at least as of 2021.)

So far, in calculating permutations, we've used all of the available objects. But what if only some of them are selected?

| Example 6 | Solving a Permutation Problem |

How many different ways can a pledge class with 20 members choose a president, vice president, and Greek Council representative? (No pledge can hold two offices.)

SOLUTION

There are 20 choices for president, 19 remaining candidates for vice president, and 18 members left to choose from for Greek Council rep. So there are $20 \cdot 19 \cdot 18 = 6,840$ different ways to assign these three offices.

Math Note

Notice that the calculation in Example 6 can also be written as $\frac{20 \cdot 19 \cdot 18 \cdot 17!}{17!} = \frac{20!}{17!}$. The factorial in the numerator is the number of people we're choosing from; the one in the denominator is the difference between the total number of people and the number being chosen.

Try This One 6

How many ways can a manager and assistant manager be selected from a department consisting of 10 employees?

In Example 6, a certain number of objects (people in this case) have been chosen from a larger pool. The order of selection is important, and no repetition is allowed. (No pledge can be both president and vice president—power trip!) We call such an arrangement of objects a **permutation of n objects taken r at a time**. In the pledge problem, n is 20 and r is 3. We will use the symbol $_nP_r$ to represent this type of permutation.

We solved Example 6 using the fundamental counting principle, but the result and the accompanying Math Note suggest the formula below:

Permutation of n Objects Taken r at a Time

Math Note

This formula shows why we were so interested in dividing factorials earlier in the section.

The arrangement of n objects in a specific order using r of those objects without replacement is called a **permutation of n objects taken r at a time**. It is written as $_nP_r$, and is calculated using the formula

$$_nP_r = \frac{n!}{(n-r)!}$$

| **Example 7** | **Solving a Permutation Problem** |

In a lottery game, 40 numbered Ping-Pong balls are put in a bin, and 4 are chosen at random, 1 at a time. To win the game, players need to match all 4 in the order in which they were drawn. How many different winning orders are there?

SOLUTION

This is a permutation problem because 4 balls are picked from 40 with no repetition, and the order is important. So we use the permutation formula with $n = 40$ and $r = 4$.

$$_{40}P_4 = \frac{40!}{(40-4)!} = \frac{40!}{36!} = \frac{40 \cdot 39 \cdot 38 \cdot 37 \cdot 36!}{36!} = 40 \cdot 39 \cdot 38 \cdot 37 = 2{,}193{,}360$$

This is why playing lotteries is not the best idea I've ever heard.

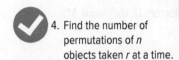

4. Find the number of permutations of n objects taken r at a time.

Try This One   7

How many six-letter passwords are there that use only lowercase letters with no letter repeated?

Using Technology: Computing Permutations

In Example 7, we computed $_{40}P_4$ using factorial division. Let's see how we could have done so more quickly using technology.

Graphing Calculator

40 **MATH**, then use the right arrow to select PRB, and press 2 to select $_nP_r$. Finally, press 4 and **ENTER** to get the result.

Spreadsheet

In a blank cell, enter =PERMUT(40,4). The first number is always the total number of objects, and the second is the number being chosen.

| **Math Note** |

There's a bit of a shortcut that can sometimes be helpful in calculating permutations. In Example 7, after using the permutation formula and simplifying, we were left with $40 \cdot 39 \cdot 38 \cdot 37$.

Notice that this is a product starting with the number of choices (40) that looks like a factorial, but has four factors, which is the number of objects being chosen.

We now have two permutation formulas: there are $n!$ permutations of n objects using all of them, and $\frac{n!}{(n-r)!}$ using only r of the objects. We should probably make sure that these formulas are consistent with each other. To check, we'll use the second formula to calculate the number of permutations when n objects are taken n at a time. The result should be $n!$ (to match the first formula). Let's see.

If we arrange five people in order for a group picture, we can think of it as a permutation of the five people chosen five at a time; that is,

$$_5P_5 = \frac{5!}{(5-5)!} = \frac{5!}{0!}$$

This will agree with the first formula if we agree to define 0! as 1. This is one of the reasons that 0! is defined to be 1: so we know that our two permutation formulas are consistent. This is good news, because it means we only need to remember the second permutation rule.

Problems involving permutation without duplicate objects can be solved using the fundamental counting rule; however, not all problems that can be solved with the fundamental counting rule can be solved using permutations.

When some of the objects are the same, a different permutation rule is used. Suppose that to solve a word puzzle, you need to find the number of permutations of the letters in the word *moon*. First label the letters as M, O_1, O_2, and N. This would be 4!, or 24, permutations. But since the O's without the subscripts are the same, the permutation M, O_1, O_2, N would be the same as M, O_2, O_1, N. The duplicates are eliminated by dividing 4! by the number of ways to arrange the O's (2!) to get 12. This leads to the next rule.

Permutation Rule When Objects Are Alike

The number of permutations of n objects in which k_1 objects are alike, k_2 objects are alike, etc., is

$$\frac{n!}{k_1! k_2! \cdots k_p!}$$

where $k_1 + k_2 + \cdots + k_p = n$

Example 8 | **Solving a Permutation Problem with Like Objects**

How many different passwords can be made using all of the letters in the word *Mississippi*?

SOLUTION

The letters can be rearranged as M IIII SSSS PP. Then $n = 11$, $k_1 = 1$, $k_2 = 4$, $k_3 = 4$, and $k_4 = 2$.

Using our newest formula, there are

$$\frac{11!}{1! 4! 4! 2!} = 34,650$$

different passwords.

5. Find the number of permutations when some objects are alike.

Try This One   **8**

Find the number of different passwords using all of the letters in the word *Massachusetts*.

In summary, the formulas for permutations are used when we're looking for the number of ways to arrange objects when the order matters, and once an object is used, it can't be used again.

Sidelight   **Win a Million or Be Struck by Lightning?**

Do you think you would be more likely to win a large lottery and become a millionaire or more likely to be struck by lightning? Of course, this depends on how often you play lotteries and how much time you spend outside, but on average it turns out that you would be quite a bit more likely to be struck by lightning.

An article in the Associated Press noted that researchers have found that the chance of winning $1 million or more is about 1 in 1.9 million. The chances of winning $1 million in a recent Pennsylvania lottery were 1 in 9.6 million. The chances of winning a $10 million prize in Publisher's Clearinghouse Sweepstakes were 1 in 200 million. In contrast, the chances of being struck by lightning are about 1 in 600,000. In other words, you're at least 3 times more likely to be struck by lightning than to win $1 million.

One way to guarantee winning a lottery is to buy all possible combinations of the winning numbers. In 1992, an Australian investment group purchased 5 million of the 7 million possible combinations of the lottery numbers in

a Virginia State Lottery. Their intent was to buy all combinations, but they ran out of time before getting all of them. Fortunately for them, they did get the winning numbers and won a jackpot of $27 million. Having bought 5 million $1 tickets, they walked away with a profit of $22 million. Not bad at all.

jerbarber/iStock/Getty Images

States have written laws to prevent this from happening today, and they are devising lottery games with many more possibilities so that it would be impossible to buy all the possible tickets to win.

The consequence is that your odds of winning are now even lower! There's an old joke among statisticians: lotteries are a tax on people who are bad at math.

Answers to Try This One

1 84   **2** (a) 125   (b) 60   **5** 87,178,291,200

3 (a) 720   (b) 11,880   (c) 30   **6** 90

4 (a) If we number them 1, 2, 3, and 4, the orders are   **7** 165,765,600
1234, 1243, 1324, 1342, 1423, 1432, 2134,
2143, 2314, 2341, 2413, 2431, 3124, 3142,   **8** 64,864,800
3214, 3241, 3412, 3421, 4123, 4132, 4213,
4231, 4312, 4321. There are 24 of them.
(b) $4 \cdot 3 \cdot 2 \cdot 1 = 24$

Exercise Set 10-2

Writing Exercises

1. What is the point of studying counting techniques in this course?
2. Explain the fundamental counting principle in your own words.
3. What do we mean by the phrase "a permutation of n distinct objects"?
4. How does a permutation of n objects differ from a permutation of n objects taken r at a time?
5. Explain what the symbols $n!$ and $_nP_r$ represent, and explain how to compute each for given values of n and r.

6. Describe the difference between finding the number of ways to choose five objects if each can only be chosen once and the number of ways if the objects can be chosen more than once.
7. We developed two permutation formulas in this section, but you really only need to know one of them. Which one and why?
8. How did we use the fundamental counting principal to develop a formula for a permutation of n objects?

Computational Exercises

Evaluate each.

9. $10!$

10. $5!$

11. $1!$

12. $0!$

13. $\frac{8!}{5!}$

14. $\frac{9!}{7!}$

15. $\frac{11!}{(11-3)!}$

16. $\frac{14!}{(14-6)!}$

17. $\frac{6!9!}{4!3!}$

18. $\frac{8!5!}{3!6!}$

19. $\frac{7!}{2!3!}$

20. $\frac{10!}{2!5!}$

21. $\frac{150!}{148!}$

22. $\frac{200!}{197!}$

23. $_8P_2$

24. $_7P_5$

25. $_{12}P_{12}$

26. $_5P_3$

27. $_6P_6$

28. $_6P_0$

29. $_{11}P_3$

30. $_6P_2$

31. $\frac{_4P_2}{_{14}P_3}$

32. $\frac{_9P_6}{_{20}P_2}$

Applications in Our World

33. How many four-letter passwords can be formed from the letters in the word *panicky* if each letter has to be distinct? What if letters are allowed to be repeated?
34. Out of a group of eight students serving on the Student Government Association, how many different ways can a president, a vice president, and a treasurer be selected?
35. How many different ID cards can be made if there are six digits on a card and no digit can be used more than once? What if digits can be repeated?
36. How many different ways can seven types of laser printer be displayed on a shelf in a computer store?
37. How many different ways can four Super Bowl raffle tickets be selected from 50 tickets if each ticket wins a different prize?

38. How many different ways can a psychology student select 5 subjects from a pool of 20 subjects and assign each one to a different experiment?
39. How many website graphics can be created by using at least three of five different bitmap images?
40. A chemistry lab group has seven experiments to choose from and five members in the group. How many different ways can the experiments be assigned if only one experiment is assigned to each group member?
41. A radio DJ has a choice of seven songs in the queue. He has to pick three different songs to play in a certain order after the commercial break. How many ways can he select the three different songs?

42. A professor has five different tasks to assign, one to each of her five teaching assistants. In how many different ways could she make the assignments?

43. A nursing student can be assigned to one of six different floors each day depending on staffing needs. How many different ways can the student be assigned during a 5-day work week?

44. A hotel manager has 12 different promotional events to choose from, and plans to run a promotion during each of the 4 weeks in February. A promotion can be repeated if it's successful. How many different ways can the promotions be chosen?

45. Out of 17 contacts in her cell phone, how many different ways can Shana set the first four speed-dial contacts?

46. In how many different ways can you visit the seven stores you like at an outlet mall?

47. How many different code words can be made from the symbols *, *, *, @, @, $, #, #, #, # if each word has 10 symbols?

48. How many different passwords can be made from the letters in the word *Alabama*?

49. A radio DJ is contractually obligated to run a commercial for Pepsi three times, a commercial for a local law firm once, and a new promo for the station twice per hour. In how many different ways can he do this?

50. A graphic designer is choosing from three red squares, two blue squares, and four white squares to design a backdrop pattern for a tech expo. How many different orders can be chosen?

51. The panel for a political forum consists of four Democrats, five Republicans, and three Independents. In how many orders can they be seated if all that matters is party, not individuals?

52. There are 16 teams in each conference in professional football, and 6 make the playoffs each year. In how many different ways can that happen? (The teams are seeded from 1 to 6.)

53. In how many ways can the panelists in Problem 51 be chosen to answer five questions posed by the moderator with no regard for party? Panelists can answer more than one question.

54. In how many ways can a team from one of the conferences be chosen (see Problem 52) to be featured on the NFL

pregame show for the first 6 weeks of the season? Teams can be featured more than once.

55. A Major League Baseball team has 25 players on the active roster. How many choices does a manager have for batting order, listing the nine starters from 1 through 9?

56. A college basketball team has 14 women on the roster. In how many ways can the coach choose a lineup featuring five different positions?

57. For the new fall season, a network president has 11 shows in development, and six openings in the prime time schedule. In how many ways can the new shows be arranged to fit into the schedule?

58. For the Orion Music and More Festival, there were eight bands scheduled on the main stage over 2 days. In how many different orders could they be scheduled? What if Metallica is scheduled to close the show both days, but every other band is scheduled only once?

While making up his schedule for spring semester, DeSean complains that he doesn't have very many choices of schedule because of the general education requirements he has to meet. His advisor tells DeSean that he has to take one course from each of English (three choices), history (five choices), math/stats (five choices), computer science (four choices), and general science (six choices). Does DeSean have a legitimate gripe? Let's examine this in Problems 59–62.

59. If every possible course is available at the time he's registering, how many possible schedules can DeSean choose from (disregarding when the classes meet)?

60. If DeSean hasn't met the prerequisites for two of the math/stat courses and three of the general science courses, by how much does this reduce his number of possible schedules?

61. When trying to schedule, DeSean finds that all but one of the English courses is closed, as are two history courses and one general science course. How many schedules does he have to choose from now?

62. In an unprecedented effort to make the general education requirements more accessible, the dean of DeSean's college decides to double the number of acceptable courses in each of those five areas. What effect does this have on the number of possible schedules?

Critical Thinking

63. A campus pizzeria offers regular crust, thin crust, or pan pizzas. You can get either white or red sauce. The owner is kind of eccentric, and only sells pizzas with one topping, chosen from pepperoni, sausage, ham, onions, and ground beef. First, use a tree diagram like the one on page 570 to diagram out all possible choices of pizza. Then show that using the fundamental counting principle yields the same answer.

64. (a) Suppose that a chef is choosing from 20 toppings to make a gourmet pizza, and he plans to choose 6 of them. How many different ways can he do so if

you're keeping track of the order in which he adds them?

   (b) Let's be honest, though . . . when you're putting toppings on a pizza, it doesn't really matter what order you choose. So how many different pizzas can the chef make? (*Hint:* How can you get rid of all of the different possible orders in your count?)

65. Show that

$$\frac{n!}{(n-r)!} = n!(n-1)!(n-2)! \cdots (n-r+1)$$

66. Explain why a combination lock should actually be called a permutation lock.

For Questions 67–70, decide if each situation is or is not a permutation, and explain your answer.

67. A regional manager has to pick three top employees to attend a trade show in Palm Springs.

68. A concert promoter picks five bands from a list of applicants and sets the bill for a day-long music festival.

69. Dave looks through his music library and picks out his five favorite bands.

70. You have to pick six numbers from 54 in a lotto game, and six numbered balls are drawn from a bin. If you match all six numbers, you become very wealthy.

Section 10-3 Combinations

✓ **LEARNING OBJECTIVES**

1. Distinguish between a combination and a permutation.

2. Find the number of combinations of *n* objects taken *r* at a time.

3. Use the combination rule in conjunction with the fundamental counting principle.

Suppose that after waiting in line overnight, you manage to snag the last three tickets to a big concert. Sweet! The bad news is that you can only take two of your four housemates. How many different ways can you choose the two friends that get to go?

This sounds a little bit like a permutation problem, but there's a key difference: the order in which you choose two friends doesn't make any difference. They either get to go or they don't. So choosing Ruth and Ama is exactly the same as choosing Ama and Ruth. When order matters in a selection, we call it a permutation, but when order is not important we call it a *combination*.

Andrey Armyagov/Shutterstock

A selection of objects without regard to order is called a **combination**.

In Example 1, we'll examine the difference between permutations and combinations.

Example 1   Comparing Permutations and Combinations

Given four housemates, Ruth, Elaine, Ama, and Jasmine, list the permutations and combinations when you are selecting two of them.

SOLUTION

We'll start with permutations, then eliminate those that have the same two people listed.

Permutations

Ruth Elaine	Elaine Ruth	Ruth Ama
Ama Ruth	Ruth Jasmine	Jasmine Ruth
Elaine Ama	Ama Elaine	Elaine Jasmine
Jasmine Elaine	Ama Jasmine	Jasmine Ama

Combinations

Ruth Elaine	Ruth Ama	Ruth Jasmine
Elaine Ama	Elaine Jasmine	Ama Jasmine

There are 12 permutations, but only 6 combinations.

Try This One   1

If you are choosing two business classes from three choices, list all the permutations and combinations.

It will be very valuable in our study of counting and probability to be able to decide if a given selection is a permutation or a combination.

| **Example 2** | **Identifying Permutations and Combinations** |

Decide if each selection is a permutation or a combination.

(a) From a class of 25 students, a group of 5 is chosen to give a presentation.
(b) A starting pitcher and catcher are picked from a 12-person intramural softball team.

SOLUTION

(a) This is a combination because there are no distinct roles for the 5 group members, so order is not important.
(b) This is a permutation because each selected person has a distinct position, so order matters.

Try This One   2

Decide if each selection is a permutation or a combination.

(a) A 5-digit passcode is chosen from the numbers 0 through 9.
(b) A gardener picks 4 vegetable plants for a garden from 10 choices.

1. Distinguish between a combination and a permutation.

| **Example 3** | **Calculating the Number of Combinations** |

(a) A college instructor gives students a list of seven projects to choose from, and each group has to do three. A week into the course, each group has to submit a list of the projects they're choosing and the order in which they plan to present those projects. Use a permutation formula to find the number of ways a group can choose.
(b) Midway through the first week, the instructor decides that groups don't need to provide an order. What would you divide your answer from part (a) by to account for this? Why?
(c) Use your line of reasoning from part (b) to write a formula for finding the number of combinations of n objects taken r at a time.

SOLUTION

(a) This is exactly what our permutation formula from Section 10-2 is used for. The number of permutations is $_7P_3 = \frac{7!}{(7-3)!} = \frac{7!}{4!} = 210$.
(b) If we were to list all the permutations, every time we listed three specific projects in different orders, we'd be listing the same three projects six times. Why six? Because that's the number of ways we can order the three projects: $3! = 6$. So to find the number of projects to choose without regard for order, we'd take the 210 permutations and divide by 6 to get 35 choices.
(c) Now we can generalize our result. To calculate the number of combinations of n objects taken r at a time, all we did was to find the number of permutations $_nP_r$, then divide by the number of permutations of r objects, which is $r!$. This is how we make order not count: by dividing out the number of possible orders. So our combination formula is $\frac{_nP_r}{r!}$.

Try This One   3

(a) Find the number of combinations of 10 objects taken 5 at a time.
(b) Write a formula for the number of combinations of n objects taken r at a time that involves only factorials.

To represent the number of combinations of n objects taken r at a time, we'll use notation similar to permutation notation: $_nC_r$. Here's the formula that we discovered in Example 3.

The Combination Rule

The number of combinations of n objects taken r at a time without replacement is denoted by $_nC_r$, and is given by the formula

$$_nC_r = \frac{_nP_r}{r!} = \frac{n!}{(n-r)!r!}$$

Example 4 — An Application of Combinations

Math Note

Some people use the terminology "10 choose 3" to describe the combination $_{10}C_3$, and it is sometimes represented using the notation $\binom{10}{3}$.

While studying abroad one semester, Tran is required to visit 10 different cities. He plans to visit 3 of the 10 over a long weekend. How many different ways can he choose the 3 to visit? Assume that distance is not a factor.

amnachphoto/123RF

SOLUTION

The problem doesn't say anything about the order in which they'll be visited, so this is a combination problem.

$$_{10}C_3 = \frac{10!}{(10-3)!3!} = \frac{10!}{7!3!} = 120$$

Try This One   4

An instructor posts a list of eight group projects to the class website. Every group is required to do four projects at some point during the semester. How many different ways can a group choose the four projects they want to do?

2. Find the number of combinations of n objects taken r at a time.

Using Technology: Computing Permutations

In Example 4, we computed $_{10}C_3$ using factorial division. Let's see how we could have done so more quickly using technology.

Graphing Calculator

10 **MATH**, then use the right arrow to select PRB, and press 3 to select $_nC_r$. Finally, press 3 and **ENTER** to get the result.

Spreadsheet

In a blank cell, enter =COMBIN(10,3). The first number is always the total number of objects, and the second is the number being chosen.

In some cases, the combination rule is used in conjunction with the fundamental counting principle. Examples 5 and 6 illustrate some specific situations.

Example 5 — Choosing a Committee

At one school, the student government consists of seven women and five men. How many different committees can be chosen with three women and two men?

SOLUTION

You can think of this problem as having two distinct stages: first choosing the three women, then choosing the two men. So it's a fundamental counting principle problem. At each stage, though, we have a combination. We can choose three women from the seven candidates in $_7C_3 = 35$ ways. We can choose two men from the five candidates in $_5C_2 = 10$ ways. Now we use the fundamental counting principle to multiply: there are $35 \cdot 10 = 350$ different potential committees.

Try This One   5

On an exam, a student has to pick 2 essay questions from 6 essay questions and 10 multiple-choice questions from 20 multiple-choice questions to answer. How many different ways can the student pick questions to answer?

Example 6   Designing a Calendar

To raise money for a charity event, a sorority plans to sell a calendar featuring pictures and bios of some of the school's most popular athletes. They will need to choose six athletes from a pool of finalists that includes nine women and six men. How many possible choices are there if they want to feature at least four women?

SOLUTION

Since we need to include at least four women, there are three possible compositions: four women and two men, five women and one man, or six women and no men.

Four women and two men:

$$_9C_4 \cdot {_6}C_2 = \frac{9!}{(9-4)!4!} \cdot \frac{6!}{(6-2)!2!} = 126 \cdot 15 = 1{,}890$$

Five women and one man:

$$_9C_5 \cdot {_6}C_1 = \frac{9!}{(9-5)!5!} \cdot \frac{6!}{(6-1)!1!} = 126 \cdot 6 = 756$$

Six women and no men:

$$_9C_6 \cdot {_6}C_0 = \frac{9!}{(9-6)!6!} \cdot \frac{6!}{(6-0)!0!} = 84 \cdot 1 = 84$$

The total number of possibilities is $1{,}890 + 756 + 84 = 2{,}730$.

Try This One   6

A four-person crew for the international space station is to be chosen from a candidate pool of 10 Americans and 12 Russians. How many different crews are possible if there must be at least 2 Russians?

3. Use the combination rule in conjunction with the fundamental counting principle.

This is a good time to talk about the significance of the words *and* and *or* when applying counting techniques. In Example 5, we needed a committee with three women AND two men; this led us to conclude that we could think of it as a two-stage process and *multiply* using the fundamental counting principle. In Example 6, we end up choosing between three possible options: four women OR five women OR six women. In this case, we found the number of possibilities for each and *added* them.

Table 10-1 summarizes all of the counting rules from Sections 10-2 and 10-3. It's important to know the formulas, but it's far more important to understand the situations that each formula is needed for.

Table 10-1	Summary of Counting Rules	
Rule	**Description**	**Formula**
Fundamental counting principle	The number of ways a sequence of n events can occur if the first event can occur in k_1 ways, the second event can occur in k_2 ways, etc. (Events are unaffected by the others.)	$k_1 \cdot k_2 \cdot k_3 \cdots k_n$
Permutation rule	The number of permutations of n objects taking r objects at a time. (Order is important.)	$\dfrac{n!}{(n-r)!}$
Permutation rule for duplicate objects	The number of permutations in which k_1 objects are alike, k_2 objects are alike, etc.	$\dfrac{n!}{k_1!k_2! \cdots k_p!}$
Combination rule	The number of combinations of r objects taken from n objects. (Order is not important.)	$\dfrac{n!}{(n-r)!r!}$

Answers to Try This One

1   Call the classes A, B, and C:

Permutations

A B   B A   A C
C A   B C   C B

Combinations

A B   A C   B C

2   (a) Permutation
(b) Combination

3   (a) 252   (b) $\dfrac{n!}{(n-r)!r!}$

4   70

5   2,771,340   **6**   5,665

Exercise Set   10-3

Writing Exercises

1. What is meant by the term combination?
2. What is the difference between a permutation and a combination?
3. Describe a real-life situation in which it would be appropriate to use combinations to count possibilities.
4. Describe a situation related to the one in Exercise 3 in which it would be appropriate to use permutations to count possibilities.
5. Describe a situation where counting would require both the combination formula and the fundamental counting principle.
6. Explain why it makes sense that $_nC_r = \frac{_nP_r}{r!}$.
7. In counting problems that use the word *and* between two different possibilities, what operation will we use? Explain.
8. In counting problems that use the word *or* between two different possibilities, what operation will we use? Explain.

Computational Exercises

For Exercises 9–22, evaluate each expression.

9. $_5C_2$
10. $_8C_3$
11. $_7C_4$
12. $_6C_2$
13. $_6C_4$

14. $_3C_0$
15. $_3C_3$
16. $_9C_7$
17. $_{12}C_2$
18. $_4C_3$

19. $_{10}C_7 \cdot {_5C_4}$
20. $_8C_5 \cdot {_6C_2} \cdot {_3C_1}$
21. $\frac{_{10}C_3 \cdot {_6C_2}}{_6C_3 \cdot {_5C_3}}$
22. $\frac{_{12}C_8 \cdot {_7C_3}}{_5C_2}$

For Exercises 23–30, find both the number of combinations and the number of permutations for the given number of objects.

23. 8 objects taken 5 at a time
24. 5 objects taken 3 at a time
25. 6 objects taken 2 at a time
26. 10 objects taken 6 at a time
27. 9 objects taken 9 at a time
28. 12 objects taken 1 at a time
29. 12 objects taken 4 at a time
30. 15 objects taken 7 at a time

For Exercises 31–40, decide whether the selection described is a combination or a permutation.

31. Ten fans at a concert are chosen to go backstage after the show.
32. From a list of 20 dishes he knows how to cook, Maurice chooses different dishes for breakfast, lunch, and dinner on his partner's birthday.
33. A state elects a governor and lieutenant governor from a pool of eight candidates.
34. A state elects two senators from a pool of 12 candidates.
35. Lupe chooses an eight-letter password from the letters of the alphabet.

36. Of the six optional community service projects in a service learning course, Haylee picks three of them.

37. When looking for a new car, you read about 10 different models and choose 4 that you would like to test drive.

38. Mark looks over the novels on his bookshelf and lists his five favorites ranked 1 through 5.

39. In planning a salad for a banquet, a chef chooses one of two types of greens, one of three types of croutons, and two of five types of vegetables.

40. Ali schedules classes for next semester from 12 different course choices, scheduling times that fit into that semester's work schedule.

Applications in Our World

41. In 5-card poker, each player is dealt 5 cards (go figure) from a standard deck of 52 cards. How many different hands can be dealt?

42. How many ways are there to select three math help websites from a list that contains six different websites?

43. How many ways can a student pick five questions from an exam containing nine questions? How many ways are there if everyone is required to answer the first question and the last question?

44. How many ways can four finalists for a job be selected from 10 interviewees?

45. A sheriff is choosing three shift commanders from 10 candidates who have expressed interest in the promotion. In how many ways can he do this?

46. How many different possible tests can be made from a test bank of 20 questions if the test consists of 5 questions? (Ignore the order of questions.)

47. The general manager of a fast-food restaurant chain must select 6 restaurants from 11 for a promotional program. How many different possible ways can this selection be done?

48. How many ways can 3 cars and 4 trucks be selected from 8 cars and 11 trucks to be tested for a safety inspection?

49. During the tryouts for a university pep band, there were 4 trumpet players, 12 drummers, and 7 saxophonists. How many ways can the jazz band be chosen so there are 2 trumpet players, 5 drummers, and 3 saxophonists?

50. There are seven men and five women in line at a salsa dance club. The bouncer can only admit two more men and two more women. How many ways can he choose from those in line? How many ways can he choose if instead he is told he can admit four people and at least two have to be women?

51. Coca-Cola comes in two low-calorie varieties: Diet Coke and Coke Zero. If a promoter has 10 cans of each, how many ways can she select 3 cans of each for a taste test at the local mall?

52. At the movies, Shana wants to get snacks for her friends. How many ways can she select three types of candy and two types of soda from the eight types of candy and five types of soda available?

53. Steve wants to download new music into an iPhone from iTunes. How many ways can Steve select 2 rock songs, 3 alternative songs, and 3 rap songs from a list of 8 rock songs, 6 alternative songs, and 10 rap songs?

54. How many ways can 2 men and 2 women be selected for a debate tournament if there are 10 male finalists and 12 female finalists?

55. A resort manager is choosing a committee of four people to discuss employment issues. They'll choose from eight housekeepers, three desk clerks, and five maintenance workers. How many possible committees are there if there have to be at least two housekeepers?

56. The California Bureau of Investigation is putting together an elite serial crimes task force with seven members. The candidates are five members of the CBI, eight members of local law enforcement agencies, and nine state patrol members. Find the number of possible task forces if there has to be at least one representative from each agency, and no more than one state patrol member.

57. An inspector with the Nuclear Regulatory Commission is tasked to visit five nuclear plants this month, randomly chosen from three in Ohio, four in New York, and five in Pennsylvania. How many different ways can he choose the plants to visit if at least three will be in Pennsylvania?

58. Eleven patients with Type 2 diabetes are being chosen for a clinical trial of an experimental medication. Twenty patients that developed the disease after age 40 have been identified as good candidates for the study, as well as 14 that developed it as children. The plan is to divide the number of patients as equally as possible among those that developed the disease early and late. Find the number of ways that patients can be chosen for the study.

Critical Thinking

59. In a class of 30 people, the professor decides that everyone should get to know each other, so everyone must have at least a 2-minute conversation with everyone else in the class. What's the least amount of total time that will be spent on these conversations?

60. (a) Refer to Exercise 59. Find the least amount of time needed if the class has 10 students, 20 students, and 40 students.

(b) Based on the amount of time calculated for 10, 20, 30, and 40 students, try to make a conjecture as to the time needed if there are 50 students, then check your answer.

61. (a) The 2017 baseball team at the Ohio State University consisted of 10 freshmen, 7 sophomores, 12 juniors, and 5 seniors. The coaches want to choose two players from each class to represent the team at a booster club banquet. How many different ways can they choose?

(b) The coaches also need to choose four players overall to visit elementary schools in the community, with each player going to a different school. How many different ways can they make this choice?

62. A state lottery offers two games in which you choose 6 numbers. In the first, there are 25 numbers to choose from, and you need to match the numbers in the order in which they are drawn. In the second, there are 50 numbers to choose from, and you need to match all 6 regardless of the order. Which one is easier to win?

63. Show that for any natural numbers n and r, $_nC_r = {_nC_{n-r}}$.

64. Using the definitions of what $_nP_r$ and $_nC_r$ mean (not the formulas for computing them), explain why both are zero if $r > n$. Then try to compute them using the formulas. What goes wrong?

Section 10-4

Tree Diagrams, Tables, and Sample Spaces

LEARNING OBJECTIVES

1. Use tree diagrams to find sample spaces and compute probabilities.

2. Use tables to find sample spaces and compute probabilities.

Photodisc/MedioImages/
Getty Images

Math Note

Recall that we used tree diagrams to illustrate the fundamental counting principle back in Section 10-2.

For centuries, people have tried a wide variety of techniques, some of them pretty bizarre, to try and influence the sex of their children. The truth is, without the aid of cutting-edge science, you don't get to choose. But that doesn't stop many young couples from planning the type of family they hope to have. Suppose that one couple would like to have three children, but they definitely want to have at least one boy and one girl. What is the probability that they'll get their wish without having to go beyond three kids?

When working with theoretical probabilities, we know that we need to decide on the sample space for an event, and then find how many individual outcomes are in that event. When situations start to get complicated, it might not always be apparent how to do so. That's where tree diagrams and tables can help.

Tree Diagrams

A **tree diagram** is a diagram consisting of branches corresponding to the outcomes of two or more probability experiments that are done in sequence.

When constructing a tree diagram, we use branches emanating from a single point to show the outcomes for the first experiment, and then add the outcomes for the second experiment using branches emanating from each branch that was used for the first experiment, and so on.

In Example 1, we'll use a tree diagram to find the sample space for our hopeful young couple.

Example 1

Using a Tree Diagram to Find a Sample Space

Use a tree diagram to find the sample space for the sexes of three children in a family.

SOLUTION

There are two possibilities for the first child, boy or girl, two for the second, boy or girl, and two for the third, boy or girl. So the tree diagram can be drawn as shown in Figure 10-3.

After a tree diagram is drawn, the outcomes can be found by tracing through

Math Note

The sexes of the children are listed in their birth order. For example, the outcome GGB means the firstborn was a girl, the second a girl, and the third a boy.

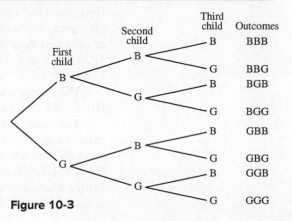

Figure 10-3

all of the branches. In this case, the sample space would be {BBB, BBG, BGB, BGG, GBB, GBG, GGB, GGG}.

Try This One   1

A soda machine dispenses both Coke and Pepsi products, in both 12-ounce cans and 20-ounce bottles. For each brand, it has a regular cola, diet cola, and lemon-lime drink. Use a tree diagram to find the sample space for the experiment of choosing one drink at random from this machine.

Once a tree diagram is drawn and the sample space is found, you can compute the probabilities for various events.

Example 2   Computing a Probability

If a family has three children, find the probability that they have at least one boy and one girl. (Assume that each child is equally likely to be a boy or girl.)

SOLUTION

The sample space we found in Example 1 has eight outcomes, and only two of them have three kids with the same sex. So six of the eight outcomes have at least one boy and one girl, making the probability $\frac{6}{8}$ or $\frac{3}{4}$.

Try This One   2

Suppose the soda machine from Try This One 1 goes berserk and starts dispensing drinks randomly. If you want a diet cola, what is the probability that you'll get one?

Example 3   Using a Tree Diagram to Compute Probabilities

A coin is flipped, and then a die is rolled. Use a tree diagram to find the probability of getting heads on the coin and an even number on the die.

SOLUTION

First, we'll use a tree diagram to find the sample space. The coin will land on either heads or tails, and there are six outcomes for the die: 1, 2, 3, 4, 5, or 6. The tree diagram is shown in Figure 10-4.

The sample space is {H1, H2, H3, H4, H5, H6, T1, T2, T3, T4, T5, T6}.

The total number of outcomes for the experiment is 12. The number of ways to get a head on the coin and an even number on the die is 3: H2, H4, or H6. So, the probability of getting a head and an even number when a coin is tossed and a die is rolled is $\frac{3}{12}$, or $\frac{1}{4}$.

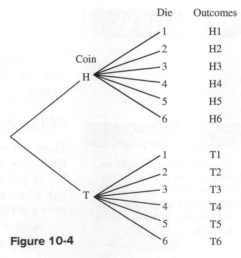

Figure 10-4

Try This One   3

In order to collect information for a student survey, a researcher classifies students according to eye color (blue, brown, green), sex (male, female), and class rank (first-year, sophomore). A folder for each classification is then made up (e.g., sophomore/female/green eyes). Find the sample space for the folders using a tree diagram. If a folder is selected at random, find the probability that

(a) It includes students with blue eyes.
(b) It includes students who are female.
(c) It includes students who are male first-year students.

1. Use tree diagrams to find sample spaces and compute probabilities.

In constructing tree diagrams, not all branches have to be the same length. For example, suppose two players, Alice and Diego, are matched up in the final round of a chess tournament, and the first one to win two games wins the tournament. The tree diagram would be like the one shown in Figure 10-5. For any game, A means that Alice wins, and D means that Diego wins.

Notice that if Alice wins the first two games, the tournament is over. So the first branch is shorter than the second one. But if Alice wins the first game and Diego wins the second game, they need to play a third game in order to decide who wins the tournament. Similar reasoning can be applied to the rest of the branches.

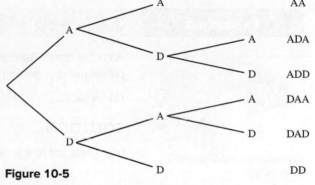

	AA
	ADA
	ADD
	DAA
	DAD
	DD

Figure 10-5

> **CAUTION**
>
> When the branches of a tree diagram are not equal in length, the outcomes are not equally likely even if the probability of every individual trial is $\frac{1}{2}$. In the tournament example above, if each player is equally likely to win any given game, the outcomes AA and DD both have probability $\frac{1}{4}$, while the outcomes ADA, ADD, DAA, and DAD all have probability $\frac{1}{8}$.

Sidelight   Math and the Draft Lottery

If you're like most people, winning the lottery sounds pretty good to you. But if you were an 18-year-old male in 1970, you would have felt a lot differently. During the Vietnam War, men were drafted for service based on their age and birthday, with older candidates chosen first. This allowed those born in the last months of the year to avoid military service, which didn't sound particularly fair to those born in January.

So in 1970, the government decided to change the system and instituted a lottery to decide who would be drafted. The days of the year were numbered 1 through 366 (for leap years) and mixed in a barrel. The numbers corresponding to birth dates were drawn one at a time, and men were drafted based on these numbers.

Sounds perfectly random, right? Everyone thought so, until mathematicians noticed that a disproportionate number of draftees were born later in the year, far more than would be expected with random chance. There's no evidence that anything nefarious was going on, however. It's most likely that the numbers later in the year were put into the barrel last, leaving them on top, and the barrel wasn't mixed well enough.

Bettmann/Getty Images

Tables

Another way of determining a sample space is by making a **table** of outcomes. Consider the sample space of selecting a card from a standard deck of 52 cards. (The cards are assumed to be shuffled to make sure that the selection occurs at random.) There are four suits—hearts, diamonds, spades, and clubs, and 13 cards of each suit consisting of the denominations ace (A), 2, 3, 4, 5, 6, 7, 8, 9, 10, and 3 picture or face cards—jack (J), queen (Q), and king (K). The sample space is shown in Figure 10-6.

	A	2	3	4	5	6	7	8	9	10	J	Q	K
♥	A♥	2♥	3♥	4♥	5♥	6♥	7♥	8♥	9♥	10♥	J♥	Q♥	K♥
♦	A♦	2♦	3♦	4♦	5♦	6♦	7♦	8♦	9♦	10♦	J♦	Q♦	K♦
♠	A♠	2♠	3♠	4♠	5♠	6♠	7♠	8♠	9♠	10♠	J♠	Q♠	K♠
♣	A♣	2♣	3♣	4♣	5♣	6♣	7♣	8♣	9♣	10♣	J♣	Q♣	K♣

Figure 10-6

Example 4　Using a Table to Compute Probabilities

Lawrence Manning/Corbis RF/
Getty Images

A card is drawn from an ordinary deck. Use the sample space shown in Figure 10-6 to find the probabilities of getting

(a) A jack.　　　　(b) The 6 of clubs.　　　　(c) A 3 or a diamond.

SOLUTION

(a) There are four jacks and 52 possible outcomes, so

$$P(\text{jack}) = \frac{4}{52} = \frac{1}{13}$$

(b) Since there is only one 6 of clubs, the probability of getting the 6 of clubs is

$$P(6 \text{ of clubs}) = \frac{1}{52}$$

(c) There are four 3s and 13 diamonds, but the 3 of diamonds is counted twice in this listing. So, there are 16 possibilities of drawing a 3 or a diamond, and

$$P(3 \text{ or diamond}) = \frac{16}{52} = \frac{4}{13}$$

Try This One　　　　4

A single card is drawn at random from a well-shuffled deck. Using the sample space shown in Figure 10-6, find the probability that the card is

(a) An ace.　　　　(c) A club.　　　　(e) The 6 or 8 of spades.
(b) A face card.　　(d) A 4 or a heart.

Example 5　Building a Table to Illustrate a Sample Space

(a) Use the fundamental counting principle to calculate the number of outcomes for the experiment of rolling two 6-sided dice.
(b) Build a table similar to Figure 10-6 that illustrates the outcomes for that experiment. Do you get the same number of possibilities?

SOLUTION

(a) There are six different outcomes for the first roll, and six for the second. So, using the fundamental counting principle, we get $6 \cdot 6 = 36$ possible outcomes.

(b) To build a table, we'll list the outcomes from the first roll vertically along the left side, and the outcomes from the second roll horizontally across the top. Just for clarity, let's pretend that the first die is red and the second is blue. We'll then list the result of each roll as an ordered pair, with the result of the red die in red and the result of the blue die in blue. The table is shown in Figure 10-7. Notice that the pair (3, 6) lives in the row for 3 on the red die and 6 for the blue die. This represents getting 3 with red and 6 with blue, for a total roll of 9.

Notice that there are 36 entries in the table, matching our answer to part (a).

	Blue die					
Red die	**1**	**2**	**3**	**4**	**5**	**6**
1	(1, 1)	(1, 2)	(1, 3)	(1, 4)	(1, 5)	(1, 6)
2	(2, 1)	(2, 2)	(2, 3)	(2, 4)	(2, 5)	(2, 6)
3	(3, 1)	(3, 2)	(3, 3)	(3, 4)	(3, 5)	(3, 6)
4	(4, 1)	(4, 2)	(4, 3)	(4, 4)	(4, 5)	(4, 6)
5	(5, 1)	(5, 2)	(5, 3)	(5, 4)	(5, 5)	(5, 6)
6	(6, 1)	(6, 2)	(6, 3)	(6, 4)	(6, 5)	(6, 6)

Figure 10-7

Try This One   5

(a) Use the fundamental counting principle to calculate the number of outcomes for the experiment of flipping a coin and then rolling a 10-sided die. (The sides of the die are numbered from 1 through 10.)
(b) Build a table that illustrates all of the outcomes. Do you get the same number you calculated in part (a)?

Example 6   **Using a Table to Compute Probabilities**

When two dice are rolled, find the probability of getting

(a) A sum of 8.
(b) Doubles (the same number on each die).
(c) A sum less than 5.

SOLUTION

Using the sample space shown in Figure 10-7, there are 36 possible outcomes.

(a) There are five ways to get a sum of 8: (2, 6), (3, 5), (4, 4), (5, 3), and (6, 2). So $n(E) = 5$, $n(S) = 36$, and

$$P(\text{sum of 8}) = \frac{n(E)}{n(S)} = \frac{5}{36}$$

(b) There are six ways to get doubles: (1, 1), (2, 2), (3, 3), (4, 4), (5, 5), and (6, 6). So $n(E) = 6$, $n(S) = 36$, and

$$P(\text{doubles}) = \frac{n(E)}{n(S)} = \frac{6}{36} = \frac{1}{6}$$

(c) A sum less than 5 means a sum of 4, 3, or 2. The number of ways this can occur is 6, as shown.

Sum of 4: (1, 3), (2, 2), (3, 1)
Sum of 3: (1, 2), (2, 1)
Sum of 2: (1, 1)
$n(E) = 6$, $n(S) = 36$, and so

$$P(\text{sum less than 5}) = \frac{6}{36} = \frac{1}{6}$$

Ingram Publishing/Alamy Stock Photo

Try This One   6

2. Use tables to find sample spaces and compute probabilities.

Two dice are rolled. Use the sample space shown in Figure 10-7 to find the probability of
(a) Getting a sum of 9.
(b) Getting a sum that is an even number.
(c) Getting a sum greater than 6.

Answers to Try This One

1   Sample space: {Coke, can, cola; Coke, can, diet cola; Coke, can, lemon-lime; Coke, bottle, cola; Coke, bottle, diet cola; Coke, bottle, lemon-lime; Pepsi, can, cola; Pepsi, can, diet cola; Pepsi, can, lemon-lime; Pepsi, bottle, cola; Pepsi, bottle, diet cola; Pepsi, bottle, lemon-lime}.

2 $\frac{1}{3}$

3   Sample space: {blue, male, first-year; blue, male, sophomore; blue, female, first-year; blue, female, sophomore; brown, male, first-year; brown, male, sophomore; brown, female, first-year; brown, female, sophomore; green, male, first-year; green,

male, sophomore; green, female, first-year; green, female, sophomore}.

(a) $\frac{1}{3}$ (b) $\frac{1}{2}$ (c) $\frac{1}{4}$

4   (a) $\frac{1}{13}$ (b) $\frac{3}{13}$ (c) $\frac{1}{4}$ (d) $\frac{4}{13}$ (e) $\frac{1}{26}$

5   (a) $2 \cdot 10 = 20$
(b)

	1	2	3	4	5	6	7	8	9	10
Heads (H)	H1	H2	H3	H4	H5	H6	H7	H8	H9	H10
Tails (T)	T1	T2	T3	T4	T5	T6	T7	T8	T9	T10

6   (a) $\frac{1}{9}$ (b) $\frac{1}{2}$ (c) $\frac{7}{12}$

Exercise Set   10-4

Writing Exercises

1. Explain how to draw a tree diagram, and how tree diagrams help to find sample spaces.
2. Think of an example, other than the one provided in the section, where a tree diagram would have branches of different lengths.
3. Could you make a table like the ones we built in this lesson for the experiment of rolling three dice and

calculating the total? Why or why not? What about a tree diagram?
4. When you roll two dice and record the total of the two rolls, in terms of sample spaces and probabilities, does it matter if you roll them one at a time or at the same time? Explain, using information from this section.

Applications in Our World

5. Three computers are chosen at random from an inventory of Dell and Acer computers for a bookstore display. Find the probability that
   (a) All three will be Acers.
   (b) Exactly two will be Dells.
   (c) At least two will be Acers.
6. In Try This One 5, you should have found the sample space for a probability experiment where a coin is flipped and a 10-sided die is rolled. This setup was used for a carnival game, and different amounts were paid out for certain outcomes. Find the probability of each outcome:
   (a) There was a head on the coin and an odd number on the die.
   (b) There was a head on the coin and a prime number on the die.

   (c) There was a tail on the coin and a number less than 5 on the die.
7. For the carnival game in Exercise 6, find the probability of these outcomes:
   (a) There was a tail on the coin and a number greater than 1 on the die.
   (b) There was a head on the coin and an even number on the die.
   (c) There was a tail on the coin and a number divisible by 3 on the die.
8. Sara came across a new website that featured a "live" fortune-teller. She typed in three questions and found (shockingly) that the fortune-teller could only answer yes or no to her questions. Draw a tree diagram for all possible answers for her three questions. Assuming that

this spooky fortune-teller is actually just choosing the responses at random, find the probability that

(a) All answers will be yes or all answers will be no.
(b) The answers will alternate (i.e., Yes–No–Yes or No–Yes–No).
(c) Exactly two answers will be yes.

9. After a late night studying, Ebony decides to grab a latte before class so she can stay awake through her morning lecture. She has only a one-dollar bill, a five-dollar bill, and a ten-dollar bill in her wallet. She pulls one out and looks at it, but then she puts it back. Distracted by a flyer for a new campus organization, she randomly hands a bill from her wallet to the clerk. Draw a tree diagram to determine the sample space and find the probability that

(a) Both bills have the same value.
(b) The second bill is larger than the first bill.
(c) The value of each of the two bills is even.
(d) The value of exactly one of the bills is odd.
(e) The sum of the values of both bills is less than $10.

10. A coin flip determines who gets the ball first at the beginning of a football game, with the visiting team calling heads or tails. The captain of one particular team always calls heads. In the first four games as visitor of a season, find the probability that the team

(a) Wins the toss three times.
(b) Loses the toss all four times.
(c) Wins the toss more than once.
(d) Loses the toss no more than twice.
(e) Loses the toss at least once.

11. Mark and Raul play three games of pool. They are equal in ability. Draw a tree diagram to determine the sample space and find the probability that

(a) Either Mark or Raul win all three games.
(b) Either Mark or Raul win two out of three games.
(c) Mark wins only two games in a row.
(d) Raul wins the first game, loses the second game, and wins the third game.

12. Frank, Sofia, Eldridge, and Jake are the four qualifiers for a charity raffle with two $500 prizes. One of their names will be drawn for the first prize then replaced, at which point the second prize winner will be drawn. Draw a tree diagram to determine the sample space and find the probability that

(a) One person wins both prizes.
(b) There are two different winners.
(c) Sofia wins at least one prize.
(d) Frank wins both prizes.
(e) The two winners are Jake and Eldridge.

13. The pool table in the student lounge is broken, and when Carrie puts a quarter into the slot, only the balls numbered 1 through 5 come down the chute. Carrie picks one without looking and puts it on the table, then picks another without looking. Draw a tree diagram to determine the sample space and find the probability that

(a) The sum of the numbers of the balls is odd.
(b) The number on the second ball is larger than the number on the first ball.
(c) The sum of the numbers on the balls is greater than 4.

14. Akira wants to purchase that all-important first new car and can select one option from each category:

Model	Engine Type	Color
Ford Fusion	Hybrid	Burnt copper
Honda Civic	E-85	Cobalt blue
Toyota Corolla		Metallic green

Draw a tree diagram and find the sample space for all possible choices. Find the probability that the car, if chosen at random,

(a) Has an E-85 engine.
(b) Is a burnt copper Ford Fusion.
(c) Is a metallic green hybrid Honda Civic.
(d) Is cobalt blue.
(e) Is either a metallic green or cobalt blue Toyota Corolla.

15. The Sleep Number bed is a mattress that adjusts to different levels of softness. After deciding that they definitely want one, a couple has a few choices to make. These choices are summarized in the table. They'll need to make one choice in each category.

Size	Top	Adjustment
King	Pillow	Sleep IQ
Queen	Memory foam	Manual
Double		

Draw a tree diagram to find the sample space and find the probability that if the couple chooses a mattress at random, it will be

(a) A memory foam-topped mattress.
(b) A queen-sized mattress with a pillow top.
(c) A memory foam-topped mattress with manual adjustment.

16. To choose the order of bands for the finals of a Battle of the Bands competition, Freddy puts a penny, a nickel, a dime, a quarter, and a half-dollar into five separate envelopes and has one band choose an envelope. Another band then chooses from the envelopes remaining. Draw a tree diagram to determine the sample space and find the probability that

(a) The amount of the first coin is less than the amount of the second coin.
(b) Neither coin is a quarter.
(c) One coin is a penny and the other coin is a nickel or a dime.
(d) The sum of the amounts of both coins is even.
(e) The sum of the amounts of both coins is less than $0.40.

17. At the beginning of a magic trick, The Great Mancini shuffles an ordinary deck of 52 cards as shown in Figure 10-6, and has the nearest person in the audience draw a single card. Using the sample space for drawing a single card from this deck, find the probability that the contestant got

(a) A 10.
(b) A club.
(c) The ace of hearts.
(d) A 3 or a 5.

(e) A 6 or a spade.
(f) A queen or a club.
(g) A diamond or a club.
(h) A red king.
(i) A black card or an 8.
(j) A red 10.

18. For one of the game booths at a town fair, contestants pick a single card from a standard deck, and payouts are based on the card chosen. Find the probability that your card is
(a) The 6 of clubs.
(b) A black card.
(c) A queen.
(d) A black 10.
(e) A red card or a 3.
(f) A club and a 6.
(g) A 2 or an ace.
(h) A club, diamond, or spade.
(i) A diamond face card.
(j) A red ace.

19. In between classes, Jade plays a game of online Monopoly. Using the sample space for rolling two dice shown in Figure 10-7, find the probability that when Jade rolls the two dice, the result is a
(a) Sum of 5.
(b) Sum of 7 or 11.
(c) Sum greater than 9.
(d) Sum less than or equal to 5.
(e) Three on one die or on both dice.
(f) Sum that is odd.
(g) Prime number on one or both dice.
(h) Sum greater than 1.

20. During a charity Las Vegas Casino Night, Rosie plays craps and gets to roll the dice. Using the sample space for rolling two dice as shown in Figure 10-7, find the probability the roll was a
(a) Sum of 8.
(b) Sum that is prime.
(c) Five on one or both dice.
(d) Sum greater than or equal to 7.
(e) Sum that is less than 3.
(f) Sum greater than 12.
(g) Six on one die and 3 on the other die.

Critical Thinking

21. An online math quiz made up of true/false questions starts every student out with two questions. If they get both wrong, they're done. If they get at least one right, they get a third question. Suppose that a student guesses on all questions. Draw a tree diagram and find the sample space for all possible results. Then explain why every outcome is not equally likely.

22. To get past security at a club, potential patrons must answer a series of four questions. If they get any of the first three wrong, they get no more questions and don't get in. Draw a tree diagram and find the sample space for all possible results. Then explain why all of the outcomes are not equally likely.

23. Consider the sample space when three dice are rolled. How many different outcomes would there be?

24. When three dice are rolled, how many ways can a sum of 6 be obtained?

25. When three dice are rolled, find the probability of getting a sum of 6.

26. Find the probability of rolling 10 with two eight-sided dice, with the sides numbered 1 through 8.

Note: Exercises 27–30 go together. Think of them as a mini-project.

27. (a) We know that when flipping a coin, the probability of getting tails is 1/2. Use a tree diagram to find the probability of getting tails twice in a row.
    (b) We also know that the probability of rolling a 5 with one die is 1/6. Use tree diagrams to find the probability of getting tails and then rolling a 5 when you flip a coin and then roll a single die.

(c) Based on the results of parts (a) and (b), how do you think you find the probability of two unconnected events occurring consecutively when you know the probability of each occurring individually? (Answer successfully, and you've discovered an important rule we'll study in Section 10-8.)

28. In the finals of an Angry Birds tournament, Bob is the number one seed, and he'll win the championship if he wins one game in the finals. The other two finalists are Fran and Julio, and they each need to win two games to claim the title.
(a) Draw a tree diagram and find the sample space for each possible result of the finals.
(b) Assuming that all three players have an equal likelihood of winning any individual game, find the probability of each outcome in the sample space. (*Hint:* Use the rule you discovered in 27[c].) Make sure that all of your probabilities add to 1!
(c) Use the results of (b) to find the probability of each contestant winning the championship.

29. Disappointed in her failure to gain the top seed for the tournament in Exercise 28, Fran resorts to using a banned substance that improves her reflexes and makes her twice as likely to win any individual game as Bob or Julio. Find the probability of each winning the championship.

30. If Fran's cheating (see Exercise 29) had made her three times as likely to win any individual match as Bob or Julio, would this have been enough to make her the most likely winner?

Section 10-5 | **Probability Using Permutations and Combinations**

LEARNING OBJECTIVES

1. Compute probabilities using combinations.

2. Compute probabilities using permutations.

Sometimes a friendly game of poker can become less friendly when someone seems just a bit too lucky. Suppose one player in such a game gets dealt all four aces in one hand. Would you suspect that something fishy was going on? What is the probability of that happening?

In Section 10-4, we used tree diagrams and tables to find sample spaces and the number of outcomes in certain events. This was a pretty good strategy, but when the number of possibilities gets larger, diagrams can get out of hand. Fortunately, we know about counting techniques that are tailor-made for answering questions about probability. If our job is to find how many ways something can happen, the combination and permutation rules from Section 10-2 will be our best friends.

George Doyle/Stockbyte/Getty Images

Our main tool will still be the formula for theoretical probability: dividing the number of outcomes that satisfy a certain event by the total number of outcomes in a sample space. But now we'll be able to answer some tougher questions using counting techniques to count outcomes.

Example 1 | **Using Counting Techniques to Compute Probability**

Fancy/SuperStock

Stacy has the option of selecting three books to read for a humanities course. The suggested book list consists of 10 biographies and five current events books. Stacy decides to pick the three books at random. Find the probability that all three books will be biographies.

SOLUTION

All we're interested in is whether or not the three books are biographies, so the order doesn't matter, making this a combination. Specifically, we want to know how many ways we can choose three books from 10 biographies: this is $_{10}C_3$.

$$_{10}C_3 = \frac{10!}{(10-3)!3!} = \frac{10!}{7!3!} = \frac{10 \cdot 9 \cdot 8 \cdot 7!}{7! \cdot 3 \cdot 2 \cdot 1} = 120 \quad \textit{120 ways to pick 3 biographies}$$

Stacy will be picking 3 books total from the 15 options, so the total number of ways to choose is $_{15}C_3$.

$$_{15}C_3 = \frac{15!}{(15-3)!3!} = \frac{15!}{12!3!} = \frac{15 \cdot 14 \cdot 13 \cdot 12!}{12! \cdot 3 \cdot 2 \cdot 1} = 455 \quad \textit{455 total choices}$$

The probability that Stacy gets three biographies if choosing at random is

$$\frac{120}{455} \approx 0.264$$

Try This One | **1**

There are 12 women and 5 men in a seminar course. If the professor chooses four-person groups at random, what is the probability that the first group chosen will consist of all women?

Now back to that poker game . . .

| Example 2 | Using Counting Techniques to Compute Probability |

What is the probability of getting four aces when drawing 5 cards from a standard deck of 52 cards?

SOLUTION

First, we'll figure out how many five-card hands have four aces. The key observation is that we need to get all four aces in the deck, and there's only one way to do that. After we have our four aces, we still have to choose a fifth card, and there are 48 other cards left in the deck. Using the fundamental counting principle, there are $1 \cdot 48$ possible hands that have four aces.

Order doesn't matter again, so the total number of five-card hands is a combination of 52 objects taken 5 at a time, or $_{52}C_5$.

$$_{52}C_5 = \frac{52!}{(52-5)!5!} = \frac{52!}{47!5!} = \frac{52 \cdot 51 \cdot 50 \cdot 49 \cdot 48 \cdot \cancel{47!}}{\cancel{47!} \cdot 5 \cdot 4 \cdot 3 \cdot 2 \cdot 1} = 2,598,960$$

The probability of getting four aces is

$$\frac{48}{2,598,960} = \frac{1}{54,145} \approx 0.0000185$$

It's certainly possible to get dealt four aces, and we would hate to accuse your friend of cheating, but to say the least that's an extremely unlikely occurrence.

> **Math Note**
>
> The probability in Example 2 means that you would expect to draw four aces on average once every 54,145 tries.

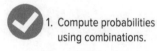
1. Compute probabilities using combinations.

| Try This One | 2 |

Suppose the deck of cards in Example 2 has all 32 cards with numbers less than 10 removed, so that only 10s, jacks, queens, kings, and aces remain. How many times more likely is it now to get 4 aces when drawing 5 cards?

Sidelight   Unlikely Events Are Very Likely to Occur

When the probability of some event is low, like say .05, it's common and reasonable to guess that event most likely will not occur. But if you look at it a certain way, events that are incredibly unlikely happen all the time. Let's use poker as an example. When you're dealt five cards, there's just a 1 in 13 chance that the first card you are dealt will be a 5, because there are 13 denominations. That's a probability of about 0.077. Now think about this: regardless of what denomination your first card ends up being, there was the same probability—just 0.077—of getting it. But of course the first card has to be *something*, so every time that first card is dealt, something with probability 0.077 is *guaranteed to happen*.

Now let's extend that out to all five cards. As we saw in Example 2, there are 2,598,960 possible hands you can be dealt. That means the probability of getting any particular hand is 0.00000038. Yet every single time that you are dealt five cards,

you have to get one possible hand, so something with that spectacularly low probability will occur with probability 1. Wild!

Finally, let's talk about the most unlikely lottery win ever, which happened to you even if you don't realize it. That's the genetic lottery, which resulted in you being born as you. Genetics tells us that the number of possible DNA combinations, which determine all of your characteristics and make you who you are, is on the order of 3 followed by 614 zeros. This is far larger than the total number of stars in the universe. Of all those possible combinations, only one of them could have resulted in you, an event so unlikely that the probability is for all intents and purposes zero. And yet somehow that happened. And continues to happen thousands of times across the world every single day. Next time you're feeling a little down, think about how incredibly lucky you were to have ever been born, and maybe you'll feel just a little special.

In Example 3, order matters.

| Example 3 | Using Permutations to Compute Probability |

(a) A combination lock has 40 numbers on it, from zero to 39. Find the probability that if the combination to unlock it consists of three numbers, it will contain the numbers 1, 2, and 3 in some order. Assume that numbers cannot be repeated in the combination.

(It's interesting to note that a combination lock should really be called a permutation lock since the order of the numbers is important when you are unlocking the lock.)

(b) How does the problem change if numbers CAN be repeated? Find the probability in that case.

SOLUTION

(a) The statement of the question was kind enough to point out that order matters when opening a combination lock, so we're dealing with permutations in this problem. The number of combinations for the lock containing 1, 2, and 3 is $_3P_3$.

$$_3P_3 = \frac{3!}{(3-3)!} = \frac{3!}{0!} = \frac{3 \cdot 2 \cdot 1}{1} = 6$$

The total number of combinations is a permutation of the 40 numbers taken 3 at a time, or $_{40}P_3$.

$$_{40}P_3 = \frac{40!}{(40-3)!} = \frac{40!}{37!} = \frac{40 \cdot 39 \cdot 38 \cdot \cancel{37!}}{\cancel{37!}} = 59{,}280$$

The probability of the combination containing 1, 2, and 3 is $\frac{6}{59{,}280} \approx 0.000101$

(b) If numbers can be repeated in the combination, this has no effect on the number of outcomes that contain 1, 2, and 3 in some order, since in that case repeats aren't possible. What this WILL affect is the total number of combinations for the lock. Now we can just use the fundamental counting principle, because there will be 40 choices at each stage of the process (choosing each digit in the combination). So we'll get $40 \cdot 40 \cdot 40 = 64{,}000$ possible combinations, and the probability is now

$$\frac{6}{64{,}000} = 0.00009375$$

Steve Sant/Alamy Stock Photo

Permutations determine the number of combinations that can open a combination lock.

2. Compute probabilities using permutations.

Try This One 3

(a) A different "permutation" lock has letters from A through L on it, and the combination consists of four letters with no repeats. What is the probability that the combination is I, J, K, and L in some order?

(b) Would a combination consisting of I, J, K, and L in some order be more or less likely if letters can be repeated? Explain.

In Example 4, we will again need to use the fundamental counting principle.

Example 4 Using Counting Techniques to Compute Probability

A store has six different fitness magazines and three different news magazines. If a customer buys three magazines at random, find the probability that he'll pick two fitness magazines and one news magazine.

SOLUTION

We want to know how many of each magazine the customer buys, not the order in which he buys them, so we're dealing with combinations. There are $_6C_2$ or 15 ways to choose two fitness magazines from six fitness magazines, as shown:

$$_6C_2 = \frac{6!}{(6-2)!2!} = \frac{6!}{4!2!} = \frac{6 \cdot 5 \cdot \cancel{4!}}{\cancel{4!} \cdot 2 \cdot 1} = 15$$

There are $_3C_1$ or three ways to choose one magazine from three news magazines:

$$_3C_1 = \frac{3!}{(3-1)!1!} = \frac{3!}{2! \cdot 1!} = \frac{3 \cdot \cancel{2!}}{\cancel{2!} \cdot 1} = 3$$

Note the use of "and"; the event in question is picking two fitness magazines AND one news magazine, so we can think of this as a two-step process and use the fundamental counting principle to multiply, giving us $15 \cdot 3 = 45$ ways to pick two fitness and one news magazine. Next, there are $_9C_3$ or 84 ways to pick three magazines from nine magazines:

$$_9C_3 = \frac{9!}{(9-3)!3!} = \frac{9!}{6!3!} = \frac{9 \cdot 8 \cdot 7 \cdot \cancel{6!}}{\cancel{6!} \cdot 3 \cdot 2 \cdot 1} = 84$$

The probability of picking two fitness magazines and one news magazine is

$$\frac{45}{84} \approx 0.536$$

Try This One   4

Find the probability that the customer in Example 4 picks at least two fitness magazines.

Now we can really see why studying counting techniques is so important in probability: tons of problems can be solved by combining them with the formula for theoretical probability.

Sidelight   **The Classical Birthday Problem**

You're in a room with 29 other people and someone says "I'll bet you fifty bucks that two people in this room have the same birthday." Would you take the bet? Most people would; since there's 365 days in a year, you'd think that the probability of 2 out of 30 having the same birthday should be pretty low. Believe it or not, the probability is actually greater than 0.7, meaning you'd have at least a 70% chance of losing that bet. With 50 people in the room, the probability is over 0.97! If you're skeptical, read on.

We can find the probability using permutation rules and the probability rule for complements: the strategy is to find the probability that everyone in the room has a different birthday, then subtract from 1. We'll need to assume that every day of the year is equally likely as a birthday, which is certainly reasonable.

Let's start with three people in the room. There are 365 possible days for the first person, then 364 for the second and 363 for the third. So the number of ways to get three different birthdays is $365 \cdot 364 \cdot 363$, which happens to be $_{365}P_3$. The total number of possible birthdays for three people is 365^3, so the probability of all three having different birthdays is

$$\frac{_{365}P_3}{365^3} \approx 0.992$$

and the probability of two having the same birthday is about $1 - 0.992 = 0.008$. That sounds totally reasonable. But

here's where things get weird. The general formula we can take from this is that the probability of k people in a room all having different birthdays is

$$\frac{_{365}P_k}{365^k}$$

which makes the probability that at least two people have the same birthday

$$1 - \frac{_{365}P_k}{365^k}$$

With 30 people, the probability is

$$1 - \frac{_{365}P_{30}}{365^{50}} \approx 0.7063$$

and with 50, it's

$$1 - \frac{_{365}P_{50}}{365^{30}} \approx 0.9704$$

It turns out that the break-even point is 23 people: for groups of 23 or more, there's a better than 50-50 chance that two have the same birthday.

It's interesting to note that two presidents, James K. Polk and Warren G. Harding, were both born on November 2. Also, John Adams and Thomas Jefferson both died on July 4, which is ironic to begin with—but it was the SAME July 4, in 1826.

Answers to Try This One

1 $495/2,380 \approx 0.208$

2   The probability is now about 0.00103, making this almost 56 times as likely.

3   (a) $1/495 \approx 0.002$

(b) It's less likely; the number of combinations with those four letters doesn't change, but there are more total combinations because we're allowing more choices for each letter other than the first.

4 $\dfrac{65}{84}$

Exercise Set   10-5

Writing Exercises

1. Explain how combinations and permutations are useful in computing probabilities.
2. What is the biggest advantage of combinations and permutations over tree diagrams when computing probability?
3. We've already asked this question, but now is a really good time for you to think about it again: how can you

decide whether to use the permutation or combination formula for a probability experiment?

4. How can you tell that you need to use the fundamental counting principle in conjunction with permutation or combination formulas?

Computational Exercises

For Exercises 5–10, perform each calculation either using factorial formulas, or permutation and combination functions on a graphing calculator or spreadsheet.

5. $\dfrac{_{12}P_2}{_{15}P_{12}}$

6. $\dfrac{_6P_4}{_9P_6}$

7. $\dfrac{_9C_5}{_{15}C_5}$

8. $\dfrac{_4C_2}{_9C_4}$

9. $\dfrac{_6C_2 \cdot _4C_3}{_{10}C_5}$

10. $\dfrac{_8P_4 \cdot _5P_4}{_{13}P_8}$

Applications in Our World

11. A student-faculty government committee of 4 people is to be formed from 20 student volunteers and 5 faculty volunteers. Find the probability that the committee will consist of the following, assuming the selection is made at random:
    (a) All faculty members.
    (b) Two students and two faculty members.
    (c) All students.
    (d) One faculty member and three students.

12. In a company there are seven executives: four women and three men. Three are chosen at random to attend a management seminar. Find these probabilities.
    (a) All three will be women.
    (b) All three will be men.
    (c) Two men and one woman will be chosen.
    (d) One man and two women will be chosen.

13. A city council consists of 10 members. Four are Republicans, three are Democrats, and three are Independents. If a committee of three is to be selected, find the probability of selecting
    (a) All Republicans.
    (b) All Democrats.
    (c) One of each party.
    (d) Two Democrats and one Independent.
    (e) One Independent and two Republicans.

14. In a class of 18 students, there are 11 men and seven women. Four students are picked to present a demonstration on the use of graphing calculators. Find the probability that the group consists of
    (a) All men.
    (b) All women.
    (c) Three men and one woman.
    (d) One man and three women.
    (e) Two men and two women.

15. A chef is choosing from 12 different entrees for an important banquet, 3 of which contain spinach. The

guests will have a choice of 4 entrees. If the chef chooses those 4 at random, find the probability that
    (a) None contain spinach.
    (b) At least one has spinach.
    (c) Three have spinach.
    (d) All 4 have spinach.

16. There are 50 tickets sold for a raffle for a Student Art Auction, and there are two prizes to be awarded. If Dionte buys two tickets, find the probability that he'll win both prizes.

17. An engineering company has four openings and the applicant pool consists of six database administrators and eight network engineers. If the hiring is done without regard for the specific qualifications of the applicants, find the probability that the four hired will be
    (a) All network engineers.
    (b) Two database administrators and two network engineers.
    (c) All database administrators.
    (d) Three database administrators and one network engineer.
    (e) One database administrator and three network engineers.

18. The list of potential parolees at a monthly parole hearing consists of eight drug offenders, five violent offenders, and two convicted of property crimes. I'd surely like to think that parolees aren't chosen at random, but if this particular board chooses three parolees randomly, find the probability that
    (a) All three are drug offenders.
    (b) Two of the three are property offenders.
    (c) All three are violent offenders.
    (d) One of each type of offender is paroled.
    (e) Two are drug offenders and one is a violent offender.

19. Find the probability of getting any triple-digit number where all the digits are the same in a lottery game that consists of selecting a three-digit number.

20. Five jurors have already been seated for a trial, and seven more need to be chosen. Attorneys on both sides are making lists of who they consider to be acceptable from the jury pool, which has eight men and nine women still eligible. If one of the attorneys has an important appointment to get to (by which I mean a golf match), and decides to pick seven ideal jurors randomly, what is the probability that he'll pick three men and four women?

21. A physical therapist is scheduling appointments for the day. They have eight worker's comp cases and four Medicare patients awaiting care. If they choose five at random for the morning schedule, find the probability that they'll pick three worker's comp cases and two Medicare patients.

22. A five-digit identification card is made. Find the probability that the card will contain the digits 0, 1, 2, 3, and 4 in any order.

23. The combination lock in Example 3 has 40 numbers from zero to 39, and a combination consists of 3 numbers in a specific order with no repeats. Find the probability that the combination consists only of even numbers.

24. Is it more or less likely that the combination for the lock in Problem 23 consists of all even numbers if it consists of four numbers in a specific order with no repeats? Explain why this answer makes sense.

In one lottery game, contestants pick five numbers from 1 through 40 and have to match all five for the big prize (in any order). Exercises 25–30 refer to this game.

25. What's the probability you'll win if you buy one ticket? (*Hint:* Your chances are NOT good.)

26. You'll get second prize in the lottery game in Problem 25 if you match four of the five numbers. Find the probability of winning second prize if you buy five tickets.

27. You'll get twice your money back if you match three of the five numbers. If you buy two tickets, what's the probability of matching three out of five numbers?

28. What's the probability of utter failure in this game, defined by yours truly to mean matching none of the numbers?

29. Find the probability that you win some money when buying one ticket. (Refer to Exercises 25–28.)

30. To celebrate Fourth of July week, the lottery commission is offering a $1 million bonus if a winner of the big prize matches the numbers in the order in which they were drawn. What's the probability of this happening if you buy one ticket?

Exercises 31–36 refer to poker hands consisting of 5 cards dealt at random from a standard deck of 52 cards. Find the probability of getting each hand.

31. A full house (three of one denomination and two of another)

32. A flush (five cards of the same suit)

33. Three of a kind (exactly three of one denomination, remaining cards are two different denominations)

34. Four of a kind (four of the same denomination)

35. A royal flush (ten, jack, queen, king, and ace of the same suit)

36. A straight flush (five cards of the same suit that are consecutive in denomination)

Critical Thinking

37. At a carnival game, the player pays a dollar, then flips a quarter, rolls two dice, and draws two cards from a standard deck. If the result is tails, 7, and a pair of spades, they get back $100. What is the probability of this happening?

38. (a) If you play the game in Problem 37 1,000 times, how many would you expect to win? (*Hint:* Remember that you can think of probability as a percent chance of something happening.)

    (b) Use your answer to part (a) to calculate how much you would expect to win or lose if you played the game 1,000 times. (Don't forget that it costs a dollar to play!)

39. Many lottery games are based on choosing some numbers from a larger group. Devise a lottery game with at least three different ways of winning, compute the probability of winning for each way, and assign payouts for each winning combination that you think are fair.

40. (a) Suppose that we choose 3 letters at random from the first 10 letters of the alphabet without repeats. Find the probability of choosing ABC in that order, then find the probability of choosing ABC in any order.

    (b) If we change the scenario in part (a) to include repeats, recompute the probabilities.

41. (a) What is the arithmetic relationship between the two probabilities in Exercise 40(a)?

    (b) Suppose the scenario in Exercise 40 is repeated, this time choosing four letters rather than three. Without actually computing either probability, what do you think the relationship between the two probabilities will be? Explain how you got your answer. (*Hint:* Think about part [a] and the effect of order mattering.)

42. Think of a real-life scenario for which it would be reasonable to compute probabilities using combinations, and one in which it would be reasonable to use permutations. (No stealing any of the scenarios from the section!)

43. (a) Think of a scenario where it would be reasonable to use combinations AND the fundamental counting principle to compute a probability.

    (b) Think of a scenario where it would be reasonable to use permutations AND the fundamental counting principle to compute a probability.

44. Make up a probability problem that is *answered* by each calculation:

    (a) $\dfrac{_5P_2}{_{20}P_5}$ (b) $\dfrac{_4C_2}{_{12}C_2}$ (c) $1 - \dfrac{_{10}P_5}{_{20}P_{10}}$

45. USING TECHNOLOGY. In the Sidelight on page 596, we learned the surprising fact that with just 30 people in a room, there's over a 70% chance that two of them will have the same birthday. This fact is so nonintuitive that a lot of people flat-out don't believe it even after computing it. It's true, I promise. Using the formula developed in that Sidelight, you can set up a spreadsheet that will calculate the probability for any number of people in the room. There's a template in the online resources for this chapter to help you get started. If you need to, review the Excel command for computing permutations in Section 10-2.

46. USING TECHNOLOGY. In this exercise, you're going to compute the probability of winning a variety of different games of chance, using Excel. In each case, you need to decide whether to use the =PERMUT command, or the =COMBIN command, then use that command to find the number of possible outcomes. You'll then compute the probability of winning in another column. A template to help you get started can be found in the online resources for this chapter, along with descriptions of the outcomes you'll be counting.

Section 10-6   Odds and Expectation

LEARNING OBJECTIVES

1. Compute the odds in favor of and against an event.
2. Compute odds from probability.
3. Compute probability from odds.
4. Compute expected value.

The Atlanta Braves won the World Series on November 2, 2021. By the time the ink dried on newspapers reporting the victory, oddsmakers in Las Vegas had listed the odds against the Braves winning it again in 2022 as 10 to 1. But what exactly does that mean? The term "odds" is used all the time in describing the likelihood of something happening, but a lot of people don't understand exactly what a given set of odds means.

MediaPunch/Shutterstock

Odds are used by casinos, racetracks, and other gambling establishments to determine the payoffs when bets are made or lottery tickets are purchased. They're also used by insurance companies in determining the amount to charge for premiums. The formulas for computing odds are similar to the formula we've been using for theoretical probability, which is both good and bad. It's good because we've been working with theoretical probability for a while now, so you should be pretty familiar with the idea and the formula. It's bad because it can be really easy to confuse odds with probability. Let's take a look at the connection between the two.

In essence, a probability is a comparison between the number of favorable outcomes in a sample space and the total number of outcomes. If we're interested in getting 2 or less when rolling a single die, there are two favorable outcomes (rolls of 1 and 2) and six total outcomes, so the probability is 2/6, or 1/3.

Odds are similar, but they provide a comparison between the number of favorable outcomes and the number of *unfavorable* outcomes. In the case of rolling a die and getting 2 or less, there are two favorable outcomes and four unfavorable outcomes. Odds can (and often are) written as fractions, but more often they're written in ratio form using a colon, which helps distinguish odds from probability in common usage. So we could say that the odds in favor of rolling 2 or less are 2:4, or 1:2.

> If an event E has a favorable outcomes and b unfavorable outcomes, then
>
> 1. The **odds in favor** of event E occurring $= \frac{a}{b}$ (also written as $a:b$)
> 2. The **odds against** event E occurring $= \frac{b}{a}$ (also written as $b:a$)

So what do odds really say? The odds of Atlanta repeating as World Series champion could be listed as 10/1, 10:1, or 10 to 1. In common usage, the phrase "the odds of" really means "the odds against"; if we are told that the odds of rolling a 12 with two dice are 35 to 1, it means that the odds against rolling 12 are 35 to 1. So, by setting Atlanta's odds at 10:1, the oddsmakers are predicting that if the season were played 11 times, Atlanta would win the championship once and not win it 10 times.

Example 1	**Computing Odds**

A card is drawn from a standard deck of 52 cards.

(a) Find the odds in favor of getting an ace.
(b) Find the odds against getting an ace.
(c) Describe what these odds mean.

SOLUTION

(a) In a deck of cards there are 52 cards and four aces, so $a = 4$ and $b = 52 - 4 = 48$. (In other words, there are 48 cards that are not aces.)

The odds in favor of an ace $= \frac{4}{48} = \frac{1}{12}$.

(b) The odds against an ace $= \frac{48}{4} = \frac{12}{1}$.

The odds in favor of an ace are 1:12 and the odds against an ace are 12:1.

(c) If the odds against are 12:1, it means that if we were to draw a single card from a full deck of 52 cards 13 times, we'd expect to get an ace once and not get an ace 12 times.

> **Math Note**
>
> Notice that if the odds in favor of an event occurring are $a{:}b$, the odds against it occurring are $b{:}a$.

1. Compute the odds in favor of and against an event.

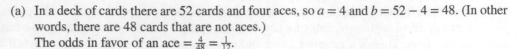

Try This One   1

(a) What are the odds in favor of rolling a prime number sum with a roll of two dice? What are the odds against? (Figure 10-7 on page 589 will help.)
(b) What do these odds mean?

When an event E has a favorable and b unfavorable outcomes, there are $a + b$ total outcomes, and the probability of E is

$$P(E) = \frac{a}{a + b}$$

The probability of E not occurring is

$$1 - P(E), \quad \text{or} \quad \frac{b}{a + b}$$

If we divide these two probabilities, we get

$$\frac{P(E)}{1 - P(E)} = \frac{\dfrac{a}{a + b}}{\dfrac{b}{a + b}} = \frac{a}{a + b} \cdot \frac{a + b}{b} = \frac{a}{b}$$

It's easy to get cross-eyed at the algebra here, but something interesting actually happened: when we divided the probability of E occurring by the probability of it not occurring, we got a/b. Why is that interesting? Remember, a is the number of favorable outcomes and b is the number of unfavorable outcomes, so a/b is exactly the odds in favor of E. This gives us a strong connection between probability and odds.

Formulas for Odds in Terms of Probability

$$\text{Odds in favor} = \frac{P(E)}{1 - P(E)}$$

$$\text{Odds against} = \frac{1 - P(E)}{P(E)}$$

where $P(E)$ is the probability that event E occurs.

Example 2

Finding Odds from Probability

The probability of getting exactly one pair in a five-card poker hand is 0.423. Find the odds in favor of getting exactly one pair, and the odds against.

SOLUTION

This is a direct application of the formula relating probability and odds. The odds in favor of getting exactly one pair are

$$\frac{P(\text{getting exactly one pair})}{1 - P(\text{getting exactly one pair})} = \frac{0.423}{1 - 0.423} = \frac{0.423}{0.577}$$

> **Math Note**
>
> The probability 0.423 in Example 2 comes from dividing the number of hands with exactly one pair by the total number of five-card hands. This uses the fundamental counting principle and the combination formula.

We can convert this into fraction form by multiplying both the numerator and denominator by 100.

$$\frac{0.423}{0.577} \cdot \frac{100}{100} = \frac{423}{577}$$

So the odds in favor of getting exactly one pair are 423:577, and the odds against are 577:423.

Try This One   2

2. Compute odds from probability.

According to the American Cancer Society, the probability of an American female developing some type of cancer at some point is about 1/3. Find the odds in favor of and against an American woman developing cancer.

There's another interesting consequence of the calculation we did after Try This One 1: we found how to find the probability of an event occurring when we know the odds.

> **Math Note**
>
> When the odds are 1:1, a game is said to be fair. That is, both parties have an equal chance of winning or losing.

Formula for Probability in Terms of Odds

If the odds in favor of an event E are $a{:}b$, then the probability that the event will occur is

$$P(E) = \frac{a}{a + b}$$

Example 3

Finding Probability from Odds

According to the National Safety Council, the odds of dying due to injury at some point in your life are about 10:237. Find the probability of dying from injury.

SOLUTION

In the formula for converting to probability, the odds in favor are $a{:}b$. In this case, those odds are 10:237, so $a = 10$ and $b = 237$.

$$P(\text{dying from injury}) = \frac{10}{10 + 237} = \frac{10}{247} \approx 0.040$$

3. Compute probability from odds.

Try This One   3

When two dice are rolled, the odds in favor of getting a sum of 9 are 1:8. Find the probability of not getting a sum of 9 when two dice are rolled.

Sidelight **Odds and Betting**

In almost any gambling enterprise, from Vegas to a church raffle, odds are used to determine what the payouts will be. Let's use the game of roulette as an example. There are 38 partitions on a roulette wheel, with a small ball that is equally likely to land in any of them. Players make various bets on where the ball will land. If you simply bet on a particular number, the odds against you are 37:1. If you win, the casino will pay 35 times what you bet. So if you bet a dollar on each spin, on average, in 38 spins, you would lose 37 times and win once, meaning you'd be two dollars in the hole. That doesn't sound like the recipe for the casino making a lot of money, but when you consider the large number of people in a typical casino, and the fact that they're placing multiple bets at frequent intervals, it starts to make more sense.

This is where it becomes clear that state lotteries are an awful bet. In the Ohio lottery's "Classic Lotto" game, for example, the odds in favor of matching 5 of 6 numbers are

1:54,021. So to make the game completely fair, the payout should be in the neighborhood of $54,000 on a $1 ticket. The actual payout? Just $1,500! Any gambler or statistician will tell you that lotteries offer the worst odds of just about any game of chance.

So why do so many people play lotteries? For one, a complete lack of understanding of probability. But the biggest reason is probably that the massive payouts grab people's attention and make it seem like it's worth the risk of losing money. In almost every case though, no matter how huge the payout is in a lottery, it's still many, many times smaller than the odds of winning would dictate. In fact, the irony of this is that if a privately owned casino tried to run a game with payouts anywhere near that low compared to the odds of winning, the states would shut them down in about 5 minutes. Yes, the same states that operate lotteries with those dreadful payouts.

Expected Value

When we described what odds mean earlier, we used the word "expect." For example, if the odds of rolling a number 2 or less with a single die are 1:2, it means that for every three rolls of the die, we'd expect to get a number 2 or less once, and 3 or more twice.

Now we're going to get a little more in-depth in terms of what we'd expect to happen in various probability experiments by studying **expectation**, or **expected value**. Expected value is used to determine the result that would be expected over the long term in some sort of gamble. It is used not only for games of chance, but in areas like insurance, management, engineering, and others. Here's the key thing to remember as we study expected value: *it only makes sense for events that have numerical outcomes.*

For example, rolling a die has a numerical outcome (1 through 6), and expected value can be used to determine what the average long-run result is likely to be. (We'll find out in Example 5.) But it doesn't make sense to ask what the long-term average of flipping a coin is because you can't average "heads" and "tails."

Example 4 will illustrate a procedure for computing the expected value of a game, and then we'll summarize that procedure for future reference.

Example 4   Developing a Procedure to Find Expected Value

You pay a dollar to roll two dice. If you roll a total of 5 or 6, you get your dollar back plus two more just like it. If not, you get nothing and like it.

(a) Find the probability of winning the game.
(b) If you play the game 100 times, on average how many times would you expect to win?
(c) What would be the total amount won or lost?

SOLUTION

(a) Using the table of outcomes for rolling two dice back in Section 10-4, there are 9 outcomes that result in a total of 5 or 6, and 36 total outcomes, so the probability of winning is $\frac{9}{36}$, or $\frac{1}{4}$.

(b) Since you have a $\frac{1}{4}$ chance of winning any game, on average you'd expect to win 25 out of every 100 games.

(c) According to the description of the game, if you lose, you lose $1, and if you win, you win $2. So in 100 games, you'd expect to win 25 times, and lose 75, so your total won or lost would be

$$25 \cdot \$2 + 75 \cdot (-\$1) = \$50 - \$75 = -\$25$$

That is, for every 100 games, you'd expect to lose on average $25. That's what we would call the expected value of playing the game 100 times.

Try This One   4

On a roulette wheel, there are 38 slots, 18 of which are colored red. If you bet $5 on red and win, you get $10 back. If red doesn't come up, you lose your $5. Use the steps from Example 4 to find the expected value of playing the game 100 times.

In Example 4, we saw that playing a dice game 100 times would result in a loss of $25 on average. That means that every time we play the game once, we can expect to lose 25 cents. That might not make sense to you at first, since you can't possibly lose 25 cents playing the game once: you either win $2 or lose $1. But that's really the point of expected value: it's about what the result of a probability experiment would be on average. It doesn't tell us what will happen on any given trial.

In any case, a close look at Example 4 allows us to develop a general procedure for finding expected value. We first found the probability of each outcome. Then we multiplied each probability by the numeric value of the outcome and added the results to get expected value.

Expected Value

The expected value for the outcomes of a probability experiment is

$$E = X_1 \cdot P(X_1) + X_2 \cdot P(X_2) + \cdots + X_n \cdot P(X_n)$$

where the X's correspond to the numerical outcomes and the $P(X)$'s are the corresponding probabilities of the outcomes.

Example 5 | Computing Expected Value

When a single die is rolled, find the expected value of the outcome, and describe what exactly that means.

SOLUTION

Since each numerical outcome, 1 through 6, has a probability of $\frac{1}{6}$, the expected value is

$$E = 1 \cdot \frac{1}{6} + 2 \cdot \frac{1}{6} + 3 \cdot \frac{1}{6} + 4 \cdot \frac{1}{6} + 5 \cdot \frac{1}{6} + 6 \cdot \frac{1}{6} = \frac{21}{6} = 3.5$$

This doesn't mean that you'd expect to get 3.5 if you roll the die one time, because that would be silly. What it means is that if you repeat that roll many, many times, you'd expect that the long-term average of the results would be 3.5.

Try This One   5

If seven cards are numbered with integers from −2 to 4, then placed into a box and picked out at random, find the expected value, and describe what exactly that means.

In gambling games, the expected value is found by multiplying the amount won, or net gain, and the amount lost by the corresponding probabilities and then finding the sum.

| **Example 6** | **Computing Expected Value** |

The prize in a raffle is a flat-screen TV valued at $350, and 1,000 tickets are sold. What's the expected value if you buy 1 ticket?

SOLUTION

We begin with two notes. First, for a win, the net gain is $349, since you don't get the cost of the ticket ($1) back. Second, for a loss, the gain is represented by a negative number, in this case, −$1.

The problem can then be set up as follows:

	Win	Lose
Gain, X	$349	−$1
Probability, $P(X)$	$\dfrac{1}{1,000}$	$\dfrac{999}{1,000}$

The expected value is

$$E(X) = \$349 \cdot \frac{1}{1,000} + (-\$1) \cdot \frac{999}{1,000} = -\$0.65$$

Try This One   6

With their house in foreclosure, a homeowner comes up with a plan to salvage the situation: they sell 10,000 raffle tickets at $50 each for the home, which is valued at $200,000. Find the expected value from buying one ticket.

In Example 6, we got an expected value of −$0.65. Again, this doesn't mean that you'd lose 65 cents from buying a ticket, because you can only lose a dollar or win a $350 prize. It means that if you were to buy tickets to a similar raffle repeatedly, in the long run you'd average a 65-cent loss for each ticket bought.

Sidelight   **Math in Slot Machines**

Today, most slot machines are electronic—really, they're video games. But early slot machines were mechanical. The first slot machines were invented by the Fey Manufacturing Company of San Francisco in 1895. There were three large wheels with different symbols on them that spun when a control handle on the side was pulled. Each of the wheels contained 20 symbols, and payouts were based on matching symbols facing forward when the wheels came to a stop. The manufacturer controlled the payouts by cleverly arranging the symbols, using probability. For example, there might be three cherries on the first wheel and six on the second, but none on the third. So if you get cherries on the first two wheels, it feels like you almost won. But since there were no cherries on the third wheel, the probability of your getting three cherries was actually zero.

Using probability theory, slot machine makers set the payouts for each winning combination. Expected value calculations will then let the owner know what long-term profits should be. The key is to let people win enough that they feel like it's worthwhile to play but not so much that the owner of the machine has to get a real job.

Vlada Photo/Shutterstock

| **Example 7** | **Computing Expected Value** |

Stan Honda/AFP/Getty Images

A stock you bought two years ago with high hopes is now selling for less than you paid, and things look grim for the company. Do you sell, or hold on and hope it will come back to the original price before you sell?

A model economists use for such situations is a game no one wants to play: suppose you have a choice: lose $100, or take a 50-50 chance between losing nothing, and losing $300. Which do you choose?

4. Compute expected value.

One thousand tickets are sold at $1 each for four prizes of $100, $50, $25, and $10. What is the expected value if you buy two tickets?

SOLUTION

First, let's find the expected value of buying one ticket.

Gain, x	$99	$49	$24	$9	−$1
Probability, $P(x)$	$\dfrac{1}{1,000}$	$\dfrac{1}{1,000}$	$\dfrac{1}{1,000}$	$\dfrac{1}{1,000}$	$\dfrac{996}{1,000}$

$$E(x) = \$99 \cdot \frac{1}{1,000} + \$49 \cdot \frac{1}{1,000} + \$24 \cdot \frac{1}{1,000} + \$9 \cdot \frac{1}{1,000} - \$1 \cdot \frac{996}{1,000} = -\$0.815$$

Now multiply by 2 since two tickets were bought.

$$-\$0.815(2) = -\$1.63$$

Try This One   7

The profit made by a small ski resort, not surprisingly, depends largely on the seasonal weather. In a season with more than 75 inches of snow, it makes an average of $250,000. If snowfall is between 40 and 75 inches, the average profit is $160,000, and if snowfall is less than 40 inches, it loses $70,000. The resort gets over 75 inches of snow 40% of years, between 40 and 75 inches 45% of years, and less than 40 inches 15% of years. Find the resort's expected yearly profit.

Math Note

In American roulette, there are 22 different types of bets you can place, but it's interesting to note that all but one of them have the exact same expected value: −$0.053 on a $1 bet. (Betting on 0, 00, 1, 2, and 3, called a five-number bet, is worse, at −$0.079.)

In gambling games, if the expected value of the gain is 0, the game is said to be fair. If the expected value of the gain of a game is positive, then the game is in favor of the player. That is, the player has a better-than-even chance of winning. If the expected value of the gain is negative, then the game is said to be in favor of the house. That is, in the long run, the players will lose money. Can you guess what the sign of the expected value is for every game in a casino?

Answers to Try This One

1 (a) In favor: 5:7; against: 7:5 (b) This means that if you rolled two dice 12 times, you would expect to get a prime number 5 times.

2 In favor: 1:2; against: 2:1   **3** $\dfrac{8}{9}$ **4** −$26.32

5 1; If you played this game repeatedly, in the long run the average card number would be 1

6 −$30   **7** $161,500

| **Exercise Set** | **10-6** |

Writing Exercises

1. Explain the difference between the odds in favor of an event and the odds against an event.
2. Explain the numerical relationship between the odds in favor of an event and the odds against an event.

3. Explain the meaning of odds in a gambling game.
4. Explain how to find the probability of an event occurring when given the odds in favor of an event.

5. Explain what is meant by the expected value of a probability experiment.

6. Why does every game in a casino have a negative expected value for the player?

Computational Exercises

In Exercises 7–12, find the odds in favor of and odds against each event given the probability.

7. $P(A) = \frac{7}{8}$

8. $P(B) = \frac{1}{9}$

9. $P(C) = \frac{5}{11}$

10. $P(D) = \frac{7}{10}$

11. $P(E) = \frac{9}{13}$

12. $P(F) = \frac{9}{14}$

In Exercises 13–18, find the probability of each event given the odds.

13. 5:8 in favor

14. 9:13 in favor

15. 6:5 against

16. 12:7 against

17. 3:7 in favor

18. 15:6 against

In Exercises 19–22, find the expected value of the probability experiment with outcomes $X_1, X_2, \ldots$

19. $X_1 = 6, X_2 = 2, X_3 = 4; P(X_1) = \frac{1}{5}, P(X_2) = \frac{2}{5}, P(X_3) = \frac{2}{5}$

20. $X_1 = \$10, X_2 = \$5, X_3 = \$1, X_4 = \$20; P(X_1) = \frac{3}{10}, P(X_2) = \frac{2}{10}, P(X_3) = \frac{1}{10}, P(X_4) = \frac{4}{10}$

21. $X_1 = \$1, X_2 = \$2, X_3 = \$3; P(X_1) = \frac{2}{7}, P(X_2) = \frac{1}{7}, P(X_3) = \frac{4}{7}$

22. $X_1 = 15, X_2 = 20, X_3 = 25, X_4 = 30, X_5 = 35; P(X_1) = \frac{2}{5}, P(X_2) = \frac{1}{15}, P(X_3) = \frac{4}{15}, P(X_4) = \frac{1}{5}, P(X_5) = \frac{1}{15}$

Applications in Our World

23. In planning a gambling booth for a charity festival, Antoine needs to know the odds of various combinations in order to decide on payouts that will be high enough that people want to play, but low enough that the charity will make money. If the player rolls two dice, find the odds
    (a) In favor of getting a sum of 10.
    (b) In favor of getting a sum of 12.
    (c) Against getting a sum of 7.
    (d) Against getting a sum of 3.
    (e) In favor of getting doubles.

24. If the game in Exercise 23 has players who roll only one die, find the odds
    (a) In favor of getting a 3.
    (b) In favor of getting a 6.
    (c) Against getting an odd number.
    (d) Against getting an even number.
    (e) In favor of getting a prime number.

25. Steve shuffled a deck of 52 cards and asked Sally to draw 1 card to start a magic trick. Find the odds
    (a) In favor of getting a queen.
    (b) In favor of getting a face card.
    (c) Against getting a club.
    (d) In favor of getting an ace.
    (e) In favor of getting a black card.

26. Monica is trying to design a coin-flipping game for a charity fair. The plan is for contestants to flip three coins. Find the odds
    (a) In favor of getting exactly three heads.
    (b) In favor of getting exactly three tails.
    (c) Against getting exactly two heads.
    (d) Against getting exactly one tail.
    (e) In favor of getting at least one tail.

27. Your friends have taken bets on whether you will pass this class. (Sounds like maybe you need new friends.) Find the probability that you will pass given these odds:
    (a) 7:4 in favor of you passing
    (b) 2:5 against you passing
    (c) 3:1 in favor of you passing
    (d) 1:4 against you passing

28. Find the probability that you will win a Wii bowling tournament given these odds:
    (a) 3:4 in favor of you winning
    (b) 1:7 against you winning
    (c) 5:4 in favor of you winning
    (d) 6:5 in favor of you winning

29. If the odds against a horse winning a race are 9:5, find the probability that the horse will win the race.

30. A game show contestant rolls two dice and wins if they throw doubles. What are the odds in favor of the event? What are the odds against the event?

A roulette wheel has 38 numbers: 1 through 36, 0, and 00. A ball is rolled, and it falls into one of the 38 slots, giving a winning number. Each bet in Problems 31–36 lists the payout for winning on a $1 bet, including the dollar that was bet. Find the odds in favor of each bet, and the expected value.

31. Betting on an individual number: $36

32. Row zero bet (0 and 00): $18

33. First dozen (1 through 12): $3

34. Even numbers from 2 through 36: $2

35. Corner (any four adjoining numbers on the betting grid): $9

36. Top line (0, 00, 1, 2, 3): $7

When 150 people were surveyed by phone about their favorite cola, the results were tabulated in the given table. Use the table to find the odds in favor of each response in Exercises 37–42.

Favorite cola	Number of responses
Coke	29
Pepsi	21
Diet Coke	37
Diet Pepsi	28
Coke Zero	19
Other	16

37. Coke

38. Diet Pepsi

39. A Pepsi product
40. A Coke product
41. Neither a Coke nor a Pepsi product
42. A low-calorie product (one of the diets or Coke Zero)

In sports betting, odds are often given in terms of the profit on a $100 bet if your team wins. For example, if a team is listed as +130, that means if you bet $100 and win, you'll get back $230: your original $100 plus a profit of $130. If the odds against a team winning are 5:1, it means that a $1 bet would return a total of $6, making a profit of $5. Use these facts in Exercises 43 and 44.

43. With five weeks left in the 2021 season, the four favorites to win the Super Bowl were Tampa Bay (+525), Kansas City (+700), Green Bay (+750), and Buffalo (+800). Find the probability of each team winning, at least in the estimation of the oddsmakers in Vegas.

44. After the first month of the 2021 NBA season, the four Las Vegas favorites to win the championship were Brooklyn (13:5 odds against), Golden State (6:1), Milwaukee (7:1), and Phoenix (9:1). According to the oddsmakers, (a) what was the probability of each team winning the championship, and (b) how would the odds have been listed in terms of profit on a $100 bet?

Baylor University won the 2021 men's NCAA basketball tournament. The table lists the heights and weights of all players on the team's roster. Use this information for Problems 45–50.

6'3"	195	6'9"	225
6'4"	195	6'10"	245
6'7"	200	6'1"	185
6'2"	205	7'0"	215
6'8"	245	6'8"	195
6'5"	250	6'1"	180
6'3"	180		

45. Find the odds in favor of a randomly selected player being over 6'1" tall.
46. Find the odds in favor of a randomly selected player being under 200 pounds.
47. Find the odds against a randomly selected player being 6'1" tall.
48. Find the odds against a randomly selected player weighing between 199 and 221 pounds.
49. Find the expected value for height if one player is chosen at random.

50. What numeric quantity does your answer from Problem 49 match?
51. A cash prize of $5,000 is to be awarded at a fundraiser. If 2,500 tickets are sold at $5 each, find the expected value of buying one ticket.
52. You start your shift as a cashier with your drawer containing ten $1 bills, five $2 bills, three $5 bills, one $10 bill, and one $100 bill. Find the expectation if one bill is chosen at random.
53. In a scratch-off game, if you scratch the two dice on the ticket and get doubles, you win $5. For the game to be fair, how much should you pay to play the game?
54. At this year's State Fair, there was a dice rolling game. If you rolled two dice and got a sum of 2 or 12, you won $20. If you rolled a 7, you won $5. Any other roll was a loss. It cost $3 to play one game with one roll of the dice. What is the expectation of the game?
55. Melinda buys one raffle ticket at the Spring Fling since there is one $1,000 prize, one $500 prize, and five $100 prizes. There were a total of 1,000 tickets sold at $3 each. What is Melinda's expectation?
56. If Melinda buys two tickets to the raffle in Exercise 55, what is the expectation?
57. For a daily lottery, players pick any three-digit number from 000 to 999. If a player pays $1, they can win $500. Find the expectation. In the same daily lottery, if a player boxes a number, they can win $80. Find the expectation if the number 123 is played for $1 and boxed. (When a number is "boxed," it wins when the digits occur in any order.)
58. If a 60-year-old buys a $1,000 life insurance policy at a cost of $60 and has a probability of 0.972 of living to age 61, find the expectation of the policy until the buyer reaches 61.
59. A new flat-screen TV comes with a 1-year warranty which completely covers any parts and repairs. At the end of the warranty period, the buyer is offered an extended warranty for 3 more years at a cost of $90. Industry records indicate that during that 3-year period, there's an 8% chance that a minor repair averaging $85 will be needed, and a 3.5% chance that a major repair averaging $370 will be needed. Find the expected value of buying the extended warranty.
60. A company that makes wireless routers has found that, on average, 1.6% of the units are defective. It makes a $38 profit from selling nondefective units but ends up losing $24 when selling a defective unit due to repair and shipping costs. Find the expected profit from selling 10,000 units.

Critical Thinking

61. You stop on the street between errands to engage in a shell game with a street vendor. The vendor shows you a two-headed penny under one shell, a two-tailed penny under the second shell, and a fair penny (one head and one tail) under the third shell. He shuffles the shells around and then you choose a shell. He shows you that under the shell is a penny with the head side up. He is willing to bet you $5 that it is the two-headed penny. He says it cannot be the two-tailed penny because a head is showing. Therefore, he says there is a 50-50 chance of it being the two-headed coin. Should you take the bet?

62. Stuck in a bad situation, you're given a choice: lose $100, or take a 50-50 chance between losing nothing and losing $300. Choose the option that sounds better to you, then find the expected value of each option over 10 trials.

63. Since expected value only applies to numerical outcomes, it takes some ingenuity to use it for flipping a coin.
    (a) Choose any two numbers at random, assigning one to heads and another to tails. Then find the expected value of flipping the coin.
    (b) Repeat part (a) for two different numbers. What can you conclude?

64. Chevalier de Mere, a famous gambler, won money when he bet unsuspecting patrons that in four rolls of a die, he could get at least one 6, but he lost money when he bet that in 24 rolls of two dice, he could get a double 6. Find the probability of each event and explain why he won the majority of the time on the first game but lost the majority of the time when playing the second game.

In many respects, investing in the stock market is just another form of gambling: you pay money for stocks, and when you sell them at some point, you can gain money, lose money, or break even. Use what you learned about expected value to answer the questions in Exercises 65 and 66 about buying and selling stock.

65. The Bui family members decide to invest their $4,000 tax refund in the stock market for 1 year. An analyst suggests that they choose between two stocks. A computer analysis of past performance predicts that if they invest in RZ Electronics, there's a 40% chance they'll make a profit of $1,600, a 40% chance they'll make $200, and a 20% chance they'll lose $2,000. For Jackson Builders, there's a 75% chance they'll make $800, and a 25% chance that they'll lose $300.
    (a) Without doing any calculations, which sounds like the better investment to you? Why?
    (b) Find the expected value of each investment.

66. You're given the option of investing $10,000 in one of three mutual funds. A prominent market analyst posts the following estimates of performance for the three funds. The Hetrick Fund: 40% chance of a 35% gain; 40% chance of a 30% loss; 20% chance of breaking even. The Abercrombie Fund: 80% chance of breaking even; 12% chance of an 18% gain; 8% chance of a 9% loss. The Goldberg Fund: 25% chance of a 90% gain; 25% chance of a 5% gain; 40% chance of breaking even; 10% chance of losing everything. Answer parts (a) and (b) without doing any calculations.
    (a) If your primary objective is to shoot for the largest possible return without regard to risk, which would you be likely to choose?
    (b) If your primary objective is the least risk of losing big, which would you be likely to choose?
    (c) Find the expected value for each fund. With these choices, who would be most likely to be successful: a timid investor or an aggressive one?

Section 10-7   The Addition Rules for Probability

LEARNING OBJECTIVES

1. Decide if two events are mutually exclusive.
2. Develop and use the addition rule for mutually exclusive events.
3. Develop and use the addition rule for events that are not mutually exclusive.

If you pay any attention at all to the news, you most likely know that ours is a country that is becoming increasingly divided by political affiliation. As the animosity between the two major parties grows, our lawmakers become more predictable. On most issues, you can have a pretty good idea of which side a politician will take just by seeing if they have a D or an R after their title. Of course, other characteristics come into play as well, and some political reporters make a living trying to forecast how our leaders are going to react to the key issues of the day.

Tupungato/iStockphoto/Getty Images

As you may know, every member of Congress declares a single party affiliation, either Democrat, Republican, or Independent. Suppose that one commentator feels like a particular issue before Congress is most likely to be supported by women and Democrats. It would be useful to find the probability that any given representative is either female or a Democrat. At first thought, it seems reasonable to add up the number of women and the number of Democrats in Congress, then divide by the total number of representatives. But if you think about it, this won't work—those that are both female and Democrats will get counted twice.

The situation would be simpler if the commentator were interested in the probability of a representative being either a Republican or an Independent. The key difference is that any individual has to be one or the other. These two events are called *mutually exclusive*, which indicates that either one or the other must occur, but not both.

Two events are **mutually exclusive** if they cannot both occur at the same time. That is, the events have no outcomes in common.

| **Example 1** | Deciding if Two Events Are Mutually Exclusive |

In drawing cards from a standard deck, decide if the two events are mutually exclusive or not. If not, list all outcomes that the events have in common.

(a) Drawing a 4, drawing a 6.
(b) Drawing a 4, drawing a heart.
(c) Drawing a red card, drawing a face card.

SOLUTION

(a) Every card has just one denomination, so a card can't be both a 4 and a 6. The events are mutually exclusive.
(b) You could draw the 4 of hearts, which is one outcome satisfying both events. The events are not mutually exclusive.
(c) These are not mutually exclusive: the common outcomes are the jack of hearts, the jack of diamonds, the queen of hearts, the queen of diamonds, the king of hearts, and the king of diamonds.

Try This One   1

If student government picks students at random to win free books for a semester, determine whether the two events are mutually exclusive or not. If not, list all outcomes that the events have in common.

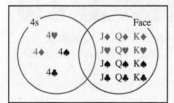
1. Decide if two events are mutually exclusive.

(a) The winner is a sophomore or a business major.
(b) The winner is a junior or a senior.
(c) The winner is a male or a member of the college cheerleading squad.

The probability of two or more events occurring can be calculated by developing addition rules, which we'll do in this section, starting with Example 2.

| **Example 2** | Developing a Probability Rule for Mutually Exclusive Events |

(a) When drawing a random card from a deck, find the probability of drawing a 4, and the probability of drawing a face card.
(b) Draw a Venn diagram that lists all outcomes that satisfy those two events.
(c) Use your Venn diagram to find the probability of drawing a 4 or a face card.
(d) What's the relationship between your answers to parts (a) and (c)?

SOLUTION

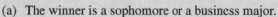

(a) There are 52 cards in the deck and four of them are 4s, so the probability is 4/52, or 1/13. There are 12 face cards (4 each of jacks, kings, and queens), so the probability of drawing a face card is 12/52, or 3/13.
(b) We'll put all of the 4s in one circle and all of the face cards in another. Since no face card is also a 4, the intersection is empty.
(c) We can get the total number of cards that are either 4s or face cards by just counting the number of cards in our diagram. There are 16, so the probability of a randomly selected card being either a 4 or a face card is 16/52, or 4/13.
(d) Now the key observation: if we add the probability of a card being a 4 to the probability of a card being a face card, we get

$$\frac{1}{13} + \frac{3}{13} = \frac{4}{13}$$

and this is the probability of a card being a 4 OR a face card.

Try This One   2

(a) If you randomly pick a day of the week to be your day to eat pizza, find the probability that the day is Tuesday, and the probability that the day is on the weekend.
(b) Draw a Venn diagram that lists all outcomes that satisfy those two events.
(c) Use your Venn diagram to find the probability that the day you pick is either Tuesday or on the weekend.
(d) How does your answer to part (c) compare to the answers from part (a)?

The method that we used in Example 2 provides our first addition rule. When two events are mutually exclusive, the intersection will always be empty, and the probability of one or the other occurring is the sum of the individual probabilities.

Addition Rule 1

When two events A and B are mutually exclusive, the probability that A or B will occur is

$$P(A \text{ or } B) = P(A) + P(B)$$

Example 3   Using Addition Rule 1

A restaurant has three pieces of apple pie, five pieces of cherry pie, and four pieces of pumpkin pie in its dessert case. If a customer selects at random one kind of pie for dessert, find the probability that it will be either cherry or pumpkin.

SOLUTION

The events are mutually exclusive. Since there is a total of 12 pieces of pie, 5 of which are cherry and 4 of which are pumpkin,

$$P(\text{cherry or pumpkin}) = P(\text{cherry}) + P(\text{pumpkin})$$
$$= \frac{5}{12} + \frac{4}{12} = \frac{9}{12} = \frac{3}{4}$$

> **Math Note**
>
> Sometimes it's a bad idea to reduce fractions in the individual probabilities when using the addition rule: you'll need a common denominator to add them. Your final answer should always be reduced if possible, though.

Try This One   3

A liberal arts math class contains 7 first-year students, 11 sophomores, 5 juniors, and 2 seniors. If the professor randomly chooses one to present a homework problem at the board, find the probability that it's either a junior or a senior.

The addition rule for mutually exclusive events can be extended to three or more events as shown in Example 4.

Example 4   Using the Addition Rule with Three Events

A card is drawn from a deck. Find the probability that it is either a club, a diamond, or a heart.

SOLUTION

In the deck of 52 cards there are 13 clubs, 13 diamonds, and 13 hearts, and any card can be only one of those suits. So

$$P(\text{club, diamond, or heart}) = P(\text{club}) + P(\text{diamond}) + P(\text{heart})$$

$$= \frac{13}{52} + \frac{13}{52} + \frac{13}{52} = \frac{39}{52} = \frac{3}{4}$$

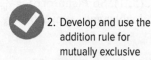

2. Develop and use the addition rule for mutually exclusive events.

Try This One   4

In rolling two dice, find the probability that the sum is 2, 3, or 4.

In Example 4, we used an extended version of addition rule 1:

$$P(A \ \text{or} \ B \ \text{or} \ C) = P(A) + P(B) + P(C)$$

What if two events aren't mutually exclusive? In this case addition rule 1 won't work, and it's helpful to think about why. Let's say that there's one outcome common to both events. When you add up the total number of outcomes in one event or the other, the outcome common to both will get counted twice, and you'll have one more outcome than you should have. To account for that, you could subtract off that single outcome so that it's only counted once. If that makes sense to you, then so will our second addition rule.

Math Note

Addition rule 2 can actually be used when events are mutually exclusive, too—in that case, $P(A \text{ and } B)$ will always be zero, and you'll get addition rule 1 back. But it's still important to distinguish between the two situations.

Addition Rule 2

When two events A and B are not mutually exclusive, the probability that A or B will occur is

$$P(A \ \text{or} \ B) = P(A) + P(B) - P(A \text{ and } B)$$

We could justify addition rule 2 using Venn diagrams, but frankly we'd rather have you do it. See Problem 46.

Example 5   Using Addition Rule 2

A single card is drawn from a standard deck of cards. Find the probability that it's a king or a red card.

SOLUTION

There are 4 kings and 26 red cards in a deck, but the events aren't mutually exclusive because the king of hearts and the king of diamonds are both. So we'll use addition rule 2 and subtract the probability of the card being a red king.

$$P(\text{king or red card}) = P(\text{king}) + P(\text{red card}) - P(\text{king and red card})$$

$$= \frac{4}{52} + \frac{26}{52} - \frac{2}{52} = \frac{28}{52} = \frac{7}{13}$$

Try This One   5

A card is drawn from an ordinary deck. Find the probability that it is a heart or a face card.

Example 6 | Using Addition Rule 2

Two dice are rolled. Find the probability of getting doubles or a sum of 6.

SOLUTION

There are six ways to get doubles: (1, 1), (2, 2), (3, 3), (4, 4), (5, 5), (6, 6). So $P(\text{doubles}) = \frac{6}{36}$.
There are five ways to get a sum of 6: (1, 5), (2, 4), (3, 3), (4, 2), (5, 1). So $P(\text{sum of } 6) = \frac{5}{36}$.
Notice that there is one way of getting doubles and a sum of 6, so $P(\text{doubles and a sum of } 6) = \frac{1}{36}$.
Finally,

$$P(\text{doubles or a sum of } 6) = P(\text{doubles}) + P(\text{sum of } 6) - P(\text{doubles and sum of } 6)$$

$$= \frac{6}{36} + \frac{5}{36} - \frac{1}{36} = \frac{10}{36} = \frac{5}{18}$$

3. Develop and use the addition rule for events that are not mutually exclusive.

Try This One   6

When two dice are rolled, find the probability that both numbers are more than 3, or that they differ by exactly 2.

In many cases, the information in probability problems can be arranged in table form in order to make it easier to compute the probabilities for various events. Example 7 uses this technique.

Example 7 | Using a Table and Addition Rule 2

In a hospital there are eight nurses and five physicians. Seven nurses and three physicians are females. If a staff person is selected, find the probability that the subject is a nurse or a male.

SOLUTION

The sample space can be written in table form.

Staff	Females	Males	Total
Nurses	7	1	8
Physicians	3	2	5
Total	10	3	13

Looking at the table, we can see that there are eight nurses and three males, and there's one person who is both a male and a nurse. The probability is

$$P(\text{nurse or male}) = P(\text{nurse}) + P(\text{male}) - P(\text{male and a nurse})$$

$$= \frac{8}{13} + \frac{3}{13} - \frac{1}{13} = \frac{10}{13}$$

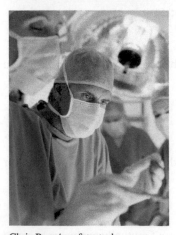

Chris Ryan/age fotostock

Try This One   7

In one class, there are 15 first-year students and 10 sophomores. Six of the first-years are education majors and four of the sophomores are education majors. If a student is selected at random, find the probability that the student is a sophomore or an education major.

Answers to Try This One

1   (a) Not mutually exclusive; a sophomore business major
       (b) Mutually exclusive
       (c) Not mutually exclusive; a male that is a member
            of the cheerleading squad

       (c) $\dfrac{3}{7}$

       (d) The answer to part (c) is the sum of the two
            answers in part (a).

2   (a) $P(\text{Tuesday}) = \dfrac{1}{7}$, $P(\text{Weekend}) = \dfrac{2}{7}$

       (b)

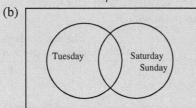

3 $\dfrac{7}{25}$ **4** $\dfrac{1}{6}$ **5** $\dfrac{11}{26}$ **6** $\dfrac{5}{12}$

7 $\dfrac{16}{25}$

Exercise Set   10-7

Writing Exercises

1. Explain how to tell if two events are mutually exclusive or not.
2. Explain the difference between the two addition rules for probability.
3. When we want to find the probability of one event or another happening and the events aren't mutually exclusive, why can't we just add the probabilities of the original events? What goes wrong?
4. Describe how we used Venn diagrams to justify addition rule 1.
5. Why does addition rule 2 in essence make addition rule 1 unnecessary? (*Hint:* Think about what would happen to addition rule 2 when the two events are mutually exclusive.)
6. Does addition rule 1 (for mutually exclusive events) only apply to two events? Explain.

For Exercises 7–14, decide if the events are mutually exclusive, and explain your answer.

7. Roll a die: get an even number, or get a number less than 3.
8. Roll a die: get a prime number (2, 3, 5), or get an odd number.
9. Roll a die: get a number greater than 3, or get a number less than 3.
10. Pick a student in your class: the student has blond hair, or the student has blue eyes.
11. Pick a student in your college: the student is a junior, or the student is a history major.
12. Pick any course: it is a calculus course, or it is an English course.
13. Pick a registered voter: the voter is a Republican, or the voter is a Democrat.
14. A chef decides to make either parmesan-crusted chicken or bacon-wrapped sirloin the daily special.

Applications in Our World

15. A young couple is arguing about what month they want to get married next year, so they decide to grab a calendar and choose a month at random. What's the probability that they get married in April or May?
16. At the animal shelter where Miguel volunteers on the weekends, there were two Siamese cats, four tabby cats, 16 mixed-breed dogs, and four iguanas in their cages. If a customer picks any of these animals at random, find the probability that the animal is either a mixed-breed dog or a Siamese cat.
17. When Milo went to register for classes at the last minute, the only classes left to take were seven math courses, five computer science courses, three statistics courses, and four science courses. He shut his eyes and picked one at random.

Find the probability that Milo selected either a science course or a math course. (The probability that Milo will ever graduate using this strategy is an interesting question, too.)
18. When Shonda looked at the favorites list on her boyfriend's smartphone, there were 10 songs she liked, 7 songs she'd heard but did not like, and 5 songs she hadn't heard. She let the phone pick a song at random. What is the probability that the selected song was one she liked or one she'd heard but did not like?
19. While conducting an experiment on perception, a psychologist randomly chooses a subject from nine adult males, six adult females, and seven children. Find the probability that the subject is
    (a) An adult male or a child.   (b) An adult.

20. In my Tae Kwon Do class, there are two black belts, three red belts, five blue belts, three green belts, and six yellow belts. If the sensei selects a student at random to lead the warm-up, find the probability that the person is
    (a) Either a green belt or a yellow belt.
    (b) Either a black belt or a blue belt.
    (c) Either a black belt, a green belt, or a yellow belt.
21. On a small college campus, there are five English professors, four math professors, two science professors, three psychology professors, and three history professors. If a professor is chosen at random, find the probability that the professor is
    (a) An English or psychology professor.
    (b) A math or science professor.
    (c) A history, science, or math professor.
    (d) An English, math, or history professor.
22. A deck of cards is randomly dealt by the computer during a game of Spider Solitaire. Find the probability the first card dealt is
    (a) A 4 or a diamond.
    (b) A club or a diamond.
    (c) A jack or a black card.
23. In a statistics class there are 18 juniors and 10 seniors; 6 of the seniors are females, and 12 of the juniors are males. If a student is chosen at random, find the probability of selecting
    (a) A junior or a female.
    (b) A senior or a female.
    (c) A junior or a senior.
24. A cell phone company gets a really good deal on 400 Samsung phones. Of them, 250 have 5″ screens and the rest have 4″ screens. Of the phones with 5″ screens, 140 are silver and the rest are black. Of the phones with 4″ screens, 80 are silver and the rest are black. If one phone is picked at random for a customer, find the probability that it will
    (a) Be silver or have a 5″ screen.
    (b) Be black or have a 4″ screen.
25. BlueFly.com, an online designer clothing marketplace, purchases items from Kenneth Cole, Michael Kors, and Vera Wang. The most recent purchases are shown here:

Product	Kenneth Cole	Michael Kors	Vera Wang
Dresses	24	18	12
Jeans	13	36	15

    If one item is chosen at random, find these probabilities:
    (a) It was purchased from Kenneth Cole or is a dress.
    (b) It was purchased from Michael Kors or Vera Wang.
    (c) It is a pair of jeans or it was purchased from Kenneth Cole.
26. In a recent campus survey, the following data were obtained in response to the question "Do you think there should be harsher penalties for underage drinking on campus?"

	Yes	No	No opinion
Males	72	81	5
Females	103	68	7

If a person is picked at random, find these probabilities:
(a) The person has no opinion.
(b) The person is a male or is against harsher penalties.
(c) The person is a female or favors harsher penalties.

27. A grocery store employs cashiers, stock clerks, and deli personnel. The distribution of employees according to marital status is shown next.

Marital status	Cashiers	Stock clerks	Deli personnel
Married	8	12	3
Not married	5	15	2

If an employee is chosen at random, find these probabilities:
(a) The employee is a stock clerk or married.
(b) The employee is not married.
(c) The employee is a cashier or is not married.

28. Students were surveyed on campus about their study habits. Some said they study in the morning, others study during the day between classes, and others study at night. Some students always study in a group and others always study alone. The distribution is shown below:

How students study	Morning	Between classes	Evening
Study in a group	2	3	1
Study alone	3	4	2

If a student who was surveyed is chosen at random, find these probabilities:
(a) The student studies in the evening.
(b) The student studies in the morning or in a group.
(c) The student studies in the evening or studies alone.

29. Three cable channels (95, 97, and 103) air quiz shows, comedies, and dramas. The numbers of shows aired are shown here.

Type of show	Channel 95	Channel 97	Channel 103
Quiz show	5	2	1
Comedy	3	2	8
Drama	4	4	2

If a show is picked at random, find these probabilities:
(a) The show is a quiz show or it is shown on Channel 97.
(b) The show is a drama or a comedy.
(c) The show is shown on Channel 103 or it is a drama.

30. A local postal carrier distributed first-class letters, advertisements, and magazines. For a certain day, the carrier distributed the following numbers of each type of item.

Delivered to	First-class letters	Ads	Magazines
Home	325	406	203
Business	732	1,021	97

If an item of mail is chosen at random, find these probabilities:

(a) The item went to a home.

(b) The item was an ad or it went to a business.

(c) The item was a first-class letter or it went to a home.

31. As part of her major in microbiology, Juanita spent 3 weeks studying the spread of a disease in the jungles of South America. When she returned, she found that she had many emails sent to her home account and her school account, and that the emails were either spam, school announcements, or messages from friends as follows:

Delivered to	School announcements	Spam	Messages from friends
Home	412	910	342
School	791	1,206	68

If an email is picked at random, find these probabilities:

(a) The email was sent to her home.

(b) The email was a school announcement or it was sent to her school account.

(c) The email was spam or it was sent to her home account.

32. Before a Walk for the Cure 10-mile walk, participants could choose a T-shirt from a box with six red shirts, two green shirts, one blue shirt, and one white shirt. When the first participant randomly picks a shirt from the box, what is the probability they will get a red shirt or a white shirt?

33. You roll two dice as part of a casino game. Find the probability of getting

(a) A sum of 6, 7, or 8.

(b) Doubles or a sum of 4 or 6.

(c) A sum of greater than 9, less than 4, or equal to 7.

34. In another dice game, you begin by rolling three dice. Find the probability that

(a) All three dice show the same number.

(b) The three numbers sum to a number less than 6.

(c) The three numbers are all odd or divisible by 3.

35. In five-card poker, a flush is a hand where all cards have the same suit, and a pair is when two cards have the same denomination. You're dealt the 3, 7, 10, and jack of clubs and the 9 of diamonds, and all of the other cards are still in the deck. If you discard the 9 of diamonds and draw one card to replace it, what's the probability of getting a flush or a pair?

36. A straight in five-card poker is five cards with consecutive denominations. If you're dealt 3, 4, 5, 6, and a queen, then discard the queen and draw one card to replace it, what's the probability of getting a straight or a pair? Assume that all other cards are still in the deck.

This table shows the top 20 highest-rated American TV shows for the 2020–2021 season. The last column is the average number of viewers in millions. You'll be using this table for all of

Exercises 37–42, so memorize it and then eat it. Or I guess you could just refer back to it. That probably makes more sense.

Rank	Program	Network	Total Viewers (in millions)
1	*NBC Sunday Night Football*	NBC	16.50
2	*Fox + NFLN Thursday Night Football*	Fox	13.42
3	*NCIS*	CBS	12.54
4	*Sunday Night NFL Pre-Kick*	NBC	11.84
5	*Equalizer*	CBS	11.78
6	*FBI*	CBS	10.97
7	*60 Minutes*	CBS	10.71
8	*Chicago Fire*	NBC	10.20
9	*Blue Bloods*	CBS	10.16
10	*The OT*	Fox	9.77
11	*Chicago Med*	NBC	9.71
12	*Chicago PD*	NBC	9.68
13	*9-1-1*	Fox	9.60
14	*Young Sheldon*	CBS	9.45
15	*This Is Us*	NBC	9.27
16	*FBI: Most Wanted*	CBS	8.78
17	*9-1-1: Lone Star*	Fox	8.69
18	*Bull*	CBS	8.54
19	*Football Night In America Pt 3*	NBC	8.41
20	*Voice - Tuesday*	NBC	8.38

37. If one of the top 20 shows is chosen at random, what is the probability that it was on NBC or CBS?

38. If you choose one top 20 show randomly, what is the probability that it was on CBS or had more than 10 million viewers on average?

39. When choosing a random top 20 show, what's more likely: that it was on CBS, or that it was on Fox or NBC? Justify by calculating probabilities, please.

40. In the rankings of all shows for the season, the probabilities of a show being on each network were as follows: Fox 0.180, NBC 0.180, ABC 0.248, CBS 0.217, the CW 0.174. Find the probability that one randomly selected network show was not on NBC or CBS.

41. There were 161 total shows listed. Fox and NBC both had 29 shows listed, while CBS had 35. Find the probability that a randomly selected show was on CBS or was ranked in the top 20.

42. Find the probability that a randomly selected show (not necessarily in the top 20) was on NBC or had more than 9 million viewers.

Critical Thinking

43. At Big Tony's Pizzas and Loans, 45% of customers order pizzas with either sausage or pepperoni. If 31% get only pepperoni and 26% only sausage, find the probability that the next person to call will order both sausage and pepperoni.

44. Use a Venn diagram to illustrate the probability calculation in Try This One 5 on page 611.

45. Use a Venn diagram to illustrate the probability calculation in Try This One 6 on page 612.

46. The Venn diagram shown can be used to show why addition rule 2 on page 611 works. Do it! Studying Example 2 should be helpful.

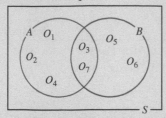

47. Use a Venn diagram to demonstrate the addition rule for probability when there are three mutually exclusive events.

48. (a) Decide if the three events are mutually exclusive when drawing a single card from a regular deck, and explain your reasoning: drawing a card with an even number, drawing a card with a number less than 4, drawing a spade.

  (b) Find the probability of each event: drawing a card with an even number, drawing a card with a number less than 4, and drawing a spade.

  (c) Do you think that the probability of drawing a card with an even number, a number less than 4, or a spade is the sum of the three answers to part (b)? Why or why not?

49. Refer to Exercise 48.
  (a) Draw a Venn diagram illustrating the three events.
  (b) Use your Venn diagram to find the probability of drawing a card with an even number, a number less than 4, or a spade.

50. Use the work you did with the Venn diagram in Exercise 49 to develop a general addition rule for three events that are not mutually exclusive.

Exercises 51 through 54 require some careful thought (and maybe some research), and may be open to interpretation. Analyze the situation and decide if the two events are or are not mutually exclusive, then write an explanation of your answer.

51. Being acquitted of a crime and spending time in jail after the trial

52. Being born outside the boundaries of the United States and being elected president of the United States

53. Being a first-year student, being a sophomore, and being a junior

54. Getting fewer votes than your opponent when running for the presidency and winning the election

---

Section 10-8   The Multiplication Rules and Conditional Probability

moodboard/age fotostock

Each day a professor picks a student at random to work out a homework problem on the board. If the student is selected regardless of who already went, the events are independent. If the student is selected from the pool of students who haven't gone yet, the events are dependent.

LEARNING OBJECTIVES

1. Find the probability of two or more independent events all occurring.

2. Find the probability of two or more dependent events all occurring.

3. Find conditional probabilities.

Even after learning about probability, a lot of people are tempted to play lottery games, looking for that one big score that will put them on easy street. Suppose that you play a daily game in which you try to match a randomly selected four-digit number. On Monday, the number is 2471. On Tuesday, the number is again 2471. Would you play 2471 on Wednesday? People are usually split on this question: some say "There's no way it's going to come up *three* days in a row!" while others say "That number's hot! Go with it!"

The logic in both of those arguments is faulty, however, assuming that the game isn't rigged and the numbers really are chosen at random. When Wednesday's numbers are being drawn, they have no idea what may have been drawn on Monday and Tuesday: that's all in the past and has no effect on Wednesday's draw.

Let's recast the question this way: What's the probability that 2471 will come up three times in a row? In this section, we'll study rules for finding the probability of consecutive events occurring. In every case, we'll have to decide if any event affects the outcome of the ones that follow.

Chip Simons/Creatas Images/Getty Images

Two events *A* and *B* are **independent** if the fact that *A* occurs has no effect on the probability of *B* occurring.

Based on our description above, the lottery numbers that are drawn on Tuesday and Wednesday are independent events.

Here are other examples of independent events:

Rolling a die and getting a 6, and then rolling a second die and getting a 3.
Drawing a card from a deck and getting a queen, replacing it, and drawing a second card and getting a queen.

On the other hand, when the occurrence of the first event changes the probability of the occurrence of the second event, the two events are said to be *dependent*. For example, suppose a card is drawn from a deck and *not* replaced, and then a second card is drawn. The probability for the second card is changed since the sample space contains only 51 cards when the first card isn't replaced.

> Two events *A* and *B* are **dependent** if the outcome of *A* has some effect on the probability of *B* occurring.

Here are some examples of dependent events:

Drawing a card from a deck, not replacing it, and then drawing a second card.

Selecting a lottery ball from a tumbler, not replacing it, and then drawing a second ball.

Parking in a no-parking zone and getting a parking ticket.

Example 1 | **Finding the Probability of Two Independent Events Both Occurring**

The events of rolling a die and getting 6, then rolling a second die and getting 3 or less are independent.

(a) Find the probability of each event. Don't reduce any fractions.
(b) Use the fundamental counting principle to find the number of ways that BOTH events can occur. Then do the same to find the number of total outcomes for this two-stage probability experiment.
(c) Multiply the two probabilities from part (a), then use your answer to part (b) to justify that this results in the probability of BOTH events occurring.

SOLUTION

(a) There's just one way to get 6 out of six possible outcomes, so the probability is 1/6. There are three ways to get 3 or less, so the probability is 3/6.
(b) With one way to get the first outcome and three ways to get the second, the total number of ways to get both is $1 \cdot 3 = 3$. There are six total outcomes for each stage, so the total number of outcomes is $6 \cdot 6 = 36$.
(c) The product of the two probabilities is

$$\frac{1}{6} \cdot \frac{3}{6} = \frac{1 \cdot 3}{6 \cdot 6} = \frac{3}{36} = \frac{1}{12}$$

Look carefully at the multiplication in the numerator: this is the calculation done, using the fundamental counting principle, to find the number of ways that both events can occur. The denominator is the calculation to find the total number of outcomes for the two-stage experiment. So what is the fraction? Exactly the probability of both events occurring. Because we multiply fractions by multiplying numerators and denominators separately, whenever we multiply the probability of two independent events, in effect we're using the fundamental counting principle and the formula for theoretical probability to find the probability that both events occurred.

Try This One | 1

The events of rolling one die and getting 5 or 6, then flipping a coin and getting tails are independent.

(a) Find the probability of each event. Don't reduce any fractions.

(b) Use the fundamental counting principle to find the number of ways that BOTH events can occur. Then do the same to find the number of total outcomes for this two-stage probability experiment.
(c) Multiply the two probabilities from part (a), then use your answer to part (b) to justify that this results in the probability of BOTH events occurring.

The result of Example 1 illustrates our first multiplication rule, which allows us to find the probability of two events both occurring when they're independent.

Multiplication Rule 1

When two events A and B are independent, the probability of both occurring is

$$P(A \text{ and } B) = P(A) \cdot P(B)$$

Example 2   Using Multiplication Rule 1

As part of a psychology experiment on perception and memory, colored balls are picked from an urn. The urn contains three red balls, two green balls, and five white balls. A ball is picked and its color is noted. Then it is replaced. A second ball is picked and its color is noted. Find the probability of each of these.

(a) Picking two green balls.
(b) Picking a green ball and then a white ball.

SOLUTION

Remember, selection is done with replacement, which makes the events independent, so multiplication rule 1 applies.

(a) The probability of picking a green ball on each trial is $\frac{2}{10}$, so

$$P(\text{green and green}) = P(\text{green}) \cdot P(\text{green}) = \frac{2}{10} \cdot \frac{2}{10} = \frac{4}{100} = \frac{1}{25}$$

(b) The probability of picking a green ball is $\frac{2}{10}$ and the probability of picking a white ball is $\frac{5}{10}$, so

$$P(\text{green and white}) = P(\text{green}) \cdot P(\text{white}) = \frac{2}{10} \cdot \frac{5}{10} = \frac{10}{100} = \frac{1}{10}$$

Try This One   2

As part of a card trick, a card is drawn from a deck and replaced; then a second card is drawn. Find the probability of getting a queen and then an ace.

Example 3   Finding Probabilities for Three Independent Events

Three cards are drawn from a deck. After each card is drawn, its denomination and suit are noted and it's mixed back into the deck before the next card is drawn.

Math Note

Multiplication rule 1 can be extended to three or more independent events using the formula

$P(A_1$ and A_2 and $A_3 \cdots$ and $A_n)$

$= P(A_1) \cdot P(A_2) \cdot P(A_3) \cdots \cdot P(A_n)$

(a) Does multiplication rule 1 apply here? Why or why not?
(b) Find the probability of getting three kings.
(c) Find the probability of getting three clubs.

SOLUTION

(a) Multiplication rule 1 does apply because the draws are independent. This is entirely because the card that was picked is mixed back into the deck. If that weren't the case, the probabilities would change for the second and third draws.
(b) Since there are four kings, the probability of getting a king on each draw is $\frac{4}{52}$ or $\frac{1}{13}$. The probability of getting three kings is

$$\frac{1}{13} \cdot \frac{1}{13} \cdot \frac{1}{13} = \frac{1}{2,197}$$

(c) Since there are 13 clubs, the probability of getting a club is $\frac{13}{52}$ or $\frac{1}{4}$. The probability of getting three clubs in a row is

$$\frac{1}{4} \cdot \frac{1}{4} \cdot \frac{1}{4} = \frac{1}{64}$$

Try This One   3

Given that the probability of rain on any given day in March in Daytona Beach is $\frac{1}{5}$, find the probability that

(a) It rains three straight days in March.
(b) It rains on March 10 and 12, but not March 11.
(c) What assumption do you have to make in order to use multiplication rule 1 for this situation?

As we've pointed out, when a card is picked from a deck and not replaced, this changes the probabilities for what will happen when a second card is drawn. That makes sense because if you draw a queen on the first card, now there are only three left, rather than the four there were originally. But what if the deck is actually 1,000 decks all mixed together? Now if you draw a queen, there are still 3,999 left in the deck, and the effect on the probability of drawing another queen isn't nearly as significant.

Math Note

I just did a quick calculation and found that a stack of 1,000 decks of cards would be about 52 feet tall. Why would I do this? Because I can. And because I wondered.

This illustrates a useful idea in computing probabilities in our world: if the sample space is really large, choosing a handful of objects without replacement technically makes the choices dependent, but the effect on probability is so small that we can pretend the choices are independent, making the calculations simpler. This is illustrated in Example 4.

Example 4	Using Multiplication Rule 1 with Large Samples

According to a report by MarketWatch, 70% of college graduates will leave college with some student debt. If four graduating seniors are chosen randomly and asked if they have student debt:

(a) Are the students' responses independent? Can we still use multiplication rule 1?
(b) Find the probability that all four will graduate with student debt.

SOLUTION

(a) The student responses are not technically independent because it would be silly to survey the same student twice. So after one has been surveyed, they're no longer in the pool of candidates, and the probabilities for later students are affected. But there are millions of college graduates every year, so excluding four of them from the total would have a miniscule effect on the probabilities, so we can still use multiplication rule 1.

(b) If 70% of students graduate with debt, then the empirical probability of any one student doing so is 0.7. Using multiplication rule 1, the probability that all four graduate with debt is $(0.7)(0.7)(0.7)(0.7) = 0.2401$.

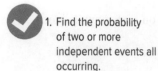

1. Find the probability of two or more independent events all occurring.

Try This One   4

MarketWatch also reported that only 37% of student borrowers were actively paying down their debt. If three student borrowers are randomly picked, find the probability that all of them are paying down their debt.

When we are interested in finding the probability of consecutive events that are dependent, we can still use the multiplication rule, but with a minor modification. For example, let's say we draw two cards at random from a standard deck. The probability of getting an ace on the first draw is $\frac{4}{52}$. But if the first card is not put back into the deck, that changes the probability of drawing another ace—it's now $\frac{3}{51}$. Using multiplication rule 1, the probability of both events occurring is

$$\frac{4}{52} \cdot \frac{3}{51} = \frac{12}{2,652} = \frac{1}{221}$$

We can summarize this procedure as multiplication rule 2.

Multiplication Rule 2

When two events are dependent, the probability of both occurring is

$$P(A \text{ and } B) = P(A) \cdot P(B \text{ given that } A \text{ has already occurred})$$

Example 5   Using Multiplication Rule 2

An appliance store gets a shipment of 25 plasma TVs, and 3 of them are defective. If two of the TVs are chosen at random, find the probability that both are defective. (The first TV is not replaced after it's tested.)

Digital Vision/SuperStock

SOLUTION

Since there are 3 defective TVs out of a total of 25, the probability of the first being defective is $\frac{3}{25}$. After the first one is found to be defective and not replaced, there are 2 defective sets left out of 24, so the probability of the second being defective given that the first one is defective is $\frac{2}{24}$. Using multiplication rule 2, the probability that both are defective is

$$P(\text{1st defective and 2nd defective}) = P(\text{1st}) \cdot P(\text{2nd given 1st})$$

$$= \frac{3}{25} \cdot \frac{2}{24} = \frac{6}{600} = \frac{1}{100}$$

Try This One   5

The 2021 NCAA women's basketball tournament field had (among 64 teams) 8 teams from the Atlantic Coast Conference, 7 from the Big Ten, 7 from the Southeastern Conference, and 6 from the Pacific 12. If you were randomly assigned two teams in a dorm pool, find the probability that

(a) Both were from the Southeastern Conference.
(b) The first was from the Atlantic Coast Conference and the second was from the Pacific 12.

Multiplication rule 2 can be extended to three or more events as shown in Example 6.

Example 6

Using Multiplication Rule 2 with Three Events

Three cards are drawn from an ordinary deck and not replaced. Find the probability of

(a) Getting three jacks.
(b) Getting an ace, a king, and a queen in order.
(c) Getting a club, a spade, and a heart in order.
(d) Getting three clubs.

SOLUTION

(a) $P(\text{three jacks}) = \frac{4}{52} \cdot \frac{3}{51} \cdot \frac{2}{50} = \frac{24}{132,600} = \frac{1}{5,525}$

One less remaining jack, and one less card to choose from at each stage

(b) $P(\text{ace and king and queen}) = \frac{4}{52} \cdot \frac{4}{51} \cdot \frac{4}{50} = \frac{64}{132,600} = \frac{8}{16,575}$

(c) $P(\text{club and spade and heart}) = \frac{13}{52} \cdot \frac{13}{51} \cdot \frac{13}{50} = \frac{2,197}{132,600} = \frac{169}{10,200}$

(d) $P(\text{three clubs}) = \frac{13}{52} \cdot \frac{12}{51} \cdot \frac{11}{50} = \frac{1,716}{132,600} = \frac{11}{850}$

2. Find the probability of two or more dependent events all occurring.

Try This One   6

When drawing four cards from a deck, what is the probability that

(a) All four are aces?   (b) All four are clubs?

Math Note

We use the term conditional probability because the condition that A has already occurred affects the probability of B.

Conditional Probability

We know that to find the probability of two dependent events occurring, it's important to find the probability of the second event occurring given that the first has already occurred. We call this the **conditional probability** of event B occurring given that event A has occurred, and denote it $P(B|A)$.

Now that we have a symbol to represent the probability of event B given that A has occurred, we can rewrite multiplication rule 2 and solve the equation for $P(B|A)$:

$$P(A \text{ and } B) = P(A) \cdot P(B|A)$$

$$P(B|A) = \frac{P(A \text{ and } B)}{P(A)}$$

This gives us a formula for conditional probability.

Formula for Conditional Probability

The probability that a second event B occurs given that a first event A has occurred can be found by dividing the probability that both events occurred by the probability that the first event has occurred. The formula is

$$P(B|A) = \frac{P(A \text{ and } B)}{P(A)}$$

Example 7 illustrates the use of this rule.

USAF/Getty Images

Military strategies (and other types of strategies) use conditional probability. If *A* occurs, how likely is *B*? How likely is *B* if *A* doesn't occur?

Example 7 — Finding a Conditional Probability

Suppose that your professor goes stark raving mad and chooses your final grade from A, B, C, D, F, or Incomplete totally at random.

(a) Find the probability of getting an A given that you get a letter grade higher than D using the conditional probability formula.
(b) Find that probability by writing out a new sample space and using the formula for theoretical probability.

SOLUTION

The goal is to find $P(A|\text{letter grade higher than D})$.

(a) The full sample space is {A, B, C, D, F, I}, which has six outcomes. Getting an A is one of them, so $P(A) = \frac{1}{6}$. Notice that this is the same as the probability of getting an A and getting a grade higher than D because obviously if you get an A, that's higher than a D. There are three outcomes matching a letter grade higher than D, so $P(\text{letter grade higher than D}) = \frac{3}{6}$. Using the conditional probability formula, we get

$$P(A|\text{letter grade higher than D}) = \frac{P(A \text{ and letter grade higher than D})}{P(\text{letter grade higher than D})} = \frac{\frac{1}{6}}{\frac{3}{6}} = \frac{1}{6} \cdot \frac{6}{3} = \frac{1}{3}$$

(b) If it's given that the grade is a letter grade higher than D, there are only three possible outcomes: A, B, or C. So in that case, $P(A|\text{letter grade higher than D}) = \frac{1}{3}$.

This example illustrates a useful point: just because you CAN memorize and use a formula, it doesn't always mean that you SHOULD. In many cases, it's simpler to compute a conditional probability by thinking about what the new sample space is under a given condition.

Try This One   7

A group of patients in a blind drug trial is assigned numbers from 1 through 8. The even numbers get an experimental drug, while the odd numbers get a placebo.

(a) Use the conditional probability formula to find the probability that one particular patient gets the experimental drug given that they weren't assigned 1, 2, or 3.
(b) Compute the same probability by writing out a new sample space and using the theoretical probability formula.
(c) Which method do you prefer? Why?

Example 8 — Finding a Conditional Probability

Hate crimes are defined to be crimes in which the victim is targeted because of one or more personal characteristics, such as race, religion, or sexual orientation. The table below lists the motivation for certain hate crimes as reported by the FBI for 2020.

Motivation	Crimes against persons	Crimes against property	Crimes against society	Total
Race	3,392	1,222	170	**4,784**
Religion	592	1,024	34	**1,650**
Sexual Orientation	1,061	324	10	**1,395**
Total	**5,045**	**2,570**	**214**	**7,829**

(a) Find the probability that a hate crime was racially motivated given that it was a crime against persons.
(b) Find the probability that a hate crime was against property given that it was motivated by the victim's sexual orientation.

SOLUTION

(a) Since we're interested only in crimes against persons, we only need to look at that column. There were 5,045 such crimes total, and 3,392 were racially motivated, so the probability is

$$\frac{3,392}{5,045} \approx 0.672$$

(b) This time we're given that the crime was motivated by sexual orientation, so we only need to look at that row. There were 1,395 such crimes total, and 324 were against property, so the probability is

$$\frac{324}{1,395} \approx 0.232$$

Try This One   8

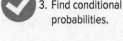

3. Find conditional probabilities.

Based on the data in the above table, find the probability that

(a) A crime was motivated by either race or religion given that it was a crime against society.
(b) A crime was against persons given that it was motivated by religion or sexual orientation.

Answers to Try This One

1 (a) $P(5 \text{ or } 6) = \frac{2}{6}$; $P(\text{tails}) = \frac{1}{2}$
   (b) Number of ways to get 5 or 6 and tails: $2 \cdot 1 = 2$; total number of outcomes: $6 \cdot 2 = 12$
   (c) $\frac{2}{6} \cdot \frac{1}{2} = \frac{2 \cdot 1}{6 \cdot 2} = \frac{2}{12} = \frac{1}{6}$; the numerator is the number of ways to get 5 or 6 and tails, while the denominator is the total number of outcomes.

2 $\frac{1}{169}$ **3** (a) $\frac{1}{125}$ (b) $\frac{4}{125}$
   (c) You have to assume that the weather on any given day is independent of what happened the day before.

4 0.051   **5** (a) $\frac{1}{96}$ (b) $\frac{1}{84}$

6 (a) $\frac{1}{270,725}$ (b) $\frac{11}{4,165}$

7 (a) $\frac{P(\text{experimental drug and not assigned 1, 2, or 3})}{P(\text{not assigned 1, 2, or 3})}$
      $= \frac{3/8}{5/8} = \frac{3}{5}$
   (b) $\frac{\text{number of ways to get even from 4, 5, 6, 7, 8}}{5 \text{ choices}} = \frac{3}{5}$
   (c) This is an opinion. Make sure you justify yours!

8 (a) $\frac{204}{214} \approx 0.953$ (b) $\frac{1,653}{3,045} \approx 0.543$

Exercise Set   10-8

Writing Exercises

1. What is the difference between independent and dependent events? Give an example of each.
2. What is meant by the term conditional probability?
3. Describe two methods for computing conditional probability.
4. Explain why the probability of two consecutive events occurring couldn't possibly be the sum of the two individual probabilities.
5. Describe how and when to use multiplication rule 1.
6. Describe how and when to use multiplication rule 2.

In Exercises 7–14, decide whether the events are independent or dependent, and explain your answer.

7. Flipping a coin and drawing a card from a deck.
8. Drawing a ball from a lottery machine, not replacing it, and then drawing a second ball.
9. Getting a raise in salary and purchasing a new car.
10. Driving on ice and having an accident.
11. Having a large shoe size and having a high IQ.
12. A father being left-handed and a daughter being left-handed.
13. Smoking regularly and having lung cancer.
14. Eating an excessive amount of ice cream and smoking an excessive amount of cigarettes.

Applications in Our World

15. On one large campus, 18% of students surveyed said that they spend less than an hour a night studying. If three students are picked at random, what's the probability that all three spend less than an hour a night studying? What's the probability that two of the three do?
16. A national study of patients who were at an unhealthy weight found that 56% also have elevated blood pressure. If two such patients are selected, find the probability that both have elevated blood pressure.
17. According to the National Highway Traffic Safety Administration, seat belt use among American drivers and front-seat passengers passed 90% for the first time in 2016. At that time, 90.1% of people used seat belts. If four people are selected at random, find the probability that all four regularly use seat belts.
18. A computer salesperson at Best Buy claims to have a 20% chance of selling a computer when helping out a customer. If this is true and the salesperson talks to four customers before lunch, find the probability that all four will buy computers.
19. If 25% of Michael's graduating class are not U.S. citizens, find the probability that two randomly selected graduating students are not U.S. citizens.
20. If two people are chosen at random, what is the probability that they were both born in December?
21. If two people are chosen at random, find the probability that they were born in the same month.
22. If three people are chosen, find the probability that all three were born in March.
23. If half of Americans believe that the federal government should take "primary responsibility" for eliminating poverty, find the probability that three randomly selected Americans will agree that it is the federal government's responsibility to eliminate poverty.
24. What is the probability that a husband, wife, and daughter have the same birthday?
25. A telecommunications company has six satellites, two of which are sending a weak signal. If two are picked at random without replacement, find the probability that both are sending a weak signal.
26. In Exercise 25, find the probability that one satellite sends a strong signal and the other sends a weak signal.

The Federal Bureau of Investigation reported the statistics in the following table for homicides in 2019. Use the data to answer Questions 27–32.

Weapon	Number	Weapon	Number
Handgun	6,368	Cutting instrument	1,476
Rifle	364	Other weapons	1,593
Shotgun	200	Hands, feet, etc.	600
Unknown firearm	3,326	Total	13,927

27. If there were three unrelated murders in Detroit in June 2019, find the probability that all three were committed with a gun.
28. Find the probability that for two randomly selected murders, neither involved a gun.
29. Find the probability that a murder was committed with a handgun given that a gun was used.
30. Find the probability that a murder was committed with a shotgun given that some type of weapon was used (other than body parts).
31. Find the probability that if four unrelated murders are studied, two involved a gun and two involved a cutting instrument.
32. Find the probability that two unrelated murders were both committed with a handgun.

The U.S. Energy Information Administration reported the following statistics for electrical energy generation in 2020. Use the data to answer Questions 33–36.

Source	Billion kilowatt hours
Fossil fuels	2,424.0
Nuclear	790.0
Hydroelectric	291.1
Other renewables	500.9

33. If three homes are picked at random, find the probability that all three got their electricity from a source other than fossil fuels.
34. If three homes are randomly chosen, find the probability that two got their electricity from fossil fuels and one from a nuclear power plant.
35. Find the probability that a home got its power from a hydroelectric source given that it didn't come from fossil fuels.
36. Find the probability that a home got its power from fossil fuels given that it didn't come from a nuclear plant.
37. In a department store there are 120 customers, 90 of whom will buy at least one item. If 5 customers are

selected at random, one by one, find the probability that all will buy at least one item.

38. During a game of online hearts, three cards are dealt, one at a time without replacement, from a shuffled, ordinary deck of cards. Find these probabilities:
    (a) All are jacks.
    (b) All are clubs.
    (c) All are red cards.

39. In a group of eight Olympic track stars, five are hurdlers. If three are selected at random without replacement, find the probability they are all hurdlers.

40. In a class consisting of 15 men and 12 women, two different homework papers were selected at random. Find the probability that both papers belonged to women.

41. I offer you the following bet: you flip a coin then roll a die, and I give you $10 if you get tails and a 5, but you give me $1 otherwise. What's the probability that you'll win given that the coin lands tails up?

42. Juan draws a black card from an ordinary deck as the first card for a game of Gin Rummy. What is the probability the card was a king? What's the probability that the next black card is also a king given that the first one was?

43. At a carnival gambling booth, two dice are rolled, and a sum of 7 wins double your money. Find the probability that the sum is 7 given that one of the numbers was a 6.

44. After Frank and Sun order a pizza, they flip a coin to see who has to go pick it up (heads and Frank goes), then roll a die to see how many dollars Sun has to pay toward the cost. Find the probability that Frank has to pick up the pizza and pay more than $3.

45. A computer randomly deals an ordinary deck of cards for a game of FreeCell, and the first card dealt is a face card. Find the probability that the card was also a diamond.

46. During a backgammon game, Kelly rolled the two dice on the board. Find the probability that the sum obtained was greater than 8 given that the number on one die was a 6.

Use this information for Exercises 47–50. Three red cards are numbered 1, 2, and 3. Three black cards are numbered 4, 5, and 6. The cards are placed in a box and one card is picked at random.

47. Find the probability that a red card was picked given that the number on the card was an odd number.

48. Find the probability that a number less than 5 was picked given that the card was a black card.

49. Find the probability that a number less than 5 was picked given that the card was red.

50. Find the probability that a black card was picked given that the number on the card was an even number.

Use the following information for Exercises 51–54. A survey of 200 college students shows the average number of text messages they send each week.

	Less than 600	600–799	800–999	1,000 or more
Men	56	18	10	16
Women	61	18	13	8

If a person is selected at random, find these probabilities:

51. The student sent less than 600 texts given that it was a woman.

52. The student sent more than 999 texts given that it was a man.

53. The student was a woman given that they sent between 600 and 799 texts.

54. The student was a man given that they sent between 600 and 999 texts.

The table below shows the number of active-duty personnel in each branch of the military in 2020, as well as the percentage that were women. Find each probability in Exercises 55–58.

Branch	Army	Navy	Marines	Air Force
Total	481,254	341,996	180,958	329,614
% Women	15.5%	20.4%	8.9%	21.1%

Source: *U.S. Department of Defense*

55. An individual on active duty was a woman given that they were in either the air force, marines, or navy.

56. An individual was in the army given that it was a man.

57. An individual was a man given that they were not in the navy.

58. An individual was not a marine given that it was a woman.

Critical Thinking

59. (a) Use the multiplication rule for two independent events to show that the rule can be used for three events as well.
    (b) How can you use the result of (a) to conclude that the rule works for any number of independent events?

60. Suppose that I roll two dice and tell you that the result is definitely even, then offer you 3 to 1 odds that the total isn't 4. Draw a diagram to illustrate the outcomes, then use the diagram to compute the conditional probability. Finally, decide if the given odds are in your favor or not.

61. In many lotto games, the player chooses 6 numbers from a pool of 40 or more, and if they match all 6 numbers

drawn, they win the jackpot. If there is more than one winner, the jackpot is split evenly among all winners.
    (a) First instinct, if you were playing a lotto game, would you choose the numbers 1, 2, 3, 4, 5, and 6? Why or why not?
    (b) After the first winning number is drawn, does it have any effect on what number is drawn next? Think carefully, then use your answer and the idea of independence of events and probability to explain why there's nothing wrong with choosing 1, 2, 3, 4, 5, and 6.
    (c) In fact, something in the statement that begins this exercise indicates that 1, 2, 3, 4, 5, and 6 might be a desirable choice. What is it?

62. Classify each statement as sensible or silly, and briefly explain your answer.
    (a) I flipped a coin five times in a row and got heads, so I'm willing to bet $100 that it will be tails on the next flip.
    (b) There's a 30% chance of rain tomorrow and there's a 50–50 chance of my only class getting cancelled, so there's a 35% chance I'll be able to go golfing without missing class. (By the way, I'm not going to go golfing if it rains.)
    (c) According to duilawblog.com, the probability of being convicted when charged with DUI in California is 0.794. If 50% of those convicted get at least 48 hours in jail, the probability of spending at least 48 hours in jail if charged with DUI in California is 1.294.
    (d) If I buy one ticket to a multistate lottery, I have a better chance of being struck by lightning than I do of winning the grand prize.

When introducing the idea of using multiplication rule 1 with large samples, we talked about the likelihood of drawing a second queen both from a deck of cards, and then from 1,000 decks. In Exercises 63–65, we'll study the numbers.

63. (a) Find the probability of drawing a queen from a regular deck.
    (b) Once that card is no longer in the deck, find the probability of drawing a second queen. How does it compare to the original probability?
64. (a) Find the probability of drawing one queen from a stack of 1,000 regular decks.
    (b) Once that card is no longer in the deck, find the probability of drawing a second queen. How does it compare to the original probability?
    (c) What can you conclude from Exercises 63 and 64?
65. USING TECHNOLOGY. Writing Exercises 63 and 64 got me to thinking: how many decks is *enough* (whatever that means) to justify treating the first and second draw as independent events? That's what we'll study now, with the help of a spreadsheet. There's a template in the online resources for this section where we'll compute the probability of drawing two consecutive queens first the "correct" way (treating the draws as dependent, like in Exercises 63 and 64), then treating the draws as independent and using our first multiplication rule. Your job is to try different numbers of decks and decide how many are necessary for the difference in probability to be negligible (whatever that means). Report on your findings.

Section 10-9 | The Binomial Distribution

LEARNING OBJECTIVES

1. Identify binomial experiments.
2. Compute probabilities of outcomes in a binomial experiment.
3. Construct a probability distribution.

Many probability problems involve situations that have only two outcomes. When a baby is born, it will be either male or female. When a coin is flipped, it will land either heads or tails. When the New York Yankees play, they either win or they lose. That intriguing classmate that sits behind you in class will either go out with you or they won't.

Other situations can be reduced to two outcomes. For example, medical procedures can be classified as either successful or unsuccessful. An answer to a multiple-choice exam question can be classified as right or wrong even though there may be four answer choices.

HBSS/Fancy Photography/Getty Images

Situations like these are called *binomial experiments*. This is the only type of probability question we'll study in this section, so we should begin by clearly defining exactly what types of questions qualify.

A **binomial experiment** is a probability experiment that satisfies the following requirements:

1. Each trial can have only two outcomes, or outcomes that can be reduced to two outcomes. These outcomes can be considered either a success or a failure.
2. The outcomes must be independent of each other.
3. There must be a fixed number of independent trials.
4. The probability of a success must remain the same for all trials of the experiment.

Example 1	Deciding if an Experiment Is Binomial

Decide whether or not each is a binomial experiment and explain your reasoning.

(a) Drawing a card from a deck and seeing what suit it is.
(b) Randomly answering the questions on a true/false test.
(c) Asking 100 people whether or not they smoke.
(d) Drawing cards at random from a deck without replacement and deciding if they are red or black cards.

SOLUTION

(a) No. There are four outcomes (the four suits), not just two, and there's no reasonable way to reduce this to two outcomes.
(b) Yes. There are only two outcomes, right and wrong. Any test has to have a certain number of questions, so there's a fixed number of trials. If you're answering randomly you have a 50-50 shot on each question, and every question is unaffected by the others.
(c) Yes. There are two outcomes (yes or no), a fixed number of trials (100), and if the people are randomly chosen, the probability for each trial is constant.
(d) No. There are two outcomes and a fixed number of trials, but the trials aren't independent, and the probability doesn't stay constant. The cards aren't being replaced in the deck, so the previously drawn cards affect the later probabilities.

Try This One   1

Decide whether or not each experiment is a binomial experiment and explain your reasoning.

(a) Picking 10 poker chips with replacement from an urn containing three chips of different colors, and seeing if the chosen chip is orange.
(b) Picking numbers from a bingo machine.
(c) Guessing on a five-question multiple-choice quiz with four choices for each question, and seeing if you got each right or wrong.
(d) Rolling a die and getting a 3.

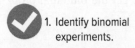

1. Identify binomial experiments.

Here's a more specific example of a binomial probability experiment. You're one of four finalists in a raffle, and each finalist is represented by a single colored ball thrown into a box. The colors are red, black, white, and green (you). A ball is picked from the box and its color is recorded, then it's mixed back into the box and a second ball is picked. If your green ball is picked twice, you win a 2-year lease on a Fiat convertible. If the first ball chosen isn't green, you don't care what color it is, so all you're interested in is whether the chosen ball is green or not. The tree diagram for this experiment is shown in Figure 10-8.

The probability of picking two green balls is $\frac{1}{16}$ since there is only one way to select two green balls, namely (G, G), and there are 16 total possible outcomes in the sample space. The probability of picking exactly one green is $\frac{6}{16}$ or $\frac{3}{8}$ since there are 6 outcomes that contain one green ball: (R, G),

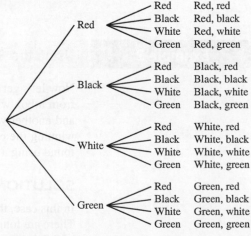

Figure 10-8

(B, G), (W, G), (G, R), (G, B), and (G, W). The probability of picking no green balls is $\frac{9}{16}$ since there are 9 outcomes that contain no green balls.

A table for the probabilities can be shown.

Outcome	Probability
0 green balls	$\frac{9}{16}$
1 green ball	$\frac{6}{16} = \frac{3}{8}$
2 green balls	$\frac{1}{16}$
Sum	$\frac{16}{16} = 1$

Math Note

There are actually four outcomes for any trial of the experiment described here (the four colors that can be picked), so from that standpoint this isn't a binomial experiment. But by focusing only on whether or not the drawn ball is green, we've reduced the number of outcomes to two (green or not green), making it a binomial experiment.

The experiment can be considered a binomial experiment since it meets the four conditions of a binomial experiment:

1. Each trial has only two outcomes: the ball chosen is either green or it isn't.

2. The outcomes are independent of each other: each ball is replaced before the second ball is selected.

3. There is a fixed number of trials. In this case, there are two draws.

4. The probability of a success remains the same in each case. In this case, since there is only one green ball and a total of four balls, $P(S) = \frac{1}{4}$ and $P(F) = 1 - \frac{1}{4} = \frac{3}{4}$. (Getting a green ball is considered a success. Getting any other colored ball is considered a failure.)

Now let's think about how we could get x successes in n trials of a binomial experiment. If p is the probability of success on any trial, the multiplication rule for independent events tells us that p^x is the probability of x successes. That leaves $n - x$ failures, and if q is the probability of failure, the probability of $n - x$ failures is q^{n-x}. We use the multiplication principle again to multiply those two probabilities. Finally, we have to choose *which* of the n trials result in success: there are $_nC_x$ ways to do this. Multiplying all the results, we get the binomial probability formula.

The Binomial Probability Formula

The probability of exactly x successes in n trials of a binomial probability experiment is

$$P(x) = {}_nC_x \cdot p^x \cdot q^{n-x}$$

where p = the probability of a success
q = the probability of a failure ($q = 1 - p$)

Example 2   Using the Binomial Probability Formula

Now let's get back to trying to win you that convertible. Remember, we were drawing a ball from a box with four balls, colored red, green, black, and white. The ball was then replaced and another was drawn. Use the binomial probability formula to find the probability of you winning the car (which means that the green ball is drawn twice). Does this match what we found using a tree diagram earlier? Why is that important?

SOLUTION

In this case, there are two trials, so $n = 2$. We're interested in two successes, so $x = 2$ as well. There are four balls and one is green, so the probability of success is $p = \frac{1}{4}$ and the probability of failure is $q = \frac{3}{4}$. Using the binomial probability formula, we get

$$_2C_2 \cdot \left(\frac{1}{4}\right)^2 \cdot \left(\frac{3}{4}\right)^{2-2} = 1 \cdot \frac{1}{16} \cdot 1 = \frac{1}{16}$$

This does match the probability we got using a tree diagram to write out the sample space. That's important for two reasons: first, it means we're right on this particular question. Better still, it provides some evidence that the formula we developed most likely works in general. It doesn't PROVE that it works, since we just did one example, but it's at least a confidence builder.

Try This One   2

If your color is picked once, no car for you! But you at least get 100 bucks and a bag of muffins, which is better than a sharp stick in the eye. What's the probability of that happening? Use the binomial probability formula, then check to see if your answer matches the probability computed using a tree diagram.

Example 3   Using the Binomial Probability Formula

Suppose that the morning after your birthday, you remember that you have a 20-question true or false quiz in your early class. Uh oh! Completely unprepared and a little woozy, you decide to guess on every question. What's the probability that you'll get 12 out of 20 right and just barely pass?

SOLUTION

Because the questions are true or false, and you're arbitrarily guessing, the probability of getting any given question right is $\frac{1}{2}$. The number of trials is $n = 20$, and the number of successes is $x = 12$. The probability of success is $p = \frac{1}{2}$, and the probability of failure is then $q = 1 - \frac{1}{2}$. Substituting these values into the binomial probability formula, we get

$$P(12 \text{ right}) = {}_nC_x \cdot p^x \cdot q^{n-x}$$

$$= {}_{20}C_{12} \cdot \left(\frac{1}{2}\right)^{12} \cdot \left(\frac{1}{2}\right)^{20-12}$$

$$= \frac{20!}{(20-12)!12!} \cdot \left(\frac{1}{2}\right)^{12} \cdot \left(\frac{1}{2}\right)^{8}$$

$$= 125{,}970 \cdot \frac{1}{4{,}096} \cdot \frac{1}{256} \approx 0.12$$

The probability is 0.12, so you have a 12% chance of that happening.

> **Math Note**
>
> Notice that because there are 20 trials in Example 3, it would be cumbersome (to say the very least!) to use a tree diagram to find the probability.

Try This One   3

If you take a 10-question multiple-choice quiz, with four choices for each question, and completely guess on every one, what's the probability of getting exactly 6 questions right? (Only one of the choices is the right answer.)

Example 4 comes with a slight twist: instead of looking for the probability of EXACTLY x successes in n trials, we'll be looking for the probability of x OR LESS successes.

Example 4   Using the Binomial Probability Formula

Of five physical therapists that work at a rehab center, three have master's degrees and two have doctorates. Each therapist is equally likely to be assigned to a patient on any given visit. If Elaine has five sessions scheduled in the next 2 weeks, find the probability that she gets a therapist with a doctorate fewer than two times.

Image Source/Alamy Stock Photo

SOLUTION

Fewer than two times means zero or one time, so we'll find the probability of each and then add the answers. (We can use the addition rule because having one therapist with a doctorate and having zero therapists with a doctorate are mutually exclusive events.)

$$P(1 \text{ doctorate}) = {}_5C_1 \cdot \left(\frac{2}{5}\right)^1 \left(\frac{3}{5}\right)^4$$

$$= \frac{5!}{(5-1)!1!} \cdot \frac{2}{5} \cdot \frac{81}{625} = 0.2592$$

$$P(0 \text{ doctorate}) = {}_5C_0 \cdot \left(\frac{2}{5}\right)^0 \left(\frac{3}{5}\right)^5$$

$$= \frac{5!}{(5-0)!0!} \cdot 1 \cdot \frac{243}{3,125} = 0.07776$$

$$P(1 \text{ or } 0 \text{ doctorate}) = 0.2592 + 0.07776 = 0.33696$$

Try This One   4

If a different patient at the rehab center in Example 4 has seven appointments scheduled, find the probability that they get a therapist with a master's degree at least five times.

Using Technology: Computing Binomial Probabilities

Graphing Calculator

You could certainly calculate directly using the combination command we already know multiplied by p^x and q^{n-x}. But graphing calculators have a built-in command for computing binomial probabilities directly. The **binompdf(*n*,*p*,*x*)** command can be found by pressing **2nd VARS**, which is the **DISTR** menu. Then scroll down to binompdf. (Don't confuse this with binomcdf, which we'll use later.) The screen shot shows how this works for finding the probability of guessing 12 out of 20 true/false questions correctly.

Notice that the inputs in parentheses correspond to the number of trials (*n*), then the probability of success (*p*), and finally the number of successes (*x*).

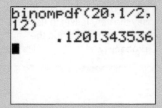

David Sobecki/McGraw Hill

Spreadsheet

The Excel command is =BINOM.DIST. When you start typing =BINOM.DIST in Excel, here's what appears, as a guide to help you:

= BINOM.DIST(number_s, trials, probability_s, cumulative), where

- number_s = number of successes (This is our *x*.)
- trials = number of trials (*n*)
- probability_s = the probability of success (*p*)
- cumulative = false; this tells Excel that you want the probability of EXACTLY *x* successes.

The sample at right also shows the calculation for guessing 12 of 20 true/false questions correctly.

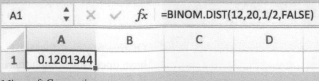

Microsoft Corporation

I'm guessing you weren't exactly blown away by the significance of Example 4. Is it really a big deal to find the probability of at most one success compared to exactly one success? It depends on the context. Think back to Example 3, when we found the probability of getting exactly 12 out of 20 questions right when guessing on a true/false test. Wouldn't it be far more reasonable to wonder about the probability of getting AT LEAST 12 right? In order to pull that off, we'd have to individually find the probability of getting 12 right, then 13, 14, 15, 16, 17, 18, 19, and 20, then add the results. Really? There has to be a better way!

Fortunately, technology can save the day. Graphing calculators and spreadsheets have functions that are specifically designed to find the probability of *x* successes OR LESS in a binomial experiment. This is known as a **cumulative probability**.

Using Technology

Graphing Calculator

The **binomcdf(*n,p,x*)** command can be found by pressing **2nd VARS**, which is the **DISTR** menu. Then scroll down to binomcdf. This calculates the probability of getting *x* successes or less in *n* trials. For a binomial experiment with five trials and a 2/5 probability of success on any trial, the screen shot shows how to find the probability of 1 or less successes.

Notice that the inputs in parentheses correspond to the number of trials (5), then the probability of success (2/5), and finally the number of successes or less that we're interested in (1).

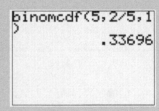

David Sobecki/McGraw Hill

Spreadsheet

To find a cumulative binomial probability, use the same command we already learned: =BINOM.DIST, followed by the number of successes, the number of trials, and the probability of success on each trial in parentheses. The only difference is that you finish the string in parentheses with TRUE rather than FALSE. This tells the formula to find the probability of *x* successes OR LESS.

A1		✕ ✓	f_x	=BINOM.DIST(1,5,2/5,TRUE)	
	A	**B**	**C**	**D**	
1	0.33696				

Microsoft Corporation

Example 5   Using a Cumulative Binomial Probability

For the test in Example 3, find the probability of passing, which would be getting 12 or more questions right.

SOLUTION

This doesn't exactly match the cumulative probability that we've learned how to compute. We're asked to find the probability of 12 or MORE successes, not 12 or less. So we'll have to be a little bit clever and use complements. The outcome of getting 12 or more questions right is the complement of getting 11 or less right, which we can find as a cumulative binomial probability. Then we'll subtract from 1 to get the probability that we want.

Using either binomcdf(20,1/2,11) on a calculator or =BINOM.DIST(11,20,1/2,TRUE), we find that the probability of getting 11 or less questions right is about 0.748. That's the probability of failing, so the probability of passing is $1 - 0.748 = 0.252$.

Try This One   5

For a 10-question multiple-choice quiz with five choices for each answer, use a cumulative binomial probability to find the probability of getting five or less right.

Use of the binomial probability formula requires a series of trials that are independent. As we saw in Section 10-8, when subjects are chosen from a sample and not replaced, the choices aren't independent, but if the sample is really large, the lack of replacement has a negligible effect on probability. So we can treat the choices as independent even though they technically aren't, which comes in handy when using the binomial probability formula.

| **Example 6** | **Using the Binomial Probability Formula** |

According to insidehighered.com, 9.2% of college undergrads don't have health insurance. If 10 undergrads are surveyed at random, find the probability that half of them will be uninsured.

SOLUTION

The population here is all college undergraduates, a group of at least 15 million individuals. When 10 undergrads are chosen, we can assume there's no replacement, but the large sample size allows us to treat the choices as independent. There are two outcomes for each trial: the student either has health insurance or not. So this is a binomial probability problem with $n = 10$, $x = 5$, $p = 0.092$, and $q = 1 - 0.092 = 0.908$.

$$P(5 \text{ uninsured}) = {}_{10}C_5 \cdot (0.092)^5 \cdot (0.908)^5$$

$$= \frac{10!}{(10-5)!5!}(0.0000066)(0.6172047) \approx 0.00103$$

Try This One   6

In a large community, it was determined that 44% of the residents use the public library at least once a year. If 10 people are picked randomly, find the probability that exactly 2 of them have used the library during the last year.

✓ 2. Compute probabilities of outcomes in a binomial experiment.

To describe all of the outcomes in a probability experiment, we can build a *probability distribution*.

A **probability distribution** consists of a list of all outcomes and the corresponding probabilities for a probability experiment.

| **Example 7** | **Constructing a Probability Distribution** |

Construct a probability distribution for the possible number of tails when you flip two coins.

SOLUTION

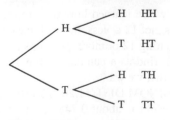

Figure 10-9

When two coins are flipped, the outcomes can be shown using a tree diagram. See Figure 10-9.
       The sample space is {HH, HT, TH, TT}. Each outcome has a probability of $\frac{1}{4}$. Notice that the outcome "No tails" is HH, and $P(HH) = \frac{1}{4}$. The outcome of one tail consists of HT and TH, and $P(HT, TH) = \frac{1}{4} + \frac{1}{4} = \frac{1}{2}$. The outcome of two tails is TT and $P(TT) = \frac{1}{4}$. Now a

Math Note

In Example 7, the number of trials was small enough that a tree diagram was more efficient than using the binomial probability formula three times.

probability distribution can be constructed by considering the outcomes to be the number of tails. The probability distribution is

Number of tails, x	0	1	2
Probability, $P(x)$	$\dfrac{1}{4}$	$\dfrac{1}{2}$	$\dfrac{1}{4}$

3. Construct a probability distribution.

Try This One   7

Three cards are numbered 1, 2, 3 and placed into a bag. Another bag is set up the same way, then one card is drawn from each bag and the numbers are added. Find all possible totals, and construct a probability distribution for the experiment.

Answers to Try This One

1   (a) Yes. One outcome (orange or not), fixed number of trials (10), and replacement means independent and constant probability.

    (b) No. Many possible outcomes.

    (c) Yes. Two outcomes (right or wrong), fixed number of trials (5), guessing on each makes choices independent and equally likely to be right or wrong.

    (d) Yes. Two outcomes (3 or not), fixed number of trials (1). With no other trials, that's all that matters.

2 $_2C_1 \cdot \left(\dfrac{1}{4}\right)^1 \cdot \left(\dfrac{3}{4}\right)^{2-1} = 2 \cdot \dfrac{1}{4} \cdot \dfrac{3}{4} = \dfrac{3}{8}$

3   About 0.016

4   About 0.420

5   0.994

6   About 0.084

7

Total	2	3	4	5	6
Probability	$\dfrac{1}{9}$	$\dfrac{2}{9}$	$\dfrac{1}{3}$	$\dfrac{2}{9}$	$\dfrac{1}{9}$

Exercise Set   10-9

Writing Exercises

1. What are the four requirements for a probability experiment to be a binomial experiment?
2. Give a brief description of how an experiment with several outcomes can be reduced to one with two outcomes.
3. What is a probability distribution?
4. In a binomial experiment, given the probability of any trial being a success, explain how to find the probability of a failure.
5. What is a cumulative binomial probability? What types of problems can it be used to solve?
6. Explain why cumulative binomial probabilities are often more useful than regular binomial probabilities.

Computational Exercises

Find the probability of each using the binary probability formula or technology.

7. $n = 3, p = 0.40, x = 2$

8. $n = 5, p = 0.80, x = 3$
9. $n = 10, p = 0.66, x = 4$
10. $n = 5, p = 0.20, x = 8$

11. $n = 20, p = 0.93, x = 12$
12. $n = 13, p = 0.16, x = 7$
13. $n = 9, p = 0.33, x = 5$
14. $n = 6, p = 0.58, x = 2$
15. $n = 4, p = 0.72, x = 0$
16. $n = 7, p = 0.25, x = 1$
17. $n = 12, p = 0.33, x = 3$ or 4
18. $n = 15, p = 0.51, x = 7$ or 10

19. $n = 9, p = 0.42, x$ is less than 3
20. $n = 16, p = 0.75, x$ is more than 13
21. $n = 50, p = 0.94, x$ is more than 40
22. $n = 35, p = 0.3, x$ is less than 6
23. $n = 25, p = 0.33, x$ is less than 16
24. $n = 18, p = 0.95, x$ is more than 16

Applications in Our World

25. During a game of Yahtzee, a player needs three sixes. She rolls the same die three times. What is the probability she got three sixes?

26. A cooler held three energy drinks. One was a Monster, another was a Red Bull, and the third was Atomic X. Simon picks an energy drink at random, puts it back, picks another, puts it back, and then picks a third drink. Find the probability that
    (a) No Red Bull drinks are picked.
    (b) Exactly one Atomic X drink is picked.
    (c) Exactly three Monster drinks are picked.

27. Of all identity thefts, 27% are solved in 1 day or less. If 10 cases are studied, find the probability that exactly 5 are solved in 1 day or less.

28. Of all people who do banking, 16% prefer to use ATMs. Of 20 people who are banking customers, find the probability that exactly 4 prefer to use the ATM.

29. According to the Bureau of Labor Statistics, among married-couple families where both the husband and wife have earnings, 26% of wives earned more than their husbands. If 18 wives are surveyed, find the probability that exactly 6 of them earn more than their husbands.

30. It is reported that 41% of people surveyed said that they eat meals while driving. If 12 drivers are picked at random, find the probability that exactly 6 will say that they eat meals while driving.

31. Approximately 45% of people who eat at fast-food places choose McDonald's. If 10 randomly selected people are surveyed, find the probability that 5 picked McDonald's.

32. About 30% of people who listen to commercial radio change the station in 1 to 2 minutes after the commercials begin. If six people are randomly selected, find the probability that two will change the station within 1 to 2 minutes after the commercials begin.

33. A survey shows that 65% of workers ages 42 to 60 would choose excellent retirement benefits over a high salary when seeking a job. If 15 people are randomly chosen, find the probability that 9 would choose excellent benefits over a high salary.

34. *Statista.com* magazine reported that at the end of 2021, 53.7% of smartphones in operation in the United States were iPhones. If this is still accurate, what's the probability that fewer than 3 of the next 10 smartphone users you see have iPhones?

35. The Bureau of Labor Statistics reported that in 2020, 63.1% of U.S. adults in the workforce were working full time. If a pollster called 20 working adults at random, what's the probability that more than 17 are working full time?

36. A Telenav survey found that one in five people surveyed would rather go shoeless than cell phone-less for a week. Find the probability that if 25 people are surveyed, between 40% and 60% of them would be more willing to give up their shoes than their phones.

37. A survey found that 55% of people said that a diet was harder to stick to than a budget. (Source: Kelton Research for Medifast.) If 22 people are randomly selected, find the probability that between 80% and 90% of them will say that a diet is harder to stick to than a budget.

38. A survey done by Harris found that 40% of the people surveyed said that their children will have between $5,001 and $20,000 in debt when they graduate from college. If 25 people are selected, find the probability that 14 or 15 will say that their children will have a debt between $5,001 and $20,000 when they graduate from college.

39. A survey done by Harris International QUICKQUERY found that 10% of women had a fear of flying. If 50 women are randomly selected, find the probability that fewer than 3 will have a fear of flying.

40. In a Harris survey, 40% of the people surveyed said that they received error messages from doing online transactions. If 18 people are randomly selected, find the probability that 5 or 6 people would have received error messages while doing online transactions.

41. A survey found that 33% of people earning between $30,000 and $75,000 said that they were very happy. If six people who earn between $30,000 and $75,000 are selected at random, find the probability that at most two would consider themselves very happy.

42. In a recent survey, 2% of the people surveyed said that they would keep their current job if they won a multi-million-dollar lottery. If 20 people are chosen randomly, find the probability that 3, 4, or 5 of them would keep their job.

43. If you take a true/false quiz with 10 questions and totally guess on every one, find the probability that you pass (get at least 60% right).

44. Repeat Problem 43 for a 10-question multiple-choice quiz where there are four choices on each question (and only one is correct).

Critical Thinking

In Exercises 45–53, decide if the experiment or question described is or is not a binomial experiment, then explain your reasoning. You don't need to find probabilities.

45. Keeping track of the number of customers in a week that get pepperoni on their pizza.

46. Recording the numerical results of rolling two dice 20 different times.

47. Playing 10 different people in golf and recording whether or not you win.

48. Drawing 12 different cards from a standard deck with replacement and recording whether or not each is a spade.

49. Surveying 20 different homeowners in your state to find out if they have life insurance.

50. Playing 25 rounds of rock, paper, scissors against the same person and recording the result. (If you're not familiar with the game, you can find the rules online very easily.)

51. According to the U.S. Department of Health and Human Services, just 3.9% of men in America are taller than 6'2". What is the probability that three out of five men chosen in your class are 6'3" or taller?

52. The Bureau of Labor Statistics reported that 7.4% of work-eligible residents in the Las Vegas metro area were unemployed in September 2021. If 30 people from that area were chosen at random, what was the probability that 5 were unemployed?

53. There are 27 people in my calc 2 class right now, and 22 of them are passing. If I pick 6 at random, what is the probability that all 6 are passing?

54. (a) Write an example of a binomial experiment in real life in which the conditions of a binomial experiment are all met and lack of replacement is not an issue.

    (b) Write an example of a binomial experiment in real life without replacement, in which case we would need a large sample to use the binomial probability formula.

Section	Important Terms	Important Ideas
10-1	Probability experiment Outcome Random Sample space Event Complement Theoretical probability Empirical probability Observed frequency Frequency distribution	**Flipping coins,** drawing cards from a deck, and rolling a die are examples of probability experiments. The set of all possible outcomes of a probability experiment is called the sample space. The two types of probability are theoretical and empirical. Theoretical probability uses sample spaces and is based on the assumption that all outcomes in the sample space are equally likely. Empirical probability uses frequency distributions and is based on observation. Probability is a number that represents how likely it is that something will occur. Probability can be zero, one, or any number in between. Probabilities closer to one mean events are more likely to occur. Probabilities closer to zero mean events are less likely to occur.
10-2	Fundamental counting principle Factorial notation Permutation Permutation rule	**Since probability** formulas completely rely on finding the number of outcomes in an event or a sample space, being able to find those numbers is a key skill. That's the point of counting techniques. 　　In order to find the total number of outcomes for a sequence of events, the fundamental counting principle or the permutation rules can be used. When the order or arrangement of the objects in a sequence of events is important, then the result is called a permutation of the objects.
10-3	Combination Combination rule	**When the** order of the objects is not important, then the result is called a combination. In this case, the combination rules can be used to count the number of possible combinations.
10-4	Tree diagram	**When sample** spaces are difficult to identify, it's often helpful to use tree diagrams or tables to identify all possible outcomes for a probability experiment.
10-5		For events and sample spaces that are more complicated, probabilities can be found by using the fundamental counting principle, the permutation rule, or the combination rule, depending on the situation. In some cases, more than one of those rules may be needed to calculate a probability.
10-6	Odds in favor Odds against Expectation (expected value)	**In order** to determine payoffs, gambling establishments give odds. There are two ways to compute odds for a game of chance: "odds in favor of an event" and "odds against the event." Odds are a ratio that compares the number of favorable and unfavorable outcomes. Probabilities, on the other hand, are ratios that compare the number of favorable outcomes to the TOTAL number of outcomes. 　　When a probability experiment has numerical outcomes, we can find the expected value, which is a long-run average of the outcomes if the experiment is repeated over and over. We find the expected value by multiplying each outcome by its probability, then adding the results.
10-7	Mutually exclusive events Addition rules	**Two events** are said to be mutually exclusive if they can't occur at the same time. If two events are mutually exclusive, the probability of one or the other occurring is the sum of the probabilities for each event. If the events are not mutually exclusive, the probability of one or the other occurring is the sum of the probabilities minus the probability that both occur.

Section	Important Terms	Important Ideas
10-8	Independent events Dependent events Multiplication rules Conditional probability	**Events can** be classified as independent or dependent. Events are said to be independent if the occurrence of the first event does not affect the probability of the occurrence of the next event. If the probability of the second event occurring is changed by the occurrence of the first event, then the events are dependent. If two events are independent, the probability of both occurring is the product of the probabilities for each event. If the probability of an event *B* occurring is affected by an event *A* occurring, then we say that a condition has been imposed on the event and the probability of event *B* occurring given that *A* has occurred is called a conditional probability. If two events are dependent, the probability of both occurring is the probability that the first occurs multiplied by the probability that the second occurs given the first. A convenient rule of thumb is that "or" goes along with addition, while "and" goes along with multiplication.
10-9	Binomial experiment Probability distribution Cumulative probability	**Many probability** experiments have two outcomes or can be reduced to two outcomes. If the trials are independent, fixed in number, and have the same probability of a success, then the experiment can be considered a binomial experiment. These problems can be solved using the binomial probability formula. For some probability experiments, a probability distribution can be constructed, listing all outcomes with the corresponding probabilities.

Math in Gambling REVISITED

Steve Allen/Stockbyte/Getty Images

1. The odds against winning the Powerball jackpot are 292 million to 1. The probability of getting struck by lightning is 1/600,000. You are WAY WAY WAY more likely to get struck by lightning. By the way, think about 292 million to one for a second. Your chance of winning Powerball with one ticket is only slightly better than your chance of having your name chosen out of a hat (a really big hat) containing the names of *every human being in the United States*.

2. The probability of getting killed in an accident if you text and drive varies by year, but estimates are that you're AT LEAST four times as likely to get killed on the road if you text. Not very smart, is it?

3. (a) Using the table for rolling two dice on page 589, the probability of rolling 2, 3, 11, or 12 is $\frac{1}{6}$, meaning the probability of losing is $\frac{5}{6}$. So the two outcomes are +$4 with probability $\frac{1}{6}$, and −$1 with probability $\frac{5}{6}$. The expected value of each trial is then

$$+4 \cdot \frac{1}{6} + (-1) \cdot \frac{5}{6} = -\frac{1}{6}$$

In 100 trials, you would expect to lose $100($\frac{1}{6}$), or $16.67.

(b) The probability of winning $35 is $\frac{1}{38}$, and the probability of losing $1 is $\frac{37}{38}$. The expected value of each spin is

$$+35 \cdot \frac{1}{38} + (-1) \cdot \frac{37}{38} = -\frac{2}{38} = -\frac{1}{19}$$

In 100 trials, you would expect to lose $100($\frac{1}{19}$) or $5.26.

(c) Using the same idea as scenarios 1 and 2, the expected value of each ticket is

$$+20,000,000 \cdot \frac{1}{175,711,536} + (-1) \cdot \frac{175,711,535}{175,711,536}$$
$$\approx -\$0.8862$$

Buying 100 tickets, you would expect to lose $100(−0.8862), or $88.62.

(d) You would win $1 with probability $\frac{1}{2}$, and lose $1 with probability $\frac{1}{2}$, so the expected value of each flip is

$$1 \cdot \frac{1}{2} + (-1) \cdot \frac{1}{2} = 0$$

In 100 flips you would expect to break even.

The best, by far, is flipping a coin, followed by roulette, the church fair dice game, and, bringing up the rear by quite a bit, the multistate lottery. Notice that under the best of circumstances, you are likely to break even!

Review Exercises

Section 10-1

1. Which of the following numbers could represent a probability?
   (a) $\frac{3}{2}$
   (b) $\frac{2}{3}$
   (c) 0.1
   (d) $-\frac{1}{2}$
   (e) 80%

2. When a die is rolled, find the probability of getting
   (a) A 5.
   (b) A 6.
   (c) A number less than 5.

3. When a card is drawn from a deck, find the probability of getting
   (a) A heart.
   (b) A 7 and a club.
   (c) A 7 or a club.
   (d) A jack.
   (e) A black card.

4. In a survey conducted at the food court in a local mall, 20 people preferred Panda Express for lunch, 16 preferred Sbarro Pizza, and 9 preferred Subway. If a shopper is chosen at random, find the probability that they prefer Sbarro Pizza.

5. If a die is rolled one time, find these probabilities:
   (a) Getting a 7.
   (b) Getting an odd number.
   (c) Getting a number less than 3.

6. In a recent survey in a college dorm that has 1,500 rooms, 450 have Play Station 2. If a room in this dorm is randomly selected, find the probability that it has Play Station 2.

7. During a Midnight Madness sale at Old Navy, 16 white cargo pants, 3 khaki cargo pants, 9 tan cargo pants, and 7 black cargo pants were sold. If a customer who made a purchase during the sale is surveyed at random, find the probability that they bought
   (a) A pair of tan cargo pants.
   (b) A pair of black cargo pants or a pair of white cargo pants.
   (c) A pair of khaki, tan, or black cargo pants.
   (d) A pair of cargo pants that was not white.

8. An urban art gallery runs an annual art competition. Among this year's finalists, there were 16 paintings, 4 metal sculptures, 3 kinetic sculptures, and 7 etchings. If the gallery decides to totally ignore ethics and chooses the winner randomly, find the probability that it's
   (a) A kinetic sculpture.
   (b) A metal sculpture or an etching.
   (c) Not a sculpture.
   (d) Not an etching.

9. When two dice are rolled, find the probability of getting
   (a) A sum of 5 or 6.
   (b) A sum greater than 9.
   (c) A sum less than 4 or greater than 9.
   (d) A sum that is divisible by 4.
   (e) A sum of 14.
   (f) A sum less than 13.

10. Two dice are rolled. Find the probability of getting a sum of 8 if the number on one die is a 5.

Section 10-2

11. Compute $\frac{14!}{11!}$.
12. Compute $_{12}P_6$.
13. A license plate consists of three letters followed by four digits. How many different plates can be made if repetitions are allowed? If repetitions are allowed in the letters but not in the digits?
14. How many different arrangements of the letters in the word *bread* are there?
15. How many different arrangements of the letters in the word *cheese* are there?
16. How many different three-digit odd numbers use only the digits 0, 1, 2, 3, 4?

Section 10-3

17. Compute $_9C_6$.
18. Find both the number of combinations and the number of permutations of 10 objects taken 4 at a time.
19. Describe the difference between combinations and permutations.
20. How many different three-digit combinations can be made by using the numbers 1, 3, 5, 7, and 9 without repetitions if the "right" combination can open a safe? Does a combination lock really use combinations?
21. How many two-card pairs (i.e., the same rank) are there in a standard deck?
22. How many ways can 5 different television programs be selected from 12 programs?
23. A quiz consists of six multiple-choice questions. Each question has three possible answer choices. How many different answer keys can be made?
24. How many different ways can a buyer pick four television models from a possible choice of six models?

Section 10-4

25. A gambler rolls an eight-sided die and then flips a coin. Draw a tree diagram and find the sample space.
26. A student can schedule one of three courses at 8:00 A.M.: English, math, or chemistry. The student can schedule either psychology or sociology at 11:00 A.M. Finally, the student can schedule either world history or economics at 1:00 P.M. (a) Draw a tree diagram and find all the different ways the student can make a schedule. (b) Repeat part (a), but include the condition that the student will take classes only at 8 A.M. and 11 A.M. if their first class is chemistry.
27. As an experiment in probability, a two-question multiple-choice quiz is given at the beginning of class, but the answers are all written in Hebrew, which none of the students can read. This forces everyone to guess. Each question has choices A, B, C, D, and E. Construct a table that displays the sample space, then use the table to find the probability that both questions in a randomly selected quiz were answered with D or E (either D-E, or E-D).

Section 10-5

28. A card is drawn from a deck. Find the probability that it is a diamond given that it is a red card.

29. An investor has six bond accounts, three stock accounts, and two mutual fund accounts. If three investments are chosen at random, find the probability that one of each type of account is selected.

30. A newspaper advertises five different movies, three plays, and two baseball games. If a couple picks three activities at random, find the probability they will attend two plays and one movie.

31. In putting together the music lineup for an outdoor spring festival, Fast Eddie can choose from 4 student bands and 12 nonstudent bands. There are five time slots for bands; the first at 3 P.M., the others on the hour until 7 P.M. If Eddie chooses the bands randomly, find the probability that no student bands will be picked.

Section 10-6

32. Find the odds in favor of an event E when $P(E) = \frac{1}{4}$.
33. Find the odds against an event E when $P(E) = \frac{5}{6}$.
34. Find the probability of an event when the odds in favor of the event are 6:4.
35. The table lists five outcomes for a probability experiment with the corresponding probabilities. Find the expected value.

Outcome	5	10	15	20	25
Probability	0.5	0.2	0.1	0.1	0.1

36. After being picked from the audience on *Let's Make a Deal,* Marlena gets to pick one of five envelopes. Each envelope has a single bill in it: $1, $10, $20, $100, or $500. Find the expected value of the game.

37. You bet $10 and get to pick one card. If it's red, you get back $5. If it's a black card with a number on it, you get back $15. A black face card gets you $20 and a black ace $30. Find the expected value of making this bet 25 times.

Section 10-7

In Exercises 38–40, decide if the two events are mutually exclusive.

38. You meet someone while out; they give you their phone number or their email address.
39. You complete a course and either pass or fail.
40. You spend a weekend in Las Vegas; you either win money, lose money, or break even.
41. If one of the 50 states is selected at random to be the site of a new nuclear power plant,
    (a) Find the probability that the state either borders Canada or Mexico.
    (b) Find the probability that the state either begins with A or ends with S.
42. There are six patients waiting at a free clinic with pain issues, three with rashes, four with fevers, and two with irregular heartbeats. If the next patient is randomly called, find the probability that they have a fever or a rash.

Section 10-8

In Exercises 43–46, decide if the two events are independent.

43. Missing 3 straight days of class and failing the next test.

44. Missing 3 straight days of class and getting overloaded with spam emails.
45. Drawing an ace from a standard deck, then drawing a second ace.
46. Drawing an ace from a standard deck, then replacing that card, shuffling, and drawing another ace.
47. In a family of three children, find the probability that all the children will be girls if it is known that at least one of the children is a girl.
48. A Gallup Poll found that 78% of Americans worry about the quality and healthfulness of their diet. If five people are selected at random, find the probability that all five worry about the quality and healthfulness of their diet.
49. Twenty-five percent of the engineering graduates of a university received a starting salary of $70,000 or more. If three of the graduates are chosen at random, find the probability that all have a starting salary of $70,000 or more.
50. Three cards are drawn from an ordinary deck *without* replacement. Find the probability of getting
    (a) All black cards.
    (b) All spades.
    (c) All queens.
51. A coin is flipped and a card is drawn from a deck. Find the probability of getting
    (a) A head and a 6.
    (b) A tail and a red card.
    (c) A head and a club.
52. The results in the table summarize responses to a survey question reported by the Pew Research Center on personal level of concern about climate change.

	Generation Z	Millennial	Generation X	Baby Boomer
A top concern	337	1,043	950	1,761
One of several important concerns	365	1,327	1,407	2,126
Not an important concern	201	758	1,126	2,187

(a) Find the probability that a randomly selected respondent feels that climate change is not an important concern given that they are a Baby Boomer.
(b) Find the probability that a randomly selected respondent feels that climate change is a top concern given that they are neither a Millennial or a Gen Xer.
(c) Find the probability that a randomly selected respondent is a member of Gen Z given that they feel that climate change is at least an important concern.
(d) Find the probability that a randomly selected respondent is not a Millennial given that they think climate change is not an important concern.

Section 10-9

53. Use the binomial probability formula to find the probability of five successes in six trials when the probability of success on each trial is 1/3.

54. A survey found that 24% of families eat at home as a family five times a week. If 10 families are surveyed, find the probability that exactly 3 will say that they eat at home as a family five times a week.

55. According to a survey, 45% of teenagers said that they have seen passengers in an automobile encouraging the driver to speed. If 16 teens are surveyed, find the probability that exactly 6 will say that they have seen passengers encouraging the driver to speed.

56. Construct a probability distribution for the possible number of heads when tossing a coin three times.

Chapter Test

In Exercises 1–3, compute the requested value.

1. $_7C_5$
2. $_{12}P_5$
3. The probability of 8 successes in 10 trials when the probability of success on each trial is 1/4.
4. If someone saw the title of Chapter 10 in this book and asked you "What is probability?" what would you say?
5. Describe a situation where combinations would be used to count possibilities and one where permutations would be used.
6. One company's ID cards consist of five letters followed by two digits. How many cards can be made if repetitions are allowed? If repetitions are not allowed?
7. How many ways can five sopranos and four altos be chosen for a university chorus from seven sopranos and nine altos?
8. When a card is drawn from a deck, find the probability of getting
   (a) A diamond.
   (b) A 5 or a heart.
   (c) A 5 and a heart.
   (d) A king.
   (e) A red card.
9. A trooper has written 12 citations for speeding this week, along with 8 for driving under the influence, 4 for reckless driving, and 7 for failure to wear a seat belt. Find the probability that the trooper writes a citation for
   (a) Reckless driving.
   (b) DUI or speeding.
   (c) Failure to wear a seat belt given that the citation isn't for speeding or reckless driving.
10. When two dice are rolled, find the probability of getting
    (a) A sum of 6 or 7.
    (b) A sum greater than 3 or greater than 8.
    (c) A sum less than 3 or greater than 8.
    (d) A sum that is divisible by 3.
    (e) A sum of 16.
    (f) A sum less than 11.
11. There are six cards numbered 1, 2, 3, 4, 5, and 6. A contestant flips a coin. If it lands heads up, they will draw a card with an odd number. If it lands tails up, they will draw a card with an even number. Draw a tree diagram and find the sample space.
12. Of the physics graduates of a university, 30% received a starting salary of $60,000 or more. If five of the graduates are chosen at random, find the probability that all had a starting salary of $60,000 or more.
13. Five cards are drawn from an ordinary deck *without* replacement. Find the probability of getting
    (a) All red cards.
    (b) All diamonds.
    (c) All aces.
14. Four coins are tossed. Find the probability of getting four heads given that two of the four coins landed heads up.
15. A coin is tossed and a die is rolled. Find the probability of getting a head on the coin if it is known that the number on the die is even.
16. Nurses at one hospital can be classified according to sex (male, female), income (low, medium, high), and rank (staff nurse, charge nurse, head nurse). Draw a tree diagram and show all possible outcomes.
17. Find the odds in favor of and odds against an event E when $P(E) = \frac{3}{8}$.
18. Find the probability of an event when the odds against the event are 3:7.
19. There are six cards placed face down in a box. Each card has a number written on it. One is a 4, one is a 5, one is a 2, one is a 10, one is a 3, and one is a 7. You pick a card. Find the expected value of the draw.
20. A gambler draws a card from an ordinary deck of cards. If it is a black card, she wins $2. If it is a red card between or including 3 and 7, she wins $10. If it is a red face card, she wins $25, and if it is a black jack, she wins an additional $100. If it is any other card, she wins nothing. Find the expectation of the game. (Careful! This is tricky.)
21. In a soda machine in the student union, there are five Diet Cokes, four Mountain Dews, and two Dr. Peppers. If a student picks three sodas at random, find the probability that the student will get one Diet Coke, one Mountain Dew, and one Dr. Pepper.
22. At Sally's orientation, there were six computer science majors, four electrical engineering majors, and three architecture majors in her group. If four students are selected at random to receive a free tote bag with the school's logo, find the probability that the selection will include two computer science majors, one electrical engineering major, and one architecture major.
23. The results of a survey revealed that 30% of the people surveyed said that they would buy home electronic equipment at post-holiday sales. If 20 people are selected, find the probability that 8 or 9 will purchase home electronic equipment after the holidays.

Projects

1. Make a set of three cards—one with the word "heads" on both sides, one with the word "tails" on both sides, and one with "heads" on one side and "tails" on the other side. With a partner, play the game described in Exercise 61 of Section 10-6 (page 607) 100 times and record how many times your partner wins. (*Note:* Do not change options during the 100 trials.)
   (a) Do you think the game is fair (i.e., does one person win approximately 50% of the time)?
   (b) If you think the game is unfair, explain what the probabilities might be and why.

2. Take a coin and tape a small weight (e.g., part of a paper clip) to one side. Flip the coin 100 times and record the results. Do you think you have changed the probabilities of the results of flipping the coin?

3. This game is called "Diet Fractions." Roll two dice and use the numbers to make a fraction less than or equal to one. Player A wins if the fraction cannot be reduced; otherwise, player B wins.
   (a) Play the game 100 times and record the results.
   (b) Decide if the game is fair or not. Explain why or why not.
   (c) Using the sample space for two dice, compute the probabilities of player A winning and player B winning. Do these agree with the results obtained in part (a)?

   Source: George W. Bright, John G. Harvey, and Margariete Montaque Wheeler, "Fair Games, Unfair Games." Chapter 8, *Teaching Statistics and Probability. NCTM 1981 Yearbook.* Reston, Virginia: The National Council of Teachers of Mathematics, Inc., 1981, p. 49.

4. Remember looking through cereal boxes for toys when you were a kid? It always seemed like you didn't get the exact one you wanted. Let's say that there was a certain toy you wanted, and five others that you could take or leave, all packed one per box at random. About how many boxes would you expect to have to buy to get the toy you wanted? Of course, you might have gotten it in the first box. Or you might have exhausted mom and dad's savings without ever getting it. These are the extremes.
   (a) You can simulate this experience using a single die, and rolling until a particular number of your choice comes up. Keep track of how many rolls it took, then repeat 99 more times, and find the average number of times it took.
   (b) If there were 10 different choices, you could simulate that by using the ace through 10 of a certain suit. Pick a certain card, shuffle the deck, and start dealing out the cards, keeping track of how many it takes to get the one you picked. Repeat 99 times and find the average.
   (c) Summarize your findings for both experiments.
   (d) Call getting the number or card you wanted a value of 2, and not getting it a value of 1. Then find the probability of each on any given roll or draw, and find the expected value for each experiment. How does it compare to the experimental results?

5. When devising the payouts for a gambling game, there are two things you have to balance: you want to make sure that players have enough chance to win that they want to keep playing, but you also want to make sure that they lose enough so that you make enough money for it to be worth your effort.

   In this project, we'll be using probability and expected value to decide how much to pay out for certain events. The players will roll three regular dice. First, decide on how much you will charge to play the game one time. Find the probability and odds against for each event listed, then decide how much you will pay out for each. (Of course, you have to pay nothing for some of the events or you'll never make any money.) After setting the payouts for each event, compute the expected value of playing your game. (Don't forget to include the amount paid to play the game in the first place!) In order for you to make money, the expected value should be negative for the players. In order for players to want to play, the expected value should be no more than 20% of the amount it costs to play the game. If that's not the case, adjust your payouts until the expected value falls within that range.

   List of Events
   Sum of the three dice: 3, 4, 5, 6, 7, 8, 9, 10, 11, 12, 13, 14, 15, 16, 17, 18
   Special rolls: triples (all three dice the same), singles (all three dice different), odds (all three odd), evens (all three even), straight (three consecutive numbers), dozen (product of all three is 12).

Design elements: Front matter, Chapter Opener, Summary and End Matter header design (random numbers background illustration): ©pixeldreams.eu/Shutterstock RF

Statistics

Photo: Helen H. Richardson/The Denver Post/Getty Images

Outline

Math in Sociology

Broadly defined, sociology is the study of human behavior within society. One important branch is criminology. This is not about investigating crimes, but rather studying patterns of criminal behavior and their effect on society. One of the main tools used by sociologists is statistics. This important area of math allows researchers to study patterns of behavior objectively, by analyzing information gathered from a mathematical perspective, not a subjective one.

It all begins with the gathering of data. Data are measurements or observations that are gathered for a study of some sort. Let's look at an example. There's an old adage in police work that violent crime increases with the temperature. A statement like "My cousin Jed is a cop, and he told me that there are more violent crimes during the summer" is not evidence, nor is it data. It's the subjective opinion of one guy. In order to study whether this phenomenon is legitimate, we would need to gather information about the number of violent crimes at different times during the year, as well as temperature information, and study those numbers objectively to see if there appears to be a connection. This is one of the most important topics in this chapter.

Data on its own isn't good for very much unless we develop methods for organizing, studying, and displaying that data. These are the methods that make up the bulk of this chapter. After discussing methods of gathering data, we'll learn effective methods for organizing and displaying data so that it can be presented in a meaningful, understandable way. We will then turn our attention to techniques of analyzing data that will help us to unlock the secrets of what sets of data may be trying to tell us.

As for math in sociology, the table shows the number of homicides in each month of 2021 in Chicago, along with the average high temperature for that month.

Month	Avg. high temp.	Homicides
1	32	54
2	37	36
3	51	44
4	57	54

Month	Avg. high temp.	Homicides
5	70	67
6	82	86
7	84	108
8	84	80
9	78	94
10	66	63
11	55	52
12	31	60

Source: *Chicago Tribune*

After finishing the chapter, you'll be able to analyze the data and answer the following questions.

1. A histogram is a graph similar to a bar graph that we'll use in this chapter. Draw a histogram for the data. Put the number of the month on the horizontal axis. What patterns do you notice? Does the data set appear to be normally distributed? Discuss.

2. There are four different numbers that might be considered an "average" for the number of homicides: mean, median, mode, and midrange. Find each. Which measure of average do you think is most revealing? Explain.

3. Draw a scatter plot for the data set. This is a graph of points of the form (average high temperature, homicides). Does there appear to be a relationship between temperature and homicides? Explain.

4. Can you conclude, using a measure called the correlation coefficient, with 95% certainty that there is in fact a relationship between temperature and homicides? What about with 99% certainty?

5. Do the results indicate that warmer temperatures cause people to behave in a more violent manner? Discuss. A one-sentence answer won't cut it here. This is a really important question.

For answers, see Math in Sociology Revisited on page 735

| **Gathering and Organizing Data**

LEARNING OBJECTIVES

1. Define data and statistics.
2. Explain the difference between a population and a sample.
3. Describe four basic methods of sampling.
4. Contrast descriptive and inferential statistics.
5. Construct a frequency distribution for a data set.
6. Draw a stem and leaf plot for a data set.

One of the big issues being discussed over the last 20 years or so in higher education is grade inflation—the perception that college students in general are getting much higher grades than they used to. But perception and reality are often two different things, so how could we decide if this is actually taking place? And if so, what are some possible reasons? Are students just getting smarter? Are professors lowering their standards?

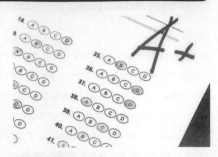

ALEX S/Shutterstock

In order to examine questions like this, it would be very valuable to gather some *data*.

Data are measurements or observations that are gathered for an event under study.

Gathering data plays an important role in many different areas. In college sports, recruiters look at past performance to help them decide which high school players will be stars in college. In public health, administrators keep track of the number of residents in a certain area who contract a new strain of the flu. In education, researchers might try to determine if a new method of teaching is an improvement over traditional methods. Media outlets report the results of Nielsen, Harris, and Gallup polls. All of these studies begin with collecting data.

Once data have been collected, in order to get anything of value from them, the data need to be organized, summarized, and presented in a form that allows observers to draw conclusions. This is what the study of statistics is all about:

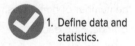

1. Define data and statistics.

Statistics is the branch of mathematics that involves collecting, organizing, summarizing, and presenting data and drawing general conclusions from the data.

Populations and Samples

In the case of grade inflation, relevant data might be average grade point averages for students in several different years. Table 11-1 shows some GPA data from the website gradeinflation.com.

Table 11-1	Average GPAs from 1983 to 2013 at Some American Universities

Year	Average Undergraduate GPA
1983	2.84
1988	2.88
1993	2.95
1998	3.01
2003	3.08
2008	3.11
2013	3.16

Data from gradeinflation.com

The numbers certainly appear to indicate that grade inflation is real. But where did the data come from? Were grades from *all* colleges included, or just some of them?

When statistical studies are performed, we usually begin by identifying the *population* for the study.

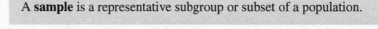

A **population** consists of all subjects under study.

In our example, the population is all colleges in the United States. If you were only interested in study habits of the students at your school, the population would be all of the students at the school. More often than not, it's not realistic to gather data from every member of a population. In that case, a smaller representative group of the population would be selected. This group is called a *sample*.

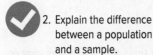
2. Explain the difference between a population and a sample.

A **sample** is a representative subgroup or subset of a population.

The GPA data in Table 11-1 were gathered from a sample of 75 schools. The reason the data are compelling is that the sample is *representative* of the population as a whole. The schools chosen vary in size, cost, and geographic location. Some are public, and some are private. In order to ensure that a sample is representative, researchers use a variety of methods.

Sampling Methods

We will study four basic sampling methods that can be used to obtain a representative sample: random, systematic, stratified, and cluster.

In order to obtain a **random sample**, each subject of the population must have an equal chance of being selected. The best way to obtain a random sample is to use a list of random numbers. Random numbers can be obtained from a table or from a computer or calculator. Subjects in the population are numbered, and then they are selected by using the corresponding random numbers.

Using a random number generator is like selecting numbers out of a hat. The difference is that when random numbers are generated by a calculator, computer, or table, there is a better chance that every number has an equally likely chance of being selected. When numbers are placed in a hat and mixed, you can never be sure that they are thoroughly mixed so that each number has an equal chance of being chosen.

A **systematic sample** is taken by numbering each member of the population and then selecting every *k*th member, where *k* is a natural number. For example, the researcher from gradeinflation.com might have numbered all of the colleges willing to release grade information and chosen every 10th one. When using systematic sampling, it's important that the starting number is selected at random.

When a population is divided into groups where the members of each group have similar characteristics (like large public schools, large private schools, small public schools, and small private schools) and members from each group are chosen at random, the result is called a **stratified sample**. The grade inflation researcher might have decided to choose five schools from each of those groups. Since 75 is a relatively small sample (there are about 5,300 colleges in the U.S.), it's possible that 40 or more came from large public schools. This would jeopardize the study because it may be the case that grade inflation is more or less likely at large public schools than schools in the other categories. So the purpose of a stratified sample is to ensure that all groups will be properly represented.

When an existing group of subjects that represent the population is used for a sample, it is called a **cluster sample**. For example, an inspector may select at random one carton of calculators from a large shipment and examine each one to determine how many are defective. The group in this carton represents a cluster. In this case, the researcher assumes that the calculators in the carton represent the population of all calculators manufactured by the company.

Samples are used in the majority of statistical studies, and if they are selected properly, the results of a study can be generalized to the population.

Mike Lockhart/U.S. Fish & Wildlife Service

Wildlife biologists capture and tag animals to study how they live in the wild. The population is all the animals of the species in this region. The tagged animals constitute the sample. If the sample is not representative of the species the data could be misleading.

Example 1	Choosing a Sample

Math Note

In the supplement, "Misuses of Statistics," at the end of this chapter, we'll study some ways that samples can be poorly chosen, either intentionally or not, which can invalidate the conclusions of a study.

A student in an education class is given an assignment to find out how late typical students at his campus stay up to study. He considered four different methods for gathering data. Describe what type of sampling method he's using in each case. Then describe whether you think he's likely to get a representative sample.

(a) Stop by the union before his 9 A.M. class and ask everyone sitting at a table how late they were up studying the night before.
(b) Ask the school registrar to give him a list of 50 students whose student ID numbers end in 4.
(c) Survey the five people sitting closest to the door in each of his five classes.
(d) Look at the campus directory and choose the first name at the top of each page.

SOLUTION

(a) Since he is choosing all students in a particular place at a particular time, this is a cluster sample. The sample is unlikely to be representative. Since he's polling people early in the morning, those that tend to stay up very late studying are less likely to be included in the sample. (Would you be at the union at 8:30 if you were up until 4 A.M. studying? Didn't think so.)
(b) This is a random sample: you'd think that the last digit of an ID number is likely to be completely random. How representative it is might depend on how big the campus is: if there are 500 students on the campus, this would probably be pretty good. If there were 50,000 students, I'd be a lot less comfortable with it.
(c) This is a stratified sample, since he's randomly picking subjects from existing groups (the classes). It sounds like a pretty good approach, depending on what kinds of classes he's taking. If they're all senior-level courses that require a lot of studying, that would probably bias the results toward people staying up later.
(d) This is a classic example of a systematic sample, and would almost certainly be a good representative sample.

Try This One   1

A marketing firm is hired to study the occupants of a medium-sized town in Ohio to see if a chain restaurant/brewery would be popular there. The company considered three different methods for gathering data. Describe what type of sampling method they're using in each case. Then describe whether you think they're likely to get a representative sample.

(a) Pass out surveys at each of the churches in town.
(b) Mail a survey to four houses on each street in town.
(c) Pay workers to conduct in-person surveys at the annual town festival.

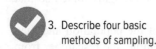 3. Describe four basic methods of sampling.

Descriptive and Inferential Statistics

There are two main branches of statistics: descriptive and inferential. Statistical techniques used to *describe* data are called **descriptive statistics**. This is based on collecting, organizing, and reporting data without using the data to draw any wide-ranging conclusions. For example, a researcher might be interested in the average age of full-time students on your campus and how many credit hours they're scheduled for this term. In that case, they could gather that information for every student, in which case the averages would be describing a definite situation. Descriptive statistics often involves gathering information about every individual in a population.

Statistical techniques used to make *inferences* are called **inferential statistics**. This is based on studying characteristics of a sample within a larger population and using them to draw conclusions about the entire population. For example, the Bureau of Labor Statistics estimates the number of people in the United States that are unemployed every month. There's just no way they can survey every person in the workforce every month, so the bureau picks a

Sidelight   Math in the Courtroom

Mathematicians are often called on to testify in court, and often it's to interpret the statistical likelihood of some event. In one case in Los Angeles, a couple was convicted of robbery based almost entirely on statistics! There were witnesses who saw the robbers leaving the scene, but they could not positively identify the suspects: just certain characteristics, like race, hair color, facial hair, and type of vehicle. A mathematician calculated that there was just one chance in 12 million of a second couple with the exact characteristics as the suspects being in the same area at that time. Despite a lack of any hard evidence, the couple was convicted.

When the conviction was appealed, the appellate judge was somewhat less impressed by the statistical argument than the jury had been. The conviction was overturned based on faulty calculations and the lack of other evidence.

Way back in Chapter 1, we discussed the fact that inductive reasoning can be used in criminal investigation to identify suspects and make it likely that the right person was arrested, but evidence gathered from deductive reasoning is far more effective in actually proving that someone is guilty. In this case, conclusions based on inferential statistics constituted the bulk of the evidence in the case, and the lack of any evidence based on deductive reasoning was enough to

Image Source Plus/Alamy Stock Photo

convince the judge that the suspects weren't guilty beyond a reasonable doubt. Inferential statistics are used to draw many useful and important conclusions, but when someone's freedom is at stake, our system dictates that inferring guilt simply isn't enough.

Janis Christie/Getty Images

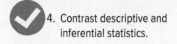

4. Contrast descriptive and inferential statistics.

sample of adults to see what percentage is unemployed. Then they use that information to estimate the unemployment rate for the entire population.

There's a connection between the mathematical reasoning we studied in Chapter 1 and the difference between descriptive and inferential statistics. When you find the average age of every student on your campus, you can use deductive reasoning to conclude that this represents a true average. But if you found the average age of a sample within the population and used that to infer an average age for the students on the entire campus, you'd be using inductive reasoning to make an educated guess as to the true average age.

As we study statistics in this chapter, from time to time we'll think about the distinction between descriptive and inferential statistics. Are we *describing* data that have been gathered? Or are we using data to draw conclusions or to make predictions? That's an important distinction to keep in mind.

Frequency Distributions

The data collected for a statistical study are called **raw data**. In order to describe situations and draw conclusions, we need to organize the data in a meaningful way. Two methods that we will use are *frequency distributions* and *stem and leaf plots*.

There are two types of frequency distributions, the categorical frequency distribution and the grouped frequency distribution. A **categorical frequency distribution** is used when the data are categorical rather than numerical. Example 2 shows how to construct a categorical frequency distribution.

| Example 2 | Constructing a Categorical Frequency Distribution |

Twenty-five volunteers for a medical research study were given a blood test to obtain their blood types. The data follow.

A	B	B	AB	O
O	O	B	AB	B
B	B	O	A	O
AB	A	O	B	A
A	O	O	O	AB

Construct a frequency distribution for the data.

SOLUTION

Step 1 Make a table as shown, with all categories represented.

Type	Tally	Frequency
A		
B		
O		
AB		

Step 2 Tally the data using the second column of your table.

Step 3 Count the tallies and place the numbers in the third column. The completed frequency distribution is shown.

Type	Tally	Frequency
A	ⅢⅡ	5
B	ⅢⅡ //	7
O	ⅢⅡ ////	9
AB	////	4

Try This One   2

A health-food store recorded the types of vitamin pills 35 customers purchased during a 1-day sale. Construct a categorical frequency distribution for the data.

C	C	C	A	D	E	C
E	E	A	B	D	C	E
C	E	C	C	C	D	A
B	B	C	C	A	A	E
E	E	E	A	B	C	B

Another type of frequency distribution that can be constructed uses numerical data and is called a **grouped frequency distribution**. In a grouped frequency distribution, the numerical data are divided into classes. For example, if you gathered data on the weights of people in your class, there's a decent chance that no two people have the exact same weight. So it would be reasonable to group people into weight ranges, like 100–119 pounds, 120–139 pounds, and so forth. In the 100–119 pound class, we call 100 the lower limit and 119 the upper limit. When deciding on classes, here are some useful guidelines:

1. Try to keep the number of classes between 5 and 15. We wouldn't want weight classes like 120–122 pounds. The result would be so many classes, we wouldn't be much better off than if we used individual weights.

2. Make sure the classes don't overlap. Notice that we didn't choose 110–120, 120–130, and so on: that would put someone that weighs 120 pounds into two different classes.

3. Don't leave out any numbers between the lowest and highest, even if nothing falls into a particular class. Even if nobody weighs between 160 and 169, we still want a class for that because it may be useful to know that nobody is in that range.

4. Make sure the range of numbers included in a class is the same for each one. (We wouldn't want the first class to go from 100 to 120 pounds and the second from 121 to 130 pounds.)

5. The beginning and ending values have to be chosen based on how the data values are rounded. If our first two classes are 100–119 and 120–139, where would someone who weighs 119.6 pounds go? If you said "nowhere," you win. If any of the weights in our data set are rounded to the nearest tenth, we'd want to make our classes things like 100–119.9, 120–139.9, and so on.

Example 3 demonstrates the steps for building a grouped frequency distribution.

Example 3 | **Constructing a Grouped Frequency Distribution**

These data represent the record high temperatures for each of the 50 states in degrees Fahrenheit. Construct a grouped frequency distribution for the data.

State	Temp.	State	Temp.	State	Temp.	State	Temp
Alabama	112	Indiana	116	Nebraska	118	South Carolina	113
Alaska	100	Iowa	118	Nevada	125	South Dakota	120
Arizona	128	Kansas	121	New Hampshire	106	Tennessee	113
Arkansas	120	Kentucky	114	New Jersey	110	Texas	120
California	134	Louisiana	114	New Mexico	122	Utah	117
Colorado	118	Maine	105	New York	108	Vermont	105
Connecticut	106	Maryland	109	North Carolina	110	Virginia	110
Delaware	110	Massachusetts	107	North Dakota	121	Washington	118
Florida	109	Michigan	112	Ohio	113	West Virginia	112
Georgia	112	Minnesota	114	Oklahoma	120	Wisconsin	114
Hawaii	100	Mississippi	115	Oregon	119	Wyoming	115
Idaho	118	Missouri	118	Pennsylvania	111		
Illinois	117	Montana	117	Rhode Island	104		

Source: *National Climatic Data Center*

SOLUTION

Step 1 Subtract the lowest value from the highest value: $134 - 100 = 34$.

Step 2 If we use a range of 5 degrees, that will give us seven classes, since the entire range (34 degrees) divided by 5 is 6.8. This is certainly not the only choice for the number of classes: if we choose a range of 4 degrees, there will be nine classes.

Step 3 Start with the lowest value and add 5 to get the lower class limits: 100, 105, 110, 115, 120, 125, 130, 135. Notice that all of the data are rounded to the nearest whole number, so that's reflected in our choices of class limits.

Step 4 Set up the classes. To find the upper limit for each, subtract 1 from the next upper limit.

Math Note

If it's convenient, you can start the first class at a value lower than the lowest data value.

Digital Vision/Getty Images

Step 5 Tally the data and record the frequencies as shown. It's a really good idea to cross out each data value as you tally it up, and an even better idea to make sure that all of the frequencies add up to the total number of data values. In this case, the following frequencies add to 50, so we can be pretty sure we didn't miss any values or record any twice.

Class	Tally	Frequency
100–104	///	3
105–109	ℍℍ ///	8
110–114	ℍℍ ℍℍ ℍℍ /	16
115–119	ℍℍ ℍℍ ///	13
120–124	ℍℍ //	7
125–129	//	2
130–134	/	1

5. Construct a frequency distribution for a data set.

Try This One   3

In one math class, the data below represent the number of hours each student spends on homework in an average week. Construct a grouped frequency distribution for the data using six classes.

1	2	6	7	12	13	2	6	9	5
18	7	3	15	15	4	17	1	14	5
4	16	4	5	8	6	5	18	5	2
9	11	12	1	9	2	10	11	4	10
9	18	8	8	4	14	7	3	2	6

Stem and Leaf Plots

Another way to organize data is to use a **stem and leaf plot** (sometimes called a stem plot). Each data value or number is separated into two parts. For a two-digit number such as 53, the tens digit, 5, is called the **stem**, and the ones digit, 3, is called its **leaf**. For the number 72, the stem is 7, and the leaf is 2. For a three-digit number, say 138, the first two digits, 13, are used as the stem, and the third digit, 8, is used as the leaf. For values rounded to the tenths place, like 8.4, you can use the value to the left of the decimal place as stem and the tenths place as leaf, as in our next example. In any case, the very last digit is used as the leaf, and what comes before is the stem. Note that we'll include a key with our plot that clarifies what the stems and leaves represent.

Example 4   Drawing a Stem and Leaf Plot

The following data are the percentage of adults that had been fully vaccinated against COVID-19 as of July 1, 2021. Draw a stem and leaf plot illustrating these data.

State	% fully vaccinated	State	% fully vaccinated	State	% fully vaccinated	State	% fully vaccinated
Alabama	33	Colorado	52	Hawaii	52	Kansas	42
Alaska	43	Connecticut	61	Idaho	36	Kentucky	44
Arizona	43	Delaware	50	Illinois	46	Louisiana	35
Arkansas	34	Florida	46	Indiana	40	Maine	62
California	50	Georgia	37	Iowa	48	Maryland	56

State	% fully vaccinated	State	% fully vaccinated	State	% fully vaccinated	State	% fully vaccinated
Massachusetts	62	New Hampshire	56	Oregon	54	Vermont	66
Michigan	47	New Jersey	55	Pennsylvania	50	Virginia	52
Minnesota	52	New Mexico	55	Rhode Island	59	Washington	55
Mississippi	30	New York State	54	South Carolina	39	West Virginia	37
Missouri	39	North Carolina	42	South Dakota	45	Wisconsin	50
Montana	43	North Dakota	39	Tennessee	35	Wyoming	35
Nebraska	48	Ohio	45	Texas	41		
Nevada	42	Oklahoma	38	Utah	37		

The Mayo Clinic

SOLUTION

The data values range from 30 to 66, so we'll use the first digits as stems and the second digits as leaves. The stems will range from 3 to 6, so we begin our setup like this:

Stems	Leaves
3	
4	
5	
6	

Now we go through the data values one-by-one, writing the appropriate leaf next to the matching stem. For 43, we put leaf 3 next to stem 4. For 52, we put leaf 2 next to stem 5. Continue in this manner until all the data values have been entered. Once again, it's a good idea to cross out the data values as you enter them on the plot, then count when finished to make sure that you have the right number of values.

Also, from looking at the plot, you can't tell if the data value with stem 4 and leaf 3 means 43, 4.3, or something else, so we put a key at the bottom of the plot to clarify.

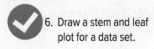

6. Draw a stem and leaf plot for a data set.

Stems	Leaves
3	3 4 7 6 5 0 9 9 8 9 5 7 7 5
4	3 3 6 6 0 8 2 4 7 3 8 2 2 5 5 1
5	0 2 0 2 6 2 6 5 5 4 4 0 9 2 5 0
6	1 2 2 6

Key: 4|3 means 43

The last step isn't strictly necessary, but many folks prefer stem and leaf plots with the data in order, so we'll rearrange the leaves corresponding to each stem from smallest to largest.

Stems	Leaves
3	0 3 4 5 5 5 6 7 7 7 8 9 9 9
4	0 1 2 2 2 3 3 3 4 5 5 6 6 7 8 8
5	0 0 0 0 2 2 2 2 4 4 5 5 5 6 6 9
6	1 2 2 6

Key: 4|3 means 43

Math Note

Remember, when drawing a stem and leaf plot with three-digit numbers, the first two digits are the stem. So for 119, the stem is 11 and the leaf is 9.

Try This One   4

It's no secret that gas prices have fluctuated wildly in the last few years, but from 1980–1999, they were surprisingly stable. According to the Energy Information Administration, the data below represent the average price (in cents) per gallon of regular unleaded gas for those years. Draw a stem and leaf plot of the data.

| 119 | 131 | 122 | 113 | 112 | 86 | 90 | 90 | 100 | 115 |
| 114 | 113 | 111 | 111 | 115 | 123 | 123 | 123 | 106 | 117 |

Deciding on stems and leaves for the vaccination data was fairly simple because there was a natural choice of stems (the first digit of each rate) that gave us a reasonable number of them (4). That's not always the case: for the high-temperature data earlier in this section, the natural choice of stem would seem to be the first two digits. But here's what the stem and leaf plot would look like in that case:

Stems	Leaves
10	0 0 4 5 5 6 6 7 8 9 9
11	0 0 0 0 1 2 2 2 2 3 3 3 3 4 4 4 4 5 5 6 7 7 7 8 8 8 8 8 8 9
12	0 0 0 0 1 1 2 5 8
13	4

Key: 13│4 means 134

This is a perfectly lovely plot, but it's not terribly useful in analyzing the data. There just aren't enough categories for this particular data set. We can easily see that high temperatures between 110 and 120 are by far most common, but other than that the plot isn't very enlightening.

In most cases, it's a good idea to choose stems so that there are between 5 and 15 categories. In the case of the high-temperature data, we'd get a better organization if we divided the categories in half, so that the first category covers 100 to 104, the second 105 to 109, and so on. So that's just what we'll do.

Example 5   Building a Stem and Leaf Plot with Divided Categories

Build a stem and leaf plot for the high temperature data in Example 3. Divide each natural category into two halves.

SOLUTION

What we're really doing here is using the same categories we did for the frequency distribution in Example 3. The main issue is that we need a way to represent this in the plot. For the values that have stem 10 in our first attempt at the plot, we'll write $10^{(0)}$ and $10^{(5)}$ to represent the values from 100 to 104, and the values from 105 to 109. Here's the resulting plot, with the leaves arranged in numerical order. Notice the appropriate additions to the key at the bottom of the plot.

Stems	Leaves
$10^{(0)}$	0 0 4
$10^{(5)}$	5 5 6 6 6 7 8 9 9
$11^{(0)}$	0 0 0 0 1 2 2 2 2 3 3 4 4 4 4
$11^{(5)}$	5 5 6 7 7 7 8 8 8 8 8 9
$12^{(0)}$	0 0 0 0 1 1 2
$12^{(5)}$	5 8
13	4

Key: $10^{(0)}$│4 means 104; $10^{(5)}$│6 means 106

Try This One 5

Below are the number of games won by the University of Connecticut women's basketball team each year starting in 1990. Draw a stem and leaf plot, dividing natural categories as appropriate.

29	23	18	30	35	34	33	34	29	36	32	39	37	31	25	32
32	36	39	39	36	33	35	40	38	38	36	36	35	29	28	

The methods that we've used for organizing and picturing data have each described all of the data available to us in a given circumstance, and so they qualify as descriptive statistics. As we continue our study of stats, we'll keep trying to think about whether we're describing data, or using it to make some sort of inference.

Answers to Try This One

1 (a) Stratified sample; subjects are being chosen from pre-existing groups. Probably not a good representative sample because it leaves out people who don't attend church, and in some cases church-goers are less likely to be favorably inclined to support a brewery.

(b) Systematic sample (but you could make a case for stratified); they would work systematically through the town by street. Should be a good representative sample.

(c) Cluster sample: they're surveying only people in one group (those that attend the festival). If the festival is really well-attended, it would probably be a fairly good representative sample.

2

Type	Tally	Frequency
A	⊦⊦⊦⊦ /	6
B	⊦⊦⊦⊦	5
C	⊦⊦⊦⊦ ⊦⊦⊦⊦ //	12
D	///	3
E	⊦⊦⊦⊦ ////	9

3

Class	Tally	Frequency
1–3	⊦⊦⊦⊦ ⊦⊦⊦⊦	10
4–6	⊦⊦⊦⊦ ⊦⊦⊦⊦ ////	14
7–9	⊦⊦⊦⊦ ⊦⊦⊦⊦	10
10–12	⊦⊦⊦⊦ /	6
13–15	⊦⊦⊦⊦	5
16–18	⊦⊦⊦⊦	5

4

Stems	Leaves
8	6
9	0 0
10	0 6
11	9 3 2 5 4 3 1 1 5 7
12	2 3 3 3
13	1

Key: 8|6 means 86

5

Stems	Leaves
$1^{(5)}$	8
$2^{(0)}$	3
$2^{(5)}$	5 8 9 9 9
$3^{(0)}$	0 1 2 2 2 3 3 4 4
$3^{(5)}$	5 5 5 6 6 6 6 7 8 8 9 9 9
$4^{(0)}$	0

Key: $1^{(5)}$|8 means 18; $2^{(5)}$|3 means 23

Exercise Set 11-1

Writing Exercises

1. What are *data*?
2. Define *statistics*.
3. Explain the difference between a population and a sample.
4. How is a random sample selected?
5. How is a systematic sample selected?
6. How is a stratified sample selected?
7. How is a cluster sample selected?
8. What are the similarities between a grouped frequency distribution and a stem and leaf plot?
9. Describe the difference between descriptive and inferential statistics.
10. How do inductive and deductive reasoning play a role in understanding descriptive and inferential statistics?

For Exercises 11–16, classify each sample as random, systematic, stratified, or cluster and explain your answer.

11. In a large school district, all teachers from two buildings are interviewed to determine whether they believe the students have less homework to do now than in previous years.

12. Every seventh customer entering a shopping mall is asked to name their favorite store.

13. Nursing supervisors are selected using random numbers in order to determine annual salaries.

14. Every hundredth hamburger manufactured is checked to determine its fat content.

15. Mail carriers of a large city are divided into four groups according to sex (male or female) and according to whether they walk or ride on their routes. Then 10 are selected from each group and interviewed to determine whether they have been bitten by a dog in the last year.

16. For the draft lottery that was conducted from 1969 to 1975, the days of the year were written on pieces of paper and put in plastic capsules. The capsules were then mixed in a barrel and drawn out one at a time, with the day recorded.

In Exercises 17–22, decide if you think the method described would result in a good random sample, and explain your answer.

17. Students on a college quad are offered $5 to participate in a survey.

18. A sociologist setting up a study of race and education level for prisoners at a state penitentiary chooses inmates whose prison ID is divisible by 11.

19. Random phone numbers are dialed in a given area code to ask people about how often they use their landline phones.

20. Random phone numbers are dialed in a given area code to survey people as to whether or not they've needed the services of a food pantry to feed their families.

21. In an attempt to study the freshness of dairy products at a supermarket, the expiration dates of the first five bottles of milk and first five cartons of eggs on each shelf are recorded.

22. All of the green M&Ms in a bag are taken out and weighed to find the average weight of each piece.

Applications in Our World

23. At a college financial aid office, students who applied for a scholarship were classified according to their class rank: FY = first year, So = sophomore, Jr = junior, Se = senior. Construct a frequency distribution for the data.

FY	FY	FY	FY	FY
Jr	FY	FY	So	FY
FY	So	Jr	So	FY
So	FY	FY	FY	So
Se	Jr	Jr	So	FY
FY	FY	FY	FY	So
Se	Se	Jr	Jr	Se
So	So	So	So	So

24. A questionnaire about how students primarily get news resulted in the following responses from 25 people. Construct a frequency distribution for the data. (I = Internet, N = newspaper, R = radio, T = TV)

T I I N I I I N I R T N I
I T R I I N I R T N I I

25. Twenty-five fans of reality TV were asked to rate four shows, and the data below reflect which one each rated highest. Construct a frequency distribution for the data. (S = *Survivor*, D = *Dancing with the Stars*, B = *Big Brother*, A = *The Amazing Race*)

S S B D D B S D A B A A S
B A D B A S A D B B S A

26. A small independent developer of apps for smartphones sells an app that marks the location of your car by GPS so you can find it later in case you forget where you parked. The number of paid downloads for the past 49 weeks is shown. Construct a frequency distribution using six classes.

373	254	237	243	308	210	266	253	201	266
239	114	224	373	286	329	236	284	247	273
198	361	416	207	243	326	251	169	360	311
215	189	344	268	363	21	270	165	240	48
150	300	207	314	197	209	210	260	327	

27. The ages of the signers of the Declaration of Independence are shown here. (Age is approximate since only the birth year appeared in the source, and one has been omitted since his birth year is unknown.) Construct a frequency distribution for the data using seven classes.

41	54	47	40	39	35	50	37	49	42	70	32
44	52	39	50	40	30	34	69	39	45	33	52
44	62	60	27	42	34	50	42	52	38	36	45
35	43	48	46	31	27	55	63	46	33	60	62
35	46	45	34	53	50	50					

Source: *The Universal Almanac*

28. The percentage of traffic fatalities in which at least one driver had a blood alcohol level over 0.08 is shown for the first 27 states alphabetically. Construct a frequency distribution using six classes.

33.0	30.1	35.9	35.0	24.5	36.0	38.0	39.9	34.2
31.3	25.8	29.0	44.4	25.8	33.4	34.0	30.3	25.7
27.1	47.7	29.6	30.8	35.0	28.2	28.7	25.7	32.2

Source: *U.S. Statistical Abstract*

29. The data shown are the cigarette taxes per pack in cents imposed by each state as of December 1, 2021. Construct a frequency distribution with first class 0–69.

68	194	320	129	351	170	200	203	153	60
200	435	57	110	200	64	435	333	62	303
200	210	298	108	304	180	45	260	141	120
115	134	100	200	68	178	44	425	170	252
287	37	136	200	17	270	160	57	308	60

Federation of Tax Administrators

30. The acreage (in thousands of acres) of the 39 U.S. National Parks is shown here. Construct a frequency distribution for the data using eight classes.

41	66	233	775	169
36	338	233	236	64
183	61	13	308	77
520	77	27	217	5

650	462	106	52	52
505	94	75	265	402
196	70	132	28	220
760	143	46	539	

Source: *The Universal Almanac*

31. The heights in feet above sea level of the major active volcanoes in Alaska are given here. Construct a frequency distribution for the data using 10 classes. Based on your distribution, what would you estimate to be the average height of an active volcano in Alaska?

4,265	3,545	4,025	7,050	11,413
3,490	5,370	4,885	5,030	6,830
4,450	5,775	3,945	7,545	8,450
3,995	10,140	6,050	10,265	6,965
150	8,185	7,295	2,015	5,055
5,315	2,945	6,720	3,465	1,980
2,560	4,450	2,759	9,430	
7,985	7,540	3,540	11,070	
5,710	885	8,960	7,015	

Source: *The Universal Almanac*

32. During the 1998 baseball season, Mark McGwire and Sammy Sosa both broke Roger Maris's home run record of 61. The distances in feet for each home run follow. Construct a frequency distribution for each player using the same eight classes. Which player do you think hit longer home runs in general? Which had more consistent distances?

McGwire				Sosa			
306	370	370	430	371	350	430	420
420	340	460	410	430	434	370	420
440	410	380	360	440	410	420	460
350	527	380	550	400	430	410	370
478	420	390	420	370	410	380	340
425	370	480	390	350	420	410	415
430	388	423	410	430	380	380	366
360	410	450	350	500	380	390	400
450	430	461	430	364	430	450	440
470	440	400	390	365	420	350	420
510	430	450	452	400	380	380	400
420	380	470	398	370	420	360	368
409	385	369	460	430	433	388	440
390	510	500	450	414	482	364	370
470	430	458	380	400	405	433	390
430	341	385	410	480	480	434	344
420	380	400	440	410	420		
377	370						

Source: *USA Today*

33. The data (in billions of dollars) are the values of the 32 National Football League franchises in 2021. Construct a frequency distribution for the data using six classes.

6.50	2.27	2.65	3.42	2.80	2.91	3.40	3.70
4.18	3.75	2.92	3.80	4.05	5.00	2.83	2.63
4.08	2.60	2.93	3.20	2.40	3.42	3.50	3.35
2.28	2.94	3.25	4.85	3.48	4.80	3.43	4.20

Source: *Forbes magazine*

34. Twenty-nine executives reported the number of telephone calls made during a randomly selected week as shown here. Construct a stem and leaf plot for the data and analyze the results.

22	14	12	9	54	12
16	12	14	49	10	14
8	21	37	28	36	22
9	33	58	31	41	19
3	18	25	28	52	

35. The National Insurance Crime Bureau reported that these data represent the number of registered vehicles per car stolen for 35 selected cities in the United States. For example, in Miami, one automobile is stolen for every 38 registered vehicles in the city. Construct a stem and leaf plot for the data and analyze the distribution. (The data have been rounded to the nearest whole number.)

38	53	53	56	69	89	94
41	58	68	66	69	89	52
50	70	83	81	80	90	74
50	70	83	59	75	78	73
92	84	87	84	85	84	89

Source: *USA Today*

36. As an experiment in a botany class, 20 plants are placed in a greenhouse, and their growth in centimeters after 20 days is recorded, with the results shown below. Construct a stem and leaf plot for the data.

20	12	39	38
41	43	51	52
59	55	53	59
50	58	35	38
23	32	43	53

37. The data shown represent the percentage of unemployed males for a sample of countries of the world. Using whole numbers as stems and the decimals as leaves, construct a stem and leaf plot.

8.8	1.9	5.6	4.6	1.5
2.2	5.6	3.1	5.9	6.6
9.8	8.7	6.0	5.2	5.6
4.4	9.6	6.6	6.0	0.3
4.6	3.1	4.1	7.7	

Source: *The Time Almanac*

38. The COVID pandemic saw air travel greatly decrease. The following data are the percentage decrease of passengers boarding flights at the 25 busiest airports in the United States in 2020 compared to 2019. Draw a stem and leaf plot for the data.

62	46	59	72	64
48	57	60	63	59
52	53	73	62	55
64	57	55	71	63
67	62	66	53	58

Source: *Federal Aviation Administration*

39. The next data set is the high temperatures for each of the last 30 days in North Port, FL. (In case you're wondering, today is March 31.) Draw a stem and leaf plot for the

data. It surely would be a good idea to divide each natural choice of category into two halves.

```
84  80  83  76  81  85
80  83  78  79  87  86
79  83  70  80  82  87
77  84  65  79  83  86
79  86  68  82  84  78
```

40. Next up, we have the closing price over the last 25 days of one share of stock in Northern Dynasty Minerals, Inc.,

which I think we can all agree is a pretty cool name for a company. Draw a stem and leaf plot for the prices, dividing the natural category choices into two halves.

```
1.45  1.38  1.22  1.23  1.46
1.45  1.39  1.34  1.30  1.59
1.40  1.42  1.20  1.28  1.54
1.45  1.59  1.29  1.36  1.59
1.35  1.37  1.34  1.40  1.57
```

Critical Thinking

For Exercises 41–46, decide whether descriptive or inferential statistics is being used.

41. A recent study showed that eating garlic can lower blood pressure.

42. The average number of students in a class at White Oak University is 22.6.

43. It is predicted that the average number of automobiles each household owns will increase next year.

44. Last year's total attendance at Long Run High School's football games was 8,325.

45. The chance that a person will be robbed in a certain city is 15%.

46. A national poll in November indicated that the presidential election would be very close.

For Exercises 47–50, have all the students in your class fill out this short survey, then compile and use the data.

Birthday (day of month)	Year in school	Credit hours this semester	Pick a number between 0 and 100

47. Draw a stem and leaf plot for the results of the birthday question.

48. Draw a frequency distribution for the year in school data.

49. Draw a grouped frequency distribution for the "number between 0 and 100" data.

50. Of the three methods for organizing data in this section, choose the one you think would be most appropriate for the credit hours data, describe why you chose that method, then use the method to organize the data.

51. In addition to the four basic sampling methods, other methods are also used. Some of these methods are *sequence sampling, double sampling,* and *multiple sampling.* Look up these methods on the Internet and explain the advantages and disadvantages of each method.

52. For the data in Exercise 33, draw a stem and leaf plot, and compare your result to the frequency distribution. Which do you think provides more useful information, and why? What are the strengths and weaknesses of each?

53. What does your result from Exercise 29 tell you about cigarette taxes by state? Explain how the frequency distribution made it easier to make this observation than the raw data.

54. What does your result from Exercise 35 tell you about car theft in the United States? How did the stem and leaf plot help you to analyze the data?

Section 11-2   Picturing Data

LEARNING OBJECTIVES

1. Draw and interpret bar graphs and pie charts.

2. Draw histograms and frequency polygons.

3. Draw time series graphs.

Now we've gathered some data, and we think maybe some of it has an interesting story to tell. How can we most effectively present that data? In Section 11-1, we displayed data in table form, using frequency distributions, and using stem and leaf plots. All are perfectly valid, but . . . they don't exactly pop.

The data in the following table, supplied by the FBI, shows the number of hate crimes reported in the U.S. for the years 2005–2020.

Year	2005	2006	2007	2008	2009	2010	2011	2012
Hate crimes	7,163	7,722	7,624	7,783	6,604	6,628	6,222	5,796

Year	2013	2014	2015	2016	2017	2018	2019	2020
Hate crimes	5,928	5,479	5,850	6,121	7,175	7,120	7,314	7,759

FBI

Now let's look at the same information presented in graphic form:

The information is the same, but with the graph, just a quick glance is enough to see that hate crimes decreased significantly from 2008 to 2014, then increased sharply starting in 2015,

Hate Crimes Rising

Source: FBI

reaching a new high in 2020. Simply put, if you want your data to really catch someone's eye, you can't do much better than a nice graphical representation of the data. In this section, we'll study several methods for accomplishing this goal.

Bar Graphs and Pie Charts

The first type of frequency distribution we studied in Section 11-1 was the categorical frequency distribution. When data are representative of certain categories, rather than numerical, we often use **bar graphs** or circle graphs (commonly known as **pie charts**) to illustrate the data. We'll start with a pie chart in Example 1. A pie chart is a circle that is divided into sections in proportion to the frequencies corresponding to the categories. The purpose of a pie chart is to show the relationship of the parts to the whole by visually comparing the size of the sections.

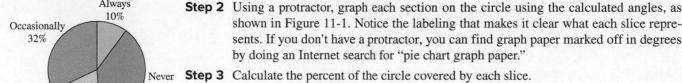

Example 1   Drawing a Pie Chart to Represent Data

The marketing firm Deloitte Retail conducted a survey of grocery shoppers. The frequency distribution below represents the responses to the survey question, "How often do you bring your own bags when grocery shopping?" Draw a pie chart to represent the data.

Response	Frequency
Always	10
Never	39
Frequently	19
Occasionally	32

SOLUTION

Step 1 Find the number of degrees corresponding to each slice using the formula

$$\text{Degrees} = \frac{f}{n} \cdot 360°$$

where f is the frequency for each class and n is the sum of the frequencies. Recall that there are 360° in a circle: f divided by n calculates the percentage of the circle that's covered by each, and multiplying that percentage by 360° gives us the number of degrees for each slice. In this case, $n = 100$, so the degree measures are:

Always $\quad \frac{10}{100} \cdot 360° = 36°$

Never $\quad \frac{39}{100} \cdot 360° = 140.4°$

Frequently $\quad \frac{19}{100} \cdot 360° = 68.4°$

Occasionally $\quad \frac{32}{100} \cdot 360° = 115.2°$

Step 2 Using a protractor, graph each section on the circle using the calculated angles, as shown in Figure 11-1. Notice the labeling that makes it clear what each slice represents. If you don't have a protractor, you can find graph paper marked off in degrees by doing an Internet search for "pie chart graph paper."

Step 3 Calculate the percent of the circle covered by each slice.

Always: $\frac{10}{100} = 10\%$ Never: $\frac{39}{100} = 39\%$

Frequently: $\frac{19}{100} = 19\%$ Occasionally: $\frac{32}{100} = 32\%$

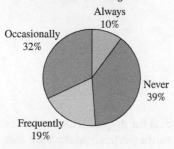

Ryan McVay/Getty Images

How Often Shoppers Bring Their Own Bags

Always 10%
Occasionally 32%
Never 39%
Frequently 19%

Figure 11-1

Label each section with the percent, as shown in Figure 11-1.

Try This One   1

The Harris poll asked 2,022 adult Americans if they have a household budget, and if they stick to it. The results are summarized below. Draw a pie chart to illustrate the data.

Response	Frequency
Have a budget, stick to it	687
Have a budget, don't stick to it	789
Don't have a budget	546

Using Technology: Creating a Pie Chart

To create a pie chart in Excel:

1. Type the category names in one column (or row).
2. Type the category values in the next column (or row).
3. Use the mouse to drag and select all the data in those two columns (or rows).
4. With the appropriate cells selected, click the **Insert** tab, then choose **Chart** and Pie chart. There are a few different styles you can experiment with, but starting with the simplest is a good idea.

You can add titles and change colors and other formatting elements by right-clicking on certain elements or by using the options on the **Charts** menu. Try some options and see what you can learn!

While a pie chart is used to compare parts to a whole, a bar graph is used for comparing parts to other parts. Let's look at some different survey results that are better suited for a bar graph.

Example 2   Drawing a Bar Graph to Represent Data

Let's revisit the marketing firm's survey of 100 grocery shoppers. Suppose the frequency distribution below represents the responses to the survey question, "Which grocery stores have you shopped at in the last month?"

(a) Draw a vertical bar graph for the results, shown in the frequency distribution.
(b) Would a pie chart make sense for these data? Why or why not?

Store	Frequency
Publix	50
Trader Joe's	40
Fareway	28
Aldi	33
Other	65

SOLUTION

(a) **Step 1**   Draw and label the axes. We were asked for a vertical bar graph, so the responses go on the horizontal axis, and the frequencies on the vertical. Make sure that your labeling on the vertical axis starts at zero.

Step 2 Draw vertical bars with heights that correspond to the frequencies. The completed graph is shown in Figure 11-2.

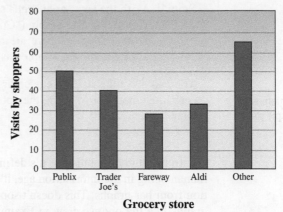

Figure 11-2

Math Note

For clarity, bar graphs are usually drawn with space between the bars.

(b) There are two good ways to decide that a pie chart would be a bad choice. One is numerical: we were told that 80 shoppers were surveyed, but the numbers for the responses add up to more than 80. So this isn't comparing distinct parts to a whole: the categories overlap. Which brings us to the other way you can tell that a pie chart is a bad choice. The nature of the survey ("Which stores have you shopped at?") pretty much guarantees that some people will choose more than one of the responses. This is a dead giveaway that the categories overlap, so a pie chart would be a bad choice.

Try This One   2

Two surveys of 100 people were done on calorie information at fast-food restaurants. The first asked how much impact posted calorie information has on buying choices. The second asked which (if any) restaurants the subjects tend to avoid because they think the food is unhealthy. Results are below. Draw a pie chart for one set of data and a bar graph for another. Choose which data set is appropriate for each type of graph.

Impact of Posted Calorie Information		Restaurants Avoided for Health Reasons	
Response	**Frequency**	**Response**	**Frequency**
Great impact	40	McDonalds	46
Some impact	42	Burger King	38
Not much impact	11	Wendy's	35
None at all	7	Subway	18

CAUTION

In drawing a vertical bar graph, it's very important that the scale on the vertical axis begins at zero and has consistent spacing. As you'll see in the supplement on misuse of statistics, an improperly labeled axis can lead to a deceptive graph.

Using Technology: Creating a Bar Graph

To create a bar graph in Excel:

Step 1: Type the category names in one column (or row).
Step 2: Type the category values in the next column (or row).

Step 3: Use the mouse to drag and select all the data in those two columns (or rows).

Step 4: With the appropriate cells selected, click the **Insert** tab, then choose **Chart** and Column or Bar graph. ("Column" gives you vertical bars, "Bar" gives you horizontal.) Again, there are a few different styles you can experiment with, but starting with the simplest is a good idea.

You can add titles and change colors and other formatting elements by right-clicking on certain elements or by using the options on the **Charts** menu. Try some options and see what you can learn!

1. Draw and interpret bar graphs and pie charts.

In the study of statistics, it's definitely a good thing to be able to draw bar graphs. As a human being in the information age, it's a great thing to be able to read and interpret information from bar graphs. This doesn't sound particularly difficult when dealing with simple bar graphs like the one we drew in Example 2. But there are bar graphs that contain a lot more information, and require some careful attention to interpret, like the one in Example 3.

Example 3   Reading a Complicated Bar Graph

As vaccines for COVID-19 became available in late 2020, the Kaiser Family Foundation began a monthly survey of adults about their intentions in regard to the vaccine. The results are summarized in the bar graph. (Note that some percentages don't add to 100 due to undecided respondents.)

(a) Describe trends among those who had already been vaccinated.

(b) Which category was most stable from December to June? What can you conclude?

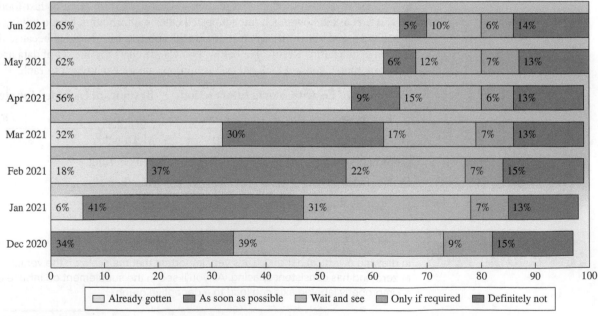

Source: Kaiser Family Foundation

SOLUTION

(a) The key at the bottom of the graph tells us that the people who had already been vaccinated is illustrated by the light green bars. No respondents had been vaccinated in December. The percentage progression by month was then 6%, 18%, 32%, 56%, 62%, 65%. This shows a clear upward trend in vaccinated respondents, with the most significant increase from March to April.

(b) There were two categories that were quite stable: those represented by the teal bars and the purple bars. These correspond to people who would choose to be vaccinated only if required, and those who would definitely not get vaccinated. It's reasonable to conclude that people opposed to vaccination had made up their minds early on and were unmoved as more information became available.

Try This One 3

For the survey illustrated by the bar graph in Example 3:

(a) What was the change in percentage of those who planned to wait and see from December to June?
(b) What does the trend in the response "As soon as possible" indicate?

Histograms and Frequency Polygons

When data are organized into grouped frequency distributions, two types of graphs are commonly used to represent them: *histograms* and *frequency polygons*.

A **histogram** is similar to a vertical bar graph in that the heights of the bars correspond to frequencies. The difference is that class limits are placed on the horizontal axis, rather than categories. A consequence of this labeling is that there's no space in between the bars. The procedure is illustrated in Example 4.

Example 4 Drawing a Histogram for Grouped Data

In Section 11-1, we analyzed and organized data representing the record high temperature in every state. It would therefore seem reasonable, if not expected, for us to illustrate that data graphically. So that's just what we'll do. Below is a grouped frequency distribution for the data. Use it to draw a histogram.

Class	Frequency
100–104	3
105–109	8
110–114	16
115–119	13
120–124	7
125–129	2
130–134	1

SOLUTION

Step 1 Write the scale for the frequencies on the vertical axis and the class limits on the horizontal axis. Make sure that your labeling on the vertical axis starts at zero.

Step 2 Draw vertical bars with heights that correspond to the frequencies for each class. See Figure 11-3.

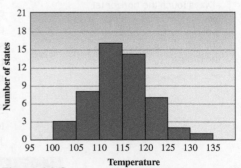

Figure 11-3

Note: Statisticians will sometimes label the horizontal axis of a histogram a bit differently to make it clearer which class a value right on a class limit falls into. For example, if you labeled 100, 105, 110, etc., on the *x* axis, it wouldn't be clear whether a temperature of 105 lives in the first bar or the second. If it's unclear at some point what values a given bar represents, refer to the frequency distribution for those data.

Math Note

When drawing a histogram, make sure that the bars touch (unlike a bar graph) and that every class is included, even if the frequency is zero.

Try This One   4

Draw a histogram for the frequency distribution to the right, which represents the number of losses by the team that won the NCAA men's basketball championship for the years from 1939–2021.

Losses	Frequency	Losses	Frequency
0–1	13	6–7	10
2–3	30	8–9	4
4–5	22	10–11	3

A **frequency polygon** is similar to a histogram, but instead of bars, a series of line segments is drawn connecting the midpoints of the classes. The heights of those points match the heights of the bars in a histogram. In Example 5, we'll draw a frequency polygon for the frequency distribution from Example 4 so you can compare the two.

Example 5   Drawing a Frequency Polygon

Math Note

Once you have the first midpoint, you can get all of the others by just adding the size of each class to the previous. For example, the second midpoint is the first, 102, plus the size of the classes, 5.

Draw a frequency polygon for the frequency distribution from Example 4. The distribution is repeated here.

Class	Frequency	Class	Frequency
100–104	3	120–124	7
105–109	8	125–129	2
110–114	16	130–134	1
115–119	13		

SOLUTION

Step 1 Find the midpoints for each class. This is accomplished by adding the upper and lower limits and dividing by 2. For the first two classes, we get:

$$\frac{100 + 104}{2} = 102$$

$$\frac{105 + 109}{2} = 107$$

The remaining midpoints are 112, 117, 122, 127, and 132.

Step 2 Write the scale for the frequencies on the vertical axis (making sure to start at zero), and label a scale on the horizontal axis so that all midpoints will be included.

Step 3 Plot points at the midpoints with heights matching the frequencies for each class, then connect those points with straight lines.

Step 4 Finish the graph by drawing a line back to the horizontal axis at the beginning and end. The horizontal distance to the axis should equal the distance between the midpoints. In this case, that distance is 5, so we extend back to 97 and forward to 137. See Figure 11-4.

Math Note

A frequency polygon should always touch the horizontal axis at both ends to indicate that the frequency is zero for any values not included on the graph.

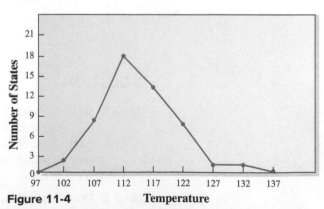

Figure 11-4

Try This One   5

Draw a frequency polygon for the frequency distribution from Try This One 4.

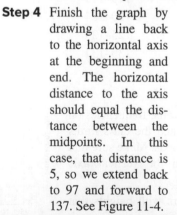
2. Draw histograms and frequency polygons.

So you may be wondering at this point: why bother with frequency polygons? Aren't they pretty much the same thing as histograms? Good question. There actually is a point, though, as we will now see.

Example 6	Identifying the Value of Frequency Polygons

These two grouped frequency distributions come from data gathered in two calculus classes that I taught recently. One is for percentage on the final exam; the other is for overall percentage in the course.

(a) Draw a frequency polygon for each distribution. Put them on the same axes.
(b) What can you learn about final exam percentage vs. overall percentage from the polygons?
(c) What would happen if you tried to draw a histogram for each on the same axes?

Final Exam Percentage

Percentage	Frequency
40–49.9	3
50–59.9	2
60–69.9	13
70–79.9	14
80–89.9	14
90–99.9	12

Overall Percentage

Percentage	Frequency
40–49.9	0
50–59.9	1
60–69.9	7
70–79.9	20
80–89.9	19
90–99.9	11

SOLUTION

(a) The midpoints for the classes are 44.95, 54.95, 64.95, 74.95, 84.95, and 94.95. The two frequency polygons are shown at right.

(b) There are lots of observations you could make from the graph. The first thing that jumps out at me is that there were a lot more students averaging in the 70s and 80s overall than on the final. There were also more low scores on the final, with the blue line living above the orange line for all of the classes below 70. The number of students averaging in the 90s looks to be about the same for the final and overall, though.

(c) In short, if we tried to draw two histograms on the same axes, we'd get a mess. Because histograms have filled-in bars, even if you use different colors, in some places the bars would obscure each other, making it almost impossible to read.

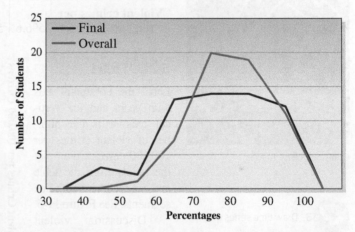

Try This One	6

The table shows two grouped frequency distributions describing the weights of 204 toddlers at age 24 months. There were 102 of each sex in the study. Draw two frequency polygons for the data on the same axes, and write any observations you can make based on your graph.

Class	Frequency (Boys)	Frequency (Girls)
9–9.9	1	3
10–10.9	7	15
11–11.9	20	30
12–12.9	34	31
13–13.9	26	17
14–14.9	11	5
15–15.9	3	1

Example 6 illustrates when frequency polygons are especially useful: for making direct comparisons of two or more data sets.

Time Series Graphs

A **time series graph** can be drawn for data collected over a period of time. This type of graph is used primarily to show trends, like prices rising or falling, for the time period. The graph illustrating ID theft complaints at the beginning of this section is a good example. There are three types of trends. Secular trends are viewed over a long period of time, such as yearly. Cyclical trends show oscillating patterns. Seasonal trends show the values of a commodity for shorter periods of the year, such as fall, winter, spring, and summer. Example 7 shows how to draw a time series graph.

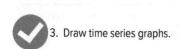

Example 7   Drawing a Time Series Graph

Comstock/Getty Images

3. Draw time series graphs.

Politicians often talk about violent crime in the United States in apocalyptic terms, but what do the data say? The table shows the number of violent crimes committed per 100,000 citizens every five years from 1980 to 2020. Draw a time series graph for the data and use it to discuss trends in violent crime.

Year	1980	1985	1990	1995	2000	2005	2010	2015	2020
Violent crimes per 100,000 citizens	596.6	556.6	729.6	684.5	506.5	469.0	404.5	372.6	398.5

SOLUTION

Label the horizontal axis with years and the vertical axis with the number of violent crimes per 100,000 citizens, then plot the points from the table and connect them with line segments. See Figure 11-5.

Discussion: violent crime rose sharply between 1985 and 1990, hitting a peak in that year. Since then, it's been declining at a pretty steady rate.

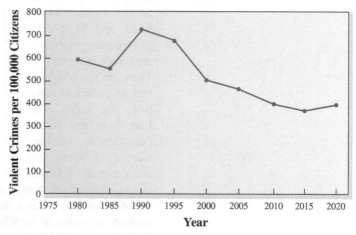

Figure 11-5

The two exceptions were a big decline from 1995 to 2000, and a slight increase from 2015 to 2020.

Try This One   7

Math Note

It's especially important to scale the vertical axis appropriately when drawing a time series graph. See Exercise 52 for some perspective on why.

The number of bankruptcy filings (in millions) in the United States for even-numbered years from 2002 to 2020 is shown in the table. Draw a time series graph for the data.

Year	2002	2004	2006	2008	2010	2012	2014	2016	2018	2020
Filings (millions)	1.58	1.60	0.62	1.12	1.50	1.22	0.94	0.79	0.78	0.76

This section provided just a sampling of the many types of charts and graphs used to picture data. A Google search for the string *picturing data* will help you find a more comprehensive list.

Answers to Try This One

1 **Does your household have a budget?**

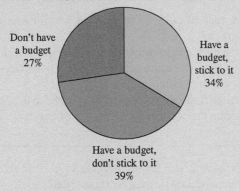

Don't have a budget 27%

Have a budget, stick to it 34%

Have a budget, don't stick to it 39%

2 **Impact of Posted Calorie Information**

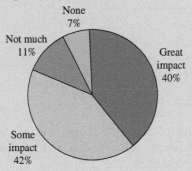

None 7%

Not much 11%

Great impact 40%

Some impact 42%

Restaurants Avoided For Health Reasons

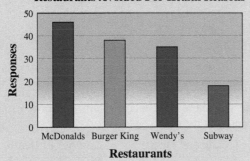

3 (a) It went down by 29%, from 39% to 10%.

(b) It went down very rapidly after March. This most likely indicates that after that point, most people who wanted the vaccine were able to get it.

4 **Losses by the NCAA Basketball Champion**

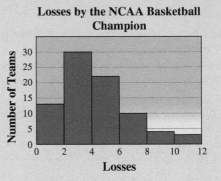

5 **Losses by the NCAA Basketball Champion**

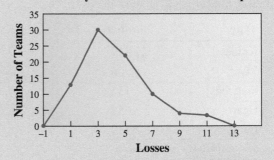

6 Observations vary, but in general this shows that boys tend to be a bit heavier than girls at 24 months.

Weights of 24-Month-Old Toddlers

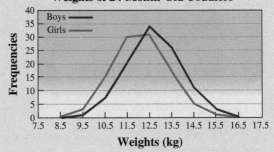

7 **Bankruptcy Filings**

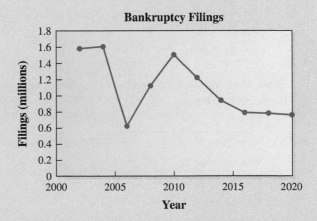

Exercise Set   11-2

Writing Exercises

1. Describe how to draw a bar graph and a pie chart given a frequency distribution.
2. Explain why bar graphs and pie charts are typically used for categorical frequency distributions, while histograms are used for grouped frequency distributions.
3. What type of categorical data are best illustrated with a pie chart?
4. What type of categorical data are best illustrated with a bar graph?
5. Describe some differences between bar graphs and histograms.
6. How are histograms and frequency polygons similar? How are they different?
7. When are frequency polygons more useful than histograms?
8. Describe the purpose of a time series graph.

Applications in Our World

9. Draw a bar graph for the number of transplants of various types performed in the United States in 2020, then use your graph to analyze the data.

Type	Number
Kidney	22,817
Liver	8,906
Pancreas	135
Heart	3,658
Lung	2,539

Source: *U.S. Department of Health and Human Services*

10. The data below show the number of students visiting a university health care center during finals week who suffered from certain conditions. Draw a bar graph to illustrate the data, then use your graph to analyze the data.

Condition	Number
Flu	48
Panic attack	36
Bronchitis	32
Headache	32
Broken bone	19
Cold/sniffles	17
Rash	16
Pneumonia	11
Ear infection	9

11. Draw a bar graph for the number of registered taxicabs in the selected cities.

City	Number
New York	13,237
Washington, D.C.	7,200
Chicago	6,650
Los Angeles	2,300
Atlanta	1,550

Source: *Wikipedia*

12. Draw a bar graph for the number of unemployed people in the selected states for November 2021.

State	Number
Texas	742,218
New York	614,325
Pennsylvania	359,349
Florida	482,551
Ohio	275,395

Source: *Bureau of Labor Statistics*

13. The number of students at one campus who had a 4.0 GPA is shown below, organized by class. Draw a pie chart for the data.

Rank	Frequency
First year	12
Sophomore	25
Junior	36
Senior	17

14. In an insurance company study of the causes of 1,000 deaths, these data were obtained. Draw a pie chart to represent the data.

Cause of Death	Number of Deaths
Heart disease	432
Cancer	227
Stroke	93
Accidents	24
Other	224
	1,000

15. In a survey of 100 college students, the numbers shown here indicate the primary reason why the students selected their majors. Draw a pie chart for the data and analyze the results.

Reason	Number
Interest in subject	62
Future earning potential	18
Pressure from parents	12
Good job prospects	8

16. A survey of the students in the school of education at a large university obtained the following data for students enrolled in specific fields. Draw a pie chart for the data and analyze the results.

Major Field	Number
Preschool	893
Elementary	605
Middle	245
Secondary	1,096

Use the graph for Questions 17–22.

Share of Non-Parents Younger than 50 Who Say They are not Likely to have Children is Up from 2018

% of non-parents ages 18 to 49 saying they are__to have children someday

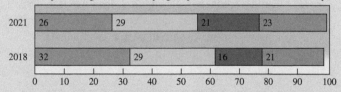

% of parents ages 18 to 49 saying they are__to have more children someday

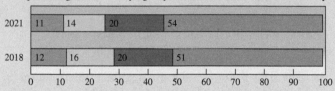

Note: Share of respondents who didn't offer an answer not shown.

Source: *Pew Research Center*

17. What percentage of all non-parents said they were likely to have children in 2018? 2021?

18. What percentage of parents said they were not likely to have children in 2018? 2021?

19. Did the percentage of respondents saying they were very likely to have children go up or down from 2018 to 2021?

20. Did the percentage of respondents saying they were not at all likely to have children go up or down from 2018 to 2021?

21. Which category of respondents is least likely to have children in the upcoming years?

22. If a random respondent is selected, are they more or less likely to plan on having children if they are currently parents?

Use the graph for Questions 23–28.

Local TV Stations are Turned to Most for Local News, Primarily Through the TV Set; Most Other Providers have Larger Digital Share

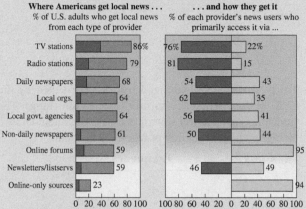

Note: For each provider type, respondents were shown the applicable non-digital platform(s), i.e., print, TV set, radio and word of mouth. These responses are all collapsed here as "analog."

Source: *Pew Research Center*

23. Which type of provider of local news is used most often?

24. Which type of provider of local news is used often by the fewest percentage of respondents?

25. Aside from the obvious sources (online forums and online-only sources), which type of provider of local news is most commonly accessed online?

26. Which type of news provider is least commonly accessed online?

27. Overall, is local news more likely to be accessed in an analog or an online format?

28. Why do you think the two bars with lengths 94 and 95 aren't length 100?

29. For 108 randomly selected college applicants, the frequency distribution shown here for entrance exam scores was obtained. Draw a histogram and frequency polygon for the data.

Class	Frequency
90–98	6
99–107	22
108–116	43
117–125	28
126–134	9

30. For 75 employees of a large department store, the distribution shown here for years of service was obtained. Draw a histogram and frequency polygon for the data.

Class	Frequency
1–5	21
6–10	25
11–15	15
16–20	0
21–25	8
26–30	6

31. Thirty American sedans and trucks were tested by the EPA for fuel efficiency in miles per gallon (mpg). The frequency distribution shown here was obtained. Draw a histogram and frequency polygon for the data.

Class	Frequency
8–12	3
13–17	5
18–22	15
23–27	5
28–32	2

32. In a study of reaction times of dogs to a specific stimulus, an animal trainer obtained these data, given in seconds. Draw a histogram and frequency polygon for the data, and analyze the results.

Class	Frequency
2.3–2.9	10
3.0–3.6	12
3.7–4.3	6
4.4–5.0	8
5.1–5.7	4
5.8–6.4	2

33. The data below show the number of families living in poverty (based on official governmental thresholds) in the United States for a selection of years from 1985 to 2020. Draw a time series graph for the data.

Year	1985	1990	1995	2000	2005	2010	2015	2020
Families in poverty (Millions)	7.2	7.1	7.5	6.4	7.7	9.4	8.6	7.3

34. The data below represent the number of laptops sold by a local computer store for the years listed. Draw a time series graph for the data.

Year	Number	Year	Number
2007	201	2014	799
2008	256	2015	873
2009	314	2016	1,012
2010	379	2017	1,105
2011	450	2018	906
2012	576	2019	899
2013	681	2020	740

35. The following data represent the number of downloads per month of an iPhone app. Draw a time series graph for the data, and use it to describe trends in the data.

Month	May	June	July	August	September
Downloads	12,413	15,160	18,201	20,206	19,143

36. The data below are the total sales for the U.S. restaurant industry in billions of dollars for a group of years. Draw a time series graph for the data, and use it to write a description of the data.

Year	1970	1980	1990	2000	2010	2020
Sales (Billion $)	45.4	126.8	253.7	401.7	586.8	555.8

37. The data below are the average daily temperatures in degrees Fahrenheit for every day of last May in Las Vegas.
   (a) Build a frequency distribution, then use it to draw a histogram for the data. You can decide on the classes, but make sure there are less than 10.
   (b) Use your histogram to write a description of the data.

77.9	74.5	83.8	90.3	93.9	68.5	81.5
78.3	77.4	85.3	84.5	90.1	75.3	
77.8	80.2	85.7	80.5	83.0	82.4	
75.3	83.0	86.5	83.1	78.5	85.0	
74.6	84.7	86.3	87.8	63.6	89.1	

38. The data below are the average daily temperatures in degrees Fahrenheit for every day of last May in Honolulu.
   (a) Build a frequency distribution, then use it to draw a histogram for the data. You can decide on the classes, but make sure there are less than 10.
   (b) Use your histogram to write a description of the data, including a comparison to the data in Problem 37.

76.2	77.0	76.0	76.4	78.3	78.2	76.6
76.0	76.6	75.4	77.9	79.5	77.7	
75.8	76.8	74.5	78.0	78.3	77.7	
75.1	76.3	73.9	77.7	77.8	77.5	
75.7	76.0	74.3	77.7	78.5	77.0	

39. The number of employed registered nurses (in thousands) in each of the 50 states is shown in the table.
   (a) Build a frequency distribution, then use it to draw a histogram for the data. You can decide on the classes, but make sure there are less than 10.
   (b) Use your histogram to write a description of the data.

45.4	10.2	87.0	15.4	11.1
5.5	11.7	89.7	176.2	60.6
44.7	123.8	56.7	91.3	184.9
23.6	61.2	29.6	9.0	19.1
250.2	33.2	67.6	124.8	6.4
42.5	28.2	8.7	28.3	62.2
36.3	44.8	19.2	31.0	54.5
10.5	40.8	16.5	130.7	18.3
164.8	14.8	13.9	12.3	56.4
65.2	49.6	78.2	42.7	4.8

40. The number of employed licensed nurse practitioners (in thousands) in each of the 50 states is shown in the table.
    (a) Build a frequency distribution, then use it to draw a histogram for the data. You can decide on the classes, but make sure there are less than 10.
    (b) Use your histogram to write a description of the data, including a comparison to the data in Problem 39.

1.9	0.9	6.8	1.2	0.5
0.7	0.7	4.2	16.0	5.8
3.6	4.5	2.9	4.0	9.7
2.7	3.1	2.7	0.5	1.4
17.0	1.3	4.0	5.4	0.5
3.2	2.2	0.6	1.1	6.6
3.4	3.0	1.0	2.5	3.8
0.7	2.4	0.7	7.5	0.9
12.7	1.1	1.5	0.7	3.6
4.9	3.5	5.5	3.7	0.4

41. Look back at the data in Exercise 31. The EPA also tested 30 imported sedans and light trucks, getting the results summarized in the next frequency distribution. Draw a frequency polygon on the same axes you used for Exercise 31, and use the results to write some observations contrasting the two data sets. If you didn't do Exercise 31, not to worry. Just draw both frequency polygons now.

Class	Frequency
8–12	1
13–17	0
18–22	17
23–27	8
28–32	4

42. Look back at the data in Exercise 32. The trainer tested reaction times to the same stimuli for cats, getting the results summarized in the next frequency distribution. Draw a frequency polygon on the same axes you used for Exercise 32, and use the results to write some observations contrasting the two data sets. If you didn't do Exercise 32, remain calm: this is fixable. Just draw both frequency polygons now. Genius!

Class	Frequency
2.3–2.9	21
3.0–3.6	15
3.7–4.3	4
4.4–5.0	0
5.1–5.7	1
5.8–6.4	1

43. Using the Internet as a resource, find the number of games won by your school's women's basketball team over the last 10 seasons and draw a table to organize the data. (If your school doesn't have a basketball team, choose a nearby school that does.) Draw a time series graph for your data, and use the graph to analyze the team's performance.

44. Using the Internet as a resource, find the number of games won by the closest professional football team to your campus over the last 20 years. (If the closest team hasn't existed for 20 years, pick another team that has.) Divide the wins into classes, and draw a frequency distribution for the data.

Critical Thinking

In Exercises 45–50, state which type of graph (bar graph, pie chart, or time series graph) would most appropriately represent the given data. Explain your answer.

45. The number of students enrolled at a local college each year for the last 5 years.
46. The budget for the student activities department at your college.
47. The number of students who get to school by automobile, bus, train, or by walking.
48. The record high temperatures of a city for the last 30 years.
49. The areas of the five lakes in the Great Lakes.
50. The amount of each dollar spent for wages, advertising, overhead, and profit by a corporation.
51. Redraw the bar graph from Example 2, but label the vertical axis so that it goes from 25 to 70. Why does this make the graph deceiving?

52. Redraw the time series graph from Example 7, but make the scale on the vertical axis range from 350 to 750 rather than from 0 to 800. Why does this make the graph deceiving?
53. Redraw the graph from Exercise 10, but make the scale on the vertical axis range from 8 to 50. Does this make the graph deceiving? How?
54. If you're going to compare data sets by drawing two frequency polygons on the same axes, what are some things that would be needed in order for the comparison to be meaningful? Think about as many as you can.
55. Why do you think we put spaces between the bars on a bar graph, but not a histogram? Looking at the horizontal axis should help.
56. Did we study descriptive or inferential statistics in this section? Explain.

57. Make up your own pie chart problem: using the Internet, a newspaper, or a library as a resource, find some data that you find interesting that would lend itself to being described by a pie chart, then draw the chart and use it to analyze the data.

58. Make up your own frequency polygon problem: using the Internet, a newspaper, or a library as a resource,

find some data that you find interesting that would lend itself to being divided into classes and described by a frequency polygon, then organize the data and draw the polygon. Finally, use your polygon to analyze the data.

Section 11-3   Measures of Average

LEARNING OBJECTIVES

1. Compute the mean of a data set.

2. Compute the median of a data set.

3. Compute the midrange of a data set.

4. Compute the mode of a data set.

5. Compute the mean for grouped data.

6. Compare the four measures of average.

Are you an average college student? What does that even mean? According to a variety of sources found on the Web, the average college student is 20 years old, lives on campus, and works off campus during the school year. She is female, talks to her parents every day, and comes from a family with an annual income between $50,000 and $100,000. She sleeps 6 hours a night, has $1,200 in credit card debt, and will leave college with $30,000 in total debt.

Ryan McVay/Photodisc/Getty Images

If this doesn't sound exactly like anyone you know, don't feel bad. As we will learn in this section, the word *average* is ambiguous—there are a variety of ways to describe an average. So what one source considers average might not match someone else's thoughts, and with so many different facets to a person, it's possible that nobody exactly meets all of the average criteria.

Our goal in Section 11-3 will be to understand the different measures of average. In casual terms, average means the most typical case, or the center of the distribution. Measures of average include the *mean, median, mode,* and *midrange.*

Let's suppose that an individual, let's just call him George, is in the market for a new job. George finds a small company, Vandelay Industries, which he's considering going to work for. While researching this company, George finds a flier on which that company advertises that its average employee makes almost $150,000 per year. This sounds pretty darn good to George, so he accepts an entry-level position at a low salary, betting on quick advancement to that alleged average.

Shortly after starting his career, while emptying the recycling bin in the copy room (just one of George's glamorous duties), he stumbles across a list of salaries for all employees at his new company. Let's investigate just how true the company's average salary claim is.

Here's the salary list for Vandelay Industries, which we'll use to study measures of average, which are more technically called **measures of central tendency**.

Employee	Salary	Employee	Salary
Jerry	$58,000	Susan	$51,000
Kramer	$65,000	Tim	$53,000
Newman	$944,000	Estelle	$55,000
George	$20,000	Frank	$50,000
Elaine	$52,000		

The Mean

There's a pretty good chance that when you think of "average" in a math class, what you're thinking of is actually called the *mean.* The mean, also known as the arithmetic average, is found by adding the values of the data and dividing by the total number of values. Here's a good way to think of the mean of a list of numbers: picture a yardstick labeled like a number line, and put a quarter down at each number on the list. Then try to balance the yardstick on your finger so that it doesn't tip in either direction. The mean of that list is where you'd have to put your finger to get it to balance. (You can actually try this, by the way. I did, and it worked.)

The Greek letter Σ (sigma) is used to represent the sum of a list of numbers. If we use the letter X to represent data values, then ΣX means to find the sum of all values in a data set. Using this notation:

> The **mean** is the sum of the values in a data set divided by the number of values. If $X_1, X_2, X_3, \ldots, X_n$ are the data values, we use $\overline{X}$ to stand for the mean, and
>
> $$\overline{X} = \frac{X_1 + X_2 + X_3 + \cdots + X_n}{n} = \frac{\Sigma X}{n}$$
>
> The symbol $\overline{X}$ is read as "X bar."

Example 1   Computing the Mean of a Data Set

Find the mean of all salaries for Vandelay Industries. Is the company's claim technically truthful? Do you think it's deceiving? Explain.

SOLUTION

The company has nine employees, so we need to add all the salaries and then divide the sum by 9.

$$\overline{X} = \frac{58 + 65 + 944 + 20 + 52 + 51 + 53 + 55 + 50}{9} = 149.8$$

All of the salaries were whole numbers of thousands, so it was easier to just add the number of thousands and divide by 9. The result of 149.8 tells us that the mean salary is 149.8 thousand dollars, or $149,800. So the claim is, in fact, truthful—provided that by "average" you mean "mean." But is it deceiving? You bet it is! There's only one person in the company that makes more than $65,000 per year—the owner (Newman) who pays himself a handsome salary of $944,000. Given that we want measures of average to describe a most typical case, $149,800 certainly doesn't fit that bill.

Try This One   1

The total areas, in thousands of square miles, of the 15 largest states in the United States are provided in the list below. Find the mean area for these 15 states. Do you think it's an accurate reflection of the most typical size? Why or why not?

665	269	164	147	122	114	111	104
98	98	97	87	85	84	82	

1. Compute the mean of a data set.

The Median

According to the National Organization of Realtors, the mean selling price for existing homes nationally in November 2021 was $372,400. The median selling price, on the other hand, was $353,900. Since both of these are measures of central tendency, how can they differ by almost $20,000? What exactly is the median? In this case, the median tells us that half of all homes sold during that month were priced above $353,900, and that half were priced below that amount.

In short, the **median** of a data set is the value in the middle if all values are arranged in order. The median will either be a specific data value in the set, or will fall in between two values. Here's how we find a median:

Steps in Computing the Median of a Data Set

Step 1   Arrange the data in order, from smallest to largest. Actually, largest to smallest will work, too. Whatever makes you happy.

Step 2   If the number of data values is odd, the median is the value in the exact middle of the list. If the number of data values is even, the median is the mean of the two middle data values.

Example 2   Finding the Median of a Data Set

(a) Find the median salary for Vandelay Industries. How does it compare to the mean?
(b) Find the mean and median if Newman's salary is left out. What can you conclude?

SOLUTION

(a) First, we need to arrange the salaries in order:

$20,000, $50,000, $51,000, $52,000, $53,000, $55,000, $58,000, $65,000, $944,000

There are nine salaries listed, and where I come from, nine is odd. So the median will be the salary right in the middle: there will be four salaries less and four more. That makes it the fifth salary on the list, which is $53,000. This is a whole lot less than the mean of $149,800, and in fact is a much more reasonable measure of average for these data.

(b) Here's the ordered list if we leave off Newman's gigantic salary:

$20,000, $50,000, $51,000, $52,000, $53,000, $55,000, $58,000, $65,000

Now there are eight salaries, so we'll need to find the mean of the two in the middle, which are $52,000 and $53,000. It would be nice if you could just figure out that the mean is halfway in between, but for the sake of completeness:

$$\frac{\$52,000 + \$53,000}{2} = \$52,500$$

So now the median is $52,500. The new mean is

$$\frac{\$20,000 + \$50,000 + \$51,000 + \$52,000 + \$53,000 + \$55,000 + \$58,000 + \$65,000}{8} = \$50,500$$

Now that's interesting. The median was almost unaffected by throwing away the largest value, but the mean changed dramatically, to say the least. This is exactly why the mean was a poor measure of average for this data set: the one very large value has a great impact on the mean, but not so much on the median.

Try This One   2

(a) Find the median of the land areas provided in Try This One 1. Compare to the mean, and discuss the difference. Try to come up with an explanation for the discrepancy.
(b) Find the mean and median if you throw out Alaska, which has the largest area by far. What does this tell you?

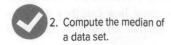
2. Compute the median of a data set.

CAUTION	When finding the median of a data set, make sure the data values are arranged in order!

The Midrange

The **midrange** is another measure of average. The advantage of the midrange is that it's very quick and easy to calculate. The disadvantage is that it totally ignores most of the data values, so it's not a particularly reliable measure. But in a pinch, it can give you a rough idea of average.

Finding the Midrange for a Data Set

$$\text{Midrange} = \frac{\text{lowest value} + \text{highest value}}{2}$$

Example 3	Finding the Midrange of a Data Set

Find the midrange of all salaries at Vandelay Industries. Is it meaningful in this case?

SOLUTION

It's not *necessary* to put a data set in order to find the midrange, but it sure doesn't hurt. All we need to know is the lowest and highest salaries, and since we already ordered the list in Example 2, it's easy to see that those are $20,000 and $944,000. So the midrange is

$$\frac{\$20,000 + \$944,000}{2} = \$482,000$$

Wow. The midrange is a whopping $482,000, which is meaningful in that it emphasizes how big Newman's salary is, but as a measure of average it's not good for much.

Purestock/Getty Images

The mean and median age for this group are different. The grandfather's age pulls up the mean, while the four children bring down the median.

 3. Compute the midrange of a data set.

Try This One	3

Find the midrange area for the top 15 largest states from Try This One 1. Do you think it's a reasonable measure of average?

The Mode

The fourth measure of average is called the *mode*. The mode is usually described to be the most typical case.

The value that occurs most often in a data set is called the **mode**.

A data set can have more than one mode or no mode at all. These situations will be shown in some of the examples that follow.

Example 4	Finding the Mode of a Data Set

These data represent the duration (in days) of the final 20 U.S. space shuttle voyages. Find the mode.

11 12 13 12 15 12 15 13 15 12 12 15 13 10 13 15 11 12 15 12

Source: *Wikipedia*

Source: NASA

SOLUTION

If we construct a frequency distribution, it will be easy to find the mode—it's simply the value with the greatest frequency. The frequency distribution for the data is shown to the right, and the mode is 12.

Days	Frequency	Days	Frequency
10	1	13	4
11	2	15	6
12	7		

Try This One   4

The data below represent the number of fatal commercial airline incidents in the United States for each year from 1998–2011. Find the mode.

1   2   3   6   0   2   2   3   2   1   2   1   0   0

Example 5   Finding the Mode of a Data Set

Math Note

When a data set has two modes, it's called *bimodal*.

The number of Atlantic hurricanes for each of the years from 1999–2021 is shown in the list. Find the mode, and describe what that tells you.

8   8   9   4   7   9   15   5   6   8   3   12   7   10   2   6   4   7   10   8   6   14   7

SOLUTION

This time, we'll find the mode without making a frequency distribution. Instead, we can just work down the list, counting the number of occurrences for each number of hurricanes. It turns out that there are two numbers that appear four times, while no others appear more than three times. Those numbers are 7 and 8, so this data set has two modes. This means that over that 22-year span, the most common number of Atlantic hurricanes was 7 and 8.

Try This One   5

The table below lists the average high temperature in degrees Fahrenheit for each month of the year on the island of Antigua. Find the mode.

Month	Jan	Feb	Mar	Apr	May	Jun	Jul	Aug	Sep	Oct	Nov	Dec
High	81	82	82	83	85	86	87	87	87	86	84	82

Example 5 shows that it's possible for a data set to have more than one mode. It's also possible for a data set to have no mode if there's no data value that occurs more than once. For an example, you can just look back at the salaries for Vandelay Industries: everyone has a different salary, so there's no mode.

The mode is probably most useful when data are classified by groups or categories. In fact, it's the only measure of central tendency that even makes sense when the data values aren't numbers, like in Example 6.

Example 6   Finding the Mode for Categorical Data

A survey of the junior class at Fiesta State University shows the following number of students majoring in each field. Find the mode.

Image Source/Getty Images

The winner of an election by popular vote is the mode.

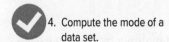

4. Compute the mode of a data set.

Business	1,425
Liberal arts	878
Computer science	632
Education	471
General studies	95

SOLUTION

You have to be a little careful here. If you focus on the numbers, you might conclude that there's no mode, since they're all different. But that would be missing the point. The mode is supposed to be the most typical case. Here, the most typical major is the one with the most students: that's business, so that's the mode.

Try This One   6

Five hundred college graduates were asked how much they donate to their alma mater on an annual basis. Find the mode of their responses, summarized below.

$500 or more	45
Between 0 and $500	150
Nothing	275
Refused to answer	30

The Mean for Grouped Data

We now return to the mean, this time in a situation that's a bit more complicated: when we have grouped data in a frequency distribution. The procedure for finding the mean for grouped data uses the midpoints and the frequencies of the classes. To see why, let's take another look at the grouped frequency distribution for the record high temperature in each state from earlier in the chapter.

Class	Frequency
100–104	3
105–109	8
110–114	16
115–119	13
120–124	7
125–129	2
130–134	1

Take a look at the third class: all we know from this frequency distribution is that there were 16 high temperatures from 110 to 114 degrees. But we have no idea what the EXACT temperatures were, so we can't add the temperatures the way we normally would for computing a mean.

Instead, we'll make an educated guess. With 16 temperatures in that range, it would be reasonable to assume that some were closer to 110, some were closer to 114, and others were right in the middle. We might even guess that the average of those 16 temperatures is likely to be 112. In that case, the sum of all the actual temperatures would be the same as the number of temperatures in that range (16) times their average (112). So we could use that number as the sum of those temperatures. We could then repeat for each class, giving us an approximate total of all temperatures. Then we'd divide by the total number as always to get the approximate mean.

Since this procedure only gives an approximate value for the mean, it's typically used in two situations. One is when the data set is very large and calculating the exact mean using all the data is impractical. The other is when the original raw data are unavailable but have been grouped by someone else.

Math Note

The symbol $\overline{X}$ is used to represent the mean of a sample within a population. The Greek lowercase letter mu (μ) is often used to represent the mean for an entire population.

Finding the Mean for Grouped Data

Step 1:   Find the midpoint of each class in the grouped data.

Step 2:   Multiply the frequency for each class by the midpoint of that class.

Step 3:   Add up all of the products from step 2.

Step 4:   Divide by the sum of all frequencies (which is the total number of data values).

If you prefer formulas to procedures:

$$\overline{X} = \frac{\Sigma(f \cdot X_m)}{n}$$

where f is the frequency for each class, X_m is the midpoint of each class, and n is the sum of all frequencies.

Example 7 | Finding the Mean for Grouped Data

Find the mean record high temperature for the 50 states.

SOLUTION

First, we'll need the midpoint for each class. Since we'll need to multiply by the frequencies, it's convenient to make a new table with the midpoints and frequencies, then multiply them. We'll also need the sum of those products and of the frequencies.

Class	Midpoint	Frequency	Midpoint × Frequency
100–104	102	3	306
105–109	107	8	856
110–114	112	16	1,792
115–119	117	13	1,521
120–124	122	7	854
125–129	127	2	254
130–134	132	1	132
Sums		50	5,715

To get our mean, we divide the sum of the products by the sum of the frequencies:

$$\overline{X} = \frac{5,715}{50} = 114.3$$

The mean state record high temperature is about 114.3°.

Try This One   7

The data at right are the number of losses by the team that won the NCAA men's basketball championship for the year from 1939–2016. Find the mean number of losses.

Losses	Frequency	Losses	Frequency
0–1	13	6–7	10
2–3	30	8–9	4
4–5	22	10–11	3

Can we find (or at least approximate) other measures of average for grouped data? We'll explore this in the Critical Thinking exercises.

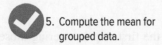

5. Compute the mean for grouped data.

Using Technology to Compute Measures of Central Tendency

Technology is certainly a useful tool in computing measure of average, so let's have a look at using spreadsheets and graphing calculators to do the heavy lifting for us. (For what it's worth, when a list of data has just a few numbers on it, it's probably quicker to find the measures of average by hand. It's the large data sets that make technology incredibly useful.)

To illustrate these features, we'll use the following data set. According to deathpenalty-info.org, there were 9 death row inmates that were exonerated and freed in 2009 (which is horrifying if you think about it). The table below shows the number freed in the 12 subsequent years.

Year	'09	'10	'11	'12	'13	'14	'15	'16	'17	'18	'19	'20	'21
Number	9	1	1	3	1	7	6	0	5	2	3	6	2

Spreadsheet

There are built-in commands for finding the mean and median in Excel. There's no midrange command, but we can build one, and ordering the list allows us to find the mode.

First, the data should be entered into a single column, as shown in Figure 11-6(a). To calculate the mean, choose an empty cell and enter "=AVERAGE(". Then use the cursor to select all of the cells containing the data, enter a close parenthesis, and hit enter. This displays the mean, shown in cell B15. To find the median, use the same routine, but enter "MEDIAN" rather than "AVERAGE". This is in cell B16.

There's no direct command for midrange, but there are commands to find the highest and lowest numbers on the list; we can embed them in a simple calculation that finds the midrange. The string we need to enter looks like

$$= (MAX(range) + MIN(range))/2$$

where "range" represents the range of cells that the data live in. Again, the simplest way to enter that range is to just use the cursor to select it after typing "MAX(" and "MIN(". The resulting midrange in shown in cell B17.

To find the mode, we'll first sort the list. Select the cells containing the data (B1 through B13 in this case), then choose "Sort..." from the "Data" menu. You can choose to sort in either ascending or descending order. The result, shown in Figure 11-6(b), makes it a simple matter to find the value or values that occur most often. Notice that (as we'd hope) sorting didn't affect the mean, median, or midrange. Figure 11-6(c) shows the commands entered for finding the mean, median, and midrange.

	A	B
1		9
2		1
3		1
4		3
5		1
6		7
7		6
8		0
9		5
10		2
11		3
12		6
13		2
14		
15	Mean	3.53846154
16	Median	3
17	Midrange	4.5

Figure 11.6 (a)

	A	B
1		0
2		1
3		1
4		1
5		2
6		2
7		3
8		3
9		5
10		6
11		6
12		7
13		9
14		
15	Mean	3.53846154
16	Median	3
17	Midrange	4.5

(b)

B15 f_x =AVERAGE(B1:B13)

B16 f_x =MEDIAN(B1:B13)

B17 f_x =(MAX(B1:B13)+MIN(B1:B13))/2

(c)

Source: Microsoft Corporation

Graphing Calculator

In order to find measures of average for a group of numbers, the first step is entering those numbers in a list. To do this, we need to use the list editor: hit the STAT key, then choose 1. Enter the numbers one at a time, hitting ENTER after each. The exoneration data that we looked at previously are shown entered in Figure 11-7(a) (most of them, anyhow). Then hit STAT again and use the right arrow to select CALC at the top of the screen, and pick choice 1: 1-Var Stats [Figure 11-7(b)]. After hitting ENTER twice, you get the 1-Variable Stats menu, shown in Figures 11-7(c and d).

There's actually quite a bit more information on this screen than we're looking for, and some of it will be discussed in later lessons. For now, the mean is the very first thing on the list: $\bar{x} = 3.538461538$. Using the down arrow to scroll down, we find the median: Med = 3 [Figure 11-7(d)]. The midrange isn't displayed, but minX = 0 and maxX = 9 provide the lowest and highest values. When you have those, it's simple to add them and divide by 2 to get the midrange of 4.5.

As with Excel, there's no specific command for finding the mode, but the calculator will order the data, which makes it a lot easier to find the mode. Hit the STAT key again and choose 2, which is SortA (for "sort ascending"). Then you need to tell it which list to sort. Since we entered the data in list L1, enter that by hitting 2nd 1 [Figure 11-8(a)]. After you press ENTER, the calculator displays Done, and the list is ordered. When you go back to the list editor, you can easily find which value or values appear most often, giving you the mode [Figure 11-8(b)].

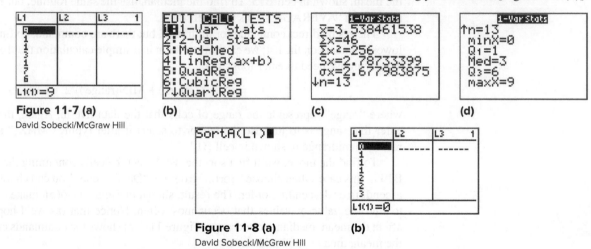

Figure 11-7 (a)   (b)   (c)   (d)
David Sobecki/McGraw Hill

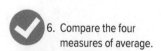

Figure 11-8 (a)   (b)
David Sobecki/McGraw Hill

Comparing Measures of Average

Since we studied four measures of average, it's a good idea to conclude the section by comparing them.

6. Compare the four measures of average.

| Example 8 | Comparing Measures of Average |

For the Vandelay Industries salary data, compare the four measures of average. Which do you think is the best description of the true average?

SOLUTION

Here's a summary of the measures of average, with the salaries repeated one more time for reference:

$20,000, \$50,000, \$51,000, \$52,000, \$53,000, \$55,000, \$58,000, \$65,000, \$944,000$

Mean: $149,800
Median: $53,000
Midrange: $482,000
Mode: None

Certainly the mode isn't helpful for this data set. In fact, the only one that could possibly be considered as a reasonable average is the median. Aside from Newman's $944k, nobody makes more than $65,000, so any "average" that's more than twice that isn't really a true reflection of the typical salary.

Try This One 8

For the land area of the top 15 states that we studied in several Try This One questions, compare the four measures of average. Which do you think is the best description of the true average?

To conclude the section, Table 11-2 summarizes some of the strengths and weaknesses of each measure of average.

Table 11-2 A Comparison of Measures of Average

Measure	Strengths	Weaknesses
Mean	• Unique—there's exactly one mean for any data set • Factors in all values in the set • Easy to understand	• Can be adversely affected by one or two unusually high or low values • Can be time-consuming to calculate for large data sets
Median	• Divides a data set neatly into two groups • Not affected by one or two extreme values	• Can ignore the effects of large or small values even if they are important to consider
Mode	• Very easy to find • Describes the most typical case • Can be used with categorical data like candidate preference, choice of major, etc.	• May not exist for a data set • May not be unique • Can be very different from mean and median if the most typical case happens to be near the low or high end of the range
Midrange	• Very quick and easy to compute • Provides a simple look at average	• Dramatically affected by extremely high or low values in the data set • Ignores all but two values in the set

Sidelight The Birth of Statistics

The study of statistics has its origin in censuses taken by the Babylonians and Egyptians almost 6,000 years ago. "Birth" is an appropriate word to use in describing the beginning of statistics, since it started with birth records. The infancy of statistics occurred much later, some time around 27 BCE, when the Roman emperor Augustus conducted surveys on births and deaths of citizens, as well as the amount of livestock each owned and the crops each had harvested. Suddenly, a large amount of data was wanted, and methods had to be developed to collect, organize, and summarize that data.

The adolescence of statistics is probably sometime around the 14th century, when people began keeping detailed records of births, deaths, and accidents in order to determine insurance rates. Over the next 600 years or so, statistics grew up, with a wide variety of methods developed to fill needs in biology, physics, astronomy, commerce, and many other areas.

The true adulthood of statistics coincides with the rise of computer technology. Today, with just a few clicks of a mouse, statisticians can process and analyze amounts of data that would have taken the pioneers in the field centuries.

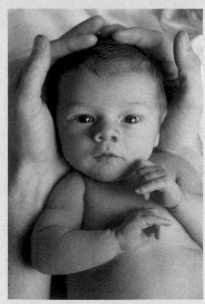

Davis Frare Photography/Brand X Pictures/ Corbis

Answers to Try This One

1   155,133 sq. mi. It seems a little high to be an accurate reflection; only three of the states have areas bigger than that, while 12 are less.

2   (a) 104,000 sq. mi. This is quite a bit lower than the mean: the one extremely large state raises the mean considerably.
    (b) Mean: 118,714 sq. mi.; median: 101,000 sq. mi.; that one large value did in fact skew the mean upward.

3   373,500 sq. mi. It's more than twice as much as the mean and median.

4   2 incidents

5   The modes are 82° and 87°.

6   The mode is donating nothing.

7   3.79 losses

8   Mean: 155,133 sq. mi.; median: 104,000 sq. mi.; midrange: 373,500 sq. mi.; mode: none. The median appears to be the best indicator of the most typical case.

Exercise Set   11-3

Writing Exercises

1. If someone asked you before covering this section what the word "average" meant, what would you have said? What would your response be now?
2. Describe how to find the mean of a data set.
3. Describe how to find the median of a data set.
4. What is meant by the term *mode of a data set*?
5. Describe how to find the midrange of a data set.
6. Which of the measures of average can be used to describe the average for categorical data? Why?

In Questions 7–10, describe some strengths and weaknesses of using the given measure of central tendency to describe the average for a data set.

7. Mean
8. Median
9. Mode
10. Midrange

Applications in Our World

For Exercises 11–22, find the mean, median, mode, and midrange. Then decide which you think is the most meaningful measure of average for that data set. For larger data sets, you should consider using technology.

11. These data are the annual in-state tuition, in thousands of dollars, for the 11 largest universities in Ohio.
    10 10 13 11 9 13 9 9 11 10 9

12. The data below show the number of attorneys employed by the 10 largest law firms in Pittsburgh.
    87   109   57   221   175   123   170   80   66   80

13. A graduate student in social work surveyed students at 10 colleges in a Midwestern state to find how many are single parents. The results are the data below.
    700 298 638 260 1,380 280 270 1,350 380 570

14. The number of wins for the teams that won college football bowl games at the end of the 2021 season:
    10 12 12 11 10 11 13 6 10 8 9 8 7 10 12 6 6
    11 8 6 8 6 7 7 11 9 6 8 6 9 10 6 8 8 10 8 8
    9 7 6

15. The average number of cigarettes smoked per person in a year in the top 10 countries for smoking are shown next.

Greece	4,313	Malta	2,668
Hungary	3,265	Bulgaria	2,574
Kuwait	3,062	Belarus	2,571
Japan	3,023	Belgium	2,428
Spain	2,779	Turkey	2,394

16. The number of existing home sales in millions for the months from April 2020 to November 2021 are shown below.
    4.4 4.0 4.8 5.9 6.0 6.4 6.7 6.6 6.7 6.7 6.2 6.0 5.9
    5.8 5.9 6.0 5.9 6.3 6.3 6.5

17. The number of bids for the five most popular products under concert tickets on eBay for a day in April is shown below.
    340   75   123   259   151

18. The number of beds at all licensed hospitals in Hillsborough County:
    398   120   493   206   73   102   183   870   76
    60   60   112   147   1,018   201

19. From 2006 to 2018, the number of deaths by suicide at U.S. schools each year was
    9 5 7 2 3 5 6 8 9 3 6 8

20. According to the website *Best Value Schools*, the nine master's degrees with the best earning potential have average salaries as listed:
$187,600 $162,800 $114,040 $138,300 $135,300
$131,800 $129,900 $128,200 $126,800

21. A data breach is an incident wherein information is stolen or taken from a system without the knowledge or authorization of the system's owner. The annual number of data breaches in the United States from 2005 to 2020 were:

Year	Data Breaches (millions)	Year	Data Breaches (millions)
2005	157	2013	614
2006	321	2014	783
2007	446	2015	784
2008	656	2016	1,106
2009	498	2017	1,632
2010	662	2018	1,257
2011	419	2019	1,473
2012	447	2020	1,001

Source: *https://www.statista.com/*

22. The 12 states with the highest unemployment rates in November 2021 and their rates of unemployment were:

State	Rate
California	6.9
Nevada	6.8
New Jersey	6.6
New York	6.6
New Mexico	6.2
Alaska	6.0
Connecticut	6.0
District of Columbia	6.0
Hawaii	6.0
Michigan	5.9
Illinois	5.7
Pennsylvania	5.7

For Exercises 23–30, find the mean of the data set.

23. For 100 students in the student union at one campus, the distribution of the students' ages was obtained as shown.

Class	Frequency	Class	Frequency
18–20	36	27–29	9
21–23	29	30–32	3
24–26	16	33–35	7

24. Forty new automobiles were tested for fuel efficiency by the EPA (in miles per gallon). This frequency distribution was obtained:

Class	Frequency	Class	Frequency
8–12	1	23–27	15
13–17	3	28–32	8
18–22	8	33–37	5

25. In a study of the time in minutes it takes an untrained mouse to run a maze, a researcher recorded these data in seconds.

Class	Frequency
2.1–2.7	5
2.8–3.4	7
3.5–4.1	12
4.2–4.8	14
4.9–5.5	16
5.6–6.2	8

26. Eighty randomly selected compact fluorescent lightbulbs were tested to determine their lifetimes (in hours). The frequency distribution was obtained as shown.

Class	Frequency
651–800	6
801–950	12
951–1,100	25
1,101–1,250	18
1,251–1,400	14
1,401–1,550	5

27. These data represent the net worth (in millions of dollars) of 45 national corporations.

Class	Frequency
10–20	2
21–31	8
32–42	15
43–53	7
54–64	10
65–75	3

28. The cost per load (in cents) of 35 laundry detergents tested by *Consumer Reports* is shown here.

Class	Frequency
13–19	2
20–26	7
27–33	12
34–40	5
41–47	6
48–54	1
55–61	0
62–68	2

29. The frequency distribution shown represents the commission earned (in dollars) by 100 salespeople employed at several branches of a large chain store.

Class	Frequency	Class	Frequency
150–158	5	186–194	20
159–167	16	195–203	15
168–176	20	204–212	3
177–185	21		

30. This frequency distribution represents the data obtained from a sample of 75 copy machine service technicians. The values represent the days between service calls for various copy machines.

Class	Frequency	Class	Frequency
16–18	14	25–27	10
19–21	12	28–30	15
22–24	18	31–33	6

Critical Thinking

31. For each of the characteristics of the average college student listed on page 670, decide which measure of average you think was used, and explain your choices.

32. For these situations, state which measure of average— mean, median, or mode—should be used.
    (a) You're interested in the most typical case.
    (b) The data are categorical.
    (c) The values are to be divided into two approximately equal groups, one group containing the larger and one containing the smaller values.

For Exercises 33–38, describe which measure of average— mean, median, or mode—was probably used in each situation.

33. Half of the factory workers make more than $12.37 per hour and half make less than $12.37 per hour.
34. The average number of children per family in Plaza Heights is 1.8.
35. Most people prefer red convertibles over any other color.
36. The average person cuts the lawn once a week.
37. The most common fear today is fear of speaking in public.
38. The average age of college professors is 42.3 years.

Exercises 39 and 40 refer back to Exercise 24. The full data set is shown here.

37	27	20	30	24	17	32	26	17	32	25	28
20	18	27	32	36	25	9	26	34	30	25	23
25	24	20	13	31	20	20	25	20	22	37	27
27	25	37	30								

39. Find the exact mean. How does it compare to the mean calculated in Exercise 24? Are you surprised? Why or why not?
40. (a) Find the median, mode, and midrange for the data.
    (b) Which measure of central tendency would a government official choose to show if they wanted to make the case that gas mileage was getting better for newer vehicles? Why?
    (c) Which measure of central tendency would an environmental lobbyist choose to show if they wanted to

make the case that gas mileage was still not very good for newer vehicles? Why?

41. Under what circumstances would the median for a data set be likely to be close to the mean? When would those two measures of average be very different?
42. We haven't talked about descriptive and inferential statistics in this section, so let's get to it. The calculations that we did to find measures of average for the state record high temperature data were most definitely descriptive statistics. Why?
43. In Exercise 26, we calculated the mean life for 80 randomly selected bulbs. Is this descriptive or inferential statistics? Does it depend on what we do with the data?
44. Based on the results of Exercises 42 and 43, describe when measures of average are part of descriptive statistics, and when they're part of inferential statistics.
45. Based on the 2020 census, the median population of a state was 4.5 million, while the mean population was 6.5 million.
    (a) What factors do you think account for the discrepancy in these measures of average?
    (b) Which of the measures of average do you think is a better indicator of what the average state population might be? Why?
46. When finding the mean for grouped data, we multiply the midpoint of each category by its frequency. Why? What are we actually doing when we use this technique to find the mean?

In Exercises 47–50, we'll think about measures of central tendency for grouped data other than the mean, which we're already experts on. We'll be referring to the grouped distribution in Exercise 26.

47. First, let's talk about the median. Talk about what you might do to estimate the median for grouped data, then execute your plan. Make note of your estimate, because we'll need it in a bit.

48. Now we turn our attention to the midrange. Devise, describe, and then execute a plan to estimate the midrange, and again make note of your estimate.

49. Now the mode: can you make an estimate of the mode based on the grouped frequency distribution? Be careful—think about the fact that there may not even BE a mode. If you can estimate, do so.

50. USING TECHNOLOGY. We know that in order to compute exact values for the measures of average, we'd need all of the data, not just the grouped distribution.

Since there are 80 data values, computing the measures of average by hand isn't exactly something you'd hope to do with your valuable time. Fortunately, we've set up a spreadsheet (available in online resources for this section) that has all of the data loaded. Your job is to use the Excel commands from this section to compute all four measures of average. Then discuss how they compare to the estimated values from Exercises 26, 47, 48, and 49.

Section 11-4 | Measures of Variation

LEARNING OBJECTIVES

1. Find the range of a data set.

2. Find the variance and standard deviation of a data set.

3. Interpret standard deviation.

Now that we know about measures of average, we'll consider the fact that there's more to the story told by a data set than just the average. For an example, consider the two pictures of dogs on this page.

If we look only at measures of average, particularly the mean, we might be fooled into thinking that the two groups are very similar, when clearly they are not. The difference, of course, is that all of the dogs in the first picture are of similar size, while those in the second picture have many different weights. Because there are some small dogs and one very large one, the mean weight in both groups is probably similar.

In this section we will study *measures of variation*, which will help to describe how the data within a set vary. The three most commonly used measures of variation are *range*, *variance*, and *standard deviation*.

Tim Davis/Corbis/VCG/Getty Images

Chris Collins/The Image Bank/ Getty Images

The two groups of dogs have about the same mean size, but the range of sizes is quite different.

Range

The *range* is the simplest of the three measures of variation that we will study.

The **range** of a data set is the difference between the highest and lowest values in the set.

$$\text{Range} = \text{Highest value} - \text{lowest value}$$

Example 1 | Finding the Range of a Data Set

The first list below is the weights of the dogs in the first picture, and the second is the weights of the dogs in the second picture. Find the mean, median, and range for each list, then describe any observations you can make based on the results.

70 73 58 60
30 85 40 125 42 75 60 55

SOLUTION

For the first list,

$$\text{Mean} = \frac{70 + 73 + 58 + 60}{4} = 65.25 \text{ lb}$$

The ordered list is 58, 60, 70, 73, so the median is the mean of 60 and 70, which is 65.

$$\text{Range} = 73 - 58 = 15 \text{ lb}$$

For the second list,

$$\text{Mean} = \frac{30 + 85 + 40 + 125 + 42 + 75 + 60 + 55}{8} = 64 \text{ lb}$$

The ordered list is 30, 40, 42, 55, 60, 75, 85, 125, so the median is the mean of 55 and 60, which is 57.5.

$$\text{Range} = 125 - 30 = 95 \text{ lb}$$

As we suspected, the means and medians are similar, but the ranges are very different. This is reflective of the fact that there's a lot more variation in size among the dogs in the second picture.

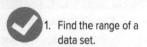

1. Find the range of a data set.

Try This One   1

The monthly high temperatures from January to December in Aruba and St. Louis are shown below. Find the median and range for each, then write any observations you make based on this information.

Aruba:	85°	85°	86°	87°	88°	89°	88°	89°	89°	89°	87°	85°
St. Louis:	38°	45°	55°	66°	77°	86°	91°	88°	81°	69°	54°	42°

Variance and Standard Deviation

The range is a limited measure of variation because it ignores all the data except the highest and lowest values. If most of the values are similar, but there's just one unusually high value, the range will make it look like there's a lot more variation than there actually is. For this reason, we will next define *variance* and *standard deviation*, which are much more reliable measures of variation.

Variance and standard deviation are a little tricky to compute. In the box below, we'll explain how to find them. Later in the section, we'll discuss their significance.

Procedure for Finding the Variance and Standard Deviation

Step 1 Find the mean.

Step 2 Subtract the mean from each data value in the data set.

Step 3 Square the differences.

Step 4 Find the sum of the squares.

Step 5 Divide the sum by $n-1$ to get the variance, where n is the number of data values.

Step 6 Take the square root of the variance to get the standard deviation.

Example 2	Finding Variance and Standard Deviation

Find the variance and standard deviation for the weights of the eight dogs in the second picture at the beginning of this section. The weights are listed again for reference.

$$30 \quad 85 \quad 40 \quad 125 \quad 42 \quad 75 \quad 60 \quad 55$$

SOLUTION

Step 1 Find the mean weight.
We found the mean of 64 lb in Example 1.

Step 2 Subtract the mean from each data value.

$30 - 64 = -34$	$85 - 64 = 21$	$40 - 64 = -24$	$125 - 64 = 61$
$42 - 64 = -22$	$75 - 64 = 11$	$60 - 64 = -4$	$55 - 64 = -9$

Step 3 Square each result.

$(-34)^2 = 1,156$	$(21)^2 = 441$	$(-24)^2 = 576$	$(61)^2 = 3,721$
$(-22)^2 = 484$	$11^2 = 121$	$(-4)^2 = 16$	$(-9)^2 = 81$

Step 4 Find the sum of the squares.

$$1,156 + 441 + 576 + 3,721 + 484 + 121 + 16 + 81 = 6,596$$

Step 5 Divide the sum by $n - 1$ to get the variance, where n is the sample size. In this case, n is 8, so $n - 1 = 7$.

$$\text{Variance} = \frac{6,596}{7} \approx 942.3$$

Step 6 Take the square root of the variance to get standard deviation.

$$\text{Standard Deviation} = \sqrt{942.3 \text{ lb}} \approx 30.7 \text{ pounds}$$

Math Note

Since it's obviously pretty involved to compute the variance and standard deviation, it shouldn't be a surprise that we'll learn how to use technology to help us out in a bit.

To organize the steps, you might find it helpful to make a table with three columns: the original data, the difference between each data value and the mean, and their squares. Then you just add the entries in the last column and divide by $n - 1$ to get the variance.

Data (X)	$X - \overline{X}$	$(X - \overline{X})^2$
30	−34	1,156
85	21	441
40	−24	576
125	61	3,721
42	−22	484
75	11	121
60	−4	16
55	−9	81

Try This One	2

Find the variance and standard deviation for the four dogs in the first picture at the beginning of the lesson. How does the standard deviation compare to the standard deviation for the dogs in the second picture? Why do you think this is reasonable?

$$70 \quad 73 \quad 58 \quad 60$$

To understand the significance of standard deviation, we'll look at the process one step at a time.

Step 1 *Compute the mean.*   Variation is a measure of how far the data vary from the mean, so it makes sense to begin there.

Step 2 *Subtract the mean from each data value.*   In this step, we are literally calculating how far away from the mean each data value is. The problem is that since some are greater than the mean and some less, their sum will always add up to zero. (Try it!) So that doesn't help much.

Step 3 *Square the differences.*   This solves the problem of those differences adding to zero—when we square them, they're all positive.

Step 4 *Add the squares.*   In the next two steps, we're getting an approximate average of the squares of the individual variations from the mean. First we add them, then . . .

Step 5 *Divide the sum by $n - 1$.*   It seems like dividing by the number of values (n) here is a good idea, but it turns out that when we're using a sample from a larger population to compute mean and variance, dividing by $n - 1$ makes the sample variance more likely to be a true reflection of the population variance. In any case, at this point we have an approximate average of the squares of the individual variations from the mean. This will be discussed in greater depth in Exercise 34.

Step 6 *Take the square root of the sum.*   This "undoes" the square we did in Step 3. It will return the units of our answer to the units of the original data, giving us a good measure of how far the typical data value varies from the mean.

> ### Math Note
> We use s to represent the variance for a sample. The lowercase Greek sigma (σ) represents the standard deviation for an entire population. We'll discuss the difference between the two in the Critical Thinking exercises.

Now we'll formally define these two key measures of variation. A lowercase s is typically used to represent standard deviation, and s^2 is used for variance.

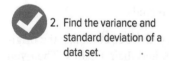

2. Find the variance and standard deviation of a data set.

The **sample variance** for a data set is an approximate average of the square of the distance between each value and the mean. If X represents individual values, $\overline{X}$ is the mean and n is the number of values:

$$s^2 = \frac{\Sigma(X - \overline{X})^2}{n - 1}$$

The **sample standard deviation** (s) is the square root of the sample variance. It provides an approximate average of the distances between data values and the mean.

Interpreting Standard Deviation

Since standard deviation measures how far typical values are from the mean, its size tells us how spread out the data are. We'll examine this idea in Examples 3 and 4.

Example 3   Interpreting Standard Deviation

The mean and standard deviation for heights of all adult males in the United States are 69.3 inches and 2.8 inches, respectively. The mean and standard deviation for the 2021–2022 Cleveland Cavaliers (a professional basketball team), on the other hand, were 77.4 inches and 3.75 inches. What can we conclude from comparing these statistics?

SOLUTION

There are two main things we can learn from these comparisons. First, the members of a professional basketball team tend to be a lot taller than average people. If you've ever watched a basketball game, this comes as no surprise. Second, the heights are spread out a bit more than the population in general. This could be due to the small sample size of 18 players, but it might also be because there are some basketball players that are good players because of speed and athleticism even though they're not that tall, while others are successful to some extent because of their extreme height. In order to draw more reliable conclusions, we'd probably need to look at a much larger sample of pro basketball players.

Try This One   3

Describe how you think the mean and standard deviations would be likely to compare for the two data sets.

Data set 1: 40-yard dash times for starting running backs in the National Football League
Data set 2: 40-yard dash times for every student in your math class

Using Technology: Finding Standard Deviation

Graphing Calculator

The 1-Var Stats screen, which we used to compute the mean and median in Section 11-3, also calculates the standard deviation. At right is another look at a screen shot from Section 11-3, where the data on death row exonerations were entered in the list editor, and then the 1-Variable Stats command was chosen. The quantities we're interested in are Sx and σx: these are the sample standard deviation (Sx) and the population standard deviation (σx).

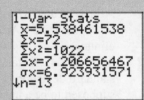

```
1-Var Stats
x̄=5.538461538
Σx=72
Σx²=1022
Sx=7.206656467
σx=6.923931571
↓n=13
```

David Sobecki/McGraw Hill

Spreadsheet

To use a spreadsheet to calculate standard deviation, the data are entered in a single column, just like we did for computing measures of average in Section 11-3. Here are sample calculations for the second group of dog weights from Example 1 in this lesson, with the formulas used to calculate each statistic illustrated to the right.

	A	B
1		30
2		85
3		40
4		125
5		42
6		75
7		60
8		55
9		
10	Range	95
11	Sample st. dev.	30.696673

Microsoft Corporation

Range

B10	▲▼	✕ ✓	*fx*	=MAX(B1:B8)-MIN(B1:B8)

Sample Standard Deviation

B11	▲▼	✕ ✓	*fx*	=STDEV(B1:B8)

Example 4   Interpreting Standard Deviation

A professor has two sections of Math 115 this semester. The 8:30 A.M. class has a mean score of 74% with a standard deviation of 3.6%. The 2 P.M. class also has a mean score of 74%, but a standard deviation of 9.2%. What can we conclude about the students' averages in these two sections?

SOLUTION

In relative terms, the morning class has a small standard deviation and the afternoon class has a large one. So even though they have the same mean, the classes are quite different. In the morning class, most of the students probably have scores relatively close to the mean, with few very high or very low scores. In the afternoon class, the scores vary more widely, with a lot of high scores and a lot of low scores that average out to a mean of 74%.

Try This One   4

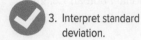

3. Interpret standard deviation.

Which data set would be likely to have the larger standard deviation? EXPLAIN.

Data set 1: Weights of all dogs at a golden retriever rescue shelter
Data set 2: Weights of all dogs for sale in a pet store

Answers to Try This One

1   Aruba, median 87.5°, range 4°; St. Louis, median 67.5°, range 53°; Aruba has a much more consistent climate.

2   Variance 54.25, standard deviation 7.37 lbs. The standard deviation is much less. Since this is supposed to be a measure of variation, you'd expect it to be smaller since the dogs are similar in size.

3   You'd certainly expect the mean to be a whole lot lower for the group of professional athletes, since they'd be much faster runners than the general

public. You'd also expect the standard deviation to get quite a bit smaller. Everyone that becomes a professional running back has to be at least reasonably fast, so you'd expect the times to be pretty consistent. But a group of students in a math class would most likely have some pretty fast runners and some dreadfully slow runners.

4   Data set 2: all of the dogs in the first set are the same breed, so you'd expect them to be reasonably similar in size.

Exercise Set   11-4

Writing Exercises

1. Name three measures of variation.
2. What is the range?
3. Why is the range not usually the best measure of variation?
4. What is the relationship between the variance and standard deviation?
5. Explain the procedure for finding the standard deviation for data.
6. Explain how the variation of two data sets can be compared by using the standard deviations.

For Exercises 7–10, discuss the relative sizes of the standard deviations for the two data sets.

7. Data set 1:   12   15   13   10   16   13   12   13
   Data set 2:   5   26   31   2   10   25   6   33
8. Data set 1:   Weights of all incoming U.S. Army recruits in February 2022
   Data set 2:   Weights of all kindergarten children in Florida
9. Data set 1:   Average monthly high temperature in Chicago
   Data set 2:   Average monthly high temperature in Los Angeles
10. Data set 1:   Amount of time spent studying per week for PhD students in math at the Ohio State University
    Data set 2:   Amount of time studying per week for all students at the Ohio State University

Applications in Our World

For Exercises 11–26, find the range, variance, and standard deviation, then for 11–18, write a sentence or two describing what the standard deviation tells you about the data. Consider using technology for the larger data sets.

11. These data are the number of junk emails Lena received for 9 consecutive days.

    61   1   1   3   2   30   18   3   7

12. The number of hospitals in the five largest hospital systems is shown here.

    340 75 123 259 151
    Source: *USA Today*

13. Ten used trail bikes are randomly selected from a bike shop, and the odometer reading of each is recorded as follows.

    1,902 103 653 1,901 788 361 216 363 223 656

14. Fifteen students were selected and asked how many hours each studied for the final exam in statistics. Their answers are recorded here.

    8 6 3 0 0 5 9 2 1 3 7 10 0 3 6

15. The weights of nine players from a college football team are recorded as follows.

    206 215 305 297 265 282 301 255 261

16. These are the numbers of stories in the 11 tallest buildings in St. Paul, Minnesota.

    37 46 32 32 33 25 17 27 21 25 33
    Source: *The World Almanac and Book of Facts*

17. The heights (in inches) of nine male army recruits are shown here.

    78 72 68 73 75 69 74 73 72

18. The number of calories in 12 randomly selected microwave dinners is shown here.

    560 832 780 650 470 920 1,090 970
    495 550 605 735

19. The table below shows the average price of a gallon of regular unleaded gas in U.S. dollars for various countries in January 2022.

Country	Price	Country	Price
Australia	4.38	Iraq	1.94
Canada	4.89	Japan	5.29
China	4.56	Mexico	4.18
France	7.31	Saudi Arabia	2.35
Germany	6.93	South Korea	5.16
Hong Kong	9.90	Taiwan	4.12
India	5.26	United Kingdom	7.49
Iran	0.19	USA	3.66

    Source: *globalpetrolprices.com*

20. The number of attorneys in 10 law firms in Pittsburgh is 87, 109, 57, 221, 175, 123, 170, 80, 66, and 80.
    Source: *Pittsburgh Tribune Review*

21. The stock prices for the top 10 rated grocery stocks according to TheStreet.com last January were $58.88, $29.75, $43.40, $110.67, $29.49, $26.21, $23.11, $11.85, $10.23, and $3.81.

22. The table reflects consumer priorities worldwide for a variety of expenditures:

Global Priority	U.S. Currency (in billions)
Cosmetics in the United States	8
Ice cream in Europe	11
Perfumes in Europe and the United States	12
Pet foods in Europe and the United States	17
Business entertainment in Japan	35
Cigarettes in Europe	50
Alcoholic drinks in Europe	105
Narcotic drugs in the world	400
Military spending in the world	780

    Source: *www.globalissues.org*

23. There were 12 outbreaks of the Ebola virus between 2007 and 2021. The number of deaths from these outbreaks were

    2,280 11,310 12 6 55 33 9 49 17 29
    37 14
    Source: *Wikipedia*

24. The five teams in baseball's American League Eastern Division had the following number of wins in 2021: 100, 92, 92, 91, 52

25. The number of workplace fatalities in the United States from 2008 to 2020 were as follows:

Year	Fatalities	Year	Fatalities
2008	5,214	2015	4,836
2009	4,551	2016	5,190
2010	4,690	2017	5,147
2011	4,693	2018	5,250
2012	4,628	2019	5,333
2013	4,585	2020	4,764
2014	4,821		

    Source: *bls.gov*

26. From 2007 to 2018, the number of deaths by suicide at U.S. schools each year was

    9 5 7 2 3 5 6 8 9 3 6 8

Critical Thinking

27. The three data sets have the same mean and range, but is the variation the same? Explain your answer.

    (a) 5 7 9 11 13 15 17
    (b) 5 6 7 11 15 16 17
    (c) 5 5 5 11 17 17 17

28. Using this set—10, 20, 30, 40, and 50,
    (a) Find the standard deviation.
    (b) Add 5 to each value and then find the standard deviation.
    (c) Subtract 5 from each value and then find the standard deviation.
    (d) Multiply each value by 5 and then find the standard deviation.
    (e) Divide each value by 5 and then find the standard deviation.
    (f) Generalize the results of (a)–(e).

29. When playing golf, in most cases the player will want to get the ball as close as possible to the hole when hitting a shot approaching the green. Suppose that two players hit 10 approach shots to a green, and their results are as shown in the two figures. The black dot is the hole, and the white dots are where each shot ended up.

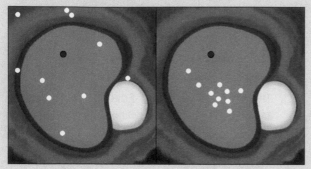

Pat's shots   Ron's shots

    (a) If you were to "average" the locations of all of the results for each player, where would you estimate the average shot would be for each?
    (b) How would you describe the measures of variation for each of the players?
    (c) Which is a better player in your opinion? What does that say about the importance of measures of average compared to measures of variation in some situations?

30. The table shows the number of class periods missed over the last 10 courses by two students.

Helena	3	4	3	2	2	4	3	2	2	3
Juanita	0	9	2	1	0	6	0	4	0	8

    (a) Find the mean number of classes missed for each student.
    (b) Without doing the calculations, how do you think the measures of variation compare for the two students?
    (c) Which is a better student in your opinion? What does that say about the importance of measures of average compared to measures of variation in some situations?

31. Think of a quantity in our world that would be likely to have a relatively small standard deviation if you compiled some data on that quantity.

32. Think of a quantity in our world that would be likely to have a relatively large standard deviation if you compiled some data on that quantity.

33. When computing standard deviation, in essence what we're finding is the average distance away from the mean for all data values. In that case, it would seem reasonable to divide a sum by n, the number of data values in the set. But, when computing what we called sample standard deviation, we instead divided by $n - 1$. Let's examine that phenomenon.
    (a) What do you think is meant by "population standard deviation" as opposed to "sample standard deviation"?
    (b) Give an example of a situation where you would be computing a population standard deviation and one where you'd be computing a sample standard deviation.

34. When we're computing a population standard deviation, we're using data from every single individual in a population. In that case, we're not leaving anything out, and in fact, we find the standard deviation as you might guess: we divide by n. But for a sample standard deviation, the data values we have represent a sample of individuals from a larger population.
    (a) If we're leaving out some (maybe many) members of the population, explain why the total amount of variation has to be no more than the total amount of variation in the population, and why in fact it's likely to be smaller.
    (b) The sum in the standard deviation formula is a measure of the total amount of variation in the data values we're using. Based on your answer to part (a), will this number be too big or too small if we're working with a sample rather than a population?
    (c) When we divide that total amount of variation by the sample size, would we want to divide by a bigger or a smaller number to correct the inaccuracy described in part (b)?
    (d) Put all the parts together to describe why we divide by n for a population standard deviation, but $n - 1$ for a sample standard deviation.

35. What does the difference between population standard deviation and sample standard deviation have to do with descriptive vs. inferential statistics?

LEARNING OBJECTIVES

1. Compute the percentile rank for a data value.

2. Find a data value corresponding to a given percentile.

3. Use percentile rank to compare values from different data sets.

4. Compute quartiles for a data set.

5. Draw a box plot.

So you managed to survive your college years and escape with that long-awaited diploma. But maybe the thought of entering the real world starts to look a little less appealing, and you realize that a couple more years of college might suit you. Next step: take the GRE (Graduate Record Examination), which is like the SAT for grad school.

When you get your score back, you see that it's 1120. How do you know if that's a good score? The quickest way is to look at the *percentile rank* on the report, which in this case would probably be 60%. Does this mean that you only got 60% of the questions right? That wouldn't be a very good score, now would it? In fact, that's not at all what it means, and you did just fine. When raw scores (or other data) are hard to interpret, it's useful to see where certain values rank within the data set, and that's what we'll study in this section.

PictureQuest/Comstock Images/Getty Images

The term percentile is used in statistics to measure the position of a data value in a data set.

> A **percentile,** or percentile rank, of a data value indicates the percent of data values in a set that are below that particular value.

Tetra Images/Getty Images

In this case, your percentile rank of 60% means that 60% of all students who took the GRE scored lower than you did. Not bad! Scores on standardized tests are one of the most common uses for percentiles because they help you to put a raw score like 1120 into context.

Percentiles are also used very commonly in health care, especially pediatrics. To monitor a child's development, doctors compare the child's height, weight, and head size at a certain age to measurements of other children at the same age. If a child's percentiles suddenly change radically, the doctor might suspect a problem in their development.

Percentiles were originally used to analyze data sets with 100 or more values. But when statistical techniques were developed for smaller data sets, the percentile concept came into use for those sets as well. There are several methods that can be used for computing percentiles, and sometimes the answers vary slightly, especially with small data sets. In this section, we'll use a basic method that works well for smaller data sets. For larger data sets, we'll rely on our good friend technology.

Example 1 | Finding the Percentile Rank of a Data Value

Suppose you score 77 on a test in a class of 10 people, with the 10 scores listed below. What percentage of the scores were lower than yours? What was your percentile rank?

93 82 64 75 98 52 77 88 90 71

SOLUTION

It's not absolutely necessary to put the data in order when computing a percentile rank, but it surely does help. So that's where we'll start. The ordered list is

52 64 71 75 77 82 88 90 93 98

Now if you focus on 77, you can see that there are exactly four scores lower than yours. Since there were 10 scores total, that means that 4/10 or 40% of the scores were lower than yours.

As for the percentile rank? That's just to see if you're paying attention. The definition of percentile rank is the percentage of data values that are lower than a given value, so we say that a score of 77 is at the 40th percentile.

Math Note

In order to keep our study of percentiles relatively simple, we'll work with data sets with no repeated values. More complicated techniques, or technology, are used to find percentiles for data sets with repeated values.

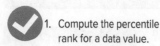
1. Compute the percentile rank for a data value.

Try This One   1

The weights in pounds for the 12 members of a college gymnastics team are below. Find the percentile rank of the gymnast who weighs 97 pounds.

101   120   88   72   75   80   98   91   105   97   78   85

Finding a Percentile Rank

The percentile rank for a given data point in a set with n total values can be calculated using this formula:

$$\text{Percentile of a value } x = \frac{\text{Number of data values less than } x}{n} \cdot 100$$

The result should be rounded to the nearest whole number.

In Example 1, we saw how to find the percentile rank for a given data value. In Example 2, we'll examine the opposite question—finding the data value that corresponds to a given percentile.

Example 2 | **Finding a Data Value Corresponding to a Given Percentile**

United States Navy

The number of words in each of the last 10 presidential inaugural addresses is listed below. Find the length that corresponds to the 30th percentile.

2,552   1,433   2,096   2,395   2,071   1,592   2,155   1,598   2,320   2,561

SOLUTION

Step 1 We're asked to find the number on the list that has 30% of the numbers below it. There are 10 numbers, and 30% of 10 is 3.

Step 2 Arrange the data in order from smallest to largest, and find the value that has 3 values below it.

1,433   1,592   1,598   2,071   2,096   2,155   2,320   2,395   2,552   2,561

The 30th percentile is the speech that consisted of 2,071 words.

Try This One   2

Comstock/PictureQuest

The average monthly rainfall in inches for St. Petersburg, Florida, is shown in the chart below. Which month is at the 75th percentile?

Jan	Feb	Mar	Apr	May	Jun	Jul	Aug	Sep	Oct	Nov	Dec
2.3	2.8	3.4	1.6	2.6	5.7	7.0	7.8	6.1	2.5	1.9	2.2

2. Find a data value corresponding to a given percentile.

The procedure in Example 2 makes perfect sense: we found a value that has 30% of the values less than it, so that's the 30th percentile, right? It turns out that there's a slight issue. If a data value is at the 30th percentile, what percentage of values should be bigger than it? Seventy percent, right? But in Example 2, the value we found at the 30th percentile has six values bigger than it, which means only 60% of the values are above it. In essence, we're "missing" 10%. It's not really missing—it's the data value itself. But it just kind of doesn't feel right to have only 60% of values above the 30th percentile.

With relatively small data sets, this issue always pops up. So statisticians developed an alternative procedure to account for it, which we'll study in the Critical Thinking exercises.

Percentile ranks are particularly useful in comparing data that come from two different sets, as shown in Example 3.

Example 3	**Using Percentiles to Compare Data from Different Sets**

Two students are competing for one remaining spot in a law school class. Miguel ranked 51st in a graduating class of 1,700, while Dustin ranked 27th in a class of 540.

(a) Without doing any calculations, which student ranked higher?
(b) Which student's percentile rank within his class was higher? What can you conclude?

SOLUTION

(a) With no calculation involved, it would be tempting to say that Dustin ranked higher. He was 27th, while Miguel was 51st. But because their class sizes were different, it's not necessarily accurate to say that Dustin ranked higher. That's why percentiles would be helpful in this situation.

(b) Miguel ranked 51st out of 1,700, so there were $1,700 - 51 = 1,649$ students ranked below him. His percentile rank is

$$\frac{1,649}{1,700} = 0.97 \text{ or } 97\%$$

Dustin had $540 - 27 = 513$ students ranked below him, so his percentile rank is

$$\frac{513}{540} = 0.95 \text{ or } 95\%$$

Both are excellent students, but Miguel's ranking is higher even though he was 51st and Dustin was 27th.

3. Use percentile rank to compare values from different data sets.

Try This One	**3**

For the 2020–2021 television season, *The Rachel Maddow Show,* broadcast on MSNBC, ranked 4th in the Nielsen ratings out of 95 news programs on cable television. The MSNBC network ranked 7th out of 159 networks (broadcast and cable combined). Which was higher rated: the show or the network it was on?

In Section 11-3, we saw that the median of a data set divides the set into equal halves. Another statistical measure we will study is the **quartile**, which divides a data set into quarters. Here's a visual that illustrates what quartiles are about:

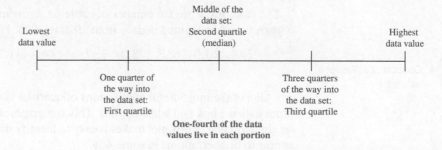

We use the symbol Q_1 to represent the first quartile, Q_2 to represent the second (which is also the median), and Q_3 for the third. This is a nice way to think of what these symbols mean:

Interpretation of Quartiles

Q_1 (first quartile): 25% of data values are less than this, and 75% are greater than it.

Q_2 (second quartile): 50% of data values are less than this, and 50% are greater than it.

Q_3 (third quartile): 75% of data values are less than this, and 25% are greater than it.

Example 4   Finding Quartiles

The data below are the percentages of total electricity generated that comes from nuclear power for the nations with the 12 largest economies in the world, listed by size of economy. Find the quartiles and describe what they mean.

19.5   2.4   0   15.8   17.2   76.9   2.9   0   3.7   18.6   16.8   0

SOLUTION

First, we'll find the median, which is also Q_2 (the second quartile). To do so, put the data values in order, from least to greatest.

0   0   0   2.4   2.9   3.7   15.8   16.8   17.2   18.6   19.5   76.9

There's an even number of values, so the median is the number halfway in between the two values in the middle, which are 3.7 and 15.8.

$$\frac{3.7 + 15.8}{2} = 9.75$$

So $Q_2 = 9.75\%$.

This divides the distribution into two halves, with six values in the lower half and six in the upper half. Next, we'll find the median of the lower half: this is Q_1 (the first quartile). The two values in the middle of the bottom half are 0 and 2.4, and halfway between them is 1.2. So $Q_1 = 1.2\%$.

Finally, we'll find the median of the upper half. The two values in the middle of the upper six values are 17.2 and 18.6, and 17.9 is halfway between. So $Q_3 = 17.9\%$.

Now let's summarize. The first quartile is 1.2%, so nations that get less than 1.2% of their power are in the bottom fourth among the world's largest economies. The second quartile is 9.75%, which tells us that nations that get less than 9.75% of their energy from nuclear are in the bottom half. The third quartile is 17.95%, so if a nation gets more than 17.95% of its energy from nuclear, it's in the top fourth in terms of nuclear generation among the world's largest economies. (*Note:* If there are an odd number of values, when finding the first and third quartiles, you will of course be finding the median of the upper and lower halves of the data set. In that case, include the median in both the upper and lower halves.)

David Wasserman/Brand X Pictures/
Getty Images

Math Note

The fact that we have repeated values on this list means we're cheating just a bit. But since all of those values are zero, and will be on the bottom fourth of the distribution under any circumstances, the quartiles aren't actually affected much.

Try This One   4

The data below are the number of cattle on farms in the United States (in millions) for each year that begins a decade from 1920 to 2020. Find the quartiles.

70.4   61.0   68.3   78.0   96.2   112.4   111.2   95.8   98.2   93.9   93.8

4. Compute quartiles for a data set.

One of the most useful applications of quartiles is using them to draw a **box plot** (sometimes called a box and whisker plot). This is a graphical way to evaluate the spread of a data set. In particular, a box plot makes it easy to identify data points that are **outliers**—those that appear to be aberrational in some way.

First, we'll need to define a new term. The distance between the first and third quartiles for a data set is called the **interquartile range**, or **IQR**. That is,

$$\text{IQR} = Q_3 - Q_1$$

Data values are considered to be outliers if they're more than 1.5 times the IQR below Q_1, or above Q_3.

Below is an example of a box plot for a group of test scores. (The labeling in purple is not typically included in a box plot: it's there to help you understand what the plot is displaying.)

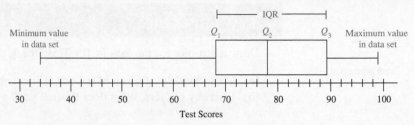

There's a lot of information about the data set here. You can see what all of the quartiles are (at least approximately): Q_1 is about 68, Q_2 is about 78, and Q_3 is about 89. The interquartile range is the distance between 89 and 68, which is 21. The lowest score was 34 and the highest was 99. Looking a little deeper, the box shows us that the middle half of all scores (which are the values between Q_1 and Q_3) fall between 68 and 89.

Using this sample as a guide, we'll draw and interpret a box plot in Example 5.

Example 5 | **Drawing and Interpreting a Box Plot**

Draw a box plot for the nuclear power data in Example 4, then use it to answer some questions about the data.

(a) What does the box plot tell us about the data set?
(b) Find any outliers in the data set.

SOLUTION

Step 1 Find the quartiles. From Example 4, we know that the quartiles are $Q_1 = 1.2$, $Q_2 = 9.75$, and $Q_3 = 17.9$.

Step 2 Draw a number line that begins before the lowest value in the data set and ends after the highest value. Locate the lowest and highest data values, and draw short vertical lines above the line at those locations.

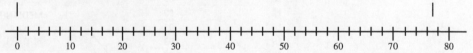

Step 3 Draw a rectangular box over the number line, beginning at Q_1 and ending at Q_3. Then draw a vertical line through the box at Q_2. Then draw horizontal lines from the lowest and highest values to the edges of the box.

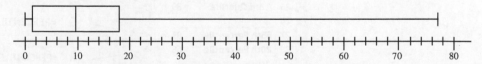

And we have a box plot! Now for the analysis.

(a) The position of the box shows that most of the values in the data set are on the low end of the distribution. Half of the values are between 1.2 and 17.9, and three fourths are less than 17.9.

(b) To decide if there are any outliers, we'll first need to multiply the interquartile range by 1.5. The IQR is $17.9 - 1.2 = 16.7$, and $1.5 \cdot \text{IQR} = 25.05$. Next, we subtract this number from the first quartile and add it to the third:

$$Q_1 - 25.05 = 1.2 - 25.05 = -23.85$$
$$Q_3 + 25.05 = 17.9 + 25.05 = 42.95$$

There are no negative data values, so there can't be any less than –23.85, but there's definitely a value greater than 42.95. Looking back at the data, there's one outlier: the maximum value of 76.9%.

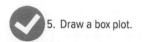

5. Draw a box plot.

Try This One   5

Draw a box plot for the data in Try This One 4, then use it to answer these questions:

(a) What does the position of the box tell you about the data?
(b) Find any outliers. What does this tell you?

Using Technology: Finding Percentiles and Quartiles

There are Excel commands that will find all of the information we studied in this lesson. It's all about knowing the syntax. The example below shows the test scores that were used for the box plot sample on page 695. The list was already sorted using the SORT command.

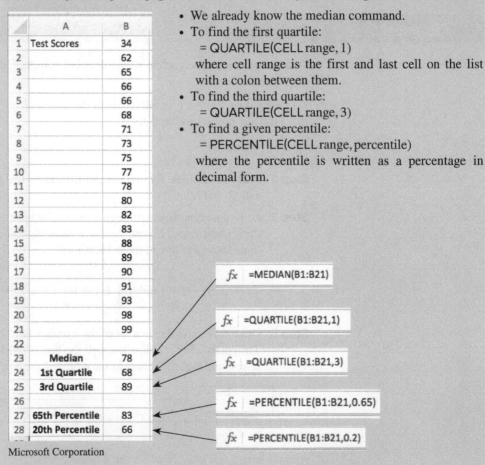

- We already know the median command.
- To find the first quartile:
  = QUARTILE(CELL range, 1)
  where cell range is the first and last cell on the list with a colon between them.
- To find the third quartile:
  = QUARTILE(CELL range, 3)
- To find a given percentile:
  = PERCENTILE(CELL range, percentile)
  where the percentile is written as a percentage in decimal form.

	A	B
1	Test Scores	34
2		62
3		65
4		66
5		66
6		68
7		71
8		73
9		75
10		77
11		78
12		80
13		82
14		83
15		88
16		89
17		90
18		91
19		93
20		98
21		99
22		
23	Median	78
24	1st Quartile	68
25	3rd Quartile	89
26		
27	65th Percentile	83
28	20th Percentile	66

fx =MEDIAN(B1:B21)

fx =QUARTILE(B1:B21,1)

fx =QUARTILE(B1:B21,3)

fx =PERCENTILE(B1:B21,0.65)

fx =PERCENTILE(B1:B21,0.2)

Microsoft Corporation

In this section, we studied two measures of position, percentile and quartile. In Section 11-6, the concept of position plays a very important role when we study a third measure of position, the z score.

Answers to Try This One

1 58th percentile

2 September

3 Neither. Both are at the 96th percentile.

4 $Q_1 = 70.4$, $Q_2 = 93.9$, $Q_3 = 98.2$

5
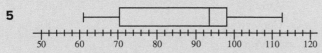

(a) The width of the box tells us the data's pretty spread out; the fact that the median falls so close to the upper end of the box means that the higher data values are clustered closer together.

(b) There are no outliers. This means that none of the data values are unusually high or low compared to the others.

Exercise Set   11-5

Writing Exercises

1. If your score in your math class puts you in the 60th percentile, what exactly does that mean?
2. Does a score in the 90th percentile mean that you got 90% of the questions right on a test? Explain.
3. Explain what quartiles are.
4. What is the connection between the second quartile and the median for a data set? Explain.
5. What does the portion inside the box in a box plot represent?
6. What is meant by the term "outlier"? How can you decide if a data set has any outliers?

Applications in Our World

7. The scores for 20 students on a 50-point math test are 42, 48, 50, 36, 35, 27, 47, 38, 32, 43, 24, 33, 39, 49, 44, 40, 29, 30, 41, and 37.
   (a) Find the percentile rank for a score of 32.
   (b) Find the percentile rank for a score of 44.
   (c) Find the percentile rank for a score of 36.
   (d) Find the percentile rank for a score of 27.
   (e) Find the percentile rank for a score of 49.
8. The heights (in inches) of the 12 students in a seminar course are 73, 68, 64, 63, 71, 70, 65, 67, 72, 66, 60, and 61.
   (a) Find the percentile rank for a height of 67 in.
   (b) Find the percentile rank for a height of 70 in.
   (c) Find the percentile rank for a height of 63 in.
   (d) Find the percentile rank for a height of 68 in.
   (e) Find the percentile rank for a height of 66 in.
9. In a class of 500 students, Carveta's rank was 125. Find the percentile rank.
10. In a class of 400 students, John's rank was 80. Find the percentile rank.
11. In a charity marathon to raise money for AIDS research, Chen finished 43rd out of 200 entrants. Find the percentile rank.
12. An investor decides to buy stock only in companies that are ranked in the 80th percentile or above in terms of dividends paid. Her broker gives her a tip on a company that ranked 80th of 1,941 companies on the New York Stock Exchange that paid dividends in the previous year. Does this qualify for her portfolio?
13. On an exam, Angela scored in the 20th percentile. If there were 50 students in the class, how many students scored lower than Angela?
14. Out of 600 applicants to a graduate program at an Ivy League school, Marissa's GRE score was in the 25th percentile. How many applicants scored lower than Marissa?
15. Eldrick finished at the 77th percentile of a men's golf league with 88 players in it. How many golfers finished above him?
16. Tina's application for a nursing position is ranked at the 83rd percentile out of 110 applications. If the top 12 applicants will get job offers, will Tina get an offer?
17. Lea's percentile rank on an exam in a class of 600 students is 60. Maurice's class rank is 220. Who is ranked higher?
18. In an English class of 30 students, Audrelia's percentile rank is 20. Maranda's class rank is 20. Whose rank is higher?
19. The University of Wisconsin men's basketball team finished the 2020 season ranked 28th out of 358 teams, while its football team was ranked 13th out of 130 teams. Which team had a better ranking?
20. A pro golfer is ranked 48th out of 230 players that played in a PGA event that year. His sister is ranked 42nd out of

193 players on the LPGA tour. Which has bragging rights as being the higher-ranked player by percentile?

21. In an evening statistics class, the ages of 20 students are as follows.

18   24   19   20   33   42   43   27   31   39
21   44   26   32   37   34   23   35   28   25

(a)   What is the percentile rank of 33?
(b)   What is the rank (from the top) of 33?
(c)   What age corresponds to the 20th percentile?

22. Twenty subjects in a psychology class experiment scored the following on an IQ test.

95   107   110   101   122   96   94   104   131   90
111   103   97   100   119   85   108   120   130   88

(a)   What is the percentile rank of 96?
(b)   What is the rank (from the top) of 96?
(c)   What score corresponds to the 40th percentile?

For Exercises 23–28, find the values for Q_1, Q_2, and Q_3.

23. Number of drive-in theaters in nine selected states:

59   20   21   34   52   48   24   29   55
Source: *National Association of Theater Owners*

24. Average cost in cents per kilowatt-hour of producing electricity using nuclear energy for the past 10 years:

2.10   2.07   2.02   1.95   1.98
2.00   2.04   2.01   2.20   2.19
Source: *Nuclear Energy Institute*

25. The average costs of a 1-hour massage in 10 large cities are:

$75 $77 $78 $90 $87 $93 $89 $93 $95 $88

26. The total number of passengers (in millions) for flights on domestic carriers for each month in 2020 were:

79.8   78.8   37.6   3.0
 7.7   15.4   21.2   23.8
27.1   29.5   29.7   31.1

27. The annual number of homicides in Milwaukee for a 10-year period is shown.

123   124   121   127   108
107   88   122   103   66

28. Following are the 10 largest salaries for CEOs in 2020 (in millions of dollars), with the associated company.

Palantir Technologies	1,098.5
DoorDash	413.7
Opendoor Technologies	370.2
Paycom Software	211.1
IAC/InterActiveCorp	189.5
Activision Blizzard	154.6
Regeneron Pharmaceuticals	135.4
Airbnb	120.1
ContextLogic	78.2
DaVita	73.4

Source: *Equilar.com*

In Problems 29–32, draw a box plot for the data, then use it to answer these questions: (a) What does the position of the box tell you about the data? (b) Find any outliers. What does this tell you about the data set?

29. The following data are the number of Internet users in 2021, in millions, in the top 15 countries for Internet usage.

1,010.7   833.7   312.3   212.4   129.2   160.0
 136.2   124.0   118.8   117.4   92.0   78.1
  77.8   73.0   68.2
Source: *Wikpedia*

30. The table shows the median price (in thousands) of all existing homes sold in the United States for the years from 2004 to 2021.

Year	04	05	06	07	08	09	10
Median Price	195.2	219.0	221.9	217.9	198.1	172.5	172.9

Year	11	12	13	14	15	16
Median Price	166.2	177.2	197.1	208.3	222.4	234.4

Year	17	18	19	20	21
Median Price	247.2	273.8	285.3	303.6	353.9

Source: *National Association of Realtors*

31. Use the data in Exercise 27.
32. Use the data in Exercise 28.

Critical Thinking

33. Is it possible to score 90% on a test and have a percentile rank less than 90th? Is it possible to score 90% and have a percentile rank of exactly 90?

34. What would you have to do in order to rank in the 100th percentile in your college graduating class?

35. In Exercise 31, you were asked to draw a box plot for the data in Exercise 27. Write a thorough analysis of anything you can learn from the box plot that wouldn't have been apparent from just finding the quartiles.

36. Draw a box plot of a data set that you gathered, then write an analysis of the data using your box plot. The data set should have at least 10 values.

37. After Example 2, we pointed out that our method for finding a data value corresponding to a certain percentile has a minor flaw when the data set is small. In that case, we saw that for a data set with 10 values, if you find the value at the 30th percentile, you'd hope to have 70% of the values bigger than it, but it's actually only 60%.

What if the data set had 100 values? If you find the value that has 30% of the values below it, what percent will be above it?

38. (a) Explain why the following claim is false. (You will most likely find it helpful to try some sample calculations on different sized data sets.) "If a data value is at the 80th percentile, then its ranking in the data set is 0.8 times the total number of values."

    (b) For what type of data set will the claim in part (a) be very close to true? Use your results from part (a).

39. The procedure used to fix the issue in Exercise 37 for small data sets is described below. Use it to find the score at the 80th percentile for this list of test scores:

    93 82 64 75 98 52 77 88 90 71
    78 55 62 99 91 83 89 61 73 80

    Step 1: Arrange the data values in order from smallest to largest.

    Step 2: Count the total number of data values, then multiply that number by the percentile rank we're looking for, written as a percentage in decimal form. We'll call this number the *index,* and represent it with the letter L.

    Step 3: If L is a whole number, then the data value corresponding to our given percentile is halfway between the value in the Lth position from the bottom and the next higher value.

    Step 4: If L is not a whole number, round it up to the next whole number. Then the data value at the given percentile is the Lth value from the bottom of the list.

40. What is unusual about the score at the 80th percentile using this procedure? How does it compare to the score at the 80th percentile using the procedure in Example 2?

41. How many scores are lower than the score you found at the 80th percentile using the new procedure? What percentage is that? What percentage of scores are above it? What can you conclude?

42. USING TECHNOLOGY. It should be pretty obvious that for large data sets, using technology to compute percentiles and quartiles is a pretty darn good idea. There's a spreadsheet in the online resources for this section that contains the net worth as of 2018 for 530 of the 535 members of the United States Congress. In Column F, use formulas to compute the statistics that are listed next to each cell in Column E. Then draw a box plot for the data by hand. Finally, write a short essay analyzing the information provided by your box plot.

43. You probably wouldn't be surprised to find that most members of Congress are really rich. But how rich? In Exercise 41, you can find the net worth of every member of the House and Senate as of 2018. Senator James Inhofe (R-OK) and Representative Earl Blumenauer (D-OR) both had a net worth of $4.5 million. Senator Inhofe ranked 21st out of 100 senators, while Congressman Blumenauer ranked 54th out of 435 members of the House. Based on percentiles, which was richer relative to their congressional body? What do you think that says about the House compared to the Senate?

Section 11-6 — The Normal Distribution

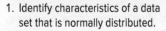

LEARNING OBJECTIVES

1. Identify characteristics of a data set that is normally distributed.

2. Apply the empirical rule.

3. Compute z scores and use them to compare values from different data sets.

4. Use a table and z scores as well as technology to find areas under the standard normal distribution.

If you grew up anywhere near maple trees, you're probably familiar with the seed pods they drop by the thousands in the spring. Like almost all living things, the individual pods vary in size—on any given tree, there's a typical size, with some pods bigger and some smaller.

On a beautiful spring day, I gathered 100 pods from the maple in my back yard, measured each, then grouped them according to their length. The smallest was 42 mm, the largest 59 mm, and in the photo they're grouped in classes 42–43, 44–45, 46–47, etc. An interesting thing happened—we see that the largest number of pods have lengths somewhere in the middle of the range, and the classes farther away from the center have less pods.

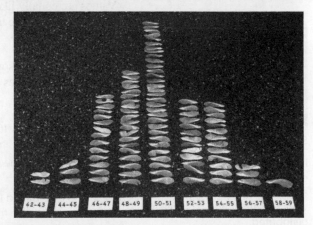

David Sobecki/McGraw Hill

A wide variety of quantities in our world, like sizes of individuals in a population, IQ scores, life spans for batteries of a certain brand, and many others, tend to exhibit this same phenomenon. In fact, it's so common that frequency distributions of this type came to be known as *normal distributions.*

Suppose a researcher selects a random sample of 100 adult women, measures their heights, and constructs a histogram. The researcher would probably get a graph similar to the one shown in Figure 11-9(a). If the researcher increases the sample size and decreases the width of the classes, the histograms will look like the ones shown in Figures 11-9(b) and 11-9(c). Finally, if it were possible to measure the exact heights of all adult females in the United States and plot them, the histogram would approach what is called the *normal distribution*, shown in Figure 11-9(d). This distribution is also known as a *bell curve* or a *Gaussian distribution*, named for the German mathematician Carl Friedrich Gauss (1777–1855) who derived its equation.

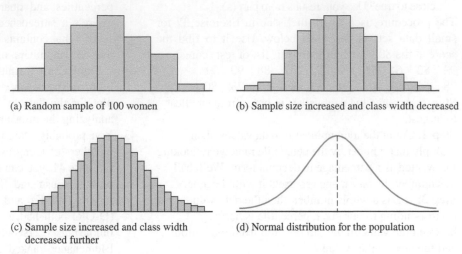

(a) Random sample of 100 women

(b) Sample size increased and class width decreased

(c) Sample size increased and class width decreased further

(d) Normal distribution for the population

Figure 11-9

To show that normal distributions are not even close to a purely mathematical topic, we're going to borrow our definition of normal distribution from an investment site called "Investopedia":

> A probability distribution that plots all of its values in a symmetrical fashion and most of the results are situated around the probability's mean is called a **normal distribution**. Values are equally likely to plot either above or below the mean. Grouping takes place at values that are close to the mean and then tails off symmetrically away from the mean.

Next, we'll highlight some of the important features of a normal distribution (which, in turn, give us the key features of a data set that is normally distributed).

Steve Mason/Photodisc/Getty Images

In a normal distribution, values cluster around a central value and fall off in both directions. Waiting times can be normally distributed.

1. Identify characteristics of a data set that is normally distributed.

Some Properties of a Normal Distribution

1. The value in the middle of the distribution, which appears most often in the sample, is the mean.

2. The distribution is symmetric about the mean. This means that the graph has two halves that are mirror images on either side of the mean value.

3. This is the key fact: the area under any portion of the curve is the percentage (in decimal form) of data values that fall between the values that begin and end that region.

4. The total area under the entire curve is 1.

Let's explore property 3 from the colored box, because it's what makes normal distributions so useful in analyzing data.

| Example 1 | Getting Information from a Normal Curve |

The graph below shows a normal distribution for heights of women in the United States. The numbers on the horizontal axis are heights in inches, and some areas are labeled for reference.

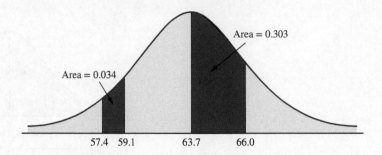

(a) What is the mean height?
(b) What percentage of women are between 57.4 and 59.1 inches tall?
(c) If there are 31,806 women at a stadium concert, how many of them would you expect to be between 63.7 and 66.0 inches tall?

SOLUTION

(a) The mean is the value in the very center of a normal distribution, and it corresponds to the value that appears most often. This would be the highest point on the graph, which is labeled 63.7. So the mean height for American women is 63.7 inches.

(b) This is what the third property in the colored box is about. The diagram indicates that the area under the normal graph between 57.4 and 59.1 is 0.034. This is the decimal form of the percentage of data values that fall in that range. Converting 0.034 to percent form by moving the decimal point two places right, we get 3.4%. So we'd expect that about 3.4% of women would have heights in that range.

(c) This is the same question with a twist. It asks for the number of women, not a percentage. But we can find that number by first finding a percentage. In this case, the area under that portion of the graph is 0.303, so we'd expect 30.3% of women to have heights between 63.7 and 66.0 inches. In particular, we'd expect 30.3% of the women at the concert to have a height in that range.

$$30.3\% \text{ of } 31,806 = 0.303 \times 31,806 = 9,637.218$$

We're talking about human beings here, not leftover body parts, so the .218 part is irrelevant. We'd expect about 9,637 women to be between 63.7 and 66.0 inches tall.

Try This One   1

Based on the normal curve in Example 1, what percentage of American women are taller than 66 inches? (*Hint:* Half of all women are taller than 63.7 inches.)

The Empirical Rule

So now we know that if we have data values that are normally distributed, and IF we can find the area under portions of the graph, we can find the percentage of data values that fall in a certain range. That leads to the big question: HOW in the world do you find areas under the graph? In this section, we'll rely on a known rule that allows us to quickly find a range in which most of the data values fall. It's known as the *empirical rule*.

The Empirical Rule

When data are normally distributed, approximately 68% of the values are within 1 standard deviation of the mean, approximately 95% are within 2 standard deviations of the mean, and approximately 99.7% are within 3 standard deviations of the mean (see Figure 11-10).

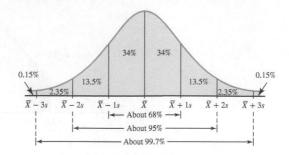

Figure 11-10 $\overline{X}$ = mean, s = standard deviation.

| **Example 2** | **Using the Empirical Rule** |

According to the website answerbag.com, the mean height for male humans is 5 feet 9.3 inches, with a standard deviation of 2.8 inches. If this is accurate, out of 1,000 randomly selected men, how many would you expect to be between 5 feet 6.5 inches and 6 feet 0.1 inch?

SOLUTION

The given range of heights corresponds to those within 1 standard deviation of the mean, so we would expect about 68% of men to fall in that range. In this case, we expect about 680 men to be between 5 feet 6.5 inches and 6 feet 0.1 inch.

Try This One   2

A standard test of intelligence is scaled so that the mean IQ is 100, and the standard deviation is 15. If there are 40,000 people in a stadium, how many would you expect to have an IQ between 70 and 130?

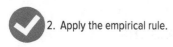
2. Apply the empirical rule.

The Standard Normal Distribution

We were able to answer the questions in Example 2 and Try This One 2 because the values we were interested in just happened to be exactly within 1 or 2 standard deviations of the mean. But what if we wanted to know how many men you'd expect to be between 5′10″ and 6′3″? We'd need to be able find the area under the graph for that exact situation, which I can guarantee you is NOT easy. Worse still, if we then try to find how many people we'd expect to have an IQ between 90 and 105, we're stuck with a similar problem, but the graph (and area calculations) will be different!

The point is that while all normally distributed data have the same *general* shape for their graph, areas under portions of the graph depend on how tall the curve is and how spread out it is. These characteristics in turn depend on what the mean and standard deviation are for that particular data set. This would require a different set of extremely difficult area calculations for every individual situation involving normally distributed data. Ouch!

Fortunately, statisticians found a way to work around this problem. The areas have been calculated for a simple normal distribution with mean 0 and standard deviation 1; then a procedure was developed to apply those areas to other distributions. This special distribution is called the *standard normal distribution*.

> The **standard normal distribution** is a normal distribution with mean 0 and standard deviation 1.

The standard normal distribution is shown in Figure 11-11. The values under the curve shown in Figure 11-11 indicate the proportion of area in each section. Hopefully you recognize that these values come from the empirical rule. For example, the area between the mean and 1 standard deviation above or below the mean is about 0.34, or 34%.

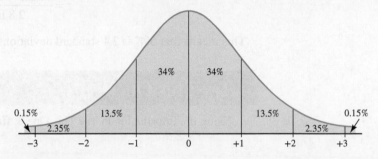

Figure 11-11

Since area calculations for normal curves are really difficult, tables have been compiled that list areas under certain portions of the standard normal curve. But they work only for data with mean 0 and standard deviation 1, so we'll apply a clever trick to change any normal distribution into one with mean 0 and standard deviation 1. To do this, we'll find *z scores* for given data values.

> For a data value from a sample with mean $\overline{X}$ and standard deviation s, the *z* **score** is
>
> $$z = \frac{X - \overline{X}}{s}$$
>
> Verbally, to find a *z* score, just subtract the mean, then divide the result by the standard deviation.

A couple of notes:

• A data point is greater than the mean if $z > 0$ and less than the mean if $z < 0$.

• *z* scores are typically rounded to two decimal places.

Here's a clever way to think about what exactly a *z* score means: what would the *z* score be for a data value equal to the mean? What about one standard deviation above the mean? For equal to the mean, we'd get

$$\frac{\overline{X} - \overline{X}}{s} = 0$$

For one standard deviation above the mean, the result is

$$\frac{(\overline{X} + s) - \overline{X}}{s} = \frac{s}{s} = 1$$

Interesting . . . a data value zero standard deviations above the mean has a *z* score of 0, a data value one standard deviation above the mean has a *z* score of 1. See the pattern?

That's exactly what *z* scores are designed to do: the *z* score for a data value is simply a measure of how many standard deviations above or below the mean that it is.

Example 3   Computing a z Score

Math Note

Notice that the units divide out, so z scores have no units.

Based on the information in Example 2, find the z score for a man who is 6 feet 4 inches tall and describe what it tells us.

SOLUTION

In Example 2 we were given that the mean height is 5 feet 9.3 inches and the standard deviation is 2.8 inches. So we can use the formula for z scores with mean 69.3 inches and standard deviation 2.8 inches. Note that we converted the heights to inches to make it easier to subtract.

$$z = \frac{76 \text{ in.} - 69.3 \text{ in.}}{2.8 \text{ in.}} \approx 2.39$$

This means that 6′4″ is 2.4 standard deviations above the mean.

Try This One   3

Using the information in Try This One 2, find the z score for a person with an IQ of 91.

Example 4 illustrates our first use for z scores: they allow us to compare data values from completely different sets.

Example 4   Using z Scores to Compare Standardized Test Scores

As you probably know, there are two main companies that offer standardized college entrance exams, ACT and SAT. Since each has a completely different scoring scale, it's really difficult to compare the scores of students that took different exams. One year the ACT had a mean score of 21.2 and a standard deviation of 5.1. That same year, the SAT had a mean score of 1498 and a standard deviation of 347. Suppose that a scholarship committee is considering two students, one who scored 26 on the ACT and another who scored 1800 on the SAT. Both are pretty good scores, but which one is better?

SOLUTION

This is not an easy question because the two tests have completely different scales. But because we know the mean and standard deviation for each, we can compute z scores. This will allow us to decide which student did better relative to the others who took the same test.

$$26 \text{ ACT: } z = \frac{26 - 21.2}{5.1} \approx 0.94 \qquad 1800 \text{ SAT: } z = \frac{1800 - 1498}{347} \approx 0.87$$

The student with 26 on the ACT did better. Their score is 0.94 standard deviations above the mean, while the student who scored 1800 on the SAT is 0.87 standard deviations above the mean.

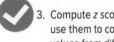

3. Compute z scores and use them to compare values from different data sets.

Try This One   4

What score is better: 2150 SAT or 139 on an IQ test?

Another way that we'll use z scores is to find areas under a normal curve using only areas under a standard normal curve, which can be read from a table, like Table 11-3. We'll also learn how to use technology to find these areas.

Table 11-3	Area Under a Normal Distribution Curve Between $z = 0$ and a Positive Value of z

Math Note

An area table with more values can be found in Appendix A. If you need the area for a z score between two values in Table 11-3, you can get an approximate area by choosing an area between areas in the table, or use the more extensive table in the appendix.

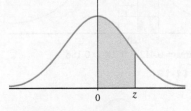

How to use the table

The value in the A column of the table is the area under the standard normal curve between $z = 0$ and each given positive value of z.

z	A	z	A	z	A
0.00	0.000	1.10	0.364	2.20	0.486
0.05	0.020	1.15	0.375	2.25	0.488
0.10	0.040	1.20	0.385	2.30	0.489
0.15	0.060	1.25	0.394	2.35	0.491
0.20	0.079	1.30	0.403	2.40	0.492
0.25	0.099	1.35	0.412	2.45	0.493
0.30	0.118	1.40	0.419	2.50	0.494
0.35	0.137	1.45	0.427	2.55	0.495
0.40	0.155	1.50	0.433	2.60	0.495
0.45	0.174	1.55	0.439	2.65	0.496
0.50	0.192	1.60	0.445	2.70	0.497
0.55	0.209	1.65	0.451	2.75	0.497
0.60	0.226	1.70	0.455	2.80	0.497
0.65	0.242	1.75	0.460	2.85	0.498
0.70	0.258	1.80	0.464	2.90	0.498
0.75	0.273	1.85	0.468	2.95	0.498
0.80	0.288	1.90	0.471	3.00	0.499
0.85	0.302	1.95	0.474	3.05	0.499
0.90	0.316	2.00	0.477	3.10	0.499
0.95	0.329	2.05	0.480	3.15	0.499
1.00	0.341	2.10	0.482	3.20	0.499
1.05	0.353	2.15	0.484	3.25*	0.499

*For z scores greater than 3.25, use $A = 0.500$

Finding Areas under the Standard Normal Distribution

In the remainder of this section we're going to focus on simply finding areas under a normal curve. It's going to be pretty abstract, but the payoff is worth it: once we're good at this skill, we'll be able to solve all sorts of problems in our world, which is the entire focus of the next section.

For now, we'll find areas corresponding to z scores using Table 11-3 and the following key facts.

Two Important Facts about the Standard Normal Curve

1. The area under any normal curve is divided into two equal halves at the mean. Each of the halves has area 0.500.

2. The area between $z = 0$ and a positive z score is the same as the area between $z = 0$ and the negative of that z score.

Each of the facts in the colored box is a consequence of the fact that normal distributions are symmetric about the mean. Examples 3, 4, and 5 will illustrate how to find areas under a normal curve.

Example 5

Finding the Area between Two z Scores

Math Note

Remember that the area under any portion of a normal distribution has to be non-negative and less than or equal to 1. If you don't get an area in that range, you must have made a mistake.

Find the area under the standard normal distribution

(a) Between $z = 1.55$ and $z = 2.25$.
(b) Between $z = -0.60$ and $z = -1.35$.
(c) Between $z = 1.50$ and $z = -1.75$.

SOLUTION

(a) Draw the picture, label z scores, and shade the requested area. See Figure 11-12.

In Table 11-3, find 2.25 in one of the z columns. The number listed next to it in the A column is 0.488. This tells us that the area between $z = 0$ and $z = 2.25$ is 0.488. The number in the A column next to $z = 1.55$ is 0.439, so the area between $z = 0$ and $z = 1.55$ is 0.439. The area we are looking for is the larger area minus the smaller:

$$0.488 - 0.439 = 0.049$$

The area between $z = 1.55$ and $z = 2.25$ is 0.049.

(b) Draw the picture, label z scores, and shade the requested area. See Figure 11-13.

All of the z scores in the table are positive, but key fact 2 on the previous page shows that the areas for negative z scores are exactly the same as for the corresponding positive z scores. So we can get the area between $z = 0$ and $z = -1.35$ by looking up $z = 1.35$ in Table 11-3, which gives us 0.412. We can then get the area for $z = -0.60$ by looking up $z = 0.60$ in the table: the result is 0.226. Again, the area we're looking for is the larger area minus the smaller:

$$0.412 - 0.226 = 0.186$$

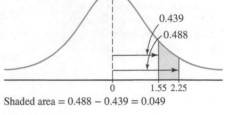

Shaded area = 0.488 − 0.439 = 0.049

Figure 11-12

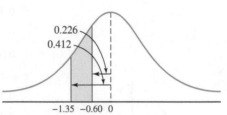

Shaded area = 0.412 − 0.226 = 0.186

Figure 11-13

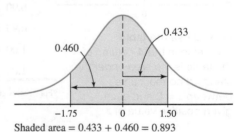

Shaded area = 0.433 + 0.460 = 0.893

Figure 11-14

The area between $z = -0.60$ and $z = -1.35$ is 0.186.

(c) Draw the picture, label the z scores, and shade the requested area. See Figure 11-14.

In this case, the values we get from the table give us the areas between $z = 0$ and the z scores of −1.75 and 1.50. The entire shaded area is the *sum* of these two areas, so we'll add the two areas we get from the table. The area corresponding to $z = -1.75$ is 0.460, and the area corresponding to $z = 1.50$ is 0.433.

$$0.460 + 0.433 = 0.893$$

The area between $z = 1.50$ and $z = -1.75$ is 0.893.

Try This One   5

Find the area under the standard normal distribution

(a) Between $z = 2.05$ and $z = 2.40$.
(b) Between $z = -3.2$ and $z = -2.0$.
(c) Between $z = -0.55$ and $z = 1.6$.

When we want to find the area to the right or left of a single z score, we will use the fact that the area under half of the curve is exactly 0.500, as in Examples 6 and 7.

| **Example 6** | Finding the Area to the Right of a z Score |

Math Note

Using a sketch and the two important facts on page 705 allows us to find areas without having to learn a variety of rules and formulas for the different possible situations.

Find the area under the standard normal distribution

(a) To the right of $z = 1.70$.   (b) To the right of $z = -0.95$.

SOLUTION

(a) Draw the picture, label the z score, and shade the area. See Figure 11-15.
    The area of the entire portion to the right of $z = 0$ is 0.500. According to the table, the area of the portion between $z = 0$ and $z = 1.70$ is 0.455. So the shaded portion is the difference between 0.500 and 0.455:

$$0.500 - 0.455 = 0.045$$

The area to the right of $z = 1.70$ is 0.045.

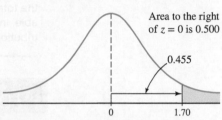

Shaded area $= 0.500 - 0.455 = 0.045$

Figure 11-15

(b) Draw the picture, label the z score, and shade the area. See Figure 11-16.
    This time, the area is the sum of 0.500 (the entire right half of the distribution) and the area between $z = 0$ and $z = -0.95$, which we find to be 0.329 using Table 11-3.

$$0.500 + 0.329 = 0.829$$

The area to the right of $z = -0.95$ is 0.829.

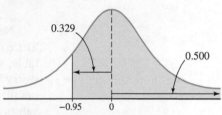

Shaded area $= 0.500 + 0.329 = 0.829$

Figure 11-16

Try This One   6

Find the area under the standard normal distribution

(a) To the right of $z = -2.40$.   (b) To the right of $z = 0.25$.

| **Example 7** | Finding the Area to the Left of a z Score |

Find the area under the standard normal distribution

(a) To the left of $z = -2.20$.   (b) To the left of $z = 1.95$.

SOLUTION

(a) Draw the picture, label the z score, and shade the area. See Figure 11-17.
    The shaded area is the entire left half (0.500) minus the area between $z = 0$ and $z = -2.2$, which is 0.486 according to Table 11-3.

$$0.500 - 0.486 = 0.014$$

The area to the left of $z = -2.2$ is 0.014.

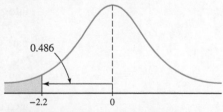

Shaded area $= 0.500 - 0.486 = 0.014$

Figure 11-17

(b) Draw the picture, label the z score, and shade the area. See Figure 11-18.

Table 11-3 gives us an area of 0.474 between $z = 0$ and $z = 1.95$. Adding to the area of the left half (0.500), we get

$$0.500 + 0.474 = 0.974$$

The area to the left of $z = 1.95$ is 0.974.

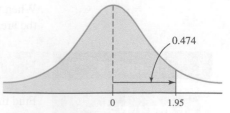

Shaded area = 0.500 + 0.474 = 0.974

Figure 11-18

CAUTION

Don't ignore your graph when writing your final answer! As long as you keep in mind that the total area under the distribution is 1, you can visually check if your answer is reasonable. In Example 7b, just from a quick glance at the graph you can see that most of the distribution is shaded, so your answer should be close to 1.

Try This One   7

Find the area under the standard normal distribution

(a) To the left of $z = -1.05$ (b) To the left of $z = 0.1$

Next we'll look at how we could use technology to find areas under a normal curve. There are a few reasons why this is really helpful. First, we don't have to look up values in a table, which is always a plus if you don't have the table handy. Second, you don't have to worry about estimating when looking up a value that's not exactly in the table. Third, it turns out that you don't even need to compute z scores: the commands allow you to enter raw data values along with the mean and standard deviation for that data set. We'll start with graphing calculators.

Using Technology: Finding Areas Beneath a Normal Curve with a Graphing Calculator

On a TI-84 calculator, the **normalcdf**(a,b,μ,σ) command will find the area bounded by a normal curve, the x axis and vertical lines at a lower bound a and an upper bound b as shown in the figure. You get to this command by pressing 2nd VARS to get the distribution menu; then it's choice 2. (Don't confuse it with choice 1, **normalpdf**, which is used for a different purpose.)

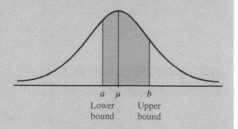

Notice that the inputs in parentheses correspond to the lower bound (a), the upper bound (b), the mean (μ), and the standard deviation (σ). This eliminates the need to calculate z scores: you just need the mean and standard deviation for the given data set.

The first screenshot below shows the area between $z = 1.50$ and $z = 1.75$, which was part (c) of Example 5. Notice that these come from a standard normal distribution, so we've input 0 and 1 as the mean and standard deviation.

If we need to compute the area to the right of a z score, we have to be a little clever. In that case, we'll use that z score as lower bound, and some really large number as upper bound. Any area left over to the right of that large upper bound is probably negligible. The second screenshot displays a sample calculation for Example 6b, which is the area to the right of $z = -0.95$.

Finally, to compute the area to the left of a *z* score, we'll enter some really large negative number as lower bound, and the given score as upper bound. The third screenshot is the area in Example 7a, to the left of *z* = 1.95. Again, in each case, the area requested was under the standard normal, so we entered 0 for mean and 1 for standard deviation.

```
normalcdf(1.5,1.
75,0,1)
        .0267481152
```
```
normalcdf(-.95,1
00000000,0,1)
        .828943888
```
```
normalcdf(-10000
000,1.95,0,1)
        .9744120102
```

David Sobecki/McGraw Hill

Excel is set up to calculate normal probabilities differently. Instead of calculating the area between two values, it calculates the area to the LEFT of any value. This works fabulously well if you're interested in the probability of a randomly selected value being LESS than some number. For other types of probability, we'll need to do some arithmetic.

Using Technology: Finding Areas Beneath a Normal Curve with a Spreadsheet

The Excel command is =NORM.DIST. When you start typing =NORM.DIST in Excel, here's what appears, as a guide to help you:
= NORM.DIST(x, mean, standard_dev, cumulative), where
- *x* is a data value.
- mean is the mean for the distribution.
- standard_dev is the standard deviation for the distribution.
- for cumulative, always enter TRUE; we won't use FALSE in this book.

This formula finds the area under a normal curve to the left of the data value *x*. So to find the area to the left of *z* = 1.95 in Example 7a, we'd use
= NORM.DIST(1.95, 0, 1, TRUE)
as shown in row 1 of the screenshots below.

If we want to find the area to the RIGHT of a data value, we'll take advantage of the fact that the area under the entire curve is always 1. The Excel command finds the area to the left; to turn that into the area to the right, just subtract from 1. So to find the area to the right of *z* = −0.95 in Example 6b, we'd enter
= 1 − NORM.DIST(−.95, 0, 1, TRUE)
as seen in row 2.

Finally, to find the area between two values, we need to subtract the area to the left of the bigger value minus the area to the left of the smaller value. To find the area between *z* = 1.50 and *z* = 1.75 in Example 5c, we can use
= NORM.DIST(1.75, 0, 1, TRUE) − NORM.DIST(1.50, 0, 1, TRUE)
as seen in row 3. This takes some thought, and a sketch really comes in handy here.

	A
1	0.97441194
2	0.828943874
3	0.026748044

	A
1	=NORM.DIST(1.95,0,1,TRUE)
2	=1-NORM.DIST(-0.95,0,1,TRUE)
3	=NORM.DIST(1.75,0,1,TRUE)-NORM.DIST(1.5,0,1,TRUE)

Microsoft Corporation

4. Use a table and *z* scores as well as technology to find areas under the standard normal distribution.

To close the section, we should point out that no data set fits the normal distribution perfectly, since the normal curve is a theoretical distribution that would come from infinitely many data values. But there are many, many quantities that vary from a true normal distribution so slightly that the normal curve can be used to study those quantities very effectively. So that's what we'll do in Section 11-7.

Answers to Try This One

1 19.7%

2 About 38,000

3 −0.6

4 SAT: $z = 1.88$; IQ: $z = 2.6$. The IQ score is better.

5 (a) 0.012
     (b) 0.022
     (c) 0.654

6 (a) 0.992
     (b) 0.401

7 (a) 0.147
     (b) 0.540

Exercise Set   11-6

Writing Exercises

1. What is the distinguishing characteristic of a quantity that is normally distributed?
2. Write as many properties of a normal curve as you can think of.
3. What does the area under a portion of a normal curve tell you about the data in the associated distribution?
4. Explain what the empirical rule says.
5. What percentage of the area under a normal curve falls to the right of the mean? To the left? Explain.
6. Why does it make sense that the total area under a normal curve is 1?

7. What is a standard normal distribution?
8. What does the standard normal distribution have to do with z scores?
9. Explain how to find the z score for a data value.
10. Why does the area table for z scores only have to include positive z scores?
11. What are z scores used for other than finding area under normal curves?
12. What exactly does the z score for a data value tell us about that value?

Computational Exercises

For Exercises 13–18, the data set described is normally distributed with the given mean and standard deviation, and with n total values. Find the approximate number of data values that will fall in the given range.

13. Mean = 12, standard deviation = 1.5, $n = 50$.
    Range: 10.5 to 13.5
14. Mean = 100, standard deviation = 8, $n = 200$.
    Range: 84 to 116
15. Mean = 400, standard deviation = 25, $n = 500$.
    Range: 325 to 475
16. Mean = −14.2, standard deviation = 1.6, $n = 120$.
    Range: −17.4 to −11
17. Mean = 36.5, standard deviation = 4.3, $n = 80$.
    Range: 32.2 to 40.8
18. Mean = 237, standard deviation = 31, $n = 750$.
    Range: 175 to 299

For Exercises 19–24, find the z score for the given data value.

19. Mean = 23, standard deviation = 3.5, data value = 21.7

20. Mean = 8, standard deviation = 2.1, data value = 7.2
21. Mean = 100, standard deviation = 15, data value = 63
22. Mean = 530, standard deviation = 42, data value = 625
23. Mean = 1.2, standard deviation = 0.25, data value = 1.31
24. Mean = 3.6, standard deviation = 0.31, data value = 2.98

For Exercises 25–46, find the area under the standard normal distribution curve.

25. Between $z = 0$ and $z = 1.95$.
26. Between $z = 0$ and $z = 0.55$.
27. Between $z = 0$ and $z = -0.5$.
28. Between $z = 0$ and $z = -2.05$.
29. To the right of $z = 1.0$.
30. To the right of $z = 0.25$.
31. To the left of $z = -0.40$.
32. To the left of $z = -1.45$.
33. Between $z = 1.25$ and $z = 1.90$.
34. Between $z = 0.8$ and $z = 1.3$.
35. Between $z = -0.85$ and $z = -0.20$.
36. Between $z = -1.55$ and $z = -1.85$.

37. Between $z = 0.25$ and $z = -1.10$.
38. Between $z = 2.45$ and $z = -1.05$.
39. To the left of $z = 1.20$.
40. To the left of $z = 2.15$.
41. To the right of $z = -1.90$.

42. To the right of $z = -0.20$.
43. To the left of $z = -0.60$.
44. To the right of $z = -1.10$.
45. Between $z = 1.90$ and $z = 1.95$.
46. Between $z = -0.1$ and $z = -0.2$.

Applications in Our World

As we know from Chapter 10, the probability of an event is the same as the percent chance that it will happen, written in decimal form. You'll use this idea in Exercises 47–50.

47. The College Board reported that SAT scores in the early 2000s were normally distributed with mean 1026 and standard deviation 210.
    (a) What percentage of students scored over 606?
    (b) Out of every 500 students, how many scored between 816 and 1236?
48. According to 9monthsafter.com, the length of human pregnancies are normally distributed with mean 268 days and standard deviation 15 days.
    (a) What is the probability that you were born more than 298 days after you were conceived?
    (b) If a child is conceived on April 19, what is the probability that it will be born before January 27?

49. According to the label, a bag of Kettle brand Tias tortilla chips contains 8 ounces of chips. Suppose that the bagging process is normally distributed with a standard deviation of 0.14 ounce. If the manager wants there to be greater than a 97.5% chance that any given bag has at least 8 ounces, and the weight they set their filling machine for provides the mean weight, how should they set the machine?
50. The final scores for students in my Math 102 classes over the last 10 years are approximately normally distributed with mean 74 and standard deviation 10.7. If I decide to assign one student's grade this semester by randomly choosing a score of some student in the last 10 years, what's the approximate probability that the student will get a score that is (a) between 63 and 85 and (b) over 53?

Critical Thinking

51. I'm a 6′4″ male, and my wife is 5′10″. We're both pretty tall, but clearly I am taller. I can send you pictures if you don't believe me. Anyhow, based on z scores, which of us is taller compared to the rest of our sex? The mean height for women in the U.S. is 63.7 inches and the standard deviation is 2.7 inches. These statistics for men can be found within the section.
52. The 1942 St. Louis Cardinals were an awesome baseball team. They won 106 games in an era when teams played only 154 games in a regular season. During the 1942 season, the mean number of games won by a National League team was 75.8 with a standard deviation of 20.5. Another great Cardinals team was the 2004 squad, which won 105 out of 162 games. In 2004, the mean number of games won by a National League team was 80.9 with a standard deviation of 13.6.
    (a) Which was the better team based strictly on percentage of games won?
    (b) Which was the better team based on z scores?
53. For this question, we'll need to assume that the finishing times for a group of runners in a marathon race are likely to be normally distributed. The mean finish time for male marathon (26.2 miles) runners is 278.42 min with a standard deviation of 39.4 min. The mean finish time for female marathon runners is 294.45 min with a standard deviation of 39.9 min. At the 2021 Boston Marathon, Benson Kipruto won the men's division with a time of 129.85 min and Diana Kipyokei won the women's division at 144.75 min. Who finished faster, relative to their sex?

54. Refer to Exercise 53. Who is more unusual within their population: a person who scores 150 on an IQ test, or a female marathon runner who finishes in 3 hours?
55. Find a z value to the right of the mean so that 67.4% of the distribution lies to the left of it.
56. Find a z value to the left of the mean so that 98.6% of the area lies to the right of it.
57. Find two z values, one positive and one negative but having the same absolute value, so that the areas in the two tails (ends) total these values.
    (a) 4%
    (b) 8%
    (c) 1.6%
58. Using the Internet as a resource, find at least five different quantities in our world that tend to be normally distributed, and discuss why you think it makes sense that each fits a normal distribution.
59. Normal distributions are often used in packaging. Packaging of most products is done by a machine, and there's going to be some variation in the amount of product in each package. Suppose that the standard deviation for one machine that fills potato chip bags is 0.4 ounces, and the package is labeled as containing 14 ounces of chips.
    (a) Would it be wise to calibrate the machine so that the mean weight of each package is 14 ounces? Why or why not?
    (b) If the plant manager wanted to be 97.5% sure that any given bag had at least the weight stated on the

package, what mean weight should they have their engineers calibrate the packaging machine for? (*Hint:* Think empirical rule.)

60. USING TECHNOLOGY. Spreadsheets can be used to efficiently do many comparisons between data values from different sets, because you can set up a formula to quickly calculate z scores. The plan here is to build a z-score calculator that allows you to input mean and standard deviation for two different data sets, then use formulas to calculate the z scores for different data values. In the online resources for this section, there's a spreadsheet with a variety of ACT and SAT scores listed. Fill in the appropriate means and standard deviations from within the lesson, then use a formula to find z scores and copy it down to find z scores for all of the test scores for comparison. There are further instructions on the spreadsheet.

Section 11-7   Applications of the Normal Distribution

LEARNING OBJECTIVES

1. Use the normal distribution to find percentages.

2. Use the normal distribution to find probabilities.

3. Use the normal distribution to find percentile ranks.

4. Recognize data that are approximately normally distributed.

Surely there must be someone out there who doesn't like Oreos, but I haven't met them. A standard package of America's favorite cookie contains 510 grams of sweet temptation. Of course, when you buy a package, you expect every one of those 510 grams. But there's variation in just about everything, so some packages will have more than the intended weight, and some will have less. The folks running Nabisco aren't dummies, though— they know that if someone decides to check the weight and finds it to be less than 510 grams, they won't be a very happy customer. So what to do?

John Flournoy/McGraw Hill

This is exactly the sort of situation where weights tend to be normally distributed. The company would likely design its production and packaging process so that the mean is somewhat larger than 510 grams, with a standard deviation that assures that the vast majority of packages will weigh at least 510 grams.

A quantity that can vary randomly from one individual trial to another, like the exact weight of a package of Oreos, is called a **random variable**. In this section, we will see how the area calculations we practiced in Section 11-6 can be used to solve problems in our world for random variables that are normally distributed. Our general strategy will be to transform specific data into z scores, then use Table 11-3 or technology to find areas. Finally, we'll interpret the area as it applies to the given data.

The reason this works is indicated by Figure 11-19. Suppose that the weight of Oreos in a package is normally distributed with mean 518 grams and standard deviation 4 grams. The distribution would look like Figure 11-19a. After using the z score formula to standardize, we get the graph in Figure 11-19b; now we can apply our area calculations, and the area will tell us the percentage of packages that will fall within a certain range.

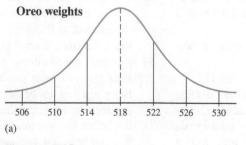
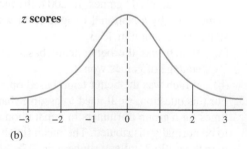

Figure 11-19

Example 1   Solving a Problem Using the Normal Distribution

If the weights of Orcos in a package are normally distributed with mean 518 grams and standard deviation 4 grams, find the percentage of packages that will weigh less than 510 grams.

Math Note

It's typical for manufacturers to make sure that the listed weight for products packaged by weight is at least two standard deviations below the mean, ensuring that over 97% of the packages have the listed weight or more.

SOLUTION

Step 1 Draw the figure and represent the area, as shown in Figure 11-20.

Step 2 Find the z score for data value 510.

$$z = \frac{\text{value} - \text{mean}}{\text{standard deviation}}$$
$$= \frac{510 - 518}{4} = -2$$

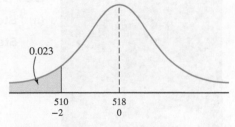

Figure 11-20

This tells us that 510 grams is 2 standard deviations below the mean.

Step 3 Find the shaded area using Table 11-3 or technology. The shaded area is the area of the left half (0.5) minus the area between $z = 0$ and $z = -2$. The area between $z = 0$ and $z = -2$ is 0.477 (using $z = +2$ from the table).

$$\text{Area} = 0.5 - 0.477 = 0.023$$

Step 4 Interpret the area. In this case, an area of 0.023 tells us that only 2.3% of packages will have weights less than 510 grams.

McGraw Hill

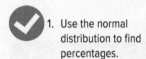

1. Use the normal distribution to find percentages.

Try This One 1

Based on data compiled by the World Health Organization, the mean systolic blood pressure in the United States is 120, the standard deviation is 16, and the pressures are normally distributed. Find each.

(a) The percent of individuals who have a blood pressure between 120 and 128
(b) The percent of individuals who have a blood pressure above 132
(c) The percent of individuals who have a blood pressure between 112 and 116
(d) The percent of individuals who have a blood pressure between 124 and 144
(e) The percent of individuals who have a blood pressure lower than 104

Our next interpretation for area under a normal distribution is tremendously important because it ties our study of statistics to the main topic of Chapter 10, probability.

Probability and Area under a Normal Distribution

The area under a normal distribution between two data values is the probability that a randomly selected data value is between those two values.

Example 2 illustrates the use of area to find probability.

Example 2 Using Area under a Normal Distribution to Find Probabilities

Based on data in an EPA study released in 2016, the average American generates 1,606 pounds of garbage per year. Let's estimate that the number of pounds generated per person is approximately normally distributed with standard deviation 200 pounds. Find the probability that a randomly selected person generates

(a) Between 1,250 and 2,050 pounds of garbage per year.
(b) More than 2,050 pounds of garbage per year.

Photodisc/Getty Images

SOLUTION

(a) **Step 1**   Draw the figure and represent the area. See Figure 11-21.

Step 2   Find the z scores for the two given weights.

$$1{,}250 \text{ pounds: } z = \frac{1{,}250 - 1{,}606}{200} = -1.78$$

$$2{,}050 \text{ pounds: } z = \frac{2{,}050 - 1{,}606}{200} = 2.22$$

Step 3   Find the shaded area. Using Table 11-3, the area between $z = 0$ and $z = 2.22$ is about 0.487, and the area between $z = 0$ and $z = -1.78$ is about 0.462. So the shaded area is

$$0.487 + 0.462 = 0.949$$

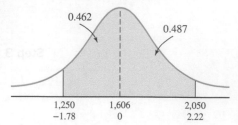

Figure 11-21

Step 4   Interpret the area. This tells us that if a person is selected at random, the probability that they generate between 1,250 and 2,050 pounds of garbage per year is 0.949.

(b) Most of the work was already done in part (a). We can adapt the graph a bit (Figure 11-22).

  The shaded area is the area between $z = 0$ and $z = 2.22$ subtracted from the area of the entire right half (0.5).

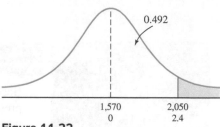

Figure 11-22

$$0.5 - 0.487 = 0.013$$

The probability that a randomly selected person generates more than 2,050 pounds of garbage per year is 0.013.

Math Note

The probability that a randomly selected value from a normal distribution is less than some value is the area to the left of that value. The probability of a randomly selected value being greater than some value is the area to the right of that value.

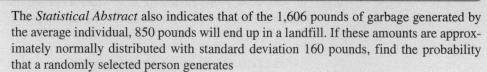

Try This One   2

The *Statistical Abstract* also indicates that of the 1,606 pounds of garbage generated by the average individual, 850 pounds will end up in a landfill. If these amounts are approximately normally distributed with standard deviation 160 pounds, find the probability that a randomly selected person generates

(a) Less than 600 pounds that end up in a landfill.
(b) Between 600 and 1,000 pounds that end up in a landfill.

2. Use the normal distribution to find probabilities.

Normal distributions can also be used to find percentiles and answer questions of "how many?" Example 3 illustrates these types of questions.

Example 3   Finding Number in a Sample and Percentile Rank

The American Automobile Association reports that the average time it takes to respond to an emergency call is 25 minutes. If the response time is approximately normally distributed and the standard deviation is 4.5 minutes:

(a) If 80 calls are randomly selected, approximately how many will have response times less than 17 minutes?
(b) In what percentile is a response time of 30 minutes?

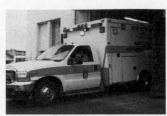

PBNJ Productions/Blend Images
LLC

SOLUTION

(a) **Step 1** Draw a figure and represent the area as shown in Figure 11-23.

Step 2 Find the z score for 17.

$$z = \frac{\text{Value} - \text{mean}}{\text{Standard deviation}}$$

$$= \frac{17 - 25}{4.5} \approx -1.78$$

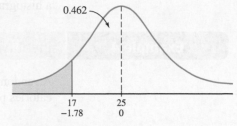

Figure 11-23

Step 3 Find the shaded area. Using Table 11-3, we find that the area between $z = 0$ and $z = -1.78$ is approximately 0.462. The shaded area is the difference of the entire left half (0.5) and 0.462.

$$\text{Area} = 0.5 - 0.462 = 0.038$$

Step 4 Interpret the area. An area of 0.038 tells us that 3.8% of calls will have a response time of less than 17 minutes. Multiply the percentage by the number of calls in our sample:

$$3.8\% \text{ of 80 calls} = (0.038)(80) = 3.04$$

About 3 calls in 80 will have response time less than 17 minutes.

(b) **Step 1** Draw a figure and represent the area as shown in Figure 11-24. Remember that to find the percentile, we are interested in the percentage of times that are less than 30 minutes.

Step 2 Find the z score for a response time of 30 minutes.

$$z = \frac{30 - 25}{4.5} \approx 1.11$$

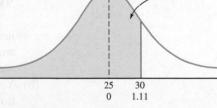

Figure 11-24

Step 3 Find the shaded area. The area to the left of $z = 1.11$ is the area of the left half (0.5) plus the area between $z = 0$ and $z = 1.11$ (about 0.366 according to Table 11-3).

$$\text{Area} = 0.5 + 0.366 = 0.866$$

Step 4 Interpret the area. An area of 0.866 means that 86.6% of calls will get a response time of less than 30 minutes, so the time of 30 minutes is in the 87th percentile.

> **Math Note**
>
> Notice that in this case (number of calls), an answer of 3.04 doesn't make sense. We need to round to the nearest whole number.

Try This One   3

The mean for a reading test given nationwide is 80, and the standard deviation is 8. The random variable is normally distributed. If 10,000 students take the test, find each.

(a) The number of students who will score above 90
(b) The number of students who will score between 78 and 88
(c) The percentile of a student who scores 94
(d) The number of students who will score below 76

3. Use the normal distribution to find percentile ranks.

The random variables we've studied so far in this section are described as being normally distributed. But how would you know that if you were gathering raw data? One way is to draw a histogram and see if it's bell-shaped, or close to it. This is illustrated in Example 4.

Example 4	Identifying Normally Distributed Data

A random sample of 25 entrées served by a college cafeteria is tested to find the number of calories per serving, with the data in the list below.

845	460	620	752	683	1,088	785
575	580	755	720	650	512	945
672	526	1,050	725	822	740	773
812	880	911	910			

(a) Draw a histogram for the data using seven classes. Do the data appear to be normally distributed?
(b) Find the mean and standard deviation for the data.
(c) What percentage of randomly selected entrées from this cafeteria has more than 700 calories?

SOLUTION

(a) The lowest number of calories is 460 and the highest is 1,088, a range of about 700 calories. A good choice of classes here is simply using the number of hundreds: 400–499, 500–599, and so on.

The histogram doesn't look exactly like a normal curve, but it's close enough that it would be reasonable to guess that the data are approximately normally distributed.

(b) You can find the mean and standard deviation by hand, but with 25 data values, using technology seems like a good idea. We used the AVERAGE and STDEV commands in Excel, but you could also use a graphing calculator. The results:

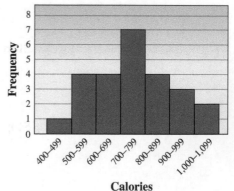

$$\overline{X} = 751.6 \quad s = 160.7$$

(c) Since we're already using technology, let's also use it to find an area under a normal curve with mean 751.6 and standard deviation 160.7. We want the area to the right of 700. For the calculator, we press 2nd VARS 2 to get the **normalcdf** command, then key in 700,1000000,751.6,160.7). For a spreadsheet, we use the command =1−NORM.DIST (700,751.6,160.7).

The screen shots show what the results look like. In each case, we see that the area is about 0.626. That tells us that there's a 62.6% chance that an entrée chosen at random has more than 700 calories.

David Sobecki/McGraw Hill

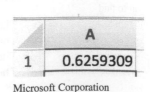

Microsoft Corporation

4. Recognize data that are approximately normally distributed.

Try This One   4

The ages of 33 randomly selected inmates at a county jail are shown below.

18	18	20	24	26	26	27	29
30	30	31	33	34	34	36	37
37	38	38	38	39	41	42	42
43	43	46	47	49	49	52	54
55							

(a) Draw a histogram to show that the ages are approximately normally distributed. Use classes 18–25, 26–33, and so on.
(b) Find the mean and standard deviation.
(c) If the prison has 1,148 inmates, about how many are under the age of 21?

Answers to Try This One

1 (a) 19.2%
  (b) 22.7%
  (c) 9.3%
  (d) 33.4%
  (e) 15.9%

2 (a) 0.059
  (b) 0.767

3 (a) 1,060
  (b) 4,400
  (c) 96th percentile
  (d) 3,080

4 (a)

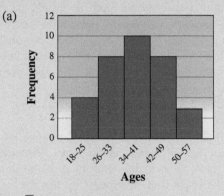

  (b) $\overline{X} = 36.5$, $s = 10.0$
  (c) 70

Exercise Set   11-7

Writing Exercises

1. Explain why the normal distribution can be used to solve many problems in our world.
2. Given a data set, how could you decide if the distribution of the data was approximately normal?
3. Describe the connection between area under a normal curve and probability.

4. When you know that a random variable is approximately normally distributed, what do you need to find in order to calculate probabilities of the variable being in some range?

Applications in Our World

5. The average hourly wage for all workers in the area of leisure and hospitality in 2021 was $19.15, according to the Bureau of Labor Statistics. If wages for this sector are normally distributed with standard deviation $3.15,

find these probabilities for a randomly selected worker in the leisure and hospitality sector:
(a) Earning more than $20.00 per hour
(b) Earning less than $13.00 per hour

6. The average cost for two people to go to a movie is $22. The standard deviation is $3. Assume the cost is normally distributed. Find the probability that at any given theater, the cost will be more than $26 for two people to go to a movie.

7. According to salary.com, the mean salary for a high school teacher in the United States in 2021 was $62,870, with a standard deviation of $15,100. Find these probabilities for a randomly selected teacher. Assume the variable is normally distributed.
   (a) The teacher earns more than $65,000.
   (b) The teacher earns less than $45,000.

8. Average sales for an online textbook distributor were $71.12 per customer per purchase. Assume the sales are normally distributed. If the standard deviation of the amount spent on textbooks is $8.42, find these probabilities for a randomly selected customer of the online textbook distributor.
   (a) They spent more than $60 per purchase.
   (b) They spent less than $80 per purchase.

9. A survey found that people keep their televisions an average of 4.8 years. The standard deviation is 0.89 year. If a person decides to buy a new TV, find the probability that they have owned the old one for the given amount of time. Assume the random variable is normally distributed.
   (a) Less than 2.5 years
   (b) Between 3 and 4 years
   (c) More than 4.2 years

10. The average age of CEOs is 56 years. Assume the random variable is normally distributed. If the standard deviation is 4 years, find the probability that the age of a randomly selected CEO will be in the given range.
   (a) Between 53 and 59 years old
   (b) Between 58 and 63 years old
   (c) Between 50 and 55 years old

11. The average life of a brand of automobile tires is 30,000 miles, with a standard deviation of 2,000 miles. If a tire is selected and tested, find the probability that it will have the given lifetime. Assume the random variable is normally distributed.
   (a) Between 25,000 and 28,000 miles
   (b) Between 27,000 and 32,000 miles
   (c) Between 31,500 and 33,500 miles
   (d) If you worked at a tire shop, would you expect to see many sets of these tires that lasted 40,000 miles? Why or why not? Make sure your answer is backed up with math!

12. The average time a person spends in each visit to an online social networking service is 62 minutes. The standard deviation is 12 minutes. If a visitor is selected at random, find the probability that they will spend the time shown on the networking service. Assume the times are normally distributed.
   (a) At least 180 minutes
   (b) At least 50 minutes

   (c) If someone logged into the service, then logged back out in 5 minutes, would you find that unusual? Why or why not? Back up your answer with math.

13. The average amount of snow per season in Trafford is 44 inches. The standard deviation is 6 inches. Find the probability that next year Trafford will receive the given amount of snowfall. Assume the random variable is normally distributed.
   (a) At most 50 inches of snow
   (b) At least 53 inches of snow

14. The average waiting time for a drive-in window at a local bank is 9.2 minutes, with a standard deviation of 2.6 minutes. When a customer arrives at the bank, find the probability that the customer will have to wait the given time. Assume the random variable is normally distributed.
   (a) Between 5 and 10 minutes
   (b) Less than 6 minutes or more than 9 minutes

15. The average time it takes a first-year college student to complete the Mason Basic Reasoning Test is 24.6 minutes. The standard deviation is 5.8 minutes. Find these probabilities. Assume the random variable is normally distributed.
   (a) It will take a student between 15 and 30 minutes to complete the test.
   (b) It will take a student less than 18 minutes or more than 28 minutes to complete the test.

16. A brisk walk at 4 miles per hour burns an average of 300 calories per hour. If the standard deviation of the distribution is 8 calories, find the probability that a person who walks 1 hour at the rate of 4 miles per hour will burn the given number of calories. Assume the random variable is normally distributed.
   (a) More than 280 calories
   (b) Less than 293 calories
   (c) Between 285 and 320 calories

17. During September, the average temperature of Laurel Lake is 64.2° and the standard deviation is 3.2°. Assume the random variable is normally distributed. For a randomly selected day in September, find the probability that the temperature will be
   (a) Above 62°.
   (b) Below 67°.
   (c) Between 65° and 68°.

18. If the systolic blood pressure for a certain group of people has a mean of 132 and a standard deviation of 8, find the probability that a randomly selected person in the group will have the given systolic blood pressure. Assume the random variable is normally distributed.
   (a) Above 130
   (b) Below 140
   (c) Between 131 and 136

19. An IQ test has a mean of 100 and a standard deviation of 15. The test scores are normally distributed. If 2,000 people take the test, find the number of people who will score

(a) Below 93.

(b) Above 120.

(c) Between 80 and 105.

(d) Between 75 and 82.

20. Based on data from the U.S. Census Bureau, I was able to estimate that the mean size in square feet of new homes sold in 2021 was 2,333, with a standard deviation of 850. Let's estimate that the sizes are normally distributed. In a sample of 500 recently built homes, find the number of homes that will

(a) Have between 1,900 and 2,000 square feet.

(b) Have more than 3,000 square feet.

(c) Have less than 2,000 square feet.

(d) Have more than 1,600 square feet.

21. Based on more data from the U.S. Census Bureau, I was able to estimate that the mean selling price of new homes built in 2020 was $336,900, with a standard deviation of $112,000. Home prices aren't exactly normally distributed, but let's estimate that they're close. If 800 home sales are selected at random, find the number of homes that cost

(a) More than $500,000.

(b) Between $200,000 and $300,000.

(c) Less than $150,000.

22. The average price of Stephen King paperbacks sold at bookstores across the country is $9.52, and the standard deviation is $1.02. Assume the price is normally distributed. If a national bookstore sells 1,000 Stephen King paperbacks during August, find the number of Stephen King paperbacks that were sold

(a) For less than $8.00.

(b) For more than $10.00.

(c) Between $9.50 and $10.50.

(d) Between $9.80 and $10.05.

23. Refer to Exercise 5. If your buddy Jamal gets a job as a hotel shift manager making $15.20 per hour, what percentile is he in?

24. Refer to Exercise 9. If you buy a TV for your dorm and keep it for your 4 years of school then sell it, what percentile does that put you in?

25. Refer to Exercise 12. Suppose that your grades have been slipping, and you impose a new rule on yourself: no more than 20 minutes at a time on social networking sites. What percentile would that put you in?

26. Refer to Exercise 20. When the one-hit wonder band The Flaming Rogers falls off the charts, their singer is forced to downsize from the 7,000 square foot home he built earlier in the year to a more modest 2,900 square foot model. What was his change in percentile rank?

USING TECHNOLOGY. In Problems 27–30, (a) construct a histogram for the data using the given number of classes to show that the data are approximately normally distributed; (b) find the mean and standard deviation using a graphing calculator, Excel, or an online calculator; (c) find the requested probabilities.

27. A researcher is studying reaction times in adult subjects who are alcohol-impaired in an effort to study the effects of impairment on driving. The amount of time 25 randomly selected adults take to react to a stimulus when they have a blood alcohol level of 0.08 was recorded and is displayed below. Use five classes; find the probability that a randomly selected driver with a blood alcohol of 0.08 will take less than 1.5 seconds to react, and the probability that they will take between 2 and 4 seconds to react.

1.2	1.5	1.6	1.6	1.7	1.9	2.0	2.1	2.1
2.2	2.2	2.2	2.3	2.4	2.4	2.5	2.5	2.7
2.8	2.8	2.8	3.0	3.1	3.3	3.5		

28. Repeat Exercise 27 for the reaction times of 25 subjects with a zero blood alcohol level:

0.7	1.1	1.1	1.2	1.4	1.4	1.4	1.5	1.5
1.5	1.6	1.6	1.7	1.7	1.7	1.7	1.7	1.8
1.9	1.9	2.1	2.2	2.3	2.5	2.6		

29. A research firm surveys 32 newly hired graphic designers in the Atlanta area and reports their annual salaries in thousands. Use seven classes; find the probability that a randomly selected new graphic designer in Atlanta makes more than $60,000, and the probability that they make between $40,000 and $50,000.

35.1	37.3	40.2	42.8	43.2	43.8	45.1	46.9
47.0	47.3	47.3	47.7	48.8	49.2	49.4	50.3
50.3	50.7	50.9	50.9	51.5	52.3	52.8	54.1
54.2	54.4	54.8	55.1	55.9	57.7	58.2	61.8

30. Repeat Exercise 29 for newly hired graphic designers in New York City:

41.3	43.8	46.4	49.4	49.9	49.7	51.5	53.1
53.0	53.2	53.3	53.7	55.3	55.8	55.2	56.4
57.0	57.1	57.0	56.8	57.7	58.5	59.3	60.3
60.9	60.4	60.9	61.0	62.2	64.4	64.3	67.6

Critical Thinking

31. Discuss whether or not you think each of the given quantities is likely to be normally distributed.

(a) Daily gas prices over a 2-year period

(b) The number of M&Ms in a 2-pound bag

(c) The heights of players on professional basketball teams

(d) The daily number of hits for Facebook

(e) The ages of all active-duty military personnel

32. The number of incoming students at two campuses of a midwestern university have historically been normally distributed. The main campus incoming class has a mean of 3,402 and a standard deviation of 425, and the regional campus incoming class has a mean of 730 and a standard deviation of 102. If there were 3,740 incoming students on the main campus and 822 on the regional campus, which had the more successful year in student recruitment based on z scores?

33. If a distribution of raw scores were plotted and then the scores were transformed into z scores, would the shape of the distribution change? Explain.

34. An instructor gives a 100-point examination in which the grades are normally distributed. The mean is 60 and the standard deviation is 10. If there are 5% A's and 5% F's, 15% B's and 15% D's, and 60% C's, find the scores that divide the distribution into those categories.

35. A researcher who is in charge of an educational study wants subjects to perform some special skill. Fearing that people who are unusually talented or unusually untalented could distort the results, he decides to use people who scored in the middle 50% on a certain test. If the mean for the population is 100 and the standard deviation is 15, find the two limits (upper and lower) for the scores that would enable a volunteer to participate in the study. Assume the scores are normally distributed.

36. While preparing for their comeback tour, The Flaming Rogers find that the average time it takes their sound tech to set up for a show is 58.6 minutes, with a standard deviation of 4.3 minutes. If the band manager decides to include only the fastest 20% of sound techs on the tour, what should the cutoff time be for concert setup? Assume the times are normally distributed.

37. For the data provided in Exercise 27, find the probabilities that were requested using the empirical probability formula from Section 10-1. How do your answers compare to the results of Exercise 27? What can you conclude?

38. Repeat Question 37 for the data in Exercise 28. How do your conclusions compare to those from Question 37?

39. Read back through all of the solved examples in this section, and write a sentence or two for each, describing whether you think the work done in that example constitutes descriptive or inferential statistics.

40. When a random variable is normally distributed, how do you think the four measures of average we've studied compare? Explain your answer in detail. Thinking about the graphs we draw to help with area calculations will be helpful.

Section 11-8   Correlation and Regression Analysis

LEARNING OBJECTIVES

1. Draw and analyze scatter plots for two data sets.

2. Calculate correlation coefficients.

3. Determine if correlation coefficients are significant.

4. Find a regression line for two data sets.

5. Use regression lines to make predictions.

As the world suffered through the first year of the COVID-19 pandemic in 2020, there seemed to be a shining light at the end of the tunnel: governments and scientists in the private sector were working at record speed toward developing a vaccine that would protect us from the most serious effects of COVID-19. As 2021 dawned, not only did these vaccines become available in the United States, but they were offered for free. Then a strange thing happened: millions of people decided not to get the vaccine, and many of them worked diligently to convince others to do the same. Not surprisingly, there was quite a bit of analysis of why this happened. In order to develop some theories, we'd want to gather some data on vaccination rates and other characteristics of certain populations and see if there's any sort of meaningful connection that may be enlightening.

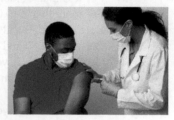
Prostock-studio/Shutterstock

Once we've found comparative data, we can then plot a graph designating one set of data as the x variable or **independent variable**, and the other as the y variable or **dependent variable**. This graph is called a *scatter plot*.

A **scatter plot** is a graph of the ordered pairs (x, y) consisting of data from two data sets.

After a scatter plot is drawn, we can analyze the graph to see if there's a noticeable pattern. If there is, like the points falling in an approximately straight line, we can conclude that the two variables might be related somehow.

If there appears to be a reasonably clear relationship between the variables, we might study that relationship further by trying to find an equation that relates the variables. This is known as *regression analysis*, which we will learn about later in the section. For now, we'll focus on building a scatter plot.

| Example 1 | Drawing and Analyzing a Scatter Plot |

Digital Vision/SuperStock

A medical researcher selects a sample of small hospitals in the state and hopes to discover if there is a relationship between the number of beds and the number of personnel employed by the hospital.

(a) Do you think there should be a relationship between these two data sets? Describe any relationship you'd expect.
(b) Draw a scatter plot for the data shown. Does it look like there's a relationship between the data sets?

No. of beds (x)	28	56	34	42	45	78	84	36	74	95
Personnel (y)	72	195	74	211	145	139	184	131	233	366

SOLUTION

(a) As a general rule, you would think that larger hospitals would have more beds in them, and you'd also think that larger hospitals would employ more personnel. So it seems reasonable to predict that less beds would go along with less personnel, and more beds would go along with more personnel.

(b) First we need to draw and label the axes. The information in the table indicates that the number of beds should go

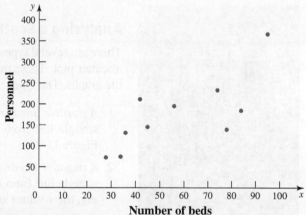

Figure 11-25

on the x axis, and the largest number of beds in the table is 95. So we'll label that axis to go from 0 up to 100. On the y axis, we'll go from 0 to 400 or so because the largest number of personnel is 366. Then we plot each point from the table, as shown in Figure 11-25.

We can see from the graph that there appears to be a connection. In general, the points progress from lower left to upper right, which confirms our prediction: lower number of beds tend to go with lower numbers of personnel, and higher numbers of beds tend to go with higher numbers of personnel.

Try This One 1

(a) Do you think there should be a relationship between the percentage of workers in a state that are unemployed and the average household income there? What kind of relationship?
(b) The table shows recent data for eight randomly selected states. Draw a scatter plot for the data. Does it look like there's a relationship between the data sets?

Unemployment %, x	6.2	6.4	2.7	4.8	3.1	4.3	4.7	5.0
Household income, y	42,278	49,875	60,730	55,173	73,397	58,080	53,875	46,140

Using Technology: Creating a Scatter Plot in Excel

To create a scatter plot in Excel:

1. Type the values that will go on the *x* axis in one column. (This would correspond to the number of beds in Example 1.)
2. Type the values that will go on the *y* axis in one column. (This would correspond to the number of personnel in Example 1.)
3. Use the mouse to drag and select all the data in those two columns.
4. With the appropriate cells selected, click the **Insert** tab, then **Charts**, and click on **Scatter**. Then choose the type of scatter diagram you want. Options include plotting only the points, connecting the points with curves, and connecting the points with line segments. For our purposes at the moment, we'll want the **Scatter** option that just plots the points.

You can add titles and change colors and other formatting elements by right-clicking on certain elements, or using the options on the **Charts** menu. Try some options and see what you can learn.

Sean Pavone/123RF

 1. Draw and analyze scatter plots for two data sets.

Analyzing a Scatter Plot

There are several types of relationships that can exist between the *x* values and the *y* values in a scatter plot. These relationships can be identified by looking at the pattern of the points on the graphs. The types of patterns and corresponding relationships are:

1. A *positive linear relationship* exists when the points fall approximately in an ascending straight line from left to right, where the *x* and *y* values increase at the same time. See Figure 11-26a.
2. A *negative linear relationship* exists when the points fall approximately in a descending straight line from left to right. See Figure 11-26b. In this case as the *x* values are increasing, the *y* values are decreasing.
3. A *nonlinear relationship* exists when the points fall in a curved line. See Figure 11-26c. The relationship is described by the nature of the curve.
4. *No relationship* exists when there is no discernible pattern to the points. See Figure 11-26d.

The relationship between the variables in Example 1 might be a positive linear relationship. It looks like as the number of beds increases, the number of personnel also increases.

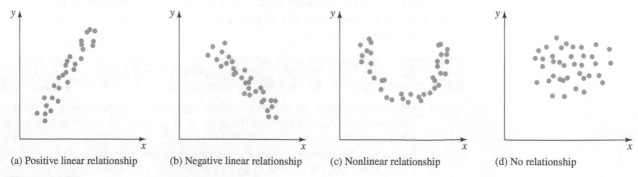

(a) Positive linear relationship   (b) Negative linear relationship   (c) Nonlinear relationship   (d) No relationship

Figure 11-26

The Correlation Coefficient

Deciding whether or not two data sets are related by simply looking at a scatter plot is a pretty subjective process, so it would be nice to have a way to quantify how strongly connected data sets are.

The **correlation coefficient** is a number that describes how close to a linear relationship there is between two data sets. Correlation coefficients range from −1 (perfect negative linear

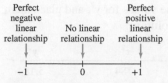

Perfect
negative
linear   No linear
relationship relationship

Perfect
positive
linear
relationship

−1   0   +1

Figure 11-27

relationship) to 1 (perfect positive linear relationship). The closer this number is to 1 in absolute value, the more likely it is that the data sets are linearly related. A correlation coefficient close to zero indicates that the data sets are most likely not linearly related. This could mean that they're related in some nonlinear way or that they're not related at all (see Figure 11-27). We use the letter r to represent the correlation coefficient. It doesn't depend on the units for the two data sets, and it also doesn't depend on which set you choose for the x variable.

Example 2   Estimating Correlation Coefficients

Based on our description of the correlation coefficient for two data sets, make an estimate of the correlation coefficient for each scatter plot in Figure 11-26.

SOLUTION

The first plot shows a positive linear relationship, so r should be close to 1. The second plot shows a negative linear relationship, so r should be close to −1. The third and fourth plots don't show linear relationships. One has a clear pattern and one doesn't, but the correlation coefficient only describes whether there's a *linear* relationship. Neither of those plots has one, so both should have r close to zero.

Try This One   2

Make an estimate of the correlation coefficient for the scatter plots in Example 1 and Try This One 1.

Calculating the Value of the Correlation Coefficient

In order to find the value of the correlation coefficient, we will use the following formula:

Formula for Finding the Value of r:

$$r = \frac{n(\sum xy) - (\sum x)(\sum y)}{\sqrt{[n(\sum x^2) - (\sum x)^2][n(\sum y^2) - (\sum y)^2]}}$$

where

n = the number of data pairs

$\sum x$ = the sum of the x values

$\sum y$ = the sum of the y values

$\sum xy$ = the sum of the products of the x and y values for each pair

$\sum x^2$ = the sum of the squares of the x values

$\sum y^2$ = the sum of the squares of the y values

Obviously, this is a pretty complicated formula, so arranging information in an orderly table is pretty much essential, as we'll see in Example 3.

Example 3   Calculating a Correlation Coefficient

Find the correlation coefficient for the data in Example 1, and discuss what you think it indicates.

SOLUTION

Step 1 Make a table as shown with the following headings: x, y, xy, x^2, and y^2.

Step 2 Find the values for the product xy, the values for x^2, the values for y^2, and place them in the columns as shown. Then find the sums of the columns.

Math Note

Since calculating the correlation coefficient is not exactly a walk in the park, technology can be a big help. The upcoming Using Technology box shows some of the ways you can get high-tech help finding r.

x	y	xy	x^2	y^2
28	72	2,016	784	5,184
56	195	10,920	3,136	38,025
34	74	2,516	1,156	5,476
42	211	8,862	1,764	44,521
45	145	6,525	2,025	21,025
78	139	10,842	6,084	19,321
84	184	15,456	7,056	33,856
36	131	4,716	1,296	17,161
74	233	17,242	5,476	54,289
95	366	34,770	9,025	133,956
$\Sigma x = 572$	$\Sigma y = 1,750$	$\Sigma xy = 113,865$	$\Sigma x^2 = 37,802$	$\Sigma y^2 = 372,814$

Step 3 Substitute into the formula and evaluate (note that there are 10 data pairs):

$$r = \frac{n(\Sigma xy) - (\Sigma x)(\Sigma y)}{\sqrt{\left[n(\Sigma x^2) - (\Sigma x)^2\right]\left[n(\Sigma y^2) - (\Sigma y)^2\right]}}$$

$$= \frac{10(113,865) - (572)(1,750)}{\sqrt{[10(37,802) - (572)^2][10(372,814) - (1,750)^2]}}$$

$$= \frac{137,650}{\sqrt{(50,836)(665,640)}}$$

$$\approx 0.748$$

The correlation coefficient for the two data sets is 0.748. This appears to confirm what our eyes tell us about the scatter plot in Example 1—it's certainly not a perfect linear relationship, but it does tend to indicate that as bed space goes up, the number of personnel does as well.

Try This One   3

Find the correlation coefficient for the data in Try This One 1, and discuss what you think it indicates.

2. Calculate correlation coefficients.

CAUTION

Based on what we've seen in Examples 1 and 3, does having more beds cause a hospital to have more staff, or does having more staff cause it to have more beds? If you answered "neither," you're very smart. All we know so far is that the *two variables appear to be related in a linear way.* This says exactly NOTHING about whether one is caused by the other. We'll discuss this very important issue later in the section.

Math Note

The paragraph to the right should make it pretty clear that this section is ALL about inferential statistics.

Next we will tackle the issue of how to interpret the correlation coefficient. In Example 2, the value of r was based on a sample of data. The population is all hospitals in a certain state with fewer than 100 beds, and a sample of such hospitals was chosen. When data from a sample is used, we can't be positive that it represents the entire population. So conclusions drawn, like the conclusion that there's a connection between numbers of beds and personnel, may not be correct.

Statisticians have traditionally agreed that when we conclude that two data sets have a relationship, we can be satisfied with that conclusion if there is either a 95% or 99% chance that we're correct. This corresponds to a 5% or 1% chance of being wrong. These percentages are called **significance levels**.

Table 11-4	Significant Values for the Correlation Coefficient					
	Sample Size	**5%**	**1%**	**Sample Size**	**5%**	**1%**
	4	.950	.990	17	.482	.606
	5	.878	.959	18	.468	.590
	6	.811	.917	19	.456	.575
	7	.754	.875	20	.444	.561
	8	.707	.834	21	.433	.549
	9	.666	.798	22	.423	.537
	10	.632	.765	23	.412	.526
	11	.602	.735	24	.403	.515
	12	.576	.708	25	.396	.505
	13	.553	.684	30	.361	.463
	14	.532	.661	40	.312	.402
	15	.514	.641	60	.254	.330
	16	.497	.623	120	.179	.234

We know that the more data we have, the more likely those data are representative of the population. Not surprisingly, the value of r needed to be reasonably sure that two data sets are correlated is higher for small sample sizes, and lower for large sample sizes. In Table 11-4, we see the minimum r values needed to have a 5% and 1% chance of being wrong when we conclude that two data sets are related.

Using Significant Values for the Correlation Coefficient

If $|r|$ is greater than or equal to the value given in Table 11-4 for either the 5% or 1% significance level, then we can reasonably conclude that the two data sets are linearly related.

The use of Table 11-4 is illustrated in Example 4.

Example 4	Deciding if a Correlation Coefficient Is Significant

Determine if the correlation coefficient $r = 0.748$ found in Example 3 is significant at the 5% level.

SOLUTION

Since the sample size is $n = 10$ and the 5% significance level is used, the value of $|r|$ has to be greater than or equal to the number found in Table 11-4 to be significant. In this case, $|r| = 0.748$; this is greater than 0.632, which we found in the table. Conclusion: there's at least a 95% chance that the data sets are significantly related.

However, we can't conclude that there is a relationship between the data sets at the 1% significance level since r would need to be greater than or equal to 0.765. In this example, $r = 0.748$, which is less than 0.765. In summary, we can be 95% sure that the data sets are linearly related, but not 99% sure.

David Buffington/Getty Images

Researchers could use correlation and regression analysis to find out if there is a correlation between the size of a family's home and the number of children in the family.

Try This One	4

Test the significance of the correlation coefficient obtained from Try This One 3. Use 5%, and then 1%.

CAUTION

Multi-symptom cold and cough relief without drowsiness

INDICATIONS: For the temporary relief of nasal congestion, minor aches, pains, headache, muscular aches, sore throat, and fever associated with the common cold. Temporarily relieves cough occurring with a cold. Helps loosen phlegm (mucus) and thin bronchial secretions to drain bronchial tubes and make coughs more productive.

DIRECTIONS: Adults and children 12 years of age and over, 2 liquid caps every 4 hours, while symptoms persist, not to exceed 8 liquid caps in 24 hours, or as directed by a doctor. Not recommended for children under 12 years of age.

WARNINGS: Do not exceed recommended dosage. If nervousness, dizziness, or sleeplessness occur, discontinue use and consult a doctor. Do not take this product for more than 10 days. A persistent cough may be a sign of a serious condition. If symptoms do not improve or if cough persists for more than 7 days, tends to recur, or is accompanied by rash, persistent headache, fever that lasts for more than 3 days, or if new symptoms occur, consult a doctor. Do not take this product for persistent or chronic cough such as occurs with smoking, asthma, chronic bronchitis, or emphysema, or where cough is accompanied by excessive phlegm (mucus) unless directed by a doctor. If sore throat is severe, persists for more than 2 days, is accompanied or followed by fever, headache, rash, nausea, or vomiting, consult a doctor promptly. Do not take this product if you have heart disease, high blood pressure, thyroid disease, diabetes, or difficulty in urination due to enlargement of the prostate gland unless directed by a doctor. As with any drug, if you are pregnant or nursing

McGraw Hill/C.P. Hammond, photographer

Medication labels help users to become aware of the effects of taking medications.

3. Determine if correlation coefficients are significant.

Remember, the result of Example 4 doesn't guarantee that the data sets are actually related—there's still a 5% chance that they're not related at all.

How should we choose whether to use the 5% or 1% significance level in a given situation? It depends on the seriousness of the situation and the importance of drawing a correct conclusion. Suppose that researchers think that a new medication helps patients with asthma to breathe easier, but that some patients have experienced side effects. A correlation study would probably be done to decide if there is a relationship between the medication and these particular side effects. If the potential side effects are serious, like heart attacks or strokes, the 1% significance level would be used. If the side effects are mild, like headache or nausea, the 5% significance level would probably be considered sufficient, since the consequences of being wrong are not as dire as in the first case.

Regression

Once we have concluded that there is a significant linear relationship between two variables, the next step is to find the *equation* of the **regression line** through the data points.

If you look back at the scatter plot in Example 1, you can see a general trend among the points from lower left to upper right. You could probably put a straightedge down and draw what seems like the closest approximation to a line by "splitting the difference" between points that don't line up exactly. That's what a regression line does, but instead of guessing, we're finding the single line with a special property: the overall distance from each point to the line is a minimum. (For that reason, the regression line is also called the *line of best fit.*)

Recall from algebra that the equation of a line in slope-intercept form is $y = mx + b$, where m is the slope and b is the y intercept. (See Section 6-3.) In statistics, the equation of the regression line is written as $y = a + bx$, where a is the y intercept and b is the slope. This is the equation that will be used here. In order to find the values for a and b by hand, we need two formulas. It's more common to use technology, which we'll learn about after Example 5.

Formulas for Finding the Values of a and b for the Equation of the Regression Line

$$b = \frac{n(\sum xy) - (\sum x)(\sum y)}{n(\sum x^2) - (\sum x)^2} \text{ slope}$$

$$a = \frac{\sum y - b(\sum x)}{n} \text{ } y \text{ intercept}$$

Example 5   Finding a Regression Line

Math Note

This procedure is very sensitive to rounding error, so using at least three decimal places is a good idea.

Find the equation of the regression line for the data in Example 1.

SOLUTION

We already calculated the values needed for each formula when we found the correlation coefficient in Example 3. Substitute into the first formula to find the value for the slope, b.

$$b = \frac{n(\sum xy) - (\sum x)(\sum y)}{n(\sum x^2) - (\sum x)^2} = \frac{10(113,865) - (572)(1,750)}{10(37,802) - (572)^2} = \frac{137,650}{50,836} \approx 2.708$$

Substitute into the second formula to find the value for the y intercept (a) when $b = 2.708$.

$$a = \frac{\sum y - b(\sum x)}{n} = \frac{(1,750) - 2.708(572)}{10} = \frac{201.02}{10} = 20.102$$

The equation of the regression line is $y = 20.102 + 2.708x$.

4. Find a regression line for two data sets.

Try This One 5

Find the equation of the regression line for the data in Try This One 1.

Using Technology: Finding the Line of Best Fit and Correlation Coefficient

Graphing Calculator

1. Press STAT ENTER to get to the list editor.
2. Enter the data set you want to use as inputs (x values) under **L1.**
3. Use the right arrow key ▶ to access **L2,** then enter the data set you want to use as outputs (y values). When entering data from a table, make sure you enter the second list in the same order as the first.
4. Turn on **Plot1** by pressing 2nd Y= 1. Set up the screen, as shown below.
5. Press Y= and if the **Y=** screen isn't blank, move the cursor over any entered equations and press CLEAR.
6. Press ZOOM 9 which is the **Zoom Stat** option; this automatically sets a graphing window that displays all of the plotted points.
7. Press STAT followed by the right arrow key to access the **STAT CALC** menu, and choose the **LinReg** (for linear regression) option, which is choice 4. Then press ENTER to calculate the line of best fit.
8. Press Y= and enter the equation of the line of best fit, then press GRAPH to display the scatter plot along with the line of best fit.
9. The correlation coefficient should display on the screen when you find the regression line. If it doesn't, go to the **CATALOG** menu (2nd 0), then scroll down and choose **DiagnosticOn.**

David Sobecki/McGraw Hill

Spreadsheet

1. Enter the data in two columns and create a scatter plot.
2. Click on one of the points on the scatter plot, then choose "Add Trendline" from the **Chart** menu.
3. In the formatting dialog box that appears, click "Options," then click the checkbox for "Display equation on chart."
4. After finding the trendline, on the format trendline menu, click the box that says "Display R-squared value on chart." This finds the square of the correlation coefficient, so you can then use the formula "=SQRT(" to find *r.* If you want to find the correlation coefficient without drawing a scatter plot and finding a line of best fit, you can use the "=CORREL" command. After entering the command in a cell, first enter the cell range for the first data set, either manually or by dragging down the column, then type a comma followed by the cell range for the second data set. Then press "enter" and the correlation coefficient will appear as if by magic. Nice.

After we find the equation of a regression line, we can see how it compares to the original data by graphing the line on the scatter plot. For the scatter plot in Example 1, we can graph the regression line by plotting two points using the equation we found in Example 5.

Choosing two *x* values between 0 and 100 (so that the points will appear on the scatter plot), if *x* = 30, then

$$y = 20.102 + 2.708x$$
$$= 20.102 + 2.708(30) \approx 101$$

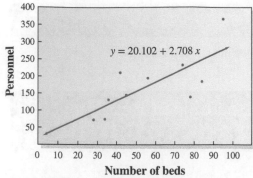

Figure 11-28

If $x = 70$, then
$$y = 20.102 + 2.708(70) \approx 210$$

So the points (30, 101) and (70, 210) are on the graph. We plot those two points and draw a line through them. See Figure 11-28.

The Relationship between r and the Regression Line

Two things should be noted about the relationship between the value of r and the regression line. First, the value of r and the value of the slope, b, always have the same sign. Second, the closer the value of r is to $+1$ or -1, the better the points will fit the line. In other words, the stronger the relationship, the better the fit. Figure 11-29 shows the relationship between the correlation coefficient and the regression line.

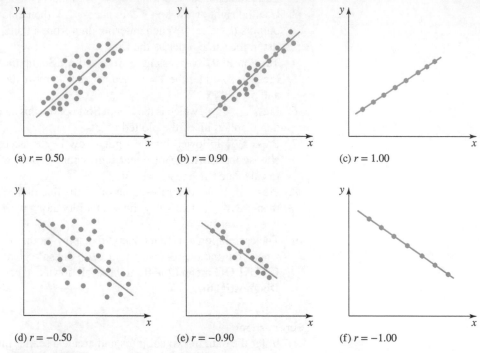

(a) $r = 0.50$

(b) $r = 0.90$

(c) $r = 1.00$

(d) $r = -0.50$

(e) $r = -0.90$

(f) $r = -1.00$

Figure 11-29

One of the most useful aspects of finding a regression line is that it can be used to make predictions for one of the variables given a value for the other. This is illustrated in Example 5.

| **Example 6** | Using a Regression Line to Make a Prediction |

Use the equation of the regression line found in Example 5 to predict the approximate number of personnel for a hospital with 65 beds.

SOLUTION

Substitute 65 for x into the equation $y = 20.120 + 2.7077x$ and find the value for y.

$$y = 20.120 + 2.7077(65) = 196.12 \text{ (which can be rounded to 196)}$$

We predict that a hospital with 65 beds will have about 196 personnel.

Try This One   6

5. Use regression lines to make predictions.

Use the equation of the regression line found in Try This One 5 to predict the household income in a state with a 5.5% unemployment rate.

CAUTION

Never use a regression line to predict anything unless you've already checked that the correlation is significant! If it isn't significant, the regression line is meaningless.

Correlation and Causation

Here's a key phrase to remember when studying correlation and regression analysis: *correlation is not the same thing as causation*. As we pointed out in a caution box earlier, knowing that two variables are related doesn't say anything about whether or not one CAUSES changes in the other. (This is discussed in the next Sidelight.) Here are some possible interpretations when two variables are significantly correlated.

Possible Relationships between Data Sets

©McGraw Hill Education/John Flournoy

1. *There is a direct cause-and-effect relationship between two variables.* That is, *x* causes *y*. For example, water causes plants to grow, poison causes death, and heat causes ice to melt.

2. *There is a reverse cause-and-effect relationship between the variables.* That is, *y* causes *x*. Suppose that we find a correlation between excessive coffee consumption and nervousness. It would be tempting to conclude that drinking too much coffee causes nervousness. But it might actually be the case that nervous people drink a lot of coffee because they think it calms their nerves.

Sidelight   Of Skiing and Killer Bedsheets

Don't ask how, but someone that's very interested in statistics discovered what appears to be a connection between the total revenue generated by skiing facilities in the United States and the number of Americans who die by becoming tangled in their bedsheets. Don't believe it? Have a look at the graphs. The first shows both of these data sets graphed by year. Note the similarity in their shapes. That's the first indication that they may be related somehow. So the second graph is a scatter plot of the data for matching years: it displays a pretty clear linear trend. How clear? The correlation coefficient, as calculated by Excel, is 0.97—close to a perfect correlation!

So should you start sleeping on the floor if ski facility revenues are up next year? If someone dies from getting tangled in their bedsheets, does everyone they know cope with their grief by going skiing? Are ski resorts stocked with killer sheets? While that would make a fine plot for a really dreadful movie, in real life it's just plain silly. This is a classic example of the fact that correlation does not in any way equate to causation. Bedsheet deaths don't cause people to go skiing, and visiting ski facilities doesn't make an individual more likely to die in bed. It's also quite a stretch to think that there's some third variable that connects these quantities. More than likely it's just a strange, cosmic coincidence. The moral of the story? Use your head. Don't be a data nerd. Never rely strictly on numbers like correlation coefficients to do your thinking for you. Statistics is a tool to help us draw conclusions, not an excuse to use our heads only to keep our hats from resting on our shoulders.

If you find this correlation interesting, do a Google search for the phrase "spurious correlations." It's HOURS of fun.

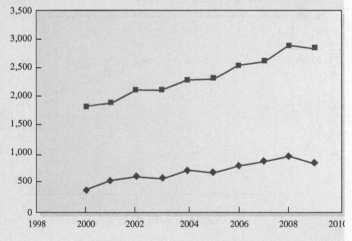

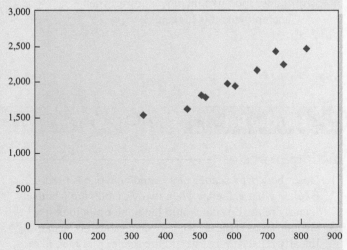

3. *The relationship between the variables may be caused by a third variable.* You could probably find a positive correlation between the amount of ice cream consumed per week and the number of drowning deaths for each week of the year. But this doesn't mean that eating ice cream causes you to drown. Both eating ice cream and swimming are more common during the summer months, and it stands to reason that more people will drown when more people are swimming. In this case, a third variable (seasonal weather) is affecting each of the original two variables.

4. *There may be a variety of complicated interrelationships among many variables.* For example, a researcher may find a significant relationship between students' high school grades and college grades. But there probably are many other variables involved, such as academic ability, hours of study, influence of parents, motivation, age, and instructors.

5. *The relationship might simply be coincidental.* For example, historians have noticed that there is a very strong correlation between the party that wins a presidential election and the result of the Washington NFL team's final game before the election. But good old common sense tells us that this relationship has to be a coincidence.

What about studying potential reasons for vaccine hesitance? We'll examine that in Exercises 25–30.

Answers to Try This One

1 (a) If unemployment is high, you'd think that income would be low, as both would tend to go with a struggling economy.

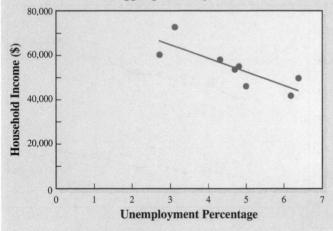

(b) It does look like income is lower when unemployment is higher.

2 Example 1: It's certainly not a perfect line, but somewhat linear. The correlation coefficient should be closer to 1 than to 0, maybe around 0.7 or so.

Try This One 1: Similar to Example 1, but the correlation appears to be negative. A guess around −0.7 is reasonable.

3 $r = -0.825$; this is close enough to −1 to conclude that it's likely there's a connection between unemployment and income.

4 Significant at the 5% level, but not 1%

5 $y = 83{,}212 - 6{,}079.2x$

6 $49{,}776.40

Exercise Set 11-8

Writing Exercises

1. Describe what a scatter plot is and how one is used.
2. Explain what is meant when two variables are positively linearly related. What would the scatter plot look like?
3. Explain what is meant when two variables are negatively linearly related. What would the scatter plot look like?
4. Explain what the value of the correlation coefficient tells you about the relationship between two data sets.
5. What is a regression line for two data sets?
6. Describe how regression lines can be used to make predictions about situations.

7. Describe the key difference between *correlation* and *causation*.

8. If we say that the relationship between two variables is significant at the 5% level, what does that actually mean?

9. Think of an example (other than the one provided in this section) of a situation where it would be reasonable to use the 5% significance level for a correlation, and one where it would be more likely that the 1% level would be used.

10. Think of an example of two data sets that you'd expect to have a negative linear relationship.

Computational Exercises

For the data sets in Exercises 11–18,

(a) Draw a scatter plot.
(b) Find the value for *r*.
(c) Test the significance of *r* at the 5% level and at the 1% level.
(d) Find the equation of the regression line and draw the line on the scatter plot, but only if *r* is significant.
(e) Describe the nature of the relationship if one exists.

11.

x	1	4	6	2	3	5	7
y	8	15	20	10	11	16	25

12.

x	21	25	24	30	36	40
y	12	8	9	5	3	2

13.

x	75	80	85	90
y	10	5	11	4

14.

x	9	12	15	11	10	13
y	50	60	71	55	53	60

15.

x	27	35	48	43	32
y	19	13	8	10	15

16.

x	31	34	37	40	46
y	3	15	2	13	5

17.

x	6.2	7.4	5.6	3.1	4.0
y	32.8	30.1	28.7	18.4	22.6

18.

x	0.5	0.9	1.6	0.7	2.8
y	15.6	9.4	5.3	11.2	6.7

Applications in Our World

For the data sets in Exercises 19–24,

(a) Draw a scatter plot.
(b) Find the value for *r*.
(c) Test the significance of *r* at the 5% level and at the 1% level.
(d) Find the equation of the regression line and draw the line on the scatter plot, but only if *r* is significant.
(e) Describe the nature of the relationship if one exists.
(f) Make the requested prediction using your regression line if possible.

19. The data represent the heights in feet and the number of stories of the tallest buildings in Cleveland.

Height, *x*	947	708	658	529	450	446	430	420	419
Stories, *y*	57	52	46	40	31	28	24	26	32

Source: *The World Almanac and Book of Facts*

Predict the number of stories in a 500-foot building.

20. A researcher hopes to determine whether the number of hours a person walks per week is related to the person's age.

Age, *x*	34	22	48	56	62
Hours, *y*	3.5	7	3.5	3	1

Predict the number of hours for a 35-year-old.

21. A study was conducted to determine if the amount a college student spends per month on recreation is related to the student's income.

Monthly income, *x*	Amount, *y*	Monthly income, *x*	Amount, *y*
$800	$160	$850	$145
$1,200	$300	$907	$190
$1,000	$260	$1,100	$250
$900	$235		

Predict the amount spent by a student with an income of $925.

22. A researcher hopes to determine if there is a relationship between the number of days an employee missed a year and the person's age.

Age, *x*	22	30	25	35	65	50	27	53	42	58
Days missed, *y*	0	4	1	2	14	7	3	8	6	4

Predict the number of days missed by a 56-year-old employee.

23. A statistics instructor plans to see if there's a relationship between the final exam score in Statistics 102 and the final exam scores of the same students who took Statistics 101.

Stat 101, *x*	87	92	68	72	95	78	83	98
Stat 102, *y*	83	88	70	74	90	74	83	99

Predict the Stat 102 score for a student who got 90 on the Stat 101 final.

24. The data shown indicate the number of wins and the number of goals scored for teams in the National Hockey League after the first month of the season.

No. of wins, x	No. of goals, y	No. of wins, x	No. of goals, y
10	23	7	21
9	22	5	16
6	15	9	12
5	15	8	19
4	10	6	16
12	26	6	16
11	26	4	11
8	26		

Source: *USA Today*

Predict the number of goals for a team with 8 wins.

USING TECHNOLOGY. In Questions 25–30, we'll be studying connections between certain data sets and the vaccination rates for each state. All data sets are provided in a spreadsheet in class resources for this section.

25. One theory is that education level might play a role in whether or not people choose to be vaccinated. On the first page of the spreadsheet, the percentage of people in each state who had received at least one dose of COVID-19 vaccine is in one column, and the percentage of adults with a college degree is in another.

(a) Find the correlation coefficient for the two data sets, and test it for significance at the 5% and 1% levels.

(b) Do you think there's a connection between education rates and vaccination rates? Explain.

26. Another theory is that political affiliation makes some people less likely to get vaccinated. Repeat Exercise 25 for the second page of the spreadsheet, which compares vaccination rates to the percentage of votes cast for the Democratic candidate in the 2020 presidential election.

27. Some people objected to the vaccine on religious grounds. Repeat Exercise 26 for the third page of the spreadsheet, which compares vaccination rates to the percentage of people who attend weekly religious services in each state.

28. Early demographic data suggested that younger Americans were less likely to get vaccinated than older. Repeat Exercise 25 for the fourth page of the spreadsheet, which compares vaccination rates to median age in each state.

29. On a hunch, I compared vaccination rates by state to teen pregnancy rates. Repeat Exercise 25 for the fifth page of the spreadsheet, which has the data. Do you think the correlation makes sense? Why or why not?

30. Based on your own thoughts and Exercises 25–30, what factors do you think played a significant role in vaccine hesitancy in the United States?

Critical Thinking

31. Find the value for r, then interchange the values for x and y and find the value of r. Explain the results.

x	1	2	3	4	5
y	3	5	7	9	11

32. (a) Draw a scatter plot of the given data and decide if the variables are related. Explain your conclusion.

x	−3	−2	−1	0	1	2	3
y	9	4	1	0	1	4	9

(b) Based on your scatter plot, make an educated guess as to what you think the correlation coefficient (r) is likely to be.

(c) Find the correlation coefficient for the data. How accurate was your guess?

33. Repeat Question 32, but this time use only the data values corresponding to $x > 0$. Why does this affect the result so much?

34. Design your own correlation problem. Think of two quantities that you think might be correlated, then gather some data, and use the techniques of this section to decide if the data sets are correlated.

35. Most people assume that there's a very strong connection between the price of crude oil and gasoline prices. But how strong is it? Do an Internet search for "oil prices historical" and "gas prices historical" and find prices for each on at least 10 dates over a span of at least 2 years. Then draw a scatter plot and calculate the correlation coefficient. Discuss how strong the connection appears to be.

36. You would think that there would be a strong correlation between a person's height and shoe size. Is that really the case? Find the heights and shoe sizes of at least 15 different people, calculate the correlation coefficient, and see if it's significant at the 5% and 1% levels. Will their sex affect the results? Can you correct for that?

In Problems 37–44, for each pair of quantities, decide if they are likely to be positively correlated, negatively correlated, or uncorrelated. Explain your reasoning.

37. Total number of years of school completed and average annual salary

38. Speed limit on a stretch of road and the probability of surviving an accident on that road

39. Number of primary schools in a town and average SAT scores of students graduating from high school in that town

40. Number of class meetings missed by college students and their average scores in the course

41. The age of a person and the probability that they wear glasses or contact lenses

42. The height of adults and the amount of money they earn relative to their coworkers

43. IQ and the probability that a person has served time in prison

44. Number of rooms in a house and how the homeowners rate their overall happiness on a scale of 1–100

45. Find the correlation coefficient for teen pregnancy rates and each of education level, percent voting for the Democratic candidate in the 2020 presidential election, median age, and percent attending weekly religious services. The data sets were provided by the spreadsheet that accompanies Exercises 25–30. List the quantities in order of how strongly they correlate with teen pregnancy rates. Then provide some theories on why the stronger correlations may exist.

46. The two graphics appeared in the *Washington Post*. Describe what you can learn from each.

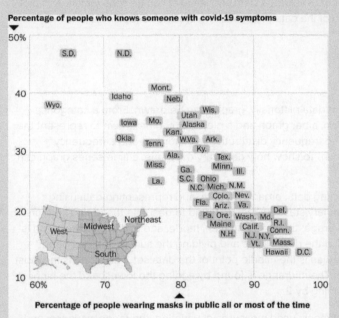

Data as of Oct. 19, 2020

Source: *Washington Post*

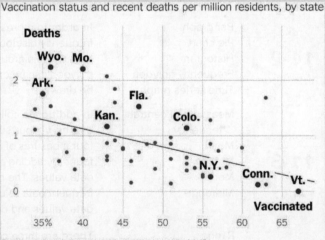

Section	Important Terms	Important Ideas
11-1	Data Statistics Population Sample Random sample Systematic sample Stratified sample Cluster sample Descriptive statistics Inferential statistics Raw data Categorical frequency distribution Grouped frequency distribution Stem and leaf plot Stem Leaf	**Statistics** is the branch of mathematics that involves collecting, organizing, summarizing, and presenting data. In addition, researchers use statistics to draw general conclusions from the data. When a study is conducted, the researcher defines a population, which consists of all subjects under study. There are two main types of statistics: descriptive and inferential. Descriptive statistics is about describing the characteristics of a population. Inferential statistics is about using characteristics of a sample to make predictions and draw conclusions about a larger population. Since populations are usually large, the researcher will select a representative subgroup of the population, called a sample, to study. There are four basic sampling methods: random, systematic, stratified, and cluster. Once data are collected, they are organized into a frequency distribution. There are two types of frequency distributions: categorical and grouped. When the data set consists of a relatively small number of values, a stem and leaf plot can be constructed. This plot illustrates the spread of the data while retaining the original data values.
11-2	Bar graph Pie chart Histogram Frequency polygon Time series graph	**In order** to represent data pictorially, graphs can be drawn. From a categorical frequency distribution, a bar graph and a pie chart can be drawn to represent the data. From a grouped frequency distribution, a histogram and a frequency polygon can be drawn. To show how data vary over time, a time series graph can be drawn.
11-3	Measures of central tendency Mean Median Mode Midrange	**In addition** to collecting data, organizing data, and representing data using graphs, various summary statistics can be found to describe the data. There are four measures of average: the mean, median, mode, and midrange. The mean is found by adding all of the data values and dividing the sum by the number of data values. The median is the middle point of the data set. The mode is the most frequent data value. The midrange is found by adding the lowest and the highest data values and dividing by 2.
11-4	Range Variance Sample standard deviation Population standard deviation	**There are** three commonly used measures of variation: the range, variance, and standard deviation. The range is found by subtracting the lowest data value from the highest data value. The standard deviation measures the spread of the data values. When the standard deviation is small, most data values are close to the mean. When the standard deviation is large, the data values are spread out farther away from the mean. The variance is the square of the standard deviation. Standard deviation can be computed either using all data values from a population (population standard deviation), or using a sample of data values (sample standard deviation).
11-5	Percentile Quartile Box plot Outlier Interquartile range (IQR)	**The position** of a data value in a data set can be determined by its percentile rank. The percentile rank of a specific data value gives the percent of data values that fall below the specific value. Quartiles divide a distribution into quarters. A box plot is a graphical representation of how data in a set are spread out. The quartiles are put on a number line with a box enclosing the middle half of values in the distribution. Any values that are above or below the box by more than 1.5 times the width of the box are considered to be outliers. These are aberrational values that should be investigated to decide if they're truly exceptional cases, or maybe a mistake in the data values.

Section	Important Terms	Important Ideas
11-6	Normal distribution Empirical rule Standard normal distribution *z* score	**Many variables** have a distribution that is bell-shaped. These variables are said to be approximately normally distributed. Statisticians use the standard normal distribution to describe these variables. The standard normal distribution has a mean of zero and a standard deviation of 1. A variable can be transformed into a standard normal variable by finding its corresponding *z* score. These *z* scores are a measure of how many standard deviations above or below the mean a given data value is. So they provide a very convenient way to compare data values that come from completely different data sets. The area under any portion of the curve is the percentage (in decimal form) of data values that fall between the values that begin and end the region. A table showing the area under the standard normal distribution can be used to find areas for various *z* scores, but graphing calculators and spreadsheets make the process considerably simpler.
11-7	Random variable	**Since many** variables in our world are approximately normally distributed, the standard normal distribution can be used to solve many applications in our world. Given a data set, we can draw a histogram to decide if the random variable is approximately normally distributed. If the histogram resembles a normal curve, then we can compute a mean and standard deviation, and use areas under the related normal curve to analyze the data.
11-8	Correlation Independent variable Dependent variable Scatter plot Correlation coefficient Significance levels Regression line	**In many situations**, it's interesting and useful to decide if two different random variables are related in some way. We begin by drawing a scatter plot, which is a graph that can be used to visualize a potential relationship. Next, we find the correlation coefficient, a number ranging from −1 to 1 that indicates whether there actually is a linear relationship. There are standards for determining how likely it is that two data sets are linearly related; if they are, we can find a regression line that best fits the data and use it to make predictions for values of one of the variables given a specific value of the other.

Math in Sociology REVISITED

Helen H. Richardson/The Denver Post/Getty Images

1. Even though there's a bit of a bell shape, with the highest values in the middle, the ends don't tail off nearly enough for us to consider this a normal distribution. But it does appear that homicides go up in the summer months and fall off in winter.

2. Mean 66.5, median 61.5, midrange 72, mode 54.

   The midrange and mode are far enough away from the other two that I wouldn't put a lot of stock in them. Also, the mean is higher than the median, but in the ballpark. This may be due to the one month that was unusually high. You could make a strong case for either being the most meaningful.

3. The scatter plot follows, with temperature as the x variable. While it's not at all clear whether there is a real correlation, it looks like there's at least a chance that the two data sets have a positive linear relationship.

4. The correlation coefficient is 0.805. Looking at the table of significant values for the correlation coefficient, we see that this is significant at both the 5% and 1% levels. So we can be 99% sure that homicides and monthly high temperature are linearly related.

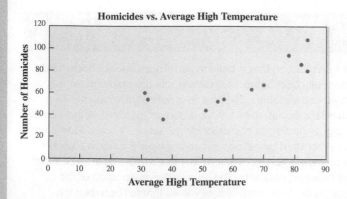

5. Absolutely not. In fact, no matter how strongly the data sets are related, we can't draw any conclusions about whether one causes the other. All we know is that for some reason, they appear to be related. There are a wide variety of theories to account for this phenomenon, but none have been proven or even widely accepted.

Review Exercises

Section 11-1

1. A sporting goods store kept a record of sales of five items for one randomly selected hour during a recent sale. Construct a frequency distribution for the data (B = baseballs, G = golf balls, T = tennis balls, S = soccer balls, F = footballs).

F	B	B	B	G	T	F
G	G	F	S	G	T	
F	T	T	T	S	T	
F	S	S	G	S	B	

2. The data set shown below represents the time in minutes spent using a computer at Kinko's Copy Center by 25 customers on a randomly selected day. Construct a stem and leaf plot for the data and analyze the results.

16	25	18	39	25	17	29	14	37
22	18	12	23	32	35	24	26	
20	19	25	26	38	38	33	29	

3. During June, a local theater company recorded the given number of patrons per day. Construct a grouped frequency distribution for the data. Use six classes.

102	116	113	132	128	117
156	182	183	171	168	179
170	160	163	187	185	158
163	167	168	186	117	108
171	173	161	163	168	182

Section 11-2

4. Construct a bar graph for the number of homicides reported for these cities.

City	Number	City	Number
New Orleans	179	Baltimore	234
Washington, D.C.	186	Atlanta	105
Chicago	509		

    Source: *Federal Bureau of Investigation*

5. Draw a pie chart for the data in Exercise 4.
6. Draw a histogram and frequency polygon for the frequency distribution obtained from the data in Exercise 3.
7. The data set shown below indicates how much Janine earned each year from a part-time job at the Otherworld Internet Cafe during the 5 years of college. Draw a time series graph for the data, then write an analysis of what the graph tells you about Janine's earnings.

Year	Amount	Year	Amount
2017	$8,973	2020	$13,877
2018	$9,388	2021	$19,203
2019	$11,271		

Section 11-3

8. The data that follow are the number of U.S. citizens (in thousands) identifying solely as Native Americans in the most recent census for the 22 states west of the Mississippi River.

363	322	297	193	171	104	72	63	61	56	53
37	33	32	31	28	27	22	21	18	13	11

    Find each of the four measures of average for the data, then discuss any similarities or differences among the four.
9. Twelve batteries were tested to see how many hours they would last. The frequency distribution is shown here.

Hours	Frequency	Hours	Frequency
1–3	1	10–12	1
4–6	4	13–15	1
7–9	5		

    Find the mean.
10. In your own words, explain the difference between mean, median, mode, and midrange. Include an explanation of why each is considered a measure of average.

Section 11-4

11. The data below are the number of beds at the eight top-ranked children's hospitals in the United States according to *U.S. News and World Report*. Find the mean, range, variance, and standard deviation. Discuss what the range and standard deviation tell you about the data set.

396	516	512	469	286	250	318	451

12. Which data set do you think would have a greater standard deviation: the ages of everyone currently working out in the gym on your campus, or the ages of everyone currently shopping at the nearest off-campus grocery store? Explain your answer.

Section 11-5

13. For the census data in Problem 8,
    (a) Find the percentile rank for the state where 193,000 citizens identified as Native American, and the state where 33,000 did.
    (b) What was the Native American population of the state at the 75th percentile?
14. Find the quartiles for the data in Problem 8, then draw a box plot and discuss what it tells you about the data.

Section 11-6

15. Find the area under the standard normal distribution curve.
    (a) Between $z = 0$ and $z = 1.95$
    (b) Between $z = 0$ and $z = 0.40$
    (c) Between $z = 1.30$ and $z = 1.80$
    (d) Between $z = -1.05$ and $z = 2.05$
    (e) Between $z = -0.05$ and $z = 0.55$
    (f) Between $z = 1.10$ and $z = -1.80$
    (g) To the right of $z = 2.00$
    (h) To the right of $z = -1.35$
    (i) To the left of $z = -2.10$
    (j) To the left of $z = 1.70$
16. A doctor has treated 45 male patients this week. Their average weight is 185 pounds, and the standard deviation is 25 pounds. If their weights are approximately normally distributed, use the empirical rule to estimate the number that weigh between 135 and 235 pounds.
17. For a data set with 90 values that is normally distributed with mean 200 and standard deviation 25, approximately how many data values will fall in each range?
    (a) The range from 190 to 210.
    (b) The range of values greater than 240.

Section 11-7

18. The most recent data available at the moment (2020 data) reports a mean SAT score of 1059 with a standard deviation of 210, and a mean ACT score of 20.8 with a standard deviation of 5.8. Which is a better score: 900 on the SAT or 17 on the ACT? Make an educated guess first, then decide by comparing z scores.
19. The mean age for IT workers in one market is 32.4, with a standard deviation of 8.1. The mean number of years of work experience is 8.2 years, with a standard deviation of 1.9. My friend Helen works in IT in this market, is 52 years old, and has 28 years of experience. Is Helen older or more experienced relative to the market in general?
20. The average number of years needed to complete a graduate degree in architecture at one major university is 3, with a standard deviation of 4 months. If the completion times are normally distributed, find the probability it will take a newly enrolled student:
    (a) More than 4 years to complete the program.
    (b) Less than 3 years to complete the program.
    (c) Between 3.8 and 4.5 years to complete the program.
    (d) Between 2.5 and 3.1 years to complete the program.
21. On the daily run of an express bus, the average number of passengers is 48. The standard deviation is 3. If the data are normally distributed, find the probability that the bus will have

(a) Between 36 and 40 passengers.
(b) Fewer than 42 passengers.
(c) More than 48 passengers.
(d) Between 43 and 47 passengers.
22. The average weight of an airline passenger's suitcase is 45 pounds. The standard deviation is 2 pounds. If the weights are normally distributed, and an airline handles 2,000 suitcases in one day, find the number that will weigh less than 43.5 pounds.
23. The average cost of Cheetah brand running shoes is $83.00 per pair, with a standard deviation of $8.00. If 90 pairs of shoes are sold, how many will cost between $80.00 and $85.00? Assume the cost is normally distributed.
24. The data below are the heights and weights of the 25 players on a major league baseball roster. Is either data set approximately normally distributed? Use a histogram to decide.

Heights (inches)

73	76	76	78	78	78	72	76	73
72	74	73	73	71	72	77	75	74
71	75	72	74	71	71	74		

Weights (pounds)

200	200	210	230	250	240	200	230	190
190	205	190	200	190	180	225	240	210
185	215	205	200	205	203	205		

Section 11-8

25. A study is done to see whether there is a relationship between a student's grade point average and the number of hours the students watches television each week. The data are shown here.

Hours, x	10	6	8	15	5	6	12
GPA, y	2.4	4	3.2	1.6	3.7	3.7	3.0

Draw a scatter plot for the data and describe the relationship. Find the value for r and determine whether or not it is significant at the 5% significance level. If the correlation is significant, find the equation of the regression line, and predict the GPA for a student that watches 9 hours of TV per week.
26. The table displays the number of homicides committed in Chicago, and the number of wins by the Chicago Bears of the NFL for each year from 2000 to 2017.

Homicides	Wins	Homicides	Wins	Homicides	Wins
628	5	452	13	516	10
666	13	442	7	441	8
647	4	509	9	432	5
598	7	458	7	492	6
448	4	449	11	762	3
449	11	440	8	650	5

(a) Without doing any calculations, discuss whether or not you think the two data sets should be related.
(b) Use technology to find the correlation coefficient for the data sets, and then see if it is significant at the 5% level. Was your prediction from part (a) right or wrong?

Chapter Test

1. A questionnaire about the last 25 peripheral devices purchased at a computer store is shown below. Construct a frequency distribution (L = laser mouse, E = external hard drive, K = wireless keyboard, W = webcam)

   L L K E E K L E W K
   W W L K W E K W L W
   E K K L L

2. Draw a bar graph for the frequency distribution obtained in Exercise 1.

3. Draw a pie chart for the data used in Exercise 2.

4. The data (in millions of dollars) are the values of the 30 Major League baseball franchises.

   (a) Construct a frequency distribution for the data using eight classes.

925	1,175	1,000	2,300	2,200	1,050
905	800	860	1,150	1,100	865
1,340	2,500	675	875	910	1,650
3,400	725	1,235	975	890	2,250
1,200	1,600	650	1,225	900	1,300

   Source: *Forbes Magazine*

   (b) Draw a histogram for the data. Does it appear to be approximately normally distributed? Explain.

   (c) The value of the St. Louis Cardinals was $1.6 billion. What was their percentile rank? (A billion is 1,000 million, by the way.)

   (d) What was the value of the franchise at the 20th percentile?

   (e) Draw a box plot for the data. Find any outliers.

5. A special aptitude test is given to job applicants. The data shown here represent the scores of 30 applicants. Construct a stem and leaf plot for the data, and summarize the results.

204	210	227	218	254
256	238	242	253	227
251	243	233	251	241
237	247	211	222	231
218	212	217	227	209
260	230	228	242	200

6. The given data represent the federal minimum hourly wage in the years shown. Draw a time series graph to represent the data, and analyze the results.

Year	Wage	Year	Wage
1960	$1.00	1990	3.80
1965	1.25	1995	4.25
1970	1.60	2000	5.15
1975	2.10	2005	5.15
1980	3.10	2010	7.25
1985	3.35	2020	7.25

7. These temperatures were recorded in Pasadena for a week in April.

   87 85 80 78 83 86 90

   Find each of these:

   (a) Mean   (c) Mode
   (b) Median   (d) Midrange

   (e) Range   (g) Standard deviation
   (f) Variance

8. The distribution of the number of errors 10 students made on a typing test is shown.

Errors	Frequency
0–2	1
3–5	3
6–8	4
9–11	1
12–14	1

   Find the mean.

9. Find the area under the standard normal distribution for each.
   (a) Between 0 and 1.50
   (b) Between 1.56 and 1.96
   (c) Between −0.06 and 0.73
   (d) To the right of $z = -1.28$
   (e) To the left of $z = 1.36$

10. The mean time it takes for a certain pain reliever to begin to reduce symptoms is 30 minutes, with a standard deviation of 4 minutes. Assuming the time is normally distributed, find the probability that it will take the medication
    (a) Between 34 and 35 minutes to begin to work.
    (b) More than 35 minutes to begin to work.
    (c) Less than 25 minutes to begin to work.
    (d) Between 35 and 40 minutes to begin to work.

11. A major brand of potato chips sells its product in a bag labeled 18 ounces. The packaging process is designed so that the mean weight is 18.35 ounces with a standard deviation of 0.12 ounce. In a production run of 10,000 bags, how many would you expect to be less than the labeled weight?

12. A study is conducted to determine the relationship between a driver's age at the beginning of the study and the number of accidents they have had over a 1-year period. The data are shown here. Draw a scatter plot for the data, and explain the nature of the relationship. Find the value for r and determine whether or not it is significant at the 5% significance level. If r is significant, find the equation of the regression line, and predict y when x is 61.

Age, x	No. of accidents, y	Age, x	No. of accidents, y
32	1	17	1
20	1	45	0
55	0	26	0
16	3	32	1
19	2	61	0

Projects

1. Survey at least 30 students on your campus to find out how many miles away from the campus they live. Construct a frequency distribution with at least five classes from the data. Draw a histogram and frequency polygon for your data, then compute the mean, median, mode, midrange, and standard deviation for the data. Write a report summarizing the results of your study, including a discussion of whether or not the data appear to be approximately normally distributed.

2. Survey at least 30 students on your campus to find out how many credit hours each is taking, and the estimated number of hours per week that they spend on schoolwork outside of class. Draw a scatter plot for the data. Then find the correlation coefficient and decide if it is significant at the 5% level and at the 1% level. Write a report summarizing your findings. Include a discussion of whether or not you were surprised by the results and why.

3. In this project, we'll study unemployment rates by state. Visit the website http://www.bls.gov/lau/, which lists the current unemployment rate for each state. Make a list of the rate for every state. (Using an Excel spreadsheet for this project would be an excellent idea due to the amount of data.)

   (a) Arrange the rates in order from smallest to largest. Without doing any calculations, make an estimate of what you think the measures of average are likely to be close to, and what you think the standard deviation will be.

   (b) Find each measure of average and the standard deviation, and discuss how accurate your forecasts were.

   (c) In what percentile is your state?

   (d) Build a box plot for the data and see if there are any outliers. If there are, do some research as to why the employment picture in that particular state is either unusually good or bad.

   (e) There is a link next to each state (at the moment it looks like a little graph; it used to look like a dinosaur, which means that it changed for the worse) that gives historical data for that state. Follow the link for your state and make a table showing the monthly rate for each month from January 2020 to January 2022. Call January 2020 month 1, February 2020 month 2, and so on. Then find the correlation coefficient to see if the

time and unemployment dates are significantly related. If so, find a regression line and use it to project what the current unemployment rate would be if the trend hadn't changed after January 2022.

4. The website salary.com provides national and regional salary statistics for just about every job you can think of in a really interesting way: a normal curve. It shows the median salary, as well as the 10th, 25th, 75th, and 90th percentiles. It turns out that this is more than enough information to find the standard deviation.

   Pick a job that you're interested in and the area in which you live. Put this information into the search field on the main page of that site. Then copy down the median salary given and the salary at the 90th percentile.

   (a) Why is the median salary the same as the mean?

   (b) Draw a diagram of a normal distribution to show that the value at the 90th percentile corresponds to an area of 0.4 between $z = 0$ and the value at the 90th percentile.

   (c) Use the table in Appendix A to find the z score that corresponds to area 0.4. This is the z score for the salary listed at the 90th percentile on the website.

   (d) Now you know the mean, the data value (salary) at the 90th percentile, and the z score it corresponds to. Use the formula for finding a z score to solve for the standard deviation.

   (e) Now that you know both the mean and standard deviation, you can analyze the salaries for that position using the techniques we learned in this chapter. Think about a salary that you'd be happy starting out at in this job. What percentage of workers in your area makes at least that much? What percentage makes between that amount and $10,000 per year more?

   (f) Think of a salary that you would like to make after gaining 5 years of experience. What percentage of workers in your area makes at least that much?

   (g) Use all of this information to write a short essay about your financial outlook if you go into this line of work. Does this change your thoughts about that job at all?

Supplement: Misuses of Statistics

Ed Araquel/TM and Copyright 20th
Century Fox Film Corp/Everett
Collection

"Trust No One" was the unofficial motto of the classic TV show *The X Files*, in which intrepid agent Fox Mulder and his faithful assistant Dr. Scully work to find the truth behind incredible events amid vast conspiracies. In the world of that show, things were rarely what they seemed, and the best course of action for the investigators (and viewers) was to assume that anything that anyone said may have been at best a stretching of the truth. In our modern world, there's a lot of wisdom in that approach. We are constantly bombarded with information and statistics, and it would be quite a leap to expect that most of the people disseminating that information have your best interests in mind. An educated adult knows that it's a good idea to view information suspiciously, and also knows how to recognize when statistics are being manipulated to tell the story that someone wants you to hear, not the truth. But, as Mulder was fond of pointing out, the truth is out there. You just have to want to find it.

Ideally, statistical research is used to provide knowledge and information to help us make intelligent decisions about our health and welfare, or maybe just to give us information about things we find interesting. But the world is a less-than-ideal place and there are people who will misuse statistics to sell us products that don't work; to attempt to prove something that isn't true; or to get our attention by using fear, shock, or outrage supported by bad statistics.

Just because we read or hear the results of a research study or an opinion poll in the media, this doesn't mean that these results are reliable or that they can be applied to any and all situations. For example, reporters sometimes leave out critical details like the size of the sample used or how the research subjects were selected. Without this information, you can't properly evaluate research and properly interpret the conclusions of a study or survey.

The purpose of this supplement is to show you some ways that statistics can be misused. The point isn't to make you reject any statistical study or result; it's to help you to recognize when information or conclusions you're being given might be presented in a biased or dishonest way.

Suspect Samples

Since there are over 300 million people in the United States (and over 7 billion in the world), the vast majority of statistical studies you run across involve sampling: this is where any study is going to begin. That's why we started our study of statistics with sampling, and it's also why that's where we'll begin our look at statistical mischief.

Unreported Sample Sizes

Plush Studios/Blend Images

People who are annoyed by
unsolicited phone calls won't
appear in a sample of people
contacted by phone.

In the first part of the 21st century, there was a chewing gum brand that ran a large campaign claiming that "Four out of five dentists surveyed" would recommend their product. That sounds like a ringing endorsement. But what does it really mean? The advertiser (of course) provided no information on how many dentists they actually surveyed. If they only surveyed five, their claim could still be technically true, but they may have stumbled across four that by coincidence would recommend their product, even if the vast majority of dentists would rather see their patients chewing razor blades. Worse still, they could easily have chosen to survey five that would be predisposed to recommend their product for one reason or another (like say a check with multiple zeroes on it).

If samples in a study are too small, the results could be completely meaningless. And if the size of a sample is conveniently left out, there's a pretty good chance that the person reporting the statistics would just as soon have you not know what that sample size was.

Samples Relying on Volunteer Participation

Not only is it important to have a sample size that is large enough, but it is also necessary to see how the subjects in the sample were selected. Studies using volunteers sometimes have a built-in bias because volunteers generally don't represent the population at large. Sometimes they are recruited from a particular socioeconomic background, and sometimes unemployed people volunteer for research studies in order to get paid. Studies that require the subjects to spend several days or weeks in an environment other than their home or workplace automatically exclude people who are employed and can't take time off from work. Sometimes college students or retirees are used. In the past, many studies have used only men but have tried to generalize the results to both men and women. Opinion polls that require a person to phone or mail in a response most often are not representative of the population in general since only those with strong feelings for or against the issue usually make the effort to call or respond by mail.

Samples Relying on Convenience

It can be difficult and time-consuming to conduct a study involving a truly random sample. It's not uncommon for people to cut corners and work with a sample that's easy to get to for one reason or another. Another type of sample that can lead to biased conclusions is the **convenience sample**. This occurs when a sample is chosen based on a preexisting group. For example, educational studies sometimes use students that are currently all in the same class because they're easy to contact all at once. But there are many factors that could lead a group of students to take the same class at the same time, so it's very possible that the group doesn't represent the population of students as a whole.

If you're trying to interpret the results of a study done using a small sample, a volunteer sample, or a convenience sample, you have to give a lot of thought to whether the characteristics of that sample make it representative of an entire population.

Ambiguous Averages

There was a time when we all thought that "average" meant one thing: the result of adding all data values and dividing by how many there are. But we know better now, don't we? When someone uses the term "average" in making some sort of claim instead of letting us know which measure of average (mean, median, mode, midrange) that they're using, it's time to trust no one and be suspicious. We know that the values of these different measures of average sometimes differ quite a lot. People who know that there are several types of average can, without lying, select the one for the data that most lends evidence to support their position.

For example, suppose the owners of a store employ four salespeople. The number of years each has been employed by the store is 22, 10, 2, and 2. The mean number of years' service is 9. The median is 6, and the mode is 2. Now if the owners wanted to advertise the fact that their employees have many years of experience, which average do you think they would use? Obviously, they would use 9. However, 9 is not very high, so since one owner sometimes doubles as a salesperson and has owned the store for 42 years, those years are added to compute the average. The mean is now 15.6 years. Much more impressive, isn't it? (Actually the midrange, 22 years, is even more impressive.) Whenever the word "average" is used instead of mean, median, mode, or midrange, ask yourself, "What average is being used?"

Convenient Use of Percentages

Another type of statistical distortion can occur when different values are used to represent the same data. For example, one political candidate who is running for reelection might say,

Stockbyte/Getty Images

In a study to determine how long it takes to fall asleep, would college students be a good sample of the overall population?

REUTERS/Jim Bourg/Alamy Stock Photo

People on one side of an issue sometimes frame statistics to support their point.

"During my administration, expenditures increased a mere 3%." The opponent, who is trying to wrest that seat away, might say, "During my opponent's administration, expenditures have increased a whopping $6,000,000." Both figures could be correct; but expressing $6,000,000 as a mere 3% makes it seem like a very small increase, whereas expressing a 3% increase as $6,000,000 makes it sound like a very large increase. Here again, ask yourself, "Which measure best represents the data?"

Detached Statistics

A claim that uses detached statistics is one in which no comparison is made. For example, you might hear a claim like "Our brand of crackers has 1/3 fewer calories." That's completely meaningless if no comparison is given. One-third fewer calories than what? Someone else's crackers? An entire cow? Another common example is time comparisons: "Our aspirin works four times faster." Than what? In some cases, if you read the fine print you'll see what the comparison is to, allowing you to decide if it's meaningful. If no comparison is available, you should consider completely disregarding the claim.

McGraw Hill/C.P. Hammond, photographer

Implied Connections

This one is a classic—if you look for it, you'll find it in advertising pretty much every day. Many claims imply that there's a connection between two variables without having any compelling evidence (or maybe any evidence at all) to back up that connection. To cover themselves legally, advertisers will use words like "may," "suggest," "some," etc., so that if the connection actually doesn't exist, they're not lying. For example, consider this statement: "Eating fish may help to reduce your cholesterol." Notice the words "may help." There is no guarantee that eating fish will definitely help you reduce your cholesterol.

"Studies suggest that using our exercise machine will reduce your weight." Here the word *suggest* is used, and again, there is no guarantee that you will lose weight using the exercise machine advertised.

Another claim might say, "Taking calcium will lower blood pressure in some people." Notice the word *some* is used. You may not be included in the group of "some" people. Be careful when drawing conclusions from claims that use words such as "may," "in some people," "might help," etc.

Misleading Graphs

Graphs are great. We love them. In general, they're a fantastic way to illustrate information, making it easier to interpret. But there's a dark side to graphs. They can often be found hanging out with shady characters and committing crimes. Well, that's an exaggeration, but an

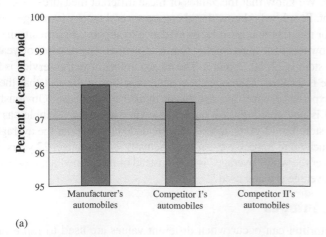

(a)

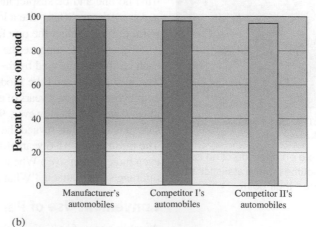

(b)

Figure 11-30

inappropriately drawn graph can (either intentionally or not) misrepresent data and lead to false or misleading conclusions. For example, a car manufacturer's ad stated that 98% of the vehicles it had sold in the past 10 years were still on the road. The ad then showed a graph similar to the one in Figure 11-30(a). The graph shows the percentage of the manufacturer's automobiles still on the road and the percentage of its competitors' automobiles still on the road. It sure looks like there's a big difference between companies because the first bar is more than twice as tall as the third. But is there really? Notice the scale on the vertical axis in Figure 11-30(a). It has been cut off (or truncated) so that it starts at 95%. When the graph is redrawn using a scale that goes from 0% to 100%, as in Figure 11-30(b), there is hardly a noticeable difference in the percentages. Changing the units at the starting point on the axis can convey a very different visual representation of the data.

It isn't necessarily wrong to truncate an axis of the graph; many times it's necessary to do so. But the reader should be aware of this fact and interpret the graph accordingly. Don't be misled if an inappropriate explanation is given.

The average lifespan for Americans has been increasing for decades (although, to be fair, we still rank 34th in the world). The data below display the average life expectancy for Americans according to the National Center for Health Statistics.

Year	1985	1990	1995	2000	2005	2010	2015
Life expectancy	74.7	75.4	75.8	76.8	77.4	78.3	78.8

If you draw a time-series graph with a scale ranging from zero to 100 years, the increase looks mild, almost negligible (Figure 11-31(a)). But when I asked Excel to chart the data, Figure 11-31(b) shows the scale it chose by default, from 74 to 80. This makes it look like immortality is just around the corner. Again, the choice of scale has a profound effect on how the information is perceived. You should also be wary if a graph appears to be leaving out some information. In this case, average life expectancy decreased to 77.3 in 2020, largely due to the pandemic. Someone interested in emphasizing the increase in life expectancy may conveniently leave out this piece of data.

Another misleading graphing technique you see sometimes is exaggerating a one-dimensional increase by showing it in two dimensions. For example, the average cost in 2017 dollars of a 30-second Super Bowl commercial increased from $1.15 million in 1995 to $5.6 million in 2021. The bar graph in Figure 11-32(a) illustrates this increase by comparing the heights of two bars

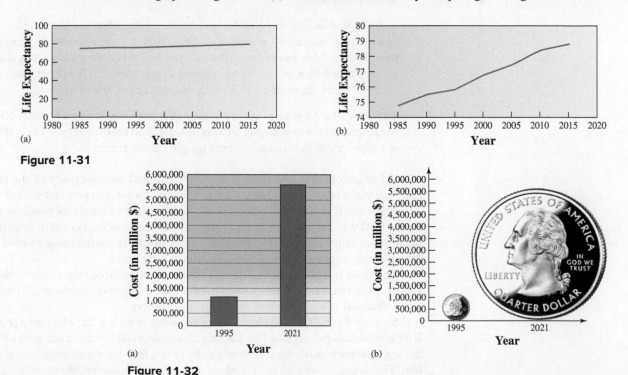

Figure 11-31

Figure 11-32

with the same width, so the difference is in one dimension. We can see that the increase is significant. The same data are shown using quarters rather than bars in Figure 11-32(b), so the difference occurs in two dimensions. It makes the change in cost look a LOT more significant because your eye is comparing the areas of the quarters, rather than the heights of the bars.

Faulty Survey Questions

Surveys and opinion polls obtain information by using questionnaires. There are two types of studies: interviews, and self-administered questionnaires. The interview survey requires a person to ask the questions either in person or by telephone. Self-administered questionnaire surveys require the participant to answer the questions by mail, computer, or in a group setting, such as a classroom. When analyzing the results of a survey using questionnaires, you should be sure that the questions are properly written, since the way questions are phrased can often influence the way people answer them. For example, when a group of people was asked, "Do you favor a waiting period before guns are sold?" 91% of the respondents said "Yes" and 6% said "No." However, when the question was rephrased as, "Do you favor a national gun-registration program costing about 20% of all dollars spent on crime control?" 37% responded "Yes" and 61% responded "No." As you can see, although the questions pertain to some form of gun control, each asks something a little different, and the responses are radically different. When reading and interpreting the results obtained from questionnaire surveys, watch out for some of these common mistakes made in the writing of the survey questions.

Photographer's Choice/SuperStock

Asking Biased Questions. By asking a question in a certain way, the researcher can lead the respondents to answer the question the way the researcher wants them to. For example, the question, "Are you going to vote for Candidate Jones, even though the latest survey shows he will lose the election?" may lead the respondent to say, "No" since many people don't want to vote for a loser, or admit it when they do.

Using Confusing Words. Using words in a survey question that are not well defined or understood can invalidate the responses. For example, a question such as, "Do you think people would live longer if they were on a diet?" would mean many different things to people since there are many types of diets, such as low-salt diets, high-protein diets, all fast-food diets, . . .

Asking Double-Barreled Questions. Sometimes two ideas are contained in one question, and the respondent may answer one or the other in their response. For example, consider the question, "Are you in favor of a national health program and do you think it should be subsidized by a special tax as opposed to other ways to finance it, such as a national lottery?" Here the respondent is really answering two questions.

Using Double Negatives. Survey questions containing double negatives often confuse the respondent. For example, what is this question really asking? "Do you feel that it is not appropriate to have areas where people cannot smoke?"

Other factors that could bias a survey would include anonymity of the participant, the time and place of the survey, and whether the questions were open-ended or closed-ended.

Participants will, in some cases, respond differently to questions based on whether or not their identity is known. This is especially true if the questions concern sensitive issues like income, sexuality, abortion, etc. Researchers try to ensure confidentiality rather than anonymity; however, many people will be suspicious in either case.

The time and place where a survey is taken can influence the results. For example, if a survey on airline safety is taken right after a major airline crash, the results may differ from those obtained in a year with no major airline disasters.

So should you "Trust No One"? We're not going to go that far: when used properly, statistics is a tremendously powerful tool for learning about important aspects of our world. But when statistics are used improperly, whether intentionally or not, they can cause confusion or outright deception. This is exactly why it's so valuable to understand the concepts and terminology of statistics.

Exercise Set Chapter 11 Supplement

1. According to a pilot study of 20 people conducted at the University of Minnesota, daily doses of a compound called arabinogalactan over a period of 6 months resulted in a significant increase in the beneficial *lactobacillus* species of bacteria. Why can't it be concluded that the compound is beneficial for the majority of people?

2. Comment on this statement taken from a magazine advertisement: "In a recent clinical study, Brand ABC* was proved to be 1,950 percent better than creatine!"

   *Actual brand will not be named.

3. In an ad for women, the following statement was made: "For every hundred women, 91 have taken the road less traveled." Comment on this statistic.

4. In many ads for weight-loss products, under the product claims and in small print, the following statement is made: "These results are not typical." What does this say about the product being advertised?

5. An article in a leading magazine stated that, "When 18 people with chronic, daily, whiplash-related headaches received steroid injections in a specific neck joint, 11% had no more headaches." Think of a possible reason why the figure 11% was used.

6. In an ad for moisturizing lotion, the following claim is made: ". . . it's the #1 dermatologist recommended brand." What is misleading about the claim?

7. An ad for an exercise product stated that, "Using this product will burn 74% more calories." What is misleading about that statement?

8. "Vitamin E is a proven antioxidant and may help in fighting cancer and heart disease." Is there anything ambiguous about this claim?

9. "Just one capsule of Brand X* can provide 24 hours of acid control." What needs to be more clearly defined in this statement?

   *Actual brand will not be named.

10. ". . . male children born to women who smoke during pregnancy run a risk of violent and criminal behavior that lasts well into adulthood." Can we infer that smoking during pregnancy is responsible for criminal behavior in people?

For Exercises 11–13, explain why the graphs are misleading.

11. A company advertises that its brand of energy pills gets into the user's blood faster than a competitor's brand and shows these two graphs to prove its claim.

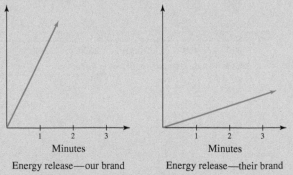

12. The graph shows the difference in sales of pumpkins during October for the years 2005 and 2020.

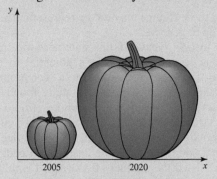

13. Explain this contradiction: the two graphs were drawn using the same data, yet the first graph shows sales remaining stable, and the second graph shows sales increasing dramatically.

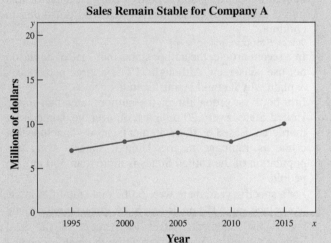

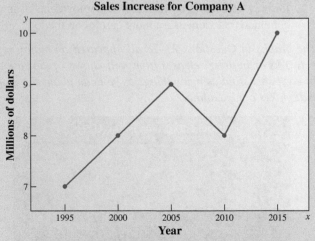

For Exercises 14–16, explain why each survey question might lead to an erroneous conclusion.

14. "How often do you run red lights?"

15. "Do you think gun manufacturers should put safety locks on all guns sold even though it would increase the cost of the gun by 20%?"

16. "Do you think that it is not important to give extra tutoring to students who are not failing?"

17. The results of a survey reported in *USA Weekend* stated that:
    "9% would drive through a toll booth without paying."
    "13% would steal cable television or inflate their resumes."
    Explain why these figures may not be representative of the population in general.

18. In an article in *USA Weekend*, this statement was made: "More serious seems to be coffee's potential to raise blood levels of homocysteine, a protein that promotes artery clogging. A recent Norwegian study found 20% higher homocysteine in heavy coffee drinkers (more than 9 cups a day) than in non-coffee drinkers." Based on this statement, should we give up our daily cup of coffee?

19. An article in a newspaper with the headline, "Lead: The Silent Killer" listed the number of confirmed childhood lead poisoning cases in Allegheny County, Pennsylvania, in 1985 as 15 and in 1997 as 124. Can you conclude that the incidence of childhood lead poisoning cases is increasing in Allegheny County? Suggest a factor that might cause an increase in reported lead poisoning in children.
    Source: *Pittsburgh Tribune Review*

20. In a recent article, the author states that 71% of adults do not use sunscreen. Although 71% is a large percentage, explain why it could be misleading.

21. In a book on probabilities, the author states that in the United States every 20 minutes, on average, someone is murdered. Based on this statement, can we conclude that crime is rampant in the United States? *Note:* the population of the United States is more than 300 million people.

22. For a specific year, there were 6,067 male fatalities in the workplace and 521 female deaths. A government official made this statement: "Over 90 percent of the fatal injuries the past year were men, although men accounted for only 54 percent of the nation's employment." Can we conclude that women are more careful on the job?

The graphs in Questions 23–28 all appeared on major news networks or in press releases from well-known organizations. In each case, find as many things as you can about the graph that makes it misleading.

23.

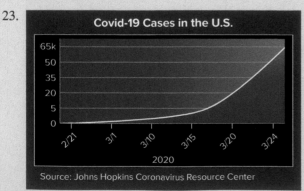

24.

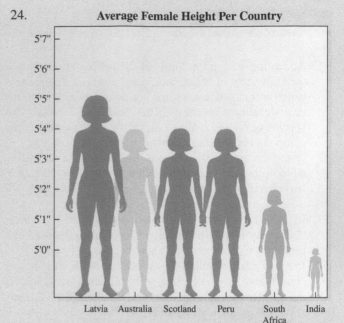

25.

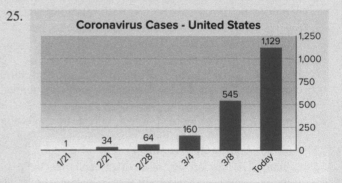

26.

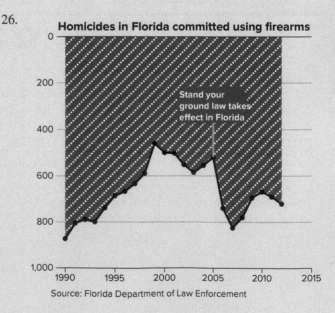

27.

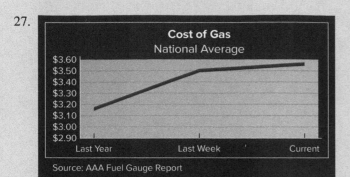

Source: AAA Fuel Gauge Report

Fatal Occupational Injuries

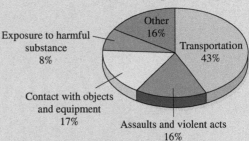

Source: *The World Almanac and Book of Facts*

28.

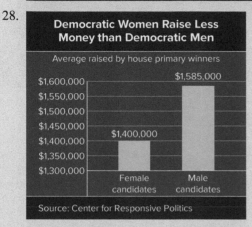

Source: Center for Responsive Politics

Fatal Occupational Injuries

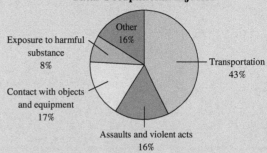

Source: *The World Almanac and Book of Facts*

29. To show that misuse of statistics can happen even with the best of intentions, compare the two pie charts shown here. The first is from the second edition of this book; the second is what the chart should have looked like. Explain why the second-edition version was misleading. (*Hint:* Look at the actual percentages carefully.)

30. I saw this claim recently in an ad for a whole grain cereal: "People who eat whole grain tend to weigh less than those who don't." The implication is that eating whole grain products (like, say, theirs) will lead to weight loss. There are at least two reasons why this claim is suspect. How many reasons can you think of?

Design elements: Front matter, Chapter Opener, Summary and End Matter header design (random numbers background illustration): ©pixeldreams.eu/Shutterstock RF

Voting Methods

Photo: Peter Joneleit/Zuma Wire Service/ZUMA Press/Alamy Stock Photo

Outline

Math in | College Football

Football and other sports have been used as examples dozens of times throughout this book because sports present excellent examples of a variety of ways that math gets used in our world. There are the obvious ways—keeping score and adding up statistics like yards gained or home runs hit. But there are also many behind-the-scenes examples of the importance of math in sports. Allocation of salaries, ticket prices, devising a schedule that meets the needs of every team in a league, assigning officials to work games . . . these are just four of many such examples.

The mathematics of voting, which we will study in Chapter 12, plays a huge role in college football. Strange, isn't it? Every game has a winner and a loser, so why should voting matter? It turns out that choice of teams that play for the national championship at the end of the season is strongly influenced by polls that are voted on by coaches, athletic directors, and members of the media. For championship contenders, the season is really one long election!

Another place where voting is important is in the awarding of the Heisman trophy, which many feel is the most prestigious individual award in amateur sports. It is given each year to the top college football player in America as determined by a voting process that we'll study in this chapter.

On the surface, voting seems like a very straightforward and simple process: everyone casts a vote for their favorite candidate, and whoever gets the most votes wins. But as we will see, there are flaws in any voting system, so a variety of systems have been developed, depending on the needs of the organization holding the election. In this chapter, we'll study five voting methods in depth and examine the strengths and weaknesses of each.

We will then turn our attention to apportionment. This is the process that is used to allocate representatives from a legislative body, like a city council or the U.S. House of Representatives, to different regions. We will see that the ideas can also be used for other practical problems of dividing resources.

By the time you finish this chapter, you should be able to answer the following questions about Heisman trophy balloting, for a year in which the vote was very controversial. A total of 902 ballots were returned, and the results for the top five candidates are summarized in the table below. The candidates are listed using letters for now so you can objectively decide who you think should have won.

The voting is done using a modified version of the Borda count method, which you will study in Section 12-2. Voters list their top three candidates in order. Any candidate not listed on a ballot gets 0 points; the candidates listed first, second, and third get 3 points, 2 points, and 1 point, respectively. The candidate with the most points wins the award.

Player	First	Second	Third
A	309	207	234
B	300	315	196
C	266	288	230
D	13	44	86
E	3	27	53

1. Who won the award?
2. Do you think the right person won the award based strictly on the ballots? Justify your response.
3. Is the winner the same if the plurality method is used?
4. How many voters chose someone other than one of the top five candidates as their first choice?
5. Discuss whether or not you think any of the fairness criteria in this chapter were violated by the election.
6. Do some online research and see if you can find what year this was, and who some of the players involved were. (*Hint:* You're probably familiar with one of the names even if you're not a big college football fan.)

For answers, see Math in College Football Revisited on page 797

Section 12-1 | Preference Tables and the Plurality Method

✓ **LEARNING OBJECTIVES**

1. Interpret the information in a preference table.

2. Determine the winner of an election using the plurality method.

3. Decide if an election violates the head-to-head comparison criterion.

Voting seems like such a simple idea: two candidates both want a position, and whichever one gets the most votes wins. But like most things in the modern world, elections rarely turn out to be as simple as they appear. The most obvious complication arises when there are more than two candidates. Should the winner just be the one who gets the most votes, even if less than half of the voters want them in office? Maybe voters should rank the candidate in order of preference . . . but then how do we decide on the winner?

We will begin our study of voting methods by examining a method for summarizing the results when candidates are ranked in order of preference by voters. We'll then study the simplest of the methods for determining the winner of an election, and begin a study of the weaknesses inherent in different voting systems.

Comstock Images/Getty Images

Preference Tables

Suppose there are three candidates running for club president. We'll call them A, B, and C. Instead of simply voting for the single candidate of your choice, you are asked to rank each candidate in order of preference. This type of ballot is called a *preference ballot*.

In this case, there are six possible ways to rank the candidates, as shown.

First choice	A	A	B	B	C	C
Second choice	B	C	A	C	A	B
Third choice	C	B	C	A	B	A

Now, suppose that the 20 club members voted as follows.

A	B	A	A	A	A	A	B	A	A	B	A	A	A
B	C	B	C	B	C	B	C	B	B	C	B	C	C
C	A	C	B	C	B	C	A	C	C	A	C	B	B

A	C	C	A	A	B
C	B	B	B	B	C
B	A	A	C	C	A

Of the 6 possible rankings, only 4 appear in the 20 ballots. Nine people voted for the candidates in order of preference ABC, five people voted ACB, four people voted BCA, and two people voted CBA.

A **preference table** can be made showing the results.

Number of voters	9	5	4	2
First choice	A	A	B	C
Second choice	B	C	C	B
Third choice	C	B	A	A

The sum of the numbers in the top row indicates the total number of voters. Also note that 9 + 5 or 14 voters picked candidate A as their first choice, 4 picked candidate B as their first choice, and 2 voters picked candidate C as their first choice.

Because no voters cast ballots ranking candidates as BAC or CAB, those possible rankings are not listed as columns in the table.

Example 1 | **Interpreting a Preference Table**

Four candidates, W, X, Y, and Z, are running for student government president. The students were asked to rank all candidates in order of preference. The results of the election are shown in the preference table.

Number of voters	86	42	19	13	40
First choice	X	W	Y	X	Y
Second choice	W	Z	Z	Z	X
Third choice	Y	X	X	W	Z
Fourth choice	Z	Y	W	Y	W

(a) How many students voted?
(b) How many people voted for candidates in the order Y, Z, X, W?
(c) How many students picked candidate Y as their first choice?
(d) How many students picked candidate W as their first choice?

Digital Vision Ltd./SuperStock

SOLUTION

(a) To find the total number of voters, find the sum of the numbers in the top row.

$$86 + 42 + 19 + 13 + 40 = 200$$

(b) The ranking Y, Z, X, W is in the third column of the table, which is headed by the number 19. This means that 19 voters chose that order.

(c) There were 19 voters who chose Y, Z, X, W (third column) and 40 that chose Y, X, Z, W (fifth column), and those are the only rankings with Y listed first. So $19 + 40 = 59$ voters listed candidate Y first.

(d) Only one ranking order has candidate W first—the one in the second column. There were 42 voters who submitted that order, so 42 people chose candidate W as their first choice.

Try This One 1

1. Interpret the information in a preference table.

The Student Activities Committee at Camden College is choosing a location for an end-of-year banquet, and they ask all members to list the four possible locations in order of preference. The choices are Airport Restaurant (A), Bob's Bar and Grill (B), The Crab Shack (C), and Dino's (D). The results are shown in the preference table.

Number of voters	19	13	12	9	4	2
First choice	C	B	C	C	A	B
Second choice	B	C	A	B	C	A
Third choice	A	D	B	D	D	D
Fourth choice	D	A	D	A	B	C

(a) How many members voted?
(b) How many members listed The Crab Shack as their first choice?
(c) How many members listed The Crab Shack and Bob's Bar and Grill in their top two?

Math Note

Plurality does not necessarily mean majority; it simply means more votes than any other candidate receives. Majority means more than 50% of the votes cast.

In the remainder of this section, and in Sections 12-2 and 12-3, we will study four common voting methods.

The Plurality Method

The simplest method of determining a winner in an election with three or more candidates is called the *plurality method*.

> In an election with three or more candidates that uses the **plurality method** to determine a winner, the candidate with the most first-place votes is the winner.

Example 2   Using the Plurality Method

The preference table for a club presidential election consisting of three candidates is shown.

(a) Using the plurality method, determine the winner.
(b) Can you make an argument as to why candidate B shouldn't win the election?

Number of votes	4	7	5	4
First choice	B	A	C	B
Second choice	C	C	A	A
Third choice	A	B	B	C

SOLUTION

(a) In this situation, only the first-place votes for each candidate are considered. Candidate A received 7 first-place votes (column 2). Candidate B received 4 + 4 or 8 first-place votes (columns 1 and 4). Candidate C received 5 first-place votes (column 3). Candidate B is the winner since that candidate received the most first-place votes.

(b) This is an important question—it's our first indication of why just calling the person with the most votes isn't necessarily the best approach. Look at the bottom row of the table: of the 20 people that voted, 12 ranked B as their LEAST favorite candidate! If more than half of those voting really really don't want that candidate to be club president, should they win?

2. Determine the winner of an election using the plurality method.

Try This One   2

An election was held for the chairperson of the Psychology Department. There were three candidates: Professor Jones (J), Professor Kline (K), and Professor Lane (L). The preference table for the ballot is shown.

Number of votes	2	4	1	3
First choice	L	J	K	L
Second choice	J	K	L	K
Third choice	K	L	J	J

(a) Who won the election if the plurality method of voting was used?
(b) Do you think this is the correct choice? Why or why not?

In Example 2, the top row consists of the number of voters who ranked the candidates in the order shown in the column. Instead of numbers in the top row, percents can also be used. That allows us to draw a pie chart illustrating the results, which we'll do in Example 3.

Example 3	Using Percentages to Summarize a Preference Table

For the preference table in Example 2, calculate the percentage of voters that chose each candidate, and rewrite the table with the percentages in place of the number of voters. Then use the results to draw a pie chart illustrating the first-place votes for each candidate.

SOLUTION

From adding the numbers along the top of the original preference table, we know that there were 20 votes cast. We can find the percentage for each ballot by dividing the number of voters by 20 and converting to percent form.

$$\text{First column: } \frac{4}{20} = 0.2 = 20\%$$

$$\text{Second column: } \frac{7}{20} = 0.35 = 35\%$$

$$\text{Third column: } \frac{5}{20} = 0.25 = 25\%$$

$$\text{Fourth column: } \frac{4}{20} = 0.2 = 20\%$$

The preference table now looks like this:

Percent of votes	20%	35%	25%	20%
First choice	B	A	C	B
Second choice	C	C	A	A
Third choice	A	B	B	C

There are 360° in a full circle, so to find the number of degrees for each portion, we find the appropriate percentage of 360°.

Candidate A: 35% of 360° = 0.35 × 360° = 126°
Candidate B: 40% of 360° = 0.40 × 360° = 144°
Candidate C: 25% of 360° = 0.25 × 360° = 90°

The pie chart is shown in Figure 12-1.

Percentage of First-Place Votes

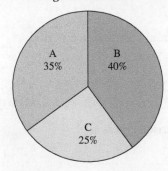

Figure 12-1

Try This One	3

Rewrite the preference table for the election in Try This One 2, replacing the number of voters with the percentage of voters for each ballot. Then draw a pie chart illustrating the first-place votes.

The plurality method is a simple way to determine the winner of an election, but it has some flaws. First, some would suggest that a candidate shouldn't win an election if less than half of the voters choose them. As we see from Figure 12-1, candidate B wins the election in Example 2, even though well less than half of the ballots listed B first. Second, the possibility of a tie exists, and it is greater when there are fewer voters. Third, the method completely ignores information about voters' preferences except for their first-place vote. Fourth, this method can sometimes violate what is called the *head-to-head comparison criterion*.

A *criterion* is a way of measuring or evaluating a situation. In this chapter, we will discuss various criteria for assessing the fairness of voting systems. The first of these is the head-to-head comparison criterion.

The **head-to-head comparison criterion** states that if a particular candidate wins all head-to-head comparisons with all other candidates, then that candidate should win the election.

Sidelight   **Some Interesting Facts About Voting**

• You might think that voting is a human creation, but it isn't. Honeybees vote, and they can't even count! When it's time to locate a new hive, scouts return from their search for a good location, and they dance. The bees that dance most vigorously manage to recruit other scouts to their side until one site has the majority of scouts dancing in its favor.

• The first election poll in U.S. history, conducted by the *Harrisburg Pennsylvanian* newspaper, correctly predicted that Andrew Jackson would win the most votes in the upcoming election. There was just one problem: Jackson lost the election in the Electoral College, and John Quincy Adams won the presidency. (Project 1 at the end of the chapter examines our system for electing a president.)

• The lever voting machine was first used in 1892, and at the time it had more moving parts than almost any other device made in America. Ironically, it was marketed as a "plain, simple" way to cast ballots.

• For the 2006 elections, the U.S. Department of Defense paid $830,000 for a Web-based system that allowed those serving in the military overseas to vote easily. Only 63 people used it.

IT Stock Free/Alamy Stock Photo

• Studies have shown that rainy days result in lower voter turnout, at about the rate of 0.8% for each inch of rain. Computer simulations indicate that if it had rained in Illinois on election day 1960, Nixon would have defeated Kennedy, and if it had been sunny in Florida on election day 2000, Gore would have defeated Bush.

Source: *Discover Magazine*

Let's see if the election in Example 2 violates the head-to-head comparison criterion.

Example 4   **The Head-to-Head Comparison Criterion**

Does the election in Example 2 violate the head-to-head comparison criterion?

SOLUTION

The idea is to compare all combinations of two candidates at a time to see which is preferred in a head-to-head matchup without the third candidate involved.

The preference table for the club president's election is reprinted here for reference.

Number of votes	4	7	5	4
First choice	B	A	C	B
Second choice	C	C	A	A
Third choice	A	B	B	C

First, compare A with B:

The second and third preference ballots have candidate A listed higher than candidate B, and there were 12 voters that chose this order. The first and fourth have candidate B listed higher, and that order was chosen by 8 candidates. So candidate A would win a head-to-head matchup with candidate B. That alone doesn't mean that the election violates the head-to-head comparison criterion: the criterion doesn't say that the winning candidate has to beat all others in a head-to-head matchup.

Next, compare A with C:

There were 11 voters who listed candidate A higher than candidate C (the second and fourth columns). There were 9 who listed candidate C higher, so candidate A also wins a head-to-head matchup with candidate C.

Without even comparing B and C, we can see that the head-to-head comparison criterion is violated: candidate A defeats both B and C head-to-head, but candidate A didn't win the election using the plurality method. (The head-to-head criterion says that any candidate who defeats all opponents should win the election.)

>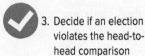
>
> The head-to-head comparison criterion doesn't say that the winner of an election has to defeat every opponent head-to-head. It says that if there is a candidate that *does* defeat all others head-to-head, that candidate should win the election.

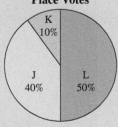

3. Decide if an election violates the head-to-head comparison criterion.

Try This One   4

Does the election in Try This One 2 violate the head-to-head comparison criterion? Why or why not?

The result of Example 4 shows that the plurality method doesn't always satisfy the head-to-head comparison criterion. This is not to say that *every* election conducted by the plurality method violates the head-to-head criterion. We have simply found that *some* do, so we say the method in general doesn't meet the criterion.

The head-to-head criterion is called a **fairness criterion**. It is one of four fairness criteria that we will study in this chapter. Political scientists have come to agree that a truly fair voting system should satisfy all of these four criteria. Think about what that means based on Example 4: we just decided that the most common and obvious method for holding an election is considered unfair! In the next two sections we'll study other voting methods and see if we can find the ultimate fair method.

Answers to Try This One

1 (a) 59   (b) 40   (c) 41

2 (a) Professor Lane
       (b) This is an opinion. Make sure you back yours up!

3

Percent of votes	20%	40%	10%	30%
First choice	L	J	K	L
Second choice	J	K	L	K
Third choice	K	L	J	J

Percentage of First-Place Votes

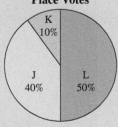

4 No. Head-to-head, J wins over K, L wins over J, and K ties L, so there is no candidate who defeats all others head-to-head.

Exercise Set   12-1

Writing Exercises

1. Explain how a preference table is used to display voting results.
2. Explain how the winner of an election is determined using the plurality voting method.
3. What is the difference between a plurality and a majority?
4. Explain the head-to-head comparison criterion.
5. What is a fairness criterion for an election?
6. Describe some weaknesses of the plurality method for elections.

Applications in Our World

7. The preference ballots for the election of a CEO by the board of directors are shown. Make a preference table for the results of the election and answer each question.
   (a) How many people voted?
   (b) How many people voted for the candidates in the order of preference XZY?
   (c) How many people voted for candidate Y as their first choice?
   (d) Using the plurality method, determine the winner of the election.

X	X	Y	Z	X	Y	Z	Z	X	Y	X	Y	X	Y
Y	Z	Z	Y	Y	Z	Y	Y	Y	Z	Z	Z	Y	Z
Z	Y	X	X	Z	X	X	X	Z	X	Y	X	Z	X

X	Y	Z	Z	X	X	Y	Y
Z	Z	Y	Y	Z	Z	Z	Z
Y	X	X	X	Y	Y	X	X

8. The Tube City Talkers Club held its annual speech contest. The preference ballots for the best speaker are shown. The candidates were Cortez (C), Lee (L), and Smith (S). Make a preference table for the results of the election and answer each question.
   (a) How many people voted?
   (b) How many people voted for the candidates in the order of preference CLS?
   (c) How many people voted for Lee for first place?
   (d) Using the plurality method, determine the winner of the election.

C	S	S	L	L	C	S	S	L	L	C	C	S	S
L	C	C	S	S	L	C	C	S	S	L	L	C	C
S	L	L	C	C	S	L	L	C	C	S	S	L	L

C	L	S	L	L
L	S	C	S	S
S	C	L	C	C

9. The preference ballots of the board of directors for the selection of a city in which to hold the next National Mathematics Instructors' Association Conference are shown. The three cities under consideration are Chicago (C), Philadelphia (P), and Miami (M). Make a preference table for the results of the election and answer each question.
   (a) How many people voted?
   (b) How many people voted for the candidates in the order of preference PMC?
   (c) How many people voted for Chicago as their first choice?
   (d) Using the plurality method, determine the winner of the election.

P	C	M	P	P	P	P	C	M	M	P	P	C
M	P	P	M	M	M	M	P	P	P	M	M	P
C	M	C	C	C	C	C	M	C	C	C	C	M

P	M	P	C	M
M	P	M	P	P
C	C	C	M	C

10. A group of club members decides to vote to select the color of its meeting room. The color choices are white (W), light blue (B), and light yellow (Y). Make a preference table for the election and answer each question.
    (a) How many club members voted?
    (b) How many club members voted for white as their first choice?
    (c) How many club members voted for the colors in the order BYW?
    (d) Using the plurality method, determine the winner.

W	B	B	Y	B	Y	W	B	Y	Y	W	B	Y
B	Y	Y	W	Y	W	B	Y	W	W	B	Y	W
Y	W	W	B	W	B	Y	W	B	B	Y	W	B

11. Students at a college were asked to rank three improvements that they would like to see at their college. The choices were build a new gymnasium (G), build a swimming pool (S), or build a baseball/football field (B). The votes are summarized in the preference table.

Number of votes	83	56	42	27
First choice	G	S	S	B
Second choice	S	G	B	S
Third choice	B	B	G	G

(a) How many students voted?
(b) What option won if the plurality method was used to determine the winner?

12. The owner of a restaurant decides to poll regular customers to choose which dish she'll submit to an annual citywide competition. The choices are lemon-crusted salmon (L), crab-stuffed chicken (C), garlic prime rib (G), and wasabi rolls (W). The results of the poll are shown in the preference table.

Number of votes	8	6	5	3	2
First choice	C	L	W	G	W
Second choice	W	W	G	C	C
Third choice	G	G	C	W	L
Fourth choice	L	C	L	L	G

(a) How many customers voted?
(b) What meal was selected as the winner if the plurality method was used to determine the winner?

13. A panel of experts is convened to decide which is the best hospital overall in a metropolitan area. The four hospitals under consideration are Regional Medical Center (R), Community General (G), Children's Hospital (C), and Derbyshire Hospital (D). The preference table is shown.

Number of votes	3	5	2	6	4
First choice	R	D	C	C	R
Second choice	G	C	R	G	C
Third choice	C	G	G	D	D
Fourth choice	D	R	D	R	G

(a) How many panelists were there?
(b) What hospital won if the plurality method was used to determine the winner?

14. The students in Dr. Lee's math class were asked to vote on the starting time for their final exam. Their choices were 8:00 A.M., 10:00 A.M., 12:00 P.M., or 2:00 P.M. The results of the election are shown in the preference table.

Number of votes	8	12	5	3	2	2
First choice	8	10	12	2	10	8
Second choice	10	8	2	12	12	2
Third choice	12	2	10	8	8	10
Fourth choice	2	12	8	10	2	12

(a) How many students voted?
(b) What time was the final exam if the plurality method was used to determine the winner?

For Exercises 15–18, rewrite the preference table for the given Exercise number with the percentage of voters in place of the number of votes, then use your results to draw a pie chart illustrating the percentage of first-place votes received by each candidate.

15. Exercise 11
16. Exercise 12
17. Exercise 13
18. Exercise 14
19. Using the election results given in Exercise 11, has the head-to-head comparison criterion been violated? Explain your answer.
20. Using the election results given in Exercise 12, has the head-to-head comparison criterion been violated? Explain your answer.
21. Using the election results given in Exercise 13, has the head-to-head criterion been violated? Explain your answer.
22. Using the election results given in Exercise 14, has the head-to-head criterion been violated? Explain your answer.

Critical Thinking

23. Using the Internet as a resource, look up the total number of votes received by presidential candidates in every election since 1960.
    (a) If victory required a majority of all votes cast, which of the elections would not have had a winner?
    (b) If the elections were decided using the plurality method, would any of the elections have turned out differently?

24. Suppose that an election has seven candidates, and 1,000 people cast a ballot. What is the smallest number of votes that someone could win with if the plurality method is used?

25. Many people think that the only method needed to decide an election is the plurality method: whoever gets the most votes wins. Based on information in this section, write a short essay justifying why other voting methods are sometimes used. Include your opinions on whether or not this is reasonable justification.

26. You probably know that in presidential elections, voters simply vote for one candidate, as opposed to ranking all candidates in order as described in this section. Discuss some potential pros and cons of having voters fill out a preference ballot for presidential elections.

Use the following information for Exercises 27 through 29.

Suppose 100 votes are cast in an election involving three candidates, A, B, and C, and 80 votes are counted so far. The results are

A 36   B 32   C 12

(Note: this is not a preference list. Voters are submitting a single name on a ballot.)

27. What is the minimum number of remaining votes candidate A needs to guarantee that they win the election using the plurality method? Explain your answer.

28. What is the minimum number of remaining votes candidate B needs to guarantee that they win the election using the plurality voting method?

29. Can candidate C win the election using the plurality voting method? Explain your answer.

30. If there are 408 votes cast in an election with four candidates, what is the smallest number of votes a candidate can win with if the plurality method is used?

31. Is it possible to have a tie if an election is held by comparing all possible combinations of two candidates at a time? If so, give an example.

32. Can you think of a circumstance under which an election decided using the plurality method is guaranteed to satisfy the head-to-head criterion?

Section 12-2

The Borda Count Method and the Plurality-with-Elimination Method

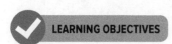

LEARNING OBJECTIVES

1. Determine the winner of an election using the Borda count method.

2. Decide if an election violates the majority criterion.

3. Determine the winner of an election using the plurality-with-elimination method.

4. Decide if an election violates the monotonicity criterion.

One of the most controversial examples of voting today comes from the world of college football. Up until the 2014 season, two teams were chosen to play in a national championship game based largely on national polls. In the current playoff system, a committee chooses the teams that make the postseason playoff, but only the most naïve observers believe that the weekly polls don't have a tremendous influence on the four teams that make the playoffs. But how exactly are the top teams in those polls determined? The answer is that coaches, athletic directors, and members of the media send in a weekly ballot listing their top 25 teams in order, and the teams receive points based on where they appear on ballots. The higher they are listed on any given ballot, the more points they get. When the points are added up, the totals determine where teams are ranked. In essence, it's an election.

John Mersits/AP Images

In this section, the first voting method we'll study is the basis for college football polls and a wide variety of other lists and awards.

The Borda Count Method

A second method of voting when there are three or more alternatives is called the Borda count method. This method was developed by a French naval captain and mathematician,

Jean-Charles de Borda. To be fair, however, the method was used to elect Roman senators about 1,600 years before Borda was born.

The **Borda count method** of voting requires the voter to rank each candidate from most favorable to least favorable then assigns 1 point to the last-place candidate, 2 points to the next-to-the-last-place candidate, 3 points to the third-from-the-last-place candidate, etc. The points for each candidate are totaled separately, and the candidate with the most points wins the election.

Example 1 illustrates how to use the Borda count method.

Example 1	Using the Borda Count Method

The preference table for a club presidential election consisting of three candidates is shown.

(a) Use the Borda count method to determine the winner.
(b) Who wins using the plurality method? Is there any real difference between the result with the two methods?

Number of votes	15	8	3	2
First choice	B	A	C	A
Second choice	C	B	A	C
Third choice	A	C	B	B

SOLUTION

(a) Since there are three candidates, we assign 1 point for the third choice, 2 points for the second choice, and 3 points for the first choice. Then multiply the number of votes by the number of the choice for each candidate to get the total points.

Candidate A:

First column: listed third, so 1 point each. The 15 at the top of that column indicates that 15 voters listed that order, so Candidate A gets 1 point per voter times 15 voters = 15 points. Second column: listed first, so 3 points each times 8 voters = 24 points. Third column: listed second, so 2 points each times 3 voters = 6 points. Fourth column: listed first, so 3 points each times 2 voters = 6 points. Candidate A has a total of $15 + 24 + 6 + 6 = 51$ points.

Candidate B:

First column: listed first, so 3 points each times 15 voters = 45 points. Second column: listed second, so 2 points each times 8 voters = 16 points. Third column: listed third, so 1 point each times 3 voters = 3 points. Fourth column: listed third, so 1 point each times 2 voters = 2 points. Candidate B has a total of $45 + 16 + 3 + 2 = 66$ points.

Candidate C:

First column: listed second, so 2 points each times 15 voters = 30 points. Second column: listed third, so 1 point each times 8 voters = 8 points. Third column: listed first, so 3 points each times 3 voters = 9 points. Fourth column: listed second, so 2 points each times 2 voters = 4 points. Candidate C has a total of $30 + 8 + 9 + 4 = 51$ points.

Candidate B has the most points and wins the election.

(b) Now all we care about is first-place votes. Candidate B got 15, candidate A got 10, and Candidate C got just 3, so again Candidate B wins. But it's interesting to note that Candidate C comes in a very distant last place using the plurality method, but ties for second using the Borda count method.

CAUTION	Make sure that you award the *most* points to the candidate listed first! It's very common to mistakenly award one point for a first-place vote, two points for second, and so on.

Sidelight   Jean-Charles de Borda (1733–1799)

Jean-Charles de Borda was a French sailor, mathematician, physicist, and engineer. As a young man he served in both the French army and navy. During the American Revolutionary War, the French and British fought for supremacy of the seas, and Borda served with "great distinction," helping the cause of American independence greatly.

Borda began his academic career as a mathematician in the army with a goal of becoming a military engineer. This training served him throughout his life, as he eventually made a wide variety of contributions to math, architecture, physics, and especially navigation. He also founded a school of naval architecture in France.

Borda was elected to the French Academy of Sciences in the 1760s, and in 1770 he devised the voting method that bears his name in order to elect officers of the academy. His method was used for 30 years. In 1801, a new president with a bit of an ego problem decided that his own method was to be used. You may have heard of him—his name was Napoleon Bonaparte. (In case you're wondering, Napoleon's method was a simple majority vote: if no candidate received more than 50 percent of the vote, then the position was left empty.)

Here are some interesting facts about Borda:

- Five French ships were named *Borda* in his honor.

- There's a crater on the moon named after him.

- His name is one of 72 inscribed on the Eiffel Tower.

Try This One   1

1. Determine the winner of an election using the Borda count method.

There are four candidates for homecoming queen at Johnsonville College: Kia (K), Latoya (L), Michelle (M), and Natalie (N). The preference table for the election is shown next:

Number of votes	232	186	95	306
First choice	M	K	M	L
Second choice	K	L	L	K
Third choice	L	N	K	N
Fourth choice	N	M	N	M

(a) Using the Borda count method, determine who won the election.
(b) Would the same person win the election if the plurality method was used? What's really unusual about the election in this case?

The Borda count method, like the plurality method, has its shortcomings. This method sometimes violates the fairness criterion called the *majority criterion*.

The **majority criterion** states that if a candidate receives a majority of first-place votes, then that candidate should be the winner of the election.

Notice that this is not called "the plurality criterion." It doesn't say that the candidate with the MOST first-place votes should always win. It says that IF there is a candidate that gets more than 50% of the votes, then that candidate should win.

Example 2 illustrates the majority criterion.

Example 2   Checking the Majority Criterion

The staff of an entertainment magazine is voting for the best new broadcast show of the 2021 fall season. The choices are *NCIS: Hawaii* (N), *Ghosts* (G), and *LaBrea* (L). The results are summarized below. If the winner is chosen using the Borda count method, does the election violate the majority criterion?

Number of votes	11	7	6	4
First choice	G	N	N	G
Second choice	N	L	G	L
Third choice	L	G	L	N

SOLUTION

First, find the winner using the Borda count method:

> *Ghosts:* $11 \cdot 3 + 7 \cdot 1 + 6 \cdot 2 + 4 \cdot 3 = 64$
> *NCIS: Hawaii:* $11 \cdot 2 + 7 \cdot 3 + 6 \cdot 3 + 4 \cdot 1 = 65$
> *LaBrea:* $11 \cdot 1 + 7 \cdot 2 + 6 \cdot 1 + 4 \cdot 2 = 39$

NCIS: Hawaii wins using the Borda count method, but of 28 ballots cast, 15 listed *Ghosts* first. This means that a majority of voters listed *Ghosts* first, and since it didn't win, the majority criterion is violated.

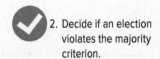

2. Decide if an election violates the majority criterion.

Try This One   2

Does the election in Try This One 1 violate the majority criterion?

In addition to its use in college football polls, the Borda count method is used to select the Heisman trophy winner, the most valuable player award in both major baseball leagues, and the country music vocalists of the year, among many others.

The Plurality-with-Elimination Method

The plurality-with-elimination method is sort of like a demolition derby, where everyone is competing against everyone else, and contestants are eliminated one at a time. (Some folks call it the "survival of the fittest" method.) It was designed specifically with the majority criterion in mind; if no candidate gets a majority of first-place votes, a series of rounds is used in which candidates are eliminated and votes are recounted.

Math Note

The plurality-with-elimination method is also commonly called *instant runoff voting*, and abbreviated IRV.

In the **plurality-with-elimination method**, the candidate with the majority of first-place votes is declared the winner. If no candidate has a majority of first-place votes, the candidate (or candidates) with the least number of first-place votes is eliminated, then the candidates who were below the eliminated candidate move up on the ballot, and the number of first-place votes is counted again. If a candidate receives the majority of first-place votes, that candidate is declared the winner. If no candidate receives a majority of first-place votes, the one with the least number of first-place votes is eliminated, and the process continues.

Example 3 shows how to use the plurality-with-elimination method.

Example 3   Using the Plurality-with-Elimination Method

Use the plurality-with-elimination method to determine the winner of the election shown in the preference table.

Number of votes	6	27	17	9
First choice	A	B	C	D
Second choice	D	A	D	B
Third choice	C	C	B	C
Fourth choice	B	D	A	A

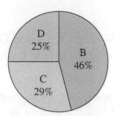

Round 1 results

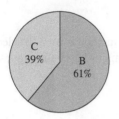

Round 2 results

Round 3 results

SOLUTION

Round 1: There were 59 votes cast, and no one received a majority of votes (30 or more), so candidate A is eliminated since he has the fewest first-place votes, 6. After candidate A is crossed out, the remaining candidates slide up to fill in the spot vacated by A.

Number of votes	6	27	17	9
First choice	~~A~~	B	C	D
Second choice	D	~~A~~	D	B
Third choice	C	C	B	C
Fourth choice	B	D	~~A~~	~~A~~

Round 2: In this round, the 6 first-place votes candidate A received go to candidate D because she moved up in the first column when candidate A was eliminated. But there is still no candidate with 30 or more first-place votes, so next candidate D is eliminated because she has the fewest first-place votes (15) in this round.

Number of votes	6	27	17	9
First choice	~~D~~	B	C	~~D~~
Second choice	C	C	~~D~~	B
Third choice	B	~~D~~	B	C

Round 3: The 6 first-place votes candidate D received in column 1 go to candidate C while the 9 first-place votes candidate D received in the fourth column go to candidate B. With A and D eliminated, the preference table looks like this.

Number of votes	6	27	17	9
First choice	C	B	C	B
Second choice	B	C	B	C

Now candidate B has 36 (27 + 9) first-place votes and candidate C has 23 (6 + 17); candidate B now has a majority and is declared the winner.

Try This One   3

The planning committee for a company's annual picnic votes for an afternoon activity. The choices are a softball game (S), a touch football game (F), a bocce game (B), and a volleyball game (V). The preference table is shown.

Number of votes	3	5	2	1	1
First choice	B	S	V	F	B
Second choice	S	F	S	S	V
Third choice	F	B	B	V	F
Fourth choice	V	V	F	B	S

Determine the winner using the plurality-with-elimination method.

Lars A. Niki

3. Determine the winner of an election using the plurality-with-elimination method.

Sidelight   Plurality-with-Elimination: A Case Study

Starting in 2018, the state of Maine changed their system of electing representatives to Congress from plurality to plurality-with-elimination. Rather than voting for one candidate, voters now rank candidates in order of preference. The intent is to ensure that a majority of voters approve of the winning candidate. And it didn't take long for this change to make a significant difference.

In Maine's second district, the incumbent, Republican Bruce Poliquin, received the most first-place votes in the 2018 election—about 2,000 more than the closest challenger, Democrat Jared Golden. But with the remaining two candidates splitting 8% of the vote, Poliquin's total was 46.3%, over 10,000 votes short of the number needed for a majority. The third- and fourth-place candidates were eliminated in one round because the last-place candidate received less than 7,000 votes, which would not have been enough to give either of the top candidates a majority.

Of the 23,427 votes that were cast for the eliminated candidates, 10,427 went to Golden and 4,747 went to Poliquin. (The remaining 8,253 did not list a ranking for either of those candidates and were ignored.) This put Golden's total at 50.62%, just enough for a majority, and he was declared the winner. The losing candidate did what anyone would do in 2018: he gave a gracious concession speech and pledged full support for the declared winner. Just kidding. He declared himself the winner and sued, a suit that proved unsuccessful. Golden was sworn into Congress on January 3, 2019. His constituents must have felt like they made the right choice, as Golden was reelected in 2020.

Plurality-with-elimination is used widely in Australia and Canada, as well as other countries. In the States, a number of cities use it, including San Francisco, Minneapolis, St. Paul, and Oakland.

The plurality-with-elimination method, like the other two methods, has some shortcomings. One shortcoming is that it sometimes fails the fairness criterion known as the *monotonicity criterion*.

The **monotonicity criterion** states that if a candidate wins an election, and a reelection is held in which the only changes in voting favor the original winning candidate, then that candidate should still win the reelection.

Consider this election:

Number of votes	7	13	11	10
First choice	X	Z	Y	X
Second choice	Z	X	Z	Y
Third choice	Y	Y	X	Z

Using the plurality-with-elimination method, candidate Y is eliminated in round 1. With Y eliminated, the preference table looks like this.

Number of votes	7	13	11	10
First choice	X	Z	Z	X
Second choice	Z	X	X	Z

Z wins with 24 first-place votes.

Now suppose the first election was declared invalid for some reason, and on a second election, the voters in column 1 change their ballots in favor of candidate Z and vote ZXY.

The new preference table will be

Number of votes	7	13	11	10
First choice	Z	Z	Y	X
Second choice	X	X	Z	Y
Third choice	Y	Y	X	Z

Here X is eliminated on the first round and the preference table becomes

Number of votes	7	13	11	10
First choice	Z	Z	Y	Y
Second choice	Y	Y	Z	Z

Now Y is the winner with 21 votes compared to 20 votes for Z.

In this case, the plurality-with-elimination method fails the monotonicity criterion. On the second election, even though candidate Z had received seven more first-place votes, they lost the election! By doing better the second time, the candidate did worse!

Example 4	**Checking the Monotonicity Criterion**

Suppose that all 6 voters from the first column of the election in Example 3 are persuaded to change their ballots to match the 17 voters in the third column. Does this violate the monotonicity criterion?

SOLUTION

First, we should point out that this change favors candidate B, since all 6 ballots move B from fourth to third choice.

Number of voters	27	23	9
First choice	B	C	D
Second choice	A	D	B
Third choice	C	B	C
Fourth choice	D	A	A

No candidate has a majority, so we eliminate the candidate with the fewest first-place votes (A). But this keeps the first-place votes the same, so we still have no majority. Next we eliminate candidate D, who now has the fewest first-place votes. The resulting preference table is

Number of voters	27	23	9
First choice	B	C	B
Second choice	C	B	C

Now B has a majority and wins. Since B was the winner of the original election, the monotonicity criterion is not violated.

Try This One   4

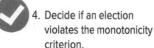

4. Decide if an election violates the monotonicity criterion.

If the one voter who listed softball last in the election in Try This One 3 changes their vote to the order B-S-F-V, does this violate the monotonicity criterion?

The plurality-with-elimination method is used in some very high profile elections. In addition to the elections mentioned in the Sidelight, the International Olympic Committee uses it to select the cities that will host the Games, and it is also used by the Academy of Motion Pictures to determine the annual Academy Award winners.

Answers to Try This One

1 (a) The winner is Kia with 2,548 points.
   (b) Michelle wins using the plurality method. Not only does Kia NOT win, she doesn't even come in second. She's a very distant third.

2 No. None of the candidates received a majority of the first-place votes.

3 The winner is softball.

4 No. Softball still wins.

Exercise Set	12-2

Writing Exercises

1. Explain how to determine the winner of an election using the Borda count method.

2. If an election violates the majority criterion, what does that mean?

3. Explain how to determine the winner of an election using the plurality-with-elimination method.

4. If an election violates the monotonicity criterion, what does that mean?

5. Is it possible for the winner of an election using the Borda count method to lose that election if the plurality method is used? How?

6. Can a candidate that gets the least first-place votes win an election held using the plurality-with-elimination method? Explain.

Applications in Our World

7. A gaming club holds a vote to decide what type of video game they'll play at the next meeting. The choices are sports (S), action (A), or role-playing (R). The preference table for the results is shown here.

Number of votes	10	6	5
First choice	R	S	S
Second choice	A	R	A
Third choice	S	A	R

Using the Borda count method of voting, determine the winner.

8. The McKees' Point Yacht Club Board of Directors wants to decide where to hold their fall business meeting. The choices are the Country Club (C), Frankie's Fine Foods (F), West Oak Golf Club (W), and Rosa's Restaurant (R). The results of the election are shown in the preference table.

Number of votes	8	6	5	2
First choice	R	W	C	F
Second choice	W	R	F	R
Third choice	C	F	R	C
Fourth choice	F	C	W	W

Determine the winner using the Borda count method of voting.

9. A local movie theater asks its patrons which movies they would like to view during next month's "Oldies but Goodies" week. The choices are *Gone with the Wind* (G), *Casablanca* (C), *Anatomy of a Murder* (A), and *Back to the Future* (B). The preference table is shown.

Number of votes	331	317	206	98
First choice	G	A	C	B
Second choice	A	C	B	G
Third choice	C	B	G	A
Fourth choice	B	G	A	C

Use the Borda count voting method to determine the winner.

10. A local police union is holding an election for a representative to the town safety board. The candidates are officers Zane (Z), Abercrombie (A), and Martinez (M). The preference table is shown.

Number of votes	15	9	5	4
First choice	Z	M	A	Z
Second choice	A	Z	M	M
Third choice	M	A	Z	A

Using the Borda count voting method, determine the winner.

11. Students at a college were asked to rank three improvements that they would like to see at their college.

Their choices were build a new gymnasium (G), build a baseball/football field (B), or build a swimming pool (S). The votes are summarized in the preference table.

Number of votes	83	56	42	27
First choice	G	S	S	B
Second choice	S	G	B	S
Third choice	B	B	G	G

(a) Using the Borda count method of voting, determine the winner.

(b) Refer to Exercise 11 in Section 12-1. Is the winner the same as the one determined by the plurality method?

12. A psychologist is performing an experiment on color perception in which the subjects are asked to order the intensities of four different color samples: pink (P), green (G), teal (T), and brown (B). The results are shown below.

Number of votes	8	6	5	3	2
First choice	P	T	T	P	P
Second choice	T	B	G	G	T
Third choice	G	G	P	T	B
Fourth choice	B	P	B	B	G

(a) Using the Borda count method, which color was rated as most intense?

(b) What percentage of the subjects rated the winner as most intense?

13. Does the election in Exercise 7 violate the majority criterion?

14. Does the election in Exercise 8 violate the majority criterion?

15. Does the election in Exercise 9 violate the majority criterion?

16. Does the election in Exercise 10 violate the majority criterion?

17. Does the election in Exercise 11 violate the majority criterion?

18. Does the election in Exercise 12 violate the majority criterion?

19. An English department is voting for a new department chairperson. The three candidates are Professor Greene (G), Professor Williams (W), and Professor Donovan (D). The results of the election are shown in the preference table.

Number of votes	10	8	7	4
First choice	D	W	G	G
Second choice	G	D	W	D
Third choice	W	G	D	W

Using the plurality-with-elimination method of voting, determine the winner.

20. The Association of Self-Employed Working Persons is picking a speaker for its next meeting. The choices for a topic are health care (H), investments (I), or advertising (A). The results of the election are shown in the preference table.

Number of votes	6	4	9	2
First choice	H	H	I	I
Second choice	I	A	H	A
Third choice	A	I	A	H

Using the plurality-with-elimination method of voting, determine the winner.

21. A panel of experts is convened to decide which is the best hospital overall in a metropolitan area. The four hospitals under consideration are Regional Medical Center (R), Community General (G), Children's Hospital (C), and Derbyshire Hospital (D). The preference table is shown.

Number of votes	3	5	2	6	4
First choice	R	D	C	C	R
Second choice	G	C	R	G	C
Third choice	C	G	G	D	D
Fourth choice	D	R	D	R	G

(a) Using the plurality-with-elimination method, determine the winner.

(b) Is the winner the same as the one determined by the plurality method used in Exercise 13 of Section 12-1?

22. The students in Dr. Lee's math class were asked to vote on the starting time for the final exam. Their choices were 8:00 A.M., 10:00 A.M., 12:00 P.M., or 2:00 P.M. The results of the election are shown in the preference table.

Number of votes	8	12	5	3	2	2
First choice	8	10	12	2	10	8
Second choice	10	8	2	12	12	2
Third choice	12	2	10	8	8	10
Fourth choice	2	12	8	10	2	12

(a) Using the plurality-with-elimination method, determine the winner.

(b) Is the winner the same as the one determined by the plurality method used in Exercise 14 of Section 12-1?

23. Suppose that all four voters from the last column of the preference table in Exercise 19 decide to change their vote to the order D-G-W (the order in the first column). Does this violate the monotonicity criterion?

24. Suppose that two of the four voters from the second column of the preference table in Exercise 20 decide to change their vote to the order H-I-A (the order in the first column). Does this violate the monotonicity criterion?

25. If two of the voters from column 1 in Exercise 21 change their vote to column 2, does this violate the monotonicity criterion?

26. If the three voters in column 4 in Exercise 22 change their vote to column 1, does this violate the monotonicity criterion?

Critical Thinking

27. Build a preference table for an election involving three candidates so that candidate A wins the election using the Borda count method, but the majority criterion is violated.

28. Build a preference table for an election involving three candidates so that candidate B wins using the plurality-with-elimination method, but the monotonicity criterion is violated.

29. Build a preference table for an election so that one candidate wins the election using the Borda count method and a different candidate wins the same election using the plurality-with-elimination method.

30. If the candidates on a preference ballot are ranked so that the lowest candidate gets 0 points, the next to the lowest candidate gets 1 point and so on, will the winner be the same using the Borda count method as the winner when the candidates are ranked the way they are explained in this section? Explain your answer using an illustration.

31. If the candidates on a preference ballot are ranked by giving the top candidate a score of 1 (first place), the next highest candidate a score of 2 (second place), and so on, can the Borda count method be used to determine a winner? Explain why or why not.

32. In an election with four candidates, how many potential different ballots can be submitted listing the candidates

from one through four? What about if there are six candidates listed from one through six?

33. Based on your answers to Exercise 32, explain why voting methods that involve a preference ballot can be difficult to tabulate when there are many candidates.

34. One way to avoid the issue described in Exercises 32 and 33 is to have ballots turned in with only the top few candidates in order of preference, as opposed to *all* of the candidates. For example, suppose that eight candidates are in the running for CEO of a company, but voters list only their top three choices in order. Discuss whether or not you think this restriction has the potential to change the outcome of an election using the Borda count method or plurality-with-elimination.

35. Let's talk about a modified Borda count method. Suppose that there are three candidates, and on each ballot the top-ranked candidate gets eight points, the second candidate gets four points, and the last candidate gets one point.

(a) Do you think this modification would change the results of some elections? Explain your answer. (You might consider reworking some of the exercises or example problems with this modification to help you decide.)

(b) If you answered yes to part (a), describe the circumstances under which certain election results would change.

Section 12-3

The Pairwise Comparison Method and Approval Voting

LEARNING OBJECTIVES

1. Determine the winner of an election using the pairwise comparison method.

2. Decide if an election violates the irrelevant alternatives criterion.

3. Describe Arrow's impossibility theorem.

4. Determine the winner of an election using approval voting.

What does it mean for an election to be fair? If everyone who wants to and is eligible can cast their vote, and all votes are properly counted, it seems like that should constitute a fair election. And all evidence indicates that election improprieties are exceedingly rare in the United States. Yet we still hear people talk about unfair elections all the time, in one case to the point where a violent mob attacked the seat of their own government.

Throughout this chapter, we've been identifying criteria that can be used to determine if an election is fair. In this section, we will examine whether any election can satisfy all these criteria.

Susan See Photography

But first, we'll study a fourth voting method, and a fourth fairness criterion.

The Pairwise Comparison Method

The pairwise comparison method uses a preference table to compare each pair of candidates. For example, if there are four candidates, A, B, C, and D, running in an election, then there would be six comparisons, as shown.

A vs. B	B vs. C
A vs. C	B vs. D
A vs. D	C vs. D

Example 1   Finding the Number of Pairwise Comparisons Needed

(a) Use the combination formula from Chapter 10 to decide how many pairwise comparisons would be needed in an election with six candidates.

(b) Develop a simple formula for finding the number of pairwise comparisons needed when there are n candidates.

SOLUTION

(a) This is a direct application of the combination formula—all we're doing is choosing two candidates from six for each comparison. So we need to know the number of combinations (order doesn't matter) of two objects chosen from six, which is $_6C_2$.

$$_6C_2 = \frac{6!}{(6-2)!2!} = \frac{6!}{4!2!} = \frac{6 \cdot 5 \cdot \cancel{4 \cdot 3 \cdot 2 \cdot 1}}{\cancel{4 \cdot 3 \cdot 2 \cdot 1} \cdot 2 \cdot 1} = \frac{6 \cdot 5}{2 \cdot 1} = \frac{30}{2} = 15$$

(b) When there are n candidates, we'll do exactly the same thing. We just need to be a little more algebraic about simplifying the expression.

$$_nC_2 = \frac{n!}{(n-2)!2!} = \frac{n \cdot (n-1) \cdot \cancel{(n-2)!}}{\cancel{(n-2)!}2!} = \frac{n(n-1)}{2}$$

This shows that if we're always choosing two candidates to compare, we don't need to deal with the combination formula anymore: the number of pairwise comparisons needed when there are n candidates is

$$\frac{n(n-1)}{2}$$

Try This One   1

Find the number of pairwise comparisons needed if eight candidates are running in an election.

Sidelight   Does History Repeat Itself?

It seems that the United States' presidents who were elected in a year with "0" at the end, which happens every 20 years, have shared some sad coincidences.

1840: William Henry Harrison died in office.

1860: Abraham Lincoln was assassinated.

1880: James A. Garfield was assassinated.

1900: William McKinley was assassinated.

1920: Warren G. Harding died in office.

1940: Franklin D. Roosevelt died in office.

1960: John F. Kennedy was assassinated.

1980: Ronald Reagan survived an assassination attempt.

Jill Braaten/McGraw Hill

George W. Bush, elected in 2000, survived 8 years in office physically unscathed, although his outgoing approval rating was the lowest recorded in the 70 years such polls have been conducted. And Joe Biden, elected in 2020, is alive and well as of this writing, but if I were him I wouldn't leave the White House without a suit of armor.

The pairwise comparisons we will make are the same ones we did to check the head-to-head comparison criterion in Section 12-1. For example, to compare candidates A and B using a preference table, add up the number of ballots on which A is preferred over B, and the number of ballots on which B is preferred over A. The candidate with the most ballots in their favor gets one point. In case of a tie, each candidate gets $\frac{1}{2}$ point. After all possible pairwise comparisons have been made, the candidate with the most points wins. In summary:

> The **pairwise comparison method** of voting requires that all candidates be ranked by the voters. Then each candidate is paired with every other candidate in a one-to-one contest. For each one-to-one comparison, the candidate who wins on more ballots gets 1 point. In case of a tie, each candidate gets $\frac{1}{2}$ point. After all possible two-candidate comparisons are made, the points for each candidate are tallied, and the candidate with the most points wins the election.

Example 2 shows how to use the pairwise comparison method.

Example 2   Using the Pairwise Comparison Method

(a) Use the pairwise comparison method to find the winner of the election whose results are shown in the following preference table.

(b) Is it the same candidate that would win using the plurality method? If the result is different, make a case that the pairwise comparison result makes more sense.

Number of votes	14	13	16	15
First choice	B	A	C	B
Second choice	C	C	A	A
Third choice	A	B	B	C

SOLUTION

(a) We will need to make three pairwise comparisons: A vs. B, A vs. C, and B vs. C. First, A vs. B:

Number of votes	14	13	16	15
First choice	Ⓑ	Ⓐ	C	Ⓑ
Second choice	C	C	Ⓐ	Ⓐ
Third choice	Ⓐ	Ⓑ	Ⓑ	C

Candidate A is ranked higher than B on 29 ballots (columns 2 and 3), and candidate B is also ranked higher on 29 ballots (columns 1 and 4). This is a tie, so each candidate gets $\frac{1}{2}$ point.

Next, compare A to C:

Number of votes	14	13	16	15
First choice	B	Ⓐ	Ⓒ	B
Second choice	Ⓒ	Ⓒ	Ⓐ	Ⓐ
Third choice	Ⓐ	B	B	Ⓒ

Candidate A is ranked higher than C in columns 2 and 4, so A gets $13 + 15 = 28$ votes.
Candidate C is ranked higher than A in columns 1 and 3, so C gets $14 + 16 = 30$ votes.
Since C has more votes, assign 1 point to C.

Finally, compare B to C:

Number of votes	14	13	16	15
First choice	Ⓑ	A	Ⓒ	Ⓑ
Second choice	Ⓒ	Ⓒ	A	A
Third choice	A	Ⓑ	Ⓑ	Ⓒ

Candidate B is ranked higher than C in columns 1 and 4, so B gets $14 + 15 = 29$ votes.
Candidate C is ranked higher than B in columns 2 and 3, so C gets $13 + 16 = 29$ votes.
Since this is a tie, assign $\frac{1}{2}$ point to B and $\frac{1}{2}$ point to C.

Now find the totals.

			Total
Candidate A	$\frac{1}{2}$	$\frac{1}{2}$	*A ties with B.*
Candidate B	$\frac{1}{2} + \frac{1}{2}$	1	*B ties with A and C.*
Candidate C	$1 + \frac{1}{2}$	$1\frac{1}{2}$	*C defeats A and ties with B.*

Candidate C has the most points and is the winner.

(b) Using the plurality method, candidate B wins easily, with 29 first-place votes out of the 58 ballots. But notice that on the two preferences that didn't have B listed first, they were listed last. So in that regard this is a better result: should a candidate win an election if half of the voters think they're the WORST choice?

> **Math Note**
>
> For the election in Example 2, candidate C wins using the Borda count method, and candidates B and C tie using the plurality-with-elimination method.

Caiaimage/Martin Barraud/OJO+/
Getty Images

1. Determine the winner of an election using the pairwise comparison method.

Try This One   2

The members of a music appreciation club vote to decide whether to attend an opera (O), a symphony (S), or a ballet (B). The results of the election are shown in the preference table.

Number of votes	13	9	6	11
First choice	O	B	S	B
Second choice	S	O	B	S
Third choice	B	S	O	O

Use the pairwise comparison voting method to determine the winning selection.

It turns out that the pairwise comparison voting method satisfies all of the fairness criteria we've studied so far: majority, head-to-head, and monotonicity. So we've found the ideal voting method, right? Not so fast, my friend. There's a fourth fairness criterion, known as the *irrelevant alternatives criterion*. As we'll see, the pairwise comparison voting method can sometimes fail this last criterion.

> The **irrelevant alternatives criterion** requires that if a certain candidate X wins an election and one of the other candidates is removed from the ballot and the ballots are recounted, candidate X still wins the election.

| Example 3 | Checking the Irrelevant Alternatives Criterion |

Does the election in Example 2 violate the irrelevant alternatives criterion?

SOLUTION

Candidate C won the election, so the two irrelevant alternatives are A and B. If A is eliminated, the preference table is now:

Number of votes	14	13	16	15
First place	B	C	C	B
Second place	C	B	B	C

Now there are $14 + 15 = 29$ voters who preferred B to C, and there are $13 + 16 = 29$ voters who preferred C to B. The result is a tie, which shows that this election violates the irrelevant alternatives criterion. Candidate C shouldn't have gone from a win to a tie because one of the losing candidates dropped out.

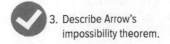

2. Decide if an election violates the irrelevant alternatives criterion.

| Try This One | 3 |

Does the election in Try This One 2 violate the irrelevant alternatives criterion?

Math Note

In all voting methods, the possibility of a tie should be considered before the votes are counted, and some way of breaking a tie should be agreed on in advance of the election.

Maybe the biggest issue with the pairwise comparison method is that it's very prone to ties. Worse still, it's possible for all of the candidates to tie. Look at this preference table:

Number of votes	8	5	7
First choice	A	C	B
Second choice	B	A	C
Third choice	C	B	A

Here A wins over B (13 to 7), B wins over C (15 to 5), and C wins over A (12 to 8); this gives us a three-way tie for first place, in which case the election accomplished absolutely nothing!

Table 12-1 summarizes the four voting methods, and the marks show which criteria are always satisfied by the voting methods.

| Table 12-1 | Fairness Criteria Satisfied by Various Voting Methods |

	Head-to-head criterion	Majority criterion	Monotonicity criterion	Irrelevant alternatives criterion
Plurality		✓	✓	
Borda count			✓	
Plurality-with-elimination		✓		
Pairwise comparison	✓	✓	✓	

Arrow's Impossibility Theorem

Table 12-1 shows that none of the four voting methods is perfectly fair. Each violates one or more of the four fairness criteria.

The obvious question now becomes: What method satisfies all of the fairness criteria? The surprising and somewhat disappointing answer is that there isn't one. In 1951, an economist named Kenneth Arrow was able to prove that there does not exist and never will exist a

3. Describe Arrow's impossibility theorem.

Sidelight **Kenneth J. Arrow (1921–2017)**

Kenneth J. Arrow was born in New York City in 1921 and studied math at the City College of New York, where he earned a B.S. He went on to study economics at Columbia University, earning master's and doctorate degrees. He was one of America's most influential economists for over 50 years before passing at the ripe age of 95 in February of 2017.

In 1951, he proved the theorem that states that no voting system will satisfy all of the four fairness criteria. The theorem became known as Arrow's impossibility theorem.

In 1972, he received the Nobel Prize in Economics for his work in social choice theory. In 1986, he won the Von Neuman Theory Prize for his contributions to decision sciences, and in 2004 he was awarded the nation's highest scientific honor, the National Medal of Science.

Jamie Rector/Bloomberg/
Getty Images

democratic voting method for three or more alternatives that satisfies all four of the fairness criteria. The result is now known as **Arrow's impossibility theorem**. (The proof is beyond the scope of our study, but don't feel bad—Arrow was a Nobel Prize winner, so he must have been awfully smart.)

So in designing an election, which option should you choose? Like many things in our world, the answer depends on a variety of factors. If you're looking for simplicity, the plurality method is the way to go. The other end of the spectrum is the pairwise comparison method, which can be a lot of work to apply in a large election but does satisfy three of the four fairness criteria. Borda count is probably least likely to result in a tie, and plurality with elimination is specifically designed to satisfy the majority criterion. After considering the number of candidates, the number of votes cast, and which fairness criteria seem most important to you, you can make an informed choice about which method meets your needs.

Approval Voting

In the late 1970s, a new voting method called *approval voting* was introduced.

> With **approval voting**, each voter gives one vote to as many candidates on the ballot as they find acceptable. The votes are counted, and the winner is the candidate who receives the most votes.

Math Note

Approval voting is so named because the voter is rating each candidate as either acceptable or unacceptable, indicating whether or not they approve of the candidate.

In this case, voters can select anything from no candidates to all of the candidates. Example 4 shows how approval voting works.

Example 4 | Using Approval Voting to Determine the Winner of an Election

Five candidates are nominated for the teacher of the year award, and 20 of their colleagues will vote using approval voting. The results are shown in the table. (For example, the first column indicates that 9 voters marked only candidates A and D acceptable.) Which candidate wins?

Number of votes	9	3	2	5	1
Candidate A	/		/		/
Candidate B		/	/	/	
Candidate C		/	/	/	
Candidate D	/	/		/	
Candidate E					/

SOLUTION

From the table, count the number of votes for each candidate:

Candidate	Votes
A	$9 + 2 + 1 = 12$
B	$3 + 2 + 5 = 10$
C	$3 + 2 + 5 = 10$
D	$9 + 3 + 5 = 17$
E	5

In this election, candidate D received 17 votes and is declared the winner.

4. Determine the winner of an election using approval voting.

Try This One   4

An election was held for an employee of the month award using approval voting. The results are shown in the table. Which candidate won?

Number of votes	20	18	12	4
Candidate F		/		/
Candidate G	/		/	/
Candidate H		/	/	/
Candidate I	/		/	
Candidate J			/	/

Several political scientists and analysts independently developed approval voting in the late 1970s. This method is now used to elect the Secretary General of the United Nations, and it is also used to elect the leaders of some academic and professional societies such as the National Academy of Sciences. The advantages of approval voting are that it is simple to use and easy to understand. The ballot is also uncomplicated.

There are some disadvantages to approval voting though. The major one is that there is no ranking or preference of the candidates. Most voters have favorite choices, but there is no way to indicate these preferences on an approval ballot.

Because of this, approval voting is also prone to violating the majority criterion. It's possible that one candidate could be considered the best choice by well more than half of voters, but if they are also left off of the other ballots, a second candidate that is considered marginally acceptable by most of the voters could win. In addition, some societies that tried approval voting dropped it when they found that most voters were only marking one candidate acceptable, in which case the election is essentially decided using the plurality method.

Tetra Images/Getty Images

Tie Breaking

Regardless of the voting method selected, the possibility of a tie between two or more candidates always exists. There are many ways to break a tie, and a fair tie-breaking method should always be decided upon in advance of an election.

In some cases, the chairperson of a committee doesn't vote on motions unless there is a tie. In this case, the chairperson would cast the tie-breaking vote. The most obvious method of breaking a tie is, of course, the age-old method of flipping a coin. In other cases, drawing a name from a hat could be used. Using a third-party judge could be considered. For example, if a tie occurs in an election of a department chairperson, the dean could decide the winner. Another possibility might be to consider some other criteria such as seniority, education, or experience of the candidates.

Since the first four methods we studied all use preference ballots, if the chosen method results in a tie, one of the other methods could be used to see if there's a winner.

Answers to Try This One

1   28

2   B wins with 2 points, compared to 1 for O and 0 for S.

3   No; if either O or S drops out, B still wins.

4   Candidate G wins with 36 votes.

Exercise Set   12-3

Writing Exercises

1. Explain how to determine the winner of an election using the pairwise comparison method.
2. Describe the irrelevant alternatives criterion. How is it different from the monotonicity criterion?
3. Describe Arrow's impossibility theorem. How is it connected to the four fairness criteria we have studied?
4. Explain how to determine the winner of an election using approval voting.
5. What are some strengths and weaknesses of approval voting?
6. Which of the five voting methods we studied do you think is easiest to apply? Which do you think is the most fair?
7. If all of the voters in an approval voting election approve of just one candidate, what method is actually being used to decide the election? Explain.
8. Fill in the table below, which summarizes our five voting methods.

Method	How are ballots cast?	How is the winner decided?	Which fairness criteria might be violated?
Plurality			
Borda count			
Plurality-with-elimination			
Pairwise comparison			
Approval			

Computational Exercises

In Exercises 9–12, find the number of pairwise comparisons that need to be made in an election with the given number of candidates.

9. 4

10. 7

11. 10

12. 9

Applications in Our World

13. A college band was invited to perform at three different shows on the same day: Real Town (R), Steel Center (S), and Temple Village (V). Since they could only perform at one show, they voted on which one they would do. The results are shown in the preference table.

Number of votes	26	19	15	6
First choice	R	T	S	R
Second choice	S	S	R	T
Third choice	T	R	T	S

Using the pairwise comparison voting method, determine the winner.

14. An incentive travel program offers customers a free weekend vacation in exchange for sitting through a sales presentation. Currently the options are Disneyland (D), Epcot Center (E), SeaWorld (S), and the Flamingo Las Vegas casino (F). Customers are asked to rate each location from most to least desirable, and the results are tabulated.

Number of voters	19	17	8	6
First choice	D	E	S	D
Second choice	E	S	D	F
Third choice	S	D	F	S
Fourth choice	F	F	E	E

If the pairwise comparison method is used to determine the most desirable location, which is it?

15. An English department is voting for a new department chairperson. The three candidates are Professor Greene (G), Professor Williams (W), and Professor Donovan (D). The results of the election are shown in the preference table.

Number of votes	10	8	7	4
First choice	D	W	G	G
Second choice	G	D	W	D
Third choice	W	G	D	W

   (a) Determine the winner using the pairwise comparison voting method.

   (b) Compare the winner with the one determined by the plurality-with-elimination method. See Exercise 19 of Section 12-2.

16. A gaming club holds a vote to decide what type of video game they'll play at the next meeting. The choices are sports (S), action (A), or role-playing (R). The preference table for the results is shown here.

Number of votes	10	6	5
First choice	R	S	S
Second choice	A	R	A
Third choice	S	A	R

   (a) Determine the winner using the pairwise comparison method.

   (b) Compare the winner with the one determined by the Borda count method used in Exercise 7 of Section 12-2.

17. The McKees' Point Yacht Club Board of Directors wants to decide where to hold their fall business meeting. The choices are the Country Club (C), Frankie's Fine Foods (F), West Oak Golf Club (W), and Rosa's Restaurant (R). The results of the election are shown in the preference table.

Number of votes	8	6	5	2
First choice	R	W	C	F
Second choice	W	R	F	R
Third choice	C	F	R	C
Fourth choice	F	C	W	W

   (a) Determine the winner using the pairwise comparison method.

   (b) Is this the same winner as the one using the Borda count method in Exercise 8 of Section 12-2?

18. The students in Dr. Lee's math class are asked to vote on the starting time for their final exam. Their choices are 8:00 A.M., 10:00 A.M., 12:00 P.M., or 2:00 P.M. The results of the election are shown in the preference table.

Number of votes	8	12	5	3	2	2
First choice	8	10	12	2	10	8
Second choice	10	8	2	12	12	2
Third choice	12	2	10	8	8	10
Fourth choice	2	12	8	10	2	12

   (a) Determine the starting time using the pairwise comparison method.

   (b) Is this the same starting time as determined by the plurality-with-elimination method used in Exercise 22 of Section 12-2? What about the plurality method in Exercise 14 of Section 12-1?

19. If Professor Donovan was unable to serve as department chairperson in the election shown in Exercise 15 and the votes are recounted, does this election violate the irrelevant alternatives criterion?

20. If the travel company from Exercise 14 loses its contract with Epcot, and the votes are recounted after Epcot is eliminated, is the irrelevant alternatives criterion violated?

21. If the West Oak Golf Club is unavailable and the votes were recounted in the election shown in Exercise 17, is the irrelevant alternatives criterion violated?

22. If a room for Dr. Lee's final exam was not available at 2:00 P.M. and the votes were recounted in the election shown in Exercise 18, is the irrelevant alternatives criterion violated?

23. A sports committee of students needs to choose a team doctor. The result of the voting is shown. The approval method will be used.

Number of votes	15	18	12	10	5
Dr. Michaels	/		/	/	/
Dr. Zhang		/	/	/	/
Dr. Philip	/	/		/	
Dr. Perez	/	/		/	

Which doctor was picked?

24. The students at a college voted on a new mascot for their sports teams. The results of the election are shown. Using the approval voting method, determine the winner.

Number of votes	235	531	436	374
Ravens	/		/	/
Panthers		/	/	/
Killer Bees	/	/	/	/
Termites		/		/

25. A nursing school committee decides to buy a van for the school. They vote on a color using the approval voting method. The results are shown.

Number of votes	1	2	1	3	2	1
White	/		/	/	/	
Blue	/	/	/			
Green		/	/	/	/	
Silver		/	/		/	

Which color was picked?

26. A research committee decides to test market a new flavor for a children's drink. The results of a survey using approval voting at a local mall are shown.

Number of votes	38	32	16	5	3
Strawberry	/		/		/
Lime	/		/	/	
Grape		/	/		
Orange		/		/	/
Bubble gum	/	/		/	

Which flavor was most popular?

27. Due to prison overcrowding, a parole board has to release one prisoner on good behavior. After hearing each case, they decide to use the approval voting method. The result is shown here.

Number of votes	1	1	1	1	1	1
Inmate W	/		/		/	/
Inmate X		/	/		/	
Inmate Y	/	/	/			
Inmate Z		/	/	/	/	/

Which inmate was released?

28. A park association committee decided to make one improvement for the local park this spring. The result of an election using approval voting is shown.

Number of votes	2	1	3	4	2
Paint benches	/		/		
Trim bushes			/	/	/
Repair snack bar			/	/	/
Patch cement walks	/	/	/	/	/

Which repair was made?

Critical Thinking

29. Build a preference table with three candidates X, Y, and Z so that the same candidate wins the election using all of the plurality, Borda count, plurality-with-elimination, and pairwise comparison methods.

30. Build a preference table so that one candidate wins using the pairwise comparison method, but that violates the irrelevant alternatives criterion.

31. Explain why Exercise 30 would be impossible if you replace "irrelevant alternatives" with "head-to-head comparison."

32. If you are voting in an election using approval voting and you approve of all of the candidates equally, discuss whether you should give every candidate one vote or every candidate no vote.

33. You may have noticed that one voting method we didn't cover is plain old "vote for one candidate and most votes wins." Or did we? How can you turn the preference ballot and table style that we studied into a "traditional" election? What method is it really?

34. Evaluate each voting method we studied if there are only two candidates in an election. Are there similarities? Differences?

35. Suppose that in an election for city council, there are nine candidates on the ballot and three will be elected.
    (a) Discuss how the approval voting method could easily be adapted to this situation.
    (b) Discuss how each of the other voting methods we studied would need to be adapted, including your thoughts on how difficult it would be to determine a winner and how fair you think the outcome would be.

36. Describe how approval voting is similar to the Borda count method with a difference in points awarded.

37. Devise a method for breaking ties when using approval voting that is based on votes cast, not something random like a coin flip.

38. Devise a variation of approval voting where every candidate is rated as "completely acceptable," "marginally acceptable," or "unacceptable." Is your method similar to any of the other methods we studied?

Section 12-4 | Apportionment

LEARNING OBJECTIVES

1. Compute standard divisors and quotas.

2. Apportion seats using Hamilton's method.

3. Apportion seats using Jefferson's method.

4. Apportion seats using Adams' method.

5. Apportion seats using Webster's method.

6. Apportion seats using the Huntington-Hill method.

You arrive at an 8 A.M. final (ugh!) and find that at least your professor was kind enough to bring donuts. Let's say she brought 4 dozen donuts and there are 24 people in class. How can she fairly distribute them?

The obvious answer is two donuts per person. This is a very simple example of *apportionment*. It's a simple example because everyone gets the same number of donuts and there are none left over. But what if some people are entitled to more donuts for one reason or another? And what if the number of students doesn't evenly divide the number of donuts? Then the situation gets more complicated.

Brand X Pictures/Getty Images

If you had to you could cut the donuts into pieces. But what if you're trying to fairly dole out computers to schools, cars to police stations, or sonogram machines to hospitals? Half of a computer won't do anyone any good. These items are called **indivisible** because they can't be divided into smaller pieces. The study of apportionment is all about coming up with fair ways to divide and assign indivisible items.

The most common example of apportionment (and, as we'll see, where the study came from in the first place) is the United States Congress. There are 100 senators and 50 states: 2 senators per state—easy! But in the House of Representatives, the number of representatives is based on the population of the state. There are 435 seats in the House, and they're divided according to population. For example, Florida has 27 seats, Pennsylvania has 18, and Rhode Island has just 2.

So why can't we just divide 435 by the population to find out how many representatives per person, then multiply by the state's population? Because it's against the law to divide politicians into pieces. (Insert politician joke here.) Almost every state would be entitled to a non-integer number of seats, which won't work.

Math Note

The term "indivisible" refers to objects that can't be split into fractional parts, like people or seats in Congress.

> **Apportionment** is a process of fairly assigning identical indivisible objects among individuals entitled to shares that may be unequal.

For the remainder of this section, we will refer to the objects being assigned as "seats." The individuals they are allocated to will be called "states" or "districts."

Standard Divisors and Quotas

A good starting point is to find how many people will be represented by each seat. This is known as the *standard divisor*.

> The **standard divisor** for an apportionment process is the average number of people per seat:
>
> $$\text{Standard divisor} = \frac{\text{Total population}}{\text{Number of seats}}$$

Example 1   Finding the Standard Divisor

A large city has 20 seats to be allocated proportionally among five districts according to the population of each district. The populations of the districts are shown in the table.

(a) If there are 20 seats and five districts, why would it be unfair to just assign 4 seats to each district?

(b) Find the standard divisor, and describe exactly what it tells us.

District	1	2	3	4	5	Total
Population (in thousands)	32	80	41	65	22	240

Aerial Archives/Alamy Stock Photo

SOLUTION

(a) While it would be easy to just divide the number of seats by the number of districts, and in this case that would actually give us a whole number, it would still be unfair because it totally ignores the sizes of the districts. If the district with 80,000 people gets the same number of seats as the district with 22,000, the people in the larger district are in essence getting about one-fourth the representation as those in the smaller district.

(b) The total population is 240,000 and there are 20 seats:

$$\text{Standard divisor} = \frac{\text{Total population}}{\text{Number of seats}} = \frac{240,000}{20} = 12,000$$

This tells us that there will be on average 1 seat for every 12,000 people.

Try This One 1

District	1	2	3	4	5	6	7	Total
Population	125	89	235	97	102	128	184	960

The table above gives the population in ten thousands of seven districts of a particular county. The county assembly has 30 seats that are to be allocated according to the population of each district.

(a) Find the standard divisor, and describe what it tells us.
(b) How many seats do you think District 4 should get? Why?

In Example 1, notice that District 2 has exactly one-third of the population. In that case, it would seem perfectly reasonable that it should get 1/3 of the seats. Unfortunately, that can't happen: there are 20 seats, which isn't divisible by 3. But it's at least a starting point on giving us an idea of roughly how many seats should go to that district. Let's take a look at another way of framing that calculation: as the fraction of the population that lives in a given district times the total number of seats available.

$$\text{Approximate number of seats for a district} = \frac{\text{Population of district}}{\text{Total population}} \cdot \text{Total number of seats}$$

Now we're going to do a little bit of algebra. First we'll rearrange the multiplication:

$$\text{Approximate number of seats} = \text{Population of district} \cdot \frac{\text{Total number of seats}}{\text{Total population}}$$

The fraction that we're multiplying by here should look vaguely familiar: it's actually the reciprocal of the standard divisor. We know that multiplying by a fraction is the same as dividing by its reciprocal, so we can rewrite this as

$$\text{Approximate number of seats} = \frac{\text{Population of district}}{\text{Standard divisor}}$$

This now becomes the most important formula in this section. It tells us roughly how many seats each district should get, so any apportionment method that doesn't give each district something close to this number probably isn't fair. This number is known as the *standard quota*.

The **standard quota** for a district in an apportionment process is the population of the district divided by the standard divisor:

$$\text{Standard quota} = \frac{\text{Population of district}}{\text{Standard divisor}}$$

Example 2 Finding Standard Quotas

Find the standard quota for each of the districts in Example 1. Based on the standard quota, how many seats do you think District 3 should get?

SOLUTION:

District	1	2	3	4	5	Total
Population (in thousands)	32	80	41	65	22	240

Math Note

The standard quota for each district is the number of seats each district would be entitled to if fractional parts were allowed.

So the sum of all the standard quotas should work out to be the total number of seats. In this case, the sum is 20.01, which differs from the number of seats (20) due to a bit of rounding error.

Refer to the population chart for the city above. From Example 1, the standard divisor is 12. Therefore the standard quota for each district is as follows:

$$\text{For District 1: Standard quota} = \frac{\text{Population of district}}{\text{Standard divisor}} = \frac{32}{12} \approx 2.67.$$

$$\text{For District 2: Standard quota} = \frac{80}{12} \approx 6.67$$

$$\text{For District 3: Standard quota} = \frac{41}{12} \approx 3.42$$

$$\text{For District 4: Standard quota} = \frac{65}{12} \approx 5.42$$

$$\text{For District 5: Standard quota} = \frac{22}{12} \approx 1.83$$

Given that the standard quota for District 3 is 3.42, it should most definitely get either three or four seats.

Try This One 2

Find the standard quota for each of the districts in Try This One 1 and describe what it tells us for each district.

As we pointed out, fractional parts of seats don't make sense, so none of the standard quotas in Example 2 can be used as seat numbers. You might guess that numbers of seats would be found using the traditional rules of rounding: round up to the next whole number if the decimal part is 0.5 or more, and round down to the previous whole number if the decimal part is less than 0.5. Let's see how that would work in Example 2:

District	1	2	3	4	5	Total
Seats	3	7	3	5	2	20

This works out perfectly! If it always worked out that way, we would be done. Since the section isn't over, you can guess that it doesn't always work out that way.

Consider the following example. Ten computers (seats) are to be apportioned to three schools (districts) based on the enrollment, which is summarized in the following table.

School	A	B	C	Total
Enrollment	561	1,015	1,324	2,900

The standard divisor is 2,900/10 = 290. The standard quotas are shown below:

School	A	B	C	Total
Standard quota	1.93	3.5	4.57	

Math Note

For a given apportionment problem, if it just so happens that all of the standard quotas work out to be whole numbers, then you have a perfectly fair apportionment and life is good.

All of the methods we study in this section are unnecessary in that case. But in most real-life situations, it would be an astronomical coincidence for this to happen.

According to the traditional rounding rule, School A gets two computers, School B gets four computers, and School C gets five computers. But, 2 + 4 + 5 = 11 and there are only 10 computers available. That's bad.

Because of this issue, other methods of apportionment will be discussed that use the **upper quota** and **lower quota**. To calculate the upper quota, round the standard quota up to the next whole number (regardless of what comes after the decimal). To calculate the lower quota, round the standard quota down to the previous whole number.

Example 3	Finding Upper and Lower Quotas

Find the upper and lower quotas for the districts in Example 2. What do they describe?

SOLUTION

Using the standard quotas from Example 2, we round up to get the upper quota and round down to get the lower quota. The results are shown in the table below.

District	1	2	3	4	5	Total
Population (in thousands)	32	80	41	65	22	240
Standard quota	2.67	6.67	3.42	5.42	1.83	20.01
Lower quota	2	6	3	5	1	17
Upper quota	3	7	4	6	2	22

In each case, the upper and lower quotas are the only reasonable choices for the number of seats assigned to a given district. It remains to be seen how to decide whether to assign a certain district its upper or lower quota.

Try This One 3

Using the information from Try This One 2, find the upper and lower quotas for each district. Find the sum of the upper and lower quotas. How does each compare to the total number of seats?

✓ 1. Compute standard divisors and quotas.

As we've seen, the upper and lower quotas are the two whole numbers closest to the standard quota (which represents how many seats each district would be entitled to if fractional parts were allowed). So it seems perfectly reasonable that every district should be assigned either its upper or lower quota. But in Example 3, if the lower quota is used, only 17 seats are filled but there were 20 to be filled. Which district will get the extra seats? If the upper quota is used, 22 seats are needed, so we are short 2 seats. Who will forfeit a seat? The entire study of apportionment, then, comes down to deciding which districts get their upper quota and which get their lower quota. A variety of methods have been developed to answer these questions. In the remainder of the section, we'll study five of them.

Hamilton's Method

This method of apportionment was suggested by Alexander Hamilton and was approved by Congress in 1791. His method, however, was subsequently vetoed by President Washington. The method was later adopted by the U.S. government in 1852 and used until 1911. Hamilton's method was also known as the "method of largest remainders" which is a pretty good description. This method starts with lower quotas, which will almost always give us too few seats. At that point, we'll need to decide which districts get extra seats. Suppose that there are only two districts: one has a standard quota of 3.98 and the other has a standard quota of 6.05. Which would be more sensible: giving four seats to the district with standard quota 3.98, or giving seven seats to the district with standard quota 6.05? If you said, "Giving four seats to the district with standard quota 3.98," then you understand Hamilton's method. Good job.

Hamilton's Method

1. Calculate the standard divisor.
2. Calculate the standard quota for each district.
3. Calculate the lower quota for each district.
4. Assign each district the number of seats according to its lower quota.
5. Assign any surplus of seats one by one to the districts in descending order of the fractional parts of their standard quotas. The largest fractional part gets the first extra seat, then the next largest, and so on, until the correct number of seats is assigned.

| Example 4 | Using Hamilton's Method |

A large city with four districts plans to select 20 council members according to the population of the districts shown in the table. Using Hamilton's method, divide the 20 seats between the four districts.

District	1	2	3	4	Total
Population (in thousands)	150	88	17	65	320

Yuri_Arcurs/Getty Images

SOLUTION:

1. Calculate the standard divisor: $\dfrac{320 \text{ (Total population)}}{20 \text{ seats}} = 16$.

2. Calculate the standard quota for each district:

   District 1: $\dfrac{150 \text{ (Population of district)}}{16 \text{ (Standard divisor)}} = 9.375$

   District 2: $\dfrac{88}{16} = 5.5$

   District 3: $\dfrac{17}{16} = 1.0625$

   District 4: $\dfrac{65}{16} = 4.0625$

3. Calculate the lower quota for each district:

   District 1: 9

   District 2: 5

   District 3: 1

   District 4: 4

4. Initially assign the districts the number of seats equal to their lower quota above.
5. The sum of the lower quotas is 19, which is 1 short of the number of seats. So we have just one extra seat to assign. We look at the fractional parts of the standard quotas, which are 0.375, 0.5, 0.0625, and 0.0625. Of these, 0.5 is the biggest, so the district with fractional part 0.5 (District 2) gets the extra seat.

District	1	2	3	4	Total
Population (in thousands)	150	88	17	65	320
Seats	9	6	1	4	20

Math Note

A useful point: remember that the population numbers are actually thousands of people, but we're working just with the numbers in the table. The good news is that this has no effect on the quotas, and it makes our calculations easier.

Sidelight   **The (Finally) Famous Alexander Hamilton**

Until the award-winning musical bearing his name came to prominence, most Americans knew Alexander Hamilton as "that dude on the ten dollar bill," if they knew him at all. But you could make a case that Hamilton was as responsible for the founding of the United States as anyone. In particular, he was the main writer of the *Federalist Papers*, which were a driving force behind ratification of the United States Constitution (and are still often consulted by modern legal scholars). Lots of people are now aware of some of Hamilton's achievements, as well as his untimely demise in a pistol duel with former Vice President Aaron Burr. Here are some lesser-known facts about Hamilton.

- He was born on the tiny Caribbean island of Nevis, and his skill as a writer inspired fellow islanders to take up a collection to send him to the colonies to attend college. He was rejected by Princeton, but the school that later became Columbia University wisely admitted him. It's believed that he lied about his age at that time, making himself two years younger than he was to appear to be a young prodigy.

- Hamilton's noble service in the Revolutionary War allowed him to eventually advance to the position of *aide de camp* to George Washington, making him essentially our first president's right-hand man.

- The last letter that Washington wrote before he died went to Hamilton.

- Hamilton basically taught himself to be a lawyer in about 6 months.

- As first Secretary of the Treasury, he pretty much founded that department, and was largely responsible for building up a banking system that allowed the fledgling United States to become an economic power.

- Hamilton's oldest son Philip also died in a duel, three years to the day before Alexander was shot by Burr. Even stranger, the same pistols were used in both duels.

Try This One   4

2. Apportion seats using Hamilton's method.

An elementary school plans to assign 35 teachers to teach each of the grades K–5. The number of students in each grade is given in the following table. Use Hamilton's method to assign each grade level the appropriate number of teachers.

Grade	K	1	2	3	4	5	Total
Enrollment	123	87	144	72	199	75	700

Jefferson's Method

This method of apportionment is attributed to Thomas Jefferson and rivaled Alexander Hamilton's method. It was used in Congress from 1791 until 1842 when it was replaced with Webster's method. Jefferson's method uses a **modified divisor**, which is determined by trial and error and is smaller than the standard divisor. If the number of seats given by the lower quota is less than the number of seats available, we "modify" the standard divisor and recalculate the lower quotas with the modified divisor until the sum equals the number of available seats. Jefferson's method is summarized below.

Jefferson's Method

1. Calculate the standard divisor.

2. Calculate the standard quota for each district.

3. Calculate the lower quota for each district.

4. (a) If the sum of the lower quotas is equal to the number of seats, assign each district the number of seats equal to its lower quota.
   (b) If the sum of the lower quotas is less than the number of seats, choose a modified divisor by trial and error until the sum of the lower quotas equals the number of seats available. The modified divisor will have to be less than the standard divisor in order to raise the number of seats assigned. Assign each district the number of seats corresponding to its modified lower quota.

If the lower quotas don't equal the number of seats, the goal is to raise some of them a bit. In finding quotas, we're dividing by the standard divisor: to make the quotas bigger, we'll need to make the divisor smaller. So the problem of finding a modified divisor is a matter of repeatedly dividing by smaller numbers until the modified lower quotas sum to the correct value. To find the modified divisor we'll start by subtracting 0.5 from the standard divisor. If the number of seats is still too few, but close, we might subtract smaller amounts, like 0.2 or 0.1; if it's not close we'll subtract larger amounts. If the number of seats is too many, then we know we've subtracted too much.

Example 5	Using Jefferson's Method

A city has 15 seats to divide among three districts according to the population shown in the table. Find the appropriate modified divisor to be used in Jefferson's method.

District	1	2	3	Total
Population (in thousands)	86	191	52	329

SOLUTION

The standard divisor is 329 divided by 15, or 21.9. The standard and lower quotas are summarized in the table:

> **Math Note**
>
> Remember, we find the standard quota for each district by dividing its population by the standard divisor, and we find the lower quota by rounding each standard quota down to the nearest whole number.

District	1	2	3	Total
Population (in thousands)	86	191	52	329
Standard Quota	3.93	8.72	2.37	
Lower Quota	3	8	2	13

Since the number of seats is too few, try using a modified divisor. Subtracting 0.5 from 21.9, we get 21.4. The modified quotas are summarized below:

District	1	2	3	Total
Population (in thousands)	86	191	52	329
Modified quota	4.02	8.93	2.43	
Modified lower quota	4	8	2	14

This added one seat, but it's still not enough, since we need 15. Try subtracting 0.5 again and use 20.9. The new modified quotas are summarized below:

District	1	2	3	Total
Population (in thousands)	86	191	52	329
Modified quota	4.11	9.14	2.49	
Modified lower quota	4	9	2	15

Since the total of the lower quotas is now the number of seats we're shooting for, we can stop here and assign seats according to the last modified lower quota.

Try This One	5

3. Apportion seats using Jefferson's method.

A county with population given in the table plans to assign 30 representatives according to population. Use Jefferson's method to calculate how many representatives each district will get.

District	1	2	3	4	Total
Population (in thousands)	92	274	79	193	638

Adams' Method

Proposed by John Quincy Adams in 1822, this method was never used in Congress. Adams' method is similar to Jefferson's method, except it uses modified upper quotas rather than lower.

Example 6	Using Adams' Method

Assign the 15 seats from Example 5 using Adams' method. Does it result in a different apportionment? Are either the same as what Hamilton's method would assign?

SOLUTION

The population information is given in the table.

District	1	2	3	Total
Population (in thousands)	86	191	52	329

The standard divisor is 21.9. The standard and upper quotas are summarized below:

District	1	2	3	Total
Population (in thousands)	86	191	52	329
Standard quota	3.93	8.72	2.37	
Upper quota	4	9	3	16

The upper quotas assign 16 seats, which is one too many. Since we need to have *fewer* seats, we will *raise* the divisor in increments of 0.5 until the modified upper quotas add to 15. All of 22.4, 22.9, and 23.4 leave the upper quotas unchanged, but for 23.9 we get the results below:

District	1	2	3	Total
Population (in thousands)	86	191	52	329
Modified quota	3.60	7.99	2.18	
Modified upper quota	4	8	3	15

Now we can assign the seats according to the modified upper quotas. This is in fact a different result than Jefferson's method, which assigned 9 seats to District 2 and 2 to District 3. The lower quotas added to 13 seats, so Hamilton's method would have assigned one more seat to the two with the larger fractional parts, which would be Districts 1 and 2. This results in the same apportionment as Jefferson's method.

Math Note

In Problems 39–42, we'll study the effects of using modified upper quotas (Adams) rather than modified lower quotas (Jefferson).

But the short answer is that Jefferson's method is more likely to assign extra seats to larger districts, while Adams' favors smaller districts.

This is the case in Examples 5 and 6: the smallest district gets one more seat using Adams' method, and the largest gets one more using Jefferson's.

Try This One	6

Assign the 30 seats from Try This One 5 using Adams' method.

✓ 4. Apportion seats using Adams' method.

Webster's Method

Daniel Webster's method of apportionment was used in Congress from 1842 until Hamilton's method was adopted in 1852. Webster's method was readopted in 1911 and was used until 1941, at which time it was replaced by the Huntington-Hill method, which was signed into law by President Roosevelt in 1941. The Huntington-Hill method is still in use in Congress today. Webster's method is a cross between Hamilton's and Adams' methods and is summarized below.

Webster's Method

1. Calculate the standard divisor.

2. Calculate the standard quota.

3. Calculate the lower and upper quotas.

4. To find the initial assignment, use standard rounding rules on the standard quota: round up if the fractional part is 0.5 or greater, and round down if it's less than 0.5.

5. Check to see if the sum of the seats is equal to the number of seats available. If so, use the assignments from step 4. If not, use a modified divisor and reassign seats based on the criteria in step 4, then repeat step 5. (If there are too many seats, try a larger divisor. If there are too few seats, try a smaller divisor.)

Example 7   **Using Webster's Method**

Use Webster's method to assign 27 seats according to population for the following city with four districts.

District	1	2	3	4	Total
Population (in thousands)	149	83	92	126	450

SOLUTION

The standard divisor is $450/27 = 16.7$. The corresponding standard quotas and initial assignments of seats are summarized in the table.

District	1	2	3	4	Total
Population (in thousands)	149	83	92	126	450
Standard Quota	8.92	4.97	5.51	7.54	
Initial Assignments	9	5	6	8	28

We have one seat too many, so we should try a larger divisor. We're pretty close, so we try a modified divisor of 16.8:

District	1	2	3	4	Total
Population (in thousands)	149	83	92	126	450
Modified quota	8.87	4.94	5.48	7.5	
Modified rounded quota	9	5	5	8	27

Now we can assign the seats according to the modified rounded quota.

Math Note

Notice that in this case, it was neither the largest nor the smallest district that had its assignment affected by the modified divisor. In that sense, Webster's method is in between Adams' and Jefferson's.

5. Apportion seats using Webster's method.

Comstock Images/Alamy Stock Photo

Try This One   **7**

A trucking company wants to assign 14 new trucks to four districts with volume of business by truckload indicated in the table. Use Webster's method to apportion the trucks to the districts according to volume of business.

District	1	2	3	4	Total
Volume (by truckload)	230	145	214	198	787

The Huntington-Hill Method

The Huntington-Hill method is the one currently used to apportion congressional seats. It was developed around 1911 by Joseph Hill, the chief statistician for the Bureau of the Census, and Edward V. Huntington, professor of mechanics and mathematics at Harvard. This method of apportionment uses the *geometric mean*.

> The **geometric mean** of two numbers x and y is given by $\sqrt{xy}$.

For example, the geometric mean of 2 and 8 is $\sqrt{2 \cdot 8} = \sqrt{16} = 4$. The geometric mean of 15 and 18 is $\sqrt{15 \cdot 18} = \sqrt{270} \approx 16.43$. The Huntington-Hill method uses the geometric mean to assign seats as follows:

> **The Huntington-Hill Method**
>
> 1. Calculate the standard divisor.
> 2. Calculate the standard, lower, and upper quotas and the geometric mean of the lower and upper quotas for each district.
> 3. Initially assign the lower quota if the standard quota is less than the geometric mean of the upper and lower quotas; assign the upper quota if the standard quota is greater than the geometric mean of the upper and lower quotas.
> 4. If the sum of the seats assigned in step 3 is equal to the number of seats available, leave seats assigned according to step 3. If the sum of the seats assigned in step 3 is not equal to the number of seats available, use a modified divisor as in the previous three methods and reassign seats accordingly until the sum equals the total number of seats available.

| **Example 8** | **Using the Huntington-Hill Method** |

A club with 400 members is to assign 10 chair positions to represent the members coming from various communities. The distribution of the members is as follows:

Community	A	B	C	Total
Number of members	153	75	172	400

Use the Huntington-Hill method to assign the 10 seats.

SOLUTION

The standard divisor is $400/10 = 40$. The standard, upper, and lower quotas are given in the table.

Community	A	B	C	Total
Number of members	153	75	172	400
Standard quota	3.825	1.875	4.3	
Lower quota	3	1	4	
Upper quota	4	2	5	

Now we compute the geometric means of the lower and upper quotas:

Community A: $\sqrt{3 \cdot 4} = \sqrt{12} \approx 3.46$
Community B: $\sqrt{1 \cdot 2} = \sqrt{2} \approx 1.41$
Community C: $\sqrt{4 \cdot 5} = \sqrt{20} \approx 4.47$

Next, compare each community's standard quota to its geometric mean. Communities A and B each have a standard quota greater than their geometric mean, so we assign the upper quotas to each: 4 seats to A and 2 seats to B. Community C's standard quota is less than its geometric mean, so we assign the lower quota of 4 to community C.

This gives us a total of $4 + 2 + 4 = 10$ seats, which is what we wanted. So there's no need to try a modified divisor.

✓ 6. Apportion seats using the Huntington-Hill method.

Try This One   8

A university math department hired five new teaching assistants to be assigned to three professors according to the total number of students each professor teaches, which is indicated in the table. Use the Huntington-Hill method to assign the five TAs to the professors.

Professor	Hunt	Tingue	Hill	Total
Number of students	200	250	145	595

So why is Huntington-Hill the preferred apportionment method used by Congress today? There's a simple answer and a long answer. The simple answer is that the previous methods were devised by politicians and statesmen using, to some extent, trial and error. In 1929, Congress became fed up with the controversies caused by apportionment (when states lose a seat, they tend to get upset) and decided to turn to mathematicians to devise a better system.

Which leads to the long answer. There is a variety of measures of fairness that have been devised to rate apportionment methods. Among them are *absolute unfairness* and *relative unfairness* (which you can learn about using a Google search if you're interested). It has been shown that the Huntington-Hill method is best for minimizing the relative unfairness in apportionment. It's interesting to note that Congress, acting with typical speed, took 12 years to finally enact Huntington-Hill as the permanent apportionment method for the House.

Maybe you're wondering what the big deal is: in the examples we've seen, there's not a whole lot of difference (if any) between the results of different apportionment methods. But keep in mind that the examples we're looking at involve a relatively small number of seats and districts. When apportionment began, with the House of Representatives in 1791, there were 14 states and 69 representatives to apportion. That's quite a bit more complex than the examples we've seen, but it's almost trivial compared to the current situation—435 seats assigned to 50 states. When the situation is that complex, and the stakes are as high as state representation in the government, changing the method can have dramatic and controversial results.

Answers to Try This One

1 (a) 320,000; this tells us that on average districts should get 1 seat for every 320,000 residents.

(b) Notice that the population of District 4 is 970,000 (which is 97 times 10,000). This is almost exactly three times as big as the standard divisor. So if there's supposed to be one seat for every 320,000 residents, that would correspond to three seats for every 960,000 residents. So it seems reasonable that District 4 should get three seats.

2

District	1	2	3	4	5	6	7
Standard quota	3.91	2.78	7.34	3.03	3.19	4	5.75

In each case, the standard quota tells us how many seats each district would be allocated if fractional parts were allowed.

3

District	1	2	3	4	5	6	7
Lower quota	3	2	7	3	3	4	5
Upper quota	4	3	8	4	4	4	6

The sum of the lower quotas is 27 and the sum of the upper quotas is 33. The lower quota sum is three less than the correct number of seats, and the upper quota sum is three more.

4

Grade	K	1	2	3	4	5
Teachers	6	4	7	4	10	4

5 Using modified divisor 19.7:

District	1	2	3	4
Representatives	4	13	4	9

6 Using modified divisor 22.9:

District	1	2	3	4
Representatives	5	12	4	9

7 Using modified divisor 56.7:

District	1	2	3	4
Trucks	4	3	4	3

8

Professor	Hunt	Tingue	Hill
TAs	2	2	1

Exercise Set 12-4

Writing Exercises

1. Describe what apportionment is in your own words. When is it needed?
2. Explain how to find the standard divisor for an apportionment process, and describe its significance.
3. Why can we not simply find the standard quota and use traditional rules of rounding to get an apportionment?
4. Describe how to find the upper and lower quotas from a standard quota.
5. Which apportionment methods use a modified divisor? How do those methods differ?

6. If an apportionment using upper or lower quotas results in too few seats, should you raise or lower the standard divisor? Explain your answer.
7. What exactly does the standard quota represent? How do we know that?
8. What happens in apportionment if all of the standard quotas work out to be whole numbers?

Computational Exercises

In Exercises 9–12, find the standard divisor for each district.

9.

District	1	2	3
Population	41	19	86

; 5 seats

10.

District	A	B	C
Population	112	89	241

; 12 seats

11.

District	A	B	C	D
Population	109	94	53	122

; 20 seats

12.

District	1	2	3	4
Population	30	19	11	12

; 8 seats

In Exercises 13–16, find the standard quota, the upper quota, and the lower quota for each district from the given exercise.

13. Exercise 9
14. Exercise 10
15. Exercise 11
16. Exercise 12

In Exercises 17–22, find the geometric mean for the pair of numbers.

17. 2 and 8
18. 5 and 20
19. 8 and 9
20. 3 and 4
21. 16 and 24
22. 19 and 30

Applications in Our World

For Exercises 23–25 find:

(a) The standard divisor.
(b) The standard quota for each district.
(c) The upper and lower quotas for each district.
(d) Assignment of seats using Hamilton's method.

23. This year 10 faculty members are to receive promotions at a community college with three campuses. The number of promotions will be apportioned according to the number of full-time faculty members as shown. Determine how many promotions there will be on each of the campuses.

Campus	South	Central	North	Total
Faculty	62	148	110	320

24. A large school district has four high schools with enrollments as shown. Forty school buses need to be allocated to the high schools. Determine how many buses are allocated to each of the high schools.

School	RRHS	CCHS	HHHS	MHS	Total
Enrollment	628	941	1,324	872	3,765

25. A county has three districts; the population of each district in thousands is shown. Nine representatives are to be appointed to the county board. Determine how many representatives each district should receive.

District	1	2	3	Total
Population (in thousands)	242	153	185	580

For Exercises 26–28 find:

(a) The standard divisor.
(b) The standard quota for each district.
(c) The upper and lower quotas for each district.
(d) Assign seats using Jefferson's method and Adams' method.
(e) Write the modified divisor you used for each method.

26. A large chain store needs to assign 12 buyers to its five stores. The number of employees for each store is shown. Determine how many buyers should be assigned to each store.

Store	1	2	3	4	5	Total
Employees	56	32	74	62	86	310

27. A freight company has four terminals. The volume of business in truckloads for each terminal is shown. The owner has purchased 12 new trucks to be allocated to the four terminals. Determine how many trucks each terminal should get based on volume of business.

Terminal	A	B	C	D	Total
Volume (in truck loads)	51	37	65	22	175

28. A large library has six branches; the number of books at each branch is shown. The library has obtained 100 new books. Determine how many books each library should get based on their current volume of books.

Branch	A	B	C	D	E	F	Total
Books	1,236	872	2,035	1,655	3,271	1,053	10,122

For Exercises 29–31, find:

(a) The standard divisor.
(b) The standard quota for each district.
(c) The upper and lower quotas for each district.
(d) Assign seats using Webster's method.
(e) Write the modified divisor you used.

29. An automotive repair company has four stores in a large city. The owner bought 10 diagnostic computers for the business. The number of cars serviced per store each month is given below. Determine how many computers each store should get based on number of cars serviced.

Store	1	2	3	4	Total
Cars serviced	119	99	186	136	540

30. A hospital has five wards; the average number of patients per ward is shown. Twelve new nurses are hired. Determine how many nurses each ward should receive based on average number of patients.

Ward	A	B	C	D	E	Total
Patients	372	465	558	619	197	2,211

31. Sixteen new police officers are to be assigned to four precincts. The populations of the precincts are shown. Determine how many of the 16 new officers each precinct should receive.

Precinct	1	2	3	4	Total
Population	3,562	8,471	2,146	5,307	19,486

For Exercises 32–34, find:

(a) The standard divisor.
(b) The standard quota for each district.
(c) The upper and lower quotas for each district.
(d) Assign seats using the Huntington-Hill method.
(e) Write the modified divisor you used.

32. A county has three clinics; the weekly number of patients seen by a nurse practitioner in each clinic is shown. The director hires eight new nurse practitioners; assign them to the three clinics based on number of patients seen.

Clinic	A	B	C	Total
Patients	456	616	573	1,645

33. A chain of physical therapy clinics has four offices; the average number of patients treated per month at each office is shown. Six new therapists are hired and apportioned to the offices according to number of patients seen. How many therapists should each office receive?

Office	1	2	3	4	Total
Patients	2,231	996	1,564	1,799	6,590

34. A department store is hiring 10 new sales associates and needs to assign them to four departments according to volume of sales in dollars. The daily sales in dollars are given. Decide how many associates each department should receive.

Department	A	B	C	D	Total
Sales	5,562	4,365	4,012	6,531	20,470

Critical Thinking

35. Describe a situation related to your campus where apportionment would be useful. Describe what things would play the roles of "total population," "seats," and "districts."

36. (a) Suppose that a company with 500 employees plans to choose representatives from each floor in its headquarters for a council to study environmental issues within the company. There will be 10 representatives total. The number of employees per floor is shown in the table. Explain why this ends up being a really, really easy apportionment problem.

Floor	1	2	3	4
# of employees	100	150	150	100

   (b) Describe a general condition under which apportionment always works out to be this easy.

37. (a) Devise your own apportionment method for dealing with the situation where each district's standard quota is not a whole number, and rework Exercises 27 and 29 using your method. Does your method provide a different apportionment than the others in either problem? (Notice that we didn't say your method had to necessarily be completely fair . . .)

   (b) Discuss whether or not you think your method is fair, giving reasons for your opinion.

38. After a year in which tax revenues are higher than budgeted, a state decides to divide the surplus among four state universities based on the enrollment at each school. Explain why this is *not* an apportionment problem.

In Problems 39–42, we'll study the effect of using upper or lower quotas with modified divisors in Adams' and Jefferson's methods.

39. (a) In Jefferson's method, the original assignment is based on lower quotas, which will almost always lead to leftover seats to be assigned. Fill in the blanks: a district with a modified quota that goes _____ more quickly is _____ likely to get another seat.

   (b) Fill in the following table and use it to decide if a large or small district is more likely to have its modified quota change a lot when modifying the divisor.

Population	20	40	60	80
Divisor 2.5				
Divisor 2.3				
Divisor 2.7				

40. Based on the results of Problem 39, explain why Jefferson's method tends to favor larger districts.

41. In Adams' method, the original assignment is based on upper quotas, which typically leads to too many seats assigned originally. Fill in the blanks: a district with a modified quota that goes _____ more quickly is _____ likely to lose a seat.

42. Use the results of Problems 39(b) and 41 to explain why Adams' method tends to favor smaller districts.

43. Of the five apportionment methods we studied, which do you think is the most fair? Explain your answer.

Now the four largest emotional hurts belong to Coalition BTO, 1, 3 and E, so these are the fractions that get the extra seat. This ends up being near-magically Hamilton's formula in the

---

Section 12-5 | Apportionment Flaws

LEARNING OBJECTIVES

1. Illustrate the Alabama paradox.
2. Illustrate the population paradox.
3. Illustrate the new states paradox.
4. Describe the quota rule.

The U.S. Constitution requires the House of Representatives to be reapportioned after each census. Except for a 4-year period after Alaska and Hawaii joined the union, the number of representatives has been fixed at 435 since 1911. But in the 1800s, that number was very fluid as the country experienced rapid population growth. Hamilton's method had been in use to apportion seats since 1852, but after the 1880 census, something funny happened. The chief clerk of the Census Bureau computed apportionment for all House sizes between 275 and 350 (by hand!), and noticed that if there were 299 seats in the House, Alabama was apportioned 8 seats. But if there were 300 seats, Alabama was only apportioned 7. In other words, adding a seat to the House would cause Alabama to *lose* a seat! This is known as a *paradox*—a conclusion that makes no sense even though it was drawn from a sound logical process. This particular result became known as the *Alabama paradox*.

Library of Congress Prints and Photographs Division [LC-DIG-pga-04078]

The Alabama Paradox

An increase in the total number of seats to be apportioned causes a district to lose a seat.

Example 1 illustrates the Alabama paradox.

| Example 1 | Illustrating the Alabama Paradox |

Creatas Images/Getty Images

A county has six districts populated as shown; 15 seats are to be apportioned according to population using Hamilton's method. The county then decides to add a representative and reapportion the seats. Does the Alabama paradox occur? Explain.

District	A	B	C	D	E	F	Total
Population (in thousands)	39	220	264	163	167	101	954

SOLUTION

First we need to find the apportionment with the original 15 seats. By Hamilton's method we get:

District	A	B	C	D	E	F	Total
Population (in thousands)	39	220	264	163	167	101	954
Standard quota	0.61	3.46	4.15	2.56	2.63	1.59	
Lower quota	0	3	4	2	2	1	12
Apportionment	1	3	4	2	3	2	15

Districts A, E, and F get the extra seats because their standard quotas have the three largest fractional parts.

When the 16th seat is added the result is:

District	A	B	C	D	E	F	Total
Population (in thousands)	39	220	264	163	167	101	954
Standard quota	0.65	3.69	4.43	2.73	2.80	1.69	
Lower quota	0	3	4	2	2	1	12
Apportionment	0	4	4	3	3	2	16

Now the four largest fractional parts belong to Districts B, D, E, and F, so those are the districts that get the extra seat. This ends up being an especially egregious example of the Alabama paradox: not only did District A lose a seat when one more seat was added, it lost its ONLY SEAT? Ouch.

Try This One 1

1. Illustrate the Alabama paradox.

A large company decided to donate 17 computers to West Hills school district. The director of the computer services decided to allot them to the three middle schools based on number of students, using Hamilton's method. However, when the computers arrived there were actually 18 computers that were then reallocated. Does the Alabama paradox occur? Explain.

School	JDM	DM	AM	Total
Number of students	1,328	1,008	384	2,720

At the time of its discovery, the Alabama paradox was overcome by choosing a number of seats for which the paradox didn't seem to come into play, not exactly a long-term solution.

Another flaw in Hamilton's method was discovered in 1900. At that time, Virginia was growing at a rate about 60% faster than Maine was, but Virginia lost a seat, while Maine gained one. This is known as the *population paradox*.

> **The Population Paradox**
>
> An increase in a district's population can cause it to lose a seat to a slower growing district.

Example 2	Illustrating the Population Paradox

A county has three districts with populations shown. There are to be 120 representatives elected to represent the three districts. A year later the population increases in two of the districts as indicated, and the seats are reapportioned. If Hamilton's method is used, does the population paradox occur?

District	A	B	C	Total
Original population	29,317	106,350	14,333	150,000
New population	29,604	106,350	14,477	150,431

SOLUTION

For the original population we have the following:

District	A	B	C	Total
Original population	29,317	106,350	14,333	150,000
Standard quota	23.45	85.08	11.47	
Lower quota	23	85	11	119
Apportionment	23	85	12	120

C has the largest fractional part.

Now let's see if the apportionment changes with the new population. If not, then no paradox occurs.

District	A	B	C	Total
New population	29,604	106,350	14,477	150,431
Standard quota	23.62	84.84	11.55	
Lower quota	23	84	11	118
Apportionment	24	85	11	120

C has the smallest fractional part.

Aha! The apportionment did change: District A gained a seat and District C lost one. To decide if the paradox occurred, we need to find what the percentage of increase was for each of those two districts. (Notice that District B didn't change population at all.)

$$\text{District A: } \frac{\text{Change in population}}{\text{Original population}} = \frac{29,604 - 29,317}{29,317} \approx 0.0098 \Rightarrow 0.98\% \text{ growth}$$

$$\text{District C: } \frac{\text{Change in population}}{\text{Original population}} = \frac{14,477 - 14,333}{14,333} \approx 0.0100 \Rightarrow 1\% \text{ growth}$$

It's close, but District C had a greater growth rate, but still lost a seat. So the population paradox did occur.

2. Illustrate the population paradox.

Try This One   2

A county with three districts has populations as shown. Six representatives are to be elected and apportioned to the districts according to population. Five years later the population increases as indicated and the six representatives are reapportioned. Did the population paradox occur using Hamilton's method?

District	A	B	C	Total
Original population (in thousands)	135	253	572	960
New population (in thousands)	173	420	927	1,520

Hamilton's method somehow managed to survive the discovery of both the Alabama and population paradoxes, but another problem was discovered in 1907 when Oklahoma joined the union. At that time there were 386 seats in the house, and it was clear based on its population that Oklahoma should get 5 seats, so the number of seats was increased to 391. The idea was Oklahoma would get the five new seats and the other states would retain the seats they had. But as you might have guessed, that didn't happen. When the 391 seats were reapportioned mathematically, Maine gained a seat and New York lost a seat. This paradox has become known as the *new states paradox*.

The New States Paradox

Adding a state with its fair share of seats can affect the number of seats due other states.

Example 3   Illustrating the New States Paradox

A company wants to distribute 10 new computers to three branch offices with number of employees as shown.

Branch	A	B	C	Total
Employees	21	146	204	371

Before the computers are delivered, the company opens a new branch office with 26 employees. Based on the apportionment of the original 10 computers, the company president decides the new branch should get one computer, so one more is ordered for that branch. If the president reapportions the 11 computers to the four branch offices using Hamilton's method, does the new states paradox occur?

SOLUTION

The original apportionment would be as follows:

Branch	A	B	C	Total
Employees	21	146	204	371
Standard quota	0.57	3.94	5.50	
Lower quota	0	3	5	8
Apportionment	1	4	5	10

The biggest fractional parts are branches A and B, so they get the extra computers.

The new apportionment with the addition of branch D is as follows:

Branch	A	B	C	D	Total
Employees	21	146	204	26	397
Standard quota	0.58	4.04	5.65	0.72	
Lower quota	0	4	5	0	9
Apportionment	0	4	6	1	11

This time the biggest fractional parts are branches C and D, so they get the extra computers. Now we can see that the new states paradox does occur: adding the extra branch and one more computer caused branch A to lose its only computer, and branch C to get an extra one.

Try This One   3

3. Illustrate the new states paradox.

A city has two districts with populations as shown. There are 100 representatives between the two districts.

District	A	B	Total
Population	1,545	8,455	10,000

(a) Apportion the representatives according to population using Hamilton's method.
(b) A new district is annexed with population 625, and correspondingly 6 new representatives are added. Reapportion the 106 seats using Hamilton's method. Does the new states paradox occur?

Just as there are fairness criteria for voting methods, there are fairness criteria for apportionment methods. In a fair method, none of the three paradoxes should occur, and the method should also satisfy the *quota rule*.

Math Note

Hamilton's method survived the first two paradoxes, but the discovery of the new states paradox in 1907 dealt it a death blow. After the census of 1910, it was replaced with Webster's method in 1911.

The Quota Rule

Every district in an apportionment should be assigned either its upper quota or its lower quota. An apportionment that violates the rule is said to *violate quota*.

The obvious question to close with is, "Is there an apportionment method that is completely fair?" The answer is no. We have seen that Hamilton's method can violate all of the paradoxes, but it satisfies the quota rule (since by definition every district gets either its upper or lower quota). It turns out that the other four apportionment methods we studied in Section 12-4 are immune from the paradoxes (which is not at all obvious), but they all use modified divisors, so they occasionally violate quota. (Very occasionally, as it turns out—none of the examples in Section 12-4 violate quota.) In fact, mathematicians Michael L. Balinski and H. Peyton Young proved in 1980 that any apportionment method that doesn't violate the quota rule must produce paradoxes, and any method that never produces paradoxes must violate the quota rule.

4. Describe the quota rule.

In summary, there is no perfect method for either voting or apportionment: it is always left to the discretion of the people involved to decide on a method that is most fair for their needs.

Answers to Try This One

1 The original apportionment is 8 for JDM, 6 for DM, and 3 for AM. The new apportionment is 9 for JDM, 7 for DM, and 2 for AM. AM lost a computer because of the district apportioning one more, so the Alabama paradox does occur.

2 The original apportionment is 1 for A, 2 for B, and 3 for C. The new apportionment is 1 for A, 1 for B,

and 4 for C. But B's growth rate was 66% compared to 62% for C, yet B lost a seat to C. The population paradox occurs.

3 The original apportionment is 15 for A, 85 for B. The new apportionment is 16 for A, 84 for B, and 6 for C. B lost a seat because C was added. The new states paradox occurred.

Exercise Set   12-5

Writing Exercises

1. What is a paradox?
2. Describe the Alabama paradox in your own words.
3. Describe the population paradox in your own words.
4. Describe the new states paradox in your own words.

5. What is the quota rule? Which apportionment methods can violate it?
6. Explain why Hamilton's method can't violate the quota rule.

Applications in Our World

7. A county library system has three branches. The number of books at each branch is shown. The head librarian bought 57 new books and apportioned them to the libraries using Hamilton's method. When the book order arrived, the librarian noticed there was an extra book added as a bonus, and she reapportioned the books with the extra book added. Determine if the reapportionment resulted in the Alabama paradox.

Branch	A	B	C	Total
Books	1,556	1,328	365	3,249

8. A school district hired nine elementary teachers to be apportioned to their four elementary schools according to enrollment, which is shown. By mistake the director called 10 of the applicants and told them they were hired! He now needs to reapportion the 10 teachers to the four schools. Determine if the Alabama paradox occurs if the teachers are assigned to schools by using Hamilton's method.

School	A	B	C	D	Total
Enrollment	876	586	758	103	2,323

9. Six computers were donated to be distributed to three local churches according to membership. When the computers arrived at the denomination's main office, there were actually seven computers. If Hamilton's method was used to decide how to distribute the computers among the churches, determine if the Alabama paradox occurs.

Church	Saint A's	Saint B's	Saint C's	Total
Membership	937	114	622	1,673

10. A city with three districts has an increase in population as indicated in the table. Using Hamilton's method, apportion the 24 seats between the three districts both before and after the population increase. Find the percent increase in population for each district, and determine if the population paradox occurs.

District	A	B	C	Total
Original population (in thousands)	988	530	2,242	3,760
New population (in thousands)	1,248	677	2,571	4,496

11. A small business has three branch offices with increase in number of employees indicated. Thirty new desks are to be apportioned according to number of employees at a given branch. Apportion the 30 desks both before and after the increase in number of employees and determine if the population paradox occurs.

Branch	1	2	3	Total
Original number of employees	161	249	490	900
New number of employees	177	275	514	966

12. Eleven legislative seats are to be apportioned to a country's three states according to population using Hamilton's method. Five years later the 11 seats are reapportioned with the indicated increase in population. Determine if the population paradox occurs when the seats are reapportioned.

State	A	B	C	Total
Original population	55	125	190	370
New population	60	150	213	423

13. The table shows the enrollment at two campuses of a community college.

Campus	North	South	Total
Enrollment	1,326	8,163	9,489

    (a) Apportion 61 new media podiums to the two campuses according to enrollment using Hamilton's method.
    (b) Suppose the college adds a third campus with enrollment of 1,070 and correspondingly adds seven podiums. Reapportion the 68 podiums using Hamilton's method.
    (c) Determine if the new states paradox occurs.

14. A state with 40 legislative seats has two districts with populations as follows.

District	A	B	Total
Population	153	787	940

    (a) Apportion the seats using Hamilton's method.
    (b) The state adds a third district with population 139 and adds 6 new seats accordingly. Reapportion the seats using Hamilton's method.
    (c) Determine if the new states paradox occurs.

15. A small island nation has 97 legislative seats to be divided among three states with population as follows.

State	A	B	C	Total
Population	2,163	9,504	8,133	19,800

(a) Apportion the seats using Hamilton's method.
(b) The country adds a state with population 2,629 and adds 13 seats accordingly; reapportion the seats using Hamilton's method.
(c) Determine if the new states paradox occurs.

Critical Thinking

16. For the city in Example 4 of Section 12-4, see if you can find an original and new number of seats that would result in the Alabama paradox.

17. Write an essay explaining why many people feel that any apportionment method that violates quota for a particular situation should be thrown out. (You should probably focus on where the upper and lower quotas come from.)

18. Which do you think is more serious: violating quota or one of the paradoxes occurring? Explain your answer.

19. Using the Internet or a library as a resource, do some research on the 1876 presidential election and describe how apportionment decided who would be president of the United States.

Section	Important Terms	Important Ideas
12-1	Preference table Plurality method Head-to-head comparison criterion	**This chapter** presented four voting methods that can be used to determine the results of an election where there are three or more choices (i.e., candidates, courses of action, etc.). These methods involve ranking the choices in order of preference on a preference ballot. A preference table can be used to summarize the results of the election. Using the plurality method, the candidate with the most first-place votes is the winner. Each method has at least one inherent flaw. That is, it fails to satisfy one of the fairness criteria for a voting method. The head-to-head criterion states that if a candidate wins all head-to-head comparisons with all other candidates, that candidate should win the election.
12-2	Borda count method Majority criterion Plurality-with-elimination method Monotonicity criterion	**In the Borda count method,** candidates are ranked on the ballot by voters. The candidate in last place on the ballot gets 1 point. The candidate in the next to the last place gets 2 points, and so on. The points are tallied and the candidate with the most points is the winner. In the plurality-with-elimination method the candidate with a majority of first-place votes is the winner. If no candidate has a majority of first-place votes, the candidate with the least number of votes is eliminated, and all candidates who were below the eliminated candidate move up. The first-place votes are counted again. The process then is repeated until a candidate receives a majority of first-place votes. The majority criterion says that if a candidate receives a majority of first-place votes, then that candidate should be the winner. The monotonicity criterion states that if a reelection is held with the same candidates and the only changes in voting favor the winner of the original election, then that winner should win the reelection.
12-3	Pairwise comparison method Irrelevant alternatives criterion Arrow's impossibility theorem Approval voting	**In the pairwise comparison method**, each candidate is ranked by the voters. Then each candidate is paired with every other candidate in a head-to-head contest by counting the number of ballots that has each listed higher. The winner of each contest gets 1 point. In case of a tie, each candidate gets $\frac{1}{2}$ point. The candidate with the most points wins the election. In the late 1970s several political scientists devised approval voting. Each voter votes for as many candidates as they find acceptable. The votes are tabulated and the candidate who receives the most votes is declared the winner. The irrelevant alternatives criterion states that if candidate A wins a certain election and if one of the other candidates is removed from the ballot and the ballots are recounted, candidate A should still win the election. The search for a perfect voting method continued until the early 1950s when an economist, Kenneth Arrow, proved that it's impossible to design a voting method for three or more candidates that doesn't violate at least one of the four fairness criteria. This fact is known as Arrow's impossibility theorem.

Section	Important Terms	Important Ideas
12-4	Indivisible Apportionment Standard divisor Standard quota Upper and lower quotas Hamilton's method Jefferson's method Modified divisor Adams' method Webster's method The Huntington-Hill method Geometric Mean	**Apportionment** is the process of fairly assigning objects among individuals that may be entitled to unequal shares. The standard quota for a district is the number of seats it would be entitled to if fractional parts were allowed. The lower quota is the standard quota rounded down to the nearest whole number, while the upper quota comes from rounding the standard quota up. Hamilton's method uses lower quotas. It initially assigns the lower quota to all districts, then assigns any leftover seats one at a time based on the size of the fractional part of the standard quota. Jefferson's method also begins with lower quotas, but if there are unassigned seats, lower quotas are recalculated using a modified divisor. Adams' method is similar, but uses upper quotas. Webster's method uses quotas rounded using traditional rounding rules, and uses a modified divisor if the number of seats apportioned is incorrect. The Huntington-Hill method is similar to Webster's method, but uses comparison to the geometric mean of the upper and lower quotas rather than traditional rounding rules.
12-5	Paradox Alabama paradox Population paradox New states paradox Quota rule	**Hamilton's method** can exhibit three different paradoxes. The Alabama paradox occurs when adding one or more seats to an apportionment causes a district to lose a seat. The population paradox occurs when a district loses a seat to another district that had a smaller percentage growth. The new states paradox occurs when adding a new district with an appropriate number of seats causes an existing district to lose a seat. The quota rule says that for an apportionment process to be fair, every district should be apportioned either its upper or lower quota. Any apportionment method that is immune to paradoxes can violate the quota rule, and any method that never violates the quota rule can exhibit one or more of the paradoxes.

Math in College Football REVISITED

Peter Joneleit/Zuma Wire Service/ZUMA Press/Alamy Stock Photo

1. Using the Borda count method, but awarding only 3 points for first place, 2 for second, and 1 for third, the results were

B	1,726 points
C	1,604 points
A	1,575 points
D	213 points
E	116 points

   Player B won the award even though he had fewer first place votes than player A.

2. I can't tell you what opinion you should have: your thoughts on whether or not you thought the election was fair are probably based on how you feel about first-place votes compared to overall favorability. A had nine more first-place votes than B, but B was in the top two on almost 100 more ballots, which is what won the award for him.

3. Since A had the most first-place votes, he would have won if the plurality method had been used. (A had the most first-place votes and didn't even finish second!)

4. The question states that 902 ballots were returned. The sum of votes in the "First" column is 891. So there were 11 voters whose first choice didn't even make the top five.

5. Because of the large number of ballots, and the many possible combinations, the voting information was not listed in preference table form, so it's difficult to determine if the fairness criteria were violated. The head-to-head comparison criterion may have been violated, since A had the most first-place votes. But the fact that he was listed third on the highest number of ballots means that one of the top two finishers may beat him head-to-head, or maybe even both. The majority criterion was definitely not violated since nobody got a majority of first-place votes. It's hard to guess if the monotonicity criterion was violated, but because of the closeness of the election, it's possible that if some ballots that had B third were changed to ballots that had him second, it could have cost him the election if a large majority of those ballots moved one of C or A to first. It's possible, and maybe even likely, that the irrelevant alternatives criterion was violated. The removal of the candidates not in the top three would have added more votes to those in the top three, and there's most likely some combination that would tip the election in favor of C or A.

6. The year was 2008, and player A, who came in third in spite of having the most first-place votes, was Tim Tebow of Florida. The winner was Oklahoma's Sam Bradford, and second place went to Colt McCoy of Texas. The odd voting pattern was strongly related to regional bias: the vast majority of voters in the southeast had Tebow first, but the other two top candidates were from the southwest, where most of the voters had them listed first or second in some order.

Review Exercises

Sections 12-1 to 12-3

Use this information for Exercises 1–4: the preference ballots for an election for the best speaker in a contest are shown. There are three candidates: Peterson (P), Quintana (Q), and Ross (R).

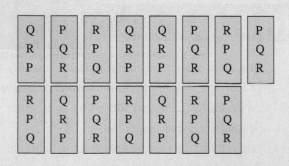

1. Build a preference table for the results of the election.
2. How many people voted in the election?
3. How many people voted for Ross as the best speaker?
4. How many people voted for the contestants in the order Quintana, Ross, Peterson?

Use this information for Exercises 5–8: a class of students decided to have a pizza party on the last day of math class. They voted to pick a pizza place. The candidates were Pizza Palace (P), Pizza Heaven (H), and Pizza City (C). The preference ballots are shown.

C	P	H	H	H	H	C	P
P	C	P	P	P	P	P	C
H	H	C	C	C	C	H	H

H	H	H	C	C	P	H	C
P	P	P	P	P	C	P	P
C	C	C	H	H	H	C	H

P	H	C	H
C	P	P	P
H	C	H	C

5. Build a preference table for the results of the election.
6. How many people voted?
7. How many people voted for Pizza City as their first choice?
8. How many people voted for the Pizza Palace, Pizza City, and Pizza Heaven in that order?

Use this information for Exercises 9–17: a large city police department has the option to choose among three styles of bulletproof vests, labeled A, B, and C. The preference table is shown.

Number of votes	26	15	10	7
First choice	A	B	C	B
Second choice	B	C	A	A
Third choice	C	A	B	C

9. How many police officers voted?
10. How many votes did the preference CAB receive?
11. Using the plurality method, which style won?
12. Using the Borda count method, which style won?
13. Using the plurality-with-elimination method, which style won?
14. Using the pairwise comparison method, which style won the election?
15. When the plurality method was used, was the head-to-head comparison criterion violated?
16. When the plurality-with-elimination method was used, was the majority criterion violated?
17. If style C was unavailable, and the votes were recounted, was the irrelevant alternatives criterion violated if the pairwise comparison voting method is used?

Use this information for Exercises 18–27. An advisory board for a company is deciding on which health insurance provider to use for the company insurance plan. Its options are Central States Medical (C), GenCare (G), Blue Shield (B), and Midwest Health Care (M). The preference table is shown here.

Number of votes	18	17	9	3
First choice	M	B	G	C
Second choice	B	G	C	G
Third choice	G	C	M	B
Fourth choice	C	M	B	M

18. How many votes did the preference GCMB receive?
19. How many board members voted in the election?
20. Using the Borda count method, which provider was selected?
21. Using the plurality method, which provider was selected?
22. Using the pairwise comparison method, which provider was selected?
23. Using the plurality-with-elimination method, which provider was selected?
24. When the Borda count method was used, was the majority criterion violated?
25. When the plurality method was used, was the head-to-head comparison criterion violated?
26. If Central States Medical was too expensive and had to be removed and the votes were recounted, was the irrelevant alternatives criterion violated if the pairwise comparison voting method is used?
27. Suppose that the plurality-with-elimination method was used, and the three members who voted CGBM decided to change their votes to MBGC. Would this violate the monotonicity criterion?
28. As a fundraiser for charity, five professors volunteer to spend a weekend in the county jail if students raised $10,000

for the local food bank. The students get to choose which professor goes behind bars, using approval voting. The results are shown below. Which professor gets locked up?

Number of voters	3	1	1	2	4	1	1
Seubert	/		/		/		/
Glass	/	/	/				
Carothers			/	/		/	/
Hern				/	/		
Nguyen		/				/	/

29. The four finalists in a regional art competition submit their favorite piece, and a committee votes on the winner using approval voting. The results of the voting are shown. Which piece won?

Number of voters	1	5	3	1	1	2
Velvet Elvis	/			/	/	/
The Drinker		/	/		/	
Vincent Van Dog	/	/	/			
The Screen		/	/	/	/	

Section 12-4

30. A community college bought 15 laptop computers to be apportioned to three departments with number of faculty as shown. Using Hamilton's method, determine how many computers each department will receive.

Department	Math	English	Science	Total
Number of faculty	25	20	15	60

31. A church has four locations with average weekly attendance as shown. Ten new assistant pastors are to be assigned to each location according to weekly attendance. Using Hamilton's method, determine how many assistant pastors each location should receive.

Location	A	B	C	D	Total
Average weekly attendance	950	265	180	450	1,845

32. Repeat Exercise 30 using Jefferson's method.
33. Repeat Exercise 31 using Adams' method.
34. Repeat Exercise 30 using Webster's method.
35. Repeat Exercise 31 using Webster's method.
36. Repeat Exercise 30 using the Huntington-Hill method.
37. Repeat Exercise 31 using the Huntington-Hill method.

Section 12-5

In Problems 38 and 39, determine if the Alabama paradox occurs if one seat is added. Use Hamilton's method.

38. Originally there were 10 seats; one seat is added to make 11.

District	A	B	C	Total
Population	32	359	433	824

39. Originally there were seven seats; one seat is added to make eight.

District	A	B	C	Total
Population	137	933	1,000	2,070

In Problems 40–41, determine if the population paradox occurs. Use Hamilton's method.

40. There are 24 seats to be assigned.

District	A	B	C	Total
Original population	529	989	2,237	3,755
New population	681	1,249	2,568	4,498

41. As part of a group project in a political science course, students have to choose how to divide up the total points earned by their group. They decide to apportion the points using Hamilton's method based on how many words each wrote for the final report. The group earns 95 points for the project, and no fractional parts of points will be assigned.

The number of words written by each student in the first draft and final report are shown below.

Student	Sinead	June	Tabitha	Cameron	Total
Words in draft	3,210	1,975	2,455	4,001	11,641
Words in final report	3,390	2,510	2,500	4,111	12,511

In Problems 42–43, determine if the new states paradox occurs. Use Hamilton's method.

42. There are 15 original seats; one new district with population 1,900 is added and correspondingly 5 seats are added.

District	A	B	Total
Population	981	4,893	5,874

43. There are 55 original seats; one new district is added with population 5,912 and correspondingly 10 seats are added.

District	A	B	C	Total
Population	6,230	16,323	10,101	32,654

Chapter Test

Use this information for Exercises 1–4: customers in a coffee shop are asked to rank three types of coffee, Arabica (A), blended (B), and Colombian (C). The preference ballots are shown.

C	B	B	C	B	A	C	B
B	A	A	B	A	B	B	A
A	C	C	A	C	C	A	C

B	C	B	B
A	B	A	A
C	A	C	C

1. Build a preference table for the ballots.
2. How many people voted?
3. How many people voted for blended as their first choice?
4. How many people selected CBA as their ranking preference?

Use this information for Exercises 5–12: a small contracting company has been invited to relocate to one of three cities: Pittsburgh (P), Baltimore (B), and Richmond (R). The employees are asked to vote on the city. The preference table is shown.

Number of votes	43	27	18	12
First choice	P	R	B	P
Second choice	R	P	R	B
Third choice	B	B	P	R

5. Using the plurality method which city won?
6. Using the Borda count method which city won?
7. Using the plurality-with-elimination method which city won?
8. Using the pairwise comparison method which city won?
9. When the plurality method was used, was the head-to-head criterion violated?
10. When the Borda count method was used, was the majority criterion violated?
11. When the plurality-with-elimination method was used, was the head-to-head criterion violated?
12. If Baltimore was ruled out and the votes were recounted, was the irrelevant alternatives criterion violated if the Borda count method was used?
13. A website devoted to criticizing TV shows holds a vote among its staff members and loyal readers for the worst reality TV show, using approval voting. A vote indicates support for that show being named worst. The results are shown. Which show was named worst?

Number of votes	247	193	52	119	220
Celebrity Bocce	/	/	/		
The Canine Bachelor		/	/	/	
Big Brother Nursing Home	/		/	/	/
Dancing with Bears				/	

14. An airline offers nonstop flights from Fort Lauderdale to three different locations. It has recently hired 12 new flight attendants and plans to assign them to the flights according

to number of passengers. The number of passengers on each flight is summarized in the table. Apportion the flight attendants using Hamilton's method.

Flight	A	B	C	Total
Number of passengers	156	264	86	506

15. Repeat Problem 14 using Jefferson's method.
16. Repeat Problem 14 using Webster's method.
17. Repeat Problem 14 using the Huntington-Hill method.
18. Repeat Problem 14 using Adams' method.
19. Determine if the Alabama paradox occurs. There were 10 original seats, and one seat is added. Use Hamilton's method.

District	A	B	C	D	Total
Population	48	393	454	258	1,153

20. The original and new populations for four districts are shown below. There are 20 seats. Determine if the population paradox occurs. Use Hamilton's method.

District	A	B	C	D	Total
Original population	1,156	2,300	3,005	1,795	8,256
New population	1,215	2,555	3,147	2,101	9,018

21. Determine if the new states paradox occurs. There were two original states as indicated in the table and 40 original seats; an additional state with population 3,400 was added and correspondingly 7 additional seats. Use Hamilton's method.

State	A	B	Total
Population	6,255	13,745	20,000

Projects

1. (a) Using the Internet or a library, write a brief report on the Electoral College system used to elect the U.S. president.
   (b) Once the Electoral College has been chosen by the individual states, what voting system does the College use to elect the president?
   (c) What voting system is used by the individual states to allocate their electoral votes?
   (d) Discuss the entire election process as it relates to the four fairness criteria we studied. Which if any of the criteria do you think may be violated by the process?
   (e) As a summary, write an opinion piece on whether you (or your group) think that the Electoral College process should be abandoned, and if so which voting system you think would be preferable.

2. (a) Print up a ballot listing five TV shows that are currently popular among college students, and ask everyone in your class to rank the shows from most favorite to least favorite. After collecting the ballots, make a preference table and use all four voting methods from Chapter 12 to determine the most popular show. Did you get the same result for all four methods?
   (b) Now have all of your classmates fill out an approval ballot, marking all of the shows that they like. Tally the results. How does the result compare to the other four voting systems?
   (c) Write a short opinion piece on which method you think provided the most reliable gauge of which show is most popular among your classmates. (If the winner was the same for all five methods, your answer will be pretty short.)

3. Using the Internet as a resource, find the most recent apportionment of seats to the U.S. House of Representatives, then find the population figures for each state. The simplest and quickest method of apportionment is Hamilton's method—use Hamilton's method to apportion the House seats to the 50 states and see if the results match the apportionment you found online. In this case, "simplest and quickest" is relative: there are 435 representatives and 50 states, so this is most definitely best done in a group setting by dividing up the work. When you're finished, write a paragraph on your thoughts on the fact that this process was done for the House of Representatives for probably 100 years with NO technology to help at all.

Design elements: Front matter, Chapter Opener, Summary and End Matter header design (random numbers background illustration): ©pixeldreams.eu/Shutterstock RF

Graph Theory

Photo: John Lund/Tiffany Schoepp/Blend Images

Outline

Math in | Road Trips

The road trip is a great American tradition. There's just something magical about the freedom of heading out to the open road and driving wherever you feel like. It usually takes about an hour for the magic to wear off, though. Then you just want to get to where you're headed as quickly as possible.

The branch of mathematics known as graph theory was created to solve problems involving the most efficient way to travel between different locations. One of the interesting things about graph theory is that we can trace its beginnings to a single problem from the early 1700s. The town of Kaliningrad in modern day Russia was founded as Königsberg, Prussia, in 1255. The Pregel river splits into two branches as it approaches the city, dividing the city into four distinct regions, as shown in the figure.

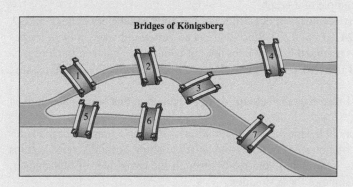

Bridges of Königsberg

The residents of Königsberg were fond of leisurely walks covering all of the bridges, but none were able to find a route that would enable them to cover all seven exactly once while beginning and ending at the same location.

At some point, someone wrote the great Swiss mathematician Leonhard Euler, who lived not far away in St. Petersburg, to ask if he could prove that such a route either was or was not possible. Although he considered the problem trivial, Euler was intrigued by the fact that it didn't appear to fit into any current branch of math. And so graph theory was born. One of the important results of this chapter bears Euler's name, and it was the one used to solve the Königsberg bridge problem. (The solution is in the Sidelight on page 819.)

From that humble beginning, graph theory expanded to cover a wide variety of practical applications, many of which we will study in this chapter. And the really amazing thing is that some of the applications we'll study are at least as important and relevant today as they've ever been, even though they can still trace their roots to the bridges of Königsberg.

And this is where we return our attention to road trips. Suppose that you're planning to visit friends at three different colleges and want to find the most efficient route to visit all three and then return home. You would probably begin by finding the distance between all of the locations. This information for one road trip is provided in the table below. When you have finished this chapter, you should be able to answer the questions below the table.

	Georgia	Georgia Tech	Auburn	Alabama
Georgia	–	70	185	271
Georgia Tech	70	–	114	200
Auburn	185	114	–	158
Alabama	271	200	158	–

1. Draw a complete weighted graph that represents the distances between these schools.

2. If a student at Georgia Tech plans on visiting friends at the other three schools and then returning home, find the route that would cover the smallest possible number of miles.

3. If the plan was to instead connect the four schools with the shortest possible path without returning to the starting point, what would that path be, and how many miles would it cover?

4. Suppose our road-tripper decides to extend the trip and include the University of Mississippi. That's 387 miles from Georgia Tech, 190 from Alabama, 304 from Auburn, and 457 from Georgia. Draw a new complete graph for the trip. Then use the nearest neighbor method and the cheapest link method to find the most efficient route starting and ending at Georgia Tech.

For answers, see Math in Road Trips Revisited on page 844

Digital Vision/Getty Images

LEARNING OBJECTIVES

1. Define basic graph theory terms.
2. Represent relationships with graphs.
3. Decide if two graphs are equivalent.
4. Recognize features of graphs.
5. Apply graph coloring.

It's time to plan that big spring break trip you've just been *living* for, and the first step is to get to South Padre Island. There's a pretty good chance that you're not going to find a flight directly from the nearest airport to Brownsville, Texas, which is the airport closest to South Padre. Instead, you'll need to make one or more intermediate connections to reach your final destination. (There's a metaphor for life in there somewhere.)

Graph theory is used to study problems like how to arrange connections that most efficiently get you to spring break nirvana. In this section we'll study the basic concepts we need to learn about this useful branch of math. The graphs that we will study in this chapter are different from the graphs of equations and functions we've worked with previously. Rather than representing relationships between two variables, these graphs will be used to illustrate relationships between things like people, cities, and streets. It begins with a new definition of the word *graph*, which is used throughout the study of graph theory.

> A **graph** consists of a finite set of points called **vertices** (singular: vertex) and line segments or curves called **edges** connecting the points.

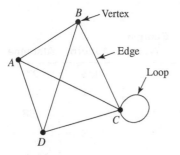

Figure 13-1

Figure 13-1 shows an example of a graph.

We will always draw vertices as black dots. In this case, they're labeled A, B, C, and D. The edges are AB, BC, CD, DA, AC, and BD. Since there is no vertex indicated at the intersection of edges AC and BD, it is assumed that the lines are not connected. You can think of it as two wires that are not joined together or two pipes that are not connected, one pipe passing over the other pipe.

The graph in Figure 13-1 also contains a **loop**. A loop is an edge that begins and ends at the same vertex.

Graphs can be used to model a wide variety of situations: cities on a map, floor plans for houses, border relationships of states or countries—even friendships or family relationships between people. In Examples 1 through 4, and really through the remainder of this chapter, we'll use graphs to illustrate some of these situations.

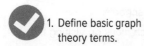

1. Define basic graph theory terms.

Example 1 | Representing Islands with a Graph

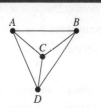

Figure 13-3

Represent the islands and the bridges shown in Figure 13-2 with a graph. Describe what you chose to be vertices and what you chose as edges, and why.

SOLUTION

In a graph, edges are lines or curves that connect the vertices, so the natural choice here would be to represent the islands with vertices and the bridges connecting them with edges. There are four islands; label them as A, B, C, and D. Next represent each bridge by an edge. For example, edge AD connects island A with island D. Each bridge connects two islands, and should be represented by an edge. The graph is shown in Figure 13-3.

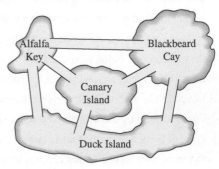

Figure 13-2

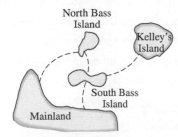

Figure 13-4

Try This One   1

Draw a graph to represent ferry service between northern Ohio and some Lake Erie islands shown in Figure 13-4. Explain your choices for vertices and edges.

Example 2   **Representing a Floor Plan with a Graph**

Represent the floor plan of the house shown in Figure 13-5 by a graph. Explain your choices for vertices and edges.

SOLUTION

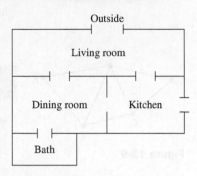

Figure 13-5

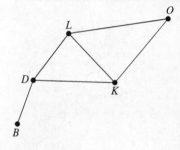

Figure 13-6

In a graph, the edges connect the vertices. So we need to think about what in the floor plan connects things. That would be the doorways, which connect rooms. So we'll make the rooms vertices, and draw an edge connecting vertices to represent doorways that connect rooms. The living room, kitchen, and dining room all have three doors; the outside is accessible only from the living room and kitchen, and the bathroom connects only to the dining room. The graph is shown in Figure 13-6.

Try This One   2

The floor plan shown in Figure 13-7 is for a two-bedroom student apartment in Oxford, Ohio. Draw a graph to represent it.

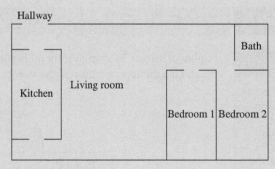

Figure 13-7

Example 3   **A Graph Representing Border Relationships**

The map shown in Figure 13-8 shows five counties in central Pennsylvania. Draw a graph that shows the counties that share a common border. Explain your choices for vertices and edges.

SOLUTION

In essence, the common borders are what connects counties: you cross a border to get from one to the other. So we should represent the counties with

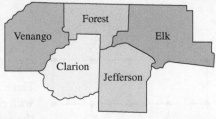

Figure 13-8

Math Note

As we can see in Example 3, a graph describing border relationships is much different from a map! The graph shows only which counties share a border. It doesn't indicate how the counties are situated.

vertices and the common borders with edges. The five counties are represented by five vertices; each is labeled with the first letter of the name of a county. Notice that Venango County and Clarion County share a common border, so connect V and C with an edge. Venango County and Forest County share a common border, so connect V and F with an edge. Since Venango County doesn't share a border with Jefferson County or Elk County, no edges connect V with J or E. Next, connect vertices F and E since Forest County and Elk County share a common border, and continue in this manner. The completed graph is shown in Figure 13-9. A good way to check your answer is to count. Look at each county one at a time on the map and count how many counties border it. Then make sure the vertex representing that county has the right number of edges hitting it.

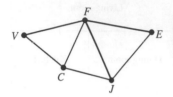

Figure 13-9

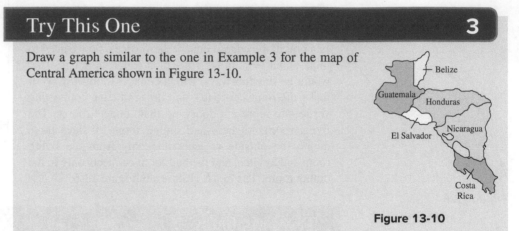

Try This One   3

Draw a graph similar to the one in Example 3 for the map of Central America shown in Figure 13-10.

Figure 13-10

Example 4   A Graph Representing City Streets

Math Note

In Example 4, we chose to represent intersections with vertices and streets with edges. In Exercise 56, you'll be asked to use vertices for the streets and edges for the intersections.

A plan of streets in a neighborhood is shown in Figure 13-11. Draw a graph for the neighborhood. Use the intersections of the streets as vertices and the streets as the edges.

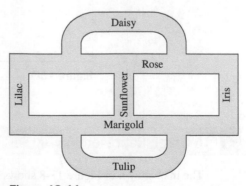

Figure 13-11

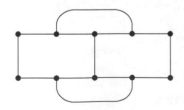

Figure 13-12

SOLUTION

Place a vertex at each point of intersection of any two or more streets. Connect the vertices with edges that follow the streets. See Figure 13-12. Since there are no obvious letters to label the vertices (which represent intersections), we decided to leave them unlabeled. If we needed to refer to particular vertices for some reason, it would be a good idea to make up a labeling system.

Try This One
4

Draw a graph for my neighborhood, shown in Figure 13-13. (I live on Henesy. Stop by—I'll make muffins.) Use vertices for the intersections and edges for the streets.

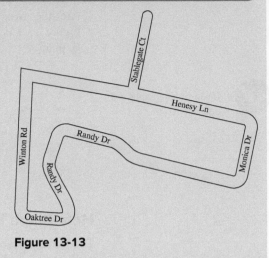

Figure 13-13

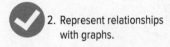

2. Represent relationships with graphs.

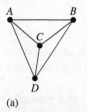 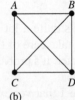

(a)   (b)

Figure 13-14

Equivalent Graphs

Glance quickly at the two graphs in Figure 13-14. Are they the same? If you answered yes, you should be very proud of yourself. Even though they physically look different, they are really the same graph, because they have the same vertices connected in the same way. The only difference is where the vertices are physically situated.

When two graphs have the same vertices connected in the same way, we call them **equivalent**.

Sidelight   Graph Theory: Making Transportation Easier

If you've ever driven or paid to park in a large urban area, you probably have a pretty good idea why public transportation is a popular alternative. The first subway system in America opened in New York City in 1870, providing an alternative to streets that had become almost impossibly congested. Since then, urban public transportation has become commonplace, in a variety of different formats.

The one thing that all of these systems have in common is that looking at an actual map of the system can be very confusing. Twists, turns, interconnections, train and bus changes—all conspire to make it difficult to figure out how to get where you're going. In the 1930s, a draftsman in London hit upon a brilliant idea: simplify the route maps by using graph theory! Each station is represented with a vertex, with the edges representing lines that connect the stations, like the example shown here of the RTA system in Cleveland.

This makes it far simpler for passengers to find their way efficiently. And ultimately, that's what graph theory was completely designed to do in the first place.

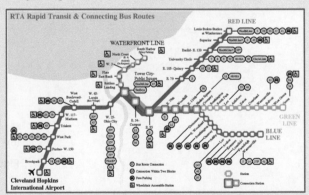

Example 5   Recognizing Equivalent Graphs

Explain why the two graphs shown in Figure 13-14 are equivalent.

SOLUTION

Each graph contains four vertices labeled A, B, C, and D. In each graph, every vertex is connected to each of the other three vertices. So the graphs have the same vertices and same connections, and they are equivalent.

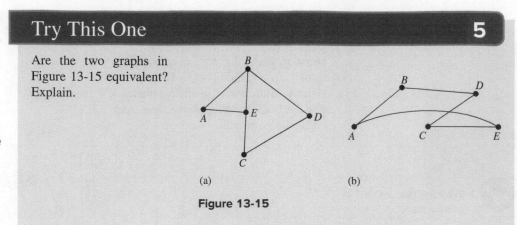

Try This One   **5**

Are the two graphs in Figure 13-15 equivalent? Explain.

3. Decide if two graphs are equivalent.

(a)   (b)

Figure 13-15

More Graph Theory Terminology

The main concepts of graph theory are fairly simple, but there's quite a lot of terminology we'll need in order to solve problems using graph theory. We'll take care of the terminology now, then use these important terms in the remainder of the chapter. It's probably a good idea to keep a running list of definitions that you can refer back to later.

The **degree** of a vertex is the number of edges that intersect that vertex. For example, each of the vertices in Figure 13-14 has degree 3 because each has three edges that intersect it. An **even vertex** has an even number of edges that intersect it, while an **odd vertex** has (you guessed it) an odd number of edges that intersect it. Loops are considered to be two edges intersecting a vertex.

Figure 13-16 shows some even and some odd vertices.

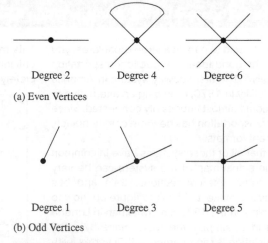

Degree 2   Degree 4   Degree 6

(a) Even Vertices

Degree 1   Degree 3   Degree 5

(b) Odd Vertices

Figure 13-16

Adjacent vertices have at least one edge connecting them. In the graph shown in Figure 13-15(a), vertex A is adjacent to vertices B and E, but is not adjacent to either vertex C or D since there is no edge connecting them.

A **path** on a graph is a sequence of adjacent vertices and the edges connecting them that uses no edge more than once. When finding a path on a graph, a vertex can be crossed more than once, but an edge can be crossed only once. Figure 13-17 shows a path on a graph. The path starts at vertex B and then goes through vertices C, E, and F in that order. The path is then described as B, C, E, F. A graph can have many paths, and a path doesn't have to include all of the vertices or edges.

A **circuit** is a path that begins and ends at the same vertex. A circuit is shown on the graph in Figure 13-18. It is given by A, B, C, E, A.

Math Note

The term *adjacent vertices* has nothing to do with whether or not vertices are actually physically next to each other. Focus on whether or not they're connected by an edge.

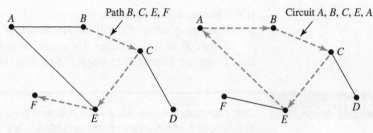

Figure 13-17   **Figure 13-18**

Math Note

Every loop is by default also a circuit, but of course a very short and simple one! Also, every circuit is a path, but not every path is a circuit since paths don't have to begin and end at the same vertex.

A graph is said to be **connected** if for any two vertices, there is at least one path that connects them. In other words, a graph is connected if you can start at any vertex, follow along the edges, and end up at any other vertex without tracing an edge more than once. If a graph is not connected, we say it is **disconnected**. Figure 13-19 shows examples of connected and disconnected graphs.

A **bridge** on a connected graph is an edge that, if removed, makes the graph disconnected. Figure 13-20 shows three graphs that have bridges. Notice that if a bridge is removed, there will be at least one pair of vertices that can no longer be connected with a path.

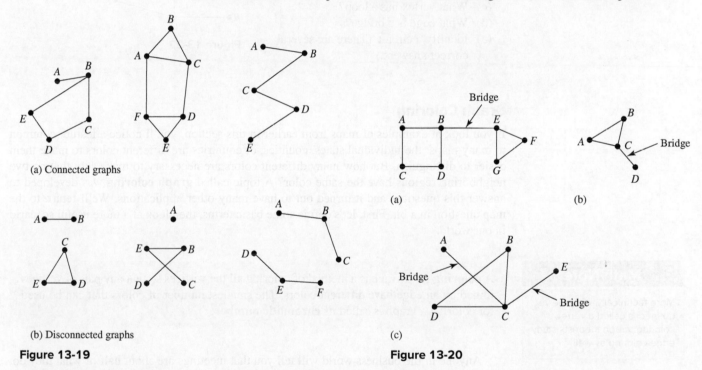

(a) Connected graphs

(b) Disconnected graphs

Figure 13-19

(a)

(b)

(c)

Figure 13-20

Example 6   Recognizing Features of a Graph

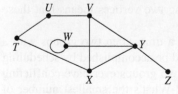

Figure 13-21

Use the graph in Figure 13-21 to answer each question.

(a) List the odd vertices.   (d) Are there any bridges?
(b) List the even vertices.   (e) Identify at least one circuit.
(c) What vertex has a loop?

SOLUTION

(a) Vertices V, W, and X have three edges touching them, so they are odd vertices. Vertex Z has just one edge touching it, so it's also odd.

(b) Vertices T and U have two edges touching them, and vertex Y has four edges touching it, so these are even vertices.

(c) There is a loop at vertex *W*.

(d) Removing edge *WY* will isolate vertex *W*, so *WY* is a bridge. Removing edge *YZ* will isolate *Z*, so it is a bridge. These are the only two bridges.

(e) There are dozens of examples of circuits. One example is *T*, *U*, *V*, *X*, *T*.

CAUTION

Don't confuse the use of the word *bridge* in graph theory with bridges in maps and diagrams. In Example 1, the bridges connecting islands are represented by edges on the graph.

Try This One   6

✔ 4. Recognize features of graphs.

Use the graph shown in Figure 13-22 to answer each question.

(a) List the odd vertices.
(b) List the even vertices.
(c) What vertex has a loop?
(d) What edge is a bridge?
(e) Identify a circuit. (There are several correct answers.)

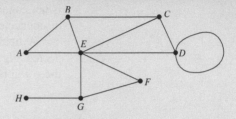

Figure 13-22

Graph Coloring

If you look at examples of maps from earlier in this section, you'll notice a feature common to many maps: the individual states, counties, or countries are different colors to make them easier to distinguish. But how many different colors are necessary to make sure that no two neighboring regions have the same color? A topic called **graph coloring** was developed to answer this question, and it turned out to have many other applications. We'll return to the map question in a bit. First, let's define some basic terms, then look at a more useful scenario in our world.

Math Note

More technically, what we've defined is called a vertex coloring: graph theorists study edge coloring as well.

A **coloring** for a graph is a method of coloring all the vertices so that any pair of vertices joined by an edge have different colors. The smallest number of colors that can be used for coloring a graph is called its **chromatic number**.

Anyone in the business world will tell you that meetings are about half of what they do, and scheduling meetings can be a real challenge. At one company, there are seven different groups that have regular meetings with many employee members of multiple groups: finance (F), strategic planning (S), human resources (H), healthcare and wellness (W), employee relations (E), union negotiations (U), and board of managers (B). The graph in Figure 13-23 shows relationships between the groups: an edge connecting two vertices means that those groups have at least one common member.

The goal is to schedule meetings so that there's never a time when two groups with a common member meet at the same time. Certainly this could be accomplished by scheduling seven different meeting times so that no two groups ever have conflicting times. But is there a more efficient way? What's the smallest number of meeting times that will work?

In terms of the graph, any two vertices that have a common edge represent groups with common members. So we can turn this into a coloring problem by using colors for vertices, making sure that no two adjacent vertices have the same color. (This would represent a meeting time when someone has two meetings to attend.)

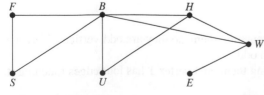

Figure 13-23

| Example 7 | Using Graph Coloring in Scheduling |

(a) For the company described earlier, find the smallest number of meeting times that guarantee no time conflicts.
(b) Which committees can be scheduled at the same times?

SOLUTION

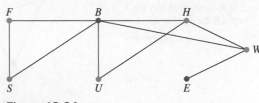

Figure 13-24

(a) The colored graph is shown in Figure 13-24. We began by coloring vertex F blue. Then B and S can't be blue because they're connected to F, and they have to be different from each other because they're connected to each other; we chose red and green. Now vertex U can be colored blue, which then requires H to be colored green. (Notice that it's connected to vertex B, which is already red.) Next, vertex W is connected to B (red) and H (green), so it has to be blue. Finally, vertex E can be either red or green. In any case, we've discovered that the graph can be colored with three colors but no less, so its chromatic number is three, and at least three different meeting times are required.

(b) The vertices that are the same color are not connected to each other by any edges, so those are the ones that do not have any common members. Vertices F, U, and W are all blue, so the finance, union negotiation, and healthcare and wellness groups can meet at the same time. Vertices B and E are both red, so the board of managers and the employee relations committees can meet at the same time. Vertices H and S are both green, so the human resources and strategic planning committees can meet at the same time.

Math Note

This isn't the only possible coloring of the graph, nor is it the only one that uses only three colors: for example, we could have chosen vertex U to have the same color as S. But what really matters is that any other coloring needs at least three colors.

Try This One   7

Suppose that you're trying to make a final exam schedule for a satellite campus that teaches six classes: anthropology 107 (A), economics 212 (E), information systems 194 (I), statistics 261 (S), history 201 (H), and geology 101 (G). Figure 13-25 shows which classes have students in common—those are connected by edges.

(a) Use graph coloring to find the chromatic number.
(b) What does the chromatic number tell us about the final exam schedule?

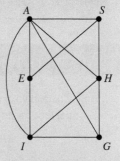

Figure 13-25

Next, we'll take a look at how graph coloring applies to map coloring.

| Example 8 | Applying Graph Coloring to Maps |

Use graph coloring to find the number of different colors that would be necessary to color the map in Figure 13-26 so that no states with a common border are the same color. Then color the map using what you learned from your graph.

SOLUTION

Using the technique we learned earlier in the section, we can draw a graph representing border relationships, shown in Figure 13-27(a). Starting with blue at Nevada, we then went to red at Oregon. If we choose different colors for Idaho and California (neither of which can be either blue or red), then we already have four colors. If we make them both green, as in Figure 13-27(b), then Arizona and Utah have to be different colors, but they can't be either blue (connected to Nevada) or green (connected to California or Idaho). This shows that at least four colors are needed. The completed

Figure 13-26

coloring shows that we can get away with just four. The colored map is shown in Figure 13-27(c). Notice that the colors for each state match the color of its vertex in the colored graph.

Math Note

When coloring a complicated graph, it's a good idea to start at the vertex that has the highest degree.

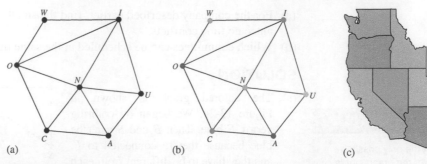

(a)   (b)   (c)

Figure 13-27

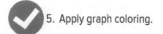

5. Apply graph coloring.

Try This One   8

Use graph coloring to find the smallest number of colors needed to color the map in Example 3 so that no counties sharing a common border are the same color.

Math Note

The four color theorem doesn't say that any *graph* can be colored with four or less colors—that's absolutely not true. It says that any *map* can be colored with four or less colors. There are complex graphs that don't represent any possible map.

To close the section, we'll look at a brief history of the map coloring problem. By the mid-19th century, it had been long noted by mapmakers that every map known could be colored using at most four colors. In 1852, an English college student named Francis Guthrie first conjectured that four colors would be sufficient to draw *any* map, a conjecture that was passed on to the well-known mathematician Augustus De Morgan (whose laws we encountered in Chapters 2 and 3). It took 27 years for someone to provide an accepted proof of this simple-sounding theorem, which came to be known as the four color theorem. But in 1890 the proof was shown to be incorrect. Believe it or not, it took until 1976 for the next widely accepted proof to appear, but even today some hard-core mathematicians don't trust the proof, as it required a tremendous amount of computer time, making it essentially impossible for a human being to check it in its entirety.

Answers to Try This One

1   The ferry service connects the islands, so that's perfect for edges, which connect vertices.

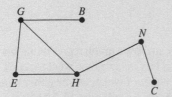

2

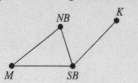

3

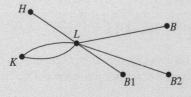

4

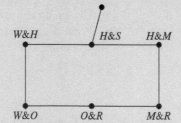

5   No. Vertex *B* connects to *E* in the first graph but not the second.

6   (a) *B*, *C*, *G*, and *H* are odd. (b) *A*, *D*, *E*, and *F* are even.
        (c) *D*   (d) *GH*
        (e) One of many possible
             answers is *A*, *B*, *C*, *D*, *E*, *A*.

7   (a) The chromatic number is
             four. One possible coloring
             is shown here.
        (b) This shows that in order to avoid
             time conflicts, at least four
             final exam times are needed.

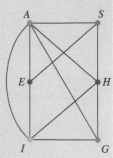

8   Three colors are required.

Exercise Set 13-1

Writing Exercises

1. Describe the difference between the meaning of the word *graph* in this chapter and the meaning we used earlier in the book.
2. What is the difference between a loop and a circuit?
3. What is the difference between a circuit and a path?
4. Draw two graphs that look physically different but are equivalent, then explain why they're equivalent.
5. How can you tell if a graph is connected?
6. Think about a graph that represents the commercial flight routes for all American cities. Do you think this graph is connected? What does it mean if it's disconnected?

7. What does it mean for two vertices in a graph to be adjacent? What would it mean in the context of Exercise 6?
8. Does every graph have a bridge? Explain.
9. What is a graph coloring?
10. What's the chromatic number of a graph?
11. If the chromatic number of a graph is five, explain why you can find a coloring that uses six colors, but not one that uses four.
12. How does graph coloring apply to maps?

Computational Exercises

Use the following graph to answer Exercises 13–24.

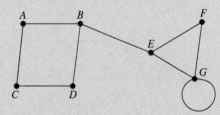

13. List the vertices of the graph.
14. How many edges does the graph have?
15. Name three vertices that are adjacent to vertex *E*.
16. List the even vertices.
17. List the odd vertices.
18. What edge is a bridge?
19. Find a path that contains vertex *E*. (There are several answers.)
20. Find a circuit that includes vertex *D*.
21. Identify a vertex that has a loop.
22. Which edges are not included in the path *A, B, E, G, F*?
23. Explain why *A, B, E, D* is a not a path.
24. Explain why *E, F, G* is not a circuit.

In Exercises 25–30, find the chromatic number of each graph and provide a coloring that uses that number of colors.

25.

26.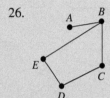

27.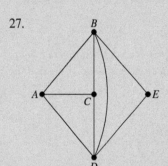

28.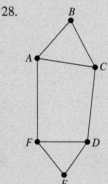

29.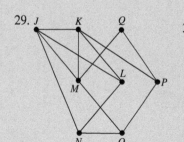

30.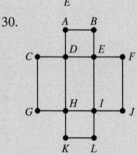

Applications in Our World

For Exercises 31–34, represent each figure using a graph. Use vertices for islands and edges for bridges.

31.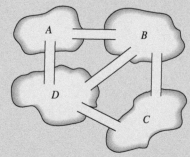

32.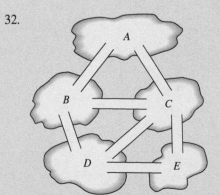

33.

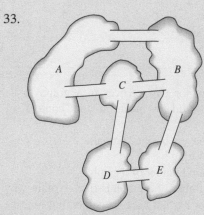

34.

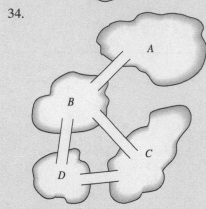

For Exercises 35–38, draw a graph to represent each map. Use vertices to represent the states and edges to represent the common borders.

35.

36.

37.

38.

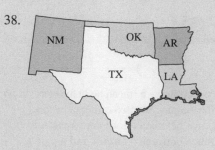

For Exercises 39–42, draw a graph that represents each floor plan. Use vertices to represent the rooms and outside area and edges to represent the connecting doors.

39.

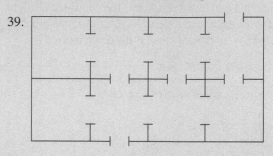

40.

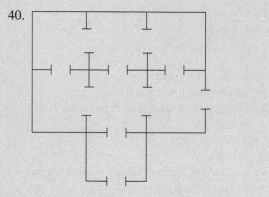

41.

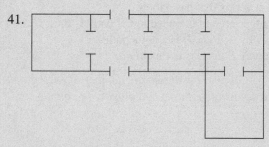

42.

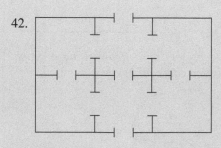

In Exercises 43–50, use graph coloring to find the smallest number of colors needed to color the map so that no regions sharing a common border are the same color.

43.

44.

49.

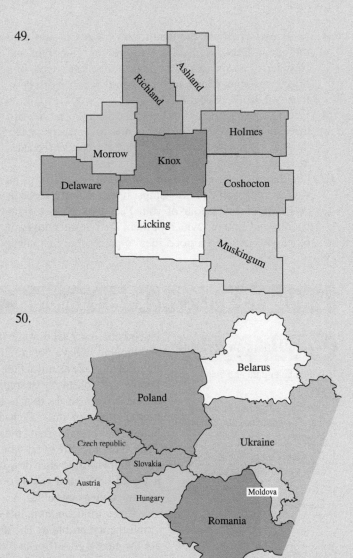

50.

45. The map in Exercise 35.
46. The map in Exercise 36.
47. The map in Exercise 37.
48. The map in Exercise 38.

In Exercises 51–54, if you wanted to paint the homes with the given floor plan so that no two adjoining rooms are the same color, find the minimum number of colors you would need. Note that you can't paint the outside. You could try, but you would not be successful.

51. Exercise 39   52. Exercise 40
53. Exercise 41   54. Exercise 42

Critical Thinking

55. Draw a circuit with five vertices, a loop, and a bridge.
56. Draw a graph that represents the street map in Example 4, but this time use vertices to represent the streets, and edges to describe which streets intersect. Is your graph different from the one in Example 4? What can you conclude?
57. Is it possible to color a graph that has a loop? Explain.
58. It's VERY common for people to object to the four color theorem based on this argument: "If there's a long country that has six others lined up on top of it, then it would need at least seven colors!" Explain why this situation does NOT violate the theorem.

59. Draw a floor plan of where you live similar to the ones in Exercises 39–42, and draw a graph that represents the floor plan. Use vertices to represent rooms and outside area, and edges to represent connecting doors. If you live in a two-story house, use the main floor. If you live in a dorm room, use a portion of the common area of your dorm with at least four rooms.
60. The city of New York consists of the boroughs of Manhattan, Brooklyn, Queens, and the Bronx. Manhattan is connected to New Jersey by several bridges. Find a map of New York City with bridges and tunnels included either online or in a library. (www.ny.com is an excellent

resource.) Then use the map to draw a graph similar to the one in Exercises 31–34, with the four boroughs and New Jersey as vertices and bridges/tunnels as edges. (If a bridge or tunnel connects more than two boroughs, draw it as two different edges.)

61. Find a map of the county you live in and the five other counties that are nearest to you, then use the map to draw a graph like the ones in Exercises 35–38. Use vertices to represent counties and edges to represent the common borders.

62. Graphs can be used to represent social relationships as well as geographic relationships. If you are on Facebook or Instagram, pick four of your Facebook or Instagram friends and find out which of them are friends with each other. (It would be a good idea to pick some that come

from different social or family groups to make things interesting.) Then draw a graph that represents the friendships. If you're not on Facebook or Instagram, ask for that information from someone you know that is. (Or join Facebook or Instagram, I guess.)

63. (a) When a graph represents the floor plan of a home as in Exercise 59, what does it mean when the graph is disconnected?
    (b) What does it mean if an edge is a loop?
    (c) What does it mean if an edge is a bridge?

64. (a) When a graph represents a map as in Exercise 61, what does it mean when the graph has a loop?
    (b) What does it mean when the graph has a bridge?
    (c) What does a circuit represent?

Section 13-2   Euler's Theorem

LEARNING OBJECTIVES

1. Define Euler path and Euler circuit.

2. Use Euler's theorem to decide if an Euler path or Euler circuit exists.

3. Use Fleury's algorithm to find an Euler path or Euler circuit.

4. Solve practical problems using Euler paths or circuits.

Nobody has ever sat down to do homework and thought "I have *got* to figure out a way to make this take as much time as possible." People are constantly searching for more efficient ways to do things, leaving more time for the fun stuff. Consider the backbone of our postal system, the mail carriers. They have to visit every house and business on their route each day, and obviously it would be helpful to do so with as little backtracking as possible.

If you think about it, this sounds like a graph theory problem. The delivery locations are the vertices, and the routes between them are the edges. An efficient route would visit every vertex without going over any edges more than once. In this section, we will study the work of Leonhard Euler, which addresses problems that involve tracing an entire graph without repeating any edges.

Don Hammond/DesignPics/Alamy Stock Photo

1. Define Euler path and Euler circuit.

> An **Euler path** is a path that passes through every edge exactly once. An **Euler circuit** is a circuit that passes through every edge exactly once.

Recall that a path is a sequence of adjacent edges and vertices that doesn't go through any edge more than once, and a circuit is a path that starts and ends at the same vertex.

The difference between a path and an Euler path is that a path can pass through any subset of edges on a graph, but an Euler path passes through *all* of the edges exactly once. Figure 13-28 shows an Euler path. As indicated by the arrows, it begins at vertex A and follows the path A, B, E, D, C, E. Figure 13-29 shows an Euler circuit that begins and ends at vertex B.

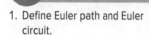

Euler path *A, B, E, D, C, E*

Figure 13-28

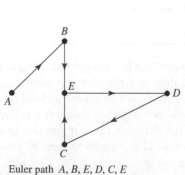

Euler circuit *B, A, E, D, C, B, E, B*

Figure 13-29

CAUTION   It's important to understand that Euler paths can (and often do) visit the same *vertex* more than once. They just can't pass through the same *edge* more than once.

Example 1 | Recognizing Euler Paths and Euler Circuits

Classify the paths shown in the graphs as Euler path, Euler circuit, or neither.

(a)

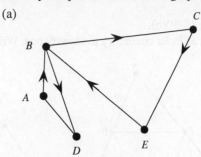

Path *A, B, C, E, B, D*

(b)

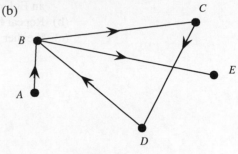

Path *A, B, C, D, B, E*

SOLUTION

(a) This path is neither. To qualify as an Euler path or circuit, every edge has to be passed through, but in this case the edge connecting *A* and *D* wasn't covered.

(b) This is an Euler path, but not an Euler circuit. All of the edges are covered exactly once, but the path begins at vertex *A* and ends at vertex *E*. To continue back to vertex *A* to make this a circuit, you'd have to pass through two different edges for a second time, and the result isn't even a path, so it can't be an Euler path or circuit. This may mean that there isn't an Euler circuit for this graph. We'll study that question shortly.

Try This One 1

Classify the paths shown in the graphs as Euler path, Euler circuit, or neither.

(a)

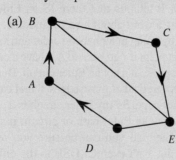

Path *B, C, E, D, A, B*

(b)

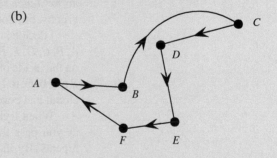

Path *A, B, C, D, E, F, A*

The question now becomes: given a graph, does it have any Euler paths or circuits? Our mail carrier would be interested in this answer—if a graph representing the route has an Euler path, they can cover every street without doubling back at all.

It turns out that some graphs have Euler paths and circuits, and some don't. Euler's theorem allows us to tell if a given graph has an Euler path or Euler circuit without first trying to find one.

Math Note

Every Euler circuit is also an Euler path, but there are Euler paths that are not Euler circuits, like the one in Figure 13-28.

Euler's Theorem

For any connected graph:

1. If all vertices are even, the graph has at least one Euler circuit (which is by definition also an Euler path). An Euler circuit can start at any vertex.

2. If exactly two vertices are odd, the graph has no Euler circuits but at least one Euler path. The path must begin at one odd vertex and end at the other odd vertex.

3. If there are more than two odd vertices, the graph has no Euler paths and no Euler circuits.

Example 2	**Using Euler's Theorem**

(a) Use Euler's theorem to decide if the graph shown in Figure 13-30(a) has an Euler path or an Euler circuit.
(b) Repeat for Figure 13-30(b).
(c) If either graph has an Euler circuit, try to identify one. Write it by listing the vertices in order.

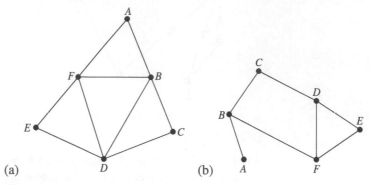

(a)   (b)

Figure 13-30

SOLUTION

(a) Vertices A, C, and E have degree 2; vertices B, D, and F have degree 4. Since all vertices have even degree, Euler's theorem guarantees that there is at least one Euler circuit, which is also an Euler path.
(b) There are four vertices with odd degrees: A has degree 1 and B, D, and F all have degree 3. That means there are no Euler paths or circuits.
(c) This is the interesting thing about Euler's theorem. It's what mathematicians would call an *existence theorem*. It tells us that there IS an Euler path or Euler circuit for certain graphs, but doesn't say a word about how to find one. Let's see what we can do.

   I decided to start by going in alphabetic order and see how that works out. So my path is A, B, C, D, E, F. At that point if I go back to A, I have a circuit, but I've left off the three edges on the inside of the graph, so it's not an Euler circuit. But if I go down to vertex D, then up to B and back to F, I've covered all of those edges, then I can continue back to A to complete the circuit and cover the last edge. So my Euler circuit is $A, B, C, D, E, F, D, B, F, A$.

   When trying to draw an Euler path or circuit, it's a good idea to mark off the edges as you pass through them so you can be sure you've covered all of them without any repeats. By the way, this isn't even CLOSE to the only Euler circuit for this graph. You might see if you can find another that starts and ends at A.

Try This One   2

✓ 2. Use Euler's theorem to decide if an Euler path or Euler circuit exists.

Use Euler's theorem to determine if the graphs shown in Figure 13-31 have an Euler path or an Euler circuit. If either graph has an Euler circuit, try to identify one. Write it by listing the vertices in order.

(a)   (b)

Figure 13-31

Finding Euler Paths and Euler Circuits

Once we know that a graph has an Euler circuit or an Euler path, it seems perfectly reasonable to try to find one. (Think of the mail carrier example—just knowing that there *is* an efficient route to take doesn't help much!) You can try the old-fashioned trial-and-error method like we did in Example 2, but this can be very challenging unless there are a small

Sidelight   **The Bridges of Königsberg**

The question in the Chapter 13 introduction about crossing all of the bridges of Königsberg is actually a question about Euler circuits. Residents in search of a path that would cross every bridge exactly once and return them to the location where they began were seeking an Euler circuit for the graph to the right. On this graph, the land masses are represented by vertices, and the bridges are represented by edges.

Notice that all four vertices are odd. Euler's theorem tells us that the residents were vainly searching for a route that didn't exist—there is no Euler circuit. Remember, this particular problem was totally responsible for the development of graph theory. It's pretty interesting that a problem that spawned an entire branch of math, which is still being studied almost 300 years later, becomes really easy to solve when you know Euler's theorem.

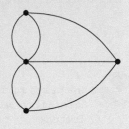

number of vertices and edges. Finding one for a mail carrier that has to visit dozens of houses would be pretty unreasonable. A procedure known as *Fleury's algorithm* has been developed to help save time.

Math Note
Before using Fleury's algorithm, make sure you've used Euler's theorem to confirm that the graph actually has an Euler path or Euler circuit.

Fleury's Algorithm

To find an Euler path or Euler circuit:

1. If a graph has no odd vertices, start at any vertex. If the graph has two odd vertices, start at either odd vertex.

2. Number the edges as you trace through the graph, making sure not to traverse any edge twice.

3. At any vertex where you have a choice of edges, choose one that is not a bridge for the part of the graph that has not yet been numbered.

The procedure, particularly point 3, is illustrated by Example 3.

Example 3	**Finding an Euler Path**

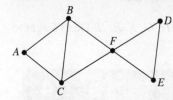

Figure 13-32

Use Fleury's algorithm to find an Euler circuit or path (if one exists) for the graph shown in Figure 13-32.

SOLUTION

First, we need to know whether or not there actually is an Euler path or circuit to look for. Vertices A, D, and E have degree 2; vertex F has degree 4; vertices B and C have degree 3. With two odd vertices, we can find an Euler path but not an Euler circuit.

Fleury's algorithm tells us to start at one of the odd vertices: we'll choose B. None of the edges have been numbered at this point, so we don't have to worry about any of the edges being a bridge for the unnumbered part. So we can go to either A, C, or F to start out. I randomly picked A. After A, we have to move on to C. At C, we can continue to either B or F since neither CB nor CF is a bridge. I picked F. Now we have three choices, but one of them won't work: with the edges we've already numbered out of the graph, edge BF is now a bridge (see Figure 13-33). So we continue to D, then E, back to F, then to B, and finally to C. The completed Euler path is B, A, C, F, D, E, F, B, C, and is shown in Figure 13-34.

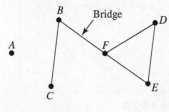

Figure 13-33

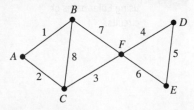

Figure 13-34

Try This One   3

Use Fleury's algorithm to find an Euler circuit or path (if one exists) for the graph shown in Figure 13-35.

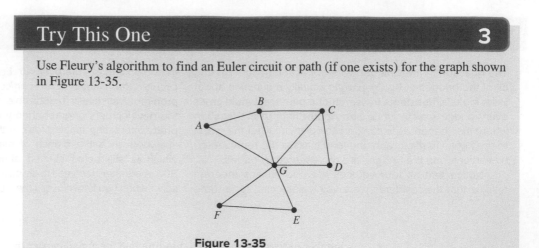

Figure 13-35

3. Use Fleury's algorithm to find an Euler path or Euler circuit.

The Euler path in Example 3 isn't unique; there are several other Euler paths that can be traced on the graph. Also notice that we started at one odd vertex *B* and ended at the other odd vertex *C*. This makes perfect sense: any time we find an Euler path or circuit, we can get another by reversing the direction. If we had ended at an even vertex, that would have to be the beginning of another Euler path, but it can't be. This can help in choosing a path or circuit when using Fleury's algorithm.

The remaining example demonstrates a practical application of Euler's theorem.

| **Example 4** | An Application of Euler's Theorem |

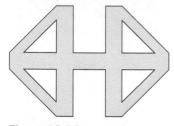

Figure 13-36

A mail carrier has the neighborhood pictured in Figure 13-36 on a daily route. The carrier wants to cover each street exactly once without retracing any street. Find an Euler path to accomplish this.

SOLUTION

Our first job is to draw a graph representing the streets as shown in Figure 13-37. There are two odd vertices, *A* and *H*, so there is an Euler path but not an Euler circuit. We should start at one of the odd vertices and finish at the

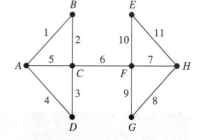

Figure 13-37

other. If we start at *A*, we can use the path *A*, *B*, *C*, *D*, *A*, *C*, *F*, *H*, *G*, *F*, *E*, *H*. Don't forget to number the edges as they are crossed so that none get repeated.

Try This One   4

A mail carrier wants to cover all streets in the neighborhood shown in Figure 13-38 without covering any street twice. Find an Euler path that will permit the carrier to do this.

4. Solve practical problems using Euler paths or circuits.

Figure 13-38

Sidelight Gimme an e!

In this chapter, we've learned that Leonhard Euler (pronounced "oiler") is the founder of graph theory. That in itself would make him pretty important among mathematicians, but graph theory was just an amusing sidelight for Euler. While Euler's theorem is one of the most important results in this chapter, if you do a Google search for "Euler's theorem," you'll find at least half a dozen different results with that name, most of them in completely different fields: geometry, number theory, physics, even economics.

To say that Euler was a prolific and important scholar is sort of like saying that the sun is a little on the warm side. A list of areas in math that he didn't make contributions to would be a whole lot shorter than a list of those that he did. In addition, he did groundbreaking work in mechanics, fluid dynamics, optics, and astronomy. Without even realizing it, you've used notation that was developed by Euler for years, because he introduced a substantial portion of the modern terminology and notation we use in math. Pierre-Simon Laplace, one of the great scientists of the late 18th and early 19th centuries, referred to Euler as "the master of us all," a title that few of his contemporaries would have disputed.

And let's face it—you have to be a superstar to have your own number! You're probably familiar with the fact that the Greek letter π is used to represent an irrational number that's close to 3.14. In the same way, the letter e is used to represent an irrational number close to 2.718. The e is in honor of (you guessed it) Euler, and the number is sometimes called Euler's number. This number plays important roles in calculus, economics, statistics, population studies, and many other areas. Interestingly, it wasn't introduced by Euler—that happened in 1690, about 17 years before Euler was born. But his pioneering work in the study of logarithms (see Section 6-5) used the constant extensively, and it eventually came to be named in his honor. To be fair, though, it was Euler himself who first started using the symbol e to represent the number. Hey, we said he was great, not humble.

Answers to Try This One

1 (a) Neither (Edge BE is not covered.)
  (b) Euler circuit

2 (a) Vertices A and C are odd; the rest are even. There is an Euler path, but no Euler circuit.
  (b) Vertices A, C, and F have degree 2; the others have degree 4. Since all vertices are even, there is an Euler circuit that is also an Euler path. One Euler circuit is $E, A, B, E, D, B, C, D, F, E$.

3 One possible Euler path is $B, A, G, F, E, G, B, C, G, D, C$. There are no Euler circuits.

4 With the diagram represented by a graph, one possible Euler path is numbered here.

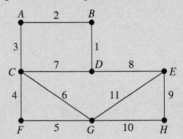

Exercise Set 13-2

Writing Exercises

1. What is the difference between a path and an Euler path?
2. Explain the difference between an Euler path and an Euler circuit.
3. Explain how you can tell if a graph has an Euler circuit.
4. What kind of graph has an Euler path but no Euler circuit?
5. What kind of graph has neither an Euler path nor an Euler circuit?
6. What is Fleury's algorithm used for?
7. What are some things that Euler paths and Euler circuits can be used for?
8. Why is it important to have an algorithm for finding Euler paths or circuits?

Computational Exercises

For Exercises 9–12, decide whether each connected graph has an Euler path, Euler circuit, or neither.

9. The graph has 6 even vertices and no odd vertices.
10. The graph has 2 odd vertices and 6 even vertices.
11. The graph has 2 even vertices and 4 odd vertices.
12. The graph has 2 odd vertices and 3 even vertices.

For Exercises 13–22,

    (a)   *State whether the graph has an Euler path, an Euler circuit, or neither*

    (b)   *If the graph has an Euler path or an Euler circuit, find one.*

13.
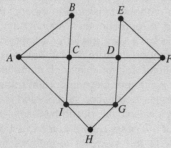

14.
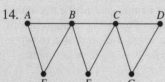

15.
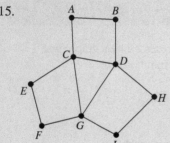

16.
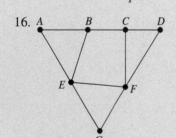

17.
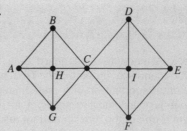

18.
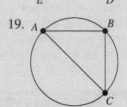

19.
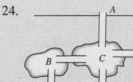

20.
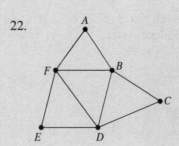

21.
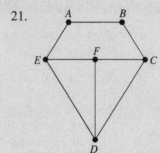

22.
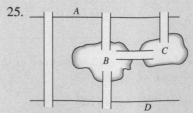

Applications in Our World

For Exercises 23–28, draw a graph for the figures using vertices for the islands and edges for the bridges. Determine if the graph has an Euler path, an Euler circuit, or neither. If it has an Euler path or Euler circuit, find one.

23.
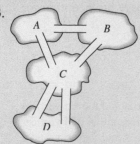

24.

25.

26.

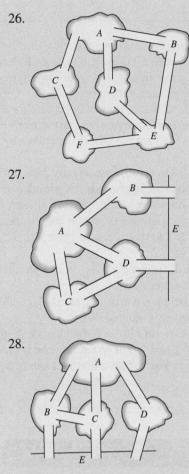

27.

28.

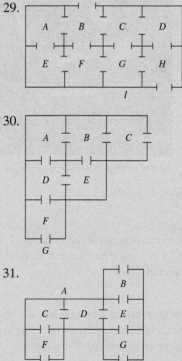

For Exercises 29–32, draw a graph for each floor plan using the rooms and the exterior area as vertices and the door openings as the edges. Determine if the graph has an Euler path, an Euler circuit, or neither. If the graph has an Euler path or an Euler circuit, find one.

29.

30.

31.

32.

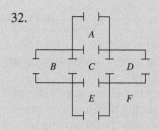

For Exercises 33 and 34, draw a graph for each map using each state as a vertex and each common border as an edge. Determine if the graph has an Euler path or an Euler circuit or neither. If an Euler path or an Euler circuit exists, find one.

33.

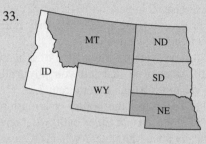

34.

For Exercises 35 and 36, determine if an Euler path or an Euler circuit exists so that a person who plows the roads does not have to pass over any street twice. If an Euler path or an Euler circuit exists, find one. (Hint: Draw a graph and state what each vertex and each edge represents.)

35.

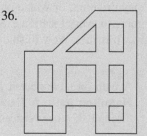

36.

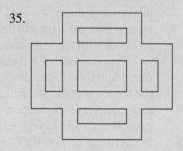

Critical Thinking

37. The study of Euler paths and circuits had its foundation in the bridges of Königsberg, as discussed on page 803 and in the Sidelight on page 819.
    (a) Try to figure out a way to remove one bridge and make it possible to cover the remaining six bridges exactly once while beginning and ending at the same location.
    (b) Repeat (a), but this time add an extra bridge.
    (c) Use Euler's theorem to explain why parts (a) and (b) can or cannot be solved.
38. Now solve the bridges of New York problem. Using the map needed for Exercise 60 in Section 13-1, restrict your graph to only the bridges. Is there any Euler path or Euler circuit? What if you include the tunnels?
39. When a graph that represents the floor plan of a house has an Euler path, what does that tell you physically about the house? What if it has an Euler circuit?
40. When a graph that represents the border relationships between a group of states has an Euler path, what does that tell you physically about the states? What if it has an Euler circuit?

41. Draw some sample graphs and use them to discuss whether or not a graph with a bridge can have an Euler circuit. Then use Euler's theorem to prove your conjecture.
42. Repeat Exercise 41 for a graph with a loop.
43. Fleury's algorithm tells us that to find an Euler path for a graph with two odd vertices, you need to start at one of them. Will you always finish at the other? Describe why your answer is correct.
44. The graph in Figure 13-30(b) doesn't have an Euler path. But there are at least two ways that you can add a single edge so that the resulting graph does have an Euler path. Find them.
45. Is there any way that you can add a single edge to the graph in Figure 13-30(b) so that it has an Euler circuit? Why or why not?
46. Explain why the word "connected" is crucial in the statements of Euler's theorem.
47. In Euler's theorem, the case where a connected graph has exactly one odd vertex is omitted. Why?
48. For the algebraic expression $(1 + 1/n)^n$, find the value if $n = 100, 200, 300, 500$, and $1,000$. What number is the result getting close to? (You'll find a hint in a sidelight within this section.)

| Section 13-3 | Hamilton Paths and Circuits |

LEARNING OBJECTIVES

1. Find Hamilton paths and Hamilton circuits on graphs.
2. Find the number of Hamilton circuits for a complete graph.
3. Solve a traveling salesperson problem using the brute force method.
4. Find an approximate optimal solution using the nearest neighbor method.
5. Find an approximate optimal solution using the cheapest link algorithm.
6. Draw a complete weighted graph based on provided information.

Every college student could use some extra money, and many of them turn to the world of pizza delivery. In business, they say that time is money. That's especially true if you're delivering pizzas: faster delivery = better tips + more deliveries. If the pizza delivery person has five deliveries to make, it would be very much to their advantage to find the most efficient route that reaches all five locations with the least amount of driving time.

This is another good example of the type of problem that graph theory was developed to solve (but to be fair, the theory was developed about a hundred years before pizza delivery). In Section 13-2, we were interested in paths that covered every edge. That's not the case here: the delivery person isn't interested in how many roads they drive on, or which ones. The deliverer's concern is going to every location exactly once in the most efficient way possible. We will study this problem with the aid of *Hamilton paths*.

Adam Crowley/Photodisc/Getty Images

A path on a connected graph that passes through every vertex exactly once is called a **Hamilton path**. A Hamilton path that begins and ends at the same vertex, but passes through all other vertices exactly once, is called a **Hamilton circuit**.

In case you're wondering, while Alexander Hamilton was an interesting guy who had his fingers in a lot of different pies, graph theory wasn't one of them. Hamilton paths and circuits are named for a different Hamilton, who you'll learn a bit about in an upcoming Sidelight.

<table>
<tr><td>

| Example 1 | Finding a Hamilton Path |

</td></tr>
</table>

Example 1 · Finding a Hamilton Path

Find a Hamilton path that begins at vertex A for the graph shown in Figure 13-39.

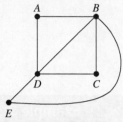

Figure 13-39

SOLUTION

There are many solutions: one of the simplest is alphabetical! The path A, B, C, D, E passes through every vertex exactly once, so it is a Hamilton path.

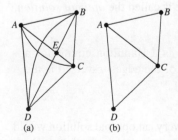

Figure 13-40

Try This One · 1

Find a Hamilton path that begins at vertex C for the graph shown in Figure 13-40.

Example 2 · Finding a Hamilton Circuit

Find a Hamilton circuit for the graph shown in Figure 13-41.

Figure 13-41

SOLUTION

Again, there are many possible answers. If we choose to start and end at vertex A, one choice is A, E, D, F, C, B, A. This path begins and ends at the same vertex and passes through every other vertex exactly once.

1. Find Hamilton paths and Hamilton circuits on graphs.

Try This One · 2

Find a Hamilton circuit that begins and ends at vertex A for the graph in Try This One 1.

Because we know Euler's theorem, we always know whether or not a given graph has an Euler path or circuit. It would be nice if there were a similar result for Hamilton paths. It turns out that there's no single result that will tell us if a graph does or does not have a Hamilton path. On the bright side, there's a condition that guarantees the existence of a Hamilton circuit.

> A **complete graph** is a graph that has an edge connecting every pair of vertices.

Figure 13-42

Figure 13-42(a) is a complete graph: every vertex has an edge that connects it to every other vertex. The graph in Figure 13-42(b) is not complete: there is no edge connecting B and D. There's a very simple way to decide if a graph is complete: a graph with n vertices is complete if it has no loops and every vertex has degree $n - 1$.

A key fact makes complete graphs important in the study of Hamilton circuits.

Complete Graphs and Hamilton Circuits

Every complete graph with more than two vertices has a Hamilton circuit. Furthermore, the number of Hamilton circuits in a complete graph with n vertices is $(n - 1)!$.

It's possible for a non-complete graph to have a Hamilton circuit—just check out Example 2 if you don't believe me. We'll study when this can be done in the Critical Thinking Exercises. But as we will see shortly, the applications we're interested in will usually result in complete graphs, so we'll restrict our attention to that case.

We'll develop the formula for finding the number of Hamilton circuits in a complete graph in Exercise 64. For now, let's think about how many Hamilton circuits we might be looking for in a graph.

Example 3	Finding the Number of Hamilton Circuits in a Graph

Math Note

Two Hamilton circuits are considered to be the same if they pass through the same vertices in the same order, regardless of the vertex where they begin and end.

How many Hamilton circuits are there in the graph in Figure 13-42a?

SOLUTION

First, since the graph is complete, we know it has Hamilton circuits. There are 5 vertices, so the graph has $(5 - 1)!$ Hamilton circuits.

$$(5 - 1)! = 4! = 4 \cdot 3 \cdot 2 \cdot 1 = 24$$

There are 24 Hamilton circuits.

 2. Find the number of Hamilton circuits for a complete graph.

Try This One	3

Find the number of Hamilton circuits for a complete graph with 13 vertices.

Hopefully, the connection between Hamilton circuits and pizza delivery is apparent: a delivery route should start and end at the pizza place and visit every destination exactly once. A problem of this nature has come to be known as a *traveling salesperson problem*. Folks who travel for business are interested in the most efficient way to visit all of their accounts. The "efficient" part could refer to time, miles, or maybe cost of travel.

The graph in Figure 13-43 shows the distances in miles between four cities that a saleswoman needs to visit in one day.

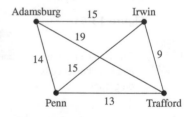

Figure 13-43

This type of graph, with the edges labeled with some quantity of interest, is called a **complete weighted graph**, and the numbers are called **weights**. In this case they are distances, but they could also be travel times, cost of flights, or something else relevant to the situation.

Looking at Figure 13-43, suppose that the saleswoman, who lives in Adamsburg, decides to visit Trafford, then Irwin, then Penn, and then return home. She would travel $19 + 9 + 15 + 14 = 57$ miles. This is found by adding the weights of the edges. Is this the shortest route in terms of miles? In this case, the shortest route is called the *optimal solution*.

Digital Vision/Digital Vision/ Getty Images

Math Note

A graph for a traveling salesperson problem is almost always complete because the salesperson would want to be able to travel directly between any pair of cities. Of course, this guarantees that there are Hamilton circuits.

The **optimal solution** for a traveling salesperson problem is a Hamilton circuit for a complete weighted graph for which the sum of the weights of the edges traversed is the smallest possible number.

Thinking back to our friends in the exciting world of pizza delivery, an optimal solution would represent the shortest and/or quickest way to hit all delivery locations and get back to pick up the next batch. The most obvious way to find an optimal solution is to find the total weights for all possible Hamilton circuits. The lack of subtlety involved justifies the name: *the brute force method*.

The Brute Force Method for Solving a Traveling Salesperson Problem

Step 1 Draw a complete weighted graph for the problem.

Step 2 List all possible Hamilton circuits.

Step 3 Find the sum of the weights of the edges for each circuit.

The circuit with the smallest sum is the optimal solution.

Example 4 | **Using the Brute Force Method**

For the traveling salesperson problem illustrated by Figure 13-43, find the optimal solution.

SOLUTION

Step 1 Draw the graph. This has been done for us.

Step 2 List all possible Hamilton circuits starting at one particular vertex. In this case, we chose vertex A. Note that there are 4 vertices, so there are $(4 - 1)!$, or 6, different Hamilton circuits. They are listed in the table below.

Step 3 Find the sum of the weights of the edges for each circuit. These are also shown in the table below.

Hamilton Circuit	Sum of Weights
A, I, T, P, A	$15 + 9 + 13 + 14 = 51$ miles
A, I, P, T, A	$15 + 15 + 13 + 19 = 62$ miles
A, T, I, P, A	$19 + 9 + 15 + 14 = 57$ miles
A, T, P, I, A	$19 + 13 + 15 + 15 = 62$ miles
A, P, T, I, A	$14 + 13 + 9 + 15 = 51$ miles
A, P, I, T, A	$14 + 15 + 9 + 19 = 57$ miles

Math Note

It makes perfect sense that there are two optimal solutions to Example 4; the second circuit is simply the first one covered in the opposite order.

There are two optimal solutions: A, I, T, P, A, and A, P, T, I, A, both covering 51 miles.

Try This One | **4**

3. Solve a traveling salesperson problem using the brute force method.

The driving times in minutes between four cities are shown in the graph in Figure 13-44. Find the optimal solution for a copy machine repair technician starting in city A who has to visit a location in each city and wants to minimize driving time.

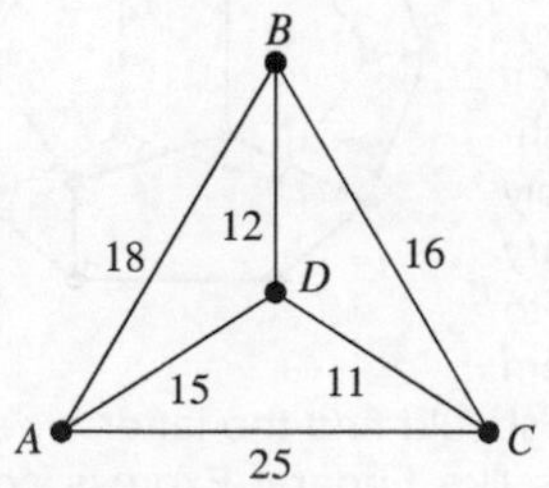

The brute force method worked very nicely in Example 4, but that's largely because there are only four vertices. If the salesperson has 10 different companies to visit, there would be 9!, or 362,880 different possibilities! Without a computer program that can both find Hamilton circuits and then add their weights, this would obviously be unreasonable. Worse still, if there are 16 locations, there are 15!, or about 1.3 *trillion* different circuits! This would take days even for a very capable computer. And the number of possible circuits grows very rapidly beyond that. Clearly, another method would be a good idea.

A second method for finding the optimal circuit is called the *nearest neighbor method*. This method doesn't always give the optimal solution, but it does give an approximation to the optimal solution: there might be a shorter route than the one given by the nearest neighbor method, but finding it by using the brute force method may be too time-consuming.

Figure 13-44

The Nearest Neighbor Method for Finding an Approximate Solution to a Traveling Salesperson Problem

Step 1   Draw a complete weighted graph for the problem.

Step 2   Starting at a designated vertex, pick the edge with the smallest weight and move to the second vertex.*

Step 3   At the next vertex, pick the edge with the smallest weight that doesn't go to a vertex already used.*

Step 4   Continue until the circuit is completed.

The sum of the weights is an approximation to the optimal solution.

*In case the weights of two edges are the same, pick either one.

| Example 5 | Using the Nearest Neighbor Method |

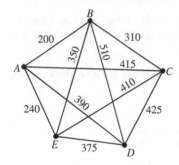

Figure 13-45

A regional manager decides that she needs to visit all of the company locations in her region during the last 2 weeks of May. She lives in city A and needs to visit all of cities B, C, D, and E, then return home. The driving times in minutes between the cities are shown in Figure 13-45. Use the nearest neighbor method to approximate the route that will give her the least amount of driving time.

SOLUTION

Starting at vertex A, the edge with the shortest driving time takes her to vertex B first. From there, the shortest driving time (without returning to A) is to vertex C. Next, the shortest driving time without returning to B is to vertex E. At that point, the only city left not visited is D, so we go there, then return to A. The circuit is A, B, C, E, D, A. The sum of the weights (driving times) is $200 + 310 + 410 + 375 + 390 = 1,685$. The approximate optimal solution is driving for 1,685 minutes, or 28 hours and 5 minutes, using the circuit A, B, C, E, D, A.

| CAUTION |

In Example 5, the nearest neighbor method doesn't find the actual optimal solution. There are two paths that have a driving time of 1,550 minutes. Remember that the nearest neighbor method finds an *approximate* optimal solution. The question is whether checking all 120 possible routes is worth saving the extra time in driving in this case. But once we get past maybe 10 vertices, it's just not very realistic to use the brute force method anymore.

Sidelight   **Hamilton's Puzzle**

Hamilton circuits are named in honor of the 19th-century physicist, astronomer, and mathematician William Rowan Hamilton. Hamilton made a wide variety of important contributions to several areas of math. He didn't spend a lot of time on graph theory, but in introducing a curious puzzle in 1857, he ushered in the study of the circuits that now bear his name.

The puzzle in question was a wooden dodecahedron, which is a three-dimensional figure made of 12 faces that are regular pentagons. There was a peg at each of the 20 vertices, and the point was to attach a string to one peg, then find a route that follows the edges of the pentagons, and has the string visit every peg exactly once.

Hamilton was able to represent the puzzle using the type of graph we've studied in this chapter, and was able to

develop a solution using algebraic methods. Unfortunately, his solution could not be generalized to other graphs.

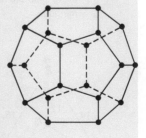

Today, over 150 years later, mathematicians are still searching for a method that can reliably find Hamilton circuits for arbitrary graphs, and the optimal solution to a traveling salesperson problem without using brute force. If you should find the latter, by all means contact delivery services like Federal Express and UPS—you would become a very, very wealthy person very, very quickly.

4. Find an approximate optimal solution using the nearest neighbor method.

Try This One 5

The owner of a retail electronics franchise has locations in four different cities. He's planning a trip to visit all four, and the one-way plane fares between the cities are shown on the graph in Figure 13-46. Use the nearest neighbor method to find the approximate optimal route in terms of cost to visit the locations starting at city *P*.

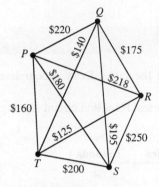

Figure 13-46

Since the nearest neighbor method isn't guaranteed to find the true optimal solution to a traveling salesperson problem, there are other methods that can be used to find approximate solutions. In some cases, you might be able to find a better solution than the one provided by the nearest neighbor method. One such method is called the *cheapest link algorithm*. It's similar to the nearest neighbor method, but instead of choosing adjacent edges to form our path, we'll piece it together from potentially different locations, making sure it becomes an actual connected Hamilton circuit at the end.

The Cheapest Link Algorithm

Step 1   Draw a complete weighted graph for the problem.

Step 2   Pick the edge with the smallest overall weight.*

Step 3   Pick the next cheapest edge that doesn't (a) enclose a smaller circuit that doesn't reach every vertex or (b) result in three chosen edges coming from the same vertex.*

Step 4   Repeat Step 2 until the Hamilton circuit is complete.

*If two qualifying edges have the same weight, pick either one.

Example 6   Using the Cheapest Link Algorithm

Math Note

When using the cheapest link algorithm, you might find it helpful to highlight or color the edges as you choose them.

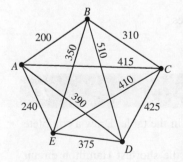

Use the cheapest link algorithm to solve the problem from Example 5.

SOLUTION

The complete weighted graph for the problem, shown previously in Figure 13-45, is repeated in the margin for reference. The smallest overall weight (200) is the edge connecting *A* and *B*, so we choose that first. The next smallest connects *A* and *E* (240), followed by the one connecting *B* and *C* (310). Now things get interesting: the edge connecting *B* and *E* has the next smallest weight (350), but if we choose that we'll have three edges coming from *B* and we'd be visiting that city twice. So we reject that side. The next smallest connects *E* and *D;* that one is fine, so we add it to our circuit. The edge with weight 390 would be next, but won't work because it results in three edges coming from *A*. The edge with weight 410 has to be rejected because it results in a circuit that doesn't include vertex *D*, while the one with weight 415 results in three edges coming from *A*. The next smallest after that (425) completes our Hamilton circuit, which is *A, B, C, D, E, A* or *A, E, D, C, B, A* depending on the direction our manager friend decides to take.

The sum of the weights (driving times) is 200 + 240 + 310 + 375 + 425 = 1,550. This improves upon the solution we got using the nearest neighbor method (and in fact happens to be the actual optimal solution). Remember, though: like the nearest neighbor method, the cheapest link algorithm is never guaranteed to find the actual optimal solution.

5. Find an approximate optimal solution using the cheapest link algorithm.

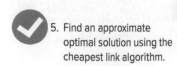

Try This One   6

Use the cheapest link algorithm to find an approximate solution to the problem in Try This One 5. How does it compare to the solution found using the nearest neighbor method?

In the examples we've looked at so far, a complete weighted graph has been provided. If we are instead given information like distances, plane fares, or driving times, we can use them to draw a complete weighted graph.

| **Example 7** | **Drawing a Complete Weighted Graph** |

(a) The distances in miles between four cities are shown in the table. Draw a complete weighted graph for the information.
(b) Use the two methods we learned for approximating the shortest Hamilton circuit starting and ending at Corning. Are the results the same?

	Corning	Mansfield	Elmira	Towanda
Corning (C)	–	30	21	54
Mansfield (M)	30	–	35	38
Elmira (E)	21	35	–	37
Towanda (T)	54	38	37	–

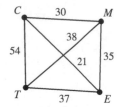

Figure 13-47

SOLUTION

(a) **Step 1** Draw and label the four vertices as C, M, E, and T.

Step 2 Connect each vertex with every other vertex using line segments (edges).

Step 3 Place the mile numbers (weights) on each segment using the information in the table. See Figure 13-47.

(b) First, we'll try the nearest neighbor method. Starting at C, the closest vertex is E. From E, the closest vertex other than C is M. Now we don't have a choice: going back to C would complete a circuit that doesn't include T, so we have to go to T next, then back to C. The sum of the weights for this circuit (C, E, M, T, C) is

$$21 + 35 + 38 + 54 = 148 \text{ miles}$$

Next, the cheapest link method. The shortest distance overall is between C and E, and the next shortest is between C and M. After that, the edge connecting M and E has the smallest remaining weight, but that would complete a circuit not containing T. The next smallest weight is the edge between T and E, followed by the edge from T to M. This completes a circuit C, E, T, M, C, with sum of weights

$$21 + 37 + 38 + 30 = 126 \text{ miles}$$

So the better of the two choices is the second circuit.

Try This One   7

(a) The distances in miles between four cities are shown in the table. Draw a complete weighted graph for the information.
(b) Use the two methods we learned for approximating the shortest Hamilton circuit starting and ending at New Stanton. Are the results the same?

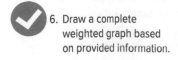

6. Draw a complete weighted graph based on provided information.

	Youngwood	Scottdale	Mt. Pleasant	New Stanton
Youngwood	–	10	12	4
Scottdale	10	–	6	8
Mt. Pleasant	12	6	–	11
New Stanton	4	8	11	–

Answers to Try This One

1 One of many possible answers is C, B, A, D, G, F, E.

2 The two possibilities are A, B, C, G, F, E, D, A and A, D, E, F, G, C, B, A.

3 479,001,600

4 Either A, B, C, D, A or A, D, C, B, A.

5 P, T, R, Q, S, P, which costs $835.

6 P, S, Q, T, R, P, which costs $858. This is a totally different circuit and costs $23 more.

7 (a)

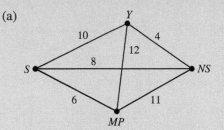

(b) Nearest neighbor: NS, Y, S, MP, NS, 31 miles
Cheapest link: NS, Y, MP, S, NS, 30 miles

Exercise Set 13-3

Writing Exercises

1. What is the difference between a Hamilton path and an Euler path?
2. What is the difference between a Hamilton path and a Hamilton circuit?
3. Give an example of a problem in our world that can be solved using a Hamilton path that's not a circuit and one that can be solved using an Euler path.
4. Give an example of a problem in our world that can be solved using a Hamilton circuit.
5. What does it mean for a graph to be complete?
6. What does it mean for a graph to be weighted?
7. Describe what a typical traveling salesperson problem entails. What is an optimal solution?
8. Describe how to use the brute force method to find an optimal solution.
9. What does it mean to find an approximate optimal solution to a traveling salesperson problem?
10. Describe two methods used to find an approximate optimal solution to a traveling salesperson problem.
11. What are some things that the weights on a graph can represent in a traveling salesperson problem?
12. Why is the brute force method not a good solution to real traveling salesperson problems, even with the use of computers to calculate circuits and weights?

Computational Exercises

For Exercises 13–20, find two different Hamilton paths.

13.

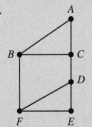

14.

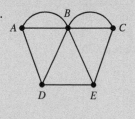

15.

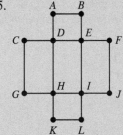

16.

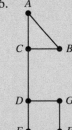

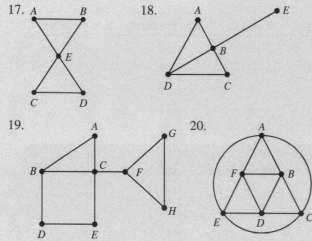

17.

18.

19.

20.

For Exercises 21–26, find two different Hamilton circuits.

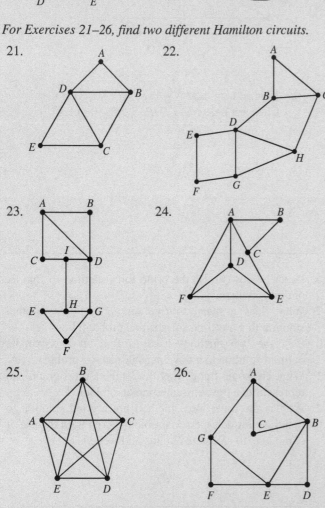

21.

22.

23.

24.

25.

26.

For Exercises 27–30, find the number of Hamilton circuits if a complete graph has the indicated number of vertices, and discuss the practicality of using the brute force method to find an optimal circuit.

27. 3

28. 6

29. 9

30. 11

For Exercises 31 and 32, use the brute force method to find the optimal solution for each weighted graph.

31.

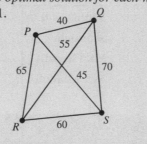

32.

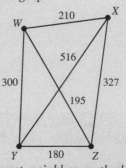

For Exercises 33–36, use the nearest neighbor method to approximate the optimal solution. Start at vertex A in each case.

33.

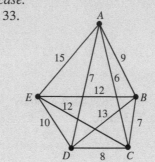

34.

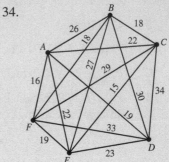

35.

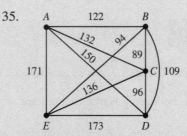

36.

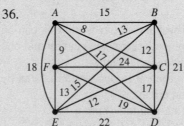

In Exercises 37–40, use the cheapest link algorithm to find an approximate optimal solution starting at vertex A for the given graph. Then compare the result to the nearest neighbor method.

37. Exercise 33

38. Exercise 34

39. Exercise 35

40. Exercise 36

Applications in Our World

For Exercises 41–44, use the information in the table shown.

Distances (in miles) between Cities

	Pittsburgh	Philadelphia	Baltimore	Washington
Pittsburgh	–	305	244	245
Philadelphia	305	–	100	136
Baltimore	244	100	–	38
Washington	245	136	38	–

41. Draw a complete weighted graph for the information in the table.
42. Find the shortest route from Pittsburgh to all other cities and back to Pittsburgh using the brute force method.
43. Use the nearest neighbor method to find an approximation of the shortest circuit starting and ending at Pittsburgh.

How do the route and mileage compare to the actual optimal solution?
44. Use the cheapest link algorithm to find an approximation of the shortest circuit starting and ending at Pittsburgh. How do the route and mileage compare to the routes from Exercises 42 and 43?

For Exercises 45–48, use the information in the table shown.

Air Fares between Cities

	New York	Cleveland	Chicago	Baltimore
New York	–	$375	$450	$200
Cleveland	$375	–	$250	$300
Chicago	$450	$250	–	$325
Baltimore	$200	$300	$325	–

45. Draw a complete weighted graph for the information in the table.
46. Find the cheapest route from Chicago to all other cities and back to Chicago using the brute force method.
47. Use the nearest neighbor method to find an approximation of the cheapest route starting and ending in Chicago.

How do the route and cost compare to the actual optimal solution?
48. Use the cheapest link algorithm to find an approximation of the cheapest route starting and ending in Chicago. How do the route and cost compare to the routes from Exercises 46 and 47?

A pizza delivery person has five prearranged deliveries scheduled for Super Bowl Sunday around 4:00 P.M. The driving times between those locations and the pizza place are shown in the table. Use this information for Exercises 49–52.

Driving Time (Minutes and Seconds)

	Home Slice Pizza	1	2	3	4	5
Home Slice Pizza	–	2:15	4:30	1:05	3:00	2:15
1	2:15	–	5:45	1:40	2:15	4:45
2	4:30	5:45	–	5:00	5:30	1:15
3	1:05	1:40	5:00	–	2:30	2:30
4	3:00	2:15	5:30	2:30	–	1:15
5	2:15	4:45	1:15	2:30	1:15	–

49. Draw a complete weighted graph for the information in the table.
50. Explain why the brute force method is not a great idea for finding the shortest circuit that begins and ends at Home Slice in terms of driving time.
51. Use the nearest neighbor method to find an approximate optimal route. What's the total driving time?

52. Use the cheapest link algorithm to find an approximate optimal route. How do the route and driving time compare to your answer from Exercise 51?

When planning routes, distance isn't always the key factor, as traffic can make a big difference in driving time. For Exercises 53–56, suppose that on a trip to Southern California, you decide that you'd like to visit the corner of Hollywood and Vine, the Rose Bowl in Pasadena, Downtown Disney in Anaheim, and Venice Beach.

53. Use Google maps to find driving distances and times between all of these sites, and draw a complete weighted graph with driving times used as weights. Use the "in traffic" times that go along with the shortest route in terms of distance.

54. Draw another complete weighted graph, this time with driving distances used as weights. Use the same routes as in Exercise 53.

55. Find the most efficient way in terms of driving time to start and finish at the Rose Bowl and visit all the other

locations using (a) the brute force method and (b) the nearest neighbor method. Did you get the same results?

56. Find the most efficient way in terms of driving distance to start and finish at the Rose Bowl and visit all the other locations using (a) the brute force method and (b) the nearest neighbor method. How do the results compare to those for driving time?

57–60. Repeat Exercises 53 through 56, choosing four sites that you might be interested in visiting, but this time use the cheapest link algorithm rather than the nearest neighbor method. They can be pretty much anywhere as long as it's reasonable to drive between them. (Honolulu and Phoenix would be silly, if not fatal.)

Critical Thinking

61. Find a road atlas (paper or online) that has a mileage chart. Pick five cities and make a table of miles between the cities. Then make a weighted graph and find an approximate optimal circuit.

62. Draw a graph with more than four vertices, then find a Hamilton circuit that is also an Euler circuit.

63. Draw a graph with more than four vertices, then find a Hamilton circuit that is not an Euler circuit.

64. Suppose that you have a complete graph with n vertices, and you choose one vertex to begin and end as many Hamilton circuits as you can find.
    (a) How many different vertices can you choose to visit after the starting vertex? What is it about the graph that ensures you can visit any of the others?
    (b) After you've visited the beginning vertex and one other, how many remaining choices of vertex do you have to visit next?
    (c) Explain how you can continue this process to show that a complete graph with n vertices has $(n - 1)!$ Hamilton circuits beginning at any vertex.

65. Is there more than one complete graph with a given number of vertices? (Think about what it means for two graphs to be equivalent.) Explain.

66. If a complete graph has n vertices, every vertex has the same degree. What is it? How can you tell if a complete graph has an Euler path or circuit?

67. After defining a complete graph, we remarked that you can tell that a graph with n vertices is complete if every vertex has degree $n - 1$. Explain why that's true.

Every complete graph has at least one Hamilton circuit. But what about graphs that aren't complete? We saw in Example 2 that it can be done, but it would be nice to know for sure whether or not it's worth the effort to try to find a Hamilton circuit. There's a result called Dirac's theorem that does just that. We'll discuss this in Exercises 68–72.

If each vertex of a connected graph with n vertices (where n > 3) is adjacent to at least n/2 vertices, then the graph has a Hamilton circuit.

68. Notice that this requires that n is more than 3. Why is this not a big deal at all when trying to decide if a graph has a Hamilton circuit?

69. None of the graphs in Exercises 13–20 is complete. Use Dirac's theorem to decide which, if any, have a Hamilton circuit.

70. Find a Hamilton circuit for any of the graphs you identified in Exercise 69 that have one.

71. Suppose that a graph representing travel by air among six cities is not complete. What would that mean in terms of travel?

72. Without using Dirac's theorem, explain why a graph with a vertex of degree 1 can't have a Hamilton circuit.

73. In this section, we used distances, travel times, and costs to represent the weights on graphs. Think of at least two other quantities that would make sense as weights in an applied problem.

74. If you were to find a simple solution to the traveling salesperson problem, list five companies you would contact to sell your solution, and explain why you would contact each one.

Section 13-4	**Trees**

According to the Federal Highway Administration, there are over 4 million miles of road in the United States. Think for a minute about how many different routes you could take to get home from school. Our road system might be incredibly convenient, but it's not exactly what you would call efficient—there are many more routes than we really need to get from one place to another.

LEARNING OBJECTIVES

1. Decide if a graph is a tree.
2. Find a spanning tree for a graph.
3. Find a minimum spanning tree for a weighted graph.
4. Apply minimum spanning trees to problems in our world.

In this section, we will look at problems of efficiency: we'll try to build graphs that connect vertices with the smallest number of edges between them. The graphs that we will construct are called *trees*.

A **tree** is a graph in which any two vertices are connected by exactly one path.

Figure 13-48(a) shows a graph that is a tree, while the graph in Figure 13-48(b) is not a tree. In the first case, if you pick any pair of vertices, there is exactly one path connecting them. In the second case, given any pair of vertices, there are two different paths connecting them. For example, A and C are connected by the paths A, B, C and A, D, C.

The colored box provides a few properties of trees that will help us to recognize when a graph is or is not a tree.

Comstock Production Department/
Comstock Images/Alamy Stock Photo

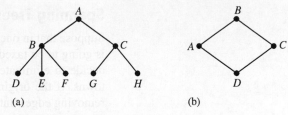
(a)   (b)

Figure 13-48

Math Note

If the tree in Figure 13-48(a) looks a tad familiar, it probably should. We used tree diagrams in Chapter 10 to help compute the number of outcomes in probability problems.

Properties of Trees

1. A tree has no circuits.
2. Trees are connected graphs.
3. Every edge in a tree is a bridge.
4. A tree with *n* vertices has exactly *n* − 1 edges.

Example 1   **Recognizing Trees**

Which of the graphs in Figure 13-49 are trees?

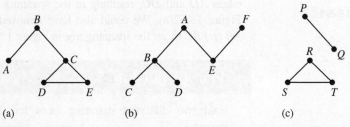
(a)   (b)   (c)

Figure 13-49

SOLUTION

The graph in Figure 13-49(a) is not a tree. The most obvious reason is that it has a circuit: C, E, D, C. It also has the same number of edges as vertices, and three of the edges (the ones that make up the circuit) aren't bridges.

The graph in Figure 13-49(b) is a tree. There is exactly one path to get from any vertex to any other. The graph is connected, has no circuits, and every edge is a bridge. Finally, there are six vertices and five edges.

The graph in Figure 13-49(c) is not a tree because it's not connected. It also has five vertices and only three edges.

1. Decide if a graph is a tree.

Try This One 1

Which of the graphs in Figure 13-50 are trees?

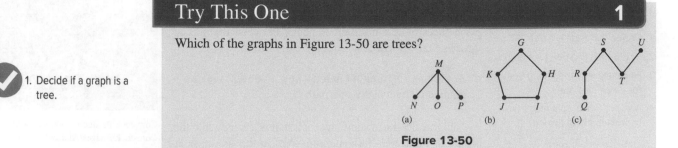

Figure 13-50

Spanning Trees

Suppose that in one county, a legislature intent on not getting reelected decides that each town is going to be taxed based on how many roads it has. Facing fiscal disaster, the town leaders decide to eliminate all roads except the minimum number necessary to travel between the towns. If the original system of roads is represented by a graph, the problem would entail removing edges until the resulting graph is a tree. Such a graph is called a *spanning tree.*

> A **spanning tree** for a graph is a tree that results from the removal of as many edges as possible from the original graph without making it disconnected.

Notice that in the definition we said "a spanning tree" not "the spanning tree." Most graphs will have more than one spanning tree.

Example 2	Finding a Spanning Tree

Figure 13-51

Find two different spanning trees for the graph shown in Figure 13-51.

SOLUTION

The graph has five vertices, so a spanning tree will have four edges. The original graph has six edges, so we need to remove two edges without making the graph disconnected. One way to accomplish this is to remove edges *AD* and *DC*, resulting in the spanning tree in Figure 13-52(a). We could also have removed edges *AB* and *BC* to get the spanning tree in Figure 13-52(b).

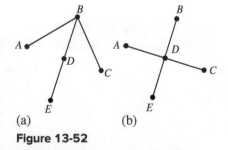

(a)   (b)

Figure 13-52

Try This One 2

Find two different spanning trees for the graph shown in Figure 13-53.

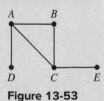

Figure 13-53

2. Find a spanning tree for a graph.

Spanning trees can be used to find the most efficient solution to a variety of problems involving cost, distance, and time. In order to do this, we'll draw a weighted graph that represents the situation. For example, suppose that a railroad company needs to connect three industrial sites with track. The distances between the sites are shown in Figure 13-54(a). This is exactly the convenience vs. efficiency issue we talked about earlier. They could lay track on all three sides, which would be convenient for travel, since you could travel from any site to either of the other two. But it would be more cost effective to find a spanning tree that connects all three sites with the least total amount of track.

Three spanning trees are shown in Figure 13-54(b).

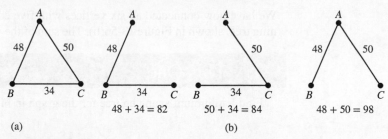

(a)   (b)

Figure 13-54

Of the three spanning trees shown, the first allows for the minimum amount of track, and is the most efficient solution. We call this a *minimum spanning tree*.

> A **minimum spanning tree** for a weighted graph is the spanning tree for that graph that has the smallest possible sum of the weights.

For a very simple weighted graph, the minimum spanning tree can be drawn by trying out all the possible choices; but when the weighted graph is more complicated, a systematic procedure is helpful. A procedure known as Kruskal's algorithm can be used.

Math Note

Don't forget that a minimum spanning tree has to have one fewer edge than the number of vertices.

Kruskal's Algorithm

To construct a minimum spanning tree for a weighted graph:

Step 1 Choose the edge with the lowest weight, and highlight it in color.

Step 2 Choose the unmarked edge with the next lowest weight that does not form a circuit with the edges already highlighted, and highlight it.

Step 3 Repeat until all vertices have been connected.

Notes: 1. If more than one edge could be chosen at any stage, pick one randomly.
2. The number of steps in any given problem depends on how many edges need to be removed.

Example 3 **Finding a Minimum Spanning Tree**

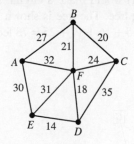

Figure 13-55

Use Kruskal's algorithm to find a minimum spanning tree for the weighted graph shown in Figure 13-55.

SOLUTION

Step 1 Highlight the lowest weighted edge, which is *ED*. See Figure 13-56(a).

Step 2 Highlight the next lowest weighted edge, which is *DF*.

Step 3 The next lowest weight is 20, which is edge *BC*. Highlighting this edge won't make a circuit with the edges already highlighted, so this is a good choice.

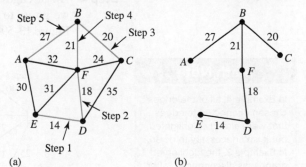

(a)   (b)

Figure 13-56

Step 4 Edge *BF* has the next lowest weight, and it also won't form a circuit if we highlight it.

Step 5 The next lowest weight is 24, but highlighting edge *FC* would form a circuit (*B*, *F*, *C*). Instead, we go to the next lowest, which is *AB*.

We have now connected all six vertices with five edges, so we have found a minimum spanning tree, shown in Figure 13-56(b). The sum of the weights is $14 + 18 + 20 + 21 + 27 = 100$.

Try This One   3

3. Find a minimum spanning tree for a weighted graph.

Find a minimum spanning tree for the graph in Figure 13-57.

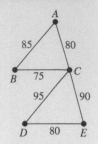

Figure 13-57

The remaining two examples are applications of minimum spanning trees.

| **Example 4** | **Spanning Trees in Agriculture** |

Pixtal/Age Fotostock

A flower farm has five greenhouses that need an irrigation system. The distances in feet between each of the greenhouses are shown in Figure 13-58. Use a minimum spanning tree to connect the greenhouses, and find the minimum amount of pipe needed.

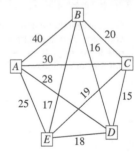

SOLUTION

Step 1 Highlight the edge with the lowest weight, *CD* (15 feet). See Figure 13-59(a).

Step 2 Highlight the edge with the next lowest weight, *DB* (16 feet).   **Figure 13-58**

Step 3 The next lowest weight is 17 feet (edge *BE*), and choosing this edge won't make a circuit with the edges already highlighted.

Step 4 All of edges *ED* (18 feet), *EC* (19 feet), and *BC* (20 feet) would make a circuit, so we go up to edge *AE* (25 feet) to complete our spanning tree. The tree is shown in Figure 13-59(b). The minimum amount of pipe needed is $15 + 16 + 17 + 25 = 73$ feet.

Math Note

In Example 4, all of the edges chosen connected to edges that were already highlighted, but it isn't necessary to do so. In Example 3, the highlighted graph was disconnected after Step 3.

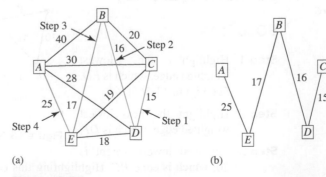

(a)   (b)

Figure 13-59

Try This One 4

A community college campus is being designed to be environmentally friendly. There will be six buildings, with the lengths in yards between them shown in Figure 13-60. The plan is to construct the minimum length of sidewalks that will connect all of the buildings. Use a minimum spanning tree to connect the buildings and find the minimum amount of sidewalk needed.

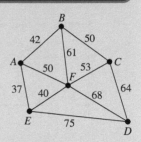

Figure 13-60

CAUTION

In any application problem involving a weighted graph, make sure you think carefully about what solves the problem: a minimum spanning tree or a Hamilton circuit. In many cases, spanning tree problems are about connecting locations, while Hamilton circuit problems are about traveling between locations and returning to a starting point.

Example 5 | Spanning Trees in Civil Engineering

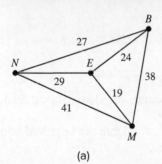

(a)

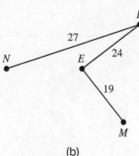

(b)

Figure 13-61

A philanthropic group is planning on building roads to connect three remote villages in Uganda with the nearest large city, Entebbe. The distances in kilometers between locations are shown in the table, and construction is estimated to cost $230,000 per kilometer. Find the lowest possible cost of construction to connect the four locations.

	Entebbe	Nbale	Makese	Bamuli
Entebbe	–	29	19	24
Nbale	29	–	41	27
Makese	19	41	–	38
Bamuli	24	27	38	–

SOLUTION

Draw a complete weighted graph and find the minimum spanning tree as shown in Figure 13-61. The minimum number of kilometers is $19 + 24 + 27 = 70$. Multiplying this by the cost of $230,000 per kilometer, we get $16.1 million.

Try This One 5

A nursing outreach program in an underdeveloped area of Appalachia has set up four medical clinics that it would like to connect with wires to share an Internet connection. The distances in miles between the locations are shown in the table, and it will cost $340 per mile to run the lines. Find the lowest possible cost of construction to connect the four centers.

	Adamsburg	Irwin	Penn	Trafford
Adamsburg	–	60	56	76
Irwin	60	–	60	36
Penn	56	60	–	52
Trafford	76	36	52	–

✓ 4. Apply minimum spanning trees to problems in our world.

Answers to Try This One

1   Graphs (a) and (c) are trees.

2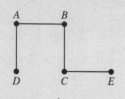

3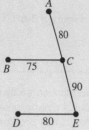

4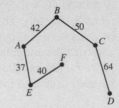

The length of sidewalk needed is 233 yards.

5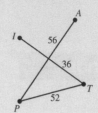

The cost is $48,960.

Exercise Set   13-4

Writing Exercises

1. Explain the difference between a graph and a tree.
2. In some sense, a tree is the opposite of a complete graph. Explain.
3. What is a spanning tree for a graph?
4. What is a minimum spanning tree for a weighted graph?
5. Describe Kruskal's algorithm for finding a minimum spanning tree.
6. Describe a problem in our world that can be solved using a minimum spanning tree.

Computational Exercises

For Exercises 7–16, decide whether or not each graph is a tree. If it isn't, explain why.

7.

8.

9.

10.

11.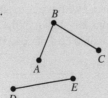

12.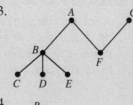

13.

14.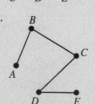

15.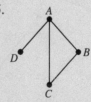

16.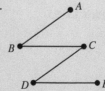

For Exercises 17–22, find a spanning tree for each. (Answers may differ from those given in the answer section.)

17.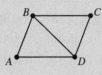

20.

18.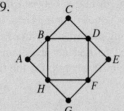

21.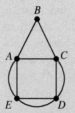

19.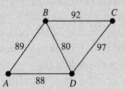

22.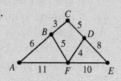

25.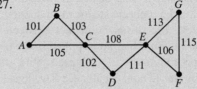

26.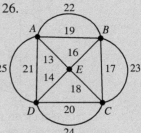

27.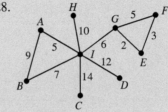

28.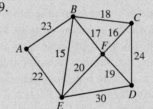

For Exercises 23–30, use Kruskal's algorithm to find a minimum spanning tree for each. Give the total weight for each one.

23.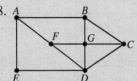

24.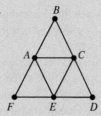

29.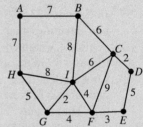

30.

Applications in Our World

31. At a convention, the manager needs to run electricity to each table. The distances in feet are shown. Use a minimum spanning tree to find the shortest amount of wire necessary. How much wire will be needed?

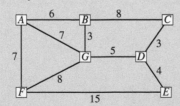

32. Eight buildings in a housing complex are connected by sidewalks. The maintenance manager wants to clear the shortest path connecting all buildings when it snows. Use a minimum spanning tree to find the shortest distance. The distances shown are in feet. Find the length of the path.

33. A committee decides to build an outdoor fitness facility with five stations in a park. Using a minimum spanning tree, find the shortest distance between all stations. If a path were to be built between the stations, find its length. The distances are given in feet.

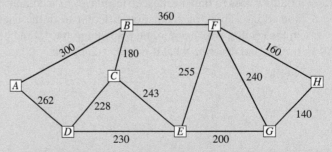

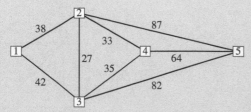

	Main	A	B	C	D
Main	–	2,200	3,250	2,370	2,960
A	2,200	–	1,450	3,820	4,050
B	3,250	1,450	–	3,700	4,100
C	2,370	3,820	3,700	–	1,110
D	2,960	4,050	4,100	1,110	–

34. As a new suburban neighborhood is being built, the local utility company needs to run underground wires to get power to all of the houses. It plans on four main distribution nodes that will in turn be connected to the main distribution center for that area. The distances in feet between the nodes and the distribution center are supplied in the table. It will cost $290 per foot to install the buried cable. Use a minimum spanning tree to connect all necessary locations at the lowest possible cost. What is the minimum cost?

35. When the floor plan of a home is represented by a graph as we saw in Section 13-1, what does it mean if the graph is a tree?

36. Two buddies decide to take a summer road trip to see baseball games in five different major and minor league ballparks: Pittsburgh, Cleveland, Columbus, Cincinnati, and Detroit. Use Google maps to find driving distances between these cities, and draw a complete weighted graph. Then use Kruskal's algorithm to find a minimum spanning tree, and use the nearest neighbor method to find an approximate optimal route for the trip starting in Pittsburgh. Do the two answers have anything in common?

Critical Thinking

37. In Exercise 73 of Section 13-3, you were asked to write two quantities that would make sense as weights for a complete weighted graph. For each of those quantities, discuss the value of finding a minimum spanning tree.

38. Look back at the graph you drew to represent the bridges and tunnels of New York in Exercise 60 of Section 13-1.
    (a) How many bridges/tunnels would you need to remove to get a spanning tree?
    (b) If we wanted to turn the graph into a weighted graph, what are some quantities we could use as weights?

39. There were a number of situations described in this section and its exercises where a minimum spanning tree could be used to find the most efficient way to accomplish something. Think of a situation different from the ones in the section where a minimum spanning tree would be useful, and describe what it would accomplish.

40. In the last two sections, we used both Hamilton circuits and minimum spanning trees to study weighted graphs. Write a description in your own words of the difference between what is accomplished by a Hamilton circuit and what is accomplished by a minimum spanning tree.

41. Using a road map, an atlas, or the Internet, find five cities and determine the distances between them. Then draw a complete graph connecting the cities. Finally, determine a minimum spanning tree for the graph.

42. Repeat Exercise 41 for five buildings on a campus of a college or university.

43. Which is more likely: an airline route being a tree, or a railway service? Explain what that would mean in each case.

44. Why is finding a minimum spanning tree so much simpler than finding the optimal solution to a traveling salesperson problem?

Section	Important Terms	Important Ideas
13-1	Graph Vertex Edge Loop Equivalent graphs Degree of a vertex Odd vertex Even vertex Adjacent vertices Path Circuit Connected graph Disconnected graph Bridge Graph coloring Chromatic number	**A graph** consists of a finite set of points called vertices and line segments called edges connecting the points. A loop is an edge that begins and ends at the same vertex. The degree of a vertex is the number of edges emanating from the vertex. Adjacent vertices have at least one edge connecting them. A path on a graph is a sequence of adjacent vertices and the edges connecting them that uses no edge more than once. A circuit is a path that begins and ends at the same vertex. A graph is said to be connected if there is at least one path that connects any two vertices; otherwise, it is said to be disconnected. A bridge on a connected graph is an edge that, if removed, makes the graph disconnected. A coloring for a graph is a method of coloring all the vertices so that any pair of adjacent vertices have different colors. The smallest number of colors that can be used to color a graph is its chromatic number. Graph coloring can be used to solve logistic problems like meeting scheduling.
13-2	Euler path Euler circuit Euler's theorem Fleury's algorithm	**Graphs** can be used as models for floor plans, neighborhoods, maps, and many other things. An Euler path is a path that passes through each edge exactly once, while an Euler circuit is a circuit that passes through each edge exactly once. Euler's theorem states that a connected graph that has all even vertices has at least one Euler circuit. If it has exactly two odd vertices, it has at least one Euler path but no Euler circuits. Finally, if a connected graph has more than two odd vertices, it has no Euler circuits or Euler paths. Fleury's algorithm can be used to find an Euler path or Euler circuit when you know that a graph has one.
13-3	Hamilton path Hamilton circuit Complete graph Complete weighted graph Traveling salesperson problem Optimal solution Brute force method Nearest neighbor method Approximate optimal solution Cheapest link algorithm	**A path** on a connected graph that passes through each vertex only once is called a Hamilton path. If a Hamilton path begins and ends at the same vertex, it is called a Hamilton circuit. A complete graph is a graph that contains an edge between any two vertices. The number of Hamilton circuits in a complete graph with n vertices is $(n - 1)!$. A complete weighted graph has weights (numbers) on all edges. The optimal solution on a weighted graph is the circuit for which the sum of the weights of the edges is the smallest. In a traveling salesperson problem, we look for a Hamilton circuit for a complete weighted graph with the smallest possible combined weights. We studied three methods for solving them: in the brute force method, we find the combined weight for all possible Hamilton circuits then pick the smallest. This finds the actual optimal solution. In the nearest neighbor method and cheapest link algorithm, we find an approximate optimal solution, which means it may or may not actually be the optimal solution. In each case, the strategy is to pick edges with the smallest weights to form a Hamilton circuit. In the nearest neighbor method, we pick connected edges in order; in the cheapest link algorithm, we pick edges that may not be connected initially but eventually form a Hamilton circuit.

843

Section	Important Terms	Important Ideas
13-4	Tree Spanning tree Minimum spanning tree Kruskal's algorithm	**A tree** is a graph in which any pair of vertices is connected by exactly one path. A spanning tree is a tree that has been created by removing edges from a graph. A minimum spanning tree for a weighted graph is the spanning tree (out of all possible spanning trees for the graph) that has the smallest sum of the weights. Kruskal's algorithm is a method that can be used to construct a minimum spanning tree for a weighted graph. Start at the lowest weighted edge and pick the next lowest weighted edge that doesn't form a circuit with previously selected edges. Continue until you have included all vertices.

Math in Road Trips REVISITED

John Lund/Tiffany Schoepp/
Blend Images

1. The graph is shown below, with weights representing the distances. Georgia is *G*, Georgia Tech is *T*, Alabama is *B*, and Auburn is *A*.

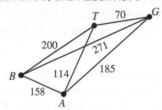

2. This is a question about Hamilton circuits. It turns out that of the six possible routes, four of them are optimal! Using the brute force method, we find that the four paths listed below all cover 613 miles.

   T, *G*, *B*, *A*, *T*; *T*, *G*, *A*, *B*, *T*; *T*, *B*, *A*, *G*, *T*; *T*, *A*, *B*, *G*, *T*

3. This is a spanning tree problem. The minimum spanning tree is shown below, and covers just 342 miles.

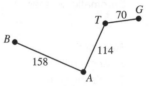

4. The nearest neighbor method gives this route: *T*, *G*, *A*, *B*, *M*, *T*, and totals 990 miles. The cheapest link method gives a different route: *T*, *A*, *B*, *M*, *G*, *T*. This route saves an entire mile! It's 989 miles.

Review Exercises

Section 13-1

Use the graph shown for Exercises 1–8.

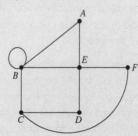

1. Identify the vertices.
2. Is vertex *E* an even or odd vertex?
3. Is vertex *C* an even or odd vertex?
4. Which vertex has a loop?
5. Which vertices are adjacent to vertex *F*?
6. Identify an edge that is a bridge.
7. Is the graph connected?
8. How many edges does the graph have?

9. Are these two graphs equivalent? Explain.

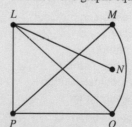

10. Draw a graph that models the region shown in the diagram. Use vertices for land masses and edges for bridges.

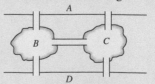

11. Draw a graph that models the floor plan. Use a vertex for each room and the outside area, and an edge for each door.

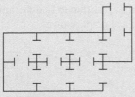

12. Draw a graph that models the border relationships among the counties shown in the map. Use vertices to represent the counties and edges to represent common borders.

13. Find the chromatic number of the graph, and find a coloring that uses that number of colors.
14. Repeat Exercise 13 for the graphs from Exercises 11 and 12.

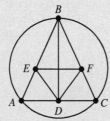

Section 13-2

In Exercises 15–17, decide if the graph has an Euler path or circuit.

15.

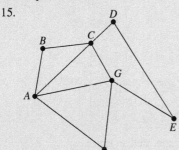

16.

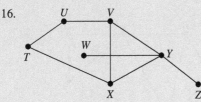

17.

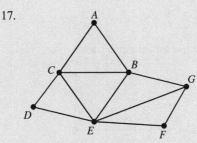

18. One of the graphs in Exercises 15–17 has an Euler circuit. Find one for that graph.
19. One of the graphs in Exercises 15–17 has an Euler path but no circuit. Find an Euler path.

20. Find an Euler circuit for the graph from Exercise 11 if possible. If not, find an Euler path. If neither exists, explain why.
21. Repeat Exercise 20 for the graph from Exercise 12.

Section 13-3

22. Describe the differences between the nearest neighbor method and the cheapest link algorithm.
23. For the given graph, find a Hamilton path.

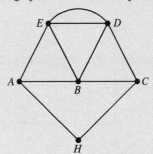

24. For the given graph, find a Hamilton circuit.

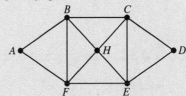

25. If a complete graph has 12 vertices, how many distinct Hamilton circuits does it have?
26. For the weighted graph shown, find an approximate optimal circuit starting at vertex *A* using the nearest neighbor method.

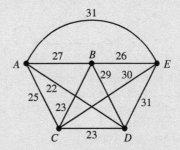

27. The table shows the one-way airfares between four cities. Draw a graph using the information in the table.

	C	D	E	F
C	–	$256	$732	$350
D	$256	–	$560	$197
E	$732	$560	–	$230
F	$350	$197	$230	–

28. Use the brute force method to find the optimal solution for the graph from Exercise 27 starting and ending at city *C*.
29. Use the cheapest link algorithm to find an approximate optimal solution to the traveling salesperson problem for the distances in the table in Exercise 27. How does it compare to the actual optimal solution?

Section 13-4

30. Is the graph a tree? Explain why each is or is not.

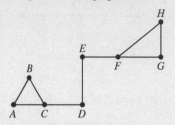

(a)   (b)

31. Find a spanning tree for the graph shown.

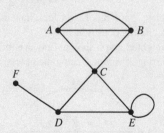

32. Use Kruskal's algorithm to find the minimum spanning tree for the weighted graph shown.

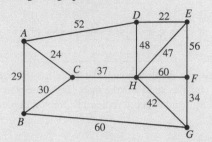

33. Explain the difference between finding a minimum spanning tree for a complete weighted graph and finding the optimal solution to a traveling salesperson problem.

34. A businessperson who travels regularly among the four cities in Exercise 27 would like to build a "flight network" from existing flights that connects the four cities with the least expensive overall cost. Find a minimum spanning tree and the associated cost.

Chapter Test

1. For the following graph:

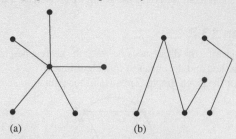

   (a) What is the degree of vertex *C?*
   (b) Which vertex has a loop?
   (c) Describe a path that starts at vertex *A* and passes through vertex *E.*
   (d) Which edge is a bridge?
   (e) Name four vertices adjacent to vertex *C.*
   (f) Which vertices are even? Which are odd?
2. Draw a graph with two bridges, and the disconnected graph that results from removing one of the bridges.
3. (a) Draw a graph for the states shown in the map. Use a vertex for each state and an edge for each common border.

   (b) Find the smallest number of colors the map could be colored with so that no states with a common border have the same color.

4. Draw a graph for the floor plan shown. Use a vertex for each room and the outside area, and an edge for each door.

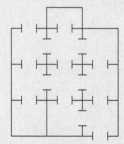

5. (a) For the given graph, find an Euler path.
   (b) Find a Hamilton path. How does it compare to the Euler path?

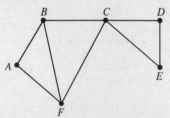

6. Make a spanning tree for the graph shown.

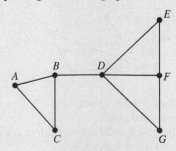

7. For the neighborhood shown in the diagram, draw a graph and find a way to plow the roads without doing any street twice.

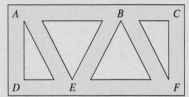

8. A salesperson who lives in Adamsburg must visit Trafford, White Oak, and Turtle Creek. The distances between the cities are as follows: Adamsburg to White Oak is 27 miles, Adamsburg to Turtle Creek is 18 miles, Adamsburg to Trafford is 20 miles, Trafford to White Oak is 14 miles, Trafford to Turtle Creek is 19 miles, and White Oak to Turtle Creek is 12 miles. Represent the distances with a complete weighted graph.

9. Use the brute force method to find the shortest distance from Adamsburg through all cities and back to Adamsburg. What is the shortest distance in miles?

10. Use the nearest neighbor method and cheapest link algorithm to approximate an optimal solution. Is either the same as the result obtained by the brute force method?

11. Find a minimum spanning tree for the graph in Exercise 8.

12. Decide whether the problem can be solved using Euler paths, Hamilton paths, or trees.

(a) A civil engineer needs to drive once over every road in a neighborhood to check for salt damage after an especially snowy winter.

(b) A group of friends goes to Epcot in Orlando on a 97 degree day and wants to visit seven different attractions while doing the least possible amount of walking.

(c) A UPS driver has 22 deliveries to make before lunch and wants to find the most time-efficient route.

(d) Two roommates plan a spring break road trip. The plan is to visit friends at four different colleges then return home while driving the shortest distance.

(e) A police officer on patrol is ordered to drive down every street on the west end of town before returning to the station to clock out.

(f) When a new campus building is being planned, the IT team is asked to wire five computer classrooms on the fifth floor to the campus network using the least amount of network cable possible.

Projects

1. As we know, graphs can be used to represent many different types of relationships. Think of at least three different types of relationships among the students in your group that you could model with a graph. Some suggestions: other classes together, other professors you've had in common, campus activities you're involved in, favorite TV shows . . . be creative! Then answer the following questions:

(a) What does each vertex represent?

(b) What does each edge represent?

(c) For each graph, find the degree of each vertex. What does that tell you about each group member?

(d) Is your graph connected or not? What does it mean if the graph is connected and if it's disconnected?

(e) Is any pair of graphs equivalent? What would that mean?

(f) Find the chromatic number of each graph and find a coloring that uses that number of colors. What does the chromatic number tell you about each graph?

2. When solving distance problems in urban areas, we have to consider the fact that you have to follow city blocks, rather than

go straight from one location to another. The figure shows a large city, with seven locations marked. A pizza delivery person based at location A has deliveries at the other six locations.

(a) Without using any graph theory, draw the path that you think would be shortest to visit all six locations and return to A.

(b) Draw a weighted graph for the figure by finding the nearest distance in blocks between each pair of locations.

(c) Using the nearest neighbor method, find an approximate optimal solution to the problem of finding the shortest circuit connecting all seven locations. Is the result shorter or longer than the one you chose in (a)?

(d) Repeat (c) using the cheapest link algorithm.

(e) See if you can find another shorter path using the brute force method. (There are 720 possible paths, so don't try to check them all!)

3. A famous problem in graph theory involves visiting every one of the state capitals in the continental United States (excluding Alaska) with the least possible amount of driving.

(a) Print a full-page map of the continental United States with all of the capitals marked, then use an atlas or the Internet to mark distances between capitals. You don't have to mark the distance from each capital to every other one—just the ones that are reasonably close to each one.

(b) Use the nearest neighbor method to find an approximate optimal solution for the graph you've drawn. Since you won't have a complete graph, it's possible that a Hamilton circuit doesn't exist. If you get stuck, you may need to go back and add some more distance measurements.

(c) Use Kruskal's algorithm to find a minimum spanning tree for your graph. Does the route differ significantly from the circuit you found in (b)?

(d) Using the Internet as a resource, find the actual optimal solution to the problem and see how it compares to the ones you found in parts (b) and (c).

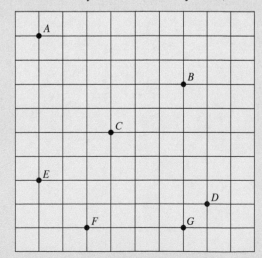

Design elements: Front matter, Chapter Opener, Summary and End Matter header design (random numbers background illustration): ©pixeldreams.eu/Shutterstock RF

Area Under the Standard Normal Distribution

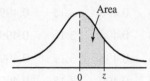

Area

The area in the A column is the area under the normal distribution between $z = 0$ and the positive value of z found in the z column.

z	A	z	A	z	A	z	A	z	A	z	A	z	A
0.00	0.000	0.25	0.099	0.50	0.192	0.75	0.273	1.00	0.341	1.25	0.394	1.50	0.433
0.01	0.004	0.26	0.103	0.51	0.195	0.76	0.276	1.01	0.344	1.26	0.396	1.51	0.435
0.02	0.008	0.27	0.106	0.52	0.199	0.77	0.279	1.02	0.346	1.27	0.398	1.52	0.436
0.03	0.012	0.28	0.110	0.53	0.202	0.78	0.282	1.03	0.349	1.28	0.400	1.53	0.437
0.04	0.016	0.29	0.114	0.54	0.205	0.79	0.285	1.04	0.351	1.29	0.402	1.54	0.438
0.05	0.020	0.30	0.118	0.55	0.209	0.80	0.288	1.05	0.353	1.30	0.403	1.55	0.439
0.06	0.024	0.31	0.122	0.56	0.212	0.81	0.291	1.06	0.355	1.31	0.405	1.56	0.441
0.07	0.028	0.32	0.126	0.57	0.216	0.82	0.294	1.07	0.358	1.32	0.407	1.57	0.442
0.08	0.032	0.33	0.129	0.58	0.219	0.83	0.297	1.08	0.360	1.33	0.408	1.58	0.443
0.09	0.036	0.34	0.133	0.59	0.222	0.84	0.300	1.09	0.362	1.34	0.410	1.59	0.444
0.10	0.040	0.35	0.137	0.60	0.226	0.85	0.302	1.10	0.364	1.35	0.412	1.60	0.445
0.11	0.044	0.36	0.141	0.61	0.229	0.86	0.305	1.11	0.367	1.36	0.413	1.61	0.446
0.12	0.048	0.37	0.144	0.62	0.232	0.87	0.308	1.12	0.369	1.37	0.415	1.62	0.447
0.13	0.052	0.38	0.148	0.63	0.236	0.88	0.311	1.13	0.371	1.38	0.416	1.63	0.449
0.14	0.056	0.39	0.152	0.64	0.239	0.89	0.313	1.14	0.373	1.39	0.418	1.64	0.450
0.15	0.060	0.40	0.155	0.65	0.242	0.90	0.316	1.15	0.375	1.40	0.419	1.65	0.451
0.16	0.064	0.41	0.159	0.66	0.245	0.91	0.319	1.16	0.377	1.41	0.421	1.66	0.452
0.17	0.068	0.42	0.163	0.67	0.249	0.92	0.321	1.17	0.379	1.42	0.422	1.67	0.453
0.18	0.071	0.43	0.166	0.68	0.252	0.93	0.324	1.18	0.381	1.43	0.424	1.68	0.454
0.19	0.075	0.44	0.170	0.69	0.255	0.94	0.326	1.19	0.383	1.44	0.425	1.69	0.455
0.20	0.079	0.45	0.174	0.70	0.258	0.95	0.329	1.20	0.385	1.45	0.427	1.70	0.455
0.21	0.083	0.46	0.177	0.71	0.261	0.96	0.332	1.21	0.387	1.46	0.428	1.71	0.456
0.22	0.087	0.47	0.181	0.72	0.264	0.97	0.334	1.22	0.389	1.47	0.429	1.72	0.457
0.23	0.091	0.48	0.184	0.73	0.267	0.98	0.337	1.23	0.391	1.48	0.431	1.73	0.458
0.24	0.095	0.49	0.188	0.74	0.270	0.99	0.339	1.24	0.393	1.49	0.432	1.74	0.459

Continued

z	A	z	A	z	A	z	A	z	A	z	A	z	A
1.75	0.460	1.97	0.476	2.19	0.486	2.41	0.492	2.63	0.496	2.85	0.498	3.07	0.499
1.76	0.461	1.98	0.476	2.20	0.486	2.42	0.492	2.64	0.496	2.86	0.498	3.08	0.499
1.77	0.462	1.99	0.477	2.21	0.487	2.43	0.493	2.65	0.496	2.87	0.498	3.09	0.499
1.78	0.463	2.00	0.477	2.22	0.487	2.44	0.493	2.66	0.496	2.88	0.498	3.10	0.499
1.79	0.463	2.01	0.478	2.23	0.487	2.45	0.493	2.67	0.496	2.89	0.498	3.11	0.499
1.80	0.464	2.02	0.478	2.24	0.488	2.46	0.493	2.68	0.496	2.90	0.498	3.12	0.499
1.81	0.465	2.03	0.479	2.25	0.488	2.47	0.493	2.69	0.496	2.91	0.498	3.13	0.499
1.82	0.466	2.04	0.479	2.26	0.488	2.48	0.493	2.70	0.497	2.92	0.498	3.14	0.499
1.83	0.466	2.05	0.480	2.27	0.488	2.49	0.494	2.71	0.497	2.93	0.498	3.15	0.499
1.84	0.467	2.06	0.480	2.28	0.489	2.50	0.494	2.72	0.497	2.94	0.498	3.16	0.499
1.85	0.468	2.07	0.481	2.29	0.489	2.51	0.494	2.73	0.497	2.95	0.498	3.17	0.499
1.86	0.469	2.08	0.481	2.30	0.489	2.52	0.494	2.74	0.497	2.96	0.499	3.18	0.499
1.87	0.469	2.09	0.482	2.31	0.490	2.53	0.494	2.75	0.497	2.97	0.499	3.19	0.499
1.88	0.470	2.10	0.482	2.32	0.490	2.54	0.495	2.76	0.497	2.98	0.499	3.20	0.499
1.89	0.471	2.11	0.483	2.33	0.490	2.55	0.495	2.77	0.497	2.99	0.499	3.21	0.499
1.90	0.471	2.12	0.483	2.34	0.490	2.56	0.495	2.78	0.497	3.00	0.499	3.22	0.499
1.91	0.472	2.13	0.483	2.35	0.491	2.57	0.495	2.79	0.497	3.01	0.499	3.23	0.499
1.92	0.473	2.14	0.484	2.36	0.491	2.58	0.495	2.80	0.497	3.02	0.499	3.24	0.499
1.93	0.473	2.15	0.484	2.37	0.491	2.59	0.495	2.81	0.498	3.03	0.499	3.25	0.499
1.94	0.474	2.16	0.485	2.38	0.491	2.60	0.495	2.82	0.498	3.04	0.499		*
1.95	0.474	2.17	0.485	2.39	0.492	2.61	0.496	2.83	0.498	3.05	0.499		
1.96	0.475	2.18	0.485	2.40	0.492	2.62	0.496	2.84	0.498	3.06	0.499		

*For z values beyond 3.25, use A = 0.500.

Design elements: Front matter, Chapter Opener, Summary and End Matter header design (random numbers background illustration): ©pixeldreams.eu/Shutterstock RF

CHAPTER 1: PROBLEM SOLVING

Exercise Set 1-1

9. 37 **11.** 10 **13.** 72

15. 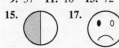 **17.**

19. $5 + 13 + 17 = 35$, which is odd.

21. $5^2 \div 2 = 12.5$

23. Conjecture: the final answer is -10.

25. Conjecture: the final answer is 18.

27. $12,345,679 \times 72 = 12,345,679 \times 9(8) = 888,888,888$

29. $999,999 \times 9 = 8,999,991$

31. $99,999 \times 99,999 = 9,999,800,001$

33. $11,111 \times 11,111 = 123,454,321$

35. When multiplied by the numbers 1–6 the digits in the answer are a permutation of the original number. But the hypothesis fails when the number is multiplied by 7 and 8.

37. The next three sums are $\frac{9}{5}, \frac{11}{6}$, and $\frac{13}{7}$.

39. g e h **41.** M J J

43. Inductive **45.** Deductive **47.** Inductive **49.** Deductive

51. Deductive **53.** Deductive **55.** Deductive **57.** Deductive

59. (a) You'd be more likely to text while driving using inductive reasoning.
(b) You'd be less likely to text while driving using deductive reasoning.

61. (a) Answers can vary, but the simplest answer is 16 and 32.
(b) Each number is twice the one before it; 2^n
(c) Add 2 to the first term, then 4 to the second, 6 to the third, and 8 to the fourth. See part (d) for formula.

(d)

n	1	2	3	4	5
$n^2 - n + 2$	2	4	8	14	22

; the formula is $n^2 - n + 2$.
There may be more than one pattern that fits a string of numbers, especially if you only have the first three.

63. Answers vary.

65. (a) The average speed always works out to be 30 miles per hour.

67. Weak **69.** Strong **71.** Strong

73. (a) 21, 28, 36 (b) 36, 49, 64 (c) 35, 51, 70 (d) 1, 6, 15, 28

Exercise Set 1-2

11. 2,900 **13.** 3,260,000 **15.** 63 **17.** 200,000 **19.** 3.67

21. 327.1 **23.** 5,460,000 **25.** 300,000 **27.** 264.9735

29. 482.60 **31.** Estimate: —44; exact value: —45.8469; 4% error

33. Estimate: —1.5; exact value: —1.8243 (to 4 decimal places); 17.8% error

35. $136; overestimate **37.** 6 hours; overestimate

39. $72; overestimate **41.** $6.00; overestimate

43. About $200; overestimate **45.** $25 per hour; overestimate

47. $54; overestimate **49.** About $50; overestimate

51. About 350 million **53.** About 1 billion **55.** 471 people

57. 249 **59.** 86% **61.** About 2,950 **63.** 350 billion

65. 1940 **67.** About 7.8 billion per year.

69. Asia; Oceania/Australia **71.** About 700 million; about 49%

73. North America must have a smaller population than several other continents.

75. Developed world: almost 70%; developing world: almost 50%

77. This is to some extent an opinion. By amount of increase, it's the developed world. But the developing world grew much more as a percentage of the 1998 value.

79. Answers vary.

81. About $27 (Answers vary depending on how you rounded.)

83. (a) May 1: 0.1 deaths per million people; August 1: 1.7 deaths per million people
(b) The deaths per million people for counties with low and high vaccination rates generally decreased between May and early July and increased from early July to August 1. However, the rate of increase in the counties with high vaccination rates was less than the rate of increase in the counties with low vaccination rates.
(c) The graphs indicate that being vaccinated decreases the chance of dying from COVID.

85. The difference between the cost of milk in 1988 and the cost in 2006 is exaggerated by the fact that the picture changed in all three dimensions, rather than just vertically.

87. 1st hour: 45 mph; 1st 2 hours: 62.5 mph; 1st 3 hours: 58.3 mph; 1st 4 hours: 53.8 mph. The average speed was in the forties for the first hour, so a lot of it probably wasn't freeway. The second hour was the fastest; the third slowed them down some, and the fourth slowed them down even more.

89. 16–17 hours: —50 mph; 17–18 hours: zero. The negative indicates that the distance away was getting smaller, so they were headed back home. They weren't moving between 17 and 18 hours. Lunch break?

91. The steepness (slope) of the graph.

Exercise Set 1-3

7. 8 and 14 **9.** 12 **11.** 43 **13.** 7 years

15. Children $40,000, grandchildren $20,000

17. 132 and 312 **19.** Barney: $1.80, Betty $3.25 **21.** 35 ladies

23. Mae earned $54.38 for working 5 hours. **25.** 8 posts

27. 48 inches wide **29.** 50 feet **31.** 21 boxes, 252 lights

33. $\frac{1}{4}$ of the original **35.** $1,093

37. There's no way to divide them evenly because 71 is a prime number.

39. $206.35

41. Mary: $1,187.50, Ji-Woo: $593.75, Claire: $296.88, Margie: $296.87

43. $465.50 **45.** 3-2-6-9-8-5-1-4-7-11-10-12

47. He cuts the bar at the 1-inch and the 3-inch marks, giving him bars of length 1, 2, and 3 inches. He pays the knight for the first day, then takes the one-inch piece back and gives him the two-inch on the second day (and continues in this fashion for six days).

49. Three, since the car in front is in front of two cars and the car at the end is behind two cars.

51. Start out with four full Jeeps and go $\frac{1}{4}$ of the way. Each will be half full. Use two of them to refill two others, then go another $\frac{1}{4}$ of the way. Each will be half full: use one to refill the other for the rest of the trip.

53. It's not possible.

55. The possibilities are: Maurice has 2 and Hani 3, Maurice has 3 and Hani 4, Maurice has 9 and Hani 8, or Maurice has 8 and Hani 7.

57. The answer is a spreadsheet that can be found in online resources.

Review Exercises

1. 18, 19, 21 **3.** q, 1,024, n **5.**

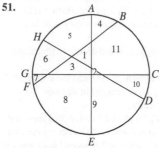

7. 5(7)(11) = 385, which is odd

9. Conjecture: the final answer is 13 more than $\frac{1}{2}$ of the original even number.

11. 337 × 12 = 4,044
337 × 15 = 5,055

13. 9 **15.** Inductive **17.** Deductive **19.** 132,000 **21.** 14.6316

23. 3,730 **25.** $340 **27.** $340

29. There are many possible combinations. If you want to spend as close to $130 as possible:

T-shirts	6	5	4	3	2	1	0
Sweatpants	0	1	2	4	5	6	7

31. 80 **33.** Some people gave more than one response. **35.** 1995

37. The graph looks steeper from 2000 to 2010. Actual rates are about $10/year for 1985–1995, and about $18/year for 2000–2010.

39. 9 **41.** 110 pounds **43.** 67 **45.** $40 **47.** 15 years old

49. 2 × 9 + 6 − 7 = 17

51.

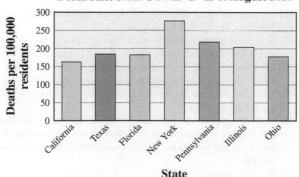

The cuts are *AE*, *BF*, *CG*, and *DH*.

53. 300 miles **55.** 4 pounds of the nature mix and 6 pounds of the soy medley

Chapter 1 Practice Test

*When using estimation, other correct answers are possible.

1. 14 10 17 **3.** 88,888,888

5. The final answer is equal to the original number plus 13.

7. (a) Ninth (b) $0.10 (c) $409.50

9. Move the last coin on the right on top of another coin or move the coin at top left below the coin at bottom left.

11. 84 **13.** 12 **15.** First person earns $36; second person earns $24.

17. (a) 90 (b) Not possible. She'd need 120%.

19. Mark is 17 years old and his mother is 49.

21. 1.38 **23.** (a) 1960: 18%; 2000: 42% (b) 2003 (c) 1.2% per year

25. Texas; it has a much bigger population.

27.

Death Rate from COVID-19 as of August 2021

Source: Statista.com

CHAPTER 2: SET THEORY

Exercise Set 2-1

9. *T* = {t, h, i, n, k, g}

11. *P* = {51, 52, 53, 54, 55, 56, 57, 58, 59}

13. *C* = {1, 2, 3, 4, 5, 6, 7, 8} **15.** *G* = {11, 12, 13, ... }

17. *Y* = {2,001, 2,002, 2,003, ... , 2,999}

19. *C* = {white, red, blue, green, gray, brown, black, yellow}

21. *L* = {medial collateral, lateral collateral, anterior cruciate, posterior cruciate}

23. True **25.** True **27.** True **29.** The set of multiples of 5

31. The set of multiples of 13 from 13 to 52

33. The set of letters in the name Steven

35. The set of natural numbers from 100 to 199

Answers can vary for the alternate descriptions.

37. {$x|x$ is a multiple of 10}; the set of positive numbers that end in zero

39. {$x|x$ is odd and x < 16}; the set of odd numbers from 1 to 15.

41. {$x|x$ is a color in an American flag} is one possible answer; the set of colors in the flag of France

43. There are no natural numbers less than zero, so $H = \varnothing$.

45. {Spring, Summer, Fall, Winter}

47. {102, 104, 106, 108, 110, 112, 114, 116, 118}

49. Well-defined **51.** Not well-defined **53.** Not well-defined

55. False **57.** True **59.** True **61.** Infinite **63.** Finite **65.** Infinite

67. Finite **69.** Equal **71.** Neither **73.** Equivalent

75. {10, 20, 30, 40}   **77.** {1, 2, 3, 4, 5, ...}
$\updownarrow$ $\updownarrow$ $\updownarrow$ $\updownarrow$ $\updownarrow$ $\updownarrow$ $\updownarrow$ $\updownarrow$ $\updownarrow$
{40, 10, 20, 30}   {4, 8, 12, 16, 20, ...}

79. 4 **81.** $n(C) = 7$ **83.** $n(E) = 1$ **85.** $n(G) = 0$

87. True **89.** False **91.** True

93. (a) {California, New York, Florida, Texas}
(b) {Illinois, Massachusetts, Virginia, Georgia, Pennsylvania}
(c) {California, New York, Florida, Texas, New Jersey}
(d) {Texas, New Jersey, Illinois}

95. (a) {Drunk driving, Injury, Assault}
(b) {Injury, Unsafe sex, Health problems}
(c) {Injury, Assault, Drunk driving}
(d) {97,000, 1,825, 150,000}
(e) No. We don't know how many students were in more than one group, so adding the numbers together probably won't be correct.

97. (a) {Government documents/benefits, loan/lease fraud, bank fraud}
(b) {60–69}
(c) {Other, credit card fraud}
(d) {14%, 16%, 20%, 13%, 6%}
(e) {Loan/lease fraud, bank fraud, utilities/phone fraud, employment/ tax fraud}
(f) 101%; some reports might fall into two categories, reports could have come from different sources, could be rounding error

99. (a) {2006, 2007, 2008, 2015, 2016, 2017, 2018, 2019, 2020, 2021}
(b) {2004, 2005, 2009, 2010, 2011, 2012, 2013, 2014}
(c) {2005, 2006, 2007, 2013, 2014, 2015, 2016, 2017, 2018, 2019, 2020, 2021}
(d) {2008, 2009, 2010, 2011, 2012}

101. No

103. (a) A appears to have more elements.
(b) {1, 2, 3, 4, 5, 6, ...}
$\updownarrow$ $\updownarrow$ $\updownarrow$ $\updownarrow$ $\updownarrow$ $\updownarrow$
{2, 4, 6, 8, 10, 12, ...}
This one-to-one correspondence shows that the two sets have the same number of elements.

105. (a) It's not clearly defined who qualifies as "an American."
(b) Who decides what a "luxury car" is?
(c) "Legitimate chance" is subjective.
(d) What exactly defines pay? Salary? Salary and benefits? Is overtime included?
(e) Biological mothers? Adopted mothers? Foster mothers?

Exercise Set 2-2

11. $A' = \{2, 3, 17, 19\}$

13. $C' = \{2, 3, 5, 7, 11\}$

15. $A' = \{1, 2, 3, 5, 7, 9, 11, ...\}$

Note: for 17–23, the list of proper subsets is the same as the list of subsets with the original set excluded.

17. {OVI, theft, fraud}, {OVI, theft}, {OVI, fraud}, {theft, fraud}, {OVI}, {theft}, {fraud}, Ø

19. {radio, TV}, {radio}, {TV}, Ø **21.** Ø

23. {fever, chills, nausea, headache}, {fever, chills, nausea}, {fever, chills, headache}, {fever, nausea, headache}, {chills, nausea, headache}, {fever, chills}, {fever, nausea}, {fever, headache}, {chills, nausea}, {chills, headache}, {nausea, headache}, {fever}, {chills}, {nausea}, {headache}, Ø

25. True   **27.** False   **29.** False   **31.** False   **33.** True

35. 8 subsets, 7 proper subsets

37. 1 subset, no proper subsets

39. 4 subsets, 3 proper subsets

41. $U = \{1, 3, 5, 7, 9, 11, 13, 15, 17, 19\}$

43. $B = \{5, 11, 13, 15\}$

45. $A \cup B = \{1, 5, 9, 11, 13, 15, 17\}$

47. $B' = \{1, 3, 7, 9, 17, 19\}$

49. $(A \cap B)' = \{1, 3, 7, 9, 13, 15, 17, 19\}$

51. $A \cup C = \{12, 14, 15, 16, 17, 19, 20\}$

53. $A' = \{11, 12, 13, 18, 19, 20\}$

55. $A' \cap (B \cup C) = \{11, 12, 13, 19, 20\}$

57. $(A \cap B)' \cap C = \{12, 20\}$

59. $(B \cup C) \cap A' = \{11, 12, 13, 19, 20\}$ **61.** $W \cap Y = Ø$

63. $W \cup X = \{2, 4, 6, 7, 8, 9, 10, 11, 12, 13, 14\}$

65. $W \cap X = \{6, 8\}$ **67.** $(X \cup Y) \cap Z = Ø$

69. $W' \cap X' = \{1, 3, 5, 15, 16, 17, 18, 19, 20, 21, 22, 23, 24\}$

71. $A \cap B = B$

73. $A \cap (B \cup C') = \{x | x$ is an odd multiple of 3 or an even multiple of 9$\}$
$= \{3, 9, 15, 18, 21, 27, 33, 36, 39, \ldots\}$

75. $C - B = \{p\}$ **77.** $B - C = \{s, u\}$ **79.** $B \cap C' = \{s, u\}$

81. $D - M = \{11, 13, 15, 17, \ldots\}$ **83.** $(D - M) - T = Ø$

85. B' **87.** $(A \cup B) - (A \cap B)$

89. {tablet, laptop, smartphone}, {tablet, laptop}, {tablet, smartphone}, {laptop, smartphone}, {tablet}, {laptop}, {smartphone}, Ø

91. $2^7 = 128$ **93.** $2^4 = 16$

95. The set of people with strong management skills, the set of people good at working as part of a team, and the set of people with 5 years' experience with a similar project

97. (a) The set of people who have been convicted of a felony
   (b) The set of people who have been convicted of a felony and have been released, or charged with a felony and found not guilty
   (c) The set of people who were charged with a felony and either found not guilty or had charges dropped before standing trial

99. (a) The set of people who have been convicted of a felony and have been released from prison
   (b) The set of people who have previously been convicted of a felony, and are currently awaiting trial on another felony charge
   (c) There's nobody in this set

101. $A \times B = \{$(chocolate, chocolate icing), (chocolate, cream cheese icing), (yellow, chocolate icing), (yellow, cream cheese icing), (red velvet, chocolate icing), (red velvet, cream cheese icing)$\}$. This is all the possible cakes that can be made choosing from chocolate, yellow, and red velvet cake, and from chocolate and cream cheese icing.

103. $n(A \times B) = n(A) \times n(B)$. In each case, we're taking each of the elements of the first set and matching them with each of the elements of the second. So you end up with the number in the first set recopied a certain number of times: the number of times corresponds to the number of items in the second set.

105–107. Answers vary.

109. For $A \cap B = A$, A has to be a subset of B. For $B \cap A = B$, B has to be a subset of A.

111. Given n elements, each has two choices: in the subset or not. So if we try to build a subset, there are $2 \cdot 2 \cdot 2 \cdots \cdot 2$ ways to choose, where there are n factors of 2. That's a long way to say 2^n.

Exercise Set 2-3

7. 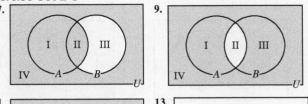 **9.**

11. 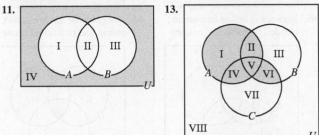 **13.**

15. 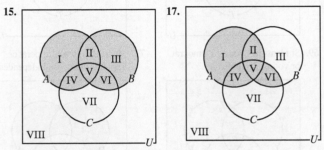 **17.**

19. 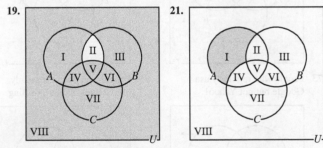 **21.**

23. 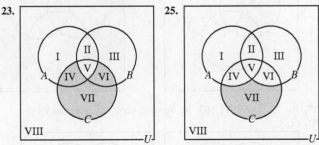 **25.**

27. 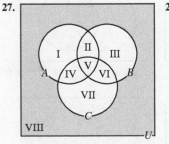 **29.**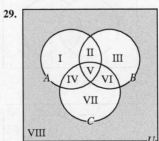

31. equal   **33.** equal   **35.** not equal   **37.** not equal   **39.** $n(A) = 10$

41. $n(A \cap B) = 4$ **43.** $n(A') = 13$ **45.** $n(A' \cap B') = 6$

47. $n(A - B) = 6$ **49.** $n(A \cap (B - A)) = 0$ **51.** $n(A) = 8$

53. $n(A \cap B) = 3$ **55.** $n(A \cap B') = 5$ **57.** $n(A') = 11$ **59.** $n(A - B) = 5$

61. People who drive an SUV or a hybrid vehicle

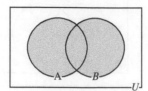

63. People who do not drive an SUV

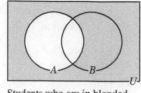

65. Students in online courses and blended or traditional courses

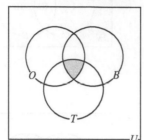

67. Students who are in blended, online, and traditional courses

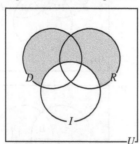

69. Students not voting Democrat or voting Republican

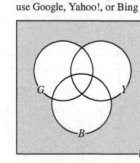

71. Students voting Democrat or Republican but not Independent

73. People who regularly use Google but not Yahoo!

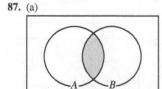

75. People who do not regularly use Google, Yahoo!, or Bing

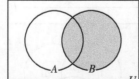

77. II   **79.** I   **81.** VI   **83.** No; Answers vary.   **85.** Answers vary.

87. (a)

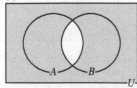

(b) Answers vary.   (c) $B \subseteq A$

89. (a)

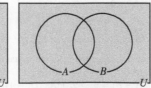

(b) Answers vary.   (c) A and B are disjoint

91. (a)

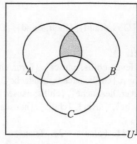

(b) Answers vary.   (c) B and C are disjoint

Exercise Set 2-4

1. (a) 10   (b) 51   (c) 3
3. (a) 22   (b) 36
5. (a) 50%   (b) 90%
7. (a) 2   (b) 9   (c) 35
9. (a) 16   (b) 7   (c) 14   (d) 3
11. (a) 192   (b) 6   (c) 87
13. (a) 18   (b) 30 in each league
15. (a) 52 listen to none, 51 to satellite.   (b) 13   (c) 30
17. The total of the eight regions is 39 but the researcher surveyed 40 people.
19. (a) We would need to know how many watch baseball but none of the other two sports.
(b) 1,000
(c) Football only: 205; basketball only: 110

Exercise Set 2-5

5. $7n$ **7.** 4^n **9.** $-3n$ **11.** $\frac{n}{4}$ **13.** $4n - 2$
15. $\frac{n+1}{n+2}$ **17.** $100n$ **19.** $-3n - 1$

For 21 through 29 we will show each set is infinite by putting it into a one-to-one correspondence with a proper subset of itself.

21. $\{3, \ 6, \ 9, \ 12, 15, \ldots, 3n, \ldots\}$
$\updownarrow \ \updownarrow \ \updownarrow \ \updownarrow \ \updownarrow \qquad \updownarrow$
$\{6, 12, \ 18, 24, 30, \ldots, 6n, \ldots\}$

23. $\{9, 18, \ 27, 36, 45, \ldots, 9n, \ldots\}$
$\updownarrow \ \updownarrow \ \updownarrow \ \updownarrow \ \updownarrow \qquad \updownarrow$
$\{18, 36, 54, 72, 90, \ldots, 18n, \ldots\}$

25. $\{2, 5, \ 8, 11, \ldots, 3n - 1, \ldots\}$
$\updownarrow \ \updownarrow \ \updownarrow \ \updownarrow \qquad \updownarrow$
$\{5, 11, 17, 23, \ldots, 6n - 1, \ldots\}$

27. $\{10, \ 100, \ldots, \ 10^n, \ldots\}$
$\updownarrow \qquad \updownarrow \qquad \updownarrow$
$\{100, 10,000, \ldots, 10^{2n}, \ldots\}$

29. $\left\{\frac{5}{1}, \frac{5}{2}, \frac{5}{3}, \ldots, \ \frac{5}{n}, \ldots\right\}$
$\updownarrow \ \updownarrow \ \updownarrow \qquad \updownarrow$
$\left\{\frac{5}{2}, \frac{5}{3}, \frac{5}{4}, \ldots, \frac{5}{n+1}, \ldots\right\}$

31. Use the correspondence $n \rightarrow 5n$.
33. Use the correspondence $n \rightarrow (n - 1)^2$. (The correspondence $n \rightarrow n^2$ shows that the natural numbers are countable, but the whole numbers include zero as well.)
35. The rational numbers can be put into a one-to-one correspondence with the natural numbers.
37. That the set in Example 2 is countable.
39. (a) Correspond every number to the number that is one less.
(b) $\aleph_0 + 1 = \aleph_0$
41. $\aleph_0$ **43.** 15   **45.** $\aleph_0$

Review Exercises

1. $D = \{52, 54, 56, 58\}$ 3. $L = \{l, e, t, r\}$
5. $B = \{501, 502, 503, \ldots\}$
7. {Buzz Aldrin, Neil Armstrong, Alan Bean, Gene Cernan, Pete Conrad, Charles Duke, James Irwin, Edgar Mitchell, Harrison Schmitt, David Scott, Alan B. Shepard, John Young}
9. $\{x | x \text{ is even and } 16 < x < 26\}$
11. $\{x | x \text{ is an odd natural number greater than } 100\}$
13. Infinite   15. Finite   17. Finite   19. Finite   21. False   23. False
25. $\varnothing$; $\{r\}$; $\{s\}$; $\{t\}$; $\{r, s\}$; $\{r, t\}$; $\{s, t\}$; $\{r, s, t\}$
27. $A \cap B = \{\text{Toyota, Honda, Lexus}\}$ 29. $(A \cap B) \cap C = \varnothing$
31. $A - B = \{\text{Chevy, BMW}\}$
33. $(A \cup B)' \cap C = \{\text{Mercedes, Acura, Dodge}\}$
35. $(B \cup C) \cap A' = \{\text{Mercedes, Acura, Hyundai, Tesla, Dodge}\}$
37. $(B' \cap C') \cup A' = \{\text{Chevy, Ford, BMW, Mercedes, Acura, Hyundai, Tesla, Dodge}\}$
39. $K \cap L = \{x | x \in E, x > 25\}$; $K \cup L = \{12, 14, 16, 18, 20, 22, 24, 26, 27, 28, 29, 30, \ldots\}$; $L - K = \{12, 14, 16, 18, 20, 22, 24\}$
41. $A - B$ 43. $B - A$ 45. $(A \cup B) - (A \cap B)$
47. 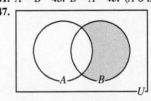   49.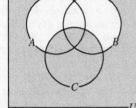

51. 20   53. VI   55. VII   57. (a) 75   (b) 20
59. (a) 3   (b) 5   (c) 6   61. $-3 - 2n$
63. Use the correspondence $n \to 12n$.

Chapter 2 Test

1. $P = \{92, 94, 96, 98\}$ 3. $X = \{1, 2, 3, 4, \ldots, 79\}$
5. $\{x | x \in E \text{ and } 10 < x < 20\}$ 7. Infinite   9. Finite
11. All subsets: $\varnothing$, {Arizona}, {Nevada}, {Oregon}, {Arizona, Nevada}, {Arizona, Oregon}, {Nevada, Oregon}, {Arizona, Nevada, Oregon}; proper subsets: all but the last one. There are 3 states that border California, so there are $2^3 = 8$ subsets.
13. $(A \cup B)' = \{c, h\}$ 15. $(A - B) - C = \{b, d, f\}$
17. {(Arizona, e), (Nevada, e), (Oregon, e), (Arizona, h), (Nevada, h), (Oregon, h) (Arizona, j), (Nevada, j), (Oregon, j)} and {(e, Arizona), (h, Arizona), (j, Arizona), (e, Nevada), (h, Nevada), (j, Nevada), (e, Oregon), (h, Oregon), (j, Oregon)}
19.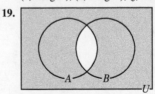

21. $n(A \cup B) = 2,300$ 23. $15n$ 25. True   27. True   29. False

CHAPTER 3: LOGIC

Exercise Set 3-1

9. Not a statement   11. Statement   13. Not a statement
15. Not a statement   17. Not a statement   19. Compound statement
21. Compound statement   23. Simple statement
25. Compound statement   27. Compound statement   29. Conjunction
31. Biconditional   33. Disjunction   35. Biconditional
37. The shirt I'm wearing to my interview is not white.
39. The hospital is full.   41. You're going to flunk this class.
43. Universal   45. Existential   47. Universal
49. Existential   51. Universal   53. Existential

55. Not all fish swim in water; Some fish do not swim in water.
57. No people who live in glass houses throw stones.
59. Not every happy dog wags its tail; Some happy dog does not wag its tail.
61. I haven't seen a four-leaf clover.
63. Somebody has survived a fall from Mt. Catherine.
65. None of my friends have a Mac laptop.
67. $p \land q$ 69. $\sim q \to p$
71. $\sim q$ 73. $q \lor \sim p$ 75. $q \leftrightarrow p$
77. $\sim q$ 79. $\sim q \to p$ 81. $p \lor \sim q$
83. $p \leftrightarrow q$ 85. $\sim p \to q$
87. The plane is on time and the sky is clear.
89. If the sky is clear, then the plane is on time.
91. The plane is not on time and the sky is not clear.
93. The plane is on time or the sky is not clear.
95. If the sky is clear, then the plane is or is not on time.
97. Water sports don't cost extra.
99. This resort is all-inclusive or water sports don't cost extra.
101. If this resort is not all-inclusive, then water sports don't cost extra.
103. This resort is all-inclusive or water sports cost extra.
105. Water sports cost extra or this resort is all-inclusive.
107. It cannot be classified as true or false.
109. (a) a is less than 20.
   (b) a is not less than 20.
   (c) $a \geq 20$ (This is if we assume that a is a real number. Otherwise, the statement would be $a \geq 20$ or a is not a real number.)
111. All are sample answers.
   (a) That movie won an Academy Award.
   (b) Statistics show that the obesity rate in the United States has been rising sharply over the last thirty years.
   (c) According to FBI data, violent crime has seen a steady decrease since the early 1990s.
   (d) The average height of a player in the National Basketball Association is 6'7".
   (e) The average high temperature in Chicago for January is 21°.
   (f) That car goes from zero to 60 mph in less than five seconds.
113. There will not be any fans at any of the games.
115. There is at least one person that likes my history professor.

Exercise Set 3-2

7. FFFT   9. FFTF   11. FTTF   13. TTFF   15. TFTT
17. TTFF   19. TTFF   21. TTFFTFFF   23. TTTTTTTT
25. TFFFTFTF   27. TTTFFFTF   29. FTFTFTTT
31. FFFTTTTT   33. FFFFTFFF   35. TFTFFFFF   37. True
39. True   41. True   43. False   45. True   47. True
49. Let p be "if you take their product daily," q be "you cut your calorie intake by 10%," and r be "you lose at least 10 pounds in the next 4 months."
51. TFTTTTTT   53. True
55. Let p be "the attendance for the following season is over 2 million," q be "he will add 20 million dollars to the payroll," and r be "the team will make the playoffs the following year."
57. TFFFTTTT   59. True
61. The truth table for $(p \land q) \lor r$ is different than the one for $p \land (q \lor r)$.
63. The statements are equivalent.
65. Answers vary.
67. Because that's exactly what it means for a biconditional statement to be true. In that case, there'd be no reason to separate the two.

Exercise Set 3-3

9. Tautology   11. Self-contradiction   13. Tautology
15. Tautology   17. Neither   19. Equivalent   21. Neither
23. Neither   25. Negations   27. Neither
29. $q \to p$; $\sim p \to \sim q$; $\sim q \to \sim p$
31. $\sim(q \land p) \to \sim p$; $p \to (q \land p)$; $(q \land p) \to p$
33. $(q \lor r) \to p$; $\sim p \to \sim(q \lor r)$; $\sim(q \lor r) \to \sim p$
35. The patient isn't septic and she's not in shock.
37. It is cold or I am not soaked.

39. I will not go to the beach or I will get sunburned.
41. The suspect is not a white male and the witness is correct.
43. It is not right and it is not wrong.
45. My grade isn't an A and it's not a B.
47. The prosecuting attorney for this case isn't experienced or prepared.
49. My friends are serious about school and prepared to work hard.
51. $p \to q$ **53.** $p \to q$ **55.** $p \to q$ **57.** $\sim p \to \sim q$

Note: for 59–63 explanations will vary.

59. *Converse*: If she did get a good job, then she graduated with a Bachelor's degree in Management Information Systems.
Inverse: If she did not graduate with a Bachelor's degree in Management Information Systems, then she will not get a good job.
Contrapositive: If she did not get a good job, she did not graduate with a Bachelor's degree in Management Information Systems.
61. *Converse*: If the patient may not survive, then we made a mistake on the dosage.
Inverse: If we don't make a mistake on the dosage, then the patient may survive.
Contrapositive: If the patient may survive, then we didn't make a mistake on the dosage.
63. *Converse*: If I go to Nassau for spring break then I will lose 10 pounds by March 1.
Inverse: If I do not lose 10 pounds by March 1 then I will not go to Nassau for spring break.
Contrapositive: If I do not go to Nassau for spring break then I did not lose 10 pounds by March 1.
65. She graduated with a Bachelor's degree in Management Information Systems and didn't get a good job.
67. We made a mistake on the dosage, and the patient will definitely survive.
69. I will lose 10 pounds by March 1 and won't go to Nassau for spring break.
71. True **73.** False **75.** False **77.** $\sim(p \to q) \equiv p \wedge \sim q$
79. Answers vary.

Exercise Set 3-4
9. Valid **11.** Invalid **13.** Invalid **15.** Valid **17.** Invalid

Note: symbolic forms for 19–23 can be found on pages 138–139.

19. Valid **21.** Valid **23.** Invalid **25.** Valid **27.** Invalid
29. Valid **31.** Valid **33.** Invalid **35.** Invalid **37.** Invalid
39. Valid **41.** Invalid **43.** Valid **45.** Invalid **47.** Valid
49. Valid **51.** Invalid **53.** Invalid **55.** Valid **57.** Valid
59. Answers vary. **61.** Answers vary.
63. The argument is valid even though the conclusion is false.

Exercise Set 3-5
5.
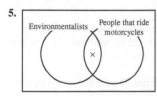
7.
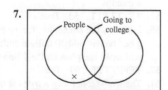
9.
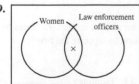
11.
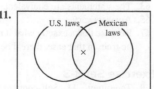
13.

15. Invalid **17.** Valid **19.** Invalid **21.** Valid **23.** Invalid
25. Invalid **27.** Invalid **29.** Valid **31.** Invalid **33.** Invalid
35. Valid **37.** Invalid **39.** Invalid **41.** Invalid **43.** All *A* is *C*.
45. No calculators can make breakfast.
47. Example 1: p = person is a politician, q = person stretches the truth, r = person takes bribes.
$(p \to q) \wedge (p \wedge r) \Rightarrow (q \wedge r)$
Example 3: p = person is a criminal, q = person is admirable, r = person is an athlete.
$(p \to \sim q) \wedge (r \wedge \sim p) \Rightarrow (q \wedge r)$

Review Exercises
1. Not a statement **3.** Not a statement
5. Not a statement **7.** Simple
9. Compound; biconditional
11. Compound; disjunction
13. The cell phone is not out of juice.
15. No failing students can learn new study methods.
17. Some SUVs are not gas guzzlers.
19. $p \wedge q$ **21.** $q \leftrightarrow p$ **23.** $\sim p \to \sim q$
25. $\sim(p \to q)$ **27.** $\sim(\sim q)$
29. It is cool or it is not cloudy.
31. It is cool if and only if it is cloudy.
33. It is not true that it is not cool or it is cloudy.
35. False **37.** TTFT **39.** TTTT
41. TFTTTTTT **43.** TTFTTTTT
45. False **47.** True
49. Tautology **51.** Neither
53. Neither **55.** Not equivalent
57. Social work is not lucrative and it's not fulfilling.
59. The signature is authentic or the check is valid.
61. Let p be the statement "I will be happy" and q be the statement "I get rich." The compound statement is $p \to q$.
63. *Converse*: If I start riding my bike to work, gas prices will go higher.
Inverse: If gas prices do not go any higher, I will not start riding my bike to work.
Contrapositive: If I do not start riding my bike to work, then gas prices will not go any higher.
65. *Converse*: If the patient gets an MRI, then the X-rays are inconclusive.
Inverse: If the X-rays aren't inconclusive then the patient won't get an MRI.
Contrapositive: If the patient doesn't get an MRI, then the X-rays aren't inconclusive.
67. Invalid **69.** Invalid **71.** Invalid **73.** Invalid **75.** Valid
77. Invalid **79.** Invalid

CHAPTER 3 Test
1. False **3.** Answers vary.
5. The image is not uploading to my online bio.
7. No students ride a bike to school.
9. $p \wedge q$ **11.** $p \leftrightarrow q$ **13.** $\sim(\sim p \wedge q)$
15. Congress is in session and my representative isn't in Aruba.
17. If Congress is in session or my representative is in Aruba, then Congress is in session.
19. FTTT **21.** TTFF **23.** Self-contradiction **25.** Tautology
27. *Converse*: If I am healthy, then I exercise regularly.
Inverse: If I do not exercise regularly, then I will not be healthy.
Contrapositive: If I am not healthy, then I do not exercise regularly.
Only the contrapositive is equivalent to the original statement.
29. Valid **31.** Invalid **33.** Invalid

CHAPTER 4: NUMERATION SYSTEMS
Exercise Set 4-1
9. 123 **11.** 20,225 **13.** 30,163 **15.** 502,111
17. ∩∩∩||||||
19. 𝟗𝟗𝟗𝟗𝟗𝟗𝟗𝟗|

21.

23.

25. **27.**

29.

31.

33. 189 **35.** 52 **37.** 713

39. 八十九 **41.** 二百八十四 **43.** 二千三百五十六

45. 4 10 7 1 **47.** 9 100 1 10 7 1 **49.** 2 10,000 3 1,000 4 100 5 10

51. (a) $1 \times 1,000 + 8 \times 100 + 5$
(b) $3 \times 10,000 + 2 \times 1,000 + 7 \times 100 + 1 \times 10 + 4$

53. (a) $1 \times 100,000 + 6 \times 10,000 + 2 \times 1,000 + 8 \times 100 + 7 \times 10 + 3$
(b) $2 \times 100,000,000 + 3 \times 100,000 + 2 \times 10,000 + 1 \times 1,000 + 4 \times 100 + 1 \times 10 + 6$

55. (a) 7,309 (b) 40,850

57. (a) 646,090 (b) 80,604,020

59. 51 **61.** 40,871 **63.** 79,893 **65.** 4,313

67.

69.

71.

73.

75. (a) 18 (b) 99 **77.** (a) 216 (b) 1,110

79. (a) 418 (b) 2,967

81. (a) 93,500 (b) 3,200,000

83. (a) LVIII (b) CXLVII

85. (a) DLXVII (b) MCCLVIII

87. (a) MCDLXII (b) $\overline{\text{MMM}}$CMIX

89. (a) $\overline{\text{LXX}}$CCCXI (b) $\overline{\text{MML}}$X

91. (a) 98 (b) 46 (c) 5,324

93. (a) $\eta\phi\,\varepsilon\alpha$ (b) $\chi\phi\,\iota\alpha$ (c) $\delta\omega\,\alpha\rho\,\gamma\alpha$

95. $\phi\phi$ **97.** $\iota\phi\,\theta\alpha$

99. (a) 277 (b) 6,852

101. (a) 3,387 (b) 113,053

103. (a) (b) **105.** (a) (b)

107. 1939 **109.** 1978

111. 2012 **113.** 423–9315 **115.** 605–12908

117. 8403–4096

119. The number of symbols needed in the Egyptian system is the sum of the digits in the Hindu-Arabic system.

121. Answers vary.

123. Because the Babylonian system is an ancient number system which is not in use today.

125. Answers vary.

127. Tally: 4; Egyptian: 4,000,000; Roman: 3,000,000; Multiplicative grouping and Chinese: 1,100,000; Hindu-Arabic: 9,999; Babylonian: 2,196,610; Mayan: 37,905

129. IIV is the same as III; XXC is the same as LXXX. Subtraction is used only when necessary to avoid writing the same symbol four times consecutively.

131. It's essentially a grouping system with two symbols, one for 5 and one for 1.

Exercise Set 4-2

3. 252 **5.** 391 **7.** 3,740 **9.** 448 **11.** 216 **13.** 253

15. 368 **17.** 374 **19.** 392 **21.** 270 **23.** 10,488

25. 166,786 **27.** 791,028 **29.** 58,625 **31.** 203,912

33. 4,707 **35.** 1,035 **37.** 5,781 **39.** 36,344 **41.** 31

43. 73 **45.** 39 **47.** Answers vary. **49.** Answers vary.

51. Multiply as if both are whole numbers: the number of decimal places in the answer is the sum of the number of decimal places in the two factors.

53. (a) $14\frac{7}{30}$ (b) $42\frac{6}{7}$ (c) $73\frac{5}{9}$

Exercise Set 4-3

7. 11 **9.** 33 **11.** 108 **13.** 106 **15.** 311 **17.** 69 **19.** 359

21. 184 **23.** 37,406 **25.** 40,530 **27.** 11111_{two} **29.** 1333_{six}

31. 22_{seven} **33.** 1017_{nine} **35.** 10110_{two} **37.** 33_{five} **39.** 939_{sixteen}

41. 1042212_{five} **43.** $B019_{\text{thirteen}}$ **45.** 100000000_{two} **47.** 111010_{two}

49. 81_{twelve} **51.** 310_{four} **53.** $1B6_{\text{twelve}}$ **55.** 15222^{7}

57. (a) 637_{eight} (b) $19F_{\text{sixteen}}$

59. (a) 2731_{eight} (b) $5D9_{\text{sixteen}}$

61. (a) 3775_{eight} (b) $7FD_{\text{sixteen}}$

63. (a) 4307_{eight} (b) $8C7_{\text{sixteen}}$

65. 1101111000_{two} **67.** $110101000100_{\text{two}}$ **69.** 110010000_{two}

71. $111010001110101_{\text{two}}$ **73.** $110101000101011_{\text{two}}$

75. 1101100010_{two} **77.** $101001011101_{\text{two}}$ **79.** $110010101001011_{\text{two}}$

81. 5 lb 7 oz **83.** 34 yd 2 ft 8 in. **85.** STOP **87.** CLASSISOVER

89. ITISRAINING **91.** 13256 **93.** 67138 **95.** 44501

97.

99.

101.

103. 67.8%, 84.7%, 90.2% **105.** 28.2%, 38.8%, 62.7%

107. 100%, 0%, 100% **109.** 3; manufacturer's **111.** 1; product

113. 7; manufacturer's **115.** one symbol; same as tally system

Note that all binary numbers in problem 117 are written without the two subscript.

117. (a)

Base 10	1	2	3	4	5	6	7	8	9
Binary	1	10	11	100	101	110	111	1000	1001

Base 10	10	11	12	13	14	15	16	17	18
Binary	1010	1011	1100	1101	1110	1111	10000	10001	10010

(b) Two digits: 2 and 3; Three digits: 4 through 7; Four digits: 8–15

(c) Conjecture: 16–31. The binary form of 32 is 100000, making it the first one with six digits.

119. *Note that base three numbers in this problem are written without the three subscript.*

(a)

Base 10	1	2	3	4	5	6	7	8	9
Base 3	1	2	10	11	12	20	21	22	100

Base 10	10	11	12	13	14	15	16	17	18
Base 3	101	102	110	111	112	120	121	122	200

Base 10	19	20	21	22	23	24	25	26	27
Base 3	201	202	210	211	212	220	221	222	1000

(b) One digit: 2; Two digits: 6; Three digits: 18

(c) Conjecture: 27–80; formula: $2 \cdot 3^{n-1}$

121. (a) b^7; add exponents (b) b^2 (c) b^n

(d) Because one is the only number that always gives back the original number when you multiply by it.

Exercise Set 4-4

5.

+	0	1	2
0	0	1	2
1	1	2	10
2	2	10	11

7.

×	0	1	2	3
0	0	0	0	0
1	0	1	2	3
2	0	2	10	12
3	0	3	12	21

9. 121_{four} **11.** $B843_{fourteen}$ **13.** 2020_{six} **15.** 121122_{nine} **17.** 1014_{five}
19. 43_{seven} **21.** 40040_{nine} **23.** 571_{twelve} **25.** 3352_{six} **27.** 56067_{nine}
29. $3976A_{twelve}$ **31.** 1501220_{six} **33.** 482_{nine} remainder 2_{nine}
35. 230_{five} remainder 3_{five} **37.** 131_{five} **39.** $1BD_{fourteen}$ **41.** 10000_{two}
43. $403_{sixteen}$ **45.** 1001_{two} **47.** $4DCA_{sixteen}$ **49.** 110010_{two}
51. $2894_{sixteen}$ **53.** 11_{two} remainder 10_{two} **55.** $B23_{sixteen}$ remainder $2_{sixteen}$
57. WORKHARD **59.** MICHAELISASPY **61.** L
63. G **65.** 01000100 01001111 01010010 01001101
67. 01010101 01001110 01001001 01001111 01001110
69. base 8
71. Yes, but only if you're adding more than two numbers. You would only carry more than 1 if the sum is at least twice the base: that can't happen if you're adding two numerals from the base number system, each of which is always less than the base.
73. ♠ = 0, ♦ = 1, ♥ = 2
75. ♣ = 0, ♥ = 1, ♠ = 2, ♦ = 3

Review Exercises

1. 1,000,221 **3.** 681 **5.** 419 **7.** 2,604 **9.** 118
11. [symbols]
13. [symbols] **15.** DIII
17. [symbols]
19. [symbols]

[symbols]

21. [symbols]
23. 1990 **25.** XXXIV **27.** 276 **29.** 756 **31.** 330 **33.** 156
35. 1,955 **37.** 365,687 **39.** 2,542 **41.** 13,965 **43.** 119
45. 17,327 **47.** 59 **49.** 19,481 **51.** 1024_{six} **53.** 2663_{nine}
55. 1313_{four} **57.** 12514_{seven}
59. (a) 733_{eight} (b) $1DB_{sixteen}$
61. (a) 1547_{eight} (b) $367_{sixteen}$
63. 111011010100_{two} **65.** 1010010110110011_{two}
67. 16.9%, 37.6%, 87.1% **69.** 251_{nine} **71.** $15AB6_{twelve}$ **73.** 331_{four}
75. 21331_{nine} **77.** $4C698_{sixteen}$ **79.** 342_{eight} remainder 6_{eight}

Chapter 4 Test

1. 2,000,312 **3.** 1966 **5.** 5,408
7. [symbols]
9. [Chinese numeral characters]
八
百
七
十
三

11. 221 **13.** 267,904 **15.** 17,375 **17.** 103 **19.** 28,474
21. 6362_{nine} **23.** 110000_{two} **25.** 621_{twelve} **27.** 331006_{eight}
29. 111011010100_{two}; $10100110110110010010_{two}$
31. $14B25_{twelve}$ **33.** 10010111_{two} **35.** 14223_{five} **37.** 121_{three}

CHAPTER 5: THE REAL NUMBER SYSTEM
Exercise Set 5-1

9. 1, 2, 4, 8, 16 **11.** 1, 2, 3, 6, 7, 9, 14, 18, 21, 42, 63, 126
13. 1, 2, 4, 8, 16, 32 **15.** 1, 2, 5, 7, 10, 14, 35, 70
17. 1, 2, 3, 4, 6, 8, 12, 16, 24, 32, 48, 96
19. 1, 17 **21.** 1, 2, 4, 8, 16, 32, 64
23. 1, 3, 5, 7, 15, 21, 35, 105 **25.** 1, 2, 7, 14, 49, 98
27. 1, 71 **29.** 3, 6, 9, 12, 15 **31.** 10, 20, 30, 40, 50
33. 15, 30, 45, 60, 75 **35.** 17, 34, 51, 68, 85
37. 1, 2, 3, 4, 5 **39.** $2^2 \times 5$ **41.** $2^4 \times 3$ **43.** 67 is prime
45. $2^3 \times 5^2$ **47.** $2 \times 3^3 \times 7$ **49.** $3 \times 5^2 \times 11$ **51.** $2^6 \times 5$
53. $3 \times 5^2 \times 29$ **55.** $2^3 \times 5 \times 11$ **57.** 5×103
59. 3 **61.** 1 **63.** 6 **65.** 21 **67.** 220 **69.** 12 **71.** 6
73. 12 **75.** 42 **77.** 10 **79.** 35 **81.** 126 **83.** 150 **85.** 630
87. 308 **89.** 120 **91.** 72 **93.** 1,040 **95.** 4 students **97.** 72 minutes
99. 6 members per team; 5 female teams, 6 male teams **101.** noon
103. 4 groups of Republicans, 5 groups of Democrats, 7 groups of Independents
105. 221 years **107.** 120 fps **109.** 150 fps
111. The only even prime number is 2.
113.

$3 + 1 = 4$	$7 + 7 = 14$
$3 + 3 = 6$	$3 + 13 = 16$
$3 + 5 = 8$	$5 + 13 = 18$
$5 + 5 = 10$	$3 + 17 = 20$
$5 + 7 = 12$	

115. Answers vary, 13.
119. We'd have to go up to $\sqrt{3,780}$, which is about 62.
121. The only limit would be the maximum number of rows allowed in a spreadsheet in whatever program you're using.

Exercise Set 5-2

9. 8 **11.** 10 **13.** 8 **15.** −10 **17.** 0 **19.** < **21.** >
23. > **25.** < **27.** < **29.** −1 **31.** 9 **33.** −11
35. −123 **37.** −33 **39.** 78 **41.** −65 **43.** 28
45. −68 **47.** −378 **49.** −24 **51.** −36 **53.** 42
55. 340 **57.** 0 **59.** 8 **61.** −5 **63.** −4 **65.** 7
67. 1 **69.** 0 **71.** Undefined **73.** 0 **75.** 111
77. 51 **79.** 36 **81.** 6 **83.** −758 **85.** −56 **87.** 59
89. −492 **91.** $1,180 **93.** 1,399 rats **95.** 8 inches
97. Puerto Rico: −439,915; West Virginia: −59,278; Mississippi: −6,018; Utah: 507,731; Idaho: 271,524; Texas: 3,999,944 **99.** 7,000 pounds
101. 2009 to 2010: 2.0; 2013 to 2014: −1.4; Difference: −3.4%
103. Biggest increase: 2.8; biggest decrease: −1.4. Difference: 4.2
105. No general statements can be made.
107. Answers vary. **109.** Answers vary.

Exercise Set 5-3

11. $\frac{1}{6}$ **13.** $\frac{7}{10}$ **15.** Already reduced **17.** $\frac{5}{6}$ **19.** $\frac{7}{8}$
21. Already reduced **23.** $\frac{5}{9}$ **25.** $\frac{7}{5}$ **27.** $\frac{15}{48}$ **29.** $\frac{38}{48}$
31. 189 **33.** 54 **35.** 3,195 **37.** $-\frac{1}{6}$ **39.** $-\frac{37}{24}$ or $-1\frac{13}{24}$
41. $\frac{7}{24}$ **43.** $\frac{7}{6}$ or $1\frac{1}{6}$ **45.** $\frac{7}{10}$ **47.** $-\frac{1}{16}$ **49.** $\frac{11}{12}$ **51.** $\frac{3}{4}$ **53.** $\frac{7}{36}$
55. $\frac{43}{64}$ **57.** $\frac{121}{1,728}$ **59.** 0.2 **61.** 0.66… or $0.\overline{6}$ **63.** 2.25
65. 0.3055… or $0.30\overline{5}$ **67.** 0.75 **69.** $0.\overline{9411764705882352}$
71. $\frac{3}{25}$ **73.** $\frac{3}{8}$ **75.** $\frac{863}{200}$ **77.** $\frac{7}{9}$ **79.** $\frac{6}{11}$ **81.** $\frac{70}{33}$ **83.** $\frac{34}{75}$
85. $-\frac{14}{39}$ **87.** $-\frac{4}{251}$ **89.** 171 miles **91.** $\frac{1}{7}$ **93.** 23 **95.** 4,740
97. 12 feet **99.** $\frac{3}{8}$ **101.** $1\frac{1}{4}$ cups of flour and $\frac{1}{3}$ cup of sugar
103. $\frac{1}{50}$ **105.** $\frac{1}{5}$ **107.** $\frac{43}{100}$ **109.** Any common denominator will do.
111. This question is a paradox, an answer of yes or no leads to a contradiction.
113. Add the fractions and divide by 2.
115. All three pairs of numbers are equal, even thought most people fill in "less than" for each in part (a).
117–119. Answers vary.

Exercise Set 5-4

9. rational **11.** irrational **13.** irrational **15.** 3 and 4 **17.** 9 and 11
19. 14 and 15 **21.** 5.477 **23.** 37.829 **25.** −6.481 **27.** $2\sqrt{6}$
29. $4\sqrt{5}$ **31.** $\sqrt{30}$ **33.** $20\sqrt{5}$ **35.** $30\sqrt{7}$ **37.** $15\sqrt{2}$

39. $-10\sqrt{7}$ **41.** $2\sqrt{5}$ **43.** $3\sqrt{30}$ **45.** $24\sqrt{3}$ **47.** $\sqrt{30}$
49. $2\sqrt{2}$ **51.** $12\sqrt{7}$ **53.** $-7\sqrt{3}$ **55.** $-2\sqrt{3}$ **57.** $4\sqrt{5}$
59. $-6\sqrt{5}$ **61.** $18\sqrt{2}$ **63.** $\sqrt{2}+8\sqrt{3}$ **65.** $10\sqrt{10}-4\sqrt{2}$
67. $7\sqrt{15}$ **69.** $-108\sqrt{6}$ **71.** $588\sqrt{2}$ **73.** $\frac{\sqrt{5}}{5}$ **75.** $\frac{\sqrt{6}}{2}$
77. $\frac{\sqrt{21}}{14}$ **79.** $\frac{21\sqrt{118}}{59}$ **81.** $\frac{\sqrt{6}}{9}$ **83.** 3 seconds   **85.** 9.2 seconds
87. 3.5 seconds, 2.5 seconds, the difference is 1 second.
89. 20 volts   **91.** 141.4 volts   **93.** 4π seconds   **95.** 0.8 foot
97. The irrational numbers are not closed under multiplication. The rational numbers are closed under multiplication.
99. You cannot compute the square root of a negative number because there is no number that, when squared, is negative.
101. (a) 3   (b) 6   (c) 2   (d) 5
103. (a) 2 and 3   (b) 3 and 4   (c) -4 and -5 (d) 4 and 5
105. (a) 4   (b) -11 (c) -16; The radical is always eliminated and the result is a rational number.
107. Answers vary.

Exercise Set 5-5

9. Integer, rational, real   **11.** Rational, real   **13.** Rational, real
15. Irrational, real   **17.** Irrational, real   **19.** Rational, real
21. Natural, whole, integer, rational, real
23. Natural, whole, integer, rational, real
25. Closure property of addition
27. Commutative property of addition
29. Commutative property of multiplication
31. Distributive property
33. Inverse property of multiplication
35. Commutative property of addition
37. Identity property of addition
39. Commutative property of multiplication
41. Distributive; commutative property of addition; distributive property
43. Addition, multiplication   **45.** Addition, subtraction, multiplication
47. None   **49.** 27.46, overweight   **51.** 21.82, normal
53. 31.66, obese   **55.** 34.98, obese   **57.** not commutative
59. commutative   **61.** associative   **63.** associative
65. The results are not the same, subtraction is not associative.
67. The results are not the same, addition does not distribute over multiplication.
69. You can write the difference as a sum: $b+(-c)$. Then a times $-c$ is the same as $-ac$.
71. $\frac{25-10}{5}=\frac{15}{5}=3$; $\frac{25}{5}-\frac{10}{5}=\frac{15}{5}=3$ "Distributing" the denominator worked. This makes sense because we can write $\frac{25-10}{5}$ as $\frac{1}{5}(25-10)$, in which case it's a multiplication and the distributive property applies.
73. You can write subtraction as addition of the opposite and the opposite of a real number is also a real number.
75. It makes it look like there are real numbers that are neither rational nor irrational: the outside rectangle shouldn't be there.

Exercise Set 5-6

9. 243   **11.** 1   **13.** 1   **15.** $\frac{1}{243}$ **17.** $\frac{1}{64}$ **19.** $3^6=729$
21. $4^7=16,384$ **23.** $3^2=9$ **25.** $12^1=12$ **27.** $5^6=15,625$
29. $\frac{1}{3^2}=\frac{1}{9}$ **31.** $\frac{1}{35^5}=\frac{1}{52,521,875}$ **33.** $\frac{1}{2^2}=\frac{1}{4}$ **35.** $\frac{1}{100^3}=\frac{1}{1,000,000}$
37. $\frac{1}{3^2}=\frac{1}{9}$ **39.** a^{20} **41.** $\frac{1}{m^8}$ **43.** a^{14} **45.** x^7 **47.** 6.25×10^8
49. 7.3×10^{-3} **51.** 5.28×10^{11} **53.** 6.18×10^{-6} **55.** 4.32×10^4
57. 59,000   **59.** 0.0000375   **61.** 2,400   **63.** 0.000003
65. -0.0000000146 **67.** $6\times10^{10}=60,000,000,000$
69. $2.67\times10^{-7}=0.000000267$ **71.** $8.8\times10^{-3}=0.0088$
73. $1.5\times10^{-9}=0.0000000015$ **75.** $2\times10^2=200$ **77.** $6\times10^0=6$
79. $6\times10^{-2}=0.06$ **81.** 2.58×10^{15} **83.** 2.4×10^1 **85.** 1×10^{-12}
87. $\frac{80}{d^2}$; it goes from 80% to 20%, 8.9% and then 5%.
89. 1.116×10^8 miles   **91.** 56%   **93.** 2.4696×10^{13} miles
95. 7.96×10^{-4} light years   **97.** 10 billion   **99.** 421 million miles
101. $79,726,027.40   **103.** 5,448,980
105. Each entry is half the previous entry, so the last three entries are $1, \frac{1}{2}$, and $\frac{1}{4}$.
107. Answers vary.
109. 66,621 miles per hour (using a 365.25 day year); it would take about 2 minutes and 13 seconds.

Exercise Set 5-7

9. (a) 1, 7, 13, 19, 25   (b) 6   (c) 67   (d) 408
11. (a) $-9, -12, -15, -18, -21$ (b) -3 (c) -42 (d) -306
13. (a) $\frac{1}{4}, \frac{5}{8}, 1, \frac{11}{8}, \frac{7}{4}$ (b) $\frac{3}{8}$ (c) $\frac{35}{8}$ (d) $\frac{111}{4}$
15. (a) $4, \frac{11}{3}, \frac{10}{3}, 3, \frac{8}{3}$ (b) $-\frac{1}{3}$ (c) $\frac{1}{3}$ (d) 26
17. (a) 5, 13, 21, 29, 37   (b) 8   (c) 93   (d) 588
19. (a) 50, 48, 46, 44, 42   (b) -2 (c) 28   (d) 468
21. (a) $\frac{1}{8}, \frac{19}{24}, \frac{35}{24}, \frac{17}{8}, \frac{67}{24}$ (b) $\frac{2}{3}$ (c) $\frac{179}{24}$ (d) $\frac{91}{2}$
23. (a) 0.6, 1.6, 2.6, 3.6, 4.6   (b) 1   (c) 11.6   (d) 73.2
25. (a) 12, 24, 48, 96, 192   (b) 2   (c) 24, 576   (d) 49,140
27. (a) $-5, -\frac{5}{4}, -\frac{5}{16}, -\frac{5}{64}, -\frac{5}{256}$ (b) $\frac{1}{4}$ (c) $-\frac{5}{4,194,304}$ (d) ≈ -6.7
29. (a) $\frac{1}{6}, -1, 6, -36, 216$ (b) -6 (c) 60,466,176   (d) $-\frac{310,968,905}{6}$
31. (a) $100, -25, \frac{25}{4}, -\frac{25}{16}, \frac{25}{64}$ (b) $-\frac{1}{4}$ (c) $-\frac{25}{1,048,576}$ (d) ≈ 80
33. (a) 4, 12, 36, 108, 324   (b) 3   (c) 708,588   (d) 1,062,880
35. (a) $\frac{1}{2}, \frac{1}{4}, \frac{1}{8}, \frac{1}{16}, \frac{1}{32}$ (b) $\frac{1}{2}$ (c) ≈ 0.000244 (d) ≈ 0.9998
37. (a) $-3, 15, -75, 375, -1,875$ (b) -5 (c) 146,484,375   (d) 122,070,312
39. (a) 1, 3, 9, 27, 81   (b) 3   (c) 177,147   (d) 265,720
41. geometric   **43.** neither   **45.** arithmetic   **47.** arithmetic
49. $3\cdot4^{n-1}$ **51.** $8n+3$ **53.** $50-10n$ **55.** $\frac{0.5}{5^{n-1}}$
57. $\frac{5\cdot3^{n-1}}{4^{n-1}}$ **59.** $4(0.6)^{n-1}$ **61.** 1,020.75   **63.** 795
65. (a) $500   (b) $13,000   **67.** no   **69.** $18,000
71. 160 ft   **73.** $3,766.11   **75.** 8 questions
77. First job pays $5,300.65 more in 10 years.
79. Yes, $2,438.81 higher.   **81.** $a_1=\frac{3}{10}$ and $r=\frac{1}{10}$. The sum is $\frac{1}{3}$.
83. Answers vary.

Review Exercises

1. 1, 2, 3, 4, 5, 6, 10, 12, 15, 20, 30, 60
3. 1, 2, 4, 5, 10, 19, 20, 38, 76, 95, 190, 380
5. 4, 8, 12, 16, 20   **7.** 9, 18, 27, 36, 45   **9.** $2^5\times3$
11. 2×5^3 **13.** 2; 30   **15.** 20; 1,200   **17.** 198 months, or 16.5 years
19. -14 **21.** -4 **23.** 44   **25.** 89   **27.** -157 **29.** $\frac{15}{19}$ **31.** $\frac{53}{6}$
33. $\frac{3}{20}$ **35.** $-\frac{25}{14}$ **37.** $\frac{51}{80}$ **39.** $\frac{21}{16}$ or $1\frac{5}{16}$ **41.** 1   **43.** $-\frac{19}{48}$
45. $0.\overline{857142}$ **47.** $\frac{11}{16}$ **49.** $\frac{23}{90}$ **51.** NFL: 12; MLB: 8; NBA and NHL: 16
53. $4\sqrt{7}$ **55.** $\frac{\sqrt{15}}{6}$ **57.** $22\sqrt{2}$ **59.** 2   **61.** 8   **63.** Rational, real
65. Irrational, real   **67.** Natural, whole, integer, rational, real
69. Commutative property of addition   **71.** Distributive property
73. 1   **75.** 117,649   **77.** 6,561   **79.** $\frac{1}{7,776}$ **81.** 2.59×10^{10}
83. 4.8×10^{-4} **85.** 2,330,000,000   **87.** 0.0000088   **89.** 6×10^{-8}
91. 1.41×10^4 seconds or about 3.9 hours
93. $-\frac{1}{5}, \frac{3}{10}, \frac{4}{5}, \frac{13}{10}, \frac{9}{5}, \frac{23}{10}; a_9=\frac{19}{5}, S_9=\frac{81}{5}$
95. $-\frac{2}{5}, \frac{1}{5}, -\frac{1}{10}, \frac{1}{20}, -\frac{1}{40}, \frac{1}{80}; a_9=-\frac{1}{640}, S_9=-\frac{171}{640}$
97. The profit for the sixth year is $25,525.63, and the total earnings for the 6 years are $136,038.26.

Chapter 5 Test

1. Integer, rational, real   **3.** Rational, real   **5.** Irrational, real
7. 14; 168   **9.** $\frac{3}{7}$ **11.** $\frac{16}{25}$ **13.** $9\sqrt{3}$ **15.** -5 **17.** $-\frac{31}{126}$
19. -8 **21.** $2\sqrt{2}$ **23.** $\frac{7}{8}$ **25.** Commutative property of addition
27. Identity property of addition
29. Associative property of multiplication
31. 4,096   **33.** 1   **35.** $\frac{1}{32}$ **37.** 2.36×10^{-3}
39. -0.00006 **41.** 3×10^3
43. $\frac{3}{4}, -\frac{1}{8}, \frac{1}{48}, -\frac{1}{288}, \frac{1}{1,728}, -\frac{1}{10,368}, \frac{1}{62,208}$
$a_{15}\approx9.57\times10^{-12}, S_{15}\approx0.643$
45. $320; $620

CHAPTER 6: TOPICS IN ALGEBRA

Exercise Set 6-1

7. $x-3$ **9.** $x+9$ **11.** $11-x$ **13.** $x-6$ **15.** $\frac{1}{2}x+x$ or $\frac{3}{2}x$
17. $\frac{3x}{6}$ or $(3x)\div6$ **19.** $\frac{x}{14}$ or $x\div14$ **21.** $3(x+\pi)$ **23.** $\frac{2x}{x+8}-3x$
25. $(5x+3y)^2$ **27.** 8   **29.** 16   **31.** 17 and 11   **33.** 12   **35.** 28
37. There are 27 students in one section and 30 students in the other section.
39. $18,295   **41.** $20.30   **43.** 18 games   **45.** $270
47. $1.25 billion on costumes and $1.93 billion on candy

49. 12   **51.** $36,000   **53.** $40   **55.** No   **57.** $563.20

59. James: 34.9%; Durant: 32.8%; Curry: 32.3%

61. $1,700 + 900x$; 3.67 fugitives per month

63. windchill = temperature − 1.5(wind speed); 3°

65. 60.3 mph

67. (a) $P = 8.775x + 11.856y$, where x is hours worked at the first job, and y is hours worked at the second.   (b) 11.5 hours

Exercise Set 6-2

9. $\frac{18}{28} = \frac{9}{14}$ **11.** $\frac{14}{32} = \frac{7}{16}$ **13.** $\frac{12}{15} = \frac{4}{5}$ **15.** $\frac{21}{8}$ **17.** $\frac{60}{30} = \frac{2}{1}$

19. $x = \frac{135}{14}$ **21.** $x = 35$ **23.** $x = 10$ **25.** $x = \frac{31}{3}$ **27.** $x = \frac{21}{4}$

29. 15 cones   **31.** $\frac{14}{3}$ inches   **33.** 4 gallons   **35.** 103 or 104

37. 37 or 38   **39.** 30 feet   **41.** 20 professors   **43.** About 15.5 million

45. 46,663   **47.** $1,200   **49.** 300 lb   **51.** 120 lb   **53.** 24 lb

55. 3 lb   **57.** 9 more students

59. To solve a proportion, you can cross multiply.

61. 25 pounds for $7.49 is a better buy.

63. 7 ounces for $1.99 is a better buy.

65. 34.5 ounces for $7.49 is a better buy.

67. (a) a varies inversely with the square of d.   (b) 250/9 or $27\frac{7}{9}$

(c) No; you can't find the individual constant of proportionality relating a and b.

69. (a) Inversely; too much time on Facebook means less time studying.

(b) Directly; bigger apartments are likely to cost more.

71. (a) Directly; as the age of a car goes up, it tends to have more mechanical problems.

(b) Directly; it is a documented fact that areas with a depressed economy have more crime.

Exercise Set 6-3

9.
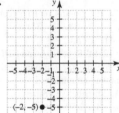

11.
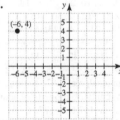

13.
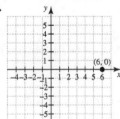

15.
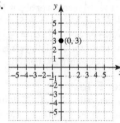

17.
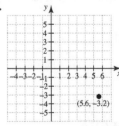

19.
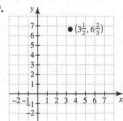

21.

23. $(-4, 2)$ **25.** $(3, 4)$

Note: answers for 27 and 29 are approximate.

27. $(-3, -3)$ **29.** $(3\frac{1}{2}, 0)$

31.
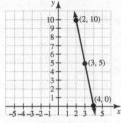

33.
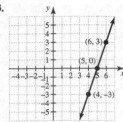

35.
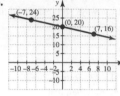

37.
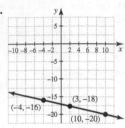

39.
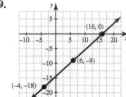

41. 1   **43.** $-\frac{7}{4}$ **45.** $-\frac{34}{19}$ **47.** $-\frac{43}{16}$

49. x intercept = (8, 0); y intercept = (0, 6)

51. x intercept = (−6, 0); y intercept = (0, −5)

53. x intercept = (72, 0); y intercept = (0, −27)

55. x intercept = (−40, 0); y intercept = (0, −56)

57. $m = -\frac{7}{5}$, y int = (0, 7)

$y = -\frac{7}{5}x + 7$

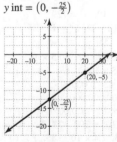

59. $m = \frac{1}{4}$, y int = (0, −4)

$y = \frac{1}{4}x - 4$

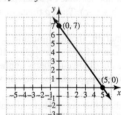

61. $y = \frac{3}{8}x - \frac{25}{2}$, $m = \frac{3}{8}$,

y int = $\left(0, -\frac{25}{2}\right)$

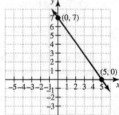

63. $y = -\frac{2}{3}x + \frac{4}{3}$, $m = -\frac{2}{3}$,

y int = $\left(0, \frac{4}{3}\right)$

65.
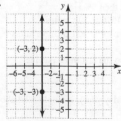

67.
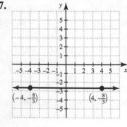

69. $y = 6.5x + 50$; (a) $69.50, (b) $82.50, (c) $115

71. $y = 0.2x + 26$; (a) $32, (b) $37, (c) $70

73. $y = 121 + 5x$, x = years after 2021; 141 messages in 2025; pass 175 in late 2031

75. $y = 28 - 1.42x$, x = years after 2012; the predicted number was very close: about 0.3% bigger).

77. $y = 36.3 + 1.1x$; 2062

79. (a) $507.32 billion in 2010, $656.50 billion in 2015, $775.84 billion in 2019 (b) 2027

(c) Actual number was $776 billion; model predicts $805.7 billion

81. (a) $y = 12,000 + 1.73x$; $13,730 (b) No; lose $740

83. $w = 160 - 3x$; 10 months   **85.** $y = 350 - 5x$; 70 years

87. (a)

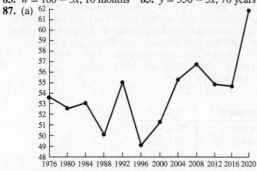

(b) Descriptions will vary.

89. (a)

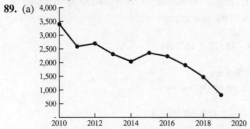

(b) Descriptions will vary, but other than small increases in 2012 and 2015, there's a clear downward trend.

91. The denominator of the fraction for the slope will always be zero.

93. $y = -\frac{a}{b}x + \frac{c}{b}$; slope = $-\frac{a}{b}$; y int = $\left(0, \frac{c}{b}\right)$

95. (a) 2011–2012: –27.1; 2012–2013: –46.0; 2013–2014: –31.4; 2014–2015: –14.0. Probably not: the rate of change varies quite a bit.

(b) Now it's out of the question. There's a change of direction.

Exercise Set 6-4

9. Function   **11.** Function   **13.** Function   **15.** Function

17. Function   **19.** Not a function   **21.** Not a function

23. $f(3) = 17; f(-2) = 2$ **25.** $f(10) = 32; f(-10) = -48$

27. $f(0) = 0; f(6) = 306$ **29.** $f(-3.6) = 5.56; f(4.5) = 45.25$

31. $f\left(\frac{1}{2}\right) = \sqrt{5}; f\left(-2\frac{1}{4}\right)$ is undefined   **33.** $f\left(-\frac{1}{2}\right) = 7; f\left(9\frac{1}{2}\right) = 25$

35. Domain $\{x | -\infty < x < \infty\}$; Range $\{y | y \geq 0\}$

37. Domain $\{x | x \neq 2\}$; Range $\{y | y \neq 1\}$

39. Domain $\{x | x \geq 0\}$; Range $\{y | y \geq 3\}$

41. Domain $\{x | -\infty < x < \infty\}$; Range $\{y | -\infty < y < \infty\}$

43. Domain $\{x | x \geq -1\}$; Range $\{y | y \leq -1\}$

45. Domain $\{x | x \leq 5\}$; Range $\{y | y \geq 7\}$

47.

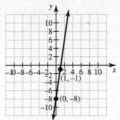

49.

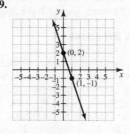

51.

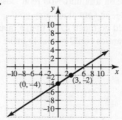

53.

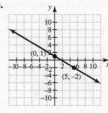

55.

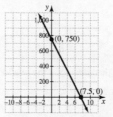

57. 42%   **59.** 1987   **61.** –6%   **63.** 1997–2015   **65.** $P(x) = 0.54x + 42$

67. $88.08 for a 4-hour shift and $98.16 for an 8-hour shift

69. No, they lose $128; $2,422   **71.** $C(d) = 45 + 42d$; $339

73. $D(G) = 3G - 10$; 65 mg   **75.** $d(t) = 80 + 65t$; 3 hours

77. $n(r) = 3r$; 135 minutes, or 2 hours and 15 minutes

79. $s(p) = 0.8p$; $95.82   **81.** $n(x) = \frac{2}{3}x + 4$; 24   **83.** $d(x) = 48 + 7.5x$; $20

85. (a) $c(p) = -20p + 1,000$

(b) No; 2nd quadrant means negative price, 4th means negative number of customers.

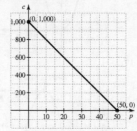

(c) 100   (d) $35   (e) (50, 0): at $50, no customers will pay.

(f) (0, 1,000); 1,000 customers will take the tour for free.

87. (a) $d(t) = 70t$ (b)   (c) 560 miles   (d) 3 hours

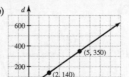

89. $f(3) = 19; f(-3) = 1$;

$f(x + 3) = x^2 + 9x + 19$;

$f(x - 3) = x^2 - 3x + 1$

91. (a) [0, 120]; we're not given any information about what happened before the oven was turned on, so we'll start at time zero. I'm guessing that it will take at most an hour after the oven was turned off for it to return to room temperature.

(b) [72, 450]; the oven was at room temperature before it was turned on, which I'm guessing is 72 degrees. The oven heats to 450 degrees and doesn't go over that, and every temperature in between will be reached at some time.

(c)

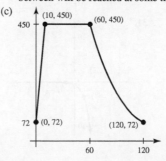

93. (a) [1, 100]; You're not going anywhere on foot until you're at least 1; after the age of 100, it's time to rest—let someone carry you.

(b) $[\frac{1}{4}, 5]$; I don't think a 1-year-old could go much more than a quarter mile, and the average man in his twenties could probably make about 5 miles.

(c)

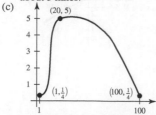

95. Function   **97.** Not a function

99. If any vertical line hits the graph in more than one point the relation graphed isn't a function.

Exercise Set 6-5

13.

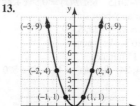

15.

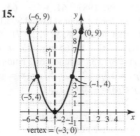

17.

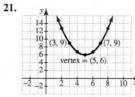

19.

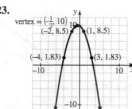

21.

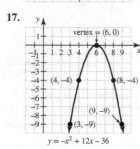

23.

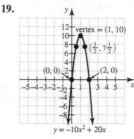

25.

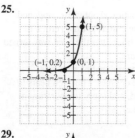

27.

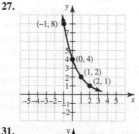

29.

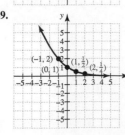

31.

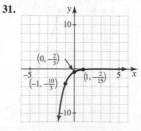

33.

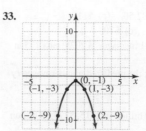

35. 3.3   **37.** 6.1   **39.** {5}   **41.** {0.6}   **43.** {3.6}

45. {−2.8}   **47.** 62 feet, 3 inches; 3.85 seconds

49. Yes, the max height is 88.6 feet.

51. (a) 4 pairs   (b) $146   (c) $930   (d) 2 or 6 pairs

53. (a) The most revenue is $411 and the number of price hikes needed is −3. This means that they need to actually decrease the price by $1.50 to maximize their revenue.   (b) $392   (c) $392

55. (a) $A = \frac{5}{2}W^2$ (b) $A = 250$ square inches and $l = 25$ inches
(c) width 4 inches; length 10 inches   (d) $A = 10w^2$

57. 4,969,507   **59.** $449.04

61. (a) $9,434.32   (b) $11,125.80

63.

Compounded	Value
Quarterly	$24,403.80
Monthly	$24,419.93
Weekly	$24,426.18
Daily	$24,427.79

More interest is paid as the number of compounding periods goes up.

65. 678 years ago, in the year 1320

67. 45,117 years old

69. 7.0   **71.** 8.7   **73.** 27.2 decibels

75. 65.1 decibels

77. There is no difference.

79. It's completely above the x axis.

81. 75.1 ft/sec

83. (a) $f(1) = 2(1) = 2, f(2) = 2(2) = 4;$
$g(1) = 2(1)^2 - 4(1) + 4 = 2; g(2) = 2(2)^2 - 4(2) + 4 = 4;$
$h(1) = 2^1 = 2; h(2) = 2^2 = 4$

x	3	4	5	6	7
$2x$	6	8	10	12	14
$2x^2 - 4x + 4$	10	20	34	52	74
$2x$	8	16	32	64	128

(b)

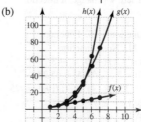

(c) Linear growth is at a steady rate. The others grow faster, with exponential growing the quickest in the long run.

85. (a) Answers vary, but most choose 1.
(b) Choice 1: $1,050; Choice 2: $1,015; Choice 3: $1,638.30
(c) Choice 1: linear; Choice 2: quadratic; Choice 3: exponential

Review Exercises

1. $8n - 4$ **3.** $6(\frac{1}{2} + x)$

5. $4\frac{4}{9}$ hours to get home and $3\frac{5}{9}$ hours to get back

7. 158 $8 seats, 79 $10 seats, and 89 $12 seats.

9. $30   **11.** $47 per couple   **13.** $\frac{16 \text{ ounces}}{$2.37}$

15. $\frac{18 \text{ minutes}}{120 \text{ minutes}} = \frac{3}{20}$ **17.** $\{\frac{389}{23}\}$ **19.** 225

21. Wife: $168,000; Son: $112,000

23. $4,320   **25.** 4.44 amps   **27.** Answers vary.

29.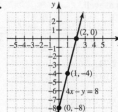

31.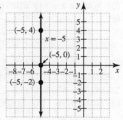

33.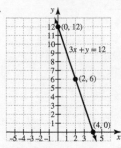

35. $\frac{2}{5}$ **37.** 0

39. $y = -3x + 12; m = -3$ **41.** $y = \frac{4}{7}x - 4; m = \frac{4}{7}$

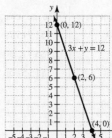 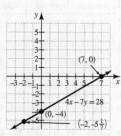

43. (a) $y = -16,333x + 81,000$ (b) $53,778
   (c) Decreasing by $16,333 per year
   (d) $-16,998$; obviously a house can't have a negative value.
45. Answers vary.
47. Function
49. Not a function
51. Domain = $\{x \mid -\infty < x < \infty\}$; range = $\{y \mid -\infty < y < \infty\}$; function
53. $f(-3) = 16, f(10) = -10$
55. $f(8) = 104, f(-5) = 65$
57. $f(x) = -\frac{3}{4}x + 4$

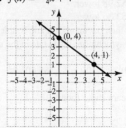

59. $G(n) = 1.8 + 0.15(n - 1)$, or $G(n) = 1.65 + 0.15n$; 3.15
61. Answers vary.
63. 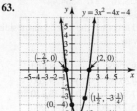   **65.**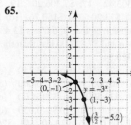

67.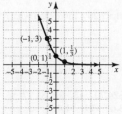

69. 4.11 **71.** Max height = 104 feet; 5.05 seconds to reach the ground

Chapter 6 Test
  1. $x = 4$
  3. $1,500 was invested at 4% and $3,500 was invested at 6%
  5. 21.9 hours **7.** 12 hours
  9. 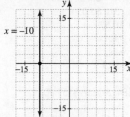   **11.**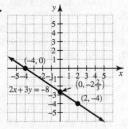

13. $y = -\frac{x}{5} + 4; m = -\frac{1}{5}$

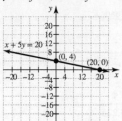

15. Not a function **17.** $f(12) = -26; f(-15) = 55$

19. 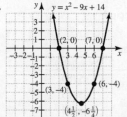   **21.**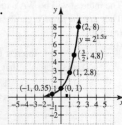

23. base width 12 inches and sides height 6 inches
25. 13.2 years

CHAPTER 7: CONSUMER MATH
Exercise Set 7-1
  7. 63% **9.** 2.5% **11.** 156% **13.** 20% **15.** $66.\overline{6}\%$ **17.** 125%
  19. 0.18 **21.** 0.06 **23.** 0.625 **25.** 3.2 **27.** $0.\overline{6}$ **29.** $\frac{6}{25}$
  31. $2\frac{9}{25}$ **33.** $\frac{1}{200}$ **35.** $\frac{1}{6}$ **37.** $15; $314.99 **39.** $15.00; $264.99
  41. 60% **43.** 24% **45.** $76.79 **47.** $126 **49.** $1,256
  51. 1996–2010: Actual 43,536, percent 50.6%, rate 3,110 per year
    2010–2019: Actual 121,732, percent 93.9%, rate 13,526 per year
  53. $800
  55. 2001–2010: Actual 125,546, percent 98.4%. 2010–2019: Actual
    142,476, percent 56.3%. Even though the actual change was similar,
    the percent change was quite a bit less for the second time period.
  57. 2010–2015: 55.1%, 11.0% per year; 2015–2020: 63.7%, 12.7% per year
  59. 25% off saves $12.25.
  61. (a) 141,210 (b) 151,685 (c) 451,515; 75,170 less
  63. Father-in-law about 31%, town about 95%
  65. 40%, 1.9%/month, about $404,000
  67. No; the total discount is 44%.

69. Total discount will be 75%, not 100%.

71. There's no comparison: 20% more what? 20% more than what?

73. $4.57   **75.** 14.3%; $4.67; 16.75%

Exercise Set 7-2

Answers for some of these exercises will vary a bit since they can be calculated in different ways.

7. $2,707.69   **9.** $1,839.60   **11.** $2,824.70   **13.** $1,681.17

15. $2,302.06   **17.** $912.50   **19.** $963.60   **21.** $1,482

23. $532.06   **25.** $990.75   **27.** $1,141.98   **29.** 32.3%

31. 61.9%   **33.** $454.57   **35.** $1,360.24

37–41. Answers vary

Exercise Set 7-3

9. $1,440   **11.** $220,340.63   **13.** 4.4%   **15.** $985

17. $8,000   **19.** $1,333.80   **21.** $497.70   **23.** 3 years

25. $960   **27.** $1,102.27   **29.** $7.88   **31.** $5.03

33. (a) $720   (b) $2,280   (c) 10.5%

35. (a) $2,025   (b) $3,975   (c) 11.3%

37. 8.43%   **39.** $141,509.43   **41.** 3 years

43. 4.25 years or 51 months   **45.** $2,096.25

47. $7,316   **49.** $92.97

51. (a) Answers vary.   (b) 10-year: $84; 5-year: $157.50
    (c) 5-year has $630 less interest

53. 13.1%   **55.** 16.4%   **57.** $6   **59.** $4.40

61. (a) $5,400   (b) $12,600   (c) 7.1%

63. The personal loan   **65.** The repair loan

67. (a) $3,000
    (b)

End of year	1	2	3	4	5
Future value ($)	10,600	11,236	11,910	12,625	13,383

    (c) $383
    (d)

End of year	$2\frac{1}{2}$	5
Future value ($)	11,500	13,225

You'd pay $225 more.

69. The interest will be WAY too high.

71. They are the same regardless of principal.

73. (a) $30
    (b)

Years	Interest earned in that year	New balance
1	$30	$780
2	$30	$810
3	$30	$840
4	$30	$870
5	$30	$900

    (c) Because they're getting 4% of the amount after 3 years, while the first investor continues to get 4% of the original $750.

Exercise Set 7-4

7. $I = \$396.20$ and $A = \$1,221.20$ **9.** $I = \$75.72$, $A = \$635.72$

11. $I = \$38.10$ and $A = \$358.10$ **13.** $10,805.48

15. $66,649.19   **17.** $574.85   **19.** $1,021.18   **21.** 6.14%

23. 6.66%   **25.** 4.5% compounded semiannually

27. 3.1% compounded quarterly   **29.** $12,175.94

31. $335,095.61   **33.** $9,507.11   **35.** $4,011.25

37. 4.45 years   **39.** 10.54 years; 10.19 years

41. $20,548.25   **43.** $27,223.76

45. (a) $24,664.64   (b) $5,153.40

47. (a) $59,556.16   (b) $19,556.16   (c) $29,398.60

49. $2,095.18   **51.** $114,717.32

53. $148.20; $1,163.29   **55.** $199.71

57. The certificate of deposit   **59.** 11.6 years

61. (a) If $u = \frac{n}{r}$, then $\frac{r}{n} = \frac{1}{u}$ and $n = ur$, so $P\left(1 + \frac{r}{n}\right)^{nt} = P\left(1 + \frac{1}{u}\right)^{urt}$. Then use a property of exponents to write as $A = P\left[\left(1 + \frac{1}{u}\right)^u\right]^{rt}$.

(b) When n tends to infinity, n/r does as well.

(c)

u	50	100	500	1,000	1,500	2,000	2,500
$\left(1 + \frac{1}{u}\right)^u$	2.692	2.705	2.716	2.717	2.717	2.718	2.718

The formula gets closer and closer to 2.718.

63. Continuously: $11,127.70; Annually: $10,955.62

65. $8,347.30   **67.** $1,235.34   **69.** 6.47%

71. (a) $40,488.76   (b) $38,902.85; $1,585.91 too low

73.

Frequency	Future value
Annually	$26,764.51
Monthly	$26,977.00
Weekly	$26,992.51
Daily	$26,996.51
Hourly	$26,997.15
Every minute	$26,997.18

Exercise Set 7-5

11. $480

13. Down payment: $56.25; installment price: $383.25

15. $43.17

	Amount financed	Total installment price	Finance charge
17.	$11,695.41	$16,011.20	$1,715.79
19.	$17,642.50	$26,915	$5,772.50
21.	$20,666	$35,018.56	$1,952.56

23. About 4.3%   **25.** 12%   **27.** 9.5%

29. $u = \$709.50$; payoff amount: $10,956.50

31. $u = \$14.54$; payoff amount: $482.81

33. $u = \$14.66$; payoff amount: $615.34

35. $180.13   **37.** $8.22   **39.** $5.18

41. (a) $10,552   (b) $6,390.48   (c) Answers vary.

43. Lender; $45.42   **45.** (a) $7,769.23   (b) $2,000 more

47. $16.65; $1,124.15   **49.** $39.49; $3,368.60   **51.** $13.32; $411.35

53. (a) $602.14   (b) $7.23   (c) $669.15

55. (a) $370.55   (b) $5.19   (c) $350.76

57. (a) $370.92   (b) $4.08   (c) $533.30

59. $892.76

61. (a) #51: $669.45; #39: $347.78; #41: $529.94
    (b) #37: unpaid balance method is $0.30 higher.
        #53: unpaid balance method is $2.98 lower.
        #55: unpaid balance method is $3.36 lower.
    (c) When purchases total a lot more than payments, unpaid balance is the better method, especially if payments are made late in the month.

63. $641.67   **65.** $3,636.36

67. The calculation ends up being $(0.99)^6 \times 2,300 = \$2,165.40$.

69. About 12.65 years

71. It'll take 133 months, which is 11 years and 1 month.

Exercise Set 7-6

9. $40.20; $2,050.20   **11.** $109.50; $5,584.50   **13.** $106.45

15. $303.42   **17.** $7,306.20   **19.** $3,053.70   **21.** $384.44

23. (a) $73.20   (b) $150.76   (c) $1,840.80

25. Subsidized loan at 6.8% is $3,434.40 less than the federal unsubsidized loan and $7,317.00 less than the private.

27. $326.84

29. (a) $21,750   (b) $123,250   (c) $871.38   (d) $138,164

31. (a) 80,000   (b) $120,000   (c) $720.00   (d) $139,200

33. (a) $32,500   (b) $295,700   (c) $1,945.71   (d) $638,240.80

35. (a) $360,000   (b) $840,000   (c) $5,779.20   (d) $547,008

37. $1,108.02; $61,970.40   **39.** $261,165.60   **41.** $1,499.44

43. $1,148.90   **45.** $4,215.22   **47.** $3,746.38

49.

Payment number	Interest	Payment on Principal	Balance of Loan
1	$718.96	$152.42	$123,097.58
2	$718.07	$153.31	$122,944.27
3	$717.17	$154.21	$122,790.06

51.

Payment number	Interest	Payment on Principal	Balance of Loan
1	$3,850	$1,929.20	$838,070.80
2	$3,841.16	$1,938.04	$836,132.76
3	$3,832.28	$1,946.92	$834,185.84

53. $255,947 **55.** $162,444
57. If you can afford the higher payment of the 20-year mortgage at 9% after making the 25% down payment, then that is the better option since the total interest paid is less. However, if you can only manage a 10% down payment, or need lower monthly payments, the 25-year mortgage at 7% would be the better choice.
59. $9,822.40, 75.0%; $138,164; 112.1%

Exercise Set 7-7
11. $37.25 **13.** $3.30 **15.** 1,022,112 **17.** $1.45 **19.** $14,428.80
21. $89.18 **23.** $0.40 **25.** 447,720 **27.** $3.99 **29.** $7.82 billion
31. $60.97 **33.** $1.96 **35.** 5,296,196 **37.** $4.68 **39.** $25,954.14
41. 24.91 **43.** 15.02 **45.** $0.34 per share **47.** $1.98 per share
49. lost $432.48 **51.** lost $2,022.63
53. The stock that pays the dividend results in $1,550 more profit.
55. $97,582.30, or $267.35 per day **57.** $1,245 **59.** $3,660.63
61. $320 **63.** $860.63 **65.** 11.5% **67.** 5.9%
69. The second investment would be the better choice.
71. (a) Curly ($3,550) (b) Curly (31.1%) (c) Moe (8.3%/yr)
73–75. Answers vary
77. (a) 13,564%; 1,214.7% per year
(b) Profit: $1,207,727.40; percent return: 15,310%, or 879% per year. Worthwhile is an opinion.

Review Exercises
1. 0.875; 87.5% **3.** $\frac{37}{20}$; 1.85 **5.** 5.75; 575%
7. 69.12 **9.** 1,100 **11.** $60 **13.** 32.7%
15. (a) The 10% is taken off the discount price, not the original.
(b) $45.05
17. $3,178.54 **19.** $2,375.54 **21.** $1,242.80 **23.** $917.67
25. $366.75 **27.** $2,322 **29.** 2.5 years **31.** 7% **33.** $1,375
35. $I = \$1,800$ and $A = \$7,800$ **37.** 1.4% **39.** $25.56
41. Discount: $1,080; Raul received $4,920
43. $I = \$603.67$ and $A = \$2,378.67$ **45.** $5.03; $50.03 **47.** 12.55%
49. $8,937.54 **51.** 8.42 years; 0.17 years, or about 62 days
53. $719.56 **55.** $15,518.75; $25,524; $1,905.25
57. 8% **59.** $1,473.96 **61.** $u = \$757.80$; payoff amount = $7,017.20
63. $9.86; $786.36 **65.** $748.82 **67.** $62.41 **69.** $275.04
71. $1,827.94 **73.** High: $59.01; low: $43.55 **75.** 15,783,230
77. $3.99 **79.** Answers vary. **81.** The buyer makes $16.55 more.

Chapter 7 Test
1. 31.25% **3.** $\frac{7}{25}$ **5.** 80% **7.** 353 **9.** $2,568
11. $2,035.18 **13.** $752.04 **15.** $124.33 **17.** 5%
19. With simple interest, the percentage is always calculated on the principal amount. With compound interest, interest is also computed on previously earned interest.
21. $33.33 **23.** $I = \$145.79$ and $A = \$645.79$
25. $345 **27.** $19,043.39 **29.** 6% **31.** $16.15
33. The finance charge is $6.63. The new balance is $387.13.
35. (a) $38.96 (b) $1,289.40 **37.** $417; $109,620
39. With a beta of 0.99, it's almost exactly as volatile as the average stock. This means the price is about as stable as an average stock.
41. $−0.65

CHAPTER 8: MEASUREMENT
Exercise Set 8-1
9. 4 yd **11.** $5\frac{2}{3}$ yd **13.** $39\frac{3}{4}$ ft **15.** $1\frac{3}{4}$ ft **17.** $\frac{1}{2}$ yd **19.** $24\frac{1}{3}$ yd
21. 800 cm **23.** 120 m **25.** 6 hm **27.** 900 dm **29.** 3.756 m
31. 0.4053 km **33.** 2,370 cm **35.** 560 mm **37.** 127,500 dm
39. 0.8 km **41.** ≈196.85 in. **43.** 406.4 mm **45.** ≈7.16 dam
47. ≈4,429.13 ft **49.** 499.53 mm **51.** ≈19.69 ft **53.** 172.21 dm
55. ≈1.46 mi **57.** ≈4.59 yd **59.** ≈8.46 m **61.** $76\frac{1}{2}$ ft; 23.32 m
63. 60 yd; 54.86 m **65.** 0.85 mi; 1.37 km **67.** $3\frac{1}{2}$ yd **69.** 2,300 mg
71. 14 km; 8.7 mi **73.** 104.65 km per hour **75.** ≈19.16 m
77. 38.1 mm by 76.2 mm **79.** 374.7 mi, 32,973.6 ft **81.** 7.92×10^9 ft
83. 573 boards **85.** 20 g **87.** 488 Canadian dollars **89.** 37.85 francs

Note: for Exercises 91–108, answers can vary: these are one guy's interpretation. The important thing is to think about the relative sizes of the different units in both systems.

91. Meters **93.** Millimeters **95.** Centimeters
97. Millimeters **99.** Yards, meters
101. Inches, centimeters **103.** Inches, millimeters
105. Yards, hectometers **107.** Inches, centimeters
109. the first runner **111.** millimeters

Exercise Set 8-2
Note: answers can vary, depending on rounding and on the conversion factors used.

7. 24 ft^2 **9.** 2.6 mi^2 **11.** $166\frac{1}{2}$ ft^2 **13.** ≈0.07 acre
15. ≈505 yd^2 **17.** ≈4.84 m^2 **19.** 43.0 km^2
21. 124.5 cm^2 **23.** 473.7 dm^2 **25.** 38,764.5 acres
27. ≈0.021 ft^2 **29.** ≈29.26 m^2 **31.** 142.7 ft^2
33. ≈2,872.32 fluid oz **35.** ≈18.44 yd^3
37. ≈6.94 ft^3 **39.** 1,570.1 in^3 **41.** 0.000097 m^3
43. 382.51 dL **45.** 87.25 L **47.** 0.00437 m^3
49. 0.5 L **51.** 920 dL **53.** 42,000 mL
55. ≈45.28 L **57.** 357,169.8 mL **59.** ≈22.31 qt
61. 46,428 gal **63.** ≈0.34 kL **65.** 45,843.75 pounds
67. 3.3 ft^3 **69.** 2 gallons **71.** 7.4 cubic yards
73. 1.56 mi^2 **75.** ≈0.67 ft^3 **77.** 634 tiles
79. 520 lb **81.** 5 bikers **83.** ≈$15\frac{1}{2}$ km^2
85. The 1970 engine was about 35% bigger.
87. Acres **89.** Square miles **91.** Square miles
93. Liters **95.** Milliliters **97.** 9 trips **99.** Answers vary.
101. Answers vary depending on your school.
103. Answers vary depending on your classroom.
105. (a) About 9,434 (b) About 110.7 lb (c) About 52 years, 117 days

Exercise Set 8-3
7. 2.0625 lb **9.** 36.8 oz **11.** 8,400 lb
13. 1.75 T **15.** ≈1.08 T **17.** 90 cg
19. 440 hg **21.** 5.15 dg **23.** 72.6 dg
25. 32.17 g **27.** 0.00543 g **29.** 22,555.6 g
31. ≈4.29 oz **33.** ≈0.106 lb **35.** ≈27,272.73 hg
37. ≈162.36 t **39.** ≈5.23 t **41.** ≈59,640 dg
43. ≈291.07 oz **45.** ≈0.03 lb **47.** 57.2°F
49. 92°F **51.** −0.4°F **53.** −15°C **55.** 35.7°C
57. ≈−23.33° C **59.** −53.85° C **61.** −7.9° F **63.** −279.1° F
65. 6.94 lb; Tricia was way off. This is less than an average baby.
67. The elevator is more than able to hold their weight.
69. She will not pack a coat. **71.** The cars are too heavy.
73. 68°F **75.** 2,240 grams **77.** Shirl lost 57.6 oz of fat.
79. They should not let them on! **81.** $12\frac{2}{3}$°F
83. Yuan did not buy the right cement.
85. The sugar should be melted. **87.** 67.5 kg
89. (a) 1,504,000 oz (b) 5.9 gal/min (c) $6,442.80
(d) $35.79 per passenger
91. Kilogram **93.** Kilogram

95. (a)

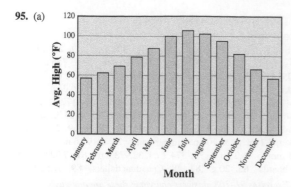

Month

The Celsius graph appears to show a greater variation in temperatures.

(b)

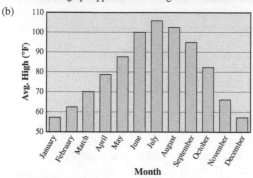

Month

Now the temperature variation looks much greater.

97. San Antonio   **99.** Football player   **101.** A delightful soak
103. Your leg would freeze solid instantly.
105. The Fahrenheit graphs shows a larger variance.

Review Exercises

1. $13\frac{1}{2}$ ft   **3.** 1.75 ft   **5.** $\approx$130.13 mi   **7.** 3.6 dm   **9.** 34.5 m
11. 290.35 in.   **13.** $\approx$33.46 in.   **15.** 1,051.56 dm   **17.** 91.44 m
19. 313.95 km/hr, 286 ft/sec   **21.** The Eiffel Tower   **23.** $\approx$39.57 ft^2
25. 68,302,080 ft^2 **27.** $\approx$8.82 mi^2 **29.** 131.13 dm^2 **31.** $\approx$0.03 ft^2
33. $\approx$6,702 oz   **35.** $\approx$1,212 gal   **37.** $\approx$1,299.47 in^3 **39.** 67,300 L
41. 2.31 L   **43.** 85 drops   **45.** The second is cheaper.
47. Acres   **49.** 4.0625 lb   **51.** 17.25 T   **53.** $\approx$18 T   **55.** 457 mg
57. 343.45 hg   **59.** 1,274 g   **61.** $\approx$0.05 lb   **63.** 55.4°F
65. 33.5°C   **67.** $-$31.45° C   **69.** $\approx$7.04 lb   **71.** 76.6° F
73. It appears the deal on the Canadian side is better, but only if they're both using the same currency!

Chapter 8 Test

1. $\approx$3.67 yd   **3.** 12.3 cm   **5.** $\approx$13.39 in.   **7.** $\approx$24.13 ft^2
9. 3.31 in^2 **11.** $\approx$6.10 in^3 **13.** 846.38 gal   **15.** 22.8125 lb
17. 59.1 cg   **19.** $\approx$0.06 lb   **21.** 29.56°C   **23.** 6 hr, 31 min
25. The first at \$1.39 per square foot is cheaper.   **27.** $\approx$3.37 hours
29. 19.44 hr   **31.** This is a waste of the agent's time.   **33.** 15.12°F

CHAPTER 9: GEOMETRY

Exercise Set 9-1

7. Ray; $\overrightarrow{AB}$ **9.** Line; l **11.** Segment; $\overline{TU}$ **13.** $\angle RST$; $\angle TSR$; $\angle S$; $\angle 3$
15. $\angle MOI$, $\angle IOW$, $\angle MOW$
17. $\angle RMU$, $\angle UMP$, $\angle PMT$, $\angle RMP$, $\angle RMT$, $\angle UMT$
19. Straight   **21.** Obtuse   **23.** Vertical angles
25. Corresponding angles   **27.** Corresponding angles
29. Alternate exterior angles   **31.** 82°   **33.** 57.6°   **35.** $71\frac{3}{4}°$
37. $(80 - x)°$ **39.** 24°   **41.** 117.5°   **43.** $68\frac{1}{6}°$ **45.** $(195 - y)°$
47. $\frac{100}{3}$ **49.** $\frac{43}{2}$ **51.** $\frac{135}{2}$ **53.** 33
55. $m\angle 1 = 37°$; $m\angle 2 = 143°$; $m\angle 3 = 37°$
57. $m\angle 1 = 90°$; $m\angle 2 = 90°$; $m\angle 3 = 90°$
59. $m\angle 1 = m\angle 3 = m\angle 5 = m\angle 7 = 15°$
$m\angle 2 = m\angle 4 = m\angle 6 = 165°$

61. $m\angle 1 = 55°$; $m\angle 2 = 55°$; $m\angle 3 = 55°$; $m\angle 4 = 125°$; $m\angle 5 = 125°$;
$m\angle 6 = 55°$; $m\angle 7 = 70°$
63. $3x = 114°$, $2x - 10 = 66°$
65. $4t - 31 = 85°$, $2t + 27 = 85°$, $x = 95°$, $y = 95°$
67. 90°   **69.** 60°   **71.** 15°   **73.** 104°   **75.** 104°   **77.** 104°   **79.** 104°
81. Answers vary.   **83.** Answers vary.   **87.** π radians $= 180°$
89. $m\angle 1 + m\angle 2 = 180°$ **91.** $m\angle 1 - m\angle 3 = 0°$
93. 3, 6; a reasonable conjecture would be 9, but yours could be different.

Exercise Set 9-2

11. Isosceles; acute   **13.** Obtuse   **15.** Scalene; acute   **17.** 70°
19. $48\frac{2}{3}°$ **21.** 40°   **23.** 34 ft   **25.** 13.8 yd   **27.** 19 mi   **29.** 36.1 ft
31. 12 ft. The first is $1\frac{1}{2}$ times as big.
33. 24 in. The second is 3 times as big.
35. All three pairs of corresponding angles are equal. $21\frac{9}{11}$ in.
37. The two angle Bs are vertical and therefore equal, so all three pairs of corresponding angles are equal. 17 m
39. Each triangle has a right angle and they share angle A, so all three pairs of corresponding angles are equal. 11 ft
41. About 127.3 feet   **43.** Yes, just barely   **45.** 16′6″
47. 14 mi   **49.** 270 ft   **51.** 10 ft   **53.** $\approx$212.2 mi
55. (a) 2.5 miles   (b) 5 miles   (c) 3.54 miles
57. 180° since the three angles form a straight angle
59. $m\angle 1 = m\angle 4$ since they are alternate interior angles between two parallel lines
61. This proves the result for every triangle.
65. If the edge is perpendicular to the ground, then drawing a line from the top left edge to the bottom right edge will form a right triangle, and the measurements will fit the Pythagorean theorem. That distance should be (to the nearest eighth of an inch) $12′10\frac{3}{8}″$.

Exercise Set 9-3

7. Octagon, 1,080°   **9.** Triangle, 180°   **11.** Hexagon, 720°
13. Rectangle   **15.** Trapezoid   **17.** 76 yd   **19.** 28 ft   **21.** 44 in.
23. 42 ft   **25.** 42 mi   **27.** 39 in.   **29.** 25 ft   **31.** 224 cm   **33.** 10 ft
35. 360 ft   **37.** 241 ft   **39.** 66 in.   **41.** 5.08 times   **43.** \$392
45. $\approx$ 1.74 mph   **47.** 118.4 yd and 191.6 yd
49. Only the longest side, at 1,325 ft
51. You can find $\angle ACB$ since it's the third side of a triangle. That angle is supplementary to $\angle BCD$.
53. The angle measures, starting with 8 sides, are 135°, 144°, 150°, 154.3°, 157.5°, 160°, and 162°. As the number of sides continues to increase, the angle measures will approach 180°.
55. (a) About 1.4 miles   (b) 400 feet less   (c) About 5.4%

Exercise Set 9-4

7. 289 in^2 **9.** 450 yd^2 **11.** 400 m^2 **13.** 120 mi^2 **15.** 467.5 in^2
17. 101.5 ft^2 **19.** 60.7 yd^2 **21.** 72 in^2 **23.** $123\frac{1}{2}$ in^2 **25.** 26.2 in^2
27. $C \approx$ 50.27 in.; $A \approx$ 201.06 in^2 **29.** $C \approx$ 50.27 m; $A \approx$ 201.06 m^2
31. $C \approx$ 131.95 km; $A \approx$ 1,385.44 km^2 **33.** 24.0 in^2
35. 148.3 ft^2 **37.** 20.0 m^2 **39.** 3.2 ft   **41.** $8\frac{1}{3}$ mi
43. 11.11 yd^2 **45.** 40 images   **47.** \$1,620
49. (a) 90.2 ft^2 (b) One gallon; 0.48 gallon left over
51. \$188.50   **53.** \$458.04   **55.** 16.28 ft^2
57. (a) That the two longest sides are parallel and the angles at the top are right angles.   (b) 16,790.6 ft^2; 0.39 acre
59. 10.6 ft   **61.** When the radius is 2.   **63.** Answers vary.   **65.** 30 in^2
67. (a) 175 sq ft
(b) 10: 300 sq ft; 15: 375 sq ft; 20: 400 sq ft
(c) Largest is 400 sq ft; there is no smallest area because you can make one of the sides as small as you like.
69. (a) 58.65 yards longer   (b) 2 minutes
71. Because the two slanted sides are parallel, as are the top and bottom, we can use theorems about parallel lines and transversals. This shows that the angle in the bottom left of the parallelogram is equal to the one in the upper right.

Exercise Set 9-5

9. 125 in³ **11.** 210 m³ **13.** 48 m³ **15.** 19,704.07 cm³
17. 1,005.31 ft³ **19.** 33,510.32 in³ **21.** 461.81 cm³ **23.** 150 in²
25. 214 m² **27.** 96 m² **29.** 4,046.4 cm² **31.** 628.3 ft²
33. 5,026.5 in² **35.** 648 ft³ **37.** 190,228,656 ft³
39. 13,684.78 in³ **41.** 141.11 in³ **43.** 55,417,694.41 mi²
45. 2,304 in³ **47.** 1,080 in² **49.** 3.4 in.
51. Two 5-gallon cans *or* one 5-gallon can and four 1-gallon cans
53. One 1-gallon can **55.** (a) 2,000 ft³ (b) 2.53 times
57. 154 in³ **59.** Area of the base is 36 m²; area of each face is 15 m²; total area is 96 m²
63. (a) 2,016 (b) 347.4 cu ft (c) 51.9% (d) 52.4% **65.** $33\frac{1}{3}$%
67. Because to have geometric meaning, raising the length of a side to the fourth power would require a four-dimensional solid, and we don't have a common name for such an object. (We don't even know what such a thing would mean, in fact!)

Exercise Set 9-6

9. $\cos A = \frac{3}{5}$; $\sin A = \frac{4}{5}$; $\tan A = \frac{4}{3}$
11. $\cos A = \frac{35}{72}$; $\sin A = \frac{\sqrt{15,836}}{144}$; $\tan A = \frac{\sqrt{15,836}}{70}$
13. $\cos A = \frac{15}{17}$; $\sin A = \frac{8}{17}$; $\tan A = \frac{8}{15}$
15. $\cos A = \frac{\sqrt{51}}{10}$; $\sin A = \frac{7}{10}$; $\tan A = \frac{7}{\sqrt{51}}$
17. 92.71 cm **19.** 76.88 in. **21.** 1.39 mm **23.** 53.82 in.
25. 1,857.41 ft **27.** 77.47 mi **29.** ≈48° **31.** 60°
33. ≈29° **35.** ≈41° **37.** c = 25.8″, A = 54.5°, B = 35.5°
39. b = 8.9 m, c = 21.9 m, B = 24°
41. 23,644.67 ft **43.** 4,980.76 m **45.** 1,218.93 ft **47.** 0° to 6.4°
49. 359.50 ft **51.** 3,932.92 ft **53.** 1.2° **55.** 6.9°
57. The angle works out to be negative.
59. Each is the ratio of one leg of a triangle to the hypotenuse. The ratios have to be greater than zero because both lengths are positive. They have to be less than 1 because the hypotenuse is the longest side, so the denominator of each ratio is larger than the numerator.
61. Height of plane = 737.6 ft and distance of car = 673.8 ft
63. Use the cosine then use the Pythagorean theorem.
65. When you know two sides, you can quickly find the third using the Pythagorean theorem. The angles in a triangle have measures that add to 180°, and you always know the 90° angle, so when you know one of the others, you can subtract from 90° to get the remaining angle.
67. Answers vary.

Exercise Set 9-7

19. Answers vary.
21. Yes; the result of the first four iterations is shown here.

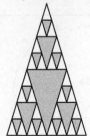

23. Yes yet again.

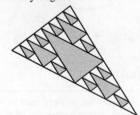

25.

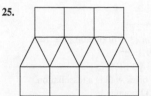

27. It can, and the pattern looks basically the same as the one in Figure 9-39(a).
29. Answers vary.
33. Only regular triangles, squares, and hexagons tessellate with themselves.
35. Answers vary.
37. The measure of each angle is $\frac{(n-2)360}{n}$, and the sum of the angles around a vertex has to be 360°, while there's already a 90° angle at each vertex from the square. So I set up and solved the equation $90 + 2\frac{(n-2)360}{n} = 360$ to find n.

Review Exercises

1. Line **3.** Line segment **5.** Trapezoid **7.** Triangle **9.** Hexagon
11. Obtuse **13.** Vertical angles **15.** Alternate exterior angles
17. (a) 63° (b) 2° **19.** $m\angle 2 = 163°$; $m\angle 3 = m\angle 1 = 17°$
21. 43° **23.** 72 ft **25.** Obtuse and scalene
27. Acute and equilateral **29.** 6 units **31.** a = 15 and b = 18
33. 15.8 ft **35.** 166.4 ft **37.** Decagon; 1,440° **39.** 16 ft
41. 28 m **43.** 1,376 cm² **45.** 9,160.88 km²
47. C = 51.84 yd; A = 213.82 yd² **49.** $49.94 **51.** 24,429.02 cm³
53. 84.82 yd³ **55.** 5,736 in² **57.** 775.71 revolutions
59. 38.79 cm³ **61.** $\sin B = \frac{3}{\sqrt{34}}$; $\cos B = \frac{5}{\sqrt{34}}$; $\tan B = \frac{3}{5}$
63. ≈2.33 yd **65.** ≈51.8° **67.** 33.04 ft **69.** 9 minutes
71.

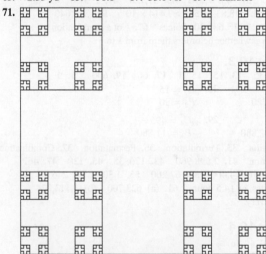

73.

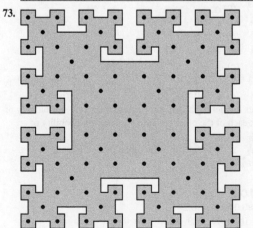

Chapter 9 Test

1. Complement: 17°; supplement: 107°

3. $m\angle 1 = m\angle 4 = m\angle 5 = 32°$; $m\angle 2 = m\angle 3 = m\angle 6 = m\angle 7 = 148°$

5. $\frac{5}{8}$ft   **7.** 1,080°   **9.** 75 ft²   **11.** 429.3 yd²

13. 282.7 cm³; 245.04 cm²   **15.** 923.63 yd³; 190 ft²

17. 33,510.32 m³; 5,026.55 m²   **19.** A brick wall is a tessellation.

21. 6,280 ft²   **23.** $41.08   **25.** 3,463.32 pounds

CHAPTER 10: PROBABILITY AND COUNTING TECHNIQUES

Exercise Set 10-1

11. Yes   **13.** No   **15.** No   **17.** Yes   **19.** Empirical   **21.** Theoretical

23. (a) $\frac{1}{6}$ (b) $\frac{1}{2}$ (c) $\frac{1}{3}$ (d) 1 (e) 1 (f) $\frac{5}{6}$ (g) $\frac{1}{6}$

25. (a) $\frac{1}{7}$ (b) $\frac{3}{7}$ (c) $\frac{2}{7}$ (d) 1 (e) 0

27. $\frac{4}{9}$ **29.** $\frac{9}{16}$ **31.** (a) ≈0.04 (b) ≈0.62 (c) ≈0.83

33. 0.84   **35.** $\frac{3}{13}$ **37.** $\frac{5}{11}$ **39.** 0.32

41. (a) 0.69 (b) 0.15 (c) 0.54   **43.** 0

45. (a) 0.239 (b) 0.761   **47.** (a) 0.42 (b) 0.90

49. (a) $\frac{6}{25}$ (b) $\frac{2}{5}$ (c) $\frac{9}{25}$ (d) $\frac{12}{25}$ (e) $\frac{1}{5}$

51. Answers vary.   **53.** (a) $\frac{1}{3}$ (b) Answers vary.

54. (a) $\frac{4}{9}$ (b) Answers vary.   **55.** Answers vary.

57. All probabilities should approach $\frac{1}{15}$, or about 0.067.

Exercise Set 10-2

9. 3,628,800   **11.** 1   **13.** 336   **15.** 990   **17.** 1,814,400   **19.** 420

21. 22,350   **23.** 56   **25.** 479,001,600   **27.** 720   **29.** 990

31. ≈0.0055   **33.** 840; 2,401   **35.** 151,200; 1,000,000   **37.** 5,527,200

39. 300   **41.** 210   **43.** 7,776   **45.** 57,120   **47.** 12,600   **49.** 60

51. 27,720   **53.** 248,832   **55.** ≈7.414 × 10¹¹   **57.** 332,640

59. 1,800   **61.** 300   **63.** 30 choices   **67.** Not a permutation

69. Depends on whether he orders them from 1 to 5

Exercise Set 10-3

9. 10   **11.** 35   **13.** 15   **15.** 1   **17.** 66   **19.** 600   **21.** 9

23. $_8C_5 = 56$ **25.** $_6C_2 = 15$
  $_8P_5 = 6,720$ $_6P_2 = 30$

27. $_9C_9 = 1$ **29.** $_{12}C_4 = 495$
  $_9P_9 = 362,880$ $_{12}P_4 = 11,880$

31. Combination   **33.** Permutation   **35.** Permutation   **37.** Combination

39. Combination   **41.** 2,598,960   **43.** 126; 35   **45.** 120   **47.** 462

49. 166,320   **51.** 14,400   **53.** 67,200   **55.** 1,302   **57.** 246

59. 870 minutes, or 14.5 hours   **61.** (a) 623,700 (b) 1,113,024

63. Answers vary.

Exercise Set 10-4

5. (a) $\frac{1}{8}$ (b) $\frac{3}{8}$ (c) $\frac{1}{2}$

7. (a) $\frac{9}{20}$ (b) $\frac{1}{4}$ (c) $\frac{3}{20}$

9. (a) $\frac{1}{3}$ (b) $\frac{1}{9}$ (c) $\frac{1}{9}$ (d) $\frac{4}{9}$ (d) $\frac{1}{3}$

11. (a) $\frac{1}{4}$ (b) $\frac{3}{4}$ (c) $\frac{1}{4}$ (d) $\frac{1}{8}$

13. (a) $\frac{3}{5}$ (b) $\frac{1}{2}$ (c) $\frac{4}{5}$

15. (a) $\frac{1}{2}$ (b) $\frac{1}{6}$ (c) $\frac{1}{4}$

17. (a) $\frac{1}{13}$ (b) $\frac{1}{4}$ (c) $\frac{1}{52}$ (d) $\frac{2}{13}$ (e) $\frac{4}{13}$ (f) $\frac{4}{13}$ (g) $\frac{1}{2}$ (h) $\frac{1}{26}$ (i) $\frac{7}{13}$

19. (a) $\frac{1}{9}$ (b) $\frac{2}{9}$ (c) $\frac{1}{6}$ (d) $\frac{5}{18}$ (e) $\frac{11}{36}$ (f) $\frac{1}{2}$ (g) $\frac{3}{4}$ (h) 1

21. The outcomes are not equally likely because there is a $\frac{1}{4}$ probability of getting WW; each of the other six outcomes has a probability of $\frac{1}{8}$.

23. 216   **25.** $\frac{5}{108}$

27. (a) $\frac{1}{4}$ (b) $\frac{1}{12}$ (c) Multiply the individual probabilities.

29. Bob: $\frac{1}{2}$; Fran: $\frac{3}{8}$; Julio: $\frac{1}{8}$

Exercise Set 10-5

5. ≈0.63   **7.** ≈0.04   **9.** ≈0.24

11. (a) 0.0004 (b) 0.15 (c) 0.383 (d) 0.45

13. (a) 0.033 (b) 0.0083 (c) 0.3 (d) 0.075 (e) 0.15

15. (a) 0.255 (b) 0.745 (c) 0.0182 (d) 0

17. (a) 0.07 (b) 0.42 (c) 0.015 (d) 0.16 (e) 0.336

19. 0.01   **21.** 0.42   **23.** 0.115   **25.** $\frac{1}{658,008}$

27. 0.018   **29.** 0.0093   **31.** 0.00144   **33.** 0.0211   **35.** 0.0000015

37. 0.0049   **39.** Answers vary.

41. (a) The probability of the letters appearing in any order is six times as great as the probability of them appearing in order.

  (b) The probability of any order should be 24 times as great. The difference between the number of possibilities is the number of permutations of the letters chosen. When it was 3 letters, any order was 3! = 6 times more likely. For 4 letters, it should be 4! = 24 times more likely.

43. Answers vary.   **45.** Answers vary.

Exercise Set 10-6

7. In favor: 7:1; against: 1:7   **9.** In favor: 5:6; against: 6:5

11. In favor: 9:4; against: 4:9   **13.** $\frac{5}{13}$ **15.** $\frac{5}{11}$ **17.** $\frac{3}{10}$ **19.** 3.6

21. $2.29   **23.** (a) 1:11 (b) 1:35 (c) 5:1 (d) 17:1 (e) 1:5

25. (a) 1:12 (b) 3:10 (c) 3:1 (d) 1:12 (e) 1:1

27. (a) $\frac{7}{11}$ (b) $\frac{5}{7}$ (c) $\frac{3}{4}$ (d) $\frac{4}{5}$ **29.** $\frac{5}{14}$ **31.** 1:37; –$0.053

33. 6:13; –$0.053   **35.** 2:17; –$0.053

37. 29:121   **39.** 49:101   **41.** 8:67

43. Tampa Bay 0.16, Kansas City 0.13, Green Bay 0.12, Buffalo 0.11

45. 11:2   **47.** 11:2   **49.** 77.6 in.   **51.** –$3.00   **53.** $0.83

55. –$1.00   **57.** –$0.50; –$0.52   **59.** –$70.25

61. No; the two-headed coin is now twice as likely.

65. (a) Answers vary. (b) RZ Electronics: +$320; Jackson Builders: +$525

Exercise Set 10-7

7. No   **9.** Yes   **11.** No   **13.** Yes   **15.** $\frac{1}{6}$ **17.** $\frac{11}{19}$

19. (a) $\frac{8}{11}$ (b) $\frac{15}{22}$ **21.** (a) $\frac{8}{17}$ (b) $\frac{6}{17}$ (c) $\frac{9}{17}$ (d) $\frac{12}{17}$

23. (a) $\frac{6}{7}$ (b) $\frac{4}{7}$ (c) 1   **25.** (a) $\frac{67}{118}$ (b) $\frac{81}{118}$ (c) $\frac{44}{59}$

27. (a) $\frac{38}{45}$ (b) $\frac{22}{45}$ (c) $\frac{2}{3}$ **29.** (a) $\frac{14}{31}$ (b) $\frac{23}{31}$ (c) $\frac{19}{31}$

31. (a) $\frac{1,664}{3,729}$ (b) $\frac{2,477}{3,729}$ (c) $\frac{2,870}{3,729}$

33. (a) $\frac{4}{9}$ (b) $\frac{1}{3}$ (c) $\frac{5}{12}$ **35.** $\frac{21}{47}$ **37.** $\frac{4}{5}$

39. CBS: 0.4; Fox or NBC: 0.6. It's more likely that it was on Fox or NBC.

41. 0.29   **43.** 0.12

45.

```
 Both > 3          Differ by 2
        (4, 4)
 (4, 5)        (1, 3)  (3, 1)
 (5, 4)  (6, 5) (4, 6) (2, 4) (4, 2)
 (5, 5)  (6, 6) (6, 4) (3, 5)
        (5, 6)        (5, 3)

The remaining 21 rolls
```

49. (a)

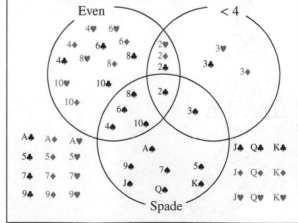

(b) There are 21 cards outside of all three circles, so there must be 31 that satisfy at least one of those events. So the probability of at least one of those events occurring is $\frac{31}{52}$.

51. Not mutually exclusive. You could be acquitted of a crime and convicted of a different one at the same time, leading to jail time after being acquitted. Or you could have been in the middle of a sentence for a different crime when tried.

53. Not mutually exclusive. There are different ways of being categorized: you could be a chronological junior (in your third year) but only have enough credits to be considered a sophomore academically.

Exercise Set 10-8

7. Independent **9.** Dependent **11.** Independent **13.** Dependent
15. (a) 0.0058 (b) 0.08 **17.** 0.66 **19.** 0.0625 **21.** $\frac{1}{12}$ **23.** $\frac{1}{8}$
25. $\frac{1}{15}$ **27.** 0.400 **29.** 0.621 **31.** 0.006 **33.** 0.062
35. 0.184 **37.** 0.2306 **39.** 0.1786 **41.** $\frac{1}{6}$ **43.** $\frac{2}{11}$
45. $\frac{1}{4}$ **47.** $\frac{2}{3}$ **49.** 1 **51.** 0.61 **53.** $\frac{1}{2}$ **55.** 0.182 **57.** 0.838
61. (a) Answers vary.
  (b) Every number drawn is independent of the previous numbers chosen, so the probability of 1, 2, 3, 4, 5, 6 is no different than the probability of any other six numbers.
  (c) That combination might be desirable because it's likely nobody else will choose it and you won't have to split the prize if it wins.
63. (a) $\frac{1}{13} \approx 0.077$ (b) $\frac{1}{17} \approx 0.059$; It's about $\frac{3}{4}$ as much.

Exercise Set 10-9

7. 0.288 **9.** 0.0616 **11.** 0.0000304 **13.** 0.0994 **15.** 0.00615
17. 0.4535 **19.** 0.1960 **21.** 0.999 **23.** 0.999 **25.** 0.00463
27. 0.0750 **29.** 0.1546 **31.** 0.2340 **33.** 0.1906 **35.** 0.008
37. 0.0080 **39.** 0.1117 **41.** 0.6870 **43.** 0.3769 **45.** Yes
47. Yes **49.** Yes **51.** Yes **53.** No

Review Exercises

1. b, c, and e **3.** (a) $\frac{1}{4}$ (b) $\frac{1}{52}$ (c) $\frac{4}{13}$ (d) $\frac{1}{13}$ (e) $\frac{1}{2}$
5. (a) 0 (b) $\frac{1}{2}$ (c) $\frac{1}{3}$ **7.** (a) $\frac{9}{35}$ (b) $\frac{23}{35}$ (c) $\frac{19}{35}$ (d) $\frac{19}{35}$
9. (a) $\frac{1}{4}$ (b) $\frac{1}{6}$ (c) $\frac{1}{4}$ (d) $\frac{1}{4}$ (e) 0 (f) $\frac{1}{4}$
11. 2,184 **13.** 175,760,000; 88,583,040 **15.** 120 **17.** 84
19. In a permutation order matters, in a combination order does not matter.
21. 78 **23.** 729
25. $S = \{$1H, 1T, 2H, 2T, 3H, 3T, 4H, 4T, 5H, 5T, 6H, 6T, 7H, 7T, 8H, 8T$\}$
27. $\frac{2}{25}$ **29.** $\frac{12}{25}$ **31.** $\frac{33}{182}$ **33.** 1:5 **35.** 10.5 **37.** $28.85
39. Mutually exclusive
41. (a) $\frac{17}{50}$ (b) $\frac{4}{25}$ **43.** Dependent
45. Dependent **47.** $\frac{1}{7}$ **49.** 0.016
51. (a) $\frac{1}{26}$ (b) $\frac{1}{4}$ (c) $\frac{1}{8}$ **53.** 0.0165 **55.** 0.1684

Chapter 10 Test

1. 21 **3.** 0.00039 **5.** Answers vary. **7.** 2,646
9. (a) $\frac{4}{31}$ (b) $\frac{20}{31}$ (c) $\frac{7}{15}$
11. $S = \{$H1, H3, H5, T2, T4, T6$\}$
13. (a) 0.025 (b) 0.000495 (c) 0 **15.** $\frac{1}{2}$
17. In favor: 3:5; against: 5:3 **19.** $5\frac{1}{6}$ **21.** $\frac{8}{33}$ **23.** 0.18

CHAPTER 11: STATISTICS

Exercise Set 11-1

11. Cluster **13.** Random **15.** Stratified
17. No; students that are well-off might be underrepresented.
19. Yes; the target group is everyone that has a phone, and everyone surveyed obviously has a phone.
21. No; the first five products on a shelf are likely to be older so that the store can get rid of older food first.

23.

Rank	Frequency
FY	18
So	12
Jr	6
Sc	4

25.

Show	Frequency
S	6
D	5
B	7
A	7

27.

Class	Frequency
27–33	7
34–40	14
41–47	14
48–54	12
55–61	3
62–68	3
69–75	2

29.

Class	Frequency
0–69	12
70–139	8
140–209	15
210–279	4
280–349	7
350–419	1
420–489	3

31.

Class	Frequency
150–1,276	2
1,277–2,403	2
2,404–3,530	5
3,531–4,657	8
4,658–5,784	7

Class	Frequency
5,785–6,911	3
6,912–8,038	7
8,039–9,165	3
9,166–10,292	3
10,293–11,419	2

The average height is an opinion question.

33.

Class	Frequency
2.20–2.99	12
3.00–3.79	11
3.80–4.59	5
4.60–5.39	3
5.40–6.19	0
6.20–6.99	1

Chosen classes can vary.

35. Most registered vehicles per car stolen are in the range of 80–89, while the least are in the 0–49 range. The most common are 84 and 89.

Stems	Leaves
3	8
4	1
5	0 0 2 3 3 6 8 9
6	6 8 9 9
7	0 0 3 4 5 8
8	0 1 3 3 4 4 4 5 7 9 9 9
9	0 2 4

Key: 3|8 means 38

37.

Stems	Leaves
0	3
1	5 9
2	2
3	1 1
4	1 4 6 6
5	2 6 6 6 9
6	0 0 6 6
7	7
8	7 8
9	6 8

Key: 1|5 means 15

39.

Stems	Leaves
$6^{(5)}$	5 8
$7^{(0)}$	0
$7^{(5)}$	6 7 8 8 9 9 9 9
$8^{(0)}$	0 0 0 1 2 2 3 3 3 3 4 4 4
$8^{(5)}$	5 6 6 6 7 7

Key $6^{(5)}$|8 means 65

41. Inferential **43.** Inferential **45.** Inferential
47. Answers vary. **49.** Answers vary. **51.** Answers vary.
53. The majority of states has taxes below $2.10, and just a handful are over $3.50. Explanations vary.

Exercise Set 11-2

9.

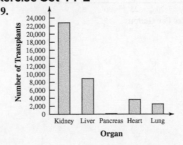

11.

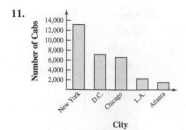

13.

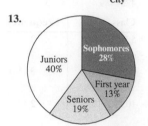

15.

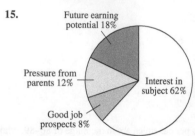

17. 2018: 61%, 2021: 55%   **19.** Down   **21.** Parents
23. TV stations   **25.** Newsletters/Listservs   **27.** Analog
29.

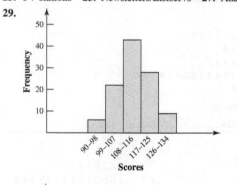

31.

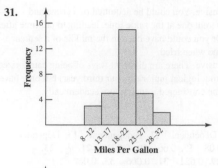

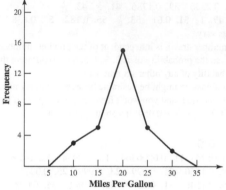

33.

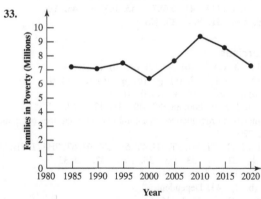

35.

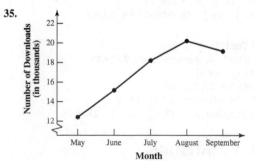

37. (a)

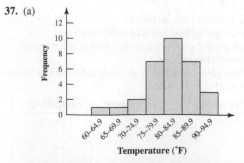

(b) The most likely high temperature in May is between 80 and 85 degrees; highs less than 75 are very unusual, and the 90s occur occasionally.

39. (a)

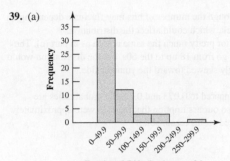

Employed RNs (in thousands)

(b) Most states have less than 50,000 employed registered nurses, and very few have more than 100,000.

41.

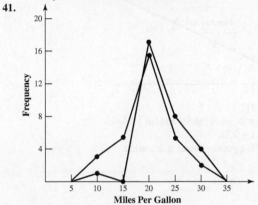

Miles Per Gallon

43. Answers vary.   **45.** Time series graph

47. Bar graph   **49.** Bar graph

51.

Grocery Store

This scale exaggerates the difference between the bar heights. In particular, it makes the number of shoppers at Fareway look negligible, but it's actually more than half of the shoppers at Publix.

53.

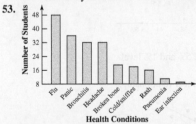

Health Conditions

In this case, changing the scale on the vertical axis had a very minimal effect.

55. In a bar graph, the labels on the horizontal axis are categories, not numerical. In a histogram, each new class begins where the previous one ended.

57. Answers vary.

Exercise Set 11-3

11. mean $\approx$ 10.4, median = 10, mode = 9, midrange = 11

13. mean = 612.6, median = 475, no mode, midrange = 820

15. mean = 2,907.7, median = 2,723.5, no mode, midrange = 3,353.5

17. mean = 189.6, median = 151, no mode, midrange = 207.5

19. mean $\approx$ 5.9, median = 6, mode = 3, 5, 6, 8, 9, midrange = 5.5

21. mean $\approx$ 766 median = 659, mode = none, midrange = 894.5

23. 23.05   **25.** mean $\approx$ 4.4 seconds

27. mean $\approx$ 42.87 or $42.87 million

29. mean = $180.28   **31.** Answers vary.   **33.** Median   **35.** Mode

37. Mode   **39.** 25.575; the approximate mean was 25.125.

41. Mean and median tend to be close when the data set doesn't have one or two terms that are unusually high or low compared to the others. But when there are outliers like that, it skews the average up or down without affecting the median much.

43. It's descriptive if we're only interested in those particular 80 bulbs. If we use it to draw a conclusion about average lifespan for all bulbs they produce, then it's inferential.

45. Answers vary.   **47.** Answers vary.   **49.** Answers vary.

Exercise Set 11-4

7. $s_1 < s_2$ **9.** $s_1 > s_2$

11. $R = 60$, variance = 406.75, $s \approx 20.17$
The number of junk emails varies pretty widely.

13. $R = 1,799$, variance = 438,113.6, $s \approx 661.90$
The odometer readings vary pretty widely.

15. $R = 99$, variance $\approx 1,288.19$, $s \approx 35.89$
The weights don't vary all that much.

17. $R = 10$, variance = 9, $s = 3$
The heights are pretty uniform.

19. $R = 9.71$, variance ≈ 5.50, $s \approx 2.35$

21. $R = 106.86$, variance ≈ 973.9, $s \approx 31.2$

23. $R = 11,304$, variance = 10,648,768.6, $s \approx 3,263.2$

25. $R = 782$, variance $\approx 80,329$, $s \approx 283$

27. The variation is not the same.

29. (a) Average would be close to the hole for Pat, close to the center of the green for Ron.
(b) Pat has a large variation, Ron has a very small variation.
(c) Answers vary, but this example shows that variation can sometimes be more meaningful than average.

31. Answers vary.

33. (a) Population standard deviation means that the data represent every member of a population. Sample standard deviation means that the data values come from a sample within a larger population.
(b) Answers vary.

35. Everything! Population standard deviation uses all data from a sample, so that's descriptive. Sample standard deviation uses a sample of data values to make a conclusion about a larger population, so that's inferential.

Exercise Set 11-5

7. (a) 20th percentile   (b) 75th percentile   (c) 35th percentile
(d) 5th percentile   (e) 90th percentile

9. 75th percentile   **11.** 79th percentile   **13.** 10   **15.** 19

17. Maurice is ranked higher.

19. Basketball 92nd percentile, football 90th

21. (a) 60th percentile   (b) 8   (c) 23 years

23. $Q_1 = 22.5$, $Q_2 = 34$, $Q_3 = 53.5$ **25.** $Q_1 = 78$, $Q_2 = 88.5$, $Q_3 = 93$

27. $Q_1 = 103$, $Q_2 = 114.5$, $Q_3 = 123$

29.

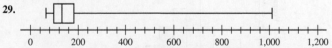

(a) Most of the data values are on the extreme low end of the distribution.
(b) There are two outliers: 833.7, and 1,010.7. This tells us that there are two countries with an unusually high number of Internet users, with the rest of the top 15 much lower.

31.

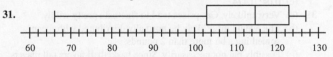

33. yes; yes   **35.** Answers vary.

37. (a) If a data set has 10 values, the value at the 80th percentile is ranked ninth, but 0.8 times 10 is 8.

(b) When the data set has a very large number of values, the claim is close to true.

39. 90.5

41. There are 16 scores below our answer, which is 80% of the scores. And there are 4 scores above it, which is 20% of the scores. This does what we had hoped for.

43. Inhofe: 79/100 gives us 79th percentile. Blumenauer: 381/435 gives us 88th percentile. Since Blumenauer ranks higher among representatives than Inhofe does among senators, and they have the same net worth, we can conclude that in general senators are richer than representatives.

Exercise Set 11-6

13. 34 **15.** 498 **17.** 54 **19.** −0.37 **21.** −2.47 **23.** 0.44
25. 0.474 **27.** 0.192 **29.** 0.159 **31.** 0.345 **33.** 0.077
35. 0.223 **37.** 0.463 **39.** 0.885 **41.** 0.971 **43.** 0.274
45. 0.003 **47.** (a) About 97.5% (b) 340 **49.** 8.28 ounces
51. My z score: 2.38. My wife's: 2.33. I am.
53. Benson: −3.77; Diana: −3.75. Benson was better. **55.** $z = +0.45$
57. (a) $z = \pm 2.05$ (b) $z = \pm 1.75$ (c) $z = \pm 2.40$
59. (a) No. That would mean that HALF of all packages would contain less than the package weight.

(b) 14.8 oz

Exercise Set 11-7

Probabilities were found using the table in Appendix A.

5. (a) 0.394 (b) 0.025 **7.** (a) 0.44 (b) 0.12
9. (a) 0.005 (b) 0.162 (c) 0.749
11. (a) 0.153 (b) 0.774 (c) 0.187

(d) No. The probability of a tire lasting more than 40,000 is less than 0.0000003.

13. (a) 0.841 (b) 0.067 **15.** (a) 0.776 (b) 0.405
17. (a) 0.755 (b) 0.811 (c) 0.284
19. (a) 638 (b) 184 (c) 1,074 (d) 136
21. (a) 58 (b) 208 (c) 38 **23.** 72nd **25.** 0th percentile
27. (a)

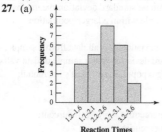

(b) Mean ≈ 2.34; $s \approx 0.58$
(c) 0.073; 0.72

29. (a)

(b) Mean ≈ 49.59; $s \approx 5.95$
(c) 0.04; 0.474

31. (a) Very unlikely. Gas prices tend to fluctuate pretty wildly.

(b) This would most likely be normally distributed with mean something a bit more than 2 pounds.

(c) Possibly but not necessarily. Since basketball favors tall players but there are still some shorter players, the heights with the largest number of players would probably be somewhat above the mean.

(d) Probably, although the number of hits may fluctuate depending on day of the week, which could affect the distribution.

(e) Probably not, for pretty much the same reason as in part (c). The ages probably go from 18 up to the 60s, but the distribution would be very strongly skewed toward the younger side.

33. No **35.** 90, 110
37. 0.04 and 0.72, compared to 0.073 and 0.72; the probabilities are reasonably close, so our assumption that the data were approximately normally distributed seems reasonable.
39. Answers vary.

Exercise Set 11-8

11. (a)

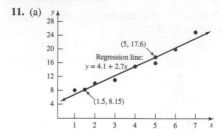

(b) $r \approx 0.977$
(c) r is significant at the 5% and the 1% level.
(d) $y = 4.1 + 2.7x$
(e) There is a positive linear relationship.

13. (a)

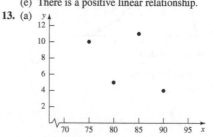

(b) $r \approx -0.441$
(c) r is not significant at 5% nor at 1% level.
(d) no regression line
(e) No relationship exists.

15. (a)

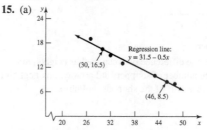

(b) $r \approx -0.983$
(c) r is significant at the 5% and 1% level.
(d) $y = 31.5 - 0.5x$
(e) There is a negative linear relationship.

17. (a)

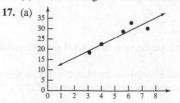

(b) $r \approx 0.909$
(c) r is significant at the 5% level.
(d) $y = 3.1x + 10.2$
(e) There is a positive linear relationship.

19. (a)

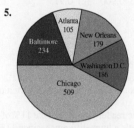

(b) $r \approx 0.942$

(c) r is significant at the 5% and 1% level.

(d) $y = 2.76 + 0.06x$

(e) There is a positive linear relationship.   (f) 33 Stories

21. (a)

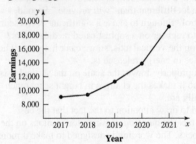

(b) $r \approx 0.896$

(c) r is significant at the 5% and 1% level.

(d) $y = -122.5 + 0.355x$

(e) There is a positive linear relationship.   (f) $205.88

23. (a)

(b) $r \approx 0.963$

(c) r is significant at the 5% and 1% level.

(d) $y = 10.25 + 0.86x$

(e) There is a positive linear relationship.

(f) 88

25. (a) 0.769; significant at both levels   (b) Answers vary.

27. (a) –0.766; significant at both levels   (b) Answers vary.

29. (a) –0.802; significant at both levels   (b) Answers vary.

31. $r = 1$ in both cases, the points lie on a line.

33. (a)

Now there appears to be a positive linear relationship.

(b) Answers vary.

(c) $r \approx 0.990$; Removing the change of direction made it possible to find a linear relationship between the remaining points.

35. Answers vary.

Some of the answers in 37–43 are open to interpretation.

37. Positive

39. None (but you could make a case that small town students might get a better education than those in large urban districts, in which case it would be negative).

41. Positive   **43.** Negative

45. From least to most, median age (–0.277), % voting for Democratic candidate (–0.662), % attending weekly church services (0.734), education level (–0.834). Discussions vary.

Review Exercises

1.

Item	Frequency
B	4
F	5
G	5
S	5
T	6

3.

Rank	Frequency
102–116	4
117–131	3
132–146	1
147–161	4
162–176	11
177–191	7

5.

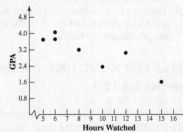

7. Janine's earning increased every year, and the rate of increase kept getting bigger.

9. mean = 7.25

11. Mean = 399.75, range = 266, variance $\approx$ 10,799.64, $s \approx 103.92$. The range of 266 tells us that the numbers vary from lowest to highest by a good amount compared to their sizes. The standard deviation tells us that overall the numbers are fairly spread out.

13. (a) 82nd; 41st   (b) 171,000

15. (a) 0.474   (b) 0.155   (c) 0.061   (d) 0.833   (e) 0.229   (f) 0.828
(g) 0.023   (h) 0.912   (i) 0.018   (j) 0.955

17. (a) 28   (b) 5

19. Helen is MUCH more experienced ($z = 10.42$) than old ($z = 2.42$).

21. (a) 0.004   (b) 0.023   (c) 0.5   (d) 0.324   **23.** 22

25. $r \approx -0.914$, r is significant at the 5% level.
$y = 4.95 - 0.21x$
$y = 3.06$ when $x = 9$

Chapter 11 Test

1.

Source	Frequency
W	6
L	7
K	7
E	5

3.

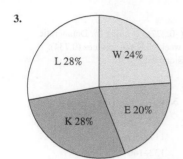

5.

Stems	Leaves
20	0 4 9
21	0 1 2 7 8 8
22	2 7 7 7 8
23	0 1 3 7 8
24	1 2 2 3 7
25	1 1 3 4 6
26	0

7. (a) mean ≈ 84.1   (b) median = 85   (c) no mode
(d) midrange = 84   (e) range = 12   (f) variance ≈ 17.1
(g) $s ≈ 4.14$
9. (a) 0.433 (b) 0.034 (c) 0.291 (d) 0.900 (e) 0.913
11. 20

Chapter 11 Supplement

1. First, 20 people is a very small sample. Second, we have no idea how those 20 were chosen. Are they representative of the general population or did they all come from a group like college students?
3. "The road less traveled" is at best vague. It could mean wildly different things to different people.
5. The writer was probably hoping that the reader would mistake 11% of patients with 11 out of 18 patients. That's a huge difference, since 11% of 18 patients is 2 patients.
7. More than what? Without a comparison, the claim is meaningless.
9. First "can provide" is a lot different than "will provide." Second, what level of acid control? Enough to cause a significant decrease in symptoms, or enough to register on a sophisticated medical exam?
11. There's no scale at all on the vertical axis, so we can't judge at all how significant the difference in energy released is.
13. The second graph is improperly drawn: the scale on the vertical axis doesn't begin at zero, so it makes the changes in height seem much greater than they actually are.
15. The question is worded to draw attention to the fact that the cost will increase, when in fact the question is asking for opinions on the importance of safety locks. The wording is designed to make it more likely that responders will say no.
17. Each of those behaviors are illegal, and people are a lot less likely to admit to doing something illegal or clearly wrong.
19. As the negative effects of lead became more widely known, it's a lot more likely that certain illnesses would be recognized as lead-related.
21. It would take some further analysis. In a country of over 300 million people, it's entirely possible that one murder every 20 minutes is a relatively low rate compared to other countries.
23. The scales on BOTH axes are inconsistent, with the y axis being particularly bad.
25. The scale on the x axis is inconsistent.
27. Both axes are a problem. The x axis has an inconsistent scale, and the y axis doesn't begin at zero.
29. The three-dimensional effect makes the "Other" category visually quite a bit smaller than the "Assaults and Violent Acts" category, but the percentages are the same.

CHAPTER 12: VOTING METHODS

Exercise Set 12-1

7. (a) 22 (b) 4 (c) 8 (d) X
9. (a) 18 (b) 9 (c) 4 (d) Philadelphia
11. (a) 208 (b) Swimming pool **13.** (a) 20 (b) Children's
15.   **17.**

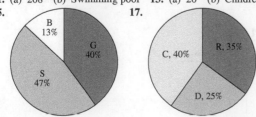

19. No **21.** No
23. (a) 1960, 1968, 1992, 1996, 2000 (b) 2000
25. Answers vary. **27.** 9 **29.** No **31.** Yes

Exercise Set 12-2

7. Role-playing (R) **9.** *Anatomy of a Murder* (A)
11. (a) Swimming pool (S) (b) Yes **13.** Yes **15.** No **17.** No
19. Professor Donovan (D) **21.** (a) Children's (b) Yes
23. Yes **25.** No **27.** Answers vary. **29.** Answers vary. **31.** Yes
33. If there are six candidates, there are 720 different ballots, making it very cumbersome to list all of the possibilities and build a preference table.
35. (a) Answers vary.
(b) It would be possible for a candidate listed last on the majority of ballots to win if he or she were listed first on enough, due to the exaggerated number of points for being listed first.

Exercise Set 12-3

9. 6 **11.** 45
13. Steel Center (S)
15. (a) There is a three-way tie. (b) The results are different.
17. (a) Rosa's Restaurant (R) (b) Yes
19. No **21.** No **23.** Dr. Zhang **25.** Green
27. Inmate Z **29.** Answers vary.
31. Using pairwise comparison puts the candidates head-to-head against each other.
33. Ignore all but the top-ranked candidate on each ballot: it's the plurality method.
35. (a) Rather than the candidate with the most approval votes winning, the candidates with the three highest approval totals are all named to council.
(b) Answers vary.
37. Answers vary.

Exercise Set 12-4

9. 29.2 **11.** 18.9
13. 1: standard 1.4, upper 2, lower 1; 2: standard 0.65, upper 1, lower 0; 3: standard 2.95, upper 3, lower 2
15. A: standard 5.77, upper 6, lower 5; B: standard 4.97, upper 5, lower 4; C: standard 2.80, upper 3, lower 2; D: standard 6.46, upper 7, lower 6
17. 4 **19.** 8.49 **21.** 19.60
23. (a) 32
(b) South: 1.9375
Central: 4.625
North: 3.4375
(c)

Campus	South	Central	North
Lower	1	4	3
Upper	2	5	4

(d)

Campus	South	Central	North
Promotions	2	5	3

25. (a) 64,444
(b) District 1: 3.755
District 2: 2.374
District 3: 2.871
(c)

District	1	2	3
Lower	3	2	2
Upper	4	3	3

(d)

District	1	2	3
Representatives	4	2	3

27. (a) 14.6
(b) Terminal A: 3.493
Terminal B: 2.534
Terminal C: 4.542
Terminal D: 1.507

(c)

Terminal	A	B	C	D
Lower	3	2	4	1
Upper	4	3	5	2

(d) Jefferson's Method:

Terminal	A	B	C	D
Trucks	4	2	5	1

Adams' Method:

Terminal	A	B	C	D
Trucks	3	3	4	2

(e) 12.7; 17.1

29. (a) 54

(b) Store 1: 2.204
Store 2: 1.833
Store 3: 3.444
Store 4: 2.519

(c)

Store	1	2	3	4
Lower	2	1	3	2
Upper	3	2	4	3

(d)

Store	1	2	3	4
Computers	2	2	3	3

(e) Use the standard divisor

31. (a) 1,217.875

(b) Precinct 1: 2.925
Precinct 2: 6.956
Precinct 3: 1.762
Precinct 4: 4.358

(c)

Precinct	1	2	3	4
Lower	2	6	1	4
Upper	3	7	2	5

(d)

Precinct	1	2	3	4
Officers	3	7	2	4

(e) Use the standard divisor

33. (a) 1,098

(b) Office 1: 2.032
Office 2: 0.907
Office 3: 1.424
Office 4: 1.638

(c)

Office	1	2	3	4
Lower	2	0	1	1
Upper	3	1	2	2

(d)

Office	1	2	3	4
Therapists	2	1	1	2

(e) 1,106

35. Answers vary. **37.** Answers vary.

39. (a) Up; more

(b)

Population	20	40	60	80	; large district
Divisor 2.5	8	16	24	32	
Divisor 2.3	8.7	17.4	26.1	34.8	
Divisor 2.7	7.4	14.8	22.2	29.6	

41. Down; more

Exercise Set 12-5

7. Alabama paradox occurred

9. Alabama paradox occurred

11. Population paradox did not occur

13. (a)

Campus	North	South
Podiums	9	52

(b)

Campus	North	South	Campus 3
Podiums	8	53	7

(c) New states paradox occurred

15. (a)

State	A	B	C
Seats	11	46	40

(b)

State	A	B	C	D
Seats	10	47	40	13

(c) The new states paradox occurred

17. Answers vary. **19.** Answers vary.

Review Exercises

1.

Number of votes	5	5	5
First choice	Q	P	R
Second choice	R	Q	P
Third choice	P	R	Q

3. 5

5.

Number of votes	6	4	10
First choice	C	P	H
Second choice	P	C	P
Third choice	H	H	C

7. 6 **9.** 58 **11.** Style A **13.** Style A

15. No **17.** No **19.** 47

21. Midwest Health Care **23.** Midwest Health Care

25. Yes **27.** No **29.** The Screen

31.

Location	A	B	C	D
Asst. Pastors	5	1	1	3

33.

Location	A	B	C	D
Asst. Pastors	5	2	1	2

35.

Location	A	B	C	D
Asst. Pastors	5	1	1	3

37.

Location	A	B	C	D
Asst. Pastors	5	2	1	2

39. Alabama paradox occurred

41. Population paradox did not occur

43. New states paradox did not occur (Note that rounding the standard divisor differently can change the answer.)

Chapter 12 Test

1.

Number of votes	1	7	4
First choice	A	B	C
Second choice	B	A	B
Third choice	C	C	A

3. 7 **5.** Pittsburgh (P) **7.** Pittsburgh (P) **9.** No

11. No **13.** *Big Brother Nursing Home*

15.

Flight	A	B	C
Flight Attendants	4	6	2

17.

Flight	A	B	C
Flight Attendants	4	6	2

19. Alabama paradox occurred

21. New states paradox occurred

CHAPTER 13: GRAPH THEORY

Exercise Set 13-1

13. *A, B, C, D, E, F, G*

15. *B, G, F* **17.** *B* and *E*

19. Answers vary; one such path is *A, B, E, F*

21. *G* **23.** There is no edge connecting *E* and *D*.

Note: for 25–29, there are other possible colorings.

25. 3;

27. 4;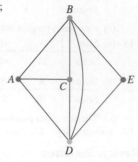

29. 3;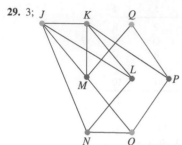

31.   **33.** 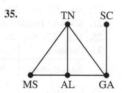 **35.**

37. 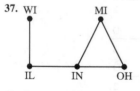 **39.**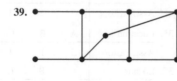

41.

43. 2 **45.** 3 **47.** 3 **49.** 4
51. 2 **53.** 2 **55.** Answers vary.
57. No; a vertex is adjacent to itself if there's a loop, so it will be adjacent to a vertex with the same color.
59. Answers vary. **61.** Answers vary.
63. (a) The house is split into distinct parts that are not connected.
  (b) There's a door that connects one part of a room to another.
  (c) There's a door that, if removed, would make it impossible to get from one part of the house to another.

Exercise Set 13-2

9. Euler circuit **11.** Neither
13. (a) Euler path   (b) $A, B, C, A, I, C, D, G, I, H, G, F, E, D, F$
15. (a) Euler circuit   (b) $A, B, D, H, I, G, D, C, G, F, E, C, A$
17. (a) Neither
19. (a) Euler circuit   (b) A, B, A, C, B, C, A
21. (a) Neither
23.   Euler circuit: A, B, C, D, C, A

25. 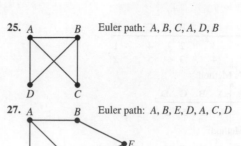 Euler path: A, B, C, A, D, B

27. 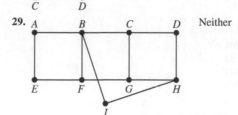 Euler path: A, B, E, D, A, C, D

29. 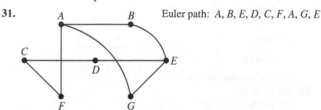 Neither

31. 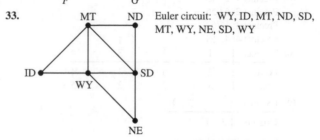 Euler path: $A, B, E, D, C, F, A, G, E$

33. 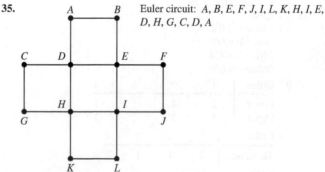 Euler circuit: WY, ID, MT, ND, SD, MT, WY, NE, SD, WY

35. Euler circuit: $A, B, E, F, J, I, L, K, H, I, E, D, H, G, C, D, A$

37. Neither can be done. Since all four vertices are odd, either removing one bridge or adding one more will make two of the vertices even and the other two odd, in which case there is still no Euler circuit.
39. Path: you can pass through every door in the house exactly once. Circuit: you can do so and end up in the room you started in.
41. Graphs will vary, but such a graph cannot have an Euler circuit because any graph with a bridge will always have at least two odd vertices.
43. Yes it will. Whether you list the path from beginning to end or end to beginning, it's the same path. So if it has to start at an odd vertex, it has to finish at one too.
45. No. It would need to have no odd vertices, but currently has four of them; no single edge can change more than two vertices.
47. It is not possible to have exactly one odd vertex in a connected graph.

Exercise Set 13-3

13. Answers vary, two possibilities are: A, B, C, D, E, F and A, C, D, E, F, B.

15. Answers vary, two possibilities are: $A, B, E, F, J, I, L, K, H, G, C, D$ and $A, B, E, D, C, G, H, K, L, I, J, F$.

17. Answers vary, two possibilities are: A, B, E, C, D and A, B, E, D, C.

19. Answers vary, two possibilities are: A, B, D, E, C, F, G, H and A, B, D, E, C, F, H, G.

21. Answers vary, two possibilities are: A, B, C, E, D, A and C, B, A, D, E, C.

23. Answers vary, two possibilities are: $A, B, D, G, F, E, H, I, C, A$ and $D, B, A, C, I, H, E, F, G, D$.

25. Answers vary, two possibilities are: A, B, C, D, E, A and A, D, B, E, C, A.

27. 2; easy to use brute force method

29. 40,320; ridiculous to use brute force method

31. P, Q, R, S, P and P, S, R, Q, P; 200

33. A, C, B, E, D, A; 42

35. A, B, C, D, E, A; 651

37. A, C, B, E, D, A; 42, the same result as the nearest neighbor method

39. A, D, C, B, E, A; 600, a better result than the nearest neighbor method

41. Let T = Pitt, L = Phil, B = Balt and W = Wash.

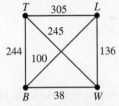

43. T, B, W, L, T; 723 miles. This route is longer than the optimal solution.

45. Let N = New York, D = Cleveland, O = Chicago, and B = Baltimore.

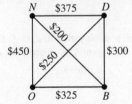

47. O, D, B, N, O; \$1,200. This route is \$50 more expensive than the optimal solution.

49.

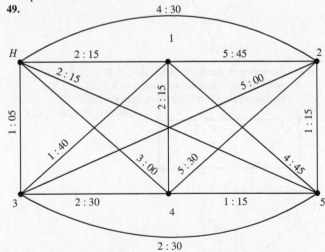

51. $H, 3, 1, 4, 5, 2, H$; 12:00

In Exercises 53–55, times and distances could vary depending on how you did the search.

53. Times are in minutes.

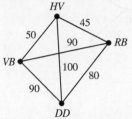

55. The results are the same: Rose Bowl – Downtown Disney – Venice Beach – Hollywood and Vine – Rose Bowl (or the reverse).

57. Answers vary.   59. Answers vary.   61. Answers vary.

63. Answers vary.   65. No; in a complete graph, every vertex is connected to every other, and there's only one way to do that.

67. To be complete, every vertex has to be adjacent to every other. If there are n vertices total, each would have to be connected to all the others, meaning each would have to be connected to $n - 1$ other vertices.

69. According to Dirac's theorem, only 20 has one.

71. In some cases, to fly from one city to another, you'll have to connect through a third city.   73. Answers vary.

Exercise Set 13-4

7. The graph is not a tree because it contains a circuit.

9. The graph is a tree.

11. The graph is not a tree because it is disconnected.

13. The graph is a tree.

15. The graph is not a tree because it contains a circuit.

17.

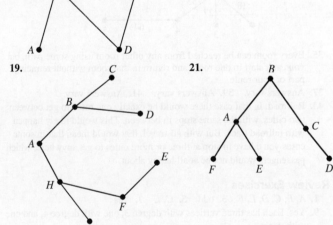

19.

21.

23. 260;

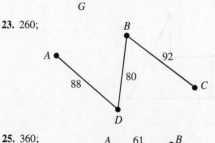

25. 360;

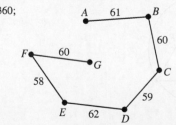

27. 633;

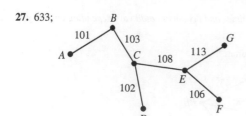

29. 89;

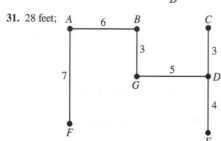

31. 28 feet;

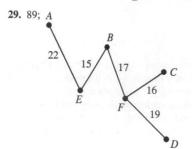

33. 162 feet;

35. Every room can be reached from any other room using some path, but you can't start in one room and return to that room without retracing part of your route.

37. Answers vary.   **39.** Answers vary.   **41.** Answers vary.

43. Railroad. In that case there would be exactly one route to get between two cities, with the same stops in between. This would likely happen with railroad track. But with air travel, that would mean that in some cases you'd have to stop at three or more cities to get anywhere, which passengers would not be at all happy about.

Review Exercises

1. A, B, C, D, E, F **3.** odd   **5.** C, E **7.** Yes
9. Yes. Each has three vertices with degree 3, one with degree 4, and one with degree 1.

11.

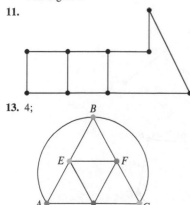

13. 4;

---

15. Euler circuit   **17.** Euler path but no Euler circuit
19. The graph from Exercise 17; one possible answer: $E, D, C, A, B, C, E, B, G, E, F, G$.
21. Euler path: C, E, W, F, V, C, W, V is one possible path.
23. Answers vary; one Hamilton path is A, E, B, D, C, H.
25. 39,916,800

27.

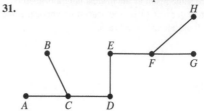

29. C, D, F, E, C or C, E, F, D, C; $1,415; the cheapest link approximate solution is $19 more than the optimal solution.

31.

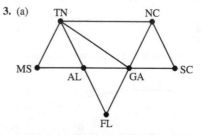

33. A minimum spanning tree connects all vertices with exactly one path that has the lowest possible overall weight. In a traveling salesperson problem, we find a circuit that visits each vertex once and has minimum weight.

Chapter 13 Test

1. (a) 4   (b) E (c) Answers vary; one possible path is A, B, C, E, D, F.
  (d) FD (e) A, B, D, E (f) C and E are even; A, B, D, and F are odd.

3. (a)

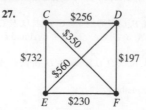

  (b) 3

5. (a) Answers vary: one Euler path is $B, A, F, B, C, D, E, C, F$.
  (b) Answers vary: one Hamilton path is A, B, F, C, E, D. The Hamilton path will typically vary from the Euler path.

7.

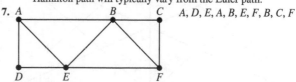

$A, D, E, A, B, E, F, B, C, F$

9. 64 miles

11.

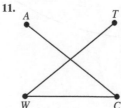